Lecture Notes in Computer Scienc

Commenced Publication in 1973
Founding and Former Series Editors:
Gerhard Goos, Juris Hartmanis, and Jan van Leeuwen

Advanced Research in Computing and Software Science

Subline of Lectures Notes in Computer Science

Kun-Mao Chao
Tsan-sheng Hsu
Der-Tsai Lee (Eds.)

Algorithms and Computation

23rd International Symposium, ISAAC 2012
Taipei, Taiwan, December 19-21, 2012
Proceedings

 Springer

Volume Editors

Kun-Mao Chao
National Taiwan University
No. 1, Sec. 4, Roosevelt Road
Taipei 106, Taiwan
E-mail: kmchao@csie.ntu.edu.tw

Tsan-sheng Hsu
Academia Sinica
128 Academia Road, Section 2
Taipei 115, Taiwan
E-mail: tshsu@iis.sinica.edu.tw

Der-Tsai Lee
National Chung Hsing University
250 Kuo Kuang Road
Taichung 402, Taiwan
E-mail: dtlee@nchu.edu.tw

ISSN 0302-9743   e-ISSN 1611-3349
ISBN 978-3-642-35260-7   e-ISBN 978-3-642-35261-4
DOI 10.1007/978-3-642-35261-4
Springer Heidelberg Dordrecht London New York

Library of Congress Control Number: 2012952397

CR Subject Classification (1998): F.2, I.3.5, E.1, C.2, G.2, F.1

LNCS Sublibrary: SL 1 – Theoretical Computer Science and General Issues

Typesetting: Camera-ready by author, data conversion by Scientific Publishing Services, Chennai, India

Printed on acid-free paper

Springer is part of Springer Science+Business Media (www.springer.com)

Preface

The papers in this volume were presented at the 23rd International Symposium on Algorithms and Computation (ISAAC 2012), held at National Taiwan University, Taipei, Taiwan, on December 19–21, 2012. In the past, ISAAC was held in Tokyo (1990), Taipei (1991), Nagoya (1992), Hong Kong (1993), Beijing (1994), Cairns (1995), Osaka (1996), Singapore (1997), Taejon (1998), Chennai (1999), Taipei (2000), Christchurch (2001), Vancouver (2002), Kyoto (2003), Hong Kong (2004), Hainan (2005), Kolkata (2006), Sendai (2007), Gold Coast (2008), Hawaii (2009), Jeju (2010), and Yokohama (2011).

The mission of the ISAAC series is to provide a top-notch forum for researchers working in algorithms and theory of computation. Papers presenting original research in all areas of algorithms and theory of computation are sought. In response to the call-for-papers, ISAAC 2012 received 174 submissions from 33 countries. The Program Committee selected 68 papers for oral presentation. "Algorithmica," "Theoretical Computer Science," and the "International Journal of Computational Geometry and Applications" will publish a special issue dedicated to selected papers from ISAAC 2012.

The best student paper awards were presented to Nanao Kita for her paper "A Partially Ordered Structure and a Generalization of the Canonical Partition for General Graphs with Perfect Matchings" and to Norie Fu for her paper "A Strongly Polynomial-Time Algorithm for the Shortest Path Problem on Coherent Planar Periodic Graphs."

We thank the invited speakers, John E. Hopcroft from Cornell University, USA, Timothy M. Chan from the University of Waterloo, Canada, and Erik D. Demaine from MIT, USA, for their distinguished lectures.

We thank all Program Committee members and external reviewers for their excellent work in the review process. We thank the Organizing Committee, chaired by Hung-Lung Wang, and conference volunteers for their extraordinary engagement in the conference preparation. We also thank the Advisory Committee, chaired by Takeshi Tokuyama, for their valuable guidance. Finally, we thank our sponsoring institutions for their support.

December 2012

Kun-Mao Chao
Tsan-sheng Hsu
Der-Tsai Lee

Organization

Symposium Chair

Der-Tsai Lee   National Chung Hsing University/Academia Sinica, Taiwan

Program Committee

Greg Aloupis	Université Libre de Bruxelles, Belgium
Lars Arge	Aarhus University, Denmark
Tetsuo Asano	JAIST, Japan
Franz Aurenhammer	Technische Universität Graz, Austria
Giorgio Ausiello	Università di Roma 'La Sapienza', Italy
Timothy M. Chan	University of Waterloo, Canada
Kun-Mao Chao	National Taiwan University, Taiwan (Co-chair)
Siu-Wing Cheng	Hong Kong University of Science and Technology, Hong Kong
Richard Cole	New York University, USA
Bhaskar DasGupta	University of Illinois at Chicago, USA
Rudolf Fleischer	GUtech, Muscat, Oman
Cyril Gavoille	Université de Bordeaux, France
Seok-Hee Hong	University of Sydney, Australia
John E. Hopcroft	Cornell University, USA
Tsan-sheng Hsu	Academia Sinica, Taiwan (Co-chair)
Hiroshi Imai	University of Tokyo, Japan
Kazuo Iwama	Kyoto University, Japan
Tao Jiang	University of California, Riverside, USA
Rolf Klein	University of Bonn, Germany
Gad M. Landau	University of Haifa, Israel and NYU-Poly, USA
Martin Middendorf	University of Leipzig, Germany
Heejin Park	Hanyang University, Korea
Peter Rossmanith	RWTH Aachen University, Germany
Xiaoming Sun	China Academy of Sciences, China
Chuan Yi Tang	Providence University, Taiwan
Takeshi Tokuyama	Tohoku University, Japan
Hsu-Chun Yen	National Taiwan University, Taiwan
Ke Yi	Hong Kong University of Science and Technology, Hong Kong
Louxin Zhang	National University of Singapore, Singapore

Organizing Committee

Kun-Mao Chao	National Taiwan University, Taiwan
Feipei Lai	National Taiwan University, Taiwan
Yuh-Dauh Lyuu	National Taiwan University, Taiwan
Hung-Lung Wang	National Taipei College of Business, Taiwan (Chair)
Yue-Li Wang	National Taiwan University of Science and Technology, Taiwan

External Reviewers

Faisal Abu-Khzam	Edith Elkind	Sung-Ryul Kim
Peyman Afshani	Irene Finocchi	Karsten Klein
Luca Castelli Aleardi	Donatella Firmani	Ton Kloks
Cyril Allauzen	Johannes Fischer	Dennis Komm
Aris Anagnostopoulos	Fabrizio Frati	Christian Komusiewicz
Sang Won Bae	Norie Fu	Matias Korman
Luca Becchetti	Toshihiro Fujito	Robin Kothari
Oren Ben-Zwi	Hiroshi Fujiwara	Richard Kralovic
Guillaume Blin	Song Gao	David Kriesel
Cecilia Bohler	Stefan Gasten	Oded Lachish
Vincenzo Bonifaci	Emilio Di Giacomo	Chi Kit Lam
Flavia Bonomo	Alexander Gilbers	Alexander Langer
Franz J. Brandenburg	Petr Golovach	Elmar Langetepe
Broňa Brejová	Joachim Gudmundsson	Kasper Green Larsen
Gerth Stølting Brodal	Vladimir Gurvich	Luigi Laura
Jin-Yi Cai	Xin Han	Francois Le Gall
Jean Cardinal	Sariel Har-Peled	Che-Rung Lee
Jou-Ming Chang	Frederic Havet	Inbok Lee
Danny Z. Chen	Danny Hermelin	Erik Jan van Leeuwen
Ho-Lin Chen	Hiroshi Hirai	Asaf Levin
Xi Chen	Wing-Kai Hon	Minming Li
Yongxi Cheng	Chien-Chung Huang	Shi Li
Otfried Cheong	Zengfeng Huang	Hongyu Liang
Mahdi Cheraghchi	Zhiyi Huang	Chung-Shou Liao
Man-Kwun Chiu	Toshimasa Ishii	Mathieu Liedloff
Sunghee Choi	Hiro Ito	Ching-Chi Lin
Morgan Chopin	Takehiro Ito	Chun-Cheng Lin
Jinhee Chun	Jiongxin Jin	Giuseppe Liotta
Paolo Codenotti	Gwenaël Joret	Chin Lung Lu
Ivan Damgaard	Hossein Jowhari	Pinyan Lu
Pooya Davoodi	Marcin Kaminski	Kazuhisa Makino
Frank Dehne	Naoki Katoh	Igor L. Markov
Camil Demetrescu	Jae-Hoon Kim	Kitty Meeks
Andrew Drucker	Jin Wook Kim	George Mertzios

Shuichi Miyazaki
Matthias Mnich
Thomas Mølhave
Gabriel Moruz
Tobias Mueller
Joong Chae Na
Mogens Nielsen
Kang Ning
Harumichi Nishimura
Pascal Ochem
Yoshio Okamoto
Hirotaka Ono
Yota Otachi
Jung-Heum Park
Vangelis Paschos
Matthew Patitz
Sheng-Lung Peng
Rainer Penninger
Marcin Pilipczuk
Michal Pilipczuk
Alexander Pilz
Sheung-Hung Poon
R. Ravi

Dror Rawitz
Liat Rozenberg
Ignaz Rutter
Toshiki Saitoh
Jayalal Sarma M.N.
Baruch Schieber
Alex Scott
Kazuhisa Seto
Devavrat Shah
Chan-Su Shin
Akiyoshi Shioura
Somnath Sikdar
Nodari Sitchinava
Michiel Smid
Christian Sommer
Hisao Tamaki
Suguru Tamaki
Ioan Todinca
Ming-Jer Tsai
Ryuhei Uehara
Kenya Ueno
Yann Vaxès
Rossano Venturini

Antoine Vigneron
Fernando S. Villaamil
Haitao Wang
Hung-Lung Wang
Yajun Wang
Zhewei Wei
Renato Werneck
Andreas Wiese
Stefanie Wuhrer
Hiroki Yanagisawa
Jonathan Yaniv
Yitong Yin
Juyoung Yon
Fang Yu
Hung-I Yu
Nengkun Yu
Tian-Li Yu
Wei Yu
Raphael Yuster
Guochuan Zhang
Qin Zhang
Yuan Zhou

Sponsoring Institutions

National Taiwan University, Taiwan
National Chung Hsing University, Taiwan
Academia Sinica, Taiwan
Institute of Information and Computing Machinery, Taiwan
National Science Council, Taiwan
Ministry of Education, Taiwan

Table of Contents

Invited Talk (I)

Future Directions in Computer Science Research 1
John E. Hopcroft

Invited Talk (II)

Combinatorial Geometry and Approximation Algorithms 2
Timothy M. Chan

Invited Talk (III)

Origami Robots and Star Trek Replicators 3
Erik D. Demaine

Graph Algorithms (I)

Strong Conflict-Free Coloring for Intervals 4
Panagiotis Cheilaris, Luisa Gargano, Adele A. Rescigno, and
Shakhar Smorodinsky

Closing Complexity Gaps for Coloring Problems on H-Free Graphs 14
Petr A. Golovach, Daniël Paulusma, and Jian Song

Randomly Coloring Regular Bipartite Graphs and Graphs with
Bounded Common Neighbors 24
Ching-Chen Kuo and Hsueh-I Lu

Reconfiguration of List $L(2,1)$-Labelings in a Graph 34
Takehiro Ito, Kazuto Kawamura, Hirotaka Ono, and Xiao Zhou

Online and Streaming Algorithms

An 8/3 Lower Bound for Online Dynamic Bin Packing 44
Prudence W.H. Wong, Fencol C.C. Yung, and Mihai Burcea

Computing k-center over Streaming Data for Small k 54
Hee-Kap Ahn, Hyo-Sil Kim, Sang-Sub Kim, and Wanbin Son

Precision vs Confidence Tradeoffs for ℓ_2-Based Frequency Estimation
in Data Streams 64
Sumit Ganguly

Competitive Design and Analysis for Machine-Minimizing Job
Scheduling Problem 75
  Mong-Jen Kao, Jian-Jia Chen, Ignaz Rutter, and Dorothea Wagner

Combinatorial Optimization (I)

A Partially Ordered Structure and a Generalization of the Canonical
Partition for General Graphs with Perfect Matchings 85
  Nanao Kita

Fast and Simple Fully-Dynamic Cut Tree Construction 95
  Tanja Hartmann and Dorothea Wagner

Green Scheduling, Flows and Matchings 106
  Evripidis Bampis, Dimitrios Letsios, and Giorgio Lucarelli

Popular and Clan-Popular b-Matchings 116
  Katarzyna Paluch

Computational Complexity (I)

Kernelization and Parameterized Complexity of Star Editing and Union
Editing 126
  Jiong Guo and Yash Raj Shrestha

On the Advice Complexity of Buffer Management 136
  Reza Dorrigiv, Meng He, and Norbert Zeh

On the Complexity of the Maximum Common Subgraph Problem
for Partial k-Trees of Bounded Degree 146
  Tatsuya Akutsu and Takeyuki Tamura

Speeding Up Shortest Path Algorithms 156
  Andrej Brodnik and Marko Grgurovič

Computational Geometry (I)

How Many Potatoes Are in a Mesh? 166
  Marc van Kreveld, Maarten Löffler, and János Pach

On Higher Order Voronoi Diagrams of Line Segments 177
  Evanthia Papadopoulou and Maksym Zavershynskyi

On the Farthest Line-Segment Voronoi Diagram 187
  Evanthia Papadopoulou and Sandeep Kumar Dey

String Algorithms

Computing the Longest Common Subsequence of Two Run-Length
Encoded Strings   197
  Yoshifumi Sakai

Efficient Counting of Square Substrings in a Tree   207
  Tomasz Kociumaka, Jakub Pachocki, Jakub Radoszewski,
  Wojciech Rytter, and Tomasz Waleń

A General Method for Improving Insertion-Based Adaptive Sorting   217
  Riku Saikkonen and Eljas Soisalon-Soininen

Computational Complexity (II)

Counting Partitions of Graphs   227
  Pavol Hell, Miki Hermann, and Mayssam Mohammadi Nevisi

Constant Unary Constraints and Symmetric Real-Weighted Counting
CSPs   237
  Tomoyuki Yamakami

Interval Scheduling and Colorful Independent Sets   247
  René van Bevern, Matthias Mnich, Rolf Niedermeier, and
  Mathias Weller

More on a Problem of Zarankiewicz   257
  Chinmoy Dutta and Jaikumar Radhakrishnan

Graph Algorithms (II)

Efficient Dominating and Edge Dominating Sets for Graphs and
Hypergraphs   267
  Andreas Brandstädt, Arne Leitert, and Dieter Rautenbach

On the Hyperbolicity of Small-World and Tree-Like Random Graphs . . .   278
  Wei Chen, Wenjie Fang, Guangda Hu, and Michael W. Mahoney

On the Neighbourhood Helly of Some Graph Classes and Applications
to the Enumeration of Minimal Dominating Sets   289
  Mamadou Moustapha Kanté, Vincent Limouzy, Arnaud Mary, and
  Lhouari Nourine

Induced Immersions   299
  Rémy Belmonte, Pim van 't Hof, and Marcin Kamiński

Computational Geometry (II)

Rectilinear Covering for Imprecise Input Points (Extended Abstract) ...   309
  Hee-Kap Ahn, Sang Won Bae, and Shin-ichi Tanigawa

Robust Nonparametric Data Approximation of Point Sets via Data
Reduction   319
  Stephane Durocher, Alexandre Leblanc, Jason Morrison, and
  Matthew Skala

Optimal Point Movement for Covering Circular Regions..............   332
  Danny Z. Chen, Xuehou Tan, Haitao Wang, and Gangshan Wu

Solving Circular Integral Block Decomposition in Polynomial Time   342
  Yunlong Liu and Xiaodong Wu

Approximation Algorithms

The Canadian Traveller Problem Revisited   352
  Yamming Huang and Chung-Shou Liao

Vehicle Scheduling on a Graph Revisited   362
  Wei Yu, Mordecai Golin, and Guochuan Zhang

A 4.31-Approximation for the Geometric Unique Coverage Problem on
Unit Disks   372
  Takehiro Ito, Shin-ichi Nakano, Yoshio Okamoto, Yota Otachi,
  Ryuhei Uehara, Takeaki Uno, and Yushi Uno

The Minimum Vulnerability Problem   382
  Sepehr Assadi, Ehsan Emamjomeh-Zadeh, Ashkan Norouzi-Fard,
  Sadra Yazdanbod, and Hamid Zarrabi-Zadeh

Graph Algorithms (III)

A Strongly Polynomial Time Algorithm for the Shortest Path Problem
on Coherent Planar Periodic Graphs   392
  Norie Fu

Cubic Augmentation of Planar Graphs............................   402
  Tanja Hartmann, Jonathan Rollin, and Ignaz Rutter

On the Number of Upward Planar Orientations of Maximal Planar
Graphs   413
  Fabrizio Frati, Joachim Gudmundsson, and Emo Welzl

Universal Point Subsets for Planar Graphs 423
  Patrizio Angelini, Carla Binucci, William Evans, Ferran Hurtado,
  Giuseppe Liotta, Tamara Mchedlidze, Henk Meijer, and
  Yoshio Okamoto

Computational Complexity (III)

Abstract Flows over Time: A First Step towards Solving Dynamic
Packing Problems 433
  Jan-Philipp W. Kappmeier, Jannik Matuschke, and Britta Peis

Extending Partial Representations of Subclasses of Chordal Graphs 444
  Pavel Klavík, Jan Kratochvíl, Yota Otachi, and Toshiki Saitoh

Isomorphism for Graphs of Bounded Connected-Path-Distance-Width .. 455
  Yota Otachi

Algorithmic Aspects of the Intersection and Overlap Numbers
of a Graph 465
  Danny Hermelin, Romeo Rizzi, and Stéphane Vialette

Graph Drawing

Linear Layouts in Submodular Systems 475
  Hiroshi Nagamochi

Segmental Mapping and Distance for Rooted Labeled Ordered Trees ... 485
  Tomohiro Kan, Shoichi Higuchi, and Kouichi Hirata

Detecting Induced Minors in AT-Free Graphs 495
  Petr A. Golovach, Dieter Kratsch, and Daniël Paulusma

Degree-Constrained Orientations of Embedded Graphs 506
  Yann Disser and Jannik Matuschke

Interval Graph Representation with Given Interval and Intersection
Lengths 517
  Johannes Köbler, Sebastian Kuhnert, and Osamu Watanabe

Data Structures

Finger Search in the Implicit Model 527
  Gerth Stølting Brodal, Jesper Sindahl Nielsen, and Jakob Truelsen

A Framework for Succinct Labeled Ordinal Trees over Large
Alphabets 537
  Meng He, J. Ian Munro, and Gelin Zhou

A Space-Efficient Framework for Dynamic Point Location 548
    Meng He, Patrick K. Nicholson, and Norbert Zeh

Selection in the Presence of Memory Faults, with Applications to
In-place Resilient Sorting 558
    Tsvi Kopelowitz and Nimrod Talmon

An Improved Algorithm for Static 3D Dominance Reporting in the
Pointer Machine 568
    Christos Makris and Konstantinos Tsakalidis

Combinatorial Optimization (II)

The Multi-Service Center Problem 578
    Hung-I Yu and Cheng-Chung Li

Computing Minmax Regret 1-Median on a Tree Network with
Positive/Negative Vertex Weights 588
    Binay Bhattacharya, Tsunehiko Kameda, and Zhao Song

Fence Patrolling by Mobile Agents with Distinct Speeds 598
    Akitoshi Kawamura and Yusuke Kobayashi

Computational Geometry (III)

Weak Visibility Queries of Line Segments in Simple Polygons 609
    Danny Z. Chen and Haitao Wang

Beyond Homothetic Polygons: Recognition and Maximum Clique....... 619
    Konstanty Junosza-Szaniawski, Jan Kratochvíl, Martin Pergel, and
    Paweł Rzążewski

Area Bounds of Rectilinear Polygons Realized by Angle Sequences 629
    Sang Won Bae, Yoshio Okamoto, and Chan-Su Shin

Randomized Algorithms

A Time-Efficient Output-Sensitive Quantum Algorithm for Boolean
Matrix Multiplication 639
    François Le Gall

On Almost Disjunct Matrices for Group Testing 649
    Arya Mazumdar

Parameterized Clique on Scale-Free Networks...................... 659
    Tobias Friedrich and Anton Krohmer

Algorithmic Game Theory

Multi-unit Auctions with Budgets and Non-uniform Valuations   669
   H.F. Ting and Xiangzhong Xiang

Efficient Computation of Power Indices for Weighted Majority
Games.   679
   Takeaki Uno

Revenue Maximization in a Bayesian Double Auction Market   690
   Xiaotie Deng, Paul Goldberg, Bo Tang, and Jinshan Zhang

Author Index   701

Future Directions in Computer Science Research

John E. Hopcroft

Cornell University
jeh@cs.cornell.edu

Over the last 40 years the computer science research was focused on making computers useful. Areas included programming languages, compilers, operating systems, data structures and algorithms. These are still important topics but with the merging of computing and communication, the emergence of social networks, and the large amount of information in digital form, focus is shifting to applications such as the structure of networks and extracting information from large data sets. This talk will give a brief vision of the future and then an introduction to the science base that needs to be formed to support these new directions.

K.-M. Chao, T.-s. Hsu, and D.-T. Lee (Eds.): ISAAC 2012, LNCS 7676, p. 1, 2012.
© Springer-Verlag Berlin Heidelberg 2012

Combinatorial Geometry and Approximation Algorithms

Timothy M. Chan

School of Computer Science, University of Waterloo
tmchan@uwaterloo.ca

In this talk, I will discuss some recent applications of combinatorial geometry to the analysis of approximation algorithms—specifically, approximation algorithms for geometric versions of *set cover*, *hitting set*, and *independent set* problems, for different types of objects such as disks and rectangles (both unweighted and weighted). These problems turn out to be related, via LP rounding, to a number of well-known combinatorial problems: ε-*nets*, *union complexity*, $(\leq k)$-*levels*, and *conflict-free coloring*. I will attempt to explain all these inter-relationships, survey some of the latest results, and mention open problems.

K.-M. Chao, T.-s. Hsu, and D.-T. Lee (Eds.): ISAAC 2012, LNCS 7676, p. 2, 2012.
© Springer-Verlag Berlin Heidelberg 2012

Origami Robots and Star Trek Replicators

Erik D. Demaine

Computer Science and Artificial Intelligence Laboratory, MIT
edemaine@MIT.EDU

Science fiction is a great inspiration for science. How can we build reconfigurable robots like Transformers or Terminator 2? How can we build replicators that mass-produce a given shape at the nano scale? Recently we've been exploring possible answers to these questions through computational geometry.

One approach to reconfigurable robots, based on computational origami design, is to build a sheet of material that can fold itself into desired shapes. We show that one pattern of hinges in the sheet lets us fold any orthogonal 3D shape up to a desired resolution–without an origamist to manually fold it. A second approach to reconfigurable robots, based on hinged dissections, is to build a chain of identical parts connected by actuated hinges. Again we show that such a chain can fold into any orthogonal 3D shape up to a desired resolution. Both of these approaches offer possible answers to programmable matter: a single piece of material that can dynamically change its shape into anything desired, in principle allowing us to download and execute geometry in the same way we download and execute software.

Going down to the nano scale, we need to work with much simpler parts, actuated by Brownian motion, van der Waal forces, etc. Tile self-assembly is one theoretical approach to manufacturing shapes within this world. A recent direction in this theory allows the use of multiple stages– operations performed by the experimenter, such as mixing two self-assembling systems together. This flexibility transforms the experimenter from a passive entity into a parallel algorithm, and vastly reduces the number of distinct parts required to construct a desired shape, possibly making the systems practical to build. The staged-assembly perspective also enables the possibility of additional operations, such as adding an enzyme that destroys all tiles with a special label. By enabling destruction in addition to the usual construction, we can perform tasks impossible in a traditional self-assembly system, such as replicating many copies of a given object's shape, without knowing anything about that shape, and building an efficient nano computer.

I will describe the algorithms and geometry underlying all of these ideas, as well as our early attempts at implementing them in practice.

K.-M. Chao, T.-s. Hsu, and D.-T. Lee (Eds.): ISAAC 2012, LNCS 7676, p. 3, 2012.
© Springer-Verlag Berlin Heidelberg 2012

Strong Conflict-Free Coloring for Intervals

Panagiotis Cheilaris[1], Luisa Gargano[2],
Adele A. Rescigno[2], and Shakhar Smorodinsky[3]

[1] Faculty of Informatics, Università della Svizzera italiana, Switzerland
[2] Dipartimento di Informatica, University of Salerno, 84084 Fisciano (SA), Italy
[3] Mathematics Department, Ben-Gurion University, Be'er Sheva 84105, Israel

Abstract. We consider the k-strong conflict-free (k-SCF) coloring of a set of points on a line with respect to a family of intervals: Each point on the line must be assigned a color so that the coloring is conflict-free in the following sense: in every interval I of the family there are at least k colors each appearing exactly once in I.

We first present a polynomial time algorithm for the general problem; the algorithm has approximation ratio 2 when $k = 1$ and $5 - \frac{2}{k}$ when $k > 1$ (our analysis is tight). In the special case of a family that contains all possible intervals on the given set of points, we show that a 2-approximation algorithm exists, for any $k \geq 1$. We also show that the problem of deciding whether a given family of intervals can be 1-SCF colored with at most q colors has a quasipolynomial time algorithm.

1 Introduction

A coloring of the vertices of a hypergraph is said to be conflict-free if every hyperedge contains a vertex whose color is unique among those colors assigned to the vertices of the hyperedge. We denote by $\mathbb{Z}^+$ the set of positive integers and by $\mathbb{N}$ the set of non-negative integers.

Definition 1. *(CF coloring) A conflict-free vertex coloring of a hypergraph $H = (V, \mathcal{E})$ is a function $C \colon V \to \mathbb{Z}^+$ such that for each $e \in \mathcal{E}$ there exists a vertex $v \in e$ such that $C(u) \neq C(v)$ for any $u \in e$ with $u \neq v$.*

Conflict-free coloring was first considered in [7]. It was motivated by a frequency assignment problem in cellular networks. Such networks consist of fixed-position *base stations*, each assigned a fixed frequency, and roaming *clients*. Roaming clients have a range of communication and come under the influence of different subsets of base stations. This situation can be modeled by means of a hypergraph whose vertices correspond to the base stations and whose hyperedges correspond to the different subsets of base stations corresponding to ranges of roaming agents. A conflict-free coloring of such a hypergraph corresponds to an assignment of frequencies to the base stations, which enables any client to connect to one of the base stations (holding the unique frequency in the client's range) without interfering with the other base stations. The goal is to minimize the number of assigned frequencies. Due to both its practical motivations and its

K.-M. Chao, T.-s. Hsu, and D.-T. Lee (Eds.): ISAAC 2012, LNCS 7676, pp. 4–13, 2012.
© Springer-Verlag Berlin Heidelberg 2012

theoretical interest, conflict-free coloring has been the subject of several papers; a survey of results in the area is given in [13].

CF-coloring also finds application in RFID (Radio Frequency Identification) networks. RFID allows a reader device to sense the presence of a nearby object by reading a tag attached to the object itself. To improve coverage, multiple RFID readers can be deployed in an area. However, two readers trying to access a tagged device simultaneously might cause mutual interference. It can be shown that CF-coloring of the readers can be used to assure that every possible tag will have a time slot and a single reader trying to access it in that time slot [13].

The notion of *k-strong* CF coloring (*k*-SCF coloring), first introduced in [2], extends that of CF-coloring. A *k*-SCF coloring is a coloring that remains conflict-free after an arbitrary collection of $k-1$ vertices is deleted from the set [1]. In the context of cellular networks, a *k*-SCF coloring implies that for any client in an area covered by at least k base stations, there always exist at least k different frequencies the client can use to communicate without interference. Therefore, up to k clients can be served at the same location, or the system can deal with malfunctioning of at most $k-1$ base stations per location. Analogously, in the RFID networks context, a *k*-SCF coloring corresponds to a fault-tolerant activation protocol, i.e., every tag can be read as long as at most $k-1$ readers are broken. A CF-coloring is just a 1-SCF coloring.

We will allow the coloring function $C\colon V \to \mathbb{Z}^+$ to be a partial function (i.e., some vertices are not assigned a color). Alternatively, we can use a special color '0' given to vertices that are not assigned any positive color and obtain a total function $C\colon V \to \mathbb{N}$. Then, we arrive at the following definition.

Definition 2. *(k-SCF coloring) Let $H = (V, \mathcal{E})$ be a hypergraph and $k \in \mathbb{Z}^+$. A coloring $C\colon V \to \mathbb{N}$ is called a k-strong conflict-free coloring if for every $e \in \mathcal{E}$ at least $\min\{|e|, k\}$ positive colors are unique in e, namely there exist $c_1, \ldots, c_{\min\{|e|, k\}} \in \mathbb{Z}^+$ such that $|\{v \mid v \in e, C(v) = c_i\}| = 1$, for $i = 1, \ldots, \min\{|e|, k\}$. The goal is to minimize the number of positive colors in the range of the k-SCF coloring function C. We denote by $\chi_k^*(H)$ the smallest number of positive colors in any possible k-SCF coloring of H.*

Remark 1. We claim that this variation of conflict-free coloring, with the partial coloring function or the placeholder color '0', is interesting from the point of view of applications. A vertex with no positive color assigned to it can model a situation where a base station is not activated at all, and therefore the base station does not consume energy. One can also think of a bi-criteria optimization problem where a conflict-free assignment of frequencies has to be found with small number of frequencies (in order to conserve the frequency spectrum) and few activated base stations (in order to conserve energy). It is not difficult to see that a partial SCF coloring with q positive colors implies always a total SCF coloring with $q + 1$ positive colors.

SCF-Coloring Points with Respect to Intervals. Several authors recently focused on the special case of CF coloring n collinear points with respect to the family of all intervals. The problem can be modeled in the hypergraph.

$H_n = ([n], \mathcal{I}^{[n]})$ with $[n] = \{1, \ldots, n\}$ and $\mathcal{I}^{[n]} = \{\{i, \ldots, j\} \mid 1 \le i \le j \le n\}$,

where each (discrete) interval is a set of consecutive points.

Conflict-free coloring for intervals models the assignment of frequencies in a chain of unit disks; this arises in approximately unidimensional networks as in the case of agents moving along a road. Moreover, it is important because it plays a role in the study of conflict-free coloring for more complicated cases, as for example in the general case of CF coloring of unit disks [7,10].

While some papers require the conflict-free property for all possible intervals on the line, in many applications good reception is needed only at some locations, i.e., it is sufficient to supply only a given subset of the cells of the arrangement of the disks [9]. In the context of channel assignment for broadcasting in a wireless mesh network, it can occur that, at some step of the broadcasting process, sparse receivers of the broadcast message are within the transmission range of a linear sequence of transmitters. In this case only part of the cells of the linear arrangements of disks representing the transmitters are involved [11,14]. In this work we consider the k-strong conflict-free coloring of points with respect to an arbitrary family of intervals. Hence, in the remainder, we consider subhypergraphs of H_n. We shall refer to these subhypergraphs of the form $H = ([n], \mathcal{I})$, where $\mathcal{I} \subseteq \mathcal{I}^{[n]}$, as *interval hypergraphs* and to H_n as the *complete interval hypergraph*.

Conflict-free coloring the complete interval hypergraph was first studied in [7], where it was shown that $\chi_1^*(H_n) = \lfloor \log n \rfloor + 1$[1]. The on-line version of the CF coloring problem for complete interval hypergraphs, where points arrive one by one and the coloring needs to remain CF all the time, has been subsequently considered in [3,4,5].

The problem of CF-coloring the points of a line with respect to an arbitrary family of intervals is studied in [9]. The k-SCF coloring problem was first considered in [2] and has since then been studied in various papers under different scenarios, we refer the reader to [13] for more details on the subject. Recently, the minimum number of colors needed for k-SCF coloring the complete interval hypergraph H_n has been studied in [6], where the exact number of needed colors for $k = 2$ and $k = 3$ has been obtained. Horev et al. show that H_n admits a k-SCF coloring with $k \log_2 n$ colors, for any k [8].

Our Results. In Section 2, we give an algorithm which outputs a k-SCF coloring of the points of the input interval hypergraph H, for any fixed value of $k \ge 1$. The algorithm has an approximation factor $5 - 2/k$ in the case $k \ge 2$ (approximation factor 2 in the case $k = 1$); moreover, it optimally uses k colors if for any $I, J \in \mathcal{I}$, interval I is not a subset of J and they differ in at least k points. In Section 3, we consider the problem of k-SCF coloring the complete interval hypergraph H_n. We give a very simple k-SCF coloring algorithm for H_n that uses $k \left(\lfloor \log \lceil \frac{n}{k} \rceil \rfloor + 1 \right)$ colors and show a lower bound of $\lceil \frac{k}{2} \rceil \lceil \log \frac{n}{k} \rceil$ colors. In section 4, we show that the decision problem whether a given interval hypergraph can be CF-colored with at most q colors has a quasipolynomial time algorithm.

---

[1] Unless otherwise specified, all logarithms are in base 2.

Notation. Through the rest of this paper we consider interval hypergraphs on n points. Given $I \in \mathcal{I}$, we denote the *leftmost* (minimum) and the *rightmost* (maximum) of the points of the interval I by $\ell(I) = \min\{p \mid p \in I\}$ and $r(I) = \max\{p \mid p \in I\}$, respectively. We will use the following order relation on the intervals of $\mathcal{I}$.

Definition 3. (Intervals ordering) *For all* $I, J \in \mathcal{I}$,

$$I \prec J \iff (r(I) < r(J)) \text{ or } (r(I) = r(J) \text{ and } \ell(I) > \ell(J)).$$

$I \in \mathcal{I}$ *is called the* i-th interval in $\mathcal{I}$ *if* $\mathcal{I} = \{I_1, \ldots, I_m\}$, $I_1 \prec I_2 \prec \cdots \prec I_m$, *and* $I = I_i$.

Given a family $\mathcal{I}$, the subfamily of intervals of $\mathcal{I}$ that are not contained in I and whose rightmost (resp. leftmost) point belongs to I is denoted by $\mathcal{L}_{\mathcal{I}}(I)$ (resp. $\mathcal{R}_{\mathcal{I}}(I)$), that is

$$\mathcal{L}_{\mathcal{I}}(I) = \{J \in \mathcal{I} \mid J \nsubseteq I, \ r(J) \in I\} \text{ and } \mathcal{R}_{\mathcal{I}}(I) = \{J \in \mathcal{I} \mid J \nsubseteq I, \ \ell(J) \in I\}$$

Clearly, $J \prec I$ (resp. $I \prec J$) for any $J \in \mathcal{L}_{\mathcal{I}}(I)$ (resp. $J \in \mathcal{R}_{\mathcal{I}}(I)$) with $J \neq I$. We denote by $\mathcal{M}_{\mathcal{I}}(I)$ the subfamily of all the intervals contained in $I \in \mathcal{I}$

$$\mathcal{M}_{\mathcal{I}}(I) = \{J \mid J \in \mathcal{I}, \ J \subset I\}.$$

2 A k-SCF Coloring Algorithm

We present an algorithm for k-SCF coloring any interval hypergraph $H = ([n], \mathcal{I})$. We prove that our algorithm achieves an approximation ratio 2 if $k = 1$ and an approximation ratio $5 - \frac{2}{k}$ if $k \geq 2$; we show that the algorithm is optimal when $\mathcal{I}$ consists of intervals differing in at least k points and not including any other interval in $\mathcal{I}$. We say that an interval $I \in \mathcal{I}$ is k-**colored** under coloring C if its points are colored with at least $\min\{|I|, k\}$ unique positive colors, where a color c is unique in I if there is exactly one point $p \in I$ such that $C(p) = c$. The k-SCF coloring algorithm, k-COLOR$(\mathcal{I})$, is given in Fig. 1. The number of colors is upper bounded by the number of iterations performed by the algorithm times $c(k)$, where $c(k) = 2k + \lceil k/2 \rceil - 1$.

At each step t of the algorithm a subset P_t of points of $[n]$ is selected (through algorithm SELECT), then $c(k)$ colors are assigned in cyclic sequence to the ordered sequence (from the minimum to the maximum) of the selected points. The intervals that are k-colored at the end of the step t are inserted in the set $\mathcal{X}_t$ and discarded. The algorithm ends when all the intervals in $\mathcal{I}$ have been discarded. At each step t a new set of $c(k)$ colors is used.

A point $p \in [n]$ can be re-colored several times during different steps of the k-COLOR algorithm; its color at the end of algorithm is the last assigned one.

The algorithm SELECT$(\mathcal{I}_t)$ considers intervals in $\mathcal{I}_t$ according to the $\prec$ relation and selects points so that P_t has at least $\min\{|I|, k\}$ points in each interval. Namely, if I is the i-th interval, then it is considered at the i-th iteration of

---

k-**COLOR**($\mathcal{I}$):

Set $t = 1$.

$\mathcal{I}_1 = \mathcal{I}$.   *[$\mathcal{I}_t$ is the set of intervals not k-colored by the beginning of step t]*

$\mathcal{X}_1 = \emptyset$.   *[$\mathcal{X}_t \subset \mathcal{I}_t$ contains the intervals that become k-colored during step t]*

while $\mathcal{I}_t \neq \emptyset$

    Execute the following **step** t

        1. Let $P_t = \{p_0, p_1, \ldots, p_{n_t}\}$ be the set returned by SELECT($\mathcal{I}_t$)

        2. **for** $i = 0$ **to** n_t

            Assign to p_i color $c_i = (t-1)c(k) + (i \mod c(k)) + 1$

        3. **for** each $I \in \mathcal{I}_t$

            if I is k-colored **then** $\mathcal{X}_t = \mathcal{X}_t \cup \{I\}$

        4. $\mathcal{I}_{t+1} = \mathcal{I}_t \setminus \mathcal{X}_t$

        5. $t = t + 1$

SELECT($\mathcal{I}_t$):

Set $P_t = \emptyset$.   *[P_t represents the set of selected points at step t]*

for each $I \in \mathcal{I}_t$ by increasing order according to relation $\prec$ *[see Def.3]*

    if $|I \cap P_t| < \min\{|I|, k\}$ **then**

        1. Let $P_t(I)$ be the set of largest $\min\{|I|, k\} - |I \cap P_t|$ points of $I \setminus P_t$

        2. $P_t = P_t \cup P_t(I)$

Return P_t

Fig. 1. The k-SCF coloring algorithm for $H = ([n], \mathcal{I})$

the **for** loop and if less than $\min\{|I|, k\}$ points of I have been already selected, then the algorithm adds the missing $\min\{|I|, k\} - |I \cap P_t|$ points of I to P_t (such points are the largest unselected ones of I).

Example 1. Consider $H = ([23], \mathcal{I})$, where $\mathcal{I}$ is the set of 13 intervals given in Fig. 2. Run k-COLOR($\mathcal{I}$) with $k = 2$; hence $c(2) = 4$ colors are used at each iteration. Initially, $\mathcal{I}_1 = \mathcal{I}$ and SELECT($\mathcal{I}_1$) returns $P_1 = \{3, 4, 7, 8, 9, 11, 12, 14, 15, 17, 18, 19, 20, 22, 23\}$ whose points are colored with c_1, c_2, c_3, c_4 in cyclic sequence. Only 3 intervals remain in $\mathcal{I}_2$; all the others are in $\mathcal{X}_1$, being 2-colored at the end of step 1. SELECT($\mathcal{I}_2$) returns $P_2 = \{14, 15, 23\}$ and these points are colored with c_5, c_6, c_7. Now $\mathcal{I}_3 = \mathcal{I}_2 \setminus \mathcal{X}_2 = \emptyset$ and the algorithm ends.

In the following, we will sketch a proof of the following theorem.

Theorem 1. *Algorithm k-COLOR($\mathcal{I}$) is a polynomial k-SCF coloring algorithm that uses less than $\frac{c(k)}{\lceil k/2 \rceil} \chi_k^*(H)$ colors on the interval hypergraph $H = ([n], \mathcal{I})$.*

2.1   Correctness of Algorithm k-COLOR

We denote by P_t the set of points returned by SELECT($\mathcal{I}_t$).

Lemma 1. *Let $I \in \mathcal{I}_t$, $t \geq 1$.*
a) $|I \cap P_t| \geq \min\{|I|, k\}$; b) $I \in \mathcal{X}_t$ if $|I \cap P_t| \leq 4k - 2$; c) $|I| \geq k$, for $t \geq 2$.

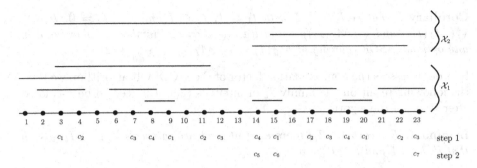

Fig. 2. Example coloring by k-COLOR for $k = 2$

Lemma 2. *If $I \in \mathcal{X}_t$ then $\mathcal{M}_{\mathcal{I}_t}(I) \subseteq \mathcal{X}_t$.*

Lemma 3. *If $\mathcal{M}_{\mathcal{I}_t}(I) = \emptyset$, then $|I \cap P_t| \leq 2k - 1$.*

With the help of the above, we can prove correctness of the algorithm.

Theorem 2. *Given interval hypergraph $H = ([n], \mathcal{I})$, algorithm k-COLOR($\mathcal{I}$) produces a k-SCF coloring of H.*

Proof. We show by induction the following statement for each $t \geq 1$: *At the end of step t of algorithm k-COLOR($\mathcal{I}$), each interval $I \in \bigcup_{i=1}^{t} \mathcal{X}_i$ is k-colored.*

For $t = 1$, the statement trivially follows. Assume the statement be true for each $i \leq t - 1$ and $t \geq 2$. We prove that it holds for t. Notice that, by c) of Lemma 1, for any $I \in \mathcal{I}_t$ it holds $\min\{|I|, k\} = k$. Clearly, if $I \in \mathcal{X}_t$, then I is k-colored by definition of $\mathcal{X}_t$. Consider then $I \in \mathcal{X}_i$ for some $i \leq t - 1$. By the inductive hypothesis I is k-colored at the end of step $t - 1$. By Lemma 2 we know that $\mathcal{M}_{\mathcal{I}_t}(I) \subseteq \mathcal{X}_i$; which implies that $\mathcal{M}_{\mathcal{I}_t}(I) = \emptyset$. Moreover, by Lemma 3, we have $|I \cap P_t| \leq 2k - 1 < c(k)$. This means that even if some points are recolored, all the assigned colors will be unique in I. $\square$

2.2   Analysis of Algorithm k-COLOR($\mathcal{I}$)

In this section we evaluate the approximation factor of the algorithm k-COLOR. We first give a lower bound tool (see also [6]). Since the vertex set $[n]$ is usually implied, we use the shorthand notation $\chi_k^*(\mathcal{I}) = \chi_k^*(([n], \mathcal{I}))$.

Theorem 3. *Let $I_1, I_2, I \in \mathcal{I}$ with $I_1, I_2 \subset I$ and $I_1 \cap I_2 = \emptyset$. Let χ_1 (resp. χ_2) be the number of colors used by an optimal k-SCF coloring of $\mathcal{M}_{\mathcal{I}}(I_1)$ (resp. $\mathcal{M}_{\mathcal{I}}(I_2)$). Then the number of colors used by any optimal k-SCF coloring of $\mathcal{M}_{\mathcal{I}}(I)$ is*

$$\chi_k^*(\mathcal{M}_{\mathcal{I}}(I)) \geq \begin{cases} \max\{\chi_1, \chi_2\} & \text{if } k \leq |\chi_2 - \chi_1|, \\ \max\{\chi_1, \chi_2\} + \left\lceil \frac{k - |\chi_2 - \chi_1|}{2} \right\rceil & \text{otherwise.} \end{cases}$$

Corollary 1. *Let $I_1, I_2, I \in \mathcal{I}$ with $I_1 \subset I$, $I_2 \subset I$ and $I_1 \cap I_2 = \emptyset$. If both $\chi_k^*(\mathcal{M}_\mathcal{I}(I_1))$ and $\chi_k^*(\mathcal{M}_\mathcal{I}(I_2))$ are at least χ, then the number of colors used in any optimal k-SCF coloring of $\mathcal{M}_\mathcal{I}(I)$ is $\chi_k^*(\mathcal{M}_\mathcal{I}(I)) \geq \chi + \lceil k/2 \rceil$.*

In order to assess the approximation factor of the k-COLOR algorithm, we need the following result on the family $\mathcal{I}_t$ of intervals that still need to be k-colored after step t of the algorithm.

Lemma 4. *For each $I \in \mathcal{I}_t$, there exist at least two intervals $I', I'' \in \mathcal{I}_{t-1}$ such that $I', I'' \subset I$ and $I' \cap I'' = \emptyset$.*

In the following we assume that there exists at least an interval $I \in \mathcal{I}$ with $|I| \geq k$. Notice that if $|I| < k$ for each $I \in \mathcal{I}$, then each interval in $\mathcal{I}$ is k-colored after the first step of the algorithm k-COLOR($\mathcal{I}$) (even using for $c(k)$ the smaller value $\max\{|I| \mid I \in \mathcal{I}\}$).

Lemma 5. *Any k-SCF coloring algorithm on $\mathcal{I}_t$ needs $k + (t - 1)\left\lceil \frac{k}{2} \right\rceil$ colors.*

We remark that the algorithm can be implemented in time $O(kn \, _2 n)$, since in each step SELECT($\mathcal{I}_t$) can be implemented in $O(kn)$ time (one does not actually need to separately consider all the intervals having the same right endpoint but only the k shortest ones) and the number of steps is upper bounded by $O(\log_2 n)$, the worst case being the complete interval hypergraph. This together with the following Theorem 4 and Theorem 2, proves the desired Theorem 1.

Theorem 4. *Consider the interval hypergraph $H = ([n], \mathcal{I})$. Then the total number of colors used by k-COLOR($\mathcal{I}$) is less than $\frac{c(k)}{\lceil k/2 \rceil} \chi_k^*(\mathcal{I})$.*

For a special class of interval hypergraphs, we show that the algorithm is optimal.

Theorem 5. *If for any $I, J \in \mathcal{I}$ such that $J \prec I$ and $I \cap J \neq \emptyset$ it holds $I \not\subseteq J$ and $|I \setminus J| \geq k$, then the algorithm k-COLOR($\mathcal{I}$), running with $c(k) = k$ on interval hypergraph $H = ([n], \mathcal{I})$, optimally uses k colors.*

3   A k-SCF Coloring Algorithm for H_n

In this section we present a k-SCF-coloring algorithm for the complete interval hypergraph $H_n = ([n], \mathcal{I}^{[n]})$. When $k = 1$ the algorithm reduces to the one in [7]. We assume that $n = hk$ for some integer $h \geq 1$. If $(h - 1)k < n < hk$ then we can add the points $n + 1, n + 2, \ldots, hk$.

A simple k-SCF-coloring algorithm for H_n can be obtained by partitioning the $n = hk$ points of V in blocks $B(1), B(2), \cdots, B(h)$ of k points and coloring their points recursively with the colors in the sets $C_1, \cdots, C_{\lfloor \log h \rfloor + 1}$, where $C_t = \{k(t - 1) + 1, \cdots, kt\}$, for $1 \leq t \leq \lfloor \log h \rfloor + 1$. The points in the median block $B(\lfloor \frac{h+1}{2} \rfloor)$ are colored with colors in C_1, then the points in the blocks $B(1), \cdots, B(\lfloor \frac{h+1}{2} \rfloor - 1)$ and in the blocks $B(\lfloor \frac{h+1}{2} \rfloor + 1), \cdots, B(h)$ are recursively colored with the same colors in the sets $C_2, \cdots, C_{\lfloor \log h \rfloor + 1}$. Formally, the algorithm is given in Fig. 3. It starts calling (k, n)-COLOR($1, h, 1$).

The proof that algorithm (k, n)-COLOR$(1, h, 1)$ provides a k-SCF coloring for H_n can be easily derived by that presented in [7,13]. Furthermore, since at each of the $\lfloor \log h \rfloor + 1$ recursive steps of algorithm (k, n)-COLOR a new set of k colors is used, we have that the number of colors is at most $k(\lfloor \log h \rfloor + 1)$. Hence, we get the following result.

Lemma 6. *At the end of algorithm (k, n)-COLOR$(1, \lceil n/k \rceil, 1)$ each $I \in \mathcal{I}$ is k-SCF colored and the number of used colors is at most $k\left(\lfloor \log \lceil \frac{n}{k} \rceil \rfloor + 1\right)$.*

(k, n)-**COLOR**(a, b, t):
if $a \leq b$ **then**
$\quad m = \lfloor \frac{a+b}{2} \rfloor$
$\quad$ Color the k points in $B(m)$ with the k colors in C_t.
$\quad (k, n)$-COLOR$(1, m - 1, t + 1)$.
$\quad (k, n)$-COLOR$(m + 1, b, t + 1)$.

Fig. 3. The k-SCF coloring algorithm for H_n

We remark that [8] shows that $\chi_k^*(H_n) \leq k \log n$ (as a specific case of a more general framework); however, we present the (k, n)-COLOR algorithm since it is very simple and gives a slightly better bound.

By Corollary 1 and considering that, for the complete interval hypergraph H_n, for each $I \in \mathcal{I}$, any of its subintervals $I' \subset I$ also belongs to $\mathcal{I}$, we get the following lower bound on $\chi^*(H_n)$.

Corollary 2. $\chi_k^*(H_n) \geq \lceil \frac{k}{2} \rceil \lceil \log \frac{n}{k} \rceil$.

Lemma 6 together with Corollary 2 proves that (k, n)-COLOR uses at most twice the minimum possible number of colors.

4   A Quasipolynomial Time Algorithm

Consider the decision problem CFSUBSETINTERVALS: *"Given an interval hypergraph H and a natural number q, is it true that $\chi_1^*(H) \leq q$?"* Notice that the above problem is non-trivial only when $q < \lfloor \log n \rfloor + 1$; if $q \geq \lfloor \log n \rfloor + 1$ the answer is always yes, since $\chi_1^*(H_n) = \lfloor \log n \rfloor + 1$.

Algorithm DECIDE-COLORS (Fig. 4) is a *non-deterministic* algorithm for CFSUBSETINTERVALS. The algorithm scans points from 1 to n, tries for every point non-deterministically every color in $\{0, \ldots, q\}$, and checks if all intervals in $\mathcal{I}$ ending at the current point have the conflict-free property. If some interval in $\mathcal{I}$ has not the conflict-free property under a non-deterministic assignment, the algorithm answers 'no'. If all intervals in $\mathcal{I}$ have the conflict-free property under some non-deterministic assignment, the algorithm answers 'yes'.

We check if an interval in $\mathcal{I}$ that ends at the current point, say t, has the conflict-free property in the following space-efficient way. For every color c in

$\{0, \ldots, q\}$, we keep track of:
(a) the closest point to t colored with c in variable p_c, and
(b) the second closest point to t colored with c in variable s_c.
Then, color c is occurring exactly one time in $[j, t] \in \mathcal{I}$ if and only if $s_c < j \le p_c$.

DECIDE-COLORS$(q, \mathcal{I})$
for $c = 0$ to q
    $s_c = 0$, $p_c = 0$.
for $t = 1$ to n
    Choose c non-deterministically from $\{0, \ldots, q\}$.
    $s_c = p_c$, $p_c = t$.
    for $j \in \{j \mid [j, t] \in I\}$
        IntervalConflict = True.
        for $c = 1$ to q
            if $s_c < j \le p_c$ **then** IntervalConflict = False
        if IntervalConflict **then return NO**
return YES

Fig. 4. A non-deterministic algorithm deciding whether $\chi_1^*(H) \le q$

Lemma 7. *The space complexity of algorithm DECIDE-COLORS is $O(\log^2 n)$.*

Proof. Since $q = O(\log n)$ and each point position can be encoded with $O(\log n)$ bits, the arrays p and s (indexed by color) take space $O(\log^2 n)$. All other variables in the algorithm can be implemented in $O(\log n)$ space. Therefore the above non-deterministic algorithm has space complexity $O(\log^2 n)$.

Theorem 6. CFSUBSETINTERVALS *has a quasipolynomial time deterministic algorithm.*

Proof. By standard computational complexity theory arguments (see, e.g., [12]), we can transform DECIDE-COLORS to a deterministic algorithm solving the same problem with time complexity $2^{O(\log^2 n)}$, i.e., CFSUBSETINTERVALS has a quasipolynomial time deterministic algorithm.

5   Conclusions, Further Work, and Open Problems

The exact complexity of computing an optimal k-SCF-coloring for an interval hypergraph remains an open problem. We have presented an algorithm with approximation ratio $5 - 2/k$ when $k \ge 2$ and 2 when $k = 1$. In a longer version of our work, we will include a proof that our analysis of the approximation ratio is tight when $k = 1$ and $k = 2$; when $k \ge 3$, we have an instance that forces the algorithm to use $(5 - 1/k)/2 > 2$ times the optimal number of colors. One might try to improve the approximation ratio, find a polynomial time approximation scheme, or even find a polynomial time exact algorithm. The last possibility is supported by the fact that the decision version of the 1-SCF problem, CFSUBSETINTERVALS, is unlikely to be NP-complete, unless NP-complete problems have quasipolynomial time algorithms. Furthermore, we have shown that the algorithm optimally uses k colors if for any $I, J \in \mathcal{I}$, interval I is not contained in J and they differ for at least k points. For the complete interval hypergraph H_n, we have presented a k-SCF coloring using at most two times the optimal number of colors. It would be interesting to close this gap.

Finally, we introduced a SCF-coloring function $C \colon V \to \mathbb{N}$, for which vertices colored with '0' can not act as uniquely-colored vertices in a hyperedge. Naturally, one could try to study the bi-criteria optimization problem, in which there two minimization goals: (a) the number of colors used, $\max_{v \in V} C(v)$ (minimization of frequency spectrum use) and (b) the number of vertices with positive colors, $|\{v \in V \mid C(v) > 0\}|$ (minimization of activated base stations).

References

1. Abam, M.A., de Berg, M., Poon, S.H.: Fault-tolerant conflict-free coloring. In: Proc. 20th Canadian Conference on Computational Geometry, CCCG (2008)
2. Abellanas, M., Bose, P., Garcia, J., Hurtado, F., Nicolas, M., Ramos, P.A.: On properties of higher order Delaunay graphs with applications. In: Proc. 21st European Workshop on Computational Geometry (EWCG), pp. 119–122 (2005)
3. Bar-Noy, A., Cheilaris, P., Olonetsky, S., Smorodinsky, S.: Online conflict-free colouring for hypergraphs. Combin. Probab. Comput. 19, 493–516 (2010)
4. Bar-Noy, A., Cheilaris, P., Smorodinsky, S.: Deterministic conflict-free coloring for intervals: from offline to online. ACM Trans. Alg. 4(4) (2008)
5. Chen, K., Fiat, A., Levy, M., Matoušek, J., Mossel, E., Pach, J., Sharir, M., Smorodinsky, S., Wagner, U., Welzl, E.: Online conflict-free coloring for intervals. SIAM J. Comput. 36, 545–554 (2006)
6. Cui, Z., Hu, Z.C.: k-conflict-free coloring and k-strong-conflict-free coloring for one class of hypergraphs and online k-conflict-free coloring. ArXiv abs/1107.0138 (2011)
7. Even, G., Lotker, Z., Ron, D., Smorodinsky, S.: Conflict-free colorings of simple geometric regions with applications to frequency assignment in cellular networks. SIAM J. Comput. 33, 94–136 (2003)
8. Horev, E., Krakovski, R., Smorodinsky, S.: Conflict-Free Coloring Made Stronger. In: Kaplan, H. (ed.) SWAT 2010. LNCS, vol. 6139, pp. 105–117. Springer, Heidelberg (2010)
9. Katz, M., Lev-Tov, N., Morgenstern, G.: Conflict-free coloring of points on a line with respect to a set of intervals. Comput. Geom. 45, 508–514 (2012)
10. Lev-Tov, N., Peleg, D.: Conflict-free coloring of unit disks. Discrete Appl. Math. 157(7), 1521–1532 (2009)
11. Nguyen, H.L., Nguyen, U.T.: Algorithms for bandwidth efficient multicast routing in multi-channel multi-radio wireless mesh networks. In: Proc. IEEE Wireless Communications and Networking Conference (WCNC), pp. 1107–1112 (2011)
12. Papadimitriou, C.: Computational Complexity. Addison Wesley (1993)
13. Smorodinsky, S.: Conflict-free coloring and its applications. ArXiv abs/1005.3616 (2010)
14. Zeng, G., Wang, B., Ding, Y., Xiao, L., Mutka, M.: Efficient multicast algorithms for multichannel wireless mesh networks. IEEE Trans. Parallel Distrib. Systems 21, 86–99 (2010)

Closing Complexity Gaps
for Coloring Problems on H-Free Graphs*

Petr A. Golovach[1], Daniël Paulusma[2], and Jian Song[2]

[1] Department of Informatics, Bergen University,
PB 7803, 5020 Bergen, Norway
petr.golovach@ii.uib.no
[2] School of Engineering and Computing Sciences, Durham University,
Science Laboratories, South Road, Durham DH1 3LE, United Kingdom
{daniel.paulusma,jian.song}@durham.ac.uk

Abstract. If a graph G contains no subgraph isomorphic to some graph H, then G is called H-free. A coloring of a graph $G = (V, E)$ is a mapping $c : V \to \{1, 2, \ldots\}$ such that no two adjacent vertices have the same color, i.e., $c(u) \neq c(v)$ if $uv \in E$; if $|c(V)| \leq k$ then c is a k-coloring. The COLORING problem is to test whether a graph has a coloring with at most k colors for some integer k. The PRECOLORING EXTENSION problem is to decide whether a partial k-coloring of a graph can be extended to a k-coloring of the whole graph for some integer k. The LIST COLORING problem is to decide whether a graph allows a coloring, such that every vertex u receives a color from some given set $L(u)$. By imposing an upper bound ℓ on the size of each $L(u)$ we obtain the ℓ-LIST COLORING problem. We first classify the PRECOLORING EXTENSION problem and the ℓ-LIST COLORING problem for H-free graphs. We then show that 3-LIST COLORING is NP-complete for n-vertex graphs of minimum degree $n - 2$, i.e., for complete graphs minus a matching, whereas LIST COLORING is fixed-parameter tractable for this graph class when parameterized by the number of vertices of degree $n - 2$. Finally, for a fixed integer $k > 0$, the LIST k-COLORING problem is to decide whether a graph allows a coloring, such that every vertex u receives a color from some given set $L(u)$ that must be a subset of $\{1, \ldots, k\}$. We show that LIST 4-COLORING is NP-complete for P_6-free graphs, where P_6 is the path on six vertices. This completes the classification of LIST k-COLORING for P_6-free graphs.

1 Introduction

Graph coloring involves the labeling of the vertices of some given graph by integers called colors such that no two adjacent vertices receive the same color. The corresponding decision problem is called COLORING and is to decide whether a graph can be colored with at most k colors for some given integer k. Because COLORING is NP-complete for any fixed $k \geq 3$, its computational complexity has been widely studied for special graph classes, see e.g. the surveys of Randerath

---

* This work has been supported by EPSRC (EP/G043434/1).

K.-M. Chao, T.-s. Hsu, and D.-T. Lee (Eds.): ISAAC 2012, LNCS 7676, pp. 14–23, 2012.

and Schiermeyer [17] and Tuza [20]. In this paper, we consider the COLORING problem together with two natural and well-studied variants, namely PRECOLORING EXTENSION and LIST COLORING for graphs characterized by some forbidden induced subgraph. Before we summarize related results and explain our new results, we first state the necessary terminology.

Terminology. We only consider finite undirected graphs $G = (V, E)$ without loops and multiple edges. The graph P_r denotes the path on r vertices. The disjoint union of two graphs G and H is denoted $G + H$, and the disjoint union of r copies of G is denoted rG. Let G be a graph and $\{H_1, \ldots, H_p\}$ be a set of graphs. We say that G is $(H_1, \ldots, H_p)$-*free* if G has no induced subgraph isomorphic to a graph in $\{H_1, \ldots, H_p\}$; if $p = 1$, we sometimes write H_1-free instead of (H_1)-free. The *complement* of a graph G denoted by $\overline{G}$ has vertex set $V(G)$ and an edge between two distinct vertices if and only if these vertices are not adjacent in G.

A *coloring* of a graph $G = (V, E)$ is a mapping $c : V \rightarrow \{1, 2, \ldots\}$ such that $c(u) \neq c(v)$ whenever $uv \in E$. We call $c(u)$ the *color* of u. A k-*coloring* of G is a coloring c of G with $1 \leq c(u) \leq k$ for all $u \in V$. The problem k-COLORING is to decide whether a given graph admits a k-coloring. Here, k is *fixed*, i.e., not part of the input. If k is part of the input, then we denote the problem as COLORING. A *list assignment* of a graph $G = (V, E)$ is a function L that assigns a list $L(u)$ of so-called *admissible* colors to each $u \in V$. If $L(u) \subseteq \{1, \ldots, k\}$ for each $u \in V$, then L is also called a k-*list assignment*. The *size* of a list assignment L is the maximum list size $|L(u)|$ over all vertices $u \in V$. We say that a coloring $c : V \rightarrow \{1, 2, \ldots\}$ *respects* L if $c(u) \in L(u)$ for all $u \in V$. The LIST COLORING problem is to test whether a given graph has a coloring that respects some given list assignment. For a fixed integer k, the LIST k-COLORING problem has as input a graph G with a k-list assignment L and asks whether G has a coloring that respects L. For a fixed integer ℓ, the ℓ-LIST COLORING problem has as input a graph G with a list assignment L of size at most ℓ and asks whether G has a coloring that respects L. In *precoloring extension* we assume that a (possibly empty) subset $W \subseteq V$ of G is precolored by a *precoloring* $c_W : W \rightarrow \{1, 2, \ldots k\}$ for some integer k, and the question is whether we can extend c_W to a k-coloring of G. For a fixed integer k, we denote this problem as k-PRECOLORING EXTENSION. If k is part of the input, then we denote this problem as PRECOLORING EXTENSION.

Note that k-COLORING can be viewed as a special case of k-PRECOLORING EXTENSION by choosing $W = \emptyset$, and that k-PRECOLORING EXTENSION can be viewed as a special case of LIST k-COLORING by choosing $L(u) = \{c_W(u)\}$ if $u \in W$ and $L(u) = \{1, \ldots, k\}$ if $u \in W \setminus V$. Moreover, LIST k-COLORING can be readily seen as a special case of k-LIST COLORING. Hence, we can make the following two observations for a graph class $\mathcal{G}$. If k-COLORING is NP-complete for $\mathcal{G}$, then k-PRECOLORING EXTENSION is NP-complete for $\mathcal{G}$, and consequently, LIST k-COLORING and hence k-LIST COLORING are NP-complete for $\mathcal{G}$. Conversely, if k-LIST COLORING is polynomial-time solvable on $\mathcal{G}$, then LIST k-COLORING is polynomial-time solvable on $\mathcal{G}$, and consequently, k-PRECOLORING EXTENSION is polynomial-time solvable on $\mathcal{G}$, and then also k-COLORING is polynomial-time solvable on $\mathcal{G}$.

Related and New Results. Král', Kratochvíl, Tuza and Woeginger [11] showed the following dichotomy for COLORING for H-free graphs.

Theorem 1 ([11]). *Let H be a fixed graph. If H is a (not necessarily proper) induced subgraph of P_4 or of $P_1 + P_3$, then COLORING can be solved in polynomial time for H-free graphs; otherwise it is NP-complete for H-free graphs.*

In Section 2 we use Theorem 1 and a number of other results from the literature to obtain the following two dichotomies, which complement Theorem 1. Theorem 3 shows amongst others that PRECOLORING EXTENSION is polynomial-time solvable on $(P_1 + P_3)$-free graphs, which contain the class of $3P_1$-free graphs, i.e., complements of triangle-free graphs. As such, this theorem also generalizes a result of Hujter and Tuza [8] who showed that PRECOLORING EXTENSION is polynomial-time solvable on complements of bipartite graphs.

Theorem 2. *Let ℓ be a fixed integer, and let H be a fixed graph. If $\ell \le 2$ or H is a (not necessarily proper) induced subgraph of P_3, then ℓ-LIST COLORING is polynomial-time solvable on H-free graphs; otherwise ℓ-LIST COLORING is NP-complete for H-free graphs.*

Theorem 3. *Let H be a fixed graph. If H is a (not necessarily proper) induced subgraph of P_4 or of $P_1 + P_3$, then PRECOLORING EXTENSION can be solved in polynomial time for H-free graphs; otherwise it is NP-complete for H-free graphs.*

In Section 3 we consider the LIST COLORING problem for $(3P_1, P_1 + P_2)$-free graphs, i.e., graphs that are obtained from a complete graph after removing the edges of some matching. We also call such a graph a *complete graph minus a matching*. Our motivation to study this graph class comes from the fact that LIST COLORING is NP-complete on almost all non-trivial graph classes, such as can be deduced from Theorem 2 and from other results known in the literature. For example, LIST COLORING is NP-complete for complete bipartite graphs [10], complete split graphs [10], line graphs of complete graphs [14], and more over, even for (not necessarily vertex-disjoint) unions of two complete graphs [9]; we refer to Table 1 in the paper by Bonomo, Durán and Marenco [1] for an overview. It is known that LIST COLORING can be solved in polynomial time for block graphs [9], which contain the class of complete graphs and trees. Our aim was to extend this positive result. However, as we show, already 3-LIST COLORING is NP-complete for complete graphs minus a matching. As a positive result, we show that LIST COLORING is fixed-parameter tractable for complete graphs minus a matching when parameterized by the number of matching edges removed.

In Section 4, we consider the LIST k-COLORING problem. As we explained, this problem is closely related to the problems k-COLORING and k-PRECOLORING EXTENSION. In contrast to COLORING and PRECOLORING EXTENSION (cf. Theorems 1 and 3), the complexity classifications of k-COLORING and k-PRECOLORING

Table 1. The complexity of k-COLORING, k-PRECOLORING EXTENSION and LIST k-COLORING on P_r-free graphs for fixed k and r. The bold entry is our new result.

	k-COLORING				k-PRECOLORING EXTENSION				LIST k-COLORING			
r	$k=3$	$k=4$	$k=5$	$k\geq 6$	$k=3$	$k=4$	$k=5$	$k\geq 6$	$k=3$	$k=4$	$k=5$	$k\geq 6$
$r\leq 5$	P	P	P	P	P	P	P	P	P	P	P	P
$r=6$	P	?	?	?	P	?	NP-c	NP-c	P	**NP-c**	NP-c	NP-c
$r=7$	?	?	?	NP-c	?	NP-c	NP-c	NP-c	?	NP-c	NP-c	NP-c
$r\geq 8$	?	NP-c	NP-c	NP-c	?	NP-c	NP-c	NP-c	?	NP-c	NP-c	NP-c

EXTENSION for H-free graphs are yet to be completed, even when H is a path. Hoàng et al. [6] showed that for any $k \geq 1$, the k-COLORING problem can be solved in polynomial time for P_5-free graphs. Randerath and Schiermeyer [16] showed that 3-COLORING can be solved in polynomial time for P_6-free graphs. These results are complemented by the following hardness results: 4-COLORING is NP-complete for P_8-free graphs [3] and 6-COLORING is NP-complete for P_7-free graphs [2]. Also the computational complexity of the LIST k-COLORING problem is still open for P_r-free graphs. Hoàng et al. [6] showed that their polynomial-time result on k-COLORING for P_5-free graphs is in fact valid for LIST k-COLORING for any fixed $k \geq 1$. Broersma et al. [2] generalized the polynomial-time result of Randerath and Schiermeyer [16] for 3-COLORING on P_6-free graphs to LIST 3-COLORING on P_6-free graphs. In addition, they showed that 5-PRECOLORING EXTENSION is NP-complete for P_6-free graphs [2], whereas 4-PRECOLORING EXTENSION is known to be NP-complete for P_7-free graphs [3]. Table 1 summarizes all existing results for these three problems restricted to P_r-free graphs. We prove that LIST 4-COLORING is NP-complete for P_6-free graphs. Because LIST 3-COLORING is polynomial-time solvable on P_6-free graphs [2], we completely characterized the computational complexity of LIST k-COLORING for P_6-free graphs. In Table 1 we indicate this result in bold. All cases marked by "?" in Table 1 are still open.

2 Classifying Precoloring Extension and 3-List Coloring

The following well-known lemma (cf. [1]) is obtained by modeling the LIST COLORING problem on n-vertex complete graphs with a k-list assignment as a maximum matching problem for an $(n + k)$-vertex bipartite graph; as such we may apply the Hopcroft-Karp algorithm [7] to obtain the bound on the running time.

Lemma 1. LIST COLORING *can be solved in* $O((n + k)^{\frac{5}{2}})$ *time on n-vertex complete graphs with a k-list assignment.*

We are now ready to state the proofs of Theorems 2 and 3.

The proof of Theorem 2. Early papers by Erdős, Rubin and Taylor [4] and Vizing [21] already observed that 2-LIST COLORING is polynomial-time solvable on general graphs. Hence, we can focus on the case $\ell \geq 3$. Because the ℓ-COLORING problem is a special case of the ℓ-LIST COLORING problem, the following results are useful. Kamiński and Lozin [13] showed that for any $k \geq 3$, the k-COLORING problem is NP-complete for the class of graphs of girth (the length of a shortest induced cycle) at least p for any fixed $p \geq 3$. Their result implies that for any $\ell \geq 3$, the ℓ-COLORING problem, and consequently, the ℓ-LIST COLORING problem is NP-complete for the class of H-free graphs whenever H contains a cycle. The proof of Theorem 4.5 in the paper by Jansen and Scheffler [10] is to show that 3-LIST COLORING is NP-complete on P_4-free graphs but as a matter of fact shows that 3-LIST COLORING is NP-complete on complete bipartite graphs, which are $(P_1 + P_2)$-free. The proof of Theorem 11 in the paper by Jansen [9] is to show that LIST COLORING is NP-complete for (not necessarily vertex-disjoint) unions of two complete graphs but as a matter of fact shows that 3-LIST COLORING is NP-complete for these graphs. As the union of two complete graphs is $3P_1$-free, this means that 3-LIST COLORING is NP-complete for $3P_1$-free graphs. This leaves us with the case when H is a (not necessarily proper) induced subgraph of P_3. By Lemma 1 we can solve LIST COLORING in polynomial time on complete graphs. This means that we can solve ℓ-LIST COLORING in polynomial time on P_3-free graphs for any $\ell \geq 1$. Hence we have proven Theorem 2.   □

The proof of Theorem 3. Let H be a fixed graph. If H is not an induced subgraph of P_4 or of $P_1 + P_3$, then Theorem 1 tells us that COLORING, and consequently, PRECOLORING EXTENSION is NP-complete for H-free graphs. Jansen and Scheffler [10] showed that PRECOLORING EXTENSION is polynomial-time solvable for P_4-free graphs. Hence, we are left with the case $H = P_1 + P_3$.

Let (G, k, c_W) be an instance of PRECOLORING EXTENSION, where G is a $(P_1 + P_3)$-free graph, k is an integer and $c_W : W \to \{1, \ldots, k\}$ for some $W \subseteq V(G)$ is a precoloring. We first prove how to transform (G, k, c_W) in polynomial time into a new instance $(G', k', c_{W'})$ with the following properties:

(i) G' is a $3P_1$-free subgraph of G, $k' \leq k$ and $c_{W'} : W' \to \{1, \ldots, k\}$ for some $W \subseteq W' \subseteq V(G)$ is a precoloring;

(ii) $(G', k', c_{W'})$ is a yes-instance if and only if (G, k, c_W) is a yes-instance.

Suppose that G is not $3P_1$-free already. Then G contains at least one triple T of three independent vertices. Let $u \in T$. Here we make the following choice if possible: if there exists a triple of three independent vertices that intersects with W, then we choose T to be such a triple and pick $u \in T \cap W$.

Let $S = V(G) \setminus (\{u\} \cup N(u))$. Because G is $(P_1 + P_3)$-free, $G[S]$ is the disjoint union of a set of complete graphs $D_1, \ldots, D_p$ for some $p \geq 2$; note that $p \geq 2$ holds, because the other two vertices of T must be in different graphs D_i and D_j. We will use the following claim.

Claim 1. Every vertex in $V(D_1) \cup \cdots \cup V(D_p)$ is adjacent to exactly the same vertices in $N(u)$.

We prove Claim 1 as follows. First suppose that w and w' are two vertices in two different graphs D_i and D_j, such that w is adjacent to some vertex $v \in N(u)$. Then w' is adjacent to v, as otherwise w' and u, v, w form an induced $P_1 + P_3$ in

G, which is not possible. Now suppose that w and w' are two vertices in the same graph D_i, say D_1, such that w is adjacent to some vertex $v \in N(u)$. Because $p \geq 2$, the graph D_2 is nonempty. Let w^* be in D_2. As we just showed, the fact that w is adjacent to v implies that w^* is adjacent to v as well. By repeating this argument with respect to w^* and w', we then find that w' is adjacent to v. Hence, we have proven Claim 1.

We now proceed as follows. First suppose that $u \in W$. By symmetry we may assume that $c_W(u) = k$. Then we assign color k to an arbitrary vertex of every D_i that does not contain a vertex colored with k already and that contains at least one vertex outside W. If $u \notin W$, then by our choice of u no vertex from $V(D_1) \cup \cdots \cup V(D_p)$ belongs to W. Either $c_W(W) = \{1, \ldots, k\}$, and we find (in polynomial time) that (G, k, c_W) is a no-instance, or $c_W(W) \subset \{1, \ldots, k\}$, and then we may assume that $c_W(W) \subseteq \{1, \ldots, k - 1\}$ by symmetry. In that case we assign color k to u and also to an arbitrary vertex of every D_i. Afterward, in both cases, we remove all vertices colored k from G. In both cases this leads to a new instance $(G', k - 1, c_{W'})$ that satisfies condition (i) except that G' may not be $3P_1$-free, and that satisfies condition (ii) due to Claim 1. We repeat this step until the graph is $3P_1$-free as claimed. Note that this takes polynomial time in total, because every step takes polynomial time and in every step the number of vertices of the graph reduces by at least 1.

Due to the above, we may assume without loss of generality that G is $3P_1$-free. We now apply the same algorithm as Hujter and Tuza [8] used for solving PRECOLORING EXTENSION on complements of bipartite graphs. Because G is $3P_1$-free, G has no three mutually nonadjacent vertices. Suppose that u and v are two nonadjacent vertices in W. Then every vertex of $V(G) \setminus \{u, v\}$ is adjacent to at least one of $\{u, v\}$. This means that we can remove u, v if they are both colored alike by c_W in order to obtain a new instance $(G - \{u, v\}, k - 1, c_{W \setminus \{u,v\}})$ that is a yes-instance of PRECOLORING EXTENSION if and only if (G, k, c_W) is a yes-instance. If u and v are colored differently by c_W, then we add an edge between them. We perform this step for any pair of non-adjacent vertices in W. Afterward, we have found in polynomial time a new instance (G^*, k^*, c_{W^*}) with the following properties. First, $|V(G^*)| \leq |V(G)|$, $k^* \leq k$ and $c_{W^*} : W^* \to \{1, \ldots, k\}$ is a precoloring defined on some clique W^* of G^*. Second, (G^*, k^*, c_{W^*}) is a yes-instance if and only if (G, k, c_W) is a yes-instance. Hence, we may consider (G^*, k^*, c_{W^*}) instead. Because W^* is a clique, we find that (G^*, k^*, c_{W^*}) is a yes-instance if and only if G^* is k^*-colorable. Because G^* is $3P_1$-free, we can solve the later problem by using Theorem 1 (which in this case comes down to computing the size of a maximum matching in the complement of G^*). This completes the proof for the case $H = P_1 + P_3$. Consequently, we have proven Theorem 3.   □

3   List Coloring for Complete Graphs Minus a Matching

We prove that 3-LIST COLORING is NP-complete for complete graphs minus a matching. In order to this we use a reduction from a variant of NOT-ALL-EQUAL 3-SATISFIABILITY with positive literals only, which we denote as

NOT-ALL-EQUAL ($\leq 3, 2/3$)-SATISFIABILITY with positive literals. The NOT-ALL-EQUAL 3-SATISFIABILITY problem is NP-complete [18] and is defined as follows. Given a set $X = \{x_1, x_2, ..., x_n\}$ of logical variables, and a set $C = \{C_1, C_2, ..., C_m\}$ of three-literal clauses over X in which all literals are positive, does there exist a truth assignment for X such that each clause contains at least one true literal and at least one false literal? The variant NOT-ALL-EQUAL ($\leq 3, 2/3$)-SATISFIABILITY with positive literals asks the same question but takes as input an instance I that has a set of variables $\{x_1, \ldots, x_n\}$ and a set of literal clauses $\{C_1, \ldots, C_m\}$ over X with the following properties. Each C_i contains either 2 or 3 literals, and these literals are all positive. Moreover, each literal occurs in at most three different clauses. One can prove that NOT-ALL-EQUAL ($\leq 3, 2/3$)-SATISFIABILITY is NP-complete by a reduction from NOT-ALL-EQUAL-3-SATISFIABILITY via a well-known folklore trick.

Let I be an arbitrary instance of NOT-ALL-EQUAL ($\leq 3, 2/3$)-SATISFIABILITY with positive literals. We let $x_1, x_2, \ldots, x_n$ be the variables of I, and we let $C_1, C_2, \ldots, C_m$ be the clauses of I. We define a graph G_I with a list assignment L of size three in the following way. We represent every variable x_i by a vertex with $L(x_i) = \{1_i, 2_i\}$ in G_I. We say that these vertices are of x-type and these colors are of 1-type and 2-type, respectively. For every clause C_p with two variables we fix an arbitrary order of its variables x_h, x_i and we introduce a set of vertices $C_p, a_{p,h}, a_{p,i}, b_{p,h}, b_{p,i}$ that have lists of admissible colors $\{3_p, 4_p\}$, $\{1_h, 3_p\}$, $\{1_i, 4_p\}$, $\{2_h, 4_p\}$, $\{2_i, 3_p\}$, respectively, and we add edges $C_p a_{p,h}$, $C_p b_{p,h}$, $C_p a_{p,i}$, $C_p b_{p,i}$, $a_{p,h} x_h$, $b_{p,h} x_h$, $a_{p,i} x_i$, $b_{p,i} x_i$. For every clause C_p with three variables we fix an arbitrary order of its variables x_h, x_i, x_j and we introduce a set of vertices $C_p, a_{p,h}, a_{p,i}, a_{p,j}, b_{p,h}, b_{p,i}, b_{p,j}$ that have lists of admissible colors $\{3_p, 4_p, 5_p\}$, $\{1_h, 3_p\}$, $\{1_i, 4_p\}$, $\{1_j, 5_p\}$, $\{2_h, 5_p\}$, $\{2_i, 3_p\}$, $\{2_j, 4_p\}$, respectively, and we add edges $C_p a_{p,h}$, $C_p b_{p,h}$, $C_p a_{p,i}$, $C_p b_{p,i}$, $C_p a_{p,j}$, $C_p b_{p,j}$, $a_{p,h} x_h$, $b_{p,h} x_h$, $a_{p,i} x_i$, $b_{p,i} x_i$, $a_{p,j} x_j$, $b_{p,j} x_j$. We say that the new vertices are of C-type, a-type and b-type, respectively. We say that the new colors are of 3-type, 4-type and 5-type, respectively. For each variable x_j that occurs in three clauses we fix an arbitrary order of the clauses C_p, C_q, C_r, in which it occurs. Then we do as follows. First, we modify the lists of $a_{p,j}$, $a_{q,j}$, $b_{p,j}$ and $b_{q,j}$. In $L(a_{p,j})$ we replace color 1_j with a new color $1'_j$. In $L(a_{q,j})$ we replace color 1_j with a new color $1''_j$. In $L(b_{p,j})$ we replace color 2_j with a new color $2'_j$. In $L(b_{q,j})$ we replace color 2_j with a new color $2''_j$. Next we introduce four vertices $a'_{p,j}, a'_{q,j}, b'_{p,j}, b'_{q,j}$ with lists of admissible colors $\{1_j, 1'_j\}$, $\{1'_j, 1''_j\}$, $\{2_j, 2'_j\}$, $\{2'_j, 2''_j\}$, respectively. We say that these vertices are of a'-type or b'-type, respectively. We say that the new colors are also of 1-type or 2-type, respectively. We add edges $a_{p,j} a'_{p,j}$, $a'_{p,j} a'_{q,j}$, $a'_{p,j} x_j$, $a_{q,j} a'_{q,j}$, $b_{p,j} b'_{p,j}$, $b'_{p,j} b'_{q,j}$, $b'_{p,j} x_j$, $b_{q,j} b'_{q,j}$. We add an edge between any two not yet adjacent vertices of G_I whenever they have no common color in their lists. In Figure 1 we give an example, where in order to increase the visibility we display the complement graph $\overline{G_I}$ of G_I instead of G_I itself.

As can be seen from Figure 1, the graph $\overline{G_I}$ is isomorphic to the disjoint union of a number of P_1s and P_2s. This means that G_I is a complete graph minus a

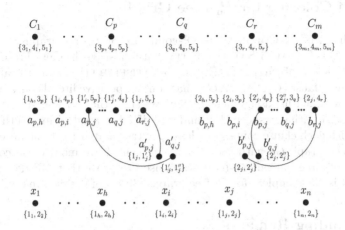

Fig. 1. An example of a graph $\overline{G_I}$ in which a clause C_p and a variable x_j are highlighted. Note that in this example C_p is a clause with ordered variables x_h, x_i, x_j, and that x_j is a variable contained in ordered clauses C_p, C_q and C_r.

matching. This leads us to Lemma 2, whereas the hardness reduction is stated in Lemma 3. The proofs of both lemmas have been omitted.

Lemma 2. *The graph G_I is a complete graph minus a matching.*

Lemma 3. *The graph G_I has a coloring that respects L if and only if I has a satisfying truth assignment in which each clause contains at least one true and at least one false literal.*

Recall that complete graphs minus a matching are exactly those graphs that are $(3P_1, P_1 + P_2)$-free, or equivalently, graphs of minimum degree at least $n - 2$, where n is the number of vertices. By observing that 3-LIST COLORING belongs to NP and using Lemmas 2 and 3, we have proven Theorem 4.

Theorem 4. *The* 3-LIST COLORING *problem is* NP-*complete for complete graphs minus a matching.*

To complement Theorem 4 we finish this section with the next result, which has as a consequence that LIST COLORING problem is fixed-parameter tractable on complete graphs minus a matching when parameterized by the number of removed matching edges, or equivalently, for n-vertex graphs G of minimum degree at least $n - 2$ when parameterized by the number of vertices of degree $n - 2$. The proof of Theorem 5 uses Lemma 1; we omit the details.

Theorem 5. *The* LIST COLORING *problem can be solved in $O(2^p(n+k)^{\frac{5}{2}})$ time on pairs (G, L) where G is an n-vertex graph with p pairs of non-adjacent vertices and L is a k-list assignment.*

4   List 4-Coloring for P_6-Free Graphs

To prove that LIST 4-COLORING is NP-complete for P_6-free graphs we reduce from NOT-ALL-EQUAL 3-SATISFIABILITY with positive literals. From an arbitrary instance I of NOT-ALL-EQUAL 3-SATISFIABILITY with variables x_1, $x_2, \ldots, x_n$ and clauses $C_1, C_2, \ldots, C_m$ that contain positive literals only, we build a graph G_I with a 4-list assignment L. Next we show that G_I is P_6-free and that G_I has a coloring that respects L if and only if I has a satisfying truth assignment in which each clause contains at least one true and at least one false literal. To obtain the graph G_I with its 4-list assignment L we modify the construction of the (P_7-free but not P_6-free) graph used to prove that 4-PRECOLORING EXTENSION is NP-complete for P_7-free graphs [3]; proof details are omitted.

5   Concluding Remarks

The main tasks are to determine the computational complexity of COLORING for AT-free graphs and to solve the open cases marked "?" in Table 1. This table shows that so far all three problems k-COLORING, k-PRECOLORING EXTENSION and LIST k-COLORING behave similarly on P_r-free graphs. Hence, our new NP-completeness result on LIST 4-COLORING for P_6-free graphs may be an indication that 4-COLORING for P_6-free graphs is NP-complete, or otherwise at least this result makes clear that new proof techniques not based on subroutines that solve LIST 4-COLORING are required for proving polynomial-time solvability.

Another open problem, which is long-standing, is to determine the computational complexity of the COLORING problem for the class of asteroidal triple-free graphs, also known as *AT-free* graphs. An *asteroidal triple* is a set of three mutually non-adjacent vertices such that each two of them are joined by a path that avoids the neighborhood of the third, and AT-free graphs are exactly those graphs that contain no such triple. We note that unions of two complete graphs are AT-free. Hence NP-completeness of 3-LIST COLORING for this graph class [9] immediately carries over to AT-free graphs. Stacho [19] showed that 3-COLORING is polynomial-time solvable on AT-free graphs. Recently, Kratsch and Müller [12] extended this result by proving that LIST k-COLORING is polynomial-time solvable on AT-free graphs for any fixed positive integer k. Marx [15] showed that PRECOLORING EXTENSION is NP-complete for proper interval graphs, which form a subclass of AT-free graphs. An *asteroidal set* in a graph G is an independent set $S \subseteq V(G)$, such that every triple of vertices of S forms an asteroidal triple. The *asteroidal number* is the size of a largest asteroidal set in G. Note that complete graphs are exactly those graphs that have asteroidal number at most one, and that AT-free graphs are exactly those graphs that have asteroidal number at most two. We observe that COLORING is NP-complete for the class of graphs with asteroidal number at most three, as this class contains the class of $4P_1$-free graphs and for the latter graph class one may apply Theorem 1.

References

1. Bonomo, F., Durán, G., Marenco, J.: Exploring the complexity boundary between coloring and list-coloring. Ann. Oper. Res. 169, 3–16 (2009)
2. Broersma, H., Fomin, F.V., Golovach, P.A., Paulusma, D.: Three Complexity Results on Coloring P_k-Free Graphs. In: Fiala, J., Kratochvíl, J., Miller, M. (eds.) IWOCA 2009. LNCS, vol. 5874, pp. 95–104. Springer, Heidelberg (2009)
3. Broersma, H.J., Golovach, P.A., Paulusma, D., Song, J.: Updating the complexity status of coloring graphs without a fixed induced linear forest. Theoretical Computer Science 414, 9–19 (2012)
4. Erdös, P., Rubin, A.L., Taylor, H.: Choosabilty in graphs. In: Proc. West Coast Conference on Combinatorics, Graph Theory and Computing, pp. 125–157 (1979)
5. Golovach, P.A., Paulusma, D., Song, J.: 4-Coloring H-Free Graphs When H Is Small. In: Bieliková, M., Friedrich, G., Gottlob, G., Katzenbeisser, S., Turán, G. (eds.) SOFSEM 2012. LNCS, vol. 7147, pp. 289–300. Springer, Heidelberg (2012)
6. Hoàng, C.T., Kamiński, M., Lozin, V., Sawada, J., Shu, X.: Deciding k-colorability of P_5-free graphs in polynomial time. Algorithmica 57, 74–81 (2010)
7. Hopcroft, J.E., Karp, R.M.: An $n^{\frac{5}{2}}$ algorithm for maximum matchings in bipartite graphs. SIAM J. Comput. 2, 225–231 (1973)
8. Hujter, M., Tuza, Z.: Precoloring extension. II. Graph classes related to bipartite graphs. Acta Math. Univ. Comenianae LXII, 1–11 (1993)
9. Jansen, K.: Complexity Results for the Optimum Cost Chromatic Partition Problem. Universität Trier, Mathematik/Informatik, Forschungsbericht, pp. 96–41 (1996)
10. Jansen, K., Scheffler, P.: Generalized coloring for tree-like graphs. Discrete Appl. Math. 75, 135–155 (1997)
11. Král', D., Kratochvíl, J., Tuza, Z., Woeginger, G.J.: Complexity of Coloring Graphs without Forbidden Induced Subgraphs. In: Brandstädt, A., Le, V.B. (eds.) WG 2001. LNCS, vol. 2204, pp. 254–262. Springer, Heidelberg (2001)
12. Kratsch, D., Müller, H.: Colouring AT-Free Graphs. In: Epstein, L., Ferragina, P. (eds.) ESA 2012. LNCS, vol. 7501, pp. 707–718. Springer, Heidelberg (2012)
13. Kamiński, M., Lozin, V.V.: Coloring edges and vertices of graphs without short or long cycles. Contributions to Discrete Math. 2, 61–66 (2007)
14. Kubale, M.: Some results concerning the complexity of restricted colorings of graphs. Discrete Applied Mathematics 36, 35–46 (1992)
15. Marx, D.: Precoloring extension on unit interval graphs. Discrete Applied Mathematics 154, 995–1002 (2006)
16. Randerath, B., Schiermeyer, I.: 3-Colorability $\in$ P for P_6-free graphs. Discrete Appl. Math. 136, 299–313 (2004)
17. Randerath, B., Schiermeyer, I.: Vertex colouring and forbidden subgraphs - a survey. Graphs Combin. 20, 1–40 (2004)
18. Schaefer, T.J.: The complexity of satisfiability problems. In: Proc. STOC 1978, pp. 216–226 (1978)
19. Stacho, J.: 3-Colouring AT-free graphs in polynomial time. Algorithmica 64, 384–399 (2012)
20. Tuza, Z.: Graph colorings with local restrictions - a survey. Discuss. Math. Graph Theory 17, 161–228 (1997)
21. Vizing, V.G.: Coloring the vertices of a graph in prescribed colors. Diskret. Analiz., no. 29, Metody Diskret. Anal. v. Teorii Kodov i Shem 101, 3–10 (1976)

Randomly Coloring Regular Bipartite Graphs and Graphs with Bounded Common Neighbors

Ching-Chen Kuo[1] and Hsueh-I Lu[2,*]

[1] Department of Computer Science and Information Engineering,
National Taiwan University, Taipei, Taiwan
`r96922074@ntu.edu.tw`
[2] Department of Computer Science and Information Engineering,
National Taiwan University, Taipei, Taiwan
`hil@csie.ntu.edu.tw`

Abstract. Let G be an n-node graph with maximum degree Δ. The *Glauber dynamics* for G, defined by Jerrum, is a Markov chain over the k-colorings of G. Many classes of G on which the Glauber dynamics mixes rapidly have been identified. Recent research efforts focus on the important case that $\Delta \geq d \log_2 n$ holds for some sufficiently large constant d. We add the following new results along this direction, where ϵ can be any constant with $0 < \epsilon < 1$.

- Let $\alpha \approx 1.645$ be the root of $(1 - e^{-1/x})^2 + 2xe^{-1/x} = 2$. If G is regular and bipartite and $k \geq (\alpha + \epsilon)\Delta$, then the mixing time of the Glauber dynamics for G is $O(n \log n)$.
- Let $\beta \approx 1.763$ be the root of $x = e^{1/x}$. If the number of common neighbors for any two adjacent nodes of G is at most $\frac{\epsilon^{1.5}\Delta}{360e}$ and $k \geq (1 + \epsilon)\beta\Delta$, then the mixing time of the Glauber dynamics is $O(n \log n)$.

1 Introduction

For any finite set S, let $|S|$ denote the cardinality of S. Let G be a simple undirected graph on a set V of n nodes. For each node v of G, let $N(v)$ consist of the neighbors of v in G. For each node subset S of G, let $N(S) = \bigcup_{v \in S} N(v)$. Let $\Delta = \max_{v \in V} |N(v)|$. Let k be a positive integer. Let $K = \{1, 2, \ldots, k\}$. A k-coloring of G is a mapping from V to a color in K such that any two adjacent nodes of G map to different colors. Let Ω consist of all k-colorings of G.

Markov Chain Monte Carlo is an important tool in sampling from complex distributions such as the uniform distribution on k-coloring. It has been successfully applied in several areas of Computer Science and more details can be found in Frieze and Vigoda [6]. Adopting terminology from statistical physics,

---

* This author also holds joint appointments in the Graduate Institute of Networking and Multimedia and the Graduate Institute of Biomedical Electronics and Bioinformatics, National Taiwan University, Taipei, Taiwan. Web: `www.csie.ntu.edu.tw/~hil`

K.-M. Chao, T.-s. Hsu, and D.-T. Lee (Eds.): ISAAC 2012, LNCS 7676, pp. 24–33, 2012.

Table 1. The currently known classes of G on which the Glauber dynamics for the k-colorings of G mixes rapidly, where α, β, and γ are some constants with $\beta > \alpha > \gamma$ and ϵ can be any constant with $0 < \epsilon < 1$. An asymptotic bound $x = \Omega(f(n))$ stands for the condition that there exists a sufficiently large constant c such that $x \geq c \cdot f(n)$ holds.

	degree Δ	girth g	number k of colors	additional constraints
Hayes et al. [11]	$\Omega(1)$	$g \geq 3$	$k = \Omega(\frac{\Delta}{\log \Delta})$	G is planar
Frieze and Vera [5]	$\Omega(\log n)$	$g \geq 3$	$k \geq (\beta + \epsilon)\Delta$	$\chi \leq \epsilon^2 \Delta/10$
Hayes and Vigoda [13]	$\Omega(\log n)$	$g \geq 4$	$k \geq (1+\epsilon)\beta\Delta$	
Hayes [9]	$\Omega(\log n)$	$g \geq 5$	$k > \beta\Delta$	
	$\Omega(\log n)$	$g \geq 6$	$k > \gamma\Delta$	
Hayes and Vigoda [7]	$\Omega(\log n)$	$g \geq 9$	$k \geq (1+\epsilon)\Delta$	
Lau and Molloy [19]	$\Omega(\log^3 n)$	$g \geq 5$	$k \geq (\alpha + \epsilon)\Delta$	
This paper	$\Omega(\log n)$	$g \geq 4$	$k \geq (\alpha + \epsilon)\Delta$	G is bipartite and regular
	$\Omega(\log n)$	$g \geq 3$	$k \geq (1+\epsilon)\beta\Delta$	$\xi \leq \frac{\epsilon^{1.5}\Delta}{360e}$

Jerrum [14] defined the *Glauber dynamics* on G as the Markov chain over Ω whose transition is (1) choosing a node v uniformly at random from V, and then (2) choosing a color uniform at random from $K \setminus K'$ for v, where K' consists of the colors of $N(v)$. The *mixing time* $\tau(G)$ for the Glauber dyanmics X on G is the time until the Glauber dynamics is close enough to its stationary distribution. We define this formally in the next section. Hayes and Sinclair [10] established a lower bound $\Omega(n \log n)$ on the mixing time $\tau(G)$ of the Glauber dynamics for G. If the Glauber dynamics mixes rapidly on G, i.e., $\tau(G)$ is bounded by a polynomial in the number n of nodes in G for sufficiently large n, then the number of k-colorings of G can be estimated in polynomial time (see, e.g., Jerrum, Valiant, and Vazirani [16].)

The classes of G on which the Glauber dynamics mixes rapidly have been extensively studied in the literature. Jerrum [14] proved $\tau = O(n \log n)$ for G with $k > 2\Delta$, whose proof was later simplified by Bubley and Dyer [1]. Vigoda [22] showed $\tau = O(n^2)$ for G with $k > 11\Delta/6$.

Dyer and Frieze [2] initiated the study for the important case that $\Delta \geq d \log n$ holds for some sufficiently large constant d. Let $\beta \approx 1.763$ be the root of $x = e^{1/x}$, Let $\alpha \approx 1.645$ be the root of $(1 - e^{-1/x})^2 + 2xe^{-1/x} = 2$. Let $\gamma \approx 1.489$ be the root of $(1 - e^{-1/x})^2 + xe^{-1/x} = 1$. Let ξ be the number of common neighbors for any adjacent nodes of G. Let χ be the average degree of the subgraph G induced by $N(v)$ over all nodes v of G. Let g be the girth of G, i.e., the length of a shortest cycle of G. Table 1 compares our results, as stated in the following theorem, with the best currently known results on the problem.

Theorem 1. *Let G be an n-node graph such that the maximum degree Δ of G is at least $d\log_2 n$ for some sufficiently large constant d. Let k be a positive* *number. The mixing time of the Glauber dynamics for the k-colorings of G is $O(n \log n)$ for the following cases, where ϵ can be any constant with $0 < \epsilon < 1$.*

1. *G is bipartite and regular and satisfies $k \geq (\alpha + \epsilon)\Delta$.*
2. *G satisfies $\xi \leq \frac{\epsilon^{1.5}\Delta}{360e}$ and $k \geq (1 + \epsilon)\beta\Delta$.*

Technical Overview. For the rest of our paper, we use X to denote the sequence X_t, with $t \geq 0$, of random variables of the Glauber dynamics over the k-colorings Ω of G. Let $[\cdot]$ denote the event indicator, i.e., $[true] = 1$ and $[false] = 0$. Let $v \in V$ and y is a k-coloring of G. We define $y(v)$ as the color assigned to v under y coloring and
$$y(S) = \{y(v) \mid v \in S\}.$$
Let $v \in V$, $c \in K$,
$$A(y, v) = K \setminus y(N(v)),$$
i.e., the set of available colors for v, and
$$T_y(v, c) = \sum_{w \in N(v)} \frac{[c \in A_v^*(y, w)]}{|A_v^*(y, w)|},$$
where $A_v^*(y, w) = K \setminus y(N(w) \setminus \{v\})$. Ignoring the color of v, $T_y(v, c)$ is the expected number of occurences of a color c in the neighborhood of v. Note that for every coloring y and vertex v,
$$\sum_{c \in K} T_y(v, c) = |N(v)|.$$

For a k-coloring y of G, suppose that we recolor $v \in V$ by $c \in A(y, v)$. Define
$$R_y(v, c) = \{w \in N(v) \mid \{y(v), c\} \subseteq y(N(w) \setminus \{v\})\}$$
as the set of v's neighbors and some of their neighbors were assigned with colors c or $y(v)$ under y coloring. Let $R_y = \min_{v \in V, c \in A(y,v)} |R_y(v, c)|$. For the first part of Theorem 1, we show that with high probability, $T_y(v, c) \approx |N(v)|/k$, for all $y \in \Omega$, $v \in V$, and $c \in K$ also holds when G is regular and bipartite. With this property, we can estimate the lower bound of $|A(y, v)|$ and R_y with high probability for any $y \in \Omega$ and $v \in V$. These two bounds help to prove the rapid mixing of X on G and we show how to improve Hayes and Vigoda [13] by less colors used. For the second part of Theorem 1, we utilize the procedure of Frieze and Vera [5] to analyze the bound of $|A(y, v)|$ and show that the colors assigned to the neighbors are nearly independent. Note that
$$|A(y, v)| = \sum_{j \in K} \prod_{w \in N(v)} (1 - [y(w) = j]).$$

In a triangle-free graph, the neighbors of v receive colors independently since each neighbor of v is not adjacent to any one else. Therefore, Vigoda and Hayes [13] can compute the expectation of $|A(y, v)|$ directly because of the independence of random variables $y(w)$, for all $w \in N(v)$. However, even if only a bit more neighbors are adjacent, the random variables $y(w)$, for all $w \in N(v)$ are no longer independent. As a result, the technique of Vigoda and Hayes cannot bound the available colors of v in this situation. In this paper, we show how to handle the above situation if any two adjacent nodes of G do not have many common neighbors.

Related Work. See [18,15,12] for more results on the Glauber dynamics for the k-colorings of G. Vigoda [22] also introduced an alternative Markov chain whose mixing time is $O(n \log n)$ for G with $k > 11\Delta/6$. Hayes [8] provided useful techniques for proving that the Glauber dynamics mixes rapidly. The Glauber dynamics is also studied on different models, such as Potts model, Ising model, and solid-on-solid model (see, e.g., [20,21]).

In this paper, we omit some proofs due to the page limit. Those omitted proofs can be found in the full version.

2   Preliminaries

Frist of all, we introduce the notion of *Markov chain*. For a finite state space Ω, a sequence of random variables(X_t) on Ω is a *Markov chain* if for all t, all $x_0, ..., x_t, y \in \Omega$,

$$Pr(X_{t+1} = y)|X_0 = x_0, X_1 = x_1, ..., X_t = x_t) = Pr(X_{t+1} = y|X_t = x_t).$$

A *Markov chain* is called *ergodic* if there exists t such that for all $x, y \in \Omega$, $P^t(x, y) > 0$, where $P^t(x, y)$ is the t-step distribution from x to y. Ergodic Markov chains are useful algorithmic tools because they eventually reach a unique stationary distribution. Therefore, we can design a approximate samplers by designing Markov chains with appropriate stationary distribution. Jerrum [14] showed that the uniform distribution over Ω is the unique stationary distribution of the Glauber dynamics. Next, we introduce several tools used in our work. For any distributions u and v, variance distance between u and v is defined as

$$d_{TV}(u, v) = \frac{1}{2} \sum_{x \in \Omega} |u(x) - v(x)|.$$

Let π be the stationary distribution of u. The mixing time of u, $\tau(\epsilon)$, is defined as

$$\tau(\epsilon) = max_{x_0 \in \Omega} min\{t : d_{TV}(P^t(x_0, x), \pi) \le \epsilon\},$$

for any $x \in \Omega$ and $\epsilon > 0$ is a sufficiently small. Consider two processes X and Y, where X_0 is chosen arbitrary and Y_0 is chosen uniformly at random from Ω. We define the set of disagreements between X_t and Y_t as

$$D_t = \{v \in V \mid X_t(v) \neq Y_t(v)\}$$

Algorithm 1. The transition of Jerrum's coupling in [17].

1. Choose a node v from V uniformly at random.
2. Choose a pair of colors (c_1, c_2) according to the joint distribution σ over $(K \setminus X_t(N(v))) \times (K \setminus Y_t(N(v)))$. The joint distribution σ should satisfy the following two conditions.
   (a) The distribution of c_1 (respectively, c_2) should be uniform over $K \setminus X_t(N(v))$ (respectively, $K \setminus Y_t(N(v))$).
   (b) σ should be chosen so as to maximize $\Pr[c_1 = c_2] = \frac{1}{\max\{|A(X_t, v))|, |A(Y_t, v)|\}}$.
3. Let $X_{t+1}(v) = c_1$ and $Y_{t+1}(v) = c_2$.

---

and the Hamming distance between X_t and Y_t is $\rho(X_t, Y_t) = |D_t|$. Let $exp(x) = e^x$. Let $P_v(x, y) = A(x, v) \setminus A(y, v)$ for any $x, y \in \Omega$.

Lemma 1 (Lau and Molloy [19, Lemma 1]). *If G is regular, then*
$\sum_{v \in V} \max\{|P_v(x, y)|, |P_v(y, x)|\} \leq \left(1 - \frac{R_y}{2\Delta}\right) \Delta\rho(x, y)$ *holds for any two colorings x and y in Ω.*

In our paper, we use Jerrum's coupling on k-colorings [17, Figure 2 of Chapter 4] and it has been a primary tool for bounding the mixing time of X for sampling k-colorings (see e.g., [6,13,5,14,9,4,2,3,19]). Starting from (X_0, Y_0), Jerrum's coupling for X moves from (X_t, Y_t) to (X_{t+1}, Y_{t+1}) by the transition shown in Algorithm 1. For $0 < \delta < 1$, we say that a pair $(x, y) \in \Omega \times \Omega$ is δ *distance decreasing* if there exist X and Y over Ω such that

$$E[\rho(X_{t+1}, Y_{t+1}) \mid X_t = x, Y_t = y] \leq (1 - \delta) \cdot \rho(x, y)$$

holds for all $t \geq 0$. The following property is useful in bounding mixing time of X.

Lemma 2 (Hayes and Vigoda [13, Theorem 1.2]). *Let $diam(\Omega) = \max_{x, y \in \Omega} \rho(x, y)$, and $0 < \epsilon, \delta < 1$. Suppose that $S \subseteq \Omega$ such that every $(x, y) \in \Omega \times S$ is δ distance decreasing, where $\frac{|S|}{|\Omega|} \geq 1 - \frac{\delta}{16diam(\Omega)}$. Then, $\tau_{mix}(\epsilon) \geq \frac{\lceil \log(32diam(\Omega)) \rceil \lceil \log(1/\epsilon) \rceil}{\delta}$.*

According to Lemm 2, we can achieve the rapid mixing time of X by finding a coupling over $\Omega \times S$ of X and the uniform stationary distribution π such that every pair $(x, y) \in \Omega \times S$ is δ distance decreasing as well as the requirement of S is satisfied.

3   Rapid Mixing on Regular Bipartite Graphs

Lemma 3. *Let y be chosen uniformly at random from Ω and $k \geq (\alpha+\epsilon)\Delta$. Let η be a number with $\eta \leq \frac{\ln(\alpha+\epsilon)}{10}$. Let δ be a number with $\delta \leq \min\{\frac{(1-e^{-\eta^2})e^{-e}}{\alpha+\epsilon}, \frac{\sqrt{5}\eta}{\epsilon}\}$.*

Let t be a number with

$$t \geq \max\left\{ 2e^{\eta}e^{\frac{-\Delta}{k}} + \frac{2}{k} \cdot \left(e^{\frac{-\Delta}{k}} - \eta + \frac{e \cdot (100 + 100\eta(\alpha + \epsilon))}{\alpha + \epsilon} \right), \frac{2}{\sqrt{5}} \cdot \frac{\delta\epsilon}{\sqrt{\alpha + \epsilon}} \right\}.$$

Let

$$m = \left(e^{\left(-\left(\frac{k-\Delta}{k-\Delta-1}-1\right)\cdot\frac{2\Delta}{k} - 2\eta\cdot\frac{k-\Delta}{k-\Delta-1}\right)} - 1 \right) \cdot e^{-2\frac{\Delta}{k}} -$$

$$\left(e^{\eta + \frac{1}{k-\Delta}} - 1 \right) \cdot 2e^{-\frac{\Delta}{k}} - \frac{\Delta - 1}{2(k - \Delta - 1)^2} \cdot 2e^{\frac{-(\Delta-1)}{k-\Delta-1}}.$$

If there exists a positive constant $d \geq 50/\delta^2\epsilon^2$ such that $\Delta \geq d\ln n$, then for all $v \in V$ and $c \in A(y, v)$, with probability at least $1 - n^{-4}$, we have

1. $||A(y, v)| - ke^{-\Delta/k}| \leq tk$,
2. $|R_y(v, c)| \geq ((1 - exp(-\frac{\Delta}{k}))^2 + m - \frac{\delta}{2})\Delta$.

3.1 Proof of the First Part of Theorem 1

Proof. Let

$$\eta = \min\left\{ e^{-100}, \frac{\ln(\alpha + \epsilon)}{10} \right\};$$

$$\delta = \min\left\{ \frac{(1 - e^{-\eta^2})e^{-e}}{\alpha + \epsilon}, \frac{\sqrt{5}\eta}{\epsilon} \right\};$$

$$t = \max\left\{ 2(e^{\eta} - 1)e^{\frac{-\Delta}{k}} + \frac{2\left(e^{\frac{-\Delta}{k}} - \eta - 1 + \frac{e\cdot(100+100\eta(\alpha+\epsilon))}{\alpha+\epsilon} \right)}{k}, \frac{2}{\sqrt{5}} \cdot \frac{\delta\epsilon}{\sqrt{\alpha + \epsilon}} \right\};$$

$$m = \left(e^{-\left(\frac{k-\Delta}{k-\Delta-1}-1\right)\cdot\frac{2\Delta}{k} - 2\eta\cdot\frac{k-\Delta}{k-\Delta-1}} - 1 \right) \cdot e^{-2\frac{\Delta}{k}} - \left(e^{\eta + \frac{1}{k-\Delta}} - 1 \right) \cdot 2e^{-\frac{\Delta}{k}}$$

$$- \frac{\Delta - 1}{2(k - \Delta - 1)^2} \cdot 2e^{\frac{-(\Delta-1)}{k-\Delta-1}}.$$

Let S consist of the colorings y in Ω such that the following conditions hold for any node $v \in V$ and any color $c \in A(y, v)$:

- $|A(y, v)| \geq ke^{-\Delta/k} - tk.$
- $|R_y(v, c)| \geq ((1 - e^{-\frac{\Delta}{k}})^2 + m - \frac{\delta}{2}\Delta.$

According to Lemma 3, if there exists a positive constant $d \geq 50/\delta^2\epsilon^2$ such that $\Delta \geq d\ln n$, then for any constant $\zeta > 0$,

$$\frac{|S|}{|\Omega|} \geq 1 - \frac{1}{n^4} \geq 1 - \frac{\zeta}{16n^2} = 1 - \frac{\frac{\zeta}{n}}{16 \cdot diam(\Omega)}$$

holds for sufficiently large n. It remains to prove that there exists a constant $\zeta > 0$ such that every pair $(x, y) \in \Omega \times S$ is $\frac{\zeta}{n}$-distance decreasing under Jerrum's coupling. Let $L_y = \min_{v \in V} |A(y, v)|$. Let $v \in V$ be the node chosen at step $t+1$. By Step 2(b) in Algorithm 1,

$$\Pr[X_{t+1}(v) = Y_{t+1}(v)] = \frac{1}{\max\{|A(X_t, v))|, |A(Y_t, v)|\}}.$$

Therefore, the probability that X and Y assign different colors to v is

$$\Pr[X_{t+1}(v) \neq Y_{t+1}(v)] = \frac{\max\{|P_v(X_t, Y_t)|, |P_v(Y_t, X_t)|\}}{\max\{|A(X_t, v))|, |A(Y_t, v)|\}},$$

where the last equality is by definition of P_v. Hence,

$$
\begin{aligned}
& E[\rho(X_{t+1}, Y_{t+1}) \mid X_t = x, Y_t = y] \\
&= \sum_{u \in V} \Pr[X_{t+1}(u) \neq Y_{t+1}(u) \mid X_t = x, Y_t = y] \\
&= \frac{n-1}{n}\rho(x, y) + \sum_{u \in V} \frac{1}{n} \cdot \frac{\max\{|P_v(x, y)|, |P_v(y, x)|\}}{\max\{|A(x, v))|, |A(y, v)|\}} \\
&\leq \frac{n-1}{n}\rho(x, y) + \frac{1}{nL_y}\left(1 - \frac{R_y}{2\Delta}\right)\Delta\rho(x, y) \qquad (1) \\
&\leq \frac{n-1}{n}\rho(x, y) + \frac{1}{n} \cdot \frac{2 - \left(1 - e^{-\frac{1}{\alpha+\epsilon}}\right)^2 - m + \frac{\delta}{2}}{2(\alpha+\epsilon)e^{\frac{-1}{\alpha+\epsilon}} - 2t(\alpha+\epsilon)}\rho(x, y),
\end{aligned}
$$

where (1) follows from Lemma 1. Recall that α is the root of $(1 - e^{-1/x})^2 + 2xe^{-1/x} = 2$ and $(1 - e^{-1/x})^2 + 2xe^{-1/x} > 2$ for all $x > \alpha$. Since $\Delta \geq d \ln n$ and $k \geq (\alpha+\epsilon)\Delta$, we have

$$2t(\alpha+\epsilon) - m + \frac{\delta}{2} < 2(\alpha+\epsilon)e^{\frac{-1}{\alpha+\epsilon}} - (2 - (1 - e^{\frac{-1}{\alpha+\epsilon}})^2),$$

where the inequality holds for all $n \geq n_0$, for some sufficiently large constant n_0. Therefore,

$$2 - (1 - e^{\frac{-1}{\alpha+\epsilon}})^2 - m + \frac{\delta}{2} < 2(\alpha+\epsilon)e^{-\frac{1}{\alpha+\epsilon}} - 2t(\alpha+\epsilon).$$

When $n = n_0$,

$$2(\alpha+\epsilon)e^{-\frac{1}{\alpha+\epsilon}} - 2t(\alpha+\epsilon) - 2 + (1 - e^{\frac{-1}{\alpha+\epsilon}})^2 + m - \frac{\delta}{2}$$

is a positive constant, and $2(\alpha+\epsilon)e^{-\frac{1}{\alpha+\epsilon}} - 2t(\alpha+\epsilon) - 2 + (1 - e^{\frac{-1}{\alpha+\epsilon}})^2 - \frac{\delta}{2}$ is independent of n and m is increasing in n. It follows that there exists a constant

$\zeta > 0$, such that

$$E[\rho(X_{t+1}, Y_{t+1}) \mid X_t = x, Y_t = y] \leq \frac{n-1}{n}\rho(x,y) + \frac{1-\zeta}{n}\rho(x,y).$$

Then,

$$E[\rho(X_{t+1}, Y_{t+1}) \mid X_t = x, Y_t = y] \leq \left(1 - \frac{\zeta}{n}\right)\rho(x,y).$$

By Lemma 2, the mixing time of X is $O(n \log n)$.

4   Rapid Mixing on Graphs with Bounded Common Neighbors

Let y be chosen from Ω uniformly at random. Let $v \in V$ be the node chosen at this step with degree $d \leq \Delta$. Let $w_1, \ldots, w_d$ be the nodes in the subgraph induced by $N(v)$. Let $Y_0 = y$, we obtain Y_i, for each $i = 1, \ldots, d$, according to the following procedure of Frieze and Vera [5].

1. Choose a color c from $A(Y_{i-1}, w_i)$ uniformly at random.
2. Let $Y_i(w_j) = Y_{i-1}(w_j)$, for all $j \neq i$.
3. Let $Y_i(w_i) = c$.

Lemma 4. *Let η be a constant with $0 < \eta < 1$. Let $G = (V, E)$ be the graph and any two adjacent nodes of G have at most ξ common neighbors with $\xi \leq \frac{\eta^{1.5}\Delta}{360e}$. Let $q = \frac{\eta}{6}$, $k \geq (1+\epsilon)\beta\Delta$, and y be chosen uniformly at random from Ω. If there exists $d \geq 6/q$ such that $\Delta \geq d\log n$, then with probability at least $1 - n^{-6}$, for all $v \in V$,*

$$|A(y,v)| \geq (1+q)\Delta.$$

4.1   Proof of the Second Part of Theorem 1

Proof. Let S consist of colorings $y \in \Omega$ with $|A(y,v)| \geq (1+q)\Delta$, for all $v \in V$. Let $0 < q < \frac{1}{6}$. According to Lemma 4, if there exists a positive constant $d \geq 6/q$ such that $\Delta \geq d\ln n$, then for any constant $\zeta > 0$,

$$\frac{|S|}{|\Omega|} \geq 1 - n^{-6} \geq 1 - \frac{\zeta}{16n^2} = 1 - \frac{\frac{\zeta}{n}}{16diam(\Omega)}$$

hold for sufficiently large n. It remains to prove that there exists a constant $\zeta > 0$ such that every pair $(x, y) \in \Omega \times S$ is $\frac{\zeta}{n}$-distance decreasing under Jerrum's coupling. Let $v \in V$ be the node chosen at step $t+1$. By Step 2(b) in Algorithm 1,

$$Pr[X_{t+1}(v) = Y_{t+1}(v)] = \frac{1}{\max\{|A(X_t, v))|, |A(Y_t, v)|\}}.$$

Therefore, the probability that X and Y assign different colors to v is

$$
\begin{aligned}
\Pr[X_{t+1}(v) \neq Y_{t+1}(v)] &= 1 - \frac{|A(X_t, v)) \cap A(Y_t, v)|}{\max\{|A(X_t, v))|, |A(Y_t, v)|\}} \\
&= \frac{\max\{|A(X_t, v))|, |A(Y_t, v)|\} - |A(X_t, v)) \cap A(Y_t, v)|}{\max\{|A(X_t, v))|, |A(Y_t, v)|\}} \\
&\leq \frac{|N(v) \cap D_t|}{\max\{|A(Y, v)|, |A(Y, v)|\}}.
\end{aligned}
$$

We have

$$
\begin{aligned}
E[\rho(X_{t+1}, Y_{t+1}) \mid X_t = x, Y_t = y] &= \sum_{u \in V} \Pr[X_{t+1}(u) \neq Y_{t+1}(u) \mid X_t = x, Y_t = y] \\
&\leq \frac{n-1}{n} \rho(X, Y) + \frac{1}{n} \sum_{v \in V} \frac{|N(v) \cap D_t|}{|A(Y, v)|} \\
&\leq \frac{n-1}{n} \rho(X_t, Y_t) + \frac{1}{n(1+q)\Delta} \sum_{v \in V} |N(v) \cap D_t| \\
&\leq \left(\frac{n-1}{n} + \frac{1}{(1+q)n}\right) \rho(X_t, Y_t) \\
&\leq \left(1 - \frac{q}{(1+q)n}\right) \rho(X_t, Y_t).
\end{aligned}
$$

Therefore, every pair $(x, y) \in \Omega \times S$ is $\frac{q}{(1+q)n}$ distance decreasing. By Lemma 2, the mixing time of X is $O(n \log n)$.

References

1. Bubley, R., Dyer, M.: Path coupling: a technique for proving rapid mixing in Markov chains. In: Proceedings of the 38th Annual IEEE Symposium on Foundations of Computer Science, pp. 223–231 (1997)
2. Dyer, M., Frieze, A.: Randomly colouring graphs with lower bounds on girth and maximum degree. In: Proceedings of the 42nd IEEE Symposium on Foundations of Computer Science, pp. 579–587 (2001)
3. Dyer, M., Frieze, A., Hayes, T.P., Vigoda, E.: Randomly coloring constant degree graphs. In: Proceedings of the 45th Annual IEEE Symposium on Foundations of Computer Science, pp. 582–589 (2004)
4. Dyer, M., Greenhill, C., Molloy, M.: Very rapid mixing of the Glauber dynamics for proper colorings on bounded-degree graphs. Random Structure and Algorithms 20(1), 98–114 (2002)
5. Frieze, A., Vera, J.: On randomly colouring locally sparse graphs. Discrete Mathematics and Theoretical Computer Science 8(1), 121–128 (2006)
6. Frieze, A., Vigoda, E.: A survey on the use of Markov chains to randomly sample colorings. In: Grimmett, G., McDiarmid, C. (eds.) Combinatorics, Complexity, and Chance — A Tribute to Dominic Welsh, ch. 4. Oxford University Press (2007)
7. Hayes, T., Vigoda, E.: A non-Markovian coupling for randomly sampling colorings. In: Proceedings of the 44th Annual IEEE Symposium on Foundations of Computer Science, pp. 618–627 (2003)

8. Hayes, T.P.: Local uniformity properties for Glauber dynamics on graph colorings (in submission)
9. Hayes, T.P.: Randomly coloring graphs of girth at least five. In: Proceedings of the 35th Annual ACM Symposium on Theory of Computing, pp. 269–278 (2003)
10. Hayes, T.P., Sinclair, A.: A general lower bound for mixing of single site dynamics on graphs. In: Proceedings of the 46th Annual IEEE Symposium on Foundations of Computer Science, pp. 511–520 (2005)
11. Hayes, T.P., Vera, J.C., Vigoda, E.: Randomly coloring planar graphs with fewer colors than the maximum degree. In: Proceedings of the 39th Annual ACM Symposium on Theory of Computing, pp. 450–458 (2007)
12. Hayes, T.P., Vigoda, E.: Variable length path coupling. In: Proceedings of the 15th Annual ACM-SIAM Symposium on Discrete Algorithms, pp. 103–110 (2004)
13. Hayes, T.P., Vigoda, E.: Coupling with stationary distribution and improved sampling for colorings and independent sets. In: Proceedings of the 16th Annual ACM-SIAM Symposium on Discrete Algorithms, pp. 971–979 (2005)
14. Jerrum, M.: A very simple algorithm for estimating the number of k-colorings of a low-degree graph. Random Structure and Algorithms 7(2), 157–166 (1995)
15. Jerrum, M., Sinclair, A.: The Markov chain Monte Carlo method: an approach to approximate counting and integration. In: Hochbaum, D.S. (ed.) Approximation Algorithms for NP-hard Problems, pp. 482–520. PWS Publishing Co. (1996)
16. Jerrum, M., Valiant, L., Vazirani, V.: Random generation of combinatorial structures from a uniform distribution. Theoretical Computer Science 43, 169–188 (1986)
17. Jerrum, M.R.: Counting, Sampling and Integrating: Algorithms and Complexity. Birkhauser Verlag, Basel (2003)
18. Kenyon, C., Mossel, E., Peres, Y.: Glauber dynamics on trees and hyperbolic graphs. In: Proceedings of the 42nd Annual IEEE Symposium on Foundations of Computer Science, pp. 568–578 (2001)
19. Lau, L.C., Molloy, M.: Randomly Colouring Graphs with Girth Five and Large Maximum Degree. In: Correa, J.R., Hevia, A., Kiwi, M. (eds.) LATIN 2006. LNCS, vol. 3887, pp. 665–676. Springer, Heidelberg (2006)
20. Martinelli, F., Sinclair, A.: Mixing time for the solid-on-solid model. In: Proceedings of the 41st Annual ACM Symposium on Theory of Computing, pp. 571–580 (2009)
21. Sly, A.: Reconstruction for the Potts model. In: Proceedings of the 41st Annual ACM Symposium on Theory of Computing, pp. 581–590 (2009)
22. Vigoda, E.: Improved bounds for sampling colorings. Journal of Mathematical Physics 41(3), 1555–1569 (2000)

Reconfiguration
of List $L(2,1)$-Labelings in a Graph

Takehiro Ito[1], Kazuto Kawamura[1], Hirotaka Ono[2], and Xiao Zhou[1]

[1] Graduate School of Information Sciences, Tohoku University,
Aoba-yama 6-6-05, Sendai, 980-8579, Japan
{takehiro,kazuto,zhou}@ecei.tohoku.ac.jp
[2] Faculty of Economics, Kyushu University,
Hakozaki 6-19-1, Higashi-ku, Fukuoka, 812-8581, Japan
hirotaka@en.kyushu-u.ac.jp

Abstract. For an integer $k \geq 0$, suppose that each vertex v of a graph G has a set $C(v) \subseteq \{0, 1, \ldots, k\}$ of labels, called a list of v. A list $L(2,1)$-labeling of G is an assignment of a label in $C(v)$ to each vertex v of G such that every two adjacent vertices receive labels which differ by at least 2 and every two vertices of distance two receive labels which differ by at least 1. In this paper, we study the problem of reconfiguring one list $L(2,1)$-labeling of a graph into another list $L(2,1)$-labeling of the same graph by changing only one label assignment at a time, while at all times maintaining a list $L(2,1)$-labeling. First we show that this decision problem is PSPACE-complete, even for bipartite planar graphs and $k \geq 6$. In contrast, we then show that the problem can be solved in linear time for general graphs if $k \leq 4$. We finally consider the problem restricted to trees, and give a sufficient condition for which any two list $L(2,1)$-labelings of a tree can be transformed into each other.

1 Introduction

Consider the graph in Fig.1 that models a wireless local area network (WLAN) in which each vertex corresponds to an access point (AP) and each edge represents the physical proximity and hence the two corresponding APs have the high potential of interference. The WLAN standard (802.11/a/b/g) divides the frequency spectrum into particular channels, and we wish to find an assignment of channels to the APs without any interference [16,17]. This kind of constraints in channel assignment have been formulated as $L(2,1)$-labelings of graphs, in which each label corresponds to a channel [3,6,14].

However, a practical issue in channel assignment requires that the formulation should be considered in more dynamic situations: in order to maintain high throughput performance, we sometimes need to change the current channel assignment to a newly found better assignment [16,17]. This reassignment must be done by individual channel changes to keep the network functionality and to prevent the need for any coordination. Furthermore, we certainly do not wish

K.-M. Chao, T.-s. Hsu, and D.-T. Lee (Eds.): ISAAC 2012, LNCS 7676, pp. 34–43, 2012.
© Springer-Verlag Berlin Heidelberg 2012

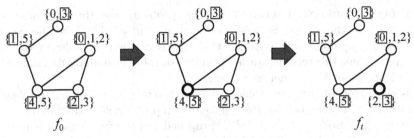

Fig. 1. A sequence of 5-list $L(2,1)$-labelings of a graph

users to be out of service during the reassignment. This situation can be formulated by the concept of reconfiguration problems that have been extensively studied in recent literature [1,2,4,5,7,8,9,10,11,12,13,15].

[$L(p, q)$-Labeling and its List Version]
For an integer $k \geq 0$, let $L = \{0, 1, \ldots, k\}$ be the *label set*. Then, for a pair of integers $p \geq 0$ and $q \geq 0$, a k-$L(p, q)$-*labeling* of a graph $G = (V, E)$ is an assignment $f : V \to L$ such that, for every two vertices x and y in V,

(i) $|f(x) - f(y)| \geq p$ if x and y are adjacent (*i.e.*, at distance 1); and
(ii) $|f(x) - f(y)| \geq q$ if x and y are at distance 2,

where the *distance* between two vertices is defined as the number of edges in a shortest path between them. Therefore, an ordinary vertex-coloring of G using $k + 1$ colors is a k-$L(1, 0)$-labeling of G. (Note that there are $k + 1$ distinct labels in L.) Given a graph G and an integer k, the k-$L(p, q)$-LABELING problem is to determine whether G has at least one k-$L(p, q)$-labeling. This problem is known to be NP-complete [3]. The k-$L(p, q)$-LABELING problem appears in several practical situations [3,6], especially in the channel assignments in WLANs (and traditionally in multi-hop radio networks), where "very close" APs must receive channels that are at least p channels apart and "close" APs must receive channels that are at least q channels apart so that they can avoid interference.

The "list" version is one of the most important and practical generalization [14], in which each vertex (AP) has a list of labels (channels) allowed to be assigned. Formally, in *list $L(p, q)$-labeling*, each vertex v of G has a set $C(v) \subseteq L$ of labels, called a *list* of v. Then, a k-$L(p, q)$-labeling f of G is called a *k-list $L(p, q)$-labeling* of G if $f(v) \in C(v)$ for each vertex $v \in V$. Figure 1 illustrates three 5-list $L(2, 1)$-labelings of the same graph with the same lists $C(v)$; the label assigned to each vertex is surrounded by a box in the list. Clearly, k-$L(p, q)$-labeling is a specialization of k-list $L(p, q)$-labeling for which $C(v) = L$ for every vertex v of G.

[Our Problem and Related Known Results]
Suppose that *two k-list $L(p, q)$-labelings* of a graph G are given as input. (For example, the leftmost and rightmost ones in Fig.1, where $k = 5$, $p = 2$ and $q = 1$.) Then, we consider the problem of determining whether we can transform one into the other via k-list $L(p, q)$-labelings of G such that each differs from the previous one in only one label assignment. We call this decision problem the

k-LIST $L(p,q)$-LABELING RECONFIGURATION problem. For the particular instance of Fig.1, the answer is "yes," as illustrated in Fig.1, where the vertex whose label assignment was changed from the previous one is depicted by a thick circle. Thus, this reconfiguration problem suitably formulates dynamic and practical situations in channel assignment [16,17].

This kind of reconfiguration problems arises when we wish to find a step-by-step transformation between two feasible solutions of a problem such that all intermediate results are also feasible. Ito *et al.* [9] proposed a framework of reconfiguration problems, and several reconfiguration problems have been studied such as SAT RECONFIGURATION [5,15], INDEPENDENT SET RECONFIGURATION [7,9,13], SUBSET SUM RECONFIGURATION [8], SHORTEST PATH RECONFIGURATION [12], etc.

In particular, reconfiguration problems for graph colorings have been intensively studied [1,2,4,10,11]. Bonsma and Cereceda [2] proved that k-$L(1,0)$-LABELING RECONFIGURATION (*i.e.*, the reconfiguration problem for $(k+1)$-vertex-colorings) is PSPACE-complete even for $k \geq 3$, while Cereceda *et al.* [4] proved that the problem is solvable in polynomial time for general graphs if $k \leq 2$. The reconfiguration problem for list edge-colorings has been studied and shown to be PSPACE-complete [10]. They [10] also gave a sufficient condition for which any two list edge-colorings of a tree can be transformed into each other, which was improved by [11].

[Our Contribution]
Among several possible settings of p and q, the k-$L(2,1)$-LABELING problem has been intensively and extensively studied due to its practical importance [3,6]. Therefore, we deal with the case where $p = 2$ and $q = 1$ in this paper, and give mainly three results for k-LIST $L(2,1)$-LABELING RECONFIGURATION. First, we show that the problem is PSPACE-complete, even for bipartite planar graphs and $k \geq 6$. In contrast, as our second result, we show that the problem can be solved in linear time for general graphs if $k \leq 4$. Third, we give a sufficient condition for which there exists a transformation between any two k-list $L(2,1)$-labelings of a tree.

These results have many implications due to the generality of "list" $L(2,1)$-labeling. For example, our proof of PSPACE-completeness can be modified so that we can prove that k-$L(2,1)$-LABELING RECONFIGURATION (*i.e.*, the non-list version) remains PSPACE-complete for bipartite planar graphs and $k \geq 8$. Furthermore, our proof for the sufficient condition is constructive, and hence yields a polynomial-time algorithm that actually finds a transformation between two given k-list $L(2,1)$-labelings of a tree.

Bonsma and Cereceda [2] also dealt with the list version, and proved that k-LIST $L(1,0)$-LABELING RECONFIGURATION is PSPACE-complete even for $k \geq 3$. Their proof can be easily modified so that we can prove that k-LIST $L(2,1)$-LABELING RECONFIGURATION is PSPACE-complete for general k. However, it is not straightforward to prove that the problem remains PSPACE-complete even for a small constant k. Notice that $k = 5$ is only the case where the complexity status remains open for k-LIST $L(2,1)$-LABELING RECONFIGURATION.

We finally remark that NP-hardness of the original problem does not imply computational hardness of its reconfiguration problem. It is interesting that k-LIST $L(2,1)$-LABELING RECONFIGURATION is solvable in linear time for $k \le 4$, although k-$L(2,1)$-LABELING is NP-complete already for $k = 4$ [3].

Due to the page limitation, we omit proofs from this extended abstract.

2   Definitions

In this section, we define some terms which will be used throughout the paper.

We may assume without loss of generality that a given graph G is simple and connected. We sometimes call a k-list $L(2,1)$-labeling of a graph simply a k-*list labeling*, and also call a k-$L(2,1)$-labeling a k-*labeling*. For two integers i and j with $0 \le i \le j$, we denote by $[i,j]$ the set of labels $i, i+1, \ldots, j$. Then, $L = [0,k]$. We say that two k-list labelings f and f' of G are *adjacent* if $|\{v \in V : f(v) \ne f'(v)\}| = 1$, that is, f' can be obtained from f by changing the label assignment of a single vertex v; we say that the vertex v is *reassigned* between f and f'. For two k-list labelings f_0 and f_t, a sequence $\langle f_0, f_1, \ldots, f_t \rangle$ is called a *reconfiguration sequence* between f_0 and f_t if $f_1, f_2, \ldots, f_t$ are k-list labelings of G such that f_{i-1} and f_i are adjacent for $i = 1, 2, \ldots, t$. We also say that two k-list labelings f and f' are *connected* if there exists a reconfiguration sequence between f and f'. Clearly, any two adjacent k-list labelings are connected. Given two k-list labelings f_0 and f_t of a graph G, the k-LIST $L(2,1)$-LABELING RECONFIGURATION problem is to determine whether f_0 and f_t are connected. We call the problem simply k-$L(2,1)$-LABELING RECONFIGURATION if $C(v) = [0,k]$ for all vertices v of G. For a reconfiguration sequence between two k-list labelings, its *length* is defined as the number of k-list labelings contained in the reconfiguration sequence.

For a graph G, we denote by $V(G)$ and $E(G)$ the vertex set and edge set of G, respectively. The maximum degree of G is denoted by $\Delta(G)$. For a vertex v, we denote by $N_1(v)$ the set of all vertices that are adjacent to v, that is, $N_1(v) = \{w \in V(G) \mid (v,w) \in E\}$. For a subgraph G' of G, we define $N_1(v; G') = N_1(v) \cap V(G')$. Note that v is not necessarily a vertex of G'. We clearly have $N_1(v; G) = N_1(v)$ and $|N_1(v)| = d(v)$, where $d(v)$ denotes the degree of v. Similarly, let $N_2(v) = \{w \in V(G) \mid \text{dist}(v,w) = 2\}$, where $\text{dist}(v,w)$ denotes the distance between v and w. For a subgraph G' of G, we define $N_2(v; G') = N_2(v) \cap V(G')$. Note that $v \notin V(G')$ may hold.

Let f be a k-list labeling of a graph G. For a vertex v of G and a subgraph G' of G, we define the subset $L_{\text{av}}(f, v; G') \subseteq C(v)$, as follows: $L_{\text{av}}(f, v; G') = C(v) \setminus (\{f(x) - 1, f(x), f(x) + 1 \mid x \in N_1(v; G')\} \cup \{f(y) \mid y \in N_2(v; G')\})$. Therefore, we can reassign v from the label $f(v)$ to any label in $L_{\text{av}}(f, v; G)$.

3   PSPACE-Completeness

The main result of this section is the following theorem.

Theorem 1. *The k-LIST $L(2,1)$-LABELING RECONFIGURATION problem is PSPACE-complete for bipartite planar graphs of maximum degree 3 and $k \ge 6$.*

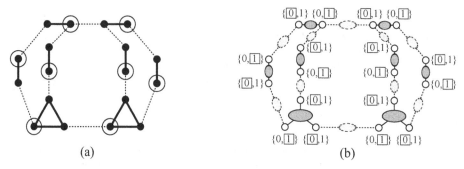

Fig. 2. (a) Graph G_s consisting of token triangles and token edges, where link edges are depicted by thin dotted lines and the vertices in a standard token configuration (namely, with tokens) are surrounded by circles, and (b) image of the corresponding graph G together with label assignments to the connectors

It is obvious that k-LIST $L(2,1)$-LABELING RECONFIGURATION can be solved in polynomial space, and hence is in PSPACE. Therefore, we prove that 6-LIST $L(2,1)$-LABELING RECONFIGURATION is PSPACE-hard.

[SLIDING TOKENS]

For a graph G_s, a subset T of $V(G_s)$ is called a *token configuration of G_s* if T forms an independent set of G_s; we may imagine that a token is placed on each vertex in T. A *move* from a token configuration to another one is to replace exactly one token from a vertex to its adjacent vertex, that is, we slide a token along an edge. Note that a move must result in a feasible token configuration. In the SLIDING TOKENS problem, we are given a graph G and two token configurations T_0 and T_t of G, both have the same number of tokens, and we are asked whether there is a sequence of moves starting from T_0 and ending in T_t. This decision problem is known to be PSPACE-complete [7].

Bonsma and Cereceda [2] showed that SLIDING TOKENS remains PSPACE-complete even for very restricted graphs and token configurations. Every vertex of a graph G_s is part of exactly one of *token triangles (i.e., copies of K_3)* and *token edges (i.e., copies of K_2)*, as illustrated in Fig.2(a). Token triangles and token edges are all mutually disjoint, and joined together by edges called *link edges*. Moreover, each vertex in a token triangle is of degree exactly 3, and G_s has a planar embedding such that every token triangle forms a face. The maximum degree of G_s is 3. We say that a token configuration T of G_s is *standard* if each of token triangles and token edges contains exactly one token (vertex) in T. Then, any move from a standard token configuration results in another standard token configuration; any token will never leave its token triangle or token edge, and will never slide along a link edge. The SLIDING TOKENS problem remains PSPACE-complete even if G_s is such a restricted graph and both T_0 and T_t are standard token configurations [2]; this restricted problem is called the STANDARD SLIDING TOKENS problem. We thus give a polynomial-time reduction from STANDARD SLIDING TOKENS to 6-LIST $L(2,1)$-LABELING RECONFIGURATION.

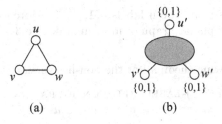

Fig. 3. (a) Token triangle and (b) image of the corresponding triangle gadget

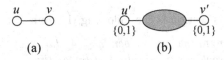

Fig. 4. (a) Token edge and (b) image of the corresponding edge gadget

[Overview of our Reduction]

We now give an overview of our reduction together with its ideas.

For a given graph G_s of STANDARD SLIDING TOKENS, we replace each token triangle (and each token edge) with a "triangle gadget" having three degree-1 vertices, called *connectors* (resp., with an "edge gadget" having two degree-1 vertices which are also called connectors.) Each connector u' in a gadget corresponds to one vertex u in the corresponding token triangle/edge; let $C(u') = \{0, 1\}$. (See Figs. 3 and 4.) Assume in our reduction that, if the label 1 is assigned to a connector u', then a token is placed on the corresponding vertex u; while a token is not placed on u if the label 0 is assigned to u'. (See Fig.2.) Therefore, for each link edge of G_s connecting two vertices in different token triangles/edges, we replace it with a "link gadget" which joins the two corresponding connectors and forbids the label 1 to be assigned to the two connectors at the same time; this ensures that the corresponding subset of vertices in G_s forms a feasible token configuration.

What about a move between two (standard) token configurations of G_s? Consider, for example, a token sliding along a token edge (u, v) in G_s, say from u to v. In the edge gadget having two corresponding connectors u' and v', this token sliding corresponds to reassigning u' and v' from $(1, 0)$ to $(0, 1)$, where (i, j) denotes a pair of labels $i \in \{0, 1\}$ and $j \in \{0, 1\}$ that are assigned to u' and v', respectively. However, since we can reassign only a single vertex at a time, such a reassignment is not allowed. Therefore, in order to simulate a token sliding along (u, v), our idea is to regard the label assignment $(1, 1)$ as also feasible; then $(1, 0)$ and $(0, 1)$ are connected via $(1, 1)$. It should be noted that this keeps the feasibility of token configurations, because no token must be placed on any vertex in $\left(N_1(u) \cup N_1(v)\right) \setminus \{u, v\}$ when we can slide the token along (u, v). Thus, for an edge gadget, we wish to forbid the assignment $(0, 0)$ only. By similar arguments, we wish to forbid the label 0 to be assigned to the three connectors of a triangle gadget at the same time. Indeed, we can construct such triangle/edge gadgets using seven labels $0, 1, \ldots, 6$. Moreover, the constructed graph G is a bipartite planar graph of maximum degree 3.

[Non-List Version]
We can give the following theorem for the non-list version.

Theorem 2. *The k-$L(2,1)$-LABELING RECONFIGURATION problem is PSPACE-complete for bipartite planar graphs of maximum degree 7 and $k \geq 8$.*

4   Linear-Time Algorithm

The main result of this section is the following theorem.

Theorem 3. *For a nonnegative integer $k \leq 4$, the k-LIST $L(2,1)$-LABELING RECONFIGURATION problem can be solved in linear time.*

To prove Theorem 3, we give several properties of k-list labelings of a graph when $k \leq 4$, based on which, we can determine in linear time whether two given k-list labelings are connected. We first remark that the following lemma holds.

Lemma 1. *For an integer $k \geq 1$, if a graph G has a k-list labeling, then $\Delta(G) \leq k - 1$.*

We first consider the case where $k \leq 3$. In this case, it suffices to consider a connected graph G consisting of a constant number of vertices, and hence the problem can be clearly solved in linear time. More specifically, if G has a k-list labeling for $k \leq 3$, then Lemma 1 implies that G is either a path or a cycle. It is known that any cycle and any path of more than four vertices has no k-list labeling if $k \leq 3$ [3]. Since G has at least two k-list labelings f_0 and f_t, the graph must consist of a constant number of vertices.

In the remainder of this section, we thus consider the case where $k = 4$. Then, by Lemma 1 it suffices to consider a graph G with $\Delta(G) \leq 3$. We now give the following lemma, which implies that we cannot reassign any vertex of degree 3 and its neighbors.

Lemma 2. *Let f be any 4-list labeling of a graph G with $\Delta(G) \leq 3$, and let u be a vertex of degree 3. Then, $L_{\mathrm{av}}(f, x; G) = \{f(x)\}$ for all vertices $x \in N_1(u) \cup \{u\}$.*

Since a given graph G is connected, we may assume without loss of generality that $\Delta(G) \geq 2$; because, if $\Delta(G) = 1$, then G consists of a single edge. We say that a vertex v is *flexible* if one of the following conditions (i) and (ii) holds:

  (i) $d(v) = 1$, and $d(z) = 2$ for the vertex z in $N_1(v)$; and
  (ii) $d(v) = 2$, and $d(x) = 1$ and $d(y) \neq 3$ for the two vertices $x, y \in N_1(v)$ with $d(x) \leq d(y)$.

The other vertices are said to be *inflexible*. We then give the following lemma, which implies that only flexible vertices are candidates for reassignments.

Lemma 3. *Let f be an arbitrary 4-list labeling of a graph G with $2 \leq \Delta(G) \leq 3$, and let v be any vertex of G. If $|L_{\mathrm{av}}(f, v; G)| \geq 2$, then v is flexible.*

Lemma 3 implies that all inflexible vertices v satisfy $|L_{av}(f, v; G)| = 1$ for any 4-list labeling f. Therefore, there is no reconfiguration sequence between two 4-list labelings f_0 and f_t if there is an inflexible vertex v in G such that $f_0(v) \neq f_t(v)$.

Let v be a vertex of degree 1, and let w be the vertex adjacent to v. Then, we call the pair (v, w) an *active pair* if v is flexible. It should be noted that w is not necessarily flexible. We now suppose that there exists a reconfiguration sequence between f_0 and f_t. Then, we say that an active pair (v, w) is *independently reconfigurable* if we can reassign the vertices v and w to their target labels without reassigning any other vertices in G. The following lemma shows that we do not care about the order of reassignments of active pairs if G has more than five vertices.

Lemma 4. *For a graph G with more than five vertices, let f_0 and f_t be any two connected 4-list labelings of G. Then, all active pairs are independently reconfigurable.*

[Proof of Theorem 3]
If a given graph G consists of at most five vertices, a simple brute-force algorithm can solve the problem. If G has more than five vertices, we first check whether $f_0(u) = f_t(u)$ for all inflexible vertices u in G. Then, by Lemmas 3 and 4, it suffices to check each active pair independently whether the two vertices in the active pair can be reassigned to their target labels. Therefore, 4-LIST $L(2,1)$-LABELING RECONFIGURATION can be solved in linear time.   □

5   Sufficient Condition for Trees

In this section, we give a sufficient condition for which any two k-list labelings of a tree are connected. Suppose that we are given a tree T together with a list $C(v)$ for each vertex v of T. The main result of this section is the following theorem.

Theorem 4. *Every two k-list labelings f and f' of a tree T are connected if*

$$|C(v)| \geq \max\{d(w) \mid w \in N_1(v)\} + 6 \tag{1}$$

for each vertex v of T. Moreover, there is a reconfiguration sequence of length $O(n^2)$ between f and f', where n is the number of vertices in T.

Remember that $L = [0, k]$, and hence L contains $k + 1$ distinct labels. Since $\Delta(T) \geq \max\{d(w) \mid w \in N_1(v)\}$ for all vertices v of a tree T, Theorem 4 immediately implies the following sufficient condition for the non-list version.

Corollary 1. *Every two k-labelings f and f' of a tree T are connected if $k \geq \Delta(T) + 5$. Moreover, there is a reconfiguration sequence of length $O(n^2)$ between f and f', where n is the number of vertices in T.*

As in Eq. (1), our concern is only the number of labels in each list $C(v)$, and we do not care about the maximum label k in the lists. Therefore, from now on, we call a k-list labeling of G simply a *list labeling*. Note that any list labeling of G is a k-list labeling of G such that $k \geq \max\{c \in C(v) \mid v \in V(G)\}$. For a given tree T, we choose an arbitrary vertex r as the root of T, and regard T as a rooted tree. For a vertex v in T, we denote by T_v the subtree of T which is rooted at v and is induced by v and all descendants of v in T. Then, $T_r = T$, and T_v consists of a single vertex v if v is a leaf of T.

We give a constructive proof, that is, we give an algorithm which actually finds a reconfiguration sequence of length $O(n^2)$ between two given list labelings f_0 and f_t. Our algorithm is outlined as follows. Let T be a tree with a list $C(v)$ satisfying Eq. (1) for each vertex v of T. We say that a vertex v in T is *fixed* if our algorithm decides not to reassign v anymore. Therefore, the target label $f_t(v)$ must be assigned to v when it is fixed. The algorithm fixes the vertices one by one, and terminates when all the vertices are fixed. More specifically, by a breadth-first search starting from the root r of T, we order all vertices $v_1, v_2, \ldots, v_n$ of T. At the i-th step, $1 \leq i \leq n$, we reassign the vertex v_i to its target label $f_t(v_i)$ via a reconfiguration (sub-)sequence $\mathcal{S}_i$ of length $O(n)$ which does not reassign any vertex v_j with $j < i$. Indeed, we reassign vertices in $V(T_{v_i})$ only, as follows.

Lemma 5. *Let v be any vertex in T, and f_{s_i} be any list labeling of T. Let c be an arbitrary label in $L_{av}(f_{s_i}, v; T \setminus T_v)$. Then, there exists a reconfiguration sequence $\mathcal{S}_i = \langle f_{s_i}, f_{s_i+1}, \ldots, f_{t_i} \rangle$ from f_{s_i} to a list labeling f_{t_i} of T such that*

(i) *$f_{t_i}(v) = c$;*
(ii) *for each list labeling $f \in \mathcal{S}_i$, $f(u) = f_{s_i}(u)$ if $u \in V(T \setminus T_v)$; and*
(iii) *$|\mathcal{S}_i| = O(|V(T_v)|)$.*

In this way, we eventually obtain the list labeling f_t. Since each reconfiguration sub-sequence $\mathcal{S}_i$ is of length $O(n)$, we can obtain a reconfiguration sequence of total length $O(n^2)$.

6   Concluding Remarks

In this paper, we analyzed complexity statuses of k-LIST $L(2,1)$-LABELING RE-CONFIGURATION with respect to the maximum label k in the lists. We remark that only the case where $k = 5$ remains open. We also remark that only three cases where $k = 5, 6, 7$ remain open for the non-list version.

Acknowledgments. We are grateful to Shinya Kumagai for interesting and fruitful discussions. This work is partially supported by JSPS KAKENHI Grant Number 21680001, 22650004, 22700001, 23500001.

References

1. Bonamy, M., Johnson, M., Lignos, I., Patel, V., Paulusma, D.: On the diameter of reconfiguration graphs for vertex colourings. Electronic Notes in Discrete Mathematics 38, 161–166 (2011)
2. Bonsma, P., Cereceda, L.: Finding paths between graph colourings: PSPACE-completeness and superpolynomial distances. Theoretical Computer Science 410, 5215–5226 (2009)
3. Calamoneri, T.: The $L(h,k)$-labelling problem: An updated survey and annotated bibliography. The Computer Journal 54, 1344–1371 (2011)
4. Cereceda, L., van den Heuvel, J., Johnson, M.: Finding paths between 3-colourings. In: Proc. of IWOCA 2008, pp. 182–196 (2008)
5. Gopalan, P., Kolaitis, P.G., Maneva, E.N., Papadimitriou, C.H.: The connectivity of Boolean satisfiability: computational and structural dichotomies. SIAM J. Computing 38, 2330–2355 (2009)
6. Hasunuma, T., Ishii, T., Ono, H., Uno, Y.: A linear time algorithm for $L(2,1)$-labeling of trees. To appear in Algorithmica, doi:10.1007/s00453-012-9657-z
7. Hearn, R.A., Demaine, E.D.: PSPACE-completeness of sliding-block puzzles and other problems through the nondeterministic constraint logic model of computation. Theoretical Computer Science 343, 72–96 (2005)
8. Ito, T., Demaine, E.D.: Approximability of the Subset Sum Reconfiguration Problem. In: Ogihara, M., Tarui, J. (eds.) TAMC 2011. LNCS, vol. 6648, pp. 58–69. Springer, Heidelberg (2011)
9. Ito, T., Demaine, E.D., Harvey, N.J.A., Papadimitriou, C.H., Sideri, M., Uehara, R., Uno, Y.: On the complexity of reconfiguration problems. Theoretical Computer Science 412, 1054–1065 (2011)
10. Ito, T., Kamiński, M., Demaine, E.D.: Reconfiguration of list edge-colorings in a graph. Discrete Applied Mathematics 160, 2199–2207 (2012)
11. Ito, T., Kawamura, K., Zhou, X.: An improved sufficient condition for reconfiguration of list edge-colorings in a tree. IEICE Trans. on Information and Systems E95-D, 737–745 (2012)
12. Kamiński, M., Medvedev, P., Milanič, M.: Shortest paths between shortest paths. Theoretical Computer Science 412, 5205–5210 (2011)
13. Kamiński, M., Medvedev, P., Milanič, M.: Complexity of independent set reconfigurability problems. Theoretical Computer Science 439, 9–15 (2012)
14. Kohl, A., Schreyer, J., Tuza, Z., Voigt, M.: List version of $L(d,s)$-labelings. Theoretical Computer Science 349, 92–98 (2005)
15. Makino, K., Tamaki, S., Yamamoto, M.: An exact algorithm for the Boolean connectivity problem for k-CNF. Theoretical Computer Science 412, 4613–4618 (2011)
16. Marias, G.F., Skyrianoglou, D., Merakos, L.: A centralized approach to dynamic channel assignment in wireless ATM LANs. In: Proc. of INFOCOM 1999, vol. 2, pp. 601–608 (1999)
17. Matsumura, Y., Kumagai, S., Obara, T., Yamamoto, T., Adachi, F.: Channel segregation based dynamic channel assignment for WLAN (preprint)

An 8/3 Lower Bound for Online Dynamic Bin Packing

Prudence W.H. Wong, Fencol C.C. Yung, and Mihai Burcea*

Department of Computer Science, University of Liverpool, UK
{pwong,m.burcea}@liverpool.ac.uk, ccyung@graduate.hku.hk

Abstract. We study the dynamic bin packing problem introduced by Coffman, Garey and Johnson. This problem is a generalization of the bin packing problem in which items may arrive and depart dynamically. The objective is to minimize the maximum number of bins used over all time. The main result is a lower bound of $8/3 \sim 2.666$ on the achievable competitive ratio, improving the best known 2.5 lower bound. The previous lower bounds were 2.388, 2.428, and 2.5. This moves a big step forward to close the gap between the lower bound and the upper bound, which currently stands at 2.788. The gap is reduced by about 60% from 0.288 to 0.122. The improvement stems from an adversarial sequence that forces an online algorithm $\mathcal{A}$ to open $2s$ bins with items having a total size of s only and this can be adapted appropriately regardless of the current load of other bins that have already been opened by $\mathcal{A}$. Comparing with the previous 2.5 lower bound, this basic step gives a better way to derive the complete adversary and a better use of items of slightly different sizes leading to a tighter lower bound. Furthermore, we show that the 2.5-lower bound can be obtained using this basic step in a much simpler way without case analysis.

1 Introduction

Bin packing is a classical combinatorial optimization problem [6,8,9]. The objective is to pack a set of items into a minimum number of unit-size bins such that the total size of the items in a bin does not exceed the bin capacity. The problem has been studied extensively both in the offline and online settings. It is well-known that the problem is NP-hard [11]. In the online setting [14,15], items may arrive at arbitrary time; item arrival time and item size are only known when an item arrives. The performance of an online algorithm is measured using competitive analysis [3]. Consider any online algorithm $\mathcal{A}$. Given an input I, let $OPT(I)$ and $\mathcal{A}(I)$ be the maximum number of bins used by the optimal offline algorithm and $\mathcal{A}$, respectively. Algorithm $\mathcal{A}$ is said to be c-competitive if there exists a constant b such that $\mathcal{A}(I) \leq c\, OPT(I) + b$ for all I.

Online Dynamic Bin Packing. Most existing work focuses on "static" bin packing in the sense that items do not depart. In some potential applications like

---

* Supported by EPSRC Studentship.

K.-M. Chao, T.-s. Hsu, and D.-T. Lee (Eds.): ISAAC 2012, LNCS 7676, pp. 44–53, 2012.
© Springer-Verlag Berlin Heidelberg 2012

warehouse storage, a more realistic model takes into consideration of dynamic arrival and departures of items. In this natural generalization, known as dynamic bin packing [7], items arrive over time, reside for some period of time, and may depart at arbitrary time. Each item has to be assigned to a bin from the time it arrives until it departs. The objective is to minimize the maximum number of bins used over all time. Note that migration to another bin is not allowed. In the online setting, the size and arrival time is only known when an item arrives and the departure time is only known when the item departs.

In this paper, we focus on online dynamic bin packing. It is shown in [7] that First-Fit has a competitive ratio between 2.75 and 2.897, and a modified first-fit algorithm is 2.788-competitive. A lower bound of 2.388 is given for any deterministic online algorithm. This lower bound has later been improved to 2.428 [4] and then 2.5 [5]. The problem has also been studied in two- and three-dimension as well as higher dimension [10,16]. Other work on dynamic bin packing considered a restricted type of items, namely unit-fraction items [2,4,12]. Furthermore, Ivkovic and Lloyd [13] studied the fully dynamic bin packing problem, which allows repacking of items for each item arrival or departure and they gave a 1.25-competitive online algorithm for this problem. Balogh et al. [1] studied the problem when a limited amount of repacking is allowed.

Our Contribution. We improve the lower bound of online dynamic bin packing for any deterministic online algorithm from 2.5 to $8/3 \sim 2.666$. This makes a big step forward to close the gap with the upper bound, which currently stands at 2.788 [7]. The improvement stems from an adversarial sequence that forces an online algorithm $\mathcal{A}$ to open $2s$ bins with items having a total size of s only and this can be adapted appropriately regardless of the load of current bins opened by $\mathcal{A}$. Comparing with the previous 2.5 lower bound, this basic step gives a better use of items of slightly different sizes leading to a tighter lower bound. Furthermore, we show in Section 3.3 that the 2.5-lower bound can be obtained using this basic step in a much simpler way without case analysis. It is worth mentioning that we consider optimal packing without migration at any time.

The adversarial sequence is composed of two operations, namely Op-Inc and Op-Comp. Roughly speaking, Op-Inc uses a load of at most s to make $\mathcal{A}$ open s bins, this is followed by some item departure such that each bin is left with only one item and the size is increasing across the bins. Op-Comp then releases items of complementary size such that for each item of size x, items of size $1 - x$ are released. The complementary size ensures that the optimal offline algorithm $\mathcal{O}$ is able to pack all these items using s bins while the sequence of arrival ensures that $\mathcal{A}$ has to pack these complementary items into separate bins.

2   Preliminaries

In dynamic bin packing, items arrive and depart at arbitrary time. Each item comes with a size. We denote by s-$item$ an item of size s. When an item arrives, it must be assigned to a unit-sized bin immediately without exceeding the bin capacity. At any time, the *load* of a bin is the total size of items currently assigned to that bin that have not yet departed. We denote by ℓ-*bin* a bin of load ℓ. Migration is not allowed, i.e., once an item is assigned to a bin, it cannot be moved to another bin. This also applies to the optimal offline algorithm. The objective is to minimize the maximum number of bins used over all time.

When we discuss how items are packed, we use the following notations:

- Item configuration ψ: y_{*z} describes a load y with $\frac{y}{z}$ items of size z, e.g., $\frac{1}{2}_{*\epsilon}$ means a load $\frac{1}{2}$ with $\frac{1}{2\epsilon}$ items of size ϵ. We skip the subscript when $y = z$.
- Bin configuration π: $(\psi_1, \psi_2, \cdots)$, e.g., $(\frac{1}{3}, \frac{1}{2}_{*\epsilon})$ means a bin has a load of $\frac{5}{6}$, with a $\frac{1}{3}$-item and an addition load $\frac{1}{2}$ with ϵ-items. In some cases, it is clearer to state the bin configuration in other ways, e.g., $(\frac{1}{2}, \frac{1}{2})$, instead of $1_{*\frac{1}{2}}$. Similarly, we will use $6 \times \frac{1}{6}$ instead of $1_{*\frac{1}{6}}$.
- Packing configuration ρ: $\{x_1{:}\pi_1, x_2{:}\pi_2, \cdots\}$ a packing where there are x_1 bins with bin configuration π_1, x_2 bins with π_2, and so on. E.g., $\{2k{:}1_{*\epsilon}, k{:}(\frac{1}{3}, \frac{1}{2}_{*\epsilon})\}$ means $2k$ bins are each packed with load 1 with ϵ-items and another k bins are each packed with a $\frac{1}{3}$-item and an addition load $\frac{1}{2}$ with ϵ-items.
- It is sometimes more convenient to describe a packing as $x{:}f(i)$, for $1 \leq i \leq x$, which means that there are x bins with different load, one bin with load $f(i)$ for each i. E.g., $k{:}\frac{1}{2}{-}i\delta$, for $1 \leq i \leq k$, means that there are k bins and one bin with load $\frac{1}{2}{-}i\delta$ for each i.

3   Op-Inc and Op-Comp

In this section, we discuss a process that the adversary uses to force an online algorithm $\mathcal{A}$ to open new bins. The adversary releases items of slightly different sizes in each stage and uses items of complementary sizes in different stages. Two operations are designed, namely, Op-Inc and Op-Comp. Op-Inc forces $\mathcal{A}$ to open some bins each with one item (of size $< \frac{1}{2}$) and the size of items is strictly increasing. Op-Comp then bases on the bins opened by Op-Inc and releases items of complementary size. This is to ensure that an item released in Op-Inc can be packed with a corresponding item released in Op-Comp into the same bin by an optimal offline algorithm. In the adversary, a stage of Op-Inc is associated with a corresponding stage of Op-Comp, but not necessarily consecutive, e.g., in one of the cases, Op-Inc is in Stage 1 and the corresponding Op-Comp is in Stage 4.

3.1   Operation Op-Inc

The aim of Op-Inc is to make $\mathcal{A}$ open at least s more bins, for some $s > 0$, such that each new bin contains one item with item size increasing over the s bins.

Pre-condition. Consider any value $0 < x < \frac{1}{2}$. Let h be the number of x-items that can be packed in existing bins.

Items to Be Involved. The items to be released have size in the range $[x, x{+}\epsilon]$, for some small ϵ, such that $x{+}\epsilon < \frac{1}{2}$. A total of $h + \lfloor \frac{s}{x} \rfloor$ items are to be released.

Outcome. $\mathcal{A}$ opens at least s more new bins with increasing load in each new bin and the load of current bins remains unchanged.

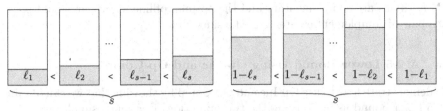

Fig. 1. Op-Comp: Assuming $k = 0$. The s bins on the left are bins created by Op-Inc. The s new bins on the right are due to Op-Comp. Note that each existing item has a complementary new item such that the sum of size is 1.

The Adversary. The adversary releases items of size x, $x + \frac{\epsilon}{s}$, $x + \frac{2\epsilon}{s}$, $\cdots$. Let $z_i = x + \frac{i\epsilon}{s}$. In each step i, the adversary releases z_i-items until $\mathcal{A}$ opens a new bin. We stop releasing items when $h + \lfloor \frac{s}{x} \rfloor$ items have been released in total. By the definition of h, s and x, $\mathcal{A}$ would have opened at least s new bins. We then let z_i-items depart except exactly one item of size z_i, for $0 \le i < s$, in the i-th new bin opened by $\mathcal{A}$.

Using Op-Inc. When we use Op-Inc later, we simply describe it as Op-Inc releasing $h + \lfloor \frac{s}{x} \rfloor$ items with the understanding that it works in phases and that items depart at the end.

3.2   Operation Op-Comp

Op-Comp is designed to work with Op-Inc and assumes that there are s existing bins each with load in the range $[x, y]$ where $x < y < \frac{1}{2}$. The outcome of Op-Comp is that $\mathcal{A}$ opens s more bins. Figure 1 gives an illustration.

Pre-condition. Consider two values $x < y < \frac{1}{2}$. Suppose $\mathcal{A}$ uses s bins with load $x = \ell_1 < \ell_2 < \cdots < \ell_s = y$. Let $\ell = \sum_{1 \le i \le s} \ell_i$. Furthermore, suppose there are some additional bins with load smaller than x. Let h be the number of $(1-y)$-items that can be packed in other existing bins with load less than x.

Items to Be Involved. The items to be released have size in the range $[1 - y, 1 - x]$. Note that $1 - x > 1 - y > \frac{1}{2}$. In each step i, for $1 \le i \le s$, the number of $(1 - \ell_{s+1-i})$-items released is at most $h + s + 2 - i$.

Outcome. $\mathcal{A}$ opens s more bins, each with an item $1 - \ell_{s+1-i}$, for $1 \le i \le s$.

The Adversary. Starting from the largest load ℓ_s, we release items of size $1-\ell_s$ until $\mathcal{A}$ opens a new bin. At most $h + s + 1$ items are needed. Then we let all $(1-\ell_s)$-items depart except the one packed in the new bin. In general, in Step i, for $1 \le i \le s$, we release items of size $1-\ell_{s+1-i}$ until $\mathcal{A}$ opens a new bin. Note that such items can only be packed in the first $s + 1 - i$ bins and so at most $h + s + 2 - i$ items are required to force $\mathcal{A}$ to open another bin. We then let all $(1-\ell_{s+1-i})$-items depart except the one packed in the new bin.

Using Op-Comp. Similar to Op-Inc, when we use Op-Comp later, we describe it as Op-Comp with h and s and the understanding is that it works in phases and there are items released and departure in between. Note that the ℓ_i- and

$(1 - \ell_i)$-items are complementary and the optimal offline algorithm would pack each pair of complementary items in the same bin.

3.3 A 2.5 Lower Bound Using Op-Inc and Op-Comp

We demonstrate how to use Op-Inc and Op-Comp by showing that we can obtain a 2.5 lower bound as in [5] using the two operations in a much simpler way.

Let k be some large even integer, $\epsilon = \frac{1}{k}$, and $\delta = \frac{\epsilon}{k+1}$. The adversary works in stages. In Stage 1, we release $\frac{k}{\epsilon}$ items of size ϵ. Any online algorithm $\mathcal{A}$ uses at least k bins. Let items depart until the configuration is $\{k{:}\epsilon\}$. In Stage 2, we aim to force $\mathcal{A}$ to use $\frac{k}{2}$ new bins. We use Op-Inc to release at most $2k$ items of size in $[\frac{1}{2}-\frac{k}{2}\delta, \frac{1}{2}-\delta]$. For each existing ϵ-bin, at most one such new items can be packed because $1-k\delta+\epsilon > 1$. The parameters for Op-Inc are therefore $x = \frac{1}{2}-\frac{k}{2}\delta$, $h = k$ and $s = \frac{k}{2}$. The configuration of $\mathcal{A}$ becomes $\{k{:}\epsilon, \frac{k}{2}{:}\frac{1}{2}-i\delta\}$, for $1 \leq i \leq \frac{k}{2}$. In Stage 3, we aim to force $\mathcal{A}$ to use $\frac{k}{2}$ new bins. We use Op-Comp to release items of size in the range $x = \frac{1}{2}+\delta$ to $y = \frac{1}{2}+\frac{k}{2}\delta$. At most one such item can be packed in the bins with an ϵ-item, i.e., $h = k$. The second $\frac{k}{2}$ bins contains items of complementary size to the items released in Stage 3, i.e., $s = \frac{k}{2}$. Note that at any time during Op-Comp, at most $\frac{3k}{2}+1$ items are released. $\mathcal{A}$ needs to open at least $\frac{k}{2}$ new bins with the configuration $\{k{:}\epsilon, \frac{k}{2}{:}\frac{1}{2}-i\delta, \frac{k}{2}{:}\frac{1}{2}+i\delta\}$, for $1 \leq i \leq \frac{k}{2}$. In the final stage, we release $\frac{k}{2}$ items of size 1 and $\mathcal{A}$ needs a new bin for each of these items. The total number of bins used by $\mathcal{A}$ becomes $\frac{5k}{2}$.

On the other hand, the optimal algorithm $\mathcal{O}$ can use $k + 2$ bins to pack all items as follows and hence the competitive ratio is at least 2.5. In Stage 1, all the ϵ-items that never depart are packed in one bin and the rest in $k - 1$ bins. In Stage 2, the new items are packed in k bins, with the $\frac{k}{2}$ bins with size $\frac{1}{2}-i\delta$, for $1 \leq i \leq \frac{k}{2}$, that never depart each packed in one bin, and the remaining $\frac{3k}{2}$ items in the remaining space. At the end of the stage, only one item is left in each of the first $\frac{k}{2}$ bins and the second $\frac{k}{2}$ bins are freed for Stage 3. In Stage 3, the complementary items that do not depart are packed in the corresponding $\frac{k}{2}$ bins, and the remaining in at most $\frac{k}{2} + 1$ bins. Finally in Stage 4, the 1-items are packed in the $\frac{k}{2}$ bins freed in Stage 3.

4   The 8/3 Lower Bound

We give an adversary such that at any time, the total load of items released and not departed is at most $6k + O(1)$, for some large integer k. We prove that any online algorithm $\mathcal{A}$ uses $16k$ bins, while the optimal offline algorithm $\mathcal{O}$ uses at most $6k + O(1)$ bins. Then, the competitive ratio of $\mathcal{A}$ is at least $\frac{8}{3}$. The adversary works in stages and uses Op-Inc and Op-Comp in pairs. Let n_i be the number of new bins used by $\mathcal{A}$ in Stage i. Let $\epsilon = \frac{1}{6k}$ and $\delta = \frac{\epsilon}{16k}$.

In Stage 0, the adversary releases $\frac{6k}{\epsilon}$ items of size ϵ, with total load $6k$. It is clear that $\mathcal{A}$ needs at least $6k$ bins, i.e., $n_0 \geq 6k$. We distinguish between two cases: $n_0 \geq 8k$ and $8k > n_0 \geq 6k$. We leave the details of the easier first case in the full paper, and we consider only the complex second case in this paper.

Case 2: $6k \leq n_0 < 8k$.

This case involves three subcases. We make two observations about the load of the n_0 bins. If less than $4k$ bins have load at least $\frac{1}{2} + \epsilon$, then the total load of all bins is at most $(4k - 1) + 4k/2 = 6k - 1$, contradicting the fact that total load of items released is $6k$. Similarly, if less than $5k$ bins have load at least $\frac{1}{4} + \epsilon$, then the total load of all bins is at most $(5k - 1) + 3k/4 < 6k$, leading to a contradiction.

Observation 1. *At the end of Stage 0 of Case 2, (i) at least $4k$ bins have load at least $\frac{1}{2} + \epsilon$; (ii) at least $5k$ bins have load at least $\frac{1}{4} + \epsilon$.*

Stage 1. We aim at $n_1 \geq 2k$. We let ϵ-items depart until the configuration of $\mathcal{A}$ becomes

$$\{4k:(\frac{1}{2}+\epsilon)_{*\epsilon}, k:(\frac{1}{4}+\epsilon)_{*\epsilon}, k:\epsilon\} ,$$

with $6k$ bins and a total load of $9k/4+O(1)$. We then use Op-Inc with $x = \frac{1}{4}+\delta$, $h = 8k$, and $s = 2k$. The first $4k$ bins can pack at most one x-item, the next k bins at most two, and the last k bins at most three, i.e., $h = 9k$. Any new bin can pack at most three items, implying that Op-Inc releases $15k = h + 3s$ items of increasing sizes, from $\frac{1}{4}+\delta$ to at most $\frac{1}{4}+15k\delta$. According to Op-Inc, $\mathcal{A}$ opens at least $2k$ bins, i.e., $n_1 \geq 2k$. We consider two subcases: $n_1 \geq 4k$ and $2k \leq n_1 < 4k$.

Case 2.1: $6k \leq n_0 < 8k$ and $n_1 \geq 4k$. In this case, we have $10k \leq n_0 + n_1$.

Stage 2. We aim at $n_2 \geq 4k$. The configuration after Op-Inc becomes

$$\{4k:(\frac{1}{4}+\epsilon)_{*\epsilon}, k:(\frac{1}{4}+\epsilon)_{*\epsilon}, k:\epsilon, 4k:\frac{1}{4}+i\delta\} , \quad \text{for } 1 \leq i \leq 4k,$$

with $10k$ bins and a total load of $9k/4+O(1)$. Note that in the last $4k$ bins, the load increases by δ from $\frac{1}{4}+\delta$ to $\frac{1}{4}+4k\delta$. We now use Op-Comp with $x = \frac{1}{4}+\delta$, $y = \frac{1}{4}+4k\delta$, $h = k$, and $s = 4k$. I.e., Op-Comp releases items of sizes from $\frac{3}{4}-4k\delta$ to $\frac{3}{4}-\delta$ and at any time, at most $5k + 1$ items are needed. None of these items can be packed in the first $5k$ bins, and only one can be packed in the next k bins, i.e., $h = k$ as said. According to Op-Comp, $\mathcal{A}$ requires $4k$ new bins.

Stage 3. We aim at $n_3 = 2k$. We let items depart until the configuration becomes

$$\{4k:\epsilon, k:\epsilon, k:\epsilon, 4k:\frac{1}{4}+i\delta, 4k:\frac{3}{4}-i\delta\} , \quad \text{for } 1 \leq i \leq 4k,$$

with $14k$ bins and a load of $4k + O(1)$. We further release $2k$ items of size 1. $\mathcal{A}$ needs to open $2k$ new bins. In total, $\mathcal{A}$ uses $6k + 4k + 4k + 2k = 16k$ bins.

We note that each item with size $\frac{1}{4}+i\delta$ has a corresponding item $\frac{3}{4}-i\delta$ such that the sum of sizes is 1. This allows the optimal offline algorithm to have a better packing. The details will be given in the full paper.

Lemma 1. *If $\mathcal{A}$ uses $[6k, 8k)$ bins in Stage 0 and at least $4k$ bins in Stage 1, then $\mathcal{A}$ uses $16k$ bins at the end while $\mathcal{O}$ uses $6k + 4$ bins.*

Case 2.2: $6k \leq n_0 < 8k$ and $2k \leq n_1 < 4k$. In this case, the Op-Inc in Stage 1 is paired with an Op-Comp in Stage 4 (not consecutively), and in between, there is another pair of Op-Inc and Op-Comp in Stages 2 and 3, respectively. Let m be the number of bins among the n_1 new bins that have been packed two items. We further distinguish two subcases: $m \geq 2k$ and $m < 2k$.

Case 2.2.1: $6k \leq n_0 < 8k$, $2k \leq n_1 < 4k$ and $m \geq 2k$. In this case, we have $8k \leq n_0 + n_1 < 10k$ and $m \geq 2k$. We make an observation about the bins containing some ϵ-items. In particular, we claim that there are at least k bins that are packed with

- either one ϵ-item and at least two $(\frac{1}{4}+i\delta)$-items,
- or one $(\frac{1}{4}+i\delta)$-item plus at least a load of $(\frac{1}{4}+\epsilon)_{*\epsilon}$.

We note that in Stage 1, $15k$ items are released, at most three items can be packed in any of the $n_1 < 4k$ new bins, i.e., at most $12k$ items. So, at least $3k$ of them have to been packed in the first $6k$ bins. Let a and b be the number of bins in the first $5k$ bins (with load at least $\frac{1}{4}+\epsilon$) that are packed at least one $(\frac{1}{4}+i\delta)$-item; z_1, z_2, z_3 be the number of bins in the next k bins (with one ϵ-item) that are packed one, two, and three $(\frac{1}{4}+i\delta)$-items, respectively. Note that $z_1 + z_2 + z_3 = k$. Since $3k$ items have to be packed in these bins, we have $a + 2b + z_1 + 2z_2 + 3z_3 \geq 3k$, hence $a + 2b + z_2 + 2z_3 \geq 2k$. The last inequality implies that $a + b + z_2 + z_3 \geq k$ and the claim holds.

Observation 2. *At the end of Stage 1 of Case 2.2.1, at least k bins are packed with either one ϵ-item and at least two $(\frac{1}{4}+i\delta)$-items, or one $(\frac{1}{4}+i\delta)$-item plus at least a load of $(\frac{1}{4}+\epsilon)_{*\epsilon}$.*

Stage 2. We aim at $n_2 \geq 2k$. Let $z = z_2 + z_3$. We let items depart until the configuration becomes

$$\{3k{:}(\tfrac{1}{2}+\epsilon)_{*\epsilon}, k{-}z{:}((\tfrac{1}{4}+\epsilon)_{*\epsilon}, \tfrac{1}{4}+i\delta), z{:}(\epsilon, \tfrac{1}{4}+i\delta, \tfrac{1}{4}+i\delta), 2k{:}\epsilon, 2k{:}(\tfrac{1}{4}+i\delta, \tfrac{1}{4}+i\delta)\} \ ,$$

with $8k$ bins and a total load of $3k + O(1)$.

Let $\delta' = \frac{\delta}{16k}$. We use Op-Inc with $x = \frac{1}{2}-6k\delta'$, $h = 2k$, and $s = 2k$. The x-items can only be packed in the $2k$ bins with load ϵ, at most one item in one bin, i.e., $h = 2k$. Any new bin can pack at most two, implying that Op-Inc releases $6k = h + 2s$ items of increasing sizes, from $\frac{1}{2}-6k\delta'$ to at most $\frac{1}{2}-\delta'$. According to Op-Inc, $\mathcal{A}$ has to open at least $2k$ new bins, i.e., $n_2 \geq 2k$.

Stage 3. In this stage, we aim at $n_3 \geq 2k$. We use Op-Comp which corresponds to Op-Inc in Stage 2. We let items depart until the configuration becomes

$$\{3k{:}(\tfrac{1}{2}+\epsilon)_{*\epsilon}, k{-}z{:}((\tfrac{1}{4}+\epsilon)_{*\epsilon}, \tfrac{1}{4}+i\delta), z{:}(\epsilon, \tfrac{1}{4}+i\delta, \tfrac{1}{4}+i\delta), 2k{:}\epsilon, 2k{:}(\tfrac{1}{4}+i\delta, \tfrac{1}{4}+i\delta), 2k{:}\tfrac{1}{2}-i\delta'\} \ ,$$

with $10k$ bins and a total load of $4k + O(1)$. We then use Op-Comp with $x = \frac{1}{2}-6k\delta'$, $y = \frac{1}{2}-5k\delta'$, $h = 2k$, and $s = 2k$. I.e., we release items of increasing size from $\frac{1}{2}+5k\delta'$ to $\frac{1}{2}+6k\delta'$, and at any time, at most $4k+1$ items are needed. The $2k$ bins of load ϵ can pack one such item. Suppose there are w, out of $2k$,

ϵ-bins that are not packed with a $\frac{1}{2}+i\delta'$-item. According to Op-Comp, $\mathcal{A}$ has to open $2k+w$ new bins.

Stage 4. In this stage, we aim at $n_4 \geq 2k-w$. We use Op-Comp which corresponds to Op-Inc in Stage 1. We let items depart until the configuration is

$$\{3k{:}(\tfrac{1}{4}+\epsilon)_{*\epsilon}, k-z{:}(\tfrac{1}{4}+\epsilon)_{*\epsilon}, z{:}(\epsilon, \tfrac{1}{4}+i\delta), 2k-w{:}(\epsilon, \tfrac{1}{2}+i\delta'), w{:}\epsilon, 2k{:}\tfrac{1}{4}+i\delta, 2k{:}\tfrac{1}{2}-i\delta', 2k+w{:}\tfrac{1}{2}+i\delta',\} ,$$

with $12k + w$ bins and a total load of $9k/2 + O(1)$. We then use Op-Comp with $x = \frac{1}{4}+\delta$, $y = \frac{1}{4}+2k\delta$, $h = w$, and $s = 2k - w$. I.e., we release items of sizes from $\frac{3}{4}-2k\delta$ to $\frac{3}{4}-\delta$ and at any time, at most $2k+1$ items are needed. Only w ϵ-bins can pack such item, i.e., $h = w$ as said. According to Op-Comp, $\mathcal{A}$ has to open $2k-w$ new bins.

Stage 5. In this final stage, we aim at $n_5 = 2k$. We let items depart until the configuration is

$$\{3k{:}\epsilon, k-z{:}\epsilon, z{:}\epsilon, 2k-w{:}\epsilon, w{:}\epsilon, 2k{:}\frac{1}{4}+i\delta, 2k{:}\frac{1}{2}-i\delta', 2k+w{:}\frac{1}{2}+i\delta', 2k-w{:}\frac{3}{4}-i\delta,\} ,$$

with $14k$ bins and a total load of $4k - \frac{w}{4} + O(1)$. Finally, we release $2k$ items of size 1 and $\mathcal{A}$ has to open $2k$ new bins. In total, $\mathcal{A}$ uses $3k + (k - z) + z + (2k - w) + w + 2k + 2k + (2k + w) + (2k - w) + 2k = 16k$. The packing of $\mathcal{O}$ will be given in the full paper.

Lemma 2. *If $\mathcal{A}$ uses $[6k, 8k)$ bins in Stage 0, $[2k, 4k)$ bins in Stage 1, and $m \geq 2k$, then $\mathcal{A}$ uses $16k$ bins at the end while $\mathcal{O}$ uses $6k + 3$ bins.*

Case 2.2.2: $6k \leq n_0 < 8k$, $2k \leq n_1 < 4k$ and $m < 2k$. We recall that in Stage 1, $15k$ items of size $\frac{1}{4}+i\delta$ are released and $\mathcal{A}$ uses $[2k, 4k)$ new bins for these items.

Observation 3. *(i) At most $8k$ items of size $\frac{1}{4}+i\delta$ can be packed to the n_1 new bins. (ii) At least k of the $\{k{:}\frac{1}{4}+i\delta, k{:}\epsilon\}$ bins have load more than $\frac{1}{2}$. (iii) At least $2k$ of the $\{4k{:}(\frac{1}{2}+\epsilon)_{*\epsilon}\}$ bins are packed with at least one $(\frac{1}{2}+i\delta)$-item.*

Let z_1 and z_2 be the number of new bins that are packed one and at least two, respectively, $(\frac{1}{4}+i\delta)$-items The following observation gives a bound on z.

Observation 4. *(i) At most $9k$ items of size $\frac{1}{4}+i\delta$ can be packed in existing bins. (ii) $z_2 \geq k$. (iii) $z_1 \geq 3(2k - z_2)$.*

Stage 2. We target $n_2 \geq z_2$. We let items depart until the configuration becomes

- $2k{:}(\frac{1}{2}+\epsilon)_{*\epsilon}$,
- $2k{:}((\frac{1}{4}+\epsilon)_{*\epsilon}, \frac{1}{4}+i\delta)$, this is possible because of Observation 3(iii),
- $x{:}(\epsilon, \frac{1}{4}+i\delta, \frac{1}{4}+i\delta)$,
- $k-x{:}((\frac{1}{4}+\epsilon)_{*\epsilon}, \frac{1}{4}+i\delta)$, this is possible because of Observation 3(ii),
- $k{:}\epsilon$,

- z_2:$(\frac{1}{4}+i\delta, \frac{1}{4}+i\delta)$, this is possible because of Observation 4(ii),
- $2(2k-z_2)$:$(\frac{1}{4}+i\delta)$, this is possible because of Observation 4(iii),

with $10k - z_2$ bins and a total load of $7k/2 + O(1)$. We then use Op-Inc with $x = \frac{1}{2}-5k\delta'$, $h = 5k - 2z_2$ and $s = z_2$. The x-items can only be packed in k of ϵ-bins and $2(2k - z_2)$ of $(\frac{1}{4}+i\delta)$-bins, i.e., $h = k + 2(2k - z_2) = 5k - 2z_2$ as said. Any new bin can pack at most two, implying that Op-Inc releases $5k = h + 2s$ items of increasing sizes from $\frac{1}{2}-5k\delta'$ to $\frac{1}{2}-\delta'$. According to Op-Inc, $\mathcal{A}$ has to open at least z_2 bins, i.e., $n_2 \geq z_2$.

Stage 3. We target $n_3 \geq z_2$. We let items depart until the configuration becomes

$$\{2k{:}(\frac{1}{2}+\epsilon)_{*\epsilon}, 2k{:}((\frac{1}{4}+\epsilon)_{*\epsilon}, \frac{1}{4}+i\delta), x{:}(\epsilon, \frac{1}{4}+i\delta, \frac{1}{4}+i\delta), k{-}x{:}((\frac{1}{4}+\epsilon)_{*\epsilon}, \frac{1}{4}+i\delta), k{:}\epsilon,$$
$$z_2{:}(\frac{1}{4}+i\delta, \frac{1}{4}+i\delta), 2(2k{-}z_2){:}\frac{1}{4}+i\delta, z_2{:}\frac{1}{2}-i\delta'\} ,$$

with $10k$ bins and a total load of $7k/2 + z_2/2 + O(1)$. We use Op-Comp with $s = z_2$ to release items of increasing size from $\frac{1}{2}+\delta'$. These items can only be packed in ϵ-bins (k of them) and $(\frac{1}{4}+i\delta)$-bins ($2(2k - z_2)$ of them). At any time, at most $(5k - z_2) + 1$ items are needed. According to Op-Comp, $\mathcal{A}$ has to open z_2 bins, i.e., $n_3 \geq z_2$.

Stage 4. We target $n_4 \geq (4k - z_2)$. We let items depart until the configuration becomes

$$\{4k{-}x{:}(\frac{1}{4}+\epsilon)_{*\epsilon}, k{+}x{:}(\epsilon, \frac{1}{4}+i\delta), k{:}\epsilon, 4k{-}z_2{:}\frac{1}{4}+i\delta, z_2{:}\frac{1}{2}-i\delta', z_2{:}\frac{1}{2}+i\delta'\} ,$$

with $10k+z_2+O(1)$ bins and a total load of $9k/4+3z_2/4$. We then use Op-Comp with $s = 4k - z_2$ and items of increasing size $\frac{3}{4}-i\delta$. Using similar ideas as before, $\mathcal{A}$ has to open $(4k - z_2)$ new bins.

Stage 5. We target $n_5 = 2k$. We let items depart until the configuration becomes

$$\{4k{-}x{:}\epsilon, k{+}x{:}\epsilon, k{:}\epsilon, 4k{-}z_2{:}\frac{1}{4}+i\delta, z_2{:}\frac{1}{2}-i\delta', z_2{:}\frac{1}{2}+i\delta', 4k{-}z_2{:}\frac{3}{4}-i\delta,\} ,$$

with $14k$ bins and a total load of $4k + O(1)$. We finally release $2k$ items of size 1 and $\mathcal{A}$ has to open $2k$ new bins. In total $\mathcal{A}$ uses $6k + 8k + 2k = 16k$ bins. The packing of $\mathcal{O}$ will be given in the full paper.

Lemma 3. *If $\mathcal{A}$ uses $[6k, 8k)$ bins in Stage 0, $[2k, 4k)$ bins in Stage 1, and $m < 2k$, then $\mathcal{A}$ uses $16k$ bins at the end while $\mathcal{O}$ uses $6k + 5$ bins.*

Theorem 5. *No online algorithm can be better than 8/3-competitive.*

5   Conclusion

We have derived a 8/3 $\sim$ 2.666 lower bound on the competitive ratio for dynamic bin packing, improving the best known 2.5 lower bound [5]. We designed two operations that release items of slightly increasing sizes and items with complementary sizes. These operations make a more systematic approach to release items: the type of item sizes used in a later case is a superset of those used in an earlier case. This is in contrast to the previous 2.5 lower bound in [5] in which rather different sizes are used in different cases. Furthermore, in each case, we use one or two pairs of Op-Inc and Op-Comp, which makes the structure clearer and the proof easier to understand. We also show that the new operations defined lead to a much easier proof for a 2.5 lower bound. An obvious open problem is to close the gap between the upper and lower bounds.

References

1. Balogh, J., Békési, J., Galambos, G., Reinelt, G.: Lower bound for the online bin packing problem with restricted repacking. SIAM J. Comput. 38, 398–410 (2008)
2. Bar-Noy, A., Ladner, R.E., Tamir, T.: Windows scheduling as a restricted version of bin packing. In: Munro, J.I. (ed.) SODA, pp. 224–233. SIAM (2004)
3. Borodin, A., El-Yaniv, R.: Online Computation and Competitive Analysis. Cambridge University Press (1998)
4. Chan, J.W.-T., Lam, T.W., Wong, P.W.H.: Dynamic bin packing of unit fractions items. Theoretical Computer Science 409(3), 172–206 (2008)
5. Chan, J.W.-T., Wong, P.W.H., Yung, F.C.C.: On dynamic bin packing: An improved lower bound and resource augmentation analysis. Algorithmica 53(2), 172–206 (2009)
6. Coffman Jr., E.G., Galambos, G., Martello, S., Vigo, D.: Bin packing approximation algorithms: Combinatorial analysis. In: Du, D.-Z., Pardalos, P.M. (eds.) Handbook of Combinatorial Optimization. Kluwer Academic Publishers (1998)
7. Coffman Jr., E.G., Garey, M.R., Johnson, D.S.: Dynamic bin packing. SIAM J. Comput. 12(2), 227–258 (1983)
8. Coffman Jr., E.G., Garey, M.R., Johnson, D.S.: Bin packing approximation algorithms: A survey. In: Hochbaum, D.S. (ed.) Approximation Algorithms for NP-Hard Problems, pp. 46–93. PWS (1996)
9. Csirik, J., Woeginger, G.J.: On-line Packing and Covering Problems. In: Fiat, A., Woeginger, G.J. (eds.) Online Algorithms 1996. LNCS, vol. 1442, pp. 147–177. Springer, Heidelberg (1998)
10. Epstein, L., Levy, M.: Dynamic multi-dimensional bin packing. Journal of Discrete Algorithms 8, 356–372 (2010)
11. Garey, M.R., Johnson, D.S.: Computers and Intractability: A Guide to the Theory of NP-Completeness. W.H. Freeman, San Francisco (1979)
12. Han, X., Peng, C., Ye, D., Zhang, D., Lan, Y.: Dynamic bin packing with unit fraction items revisited. Information Processing Letters 110, 1049–1054 (2010)
13. Ivkovic, Z., Lloyd, E.L.: A fundamental restriction on fully dynamic maintenance of bin packing. Inf. Process. Lett. 59(4), 229–232 (1996)
14. Seiden, S.S.: On the online bin packing problem. J. ACM 49(5), 640–671 (2002)
15. van Vliet, A.: An improved lower bound for on-line bin packing algorithms. Information Processing Letters 43(5), 277–284 (1992)
16. Wong, P.W.H., Yung, F.C.C.: Competitive Multi-dimensional Dynamic Bin Packing via L-Shape Bin Packing. In: Bampis, E., Jansen, K. (eds.) WAOA 2009. LNCS, vol. 5893, pp. 242–254. Springer, Heidelberg (2010)

Computing k-center over Streaming Data for Small $k^\star$

Hee-Kap Ahn, Hyo-Sil Kim, Sang-Sub Kim, and Wanbin Son

Pohang University of Science and Technology, Korea
{heekap,allisabeth,helmet1981,mnbiny}@postech.ac.kr

Abstract. The Euclidean k-center problem is to compute k congruent balls covering a given set of points in $\mathbb{R}^d$ such that the radius is minimized. We consider the k-center problem in $\mathbb{R}^d$ for $k = 2, 3$ in a single-pass streaming model, where data is allowed to be examined once and only a small amount of information can be stored in a device. We present two approximation algorithms whose space complexity does not depend on the size of the input data. The first algorithm guarantees a $(2+\varepsilon)$-factor using $O(d/\varepsilon)$ space in arbitrary dimensions, and the second algorithm guarantees a $(1+\varepsilon)$-factor using $O(1/\varepsilon^d)$ space in constant dimensions. The same algorithms can be used to compute a k-center under any L_p metric for $k = 2, 3$.

1 Introduction

A *clustering*, one of the fundamental problems in computer science, is to partition a given set into subsets, called *clusters*, subject to various objective functions. In the past, most of the clustering problems are considered in the static setting (off-line), that is, the data is known in advance. In recent decades, massive data set made a *streaming model* extremely popular as the size of memory is much smaller than the size of data. In this paper, we consider a *single-pass streaming model* [4], where data is allowed to be examined once and only a limited amount of information can be stored in a device. In this model, it is important to develop an algorithm whose space complexity does not depend on the size of input, since the input size in data streams is typically huge.

Among various clustering problems, we consider the *minmax radius clustering*, also known as the (metric) *k-center problem*: Given a set P of n points in d-dimensional metric space, find k points called *centers* such that the maximum distance between a point in P and its nearest center is minimized. The metric k-center problem can be formulated to find k congruent balls covering P such that the radius is minimized. The *discrete* k-center problem requires that obtained k centers be a subset of P.

In this paper, we study the Euclidean k-center problem in a single-pass streaming model for $k = 2, 3$. We show that our approach can be used to compute a k-center under any L_p metric for $k = 2, 3$.

---

* This research is supported by the NRF grant 2011-0030044 (SRC-GAIA) funded by the government of Korea.

K.-M. Chao, T.-s. Hsu, and D.-T. Lee (Eds.): ISAAC 2012, LNCS 7676, pp. 54–63, 2012.

Previous Work in the Static Setting. If k is a part of input, the k-center problem is NP-hard [11], even in the plane [17]. In fact, it is known to be NP-hard to approximate within factor 2 for arbitrary metric spaces, and within factor 1.822 for the Euclidean space [5]. For fixed k and d, Agarwal and Procopiuc [2] gave an exact algorithm that runs in $n^{O(k^{1-1/d})}$ time for any L_p metric.

If k and d are small, there exist more efficient algorithms. For the Euclidean 1-center problem (computing the smallest ball enclosing P), there exists a linear time algorithm for any fixed dimension [10]. For the Euclidean 2-center problem in the plane, the best known algorithm is given by Chan [6] and it runs deterministically in $O(n \log^2 n \log^2 \log n)$ time using $O(n)$ space.

Previous Work on Data Streams. The *coreset* framework is one of the fundamental tools for designing streaming algorithms since it captures an approximate "shape" of input in small size; see [1,7,19]. For the k-center problem, Zarrabi-Zadeh [14] shows that one can maintain an ε-coreset using $O(k/\varepsilon^d)$ space under any L_p metric, which is, to authors' knowledge, the only known result of obtaining ε-coresets for the k-center problem whose space does not depend on n.

While the coreset framework guarantees an ε-approximate solution to the problem, its exponential dependency on d is not so attractive in high dimensions. Some work which uses polynomial space in d has been presented independently; Charikar et al. [9] gave an 8-approximation algorithm using $O(dk)$ space and Guha [12] gave a $(2+\varepsilon)$-approximation algorithm using $O((dk/\varepsilon) \log(1/\varepsilon))$ space to the k-center problem for any metric space. (see also [15] for an algorithm similar to Guha [12]).

For small k, especially for the Euclidean 1-center, the problem is well studied. In fixed dimensions, one can devise a $(1 + \varepsilon)$-approximation algorithm using $O(1/\varepsilon^{(d-1)/2})$ space by maintaining extreme points along a number of different directions (a generalization of the algorithm given by Hershberger and Suri [13]). For arbitrary dimensions, Zarrabi-Zadeh and Chan [20] gave a 1.5-approximation algorithm maintaining only one center and one radius, which is the minimal amount of storage to compute a 1-center. Agarwal and Sharathkumar [3] developed a $(1 + \sqrt{3})/2 + \varepsilon \approx 1.3661$-approximation algorithm using $O((d/\varepsilon^3) \log(1/\varepsilon))$ space, showing that any algorithm in the one-pass streaming algorithm with space polynomially bounded in d cannot approximate an optimal 1-center within factor $(1 + \sqrt{2})/2 > 1.207$. Chan and Pathak [8] improved the approximation factor to 1.22 by analysing their algorithm more carefully.

For the Euclidean 2-center problem, Poon and Zhu [18] propose a 5.708-approximation algorithm for $d > 1$ (for $d = 1$, the same algorithm guarantees a factor of 2) using the minimum space, that is, two centers and one radius.

We are not aware of any other result on the k-center problem for the special case of $k > 2$.

Our Results. We present two approximation algorithms for computing a Euclidean 2-center in a single-pass streaming model. Both of our algorithms improve the previous known results for the k-center problem for $k = 2$ in approximation factor, space or update time.

The first algorithm MELP guarantees a $(2 + \varepsilon)$-approximation that works in arbitrary dimensions using $O(d/\varepsilon)$ space and $O(d/\varepsilon)$ update time. In fact, one can easily show that MELP can be used in any metric space using *oracle distance model* and also in the discrete setting. While Guha's algorithm [12], when plugged in $k = 2$, uses $O(d/\varepsilon \log(1/\varepsilon))$ space and $O((dn/\varepsilon) \log(1/\varepsilon) + (1/\varepsilon) \log r^*)$ time in total, where r^* is the optimal radius of the two clusters, MELP uses less space and less running time. Especially, note that the running time of our algorithm does not depend on the optimal radius. Compared to Poon and Zhu [18], MELP improves their approximation factor by relaxing the (too strict) restriction that maintains only two centers and one radius.

Our second algorithm MEDG guarantees a $(1 + \varepsilon)$-approximation to the Euclidean 2-center problem using $O(1/\varepsilon^d)$ space and $O(1)$ amortized update time (through a certain *lazy update* procedure) for constant dimensions. The running time and space complexity of MEDG is asymptotically same as the one by Zarrabi-Zadeh [14], where the proof for the time complexity is not given clearly. In this paper, we provide a complexity analysis and a correctness proof of our algorithm.

In the full version of this paper, we show that our algorithms can be applied to the Euclidean 3-center problem. Its extension to the L_p-metric for $p \geqslant 1$ is rather straightforward and we omit the proof. See Table 1 for summary.

Table 1. Our results for computing a k-center under L_p for $k = 2, 3$

	approx. factor	space	update time	remarks
MELP	$2 + \varepsilon$	$O(d/\varepsilon)$	$O(d/\varepsilon)$	in arbitrary dimensions
MEDG	$1 + \varepsilon$	$O(1/\varepsilon^d)$	$O(1)$ amortized	in constant dimensions

2 Preliminaries

Let P be a set of n points in d-dimensional Euclidean space $\mathbb{R}^d$. In a single-pass streaming model, a point in P is arriving one by one, and is allowed to be examined only once. The points in P are labeled in order of their arrivals. That is, p_i is a point in P that has arrived at the i-th step. We denote by P_i a subset of points in P that have arrived until the i-th step, that is, $P_i = \{p_1, p_2, \ldots, p_i\}$.

An *optimal solution* is a solution of the 2-center problem in $\mathbb{R}^d$, when all the points in P are assumed to be known in advance.

Let B_1^* and B_2^* denote the two congruent balls in an optimal solution, c_1^* and c_2^* denote their centers, r^* denote their radius, and δ^* denote the distance between B_1^* and B_2^*, that is, $\|c_1^* c_2^*\| = 2r^* + \delta^*$. Let P_1^* and P_2^* be subsets of P such that $P_1^* = P \cap B_1^*$ and $P_2^* = P \cap B_2^*$. If B_1^* and B_2^* are disjoint, so are P_1^* and P_2^*. If the notations are used without *, they denote an approximate solution generated by our algorithms.

Let $B(c, r)$ denote a ball of radius r centered at c, and $r(B)$ denote the radius of a ball B.

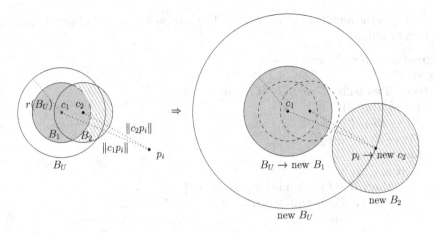

Fig. 1. When p_i arrives, MERGE occurs if $r(B_U) \leqslant \min\{\|c_1 p_i\|, \|c_2 p_i\|\}$

Overview. The rest of the paper is organized as follows. In Section 3 we consider when the point set P is well-separated, that is, δ^* is relatively big compared to the optimal radius r^*. We consider $\delta^* > 2r^*$ and devise an algorithm *MergeExpand* which partitions P optimally. We prove that this algorithm guarantees a 2-factor to the Euclidean 2-center problem using $O(d)$ space for $\delta^* > 2r^*$. By maintaining each partition using $O(1/\varepsilon^d)$ grid cells, the algorithm guarantees a $(1+\varepsilon)$-factor using $O(1/\varepsilon^d)$ space in fixed dimensions.

In Section 4 we consider the case $\delta^* \leqslant 2r^*$. Unlike $\delta^* > 2r^*$, it is tricky to partition P optimally in the single-pass streaming setting. We present two algorithms. In Section 4.1 we develop an algorithm *LayerPartition* which maintains $O(1/\varepsilon)$ partitions and guarantees a $(2+\varepsilon)$-factor to the Euclidean 2-center problem. To obtain a $(1+\varepsilon)$-approximation, in Section 4.2, we introduce another algorithm *DoublingGrid* which maintains $O(1/\varepsilon^d)$ grid cells and computes a 2-center over those grid cells when needed[1].

3   The Case $\delta^* > 2r^*$

We here introduce an algorithm *MergeExpand* that can be used for $\delta^* > 2r^*$.

MergeExpand always maintains a ball B_U centered at p_1 and whose radius is determined by the distance between p_1 and its farthest point in P_{i-1}. The basic procedure of *MergeExpand* is as follows: When a new point p_i arrives, if p_i is contained in one of the current two balls B_1 and B_2, we just update B_U if $p_i \notin B_U$. Otherwise, we enter an update-stage where either MERGE- or EXPAND-operation occurs. MERGE replaces the current two balls B_1 and B_2 into B_U and $B(p_i, r(B_U))$, respectively. See Figure 1. EXPAND replaces B_1 and B_2 to two new balls such that the centers remain the same but the radius becomes $\min\{\|c_1 p_i\| \|c_2, p_i\|\}$. Then we update B_U if $p_i \notin B_U$. We choose the operation

---

[1] Note that *DoublingGrid* works only for constant dimensions, since the 2-center problem is NP-complete if d is not fixed [16].

which makes the updated radius smaller. The precise description of the algorithm is given as follows:

Algorithm *MergeExpand*
Input: P
Output: Two balls B_1 and B_2 of radius r
1. $B_1 \leftarrow B(p_1, 0)$
2. $B_2 \leftarrow B(p_2, 0)$
3. $B_U \leftarrow B(p_1, \|p_1 p_2\|)$
4.   **for** $i \leftarrow 3$ **to** n
5.   **if** $p_i \notin B_1 \cup B_2$
6.   **then if** $r(B_U) \leqslant \min\{\|c_1 p_i\|, \|c_2 p_i\|\}$
7.   **then** MERGE $B_1 \leftarrow B_U, B_2 \leftarrow B(p_i, r(B_U))$.
8.   **else** EXPAND $B_1 \leftarrow B(c_1, \|c_j p_i\|)$ and $B_2 \leftarrow B(c_2, \|c_j p_i\|)$, where $j = 1$ if $\|c_1 p_i\| \leqslant \|c_2 p_i\|$ and $j = 2$ otherwise.
9. $B_U \leftarrow B(p_1, \|p_1 p_i\|)$ if $p_i \notin B_U$.
10. **return** B_1 and B_2

Note that *MergeExpand* computes *discrete* two centers, which means that the two centers of B_1 and B_2 are the input points of P. Especially, B_1 and B_U always take p_1 as their centers. The algorithm sets the radii of B_1 and B_2 the same. The space that the algorithm maintains is then the coordinates of the centers of B_1 and B_2 and the radii of B_1 and B_U.

We now show that for $\delta^* > 2r^*$, *MergeExpand* guarantees an *optimal partition* of P, which means that $P_1^* \subset B_1$, $P_2^* \subset B_2$, $P_1^* \cap B_2 = \emptyset$, and $P_2^* \cap B_1 = \emptyset$.

Lemma 1. *For $\delta^* > 2r^*$, MergeExpand guarantees an optimal partition.*

Proof. Without loss of generality, assume that $P_{i-1} \subset B_1^*$, and p_i is the first point that lies in B_2^*. We claim that this moment is the last moment when a MERGE-operation occurs. Indeed, at the ith-step, MERGE occurs since $r(B_U) \leqslant 2r^*$ and $\|c_j p_i\| \geqslant \delta^* > 2r^*$ for $j = 1, 2$. For any $l > i$, we have $r(B_U) > \min\{\|c_1 p_l\|, \|c_2 p_l\|\}$ since $r(B_U) \geqslant \delta^*$ and $\|c_j p_l\| \leqslant 2r^* < \delta^*$ for at least one of $j = 1, 2$, and the claim follows.

Since after ith-step only EXPAND-operations occur, this proves lemma.   □

Corollary 1. *For $\delta^* > 2r^*$, MergeExpand computes a 2-approximation to the 2-center problem using $O(d)$ space and update time.*

Proof. By Lemma 1, $P_1^* \subset B_1$ and $P_2^* \subset B_2$, $P_1^* \cap B_2 = \emptyset$, and $P_2^* \cap B_1 = \emptyset$. As noted, c_1 and c_2 are the input points of P and so $c_1 \in P_1^*$ and $c_2 \in P_2^*$. Since the distance between any two points lying in B_1^* is at most $2r^*$ and the radius of B_1 is determined by one of those distances, we have $r(B_1) \leqslant 2r^*$. The same holds for B_2. An update takes $O(d)$ to compute the distances and the corollary follows.   □

MergeExpand can be modified to use $O(1/\varepsilon^d)$ grid cells to maintain each partition of P using a technique called a *doubling*, which will be described in Section 4.2. Let us call it *MergeExpand$_\varepsilon$*. Due to the space limit we omit the proof and only state the result.

Corollary 2. *For $\delta^* > 2r^*$, MergeExpand$_\varepsilon$ computes a $(1 + \varepsilon)$-approximation to the 2-center problem using $O(1/\varepsilon^d)$ space and $O(1)$ amortized time per update in fixed dimensions under any L_p metric.*

4   The Case $\delta^* \leqslant 2r^*$

For the case $\delta^* \leqslant 2r^*$, it is not so easy to partition the point set optimally, since the optimal two balls are too close and we cannot keep enough information in the single-pass streaming setting. Indeed, one can show that any algorithm that assigns the partition of a point p_i definitively (like *MergeExpand*) cannot guarantee an approximation factor within 2. In this section, we suggest two algorithms that choose the best partition upon request.

4.1   LayerPartition

The first algorithm *LayerPartition* always maintains *layered* $m = \lceil 12/\varepsilon \rceil$ balls centered at p_1. Let $\mathcal{B}_i = \{b_1, b_2, \ldots, b_m\}$ denote such m balls for P_i. For each ball, we define a partition which divides the point set into two, one lying in b_j and the other lying in $\mathbb{R}^d \setminus b_j$. For each partition of b_j, we maintain two balls B_{j1} and B_{j2} of *discrete* centers to enclose the partitioned points. See Figure 2.

   The algorithm starts from two input points p_1 and p_2. Initially it creates layered m balls centered at p_1 and whose radii are $(j/m) \cdot \|p_1 p_2\|$ for $j = 1, 2, \ldots, m$. When a new point $p_i \in P$ arrives, if it is contained in b_m then we update each partition by expanding B_{j1} if $p_i \in b_j$ or B_{j2} if $p_i \in \mathbb{R}^d \setminus b_j$ for $j = 1, \ldots, m$. If p_i is not contained in b_m, we create new m layered balls. Let x be the smallest integer satisfying $2^x r(b_m) \geqslant \|p_1 p_i\| > 2^{x-1} r(b_m)$. The new balls b'_j have radius $(j/m) \cdot 2^x r(b_m)$ for $j = 1, \ldots, m$. If the new ball b'_j coincides with the old ball, we keep its partitioning information and update for p_i. Otherwise we create the new partitions; B'_{j1} has p_1 as its center. B'_{j2} has p_i as its center if p_i lies in $\mathbb{R}^d \setminus b'_j$ and otherwise we let $B_{j2} = \emptyset$. Let $r(j)$ denote $\max\{r(B_{j1}), r(B_{j2})\}$.

Algorithm *LayerPartition*
Input: P, ε
Output: Two balls B_1 and B_2 of radius r
1.   Accept two input points p_1, p_2.
2. $m \leftarrow \lceil 12/\varepsilon \rceil$
3.   Make m layered balls b_j whose radii are $(j/m) \cdot \|p_1 p_2\|$ for $j = 1, 2, \ldots, m$.
4.   Make a partition of b_j, $\{B_{j1} = B(p_1, 0), B_{j2} = B(p_2, 0)\}$ for $j = 1, 2, \ldots, m$.
5.   **for** $i \leftarrow 3$ **to** n
6.   **if** $p_i \in b_m$
7.   **then** Update each partition of b_j for $j = 1, \ldots, m$
8.   **else** Make new m layered balls whose radii are $(j/m) \cdot 2^x r(b_m)$ for $j = 1, 2, \ldots, m$, and update information.
9.   Take j such that $r(j)$ is the smallest among m partitions.
10. **return** B_{j1} and B_{j2}

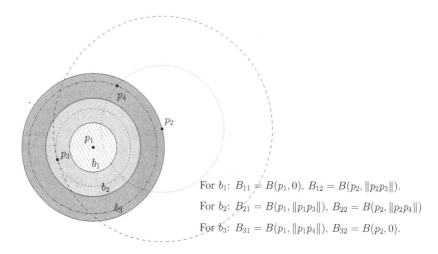

For b_1: $B_{11} = B(p_1, 0)$, $B_{12} = B(p_2, \|p_2p_3\|)$.

For b_2: $B_{21} = B(p_1, \|p_1p_3\|)$, $B_{22} = B(p_2, \|p_2p_4\|)$

For b_3: $B_{31} = B(p_1, \|p_1p_4\|)$, $B_{32} = B(p_2, 0)$.

Fig. 2. Three layered balls and their corresponding partitions

Lemma 2. *For $\delta^* \leqslant 2r^*$, LayerPartition guarantees a $(2 + \varepsilon)$-approximation to the 2-center problem using $O(d/\varepsilon)$ space and update time.*

Proof. Let B_1^* be the ball in an optimal solution which contains p_1. We look at $\mathcal{B}_n$, the final layered m balls. To bound the radius of $b_j \in \mathcal{B}_n$ for $j = 1, \ldots, m$, we think about the moment when p_i, which causes creating $\mathcal{B}_n$, arrives. Note that we have $\mathcal{B}_{i-1} \neq \mathcal{B}_i$ and $\mathcal{B}_i = \mathcal{B}_{i+1} = \cdots = \mathcal{B}_n$.

Let us denote the balls in $\mathcal{B}_{i-1}$ by $\overline{b}_j$ for $j = 1, \ldots, m$. By definition $r(b_m) = 2^x r(\overline{b_m})$, where x satisfies $2^{x-1} r(\overline{b_m}) < \|p_1 p_i\|$. Since $\delta^* \leqslant 2r^*$, any pairwise distance in P cannot exceed $6r^*$, and thus $\|p_1 p_i\| \leqslant 6r^*$ and $r(b_m) = 2^x r(\overline{b_m}) < 2\|p_1 p_i\| \leqslant 12r^*$. Since $m = \lceil 12/\varepsilon \rceil$, $r(b_j) = (j/m) \cdot r(b_m) < j \cdot \varepsilon r^*$ for $j = 1, \ldots, m$.

Consider now the smallest ball b_j that contains $P_1^* = B_1^* \cap P$. If $j = 1$, $r(b_1) < \varepsilon r^*$, and b_1 gives a $(2 + \varepsilon)$-approximation. Otherwise, look at b_{j-1}. Note that b_{j-1} contains a proper subset of P_1^*. Since the distance between any pair of points in P_1^* is at most $2r^*$ and p_1 is the center of b_{j-1}, $r(b_{j-1}) < 2r^*$. Therefore,

$$r(b_j) = (j/m) \cdot r(b_m) = r(b_{j-1}) + (1/m)r(b_m) < 2r^* + \varepsilon r^* = (2 + \varepsilon)r^*.$$

Since b_j contains all the points in P_1^*, the points lying in $\mathbb{R}^d \backslash b_j$ must be contained in P_2^* and thus have radius at most r^*. Since the center of B_{j2} is one of the input points, its radius can be at most $2r^*$, which proves the lemma. $\quad\square$

By running *MergeExpand* and *LayerPartition* together and taking the minimum radius of the balls obtained by these subroutines, we have an algorithm MELP:

Theorem 1. MELP *works in arbitrary dimensions and guarantees a $(2 + \varepsilon)$-approximation to the 2-center problem using $O(d/\varepsilon)$ space and update time.*

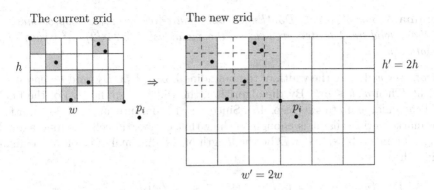

Fig. 3. Create a new grid by doubling the width and the height of the current grid upon arrival of p_i; gray grid cells are marked and do not contain any actual point

4.2 DoublingGrid

To obtain a $(1 + \varepsilon)$-approximation factor, we develop an algorithm called *DoublingGrid*. We first explain the algorithm in the plane and its extension to the general dimensions is straightforward.

It initially makes a $\lceil c/\varepsilon \rceil \times \lceil c/\varepsilon \rceil$ grid for the bounding box of p_1 and p_2, where $c = 24\sqrt{2}$. If a new point p_i is contained in the current grid, we mark the cell which contains p_i, and otherwise, we perform an operation called a *doubling*: we make a new grid by doubling the width or the height or both of the current grid so that the new grid contains p_i. See Figure 3. Let w and h be the width and the height of the current grid, respectively. Then the width w' and the height h' of the new grid is $2^j w$ and $2^k h$, where j and k is the smallest integer such that the new grid contains p_i.

Algorithm *DoublingGrid2D*
Input: P, ε
Output: Two balls B_1 and B_2 of radius r
1.   Make a bounding box of p_1 and p_2
2. $c \leftarrow 24\sqrt{2}$
3.   Make a $\lceil c/\varepsilon \rceil \times \lceil c/\varepsilon \rceil$ grid G on the bounding box of p_1 and p_2.
4.   Mark the cells that contains p_1 and p_2.
5.   **for** $i \leftarrow 3$ **to** n
6.   **if** $p_i \in G$
7.   **then** Mark the cell containing p_i.
8.   **else** Make a new grid by doubling G such that it contains p_i.
9.   Mark cells in the new grid if it contains points.
10. Compute an exact 2-center for marked grid cells.
11. **return** B_1 and B_2

Computing an exact 2-center in the plane can be done in $O((1/\varepsilon^2) \log^2(1/\varepsilon) \log^2 \log(1/\varepsilon))$ time by Chan [6].

Lemma 3. *For $\delta^* \leqslant 2r^*$, DoublingGrid2D guarantees a $(1 + \varepsilon)$-approximation to the Euclidean 2-center problem in the plane using $\lceil (24\sqrt{2}/\varepsilon) \rceil^2$ space and update time.*

Proof. Let w denote the width of the bounding box of $B_1^* \cup B_2^*$, and h denote its height. Then $w, h \leqslant 6r^*$. By the algorithm, the width (resp. height) of the final grid can enlarge up to $4w$ (resp. $4h$). Since we will compute an exact 2-center for the marked grid cells, it is enough to show that every grid cell has its diagonal length bounded by εr^*. Let D be the length of the diagonal of a cell in the final grid. Then

$$D = \left(4w/\lceil c/\varepsilon \rceil \right)^2 + (4h/\lceil c/\varepsilon \rceil)^2 \right)^{1/2} \leqslant (4\varepsilon/c)(w^2 + h^2)^{1/2} \leqslant \varepsilon r^*,$$

since $(w^2 + h^2)^{1/2} \leqslant 6\sqrt{2}r^*$ and $c = 24\sqrt{2}$. For each update, we may need to update every cell, which yields the time complexity. □

Note that *DoublingGrid2D* works for any L_p metric by replacing $c = 24\sqrt{2}$ by $c = 24 \cdot 2^{1/p}$. Note also that since we use $O(1/\varepsilon^2)$ space anyway, we can *postpone* an update of the grid until we have $O(1/\varepsilon^2)$ new points. We know that which point will cause the last doubling so we can compute the side-length of the new grid and update the cell information *once* for the $O(1/\varepsilon^2)$ points, in which we spend $O(1)$ amortized time per update.

The generalization of *DoublingGrid2D* to a d-dimensional L_p-metric space is straightforward, which we call *DoublingGrid* (here, we can use Agarwal and Procopiuc's algorithm to compute an exact 2-center [2]).

By running *DoublingGrid* and *MergeExpand$_\varepsilon$*, we obtain an algorithm MEDG:

Theorem 2. MEDG *guarantees a $(1+\varepsilon)$-approximation to the Euclidean 2-center problem by using $O(1/\varepsilon^d)$ space and $O(1)$ amortized time per update in fixed dimensions under any L_p metric.*

Due to the limit of space, we only state the results on the 3-center problem and postpone the proofs in the full version of this paper.

Theorem 3. *The streaming 3-center problem can be approximated by a factor of $(2 + \varepsilon)$ using $O(d/\varepsilon)$ space and update time in arbitrary dimensions, and by a factor of $(1 + \varepsilon)$ using $O(1/\varepsilon^d)$ space and $O(1)$ amortized time per update in fixed dimensions under any L_p metric.*

5   Conclusions

In this paper, we mainly consider computing a Euclidean 2-center in a single-pass streaming setting. Because of the limitation that we are not allowed to see the streaming data more than twice, it is not so easy to devise algorithms that guarantees a good approximation factor using as small space as possible. Nevertheless, we have provided two algorithms: for any given constant $\varepsilon > 0$, MELP guarantees a $(2 + \varepsilon)$-factor using $O(d/\varepsilon)$ space for arbitrary dimensions and MEDG guarantees a $(1 + \varepsilon)$-factor using $O(1/\varepsilon^d)$ in fixed dimensions.

We do not know any better lower bound than $(1 + \sqrt{2})/2 \approx 1.207$ for the worst-case approximation ratio of streaming algorithms using space polynomially bounded in d (this lower bound can be achieved from Agarwal and Shrathkumar's [3]); we suspect that the tight lower bound is higher. Another interesting question is whether we really need $\Omega(1/\varepsilon^d)$ space to obtain a $(1 + \varepsilon)$-factor.

References

1. Agarwal, P.K., Har-Peled, S., Varadarajan, K.R.: Approximating extent measures of points. Journal of the ACM 51(4), 606–635 (2004)
2. Agarwal, P.K., Procopiuc, C.M.: Exact and approximation algorithms for clustering. Algorithmica 33, 201–226 (2002)
3. Agarwal, P.K., Sharathkumar, R.: Streaming algorithms for extent problems in high dimensions. In: Proc. of the 21st ACM-SIAM Sympos. Discrete Algorithms, pp. 1481–1489 (2010)
4. Aggarwal, C.C.: Data streams: models and algorithms. Springer (2007)
5. Bern, M., Eppstein, D.: Approximation algorithms for geometric problems. In: Approximation Algorithms for NP-Hard Problems. PWS Publishing Co. (1996)
6. Chan, T.M.: More planar two-center algorithms. Computational Geometry 13(3), 189–198 (1999)
7. Chan, T.M.: Faster core-set constructions and data-stream algorithms in fixed dimensions. Computational Geometry 35, 20–35 (2006)
8. Chan, T.M., Pathak, V.: Streaming and Dynamic Algorithms for Minimum Enclosing Balls in High Dimensions. In: Dehne, F., Iacono, J., Sack, J.-R. (eds.) WADS 2011. LNCS, vol. 6844, pp. 195–206. Springer, Heidelberg (2011)
9. Charikar, M., Chekuri, C., Feder, T., Motwani, R.: Incremental clustering and dynamic information retrieval. SIAM J. Comput. 33(6), 1417–1440 (2004)
10. Chazelle, B., Matoušek, J.: On linear-time deterministic algorithms for optimization problems in fixed dimension. Journal of Algorithms 21, 579–597 (1996)
11. Garey, M.R., Johnson, D.S.: Computers and Intractability: A Guide to the Theory of NP-Completeness. W.H. Freeman, New York (1979)
12. Guha, S.: Tight results for clustering and summarizing data streams. In: Proc. of the 12th Int. Conf. on Database Theory, pp. 268–275. ACM (2009)
13. Hershberger, J., Suri, S.: Adaptive sampling for geometric problems over data streams. Computational Geometry 39(3), 191–208 (2008)
14. Zarrabi-Zadeh, H.: Core-preserving algorithms. In: Proc. of 20th Canadian Conf. on Comput. Geom. (CCCG), pp. 159–162 (2008)
15. McCutchen, R.M., Khuller, S.: Streaming Algorithms for k-Center Clustering with Outliers and with Anonymity. In: Goel, A., Jansen, K., Rolim, J.D.P., Rubinfeld, R. (eds.) APPROX and RANDOM 2008. LNCS, vol. 5171, pp. 165–178. Springer, Heidelberg (2008)
16. Megiddo, M.: On the complexity of some geometric problems in unbounded dimension. J. Symbolic Comput. 10, 327–334 (1990)
17. Megiddo, M., Supowit, K.J.: On the complexity of some common geometric location problems. SIAM J. Comput. 13(1), 182–196 (1984)
18. Poon, C.K., Zhu, B.: Streaming with Minimum Space: An Algorithm for Covering by Two Congruent Balls. In: Lin, G. (ed.) COCOA 2012. LNCS, vol. 7402, pp. 269–280. Springer, Heidelberg (2012)
19. Zarrabi-Zadeh, H.: An almost space-optimal streaming algorithm for coresets in fixed dimensions. Algorithmica 60, 46–59 (2011)
20. Zarrabi-Zadeh, H., Chan, T.M.: A simple streaming algorithm for minimum enclosing balls. In: Proc. of 18th CCCG, pp. 139–142 (2006)

Precision vs Confidence Tradeoffs for ℓ_2-Based Frequency Estimation in Data Streams

Sumit Ganguly

Indian Institute of Technology

Abstract. We consider the data stream model where an n-dimensional vector x is updated coordinate-wise by a stream of updates. The frequency estimation problem is to process the stream in a single pass and using small memory such that an estimate for x_i for any i can be retrieved. We present the first algorithms for ℓ_2-based frequency estimation that exhibit a tradeoff between the precision (additive error) of its estimate and the confidence on that estimate, for a range of parameter values. We show that our algorithms are optimal for a range of parameters for the class of matrix algorithms, namely, those whose state corresponding to a vector x can be represented as Ax for some $m \times n$ matrix A. All known algorithms for ℓ_2-based frequency estimation are matrix algorithms.

1 Introduction

The problem of estimating frequencies is one of the most basic problems in data stream processing. It is used for tracking heavy-hitters in low space and real time, for example, finding popular web-sites accessed, most frequently accessed terms in search-engines, popular sale items in supermarket transaction database, etc.. In the general *turnstile* data streaming model, an n-dimensional vector x is updated by a sequence of update entries of the form (i, v). Each update (i, v) transforms $x_i \leftarrow x_i + v$. The frequency estimation problem is to design a data structure and an algorithm $\mathcal{A}$ that (i) processes the input stream *in a single pass using as little memory as possible*, and, (ii) given any $i \in [n]$, uses the structure to return an estimate $\hat{x}_i$ for x_i satisfying, $|\hat{x}_i - x_i| \leq Err_{\mathcal{A}}$, with confidence $1 - \delta$, where, C is a space parameter of $\mathcal{A}$ and $Err_{\mathcal{A}}$ denotes the *precision* or the additive error of the estimation. We consider frequency estimation algorithms whose error guarantees are in terms of the ℓ_2-norm. The COUNTSKETCH algorithm by Charikar et. al. [1] is the most well-known ℓ_2-based frequency estimation and has precision $Err_{\mathsf{CSK}} = \|x^{\mathrm{res}(C)}\|_2/\sqrt{C}$ and confidence $1 - n^{-\Omega(1)}$. Here, $\|x^{\mathrm{res}(C)}\|_2$ is the second norm of x calculated after removing the top-C absolute frequencies from it. The residual norm is often smaller than the standard norm, since in many scenarios, much of the energy of x may concentrate in the top few frequencies.

Precision-Confidence Trade-offs. Let us associate with a randomized estimation algorithm $\mathcal{A}$ running on an input x, a pair of numbers namely, (1) its

K.-M. Chao, T.-s. Hsu, and D.-T. Lee (Eds.): ISAAC 2012, LNCS 7676, pp. 64–74, 2012.
© Springer-Verlag Berlin Heidelberg 2012

Work	Precision	Failure Probability	Space O(words)	Update time $O(\cdot)$	Estimation Time $O(\cdot)$
COUNTSKETCH[1]	$\|x^{\text{res}(C)}\|_2/\sqrt{C}$	$n^{-\Omega(1)}$	$C\log n$	$\log n$	n
ACSK-I $4 \le d \le O(\log n)$	$\|x^{\text{res}(C)}\|_2$ $\times\sqrt{d/(C\log n)}$	$n^{-\Omega(1)}$ $+2^{-d}$	$(C+\log n)$ $\times \log n$	$\log^2 n$	n
ACSK-II $4 \le d \le O(\log(n/C))$	$\|x^{\text{res}(C)}\|_2$ $\times\sqrt{d/(C\log(n/C))}$	$1/16$ $+2^{-d}$	$(C+\log n)$ $\times \log \frac{n}{C}$	$\log^{O(1)} n$	$C\times$ $\log^{O(1)} n$

Fig. 1. Precision-Confidence tradeoffs for ℓ_2-based frequency estimation. For ACSK-I and ACSK-II , the parameter $d \ge 4$ controls the precision-confidence tradeoff.

precision $Err_{\mathcal{A}}(x)$, and (2) the confidence denoted $1 - \delta_{\mathcal{A}}$ with which the precision holds. We say that $\mathcal{A}$ exhibits a precision-confidence tradeoff if for each fixed input x, the set of feasible non-dominating $(Err_{\mathcal{A}}(x), \delta_{\mathcal{A}})$ pairs is at least 2 and preferably, is a large set. A point $(Err_{\mathcal{A}}(x), \delta_{\mathcal{A}})$ dominates $(Err'_{\mathcal{A}}(x), \delta'_{\mathcal{A}})$ if $Err_{\mathcal{A}}(x) < Err'_{\mathcal{A}}(x)$ and $\delta_{\mathcal{A}}(x) < \delta'_{\mathcal{A}}(x)$. For example, COUNTSKETCH has the single point $(\|x^{\text{res}(C)}\|_2/\sqrt{C}, 1 - n^{-\Omega(1)})$ and does not exhibit a tradeoff. Why are algorithms with precision-confidence tradeoffs useful? To illustrate, suppose that an application requires frequency estimation of items in some input set H of a-priori unknown size t with high constant probability. Using Algorithm ACSK-I (see Figure 1) with $d = \log(t) + O(1)$ gives a precision of $\|x^{\text{res}(C)}\|_2\sqrt{\log t/(C\log n)}$ and confidence of $1 - t2^{-c\log t} = 1 - t^{1-c}$. If $t = O(1)$, the precision is superior to that of COUNTSKETCH by a factor of $\sqrt{\log n}$. If $t = n$ this matches the COUNTSKETCH guarantees. The important property is that *no changes or re-runs of the algorithm are needed*. The same output simultaneously satisfies all the precision-confidence pairs in its tradeoff set.

Contributions. We present a frequency estimation algorithm ACSK-I (Averaged CountSketch-I) that has precision $O(\|x^{\text{res}(C)}\|_2\sqrt{d/(C\log n)})$ and confidence $1 - 2^{-d}$, where, $4 \le d \le \Theta(\log n)$. A second frequency estimation algorithm ACSK-II has precision $O(\|x^{\text{res}(C)}\|_2\sqrt{d/(C\log(n/C))})$ and confidence $1 - 2^{-d}$. Both algorithms show precision-confidence tradeoff by tuning the value of d in the allowed range. Figure 1 compares the algorithms along different measures. We also show that the algorithms are optimal up to constant factors for a wide range of the parameters among the class of algorithms whose state on input x can be represented as Ax, for some $m \times n$ matrix A.

Summary. We build on the COUNTSKETCH algorithm of Charikar et.al. in [1]. Instead of taking the median of estimates for x_i from the individual tables, we take the averages over the estimates for x_i from those tables where a set of heavy-hitters do not collide with i. The analysis uses the $2d$th moment method which requires $O(d)$-wise independence of the random variables. This degree of independence d parameterizes the precision-confidence tradeoff.

2 The ACSK Algorithms

Notation. Let $\textsc{CountSketch}(C, s)$ denote the structure consisting of s hash tables $T_1, \ldots, T_s$, each having $8C$ buckets, using independently chosen pair-wise independent hash functions $h_1, \ldots, h_s$ respectively. The bucket $T_l[b]$ is the sketch: $T_l[b] = \sum_{h_l(i)=b} x_i \xi_{il}$, where the family $\{\xi_{il}\}_{i \in [n]}$ for each $l \in [s]$ is four-wise independent and the families use independent seeds across the tables. The estimated frequency is the median of the table estimates, that is, $\hat{x}_i = median_{l=1}^s T_l[h_l(i)] \xi_{il}$. Then, $|\hat{x}_i - x_i| \le \|x^{\text{res}(C)}\|_2/\sqrt{C}$, with probability $1 - 2^{-\Omega(s)}$.

For an n-dimensional vector x and $H \subset [n]$, let x_H denote the sub-vector of x with coordinates in H.

The ACSK-I (C, s_0, s, d) structure with space parameter C, number of tables parameters s_0 and s, and degree of independence parameter d, maintains two structures, namely, (1) $\textsc{CountSketch}(2C, s_0)$, where, $s_0 = c \log n$ for some constant $c > 0$, and, (2) $\textsc{CountSketch}(C', s)$, where, $C' = \lceil 3eC \rceil$, that uses (a) $2d + 1$-wise independent Rademacher families $\{\xi_{il}\}_{i \in [n]}$ for each $l \in [s]$, and, (b) the hash functions $h_1, \ldots, h_s$ corresponding to the tables $T_1, \ldots, T_s$ are independently drawn from a $d + 3$-wise independent hash family that maps $[n]$ to $[C']$. Both structures are updated as in the classical case. The frequency estimation algorithm is as follows.

1. Use the first $\textsc{CountSketch}$ structure to obtain a set H of the top-$2C$ items by absolute values of their estimated frequencies (by making a pass over $[n]$).
2. Let $S(i, H)$ be the set of table indices in the second $\textsc{CountSketch}$ structure where i does not collide with any item in $H \setminus \{i\}$. Return the average of the estimates for x_i obtained from the tables in $S(i, H)$.

$$\hat{x}_i = \text{average}_{l \in S(i,H)} T_l[h_l(i)] \cdot \xi_{il} \ .$$

Analysis. Let x_i' denote the estimated frequency obtained from the first structure. By property of precision of $\textsc{CountSketch}[1]$ we have, $|x_i' - x_i| \le \Delta$, where, $\Delta = \|x^{\text{res}(2C)}\|_2/\sqrt{2C}$. Let GoodH denote the event GoodH $\equiv \forall i \in [n], |x_i' - x_i| \le \Delta$. So by union bound, $\Pr[\text{GoodH}] \ge 1 - n2^{-\Omega(s_0)}$. We first prove simple upper bounds for (a) the maximum frequency of an item in $\bar{H}$, and, (b) $\|x^{\text{res}(H)}\|_2^2 = \sum_{j \in [n] \setminus H} x_j^2$. Let T_H denote the maximum absolute frequency of an item not in H. Lemma 1 (a) is proved in Appendix A. Lemma 1 (b) follows variants proved in [3,2].

Lemma 1. *Conditional on GoodH, (a) $T_H \le (1 + \sqrt{2})\|x^{\text{res}(C)}\|_2/\sqrt{C}$, and, (b) $\|x^{\text{res}(H)}\|_2^2 \le 9\|x^{\text{res}(2C)}\|_2^2$.*

Consider the second $\textsc{CountSketch}$ structure of ACSK-I. Let $p = 1/(8C') = 1/(8\lceil 3eC \rceil) \le 1/(24eC)$, which is the probability that a given item maps to a given bucket in a hash table. For $i, j \in [n], j \ne i$ and table index $l \in [s]$, let χ_{ijl} be 1 if $h_l(i) = h_l(j)$ and 0 otherwise. Lemma 2 shows that given sufficient independence of the hash functions, $S(i, H) = \Theta(s)$ with high probability.

Lemma 2. *Suppose the hash functions $h_1, \ldots, h_s$ of a* COUNTSKETCH *structure are each chosen from a pair-wise independent family. Let $C' \geq \lceil 1.5et \rceil + 1$. Then, for any given set H with $|H| = t$, $|S(i, H)| \geq 3s/5$ with probability $1 - e^{-s/3}$.*

Lemma 3 presents an upper bound on the $2d$th moment for the sum of $2d$-wise independent random variables, each with support in the interval $[-1, 1]$ and having a symmetric distribution about 0. Its proof, given in the Appendix, uses ideas from the proof of Theorem 2.4 in [6] but gives a slightly stronger result in comparison.

Lemma 3. *Suppose $X_1, X_2, \ldots, X_n$ are $2d$-wise independent random variables such that the X_i's have support in the interval $[-1, 1]$ and have a symmetric distribution about 0. Let $X = X_1 + X_2 + \ldots + X_n$. Then,*

$$E\left[X^{2d}\right] \leq \sqrt{2} \left(\frac{2d\mathsf{Var}\,[X]}{e}\right)^d \left(1 + \frac{d}{\mathsf{Var}\,[X]}\right)^{d-1}.$$

For a suitable normalization value T_1 and $j \in [n] \setminus (H \cup \{i\})$, let $X_{ijl} = (x_j/T_1)\xi_{jl}\xi_{il}\chi_{ijl}$ and let

$$X_i = (\hat{x}_i - x_i)|S(i, H)|/T_1 = \sum_{l \in S(i,H)} \sum_{j \notin H \cup \{i\}} X_{ijl}.$$

We wish to calculate $\mathbb{E}\left[X_i^{2d}\right]$ and use it to obtain a concentration of measure for X_i. However, the X_{ijl}'s contributing to X_i are conditioned on the event that $l \in S(i, H)$, a direct application of Lemma 3 is not possible. Lemma 4 gives an approximation for $\mathbb{E}\left[X_i^{2d}\right]$ in terms of $E[X_i^{2d}]$, where, $E[X_i^{2d}]$ is the $2d$th moment of the same random variable but under the assumption that the ξ_{jl}'s and the hash functions h_l's for each l are fully independent.

Lemma 4. *Let $C' = \lceil 3eC \rceil$, the h_l's be $d + 1 + t$-wise independent, $t \geq 2$ and $\{\xi_{il}\}_{i \in [n]}$ be $2d + 1$-wise independent. Then $\mathbb{E}\left[X_i^{2d}\right] \leq (1 + 8(12t)^{-t})^d E[X_i^{2d}]$.*

The proof of Lemma 4 requires the following Lemma 5, which is an application of the principle of inclusion-exclusion and Bayes' rule.

Lemma 5. *For any $s \geq 1$ and $t \geq 2$, let $X_1, \ldots, X_n$ be $s + t$-wise independent and identically distributed Bernoulli (i.e., 0/1) random variables with $t \geq 2$ and $p = \Pr[X_i = 1] \leq 1/(12e)$. Then, for disjoint sets $S, H \subset [n]$, with $|S| = s$ and $|H| \leq 1/(12pe)$, $\left|\Pr[\forall j \in S, X_j = 1 \mid \forall j \in H, X_j = 0] - p^s\right| \leq 8(12t)^{-t}$.*

The proof of Lemma 5 is given in the Appendix. We can now prove Lemma 4.

Proof (Of Lemma 4.).

$$\mathbb{E}\left[X_i^{2d}\right] = \mathbb{E}\left[\left(\sum_{l \in S(i,H), j \neq i} (x_j/T_1)\xi_{jl}\xi_{il}\chi_{ijl}\right)^{2d}\right]$$

$$= \sum_{\substack{\sum_{l \in S(i,H), j \neq i} e_{jl} = 2d \\ e_{jl}\text{'s even}}} \binom{2d}{e_{11}, \ldots, e_{ns}} \prod_{l \in S(i,H)} \mathbb{E}\left[\prod_{j: e_{jl} > 0} (x_j/T_1)^{e_{jl}} \chi_{ijl} \mid l \in S(i, H)\right]$$

Let e denote the vector $(e_{11}, \ldots, e_{ns})$ that satisfies the constraints in the summation, that is, (1) $\sum_{l \in S(i,H), j \neq i} e_{jl} = 2d$, (2) $e_{jl} = 0$ for each $l \in [s] \setminus S(i, H), j \in [n]$, and, (3) each e_{jl} is even. Let $S_{ile} = \{j : e_{jl} > 0\}$. Define the events:

$$E_1(i, l, e) : \quad \forall j \in S_{ile}, \chi_{ijl} = 1 \quad \text{and} \quad E_2(i, l, H) : \quad \forall j \in H \setminus \{i\}, \chi_{ijl} = 0.$$

Then,

$$\mathbb{E}\left[\prod_{j:e_{jl}>0} \chi_{ijl} \mid l \in S(i, H)\right] = \Pr\left[E_1(i, l, e) \mid E_2(i, l, H)\right] .$$

Since the product is taken over positive e_{jl}'s, for each such l, S_{ile} is non-empty. A bound on $\Pr\left[E_1(i, l, e)) \mid E_2(i, l, H)\right]$ can now be obtained using Lemma 5, where, $p = \Pr\left[\chi_{ijl} = 1\right] = 1/C' \leq 1/(24eC)$. Further, $|S_{ile}| \leq d$ and $|H| = 2C \leq 1/(12pe)$. So the premises of Lemma 5 are satisfied. Also, since the hash function h_l is drawn from a $d + 1 + t$-wise independent family, the family of random variables $\{\chi_{ijl} : j \in [n], j \neq i\}$, for each fixed i and l, is $d + t$-wise independent, and across the l's is fully independent. Applying Lemma 5, we obtain $\Pr\left[E_1(i, l, e) \mid E_2(i, l, H)\right] \in p^{|S_{ile}|}(1 \pm 8(12t)^{-t})$. Hence,

$$\mathbb{E}\left[X_i^{2d}\right]$$

$$\leq \sum_{\substack{\sum_{l \in S(i,H), j \neq i} e_{jl} = 2d \\ e'_{jl}s \text{ even}}} \binom{2d}{e_{11} \ldots e_{ns}} \prod_{l \in S(i,H)} \left[(p^{|S_{ile}|}(1 + 8(12t)^{-t})) \prod_{j:e_{jl}>0} (x_j/T_1)^{e_{jl}}\right]$$

$$\leq (1 + 8(12t)^{-t})^d \sum_{\substack{\sum_{l \in S(i,H), j \neq i} e_{jl} = 2d \\ e_{jl}\text{'s even}}} \binom{2d}{e_{11} \ldots e_{ns}} \prod_{l \in S(i,H)} p^{|S_{ile}|} \prod_{j:e_{jl}>0} (x_j/T_1)^{e_{jl}}$$

$$\leq (1 + 8(12t)^{-t})^d E[X_i^{2d}]$$

since, the *RHS*, discounting the multiplicative factor of $(1 + 8(12t)^{-t})^d$, is the expansion of $E[X_i^{2d}]$. □

We now prove the main theorem regarding the ACSK-I algorithm.

Theorem 6. *For $C \geq 2$, $s_0 = \Theta(\log n)$ and $s \geq 20d$, there is an algorithm that for any $i \in [n]$ returns $\hat{x}_i$ satisfying $|\hat{x}_i - x_i| \leq \|x^{res(C)}\|_2 \sqrt{3d/(sC)}$ with probability at least $1 - 2^{-\Omega(s)} - 2^{-d} - n2^{-s_0}$. Moreover, $\mathbb{E}[\hat{x}_i] = x_i$. The algorithm uses space $O(C(s + s_0))$ words.*

Proof. Consider the ACSK-I algorithm. For $l \in [s]$, $\mathbb{E}[T_l[h_l(i)] \cdot \xi_{il}] = x_i$. Hence the average of $T_l[h_l(i)] \cdot \xi_{il}$'s over some subset of the l's has the same expectation.

Fix $i \in [n]$. Let $T_1 \geq T_H$ which will be chosen later. Recall that for $j \in [n] \setminus (H \cup \{i\})$ and $l \in S(i, H)$, $X_{ijl} = (x_j/T_1)\xi_{il}\xi_{jl}\chi_{ijl}$. Since, $j \notin H$, $|X_{ijl}| \leq 1$ and X_{ijl} has 3-valued support $\{-x_j/T_1, 0, x_j/T_1\}$ with a symmetric distribution over it. Let $p = \Pr\left[\chi_{ijl} = 1\right] = 1/(8C') = 1/(24eC)$. By direct calculation,

$$\mathsf{Var}\left[X_i\right] = \sum_{l \in S(i,H)} \sum_{j \neq i} \left(\frac{x_j}{T_1}\right)^2 p = |S(i, H)| \frac{\|x^{res(H \cup \{i\})}\|_2^2}{24eCT_1^2} \tag{1}$$

By Lemma 3 and assuming full independence we have,

$$E[X_i^{2d}] \leq \sqrt{2}\left(\frac{2d\mathsf{Var}\,[X_i]}{e}\right)^d\left(1+\frac{2d}{9\mathsf{Var}\,[X_i]}\right)^{d-1}.$$

Let $t = 2$. Sine the hash functions are $d + 3 = d + t + 1$-wise independent and the Rademacher variables are $2d + 1$-wise independent, by Lemma 4 we have,

$$\mathbb{E}\left[X_i^{2d}\right] \leq (1 + 8(12t)^{-t})^d E[X_i^{2d}] \leq (1 + 1/72)^d E[X_i^{2d}], \quad \text{for } t = 2.$$

By $2d$th moment inequality, $\mathsf{Pr}\left[|X_i| > \sqrt{2}(\mathbb{E}[X_i^{2d}])^{1/(2d)}\right] \leq 2^{-d}$. Therefore,

$$\mathsf{Pr}\left[|X_i| > \sqrt{2(1+1/72)}\left(\frac{2d\mathsf{Var}\,[X_i]}{e}\left(1+\frac{d}{\mathsf{Var}\,[X_i]}\right)\right)^{1/2}\right] \leq 2^{-d} \quad (2)$$

Let $E_{d,i}$ denote the event whose probability is given in (2). Consider the intersection of the following three events: (1) GoodH, (2) $|S(i, H)| \geq 3s/5$, and, (3) $E_{d,i}$. By union bound, the above three events hold with probability $1 - n2^{-\Omega(s_0)} - e^{-s/3} - 2^{-d} = 1 - \delta$ (say). Since, GoodH holds, we can choose $T_1 = (1 + \sqrt{2})\|x^{\text{res}(C)}\|_2/\sqrt{C}$. Then, (1) $T_H \leq T_1$, by Lemma 1, and, (2) $\|x^{\text{res}(H\cup\{i\})}\|_2^2 \leq 9\|x^{\text{res}(2C)}\|_2^2$, by Lemma 1 (b). Substituting in (1),

$$\mathsf{Var}\,[X_i] \leq \frac{|S(i,H)|\|x^{\text{res}(H\cup\{i\})}\|_2^2}{(24eC)T_1^2} \leq \frac{s \cdot 9\|x^{\text{res}(2C)}\|_2^2}{(24eC)(1+\sqrt{2})^2(\|x^{\text{res}(C)}\|_2^2/C)} \leq \frac{s}{20} \quad (3)$$

The deviation for $|X_i|$ in (2) is an increasing function of $\mathsf{Var}\,[X]$. Hence, replacing $\mathsf{Var}\,[X_i]$ by its upper bound gives us an upper bound on the deviation for the same tail probability. Hence, with probability $1 - \delta$, we have from (2) that

$$|X_i| \leq \sqrt{2.5}\left(\tfrac{2ds}{20e}\left(1+\tfrac{20d}{s}\right)\right)^{1/2} \leq \sqrt{\tfrac{ds}{2e}}$$

since, $s \geq 20d$. Since, $|\hat{x}_i - x_i| = |X_i|T_1/|S(i, H)|$, we have,

$$|\hat{x}_i - x_i| \leq \sqrt{\frac{ds}{2e}} \cdot \frac{(1+\sqrt{2})\|x^{\text{res}(C)}\|_2}{\sqrt{C}} \cdot \frac{1}{(3s/5)} \leq \sqrt{\frac{3d}{sC}}\|x^{\text{res}(C)}\|_2 . \qquad \square$$

Precision-Confidence Tradeoff. Theorem 6 can be applied using *any value of d in the range $4 \leq d \leq s/4 = \Theta(\log n)$* (even after the estimate has been obtained). One can choose d to match the confidence to the desired level and minimize the precision (for e.g., choose $d = O(\log r)$, where r is the number of estimates taken).

The ACSK-II Algorithm. The ACSK-II algorithm uses the heavy-hitter algorithm by Gilbert et. al. in [4], denoted by HH^{GLPS}, to find the heavy hitters.

Theorem 7 ([4]). *There is an algorithm and distribution on matrices Φ such that, given Φx and a concise description of Φ, the algorithm returns $\hat{x}$ such that $\|x - \hat{x}\|_2^2 \leq (1 + \epsilon)\|x^{res(C)}\|_2^2$ holds with probability $3/4$. The algorithm runs in time $C \log^{O(1)} n$ and Φ has $O((C/\epsilon) \log(n/C))$ rows.*

The only difference in the ACSK-II (C, s) algorithm is that it uses an $\text{HH}^{\text{GLPS}}(2C, 1/2)$ structure to obtain a set H of heavy-hitters. The second COUNT-SKETCH(C', s) structure of ACSK-I , and the estimation algorithm is otherwise identical. Here, $C' = \lceil 6eC \rceil$ and $s = O(\log(n/C))$. ACSK-II has significantly faster estimation time than ACSK-I due to the efficiency of Gilbert et. al.'s algorithm. However its guarantee holds only with high constant probability. We have the following theorem.

Theorem 8. *For each $C \geq 2$, $s \geq 20d$ and $r \geq 1$, there is an algorithm that given any set of distinct indices $i_1, \ldots, i_r$ from $[n]$, returns $\hat{x}_{i_j}$ corresponding to x_{i_j} satisfying $|\hat{x}_{i_j} - x_{i_j}| \leq \|x^{res(C)}\|_2\sqrt{2d/(C \log(n/C))}$ for all $j \in [r]$, with probability $15/16 - r2^{-d}$. Moreover, $\mathbb{E}[\hat{x}_{i_j}] = x_{i_j}$, $j \in [r]$. The algorithm uses space $O(C \log(n/C))$ words and has update time $O(\log^{O(1)} n)$. The estimation time is $O(C \log^{O(1)}(n) + rCd \log(n))$.*

Proof. It follows from Theorem 7 that $\|x^{res(H)}\|_2^2 \leq (1 + 1/2)\|x^{res(C)}\|_2^2$. Further, the Loop Invariant in [4] ensures that upon termination, (a) the largest element not in H has frequency at most $T_H^2 < \|x^{res(C)}\|_2^2/C$, and, (b) $|H| = \|\hat{x}\|_0 \leq 4C$. We have upper bounds on all the parameters as needed, and the proof of Theorem 6 can be followed. □

3 Lower Bound on Frequency Estimation

We say that a streaming algorithm has a *matrix representation with m rows* if the state of the structure on any input vector x can always be represented as Ax, where, A is some $m \times n$ matrix. All known data streaming algorithms for ℓ_2-based frequency estimation have a matrix representation. We show a lower bound on the number of rows in the matrix representation of a frequency estimation algorithm.

Theorem 9. *Suppose that a frequency estimation algorithm has a matrix representation with m rows. Let it have precision $\|x^{res(C)}\|_2\sqrt{d/(C \log(n/C))}$ such that for any number r of estimations, all the estimates satisfy the precision with probability $15/16 - r \cdot 2^{-d}$. Then, for $d = \Omega(1)$, $2 + \log C \leq d \leq \log \frac{n}{C}$ and $n = \Omega(C \log(\frac{n}{C}) \log(C \log \frac{n}{C}))$, $m = \Omega(C \log(\frac{n}{C}) \cdot (1 - \frac{\log C}{d}))$.*

Proof. Let $D = [2^{d-3}]$ and $C = 4k$. Given a vector x with coordinates in D we make a pass over D and obtain the estimated frequency vector $\hat{x}$. Let H be the set of the top-$2k$ coordinates by absolute values of estimated frequency. Then, $\forall i \in D$, $|\hat{x}_i - x_i| \leq \|x^{res(4k)}\|_2\sqrt{\frac{d}{4k \log(n/C)}}$ holds with probability

$15/16 - 2^{d-3}2^{-d} > 2/3$. Following the proof of Theorem 3.1 in [5]), the resulting vector satisfies $\|x - \hat{x}_H\|_2^2 \le \left(1 + \frac{d}{\log(n/C)}\right)\|x^{\mathrm{res}(k)}\|_2^2$. Thus we have an ℓ_2/ℓ_2 k-sparse recovery algorithm with approximation factor $1 + d/\log(n/C)$ that succeeds with probability $2/3$. Since, $n = \Omega(C\log(\frac{n}{C})\log(C\log\frac{n}{C}))$ and $n = \Omega(C\log^2(n/C)(\frac{1}{d} - \frac{\log(C)}{d}))$, by the Price-Woodruff lower bound for $(1+\epsilon)$-approximate k-sparse recovery [5], such a matrix A has number of rows

$$m = \Omega\left(\frac{k}{\epsilon}\log\frac{2^{d-3}}{k}\right) = \Omega\left(C\log\left(\frac{n}{C}\right)\cdot\left(1 - \frac{\log C}{d}\right)\right) . \qquad \square$$

Clearly, both ACSK algorithms have a matrix representation. Also ACSK-II satisfies the premise regarding precision and confidence of Theorem 9 and uses $O(C\log(n/C))$ rows. ACSK-I does too provided $C = n^{1-\Omega(1)}$. Hence, they are optimal up to constant factors in the range $\frac{d}{100} \le \log C \le d - 2$ and $d \le \log\frac{n}{C}$ along with the other constraints of Theorem 9 on d, n and C.

References

1. Charikar, M., Chen, K., Farach-Colton, M.: Finding frequent items in data streams. Theoretical Computer Science 312(1), 3–15 (2004)
2. Cormode, G., Muthukrishnan, S.: Combinatorial Algorithms for Compressed Sensing. In: Flocchini, P., Gąsieniec, L. (eds.) SIROCCO 2006. LNCS, vol. 4056, pp. 280–294. Springer, Heidelberg (2006)
3. Ganguly, S., Kesh, D., Saha, C.: Practical Algorithms for Tracking Database Join Sizes. In: Ramanujam, R., Sen, S. (eds.) FSTTCS 2005. LNCS, vol. 3821, pp. 297–309. Springer, Heidelberg (2005)
4. Gilbert, A.C., Li, Y., Porat, E., Strauss, M.J.: Approximate sparse recovery: optimizing time and measurements. In: Proceedings of ACM Symposium on Theory of Computing, STOC, pp. 475–484 (2010)
5. Price, E., Woodruff, D.: $(1 + \epsilon)$-approximate Sparse Recovery. In: Proceedings of IEEE Foundations of Computer Science (FOCS) (2011)
6. Schmidt, J., Siegel, A., Srinivasan, A.: Chernoff-Hoeffding Bounds with Applications for Limited Independence. In: Proceedings of ACM Symposium on Discrete Algorithms (SODA), pp. 331–340 (1993)

A   Proofs

Proof (Of Lemma 1). Assume GoodH holds. Let $|x_i| = T_H = \max_{j\notin H}|x_j|$. So if $|x_j| < T_H - 2\Delta$, then, $j \notin H$. Hence, $H \subset J = \{j : x_j \ge T_H - 2\Delta\}$. Now, $|J \setminus \mathrm{Top}(C)| \ge |H \setminus \mathrm{Top}(C)| \ge C$. Thus,

$$\|x^{\mathrm{res}(C)}\|_2^2 \ge \sum_{j\in J\setminus\mathrm{Top}(C)} x_j^2 \ge |J \setminus \mathrm{Top}(C)|(T_H - 2\Delta)^2 \ge C(T_H - 2\Delta)^2$$

or, $T_H \le \left(\frac{\|x^{\mathrm{res}(C)}\|_2^2}{C}\right)^{1/2} + 2\Delta = (1 + \sqrt{2})\|x^{\mathrm{res}(C)}\|_2/\sqrt{C}.$ $\qquad \square$

Proof (Of Lemma 2.). Assume $t > 0$, otherwise the lemma trivially holds. Since $8C' \geq 8\lceil 1.5et \rceil \geq 12et$, we have, $\Pr[\chi_{ijl} = 1] = p = 1/(8C') \leq 1/(12et)$. Let $w = |H \setminus \{i\}|$. Denote by $Pr[\cdot]$ the probability measure under the assumption that the hash functions are fully independent. By inclusion-exclusion applied for $\Pr[\bigvee_j (\chi_{ijl} = 1)]$ and $Pr[\bigvee_j (\chi_{ijl} = 1)]$ respectively, where, j runs over $H \setminus \{i\}$), $d + 1$-wise independence of the hash function h_l for $\Pr[\cdot]$ and using triangle inequality once, we have, $\left| \Pr[\bigvee_j \chi_{ijl} = 1] - Pr\{\bigvee_j \chi_{ijl} = 1\} \right| \leq 2\binom{w}{d}p^d$.

Since, $\Pr[\bigwedge_j (\chi_{ijl} = 0)] = 1 - \Pr[\bigvee_j (\chi_{ijl} = 1)]$, and $Pr[\bigwedge_j (\chi_{ijl} = 0)] = (1 - p)^w$, we have, $\left| \Pr[\bigwedge_j \chi_{ijl} = 0] - (1 - p)^w \right| \leq 2\binom{w}{d}p^d$. Further since $w \leq t$, we have, $\binom{w}{d}p^d \leq (pet/d)^d \leq (12d)^{-d}$. Also $(1 - p)^w \geq 1 - tp \geq 1 - 1/(12e)$.

Therefore, $\Pr[\bigwedge_j \chi_{ijl} = 0] \geq 1 - 1/(12e) - 2(12d)^{-d} \geq 24/25$, for $d \geq 2$. Since the hash functions are independent across the tables, applying Chernoff's bounds, we have, $\Pr[|S(i, H)| \geq (3/5)s] \geq 1 - \exp\{-s/3\}$.   □

Proof (of Lemma 3.). We have, $X_i^{2j} \leq X_i^2$ and so $\mathbb{E}[X_i^{2j}] \leq \mathbb{E}[X_i^2]$. Also $\mathsf{Var}[X] = \sum_{j=1}^n \mathbb{E}[X_i^2]$. So for $X = X_1 + \ldots + X_n$, and since all odd moments of X_i's are 0, by symmetry of the individual distributions, we have,

$$\mathbb{E}[X^{2d}] = \sum_{r=1}^d \sum_{\substack{t_1+\ldots+t_r=d \\ t_j's > 0}} \binom{2d}{2t_1, 2t_2, \ldots, 2t_r} \sum_{1 \leq j_1 < \ldots < j_r \leq n} \prod_{u=1}^r \mathbb{E}[X_{j_u}^{2t_u}]$$

$$= \sum_{r=1}^d \sum_{\substack{t_1+\ldots+t_r=d \\ t_j's > 0}} \binom{2d}{2t_1, 2t_2, \ldots, 2t_r} \sum_{1 \leq j_1 < \ldots < j_r \leq n} \prod_{u=1}^r \mathbb{E}[X_{j_u}^2]$$

$$\leq \sum_{r=1}^d \sum_{\substack{t_1+\ldots+t_r=d \\ t_j's > 0}} \binom{2d}{2t_1, 2t_2, \ldots, 2t_r} \frac{(\mathsf{Var}[X])^r}{r!}$$

$$= \sum_{l=0}^{d-1} T_l, \text{ where, } T_l = \sum_{t_1+\ldots+t_{d-l}=d, \, t_j's>0} \binom{2d}{2t_1, 2t_2, \ldots, 2t_{d-l}} \frac{(\mathsf{Var}[X])^{d-l}}{(d-l)!}$$

Since $\binom{2d}{2t_1, 2t_2, \ldots, 2t_{d-l}} \leq \binom{2d}{2, 2, \ldots, 2}$, we have,

$$T_l \leq \sum_{t_1+\ldots+t_{d-l}=d, \, t_j's>0} \binom{2d}{2, 2, \ldots, 2} \frac{(\mathsf{Var}[X])^{d-l}}{(d-l)!} \leq \binom{d-1}{d-l-1}\binom{2d}{2, 2, \ldots, 2} \frac{(\mathsf{Var}[X])^{d-l}}{(d-l)!}$$

$$= \binom{d-1}{d-l-1}\left(\frac{1}{\mathsf{Var}[X]}\right)^l \frac{d!}{(d-l)!} T_0$$

since, there are $\binom{d-1}{d-l-1}$ assignments for $t_1, \ldots, t_{d-l}$, all positive with sum d. Therefore,

$$\mathbb{E}[X^{2d}] \leq \sum_{l=0}^{d-1} T_l \leq \sum_{l=0}^{d-1} \binom{d-1}{d-l-1}\left(\frac{1}{\mathsf{Var}[X]}\right)^l \frac{d!}{(d-l)!} T_0$$

$$\leq T_0 \sum_{l=0}^{d-1} \binom{d-1}{l}\left(\frac{1}{\mathsf{Var}[X]}\right)^l d^l = T_0\left(1 + \frac{d}{\mathsf{Var}[X]}\right)^{d-1}$$

Since,

$$T_0 = \binom{2d}{2, 2, \ldots, 2} \frac{(\mathsf{Var}[X])^d}{d!} = \frac{(2d)!}{2^d d!}(\mathsf{Var}[X])^d \leq \frac{2^{d+1/2} d^d}{e^d}(\mathsf{Var}[X])^d$$

by Stirling's approximation, we have,

$$\mathbb{E}[X^{2d}] \leq \sqrt{2}\left(\frac{2d\mathsf{Var}[X]}{e}\right)^d \left(1 + \frac{d}{\mathsf{Var}[X]}\right)^{d-1}.$$ □

Proof (Of Lemma 5.). Define events $E_1 \equiv \forall j \in S, X_j = 1$ and $E_2 \equiv \forall j \in H, X_j = 0$. We have to bound the probability $\Pr[E_1 \mid E_2]$. Let $|H| = w$. Since, $|S| = s$, $\Pr[E_1] = p^s$. By inclusion and exclusion,

$$\left| \Pr[\exists j \in H, X_j = 1 \mid E_1] - \sum_{r=1}^{t-1} (-1)^{r-1} \sum_{\substack{j_1,\ldots,j_r \in H \\ j_1 < \ldots < j_r}} \Pr[X_{j_1} = 1 \wedge \ldots \wedge X_{j_r} = 1 \mid E_1] \right|$$

$$\leq \sum_{\substack{j_1,\ldots,j_t \in H \\ j_1 < \ldots < j_t}} \Pr[X_{j_1} = 1 \wedge \ldots X_{j_t} = 1 \mid E_1]$$

Since the X_j's are $s + t$-wise independent and the event E_1 is a property of the X_j's for $j \in S$ and $|S| = s$, we have for distinct elements $j_1, \ldots, j_r$ from H (given $H \cap S$ is empty) and $1 \leq r \leq t$, $\Pr[X_{j_1} = 1 \wedge \ldots X_{j_r} = 1 \mid E_1] = \Pr[X_{j_1} = 1] \cdot \ldots \cdot \Pr[X_{j_r} = 1] = p^r$. Let $|H| = w$. The above equation is equivalently,

$$\left| \Pr[\exists j \in H, X_j = 1 \mid E_1] - \sum_{r=1}^{t-1} (-1)^{r-1} \binom{w}{r} p^r \right| \leq \binom{w}{t} p^t \qquad (4)$$

Suppose we denote by $Pr[E]$ the probability of an event $E = E(X_1, \ldots, X_n)$ assuming that the X_j's are fully independent. Then, by inclusion-exclusion, we have

$$\left| Pr[\exists j \in H, X_j = 1 \mid E_1] - \sum_{r=1}^{t-1} (-1)^{r-1} \binom{w}{r} p^r \right| \leq \binom{w}{t} p^t \qquad (5)$$

Since, $\Pr[X_j = 1] = Pr[X_j = 1] = p$, combining (4) and (5), we have by triangle inequality,

$$\left| \Pr[\exists j \in H, X_j = 1 \mid E_1] - Pr[\exists j \in H, X_j = 1 \mid E_1] \right| \leq 2 \binom{w}{t} p^t$$

Also, $Pr[E_2 \mid E_1] = 1 - \Pr[\exists j \in H, X_j = 1 \mid E_1]$ and $Pr[E_2 \mid E_1] = 1 - Pr[\exists j \in H, X_j = 1 \mid E_1] = (1 - p)^w$. Hence,

$$\left| \Pr[E_2 \mid E_1] - (1 - p)^w \right| \leq 2 \binom{w}{t} p^t \qquad (6)$$

Further, $\Pr[E_1] = Pr[\forall j \in S, X_j = 1] = p^s$. Using $s + t$-wise independence of the X_j's for $j \in H$, we can show similarly that

$$\left| \Pr[E_2] - (1 - p)^w \right| \leq 2 \binom{w}{s + t} p^{s+t} \ .$$

Combining,

$$\Pr[E_1 \mid E_2] = \frac{\Pr[E_2 \mid E_1] \Pr[E_1]}{\Pr[E_2]} \in p^s \left(1 \pm \frac{2\binom{w}{t} p^t + 2\binom{w}{s+t} p^{s+t}}{(1 - p)^w - 2\binom{w}{s+t} p^{s+t}} \right) \qquad (7)$$

Since, $pw \leq 1/(12e)$, $(1-p)^w \geq 1 - wp \geq 1 - 1/(12e)$, $\binom{w}{t}p^t \leq (wep/t)^t \leq 1/(12t)^t$ and $\binom{w}{s+t}p^{s+t} \leq 1/(12(s+t))^{s+t}$. Thus, for $t \geq 2$, we have,

$$\frac{2\binom{w}{t}p^t + 2\binom{w}{s+t}p^{s+t}}{(1-p)^w - 2\binom{w}{s+t}p^{s+t}} \leq \frac{2(12t)^{-t} + 2(12(s+t))^{-t-s}}{(1 - 1/(12e)) - 2(12(s+t))^{-s-t}} \leq 8(12t)^{-t} \; .$$

since $t \geq 2$. Hence, (7) becomes

$$\Pr[E_1 \mid E_2] \in p^s\left[1 \pm 8(12t)^{-t}\right] \; . \qquad \qquad \square$$

Competitive Design and Analysis for Machine-Minimizing Job Scheduling Problem*

Mong-Jen Kao[2], Jian-Jia Chen[1], Ignaz Rutter[1], and Dorothea Wagner[1]

[1] Faculty for Informatics, Karlsruhe Institute of Technology (KIT), Germany
[2] Research Center for Infor. Tech. Innovation, Academia Sinica, Taiwan
mong@citi.sinica.edu.tw, {j.chen,rutter,dorothea.wagner}@kit.edu

Abstract. We explore the machine-minimizing job scheduling problem, which has a rich history in the line of research, under an online setting. We consider systems with arbitrary job arrival times, arbitrary job deadlines, and unit job execution time. For this problem, we present a lower bound 2.09 on the competitive factor of *any* online algorithms, followed by designing a 5.2-competitive online algorithm. We would also like to point out a false claim made in an existing paper of Shi and Ye regarding a further restricted case of the considered problem. To the best of our knowledge, what we present is the first concrete result concerning online machine-minimizing job scheduling with arbitrary job arrival times and deadlines.

1 Introduction

Scheduling jobs with interval constraints is one of the most well-known models in classical scheduling theory that provides an elegant formulation for numerous applications and which also has a rich history in the line of research that goes back to the 1950s. For example, assembly line placement of circuit boards [12, 17], time-constrained communication scheduling [1], adaptive rate-controlled scheduling for multimedia applications [15, 18], etc.

In the basic framework, we are given a set of jobs, each associated with a set of time intervals during which it can be scheduled. Scheduling a job means selecting one of its associated time interval. The goal is to schedule all the jobs on a minimum number of machines such that no two jobs assigned to the same machine overlap in time. Two variations have been considered in the literature, differing in the way how the time intervals of the jobs are specified. In the *discrete machine minimization*, the time intervals are listed explicitly as the input, while in the *continuous* version, the set of time intervals for each job is specified by a release time, a deadline, and an execution time.

---

* This work was supported in part by National Science Council (NSC), Taiwan, under Grants NSC99-2911-I-002-055-2, NSC98-2221-E-001-007-MY3, and Karlsruhe House of Young Scientists (KHYS), KIT, Germany, under a Grant of Visiting Researcher Scholarship.

K.-M. Chao, T.-s. Hsu, and D.-T. Lee (Eds.): ISAAC 2012, LNCS 7676, pp. 75–84, 2012.
© Springer-Verlag Berlin Heidelberg 2012

In terms of problem complexity, it is known that deciding whether one machine suffices to schedule all the jobs is already strongly NP-complete [13]. Raghavan and Thompson [14] gave an $O(\log n/\log\log n)$-approximation via randomized rounding of linear programs for both versions. This result is also the best known approximation for the discrete version. An $\Omega(\log\log n)$-approximation lower-bound is given by Chuzhoy and Naor [9]. For the continuous machine minimization, Chuzhoy et al. [8] improved the factor to $O(\sqrt{\log n})$. When the number of machines used by the optimal schedule is small, they provided an $O(k^2)$-approximation, where k is the number of machines used by the optimal schedule. Recently, Chuzhoy et al. [7] further improved their previous result to a (large) constant.

In addition, results regarding special constraints have been proposed as well. Cieliebak et al. [11] studied the situation when the lengths of the time intervals during which the jobs can be scheduled are small. Several exact algorithms and hardness results were presented. Yu and Zhang [19] considered two special cases. When the jobs have equal release times, they provided a 2-approximation. When the jobs have equal execution time, they showed that the classical greedy best-fit algorithm achieves a 6-approximation.

From the perspective of utilization-enhancing, a problem that can be seen as dual to machine minimization is the *throughput maximization* problem, whose goal is to maximize the number of jobs that can be scheduled on a single machine. Chuzhoy et al. [10] provided an $O\left(\frac{e}{e-1}+\epsilon\right)$-approximation for any $\epsilon > 0$ for both discrete and continuous settings, where e is the Euler's number. Spieksma [17] proved that the discrete version of this problem is MAX-SNP hard, even when the set of time intervals for each job has cardinality two.

Several natural generalizations of this problem have been considered. Bar-Noy et al. [4] considered the weighted throughput maximization problem, in which the objective is to maximize the weighted throughput for a set of weighted jobs, and presented a 2-approximation. Furthermore, when multiple jobs are allowed to share the time-frame of the same machine, i.e., the concept of context switch is introduced to enhance the throughput, Bar-Noy et al. [3] presented a 5-approximation and a $\left(\frac{2e-1}{e-1}+\epsilon\right)$-approximation for both weighted and unweighted versions. When the set of time intervals for each job has cardinality one, i.e., only job selection is taken into consideration to maximize the weighted throughput, Calinescu et al. [6] presented a $(2+\epsilon)$-approximation while Bansal et al. [2] presented a quasi-PTAS.

Our Focus and Contribution. This paper explores the design and competitive analysis for the *continuous* machine-minimizing job scheduling problem. We consider a real-time system in which we do not have prior knowledge on the arrival of a job until it arrives to the system, and the scheduling decisions have to be made online. As an initial step to exploring the general problem complexity, we consider the case for which all the jobs have *unit* execution time.

In this paper, we provide for this problem:
- a lower bound of 2.09 on the competitive factor of *any* online algorithm, and
- a 5.2-competitive online algorithm.

To the best of our knowledge, this is the first result presented under the concept of real-time machine minimization with arbitrary job arrival times and deadlines.

We would also like to point out a major flaw in a previous paper [16] in which the authors claimed to have an optimal 2-competitive algorithm for a restricted case where the jobs have a universal deadline. In fact, our lower bound proof is built exactly under this restricted case, thereby showing that even when the jobs have a universal deadline, any *feasible* online algorithm has a competitive factor no less than 2.09.

2   Notations and Problem Model

This section describes the job scheduling model adopted in this paper, followed by a formal problem definition.

Job Model. We consider a set of real-time jobs, arriving to the system dynamically. When a job j arrives to the system, say, at time t, its *arrival time* a_j is defined to be t and the job is put into the *ready queue*. The *absolute deadline*, or, deadline for simplicity, for which j must finish its execution is denoted by d_j. The amount of time j requires to finish its execution, also called the *execution time* of j, is denoted by c_j. We consider systems with *discretized timing line* and *unit jobs*, i.e., a_j and d_j are non-negative integers, and $c_j = 1$. When a job finishes, it is removed from the ready queue. For notational brevity, for a job j, we implicitly use a pair $j = (a_j, d_j)$ to denote the corresponding properties.

Job Scheduling. A schedule $\mathbf{S}$ for a set of jobs $\mathcal{J}$ is to decide for each job $j \in \mathcal{J}$ the time at which j starts its execution. $\mathbf{S}$ is said to be *feasible* if each job starts its execution no earlier than its arrival and has its execution finished at its deadline. Moreover, we say that a schedule $\mathbf{S}$ follows the earliest-deadline-first (EDF) principle if whenever there are multiple choices on the jobs to schedule, it always gives the highest priority to the one with earliest deadline.

Let $\#_{\mathbf{S}}(t)$ be the number of jobs which are scheduled for execution at time t in schedule $\mathbf{S}$. The number of machines $\mathbf{S}$ requires to finish the execution of the entire job set, denoted by $M(\mathbf{S})$, is then $\max_{t \geq 0} \#_{\mathbf{S}}(t)$. For any $0 \leq \ell < r$, let

$$\mathcal{J}(\ell, r) = |\{j : j \in \mathcal{J}, \ell \leq a_j, d_j \leq r\}|$$

denote the total amount of workload, i.e., the total number of jobs due to the unit execution time of the jobs in our setting, that arrives and has to be done within the time interval $[\ell, r]$. The following lemma provides a characterization of a feasible EDF schedule for any job set.

Lemma 1. *For any set $\mathcal{J}$ of unit jobs, a schedule $\mathbf{S}$ following the earliest-deadline-first principle is feasible if and only if for any $0 \leq \ell < r$,*

$$\sum_{\ell \leq t < r} \#_{\mathbf{S}}(t) \geq \mathcal{J}(\ell, r).$$

In the *offline machine-minimizing job scheduling* problem, we wish to find a schedule $\mathbf{S}_\mathcal{J}$ for a given set of jobs $\mathcal{J}$ such that $M(\mathbf{S}_\mathcal{J})$ is minimized. For a better depiction of this notion, for any $0 \le \ell < r$, let

$$\rho(\mathcal{J}, \ell, r) = \frac{\mathcal{J}(\ell, r)}{r - \ell}$$

denote the density of workload $\mathcal{J}(\ell, r)$, and let $OPT(\mathcal{J})$ denote the number of machines required by an optimal schedule for $\mathcal{J}$. The following lemma shows that, when the jobs have unit execution time, there is a direct link between $OPT(\mathcal{J})$ and the density of workload.

Lemma 2. *For any set of jobs $\mathcal{J}$, we have $OPT(\mathcal{J}) = \lceil \max_{0 \le \ell < r} \rho(\mathcal{J}, \ell, r) \rceil$.*

Online Job Scheduling. We consider the case where the jobs are arriving in an online setting, i.e., at any time t, we only know the job arrivals up to time t, and the scheduling decisions have to be made without prior knowledge on future job arrivals. To be more precise, let $\mathcal{J}(t) = \{j : j \in \mathcal{J}, a_j \le t\}$ be the subset of $\mathcal{J}$ which contains jobs that have arrived to the system up to time t. In the *online machine-minimizing job scheduling* problem, we wish to find a feasible schedule for a given set of jobs $\mathcal{J}$ such that the number of machines required up to time t is small with respect to $OPT(\mathcal{J}(t))$ for any $t \ge 0$.

Definition 1 (Competitive Factor of an Online Algorithm [5]). *An online algorithm Γ is said to be c-competitive for an optimization problem Π if for any instance $\mathcal{I}$ of Π, we have $\Gamma(\mathcal{I}) \le c \cdot Opt(\mathcal{I}) + x$, where $\Gamma(\mathcal{I})$ and $OPT(\mathcal{I})$ are the values computed by Γ and the optimal solution for $\mathcal{I}$, respectively, and x is a constant. The* asymptotic *competitive factor of Γ is defined to be*

$$\limsup_{n \to \infty} \left\{ \frac{\Gamma(\mathcal{I})}{n} : \mathcal{I} \text{ is an instance of } \Pi \text{ such that } Opt(\mathcal{I}) = n. \right\}.$$

Other Notations and Special Job Sets. Let $\mathcal{J}$ be a set of jobs. For any $t \ge 0$, we use $\hat{\rho}(\mathcal{J}, t)$ to denote the maximum density among those time intervals containing t with respect to $\mathcal{J}$, i.e.,

$$\hat{\rho}(\mathcal{J}, t) = \max_{0 \le \ell \le t < r} \rho(\mathcal{J}, \ell, r).$$

Furthermore, we use $\mathrm{def}(\mathcal{J}, t) = [\ell(\mathcal{J}, t), r(\mathcal{J}, t)]$ to denote the specific time interval that achieves the maximum density in the defining domain of $\hat{\rho}(\mathcal{J}, t)$. If there are more than one such an interval, $\mathrm{def}(\mathcal{J}, t)$ is defined to be the one with the smallest left-end. We call $\mathrm{def}(\mathcal{J}, t)$ the *defining interval* of $\hat{\rho}(\mathcal{J}, t)$. In addition, we define $\hat{\varrho}(\mathcal{J}) = \max_{t \ge 0} \hat{\rho}(\mathcal{J}, t)$ to denote the maximum density for a job set $\mathcal{J}$. Notice that, by Lemma 2, $\hat{\varrho}(\mathcal{J})$ is an alternative definition of $OPT(\mathcal{J})$.

For ease of presentation, throughout this paper, we use a pair (d, σ) to denote a special problem instance for which the jobs have a universal deadline d, where $\sigma = (\sigma(0), \sigma(1), \sigma(2), \ldots, \sigma(d-1))$ is a sequence of length d such that $\sigma(t)$ is the number of jobs arriving at time t. Note that, under this setting, every defining interval has a right-end d.

3   Problem Complexity

This section presents a lower bound of the competitive factor for the studied problem. We consider a special case for which the jobs have a universal deadline, which will later serve as a basis to our main algorithm. In §3.1, we show why the online algorithm, provided in [16] for this special case and claimed to be optimally 2-competitive, fails to produce feasible schedules. Built upon the idea behind the counter-example, we then prove a lower bound of 2.09 for the competitive factor of any online algorithm in §3.2. We begin with the following lemma, which draws up the curtain on the difficulty of this problem led by unknown job arrivals.

Lemma 3 (2-*competitivity lower bound* [16]). *Any online algorithm for the machine-minimizing job scheduling problem with unit jobs and a universal deadline has a competitive factor of at least 2.*

Proof. Below we sketch the proof provided in [16]. Let $d \geq 1$ be an arbitrary integer. Consider the job set $\mathcal{J}_2^* = (d, \sigma_2^*)$, where $\sigma_2^* = (d, d, d, \ldots, d)$ is a sequence containing d elements of value d. Notice that, we have $\text{OPT}(\mathcal{J}_2^*(t)) = t + 1$ for all $0 \leq t < d$. Therefore any online algorithm with competitive factor R uses at most $R \cdot (t + 1)$ machines at time t. Taking the summation over $0 \leq t < d$, we know that any online algorithm with competitive factor R can schedule at most $\sum_{0 \leq t < d} R \cdot (t + 1) = \frac{1}{2}d(d + 1)R$ jobs. Since there are d^2 jobs in total, we conclude that $R \geq 2$ when d goes to infinity. $\square$

3.1   Why the Known Algorithm Fails to Produce Feasible Schedules

This section presents a counter-example for the online algorithm provided in [16], which we will also refer to as the *packing-via-density* algorithm in the following. Given a problem instance $\mathcal{J} = (d, \sigma)$, *packing-via-density* works as follows.

   At any time t, $t \geq 0$, the algorithm computes the maximum density with respect to the current job set $\mathcal{J}(t)$. More precisely, it computes $\hat{\rho}(\mathcal{J}(t), t)$. Then the algorithm assigns $2 \cdot \lceil \hat{\rho}(\mathcal{J}(t), t) \rceil$ jobs for execution.

   Intuitively, in the computation of density, the workload of each job is equally distributed to the time interval from its arrival till its deadline, or, possibly to a larger super time interval containing it if this leads to a higher density. The algorithm uses another factor of $\hat{\rho}(\mathcal{J}(t), t)$ in order to cover the unknown future job arrivals, which seems to be a good direction for getting a feasible scheduling.

   However, in [16], the authors claimed that the job set $\mathcal{J}_2^*$ represents one of the worst case scenarios, followed by sketching the feasibility of *packing-via-density* on $\mathcal{J}_2^*$. Although from intuition this looks promising, and, by suitably defining the potential function, one can indeed prove the feasibility of *packing-via-density* for $\mathcal{J}_2^*$ and other similar job sets, the job set $\mathcal{J}_2^*$ they considered is in fact not a worst scenario. The main reason is that, for each t with $0 \leq t < d$, the left-end of the defining interval for t is always zero. That is, we have $\text{def}(\mathcal{J}_2^*(t), t) = [0, d]$ for all $0 \leq t < d$. This implicitly takes all job arrivals into consideration when computing the densities. When the sequence is more complicated and the defining intervals change over time, using $2 \cdot \lceil \hat{\rho}(\mathcal{J}(t), t) \rceil$ machines is no longer

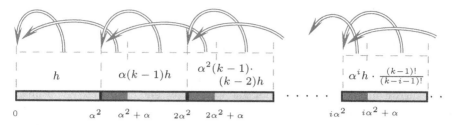

Fig. 1. An illustration for the arrival sequence $\sigma_{k,\alpha}^*$ of $\mathcal{J}_{k,\alpha}^*$ and the changes of the left-ends of defining intervals over time

able to cover the *unpaid debt* created before $\ell(\mathcal{J}(t), t)$, i.e., the jobs which are not yet finished but no longer contributing to the computation of $\hat{\varrho}(\mathcal{J}(t))$. This is illustrated by the following example.

Consider the job set $\mathcal{J}^* = (32, \sigma^*)$, where σ^* is defined as

$$
\sigma^* = \left(\underbrace{75, 75, \ldots, 75}_{0 \sim 15}, \, 1200, \, 0, 0, 0, \, \underbrace{300, 300, \ldots, 300}_{20 \sim 31} \right)
$$

Notice that, for $0 \le t \le 19$, we have $\mathrm{def}(\mathcal{J}^*(t), t) = [0, 32]$, and for $20 \le t \le 31$, we have $\mathrm{def}(\mathcal{J}^*(t), t) = [16, 32]$. A direct calculation shows that *packing-via-density* results in deadline misses of 10 jobs.

3.2   Lower Bound on the Competitive Factor

In fact, by further generalizing the construction of $\mathcal{J}^*$, we can design an *online adversary* that proves a lower bound strictly greater than 2 for the competitive factor of any online algorithm. In the job set $\mathcal{J}^*$, the debt is created by making a one-time change of the defining interval at time 20. Below, we construct an example whose defining intervals can alter for arbitrarily many times, thereby creating sufficiently large debts. Then, we present our online adversary.

Let $h, k, \alpha \in \mathbb{N}$ be three constants to be decided later. We define the job set $\mathcal{J}_{k,\alpha}^* = \left(k\alpha^2, \sigma_{k,\alpha}^* \right)$ as follows. For each $0 \le i < k$ and each $0 \le j < \alpha^2$,

$$
\sigma_{k,\alpha}^* \left(i\alpha^2 + j \right) = \begin{cases} h, & \text{if } i = 0, \\ \alpha(k - i) \cdot \sigma_{k,\alpha}^* \left((i-1)\alpha^2 + j \right), & \text{otherwise.} \end{cases}
$$

Also refer to Fig. 1 for an illustration of the sequence. The following lemma shows the changes of the defining intervals over time.

Lemma 4. *For each time* $t = i\alpha^2 + j$, *where* $0 \le i < k$, $0 \le j < \alpha^2$, *we have*

$$
\ell\left(\mathcal{J}_{k,\alpha}^*(t), t\right) = \begin{cases} 0, & \text{for } i = 0, \\ (i-1)\alpha^2, & \text{for } i > 0 \text{ and } 0 \le j < \alpha, \\ i\alpha^2, & \text{for } i > 0 \text{ and } \alpha \le j < \alpha^2. \end{cases}
$$

Below we present our online adversary, which we denote by $\mathcal{A}^*(c)$, where $c > 0$ is a constant. Let Γ be an arbitrary feasible online scheduling algorithm for this problem. The adversary works as follows. At each time t with $0 \leq t < k\alpha^2$, $\mathcal{A}^*(c)$ releases $\sigma_{k,\alpha}^*(t)$ jobs with deadline $k\alpha^2$ for algorithm Γ and observes the behavior of Γ. If Γ uses more than $c \cdot \hat{\varrho}\left(\mathcal{J}_{k,\alpha}^*(t)\right)$ machines, then $\mathcal{A}^*(c)$ terminates immediately. Otherwise, $\mathcal{A}^*(c)$ proceeds to time $t + 1$ and repeats the same procedure. This process continues till time $k\alpha^2$.

Theorem 1. *Any feasible online algorithm for machine-minimizing job scheduling problem has a competitive factor at least 2.09, even for the case when the jobs have a universal deadline.*

4   5.2-Competitive Packing-via-Density

As indicated in Lemma 2, to come up with a good scheduling algorithm for online machine-minimizing job scheduling with unit jobs, it suffices to compute a good approximation of the offline density for the entire job set, as this corresponds directly to the number of machines required by any optimal schedule. From the proofs for the problem complexity in Section 3.2, for any job set $\mathcal{J}$ and $t \geq 0$, the gap between $\hat{\varrho}(\mathcal{J})$, which is the maximum offline density for the entire job set, and $\hat{\varrho}(\mathcal{J}(t))$, which is the maximum density the online algorithm for the jobs arrived before and at time t, can be arbitrarily large. For instance, in the simple job set $\mathcal{J}_2^*$, we have $\hat{\varrho}(\mathcal{J}_2^*) = d$ while $\hat{\varrho}(\mathcal{J}_2^*(t)) = t + 1$ for all $0 \leq t < d$.

In [16], the authors proved that, simply using $\lceil \hat{\varrho}(\mathcal{J}(t)) \rceil$ to approximate the offline density as suggested in the classical mainstream *any fit* algorithms, such as *best fit*, *first fit*, etc., can results in the deadline misses of $\Theta(\log n)$ jobs even for the job set $\mathcal{J}_2^*$, meaning that we will have to use $\Theta(\log n)$ machines in the very last moment in order to prevent deadline misses if we apply these classical packing algorithms. Therefore, additional space sparing at each moment is necessary for obtaining a better approximation guarantee on the offline density in later times. One natural question to ask is:

> Is there a constant c such that $c \cdot \hat{\varrho}(\mathcal{J}(t))$ is an approximation of $\hat{\varrho}(\mathcal{J})$ for all $t \geq 0$?

In this section, we give a positive answer to the above question in a slightly more general way. We show that, with a properly chosen constant c, using $\lceil c \cdot \hat{\varrho}(\mathcal{J}(t)) \rceil$ machines at all times gives a feasible scheduling which is also c-competitive for the machine-minimizing job scheduling with unit job execution time and *arbitrary job deadlines*.

The packing-via-density(c) *algorithm.* At time t, $t \geq 0$, the algorithm computes the maximum density it has seen so far, i.e., $\hat{\varrho}(\mathcal{J}(t))$. Then the algorithm assigns $\lceil c \cdot \hat{\varrho}(\mathcal{J}(t)) \rceil$ jobs with earliest deadlines form the ready queue for execution.

Since $\hat{\varrho}(\mathcal{J}(t)) \leq \hat{\varrho}(\mathcal{J})$ for all $t \geq 0$, by Lemma 2, we have $\hat{\varrho}(\mathcal{J}(t)) \leq \text{OPT}(\mathcal{J})$ for all $t \geq 0$. Hence, we know that the schedule produced by *packing-via-density(c)* is c-competitive as long as it is feasible. In the following, we show that, for a properly chosen constant c, *packing-via-density(c)* always produces a feasible schedule for any upcoming job set. To this end, for any job set, we present two reductions to obtain a sequence whose structure is relatively simple in terms of the altering of defining intervals, followed by providing a direct analysis on the potential function of that sequence.

Feasibility of packing-via-density(c). For any job set $\mathcal{J}$, any t_1, t_2 with $0 \leq t_1 < t_2$, and any $c > 0$, consider the potential function $\Phi_c(\mathcal{J}, t_1, t_2)$ defined as

$$\Phi_c(\mathcal{J}, t_1, t_2) = \left(\sum_{t_1 \leq t < t_2} c \cdot \hat{\varrho}(\mathcal{J}(t)) \right) - \mathcal{J}(t_1, t_2).$$

Literally, in this potential function we consider the sum of maximum densities over each moment between time t_1 and time t_2, subtracted by the total amount of workload which arrives and has to be done within the time interval $[t_1, t_2]$. Since *packing-via-density(c)* assigns $\lceil c \cdot \hat{\varrho}(\mathcal{J}(t)) \rceil$ jobs for execution for any moment t, by Lemma 1, we have the feasibility of this algorithm if and only if $\Phi_c(\mathcal{J}, t_1, t_2) \geq 0$ for all $0 \leq t_1 < t_2$.

Below, we present our first reduction and show that, it suffices to prove the non-negativity of this potential function for any job set with a universal deadline. For any $d \geq 0$, consider the job set $\mathcal{J}_d$ defined as follows. For each $(i, j) \in \mathcal{J}$ such that $j \leq d$, we create a job (i, d) and put it into $\mathcal{J}_d$.

Lemma 5 (Reduction to the case of equal deadlines). *For any $c \geq 0$, $d \geq 0$, and $0 \leq t < d$, we have $\Phi_c(\mathcal{J}_d, t, d) \leq \Phi_c(\mathcal{J}, t, d)$.*

As the mapping from $\mathcal{J}$ to $\mathcal{J}_d$ is well-defined for each $d \geq 0$, by Lemma 5, the non-negativity of the potential function with respect to *any job set with a universal deadline* will in turn imply the non-negativity of that with respect to the given job set $\mathcal{J}$. Therefore, it suffices to show that, for any job set $\mathcal{J}_d = (d, \sigma_d)$, we have $\Phi_c(\mathcal{J}_d, 0, d) \geq 0$.

Next, we make another reduction and show that, it suffices to consider job sets with non-decreasing arrival sequences: (i) If σ_d is already a non-decreasing sequence, then there is nothing to argue. (ii) Otherwise, let k, $0 \leq k < d - 1$, be the largest integer such that $\sigma_d(k) > \sigma_d(k+1)$. Furthermore, let m, $k < m < d$, be the largest integer such that $\sigma_d(m)$ is under the average of $\sigma_d(k), \sigma_d(k + 1), \ldots, \sigma_d(m)$, i.e.,

$$\sigma_d(m) < \frac{\sum_{k \leq i \leq m} \sigma_d(i)}{m - k + 1}, \text{ and, } \sigma_d(m+1) \geq \frac{\sum_{k \leq i \leq m+1} \sigma_d(i)}{m - k + 2} \text{ if } m < d - 1.$$

Consider the job set $\mathcal{J}_{d,k,m} = (d, \sigma_{d,k,m})$ defined as follows. For each i with $0 \leq i < k$ or $m < i < d$, we set $\sigma_{d,k,m}(i) = \sigma_d(i)$. Otherwise, for i with $k \leq i \leq m$, we set $\sigma_{d,k,m}(i) = \left(\sum_{k \leq i \leq m} \sigma_d(i) \right) / (m - k + 1)$. The following lemma shows that the effect of this change on the potential of the resulting job set is non-increasing.

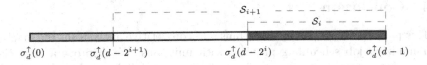

Fig. 2. The subsequence $\mathcal{S}_i$, $0 \leq i \leq \lfloor \log_2 d \rfloor$

Lemma 6 (*Reduc. to non-decreasing seq.*). $\Phi_c(\mathcal{J}_{d,k,m}, 0, d) \leq \Phi_c(\mathcal{J}_d, 0, d)$.

Repeating the process described in (ii) above, we get a non-decreasing sequence $\sigma_d^\uparrow$ for the job set $\mathcal{J}_d$. By Lemma 6, we know that the non-negativity of the potential function with respect to $(d, \sigma_d^\uparrow)$ will in turn imply the non-negativity of that with respect to $\mathcal{J}_d$.

Now, we consider the job set $\mathcal{J}_d^\uparrow = (d, \sigma_d^\uparrow)$ and provide a direct analysis on the value of $\Phi_c(\mathcal{J}_d^\uparrow, 0, d)$. To take the impact of unknown job arrivals and the unpaid-debt excluded implicitly in the computation of $\hat{\varrho}(\mathcal{J}_d^\uparrow(t))$, we exploit the non-decreasing property of the sequence and use a backward analysis. More precisely, starting from the tail, for each $0 \leq i \leq \lfloor \log_2 d \rfloor$, we consider the sequence

$$\mathcal{S}_i = \left(\sigma_d^\uparrow(d - 2^i), \sigma_d^\uparrow(d - 2^i + 1), \ldots, \sigma_d^\uparrow(d - 1) \right),$$

whose length grows exponentially as i increases. Also refer to Fig. 2 for an illustration. The following lemma shows that a simple argument, which takes eight times the sum of densities over $\mathcal{S}_i$ to cover the total workload in $\mathcal{S}_{i+1}$, already asserts the feasibility of *packing-via-density*(8).

Lemma 7 (*Feasibility of packing-via-density*(8)). $\Phi_8\left(\mathcal{J}_d^\uparrow, 0, d\right) \geq 0$.

The basic argument gives a hint on the reason why the potential function can be made positive-definite by a carefully chosen constant c. Below we show that, a further generalized approach gives a better bound. The idea is to further exploit the densities generated by $\mathcal{S}_{i+1}$ itself: when the amount of workload in $\mathcal{S}_{i+1}$ is relatively low, then a smaller factor from $\mathcal{S}_i$ suffices, and when the amount of workload in $\mathcal{S}_{i+1}$ becomes higher, then most of the workload can be covered by the densities $\mathcal{S}_{i+1}$ itself generates. In addition, we use the exponential base of 3 to define the subsequences instead of 2.

Lemma 8.
$$\Phi_{5.2}\left(\mathcal{J}_d^\uparrow, 0, d\right) \geq 0.$$

We conclude our result by the following theorem.

Theorem 2. *The packing-via-density(5.2) algorithm computes a feasible 5.2-competitive schedule for the machine-minimizing job scheduling problem with unit job execution time.*

5 Conclusion

This paper presents online algorithms and competitive analysis for the machine-minimizing job scheduling problem with unit jobs. We disprove a false claim made by a previous paper regarding a further restricted case. We also provide a lower bound on the competitive factor of any online algorithm for this problem.

References

1. Adler, M., Sitaraman, R., Rosenberg, A., Unger, W.: Scheduling time-constrained communication in linear networks. In: SPAA 1998, NY, pp. 269–278 (1998)
2. Bansal, N., Chakrabarti, A., Epstein, A., Schieber, B.: A quasi-ptas for unsplittable flow on line graphs. In: STOC 2006, pp. 721–729. ACM, NY (2006)
3. Bar-Noy, A., Bar-Yehuda, R., Freund, A., Naor, J.S., Schieber, B.: A unified approach to approximating resource allocation and scheduling. J. ACM 48, 1069–1090 (2001)
4. Bar-Noy, A., Guha, S.: Approximating the throughput of multiple machines in real-time scheduling. SIAM J. Comput. 31, 331–352 (2002)
5. Borodin, A., El-Yaniv, R.: Online computation and competitive analysis. Cambridge University Press, New York (1998)
6. Calinescu, G., Chakrabarti, A., Karloff, H., Rabani, Y.: An improved approximation algorithm for resource allocation. ACM Trans. Algorithms 7, 48:1–48:7 (2011)
7. Chuzhoy, J., Codenotti, P.: Resource Minimization Job Scheduling. In: Dinur, I., Jansen, K., Naor, J., Rolim, J. (eds.) APPROX and RANDOM 2009. LNCS, vol. 5687, pp. 70–83. Springer, Heidelberg (2009)
8. Chuzhoy, J., Guha, S., Khanna, S., Naor, J.S.: Machine minimization for scheduling jobs with interval constraints. In: FOCS 2004, Washington, pp. 81–90 (2004)
9. Chuzhoy, J., Naor, J.S.: New hardness results for congestion minimization and machine scheduling. J. ACM 53, 707–721 (2006)
10. Chuzhoy, J., Ostrovsky, R., Rabani, Y.: Approximation algorithms for the job interval selection problem and related scheduling problems. Math. Oper. Res. 31, 730–738 (2006)
11. Cieliebak, M., Erlebach, T., Hennecke, F., Weber, B., Widmayer, P.: Scheduling with release times and deadlines on a minimum number of machines. IFIP, vol. 155, pp. 209–222. Springer, Boston
12. Crama, Y., Flippo, O., Klundert, J., Spieksma, F.: The assembly of printed circuit boards: A case with multiple machines and multiple board types (1998)
13. Garey, M.R., Johnson, D.S.: Computers and Intractability: A Guide to the Theory of NP-Completeness. W.H. Freeman & Co., New York (1979)
14. Raghavan, P., Tompson, C.D.: Randomized rounding: a technique for provably good algorithms and algorithmic proofs. Combinatorica 7, 365–374 (1987)
15. Rajugopal, G., Hafez, R.: Adaptive rate controlled, robust video communication over packet wireless networks. Mob. Netw. Appl. 3, 33–47 (1998)
16. Shi, Y., Ye, D.: Online bin packing with arbitrary release times. Theoretical Computer Science 390, 110–119 (2008)
17. Spieksma, F.: On the approximability of an interval scheduling problem. Journal of Scheduling 2, 215–225 (1999)
18. Yau, D., Lam, S.: Adaptive rate-controlled scheduling for multimedia applications. In: MULTIMEDIA 1996, pp. 129–140. ACM, New York (1996)
19. Yu, G., Zhang, G.: Scheduling with a minimum number of machines. Operations Research Letters 37, 97–101 (2009)

A Partially Ordered Structure
and a Generalization of the Canonical Partition
for General Graphs with Perfect Matchings

Nanao Kita

Keio University, Yokohama, Japan
kita@a2.keio.jp

Abstract. This paper is concerned with structures of general graphs with perfect matchings. We first reveal a partially ordered structure among elementary components of general graphs with perfect matchings. Our second result is a generalization of Kotzig's canonical partition to a decomposition of general graphs with perfect matchings. It contains a short proof for the theorem of the canonical partition. These results give decompositions which are canonical, that is, unique to given graphs. We also show that there are correlations between these two and that these can be computed in polynomial time.

1   Introduction

This paper is concerned with matchings on graphs. For general accounts on matching theory we refer to Lovász and Plummer's book [1].

A *matching* of a graph G is a set of edges $F \subseteq E(G)$ no two of which have common vertices. A matching of cardinality $|V(G)|/2$ (resp. $|V(G)|/2 - 1$) is called a *perfect matching* (resp. a *near-perfect matching*). We call a graph with a perfect matching *factorizable*. An edge of a factorizable graph is called *allowed* if it is contained in a perfect matching. Generally, in a factorizable graph, the subgraph induced by the union of all the allowed edges has several components, and these are called *elementary components*. In this paper, we denote the family of elementary components of a factorizable graph G as $\mathcal{G}(G)$. A factorizable graph that has only one elementary component is called an *elementary graph.*

Matching theory is of central importance in graph theory and combinatorial optimization [2], with numerous practical applications [3]. Structure theorems that give decompositions which are canonical, namely, unique to given graphs, play important roles in matching theory. Only three theorems, i.e. the canonical partition [4–6], the Dulmage-Mendelsohn decomposition [1], and the Gallai-Edmonds structure theorem [1] have been known as such. The first two are not applicable for general graphs with perfect matchings, and the last one treats them as irreducible and does not decompose them properly, which means nothing has been known that tells non-trivial canonical structures of general graphs with perfect matchings. Therefore, in this paper, we give new canonical structure theorems for them.

K.-M. Chao, T.-s. Hsu, and D.-T. Lee (Eds.): ISAAC 2012, LNCS 7676, pp. 85–94, 2012.
© Springer-Verlag Berlin Heidelberg 2012

By the definitions, we can view factorizable graphs as being "built" up by combining elementary components with additional edges. However it does not mean that all combinations result in graphs with desired elementary components. Thus the family of elementary components must have a certain non-trivial structure. For bipartite factorizable graphs, the *Dulmage-Mendelsohn decomposition* (in short, the *DM-decomposition*) reveals the ordered structure of their elementary components. However, as for non-bipartite graphs, no counterpart has been known.

In this paper, as our first contribution, we reveal a partially ordered structure between elementary components of general graphs with perfect matchings. It has some similar natures to the DM-decomposition, however they are distinct.

The second contribution is a generalization of the *canonical partition* [4–6]; see also [1], which is originally a decomposition of elementary graphs. Kotzig [4–6] first investigated the canonical partition of elementary graphs as the quotient set of a certain equivalence relation, and later, Lovász redefined it from the point of view of maximal barriers [1]. In this paper we generalize the canonical partition to a decomposition of general graphs with perfect matchings, based on Kotzig's way. It contains a short proof for the theorem of the canonical partition.

Note that these two results of us give canonical decompositions of graphs. We also show that there are correlations between these two and that these can be computed in polynomial time.

Any of the three existing canonical structure theorems plays significant roles in combinatorics including matching theory. The canonical partition plays a crucial role in matching theory, especially from the polyhedral point of view, that is, in the study of the matching polytope and the matching lattice [7–9]. The Dulmage-Mendelsohn decomposition is known for its application to the efficient solution of linear equations determined by large sparse matrices [1]. Additionally, it is an origin of a series of studies on submodular functions, that is, the field of the principal partition [10,11]. The Gallai-Edmonds structure theorem is essential to the optimality of the maximum matching [1,12]. Thus it also underlies reasonable generalizations of maximum matching problem [13,14].

By combining the results in this paper with the Gallai-Edmonds structure theorem, we can easily obtain a refinement of the Gallai-Edmonds structure theorem, which gives a consistent view of graphs, whether they are factorizable or not, or, elementary or not [15]. Hence, we are sure that our structure theorems should be powerful tools in matching theory. In fact, the cathedral theorem [1] can be obtained from our results in a quite natural way [15].

2   Preliminaries

In this section, we list some standard definitions and well-known properties. Basics on sets, graphs, digraphs, and algorithms mostly conform to [2].

Let G be a graph and $X \subseteq V(G)$. The subgraph of G induced by X is denoted by $G[X]$. $G - X$ means $G[V(G) \setminus X]$. Given $F \subseteq E(G)$, we define the *contraction* of G by F as the graph obtained from contracting all the edges in

F, and denote as G/F. Additionally, We define the *contraction* of G by X as $G/X := G/E(G[X])$. We say $H \subseteq G$ if H is a subgraph of G. If it is clear from the context, we sometimes regard a subgraph $H \subseteq G$ as the vertex set $V(H)$, a vertex v as a graph $(\{v\}, \emptyset)$.

The set of edges that has one end vertex in $X \subseteq V(G)$ and the other vertex in $Y \subseteq V(G)$ is denoted as $E_G[X, Y]$. We denote $E_G[X, V(G) \setminus X]$ as $\delta_G(X)$. We define the *set of neighbors* of X as the set of vertices in $V(G) \setminus X$ that are adjacent to vertices in X, and denote as $N_G(X)$. We sometimes denote $E_G[X, Y]$, $\delta_G(X)$, $N_G(X)$ as just $E[X, Y]$, $\delta(X)$, $N(X)$ if they are apparent from the context.

For two graphs G_1 and G_2, $G_1 + G_2 := (V(G_1) \cup V(G_2), E(G_1) \cup E(G_2))$ is called the *union* of them, and $G_1 \cap G_2 := (V(G_1) \cap V(G_2), E(G_1) \cap E(G_2))$ the *intersection* of them.

Let $\hat{G}$ be a graph such that $G \subseteq \hat{G}$. For $e = uv \in E(\hat{G})$, $G + e$ means $(V(G) \cup \{u, v\}, E(G) \cup \{e\})$, and $G - e$ means $(V(G), E(G) \setminus \{e\})$. For a set of edges $F = \{e_i\}_{i=1}^{k}$, $G + F$ and $G - F$ means respectively $G + e_1 + \cdots + e_k$ and $G - e_1 - \cdots - e_k$.

For a path P and $x, y \in V(P)$, xPy means the subpath on P between x and y. For a circuit C with an orientation that makes it a dicircuit, and $x, y \in V(C)$ where $x \neq y$, xCy means the subpath in C that can be regarded as a dipath from x to y.

A vertex $v \in V(G)$ satisfying $\delta(v) \cap M = \emptyset$ is called *exposed* by M. For a matching M of G and $u \in V(G)$, u' denote the vertex to which u is matched by M. For $X \subseteq V(G)$, M_X denotes $M \cap E(G[X])$.

Let M be a matching of G. For $Q \subseteq G$, which is a path or circuit, we call Q *M-alternating* if $E(Q) \setminus M$ is a matching of Q. Let $P \subseteq G$ be an M-alternating path with end vertices u and v. If P has an even number of edges and starts with an edge in M if it is traced from u, we call it an *M-balanced path* from u to v. We regard a trivial path, that is, a path composed of one vertex and no edges as an M-balanced path. If P has an odd number of edges and $M \cap E(P)$ (resp. $E(P) \setminus M$) is a perfect matching of P, we call it *M-saturated* (resp. *M-exposed*).

Let $H \subseteq G$. We say a path $P \subseteq G$ is an *ear relative to* H if both end vertices of P are in H while internal vertices are not. So do we to a circuit if exactly one vertex of it is in H. For simplicity, we call the vertices of $V(P) \cap V(H)$ *end vertices* of P, even if P is a circuit. For an ear $R \subseteq G$ relative to H, we call it an *M-ear* if $P - V(H)$ is an M-saturated path.

A graph is called *factor-critical* if any deletion of its single vertex leaves a factorizable graph. A subgraph $G' \subseteq G$ is called *nice* if $G - V(G')$ is factorizable. The next two propositions are well-known and might be regarded as folklores.

Proposition 1. *Let M be a near-perfect matching of a graph G that exposes $v \in V(G)$. Then, G is factor-critical if and only if for any $u \in V(G)$ there exists an M-balanced path from u to v.*

Proposition 2. *Let G be a graph. Then G is factor-critical if and only if each block of G is factor-critical.*

Proposition 3 (implicitly stated in [16]). *Let G be a factor-critical graph, $v \in V(G)$, and M be a near-perfect matching that exposes v. Then for any non-loop edge $e = vu \in E(G)$, there is a nice circuit C of G which is an M-ear relative to v and contains e.*

Theorem 1 (implicitly stated in [16]). *Let G be a factor-critical graph. For any nice factor-critical subgraph G' of G, G/G' is factor-critical.*

Let us denote the number of odd components (i.e. connected components with odd numbers of vertices) of a graph G as $oc(G)$, and the cardinality of a maximum matching of G as $\nu(G)$. It is known as the *Berge formula* [1] that for any graph G, $|V(G)| - 2\nu(G) = \max\{oc(G - X) - |X| : X \subseteq V(G)\}$. A set of vertices that attains the maximum in the right side of the equation is called a *barrier*.

The *canonical partition* is a decomposition for elementary graphs and plays a crucial role in matching theory. First Kotzig introduced the canonical partition as a quotient set of a certain equivalence relation [4–6], and later Lovász redefined it from the point of view of barriers [1]. In fact, these are equivalent. For an elementary graph G and $u, v \in V(G)$, we say $u \sim v$ if $u = v$ or $G - u - v$ is not factorizable.

Theorem 2 (**Kotzig [4–6], Lovász [1]**). *Let G be an elementary graph. Then $\sim$ is an equivalence relation on $V(G)$ and the family of equivalence classes is exactly the family of maximal barriers of G.*

The family of equivalence classes of $\sim$ is called the *canonical partition* of G, and denoted by $\mathcal{P}(G)$. From this theorem, following is derived for elementary graphs: For an arbitrary perfect matching M of an elementary graph G, there is a u-v M-saturated path if and only if $u \not\sim v$. Thus, $uv \in E(G)$ is allowed if and only if $u \not\sim v$.

An *ear-decomposition* of graph G is a sequence of subgraphs $G_0, \subseteq, \cdots, \subseteq G_k = G$ such that $G_0 = (\{r\}, \emptyset)$ for some $r \in V(G)$ and for each $i \geq 1$, G_i is obtained from G_{i-1} by adding an ear P_i relative to G_{i-1}. We sometimes regard an ear-decomposition as a family of ears $\mathcal{P} = \{P_1, \ldots, P_k\}$. An ear-decomposition is called *odd* if any of its ears has an odd number of edges.

Theorem 3 (**Lovász [16]**). *A graph is factor-critical if and only if it has an odd ear-decomposition.*

For a factor-critical graph G and its near-perfect matching M, we call an ear-decomposition *alternating with respect to M*, or just M-alternating, if each ear is an M-ear.

Proposition 4 (**Lovász [16]**). *Let G be a factor-critical graph. Then for any near-perfect matching M of G, there is an M-alternating ear-decomposition of G.*

Later on this paper, note the following fundamental properties; for a factorizable graph G and its perfect matching M, $e \in E(G)$ is allowed if and only if there is an M-alternating circuit containing e; for $u, v \in V(G)$, $G - u - v$ is factorizable if and only if there is an M-saturated path between u and v; for two M-alternating path P and Q, a segment of $P \cap Q$ is a M-saturated paths if it contains no end vertices of P nor Q.

3   A Partially Ordered Structure in Factorizable Graphs

In this section we show our first result: a partially ordered structure among elementary components of factorizable graphs. For a factorizable graph G and $G_1 \in \mathcal{G}(G)$, we call $X \subseteq V(G)$ a *separable set for G_1* if each $H \in \mathcal{G}(G)$ satisfies $V(H) \subseteq X$ or $V(H) \cap X = \emptyset$, and $G[X]/G_1$ is factor-critical.

Definition 1. *Let G be a factorizable graph and $G_1, G_2 \in \mathcal{G}(G)$. We say $G_1 \triangleright G_2$ if there exists a separable set $X \subseteq V(G)$ for G_1 that contains G_2.*

Lemma 1. *Let G be a factorizable graph and M be a perfect matching of G. Let $G_1, G_2 \in \mathcal{G}(G)$ such that $G_1 \triangleright G_2$ and $X \subseteq V(G)$ be a separable set for G_1 that contains G_2. Then for any $u \in V(G_2)$ there exists $v \in V(G_1)$ such that there is an M-balanced path from u to v whose vertices except v are contained in $X \setminus V(G_1)$.*

Proof. Since $M_{X \setminus V(G_1)}$ is a near-perfect matching of $G[X]/G_1$ that exposes the contracted vertex g_1 corresponding to G_1, we are done by Proposition 1.   □

Proposition 5. *Let G be an elementary graph and M be a perfect matching of G. Then for any two vertices $u, v \in V(G)$ there is an M-saturated path between u and v, or an M-balanced path from u to v.*

Let G be a factorizable graph and M be a perfect matching of G. We call a sequence of elementary components $H_0, \ldots, H_k \in \mathcal{G}(G)$ an *M-ear sequence from H_0 to H_k* if there is an M-ear relative to H_{i-1} and through H_i for each $i = 1, \ldots, k-1$. The next theorem gives an equivalent definition of $\triangleright$.

Theorem 4. *Let G be a factorizable graph, M be a perfect matching of G, and $G_1, G_2 \in \mathcal{G}(G)$. Then, $G_1 \triangleright G_2$ if and only if there exists an M-ear sequence from G_1 to G_2.*

Theorem 5. *$\triangleright$ is a partial order.*

Proof. The reflexivity is obvious from the definition. The transitivity obviously follows from Theorem 4. Hence, we will prove the antisymmetry. Let $G_1, G_2 \in \mathcal{G}(G)$ be such that $G_1 \triangleright G_2$ and $G_2 \triangleright G_1$. Suppose the claim fails, that is, $G_1 \neq G_2$. Let $X \subseteq V(G)$ be a separable set for G_1 that contains G_2, and let M be a perfect matching of G. Let g_1 be the contracted vertex of $G[X]/G_1$, and take an M-ear P relative to g_1 in $G[X]/G_1$ by Proposition 3, whose corresponding end vertices in G are p and q. Since $G_2 \triangleright G_1$, by Lemma 1, there is an M-balanced path Q from p to some vertex in $V(G_2)$. Trace Q from p and let x be the first vertex we encounter that is in $X \setminus V(G_1)$. Since $G[X]/G_1$ is factor-critical, there is an M-balanced path R from x to q whose vertices except q are contained in $X \setminus V(G_1)$. Trace R from x and let y be the first vertex we encounter that is on P. If pPy has an even number of edges, $pQx + xRy + yPp$ is an M-alternating circuit containing non-allowed edges, a contradiction.

Hence hereafter we assume pPw has an odd number of edges. By Proposition 5, there is an M-saturated or balanced path L from q to p which is contained in G_1. Trace L from q and let w be the first vertex on Q. If pQw has an odd number of edges, then $wQp + P + qLw$ is an M-alternating circuit, a contradiction. If pQw has an even number of edges, then $qLw + wQx + xRy + yPq$ is an M-alternating circuit, which is also a contradiction. Thus we get $G_1 = G_2$, and the claim follows.   □

4   A Generalization of the Canonical Partition

For non-elementary graphs, the family of maximal barriers never gives a partition of its vertex set [1]. Therefore, to analyze the structures of general graphs with perfect matchings, we generalized the canonical partition based on Kotzig's way [4–6].

Definition 2. *Let G be a factorizable graph and $H \in \mathcal{G}(G)$. For $u, v \in V(H)$, we say $u \sim_g v$ if $u = v$ or $G - u - v$ is not factorizable.*

Theorem 6. $\sim_g$ *is an equivalence relation.*

Proof. Since the reflexivity and the symmetry are obvious from the definition, we prove the transitivity. Let $u, v, w \in V(H)$ be such that $u \sim_g v$ and $v \sim_g w$. If any two of them are identical, clearly the claim follows. Therefore it suffices to consider the case that they are mutually distinct. Suppose that the claim fails, that is, $u \not\sim_g w$. Then there is an M-saturated path P between u and w. By Proposition 5, there is an M-balanced path Q from v to u. Trace Q from v and let x be the first vertex we encounter that in $V(Q) \cap V(P)$. If uPx has an odd number of edges, $vQx + xPu$ is an M-saturated path between u and v, a contradiction. If uPx has an even number of edges, then xPw has an odd number of edges, and by the same argument we have a contradiction.   □

We call the family of equivalence classes of $\sim_g$ as the *generalized canonical partition* and denote as $\mathcal{P}_G(H)$ for each elementary component $H \in \mathcal{G}(G)$ of a factorizable graph G. Note that the notions of the canonical partition and the generalized one are coincident for an elementary graph. Thus we denote the union of equivalence classes of all the elementary components of G as $\mathcal{P}(G)$, and call it just as the *canonical partition*. Moreover our proof for Theorem 6 contains a short proof for the existence of the canonical partition. Kotzig takes three papers to prove it, thus to prove that $\sim$ is an equivalence relation "from scratch" is considered to be hard [1]. However, in fact, it can be shown in a simple way even without the premise of the Gallai-Edmonds structure theorem nor the notion of barriers. Note also that the generalized canonical partition $\mathcal{P}_G(H)$ is a refinement of $\mathcal{P}(H)$ for each $H \in \mathcal{G}(G)$.

5   Correlations between $\triangleright$ and $\sim_g$

In this section we further analyze properties of factorizable graphs. We denote all the upper bounds of $H \in \mathcal{G}(G)$ in $(\mathcal{G}(G), \triangleright)$ as $up_G^*(H)$ and define $up_G(H)$

as $up_G^*(H) \setminus \{H\}$. We sometimes omit the subscripts if they are apparent from the context. For simplicity, we sometimes denote the subgraph induced by the vertices in $up(H)$ (resp. $up^*(H)$) just as $G[up(H)]$ (resp. $G[up^*(H)]$).

Lemma 2. *Let G be a factorizable graph, M be a perfect matching of G, and $H \in \mathcal{G}(G)$. Let P be an M-ear relative to H with end vertices $u, v \in V(H)$. Then $u \sim_g v$.*

Proof. Suppose the claim fails, that is, $u \neq v$ and there is an M-saturated path Q between u and v. Trace Q from u and let x be the first vertex we encounter that is on $P - u$. If uPx has an even number of edges, $uQx + xPu$ is an M-alternating circuit containing non-allowed edges, a contradiction. Hence we suppose uPx has an odd number of edges. Let $I \in \mathcal{G}(G)$ be the elementary component such that $x \in V(I)$. Then one of the components of $uQx + xPu - V(I)$ is an M-ear relative to I and through H, a contradiction by Theorem 4. $\square$

Theorem 7. *Let H be an elementary component of a factorizable graph G. Then for each connected component K of $G[up(H)]$, there exists $S \in \mathcal{P}_G(H)$ such that $N(K) \cap V(H) \subseteq S$.*

By Theorem 7, we can see that upper bounds of an elementary component are each "attached" to an equivalence class of the generalized canonical partition.

Remark 1. There are factorizable graphs where $\triangleright$ does not hold for any two elementary components, in other words, where all the elementary components are minimal in the poset. For example, we can see by Theorem 4 and Theorem 7 that bipartite factorizable graphs are such, which means Theorem 5 is not a generalization of the DM-decomposition, even though they have similar natures.

The following theorem shows that most of the factorizable graphs with $|\mathcal{G}(G)| \geq 2$, in a sense, have non-trivial structures as posets.

Theorem 8. *Let G be a factorizable graph, $G_1, G_2 \in \mathcal{G}(G)$ be elementary components for which $G_1 \triangleright G_2$ does not hold, and let G_1 be minimal in the poset $(\mathcal{G}(G), \triangleright)$. Then there are possibly identical complement edges e, f of G between G_1 and G_2 such that $\mathcal{G}(G + e + f) = \mathcal{G}(G)$ and $G_1 \triangleright G_2$ in $(\mathcal{G}(G + e + f), \triangleright)$.*

6 Algorithmic Result

In this section, we discuss the algorithmic aspects of the partial order and the generalized canonical partition. We denote by n and m respectively the number of vertices and edges of input graphs. As we work on factorizable graphs and graphs with near-perfect matchings, we can assume $m = \Omega(n)$.

We start with some materials from Edmonds' maximum matching algorithm [12], referring mainly to [1,17]. For a tree T with a specified root vertex r, we call a vertex $v \in V(T)$ *inner* (resp. *outer*) if the unique path in T from r to v has an odd (resp. even) number of edges. Let G be a graph and M be a matching of G. A tree $T \subseteq G$ is called *M-alternating* if exactly one vertex of it, the root, is exposed by M in G, and each inner vertex $v \in V(T)$ satisfies $|\delta(v) \cap E(T)| = 2$ and one of the edges of $\delta(v) \cap E(T)$ is contained in M.

A subgraph $S \subseteq G$ is called a *special blossom tree with respect to M* (M-SBT) if there is a partition $V(C_1) \dot\cup \cdots \dot\cup V(C_k) = V(S)$ such that

1. $S' := S/C_1/\cdots/C_k$ is an M-alternating tree,
2. M_{C_i} is a near-perfect matching of C_i,
3. C_i is a maximal factor-critical subgraph of G if it corresponds to an outer vertex of S', and called an *outer blossom*, and
4. $|V(C_i)| > 1$ only if C_i is an outer blossom, for each $i = 1, \ldots, k$.

Edmonds' maximum matching algorithm tells us the following facts. Let G be a graph, M be a near-perfect matching of G, and $r \in V(G)$ be the vertex exposed by M. Then an M-SBT S, with root r, can be computed, if it is carefully implemented [18,19], in $O(m)$ time. Additionally, the set of vertices from which r can be reached by an M-balanced path is exactly the set of vertices contained in the outer blossoms of S.

Thus, due to an easy reduction of the above facts, the following proposition holds; they can be regarded as a folklore. See [3]. (In [3] they are presented as those for elementary graphs, but in fact, they can be applicable for general factorizable graphs.)

Proposition 6. *Let G be a factorizable graph, M be a perfect matching of G, and $u \in V(G)$.*

1. *The set of vertices that can be reached from u by an M-saturated path can be computed in $O(m)$ time.*
2. *All the allowed edges adjacent to u can be computed in $O(m)$ time.*
3. *All the elementary components of G can be computed in $O(nm)$ time.*

Proposition 7. *Given a factorizable graph G, one of its perfect matchings M and $\mathcal{G}(G)$, we can compute the generalized canonical partition of G in $O(nm)$ time.*

Let G be a factorizable graph and M be a perfect matching of G. We say two distinct elementary components G_1, G_2 of G with $G_1 \triangleright G_2$ are *non-refinable* if $G_1 \triangleright H \triangleright G_2$ yields $G_1 = H$ or $G_2 = H$ for any $H \in \mathcal{G}(G)$. Note that if G_1 and G_2 are non-refinable, then there is an M-ear relative to G_1 and through G_2 by Theorem 4. Note also that the converse of the above fact does not hold.

Lemma 3. *Let G be a factorizable graph, M be a perfect matching of G, and $H \in \mathcal{G}(G)$. Let S be a maximal M-SBT in G/H and let C be the blossom of T containing the contracted vertex h corresponding to H. Then any non-refinable upper bound of H in $(\mathcal{G}(G), \triangleright)$ has common vertices with C. Additionally, if an elementary component $I \in \mathcal{G}(G)$ has some common vertices with C, then $H \triangleright I$.*

Proposition 8. *Given a factorizable graph G, its perfect matching M, and $\mathcal{G}(G)$, we can compute the poset $(\mathcal{G}(G), \triangleright)$ in $O(nm)$ time.*

Proof. It is sufficient to list all the non-refinable upper bounds for each elementary component of G by the following procedure:

1: $D := (\mathcal{G}(G), \emptyset)$; $A := \emptyset$;
2: **for all** $H \in \mathcal{G}(G)$ **do**
3:      compute a maximal M-SBT T; let C be the blossom of T corresponding to its root;
4:      **for all** $x \in V(C)$, which satisfies $x \in V(I)$ for some $I \in \mathcal{G}(G)$ **do**
5:          $A := A \cup \{(H, I)\}$;
6:      **end for**
7: **end for**
8: $D := (\mathcal{G}(G), A)$; STOP.

By Lemma 3, the partial order on $V(D)$ determined by the reachability corresponds to $\triangleright$ after the above procedure. That is, if we define a binary relation $\prec$ on $V(D)$ so that $H' \triangleright I'$ if there is a dipath from H' to I' in D, then $\prec$ and $\triangleright$ coincide. For each $H \in \mathcal{G}(G)$, the above procedure costs $O(m)$ time, thus it costs $O(nm)$ time over the whole computation. $\square$

Remark 2. Given the digraph D after the procedure in Proposition 8, we can compute all the upper bounds of an elementary component in $O(n^2)$ time. Thus, an efficient data structure that represents the poset, for example, a boolean-valued matrix L where $L[i, j]$ = true if and only if $G_i \triangleright G_j$, can be obtained in additional $O(n^3)$ time.

As a maximum matching of a graph can be computed in $O(\sqrt{n}m)$ time [20, 21], we have the following, combining Propositions 6, 7, and 8.

Theorem 9. *Let G be a factorizable graph. Then the poset $(\mathcal{G}(G), \triangleright)$ and the generalized canonical partition $\mathcal{P}(G)$ can be computed in $O(nm)$ time.*

Acknowlegements. The author is grateful to Yusuke Kobayashi and Richard Hoshino for carefully reading the paper, and Akihisa Tamura for useful discussions.

References

1. Lovász, L., Plummer, M.D.: Matching Theory. Elsevier Science (1986)
2. Schrijver, A.: Combinatorial Optimization: Polyhedra and Efficiency. Springer (2003)
3. Carvalho, M.H., Cheriyan, J.: An $O(VE)$ algorithm for ear decompositions of matching-covered graphs. ACM Transactions on Algorithms 1(2), 324–337 (2005)
4. Kotzig, A.: Z teórie konečných grafov s lineárnym faktorom. I. Mathematica Slovaca 9(2), 73–91 (1959) (in Slovak)
5. Kotzig, A.: Zteórie konečných grafov s lineárnym faktorom. II. Mathematica Slovaca 9(3), 136–159 (1959) (in Slovak)
6. Kotzig, A.: Z teórie konečných grafov s lineárnym faktorom. III. Mathematica Slovaca 10(4), 205–215 (1960) (in Slovak)
7. Edomonds, J., Lovász, L., Pulleyblank, W.R.: Brick decompositions and the matching rank of graphs. Combinatorica 2(3), 247–274 (1982)

8. Lovász, L.: Matching structure and the matching lattice. Journal of Combinatorial Theory, Series B 43, 187–222 (1987)
9. Carvalho, M.H., Lucchesi, C.L., Murty, U.S.R.: The matching lattice. In: Reed, B., Sales, C.L. (eds.) Recent Advances in Algorithms and Combinatorics. Springer (2003)
10. Nakamura, M.: Structural theorems for submodular functions, polymatroids and polymatroid intersections. Graphs and Combinatorics 4, 257–284 (1988)
11. Fujishige, S.: Submodular Functions and Optimization, 2nd edn. Elsevier Science (2005)
12. Edmonds, J.: Paths, trees and flowers. Canadian Journal of Mathematics 17, 449–467 (1965)
13. Pap, G., Szegő, L.: On the maximum even factor in weakly symmetric graphs. Journal of Combinatorial Theory, Series B 91(2), 201–213 (2004)
14. Spille, B., Szegő, L.: A gallai-edmonds type structure theorem for path-matchings. Journal of Graph Theory 46(2), 93–102 (2004)
15. Kita, N.: Another proof for Lovász's cathedral theorem (preprint)
16. Lovász, L.: A note on factor-critical graphs. Studia Scientiarum Mathematicarum Hungarica 7, 279–280 (1972)
17. Korte, B., Vygen, J.: Combinatorial Optimization; Theory and Algorithms, 4th edn. Springer (2007)
18. Tarjan, R.E.: Data Structures and Network Algorithms. Society for Industrial and Applied Mathematics (1983)
19. Gabow, H.N., Tarjan, R.E.: A linear-time algorithm for a special case of disjoint set union. Journal of Computer and System Sciences 30, 209–221 (1985)
20. Micali, S., Vazirani, V.V.: An $O(\sqrt{|v|} \cdot |E|)$ algorithm for finding maximum matching in general graphs. In: Proceedings of the 21st Annual IEEE Symposium on Foundations of Computer Science, pp. 17–27 (1980)
21. Vazirani, V.V.: A theory of alternating paths and blossoms for proving correctness of the $O(\sqrt{V}E)$ general graph maximum matching algorithm. Combinatorica 14, 71–109 (1994)

Fast and Simple Fully-Dynamic Cut Tree Construction[*]

Tanja Hartmann and Dorothea Wagner

Department of Informatics, Karlsruhe Institute of Technology (KIT)
{t.hartmann,dorothea.wagner}@kit.edu

Abstract. A cut tree of an undirected weighted graph $G = (V,E)$ encodes a minimum s-t-cut for each vertex pair $\{s,t\} \subseteq V$ and can be iteratively constructed by $n-1$ maximum flow computations. They solve the multiterminal network flow problem, which asks for the all-pairs maximum flow values in a network and at the same time they represent $n-1$ non-crossing, linearly independent cuts that constitute a minimum cut basis of G. Hence, cut trees are resident in at least two fundamental fields of network analysis and graph theory, which emphasizes their importance for many applications. In this work we present a fully-dynamic algorithm that efficiently maintains a cut tree for a changing graph. The algorithm is easy to implement and has a high potential for saving cut computations under the assumption that a local change in the underlying graph does rarely affect the global cut structure. We document the good practicability of our approach in a brief experiment on real world data.

1 Introduction

A *cut tree* is a weighted tree $T(G) = (V,E_T,c_T)$ on the vertices of an undirected (weighted) graph $G = (V,E,c)$ (with edges not necessarily in G) such that each $\{u,v\} \in E_T$ induces a minimum u-v-cut in G (by decomposing $T(G)$ into two connected components) and such that $c_T(\{u,v\})$ is equal to the cost of the induced cut. The cuts induced by $T(G)$ are non-crossing and for each $\{x,y\} \subseteq V$ each cheapest edge on the path $\pi(x,y)$ between x and y in $T(G)$ corresponds to a minimum x-y-cut in G. If G is disconnected, $T(G)$ contains edges of cost 0 between connected components.

Cut trees were first introduced by Gomory and Hu [1] in 1961 in the field of multiterminal network flow analysis. Shortly afterwards, in 1964, Elmaghraby [3] already studied how the values of multiterminal flows change if the capacity of an edge in the network varies. Elmaghraby established the *sensitivity analysis of multiterminal flow networks*, which asks for the all-pairs maximum flow values (or all-pairs minimum cut values) in a network considering any possible capacity of the varying edge. According to Barth et al. [4] this can be answered by constructing two cut trees. In contrast, the *parametric maximum flow problem* considers a flow network with only two terminals s and t and with several parametric edge capacities. The goal is to give a maximum s-t-flow (or minimum s-t-cut) regarding all possible capacities of the parametric edges. Parametric maximum flows were studied, e.g., by Gallo et al. [5] and Scutellà [6].

However, in many applications we are neither interested in *all-pairs* values nor in one minimum s-t-cut regarding *all possible* changes of varying edges. Instead we face

---

[*] This work was partially supported by the DFG under grant WA 654/15-2 and by the Concept for the Future of Karlsruhe Institute of Technology within the German Excellence Initiative.

K.-M. Chao, T.-s. Hsu, and D.-T. Lee (Eds.): ISAAC 2012, LNCS 7676, pp. 95–105, 2012.

a concrete change on a concrete edge and need all-pairs minimum cuts regarding this single change. This is answered by *dynamic cut trees*, which thus bridge the two sides of sensitivity analysis and parametric maximum flows.

Contribution and Outline. In this work we develop the first algorithm that efficiently and dynamically maintains a cut tree for a changing graph allowing arbitrary atomic changes. To the best of our knowledge no fully-dynamic approach for updating cut trees exists. Coming from sensitivity analysis, Barth et al. [4] state that after the capacity of an edge has increased the path in $T(G)$ between the vertices that define the changing edge in G is the only part of a given cut tree that needs to be recomputed, which is rather obvious. Besides they stress the difficulty for the case of decreasing edge capacities.

In our work we formulate a general condition for the (re)use of given cuts in an (iterative) cut tree construction, which directly implies the result of Barth et al. We further solve the case of decreasing edge capacities showing by an experiment that this has a similar potential for saving cut computations like the case of increasing capacities. In the spirit of Gusfield [2], who simplified the pioneering cut tree algorithm of Gomory and Hu [1], we also allow the use of crossing cuts and give a representation of intermediate trees (during the iteration) that makes our approach very easy to implement.

We give our notational conventions and a first folklore insight in Sec. 1. In Sec. 2 we revisit the static cut tree algorithm [1] and the key for its simplification [2], and construct a first intermediate cut tree by reusing cuts that obviously remain valid after a change in G. We also state several lemmas that imply techniques to find further reusable cuts in this section. Our update approach is described in Sec. 3. In Sec. 4 we finally discuss the performance of our algorithm based on a brief experiment. Proofs omitted due to space constraints can be found in the full paper [7].

Preliminaries and Notation. In this work we consider an undirected, weighted graph $G = (V, E, c)$ with vertices V, edges E and a positive edge cost function c, writing $c(u, v)$ as a shorthand for $c(\{u, v\})$ with $\{u, v\} \in E$. We reserve the term *node* for compound vertices of abstracted graphs, which may contain several basic vertices of a concrete graph; however, we identify singleton nodes with the contained vertex without further notice. *Contracting* a set $N \subseteq V$ in G means replacing N by a single node, and leaving this node adjacent to all former adjacencies u of vertices of N, with an edge cost equal to the sum of all former edges between N and u. Analogously we contract a set $M \subseteq E$ or a subgraph of G by contracting the corresponding vertices.

A *cut* in G is a partition of V into two *cut sides* S and $V \setminus S$. The cost $c(S, V \setminus S)$ of a cut is the sum of the costs of all edges *crossing* the cut, i.e., edges $\{u, v\}$ with $u \in S$, $v \in V \setminus S$. For two disjoint sets $A, B \subseteq V$ we define the cost $c(A, B)$ analogously. Note that a cut is defined by the edges crossing it. Two cuts are *non-crossing* if their cut sides are pairwise nested or disjoint. Two vertices $u, v \in V$ are *separated* by a cut if they lie on different cut sides. A minimum u-v-cut is a cut that separates u and v and is the cheapest cut among all cuts separating these vertices. We call a cut a *minimum separating cut* if there exists an arbitrary vertex pair $\{u, v\}$ for which it is a minimum u-v-cut; $\{u, v\}$ is called a *cut pair* of the minimum separating cut. We further denote the *connectivity* of $\{u, v\} \subseteq V$ by $\lambda(u, v)$, describing the cost of a minimum u-v-cut.

Since each edge in a tree $T(G)$ on the vertices of G induces a unique cut in G, we identify tree edges with corresponding cuts without further notice. This allows for saying that a vertex is *incident* to a cut and an edge *separates* a pair of vertices. We consider the path $\pi(u,v)$ between u and v in $T(G)$ as the set of edges or the set of vertices on it, as convenient.

A change in G either involves an edge $\{b,d\}$ or a vertex b. If the cost of $\{b,d\}$ in G descreases by $\Delta > 0$ or $\{b,d\}$ with $c(b,d) = \Delta > 0$ is deleted, the change yields $G^{\ominus}$. Analogously, inserting $\{b,d\}$ or increasing the cost yields $G^{\oplus}$. We denote the cost function after a change by $c^{\ominus}$ and $c^{\oplus}$, the connectivity by $\lambda^{\ominus}$ and $\lambda^{\oplus}$, respectively. We assume that only degree-0 vertices can be deleted from G. Hence, inserting or deleting b changes neither the cost function nor the connectivity. We start with a fundamental insight on the reusability of cuts. Recall that $T(G) = (V, E_T, c_T)$ denotes a cut tree.

Lemma 1. *If $c(b,d)$ changes by $\Delta > 0$, then each $\{u,v\} \in E_T$ remains a minimum u-v-cut (i) in $G^{\oplus}$ with cost $\lambda(u,v)$ if $\{u,v\} \notin \pi(b,d)$, (ii) in $G^{\ominus}$ with cost $\lambda(u,v) - \Delta$ if $\{u,v\} \in \pi(b,d)$.*

2 The Static Algorithm and Insights on Reusable Cuts

The Static Algorithm. As a basis for our dynamic approach, we briefly revisit the static construction of a cut tree [1,2]. This algorithm iteratively constructs $n - 1$ non-crossing minimum separating cuts for $n - 1$ vertex pairs, which we call *step pairs*. These pairs are chosen arbitrarily from the set of pairs not separated by any of the cuts constructed so far. Algorithm 1 briefly describes the cut tree algorithm of Gomory and Hu.

Algorithm 1. CUT TREE

Input: Graph $G = (V, E, c)$
Output: Cut tree of G

1 Initialize tree $T_* := (V_*, E_*, c_*)$ with $V_* \leftarrow \{V\}, E_* \leftarrow \emptyset$ and c_* empty
2 **while** $\exists S \in V_*$ *with* $|S| > 1$ **do**   // unfold all nodes
3    $\{u,v\} \leftarrow$ arbitrary pair from $\binom{S}{2}$
4   **forall the** S_j *adjacent to* S *in* T_* **do** $N_j \leftarrow$ subtree of S in T_* with $S_j \in N_j$
5    $G_S \leftarrow (V_S, E_S, c_S) \leftarrow$ in G contract each N_j to $[N_j]$ // contraction
6    $(U, V \setminus U) \leftarrow$ min-u-v-cut in G_S, cost $\lambda(u,v)$, $u \in U$
7    $S_u \leftarrow S \cap U$ and $S_v \leftarrow S \cap (V_S \setminus U)$ // split $S = S_u \cup S_v$
8    $V_* \leftarrow (V_* \setminus \{S\}) \cup \{S_u, S_v\}, E_* \leftarrow E_* \cup \{\{S_u, S_v\}\}, c_*(S_u, S_v) \leftarrow \lambda(u,v)$
9   **forall the** *former edges* $e_j = \{S, S_j\} \in E_*$ **do**
10   **if** $[N_j] \in U$ **then** $e_j \leftarrow \{S_u, S_j\}$;   // reconnect S_j to S_u
11   **else** $e_j \leftarrow \{S_v, S_j\}$;   // reconnect S_j to S_v

12 **return** T_*

The *intermediate* cut tree $T_* = (V_*, E_*, c_*)$ is initialized as an isolated, edgeless node containing all original vertices. Then, until each node of T_* is a singleton node, a node $S \in V_*$ is *split*. To this end, nodes $S' \neq S$ are dealt with by contracting in G whole subtrees N_j of S in T_*, connected to S via edges $\{S, S_j\}$, to single nodes $[N_j]$ before cutting, which yields G_S. The split of S into S_u and S_v is then defined by a minimum u-v-cut (*split cut*) in G_S, which does not cross any of the previously used cuts due to the contraction technique. Afterwards, each N_j is reconnected, again by S_j, to either S_u

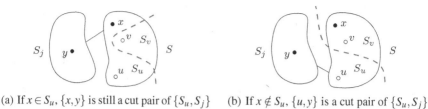

(a) If $x \in S_u$, $\{x,y\}$ is still a cut pair of $\{S_u, S_j\}$ (b) If $x \notin S_u$, $\{u,y\}$ is a cut pair of $\{S_u, S_j\}$

Fig. 1. Situation in Lemma 2. There always exists a cut pair of the edge $\{S_u, S_j\}$ in the nodes incident to the edge, independent of the shape of the split cut (dashed).

or S_v depending on which side of the cut $[N_j]$ ended up. Note that this cut in G_S can be proven to induce a minimum u-v-cut in G.

The correctness of CUT TREE is guaranteed by Lemma 2, which takes care for the *cut pairs* of the reconnected edges. It states that each edge $\{S, S'\}$ in T_* has a cut pair $\{x,y\}$ with $x \in S$, $y \in S'$. An intermediate cut tree satisfying this condition is *valid*. The assertion is not obvious, since the nodes incident to the edges in T_* change whenever the edges are reconnected. Nevertheless, each edge in the final cut tree represents a minimum separating cut of its incident vertices, due to Lemma 2. The lemma was formulated and proven in [1] and rephrased in [2]. See Figure 1.

Lemma 2 (Gus. [2], Lem. 4). *Let $\{S, S_j\}$ be an edge in T_* inducing a cut with cut pair $\{x,y\}$, w.l.o.g. $x \in S$. Consider step pair $\{u,v\} \subseteq S$ that splits S into S_u and S_v, w.l.o.g. S_j and S_u ending up on the same cut side, i.e. $\{S_u, S_j\}$ becomes a new edge in T_*. If $x \in S_u$, $\{x,y\}$ remains a cut pair for $\{S_u, S_j\}$. If $x \in S_v$, $\{u,y\}$ is also a cut pair of $\{S_u, S_j\}$.*

While Gomory and Hu use contractions in G to prevent crossings of the cuts, as a simplification, Gusfield introduced the following lemma showing that contractions are not necessary, since any arbitrary minimum separating cut can be bent along the previous cuts resolving any potential crossings. See Figure 2.

Lemma 3 (Gus. [2], Lem. 1). *Let $(X, V \setminus X)$ be a minimum x-y-cut in G, with $x \in X$. Let $(H, V \setminus H)$ be a minimum u-v-cut, with $u, v \in V \setminus X$ and $x \in H$. Then the cut $(H \cup X, (V \setminus H) \cap (V \setminus X))$ is also a minimum u-v-cut.*

We say that $(X, V \setminus X)$ *shelters* X, meaning that each minimum u-v-cut with $u, v \notin X$ can be reshaped, such that it does no longer split X.

Representation of Intermediate Trees. In the remainder of this work we represent each node in T_*, which consists of original vertices in G, by an arbitrary tree of *thin* edges connecting the contained vertices in order to indicate their membership to the node. An edge connecting two nodes in T_* is represented by a *fat* edge, which we connect to an arbitrary vertex in each incident compound node. Fat edges represent minimum separating cuts in G. If a node contains only one vertex, we color this vertex black. Black vertices are only

Fig. 2. Depending on x Lem. 3 bends the cut $(H, V \setminus H)$ upwards or downwards

incident to fat edges. The vertices in non-singleton nodes are colored white. White vertices are incident to at least one thin edge. In this way, T_* becomes a tree on V with two types of edges and vertices. For an example see Figure 3.

Conditions for Reusing Cuts. Consider a set K of $k \leq n-1$ cuts in G for example given by a previous cut tree in a dynamic scenario. The following theorem states sufficient conditions for K, such that there exists a valid intermediate cut tree that represents exactly the cuts in K. Such a tree can then be further processed to a proper tree by CUT TREE, saving at least $|K|$ cut computations compared to a construction from scratch.

Theorem 1. *Let K denote a set of non-crossing minimum separating cuts in G and let F denote a set of associated cut pairs such that each cut in K separates exactly one pair in F. Then there exists a valid intermediate cut tree representing exactly the cuts in K.*

Proof. Theorem 1 follows inductively from the correctness of CUT TREE. Consider a run of CUT TREE that uses the elements in F as step pairs in an arbitrary order and the associated cuts in K as split cuts. Since the cuts in K are non-crossing each separating exactly one cut pair in F, splitting a node neither causes reconnections nor the separation of a pair that was not yet considered. Thus, CUT TREE reaches an intermediate tree representing the cuts in K with the cut pairs located in the incident nodes.   □

With the help of Theorem 1 we can now construct a valid intermediate cut tree from the cuts that remain valid after a change of G according to Lemma 1. These cuts are non-crossing as they are represented by tree edges, and the vertices incident to these edges constitute a set of cut pairs as required by Theorem 1. The resulting tree for an inserted edge or an increased edge cost is shown in Figure 3(a). In this case, all but the edges on $\pi(b,d)$ can be reused. Hence, we draw these edges fat. The remaining edges are thinly drawn. The vertices are colored according to the compound nodes indicated by the thickness of the edges. Vertices incident to a fat edge correspond to a cut pair.

For a deleted edge or a decreased edge cost, the edges on $\pi(b,d)$ are fat, while the edges that do not lie on $\pi(b,d)$ are thin (cp. Figure 3(b)). Furthermore, the costs of the fat edges decrease by Δ, since they all cross the changing edge $\{b,d\}$ in G. Compared to a construction from scratch, starting the CUT TREE routine from these intermediate trees already saves $n - 1 - |\pi(b,d)|$ cut computations in the first case and $|\pi(b,d)|$ cut computations in the second case, where $|\pi(b,d)|$ counts the edges on $\pi(b,d)$. Hence, in scenarios with only little varying path lengths and a balanced number of increasing and decreasing costs, we can already save about half of the cut computations. We further remark that the result of Barth et. al. [4], who costly prove the existence of the intermediate cut tree in Figure 3(a), easily follows by Theorem 1 applied to the cuts in Lemma 1 as seen above. In the following we want to use even more information from the previous cut tree $T(G)$ when executing CUT TREE unfolding the intermediate tree to a proper cut tree of $(n-1)$ fat edges. The next section lists further lemmas that allow the reuse of cuts already given by $T(G)$.

Further Reusable Cuts. In this section we focus on the reuse of those cuts that are still represented by thin edges in Figure 3. If $\{b,d\}$ is inserted or the cost increases, the following corollary obviously holds, since $\{b,d\}$ crosses each minimum b-d-cut.

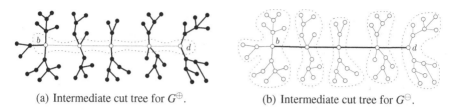

(a) Intermediate cut tree for $G^{\oplus}$.   (b) Intermediate cut tree for $G^{\ominus}$.

Fig. 3. Intermediate cut trees in dynamic scenarios. Fat edges represent valid minimum cuts, thin edges indicate compound nodes. Contracting the thin edges yields nodes of white vertices (indicated by dotted lines). Black vertices correspond to singletons.

Corollary 1. *If $\{b,d\}$ is newly inserted with $c^{\oplus}(b,d) = \Delta$ or $c(b,d)$ increases by Δ, any minimum b-d-cut in G remains valid in $G^{\oplus}$ with $\lambda^{\oplus}(b,d) = \lambda(b,d) + \Delta$.*

Note that reusing a valid minimum b-d-cut as split cut in CUT TREE separates b and d such that $\{b,d\}$ cannot be used again as step pair in a later iteration step. This is, we can reuse only one minimum b-d-cut, even if there are several such cuts represented in $T(G)$. Together with the following corollary, Corollary 1 directly allows the reuse of the whole cut tree $T(G)$ if $\{b,d\}$ is an existing bridge in G (with increasing cost).

Corollary 2. *An edge $\{u,v\}$ is a bridge in G iff $c(u,v) = \lambda(u,v) > 0$. Then $\{u,v\}$ is also an edge in $T(G)$ representing the cut that is given by the two sides of the bridge.*

While the first part of Corollary 2 is obvious, the second part follows by the fact that a bridge induces a minimum separating cut for all vertices on different bridge sides, while it does not cross any minimum separating cut of vertices on a common side. If G is disconnected and $\{b,d\}$ is a new bridge in $G^{\oplus}$, reusing the whole tree is also possible by replacing a single edge. Such bridges can be easily detected having the cut tree $T(G)$ at hand, since $\{b,d\}$ is a new bridge if and only if $\lambda(b,d) = 0$. New bridges particularly occur if newly inserted vertices are connected for the first time.

Lemma 4. *Let $\{b,d\}$ be a new bridge in $G^{\oplus}$. Then replacing an edge of cost 0 by $\{b,d\}$ with cost $c^{\oplus}(b,d)$ on $\pi(b,d)$ in $T(G)$ yields a new cut tree $T(G^{\oplus})$.*

If $\{b,d\}$ is deleted or the cost decreases, handling bridges (always detectable by Corollary 2) is also easy.

Lemma 5. *If $\{b,d\}$ is a bridge in G and the cost decreases by Δ (or $\{b,d\}$ is deleted), decreasing the edge cost on $\pi(b,d)$ in $T(G)$ by Δ yields a new cut tree $T(G^{\ominus})$.*

If $\{b,d\}$ is no bridge, at least other bridges in G can still be reused if $\{b,d\}$ is deleted or the edge cost decreases. Observe that a minimum separating cut in G only becomes invalid in $G^{\ominus}$ if there is a cheaper cut in $G^{\ominus}$ that separates the same vertex pair. Such a cut necessarily crosses the changing edge $\{b,d\}$ in G, since otherwise it would have been already cheaper in G. Hence, an edge in E_T corresponding to a bridge in G cannot become invalid, since any cut in $G^{\ominus}$ that crosses $\{b,d\}$ besides the bridge would be more expensive. In particular, this also holds for zero-weighted edges in E_T.

Corollary 3. *Let $\{u,v\}$ denote an edge in $T(G)$ with $c_T(u,v) = 0$ or an edge that corresponds to a bridge in G. Then $\{u,v\}$ is still a minimum u-v-cut in $G^{\ominus}$.*

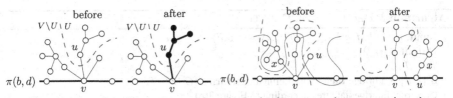

(a) Edges in U remain valid, cp. Lemma 6.   (b) Reshaping new cut by reconnecting edges.

Fig. 4. (a) cut $\{u,v\}$ remains valid, subtree U can be reused. (b) new cheaper cut for $\{u,v\}$ (black) can be reshaped by Theo. 2, Lem. 3 (dashed), $\{u,v\}$ becomes a fat edge.

Lemma 6 shows how a cut that is still valid in $G^\ominus$ may allow the reuse of all edges in E_T that lie on one cut side. Figure 4(a) shows an example. Lemma 7 says that a cut that is cheap enough, cannot become invalid in $G^\ominus$. Note that the bound considered in this context depends on the current intermediate tree.

Lemma 6. *Let $(U, V \setminus U)$ be a minimum u-v-cut in $G^\ominus$ with $\{b,d\} \subseteq V \setminus U$ and $\{g,h\} \in E_T$ with $g, h \in U$. Then $\{g,h\}$ is a minimum separating cut in $G^\ominus$ for all its previous cut pairs within U.*

Lemma 7. *Let $T_* = (V, E_*, c_*)$ denote a valid intermediate cut tree for $G^\ominus$, where all edges on $\pi(b,d)$ are fat and let $\{u,v\}$ be a thin edge with v on $\pi(b,d)$ such that $\{u,v\}$ represents a minimum u-v-cut in G. Let N_π denote the set of neighbors of v on $\pi(b,d)$. If $\lambda(u,v) < \min_{x \in N_\pi}\{c_*(x,v)\}$, then $\{u,v\}$ is a minimum u-v-cut in $G^\ominus$.*

3   The Dynamic Cut Tree Algorithm

In this section we introduce one update routine for each type of change: inserting a vertex, deleting a vertex, increasing an edge cost or inserting an edge, decreasing an edge cost or deleting an edge. These routines base on the static iterative approach but involve the lemmas from Sec. 2 in order to save cut computations. We again represent intermediate cut trees by fat and thin edges, which simplifies the reshaping of cuts.

We start with the routines for vertex insertion and deletion, which trivially abandon cut computations. We leave the rather basic proofs of correctness to the reader. A vertex b inserted into G forms a connected component in $G^\oplus$. Hence, we insert b into $T(G)$ connecting it to the remaining tree by an arbitrary zero-weighted edge. If b is deleted from G, it was a single connected component in G before. Hence, in $T(G)$ b is only incident to zero-weighted edges. Deleting b from $T(G)$ and reconnecting the resulting subtrees by arbitrary edges of cost 0 yields a valid intermediate cut tree for $G^\ominus$.

The routine for increasing an edge cost or inserting an edge first checks if $\{b,d\}$ is a (maybe newly inserted) bridge in G. In this case, it adapts $c_T(b,d)$ according to Corollary 1 if $\{b,d\}$ already exists in G, and rebuilds $T(G)$ according to Lemma 4 otherwise. Both requires no cut computation. If $\{b,d\}$ is no bridge, the routine constructs the intermediate cut tree shown in Figure 3(a), reusing all edges that are not on $\pi(b,d)$. Furthermore, it chooses one edge on $\pi(b,d)$ that represents a minimum b-d-cut in $G^\oplus$ and draws this edge fat (cp. Corollary 1). The resulting tree is then further processed by CUT TREE, which costs $|\pi(b,d)| - 1$ cut computations and is correct by Theorem 1.

Algorithm 2. DECREASE OR DELETE

Input: $T(G)$, b,d, $c(b,d)$, $c^\ominus(b,d)$, $\Delta := c(b,d) - c^\ominus(b,d)$
Output: $T(G^\ominus)$

1 $T_* \leftarrow T(G)$
2 **if** $\{b,d\}$ is a bridge **then** apply Lemma 5; **return** $T(G^\ominus) \leftarrow T_*$
3 Construct intermediate tree according to Figure 3(b)
4 $Q \leftarrow$ thin edges non-increasingly ordered by their costs
5 **while** $Q \neq \emptyset$ **do**
6 | $\{u,v\} \leftarrow$ most expensive thin edge with v on $\pi(b,d)$
7 | $N_\pi \leftarrow$ neighbors of v on $\pi(b,d)$; $L \leftarrow \min_{x \in N_\pi}\{c_*(x,v)\}$
8 | **if** $L > \lambda(u,v)$ or $\{u,v\} \in E$ with $\lambda(u,v) = c(u,v)$ **then**   // Lem. 7 and Cor. 3
9 | | draw $\{u,v\}$ as a fat edge
10 | | consider the subtree U rooted at u with $v \notin U$,   // Lem. 6 and Fig. 4(a)
11 | | draw all edges in U fat, remove fat edges from Q
12 | | continue loop
13 | $(U, V \setminus U) \leftarrow$ minimum u-v-cut in $G^\ominus$ with $u \in U$
14 | draw $\{u,v\}$ as a fat edge, remove $\{u,v\}$ from Q
15 | **if** $\lambda(u,v) = c^\ominus(U, V \setminus U)$ **then** goto line 10   // old cut still valid
16 | $c_*(u,v) \leftarrow c^\ominus(U, V \setminus U)$ // otherwise
17 | $N \leftarrow$ neighbors of v
18 | **forall the** $x \in N$ **do**   // bend split cut by Theo. 2 and Lem. 3
19 | | **if** $x \in U$ **then** reconnect x to u
20 **return** $T(G^\ominus) \leftarrow T_*$

The routine for decreasing an edge cost or deleting an edge is given by Algorithm 2. We assume G and $G^\ominus$ to be available as global variables. Whenever the intermediate tree T_* changes during the run of Algorithm 2, the path $\pi(b,d)$ is implicitly updated without further notice. Thin edges are weighted by the old connectivity, fat edges by the new connectivity of their incident vertices. Whenever a vertex is reconnected, the newly occurring edge inherits the cost and the thickness from the disappearing edge.

Algorithm 2 starts by checking if $\{b,d\}$ is a bridge (line 2) and reuses the whole cut tree $T(G)$ with adapted cost $c_T(b,d)$ (cp. Lemma 5) in this case. Otherwise (line 3), it constructs the intermediate tree shown in Figure 3(b), reusing all edges on $\pi(b,d)$ with adapted costs. Then it proceeds with iterative steps similar to CUT TREE. However, the difference is, that the step pairs are not chosen arbitrarily, but according to the edges in $T(G)$, starting with those edges that are incident to a vertex v on $\pi(b,d)$ (line 6). In this way, each edge $\{u,v\}$ which is found to remain valid in line 8 or line 15 allows to retain a maximal subtree (cp. Lemma 6), since $\{u,v\}$ is as close as possible to $\pi(b,d)$. The problem however is that cuts that are no longer valid, must be replaced by new cuts, which not necessarily respect the tree structure of $T(G)$. This is, a new cut possibly separates adjacent vertices in $T(G)$, which hence cannot be used as a step pair in a later step. Thus, we potentially miss valid cuts and the chance to retain further subtrees.

We solve this problem by reshaping the new cuts in the spirit of Gusfield. Theorem 2 shows how arbitrary cuts in $G^\ominus$ (that separate b and d) can be bend along old minimum separating cuts in G without becoming more expensive (see Figure 5).

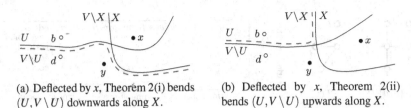

(a) Deflected by x, Theorem 2(i) bends $(U, V \setminus U)$ downwards along X.

(b) Deflected by x, Theorem 2(ii) bends $(U, V \setminus U)$ upwards along X.

Fig. 5. Situation of Theorem 2. Reshaping cuts in $G^{\ominus}$ along previous cuts in G.

Theorem 2. *Let $(X, V \setminus X)$ denote a minimum x-y-cut in G with $x \in X$, $y \in V \setminus X$ and $\{b, d\} \subseteq V \setminus X$. Let further $(U, V \setminus U)$ denote a cut that separates b, d. If (i) $(U, V \setminus U)$ separates x, y with $x \in U$, then $c^{\ominus}(U \cup X, V \setminus (U \cup X)) \leq c^{\ominus}(U, V \setminus U)$. If (ii) $(U, V \setminus U)$ does not separate x, y with $x \in V \setminus U$, then $c^{\ominus}(U \setminus X, V \setminus (U \setminus X)) \leq c^{\ominus}(U, V \setminus U)$.*

Since any new cheaper cut found in line 13 needs to separate b and d, we can apply Theorem 2 to this cut regarding the old cuts that are induced by the other thin edges $\{x, v\}$ incident to v. As a result, the new cut gets reshaped without changing its cost such that each subtree rooted at a vertex x is completely assigned to either side of the reshaped cut (line 19), depending on if the new cut separates x and v (cp. Figure 4(b)). Furthermore, Lemma 3 allows the reshaping of the new cut along the cuts induced by the fat edge on $\pi(b, d)$ that are incident to v. This ensures that the new cut does not cross parts of T_* that are beyond these flanking fat edges. Since after the reshaping exact one vertex adjacent to v on $\pi(b, d)$ ends up on the same cut side as u, u finally becomes a part of $\pi(b, d)$.

It remains to show that after the reconnection the reconnected edges are still incident to one of their cut pairs in $G^{\ominus}$ (for fat edges) and G (for thin edges), respectively. For fat edges this holds according to Lemma 2. For thin edges the order in line 4 guarantees that an edge $\{x, v\}$ that will be reconnected to u in line 19 is at most as expensive as the current edge $\{u, v\}$, and thus, also induces a minimum u-x-cut in G. This allows applying Lemma 6 and 7 as well as the comparison in line 15 to reconnected thin edges, too. Observe that an edge corresponding to a bridge never crosses a new cheaper cut, and thus, gets never reconnected. In the end all edges in T_* are fat, since each edge is either a part of a reused subtree or was considered in line 6. Note that reconnecting a thin edge makes this edge incident to a vertex on $\pi(b, d)$ and decrements the hight of the related subtree.

4   Performance of the Algorithm

Unfortunately we cannot give a meaningful guarantee on the number of saved cut computations. The saving depends on the length of the path $\pi(b, d)$, the number of $\{u, v\} \in E_T$ for which the connectivity $\lambda(u, v)$ changes, and the shape of the cut tree. In a star, for example, there exist no subtrees that could be reused by Lemma 6 (see Figure 6 (left) for a bad case example for edge deletion). Nevertheless, a first experimental proof of concept promises high practicability, particularly on graphs with less regular cut structures. The instance we use is a network of e-mail communications within the

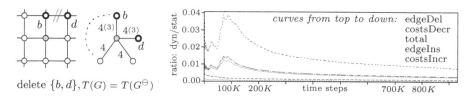

delete $\{b,d\}, T(G) = T(G^{\ominus})$

Fig. 6. left: $T(G)$ could be reused (new cost on $\pi(b,d)$ in brackets), but Alg. 2 computes $n-3$ cuts. right: Cumulative ratio of dynamic and static cut computations.

Department of Informatics at KIT [8]. Vertices represent members, edges correspond to e-mail contacts, weighted by the number of e-mails sent between two individuals during the last 72 hours. We process a queue of 924 900 elementary changes, which indicate the time steps in Figure 6 (right), and 923 031 of which concern edges. We start with an empty graph, constructing the network from scratch. Figure 6 shows the ratio of cuts computed by the update algorithm and cuts needed by the static approach until the particular time step. The ratio is shown in total, and broken down to edge insertions (151 169 occurrences), increasing costs (310473), edge deletions (151 061) and decreasing costs (310 328). The trend of the curves follows the evolution of the graph, which slightly densifies around time step 100000 due to a spam-attack; however, the update algorithm needs less than 4% of the static computations even during this period. We further observe that for decreasing costs, Theorem 2 together with Lemma 3 allows to contract all subtrees incident to the current vertex v on $\pi(b,d)$, which shrinks the underlying graph to $deg_*(v)$ vertices, with $deg_*(v)$ the degree of v in T_*. Such contractions could further speed up the single cut computations. Similar shrinkings can obviously be done for increasing costs, as well.

5 Conclusion

We introduced a simple and fast algorithm for dynamically updating a cut tree for a changing graph. In a first prove of concept our approach allowed to save over 96% of the cut computations and it provides even more possibilities for effort saving due to contractions. A more extensive experimental study is given in the full paper [7]. Recently, we further succeeded in improving the routine for an inserted edge or an increased cost such that it guarantees that each cut that remains valid is also represented by the new cut tree. This yields a high temporal smoothness, which is desirable in many applications. Note that the routine for a deleted edge or a decreased cost as presented in this work already provides this temporal smoothness.

References

1. Gomory, R.E., Hu, T.: Multi-terminal network flows. Journal of the Society for Industrial and Applied Mathematics 9(4), 551–570 (1961)
2. Gusfield, D.: Very simple methods for all pairs network flow analysis. SIAM Journal on Computing 19(1), 143–155 (1990)

3. Elmaghraby, S.E.: Sensitivity Analysis of Multiterminal Flow Networks. Operations Research 12(5), 680–688 (1964)
4. Barth, D., Berthomé, P., Diallo, M., Ferreira, A.: Revisiting parametric multi-terminal problems: Maximum flows, minimum cuts and cut-tree computations. Discrete Optimization 3(3), 195–205 (2006)
5. Gallo, G., Grigoriadis, M.D., Tarjan, R.E.: A fast parametric maximum flow algorithm and applications. SIAM Journal on Computing 18(1), 30–55 (1989)
6. Scutellà, M.G.: A note on the parametric maximum flow problem and some related reoptimization issues. Annals of Operations Research 150(1), 231–244 (2006)
7. Hartmann, T., Wagner, D.: Fast and Simple Fully-Dynamic Cut Tree Construction. Karlsruhe Reports in Informatics 2012-18, KIT Karlsruhe Institute of Technology (2012), http://digbib.ubka.uni-karlsruhe.de/volltexte/1000030004
8. Görke, R., Holzer, M., Hopp, O., Theuerkorn, J., Scheibenberger, K.: Dynamic network of email communication at the Department of Informatics at Karlsruhe Institute of Technology, KIT (2011), http://i11www.iti.kit.edu/projects/spp1307/emaildata

Green Scheduling, Flows and Matchings*

Evripidis Bampis[1], Dimitrios Letsios[1,2], and Giorgio Lucarelli[1,2]

[1] LIP6, Université Pierre et Marie Curie, France
{Evripidis.Bampis,Giorgio.Lucarelli}@lip6.fr
[2] IBISC, Université d' Évry, France
dimitris.letsios@ibisc.univ-evry.fr

Abstract. Recently, optimal combinatorial algorithms have been presented for the energy minimization *multi-processor speed scaling problem with migration* [Albers et al., SPAA 2011], [Angel et al., Euro-Par 2012]. These algorithms are based on repeated maximum-flow computations allowing the partition of the set of jobs into subsets in which all the jobs are executed at the same speed. The optimality of these algorithms is based on a series of technical lemmas showing that this partition and the corresponding speeds lead to the minimization of the energy consumption. In this paper, we show that both the algorithms and their analysis can be greatly simplified. In order to do this, we formulate the problem as a convex cost flow problem in an appropriate flow network. Furthermore, we show that our approach is useful to solve other problems in the dynamic speed scaling setting. As an example, we consider the *preemptive open-shop speed scaling problem* and we propose a polynomial-time algorithm for finding an optimal solution based on the computation of convex cost flows. We also propose a polynomial-time algorithm for minimizing a linear combination of the sum of the completion times of the jobs and the total energy consumption, for the *multi-processor speed scaling problem without preemptions*. Instead of using convex cost flows, our algorithm is based on the computation of a minimum weighted maximum matching in an appropriate bipartite graph.

1 Introduction

In the last few years, a series of papers deal with the minimization of the energy consumption in the area of scheduling (see the recent surveys [2] and [3] and the references therein). One of the most studied models in this context is the speed scaling model in which a set of tasks has to be executed on one or more processors whose speed may change dynamically during the schedule. Hence, the scheduler has to decide not only the job to execute at any given time, but also the speed of the processor(s) in order to satisfy some level of Quality of Service (QoS), while at the same time to minimize the overall energy consumption. In speed scaling, power is usually defined as a convex function of the speed and

---

* Research supported by the French Agency for Research under the DEFIS program TODO, ANR-09-EMER-010, by GDR-RO of CNRS, and by THALIS-ALGONOW.

K.-M. Chao, T.-s. Hsu, and D.-T. Lee (Eds.): ISAAC 2012, LNCS 7676, pp. 106–115, 2012.
© Springer-Verlag Berlin Heidelberg 2012

the energy is power integrated over time. Intuitively, the higher is the speed, the better is the performance in terms of QoS, but the higher is the consumption of energy.

The first theoretical result in this area has been proposed in the seminal paper of Yao et al. [14], where the authors considered the energy minimization problem when a set of jobs, each one specified by its processing volume (work), its release date and its (strict) deadline, has to be scheduled on a single speed-scalable processor. They proposed an algorithm that solves the problem optimally in polynomial time, when the preemption of the jobs, i.e. the possibility to interrupt the execution of a job and resume it later, is allowed. Since then, different problems have been studied taking into account the energy consumption, mainly in the single processor case (e.g., [5,13]), but more recently in the multiprocessor case as well (e.g. [4,6,7,11]). Different algorithmic techniques have been used in order to optimally solve different speed-scaling scheduling problems, including the use of greedy algorithms, dynamic programming, convex programming, and more recently, maximum flows.

In this paper, we show that the use of convex cost flow computations may lead to polynomial-time algorithms for basic speed scaling scheduling problems. This adds a new tool for solving speed scaling scheduling problems. More precisely, we first revisit the *multiprocessor speed scaling with migration* problem, studied in [4,6,7], and we show that it can be solved easily using a convex cost flow formulation, simplifying both the existing algorithms and their proofs of optimality. This problem is the same as the one considered in [14], except that now there are m processors on which the jobs have to be executed and that the execution of an interrupted job maybe continued on the same or on another processor (i.e. the migration of jobs is allowed).

We also consider the *preemptive open-shop speed scaling* problem and we show how it can be solved using a series of convex cost flow computations. In this problem, there are m speed-scalable processors and n jobs, but every job is composed by a set of operations, at most one per processor, and every operation is characterized by its processing volume. There is no order in the execution of the operations and two operations of the same job cannot be executed in parallel. The jobs are all available at the same time and there is a common deadline. This is the first attempt to study a speed scaling problem in a shop scheduling environment.

Finally, we propose a polynomial-time algorithm for minimizing a linear combination of the sum of the completion times of the jobs and the total energy consumption, for the *multi-processor speed scaling problem without preemptions*. Here, we are given m processors and n jobs, each one characterized by its processing volume, while the preemption of the jobs is not allowed. The proposed algorithm is based on the computation of a minimum weighted maximum matching in an appropriate bipartite graph. Notice that in [13] the complexity of the single-processor speed scaling problem with preemptions where the jobs are subject to release dates has been left open. Our result makes progress towards answering this challenging question.

In Section 1.1 we present the notation concerning the convex cost flow and the minimum weighted maximum matching problems. In Sections 2, 3 and 4 we deal with the three speed-scaling scheduling problems mentioned above. In each section, we formally define the studied problem and we give the related work and our approach for it. Due to space constraints the proofs are omitted.

1.1   Preliminaries

An instance of the convex cost flow problem consists of a network $N = (V, A)$, where V is a set of nodes and $A \subseteq V \times V$ is a set of arcs between the nodes. Each arc $(u, v) \in A$ is associated with a capacity $c_{u,v} \geq 0$ and a cost function $\kappa_{u,v}(f) \geq 0$, where $f \geq 0$. The function $\kappa_{u,v}(f)$ is convex w.r.t. f and it represents the cost incurred if f units of flow pass through the arc (u, v). Moreover, we are given an amount of flow $\mathcal{F}$, a source node $s \in V$ and a destination node $t \in V$. The objective is to route the amount of flow $\mathcal{F}$ from s to t so that the total cost is minimized and the amount of flow that crosses each edge (u, v) does not exceed the capacity $c_{u,v}$, for each $(u, v) \in A$. The convex cost flow problem can be efficiently solved in $O(|A| \log(\max\{\mathcal{F}, c_{max}\})(|A| + |V| \log |V|))$ time, where $c_{max} = \max_{(u,v) \in A}\{c_{u,v}\}$ (see for example [1]).

An instance of a minimum weighted maximum bipartite matching problem consists of a bipartite graph $G = (V, U; E)$, where each edge $e \in E$ has a weight $w_e \geq 0$. A matching M in G is a subset of edges, i.e. $M \subseteq E$, such that no two edges in M have a common endpoint, while the weight of the matching M is equal to $\sum_{e \in M} w_e$. A matching of maximum cardinality is a matching that contains the maximum number of edges among all the possible matchings in G. The objective is to find the matching with the minimum weight among the matchings of maximum cardinality. There exists an algorithm for finding such a matching in $O(|V|(|E| + |V| \log |V|))$ time (see for example [1]).

Using a standard exchange argument and based on the convexity of the speed-to-power function, it can be shown that each job/operation runs at a constant speed in any optimal schedule for the considered scheduling problems.

2   Energy Minimization on Parallel Processors

The problem. We consider the scheduling problem of minimizing the energy consumption of a set of n jobs that have to be executed on m parallel processors, where each job J_j is characterized by a processing volume (or work) w_j, a release date r_j and a deadline d_j. In this setting, preemption and migration of jobs are allowed, i.e., a job may suspend its execution and continue on the same or another processor, later from the point of suspension. Moreover, if any processor operates at a speed s, then its energy consumption rate is equal to $P(s)$, where P is a convex function of the speed s. By extending the Graham's classical three-field notation [10], we denote this problem as $S|pmtn, r_j, d_j|E$.

Previous Results. This problem is an extension in the speed scaling setting of one basic problem in scheduling theory, the well-known $P|pmtn, r_j, d_j|-$ problem. In this problem, we are given a set of n jobs that have to be executed on a set of m parallel identical processors, while preemption and migration of jobs are permitted. Each job J_j has a processing time p_j, a release date r_j and a deadline d_j. The objective is to either construct a feasible schedule in which every job J_j is executed during its interval $[r_j, d_j]$, or decide that such a schedule does not exist. The $P|pmtn, r_j, d_j|-$ problem can be solved in polynomial time (see [8]).

Polynomial-time algorithms for finding an optimal solution for $S|pmtn, r_j, d_j|E$, known as the *multi-processor speed scaling with migration* problem, have been proposed by Bingham and Greenstreet [7], Albers et al. [4] and Angel et al. [6]. The algorithm in [7] is based on the use of the Ellipsoid method. As the complexity of the Ellipsoid algorithm is high for practical applications, [4] and [6] proposed purely combinatorial algorithms. These algorithms use repeated computations of maximum flows in appropriate flow networks, in order to determine a partition of the set of jobs into subsets in which all the jobs are executed with the same speed. When the speed of such a subset of jobs is determined, these jobs as well as the corresponding time-intervals and processors are removed from the flow network and the process continuous until no job remains unscheduled. At the end, every job is associated with a unique speed and thus an execution time. The final schedule can be produced by applying the algorithm of McNaughton [12]. The optimality of the algorithms in [4] and [6] is based on a series of technical lemmas showing that this partition and the corresponding speeds lead to the minimization of the energy consumption.

Our Approach. The rough idea of our algorithm is the following: we first formulate $S|pmtn, r_j, d_j|E$ as a convex cost flow problem. An optimal convex cost flow allows us to get the optimal speed s_j for every job J_j, and thus its total execution time $t_j = \frac{w_j}{s_j}$. Then, given the execution times of the jobs, the algorithm constructs a feasible schedule by applying a polynomial-time algorithm for $P|pmtn, r_j, d_j|-$.

Convex Cost Flow Formulation. We consider that the time is partitioned into intervals defined by the release dates and the deadlines of jobs. That is, we define the time points $t_0, t_1, \ldots, t_k$, in increasing order, where each t_i corresponds to either a release date or a deadline, so that for each release date and deadline of job there is a corresponding t_i. Then, we define the intervals $I_i = [t_{i-1}, t_i]$, for $1 \le i \le k$, and we denote by $|I_i|$ the length of I_i. We call a job J_j alive in a given interval I_i, if $I_i \subseteq [r_j, d_j]$. The number of alive jobs in interval I_i is denoted by $A(I_i)$.

Then, in the flow network N_s for $S|pmtn, r_j, d_j|E$, we introduce a source node s, a destination node t, a node for each job J_j, $1 \le j \le n$, and a node for each interval I_i, $1 \le i \le k$. For each j, $1 \le j \le n$, we add an arc (s, J_j) and, for each i, $1 \le i \le k$, we add an arc (I_i, t). If the job J_j, $1 \le j \le n$, is alive during the interval I_i, $1 \le i \le k$, we introduce an arc from the node J_j to the node I_i. The capacity of the arc (u, v) is

$$c_{u,v} = \begin{cases} +\infty & \text{if } u = s \text{ and } v = J_j \\ |I_i| & \text{if } u = J_j \text{ and } v = I_i \\ m|I_i| & \text{if } u = I_i \text{ and } v = t \end{cases}$$

If an amount of flow $f_{u,v}$ passes through the arc (u,v) of N_s, then the cost function of the arc is defined as

$$\kappa_{u,v}(f_{u,v}) = \begin{cases} f_{u,v} \cdot P(\frac{w_j}{f_{u,v}}) & \text{if } u = s \text{ and } v = J_j \\ 0 & \text{if } u = J_j \text{ and } v = I_i \\ 0 & \text{if } u = I_i \text{ and } v = t \end{cases}$$

In the network N_s, if an amount $f_{u,v}$ of flow passes through the arc $(u,v) = (s, J_j)$, then $f_{u,v}$ corresponds to the execution time of job J_j, $\frac{w_j}{f_{u,v}}$ corresponds to the speed of J_j and $f_{u,v} \cdot P(\frac{w_j}{f_{u,v}})$ is the energy consumed for the execution of J_j. Furthermore, the flow passing through an edge (J_j, I_i) (resp. an edge (I_i, t)) represents the execution time of the job J_j (resp. the execution time of all jobs) during the interval I_i. Hence, the total flow that leaves the source node corresponds to the total execution time of all jobs. In [4], it was shown that the total execution time of all jobs in an optimal schedule for $S|pmtn, r_j, d_j|E$ can be easily computed using the following lemma, whose proof is based on the convexity of the speed-to-power function.

Lemma 1. [4] In an optimal schedule for $S|pmtn, r_j, d_j|E$, where each job J_j is executed with speed s_j, $1 \le j \le n$, the total execution time, $\mathcal{T}$, of all jobs is

$$\mathcal{T} = \sum_{j=1}^{n} \frac{w_j}{s_j} = \sum_{i=1}^{k} \left(\min\{m, A(I_i)\} \cdot |I_i| \right)$$

The above lemma gives the total amount of flow that has to be sent from the source node to the destination node, concluding the formulation of $S|pmtn, r_j, d_j|E$ as a convex cost flow problem.

The Algorithm and Its Optimality. Our algorithm for $S|pmtn, r_j, d_j|E$ can be summarized as follows.

ALGORITHM. GREEN-PP

1: Construct the flow network N_s;
2: Find a convex cost flow $\mathcal{F}$ of value $\sum_{i=1}^{k}(\min\{m, A(I_i)\} \cdot |I_i|)$ in N_s;
3: Determine the execution time of each job;
4: Apply a polynomial-time algorithm for $P|pmtn, r_j, d_j|-$ to find a feasible schedule;

Theorem 1. ALGORITHM GREEN-PP *finds an optimal schedule for* $S|pmtn, r_j, d_j|E$ *in* $O(n^4 m \log L)$ *time, where* $L = \max_{1 \le j \le n}\{d_j\}$.

3   Energy Minimization in an Open Shop

The Problem. We consider the scheduling problem of minimizing the energy consumed in an open shop setting. We are given a set of n jobs that have to be executed in a prespecified time interval $[0, d]$ on a set of m parallel processors. Each job J_j consists of m operations $O_{1,j}, O_{2,j}, \ldots, O_{m,j}$. Each operation $O_{i,j}$, $1 \leq i \leq m$ and $1 \leq j \leq n$, has an amount of work $w_{i,j} \geq 0$ and can only be executed on the processor M_i. The preemption of the operations is allowed, while the parallel execution of operations of the same job is not permitted. A convex speed-to-power function $P(s)$ defines the energy consumption rate of any processor running at speed $s \geq 0$. The goal is to schedule the jobs within the interval $[0, d]$ so that the total energy consumption is minimized. We denote this problem as $OS|pmtn, r_j = 0, d_j = d|E$.

Previous Results. This problem is an extension of the preemptive open shop problem $O|pmtn, r_j = 0, d_j = d|-$ [8] in the speed scaling setting. In this problem, we are given a set of n jobs and a set of m processors. Each job J_j consists of a set of m operations, where the processing time of the operation $O_{i,j}$, $1 \leq i \leq m$ and $1 \leq j \leq n$, is $p_{ij} \geq 0$. The open shop constraint enforces that no pair of operations of a job are executed at the same time. The goal is to find a feasible schedule such that all operations are preemptively scheduled during the interval $[0, d]$ or decide that such a schedule does not exist. A polynomial algorithm for this problem can be found in [8].

Up to the best of our knowledge, no results were known for this problem in the speed scaling setting until now.

Our Approach. As in the previous section, we first give a formulation of $OS|pmtn, r_j = 0, d_j = d|E$ as a convex cost flow problem. In order to compute the total execution time of all operations in an optimal schedule for $OS|pmtn$, $r_j = 0, d_j = d|E$, we use a search algorithm that repeatedly computes convex cost flows. Given the optimal value of the total execution time of all operations, an optimal convex cost flow gives the speeds, and hence the execution times, of the operations in an optimal schedule for $OS|pmtn, r_j = 0, d_j = d|E$. To compute a feasible schedule, we solve the corresponding instance of $O|pmtn, r_j = 0, d_j = d|-$ (see for example [8]).

Convex Cost Flow Formulation. We construct the flow network N_s which consists of a source node s, a destination node t, a job-node J_j, for each $1 \leq j \leq n$, and a processor-node M_i, for each $1 \leq i \leq m$. The network N_s contains an arc (s, J_j) for each job J_j, $1 \leq j \leq n$, an arc (M_i, t) for each processor M_i, $1 \leq i \leq m$, and an arc (J_j, M_i), $1 \leq j \leq n$ and $1 \leq i \leq m$ if $w_{i,j} > 0$. The capacity of the arc (u, v) is

$$c_{u,v} = \begin{cases} d & \text{if } u = s \text{ and } v = J_j \\ +\infty & \text{if } u = J_j \text{ and } v = M_i \\ d & \text{if } u = M_i \text{ and } v = t \end{cases}$$

Assuming that flow $f_{u,v}$ passes through the arc (u,v), the cost of this arc is

$$\kappa_{u,v}(f_{u,v}) = \begin{cases} 0 & \text{if } u = s \text{ and } v = J_j \\ f_{J_j,M_i} \cdot P(\frac{w_j}{f_{J_j,M_i}}) & \text{if } u = J_j \text{ and } v = M_i \\ 0 & \text{if } u = M_i \text{ and } v = t \end{cases}$$

As in the network for $S|pmtn, r_j, d_j|E$ presented in the previous section, in the network N_s for $OS|pmtn, r_j = 0, d_j = d|E$, the flow traversing the arcs corresponds to execution time. More specifically, if an amount $f_{u,v}$ of flow passes through the arc $(u,v) = (J_j, M_i)$, then $f_{u,v}$ corresponds to the execution time of operation $O_{i,j}$, $\frac{w_{i,j}}{f_{u,v}}$ corresponds to the speed of $O_{i,j}$ and $f_{u,v} \cdot P(\frac{w_{i,j}}{f_{u,v}})$ is the energy consumed for the execution of $O_{i,j}$. Furthermore, the flow passing through an edge (s, J_j) (resp. an edge (M_i, t)) represents the total execution time of the job J_j (resp. the total time that M_i operates). Hence, the total flow that leaves the source node corresponds to the total execution time of all operations. However, the total execution time of all operations in an optimal schedule for $OS|pmtn, r_j = 0, d_j = d|E$, and thus the total amount of flow that has to be sent from the source node to the destination node, cannot be easily computed as in the previous section. At the end of this section, we describe how to compute it in polynomial time.

The Algorithm and Its Correctness. Our algorithm for $OS|pmtn, r_j = 0, d_j = d|E$ can be summarized as follows.

ALGORITHM. GREEN-OS

  1: Construct the flow network N_s;
  2: Determine the total execution time of all operations $\mathcal{T}$ in an optimal schedule;
  3: Find a convex cost flow $\mathcal{F}$ of value $\mathcal{T}$ in N_s;
  4: Determine the execution time of each operation;
  5: Apply a polynomial-time algorithm for $O|pmtn, r_j = 0, d_j = d|-$ to find a feasible schedule;

We will initially assume that the total execution time of all operations $\mathcal{T}$ in an optimal schedule for $OS|pmtn, r_j = 0, d_j = d|E$ can be computed in polynomial time. Then, the proof of the following theorem is similar with the proof of Theorem 1.

Theorem 2. ALGORITHM GREEN-OS *finds an optimal schedule for* $OS|pmtn, r_j = 0, d_j = d|E.$

Computing the Total Execution Time of Operations. It remains to show how we algorithmically determine the total execution time of all operations in an optimal schedule for $OS|pmtn, r_j = 0, d_j = d|E$.

First, we introduce some additional notation. Henceforth, we denote by T^* the sum of execution times of all operations in an optimal schedule for $OS|pmtn,$

$r_j = 0, d_j = d|E$. Let $\mathcal{S}$ be any feasible schedule for the problem and assume that $t_{i,j}$ is the execution time of the operation $O_{i,j}$ in $\mathcal{S}$, $1 \leq i \leq m$ and $1 \leq j \leq n$. We denote by $\boldsymbol{t} = (t_{1,1}, t_{2,1}, \ldots, t_{m,1}, t_{1,2}, \ldots, t_{m,n})$ the vector that contains the execution times of all the operations. Then, let $T(\boldsymbol{t}) = \sum_{i=1}^{m} \sum_{j=1}^{n} t_{ij}$ and $E(\boldsymbol{t}) = \sum_{i=1}^{m} \sum_{j=1}^{n} t_{ij} \cdot P(\frac{w_{ij}}{t_{ij}})$ be the functions that map any vector of execution times $\boldsymbol{t}$ to the total execution time and the total energy consumption of the schedule $\mathcal{S}$. Note that, $E(\boldsymbol{t})$ is convex w.r.t. the vector $\boldsymbol{t}$ as a sum of convex functions. Furthermore, we define the function $E^*(T) = \min\{E(\boldsymbol{t}) : T(\boldsymbol{t}) = T\}$ which indicates the minimum energy consumption when the sum of execution times of all operations has to be equal to T.

Proposition 1. $E^*(T)$ is convex w.r.t. T.

Next, we give the search algorithm that finds the value $T^* = \arg\min_T \{E^*(T)\}$ with accuracy $1/\epsilon$. Consider any $T_1, T_2, T_3 > 0$ such that $T_1 < T_2 = \frac{T_1 + T_3}{2} < T_3$. As $E^*(T)$ is convex, we have that $E^*(T_2) \leq \frac{E^*(T_1) + E^*(T_3)}{2}$. Therefore, it follows that either $E^*(T_2) \leq E^*(T_1)$ or $E^*(T_2) \leq E^*(T_3)$ (or both). If only the first is true, then we reduce our search space to $[T_2, T_3]$. Accordingly, if only the second is true, then we reduce our search space to $[T_1, T_2]$. Finally, if both are true, then we reduce our search space to one of the following intervals: $[T_1, T_2]$, $[T_2, T_3]$ or $[\frac{T_1 + T_2}{2}, \frac{T_2 + T_3}{2}]$. If $E^*(\frac{T_1 + T_2}{2}) \leq E^*(T_2)$, then the search space is reduced to $[T_1, T_2]$. If $E^*(\frac{T_2 + T_3}{2}) \leq E^*(T_2)$, then the search space is reduced to $[T_2, T_3]$. Finally, if $E^*(\frac{T_1 + T_2}{2}) > E^*(T_2)$ and $E^*(\frac{T_2 + T_3}{2}) > E^*(T_2)$, then the search space is reduced to $[\frac{T_1 + T_2}{2}, \frac{T_2 + T_3}{2}]$. The correctness of all the cases is based on the fact that $E^*(T)$ is convex. We call this procedure ALGORITHM FIND-FLOW and we initialize T_1, T_2 and T_3 with 0, $\frac{\mathcal{T}}{2}$ and $\mathcal{T}$, respectively, where $\mathcal{T}$ is an upper bound on the sum of execution times for all operations, i.e., $\mathcal{T} = m \cdot d$.

Lemma 2. ALGORITHM FIND-FLOW returns a value T^* such that the term $E^*(T^*)$ is minimized among all $T^* > 0$. The complexity of the algorithm is $O(nm \log(md)(nm + (n + m) \log(n + m))(\log md + \log(1/\epsilon))$, where ϵ is the inverse of the desired accuracy.

4   Mean Completion Time Plus Energy Minimization on Parallel Processors

The Problem. We consider the multiprocessor scheduling problem of minimizing a linear combination of the sum of completion times of a set of n jobs and their total energy consumption. The jobs have to be executed by a set of m parallel processors where the preemption and migration of jobs are not allowed. Each job J_j, $1 \leq j \leq n$, has an amount of work w_j to accomplish and all jobs are released at time $t = 0$. We denote by C_j the completion time of job J_j, $1 \leq j \leq n$. A convex speed-to-power function $P(s)$ defines the energy consumption rate when a job is executed with speed $s > 0$ on any processor. The goal is to minimize the sum of completion times of all the jobs plus β times their total energy consumption. The parameter $\beta > 0$ is used to specify the relevant importance of the mean completion time criterion versus the total energy consumption criterion. We denote this problem as $S||\sum C_j + \beta \cdot E$.

Known Results. This problem is an extension in the speed scaling setting of the problem $P||\sum C_j$ of scheduling non-preemptively a set of n jobs, each one characterized by its processing time p_j, on a set of m machines such that the sum of the completion times of all jobs is minimized. A polynomial-time algorithm has been proposed for $P||\sum C_j$ [9].

The single-processor speed scaling problem without preemptions of minimizing the jobs' mean completion time has been studied by Albers et al. [5] and Pruhs et al. [13] in the presence of release dates and unit work jobs. In [13] the objective is the minimization of the sum of the flow times of the jobs[1] under a given budget of energy, while in [5] the goal is to minimize the sum of the flow times of the jobs and of the consumed energy.

Our Approach. The main idea is to formulate $S||\sum C_j + \beta \cdot E$ as a problem of searching for a minimum weighted maximum matching in an appropriate bipartite graph. This formulation is based on two observations. Firstly, the fact that preemption and migration of jobs is not allowed means that there is an order of the jobs executed by any processor in any feasible schedule. Given such a schedule, if ℓ jobs are executed by the processor M_i, then we can consider that there are ℓ available positions on M_i, one for the execution of each of the ℓ jobs. If the job J_j is executed in the k-th position of the processor M_i, then $k-1$ jobs precede J_j and $\ell - k$ jobs succeed J_j. Clearly, there can be at most n such positions for each processor. Secondly, the contribution of a job J_j to the objective function depends only on its position on the processor by which it is executed and it is independent of where the other jobs are executed. Overall, our problem reduces to assigning every job to a position of a machine so that our objective is minimized.

Minimum Weighted Maximum Matching Formulation. In order to formulate the $S||\sum C_j + \beta \cdot E$ problem as a minimum weighted maximum matching problem, we define a bipartite graph G_s whose edges are weighted. The following lemma is our guide for assigning weights on the edges of G_s and fixes the cost of executing a job J_j to the k-th position of any processor.

Lemma 3. *Assume that in an optimal schedule for $S||\sum C_j + \beta \cdot E$ the job J_j, $1 \leq j \leq n$, is executed at speed s_j on processor M_i in the k-th position from the end of M_i. Then, the total contribution of J_j to the objective function is minimized if it holds that $s_j P'(s_j) - P(s_j) = \frac{k}{\beta}$, where $P'(s)$ is the derivative of the power function P w.r.t. the speed s.*

Based on the above lemma, we create the complete bipartite graph $G_s = (V, U; E)$ as follows: (i) for each job J_j, $1 \leq j \leq n$, we add a vertex in V, (ii) for each processor M_i, $1 \leq i \leq m$, and each position k, $1 \leq k \leq n$, (counting from the

---

[1] The flow time of a job is defined as the amount of time that the job spends in the system, i.e., the difference between its completion time and its release date.

end) we add a vertex in U, and (iii) for each edge $(J_j, (M_i, k))$, we set its weight $c_{i,j,k} = k \cdot \frac{w_j}{s_j} + \beta \cdot \frac{w_j}{s_j} P(s_j)$ where s_j is computed according to Lemma 3.

The Algorithm and Its Optimality. Recall that each job J_j runs at a constant speed s_j in any optimal schedule for $S||\sum C_j + \beta \cdot E$. Moreover, based on a similar argument, it holds that there is no idle period on any processor between the common release date of the jobs and the date at which the last job completes its execution, in any optimal schedule. A description of our algorithm follows.

ALGORITHM. GREEN-(F+E)

1: Construct the bipartite graph G_s;
2: Find a minimum weighted maximum matching M in G_s;
3: **for each** $(J_j, (M_i, k)) \in M$ **do**
4:     Schedule J_j to the position k of M_i with speed s_j such that $s_j P'(s_j) - P(s_j) = \frac{k}{\beta}$;

Theorem 3. $S||\sum C_j + \beta \cdot E$ *can be solved in* $O(n^3 m^2)$ *time.*

References

1. Ahuja, R.K., Magnanti, T.L., Orlin, J.B.: Network flows: theory, algorithms and applications. Prentice Hall (1993)
2. Albers, S.: Energy-efficient algorithms. Commun. ACM 53, 86–96 (2010)
3. Albers, S.: Algorithms for dynamic speed scaling. In: STACS. LIPIcs, vol. 9, pp. 1–11. Schloss Dagstuhl - Leibniz-Zentrum fuer Informatik (2011)
4. Albers, S., Antoniadis, A., Greiner, G.: On multi-processor speed scaling with migration: extended abstract. In: SPAA, pp. 279–288. ACM (2011)
5. Albers, S., Fujiwara, H.: Energy-efficient algorithms for flow time minimization. ACM Trans. on Algorithms 3(4), Article 49 (2007)
6. Angel, E., Bampis, E., Kacem, F., Letsios, D.: Speed Scaling on Parallel Processors with Migration. In: Kaklamanis, C., Papatheodorou, T., Spirakis, P.G. (eds.) Euro-Par 2012. LNCS, vol. 7484, pp. 128–140. Springer, Heidelberg (2012)
7. Bingham, B.D., Greenstreet, M.R.: Energy optimal scheduling on multiprocessors with migration. In: ISPA, pp. 153–161. IEEE (2008)
8. Brucker, P.: Scheduling algorithms, 4th edn. Springer (2004)
9. Bruno, J., Coffman Jr., E.G., Sethi, R.: Scheduling independent tasks to reduce mean finishing time. Commun. ACM 17, 382–387 (1974)
10. Graham, R.L., Lawler, E.L., Lenstra, J.K., Rinnooy Kan, A.H.G.: Optimization and approximation in deterministic sequencing and scheduling. Annals of Discrete Mathematics 5, 287–326 (1979)
11. Lam, T.W., Lee, L.-K., To, I.K.-K., Wong, P.W.H.: Competitive non-migratory scheduling for flow time and energy. In: SPAA, pp. 256–264 (2008)
12. McNaughton, R.: Scheduling with deadlines and loss functions. Management Science 6, 1–12 (1959)
13. Pruhs, K., Uthaisombut, P., Woeginger, G.: Getting the best response for your erg. ACM Trans. on Algorithms 4(3), Article 38 (2008)
14. Yao, F., Demers, A., Shenker, S.: A scheduling model for reduced CPU energy. In: FOCS, pp. 374–382 (1995)

Popular and Clan-Popular b-Matchings

Katarzyna Paluch*

Institute of Computer Science, University of Wrocław

Abstract. Suppose that each member of a set of agents has a preference list of a subset of houses, possibly involving ties, and each agent and house has their capacity denoting the maximum number of houses/agents (respectively) that can be matched to him/her/it. We want to find a matching M, called popular, for which there is no other matching M' such that more agents prefer M' to M than M to M', subject to a suitable definition of "prefers". In the above problem each agent uses exactly one vote to compare two matchings. In the other problem we consider in the paper each agent has a number of votes equal to their capacity. Given two matchings M and M', an agent compares their best house in matching $M\backslash(M\cap M')$ to their best house in matching $M'\backslash(M\cap M')$ and gives one vote accordingly, then their second best houses and so on. A matching M for which there is no matching M' such that M' gets a bigger number of votes than M, when M and M' are compared in the way described above, is then called clan-popular. Popular matchings have been studied quite extensively, especially in the one-to-one setting. In the many-to-many setting we provide a characterisation of popular and clan-popular matchings, show NP-hardness results for very restricted cases of the above problems and for certain versions describe novel polynomial algorithms. The given characterisation is also valid for popular matchings in the one-to-one setting.

1   Introduction

In the paper we study popular and clan-popular b-matchings, which in other words are popular many-to-many matchings. The problems can be best described in graph terms: We are given a bipartite graph $G = (A \cup H, E)$, a capacity function on vertices $b : A \cup H \to N$ and a rank function on edges $r : E \to N$. A stands for the set of agents and H for the set of houses. For $a \in A$ and $h \in H$ $r((a,h)) = i$ means that house h belongs to (one of) agent a's ith choices. We say that a **prefers** h_1 to h_2 (or **ranks** h_1 **higher than** h_2) if $r((a,h_1)) < r((a,h_2))$. If $r(e_1) < r(e_2)$ we say that e_1 has a **higher rank** than e_2. If there exist $a \in A$ and $h_1, h_2 \in H_a, h_1 \neq h_2$ such that $r(e_1) = r((a,h_1)) = r(e_2) = r((a,h_2))$, then we say that e_1, e_2 belong to a **tie** and graph G contains ties. Otherwise we say that G does not contain ties. A b-**matching** M of G is such a subset of edges that $deg_M(v) \leq b(v)$ for every $v \in A \cup H$, meaning that every vertex v has at most $b(v)$ edges of M incident with it. For every $v \in A \cup H$ by $M(v)$ we mean

---

* Supported by MNiSW grant number N N206 368839, 2010-2013.

K.-M. Chao, T.-s. Hsu, and D.-T. Lee (Eds.): ISAAC 2012, LNCS 7676, pp. 116–125, 2012.

the set $\{w \in A \cup H : (v, w) \in M\}$. With each agent a and each b-matching M we associate a **signature** denoted as $sig_M(a)$ defined as follows.

Definition 1. *Let s denote the greatest rank (i.e. the largest number) given to any edge of E. Let $(y_1, y_2, \ldots, y_t)$ denote the ranks of all edges (a, h) such that $h \in M(a)$ sorted non-decreasingly. By $sig_M(a)$ we will denote a $b(a)$-tuple $(x_1, x_2, \ldots x_{b(a)})$ such that $x_i = y_i$ for $1 \le i \le t$ and $x_i = s + 1$ for $i > t$.*

We introduce a **lexicographic order** $\succ$ **on signatures** as follows. We will say that $(x_1, x_2, \ldots, x_d) \succ (y_1, y_2, \ldots, y_d)$ if there exists j such that $1 \le j \le d$ and for each $1 \le i \le j - 1$ we have $x_i = y_i$ and $x_j < y_j$. We say that an agent a **prefers** b-matching M to M' if $sig_M(a) \succ sig_{M'}(a)$. M' is **more popular** than M, denoted by $M' \succ M$, if the number of agents that prefer M' to M exceeds the number of agents that prefer M to M'.

Definition 2. *A b-matching M is **popular** if there exists no b-matching M' that is more popular than M. The **popular b-matching problem** is to determine if a given triple (G, b, r) admits a popular b-matching and find one if it exists.*

The popular b-matching problem has applications in one-sided markets, where agents are allowed to have more than one house and one house can be owned by more than one agent. The problem is a generalisation of the one considered in [19] in which each agent is allowed to have at most one house, but each house can be owned by more than one agent. Clearly the version we consider is no less natural than the one in [19].

In the popular b-matching problem we assume that each agent has one vote and given two b-matchings M and M', they vote $+1, -1$ or 0 depending on the case being respectively that they prefer M to M' or M' to M or are indifferent between M and M'.

In the **clan-popular b-matching** problem we will assume that each agent has a number of votes equal to their capacity. Given two b-matchings M and M' agent a compares their best house in matching $M \setminus (M \cap M')$ to their best house in matching $M' \setminus (M \cap M')$ and gives one vote accordingly $(1, -1$ or $0)$, then their second best houses and so on. More formally, for two b-matchings M and M' and agent a we define $vote_a(M, M')$ as $vote_a(M, M') = \sum_{i=1}^{b(a) - |M(a) \cap M'(a)|} signum(sig_{M' \setminus (M \cap M')}(a)_i - sig_{M \setminus (M \cap M')}(a)_i)$, where $signum(x) = 1, 0$ or -1 if correspondingly $x > 0, x = 0$ or $x < 0$ and $sig_M(a)_i$ denotes the i-th position in $sig_M(a)$. We will say that a b-matching M is **more clan-popular** than M' if $\sum_{a \in A} vote_a(M, M') > 0$. A b-matching M will be called **clan-popular** if there does not exist a b-matching M' that is more clan-popular than M.

Another application of the clan-popular b-matching problem is as follows. Suppose we are given a set of agents and a set of houses. Each agent has capacity 1 and each house has an arbitrary capacity. Agents are partitioned into clans and all agents from the same clan have the same preferences over houses. In graph terms it means we are given a graph $G = (A \cup H, E)$, a capacity function b and a rank function r. Before voting between two b-matchings M and M', one of the b-matchings, say M', is changed within clans as follows. If $h \in V(M \cap M')$, then $M'(h) = M(h)$. (The same house is given to the same agent. $V(M \cap M')$ denotes the set of end vertices of edges of $M \cap M'$.) Also b-matchings $M \setminus (M \cap M')$ and $M' \setminus (M \cap M')$ are such that the person who gets the best house in $M \setminus (M \cap M')$, gets also the best house in $M' \setminus (M \cap M')$ and the person who gets the second best house in $M \setminus (M \cap M')$, gets also the second best house in $M' \setminus (M \cap M')$ and so on. The ties are broken arbitrarily. The task is to find the popular matching in the standard sense, if it exists.

Previous Work. The notion of popularity was first introduced by Gärdenfors [4] in the one-to-one and two-sided context, where two-sided means that both agents and houses express their preferences over the other side and a matching M is popular if there is no other matching M' such that more participants (i.e. agents plus houses) prefer M' to M than prefer M to M'. (He used the term of a *majority assignment*.) He proved that every stable matching is a popular matching if there are no ties.

One-sided popular matchings were first studied in the one-to-one setting by Abraham et al. in [1]. They proved that a popular matching need not exist and described fast polynomial algorithms to compute a popular matching, if it exists. Manlove and Sng in [14] extended an algorithm from [1] to the one-to-many setting (notice that this not equivalent to the many-to-one setting.) Other results concerning popular matchings appeared in [2], [3], [9], [13], [15], [16], [17].

Our Contribution. We provide a characterisation of popular b-matchings and prove that the popular b-matching problem is NP-hard even when agents use only two ranks and have capacity at most 2 and houses have capacity 1. This in particular answers the question about many-to-one popular matchings asked in [14]. Next we modify the notion of popularity and consider clan-popular b-matchings. We give their characterisation and show that finding a clan-popular b-matching or reporting that it does not exist is NP-hard even if all agents use at most three ranks, there are no ties and houses have capacity 1. The given characterisations allow checking whether a given b-matching is popular or corresp. clan-popular in time polynomial in the number of edges. The characterisation remains valid in the one-to-one setting (both characterisations denote the same thing then) and thus provides an alternative way of checking whether a given matching is popular, moreover in linear (in the number of edges) time. (In the presence of ties an algorithm in [1] computing a popular matching, if it exists, runs in $O(\sqrt{n}m)$ time, where n, m denote the number of vertices and edges resp..) Algorithmically, the most interesting (in our opinion) part concerns polynomial algorithms for clan-popular b-matchings. We construct a novel polynomial algorithm computing a clan-popular b-matching, if it exists, for the version in which agents have capacity 2 and use two ranks, houses have arbitrary capacities and there are no ties.

2 Characterisations

First we introduce some terminology and recall a few facts from the matching theory.

By a path P we will mean a sequence of edges. Usually a path P will be denoted as $(v_1, v_2, \ldots, v_k)$, where $v_1, \ldots, v_k$ are vertices from the graph $G = (V, E)$, not necessarily all different, and for each i $(1 \le i \le k - 1)$ $(v_i, v_{i+1}) \in E$ and no edge of G occurs twice in P. We will sometimes treat a path as a sequence of edges and sometimes as a set of edges.

Let M be a b-matching. Then we will say that v is **unsaturated**, if $deg_M(v) < b(v)$ and we will say that v is **saturated** if $deg_M(v) = b(v)$. If $b(v) = 1$, then we will also use the terms **matched** and **unmatched** instead of saturated and unsaturated. If $e \in M$ we will call e an M-**edge** and otherwise – a non-M-**edge**. A path is said to be **alternating** (with respect to M) or M-**alternating** if its edges are alternately M-edges and non-M-edges. An alternating path is said to be $(M$-**)augmenting** if its end vertices are unsaturated and both its first and last edge is a non-M-edge.

For two sets Z_1, Z_2 a symmetric difference $Z_1 \oplus Z_2$ is defined as $(Z_1 \setminus Z_2) \cup (Z_2 \setminus Z_1)$. If M is a b-matching and P is an alternating path with respect to M such that (1) its beginning edge (v_1, v_2) is an M-edge or v_1 is unsaturated and (2) its ending edge (v_{k-1}, v_k) is an M-edge or v_k is unsaturated, then $M \oplus P$ is a also a b-matching. If P is M-augmenting, then $M \oplus P$ has more edges than M. A b-matching of maximum size is called a **maximum b-matching**.

We will also need a notion of an **equal path**: a path $(v_1, v_2, \ldots, v_k)$ is defined to be equal if it is alternating, (v_1, v_2) is a non-M-edge and for each $i, 1 < i < k$ such that $v_i \in A$ edges $(v_{i-1}, v_i), (v_i, v_{i+1})$ have the same rank. An **equal cycle** is an equal path of the form $(a_1, h_1, a_2, h_2, \ldots, h_k, a_1)$ and such that edges $(h_k, a_1), (a_1, h_1)$ have the same rank. We will call an equal path $(v_1 = a_1, v_2, \ldots, v_k)$ **smooth** if a_1 is unsaturated or there exists a house $h \in M(a_1)$ such that $r((a_1, h)) > r((a_1, v_2))$.

Theorem 1. *A b-matching M is popular iff graph G does not contain a path of one of the following four types:*

1. $(a_1, \ldots, h_{k-1}, a_k, h_k, a_{k+1})$, *where (a) the path is alternating, (b) a path $(a_1, \ldots, h_{k-1})$ is smooth, (c) $r((h_{k-1}, a_k)) > r((a_k, h_k))$ and (d) agents a_1, a_k, a_{k+1} are pairwise different or $a_1 \ne a_k, a_1 = a_{k+1}$ and $r((a_1, h_1)) < r((a_1, h_k))$),*
2. $(a_1, h_1, \ldots, h_k, a, h'_{k'}, a'_{k'}, h'_{k'-1}, \ldots, a'_1)$, *where paths $(a_1, h_1, \ldots, h_k, a)$ and $(a'_1, h'_1, \ldots, h'_k, a)$ are smooth and agents a_1, a, a'_1 are pairwise different,*
3. $(a_1, h_1, \ldots, h_k)$, *where the path is smooth and h_k is unsaturated,*
4. $(a_1, h_1, \ldots, h_k, a_1)$, *where the path $(a_1, h_1, \ldots, h_k)$ is equal and $r((h_k, a_1)) > r((a_1, h_1))$.*

Proof. Let M be a popular b-matching. Suppose the graph contains a path P_1 of type (1). If a_1 is saturated let $P'_1 = P_1 \cup e$, where e denotes edge $(a_1, h) \in M \setminus M'$ having lower rank than (a_1, h_1) (notice that $e \notin P_1$), otherwise let $P'_1 = P_1$.

$M' = M \oplus P'_1$ is a b-matching such that $sig_{M'}(a_1) \succ sig_M(a_1), sig_{M'}(a_k) \succ sig_M(a_k), sig_M(a_{k+1}) \succ sig_{M'}(a_{k+1})$ and for each a different from a_1, a_k, a_{k+1} we have $sig_{M'}(a) = sig_M(a)$. Therefore M' is more popular than M.

Suppose the graph contains a path P_3 of type (3). If a_1 is saturated let $P'_3 = P_3 \cup e$, where again e denotes edge $(a_1, h) \in M \setminus M'$ having lower rank than (a_1, h_1), otherwise let $P'_3 = P_3$. $M' = M \oplus P'_3$ is a b-matching such that $sig_{M'}(a_1) \succ sig_M(a_1)$ and for each a different from a_1 we have $sig_{M'}(a) = sig_M(a)$. Therefore M' is more popular than M.

If the graph contains a path of type (2) or (4), we proceed analogously.

For the other direction suppose now that M is not popular. There exists then a b-matching M' which is more popular than M. This means that the set A_1 of agents who prefer M' to M outnumbers the set A_2 of agents who prefer M to M'.

As a preprocessing step, we remove from $M \oplus M'$ all equal cycles. Therefore from now on, we will assume that $M \oplus M'$ does not contain equal cycles. For each $a \in A_1$ we build a path P_a in the following way. We will use only edges of $M \oplus M'$. We start with an edge $(a, h_1) \in M'$ having the highest possible rank (i.e. lowest possible number) and such that it has not been used in any other path $P_{a'}$ ($a \neq a'$, $a' \in A_1$). Now assume that our so far built path P_a ends on h_i. If h_i is unsaturated in M, we end. Otherwise we consider edges (h_i, a_i) belonging to M and not already used by other paths P'_a ($a' \in A_1$). (The set of such unused edges is nonempty as h_i is saturated in M and thus there are at least as many M-edges as M'-edges incident with h_i and each time we arrive at h_i while building some path P_a ($a \in A_1$) we use one M-edge and one M'-edge.) If among these edges, there is such one that $a_i \in A_1$, we proceed as follows. If there exists an edge $(a_i, h_{i+1}) \in M'$ (not necessarily unused) having higher rank than edge (h_i, a_i), we add edge (h_i, a_i) to P_a and stop; otherwise since $a_i \in A_1$, there exists an unused edge (a_i, h_{i+1}) of the same rank as (h_i, a_i) and we add edges (h_i, a_i) and (a_i, h_{i+1}) to P_a. Otherwise if among unused edges (h_i, a_i) there is such one that $a_i \in A_2$, we add it to P_a and stop. Otherwise consider any unused edge $(h_i, a_i) \in M$. a_i belongs neither to A_1 nor to A_2 and therefore has the same signature in M and in M'. There exists then an unused edge $(a_i, h_{i+1}) \in M'$ having the same rank as (h_i, a_i) (it is so because the number of edges of a given rank incident with a_i is the same in M and in M') and we add any such edge (a_i, h_{i+1}) as well as edge (h_i, a_i) to P_a.

Clearly we stop building P_a at some point because we either arrive at an unsaturated vertex $h \in H$ or at a vertex of $A_1 \cup A_2$, which may be a itself. Suppose there exists $a \in A_1$ such that P_a ends on a. Since P_a starts with an edge e of M' having the highest rank and $M \oplus M'$ does not contain equal cycles, the ending edge of P_a must have a lower rank than e, hence P_a is a path of type (4). If there exists a path P_a ending on an unsaturated vertex, it is of type (3). If there exists $a \in A_1$ such that P_a ends on $a_1 \in A_1, a_1 \neq a$, then let $e = (h', a_1)$ denote the ending edge of P_a and let $e_0 = (a, h_1)$ denote the starting edge of P_a. Because we have stopped building P_a on e, it means that there exists edge $e' = (a_1, h) \in M'$ having higher rank than e. If e' already belongs to P_a or h

is already on P_a, then $P_a \cup e'$ contains a path of type (4). Otherwise if there exists an edge $e_3 = (h, a) \in M$, $P_a \cup e' \cup e_3$ forms a path of type (1) or (4) depending on whether e_3 has the same rank as e_0 or higher. Otherwise there exists an edge $e_3 = (h, a_2) \in M$, where $a_2 \neq a$ and of course $a_2 \neq a_1$ and path $P_a \cup e' \cup e_3$ forms a path of type (1). If none of the paths P_a ends on un unsaturated vertex or on a vertex in A_1, each P_a ends on some agent $a_2 \in A_2$. Because A_2 outnumbers A_1 there exist $a_1, a_1' \in A_1$, $a_1 \neq a_1'$ and $a_2 \in A_2$ such that P_{a_1} and $P_{a_1'}$ both end on a_2. These paths are edge-disjoint and together form a path of type (2).   □

The second part of the above proof indicates a polynomial time algorithm for checking the popularity of a given b-matching. In case of popular matchings (i.e. 1-matchings) the algorithm runs in time linear in the number of edges.

Next we give a characterisation of clan-popular b-matchings.

Theorem 2. *A b-matching M is clan-popular iff graph G does not contain a path of type (3) or (4) from Theorem 1 or a path of type (1'), which is any path $(a_1, \ldots, h_{k-1}, a_k, h_k, a_{k+1})$, where (a)   the path is alternating, (b)   path $(a_1, \ldots, h_{k-1})$ is smooth, (c) $r((h_{k-1}, a_k)) > r((a_k, h_k))$ and (d)   if $a_{k+1} = a_1$, then $r((h_k, a_{k+1})) > r((a_1, h_1))$.*

The proof is similar to that of Theorem 1.

Notice that a popular b-matching need not be clan-popular.

3   Polynomial Algorithms for Clan-Popular b-matchings

First we state the following NP-hardness results.

Theorem 3. *The problem of deciding whether a given triple (G, b, r) has a popular b-matching is NP-hard, even if all edges are of one of two ranks, each agent $a \in A$ has capacity at most 2 and each house $h \in H$ has capacity 1.*

Theorem 4. *The problem of deciding whether a given triple (G, b, r) admits a clan-popular b-matching is NP-hard, even if all edges are of one of three ranks, each agent $a \in A$ has capacity at most 3, each house $h \in H$ has capacity 1 and there are no ties.*

The proofs are omitted due to space constraints.

We will now give an algorithm for the version in which there are no ties, each agent uses two ranks and has capacity 2, and houses have arbitrary capacities. (This algorithm can be easily extended to the one where we allow agents to have capacity 1 and use only one rank.) The underlying graph is $J = (A \cup H, E)$. Without loss of generality we can assume that agents use rank 1 and rank 2 edges.

Let J_1 denote $(A \cup H, E_1)$ and $b_1 : A \cup H \to N$ be defined as $b_1(a) = 1$ for $a \in A$ and $b_1(h) = b(h)$. Let $J_2 = (A \cup H, E_2)$ and b_2 be defined as $b_2(a) = 1$ for $a \in A$ and $b_2(h) = b(h) - deg_{E_1}(h)$ for $h \in H$.

First we observe the following.

Lemma 1. *If J contains a clan-popular b-matching M, then J contains a clan-popular b-matching M' that contains some maximum b_1-matching of J_1.*

Proof. Assume, that a clan-popular b-matching M restricted to rank 1 edges (called M_1) is not a maximum b_1-matching of J_1. Then in J_1 there exists an M_1-augmenting path, which must be of the form (a, h), where a is not matched in M_1. Since M is clan-popular, h is saturated in M. Therefore there exists a' such that $(a', h) \in M$ and $(a', h) \in E_2$. It is not difficult to see that $M \setminus \{(a', h)\} \cup \{(a, h)\}$ must also be a clan-popular b-matching of J. We can proceed in this way until we have a clan-popular b-matching of J that contains a maximum b_1-matching of J_1. □

From Lemma 1 we know that if J contains a clan-popular b-matching, then there exists a clan-popular b-matching $M = M_1 \cup M_2$ of J such that M_1 is a maximum b_1-matching of J_1 and M_2 is a maximum b_2-matching of J_2.

The key property of clan-popular b-matchings is stated in

Lemma 2. *If J has a clan-popular b-matching, then there exists such a clan-popular b-matching $M = M_1 \cup M_2$ that M_1 is a maximum b_1-matching of J_1 and M_2 is a b_2-matching of J_2 and M has the following property. Let $h \in H$ be such that $deg_{E_2}(h) > b_2(h)$ and $N_2(h) = \{a \in A : (a, h) \in E_2\}$. Then if $a \in N_2(h)$ is matched in M_2, then a is also matched in M_1.*

Proof. The first part of the lemma follows from Lemma 1. Suppose that some $a \in N_2(h)$ is matched in M_2 and unmatched in M_1. Since $deg_{E_2}(h) > b_2(h)$, there exists some $a' \in N_2(h)$ that is unmatched in M_2. Let h', a'' be such that $(a, h') \in E_1$ and $(h', a'') \in M_1$ (such (h', a'') exists because M_1 has maximum cardinality.) If $a' \neq a''$, then (a', h, a, h', a'') forms a path of type $(1')$ and thus M is not clan-popular. Therefore $a' = a''$. But we can remove from M edge (a', h') and add (a, h') and obtain another clan-popular b-matching of J. This way we diminish the number of agent vertices in $N_2(h)$, who are matched in M_2 but not in M_1. Proceeding in this way we obtain the wanted b-matching. □

We are going to use the algorithm for the following problem. In the **two level-matching problem** we are given a graph G consisting of two bipartite graphs $G_1 = (A \cup B_1, E_1), G_2 = (A \cup B_2, E_2)$, where sets B_1, B_2 are disjoint. We want to find M_1, M_2, where M_1 is a matching of G_1, M_2 is a matching of G_2 such that if $a \in A$ is matched in M_2, it is also matched in M_1. Such two matchings M_1, M_2 are called a **two-level matching** of G. In the **maximum two-level matching problem** we want to find a two-level matching of maximum cardinality, i.e. such one that $|M_1 \cup M_2|$ is maximised. The algorithm for computing a maximum two-level matching is given in [18].

Fact 1. *If M_1, M_2 is a maximum two-level matching of G, then M_1 is a maximum cardinality matching of G_1.*

(The above algorithm extends in an obvious way to b-matchings i.e. to the version in which each vertex of B_1, B_2 can be incident with more than one edge.)

Because of Lemma 1 we can assume that J_1, J_2 are such that $J_1 = (A \cup H_1, E_1), J_2 = (A \cup H_2, E_2)$ and H_1, H_2 are disjoint. (We split each $h \in H$ into h_1 and h_2, h_1 goes to H_1, h_2 goes to H_2 and $b(h_1) = b_1(h)$, $b(h_2) = b_2(h)$.)

Let J_2^+ be a subgraph of J_2 induced on vertices $A^+ \cup H_2^+$, where $h \in H_2^+$ iff $deg_{E_2}(h) > b_2(h)$ and $A^+ = N(H_2^+)$ contains neighbours of H_2^+ in J_2.

Theorem 5. *A b-matching of J of the form $M = M' \cup M''$, where M'' consists of all edges of E_2 incident with $A \setminus A^+$ and M' is a maximum two-level matching of J_1, J_2^+ such that each $h \in H_2^+$ is saturated in M' is clan-popular. Moreover, if J contains a clan-popular b-matching, then it contains one of the above form.*

Proof. Suppose that $M = M' \cup M''$ is such that M'' consists of all edges of E_2 incident with $A \setminus A^+$ and M' is a maximum two-level matching of J_1, J_2^+ such that each $h \in H_2^+$ is saturated in M'. We prove that M is a clan-popular b-matching of J.

First we show that J does not contain a path of type $(1')$. Suppose to the contrary that J contains some path $P = (a_1, h_1, a_2, h_2, a_3)$ of type $(1')$. Then edges $(a_1, h_1), (h_1, a_2)$ must be of rank 2 and edges $(a_2, h_2), (h_2, a_3)$ of rank 1. Therefore a_1 is unsaturated in M. Also $a_1 \neq a_2$ (since a_2 has a rank 2 edge incident with him/her and a_1 does not). Since $a_1 \neq a_2$ and h_1 is saturated (otherwise P would be a path of type (3)), we get that $a_1, a_2 \in A^+$. However from the fact that M' is a two-level matching we know that since a_2 has a rank 2 edge incident with him/her, he/she has also a rank 1 edge incident with him/her. Therefore no path of type $(1')$ occurs in the graph.

J cannot contain a path of type (3) because each $h \in H_2^+$ is saturated in M' and a path of type (3) ending on some $h \in (H_1 \cup H_2) \setminus H_2^+$ would have to be contained in J_1. This would mean that $M' \cap E_1$ is not a maximum cardinality b_1-matching of J_1 and this that M' is not a maximum two-level matching of J.

Because there are no ties, a path of type (4) does not exist in the graph either.

For the other part of the lemma, suppose J contains a clan-popular b-matching M. Let M'' denote all edges of E_2 incident with $A \setminus A^+$ and let M' denote $M \setminus M''$. If some $h \in H_2^+$ is unsaturated in M' or an edge of E_2 incident with some vertex of $A \setminus A^+$ does not belong to M, then J contains a path of type (3) beginning with some unsaturated $a \in A$ and M is not clan-popular. By Lemma 2 we can assume that M' is a two-level matching. If M' were not a maximum two-level matching, it would mean that $M' \cap E_1$ is not a maximum b_1-matching of J_1. It is so because of the following. Since each $h \in H_2^+$ is saturated in M', we cannot increase the number of E_2-edges in M' and thus if M' is not a maximum two-level matching, then we can increase the number of E_1-edges in M'. If $M' \cap E_1$ is not a maximum b_1-matching of J_1, then M is not a clan-popular b-matching of J, because J contains a path of type (3). $\qquad \square$

Algorithm CPBM *(short for Clan-Popular b-Matching)*

Input: graph J consisting of two bipartite graphs $J_1 = (A \cup H_1, E_1), J_2 = (A \cup H_2, E_2)$ such that H_1, H_2 are disjoint, function $b : A \cup H_1 \cup H_2 \to N$

Output: a clan-popular b-matching M of J or a report that it does not exist

Compute a maximum two-level b-matching M' of J_1, J_2^+.
If M' does not saturate each $h \in H_2^+$, write "does not exist" and halt.
Otherwise let M'' denote all edges of E_2 incident with $A \setminus A^+$.
Output $M = M' \cup M''$.

Theorem 6. *Algorithm CPBM solves the clan-popular b-matching problem for the cases when each agent has capacity 2 and uses two ranks, houses have arbitrary capacities and there are no ties.*

Proof. If the algorithm outputs a b-matching M, by Theorem 5 we know that it is clan-popular. If the algorithm does not output a b-matching, it is because a computed maximum two-level b-matching M' does not saturate all houses in H_2^+. Let M_1 denote $M' \cap E_1$. By Fact 1 M_1 is a maximum cardinality b_1-matching of J_1. A maximum cardinality two-level matching has cardinality $|M_1|$ plus the number of edges incident with H_2^+. Therefore if a maximum cardinality two-level matching does not saturate each $h \in H_2^+$, then there does not exist a two-level matching which does. Hence by Theorem 5 a clan-popular b-matching does not exist in J.   □

Algorithm CPBM runs in $O(\sqrt{n}n)$ time, where n is the number of vertices in the graph. (The number of edges is $O(n)$ since there are no ties.)

In the extended version of the paper we present also the algorithm for the setting in which ties among rank 2 edges are allowed.

References

1. Abraham, D.J., Irving, R.W., Kavitha, T., Mehlhorn, K.: Popular matchings. SIAM Journal on Computing 37, 1030–1045 (2007)
2. Abraham, D.J., Kavitha, T.: Voting Paths. SIAM J. Discrete Math. 24(2), 520–537 (2010)
3. Biró, P., Irving, R.W., Manlove, D.F.: Popular Matchings in the Marriage and Roommates Problems. In: Calamoneri, T., Diaz, J. (eds.) CIAC 2010. LNCS, vol. 6078, pp. 97–108. Springer, Heidelberg (2010)
4. Gärdenfors, P.: Match making: Assignments based on bilateral preferences. Behavioural Sciences 20, 166–173 (1975)
5. Cornuejols, G.: General factors of graphs. J. Comb. Theory, Ser. B 45(2), 185–198 (1988)
6. Dyer, M.E., Friese, A.M.: Planar 3DM is NP-complete. J. Algorithms 7, 174–184 (1986)

7. Huang, C.-C., Kavitha, T., Michail, D., Nasre, M.: Bounded Unpopularity Matchings. In: Gudmundsson, J. (ed.) SWAT 2008. LNCS, vol. 5124, pp. 127–137. Springer, Heidelberg (2008)
8. Irving, R.W., Kavitha, T., Mehlhorn, K., Michail, D., Paluch, K.: Rank-maximal matchings. ACM Transactions on Algorithms 2(4), 602–610 (2006)
9. Kavitha, T., Nasre, M.: Optimal popular matchings. Discrete Applied Mathematics 157(14), 3181–3186 (2009)
10. Kavitha, T., Shah, C.D.: Efficient Algorithms for Weighted Rank-Maximal Matchings and Related Problems. In: Asano, T. (ed.) ISAAC 2006. LNCS, vol. 4288, pp. 153–162. Springer, Heidelberg (2006)
11. Kavitha, T., Mestre, J., Nasre, M.: Popular Mixed Matchings. In: Albers, S., Marchetti-Spaccamela, A., Matias, Y., Nikoletseas, S., Thomas, W. (eds.) ICALP 2009, Part I. LNCS, vol. 5555, pp. 574–584. Springer, Heidelberg (2009)
12. Lovasz, L., Plummer, M.D.: Matching Theory. Ann. Discrete Math., vol. 29. North-Holland, Amsterdam (1986)
13. Mahdian, M.: Random popular matchings. In: Proceedings of EC 2006: the 7th ACM Conference on Electronic Commerce, pp. 238–242. ACM (2006)
14. Manlove, D.F., Sng, C.T.S.: Popular Matchings in the Capacitated House Allocation Problem. In: Azar, Y., Erlebach, T. (eds.) ESA 2006. LNCS, vol. 4168, pp. 492–503. Springer, Heidelberg (2006)
15. McCutchen, R.M.: The Least-Unpopularity-Factor and Least-Unpopularity-Margin Criteria for Matching Problems with One-Sided Preferences. In: Laber, E.S., Bornstein, C., Nogueira, L.T., Faria, L. (eds.) LATIN 2008. LNCS, vol. 4957, pp. 593–604. Springer, Heidelberg (2008)
16. McDermid, E., Irving, R.W.: Popular Matchings: Structure and Algorithms. In: Ngo, H.Q. (ed.) COCOON 2009. LNCS, vol. 5609, pp. 506–515. Springer, Heidelberg (2009)
17. Mestre, J.: Weighted Popular Matchings. In: Bugliesi, M., Preneel, B., Sassone, V., Wegener, I. (eds.) ICALP 2006, Part I. LNCS, vol. 4051, pp. 715–726. Springer, Heidelberg (2006)
18. Paluch, K.: Balanced Matchings, Unbalanced Ones and Related Problems (2011) (manuscript)
19. Sng, C.T.S., Manlove, D.F.: Popular matchings in the weighted capacitated house allocation problem. J. Discrete Algorithms 8(2), 102–116 (2010)

Kernelization and Parameterized Complexity of Star Editing and Union Editing

Jiong Guo* and Yash Raj Shrestha**

Universität des Saarlandes,
Campus E 1.7, D-66123 Saarbrücken, Germany
{jguo,yashraj}@mmci.uni-saarland.de

Abstract. The NP-hard STAR EDITING problem has as input a graph $G = (V, E)$ with edges colored red and black and two positive integers k_1 and k_2, and determines whether one can recolor at most k_1 black edges to red and at most k_2 red edges to black, such that the resulting graph has an induced subgraph whose edge set is exactly the set of black edges. A generalization of STAR EDITING is UNION EDITING, which, given a hypergraph H with the vertices colored by red and black and two positive integers k_1 and k_2, determines whether one can recolor at most k_1 black vertices to red and at most k_2 red vertices to black, such that the set of red vertices becomes exactly the union of some hyperedges. STAR EDITING is equivalent to UNION EDITING when the maximum degree of H is bounded by 2. Both problems are NP-hard and have applications in chemical analytics. Damaschke and Molokov [WADS 2011] introduced another version of STAR EDITING, which has only one integer k in the input and asks for a solution of totally at most k recolorings, and proposed an $O(k^3)$-edge kernel for this new version. We improve this bound to $O(k^2)$ and show that the $O(k^2)$-bound is basically tight. Moreover, we also derive a kernel with $O((k_1 + k_2)^2)$ edges for STAR EDITING. Fixed-parameter intractability results are achieved for STAR EDITING parameterized by any one of k_1 and k_2. Finally, we extend and complete the parameterized complexity picture of UNION EDITING parameterized by $k_1 + k_2$.

1 Introduction

In this paper we study the STAR EDITING problem and its general cases. The problem is defined as follows.

Problem: STAR EDITING(SE)

Input: A graph $G = (V, E)$ with edges colored red and black and two positive integers k_1 and k_2.

Parameter: $k_1 + k_2$

Question: Can we recolor at most k_1 black edges to red and at most k_2 red edges to black such that the resulting graph has an induced subgraph whose edge set is exactly the set of black edges?

---

* Supported by the DFG Excellence Cluster MMCI.
** Supported by the DFG research project DARE GU 1023/1.

K.-M. Chao, T.-s. Hsu, and D.-T. Lee (Eds.): ISAAC 2012, LNCS 7676, pp. 126–135, 2012.
© Springer-Verlag Berlin Heidelberg 2012

A generalization of STAR EDITING is UNION EDITING which can be defined as follows:

Problem: UNION EDITING(UE)

Input: A hypergraph $H = (V, S)$ whose vertices are colored red and black, and two positive integers k_1 and k_2

Parameter: $k_1 + k_2$

Question: Can we recolor at most k_1 black vertices to red and at most k_2 red vertices to black such that the set of red vertices becomes exactly the union of some hyperedges in S?

STAR EDITING and UNION EDITING arise from chemical analytics; for more details of the applications we refer to [6,2]. STAR EDITING is equivalent to UNION EDITING when each vertex in V appears in at most 2 hyperedges [2]. Another equivalent formulation of the question of STAR EDITING concerning the red edges is as follows: We say that there is a *red star* at a vertex v, if all edges incident to v are red. Then, the STAR EDITING problem asks for recoloring at most k_1 black edges and at most k_2 red edges such that in the resulting graph, all red edges are contained in red stars. Our algorithms are mainly based on this red star formulation.

Damaschke and Molokov [1,2] initialized the study of the computational complexity of STAR EDITING and UNION EDITING. They proved that both problems are NP-hard and revealed the relation between STAR EDITING and some generalizations of VERTEX COVER. Moreover, they also studied STAR EDITING from the viewpoint of parameterized algorithms. However, the algorithms given in [1,2] are not for STAR EDITING, but for a slightly modified version of STAR EDITING, which we call STAR EDITING WITH TOTAL RECOLORING defined as follows:

Problem: STAR EDITING WITH TOTAL RECOLORING (SETR)

Input: A graph $G = (V, E)$ with edges colored red and black and a positive integer k.

Parameter: k

Question: Can we recolor at most k edges such that all red edges are contained in red stars?

With k as parameter, Damaschke and Molokov [1,2] showed that SETR can be solved in $O^*(1.84^k)$ time[1] by a bounded search tree algorithm. Moreover, a problem kernel with $O(k^3)$ edges is proposed for SETR [1]. However, these algorithms do not work for STAR EDITING [1,2]; for instance, the search tree algorithm could end up with a solution with at most $k_1 + k_2$ total recolorings which recolors more than k_1 black edges.

We first improve the problem kernel for SETR to $O(k^2)$ edges and prove that this bound is basically tight. Moreover, we study STAR EDITING and derive a problem kernel with $O((k_1 + k_2)^2)$ edges. We then present intractability results for STAR

---

[1] $O^*()$ notation hides factors that are polynomial in the input size.

EDITING: STAR EDITING remains NP-hard even with $k_2 = 0$; this means the parameterization with k_2 alone is not in XP. With only k_1 as parameter, this problem becomes W[1]-hard. Finally, we extend and complete the parameterized complexity picture of UNION EDITING. If the maximum degree of the hypergraph H is not bounded, UNION EDITING becomes W[2]-hard with $k_1 + k_2$ as parameter, even with all hyperedges containing at most 2 vertices. Moreover, if the maximum degree of the hypergraph H is bounded, UNION EDITING becomes fixed-parameter tractable with $k_1 + k_2$ as parameter.

2   Preliminaries

Parameterized complexity theory is a two-dimensional framework for studying the computational complexity of problems [3,7]. A core tool in the development of fixed-parameter algorithms is polynomial-time preprocessing by *data reduction rules*, often yielding a reduction to a *problem kernel (kernelization)*. Herein, the goal is, given any problem instance x with parameter k, to transform it in polynomial time into a new instance x' with parameter k' such that the size of x' is bounded from above by some function only depending on k, $k' \leq k$, and (x, k) is a yes-instance iff (x', k') is a yes-instance.

We mostly deal with vertex-colored or edge-colored, simple, and undirected graphs. Given a graph G with edges colored red and black, let E_b denote the set of black edges and E_r denote the set of red edges. We use $\deg_{b,G}(v)$ and $\deg_{r,G}(v)$ to denote the numbers of black and red edges in G incident to v, respectively. Then, the degree $\deg_G(v)$ of v is equal to $\deg_{b,G}(v) + \deg_{r,G}(v)$. We say that a vertex v is degree-1 if $\deg_G(v) = 1$. An edge is called a degree-1 edge if it is incident to at least one degree-1 vertex. For a subset of vertices, the subgraph of G induced by V', denoted by $G[V']$, has the vertex set V' and the edge set $\{(u,v) \in E \mid u, v \in V'\}$. *Breaking* a black edge (u,v) means to delete the edge (u,v) and to add two degree-1 vertices u' and v' and two black edges (u, u') and (v, v'). If all edges incident to a vertex v are black (or red), we say that there is a *black (or red) star* at v. We *break* the black star at v, if we delete v and add to each of v's neighbours a degree-1 neighbour with a black edge between them. Given a set S of recolorings, we use $G \odot S$ to denote the graph resulting by applying S to G. We say S creates a *necessary* red star at $v \in V$, if in $G \odot S$, there is a red star at v and there is a red edge which is contained only in the red star at v.

3   Improved Kernel for STAR EDITING WITH TOTAL RECOLORING

Damaschke and Molokov [1] proved that STAR EDITING WITH TOTAL RECOLORING with the total number of recolorings k as parameter has a problem kernel with $O(k^3)$ edges. In this section we improve this bound to $O(k^2)$. We derive a completely different approach than in [1]. We apply the following six rules to an instance $(G = (V, E), k)$ in the given order.

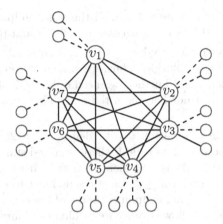

Fig. 1. A subset of vertices to which Rule 6 can be applied (black edges are denoted by solid lines and red edges are denoted by dashed lines)

Rule 1. If there is a red star at $v \in V$, then remove v.

Rule 2. Remove all black edges which are not adjacent to any red edge.

Rule 3. Break all black stars.

Rule 4. If there is a $v \in V$ with $\deg_{b,G}(v) > k$, then break all black edges incident to v and remove the newly added degree-1 black edges until only $k + 1$ remain.

Rule 5. If there is a $v \in V$ such that $\deg_{r,G}(v) > 2k$, then remove v and decrease k by $\deg_{b,G}(v)$.

Rule 6. If there is a non-empty set $V' \subseteq V$ such that $\deg_{b,G[V']}(v) \geq 2\deg_{r,G}(v)$ for all $v \in V'$, then break all black edges in $G[V']$ as shown in Fig 1.

The correctness of Rules 1 and 2 is obvious, as edges in red stars and black edges which are not incident to any red edges will never be recolored by any optimal solution. Clearly, these two rules can be applied in $O(|E| + |V|)$ time. To prove the correctness of other rules, we need the following lemmas:

Lemma 1. [*][2] *Let (u, v) be a black edge in G. If there exists an optimal solution S for G which does not create a necessary red star at v, then we can safely break (u, v). That is, G has a solution of size at most k iff G' has such a solution, where G' is the graph resulting by breaking (u, v).*

Lemma 2. *Rules 3, 4, and 5 are correct and executable in $O(|E| + |V|)$ time.*

Proof. **Rule 3.** Breaking a black star at a vertex v can be seen as breaking all incident black edges to v and then removing the isolated black star at v. Then, by Lemma 1, it suffices to prove that there is an optimal solution S that does not create a necessary red star at v. If an optimal solution S creates a red star at v, then it has to recolor all edges incident to v to red. The only reason for S

---

[2] Due to lack of space proofs of results marked [*] are given in the full version of the paper.

to recolor these edges is to create red stars at the other endpoints of these edges. Therefore, the red star at v is not a necessary red star and Rule 3 is correct.

Rule 4. If there is a vertex v with $\deg_{b,G}(v) > k$, then it will never be converted to a red star by a size-$\leq k$ solution. Thus, we can safely break the black edges incident to v due to Lemma 1. Further, we need only $k+1$ black edges incident to v to enforce that v will not become a red star. Thus, other black edges can be safely removed.

Rule 5. Suppose there is a vertex v with $\deg_{r,G}(v) > 2k$. Since we already removed all red stars according to Rule 1, there is no red star at v. To satisfy the required condition for the red edges incident to v, we either need to convert v into a red star or we need to recolor $k' \leq k$ red edges incident to v and recolor all the black edges incident to the other endpoints than v of the remaining $\deg_{r,G}(v) - k'$ red edges. The latter is not allowed, since we would need more than k recolorings in this case. Hence, the only feasible way is to recolor the black edges incident to v and decrease k by $\deg_{b,G}(v)$.

It is easy to see that all these rules can be applied in $O(|E| + |V|)$ time. □

Lemma 3. *If there is a non-empty set $V' \subseteq V$ such that $\deg_{b,G[V']}(v) \geq 2\deg_{r,G}(v)$ for all $v \in V'$, then it is safe to break all black edges in $G[V']$. We can identify such a set V' and break all black edges in $G[V']$ in $O(|E| \cdot |V|)$ time.*

Proof. Suppose that there is a set of vertices $V' = \{v_1, v_2, \ldots, v_m\}$ such that for each $v_i \in V'$, $\deg_{b,G[V']}(v_i) \geq 2\deg_{r,G}(v_i)$. By Lemma 1, it suffices to show that there exists an optimal solution S that creates no necessary red star at vertices in V'. Let S be an arbitrary optimal solution which creates some necessary red stars at vertices $v_1, v_2, \ldots, v_t \in V'$. Then S needs to recolor at least $\sum_{i=1}^{t} \deg_{b,G[V']}(v_i)/2$ many black edges. Let X denote the set of the red edges in G which are contained only in the necessary red stars at v_i's. Clearly $|X| \leq \sum_{i=1}^{t} \deg_{r,G}(v_i)$. Then we can construct another optimal solution S' from S by recoloring the red edges in X and keeping the edges in $G[V']$ black. For each of the red edges which are not in X, there is a red star in $G \odot S$ at a vertex outside of V' containing this red edge. Since the red stars outside of V' cannot be affected by recoloring edges in $G[V']$, we have S' as a solution. Obviously, $|S'| \leq |S|$.

Now, we present an algorithm which can find a vertex subset V' satisfying the condition of Rule 6. First, it computes $\deg_{r,G}(v)$ for each vertex v and stores the value in x_v. Then, the algorithm iteratively removes all vertices v having less than $2x_v$ many black edges incident to it in the current graph. Finally, it outputs the remaining vertices as V'. Note that x_v's remain unchanged during the whole iteration but after some iterations, some vertices may have less incident black edges than in G. On the one side, the output of the algorithm clearly satisfies the condition of Rule 6. On the other side, if there exists some non-empty vertex subset satisfying the condition of Rule 6, then the algorithm can find one. To show this, observe that, if a vertex subset V' satisfies the condition of Rule 6, then the vertices v in V' have $\deg_{b,G[V'']}(v) \geq 2\deg_{r,G}(v)$ for all sets V'' with $V' \subseteq V''$. This implies that, if there are two vertex subsets V' and V'' that both satisfy

the condition of Rule 6, then $V' \cup V''$ satisfies the condition of Rule 6 as well. In other words, there is exactly one maximal vertex subset V' satisfying the condition of Rule 6 (V' could be empty). Let V' be the maximal subset of the input graph G. The algorithm starts with the whole vertex set V and iteratively removes vertices. And only those vertices are removed, which cannot satisfy the condition of Rule 6 in the supersets of V'. Thus, the algorithm will never remove vertices in V'. Due to the uniqueness of V', the algorithm terminates with the output V'. Finally, it is easy to see that this algorithm runs in $O(|E| \cdot |V|)$ time
□

From the algorithm in the proof of Lemma 3, we observe the following lemma, which is crucial for upper-bounding the number of black edges in the problem kernel.

Lemma 4. *If $|E_b| \geq 4|E_r|$, then there exists a non-empty set $V' \subseteq V$ such that $deg_{b,G[V']}(v) \geq 2deg_{r,G}(v)$ for all $v \in V'$.*

Proof. Each iteration of the algorithm in the proof of Lemma 3 removes a vertex v which has less than $2deg_{r,G}(v)$ incident black edges in the current graph. By deleting this vertex, we remove also the incident black edges. Then the total number of black edges removed in all iterations is less than $2 \cdot \sum_{v \in V \setminus V'} deg_{r,G}(v)$, where V' is the set of remaining vertices after all iterations. If G has no non-empty subset V' satisfying the condition of Rule 6, then, by the proof of Lemma 3, the algorithm will remove all vertices, i.e., $|V'| = 0$, and then, G has less than $2 \sum_{v \in V} deg_{r,G}(v) = 4|E_r|$ black edges, contradicting $|E_b| \geq 4|E_r|$. □

In the following, we call an instance *reduced* if none of the above rules can be applied.

Theorem 1. STAR EDITING WITH TOTAL RECOLORING *has a problem kernel with $O(k^2)$ edges parameterized by the total number k of allowed recolorings.*

Proof. If a reduced instance $G = (V, E)$ has a solution S with at most k recolorings, then let $V(S)$ be the set of vertices that are incident to the recolored edges. Let V_1 be the set of vertices in $V \setminus V(S)$ that are adjacent to $V(S)$ and $V_2 := V \setminus (V(S) \cup V_1)$. By Lemma 4, it suffices to show that the number of red edges in G can be upper-bounded by $O(k^2)$. It is clear that all red edges in $G \odot S$ will have at least one of its endpoints in $V(S)$, since all red stars in G have been removed by Rule 1. Clearly, $|V(S)| \leq 2k$ and there are $O(k^2)$ edges in $G[V(S)]$. Next consider the red edges between $V(S)$ and V_1. Rule 5 implies that there exists no vertex v such that $deg_{r,G}(v) \geq 2k$. Hence, we have at most $4k^2$ red edges between $V(S)$ and V_1. This gives a bound of $O(k^2)$ on the number of red edges and completes the proof. □

4 Kernelization for STAR EDITING

In this section we derive a quadratic kernel for STAR EDITING with the numbers of recolorings k_1 and k_2 as parameters. The kernelization in Section 3 cannot be applied directly to STAR EDITING, since the correctness of some rules and the proof of the kernel size are heavily based on the argument that in order to create red stars for l red edges, we never recolor more than l black edges. However, this argument is no longer true for STAR EDITING as we have two separate bounds on the two types of recolorings. The first three rules from Section 3 can be directly applied here. The other three rules need some modifications. Additionally, we introduce a new reduction rule. To present the rules, we need some new notations. We define $N_{r,G}(v) := \{x \mid (x, v) \in E_r\}$ and $N_{b,G}(v) := \{x \mid (x, v) \in E_b\}$ as sets of red and black neighbours of v respectively. Moreover, we define $E_{r,G}(v) := \{(x, y) \in E_b \mid (v, x) \in E_r \text{ or } (v, y) \in E_r\}$. Let $N^1_{b,G}(v) = \{x \mid x \in N_{b,G}(v) \text{ and } \deg_G(x) = 1\}$ denote th set of degree-1 black neighbours of v and $\deg^1_{b,G}(v) := |N^1_{b,G}(v)|$. We have the following seven rules which must be applied in the given order:

Rule 1. If there is a red star at v, then remove v.

Rule 2. Remove all black edges which are not adjacent to any red edge.

Rule 3. Break all black stars.

Rule 4. If there is a $v \in V$ with $\deg_{b,G}(v) > k_1$, then break all black edges incident to v and remove the newly added degree-1 black edges until only $k_1 + 1$ remain.

Rule 5. If there is a $v \in V$ with $\deg_{r,G}(v) > 2k_1 + k_2$, then remove v and reduce k_1 by $\deg_{b,G}(v)$.

Rule 6. If there is a non-empty vertex set $V' \subseteq V$ such that for each vertex $v \in V'$, it holds that $\deg_{b,G[V']}(v) \geq 2|E_{r,G}(v)|$, then break all black edges in $G[V']$.

Rule 7. If there is a vertex v with $\deg^1_{b,G}(v) > |E_{r,G}(v)| + 1$, then keep arbitrary $|E_{r,G}(v)| + 1$ degree-1 black edges incident to v and remove the others.

The first two rules are clearly correct and run in linear time. It is easy to verify that Lemma 1 holds also for STAR EDITING. Therefore, the correctness of Rules 3 and 4 follows directly from the correctness proofs in Section 3. The following three lemmas prove the correctness of Rules 5 to 7.

Lemma 5. [∗] *Rule 5 is correct and executable in $O(|E| + |V|)$ time.*

Lemma 6. [∗] *If there is a non-empty vertex set $V' \subseteq V$ such that for each vertex $v \in V'$, it holds that $\deg_{b,G[V']}(v) \geq 2|E_{r,G}(v)|$, then it is safe to break all black edges in $G[V']$. We can identify such V' and break all the black edges in $G[V']$ in $O(|E| \cdot |V|)$ time.*

Lemma 7. [∗] *Rule 7 is correct and can be applied in $O(|E| \cdot |V|)$ time.*

Next we prove the problem kernel.

Theorem 2. STAR EDITING *has a problem kernel with $O((k_1 + k_2)^2)$ edges, parameterized by $k_1 + k_2$ where k_1 and k_2 denote the number bounds for the recolorings of black and red edges, respectively.*

Proof. Again, we use $V(S)$ to denote the set of endpoints of the edges recolored by a solution S satisfying both bounds k_1 and k_2. The set V_1 contains all vertices in $V \setminus V(S)$ that are adjacent to $V(S)$ and $V_2 := V \setminus (V(S) \cup V_1)$. We bound the numbers of red and black edges in the reduced instance, separately. By Rule 1, red edges can only be in $G[V(S)]$ or between $V(S)$ and V_1. Clearly, as in Theorem 1, there are totally $O((k_1 + k_2)^2)$ edges in $G[V(S)]$. Moreover, We know that $|V(S)| \leq 2 \cdot (k_1 + k_2)$ and in particular, there are at most $2k_1$ vertices incident to recolored black edges. Rule 5 implies that there exists no vertex v such that $\deg_{r,G}(v) > 2k_1 + k_2$. Hence, since each red edge between $V(S)$ and V_1 must be adjacent to at least one recolored black edge, we have at most $2 \cdot k_1 \cdot (2k_1 + k_2) = O((k_1 + k_2)^2)$ red edges between S and V_1.

We partition the black edges into four different subsets and bound their sizes individually. As already observed, the number of black edges in $G[V']$ is clearly bounded by $O((k_1 + k_2)^2)$. Consider the black edges between $V(S)$ and V_1. We know that $|V(S)| \leq 2 \cdot (k_1 + k_2)$. Rule 4 implies that there exists no vertex v such that $\deg_{b,G}(v) > k_1 + 1$. Hence, we have at most $2 \cdot (k_1 + k_2) \cdot (k_1 + 1)$ black edges between $V(S)$ and V_1. Due to Rules 1 to 3, there is no edge in $G[V_2]$, all vertices in V_2 are degree-1 vertices, and they are directly connected to V_1 by black edges. Therefore, the remaining black edges are either between V_1 and V_2 or in $G[V_1]$. Consider the black edges between V_2 and V_1 which are all degree-1 edges. By Rule 7, we know $N_{b,G}^1(v) \leq |E_{r,G}(v)| + 1$ for each vertex $v \in V_1$. Thus, the number of the black edges between V_1 and V_2 is bounded by $\sum_{v \in V_1}(|E_{r,G}(v)| + 1)$. Moreover, since $v \in V_1$, all black edges in $E_{r,G}(v)$ have been recolored by S. A recolored black edge can be in $E_{r,G}(v)$ of at most $2(2k_1 + k_2)$ vertices $v \in V_1$, since each of its two endpoints can be adjacent to at most $2k_1 + k_2$ many red edges between $V(S)$ and V_1. Therefore, $\sum_{v \in V_1}(|E_{r,G}(v)| + 1) \leq 4k_1(2k_1 + k_2)$ and thus, there are $O((k_1 + k_2)^2)$ black edges between V_1 and V_2. Finally, we bound the number of the black edges in $G[V_1]$. By Rule 6, there exists no set of vertices V' such that $\deg_{b,G[V']}(v) \geq 2|E_{r,G}(v)|$ for each vertex $v \in V'$. Then, there exists in V_1 a vertex v_1 with $\deg_{b,G[V_1]}(v_1) < 2|E_{r,G}(v_1)|$. Moreover, this arguments also applies for every subset of V_1. This means that there exists $v_2 \in V_1$ with $\deg_{b,G[V_1 \setminus \{v_1\}]}(v_2) \leq 2|E_{r,G}(v_2)|$ and so on, until all vertices in V_1 have been considered. Then, the number of edges in $G[V_1]$ is bounded by $\sum_{v \in V_1} 2|E_{r,G}(v)|$. As for the edges between V_1 and V_2, the black edges in $E_{r,G}(v)$ for all vertices $v \in V_1$ are recolored by S. And each recolored black edge can be in $E_{r,G}(v)$ for at most $2(2k_1 + k_2)$ many vertices $v \in V_1$. Therefore, $\sum_{v \in V_1} 2|E_{r,G}(v)| \leq 4k_1(2k_1 + k_2)$, which completes the proof of the bound on black edges. $\square$

5   Hardness Results for STAR EDITING

In this section we present several hardness results concerning the parameterized complexity of STAR EDITING. First, we show that STAR EDITING becomes fixed-parameterized intractable, parameterized by only one of k_1 and k_2. More precisely, we show that the parameterization with k_2 of STAR EDITING is not in XP and the one with k_1 is W[1]-hard.

Theorem 3. STAR EDITING *is NP-hard even when* $k_2 = 0$.

Proof. To this end, we reduce VERTEX COVER to STAR EDITING with $k_2 = 0$. VERTEX COVER (VC) asks for a given undirected graph $G = (V, E)$ and an integer $l \geq 0$, whether there is a subset $C \subseteq V$ with $|C| \leq l$ such that each edge $e \in E$ has at least one endpoint in C. VERTEX COVER is NP-complete [4]. For each vertex $v \in V$ of a given VC-instance $(G = (V, E), l)$, we create a black edge (v, v') in the corresponding STAR EDITING-instance G' and for each edge $(u, v) \in E$ we add two red edges (u, v) and (u', v'). Then, set $k_2 = 0$ and $k_1 = l$. We claim that G' has a solution recoloring at most k_1 black edges if and only if $G = (V, E)$ has a size-$\leq l$ vertex cover. Let $C \subseteq V$ be the solution of VERTEX COVER on $G = (V, E)$. It is clear to see that recoloring the black edges in G' corresponding to vertices in C gives a solution for STAR EDITING with $k_2 = 0$. For the reversed direction, STAR EDITING with $k_2 = 0$ is equivalent to finding the minimum-size subset of black edges that covers all red edges, which in turn is equivalent to finding a minimum vertex cover in the graph obtained by contracting all black edges, i.e., finding a minimum vertex cover in G. □

The reduction given in the proof of Theorem 3 implies also a lower bound of the kernel size for STAR EDITING WITH TOTAL RECOLORING.

Corollary 1. [∗] STAR EDITING WITH TOTAL RECOLORING *does not have kernels with* $O(k^{2-\epsilon})$ *edges for any* $\epsilon > 0$ *unless* coNP $\subseteq$ NP/poly, *where* k *denotes the number of allowed recolorings.*

Theorem 4. STAR EDITING *is W[1]-hard with respect to* k_1.

Proof. We prove the theorem by reducing from the PARTIAL VERTEX COVER (PVC) problem, which, given an undirected graph $G = (V, E)$ and two integers $l \geq 0$ and $t \geq 0$, asks for a subset $C \subseteq V$ such that $|C| \leq l$ and C covers at least t edges. PVC is W[1]-hard with l as parameter [5]. Given an input instance $(G = (V, E), l, t)$ of PVC, we construct an instance $G' = (V', E')$ for STAR EDITING as follows. First, set $G' := G$ and color all edges in G' with red. Then for each vertex v add a new degree-1 vertex v' with the black edge (v, v') in G'. Finally, set $k_1 = l$ and $k_2 = |E| - t$. It is obvious that a size-$\leq l$ vertex set in G which covers $\geq t$ edges one-to-one corresponds to a solution for the STAR EDITING instance. □

6   Union Editing

In this section, we consider the UNION EDITING problem and extend and complete its parameterized complexity picture. A hypergraph $H = (V, S)$ is a vertex set H equipped with a family S of subsets of vertices called hyperedges, i.e., $S = \{S_1, \ldots, S_m\}$. The degree $\deg(v)$ of a vertex v is the number of hyperedges it belongs to, and the degree $\deg(H)$ of a hypergraph is the maximum vertex degree. UNION EDITING with the maximum degree $\deg(H)$ bounded by 2 and $|S_i| \leq 2$ for all $1 \leq i \leq m$ is clearly solvable in polynomial time. STAR EDITING is the special case of UNION EDITING with $\deg(H) \leq 2$. First we consider UNION EDITING with bounded $|S_i|$ for all i's.

Theorem 5. [∗] *If deg(H) is unbounded, then* UNION EDITING *is W[2]-hard with respect to the total number $k_1 + k_2$ of recolorings even with all hyperedges satisfying $|S_i| \le c$ for a constant $c \ge 2$.*

Next, we consider UNION EDITING with bounded maximum degree deg(H).

Theorem 6. *If deg(H) is bounded by some constant c, then* UNION EDITING *with respect to the total number $k_1 + k_2$ of recolorings is fixed-parameter tractable.*

Proof. Given a hypergraph with its maximum degree bounded by a constant c, we can find the solution for UNION EDITING by a simple branching algorithm. Here, we use *red hyperedge* to denote a hyperedge which contains only red vertices. As long as there is a red vertex v that is not contained in any red hyperedge, branch into the following cases: The first case is that if $k_2 > 0$, then recolor v to black and decrease k_2 by one. Then for each hyperdege e containing v, recolor all black vertices in e and decrease k_1 by the number of such recolored black vertices as long as $k_1 \ge 0$. The algorithm is correct, since it considers all feasible possibilities creating red hyperedge containing a red vertex. The number of children of each node in the search tree is clearly bounded by $c + 1$. Moreover, the height of the tree is bounded by $k_1 + k_2$ as in each level we decrease it by at least one, and hence the search tree size is of $O^*((c + 1)^{(k_1 + k_2)})$. We can further observe that UNION EDITING remains fixed-parameter tractable when in addition to $k_1 + k_2$ the maximum degree of the hypergraph deg(H) is also taken as parameter. □

References

1. Damaschke, P., Molokov, L.: Parameterized reductions and algorithms for a graph editing problem that generalizes vertex cover (2011), Extended version of [2], http://www.cse.chalmers.se/~ptr/vcgj.pdf
2. Damaschke, P., Molokov, L.: Parameterized Reductions and Algorithms for Another Vertex Cover Generalization. In: Dehne, F., Iacono, J., Sack, J.-R. (eds.) WADS 2011. LNCS, vol. 6844, pp. 279–289. Springer, Heidelberg (2011)
3. Downey, R.G., Fellows, M.R.: Parameterized Complexity. Springer (1999)
4. Garey, M.R., Johnson, D.S.: Computers and Intractability, A Guide to the Theory of NP-Completeness. W.H. Freeman and Company, San Francisco (1979)
5. Guo, J., Niedermeier, R., Wernicke, S.: Parameterized complexity of vertex cover variants. Theory Comput. Syst. 41(3), 501–520 (2007)
6. Molokov, L.: Application of combinatorial methods to protein identification in peptide mass fingerprinting. In: KDIR, pp. 307–313. SciTePress (2010)
7. Niedermeier, R.: Invitation to Fixed-Parameter Algorithms. Oxford University Press (2006)

On the Advice Complexity of Buffer Management

Reza Dorrigiv*, Meng He**, and Norbert Zeh***

Faculty of Computer Science, Dalhousie University,
Halifax, NS, B3H 1W5, Canada
{rdorrigiv,mhe,nzeh}@cs.dal.ca

Abstract. We study the advice complexity of online buffer management. Advice complexity measures the amount of information about the future that an online algorithm needs to achieve optimality or a good competitive ratio. We study the 2-valued buffer management problem in both preemptive and nonpreemptive models and prove lower and upper bounds on the number of bits required by an optimal online algorithm in either model. We also provide results that shed light on the ineffectiveness of advice to improve the competitiveness of the best online algorithm for nonpreemptive buffer management.

1 Introduction

Buffer management is an important online problem with applications in network communication [13]. It models admission policies for the buffers of packet switches in networks that support the QoS (Quality of Service) feature. In this problem, packets have different *values* corresponding to their importance or priority, and the packet switch has a FIFO buffer of size B. Packets arrive at arbitrary times and are placed at the end of the buffer. In each time unit, the switch retrieves a packet from the front of the buffer and transmits it, unless the buffer is empty. Multiple packets may arrive in the same time unit. If the buffer is full when a packet arrives, the switch's buffer management algorithm must reject the packet. Otherwise it may choose to accept or reject the packet. The goal of the algorithm is to maximize the total value of the transmitted packets, which, in the absence of preemption, are exactly the packets it accepts. We consider the *2-valued model* here, where there are two types of packets: low-priority packets (L-packets) of value 1 and high-priority packets (H-packets) of value $\alpha > 1$. We use $p_1, p_2, \ldots, p_n$ to denote the sequence of packets, and we assume for simplicity that transmission occurs at integral times, no two packets arrive at the same time, and no packets arrive at integral times. Time starts at 0. For an integer $j \geq 0$, we define *time unit j* to be the time interval $[j, j+1)$.

---

* Research supported by an NSERC postdoctoral fellowship.
** Research supported by NSERC.
*** Research supported by NSERC and the Canada Research Chairs programme.

K.-M. Chao, T.-s. Hsu, and D.-T. Lee (Eds.): ISAAC 2012, LNCS 7676, pp. 136–145, 2012.

There are two main models for buffer management. In the *nonpreemptive model* all packets accepted into the buffer are eventually transmitted, that is, accepted packets cannot be dropped at a later time. In the *preemptive model*, accepted packets can be dropped as long as they have not been transmitted yet. Online buffer management algorithms are usually analyzed using competitive analysis [19]: Let OPT be an optimal offline buffer management algorithm, and $\mathcal{A}$ an online algorithm. We use $OPT(S)$ and $\mathcal{A}(S)$ to denote the solutions produced by these algorithms on an input sequence S, or the total values of the packets in these solutions. Which will be clear from the context. Algorithm $\mathcal{A}$ has *competitive ratio* c if $OPT(S) \leq c \cdot \mathcal{A}(S)$, for every request sequence S.

Competitive analysis of buffer management algorithms was initiated by Aiello et al. [1], Mansour et al. [17], and Kesselman et al. [16]. Since then, various algorithms have been proposed. We briefly review the results most relevant to our work and refer the reader to surveys by Azar [5], Epstein and van Stee [12], and Goldwasser [13] for a more comprehensive coverage. Aiello et al. [1] introduced the nonpreemptive 2-valued model and proved a lower bound of $(2 - \frac{1}{\alpha})$ on the competitiveness of any deterministic or randomized online algorithm in this model. The RATIO PARTITION algorithm by Andelman, Mansour, and Zhu [4] achieves this bound. This algorithm accepts each H-packet if possible and, for each accepted H-packet, marks the earliest $\frac{\alpha}{\alpha-1}$ unmarked L-packets in the buffer. It accepts an L-packet only if, after accepting it, the number of unmarked L-packets in the buffer is at most $\frac{\alpha}{\alpha-1}$ times the number of empty buffer slots. Mansour et al. introduced the preemptive model in the context of video streaming [17]. Kesselman et al. proved a lower bound of 1.282 on the competitiveness of any deterministic preemptive online algorithm [16]. The ACCOUNT STRATEGY algorithm by Englert and Westermann [4] achieves this bound.

These results completely characterize the competitiveness achievable for nonpreemptive or preemptive online buffer management without information about future requests. They do not, however, provide any insight into how much an algorithm may benefit from partial information about future requests that may be available in some applications. Various models have been proposed to facilitate the analysis of online algorithms that have access to such partial information, e.g., the finite lookahead model [15,14,2,9]. A more recent model, and the one we adopt in this paper, is *advice complexity* [10,11,7]. It measures the amount of information about the future that an online algorithm needs in order to achieve optimality or a certain competitive ratio. Two variants of this model have been proposed [11,7]. In the model by Böckenhauer et al. [7], the online algorithm $\mathcal{A}$ has access to a tape of advice bits produced by an oracle. The oracle has unlimited computational power and has access to the whole input. No restrictions are placed on how $\mathcal{A}$ uses the advice bits. The performance of $\mathcal{A}$ is expressed as a combination of its competitive ratio and the number of advice bits it uses on an input of size n. In the model by Emek et al. [11], the oracle provides a fixed number of advice bits with each request, and the algorithm has access only to the advice bits associated with the requests that have arrived so far. Previous work on advice complexity of online algorithms has focused on paging [10,11], ski rental [10], metrical task systems [11], the k-server problem [11,6,18], job shop scheduling and routing [7], and the knapsack problem [8]. To the best of our knowledge, the advice complexity of online buffer management has not been studied so far. This is the focus of this paper.

It is trivial even for a nonpreemptive buffer management algorithm to achieve optimality with one bit of advice per request: the oracle runs an offline optimal buffer management algorithm on the given request sequence and tells the online algorithm for each packet whether the optimal algorithm accepts or rejects this packet. Since one bit of advice per request is the minimum possible in the model of [11], online buffer management is not interesting in this model, and we adopt the model of [7]. Our main result is that $\Theta((n/B)\log B)$ bits of advice are necessary and sufficient for a preemptive or nonpreemptive online buffer management algorithm to produce an optimal solution. In this paper all logarithms are base 2. We also prove that a generalization of the RATIO PARTITION algorithm, which uses advice to choose the optimal ratio between unmarked L-packets and empty buffer slots, cannot outperform RATIO PARTITION without advice. We conjecture that an algorithm that chooses different ratios for different parts of the input *can* outperform RATIO PARTITION, but we were unable to prove this.

2 Optimal Preemptive Online Buffer Management

We focus on preemptive buffer management first, as our results for the nonpreemptive case are extensions of the ones for the preemptive case. We prove that, for a preemptive buffer management algorithm, $\Theta((n/B)\log B)$ bits of advice are sufficient and necessary to achieve optimality.

2.1 The Lower Bound

Theorem 1. *Any optimal preemptive online buffer management algorithm requires at least $(n/(3B))\log(B+1)$ bits of advice.*

Proof. Consider the following family of request sequences of length between $2B$ and $3B$: In time unit 0, B L-packets arrive. Between times 1 and $B + 1$, one H-packet arrives per time unit. In time unit $B + 1$, k H-packets arrive, where $0 \le k \le B$. In the next B time units, no packets arrive. The optimal solution accepts $B - k$ of the L-packets that arrive in time unit 0 and rejects the other L-packets. It then accepts all H-packets that arrive in subsequent time units.

The online algorithm also has to accept exactly $B - k$ L-packets in time unit 0. To see this, observe that, no matter how many of the L-packets are accepted, the algorithm will not preempt them. This is true because between times 1 and B, the buffer does not overflow, which implies that by time $B + 1$, all L-packets have been transmitted and the buffer contains h H-packets, where h is the number of L-packets we accepted in time unit 0. Now, if $h > B - k$, we are forced to reject one of the k H-packets that arrive in time unit $B + 1$, which leads to a suboptimal solution. If $h < B - k$, we could have accepted at least one more L-packet in time unit 0 without forcing us to reject an H-packet in time unit $B + 1$, which is again suboptimal.

Since at time 0 the algorithm needs to know the number, k, of H-packets that arrive in time unit $B+1$, and k can assume any value between 0 and B, we need $\log(B + 1)$ bits of advice for this sequence of length between $2B$ and $3B$.

We can construct arbitrarily long inputs of this type by concatenating a number, q, of such request sequences $S_1, S_2, \ldots, S_q$. The last B time units of each sequence S_i during which no packets arrive guarantee that the algorithm needs to behave on each S_i as if S_i were the whole request sequence. Thus, at least $\log(B + 1)$ bits of advice are needed for each S_i, for a total of $q \log(B + 1)$ bits. The length of the entire sequence $S_1 S_2 \ldots S_q$ is $2qB \le n \le 3qB$. Thus, we need at least $(n/(3B)) \log(B + 1)$ bits of advice.   □

2.2   The Upper Bound

We now describe an optimal preemptive online buffer management algorithm that matches the lower bound of Theorem 1 up to a constant factor.

Theorem 2. *There exists an optimal preemptive online buffer management algorithm that uses* $\lceil n/B \rceil \lceil \log(B + 1) \rceil$ *bits of advice.*

To prove Theorem 2, we consider a request sequence S and a "canonical" optimal solution $\text{OPT}_C(S)$ for S, and we propose an online algorithm $\mathcal{A}$ that uses $\lceil n/B \rceil \lceil \log(B + 1) \rceil$ bits of advice to produce this solution. Note that an optimal offline algorithm cannot benefit from preemption because it can immediately reject any packets it would preempt later. Thus, we define $\text{OPT}_C(S)$ by describing a nonpreemptive optimal offline algorithm that produces $\text{OPT}_C(S)$.

We divide S into contiguous subsequences $U_0, U_1, \ldots, U_T$, where U_t is the subsequence of requests that arrive in time unit t. Let H_t and L_t respectively be the numbers of H- and L-packets in U_t; let H'_t and L'_t respectively be the numbers of H- and L-packets in U_t that we accept; and let $Y_t = 1$ if we transmit a packet at time t, and $Y_t = 0$ otherwise. Since the buffer has capacity B, any feasible solution satisfies

$$\sum_{t=0}^{t'}(H'_t + L'_t) \le B + \sum_{t=0}^{t'} Y_t, \tag{1}$$

for all $0 \le t' \le T$. Since we can transmit a packet only if we have not already transmitted all the accepted packets, a feasible solution must also satisfy

$$\sum_{t=0}^{t'-1}(H'_t + L'_t) \ge \sum_{t=0}^{t'} Y_t, \tag{2}$$

for all $0 \le t' \le T$. Conversely, any set of values of H'_t, L'_t and Y'_t, $0 \le t \le T$, that satisfy (1) and (2) (as well as the trivial constraints that $0 \le H'_t \le H_t$, $0 \le L'_t \le L_t$, and $Y_t \in \{0, 1\}$, for all $0 \le t \le T$) yields a feasible solution.

The construction of $\mathrm{OPT_C}(S)$ starts with $Y_t = H'_t = L'_t = 0$, for all $0 \le t \le T$. Next we greedily increase the number of H-packets accepted in each time unit and then greedily increase the number of L-packets accepted in each time unit, given the set of accepted H-packets. More precisely, we proceed in two rounds. In the first round, we iterate over $t' := 0, 1, \ldots, T$ and set $Y_{t'} := 1$ if $\sum_{t=0}^{t'-1} H'_t > \sum_{t=0}^{t'-1} Y_t$, and $Y_{t'} := 0$ otherwise; then we set $H'_{t'} := \min(H_{t'}, B + \sum_{t=0}^{t'} Y_t - \sum_{t=0}^{t'-1} H'_t)$. Before the second round, we set $Y_0 := 0$ and $Y_t := 1$, for all $1 \le t \le T$. Now we iterate over $t' := 0, 1, \ldots, T$ again. For time t', we update $Y_{t'}$ so that $Y_{t'} := 1$ if $\sum_{t=0}^{t'-1}(H'_t + L'_t) > \sum_{t=0}^{t'-1} Y_t$, and $Y_{t'} := 0$ otherwise. Next we choose $L'_{t'}$ maximally so that $L'_{t'} \le L_{t'}$ and $\sum_{t=0}^{t''}(H'_t + L'_t) \le B + \sum_{t=0}^{t''} Y_t$, for all $t' \le t'' \le T$. $\mathrm{OPT_C}(S)$ accepts the first H'_t H-packets and the first L'_t L-packets in each subsequence U_t, and rejects all other packets. It transmits a packet at time t if and only if $Y_t = 1$. The proof of the following lemma is omitted due to lack of space.

Lemma 1. $\mathrm{OPT_C}(S)$ *is an optimal solution for the request sequence S.*

Next we describe a preemptive online algorithm with advice, $\mathcal{A}$, that computes $\mathrm{OPT_C}(S)$. We divide the request sequence S into $q := \lceil n/B \rceil$ subsequences $S_1, S_2, \ldots, S_q$, which we call *phases*. For $1 \le i < q$, $|S_i| = B$. S_q contains the remaining $n - (q-1)B$ requests. For each phase S_i, the advice given to the algorithm is the number, a_i, of L-packets $\mathrm{OPT_C}(S)$ accepts from S_i. Since $|S_i| \le B$, this requires $\lceil \log(B+1) \rceil$ bits of advice for each S_i, and thus $\lceil n/B \rceil \lceil \log(B+1) \rceil$ bits in total. The online algorithm now processes the packets in S one by one. While processing the requests in S_i, it keeps a count, c_i, of the number of L-packets in S_i that it has accepted and not preempted. Immediately before processing the first request in S_i, we set $c_i := 0$. For each L-packet in S_i, if the buffer is not full and $c_i < a_i$, we accept the packet and increase c_i by one; otherwise we reject the packet. For each H-packet in S_i, if the buffer is not full, we accept the packet. If the buffer is full but contains an L-packet, we preempt the *most recently* queued L-packet, decrease c_i by one, and accept the H-packet. If the buffer is full and contains only H-packets, we reject the packet. Let $\mathcal{A}(S)$ be the solution this algorithm produces on input S. Together with Lemma 1, the next lemma implies that $\mathcal{A}(S)$ is an optimal solution.

Lemma 2. $\mathcal{A}(S) = \mathrm{OPT_C}(S)$.

Proof sketch. We use induction on i to prove: (i) The set of packets from S_i accepted and not preempted by $\mathcal{A}(S)$ by the end of phase S_i is the set of packets $\mathrm{OPT_C}(S)$ accepts (and does not preempt) from S_i. (ii) While processing the packets in S_i, $\mathcal{A}(S)$ does not preempt any packets from $S_1 S_2 \ldots S_{i-1}$. These two claims together imply that $\mathcal{A}(S)$ and $\mathrm{OPT_C}(S)$ transmit the same set of packets.

For $i = 0$, the two claims hold vacuously, so assume $i > 0$ and the two claims hold for phases $S_0, S_1, \ldots, S_{i-1}$. Then the buffer states of $\mathcal{A}(S)$ and $\mathrm{OPT_C}(S)$ at the beginning of phase S_i are identical. First we prove that $\mathcal{A}(S)$ and $\mathrm{OPT_C}(S)$ accept the same H-packets from S_i. Since $\mathrm{OPT_C}(S)$ and $\mathcal{A}(S)$ accept H-packets greedily, this follows if no L-packet accepted by $\mathcal{A}(S)$ forces it to reject an H-packet accepted by $\mathrm{OPT}_C(S)$. Any L-packet accepted but not yet transmitted by $\mathcal{A}(S)$ can be preempted if necessary to make room for an H-packet and thus does not force the rejection of an H-packet. After transmitting the first L-packet p_j from S_i and before processing the last packet from S_i, the buffer contains a subset of the packets $p_{j+1}, p_{j+2}, \ldots, p_{iB}$ and thus is not full. Thus, after transmitting p_j, no H-packet from S_i is rejected by $\mathcal{A}(S)$.

Next we show that $\mathcal{A}(S)$ does not preempt any packets from $S_0, S_1, \ldots, S_{i-1}$ while processing the packets in S_i. Let p_j once again be the first L-packet from S_i transmitted by $\mathcal{A}(S)$. We have just proved that $\mathcal{A}(S)$ and $\mathrm{OPT}_C(S)$ accept the same set of H-packets from S_i. We also argued that no H-packet p_k in S_i that succeeds p_j forces any preemption because $\mathcal{A}(S)$'s buffer cannot be full when p_k arrives. If p_k precedes p_j, it can force a preemption only if $\mathcal{A}(S)$'s buffer is full when p_k arrives. Since the buffer states of $\mathcal{A}(S)$ and $\mathrm{OPT}_C(S)$ are identical at the beginning of the ith phase and $\mathrm{OPT}_C(S)$ can accept p_k without preempting any packet, $\mathcal{A}(S)$'s buffer must contain an L-packet p_h from S_i that is not in $\mathrm{OPT}_C(S)$'s buffer. Thus, $\mathcal{A}(S)$ preempts $p_h \in S_i$ to make room for p_k.

It remains to prove that, by the end of phase S_i, the set of L-packets from S_i accepted but not preempted by $\mathcal{A}(S)$ is the same as the set of L-packets $\mathrm{OPT}_C(S)$ accepts from S_i. Since none of these packets are preempted in subsequent phases, these are exactly the L-packets from S_i *transmitted* by $\mathcal{A}(S)$. We prove here that $\mathcal{A}(S)$ and $\mathrm{OPT}_C(S)$ transmit the same *number* of L-packets from S_i. This is the first part of the proof. The second part of the proof uses this fact to prove that $\mathcal{A}(S)$ and $\mathrm{OPT}_C(S)$ transmit the same *set* of L-packets from S_i. The proof of this second part is omitted due to lack of space.

$\mathcal{A}(S)$ cannot transmit more L-packets than $\mathrm{OPT}_C(S)$ because it bounds the number of L-packets from S_i it has accepted and not preempted by a_i. Next we prove that, after processing each packet p_j in S_i, the number of L-packets from S_i that $\mathcal{A}(S)$ has accepted and not preempted so far is no less than the number of L-packets $\mathrm{OPT}_C(S)$ has accepted up to that point. We use induction on the position of p_j in S_i. Before processing the first packet in S_i, the claim holds. If $\mathcal{A}(S)$ accepts p_j without preempting any other packet or if it rejects p_j because $c_i = a_i$, the invariant is maintained. If p_j is an H-packet and $\mathcal{A}(S)$ preempts an L-packet in favour of p_j or p_j is an L-packet and $\mathcal{A}(S)$ rejects it, then $\mathcal{A}(S)$'s buffer is full when p_j arrives. Since $\mathcal{A}(S)$ had accepted and not preempted at least as many packets as $\mathrm{OPT}_C(S)$ after processing each of the packets $p_1, p_2, \ldots, p_{j-1}$, $\mathcal{A}(S)$ also transmitted at least as many packets as $\mathrm{OPT}_C(S)$ by the time each of these packets was processed. The number of packets any algorithm can accept up to a certain point is B plus the number of packets it has transmitted so far. Since $\mathcal{A}(S)$'s buffer is full after processing p_j, $\mathcal{A}(S)$ has accepted and not preempted exactly this number of packets so far, while $\mathrm{OPT}_C(S)$ cannot have accepted more packets by this time. Since we already proved that $\mathcal{A}(S)$ and $\mathrm{OPT}_C(S)$ accept the same set of H-packets, this shows that the number of L-packets accepted and not preempted by $\mathcal{A}(S)$ by the time p_j has been processed is at least the number of L-packets accepted by $\mathrm{OPT}_C(S)$ by this time. $\qquad\square$

3   Optimal Nonpreemptive Online Buffer Management

In this section, we prove the somewhat surprising result that, up to constant factors, the same number of bits of advice required for an optimal preemptive online buffer management algorithm suffice for a nonpreemptive online buffer management algorithm to achieve optimality. Our lower bound for nonpreemptive online buffer management is slightly stronger than for the preemptive case.

Theorem 3. *There exists an optimal nonpreemptive online buffer management algorithm that achieves optimality using $\lceil n/B \rceil (3\lfloor \log B \rfloor + 4)$ bits of advice. Moreover, any optimal nonpreemptive online buffer management algorithm requires at least $(n \log(B + 1))/(2B)$ bits of advice.*

The lower bound proof is similar to the lower bound proof for the preemptive case and is thus omitted. In the remainder of this subsection, we prove the upper bound. We describe an online algorithm $\mathcal{A}$ that accepts the same number of H-packets and L-packets as the canonical optimal solution $\mathrm{OPT}_C(S)$ from Section 2, even though the sets of accepted packets may differ.

$\mathcal{A}$ accepts H-packets greedily, that is, it accepts each H-packet when it arrives unless the buffer is full. This does not require any advice. To define the advice that determines the behaviour of $\mathcal{A}$ for L-packets, we divide the request sequence S into $q := \lceil n/B \rceil$ phases $S_1, S_2, \ldots, S_q$ as in Section 2. For $1 \le i \le q$, let s_i and e_i be the time units during which the first and last packets in S_i arrive, respectively, and let f_i be the number of packets in $\mathrm{OPT}_C(S)$'s buffer just before the arrival of the first packet in S_i. Observe that $f_0 = 0$. We distinguish between three different cases depending on the behaviour of $\mathrm{OPT}_C(S)$ in phase S_i.

- If $\mathrm{OPT}_C(S)$'s buffer is never full during phase S_i, we call S_i a type-I phase. In this case, $\mathcal{A}$'s advice consists of the number, a_i, of L-packets in S_i $\mathrm{OPT}_C(S)$ accepts. $\mathcal{A}(S)$ accepts the first a_i L-packets in S_i it can accept and rejects the remaining L-packets in S_i.
- If $\mathrm{OPT}_C(S)$'s buffer is full at least once during phase S_i and all packets in S_i arrive over at most f_i time units (i.e., $e_i - s_i + 1 \le f_i$), we call S_i a type-II phase. In this case, $\mathcal{A}$'s advice consists of the number, r_i, of L-packets in S_i $\mathrm{OPT}_C(S)$ rejects. $\mathcal{A}(S)$ rejects the first r_i packets in S_i and accepts the remaining L-packets in S_i. (We prove below that it can accept these packets.)
- If $\mathrm{OPT}_C(S)$'s buffer is full at least once during phase S_i and the packets in S_i arrive over more than f_i time units (i.e., $e_i - s_i + 1 > f_i$), we call S_i a type-III phase. Let t_i be the last time unit in S_i when $\mathrm{OPT}_C(S)$'s buffer is full. We divide S_i into two subphases: S_i^1 contains the packets in S_i that arrive between time units s_i and t_i, inclusive, and S_i^2 contains the remaining packets in S_i. $\mathcal{A}$'s advice consists of t_i, the number, r_i', of L-packets in S_i^1 rejected by $\mathrm{OPT}_C(S)$, and the number, a_i', of L-packets in S_i^2 accepted by $\mathrm{OPT}_C(S)$. $\mathcal{A}(S)$ then treats S_i^1 as a type-II phase and S_i^2 as a type-I phase.

To encode this advice, we use one bit per phase to indicate whether it is a type-III phase. If it is, then the next $3(\lfloor \log B \rfloor + 1)$ bits represent x_i, r_i' and a_i'. If not, we use another bit to indicate whether this phase is of type I or II and accordingly encode a_i or r_i using $\lfloor \log B \rfloor + 1$ bits. In the worst case, we use $1 + 3(\lfloor \log B \rfloor + 1) = 3\lfloor \log B \rfloor + 4$ bits per phase and $\lceil n/B \rceil (3\lfloor \log B \rfloor + 4)$ bits for the whole sequence. It remains to show that $\mathcal{A}(S)$ is an optimal solution.

Lemma 3. $\mathcal{A}(S)$ *is an optimal solution.*

Proof. It suffices to prove the following three claims for every phase S_i: (i) The buffers of $\mathcal{A}(S)$ and $\mathrm{OPT}_\mathrm{C}(S)$ contain the same number of packets at the end of phase S_i. (ii) $\mathcal{A}(S)$ and $\mathrm{OPT}_\mathrm{C}(S)$ accept the same number of H-packets in S_i. (iii) $\mathcal{A}(S)$ and $\mathrm{OPT}_\mathrm{C}(S)$ accept the same number of L-packets in S_i.

For $i = 0$, these claims hold vacuously, so assume $i > 0$ and (i)–(iii) hold for phases $S_0, S_1, \ldots, S_{i-1}$. By the induction hypothesis, $\mathcal{A}(S)$'s and $\mathrm{OPT}_\mathrm{C}(S)$'s buffers both contain f_i packets at the beginning of phase S_i. We prove that (i)–(iii) hold for phase S_i by considering the three possible types of S_i.

If S_i is of type I, $\mathrm{OPT}_\mathrm{C}(S)$ does not reject any H-packet in S_i because $\mathrm{OPT}_\mathrm{C}(S)$'s buffer is never full during S_i. Moreover, the L-packets in S_i accepted by $\mathrm{OPT}_\mathrm{C}(S)$ are exactly the first a_i L-packets in S_i. To see this, observe that $\mathrm{OPT}_\mathrm{C}(S)$ rejects an L-packet p_j in S_i and accepts a subsequent L-packet p_k in S_i only if its buffer is full when p_j arrives or accepting p_j forces $\mathrm{OPT}_\mathrm{C}(S)$ to reject an H-packet p_l in S_i. (If accepting p_j forces the rejection of an H-packet after phase S_i, so does accepting p_k. Since $\mathrm{OPT}_\mathrm{C}(S)$ never accepts an L-packet that forces the rejection of an H-packet, accepting p_j can only force the rejection of an H-packet in S_i.) Then, however, $\mathrm{OPT}_\mathrm{C}(S)$'s buffer must be full either when p_j arrives or when p_l arrives, a contradiction because S_i is of type I. Now, since $\mathcal{A}(S)$ and $\mathrm{OPT}_\mathrm{C}(S)$ have the same number of packets in their buffers at the beginning of S_i, $\mathcal{A}(S)$ can also accept the first a_i L-packets and all the H-packets in this phase without filling its buffer. Therefore, $\mathcal{A}(S)$ and $\mathrm{OPT}_\mathrm{C}(S)$ accept the same set of packets in S_i and claims (i)–(iii) hold.

Next consider the case when S_i is of type II. Since all packets of S_i arrive over at most f_i time units and the buffers of $\mathcal{A}(S)$ and $\mathrm{OPT}_\mathrm{C}(S)$ contain f_i packets at the beginning of S_i, $\mathcal{A}(S)$'s and $\mathrm{OPT}_\mathrm{C}(S)$'s buffers are never empty during this phase. Thus, $\mathcal{A}(S)$ and $\mathrm{OPT}_\mathrm{C}(S)$ transmit the same number of packets during phase S_i, and claim (i) follows if we can prove claims (ii) and (iii).

First (ii). We observe that, at any time during S_i, the number of L-packets from S_i that $\mathcal{A}(S)$ has accepted so far cannot be larger than the number of L-packets from S_i that $\mathrm{OPT}_\mathrm{C}(S)$ has accepted so far. This is true because $\mathcal{A}(S)$ rejects the first r_i L-packets it receives during S_i, while $\mathrm{OPT}_\mathrm{C}(S)$ rejects a total of r_i L-packets from S_i. Since $\mathcal{A}(S)$'s buffer and $\mathrm{OPT}_\mathrm{C}(S)$'s buffer contain the same number of packets at the beginning of S_i and $\mathcal{A}(S)$ accepts H-packets greedily, this implies that, at any time during S_i, the number of H-packets from S_i accepted by $\mathcal{A}(S)$ so far is no less than the number of such packets accepted by $\mathrm{OPT}_\mathrm{C}(S)$ so far. Conversely, $\mathrm{OPT}_\mathrm{C}(S)$ accepts H-packets greedily and accepts an L-packet only if this does not prevent it from accepting an H-packet it could otherwise have accepted. Thus, the number of H-packets from S_i accepted by $\mathcal{A}(S)$ up to some point during S_i is no greater than the number of such packets accepted by $\mathrm{OPT}_\mathrm{C}(S)$ up to this point.

To show that $\mathcal{A}(S)$ and $\mathrm{OPT}_C(S)$ accept the same number of L-packets, we prove that $\mathcal{A}(S)$ rejects only the first r_i L-packets in S_i. Assume the contrary. Among the L-packets after the first r_i L-packets in S_i, let p_j be the first L-packet rejected by $\mathcal{A}(S)$. This means that the buffer of $\mathcal{A}(S)$ is full when p_j arrives. As shown in the previous paragraph, we know that up to this point, $\mathcal{A}(S)$ and $\mathrm{OPT}_C(S)$ have accepted the same number of H-packets from S_i, and $\mathcal{A}(S)$ has accepted no more L-packets from S_i than $\mathrm{OPT}_C(S)$ has. This implies that $\mathrm{OPT}_C(S)$'s buffer is also full when p_j arrives, and $\mathrm{OPT}_C(S)$ has to reject p_j. Since both algorithms have f_i packets in their buffers at the beginning of S_i, they transmit the same number of packets up to the arrival of p_j, and both algorithms' buffers are full when p_j arrives, they must both have rejected the same number of L-packets from S_i before the arrival of p_j. Since $\mathcal{A}(S)$ rejects r_i L-packets from S_i before p_j, so does $\mathrm{OPT}_C(S)$, and p_j is the $(r_i + 1)$st L-packet from S_i rejected by $\mathrm{OPT}_C(S)$, which is a contradiction.

Finally, consider the case when S_i is a type-III phase. Since there are at most B packets in S_i and $\mathrm{OPT}_C(S)$'s buffer is full at time t_i, the buffer can never run empty during subphase S_i^1. Thus, $\mathrm{OPT}_C(S)$ transmits a packet in each time unit between s_i and t_i, and $\mathrm{OPT}_C(S)$'s buffer contains at most $f_i + |S_i^1| - (t_i - s_i + 1) \le f_i + B - (t_i - s_i + 1)$ packets at the end of time unit t_i. Since $\mathrm{OPT}_C(S)$'s buffer is full at the end of time unit t_i, this implies that $t_i - s_i + 1 \le f_i$. Thus, the argument for type-II phases shows that $\mathcal{A}(S)$ and $\mathrm{OPT}_C(S)$ accept the same number of packets from S_i^1, and the argument for type-I phases shows that they accept the same number of packets from S_i^2. This completes the proof.   □

4   Advice Does Not Help Ratio Partition

The final question we investigate is whether using advice to adjust the ratio in the RATIO PARTITION algorithm helps. More precisely, we consider a class of algorithms $\Gamma(\tau)$. For a fixed parameter τ, $\Gamma(\tau)$ accepts each H-packet whenever possible and marks the τ earliest unmarked L-packets in the buffer. It accepts an L-packet only if after accepting it, the number of unmarked L-packets in the buffer is at most τ times the number of empty slots in the buffer. RATIO PARTITION is the same as $\Gamma(\frac{\alpha}{\alpha-1})$. Let BEST-THRESHOLD be an algorithm that uses advice to choose the best possible threshold τ for the given input and then runs $\Gamma(\tau)$. The following result shows that this use of advice is ineffective, that is, that BEST-THRESHOLD is no better than RATIO PARTITION.

Theorem 4. *The competitive ratio of* BEST-THRESHOLD *is at least* $2 - 1/\alpha$.

To prove Theorem 4, we ask the reader to verify that $\Gamma(\tau)$, for *any* $\tau \in [0, \infty)$, achieves a competitive ratio of exactly $2 - 1/\alpha$ on the following input S. S consists of α subsequences, each spanning $B + 1$ time units. In the first subsequence, B L-packets arrive in time unit 0, followed immediately by B H-packets in the same time unit. No further packets arrive in the remaining B time units of this subsequence. For each of the remaining $\alpha - 1$ subsequences, B L-packets arrive in time unit 0, and no further packets arrive in the remaining B time units. The key to this proof is that any threshold that is good for the first subsequence of

S is bad for the remaining α subsequences and vice versa. We conjecture that an *adaptive* threshold algorithm, which chooses different thresholds for different portions of the input, can achieve a better competitive ratio.

Acknowledgements. We would like to thank Marc Renault and two anonymous reviewers for noticing an error in the original lower bound proof for nonpreemptive online buffer management and for suggesting a way to correct it.

References

1. Aiello, W., Mansour, Y., Rajagopolan, S., Rosen, A.: Competitive queue policies for differentiated services. In: INFOCOM, pp. 414–420 (2000)
2. Albers, S.: On the influence of lookahead in competitive paging algorithms. Algorithmica 18(3), 283–305 (1997)
3. Albers, S.: A competitive analysis of the list update problem with lookahead. Theoretical Computer Science 197(1-2), 95–109 (1998)
4. Andelman, N., Mansour, Y., Zhu, A.: Competitive queueing policies for QoS switches. In: SODA, pp. 761–770 (2003)
5. Azar, Y.: Online Packet Switching. In: Persiano, G., Solis-Oba, R. (eds.) WAOA 2004. LNCS, vol. 3351, pp. 1–5. Springer, Heidelberg (2005)
6. Böckenhauer, H.-J., Komm, D., Královič, R., Královič, R.: On the Advice Complexity of the k-Server Problem. In: Aceto, L., Henzinger, M., Sgall, J. (eds.) ICALP 2011, Part I. LNCS, vol. 6755, pp. 207–218. Springer, Heidelberg (2011)
7. Böckenhauer, H.-J., Komm, D., Královič, R., Královič, R., Mömke, T.: On the Advice Complexity of Online Problems. In: Dong, Y., Du, D.-Z., Ibarra, O. (eds.) ISAAC 2009. LNCS, vol. 5878, pp. 331–340. Springer, Heidelberg (2009)
8. Böckenhauer, H.-J., Komm, D., Královič, R., Rossmanith, P.: On the Advice Complexity of the Knapsack Problem. In: Fernández-Baca, D. (ed.) LATIN 2012. LNCS, vol. 7256, pp. 61–72. Springer, Heidelberg (2012)
9. Breslauer, D.: On competitive on-line paging with lookahead. Theoretical Computer Science 209(1-2), 365–375 (1998)
10. Dobrev, S., Královic, R., Pardubská, D.: Measuring the problem-relevant information in input. ITA 43(3), 585–613 (2009)
11. Emek, Y., Fraigniaud, P., Korman, A., Rosén, A.: Online Computation with Advice. In: Albers, S., Marchetti-Spaccamela, A., Matias, Y., Nikoletseas, S., Thomas, W. (eds.) ICALP 2009, Part I. LNCS, vol. 5555, pp. 427–438. Springer, Heidelberg (2009)
12. Epstein, L., van Stee, R.: Buffer management problems. ACM SIGACT News 35(3), 58–66 (2004)
13. Goldwasser, M.H.: A survey of buffer management policies for packet switches. SIGACT News 41(1), 100–128 (2010)
14. Grove, E.F.: Online bin packing with lookahead. In: SODA, pp. 430–436 (1995)
15. Kao, M., Tate, S.R.: Online matching with blocked input. Information Processing Letters 38(3), 113–116 (1991)
16. Kesselman, A., Lotker, Z., Mansour, Y., Patt-Shamir, B., Schieber, B., Sviridenko, M.: Buffer overflow management in QoS switches. In: STOC, pp. 520–529 (2001)
17. Mansour, Y., Patt-Shamir, B., Lapid, O.: Optimal smoothing schedules for real-time streams (extended abstract). In: PODC, pp. 21–29 (2000)
18. Renault, M.P., Rosén, A.: On Online Algorithms with Advice for the k-Server Problem. In: Solis-Oba, R., Persiano, G. (eds.) WAOA 2011. LNCS, vol. 7164, pp. 198–210. Springer, Heidelberg (2012)
19. Sleator, D.D., Tarjan, R.E.: Amortized efficiency of list update and paging rules. Communications of the ACM 28(2), 202–208 (1985)

On the Complexity of the Maximum Common Subgraph Problem for Partial k-Trees of Bounded Degree

Tatsuya Akutsu[*] and Takeyuki Tamura[**]

Bioinformatics Center, Institute for Chemical Research, Kyoto University,
Gokasho, Uji, Kyoto 611-0011, Japan
{takutsu,tamura}@kuicr.kyoto-u.ac.jp

Abstract. This paper considers two versions of the maximum common subgraph problem for vertex-labeled graphs: the maximum common connected edge subgraph problem and the maximum common connected induced subgraph problem. The former is to find a connected graph with the maximum number of edges that is isomorphic to a subgraph of each of the two input graphs. The latter is to find a common connected induced subgraph with the maximum number of vertices. This paper shows that both problems are NP-hard even for labeled partial k-trees of bounded degree. It also presents some exponential-time algorithms for both problems.

Keywords: maximum common subgraph, partial k-tree, treewidth, NP-hard.

1 Introduction

Graph isomorphism and subgraph isomorphism are fundamental concepts in computer science and thus have been extensively studied. As a related problem, the *maximum common subgraph problem* is also important because it has applications in pattern recognition [6,19] and chemistry [17]. Although there exist several variants, this paper considers the *maximum common connected edge subgraph* problem (MCCES) and the *maximum common connected induced subgraph* problem (MCCIS) because these two variants have been well studied.

Due to its importance in pattern recognition and chemistry, many practical algorithms have been developed for various variants of the maximum common subgraph problem [6,17,19]. Some exponential-time algorithms have also been developed for the *maximum common induced subgraph* problem (MCIS) [1,11]. Kann studied the approximability of the maximum common subgraph problem and related problems [13].

Polynomially solvable subclasses of graphs have also been studied for MCCES and MCCIS. It is well-known that if input graphs are trees, both MCCES

---

[*] Partially supported by JSPS, Japan: Grant-in-Aid 22650045.
[**] Partially supported by JSPS, Japan: Grant-in-Aid 23700017.

K.-M. Chao, T.-s. Hsu, and D.-T. Lee (Eds.): ISAAC 2012, LNCS 7676, pp. 146–155, 2012.

and MCCIS can be solved in polynomial time using maximum weight bipartite matching [8].[1] Akutsu showed that MCCES can be solved in polynomial time if input graphs are almost trees of bounded degree whereas it remains NP-hard for almost trees of unbounded degree [2], where a graph is called almost tree if it is connected and the number of edges in each biconnected component is bounded by the number of vertices plus some constant. Yamaguchi et al. developed a polynomial-time algorithm for MCCIS and MCCES between a degree bounded partial k-tree and a graph with a polynomially bounded number of spanning trees, where k is a constant [21]. However, the latter condition seems too strong. Schietgat et al. developed a polynomial-time algorithm for outerplanar graphs under the block-and-bridge preserving subgraph isomorphism [18]. However, they modified the definition of the problem by this restriction. Although it was announced that MCCIS can be solved in polynomial time if input graphs are partial k-trees and the common subgraph must be k-connected (for example, see [5]), the restriction that subgraphs are k-connected is too strict from a practical viewpoint. On the subgraph isomorphism problem, polynomial-time algorithms have been developed for biconnected outerplanar graphs [14,20] and for partial k-trees with some constraints as well as their extensions [9,16]. We recently showed that MCCES for outerplanar graphs of bounded degree can be solved in polynomial time [3].

Since it is known that subgraph isomorphism problems can be solved in polynomial time for partial k-trees of bounded degree, it is reasonable to try to develop polynomial-time algorithms for MCCES and MCCIS for partial k-trees of bounded degree. Surprisingly, despite its practical importance,[2] the complexity of MCCES or MCCIS for partial k-trees of bounded degree has not been known. In this paper, we clarify the complexity of both problems: we show that both MCCES and MCCIS are NP-hard for vertex-labeled partial k-trees of bounded degree. This result implies under the assumption of P$\neq$NP that we cannot develop fixed-parameter algorithms for MCCES or MCCIS when the treewidth is used as the parameter. It also suggests that the maximum common subgraph problems are harder than the subgraph isomorphism problems. As positive results, we present exponential-time algorithms for MCCES and MCCIS for partial k-trees of bounded degree. In particular, we show that MCCIS can be solved in $O((2-\epsilon)^{\min(n_1,n_2)} poly(n_1, n_2))$ time for partial k-trees of bounded degree, where ϵ is a constant depending on the maximum degree (and $\epsilon = 0$ for the case of MCIS), and n_1 and n_2 denote the number of vertices of the two input graphs. It is much better (when c is not small) than an existing algorithm for MCIS [1] that works in $O(3^{\max(n_1,n_2)/3}(\max(n_1, n_2) + 1)^c)$ time where c is the size of smaller minimum vertex cover between the two input graphs. Therefore, this result suggests that the maximum common subgraph problems may be solved much faster for partial k-trees of bounded degree than for general graphs.

---

[1] For trees, MCCES and MCCIS become the same problem.

[2] The maximum degree of almost all chemical compounds is bounded by a constant (e.g., 8). It is reported that most chemical compounds have treewidth at most 3 [10,21].

2   Preliminaries

If a graph $G_c(V_c, E_c)$ is isomorphic to a subgraph of G_1 and a subgraph of G_2, we call G_c a *common subgraph* of G_1 and G_2. In addition, if a graph $G_c(V_c, E_c)$ is isomorphic to an induced subgraph of G_1 and an induced subgraph of G_2, we call G_c a *common induced subgraph* of G_1 and G_2.

In this paper, we consider the following problems.

Maximum Common Connected Edge Subgraph Problem (MCCES). Given two undirected graphs G_1 and G_2, find a common connected subgraph G_c with the maximum number of edges.

Maximum Common Connected Induced Subgraph Problem (MCCIS). Given two undirected graphs G_1 and G_2, find a common connected induced subgraph G_c with the maximum number of vertices.

In the following, we use MCS to denote the maximum common connected subgraph in both MCCES and MCCIS, where the distinction (i.e., for MCCES or MCCIS) is clear from the context.

Since we mainly consider partial k-trees, or equivalently trees of treewidth k, we briefly review it (see also Fig. 1). To define treewidth, we need the notion of tree decomposition [7]. A *tree decomposition* of a graph $G(V, E)$ is a pair $\langle \mathcal{T}(\mathcal{V}_T, \mathcal{E}_T), (B_t)_{t \in \mathcal{V}_T} \rangle$, where $\mathcal{T}(\mathcal{V}_T, \mathcal{E}_T)$ is a rooted tree and $(B_t)_{t \in \mathcal{V}_T}$ is a family of subsets of V such that (see Fig. 1).

- for every $v \in V$, $B^{-1}(v) = \{t \in \mathcal{V}_T | v \in B_t\}$ is nonempty and connected in $\mathcal{T}$, and
- for every edge $\{u, v\} \in E$, there exists $t \in \mathcal{V}_T$ such that $u, v \in B_t$.

The *width* of the decomposition is defined as $\max_{t \in \mathcal{V}_T}(|B_t| - 1)$ and the *treewidth of G* is the minimum of the widths among all the tree decompositions of G. Graphs with treewidth at most k are also known as *partial k-trees* [7].

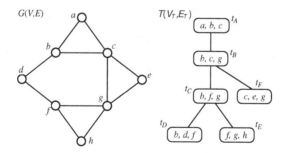

Fig. 1. Example of tree decomposition with treewidth 2

3   Hardness Results

In this section, we show that both MCCES and MCCIS are NP-hard even for partial k-trees of bounded degree.[3] For both MCCES and MCCIS, we use reductions from the maximum clique problem.

These reductions are based on a reduction from the maximum clique problem to the longest common subsequence problem (LCS) [12] although we consider the following variant of LCS (Even-Odd-LCS) here: given an integer h and n strings $s_1, \ldots, s_n$ of length $2m$, find subsequences $s'_1, \ldots, s'_n$ of these strings such that $s'_1 = s'_2 = \cdots = s'_n$, $|s'_i| \geq h$, and each s'_i consists of either letters in odd positions of s_i or letters in even positions of s_i, where s'_i is called an *odd subsequence* in the former case and an *even subsequence* in the latter case.

Proposition 1. *Even-Odd-LCS is NP-hard.*

Proof. Let $G(V, E)$ be an instance (i.e., an undirected graph) of the maximum clique problem, where $V = \{v_1, \ldots, v_n\}$.

We construct from this graph n strings $s_1, \ldots, s_n$ of length $2n$ over $\Sigma = \{a_1, \ldots, a_n\} \cup \{b_1, \ldots, b_n\}$ as follows (see Fig. 2):

$$s_i[2j - 1] = \begin{cases} a_j, & \text{if } \{v_i, v_j\} \in E \text{ or } j = i, \\ b_j, & \text{otherwise}, \end{cases}$$

$$s_i[2j] = \begin{cases} a_j, & \text{if } j \neq i, \\ b_j, & \text{otherwise}, \end{cases}$$

where $j = 1, \ldots, n$. From the definition of Even-Odd-LCS, each s'_i is a subsequence of either $s_i[1]s_i[3] \ldots s_i[2n - 1]$ (odd subsequence) or $s_i[2]s_i[4] \ldots s_i[2n]$ (even subsequence).

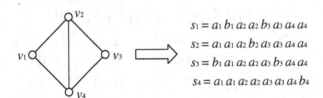

$$s_1 = a_1\, b_1\, a_2\, a_2\, b_3\, a_3\, a_4\, a_4$$
$$s_2 = a_1\, a_1\, a_2\, b_2\, a_3\, a_3\, a_4\, a_4$$
$$s_3 = b_1\, a_1\, a_2\, a_2\, a_3\, b_3\, a_4\, a_4$$
$$s_4 = a_1\, a_1\, a_2\, a_2\, a_3\, a_3\, a_4\, b_4$$

Fig. 2. Reduction from maximum clique to Even-Odd-LCS

We consider the following correspondence: v_i is in a clique $\iff$ s'_i is a subsequence of $s_i[1]s_i[3] \ldots s_i[2n - 1]$. Then, we can see that there exists a clique (i.e., a complete subgraph) of size h in $G(V, E)$ if and only if there exists a common subsequence s' of length h. Since all the constructions can be done in polynomial time, the proposition holds.   □

---

[3] Although it seems that the same NP-hardness results hold for unlabeled graphs by using appropriate encoding, we are not yet successful to get proofs.

Theorem 1. *MCCIS is NP-hard even for vertex-labeled graphs of treewidth at most 11 and maximum degree 6.*

Proof. We simulate the reduction given in the proof of Proposition 1 using MCCIS.

Let $G(V, E)$ be an instance of the maximum clique problem, where $V = \{v_1, \ldots, v_n\}$ and we assume without loss of generality that n is even. We construct gadgets $F_0, F_1, F_2, \ldots, F_n, F_{n+1}$ (see Fig. 3). Let A, B, and C be paths respectively consisting of $c_1 n^4$, $c_2 n^3$, and $c_3 n^2$ vertices of label e.[4] In the following, X^i and v_X^i respectively denote a copy of X and one of the endpoints of the path of the copy where $X \in \{A, B, C\}$, and edges in X^is are omitted in the description.

F_0 is very simple and is defined as follows.

$$V(F_0) = \{v_j^0 \mid j = 1, \ldots, n+1\} \cup A^0,$$
$$E(F_0) = \{\{v_j^0, v_{j+1}^0\} \mid j = 1, \ldots, n\} \cup \{\{v_1^0, v_A^0\}\},$$
$$\ell(v_j^0) = c \text{ for } j = 1, \ldots, n, \quad \ell(v_{n+1}^0) = e.$$

F_1 is defined as follows.

$$V(F_1) = \{v_{k,j}^1 \mid k = 1, \ldots, 6, \; j = 1, \ldots, n\} \cup \{v_{k,n+1}^1 \mid k = 1, 2, 3\}$$
$$\cup A^1 \cup B^1 \cup C^1,$$
$$E(F_1) = \{\{v_{k,j}^1, v_{k,j+1}^1\} \mid k = 1, \ldots, 6, \; j = 1, \ldots, n-1\}$$
$$\cup \{\{v_{k,j}^1, v_{k+1,j+1}^1\} \mid k = 1, 3, 5, \; j = 1, \ldots, n-1\}$$
$$\cup \{\{v_{k+1,j}^1, v_{k,j+1}^1\} \mid k = 1, 3, 5, \; j = 1, \ldots, n-1\}$$
$$\cup \{\{v_{1,j}^1, v_{4,j}^1\} \mid j = 1, \ldots, n\} \cup \{\{v_{2,j}^1, v_{3,j}^1\} \mid j = 1, \ldots, n\}$$
$$\cup \{\{v_{3,j}^1, v_{6,j}^1\} \mid j = 1, \ldots, n\} \cup \{\{v_{4,j}^1, v_{5,j}^1\} \mid j = 1, \ldots, n\}$$
$$\cup \{\{v_{1,1}^1, v_A^1\}, \{v_{2,1}^1, v_A^1\}, \{v_{3,1}^1, v_B^1\}, \{v_{4,1}^1, v_B^1\}\}$$
$$\cup \{\{v_{5,1}^1, v_C^1\}, \{v_{6,1}^1, v_C^1\}, \{v_{1,n+1}^1, v_{2,n+1}^1\}, \{v_{2,n+1}^1, v_{3,n+1}^1\}\}$$
$$\cup \{\{v_{2k-1,n}^1, v_{k,n+1}^1\}, \{v_{2k,n}^1, v_{k,n+1}^1\} \mid k = 1, 2, 3\},$$
$$\ell(v_{k,j}^1) = c \text{ for } k = 1, 2, 3, \; j = 1, \ldots, n, \quad \ell(v_{4,j}^1) = d \text{ for } j = 1, \ldots, n,$$
$$\ell(v_{5,j}^1) = a \text{ if } \{v_1, v_j\} \in E \text{ or } j = 1, \text{ otherwise } b,$$
$$\ell(v_{6,j}^1) = a \text{ if } j \neq i, \text{ otherwise } b, \quad \ell(v_{k,n+1}^1) = e \text{ for } k = 1, 2, 3.$$

Each F_i $(i = 2, \ldots, n)$ is defined as follows.

$$V(F_i) = \{v_{k,j}^i \mid k = 1, \ldots, 7, \; j = 1, \ldots, n\} \cup \{v_{k,n+1}^i \mid k = 1, \ldots, 4\}$$
$$\cup B^{2i-2} \cup C^{2i-2} \cup B^{2i-1} \cup C^{2i-1},$$
$$E(F_i) = \{\{v_{k,j}^i, v_{k,j+1}^i\} \mid k = 1, \ldots, 7, \; j = 1, \ldots, n-1\}$$
$$\cup \{\{v_{k,j}^i, v_{k+1,j+1}^i\} \mid k = 1, 4, 6, \; j = 1, \ldots, n-1\}$$

---

[4] We can use $c_1 = 1000000, c_2 = 10000, c_3 = 100$ where these are too large estimates.

$$\cup \{\{v_{k+1,j}^i, v_{k,j+1}^i\} \mid k = 1, 4, 6, \ j = 1, \ldots, n-1\}$$
$$\cup \{\{v_{1,j}^i, v_{5,j}^i\} \mid j = 1, \ldots, n\} \ \cup \ \{\{v_{2,j}^i, v_{4,j}^i\} \mid j = 1, \ldots, n\}$$
$$\cup \{\{v_{4,j}^i, v_{7,j}^i\} \mid j = 1, \ldots, n\} \ \cup \ \{\{v_{5,j}^i, v_{6,j}^i\} \mid j = 1, \ldots, n\}$$
$$\cup \{\{v_{1,1}^i, v_B^{2i-2}\}, \{v_{2,1}^i, v_B^{2i-2}\}, \{v_{3,1}^i, v_C^{2i-2}\}\}$$
$$\cup \{\{v_{4,1}^i, v_B^{2i-1}\}, \{v_{5,1}^i, v_B^{2i-1}\}, \{v_{6,1}^i, v_C^{2i-1}\}, \{v_{7,1}^i, v_C^{2i-1}\}\}$$
$$\cup \{\{v_{1,n+1}^i, v_{2,n+1}^i\}, \{v_{2,n+1}^i, v_{3,n+1}^i\}, \{v_{3,n+1}^i, v_{4,n+1}^i\}\}$$
$$\cup \{\{v_{1,n}^i, v_{1,n+1}^i\}, \{v_{2,n}^i, v_{1,n+1}^i\}, \{v_{3,n}^i, v_{2,n+1}^i\}\}$$
$$\cup \{\{v_{4,n}^i, v_{3,n+1}^i\}, \{v_{5,n}^i, v_{3,n+1}^i\}, \{v_{6,n}^i, v_{4,n+1}^i\}, \{v_{7,n}^i, v_{4,n+1}^i\}\},$$

$\ell(v_{k,j}^i) = c$ for $k = 1, 4, \ j = 1, \ldots, n, \quad \ell(v_{3,j}^i) = a$ for $j = 1, \ldots, n,$

$\ell(v_{k,j}^i) = d$ for $k = 2, 5, \ j = 1, \ldots, n,$

$\ell(v_{6,j}^i) = a$ if $\{v_i, v_j\} \in E$ or $j = i$, otherwise b ,

$\ell(v_{7,j}^i) = a$ if $j \neq i$, otherwise b , $\quad \ell(v_{k,n+1}^i) = e$ for $k = 1, \ldots, 4.$

F_{n+1} is a subgraph of the above F_i $(2 \leq i \leq n)$ induced by $v_{k,j}^{n+1}$s $(k = 1, 2, 3)$, B^{2n}, C^{2n}, $v_{1,n+1}^{n+1}$, and $v_{2,n+1}^{n+1}$.

We construct $G_1(V_1, E_1)$ by connecting $F_1, F_3, \ldots, F_{n-1}$, and F_{n+1}, and construct $G_2(V_2, E_2)$ by connecting $F_0, F_2, \ldots, F_{n-2}$, and F_n, as in Fig. 3. Clearly, this construction can be done in polynomial time.

The purpose of this construction is to establish the following correspondence in the optimal solutions among MCCIS, Even-Odd-LCS, and the maximum clique problem:

- for $i = 1, \ldots, n$, $F_i \iff s_j[2i-1]s_j[2i]$ $(j = 1, \ldots, n) \iff v_i$,
- $v_{1,j}^1$, $v_{3,j}^1$ and $v_{5,j}^1$ appear in MCS $\iff s_j'$ is an odd subsequence $\iff v_j$ is in the clique,
- $v_{2,j}^1$, $v_{4,j}^1$ and $v_{6,j}^1$ appear in MCS $\iff s_j'$ is an even subsequence, $\iff v_j$ is not in the clique,
- for $i = 2, \ldots, n$, $v_{1,j}^i$, $v_{4,j}^i$ and $v_{6,j}^i$ appear in MCS $\iff s_j'$ is an odd subsequence $\iff v_j$ is in the clique,
- for $i = 2, \ldots, n$, $v_{2,j}^i$, $v_{5,j}^i$ and $v_{7,j}^i$ appear in MCS $\iff s_j'$ is an even subsequence $\iff v_j$ is not in the clique,
- all A_is and B_is are included in MCS,
- C_1 and C_2 are included in MCS $\iff a_1$ appears in LCS $\iff v_1$ is in the clique,
- for $i = 2, \ldots, n$, C_{2i-1} and C_{2i} are included in MCS $\iff a_i$ appears in LCS $\iff v_i$ is in the clique.

With the above correspondence in mind, we show that $G(V, E)$ has a clique of size at least h if and only if the number of vertices in MCS is at least

$$c_1 n^4 + n c_2 n^3 + h(c_3 n^2 + n) + n^2 + 3n + 1.$$

Since 'only if' part is not difficult to prove, we prove 'if' part. Observe that A must be included in MCS because the size of A is very large. Then, for each

$j = 1, \ldots, n$, either $v_{1,j}^1$ or $v_{2,j}^1$ must correspond to v_j^0. Since Bs are very large, Bs must be included in MCS too. Therefore, since MCS must be an induced subgraph, either $v_{1,j}^i$ or $v_{2,j}^i$ (resp. $v_{4,j}^i$ or $v_{5,j}^i$, $v_{6,j}^i$ or $v_{7,j}^i$) can be included in MCS for each i, j (except for $i = 0, 1, n+1$). Furthermore, if $v_{1,j}^1$ (resp. $v_{2,j}^1$) is selected for MCS, $v_{1,j}^i$, $v_{4,j}^i$, $v_{6,j}^i$ (resp. $v_{2,j}^i$, $v_{5,j}^i$, $v_{7,j}^i$) must be selected for MCS for all i. That is, in order to match all As and Bs, matching paths from As and Bs to bottom es are determined uniquely once a path P from A to the leftmost bottom e in F_1 is determined, where P corresponds to choices of even/odd in the proof of Proposition 1. As a result, matching paths from bottom es to Cs are also uniquely determined, where some paths may not reach Cs. Then, there exist $c_1 n^4$ pairs of matching vertices in copies of A, $c_2 n^3$ pairs of matching vertices in copies of B, $(n+1)n$ pairs of matching vertices with label c or d, $2n+1$ pairs of matching bottom vertices with label e. Therefore, if the number of vertices in MCS is at least $c_1 n^4 + n c_2 n^3 + h(c_3 n^2 + n) + n^2 + 3n + 1$, there exist h pairs of matching paths from bottom es to Cs, which means that there exists a clique of size at least h in $G(V, E)$.

Finally, we can see that G_1 and G_2 constructed above have treewidth at most 11 by making each connecting component in consecutive two layers of a gadget as a bag (B_t). Since the maximum degree of these graphs is bounded by 6, the theorem holds.   □

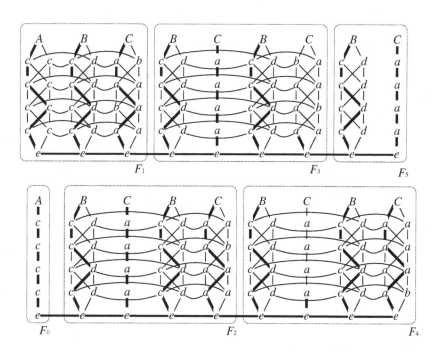

Fig. 3. Reduction from maximum clique (shown in Fig. 2) to MCCIS. Bold lines correspond to MCS.

We can also prove the following, where the proof is omitted here.

Theorem 2. *MCCES is NP-hard even for vertex-labeled graphs of treewidth at most 11 and maximum degree 5.*

4   Exponential-Time Algorithms

In this section, we present several exponential-time algorithms for MCCIS and MCCES. Although the following result is almost trivial and may be a folklore, it has a better worst case time complexity (measured by the size of the two input graphs) than an existing one (for MCIS) [1].

Proposition 2. *Both MCCIS and MCIS can be solved in* $2^{n_1+n_2+\sqrt{O(\min(n_1,n_2)\log\min(n_1,n_2))}}$ *time for general graphs, and in* $O(2^{n_1+n_2}poly(n_1,n_2))$ *time for graphs of bounded degree.*

Proof. It is known that isomorphism of graphs with n vertices can be tested in $2^{\sqrt{O(n\log n)}}$ time [4,22]. Since an induced subgraph is uniquely determined from a subset of vertices, we can solve both MCCIS and MCIS by examining all combinations of 2^{n_1} induced subgraphs from G_1 and 2^{n_2} induced subgraphs from G_2. For graphs of bounded degree, we can employ a polynomial-time isomorphism algorithm for graphs of bounded degree [15].   □

It is known that the induced subgraph isomorphism problem can be solved in polynomial time if both treewidth and degree are bounded by a constant [9,16]. Therefore, we can solve MCCIS by examining all 2^{n_1} induced subgraphs of G_1 or all 2^{n_2} induced subgraphs of G_2.

Proposition 3. *MCCIS can be solved in* $O(2^{\min(n_1,n_2)}poly(n_1,n_2))$ *time for graphs of bounded degree and bounded treewidth.*

We can slightly improve this result. The basic idea is that each vertex v has at most d neighbors and thus we need not consider the case where d neighbors are not selected in MCS but v is selected in MCS. Then, it is enough to examine $2^{d+1}-1$ assignments instead of 2^{d+1} assignments. Repeated application of this procedure leads to an $o(2^{\min(n_1,n_2)}poly(n_1,n_2))$ time algorithm as shown below.

Theorem 3. *For graphs of bounded degree d and bounded treewidth, MCCIS can be solved in* $O((2^{d+1}-1)^{n/(d^2+1)} \cdot 2^{n(1-(d+1)/(d^2+1))}poly(n_1,n_2))$ *time where* $n = \min(n_1,n_2)$.

Proof. As mentioned above, if we choose a vertex v (we call it as a *star-center*) and its neighboring vertices, we only need to examine $2^{d+1}-1$ assignments. In order to repeat this procedure, some special care is needed because the next star-center should not be adjacent to those neighbors. That is, the next star-center should be chosen from the vertices that are not neighbors of neighbors of v. Then, the number h of possible star-centers is given by

$$h \geq \lceil \frac{n}{d(d-1)+d+1} \rceil = \lceil \frac{n}{d^2+1} \rceil.$$

Since we will discuss the order of the time complexity, we assume without loss of generality that $\frac{n}{d^2+1}$ is an integer.

Let $v_{i_1}, v_{i_2}, \ldots, v_{i_h}$ be the star-centers selected in this order. Let d_j be the number of neighbors of v_{i_j}. Then, the number of assignments to be examined is bounded by

$$\left(\prod (2^{d_i+1} - 1)\right) \cdot 2^{n-\sum(d_i+1)} \leq (2^{d+1} - 1)^h \cdot 2^{n-h(d+1)}$$

$$\leq (2^{d+1} - 1)^{n/(d^2+1)} \cdot 2^{n(1-(d+1)/(d^2+1))}.$$

$\square$

We can see that the above number is $O((2 - \epsilon)^n poly(n_1, n_2))$ for some $\epsilon > 0$ by taking logarithm (with base 2) of $(2^{d+1} - 1)^h \cdot 2^{n-h(d+1)}$ as follows:

$$\log((2^{d+1} - 1)^h \cdot 2^{n-h(d+1)}) = n + h \left[\log(2^{d+1} - 1) - (d+1)\right]$$

$$\leq n \left[1 - \frac{1}{d^2+1} \cdot (d+1 - \log(2^{d+1} - 1))\right] < n.$$

For example, the complexity is $O(1.9872^n poly(n_1, n_2))$ for $d = 3$.

For MCCES, we only have the following results as in Propositions 2 and 3.

Proposition 4. *MCCES can be solved in $O(2^{|E_1|+|E_2|} poly(n_1, n_2))$ time for graphs of bounded degree, and in $O(2^{\min(|E_1|,|E_2|)} poly(n_1, n_2))$ time for graphs of bounded degree and bounded treewidth.*

5   Concluding Remarks

In this paper, we have shown that both the maximum common connected edge subgraph problem and the maximum common connected induced subgraph problem are NP-hard even for partial k-trees of bounded degree where $k = 11$. On the other hand, it is known that the former problem is solved in polynomial time for outerplanar graphs of bounded degree [3]. Since outerplanar graphs have treewidth 2 and most chemical compounds have treewidth at most 3 [10,21], it remains as an interesting open problem to decide whether the problems are NP-hard even for $k = 3$. It is also interesting to study whether the problems are NP-hard for unlabeled partial k-trees of bounded degree.

Although we have presented several exponential-time algorithms, it seems that there are many rooms for improvement. Therefore, faster exponential-time algorithms should be developed.

References

1. Abu-Khzam, F.N., Samatova, N.F., Rizk, M.A., Langston, M.A.: The maximum common subgraph problem: faster solutions via vertex cover. In: Proc. 2007 IEEE/ACS Int. Conf. Computer Systems and Applications, pp. 367–373. IEEE (2007)

2. Akutsu, T.: A polynomial time algorithm for finding a largest common subgraph of almost trees of bounded degree. IEICE Trans. Fundamentals E76-A, 1488–1493 (1993)
3. Akutsu, T., Tamura, T.: A Polynomial-Time Algorithm for Computing the Maximum Common Subgraph of Outerplanar Graphs of Bounded Degree. In: Rovan, B., Sassone, V., Widmayer, P. (eds.) MFCS 2012. LNCS, vol. 7464, pp. 76–87. Springer, Heidelberg (2012)
4. Babai, L.: Luks. E. M.: Canonical labeling of graphs. In: Proc. 15th ACM Symp. Theory of Computing, pp. 171–183. ACM Press (1983)
5. Bachl, S., Brandenburg, F.-J., Gmach, D.: Computing and drawing isomorphic subgraphs. J. Graph Algorithms and Applications 8, 215–238 (2004)
6. Conte, D., Foggia, P., Sansone, C., Vento, M.: Thirty years of graph matching in pattern recognition. Int. J. Pattern Recognition and Artificial Intelligence 18, 265–298 (2004)
7. Flum, J., Grohe, M.: Parameterized Complexity Theory. Springer, Berlin (2006)
8. Garey, M.R., Johnson, D.S.: Computers and Intractability. Freeman, New York (1979)
9. Hajiaghayi, M., Nishimura, N.: Subgraph isomorphism, log-bounded fragmentation and graphs of (locally) bounded treewidth. J. Comput. Syst. Sci. 73, 755–768 (2007)
10. Horváth, T., Ramon, J.: Efficient frequent connected subgraph mining in graphs of bounded tree-width. Theoret. Comput. Sci. 411, 2784–2797 (2010)
11. Huang, X., Lai, J., Jennings, S.F.: Maximum common subgraph: some upper bound and lower bound results. BMC Bioinformatics 7(suppl. 4), S-4 (2006)
12. Jiang, T., Li, M.: On the approximation of shortest common supersequences and longest common subsequences. SIAM J. Comput. 24, 1122–1139 (1995)
13. Kann, V.: On the Approximability of the Maximum Common Subgraph Problem. In: Finkel, A., Jantzen, M. (eds.) STACS 1992. LNCS, vol. 577, pp. 375–388. Springer, Heidelberg (1992)
14. Lingas, A.: Subgraph isomorphism for biconnected outerplanar graphs in cubic time. Theoret. Comput. Sci. 63, 295–302 (1989)
15. Luks, E.M.: Isomorphism of graphs of bounded valence can be tested in polynomial time. J. Comput. Syst. Sci. 25, 42–65 (1982)
16. Matoušek, J., Thomas, R.: On the complexity of finding iso- and other morphisms for partial k-trees. Discrete Math. 108, 343–364 (1992)
17. Raymond, J.W., Willett, P.: Maximum common subgraph isomorphism algorithms for the matching of chemical structures. J. Computer-Aided Molecular Design 16, 521–533 (2002)
18. Schietgat, L., Ramon, J., Bruynooghe, M.: A polynomial-time metric for outerplanar graphs. In: Proc. Workshop on Mining and Learning with Graphs (2007)
19. Shearer, K., Bunke, H., Venkatesh, S.: Video indexing and similarity retrieval by largest common subgraph detection using decision trees. Pattern Recognition 34, 1075–1091 (2001)
20. Syslo, M.M.: The subgraph isomorphism problem for outerplanar graphs. Theoret. Comput. Sci. 17, 91–97 (1982)
21. Yamaguchi, A., Aoki, K.F., Mamitsuka, H.: Finding the maximum common subgraph of a partial k-tree and a graph with a polynomially bounded number of spanning trees. Inf. Proc. Lett. 92, 57–63 (2004)
22. Zemlyachenko, V.M., Kornienko, N.M., Tyshkevich, R.I.: Graph isomorphism problem. J. Soviet Math. 29, 1426–1481 (1985)

Speeding Up Shortest Path Algorithms

Andrej Brodnik[1,2] and Marko Grgurovič[1]

[1] University of Primorska, Department of Information Science and Technology, Slovenia
andrej.brodnik@upr.si, marko.grgurovic@student.upr.si
[2] University of Ljubljana, Faculty of Computer and Information Science, Slovenia

Abstract. Given an arbitrary, non-negatively weighted, directed graph $G = (V, E)$ we present an algorithm that computes all pairs shortest paths in time $\mathcal{O}(m^*n + m \lg n + nT_\psi(m^*, n))$, where m^* is the number of different edges contained in shortest paths and $T_\psi(m^*, n)$ is a running time of an algorithm to solve a single-source shortest path problem (SSSP). This is a substantial improvement over a trivial n times application of ψ that runs in $\mathcal{O}(nT_\psi(m, n))$. In our algorithm we use ψ as a black box and hence any improvement on ψ results also in improvement of our algorithm.

Furthermore, a combination of our method, Johnson's reweighting technique and topological sorting results in an $\mathcal{O}(m^*n + m \lg n)$ all-pairs shortest path algorithm for arbitrarily-weighted directed acyclic graphs.

In addition, we also point out a connection between the complexity of a certain sorting problem defined on shortest paths and SSSP.

Keywords: all pairs shortest path, single source shortest path.

1 Introduction

Let $G = (V, E)$ denote a directed graph where E is the set of edges and V is the set of vertices of the graph and let $\ell(\cdot)$ be a function mapping each edge to its length. Without loss of generality, we assume G is strongly connected. To simplify notation, we define $m = |E|$ and $n = |V|$. Furthermore, we define $d(u, v)$ for two vertices $u, v \in V$ as the length of the shortest path from u to v. A classic problem in algorithmic graph theory is to find shortest paths. Two of the most common variants of the problem are the single-source shortest path (SSSP) problem and the all-pairs shortest path problem (APSP). In the SSSP variant, we are asked to find the path with the least total length from a fixed vertex $s \in V$ to every other vertex in the graph. Similarly, the APSP problem asks for the shortest path between every pair of vertices $u, v \in V$. A common simplification of the problem constrains the edge length function to be non-negative, i.e. $\ell : E \to \mathbb{R}^+$, which we assume throughout the rest of the paper, except where explicitly stated otherwise. Additionally, we define $\forall(u, v) \notin E : \ell(u, v) = \infty$.

It is obvious that the APSP problem can be solved by n calls to an SSSP algorithm. Let us denote the SSSP algorithm as ψ. We can quantify the asymptotic time bound of such an APSP algorithm as $\mathcal{O}(nT_\psi(m, n))$ and the asymptotic

K.-M. Chao, T.-s. Hsu, and D.-T. Lee (Eds.): ISAAC 2012, LNCS 7676, pp. 156–165, 2012.

space bound as $\mathcal{O}(S_\psi(m,n))$, where $T_\psi(m,n)$ is the time required by algorithm ψ and $S_\psi(m,n)$ is the space requirement of the same algorithm. We assume that the time and space bounds can be written as functions of m and n only, even though this is not necessarily the case in more "exotic" algorithms that depend on other parameters of G. Note, that if we are required to store the computed distance matrix, then we will need at least $\Theta(n^2)$ additional space. If we account for this, then the space bound becomes $\mathcal{O}(S_\psi(m,n)+n^2)$.

In this paper we are interested in the following problem: what is the best way to make use of an SSSP algorithm ψ when solving APSP? There exists some prior work on a very similar subject in the form of an algorithm named the Hidden Paths Algorithm [1]. The Hidden Paths Algorithm is essentially a modification of Dijkstra's algorithm [2] to make it more efficient when solving APSP. Solving the APSP problem by repeated calls to Dijkstra's algorithm requires $\mathcal{O}(mn+n^2 \lg n)$ time using Fibonacci heaps [3]. The Hidden Paths Algorithm then reduces the running time to $\mathcal{O}(m^*n + n^2 \lg n)$. The quantity m^* represents the number of edges $(u,v) \in E$ such that (u,v) is included in at least one shortest path. In the Hidden Paths Algorithm this is accomplished by modifying Dijkstra's algorithm, so that it essentially runs in parallel from all vertex sources in G, and then reusing the computations performed by other vertices. The idea is simple: we can delay the inclusion of an edge (u,v) as a candidate for forming shortest paths until vertex u has found (u,v) to be the shortest path to v. However, the Hidden Paths Algorithm is limited to Dijkstra's algorithm, since it explicitly sorts the shortest path lists by path lengths, through the use of a priority queue. As a related algorithm, we also point out that a different measure $|UP|$ related to the number of so-called uniform paths has also been exploited to yield faster algorithms [4].

In Sections 3, 4 and 5 we show that there is a method for solving APSP which produces the shortest path lists of individual vertices in sorted order according to the path lengths. The interesting part is that it can accomplish this without the use of priority queues of any form and requires only an SSSP algorithm to be provided. This avoidance of priority queues permits us to state a time complexity relationship between a sorted variant of APSP and SSSP. Since it is very difficult to prove meaningful lower bounds for SSSP, we believe this connection might prove useful.

As a direct application of our approach, we show that an algorithm with a similar time bound to the Hidden Paths Algorithm can be obtained. Unlike the Hidden Paths Algorithm, the resulting method is general in that it works for any SSSP algorithm, effectively providing a speed-up for arbitrary SSSP algorithms. The proposed method, given an SSSP algorithm ψ, has an asymptotic worst-case running time of $\mathcal{O}(m^*n + m \lg n + nT_\psi(m^*,n))$ and space $\mathcal{O}(S_\psi(m,n)+n^2)$. We point out that the m^*n term is dominated by the $nT_\psi(m^*,n)$ term, but we feel that stating the complexity in this (redundant) form makes the result clearer to the reader. For the case of ψ being Dijkstra's algorithm, this is asymptotically equivalent to the Hidden Paths Algorithm. However, since the algorithm ψ is arbitrary, we show that the combination of our method, Johnson's reweighting technique [5] and topological sorting gives an $\mathcal{O}(m^*n + m \lg n)$ APSP algorithm for arbitrarily-weighted directed acyclic graphs.

2   Preliminaries

Throughout the paper and without loss of generality, we assume that we are not interested in paths beginning in v and returning back to v. We have previously defined the edge length function $\ell(\cdot)$, which we now extend to the case of paths. Thus, for a path π, we write $\ell(\pi)$ to denote its length, which corresponds to the sum of the length of its edges.

Similar to the way shortest paths are discovered in Dijkstra's algorithm, we rank shortest paths in nondecreasing order of their lengths. Thus, we call a path π the k-th shortest path if it is at position k in the length-sorted shortest path list. The list of paths is typically taken to be from a single source to variable target vertices. In contrast, we store paths from variable sources to a single target. By reversing the edge directions we obtain the same lists, but it is conceptually simpler to consider the modified case. Thus, the k-th shortest path of vertex v actually represents the k-th shortest incoming path into v. We will now prove a theorem on the structure of shortest paths, which is the cornerstone of the proposed algorithm.

Definition 1. *(Ordered shortest path list P_v)*
Let $P_v = (\pi_1, \pi_2, ..., \pi_{n-1})$ denote the shortest path list for each vertex $v \in V$. Then, let $P_{v,k}$ denote the k-th element in the list P_v. The shortest path lists are ordered according to path lengths, thus we have $\forall i, j : 0 < i < j < n \Rightarrow \ell(\pi_i) \le \ell(\pi_j)$.

Theorem 1. *To determine $P_{v,k}$ we only need to know every edge $\{(u,v) \in E \mid \forall u \in V\}$ and the first k elements of each list P_u, where $(u,v) \in E$.*

Proof. We assume that we have found the first k shortest paths for all neighbors of v, and are now looking for the k-th shortest path into v, which we denote as π_k. There are two possibilities: either π_k is simply an edge (u,v), in which case we already have the relevant information, or it is the concatenation of some path π and an edge (u,v). The next step is to show that π is already contained in $P_{u,i}$ where $i \le k$.

We will prove this by contradiction. Assume the contrary, that π is either not included in P_u, or is included at position $i > k$. This would imply the existence of some path π' for which $\ell(\pi') \le \ell(\pi)$ and which is contained in P_u at position $i \le k$. Then we could simply take π_k to be the concatenation of (u,v) and π', thereby obtaining a shorter path than the concatenation of (u,v) and π. However, this is not yet sufficient for a contradiction. Note that we may obtain a path that is shorter, but connects vertices that have an even shorter path between them, i.e. the path is not the shortest path between the source s and target v.

To show that it does contradict our initial assumption, we point out that P_u contains k shortest paths, therefore it contains shortest paths from k unique sources. In contrast, the list P_v contains at most $k - 1$ shortest paths. By a counting argument we have that there must exist a path π', stored in P_u with an index $i \leq k$, which originates from a source vertex s that is not contained in P_v, thereby obtaining a contradiction. $\qquad\qquad$ $\square$

3   The Algorithm

Suppose we have an SSSP algorithm ψ and we can call it using $\psi(V, E, s)$ where V and E correspond to the vertex and edge sets, respectively and s corresponds to the source vertex. The method we propose works in the fundamental comparison-addition model and does not assume a specific kind of edge length function, except the requirement that it is non-negative. However, the algorithm ψ that is invoked can be arbitrary, so if ψ requires a different model or a specific length function, then implicitly by using ψ, our algorithm does as well.

First we give a simpler variant of the algorithm, resulting in bounds $\mathcal{O}(mn + nT_\psi(m^*, n))$. We limit our interaction with ψ only to execution and reading its output. To improve the running time we construct a graph $G' = (V', E')$ on which we run ψ. There are two processes involved: the method for solving APSP which runs on G, and the SSSP algorithm ψ which runs on G'. Let $n' = |V'|$ and $m' = |E'|$. We will maintain $m' \leq m^* + n$ and $n' = n + 1$ throughout the execution. There are $n - 1$ phases of the main algorithm, each composed of three steps: (1) Prepare the graph G'; (2) Run ψ on G'; and (3) Interpret the results of ψ.

Although the proposed algorithm effectively works on $n - 1$ new graphs, these graphs are similar to one another. Thus, we can consider the algorithm to work only on a single graph G', with the ability to modify edge lengths and introduce new edges into G'. Initially we define $V' = V \cup \{i\}$, where i is a new vertex unrelated to the graph G. We create n new edges from i to every vertex $v \in V$, i.e. $E' = \bigcup_{v \in V}\{(i, v)\}$. We set the cost of these edges to some arbitrary value in the beginning.

Definition 2. (Shortest path list for vertex v, S_v) *The shortest path list of some vertex $v \in V$ is denoted by S_v. The length of S_v is at most $n + 1$ and contains pairs of the form (a, δ) where $a \in V \cup \{null\}$ and $\delta \in \mathbb{R}^+$. The first element of S_v is always $(v, 0)$, the last element plays the role of a sentinel and is always $(null, \infty)$. For all inner (between the first and the last element) elements (a, δ), we require that $\delta = d(a, v)$. A list with $k \leq n - 1$ inner elements:*

$$S_v = ((v, 0), (a_1, \delta_1), (a_2, \delta_2), ..., (a_k, \delta_k), (null, \infty)).$$

Next we describe the data structures. Each vertex $v \in V$ keeps its shortest path list S_v, which initially contains only two pairs $(v, 0)$ and $(null, \infty)$. For each edge $(u, v) \in E$, vertex v keeps a pointer $p[(u, v)]$, which points to some element in the shortest path list S_u. Initially, each such pointer $p[(u, v)]$ is set to point to the first element of S_u.

Definition 3. (Viable pair for vertex v) *A pair (a, δ) is viable for a vertex $v \in V$ if $\forall (a', \delta') \in S_v : a \neq a'$. Alternatively, if $a = null$ we define the pair as viable.*

Definition 4. (Currently best pair for vertex v, (a_v, δ_v)) *A pair $(a_v, \delta_v) \in S_w$, where $(w, v) \in E$ is the currently best pair for vertex v if and only if (a_v, δ_v) is viable for v and: $\forall (u, v) \in E : \forall (a', \delta') \in S_u : (a', \delta')$ viable for v and $\delta' + \ell(u, v) \geq \delta_v + \ell(w, v)$.*

We now look at the first step taken in each phase of the algorithm: preparation of the graph G'. In this step, each vertex v finds the currently best pair (a_v, δ_v). To determine the currently best pair, a vertex v inspects the elements pointed to by its pointers $p[(u, v)]$ for each $(u, v) \in E$ in the following manner: For each pointer $p[(u, v)]$, vertex v keeps moving the pointer to the next element in the list S_u until it reaches a viable pair, and takes the minimum amongst these as per Definition 4. We call this process *reloading*.

Once reloaded we modify the edges in the graph G'. Let $(a_v, \delta_v) \in S_w$ where $(w, v) \in E$ be the currently best pair for vertex v, then we set $\ell(i, v) \leftarrow \delta_v + \ell(w, v)$. Now we call $\psi(V', E', i)$. Suppose the SSSP algorithm returns an array $\Pi[\ ]$ of length n. Let each element $\Pi[v]$ be a pair (c, δ) where δ is the length of the shortest path from i to v, and c is the first vertex encountered on this path. When determining the first vertex on the path we exclude i, i.e. if the path is $\pi_v = \{(i, v)\}$ then $\Pi[v].c = v$. The inclusion of the first encountered vertex is a mere convenience, and can otherwise easily be accomodated by examining the shortest path tree returned by the algorithm. For each vertex $v \in V$ we append the pair $(a_{\Pi[v].c}, \Pi[v].\delta)$ to its shortest path list. Note, that the edges $(i, v) \in E'$ are essentially shorthands for paths in G. Thus, $a_{\Pi[v].c}$ represents the source of the path in G. We call this process *propagation*.

After propagation, we modify the graph G' as follows. For each vertex $v \in V$ such that $\Pi[v].c = v$, we check whether the currently best pair $(a_v, \delta_v) \in S_u$ that was selected during the reloading phase is the first element of the list S_u. If it is the first element, then we add the edge (u, v) into the set E'. This concludes the description of the algorithm. We formalize the procedure in pseudocode and obtain Algorithm 1. To see why the algorithm correctly computes the shortest paths, we prove the following two lemmata.

Lemma 1. *For each vertex $v \in V$ whose k-th shortest path was found during the reloading step, $\psi(V', E', i)$ finds the edge (i, v) to be the shortest path into v.*

Proof. For the case when the k-th shortest path depends only on a path at position $j < k$ in a neighbor's list, the path is already found during the reloading step. What has to be shown is that this is preserved after the execution of the SSSP algorithm. Consider a vertex $v \in V$ which has already found the k-th shortest path during the reloading step. This path is represented by the edge (i, v) of the same length as the k-th shortest path. Now consider the case that some path, other than the edge (i, v) itself, would be found to be a better path to v by the SSSP algorithm. Since each of the outgoing edges of i represents a path in G, this would mean that taking this path and adding the remaining edges used to reach v would consistute a shorter path than the k-th shortest path of v. Let us denote the path obtained by this construction as π'. Clearly

Algorithm 1. All-pairs shortest path

 1: **procedure** APSP(V, E, ψ)
 2:     $V' := V \cup \{i\}$
 3:     $E' := \bigcup_{\forall v \in V} \{(i, v)\}$
 4:     $best[\,] :=$ new array $[n]$ of pairs (a, δ)
 5:     $solved[\,][\,] :=$ new array $[n][n]$ of boolean values
 6:     Initialize $solved[\,][\,]$ to $false$
 7:     **for all** $v \in V$ **do**
 8:         $S_v.append(\,(v, 0)\,)$
 9:     **end for**
10:     **for** $k := 1$ **to** $n - 1$ **do**
11:         **for all** $v \in V$ **do**   ▷ Reloading
12:             $best[v] := (null, \infty)$
13:             **for all** $u \in V$ s.t. $(u, v) \in E$ **do**
14:                 **while** $solved[v][p[(u, v)].a]$ **do**
15:                     $p[(u, v)].next()$ ▷ An end-of-list element is always viable
16:                 **end while**
17:                 **if** $p[(u, v)].\delta + \ell(u, v) < best[v].\delta$ **then**
18:                     $best[v].a := p[(u, v)].a$
19:                     $best[v].\delta := p[(u, v)].\delta + \ell(u, v)$
20:                 **end if**
21:             **end for**
22:             $\ell(i, v) := best[v].\delta$ ▷ Considering only $k - 1$ neighboring paths
23:         **end for**
24:         $\Pi[\,] := \psi(V', E', i)$
25:         **for all** $v \in V$ **do**   ▷ Propagation
26:             $S_v.append(\,(best[\Pi[v].c].a,\ \Pi[v].\delta)\,)$
27:             $solved[v][best[\Pi[v].c].a] := true$
28:             **if** $\Pi[v].c = v$ and $best[v]$ was the first element of some list S_u **then**
29:                 $E' := E' \cup (u, v)$
30:             **end if**
31:         **end for**
32:     **end for**
33: **end procedure**

---

this is a contradiction unless π' is not the k-th shortest path, i.e. a shorter path connecting the two vertices is already known.

Without loss of generality, assume that $\pi' = \{(i, u), (u, v)\}$. However, $\ell(\pi')$ can only be shorter than $\ell(i, v)$ if v could not find a viable (non-*null*) pair in the list S_u, since otherwise a shorter path would have been chosen in the reloading phase. This means that all vertex sources (the a component of a pair) contained in the list S_u are also contained in the list S_v. Therefore a viable pair for u must also be a viable pair for v. This concludes the proof by contradiction, since the path obtained is indeed the shortest path between the two vertices.   □

Lemma 2. $\psi(V', E', i)$ *correctly computes the k-th shortest paths for all vertices* $v \in V$ *given only $k - 1$ shortest paths for each vertex.*

Proof. The case when the k-th path requires only $k - 1$ neighboring paths to be known has already been proven by the proof of Lemma 1. We now consider the case when the k-th path depends on a neighbor's k-th path. If the k-th path of vertex v requires the k-th path from the list of its neighbor u, then we know the k-th path of u must be the same as that of v except for the inclusion of the edge (u, v). The same argument applies to the dependency of vertex u on its neighbor's list. Thus, the path becomes shorter after each such dependency, eventually becoming dependent on a path included at position $j < k$ in a neighbor's list (this includes edges), which has already been found during the reloading step and is preserved as the shortest path due to Lemma 1.

We now proceed in the same way that we obtained the contradiction in the proof of Lemma 1, except it is not a contradiction in this case. What follows is that any path from i to v in G' which is shorter than $\ell(i, v)$ must represent a viable pair for v. It is easy to see, then, that the shortest among these paths is the k-th shortest path for v in G and also the shortest path from i to v in G'.   □

3.1   Time and Space Complexity

First, we look at the time complexity. The main loop of Algorithm 1 (lines 7–29) performs $n - 1$ iterations. The reloading loop (lines 8–20) considers each edge $(u, v) \in E$ which takes m steps. This amounts to $\mathcal{O}(mn)$. Since each shortest path list is of length $n + 1$, each pointer is moved to the next element n times over the execution of the algorithm. There are m pointers, so this amounts to $\mathcal{O}(mn)$. Algorithm ψ is executed $n - 1$ times. In total, the running time of Algorithm 1 is $\mathcal{O}(mn + nT_\psi(m^*, n))$.

The space complexity of Algorithm 1 is as follows. Each vertex keeps track of its shortest path list, which is of size $n + 1$ and amounts to $\Theta(n^2)$ space over all vertices. Since there are exactly m pointers in total, the space needed for them is simply $\mathcal{O}(m)$. On top of the costs mentioned, we require as much space as is required by algorithm ψ. In total, the combined space complexity for Algorithm 1 is $\mathcal{O}(n^2 + S_\psi(m^*, n))$.

3.2   Implications

We will show how to further improve the time complexity of the algorithm in Section 4, but already at its current stage, the algorithm reveals an interesting relationship between the complexity of non-negative SSSP and a stricter variant of APSP.

Definition 5. (Sorted all-pairs shortest path SAPSP)
The problem SAPSP(m, n) is that of finding shortest paths between all pairs of vertices in a non-negatively weighted graph with m edges and n vertices in the form of P_v for each $v \in V$ (see Definition 1).

Theorem 2. *Let T_{SSSP} denote the complexity of the single-source shortest path problem on non-negatively weighted graphs with m edges and n vertices. Then the complexity of SAPSP is at most $\mathcal{O}(nT_{SSSP})$.*

Proof. Given an algorithm ψ which solves SSSP, we can construct a solution to SAPSP in time $\mathcal{O}(nT_\psi(m, n))$ according to Algorithm 1, since the lists S_v found by the algorithm are ordered by increasing distance from the source. $\quad\square$

What Theorem 2 says is that when solving APSP, either we can follow in the footsteps of Dijkstra and visit vertices in increasing distance from the source without worrying about a sorting bottleneck, or that if such a sorting bottleneck exists, then it proves a non-trivial lower bound for the single-source case.

4   Improving the Time Bound

The algorithm presented in the previous section has a running time of $\mathcal{O}(mn + nT_\psi(m^*, n))$. We show how to bring this down to $\mathcal{O}(m^*n + m \lg n + nT_\psi(m^*, n))$. We sort each set of incoming edges $E_v = \bigcup_{(u,v)\in E}\{(u, v)\}$ by edge lengths in non-decreasing order. By using any off-the-shelf sorting algorithm, this takes $\mathcal{O}(m \lg n)$ time.

We only keep pointers $p[(u, v)]$ for the edges which are shortest paths between u and v, and up to one additional edge per vertex for which we do not know whether it is part of a shortest path. Since edges are sorted by their lengths, a vertex v can ignore an edge at position t in the sorted list E_v until the edge at position $t - 1$ is either found to be a shortest path, or found *not* to be a shortest path. For some edge (u, v) the former case simply corresponds to using the first element, i.e. u, provided by $p[(u, v)]$ as a shortest path. The latter case on the other hand, is *not* using the first element offered by $p[(u, v)]$, i.e. finding it is not viable during the reloading phase. Whenever one of these two conditions is met, we include the next edge in the sorted list as a pointer, and either throw away the previous edge if it was found not to be a shortest path, or keep it otherwise. This means the total amount of pointers is at most $m^* + n$ at any given time, which is $\mathcal{O}(m^*)$, since m^* is at least n. The total amount of time spent by the algorithm then becomes $\mathcal{O}(m^*n + m \lg n + nT_\psi(m^*, n))$.

Theorem 3. *Let ψ be an algorithm which solves the single-source shortest path problem on non-negatively weighted graphs. Then, the all-pairs shortest path problem on non-negatively weighted graphs can be solved in time $\mathcal{O}(m^*n + m \lg n + nT_\psi(m^*, n))$ and space $\mathcal{O}(n^2 + S_\psi(m^*, n))$ where $T_\psi(m, n)$ is the time required by algorithm ψ on a graph with m edges and n nodes and $S_\psi(m, n)$ is the space required by algorithm ψ on the same graph.*

Proof. See discussion above and in Section 3.   □

5   Directed Acyclic Graphs

A combination of a few techniques yields an $\mathcal{O}(m^*n + m \lg n)$ APSP algorithm for arbitrarily weighted directed acyclic graphs (DAGs). The first step is to transform the original (possibly negatively-weighted) graph into a non-negatively weighted graph through Johnson's [5] reweighting technique. Instead of using Bellman-Ford in the Johnson step, we visit nodes in their topological order, thus obtaining a non-negatively weighted graph in $\mathcal{O}(m)$ time. Next, we use the improved time bound algorithm as presented in Section 4. For the SSSP algorithm, we again visit nodes according to their topological order. Note that if the graph G is a DAG then G' is also a DAG. The reasoning is simple: the only new edges introduced in G' are those from i to each vertex $v \in V$. But since i has no incoming edges, the acyclic property of the graph is preserved. The time bounds become $\mathcal{O}(m)$ for Johnson's step and $\mathcal{O}(m^*n + m \lg n + nT_\psi(m^*, n))$ for the APSP algorithm where $T_\psi(m^*, n) = \mathcal{O}(m^*)$. Thus, the combined asymptotic running time is $\mathcal{O}(m^*n + m \lg n)$. The asymptotic space bound is simply $\Theta(n^2)$.

Theorem 4. *All-pairs shortest path on directed acyclic graphs can be solved in time $\mathcal{O}(m^*n + m \lg n)$ and $\Theta(n^2)$ space.*

Proof. See discussion above.   □

6   Discussion

In this paper we have shown that the "standard" approach to solving APSP via independent SSSP computations can be improved upon even if we know virtually nothing about the SSSP algorithm itself. However, we should mention that in recent years, asymptotically efficient algorithms for APSP have been formulated in the so-called component hierarchy framework. These algorithms can be seen as computing either SSSP or APSP. Our algorithm is only capable of speeding up SSSP hierarchy algorithms, such as Thorup's [6], but not those which reuse the hierarchy, such as Pettie's [7], Pettie-Ramachandran [8] or Hagerup's [9] since our SSSP reduction requires modifications to the graph G'. These modifications would require the hierarchy to be recomputed, making the algorithms prohibitively slow. This raises the following question: is there a way to avoid recomputing the hierarchy at each step, while keeping the number of edges in the hierarchy $\mathcal{O}(m^*)$?

Further, if there exists an $o(mn)$ algorithm for the arbitrarily-weighted SSSP problem, then by using Johnson's reweighting technique, our algorithm might become an attractive solution for that case. For the general case, no such algorithms are known, but for certain types of graphs, there exist algorithms with an $o(mn)$ asymptotic time bound [10,11].

Furthermore, we can generalize the approach used on DAGs. Namely, in Algorithm 1 we can use an SSSP algorithm ψ that works on a specialized graph G, as long our constructed graph G' has these properties. Therefore, our algorithm can be applied to undirected graphs, integer-weighted graphs, etc., but it cannot be applied, for example, to planar graphs, since G' is not necessarily planar.

Finally, we have shown a connection between the sorted all-pairs shortest path problem and the single-source shortest path problem. If a meaningful lower bound can be proven for SAPSP, then this would imply a non-trivial lower bound for SSSP. Alternatively, if SAPSP can be solved in $O(mn)$ time, then this implies a Dijkstra-like algorithm for APSP, which visits vertices in increasing distance from the source.

References

1. Karger, D., Koller, D., Phillips, S.J.: Finding the hidden path: time bounds for all-pairs shortest paths. SIAM Journal on Computing 22(6), 1199–1217 (1993)
2. Dijkstra, E.W.: A note on two problems in connexion with graphs. Numerische Mathematik 1, 269–271 (1959)
3. Fredman, M.L., Tarjan, R.E.: Fibonacci heaps and their uses in improved network optimization algorithms. J. ACM 34(3), 596–615 (1987)
4. Demetrescu, C., Italiano, G.F.: Experimental analysis of dynamic all pairs shortest path algorithms. ACM Transactions on Algorithms 2(4), 578–601 (2006)
5. Johnson, D.B.: Efficient algorithms for shortest paths in sparse networks. J. ACM 24(1), 1–13 (1977)
6. Thorup, M.: Undirected single-source shortest paths with positive integer weights in linear time. J. ACM 46(3), 362–394 (1999)
7. Pettie, S.: A new approach to all-pairs shortest paths on real-weighted graphs. Theor. Comput. Sci. 312(1), 47–74 (2004)
8. Pettie, S., Ramachandran, V.: A shortest path algorithm for real-weighted undirected graphs. SIAM J. Comput. 34(6), 1398–1431 (2005)
9. Hagerup, T.: Improved Shortest Paths on the Word RAM. In: Welzl, E., Montanari, U., Rolim, J.D.P. (eds.) ICALP 2000. LNCS, vol. 1853, pp. 61–72. Springer, Heidelberg (2000)
10. Goldberg, A.V.: Scaling algorithms for the shortest paths problem. In: Proceedings of the Fourth Annual ACM-SIAM Symposium on Discrete Algorithms, SODA 1993, pp. 222–231. Society for Industrial and Applied Mathematics, Philadelphia (1993)
11. Gabow, H.N., Tarjan, R.E.: Faster scaling algorithms for network problems. SIAM J. Comput. 18(5), 1013–1036 (1989)

How Many Potatoes Are in a Mesh?*

Marc van Kreveld[1], Maarten Löffler[1], and János Pach[2]

[1] Dept. of Information and Computing Sciences, Utrecht University, The Netherlands
[2] Ecole Polytechnique Féderale de Lausanne and Rényi Institute, Budapest

Abstract. We consider the combinatorial question of how many convex polygons can be made at most by using the edges taken from a fixed triangulation of n vertices. For general triangulations, there can be exponentially many: $\Omega(1.5028^n)$ and $O(1.62^n)$ in the worst case. If the triangulation is fat (every triangle has its angles lower-bounded by a constant $\delta > 0$), then there can be only polynomially many: $\Omega(n^{\frac{1}{2}\lfloor\frac{2\pi}{\delta}\rfloor})$ and $O(n^{\lceil\frac{\pi}{\delta}\rceil})$. If we count convex polygons with the additional property that they contain no vertices of the triangulation in their interiors, we get the same exponential bounds in general triangulations, and $\Omega(n^{\lfloor\frac{2\pi}{3\delta}\rfloor})$ and $O(n^{\lfloor\frac{2\pi}{3\delta}\rfloor})$ in fat triangulations.

1 Introduction

It is a common task in combinatorial geometry to give lower and upper bounds for the number of occurrences of a certain subconfiguration in a geometric structure. Well-known examples are the number of vertices in the lower envelope or single face in an arrangement of line segments, the number of triangulations that have a given set of points as their vertices, etc. [10].

In this paper we analyze how many convex polygons (potatoes) can be constructed by taking unions of triangles from a fixed triangulation (mesh) M with n vertices. Equivalently, we analyze how many convex polygon boundaries can be made using the edges of a fixed triangulation, see Figure 1. For general triangulations there can be exponentially many. However, the lower-bound examples use many triangles with very small angles. When $n \to \infty$, the smallest angles tend to zero. To understand if this is necessary, we also study the number of convex polygons in a triangulation, where all angles are bounded from below by a fixed constant. It turns out that the number of convex polygons is polynomial in this case. We also study the same questions when the convex polygons cannot have vertices of M interior to them (carrots). This is the same as requiring that the submesh bounded by the convex polygon is outerplanar.

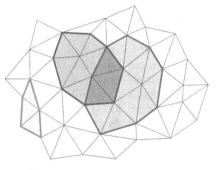

Fig. 1. A mesh M. Three convex polygons that respect M are marked.

---

* A preliminary version of this work was presented at EuroCG 2012. A full version is available on arXiv under number 1209.3954. http://arxiv.org/abs/1209.3954

K.-M. Chao, T.-s. Hsu, and D.-T. Lee (Eds.): ISAAC 2012, LNCS 7676, pp. 166–176, 2012.
© Springer-Verlag Berlin Heidelberg 2012

Table 1. Results in this paper; open spaces are directly implied by other bounds

input mesh	output vegetable	lower bound	upper bound	source
general	fat carrots	$\Omega(1.5028^n)$		Section 3
general	any potato		$O(1.62^n)$	Section 3
δ-fat	fat potatoes	$\Omega(n^{\frac{1}{2}\lfloor\frac{2\pi}{\delta}\rfloor})$		Section 4
δ-fat	any potato		$O(n^{\lceil\frac{\pi}{\delta}\rceil})$	Section 4
δ-fat	fat carrots	$\Omega(n^{\lfloor\frac{2\pi}{3\delta}\rfloor})$		Section 5
δ-fat	any carrot		$O(n^{\lfloor\frac{2\pi}{3\delta}\rfloor})$	Section 5
compact fat	any carrot	$\Omega(n^2)$	$O(n^2)$	Full version
compact fat	fat carrots	$\Omega(n)$	$O(n)$	Full version

Related Work. This paper is motivated by the *potato peeling problem*: Find a maximum area convex polygon whose vertices and edges are taken from the triangulation of a given point set [2] or a given polygon [4,7].

In computational geometry, *realistic input models* have received considerable attention in the last two decades. By making assumptions on the input, many computational problems can be solved provably faster than what is possible without these assumptions. One of the early examples concerned fat triangles: a triangle is δ-fat if each of its angles is at least δ, for some fixed constant $\delta > 0$. Matousek *et al.* [8] show that the union of n δ-fat triangles has complexity $O(n \log \log n)$ while for n general triangles this is $\Omega(n^2)$. As a consequence, the union of fat triangles can be computed more efficiently as well.

In [1,5,6,9], *fat triangulations* were used as a realistic input model motivated by polyhedral terrains, sometimes with extra assumptions. Fat triangulations are also related to the meshes computed in the area of high-quality mesh generation. The smallest angle of the elements of the mesh is a common quality measure [3]. In graph drawing, an embedded planar straight-line graph is said to have constant *angular resolution* if any two edges meeting at a vertex make at least a constant angle. Hence, fatness and constant angular resolution are the same for triangulations. The original definition of realistic terrains applied to meshes has stronger assumptions than fatness [9]. It also assumes that any two edges in the triangulation differ in length by at most a constant factor, and the outer boundary of the triangulation is a fat convex polygon.

Results. We present lower and upper bounds on the maximum number of convex polygons in a mesh in several settings. The input can be either a *general* mesh, a *fat* mesh (where every angle of each triangle is at least δ), or a *compact fat* mesh (where additionally, the ratio between the shortest and longest edge is at most ρ). The output can be either a *potato* (general convex submesh) or a *carrot* (outerplanar convex submesh, that is, one that contains no vertex of the underlying mesh in its interior), and each can additionally be required to be *fat* (where the ratio between the largest inscribed disk and the smallest containing disk is at most γ). Table 1 summarizes our results. Note that ρ and γ do not show up; the bounds hold for any constant values of ρ and γ.

2   Preliminaries

A *mesh* is a plane straight-line graph with a finite set of vertices, such that all bounded faces are triangles, the interiors of all triangles are disjoint and the intersection of any pair of triangles is either a vertex or a shared edge. We also denote the set of vertices of a graph G by $V(G)$ and the set of edges by $E(G)$, and say the *size* of G is $n = |V(G)|$. We say a mesh M is *maximal* if its triangles completely cover the convex hull of its vertices.[1] A polygon P is said to *respect* a graph G if all of its edges belong to G.

We assume a mesh M is given. We call M δ-*fat*, for some $\delta \in (0, \frac{2}{3}\pi]$, if every angle of every triangle of M is at least δ.

Let $S = [0, 2\pi)$. We define cyclic addition and subtraction $(+, -) : S \times \mathbb{R} \to S$ in the usual way, modulo 2π. We call the elements of S *directions* and implicitly associate an element $s \in S$ with the vector $(\sin s, \cos s)$.

3   Potatoes in General Meshes

Lower Bound. Let Q be a set of m points evenly spaced on the upper half of a circle. Assume $m = 2^k + 1$ for some integer k, and let the points be $v_0, \ldots, v_{m-1}$, clockwise. Let M consist of the convex hull edges, then connect v_0 and v_{m-1} to $v_{(m-1)/2}$, and recursively triangulate the subpolygons by always connecting the furthest pair to the midpoint. Figure 2 illustrates the construction.

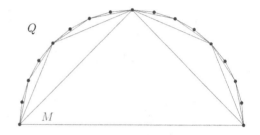

Fig. 2. A set Q of n points on a half-circle, triangulated such that the dual tree is a balanced binary tree

Let $N(k)$ be the number of different convex paths in M from v_0 to v_{m-1}. Then we have $N(k) = 1 + (N(k-1))^2$, $N(0) = 1$, because we can combine every path from v_0 to $v_{(m-1)/2}$ with every path from $v_{(m-1)/2}$ to v_{m-1}, and the extra path is $\overline{v_0, v_{m-1}}$ itself. Using this recurrence, we can relate the number m of vertices used to the number $P(m)$ of convex paths obtained: $P(3) = 2$; $P(5) = 5$; $P(9) = 26$; $P(17) = 677$; etc.

Now we place n points evenly spaced on the upper half of a circle. We triangulate $v_0, \ldots, v_{16}$ as above, and also $v_{16} \ldots, v_{32}$, and so on. We can make $n/16$ groups of 17 points where the first and last point of each group are the same. Each group is triangulated to give 677 convex paths; the rest is triangulated arbitrarily. In total we get $677^{n/16} = \Omega(1.5028^n)$ convex paths from v_0 to v_{n-1}. We omit the one from v_0 directly to v_{n-1}, and use this edge to complete every convex path to a convex polygon. The number of convex polygons is $\Omega(1.5028^n)$.[2]

---

[1] A maximal mesh is also called a *triangulation*.

[2] We can, of course, make larger groups of vertices to slightly improve the lower bound, but this does not appear to affect the given 4 significant digits.

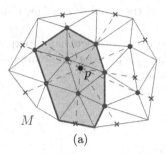 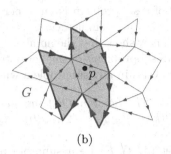

(a)   (b)

Fig. 3. (a) We project each interior vertex of M from p onto the next edge. An example potato is marked in blue. (b) The graph G obtained by removing the marked edges and orienting the others around p. The potato becomes a cycle.

Theorem 1. *There exists a mesh M with n vertices such that the number of convex polygons that respect M is $\Omega(1.5028^n)$. This is true even if M is the Delaunay triangulation of its vertices.*

Upper Bound. First, fix a point p inside some triangle of M, not collinear with any pair of vertices of M. We count only the convex polygons that contain p for now.

For every vertex v of M, let e_v be the first edge of the mesh beyond v that is hit by a ray from p and through v. Let G be the graph obtained from M by removing all such edges e_v, $v \in V(M)$. Figure 3 shows an example. We turn G into a directed graph by orienting every edge such that p lies to the left of its supporting line. We are interested in the number of simple cycles that respect G. Note that G has exactly $2n-3$ edges, since every vertex not on the convex hull causes one edge to disappear.

Lemma 1. *The number of convex polygons in M that have p in their interior is bounded from above by the number of simple cycles in G.*

Proof. With each convex polygon, we associate a cycle by replacing any edges e_v that were removed by the two edges via v, recursively. This results in a proper cycle because the convex polygon was already a monotone path around p, and this property is maintained. Each convex polygon results in a different cycle because the angle from the vertices of e_v via v is always concave.   □

Observation 1. *The complement of the outer face of G is star-shaped with p in its kernel.*

Observation 2. *Let e be an edge on the outer face of G from u to v. Then u has outdegree 1, or v has indegree 1 (or both).*

If $F \subset E(G)$ is a subset of the edges of G, we also consider the subproblem of counting all simple cycles in G that use all edges in F, the *fixed* edges. For a triple (M, G, F), we define the *potential* ρ to be the number of vertices of M (or G) minus the number of edges in F, i.e., $\rho(M, G, F) = |V(G)| - |F|$. Clearly, the potential of a subproblem is an upper bound on the number of edges that can still be used in any simple cycle.

We will now show that the number of cycles in a subproblem can be expressed in terms of subproblems of smaller potential. Let $Q(k)$ be the maximum number of simple cycles in any subproblem with potential k.

Lemma 2. *The function* $Q(\cdot)$ *satisfies*
$Q(k) \leq Q(k-1) + Q(k-2)$, $Q(0) = Q(1) = 1$.

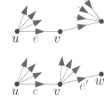

Fig. 4. Two cases for e

Proof. Let (M, G, F) be a subproblem and let $k = \rho(M, G, F)$. If $k = 1$ then $|F| = |V(G)| - 1$, so the number of fixed edges on the cycle is one less than the number of vertices available. Therefore the last edge is also fixed, if any cycle is possible. If $k = 0$, all edges are fixed.

For the general case, suppose all edges on the outer face of G are fixed. Then there is only one possible cycle. If any vertex on the outer face has degree 2 and only one incident edge fixed, we fix the other incident edge too. Suppose there is at least one edge, $e = \overline{uv}$, on the outer face that is not fixed. By Observation 2, one of its neighbors must have degree 1 towards e. Assume without loss of generality that this is v. We distinguish two cases, see Figure 4.

(i) The degree of v is 2. Any cycle in G either uses v or does not use v. If it does not use v we have a subproblem of potential $k - 1$. If it uses v, it must also use its two incident edges, so we can include these edges in F to obtain a subproblem of potential $k-2$. So, the potential $\rho(M, G, F) \leq Q(k-1) + Q(k-2)$.

(ii) The degree of v is larger than 2. Any cycle in G either uses e or does not use e. If it uses e, we can add e to F to obtain a subproblem of potential $k - 1$. If it does not use e, then consider v and the edge $e' = \overline{vw}$ that leaves v on the outer face. Since v has indegree 1 but total degree greater than 2, it must have outdegree greater than 1. Therefore, by Observation 2, w must have indegree 1. Therefore, w will not be used by any cycle in G that does not use e, and we can remove v and w to obtain a smaller graph. We also remove all incident edges; if any of them was fixed we have no solutions. We obtain a subproblem of potential $k - 2$ in this case. Again, the potential $\rho(M, G, F) \leq Q(k - 1) + Q(k - 2)$. □

This expression grows at a rate of the root of $x^2 - x - 1 = 0$, which is approximately 1.618034.

Because every convex polygon must contain at least one triangle of M, we just place p in each triangle and multiply the bound by $2n$. Since 1.62 is a slight overestimate (by rounding) of the root, we can ignore the factor $2n$ in the bound.

Theorem 2. *Any mesh M with n vertices has $O(1.62^n)$ convex polygons that respect M.*

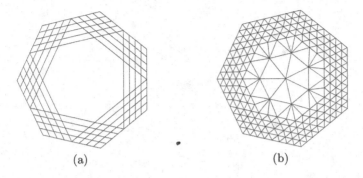

Fig. 5. (a) Essential part of the construction, allowing l^k convex polygons. (b) Final mesh.

4   Potatoes in Fat Meshes

Lower Bound. Let $k = \lfloor \frac{2\pi}{\delta} \rfloor$, and let $l = \sqrt{\frac{n}{2k}}$. Let Q be a regular k-gon, and for each edge e of Q consider the intersection point of the supporting lines of the neighboring edges. Let Q' be a scaled copy of Q that goes through these points. Now, consider a sequence $Q = Q_1, Q_2, \ldots, Q_l = Q'$ of l scaled copies of Q such that the difference in the radii of consecutive copies is equal. We extend the edges of each copy until they touch Q'. Figure 5(a) illustrates the construction.[3]

Observation 3. *The constructed graph has at least l^k different convex polygons.*

We now add vertices and edges to build a δ-fat mesh. We use $\binom{l-1}{2}$ more vertices per sector, placing $l - i$ vertices on each edge of Q_i to ensure that all angles are bounded by δ. We need $O(lk)$ vertices to triangulate the interior using some adaptive mesh generation method. The final mesh can be seen in Figure 5(b). The construction uses $\frac{3}{2}kl^2 + O(kl)$ vertices, and since we have $l = \sqrt{\frac{n}{2k}}$, there are $\frac{3}{2}kl^2 + O(kl) = \frac{3}{2}k\frac{n}{2k} + O(k\sqrt{\frac{n}{2k}}) = \frac{3}{4}n + O(\sqrt{nk}) \leq n$ vertices in total.

Observe that the triangles of the outer ring are Delaunay triangles. The inner part can also be triangulated with Delaunay triangles, since the Delaunay triangulation maximizes the smallest angle of any triangle.

Theorem 3. *There exists a δ-fat mesh M of size n such that the number of convex polygons that respect M is $\Omega(n^{\frac{1}{2}\lfloor \frac{2\pi}{\delta} \rfloor})$. This is true even if M is required to be the Delaunay triangulation of its vertices.*

Upper Bound. We consider paths in M that have roughly consistent directions.

Lemma 3. *Let $u, v \in V(M)$ be two vertices, and let $c, d \in S$ be two directions such that $d - c \leq 2\delta$. Then there is at most one convex path in M from u to v that uses only directions in $[c, d)$.*

---

[3] Our lower bound constructions use collinear points. We show in the full version that this is not essential, and the same bounds apply to "strictly convex" potatoes.

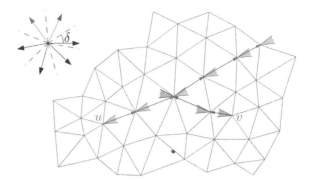

Fig. 6. Two vertices u and v that need to be extreme in two directions that differ by at most 2δ (indicated by red and blue) define a unique potential convex chain since there can be at most one edge in each sector

Proof. Let $m = c + \frac{1}{2}(d - c)$ be the direction bisecting c and d. Because M is δ-fat, for any vertex in $V(M)$ there is at most one incident edge with outgoing direction in $[c, m)$, and also at most one with direction in $[m, d)$. Because the path needs to be convex, it must first use only edges from $[c, m)$ and then switch to only edges from $[m, d)$. We can follow the unique path of edges with direction in $[c, m)$ from u and the unique path of edges with direction in $[m + \pi, d + \pi)$ from v. If these paths intersect, the concatenation may be a unique convex path from u to v as desired (clearly, the path is not guaranteed to be convex, but for an upper bound this does not matter). Figure 6 illustrates this. □

Given a convex polygon P that respects M, a vertex v of P is *extreme* in direction $s \in S$ if there are no other vertices of P further in that direction, that is, if P lies to the left of the line through v with direction $s + \frac{1}{2}\pi$.

Let $\Gamma_\delta = \{0, 2\delta, 4\delta, \ldots, 2\pi\}$ be a set of directions. As an easy corollary of Lemma 3, the vertices of a convex polygon P respecting M that are extreme in the directions of Γ_δ uniquely define P. There are at most n choices for each extreme vertex, so the number of convex polygons is at most $n^{|\Gamma_\delta|}$. Substituting $|\Gamma_\delta| = \lceil \frac{\pi}{\delta} \rceil$ we obtain the following theorem.

Theorem 4. *Any δ-fat mesh M of size n has at most $O(n^{\lceil \frac{\pi}{\delta} \rceil})$ convex polygons that respect M.*

5   Carrots in Fat Meshes

Recall that carrots are potatoes that have no interior vertices from the mesh. So we expect fewer carrots than potatoes. However, our lower bound construction for general meshes only has potatoes that are also carrots. In this section we therefore consider carrots in fat meshes.

Lower Bound. Let $k = \lfloor 2\pi/3\delta \rfloor$, and consider a regular k-gon Q. On each edge of Q, we place a triangle with angles δ, 2δ, and $\pi - 3\delta$. Then, we subdivide

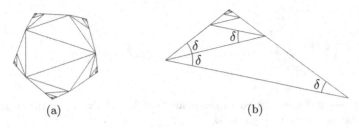

Fig. 7. (a) An example of a δ-fat mesh obtained from a k-gon ($k = 5$), which has $\Omega(n^k)$ carrots. (b) A tower of δ-δ-($\pi - 2\delta$) triangles.

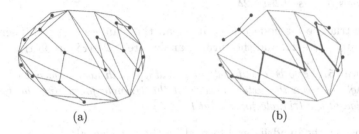

Fig. 8. (a) A carrot and its dual tree. (b) The skeleton (shown bold) of the dual tree is the spanning tree of all vertices of degree 2.

each such triangle into $\frac{n-k}{k}$ smaller triangles with angles δ, δ, and $\pi - 2\delta$, as illustrated in Figure 7(b). Finally, we triangulate the internal region of Q in any way we want, giving a mesh M.

Lemma 4. *M is convex, δ-fat, and contains $\Omega(n^{\lfloor \frac{2\pi}{3\delta} \rfloor})$ carrots.*

Proof. M is convex because $\delta + 2\delta \leq \frac{2\pi}{k}$. Every angle in the triangles outside Q is at least δ, and the angles in the interior of Q are multiples of $\frac{\pi}{k} > \delta$. Therefore, every connected subset of M is a carrot. The dual tree T of M has a central component consisting of k vertices, and then k paths of length $\frac{n}{k} - 1$. Hence, the number of subtrees of T is at least $(\frac{n}{k} - 1)^k$, which is $\Omega(n^{\lfloor \frac{2\pi}{3\delta} \rfloor})$. □

Theorem 5. *There exists a δ-fat mesh M of size n such that the number of convex outerplanar polygons that respect M is $\Omega(n^{\lfloor \frac{2\pi}{3\delta} \rfloor})$.*

Upper Bound. We will next show that given any δ-fat mesh M, the number of carrots that respect M can be at most $O(n^{\lfloor \frac{2\pi}{3\delta} \rfloor})$.

Consider any carrot. We inspect the dual tree T of the carrot and make some observations. Each node of T is either a *branch node* (if it has degree 3), a *path node* (if it has degree 2), or a *leaf* (if it has degree 1). Path nodes have one edge on the boundary of the carrot, and leaves have two edges on the boundary of the carrot. Figure 8(a) shows an example.

Fig. 9. (a) Every leaf gives rise to a turning angle of 2δ. (b) Every leaf that is an only child gives rise to a turning angle of 3δ.

Observation 4. *Let v be a leaf node of T. The turning angle between the two external edges of v is at least 2δ.*

Proof. The triangle for node v is δ-fat, so all three angles are $\geq \delta$. Therefore, the angles are $\leq \pi - 2\delta$, and the turning angles are $\geq 2\delta$ (Figure 9(a)).   □

Observation 5. *Let v be a leaf node of T and u a path node adjacent to v. The turning angle between the external edge of the triangle for u and the furthest external edge of the triangle for v is at least 3δ.*

Proof. Consider the quadrilateral formed by the two triangles of u and v. The edge in M separating u from the rest of T has two δ-fat triangles incident to one of its endpoints, and one to its other endpoint. This means that the turning angle between the edges in the observation is $\geq 3\delta$ (Figure 9(b)).   □

By Observation 4, the number of leaves in a carrot is bounded by $\lfloor \frac{\pi}{\delta} \rfloor$, and therefore, also the number of branch nodes is bounded by $\lfloor \frac{\pi}{\delta} \rfloor - 2$. However, the number of path nodes can be unbounded. Consider subtree S of T that is the spanning tree of all the path nodes. We call S the *skeleton* of the carrot. Figure 8(b) shows an example. By Observation 5, the number of leaves of S is bounded by $\lfloor \frac{2\pi}{3\delta} \rfloor$.

We will charge the carrot to the set of leaves of S, and we will argue that every set of $\lfloor \frac{2\pi}{3\delta} \rfloor$ triangles in M is charged only constantly often (for constant δ).

Observation 6. *Let Δ be any set of triangles of M. If there exists a carrot that contains all triangles in Δ, then there is a unique smallest such carrot.*

Lemma 5. *Let Δ be any set of triangles in M. The number of carrots that charge Δ is at most $2^{\lfloor \frac{2\pi}{\delta} \rfloor}$.*

Proof. Consider the tree S that is the dual of the unique smallest carrot that contains Δ, as per Observation 6. Any carrot that charges Δ has S as its skeleton. First, we argue that the set of path nodes in any carrot that charges Δ is a subset of S. Indeed, if there was any path node in T outside S, then there would be at least one leaf component of T that is disconnected from S, and there would be an edge outside Δ that gets charged by the carrot of T. Therefore, only branch nodes and leaves can still be added to S to obtain a carrot that charges Δ.

Then, we argue that there are at most $2^{\lfloor \frac{2\pi}{\delta} \rfloor}$ other nodes that can be part of a carrot that charges Δ. We can augment S by adding on components consisting of only k leaves and $k-1$ branch nodes. By Observation 4, each such component consumes a turning angle of $2k\delta$. Therefore, they can only be added on edges of S which have a cap angle of at least $2k\delta$. Therefore, there can be at most $2\pi/\delta$ potential leaves, leading to $2^{\lfloor \frac{2\pi}{\delta} \rfloor}$ choices.[4] $\square$

Theorem 6. *Any δ-fat mesh M of size n has at most $O(n^{\lfloor \frac{2\pi}{3\delta} \rfloor})$ convex outerplanar polygons that respect M.*

When the mesh is not only fat, but the edge length ratio is also bounded by a constant, we can prove better bounds. We call such meshes *compact fat*. We state the results here but defer the proofs to the full version.

Theorem 7. *Any compact fat mesh M of size n has at most $O(n)$ convex fat outerplanar polygons that respect M.*

Theorem 8. *Any compact fat mesh M of size n has at most $O(n^2)$ convex outerplanar polygons that respect M.*

Acknowledgements. We thank Stefan Langerman and John Iacono for detecting an error in an earlier version of this paper.

M.L. was supported by the Netherlands Organisation for Scientific Research (NWO) under grant 639.021.123. J.P. was supported by NSF Grant CCF-08-30272, by NSA, by OTKA under EUROGIGA project GraDR 10-EuroGIGA-OP-003, and by Swiss National Science Foundation Grant 200021-125287/1.

References

1. Aronov, B., de Berg, M., Thite, S.: The Complexity of Bisectors and Voronoi Diagrams on Realistic Terrains. In: Halperin, D., Mehlhorn, K. (eds.) ESA 2008. LNCS, vol. 5193, pp. 100–111. Springer, Heidelberg (2008)
2. Aronov, B., van Kreveld, M., Löffler, M., Silveira, R.I.: Peeling meshed potatoes. Algorithmica 60(2), 349–367 (2011)
3. Bern, M., Eppstein, D., Gilbert, J.: Provably good mesh generation. J. Comput. Syst. Sci. 48(3), 384–409 (1994)
4. Chang, J.S., Yap, C.K.: A polynomial solution for the potato-peeling problem. Discrete Comput. Geom. 1, 155–182 (1986)
5. de Berg, M., Cheong, O., Haverkort, H.J., Lim, J.-G., Toma, L.: The complexity of flow on fat terrains and its I/O-efficient computation. Comput. Geom. 43(4), 331–356 (2010)
6. de Berg, M., van der Stappen, A.F., Vleugels, J., Katz, M.J.: Realistic input models for geometric algorithms. Algorithmica 34(1), 81–97 (2002)
7. Goodman, J.E.: On the largest convex polygon contained in a non-convex n-gon or how to peel a potato. Geom. Dedicata 11, 99–106 (1981)

---

[4] Not all potential leaves can be chosen independently, but we ignore this issue since the factor is dominated by the dependency on n anyway.

8. Matousek, J., Pach, J., Sharir, M., Sifrony, S., Welzl, E.: Fat triangles determine linearly many holes. SIAM J. Comput. 23(1), 154–169 (1994)
9. Moet, E., van Kreveld, M., van der Stappen, A.F.: On realistic terrains. Comput. Geom. 41(1-2), 48–67 (2008)
10. Pach, J., Sharir, M.: Combinatorial Geometry and Its Algorithmic Applications: The Alcala Lectures. Mathematical Surveys and Monographs. AMS (2009)

On Higher Order Voronoi Diagrams
of Line Segments*

Evanthia Papadopoulou and Maksym Zavershynskyi

Faculty of Informatics, Università della Svizzera Italiana, Lugano, Switzerland
{evanthia.papadopoulou,maksym.zavershynskyi}@usi.ch

Abstract. We analyze structural properties of the order-k Voronoi diagram of line segments, which surprisingly has not received any attention in the computational geometry literature. We show that order-k Voronoi regions of line segments may be disconnected; in fact a single order-k Voronoi region may consist of $\Omega(n)$ disjoint faces. Nevertheless, the structural complexity of the order-k Voronoi diagram of non-intersecting segments remains $O(k(n-k))$ similarly to points. For intersecting line segments the structural complexity remains $O(k(n-k))$ for $k \geq n/2$.

Keywords: computational geometry, Voronoi diagrams, line segments, higher order Voronoi diagrams.

1 Introduction

Given a set of n simple geometric objects in the plane, called sites, the order-k Voronoi diagram of S is a partitioning of the plane into regions, such that every point within a fixed order-k region has the same set of k nearest sites. For $k = 1$ this is the *nearest-neighbor Voronoi diagram*, and for $k = n - 1$ the *farthest-site Voronoi diagram*. For n point sites in the plane, the order-k Voronoi diagram has been well studied, see e.g [9,1,4,6]. Its structural complexity has been shown to be $O(k(n-k))$ [9]. Surprisingly, order-k Voronoi diagrams of more general sites, including simple line segments, have been largely ignored. The farthest line segment Voronoi diagram was only recently considered in [3], showing properties surprisingly different than its counterpart for points. The nearest neighbor Voronoi diagram of line segments has received extensive attention, see e.g. [10,14,8] or [4] for a survey.

In this paper, we analyze the structural properties of the order-k Voronoi diagram of line segments. We first consider disjoint line segments and then extend our results to line segments that may share endpoints, such as line segments forming simple polygons or line segments forming a planar straight-line graph, and intersecting line segments. Unlike points, order-k Voronoi regions of line segments may be disconnected; in fact a single order-k Voronoi region may disconnect to $\Omega(n)$ disjoint faces. However, the structural complexity of the order-k

---

* Supported in part by the Swiss National Science Foundation grant 200021-127137 and the ESF EUROCORES program EuroGIGA/VORONOI, SNF 20GG21-134355.

K.-M. Chao, T.-s. Hsu, and D.-T. Lee (Eds.): ISAAC 2012, LNCS 7676, pp. 177–186, 2012.

line segment Voronoi diagram remains $O(k(n - k))$, assuming non-intersecting line segments, similarly to points, despite the disconnected regions. For intersecting line segments the dependency of the structural complexity on the number of intersections reduces as k increases and it remains $O(k(n-k))$ for $k \geq n/2$. The case of line segments involving polygonal objects is important for applications such as [11] that motivated our study.

For points, the derivation of the $O(k(n - k))$ bound relies on three facts: 1. an exact formula in [9] that relates F_k, the total number of faces on the order-k Voronoi diagram, with n, k and the number of unbounded faces in previous diagrams, 2. a symmetry property stating that $S_k = S_{n-k}$, where S_k denotes the number of unbounded faces in the order-k Voronoi diagram, and 3. an upper bound result from k-set theory [2,6]. In the case of line segments we first show that the formula of [9] remains valid, despite the presence of disconnected regions. However, the symmetry property no longer holds and results available from k-set theory are not directly applicable. Thus, a different approach has to be derived.

2   Preliminaries

Let $S = \{s_1, s_2, \ldots, s_n\}$ be a set of n line segments in $\mathbb{R}^2$. Each segment consists of three *elementary sites*: two endpoints and an open line segment. We make a *general position assumption* that no more than three elementary sites can touch the same circle.

The Euclidean distance between two points p, q is denoted as $d(p, q)$. The distance between point p and a line segment s is the minimum Euclidean distance $d(p, s) = \min_{q \in s} d(p, q)$. The bisector of two segments s_i and s_j is the locus of points equidistant from both, i.e., $b(s_i, s_j) = \{x \mid d(x, s_i) = d(x, s_j)\}$. If s_i and s_j are disjoint their bisector is a curve that consists of a constant number of line segments, rays and parabolic arcs. If segments intersect at point p the bisector consists of two such curves intersecting at point p. If segments share a common endpoint the bisector contains a two-dimensional region. In the following, we assume that segments are disjoint. We deal with segments that share endpoints in Section 5 and segments that intersect in Section 6.

Let $H \subset S$. The generalized Voronoi region of H, $\mathcal{V}(H, S)$ is the locus of points that are closer to all segments in H than to any segment not in H.

$$\mathcal{V}(H, S) = \{x \mid \forall s \in H, \forall t \in S \setminus H \ d(x, s) < d(x, t)\} \tag{1}$$

For $|H| = k$, $\mathcal{V}(H, S)$ is the order-k Voronoi region of H, denoted $\mathcal{V}_k(H, S)$.

$$\mathcal{V}_k(H, S) = \mathcal{V}(H, S) \text{ for } |H| = k \tag{2}$$

The order-k Voronoi diagram of S, $V_k(S)$, is the partitioning of the plane into order-k Voronoi regions. A maximal interior-connected subset of a region is called a face. The farthest Voronoi diagram of S is denoted as $V_f(S)$ $(V_f(S) = V_{n-1}(S))$ and a farthest Voronoi region as $\mathcal{V}_f(s, S)$ $(\mathcal{V}_f(s, S) = \mathcal{V}_{n-1}(S \setminus \{s\}, S))$.

An order-k Voronoi region $\mathcal{V}_k(H, S)$ can be interpreted as the locus of points closer to H than to any other subset of S of size k, where the distance between a point x and a set H is measured as the farthest distance $d(x, H) = \max_{s \in H} d(x, s)$.

The following lemma is a simple generalization of [3] for $1 \le k \le n - 1$.

Lemma 1. *Consider a face F of region $\mathcal{V}_k(H, S)$. F is unbounded (in the direction r) iff there exists an open halfplane (normal to r) that intersects all segments in H but no segment in $S \setminus H$.*

Corollary 1. *There is an unbounded Voronoi edge separating regions $\mathcal{V}_k(H \cup \{s_1\}, S)$ and $\mathcal{V}_k(H \cup \{s_2\}, S)$ iff a line through the endpoints of s_1 and s_2 induces an open halfplane $r(s_1, s_2)$ such that $r(s_1, s_2)$ intersects all segments in H but no segment in $S \setminus H$.*

3 Disconnected Regions

The order-k line segment Voronoi diagram may have disconnected regions, unlike its counterpart of points, see e.g., Fig. 1. This phenomenon was first pointed out in [3] for the farthest line segment Voronoi diagram, where a single region was shown possible to be disconnected in $\Theta(n)$ faces.

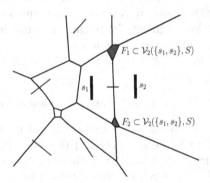

Fig. 1. $V_2(S)$ with two disconnected faces, induced by the same pair of sites

Lemma 2. *An order-k region of $V_k(S)$ can have $\Omega(n)$ disconnected faces, in the worst case, for $k > 1$.*

Proof. We describe an example where an order-k Voronoi region is disconnected in $\Omega(n - k)$ bounded faces. Consider k almost parallel long segments H. These segments induce a region $\mathcal{V}_k(H, S)$. Consider a minimum disk, that intersects all segments in H, and moves along their length. We place the remaining $n - k$ segments of $S \setminus H$ in a such way that they create obstacles for the disk. While the disk moves along the tree of $V_f(H)$ it intersects the segments of $S \setminus H$ one by one, and creates $\Omega(n - k)$ disconnectivities (see Fig. 2 (a)).

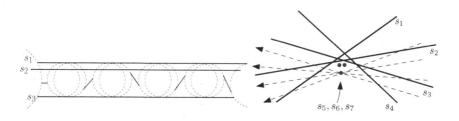

Fig. 2. (a) While the circle moves it encounters 5 obstacles, which induce $\Omega(n-k)$ disconnectivities of the region $\mathcal{V}_3(H,S)$; (b) $\mathcal{V}_4(\{s_1,s_2,s_3,s_4\},S)$ has $k=4$ disconnected unbounded faces. The dashed arrows represent the rotation of the directed line g.

We now follow [3] and describe an example where an order-k Voronoi region is disconnected in $\Omega(k)$ unbounded faces. Consider $n-k$ segments in $S \setminus H$ degenerated into points placed close to each other. The remaining k non-degenerate segments in H are organized in a cyclic fashion around them (see Fig. 2 (b)). Consider a directed line g through one of the degenerate segments s'. Rotate g around s' and consider the open halfplane to the left of g. During the rotation, the positions of g in which the halfplane intersects all k segments, alternate with the positions in which it does not. The positions in which the halfplane touches endpoints of non-degenerate segments, correspond to unbounded Voronoi edges. Each pair of consecutive unbounded Voronoi edges bounds a distinct unbounded face. Following [3], the line segments in H can be untangled into non-crossing segments while the same phenomenon remains.

Note that for small k, $1 < k \leq n/2$, $\Omega(n-k) = \Omega(n)$, while for large k, $n/2 \leq k \leq n-1$, $\Omega(k) = \Omega(n)$. $\square$

Lemma 3. *An order-k region $\mathcal{V}_k(H,S)$ has $O(k)$ unbounded disconnected faces.*

Proof. We show that an endpoint p of a segment $s \in H$ may induce at most two unbounded Voronoi edges bordering $\mathcal{V}_k(H,S)$ (see Fig. 3). Consider two such unbounded Voronoi edges. By Corollary 1 there are open halfplanes $r(s,t_1)$, $r(s,t_2)$, for $s \in H$ and $t_1, t_2 \in S \setminus H$, that intersect all segments in H but no segments in $S \setminus H$. Thus, any halfplane $r(s,t_3)$, $t_3 \in S \setminus H$ must intersect either t_1 or t_2. Since $|H| = k$ and a segment has two endpoints, the lemma follows. $\square$

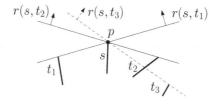

Fig. 3. Every endpoint of a segment $s \in H$ can induce at most 2 halfplanes

4   Structural Complexity

In this section we prove that the structural complexity of the order-k Voronoi diagram of n disjoint line segments is $O(k(n-k))$ despite disconnected regions. We first prove Theorem 1 which is a generalization of [9] for line segments exploiting the fact that the farthest line segment Voronoi diagram remains a tree structure [3]. Then in Lemma 7 we analyze the number of unbounded faces of the order-k Voronoi diagram in dual setting using results on arrangements of wedges [3,7] and ($\leq k$)-level in arrangements of Jordan-curves [13]. Combining Theorem 1 and Lemma 7 we derive the $O(k(n-k))$ bound.

Voronoi vertices in $V_k(S)$ are classified into *new* and *old*. A Voronoi vertex of $V_k(S)$ is called *new* (respectively *old*) if it is the center of a disk that touches 3 line segments and its interior intersects exactly $k-1$ (respectively $k-2$) segments. By the definition of the order-k Voronoi diagram we have the following properties:

1. Every Voronoi vertex of $V_k(S)$ is either *new* or *old*.
2. A *new* Voronoi vertex in $V_k(S)$ is an *old* Voronoi vertex in $V_{k+1}(S)$.
3. Under a general position assumption, an *old* Voronoi vertex in $V_k(S)$ is a *new* Voronoi vertex in $V_{k-1}(S)$.

Lemma 4. *Consider a face F of the region $\mathcal{V}_{k+1}(H, S)$. The portion of $V_k(S)$ enclosed in F is exactly the farthest Voronoi diagram $V_f(H)$ enclosed in F.*

Proof. Let x be a point in F. Suppose that among all segments in H, x is farthest from s_i. Then $x \in \mathcal{V}_f(s_i, H)$. Let $H_i = H \setminus \{s_i\}$. Since $x \in \mathcal{V}_{k+1}(H, S)$, x is farthest from s_i among all segments in H, and $|H_i| = k$, $x \in \mathcal{V}_k(H_i, S)$.   □

Regions in the farthest line segment Voronoi diagram have the following *visibility property*: Let x be a point in $\mathcal{V}_f(s, H)$ of $V_f(H)$ for a set of segments H. Let $r(s, x)$ be the ray realizing the distance $d(s, x)$, emanating from point $p \in s$ such that $d(p, x) = d(s, x)$, extending to infinity (see Fig. 4). Ray $r(s, x)$ intersects the boundary of $\mathcal{V}_f(s, H)$ at a point a_x and the part of the ray beyond a_x is entirely in $\mathcal{V}_f(s, H)$. Using this property we derive the following lemma.

Lemma 5. *Let F be a face of region $\mathcal{V}_{k+1}(H, S)$ in $V_{k+1}(S)$. The graph structure of $V_k(S)$ enclosed in F is a connected tree that consists of at least one edge. Each leaf of the tree ends at a vertex of face F (see Fig. 4).*

Proof. (Sketch) Assume that the tree of $V_f(H)$ in F is disconnected. Consider a point x on the path that connects two disconnected parts and bounds $\mathcal{V}_f(s, H)$ such that $x \notin F$. Using the *visibility property* we derive a contradiction.   □

Corollary 2. *Consider a face F of the Voronoi region $\mathcal{V}_{k+1}(H, S)$. Let m be the number of Voronoi vertices of $V_k(S)$ enclosed in its interior. Then F encloses $e = 2m + 1$ Voronoi edges of $V_k(S)$.*

Let F_k, E_k, V_k and S_k denote respectively the number of faces, edges, vertices and unbounded faces in $V_k(S)$. By Euler's formula we derive

$$E_k = 3(F_k - 1) - S_k, \quad V_k = 2(F_k - 1) - S_k \tag{3}$$

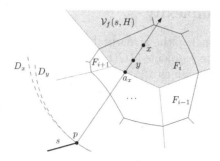

Fig. 4. The part of the ray $r(s, x)$ beyond a_x entirely belongs to $\mathcal{V}_f(s, H)$

Lemma 6. *The total number of unbounded faces in the order-k Voronoi diagram for all orders is* $\sum_{i=1}^{n-1} S_i = n(n-1)$

Theorem 1. *The number of Voronoi faces in order-k Voronoi diagram of n disjoint line segments is:*

$$F_k = 2kn - k^2 - n + 1 - \sum_{i=1}^{k-1} S_i, \quad F_k = 1 - (n-k)^2 + \sum_{i=k}^{n-1} S_i \quad (4)$$

Proof. (Sketch) Corollary 2 implies that $E_{k+1} = 2V_k' + F_{k+2}$, where V_k' is the number of new Voronoi vertices in $V_k(S)$. Combining Corollary 2 and (3) we obtain a recursive formula $F_{k+3} = 2F_{k+2} - F_{k+1} - 2 - S_{k+2} + S_{k+1}$. Using as base cases $F_1 = n$ and $F_2 = 3(n-1) - S_1$ we prove the first part of (4). Lemma 6 implies $\sum_{i=1}^{k-1} S_i + \sum_{i=k}^{n-1} S_i = \sum_{i=1}^{n-1} S_i = n(n-1)$. Combining this with the first part of (4) we derive the second part. □

Lemma 7. *For a given set of n segments,* $\sum_{i=k}^{n-1} S_i$ *is* $O(k(n-k))$, *for* $k \geq n/2$.

Proof. Following [3], we use the well-known point-line duality transformation T, which maps a point $p = (a, b)$ in the primal plane to a line $T(p) : y = ax - b$ in the dual plane, and vice versa. We call the set of points above both lines $T(p)$ and $T(q)$ the *wedge* of $s = (p, q)$. Consider a line l and a segment $s = (p, q)$. Segment s is above line l iff point $T(l)$ is strictly above lines $T(p)$ and $T(q)$ [3].

Consider the arrangement W of wedges w_i, $i = 1, \ldots, n$, as defined by the segments of $S = \{s_1, \ldots, s_n\}$. For our analysis we need the notions of r-level and $(\leq r)$-level. The r-level of W is a set of edges such that every point on it is above r wedges. The r-level shares its vertices with the $(r-1)$-level and the $(r+1)$-level. The $(\leq r)$-level of W is the set of edges such that every point on it is above at most r wedges. The complexity of the r-level and the $(\leq r)$-level is the number of their vertices, excluding the wedge apices. We denote the maximum complexity of the r-level and the $(\leq r)$-level of n wedges as $g_r(n)$ and $g_{\leq r}(n)$, respectively.

Claim: The number of unbounded Voronoi edges of $V_k(S)$, unbounded in direction $\phi \in [\pi, 2\pi]$, is exactly the number of vertices shared by the $(n-k-1)$-level and the $(n-k)$-level of W. Thus $S_k = O(g_{n-k}(n))$.

Proof of claim: Consider a vertex p (see Fig. 5) of the r-level and $(r+1)$-level. Let w_i, w_j be the wedges, that intersect at p, and let s_i, s_j be their corresponding segments. Let W_p, $|W_p| = n - r - 2$, be the set of wedges strictly above p, and let S_p be the set of the corresponding segments. Then $T(p)$ induces the unbounded Voronoi edge that separates regions $V_{n-r-1}(S_p \cup \{s_i\}, S)$ and $V_{n-r-1}(S_p \cup \{s_j\}, S)$ of $V_{n-r-1}(S)$. Let $r = n - k - 1$ to derive the claim.

The claim implies the following.

$$\sum_{i=k}^{n-1} S_i = O(g_{\leq n-k}(n)) \tag{5}$$

Since the arrangement of wedges is a special case of arrangements of Jordan curves, we use the formula from [13] to bound the complexity of the $(\leq r)$-level:

$$g_{\leq r}(n) = O\left((r+1)^2 g_0\left(\left\lfloor \frac{n}{r+1} \right\rfloor\right)\right) \tag{6}$$

It is known that the complexity of the lower envelope of such wedges is $g_0(x) = O(x)$ [7,3]. (Note that [13] implies a weaker $g_0(x) = O(x \log x)$). Therefore, $g_{\leq r}(n) = O(n(r+1))$. Substituting into formula (5) we obtain $\sum_{i=k}^{n-1} S_i = O(n(n-k))$. Since $n/2 \leq k \leq n-1$, $\sum_{i=k}^{n-1} S_i = O(k(n-k))$. $\quad\square$

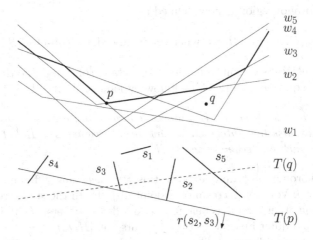

Fig. 5. (a) In the dual plane point p belongs to the 2-level and the 3-level of the arrangement W. (b) In the primal plane the halfplane $r(s_2, s_3)$ below $T(p)$ defines the unbounded Voronoi edge that separates $V_2(\{s_2, s_4\}, S)$ and $V_2(\{s_3, s_4\}, S)$.

Combining Lemma 7 and Theorem 1 we obtain the following theorem.

Theorem 2. *The number of Voronoi faces in order-k Voronoi diagram of n disjoint line segments is $F_k = O(k(n-k))$.*

5   Line Segments Forming a Planar Straight-Line Graph

In this section we consider line segments that may touch at endpoints, such as line segments forming a simple polygon, more generally line segments forming a planar straight-line graph. This is important for applications involving polygonal shapes such as [11].

When line segments share endpoints, bisectors may contain 2-dimensional portions. The standard approach to avoid this issue for $k = 1$ is to consider S as a set of distinct elementary sites. Our goal is to extend this notion for the order-k Voronoi diagram without altering the structure of the order-k Voronoi diagram for disjoint line segments. Note that we cannot simply consider elementary sites as distinct when defining an order-k Voronoi region as this will lead to a different type of order-k Voronoi diagram for disjoint segments that is not very interesting. We first extend the notion of a subset of S of cardinality k as follows.

Definition 1. *A set H, $H \subseteq S$, is called an order-k subset iff*

1. $|H| = k$ *(type 1) or*
2. $H = H' \cup I(p)$, *where $H' \subseteq S$, $|H'| < k$, p is a segment endpoint incident to set of segments $I(p)$, and $|H' \cup I(p)| > k$ (type 2). Set $rep(H) = \{p\} \cup \{H \setminus I(p)\}$ is called the* representative *of the order-k subset.*

An order-k Voronoi region is now defined as

$$\mathcal{V}_k(H, S) = \mathcal{V}(H, S), \text{ where } H \text{ is an order-}k \text{ subset of } S$$

Note that for disjoint segments all order-k subsets are of type 1 and the definition of $\mathcal{V}_k(H, S)$ is equivalent to (2). The following lemma clarifies Def. 1.

Lemma 8. *An order-k subset H induces a non-empty order-k Voronoi region iff there exists a disk that intersects or touches all segments in H but it does not intersect nor touch any segment in $S \setminus H$.*

The order-k Voronoi diagram defined in this way has some differences from the standard order-k Voronoi diagram of disjoint objects. In the standard case, any two neighboring order-k regions belong to two order-k subsets, H_1 and H_2, which are of type 1 and differ by exactly two elements, i.e. $|H_1 \triangle H_2| = 2$, where $\triangle$ denotes the symmetric difference. The bisector of these two elements defines exactly the Voronoi edge separating the two regions. Here, two neighboring regions of order-k subsets H_1 and H_2 that are not both type 1, may differ as $|H_1 \triangle H_2| \geq 1$, see e.g., regions $V(6, 5)$ and $V(3, 4)$ or regions $V(5, 7, 8)$ and $V(7, 8)$ in Fig. 6. However, the representatives of H_1 and H_2 may differ in exactly one or two elements. If $|rep(H_1) \triangle rep(H_2)| = 1$ then the bisector bounding the two regions is $b(p, h)$, where $H_1 = H_1' \cup I(p)$, $h \in H_2 \setminus H_1$, and $|H_2 \setminus H_1| = 1$. (For a type 1 subset, $rep(H) = H$). Voronoi edges bounding the regions of order-k subsets, which are not both of type 1, may remain in the order-k Voronoi diagram for severaf orders, while $|H_1 \cup H_2| > k$.

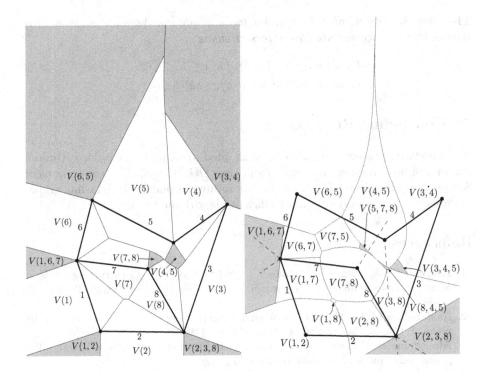

Fig. 6. (a) $V_1(S)$ of a planar straight-line graph. The bold regions are induced by order-k subsets of type-2; (b) $V_2(S)$ of a planar-straight line graph. For brevity we use $V(s_1, \ldots, s_m)$ notation instead of $\mathcal{V}_k(\{s_1, \ldots, s_m\}, S)$.

6   Intersecting Line Segments

Let S be a set of line segments that may intersect in a total I intersection points. For simplicity we assume that no two segments share an endpoint and that no more than two segments intersect at a common point. Intuitively, intersections influence Voronoi diagrams of small order and the influence grows weaker as k increases. Recall that the number of faces, edges and vertices of $\mathcal{V}_k(S)$ are denoted F_k, E_k and V_k, respectively.

Lemma 9. *The total number of unbounded faces in all orders is* $\sum_{i=1}^{n-1} S_i = n(n-1) + 2I$

Lemma 10. $F_1 = n + 2I$, $F_2 = 3n - 3 - S_1 + 2I$, and $F_3 = 5n - 8 - S_1 - S_2 + 2I$.

Following the induction scheme of Theorem 1 and using Lemma 10 as the base case we derive

$$F_k = 2kn - k^2 - n + 1 - \sum_{i=1}^{k-1} S_i + 2I, \quad F_k = 1 - (n-k)^2 + \sum_{i=k}^{n-1} S_i$$

Lemma 7 is valid for arbitrary segments including intersecting ones.

Theorem 3. *The number of Voronoi faces in order-k Voronoi diagram of n intersecting line segments with I intersections is*

$$F_k = O(k(n - k) + I), \text{ for } 1 \le k < n/2$$
$$F_k = O(k(n - k)), \text{ for } n/2 \le k \le n - 1$$

7 Concluding Remarks

Any standard iterative approach can be adapted to compute the order-k Voronoi diagram of non-crossing line segments in time $O(k^2 n \log n)$. The conventions of Section 5 are important for line segments forming a planar straight-line graph. We are currently considering more efficient algorithmic techniques.

References

1. Agarwal, P., de Berg, M., Matousek, J., Schwarzkopf, O.: Constructing levels in arrangements and higher order Voronoi diagrams. SIAM J. Comput. 27(3), 654–667 (1998)
2. Alon, N., Györi, E.: The number of small semispaces of a finite set of points in the plane. J. Comb. Theory, Ser. A 41(1), 154–157 (1986)
3. Aurenhammer, F., Drysdale, R., Krasser, H.: Farthest line segment Voronoi diagrams. Inf. Process. Lett. 100(6), 220–225 (2006)
4. Aurenhammer, F., Klein, R.: Voronoi Diagrams. In: Sack, J.-R., Urrutia, J. (eds.) Handbook of Computational Geometry. North-Holland Publishing Co. (2000)
5. Boissonnat, J.-D., Devillers, O., Teillaud, M.: A Semidynamic Construction of Higher-Order Voronoi Diagrams and Its Randomized Analysis. Algorithmica 9(4), 329–356 (1993)
6. Edelsbrunner, H.: Algorithms in combinatorial geometry. EATCS Monographs on Theoretical Computer Science, ch. 13.4. Springer (1987)
7. Edelsbrunner, H., Maurer, H.A., Preparata, F.P., Rosenberg, A.L., Welzl, E., Wood, D.: Stabbing Line Segments. BIT 22(3), 274–281 (1982)
8. Karavelas, M.I.: A robust and efficient implementation for the segment Voronoi diagram. In: Proc. 1st Int. Symp. on Voronoi Diagrams in Science and Engineering, Tokyo, pp. 51–62 (2004)
9. Lee, D.T.: On k-Nearest Neighbor Voronoi Diagrams in the Plane. IEEE Trans. Computers 31(6), 478–487 (1982)
10. Lee, D.T., Drysdale, R.L.S.: Generalization of Voronoi Diagrams in the Plane. SIAM J. Comput. 10(1), 73–87 (1981)
11. Papadopoulou, E.: Net-Aware Critical Area Extraction for Opens in VLSI Circuits Via Higher-Order Voronoi Diagrams. IEEE Trans. on CAD of Integrated Circuits and Systems 30(5), 704–717 (2011)
12. Shamos, M.I., Hoey, D.: Closest-point problems. In: Proc. 16th IEEE Symp. on Foundations of Comput. Sci., pp. 151–162 (1975)
13. Sharir, M., Agarwal, P.: Davenport-Schinzel Sequences and their Geometric Applications, ch. 5.4. Cambridge University Press (1995)
14. Yap, C.-K.: An $O(n \log n)$ Algorithm for the Voronoi Diagram of a Set of Simple Curve Segments. Discrete & Computational Geometry 2, 365–393 (1987)

On the Farthest Line-Segment Voronoi Diagram

Evanthia Papadopoulou* and Sandeep Kumar Dey

Faculty of Informatics, USI - Università della Svizzera Italiana, Lugano, Switzerland
{evanthia.papadopoulou,deys}@usi.ch

Abstract. The farthest line-segment Voronoi diagram shows properties surprisingly different from the farthest point Voronoi diagram: Voronoi regions may be disconnected and they are not characterized by convex-hull properties. In this paper we introduce the *farthest line-segment hull* and its *Gaussian map*, a closed polygonal curve that characterizes the regions of the farthest line-segment Voronoi diagram similarly to the way an ordinary convex hull characterizes the regions of the farthest-point Voronoi diagram. We also derive tighter bounds on the (linear) size of the farthest line-segment Voronoi diagram. With the purpose of unifying construction algorithms for farthest-point and farthest line-segment Voronoi diagrams, we adapt standard techniques for the construction of a convex hull to compute the farthest line-segment hull in $O(n \log n)$ or output-sensitive $O(n \log h)$ time, where n is the number of segments and h is the size of the hull (number of Voronoi faces). As a result, the farthest line-segment Voronoi diagram can be constructed in output sensitive $O(n \log h)$ time.

1 Introduction

Let S be a set of n simple geometric objects in the plane, such as points or line segments, called sites. The *farthest-site Voronoi diagram* of S is a subdivision of the plane into regions such that the region of a site s is the locus of points farther away from s than from any other site. Surprisingly, the farthest line-segment Voronoi diagram illustrates properties different from its counterpart for points [1]. For example, Voronoi regions are not characterized by convex-hull properties and they may be disconnected; a Voronoi region may consist of $\Theta(n)$ disconnected faces. Nevertheless, the graph structure of the diagram remains a tree and its structural complexity is $O(n)$. An abstract framework on the farthest-site Voronoi diagram (which does not include the case of inter-secting line-segments) was given in [11]. Related is the farthest-polygon Voronoi diagram, later addressed in [4].

In this paper we further study the structural properties of the farthest line-segment Voronoi diagram. We introduce the *farthest line-segment hull* and its

* Research supported in part by the Swiss National Science Foundation, grant 200021-127137 and the ESF EUROCORES program EuroGIGA/VORONOI, SNF 20GG21-134355.

K.-M. Chao, T.-s. Hsu, and D.-T. Lee (Eds.): ISAAC 2012, LNCS 7676, pp. 187–196, 2012.

Gaussian map, a closed polygonal curve that characterizes the regions of the farthest line-segment Voronoi diagram similarly to the way an ordinary convex hull characterizes the regions of the farthest-point Voronoi diagram. Using the farthest line-segment hull we derive tighter upper and lower bounds on the (linear) structural complexity of the diagram improving the bounds in [1] by a constant factor. We provide $O(n \log n)$ and output sensitive $O(n \log h)$-time algorithms for the construction of the farthest line-segment hull, where h, $h \in O(n)$, is the size of the hull, by adapting standard approaches for the construction of an ordinary convex hull. Then the farthest line-segment Voronoi diagram can be constructed in additional $O(h \log h)$ time as given in [1] or in additional expected-$O(h)$ time by adapting the randomized incremental construction for points in [5]. The concept of the farthest hull is applicable to the entire L_p metric, $1 \leq p \leq \infty$, and it is identical for $1 < p < \infty$.

The farthest line-segment Voronoi diagram finds applications in computing the smallest disk that overlaps all given line-segments. It is necessary in defining and computing the *Hausdorff Voronoi diagram* of clusters of line segments, which finds applications in VLSI design automation, see e.g., [12] and references therein.

2   Definitions and the Farthest Hull

Let $S = \{s_1, \ldots, s_n\}$ be a set of n arbitrary line-segments in the plane. Line-segments may intersect or touch at single points. The distance between a point q and a line-segment s_i is $d(q, s_i) = \min\{d(q, y), \forall y \in s_i\}$, where $d(q, y)$ denotes the ordinary distance between two points q, y in the L_p metric $1 \leq p \leq \infty$. The farthest Voronoi region of a line-segment s_i is

$$freg(s_i) = \{x \in \mathbb{R}^2 \mid d(x, s_i) \geq d(x, s_j), 1 \leq j \leq n\}$$

The collection of all farthest Voronoi regions, together with their bounding edges and vertices, constitute the *farthest line-segment Voronoi diagram* of S, denoted as *FVD(S)* (see Fig. 1). Any maximally connected subset of a region in *FVD(S)* is called a *face*.

Any Voronoi edge bounding two neighboring regions, $freg(s_i)$ and $freg(s_j)$, is portion of the bisector $b(s_i, s_j)$, which is the locus of points equidistant from s_i and s_j. For line-segments in general position that are non-intersecting, $b(s_i, s_j)$ is an unbounded curve that consists of a constant number of simple pieces as induced by elementary bisectors between the endpoints and open portions of s_i, s_j. If segments intersect at a point p the bisector consists of two such curves intersecting at point p. If segments share a common endpoint the bisector may contain two dimensional regions in which case standard conventions get applied. Typically, a single line-segment is treated as three entities: two endpoints and an open line-segment; the entire equidistant area is assigned to the common endpoint. For more information on line-segment bisectors, see e.g., [9,10] for the L_2 metric, [14] for the L_∞ metric, and [8] for points in L_p.

The *farthest line-segment hull*, for brevity the *farthest hull*, is a closed polygonal curve that encodes the unbounded bisectors of *FVD(S)* maintaining their

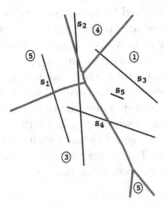

Fig. 1. A farthest line-segment Voronoi diagram in the Euclidean plane

cyclic order. In the following, we assume the ordinary Euclidean distance, however, definitions remain identical in the entire L_p, $1 < p < \infty$, metric. For the L_∞ (L_1) version of the farthest hull see [6].

Definition 1. *A* line l *through the endpoint* p *of a line-segment* s, $s \in S$, *is called a* supporting line *of* S *if and only if an open halfplane induced by* l, *denoted* $H(l)$, *intersects all segments in* S, *except* s *(and possibly except additional segments incident to* p*). Point* p *is said to* admit *a supporting line and it induces a vertex on the farthest hull. The unit normal of* l *pointing away from* $H(l)$, *is called the* unit vector *of* l *and is denoted* $\nu(l)$. *A line-segment* s, $s \in S$, *such that the line* l *through* s *is a supporting line of* S *and* $H(l)$ *intersects all segments in* $S \setminus \{s\}$, *is called a* hull segment; *the unit vector of* l *is also denoted* $\nu(s)$.

A single line-segment s may result in two hull segments of two opposite unit vectors if the supporting line through s intersects all segments in S.

Definition 2. *The line segment* $\overline{pq}$ *joining the endpoints* p, q, *of two line segments* $s_i, s_j \in S$ *is called a* supporting segment *if and only if an open halfplane induced by the line* l *through* $\overline{pq}$, *denoted* $H(\overline{pq})$, *intersects all segments in* S, *except* s_i, s_j *(and possibly except additional segments incident to* p, q*). The unit normal of* $\overline{pq}$ *pointing away from* $H(\overline{pq})$ *is called the* unit vector *of* $\overline{pq}$, $\nu(\overline{pq})$.

As shown in [1], a segment s has a non-empty Voronoi region in *FVD(S)* if and only if s or an endpoint of s admits a supporting line. The unbounded bisectors of *FVD(S)* correspond exactly to the supporting segments of S.

Theorem 1. *Let* *f-hull(S)* *denote the sequence of the hull segments and the supporting segments of* S, *ordered according to the angular order of their unit vectors.* *f-hull(S)* *forms a closed, possibly self intersecting, polygonal curve, which is called the* farthest line-segment hull *of* S *(for brevity, the* farthest hull*).*

Proof. (Sketch). The unit vector of any supporting line, hull segment, or supporting segment of S must be unique (by Definitions 1, 2). Thus, *f-hull(S)* admits a well defined ordering as obtained by the angular order of the unit vectors of its edges. Given a farthest hull edge e_i of unit vector $\nu_i = \nu(e_i)$, let $\nu_{i+1} = \nu(e_{i+1})$ be the unit vector following ν_i in a clockwise traversal of the circular list of all unit vectors of S. It is not hard to argue that e_{i+1} must be incident to an endpoint of e_i, m, by considering a supporting line through m, l_m, which starts as the line through e_i and rotates clockwise around m until it hits e_{i+1}. During the rotation l_m always remains a supporting line of S. Vertex m is chosen between the endpoints of e_i according to whether e_i is a hull or a supporting segment.

Thus, f-$hull(S)$ forms a polygonal chain, which may self intersect and may visit a vertex multiple times. The uniqueness of unit vectors implies that edges are visited only once, hence the polygonal chain must be closed.   □

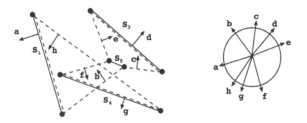

Fig. 2. The farthest line-segment hull for Fig. 1 and its Gaussian map

Fig. 2 illustrates the farthest hull and its angular ordering for the Voronoi diagram of Fig. 1. If line-segments in S degenerate to points, the farthest hull corresponds exactly to the convex hull of S. The vertices of the farthest hull are exactly the endpoints of S that admit a supporting line. The edges are of two types: supporting segments and hull segments. A supporting line of S is exactly a supporting line of the farthest hull. Any maximal chain of supporting segments between two consecutive hull segments must be convex.

Consider the *Gaussian map* of the farthest hull of S, for short *Gmap*, denoted *Gmap(S)*, which is a mapping of the farthest hull onto the unit circle K_o, such that every edge e is mapped to a point on the circumference of K_o as obtained by its unit vector $\nu(e)$, and every vertex is mapped to one or more arcs as delimited by the unit vectors of the incident edges (see Fig. 2). For more information on the Gaussian map see e.g., [3,13]. The Gaussian map can be viewed as a cyclic sequence of vertices of the farthest hull, each represented as an arc along the circumference of K_0. Each point along an arc of K_0 corresponds to the unit vector of a supporting line through the corresponding hull-vertex. The Gaussian map provides an encoding of all the supporting lines of the farthest hull as well as an encoding of the unbounded bisectors and the regions of the farthest line-segment Voronoi diagram. The portion of the Gaussian map above (resp. below) the horizontal diameter of K_o is referred to as the upper (resp. lower) Gmap. Thus, we can define the upper (resp. lower) farthest hull as the portion that corresponds to the upper (resp. lower) Gmap, similarly to an ordinary upper (resp. lower) convex hull.

Corollary 1. *FVD(S) has exactly one unbounded bisector for every supporting segment s of f-hull(S), which is unbounded in the direction opposite to $\nu(s)$. Unbounded bisectors in FVD(S) are cyclically ordered following exactly the cyclic ordering of Gmap(S).*

In [1] the unbounded bisectors of *FVD(S)* are identified through the point-line duality transformation T, which maps a point $p = (a, b)$ in the primal plane to a line $T(p) : y = ax - b$ in the dual plane, and vice versa. A segment $s_i = uv$ is sent into the wedge w_i that lies below (resp. above) both lines $T(u)$ and $T(v)$ (see Fig. 3 of [1]), referred to as the lower (resp. upper) wedge. Let E (resp. E') be the boundary of the union of the lower (resp. upper) wedges $w_1, \ldots, w_n$. As shown in [1], the edges of E (in x-order) correspond to the faces of *FVD(S)*, which are unbounded in directions 0 to π (in cyclic order). Respectively for the edges in E' and the Voronoi faces unbounded in directions π to 2π. In this paper we point out the equivalence between E and the lower Gmap. The edges of E in increasing x-order correspond exactly to the arcs of the lower Gmap in counterclockwise order; the vertices of E are exactly the unit vectors of the Gmap; the apexes of wedges in E are the unit vectors of hull segments. Respectively for E' and the upper Gmap. As pointed out in [1], E forms a Davenport-Schinzel sequence of order 3, thus the same holds for the lower Gmap. This observation does not imply a linear complexity bound, however. A $4n + 2$ upper bound on the size of E was shown in [1] based on [7]. For non-crossing segments, the order of the sequence is 2, which directly implies a linear complexity bound (see also [4]).

3   Improved Combinatorial Bounds

In this section we give tighter upper and lower bounds on the number of faces of the farthest line-segment Voronoi diagram for arbitrary line segments. Let a *start-vertex* and an *end-vertex* respectively stand for the right and the left endpoint of a line segment. Let an *interval* $[a_i, a_{i+1}]$ denote the portion of the lower Gmap between two consecutive (but not adjacent) occurrences of arcs for segment $s_a = (a', a)$, where a, a' denote the start-vertex and end-vertex of s_a respectively. Interval $[a_i, a_{i+1}]$ is assumed to be *non-trivial* i.e., it contains at least one arc in addition to a, a'. The following lemma is easy to derive using the duality transformation of [1].

Lemma 1. *Let $[a_i, a_{i+1}]$ be a non-trivial interval of segment $s_a = (a', a)$ on lower Gmap(S). We have the following properties: 1. The vertex following a_i (resp. preceding a_{i+1}) in $[a_i, a_{i+1}]$ must be a start-vertex (resp. end-vertex). 2. If a_i is a start-vertex (resp. a_{i+1} is end-vertex), no other start-vertex (resp. end-vertex) in the interval $[a_i, a_{i+1}]$ can appear before a_i or past a_{i+1} on the lower Gmap, and no end-vertex (resp. start-vertex) in $[a_i, a_{i+1}]$ can appear before a_i (resp. past a_{i+1}) on the lower Gmap.*

We use the following charging scheme for a non-trivial interval $[a_i, a_{i+1}]$: If a_i is a start-vertex, let u be the vertex immediately following a_i in $[a_i, a_{i+1}]$; the appearance of a_{i+1} is charged to u, which must be a start-vertex by Lemma 1. If a_{i+1} is an end-vertex, let u be the vertex in $[a_i, a_{i+1}]$ immediately preceding a_{i+1}; the appearance of a_i is charged to u, which by Lemma 1, must be an end-vertex.

Lemma 2. *The re-appearance along the lower Gmap of any endpoint of a segment s_a is charged to a unique vertex u of the lower f-hull, such that no other re-appearance of a segment endpoint on the lower Gmap can be charged to u.*

Proof. Let $[a_i, a_{i+1}]$ be a non-trivial interval of the lower Gmap and let u be the vertex charged the re-appearance of a_{i+1} (or a_i) as described in the above charging scheme. By Lemma 1, all occurrences of u are in $[a_i, a_{i+1}]$. Suppose (for a contradiction) that u can be charged the re-appearance of some other vertex c. Assuming that u is a start vertex, c must be a start vertex and an interval $[cu...c]$ must exist such that $cu \in [a_i, a_{i+1}]$, and thus $[cu...c] \in [a_i, a_{i+1}]$. But then u could not appear outside $[cu...c]$, contradicting the fact that u has been charged the reappearance of a_{i+1}. Similarly for an end vertex u.   □

By Corollary 1, the number of faces of $FVD(S)$ equals the number of supporting segments of the farthest hull, which equals the number of *maximal arcs* along the Gmap between consecutive pairs of unit vectors of supporting segments. There are two types of maximal arcs: *segment arcs*, which consist of a segment unit vector and its two incident arcs of the segment endpoints, and *single-vertex* arcs, which are single arcs bounded by the unit vectors of the two incident supporting segments.

Lemma 3. *The number of maximal arcs along the lower (resp. upper) Gmap, and thus the number of faces of FVD(S) unbounded in directions 0 to π (resp. π to 2π), is at most 3n-2. This bound is tight.*

Proof. Consider the sequence of all occurrences of a single segment $s_a = (a', a)$ on the lower Gmap. It is a sequence of the form

$$...a...aa'...a'... \quad \text{or} \quad ...a...a...a'...a'...$$

The number of maximal arcs involving s_a is exactly one plus the number of (non-trivial) intervals involving the endpoints of s_a. Summing over all segments, the total number of maximal arcs on the lower Gmap is at most n plus the total number of vertices that may get charged due to a non-trivial interval (i.e., the re-appearance of a vertex). By Lemma 2, a vertex can be charged at most once and there are 2n vertices in total. However, the first and last vertex along the lower Gmap cannot be charged at all. Thus, in total $3n - 2$. Similarly for the upper Gmap.

Figure 3a illustrates an example, in dual space, of n line segments (lower wedges) whose lower Gmap (boundary of the wedge union) consists of 3n-2 maximal arcs. There are exactly 2n-2 charges for vertex (wedge) re-appearances and exactly n hull segments.   □

Theorem 2. *The total number of faces of the farthest line-segment Voronoi diagram of a set S of n arbitrary line segments is at most 6n−6. A corresponding lower bound is 5n-4.*

Proof. The upper bound is derived by Lemma 3 and Corollary 1. A lower bound of 5n-4 faces can be derived by the example, in dual space, of Figure 3b. In Figure 3b, n line segments are depicted as lower and upper wedges using the point-line duality. There are $2n$ hull segments, $2(n − 2) + 1$ charges for vertex reappearances on lower wedges, and $n − 1$ charges on upper wedges.   □

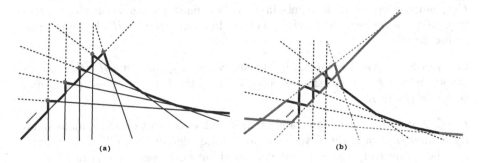

(a)   (b)

Fig. 3. (a)Lower Gmap of 3n-2 arcs. (b)Gmap of 5n-4 arcs.

Theorem 2 improves the $8n + 4$ upper bound and the $4n − 4$ lower bound on the number of faces of the farthest line-segment Voronoi diagram given in [1], as based on [7]. For disjoint segments the corresponding bound is $2n − 2$ [4].

4   Algorithms for the Farthest Line-Segment Hull

Using the Gmap, we can adapt most standard techniques to compute a convex hull with the ability to compute the farthest line-segment hull, within the same time complexity. Our goal is to unify techniques for the construction of farthest-point and farthest line-segment Voronoi diagrams. By adapting Chan's output sensitive algorithm we derive an $O(n \log h)$ output-sensitive algorithm to compute the farthest line-segment Voronoi diagram. A two-phase $O(n \log n)$ algorithm in dual space, which is based on divide and conquer paired with plane sweep is outlined in [1].

4.1   Divide and Conquer or Incremental Constructions

We list properties that lead to a linear-time merging scheme for the farthest hulls of two disjoint sets of segments, L, R, and their Gaussian maps. The merging scheme leads to an $O(n \log n)$ divide-and-conquer approach and to an $O(n \log n)$ two-stage incremental construction. Recall that there may be $\Theta(|L| + |R|)$ supporting segments between the two farthest hulls. Given a unit vector $\nu(e)$ of

Gmap(L), let its *R*-vertex be the vertex v in R such that $\nu(e)$ falls along the arc of v in *Gmap(R)*. Respectively for *Gmap(R)*. A supporting line, an edge or a vertex of the farthest hull of L (resp. R) and their corresponding unit vector or arc are called *valid* if they remain in the farthest hull of $L \cup R$.

Lemma 4. *An edge e of f-hull(L) and its unit vector $\nu(e)$ remain valid if and only if the R-vertex of $\nu(e)$ in Gmap(R) lies in $H(e)$.*

Proof. Let q, in segment s_q, be the *R*-vertex of $\nu(e)$. The line parallel to e passing through q, $l_q(e)$, must be a supporting line of *f-hull(R)*. $H(e)$ must intersect all segments in L except those inducing e. If $q \in H(e)$ then $l_q(e) \in H(e)$ and thus $H(e)$ must intersect all segments in R, thus e must remain valid. If $q \notin H(e)$ then $l_q(e)$ lies entirely outside $H(e)$ and s_q does not intersect $H(e)$, thus e must be invalid.   □

Lemma 5. *A vertex m of f-hull(L) remains valid if and only if either an incident farthest hull edge remains valid, or m is the L-vertex of an invalid unit vector in Gmap(R).*

Proof. A vertex incident to a valid farthest hull edge must clearly remain valid. By Lemma 5, if m is the *L*-vertex of an invalid edge in *f-hull(R)* then m must be valid. Conversely, suppose that m is valid but both edges e_i, e_{i+1} in *f-hull(L)* incident to m are invalid. Since m is valid, there is a supporting line of $L \cup R$ passing through m, denoted l_m, such that $\nu(l_m)$ is between $\nu(e_i)$ and $\nu(e_{i+1})$. Let u be the *R*-vertex of $\nu(l_m)$, where $\nu(l_m)$ is between $\nu(f_j)$ and $\nu(f_{j+1})$ in *Gmap(R)*. Since l_m is valid, u is in $H(l_m)$. Since $\nu(l_m)$ is between $\nu(f_j)$ and $\nu(f_{j+1})$, m cannot belong in $H(f_j) \cap H(f_{j+1})$. This implies that at least one of f_j and f_{j+1} must be invalid.   □

Let *Gmap(LR)* denote the circular list of unit vectors and arcs derived by superimposing *Gmap(L)* and *Gmap(R)*. After determining valid and invalid unit vectors using Lemmas 4 and 5, *Gmap(LR)* represents a circular list of the vertices in *f-hull(L∪R)* (by Lemma 5). Thus, supporting segments between the two hulls can be easily derived by the pairs of consecutive valid vertices in *Gmap(LR)* that belong to different sets. In particular, to derive *Gmap(L∪R)* from *Gmap(LR)* do the following: For any two consecutive valid vectors, one in *Gmap(L)* and one in *Gmap(R)*, insert a new unit vector (see Fig. 4). For any valid vertex m of *Gmap(L)* (resp. *Gmap(R)*) between two consecutive valid vectors of *Gmap(R)* (resp. *Gmap(L)*), insert new unit vectors for the corresponding supporting segments incident to m.

For $R = \{s\}$, the merging process can be refined as follows: Let $\nu_1(s)$ be the unit vector of s in its upper Gmap. 1. Perform binary search to locate $\nu_1(s)$ in *upper-Gmap(L)*. 2. Sequentially move counterclockwise along *upper-Gmap(L)* to test the validity of the encountered unit vectors until either a valid start-vertex of x-coordinate smaller than the start-vertex of s is found or the beginning of *upper-Gmap(L)* is reached. 3. Move clockwise along *upper-Gmap(L)* until either a valid end-vertex of x-coordinate larger than the end-vertex of s is found or the end of *upper-Gmap(L)* is reached. In the process, all relevant supporting

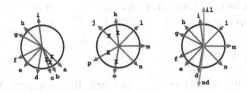

Fig. 4. Merging of two Gmaps

segments in the upper Gmap are identified. Similarly for the lower Gmap. This insertion procedure can give an alternative $O(n \log n)$ algorithm to compute the farthest hull by independently considering start-vertices and end-vertices in increasing x-coordinate followed by a merging step for the two Gmaps: First compute a partial Gmap of start vertices by inserting start-vertices in order of increasing x-coordinate. (Recall that insertion starts with a binary search). Then compute a partial Gmap of end-vertices by considering end-vertices in decreasing x-coordinate. Finally, merge the two partial Gmaps to obtain the complete Gmap. Because of the order of insertion, any portion of the partial Gmap traversed during an insertion phase gets deleted as invalid, thus the time complexity bound is achieved.

4.2   Output Sensitive Approaches

Using the Gmap, Jarvis march and quick hull are simple to generalize to construct the farthest hull within $O(nh)$ time, where h is the size of the farthest hull. For Jarvis march, unit vectors are identified one by one in say counterclockwise order starting at a vertex in some given direction, e.g., the bottommost horizontal supporting line. By combining an $O(n \log n)$ construction algorithm and Jarvis march as detailed in [2], we can obtain an $O(n \log h)$ output sensitive algorithm to construct the farthest hull. There is one point in [2] that needs modification to be applicable to the farthest hull, namely computing the tangent (here a supporting segment) between a given point and a convex hull (here a farthest hull, $f\text{-}hull(L)$). Note that unlike the ordinary convex hull, $\Theta(|L|)$ such supporting segments may exist, complicating a binary search. However, sequential search to compute those tangents can work within the same overall time complexity for one *wrapping phase* of the Jarvis march (see [2]). During a *wrapping phase*, a set T of at most hr tangents are computed, where r is the number of groups that partition the initial set of n points (here segments), each group being of size at most m, $m = \lceil n/r \rceil$. The set of tangents T can be computed in $O(n)$ time. In particular, given $Gmap(S_i)$, $S_i \subseteq S$, a hull vertex p incident to segment s, and a supporting segment of unit vector ν, we can compute the next unit vector ν_{next} in $Gmap(S_i \cup \{s\})$ in counterclockwise order by sequentially scanning $Gmap(S_i)$ starting at ν, applying the criteria of Lemmas 4 and 5 until ν_{next} is encountered. No portion of $Gmap(S_i)$ between ν and ν_{next} needs to be encountered again during this wrapping phase. Thus, the $O(hr)$ supporting segments of one wrapping phase can be computed in $O(n)$ time. Note that if s is a hull segment, a supporting line through s may be included in T.

5 Concluding Remarks

Once the farthest hull is derived, the farthest line segment Voronoi diagram can be computed similarly to its counterpart for points. For example, one can use the simple $O(h \log h)$ algorithm of [1] or adapt the randomized incremental construction for the farthest-point Voronoi diagram of [5] in expected $O(h)$-time. The randomized analysis in [5] remains valid simply by substituting points along a convex hull with the elements of the farthest hull, where elements of the farthest hull are independent objects even when they refer to the same line segment. Adapting Chan's algorithm [2] for the construction of the farthest hull results in an output sensitive $O(n \log h)$ algorithm for the farthest line-segment Voronoi diagram.

References

1. Aurenhammer, F., Drysdale, R.L.S., Krasser, H.: Farthest line segment Voronoi diagrams. Information Processing Letters 100(6), 220–225 (2006)
2. Chan, T.M.: Optimal output-sensitive convex-hull algorithms in two and three dimensions. Discrete and Computational Geometry 16, 361–368 (1996)
3. Chen, L.L., Chou, S.Y., Woo, T.C.: Parting directions for mould and die design. Computer-Aided Design 25(12), 762–768 (1993)
4. Cheong, O., Everett, H., Glisse, M., Gudmundsson, J., Hornus, S., Lazard, S., Lee, M., Na, H.S.: Farthest-Polygon Voronoi Diagrams. arXiv:1001.3593v1 (cs.CG) (2010)
5. de Berg, M., Cheong, O., van Kreveld, M., Overmars, M.: Computational Geometry: Algorithms and Applications, 3rd edn. Springer (2008)
6. Dey, S.K., Papadopoulou, E.: The L_∞ farthest line segment Voronoi diagram. In: Proc. 9th Int. Symposium on Voronoi Diagrams in Science and Engineering (2012)
7. Edelsbrunner, H., Maurer, H.A., Preparata, F.P., Rosenberg, A.L., Welzl, E., Wood, D.: Stabbing Line Segments. BIT 22(3), 274–281 (1982)
8. Lee, D.T.: Two-dimensional Voronoi diagrams in the L_p metric. J. ACM 27(4), 604–618 (1980)
9. Lee, D.T., Drysdale, R.L.S.: Generalization of Voronoi Diagrams in the Plane. SIAM J. Comput. 10(1), 73–87 (1981)
10. Karavelas, M.I.: A robust and efficient implementation for the segment Voronoi diagram. In: Proc. 1st. Int. Symposium on Voronoi Diagrams in Science and Engineering, pp. 51–62 (2004)
11. Mehlhorn, K., Meiser, S., Rasch, R.: Furthest site abstract Voronoi diagrams. Int. J. of Comput. Geometry and Applications 11(6), 583–616 (2001)
12. Papadopoulou, E.: Net-aware critical area extraction for opens in VLSI circuits via higher-order Voronoi diagrams. IEEE Trans. on CAD 30(5), 704–716 (2011)
13. Papadopoulou, E., Lee, D.T.: The Hausdorff Voronoi diagram of polygonal objects: A divide and conquer approach. Int. J. of Computational Geometry and Applications 14(6), 421–452 (2004)
14. Papadopoulou, E., Lee, D.T.: The L_∞ Voronoi Diagram of Segments and VLSI Applications. Int. J. Comp. Geom. and Applications 11(5), 503–528 (2001)

Computing the Longest Common Subsequence of Two Run-Length Encoded Strings

Yoshifumi Sakai

Graduate School of Agricultural Science, Tohoku University,
1-1 Amamiya-machi, Tsutsumidori, Aoba-ku, Sendai 981-8555, Japan
sakai@biochem.tohoku.ac.jp

Abstract. The present article reveals that the problem of finding the longest common subsequence of two strings given in run-length encoded form can be solved in $O(mn \log \log \min(m, n, M/m, N/n, X))$ time, where one input string is of length M with m runs, the other is of length N with n runs, and X is the average difference between the length of a run from one input string and that of a run from the other.

1 Introduction

A maximal-length substring of a string that consists of common symbols is referred to as a run of the string. A sting is run-length encoded if the string is decomposed into a sequence of runs, and each run is represented as a pair of its common symbol and length. This form of a string can compactly represent the string, if the string consists of a small number of long runs.

Let A and B be strings of length M with m runs, and of length N with n runs, respectively. The problem of computing the similarity between A and B, which is given in run-length encoded form, has been investigated with respect to various scoring metrics. This problem was first shown to be solvable in $O(mn \max(M/m, N/n))$ time for the longest common subsequence (LCS) metric by Bunke and Csirik [3]. The same time bound was then achieved for more relaxed scoring metrics, such as the weighted edit distance metric [5,9] and the affine gap penalty metric [7]. Basically, these algorithms process any pair of runs, one from A and the other from B, in $O(\max(M/m, N/n))$ amortized time, where M/m and N/n are the average run lengths of A and B, respectively. For restricted metrics, algorithms that perform each run pair in time depending only on the number of runs were also proposed. The recent algorithm due to Chen and Chao [4], which processes each run pair in $O(\max(m, n))$ amortized time for the edit distance metric, is one such algorithm. This algorithm performs in $O(mn \max(m, n))$ time.

For the LCS metric, further faster algorithms have been developed for the problem. Apostolico et al. [2] and Mitchell [10] proposed algorithms that perform in time $O(mn \log \max(m, n))$. In particular, the algorithm of Apostolico et al. [2] processes each run pair in $O(\log \max(m, n))$ amortized time using a balanced binary search tree. This process time can be improved to $O(\log \log \max(m, n))$

K.-M. Chao, T.-s. Hsu, and D.-T. Lee (Eds.): ISAAC 2012, LNCS 7676, pp. 197–206, 2012.

amortized time if the balanced binary search tree is replaced by a data structure given by van Emde Boas [11] after a certain preprocessing based on the integer sorting due to Han [6]. According to this observation, the algorithm of Apostolico et al. [2] can immediately be modified so as to perform in $O(mn \log \log \max(m, n))$ time without changing its outline. About a decade later, another best previous time bound was achieved by Liu et al. [8] and Ann et al. [1]. They improved the amortized process time for each run pair from $O(\max(M/m, N/n))$ due to Bunke and Csirik [3] to $O(\min(M/m, N/n))$, and hence, their algorithms perform in $O(mn \min(M/m, N/n))$ time. Whether the problem can be solved in $O(mn)$ time remains an open question.

Although Apostolico et al. [2] and Mitchell [10] have achieved an almost quadratic-time computation of the problem for the LCS metric, the asymptotic time bound given by Liu et al. [8] and Ann et al. [1] naturally poses two theoretical issues about process time of each run pair: whether each run pair can be processed in amortized time depending only on the number of runs in any of the input strings, and whether each run pair can be processed in amortized time logarithmic or further less in the average run length of any of the input strings. The present article settles these issues positively, by showing that a variant of the algorithm of Apostolico et al. [2] with the modification mentioned earlier can process each run pair in $O(\log \log \min(m, n, M/m, N/n, X))$ amortized time, where X is the average difference between the length of a run from A and that of a run from B. The execution time of the proposed variant is hence $O(mn \log \log \min(m, n, M/m, N/n, X))$, improving the previous best time bounds of the problem for the LCS metric.

2   Preliminaries

Let Σ be a finite set of symbols. For any string A, let $A[I]$ denote the Ith symbol of A. For any strings A and B, let AB denote the concatenation of A followed by B. For any symbol a and any positive integer K, let a^K denote the string $aa \cdots a$ of length K. The *run-length encoded (RLE) form* representing string A is the shortest sequence $(\alpha(1), M(1))(\alpha(2), M(2)) \cdots (\alpha(m), M(m))$ such that $A = \alpha(1)^{M(1)} \alpha(2)^{M(2)} \cdots \alpha(m)^{M(m)}$, where $\alpha(i)$ and $M(i)$ are a symbol in Σ and a positive integer, respectively. Each substring $\alpha(i)^{M(i)}$ is referred to as the ith *run* of A. The ith run of A is denoted by $A(i)$. Let $A[G..I]$ and $A(g..i)$ denote strings $A[G]A[G+1] \cdots A[I]$ and $A(g)A(g+1) \cdots A(i)$, respectively.

A *subsequence* of string A is a string obtained from A by deleting zero or more symbols at any position. A *longest common subsequence (LCS)* of strings A and B is one of the longest subsequences common to A and B. Given strings A and B over Σ, the *LCS problem* consists of finding an arbitrary LCS of A and B. The present article considers the LCS problem for the case in which two input strings, A of length M and B of length N, are given in run-length encoded form as $(\alpha(1), M(1)) \cdots (\alpha(m), M(m))$ and $(\beta(1), N(1)) \cdots (\beta(n), N(n))$, respectively. The solution of this problem is the RLE form representing an arbitrary LCS of A and B. This problem is referred to as the *LCS problem in RLE form*.

For any symbol a in Σ, let m_a be the number of indices i with $1 \leq i \leq m$ such that $\alpha(i) = a$, and let $i_a(u)$ denote the uth least such index i. For convenience, let $i_a(0) = 0$. Let $M_a(u)$ denote the value $M(i_a(u))$, and let $M_a(w..u)$ denote the sum $M_a(w) + M_a(w+1) + \cdots + M_a(u)$. Note that $M_a(w..u) = M_a(1..u) - M_a(1..w-1)$. For simplicity, $M_a(1..m_a)$ is denoted by M_a. Define n_a, $j_a(v)$, $N_a(v)$, $N_a(x..v)$, and N_a analogously with respect to B. Let $R_a(u,v)$ denote the difference $N_a(1..v) - M_a(1..u)$. Hence, for example, $R_a(w-1, x-1) \leq R_a(u,v)$ if and only if $M_a(w..u) \leq N_a(x..v)$. The present article assumes that any algorithm has a constant-time access to any value introduced above, which can be achieved by a straightforward $O(m + n)$-time preprocessing.

Let $L_a(u,v) = \min(M_a(u), |N_a(v) - M_a(u)|, N_a(v))$, and let L_a be the sum of $L_a(u,v)$, where (u,v) ranges over all index pairs such that $1 \leq u \leq m_a$ and $1 \leq v \leq n_a$. Let L be the sum of L_a with a ranging over all symbols in Σ.

3   Algorithm

This section proposes an $O(mn \log \log \min(m, n, L/(mn)))$-time algorithm for the LCS problem in RLE form based on the dynamic programming technique. This algorithm should be regarded as a variant of the algorithm of Apostolico et al. [2], because the algorithm performs well depending on a recurrence similar to (but not exactly the same as) that used in their algorithm.

3.1   Outline of the Algorithm

This subsection presents an outline of the proposed algorithm. (Due to limitation of space, the present article does not present proofs of the lemmas introduced in this subsection. The proofs will be presented in the full paper.)

For any index pair (I, J) with $0 \leq I \leq M$ and $0 \leq J \leq N$, let $C[I, J]$ denote an arbitrary LCS of prefixes $A[1..I]$ and $B[1..J]$. Similarly, for any index pair (i, j) with $0 \leq i \leq m$ and $0 \leq j \leq n$, let $C(i, j)$ denote an arbitrary LCS of $A(1..i)$ and $B(1..j)$. It is easy to verify that concatenation $C[I-1, J-1]a$ is an LCS of $A[1..I]$ and $B[1..J]$ if both $A[I]$ and $B[J]$ are identical to a; otherwise, at least one of $C[I, J-1]$ or $C[I-1, J]$ is an LCS of $A[1..I]$ and $B[1..J]$. Using the dynamic programming technique based on this simple recurrence, the LCS problem can be solved in $O(MN)$ time by constructing a table of the length of strings $C[I, J]$ for all index pairs (I, J) with $0 \leq I \leq M$ and $0 \leq J \leq N$ in $O(MN)$ time, and then performing a traceback process in $O(M + N)$ time [12]. The proposed algorithm for the LCS problem in RLE form also follows this outline, but instead uses a table of the length of strings $C(i, j)$ for all index pairs (i, j) with $0 \leq i \leq m$ and $0 \leq j \leq n$.

Let any index pair (I, J) with $A[I] = B[J]$ be referred to as a *match*. Let any match (I, J) such that both $C[I, J-1]$ and $C[I-1, J]$ are shorter than $C[I, J]$ be referred to as *dominant*. In other words, a match (I, J) is dominant if and only if $C[I, J]$ is a subsequence of neither $A[1..I-1]$ nor $B[1..J-1]$. One important property of dominant matches is that, for any match (G, H), there exists a dominant match (I, J) with $I = G$ or $J = H$ such that $C[I, J]$ is an LCS of $A[1..G]$ and $B[1..H]$. Due to this property, for any index pair (i, j) with $\alpha(i) = \beta(j)$, at least one of $C[M(1..i), J(i, j)]$ or $C[I(i, j), N(1..j)]$ is an LCS of $A(1..i)$ and $B(1..j)$, where $J(i, j)$ is the greatest index J less than or equal to $N(1..j)$ such that $(M(1..i), J)$ is dominant, and $I(i, j)$ is the greatest index I less than or equal to $M(1..i)$ such that $(I, N(1..i))$ is dominant. Furthermore, in the above observation, $C[M(1..i), J(i, j)]$ can be replaced by $C(i, j - 1)$ if there exist no such indices J with $N(1..j - 1) < J$, and similarly, $C[I(i, j), N(1..j)]$ can be replaced by $C(i - 1, j)$ if there exist no such indices I with $M(1..i - 1) < I$. The proposed algorithm constructs the table of the length of strings $C(i, j)$ based on these observations.

The (i, j)th entry of the table for each index pair (i, j) is determined in row-by-row order (i.e., in ascending order with respect to $(i-1)n+j$). If $\alpha(i) \neq \beta(j)$, then at least one of $C(i, j - 1)$ or $C(i - 1, j)$ is an LCS of $A(1..i)$ and $B(1..j)$. Hence, the (i, j)th entry of the table can be determined in constant time by referring to the $(i, j - 1)$th and the $(i - 1, j)$th entries. In contrast, the (i, j)th entry with $\alpha(i) = \beta(j)$ is determined, after obtaining a list $\mathcal{L}(i, j)$ of dominant matches defined as follows. Let $\mathcal{J}(i, j)$ be the list of all dominant pairs $(M(1..i), J)$ with $N(1..j - 1) < J < N(1..j)$ in ascending order with respect to J. Let $\mathcal{K}(i, j)$ be the list consisting of a single element $(M(1..i), N(1..j))$ if it is dominant; otherwise, let $\mathcal{K}(i, j)$ be the empty list. Let $\mathcal{I}(i, j)$ be the list of all dominant pairs $(I, N(1..j))$ with $M(1..i - 1) < I < M(1..i)$ in descending order with respect to I. Then, $\mathcal{L}(i, j)$ is defined as the concatenation of $\mathcal{J}(i, j)$, $\mathcal{K}(i, j)$, and $\mathcal{I}(i, j)$ in this order. It follows from this definition that, if $\mathcal{L}(i, j)$ is non-empty, then $C[M(1..i), N(1..j)]$ is an LCS of $A(1..i)$ and $B(1..j)$. Otherwise, at least one of $C(i, j - 1)$, $C[M(1..i), J]$, $C[I, N(1..j)]$, or $C(i - 1, j)$ is an LCS of $A(1..i)$ and $B(1..j)$, where $(M(1..i), J)$ is the last element of $\mathcal{J}(i, j)$, and $(I, N(1..j))$ is the first element of $\mathcal{I}(i, j)$.

The length of $C(i, j)$ and the entry of the table to which the backward pointer from the (i, j)th entry points are determined according to the following lemma based on the observations presented by Apostolico et al. [2]. For any dominant match (I, J), let $p(I, J)$ denote the index pair $\arg\max_{(w,x)} w$ (hence, $\arg\max_{(w,x)} x$), where $M(1..i_a(u) - 1) < I \leq M(1..i_a(u))$, $N(1..j_a(v) - 1) < J \leq N(1..j_a(v))$, and (w, x) ranges over all index pairs such that $1 \leq w \leq u$, $1 \leq x \leq v$, and $R_a(w-1, x-1) = R_a(u, v) - (N(1..j_a(v)) - J) + (M(1..i_a(u)) - I)$ (i.e., $M_a(w..u) - (M(1..i_a(u)) - I) = N_a(x..v) - (N(1..j_a(v)) - J))$. The lemma claims that $p(I, J)$ recursively specifies an LCS of $A[1..I]$ and $B[1..J]$.

Lemma 1 (Apostolico et al. [2]). *For any dominant match* (I, J), $C(i_a(w) - 1, j_a(x) - 1)a^K$ *is an LCS of* $A[1..I]$ *and* $B[1..J]$, *where* $(w, x) = p(I, J)$, $M(1..i_a(u) - 1) < I \leq M(1..i_a(u))$, $N(1..j_a(v) - 1) < J \leq N(1..j_a(v))$, *and* $K = M_a(w..u) - (M(1..i_a(u)) - I) = N_a(x..v) - (N(1..j_a(v)) - J)$.

Based on this lemma, it is useful to introduce the following lists. Let $\mathcal{J}_a(u, v)$, $\mathcal{K}_a(u, v)$, $\mathcal{I}_a(u, v)$, and $\mathcal{L}_a(u, v)$ be the lists obtained from $\mathcal{J}(i, j)$, $\mathcal{K}(i, j)$, $\mathcal{I}(i, j)$, and $\mathcal{L}(i, j)$ with $i = i_a(u)$ and $j = j_a(v)$, respectively, by replacing each element (I, J) by the index pair $p(I, J)$. It follows from Lemma 1 that, if $\mathcal{K}_a(u, v)$

is non-empty, then $C(i_a(w) - 1, j_a(x) - 1)a^K$ is an LCS of $A(1..i_a(u))$ and $B(1..j_a(v))$, and the entry to which the backward pointer points is set to the $(i_a(w) - 1, j_a(x) - 1)$th entry, where (w, x) is the only element belonging to $\mathcal{K}_a(u, v)$, and $K = M_a(w..u) = N_a(x..v)$. Otherwise, if the longest one of $C(i_a(w) - 1, j_a(x) - 1)a^{M_a(w..u)}$ for the last element (w, x) of $\mathcal{J}_a(u, v)$, or $C(i_a(w) - 1, j_a(x) - 1)a^{N_a(x..v)}$ for the last element (w, x) of $\mathcal{I}_a(u, v)$ is longer than both $C(i, j - 1)$ and $C(i - 1, j)$, then this string is an LCS of $A(1..i_a(u))$ and $B(1..j_a(v))$, and the entry to which the backward pointer points is set to the $(i_a(w) - 1, j_a(x) - 1)$th entry.

When determining the $(i_a(u), j_a(v))$th entry of the table, instead of explicitly obtaining $\mathcal{L}(i_a(u), j_a(v))$, the algorithm obtains $\mathcal{L}_a(u, v)$, because of the existence of a simple recurrence, which is presented in the following two lemmas.

Lemma 2. *For any symbol a in Σ and any index pair (u, v) with $1 \le u \le m_a$ and $1 \le v \le n_a$, $\mathcal{L}_a(u, v)$ is identical to the concatenation of*

- *the prefix of $\mathcal{I}_a(u, v - 1)$ with $v \ge 2$ consisting of all elements (w, x) such that $C(i_a(w) - 1, j_a(x) - 1)a^{N_a(x..v-1)}$ is longer than $C(i_a(u) - 1, j_a(v) - 1)$,*
- *(u, v), if (u, v) belongs to $\mathcal{L}_a(u, v)$, and*
- *the suffix of $\mathcal{I}_a(u, v - 1)$ with $u \ge 2$ consisting of all elements (w, x) such that $C(i_a(w) - 1, j_a(x) - 1)a^{M_a(w..u-1)}$ is longer than $C(i_a(u) - 1, j_a(v) - 1)$,*

in this order.

Lemma 3. *For any symbol a in Σ, any index pair (u, v) with $1 \le u \le m_a$ and $1 \le v \le n_a$ does not belong to $\mathcal{L}_a(u, v)$ if and only if*

1. *$C(i_a(u), j_a(v) - 1)$ is exactly $M_a(u)$ longer than $C(i_a(u) - 1, j_a(v) - 1)$,*
2. *$C(i_a(w) - 1, j_a(x) - 1)a^{M_a(w..u-1)}$ is as long as $C(i_a(u) - 1, j_a(v) - 1)$, where (w, x) is the last element of non-empty $\mathcal{I}_a(u, v - 1)$ with $v \ge 2$,*
3. *$C(i_a(u) - 1, j_a(v))$ is exactly $N_a(v)$ longer than $C(i_a(u) - 1, j_a(v) - 1)$, or*
4. *$C(i_a(w) - 1, j_a(x) - 1)a^{N_a(x..v-1)}$ is as long as $C(i_a(u) - 1, j_a(v) - 1)$, where (w, x) is the first element of non-empty $\mathcal{J}_a(u - 1, v)$ with $u \ge 2$.*

It follows from the outline and lemmas given above that, in order to present the proposed algorithm, it suffices to show how to implement $\mathcal{J}_a(u, v)$, $\mathcal{K}_a(u, v)$, and $\mathcal{I}_a(u, v)$ so that they can be obtained from $\mathcal{I}_a(u, v - 1)$ and $\mathcal{J}_a(u - 1, v)$ according to Lemmas 2 and 3 in $O(\log \log \min(m_a, n_a, L_a/(m_a n_a)))$ amortized time. The implementation is presented in the following three subsections. Section 3.2 introduces a simple implementation that allows the algorithm to obtain $\mathcal{J}_a(u, v)$, $\mathcal{K}_a(u, v)$, and $\mathcal{I}_a(u, v)$ in $O(L_a(u, v))$ amortized time. Section 3.3 modifies this implementation so that the algorithm can obtain these data structures in $O(\log \log \min(m_a, n_a, L_a/(m_a n_a)))$ amortized time, if a certain preprocessing has been performed. The method by which to perform the preprocessing in $O(m_a n_a \log \log \min(m_a, n_a, L_a/(m_a n_a)))$ time is then presented in Sect. 3.4. Based on these considerations, the following theorem immediately holds.

Theorem 1. *The longest common subsequence problem in run-length encoded form can be solved in $O(mn \log \log \min(m, n, L/(mn)))$ time.*

1: Examine whether (u, v) belongs to $\mathcal{L}_a(u, v)$, based on Lemma 3;
2: delete the last element (w, x) from $\mathcal{I}_a(u)$, while $\mathcal{I}_a(u)$ is non-empty, and
    $C(i_a(w)-1, j_a(x)-1)a^{N_a(x..v-1)}$ is not longer than $C(i_a(u)-1, j_a(v)-1)$;
3: delete the first element (w, x) from $\mathcal{J}_a(v)$, while $\mathcal{J}_a(v)$ is non-empty, and
    $C(i_a(w)-1, j_a(x)-1)a^{M_a(w..u-1)}$ is not longer than $C(i_a(u)-1, j_a(v)-1)$;
4: construct $\mathcal{L}_a(u, v)$ by concatenating $\mathcal{I}_a(u)$, (u, v) if it belongs to $\mathcal{L}_a(u, v)$, and
    $\mathcal{J}_a(v)$ in this order;
5: let $\mathcal{J}_a(v)$ be the prefix of $\mathcal{L}_a(u, v)$ consisting of all index pairs (w, x) such that
    $R_a(w - 1, x - 1) < R_a(u, v)$, and let $\mathcal{I}_a(u)$ be the suffix of $\mathcal{L}_a(u, v)$
    consisting of all index pairs (w, x) such that $R_a(u, v) < R_a(w - 1, x - 1)$.

Fig. 1. Procedure that updates $\mathcal{I}_a(u)$ and $\mathcal{J}_a(v)$

3.2   Simple Implementation

This subsection introduces a simple implementation that allows the algorithm
to obtain $\mathcal{J}_a(u, v)$, $\mathcal{K}_a(u, v)$, and $\mathcal{I}_a(u, v)$ from $\mathcal{I}_a(u, v - 1)$ and $\mathcal{J}_a(u - 1, v)$ in
$O(L_a(u, v))$ amortized time.

In the implementation, the algorithm uses $m_a + n_a$ doubly-linked lists,
$\mathcal{I}_a(1), \ldots, \mathcal{I}_a(m_a)$ and $\mathcal{J}_a(1), \ldots, \mathcal{J}_a(n_a)$. These lists initially contain no elements, and when determining the $(i_a(u), j_a(v))$th entry of the dynamic pro-
gramming table, $\mathcal{I}_a(u)$ and $\mathcal{J}_a(v)$ are updated so as to contain $\mathcal{I}_a(u, v)$ and
$\mathcal{J}_a(u, v)$, respectively. It can be verified, by induction, that, just before being
updated, $\mathcal{I}_a(u)$ and $\mathcal{J}_a(v)$ contain $\mathcal{I}_a(u - 1, v)$ and $\mathcal{J}_a(u, v - 1)$, respectively.
Therefore, according to Lemma 2, updating $\mathcal{I}_a(u)$ and $\mathcal{J}_a(v)$ from $\mathcal{I}_a(u, v - 1)$
and $\mathcal{J}_a(u - 1, v)$ to $\mathcal{I}_a(u, v)$ and $\mathcal{J}_a(u, v)$ can be performed by executing the
procedure given in Fig. 1. Statement 5 is valid because all elements (w, x) in
$\mathcal{L}_a(u, v)$ are in ascending order with respect to $R_a(w - 1, x - 1)$.

The amortized execution time of the procedure for any index pair (u, v) is
estimated as follows. Since no index pair is added to the $m_a + n_a$ lists more than
once, the total number of index pairs deleted from the lists by statements 2 and
3 throughout the entire algorithm is $O(m_a n_a)$. This implies that statements 1
through 4 for each execution of the procedure can be performed in a constant
amortized time. Just before the execution of statement 5, $\mathcal{I}_a(u)$ and $\mathcal{J}_a(v)$ contain at most $M_a(u)$ and $N_a(v)$ elements, respectively. Hence, if $M_a(u) \geq N_a(v)$,
then statement 5 can be executed in $O(\min(N_a(v), M_a(u) - N_a(v)))$ time, by performing a linear search of $\mathcal{I}_a(u)$ from the first element, if $N_a(v) \leq M_a(u) - N_a(v)$,
or from the last element, otherwise, for the last element of $\mathcal{J}_a(u, v)$. Otherwise,
statement 5 can be executed in $O(\min(M_a(u), N_a(v) - M_a(u)))$ time analogously.
Thus, the procedure in Fig. 1 can be performed in $O(L_a(u, v))$ amortized time.

3.3   Proposed Implementation

This subsection proposes an implementation that allows the algorithm to obtain
$\mathcal{J}_a(u, v)$, $\mathcal{K}_a(u, v)$, and $\mathcal{I}_a(u, v)$ in $O(\log \log \min(m_a, n_a, L_a/(m_a n_a)))$ amortized
time, if a certain preprocessing has been performed.

In order to present the implementation, it is necessary to introduce a data structure that can efficiently maintain a set of integers in a short interval and integers in a short interval that preserve a certain property of integers $R_a(w, x)$. A *van Emde Boas data structure* [11] represents a set S of integers between 0 and s and supports any of the following operations in $O(\log \log s)$ time: inserting r in S, deleting r from S, reporting the greatest integer in S that is less than r, and reporting the least integer in S that is greater than r, where r is an arbitrary integer between 0 and s. For any index pair (w, x) with $0 \le w \le m_a$ and $0 \le x \le n_a$, let $r_a(w, x)$ denote the number of integers R less than $R_a(w, x)$ such that $R = R_a(y, z)$ for some index pair (y, z) with $0 \le y \le m_a$ and $0 \le z \le n_a$, so that $0 \le r_a(w, x) < (m_a + 1)(n_a + 1)$. It is easy to verify that, for any index pairs (w, x) and (y, z), $R_a(w, x) \le R_a(y, z)$ if and only if $r_a(w, x) \le r_a(y, z)$.

The proposed implementation is strictly follows the outline of that introduced in the previous subsection. However, the proposed implementation also exploits the technique used in the algorithm of Apostolico et al. [2] (with the modification mentioned in Sect. 1) to efficiently search $\mathcal{I}_a(u)$ for the last element of $\mathcal{J}_a(u, v)$, or $\mathcal{J}_a(v)$ for the first element of $\mathcal{I}_a(u, v)$, when executing statement 5 of the procedure in Fig. 1. Ignoring the details, the algorithm of Apostolico et al. [2] can be thought of as the algorithm that obtains $\mathcal{J}_a(u, v)$, $\mathcal{K}_a(u, v)$, and $\mathcal{I}_a(u, v)$, without separately maintaining the $m_a + n_a$ lists, $\mathcal{I}_a(1), \ldots, \mathcal{I}_a(m_a)$ and $\mathcal{J}_a(1), \ldots, \mathcal{J}_a(n_a)$. Instead, access to any element of the lists is provided by searching a van Emde Boas data structure that represents a set of integers $r_a(w - 1, x - 1)$ labeled by (w, x) for all index pairs (w, x) in the $m_a + n_a$ lists for that in $O(\log \log(m_a n_a))$ time.

Recall that statements 1 through 4 of the procedure in Fig. 1 can be executed in constant amortized time, while statement 5 takes $O(L_a(u, v))$ time due to the linear search of $\mathcal{L}_a(u, v)$. The main idea underlying the proposed implementation is to treat the $m_a + n_a$ lists as consisting of sets of index pairs so that each linear search can be performed by traversing at most a constant number of elements on average. The details of the proposed implementation are presented in the proof of the following lemma. Let $l_a(u, v) = \min(\rho_* - r_a(u, v-1), \rho^* - \rho_*, r_a(u-1, v) - \rho^*)$, where $\rho_* = \min(r_a(u-1, v-1), r_a(u, v))$ and $\rho^* = \max(r_a(u-1, v-1), r_a(u, v))$, and let l_a be the sum of $l_a(u, v)$, where (u, v) ranges over all index pairs such that $1 \le u \le m_a$ and $1 \le v \le n_a$. It is not difficult to verify that $l_a/(m_a n_a) = O(\min(m_a, n_a, L_a/(m_a n_a)))$.

Lemma 4. *If the value $r_a(w, x)$ is available for any symbol a in Σ, and any index pair (w, x) with $0 \le w \le m_a$ and $0 \le x \le n_a$, then $\mathcal{I}_a(u)$ and $\mathcal{J}_a(v)$ are updated from $\mathcal{I}_a(u, v-1)$ and $\mathcal{J}_a(u-1, v)$ to $\mathcal{I}_a(u, v)$ and $\mathcal{J}_a(u, v)$, respectively, in $O(\log \log(l_a/(m_a n_a)))$ amortized time when determining the $(i_a(u), j_a(v))$th entry of the dynamic programming table.*

Proof. Let s be the least power of two that is greater than or equal to $l_a/(m_a n_a)$. This value can be determined in $O(m_a n_a)$ time.

The proposed implementation can be obtained by modifying the procedure in Fig. 1 so that both statements 1 through 4 and statement 5 can be performed in $O(\log \log s)$ amortized time. To do this, each of the $m_a + n_a$ lists,

$\mathcal{I}_a(1), \ldots, \mathcal{I}_a(m_a)$ and $\mathcal{J}_a(1), \ldots, \mathcal{J}_a(n_a)$, is represented as a list of van Emde Boas data structures as follows. For any index k with $0 \le k \le \lfloor (m_a + 1)(n_a + 1)/s \rfloor$, let $T_a(k)$ be a van Emde Boas data structure that maintains at most s integers $r_a(w - 1, x - 1) - ks$, each of which is labeled by (w, x), for all index pairs (w, x) with $ks \le r_a(w - 1, x - 1) < (k + 1)s$ that belong to union $\mathcal{I}_a(1) \cdots \mathcal{I}_a(m_a)\mathcal{J}_a(1) \cdots \mathcal{J}_a(n_a)$. Let $\hat{\mathcal{I}}_a(y)$ be the list of all indices k such that $r_a(w - 1, x - 1)$ belongs to $T_a(k)$ for some index pair (w, x) in $\mathcal{I}_a(y)$ in ascending order and define $\hat{\mathcal{J}}_a(z)$ analogously with respect to $\mathcal{J}_a(z)$. Simulating the procedure in Fig. 1 using lists $\hat{\mathcal{I}}_a(1), \ldots, \hat{\mathcal{I}}_a(m_a)$ and $\hat{\mathcal{J}}_a(1), \ldots, \hat{\mathcal{J}}_a(n_a)$, together with van Emde Boas data structures $T_a(k)$, in a straightforward manner, statements 1 through 4 of the procedure can be performed in $O(\log \log s)$ amortized time. Furthermore, since $s = O(\min(m_a, n_a, L_a/(m_a n_a)))$, statement 5 can also be executed in $O(\log \log s)$ amortized time. □

3.4   Preprocessing

In order to complete the proposed algorithm, this subsection presents a method to determine values $r_a(w, x)$ for all index pairs (w, x) with $0 \le w \le m_a$ and $0 \le x \le n_a$ in $O(m_a n_a \log \log \min(m_a, n_a, L_a/(m_a n_a)))$ time.

Let $R'_a(w, x)$ denote an integer $2^{\lceil \log_2 (m_a+1)(n_a+1) \rceil}(R_a(w, x) + M_a) + 2^{\lceil \log_2 (n_a+1) \rceil}w + x$ for any index pair (w, x), so that integers $R'(w, x)$ preserve the order of integers $R_a(w, x)$, and so that (w, x) can be recovered from $R'_a(w, x)$ in constant time. A naive use of the integer sorting algorithm of Han [6] that takes integers $R'_a(w, x)$ for all index pairs (w, x) with $0 \le w \le m_a$ and $0 \le x \le n_a$ in an arbitrary order as input yields only an $O(m_a n_a \log \log \max(m_a, n_a))$-time method. However, exploiting facts that $R_a(w, 0) < R_a(w, 1) < \cdots < R_a(w, n_a)$ for any index w with $0 \le w \le m_a$, and that $R_a(m_a, x) < R_a(m_a + 1, x) < \cdots < R_a(0, x)$ for any index x with $0 \le x \le n_a$, an $O(m_a n_a \log \log \min(m_a, n_a, L_a/(m_a n_a)))$-time method can also be obtained as shown in the following lemmas.

Lemma 5. *The values $r_a(w, x)$ for all index pairs (w, x) with $0 \le w \le m_a$ and $0 \le x \le n_a$ can be determined in $O(m_a n_a \log \log \min(m_a, n_a))$ time.*

Proof. By symmetry, it suffices to present an $O(m_a n_a \log \log m_a)$-time method that can obtain a list of all index pairs (w, x) with $0 \le w \le m_a$ and $0 \le x \le n_a$ in ascending order with respect to $R_a(w, x)$.

An outline of the method is as follows. Let Q_a be a set initially containing all index pairs (w, x) with $0 \le x \le m_a$ and $0 \le x \le n_a$. The method iteratively executes the following two steps until Q_a becomes empty:

1. extract all elements (w, x) such that $R_a(w, x)$ is less than or equal to a certain threshold R from Q_a in $O(s)$ time, and
2. execute the algorithm of Han [6] to obtain a list of the extracted elements (w, x) in ascending order with respect to $R_a(w, x)$ in $O(s \log \log s)$ time,

where s is the number of extracted elements. Since the threshold R is set such that s is polynomial in m_a, this method performs in $O(m_a n_a \log \log m_a)$ time.

The details of how to execute step 1 of the above method is described below. Let H_a be a list of all index pairs (w, x) in Q_a such that $(w, x - 1)$ does not belong to Q_a. Since $R_a(w, 0) < R_a(w, 1) < \cdots < R_a(w, n_a)$ for any index w with $0 \leq w \leq m_a$, H_a strictly represents Q_a. Hence, the method uses this list to maintain Q_a. In each iteration of the method, the extracted elements from Q_a are as follows. Let t be the number of elements in H_a, which is less than or equal to $m_a + 1$. If the number of elements in Q_a is less than t^2, then let $R = \infty$. Otherwise, let R be the minimum value of $R_a(w, x+t)$, where (w, x) ranges over all index pairs in H_a such that $x+t \leq n_a$. Recall that s is the number of elements (w, x) in Q_a such that $R_a(w, x) \leq R$. It follows from the above definitions that $t \leq s \leq t^2$. After determining threshold R in $O(t)$ time, all elements (w, x) with $R_a(w, x) \leq R$ can successively be extracted from Q_a in $O(s)$ time. □

Lemma 6. *The values $r_a(w, x)$ for all index pairs (w, x) with $0 \leq w \leq m_a$ and $0 \leq x \leq n_a$ can be determined in $O(m_a n_a \log \log(L_a/(m_a n_a)))$ time.*

Proof. The lemma is proven by presenting a method that lists all index pairs (w, x) in ascending order with respect to $R_a(w, x)$.

An outline of the method is as follows. Let s be the least power of two greater than or equal to $L_a/(m_a n_a)$. The method classifies all index pairs (w, x) into $O(m_a n_a)$ groups $Q_a(k)$, each of which consists of all index pairs (w, x) such that $ks \leq R_a(w, x) + M_a < (k + 1)s$. All index pairs (w, x) in each group $Q_a(k)$ are then listed in ascending order with respect to $R_a(w, x)$ in $O(t \log \log s)$ time using the algorithm of Han [6], if $t < s$, and, otherwise, in $O(t)$ time using the bucket sort algorithm, where t is the number of index pairs belonging to $Q_a(k)$.

In order to classify index pairs into $O(m_a n_a)$ groups $Q_a(k)$, a naive use of the bucket sort algorithm requires $O((M_a + N_a)/s)$ space, because the values $R_a(w, x)$ vary between $-M_a$ and N_a. The method reduces the required space to $O(m_a n_a)$ based on a graph introduced below. Let G_a be an undirected graph having all index pairs (w, x) with $0 \leq w \leq m_a$ and $1 \leq x \leq n_a$ as its vertices. The set of edges in G_a consists of

- edges between $(w - 1, x - 1)$ and $(w, x - 1)$, and edges between $(w - 1, x)$ and (w, x), for all index pairs (w, x) with $M_a(w) = L_a(w, x)$,
- edges between $(w-1, x-1)$ and (w, x), for all index pairs (w, x) with $|N_a(x) - M_a(w)| = L_a(w, x)$, and edges between $(w - 1, x - 1)$ and $(w - 1, x)$, and
- edges between $(w, x - 1)$ and (w, x), for all index pairs (w, x) with $N_a(w) = L_a(w, x)$,

where $1 \leq w \leq m_a$ and $1 \leq x \leq n_a$. Based on this definition, it is not difficult to verify that any index pair (w, x) is connected to at least one of $m_a + n_a + 1$ vertices, $(m_a, 0), (m_a - 1, 0), \ldots, (0, 0), (0, 1), \ldots, (0, n_a)$. Therefore, G_a can be partitioned into at most $m_a + n_a + 1$ connected subgraphs, each of which contain one of the $m_a + n_a + 1$ vertices. Let $G_a(1), G_a(2), \ldots, G_a(f)$ be a list of all such subgraphs of G_a in ascending order with respect to $R_a^{\min}(e)$, and hence, with respect to $R_a^{\max}(e)$, where $R_a^{\min}(e)$ and $R_a^{\max}(e)$ denote the minimum value and the maximum value of $R_a(w, x)$, respectively, with (w, x) ranging over all vertices of $G_a(e)$. By visiting all vertices in G_a in breadth-first order, subgraphs $G_a(e)$, together with values $R_a^{\min}(e)$ and $R_a^{\max}(e)$, for all indices e from 1 to f can be obtained in $O(m_a n_a)$ time. Recall that any vertex (w, x) in $G_a(e)$ is classified so as to belong to $Q_a(\lfloor (R_a(w, x) + M_a)/s \rfloor)$, which lies in a series of $O((R_a^{\max}(e) - R_a^{\min}(e))/s)$ groups, $Q_a(\lfloor (R_a^{\min}(e) + M_a)/s \rfloor), \ldots, Q_a(\lfloor (R_a^{\max}(e) + M_a)/s \rfloor)$. Therefore, if the sum of differences between $R_a^{\min}(e)$ and $R_a^{\max}(e)$ for all indices e from 1 to f is $O(L_a)$, then all index pairs (w, x) can be classified into groups $Q_a(k)$ in $O(m_a n_a)$ time and $O(m_a n_a)$ space in a straightforward manner using the bucket sort algorithm, because $L_a/s = O(m_a n_a)$ due to the setting of s. Based on the definition of the set of edges in G_a, the sum of differences between $R_a(w, x)$ and $R_a(y, z)$ is $O(L_a)$, where $((w, x), (y, z))$ ranges over all vertex pairs such that G_a has an edge between (w, x) and (y, z). This implies that the sum of differences between $R_a^{\min}(e)$ and $R_a^{\max}(e)$ for all indices e from 1 to f is $O(L_a)$, because $G_a(e)$ is a connected graph.   □

References

1. Ann, H.Y., Yang, C.B., Tseng, C.T., Hor, C.Y.: A fast and simple algorithm for computing the longest common subsequence of run-length encoded strings. Inform. Process. Lett. 108, 360–364 (2008)
2. Apostolico, A., Landau, G.M., Skiena, S.: Matching for run-length encoded strings. J. Complexity 15, 4–16 (1999)
3. Bunke, H., Csirik, J.: An improved algorithm for computing the edit distance of run-length coded strings. Inform. Process. Lett. 54, 93–96 (1995)
4. Chen, K.-Y., Chao, K.-M.: A Fully Compressed Algorithm for Computing the Edit Distance of Run-Length Encoded Strings. In: de Berg, M., Meyer, U. (eds.) ESA 2010, Part I. LNCS, vol. 6346, pp. 415–426. Springer, Heidelberg (2010)
5. Crochemore, M., Landau, G.M., Ziv-Ukelson, M.: A subquadratic sequence alignment algorithm for unrestricted scoring matrices. SIAM J. Comput. 32, 1654–1673 (2003)
6. Han, Y.: Deterministic sorting in $O(n \log \log n)$ time and linear space. J. Algorithms 50, 96–105 (2004)
7. Kim, J.W., Amir, A., Landau, G.M., Park, K.: Computing similarity of run-length encoded strings with affine gap penalty. Theor. Comput. Sci. 395, 268–282 (2008)
8. Liu, J.J., Wang, Y.L., Lee, R.C.T.: Finding a longest common subsequence between a run-length-encoded string and an uncompressed string. J. Complexity 24, 173–184 (2008)
9. Mäkinen, V., Navarro, G., Ukkonen, E.: Approximate Matching of run-length compressed strings. Algorithmica 35, 347–369 (2003)
10. Mitchell, J.: A geometric shortest path problem, with application to computing a longest common subsequence in run-length encoded strings. Technical Report, Department of Applied Mathematics, Suny Stony Brook, NY (1997)
11. van Emde Boas, P.: Preserving order in a forest in less than logarithmic time and linear space. Inform. Process. Lett. 6, 80–82 (1977)
12. Wagner, R.A., Fischer, M.J.: The string to string correction problem. J. ACM 21, 168–173 (1974)

Efficient Counting of Square Substrings in a Tree

Tomasz Kociumaka[1], Jakub Pachocki[1], Jakub Radoszewski[1],
Wojciech Rytter[1,2,*], and Tomasz Waleń[3,1]

[1] Faculty of Mathematics, Informatics and Mechanics,
University of Warsaw, Warsaw, Poland
{kociumaka,jrad,pachocki,rytter,walen}@mimuw.edu.pl
[2] Faculty of Mathematics and Computer Science,
Nicolaus Copernicus University, Toruń, Poland
[3] Laboratory of Bioinformatics and Protein Engineering,
International Institute of Molecular and Cell Biology in Warsaw, Poland

Abstract. We give an algorithm which in $O(n \log^2 n)$ time counts all
distinct squares in labeled trees. There are two main obstacles to overcome. Crochemore et al. showed in 2012 that the number of such squares
is bounded by $\Theta(n^{4/3})$. This is substantialy different from the case of
classical strings, which admit only a linear number of distinct squares.
We deal with this difficulty by introducing a compact representation of
all squares (based on maximal cyclic shifts) that requires only $O(n \log n)$
space. The second obstacle is lack of adequate algorithmic tools for labeled trees. Consequently we develop several novel techniques, which
form the most complex part of the paper. In particular we extend Imre Simon's implementation of the failure function in pattern matching
machines.

1 Introduction

Various types of repetitions play an important role in combinatorics on words
with particular applications in pattern matching, text compression, computational biology etc., see [3]. The basic type of repetitions are squares: strings of
the form ww. Here we consider square substrings corresponding to simple paths
in labeled unrooted trees. Squares in trees and graphs have already been considered e.g. in [2]. Recently it has been shown that a tree with n nodes can contain
at most $\Theta(n^{4/3})$ distinct squares, see [4], while the number of distinct squares
in a string of length n does not exceed $2n$, as shown in [7]. This paper can be
viewed as an algorithmic continuation of [4].

Despite the linear upper bound, enumerating squares in ordinary strings is
already a difficult problem. Complex $O(n)$ time solutions using suffix trees [8]
or runs [5] are known.

Assume we have a tree T whose edges are labeled with symbols from an
integer alphabet Σ. If u and v are two nodes of T, then let $val(u,v)$ denote the
sequence of labels of edges on the path from u to v (denoted as $u \rightsquigarrow v$). We

---

* The author is supported by grant no. N206 566740 of the National Science Centre.

K.-M. Chao, T.-s. Hsu, and D.-T. Lee (Eds.): ISAAC 2012, LNCS 7676, pp. 207–216, 2012.

call $val(u,v)$ a *substring* of T. (Note that a substring is a string, not a path.) Also let $dist(u,v) = |val(u,v)|$. Fig. 1 presents square substrings in a sample tree. We consider only simple paths, that is the vertices of a path do not repeat. Denote by $sq(T)$ the set of different square substrings in T. Our main result is computing $|sq(T)|$ in $O(n \log^2 n)$ time.

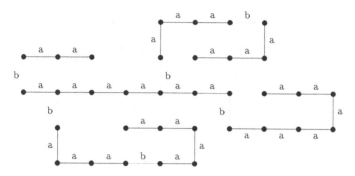

Fig. 1. We have here $|sq(T)| = 31$. There are 10 groups of cyclically equivalent squares, the representatives (maximal cyclic shift of of a square half) are: a, a^2, a^3, ba, ba^2, ba^3, ba^4, ba^5, ba^6, $(ba^3)^2$. For example, the equivalence class of $u^2 = (ba^6)^2$ contains the strings $rot(u,q)^2$ for $q \in [0,3] \cup [5,6]$, this is a single cyclic interval modulo 7 (see Section 3).

2 Algorithmic Toolbox for Trees

In this section we apply several well known concepts to design algorithms and data structures for labeled trees.

Navigation in Trees. Recall two widely known tools for rooted trees: the LCA queries and the LA queries. The LCA query given two nodes x,y returns their *lower common ancestor* $LCA(x,y)$. The LA query given a node x and an integer $h \geq 0$ returns the *ancestor* of x at *level* h, i.e. with distance h from the root. After $O(n)$ preprocessing both types of queries can be answered in $O(1)$ time [9,1]. We use them to efficiently navigate also in unrooted trees. For this purpose, we root the tree in an arbitrary node and split each path $x \rightsquigarrow y$ in $LCA(x,y)$. This way we obtain the following result:

Fact 1. *Let T be a tree with n nodes. After $O(n)$ time preprocessing we answer the following queries in constant time:*
(a) for any two nodes x,y compute $dist(x,y)$,
(b) for any two nodes x,y and a nonnegative integer $d \leq dist(x,y)$ compute $jump(x,y,d)$ — the node z on the path $x \rightsquigarrow y$ with $dist(x,z) = d$.

Dictionary of Basic Factors. The *dictionary of basic factors* (DBF, in short) is a widely known data structure for comparing substrings of a string. For a string w of length n it takes $O(n \log n)$ time and space to construct and enables lexicographical comparison of any two substrings of w in $O(1)$ time, see [6]. The

DBF can be extended to arbitrary labeled trees. Due to the lack of space the proof of the following (nontrivial) fact will be presented in the full version of the paper.

Fact 2. *Let T be a labeled tree with n nodes. After $O(n \log n)$ time preprocessing any two substrings $val(x_1, y_1)$ and $val(x_2, y_2)$ of T of the same length can be compared lexicographically in $O(1)$ time (given x_1, y_1, x_2, y_2).*

Centroid Decomposition. The centroid decomposition enables to consider paths going through the root in rooted trees instead of arbitrary paths in an unrooted tree. Let T be an unrooted tree with n nodes and $T_1, T_2, \ldots, T_k$ be the connected components obtained after removing a node r from T. The node r is called a *centroid* of T if $|T_i| \leq n/2$ for each i. The *centroid decomposition* of T, $CDecomp(T)$, is defined recursively:

$$CDecomp(T) = \{(T, r)\} \cup \bigcup_{i=1}^{k} CDecomp(T_i).$$

The centroid of a tree can be computed in $O(n)$ time. The recursive definition of $CDecomp(T)$ implies an $O(n \log n)$ bound on its total size.

Fact 3. *Let T be a tree with n nodes. The total size of all subtrees in $CDecomp(T)$ is $O(n \log n)$. The decomposition $CDecomp(T)$ can be computed in $O(n \log n)$ time.*

Determinization. Let T be a tree rooted at r. We write $val(v)$ instead of $val(r, v)$, $val^R(v)$ instead of $val(v, r)$ and $dist(v)$ instead of $dist(r, v)$.

We say that T is *deterministic* if $val(v) = val(w)$ implies that $v = w$. We say that T is *semideterministic* if $val(v) = val(w)$ implies that $v = w$ unless $r \rightsquigarrow v$ and $r \rightsquigarrow w$ are disjoint except r. Informally, T is semideterministic if it is "deterministic anywhere except for the root".

For an arbitrary tree T an "equivalent" deterministic tree $dtr(T)$ can be obtained by identifying nodes v and w if $val(v) = val(w)$. If we perform such identification only when the paths $r \rightsquigarrow v$ and $r \rightsquigarrow w$ share the first edge, we get a semideterministic tree $semidtr(T)$. This way we also obtain functions φ mapping nodes of T to corresponding nodes in $dtr(T)$ (in $semidtr(T)$ respectively). Additionally we define $\psi(v)$ as an arbitrary element of $\varphi^{-1}(v)$ for $v \in dtr(T)$ ($v \in semidtr(T)$ respectively). Note that φ and ψ for $semidtr(T)$ preserve the values of paths going through r; this property does not hold for $dtr(T)$.

The details of efficient implementation are left for the full version of the paper:

Fact 4. *Let T be a rooted tree with n nodes. The trees $dtr(T)$ and $semidtr(T)$ together with the corresponding pairs of functions φ and ψ can be computed in $O(n)$ time.*

3   Compact Representations of Sets of Squares

For a string u, let $rot(u)$ denote the string u with its first letter moved to the end. For an integer q, let $rot(u, q)$ denote $rot^q(u)$, i.e., the result of q iterations of the rot operation on the string u. If $v = rot(u, c)$ then u and v are called cyclically equivalent, we also say that v is a cyclic rotation of u and vice-versa. Let $maxRot(u)$ denote the lexicographically maximal cyclic rotation of u. Let T be a labeled tree and let x, y be nodes of T such that $val(x, y) = maxRot(val(x, y))$. Moreover let I be a cyclic interval of integers modulo $dist(x, y)$. Define a *package* as a set of cyclically equivalent squares:

$$package(x, y, I) = \{rot(val(x, y), q)^2 : q \in I\}.$$

A family of packages which altogether represent the set of square substrings of T is called a *cyclic representation* of squares in T. Such a family is called *disjoint* if the packages represent pairwise disjoint sets of squares.

Anchored Squares. Let v be a node of T. A square in T is called *anchored* in v if it is the value of a path passing through v. Let $sq(T, v)$ denote the set of squares anchored in v. Assume that T is rooted at r and let $v \neq r$ be a node of T with $dist(v) = p$. Let $sq(T, r, v)$ denote the set of squares of length $2p$ that have an occurrence passing through both r and v. Note that each path of length $2p$ passing through r contains a node v with $dist(v) = p$. Hence $sq(T, r)$ is the sum of $sq(T, r, v)$ over all nodes $v \neq r$.

We introduce two tables, defined for all $v \neq r$, similar to the tables used in the Main-Lorenz square-reporting algorithm for strings [11]. In Section 5 we sketch algorithms computing these tables in linear time (for $PREF$ under the additional assumption that the tree is semideterministic).

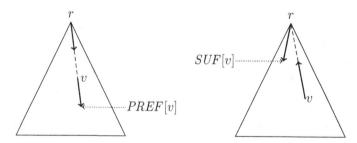

- **[Prefix table]** $PREF[v]$ is a lowest node x in the subtree rooted at v such that $val(v, x)$ is a prefix of $val(v)$, see figure above.
- **[Suffix table]** $SUF[v]$ is a lowest node x in T such that $val(x)$ is a prefix of $val^R(v)$ and $LCA(v, x) = r$.

We say that a string $s = s_1 \ldots s_k$ has a period p if $s_i = s_{i+p}$ for all $i = 1, \ldots, k-p$. Let x and y be nodes of T. A triple (x, y, p) is called a *semirun* if $val(x, y)$ has a period p and $dist(x, y) \geq 2p$. All substrings of $x \rightsquigarrow y$ and $y \rightsquigarrow x$ of length $2p$ are squares. We say that these squares are *induced* by the semirun. Let us fix

$v \neq r$ with $dist(v) = p$. Note that $val(PREF[v], SUF[v])$ is periodic with period p. By the definitions of $PREF[v]$ and $SUF[v]$, if $dist(PREF[v], SUF[v]) < 2p$ then $sq(T, r, v) = \emptyset$. Otherwise $(PREF[v], SUF[v], p)$ is a semirun *anchored* in r and the set of squares it induces is exactly $sq(T, r, v)$, see also Figure 2.

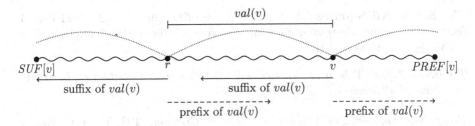

Fig. 2. The semirun $(SUF[v], PREF[v], |val(v)|)$ induces $sq(T, r, v)$

For a set of semiruns S, let $sq(S)$ denote the set of squares induced by at least one semirun in S. The following lemma summarizes the discussion on semiruns.

Lemma 5. *Let T be a tree of size n rooted at r. There exists a family S of $O(n)$ semiruns anchored in r such that $sq(S) = sq(T, r)$.*

Packages and Semiruns. Semiruns can be regarded as a way to represent sets of squares. Nevertheless, this representation cannot be directly used to count the number of different squares and needs to be translated to a cyclic representation (packages). The key tools for performing this translation are the following two tables defined for any node v of T.

1. **[Shift Table]** $SHIFT[v]$ is the smallest nonnegative integer r such that $rot(val(v), r) = maxRot(val(v))$.
2. **[Reversed Shift Table]** $SHIFT^R[v]$ is the smallest nonnegative integer r such that $rot(val^R(v), r) = maxRot(val^R(v))$.

In Section 5 we sketch algorithms computing these tables for a tree with n nodes in $O(n \log n)$ time.

Using these tables and the *jump* queries (Fact 1) we compute the cyclic representation of the set of squares induced by a family of semiruns.

Lemma 6. *Let T a be tree of size n rooted at r and let S be a family of semiruns (x, y, p) anchored in r. There exists a cyclic representation of the set of squares induced by S that contains $O(|S|)$ packages and can be computed in $O(n \log n + |S|)$ time.*

Proof. Let $(x, y, p) \in S$. We have $LCA(x, y) = r$ and $dist(x, y) \geq 2p$, consequently there exists a node z on $x \leadsto y$ such that $dist(z) = p$ and all squares induced by (x, y, p) are cyclic rotations of $val(z)^2$ and $val^R(z)^2$. Observe that $maxRot(val(z))$ and $maxRot(val^R(z))$, as any cyclic rotation of $val(z)$ or $val^R(z)$, occur on the path $x \rightsquigarrow y$. The $SHIFT$ and $SHIFT^R$ tables can be used to locate these occurrences, then *jump* queries allow to find their exact endpoints, nodes x_1, y_1 and x_2, y_2 respectively. This way we also obtain the cyclic intervals I_1 and I_2 that represent the set of squares induced by (x, y, p) as $package(x_1, y_1, I_1)$ and $package(x_2, y_2, I_2)$. □

The Set of All Squares. As a consequence of Lemmas 5 and 6 and Fact 3 (centroid decomposition) we obtain the following combinatorial characterization of the set of squares in a tree:

Theorem 7. *Let* **T** *be a labeled tree with* n *nodes. There exists a cyclic representation of all squares in* **T** *of* $O(n \log n)$ *size.*

Proof. Note that $sq(\mathbf{T}) = \bigcup\{sq(T, r) : (T, r) \in CDecomp(\mathbf{T})\}$. The total size of trees in $CDecomp(\mathbf{T})$ is $O(n \log n)$ and for each of them the squares anchored in its root have a linear-size representation. This gives a representation of all squares in **T** that contains $O(n \log n)$ packages. □

4   Main Algorithm

Computing Semiruns. Let $T' = semidtr(T)$. The following algorithm computes the set $S = \{(\psi(x), \psi(y), p) : (x, y, p) \in semiruns(T', r)\}$. This set of semiruns induces $sq(T, r)$, since $sq(T, r) = sq(T', r)$.

Algorithm 1. Compute $semiruns(T, r)$

$S := \emptyset;\ T' := semidtr(T)$
Compute the tables $PREF(T'), SUF(T')$
foreach $v \in T' \setminus \{r\}$ **do**
  $x := PREF[v];\ y := SUF[v]$
  if $dist(x, y) \geq 2 \cdot dist(v)$ **then**
    $S := S \cup \{(\psi(x),\ \psi(y),\ dist(v))\}$
return S

Computing a Disjoint Representation of Packages. In this phase we compute a compact representation of distinct squares. For this, we group packages (x, y, I) according to $val(x, y)$, which is done by sorting them using Fact 2 to implement the comparison criterion efficiently. Finally in each group by elementary computations we turn a union of arbitrary cyclic intervals into a union of pairwise disjoint intervals. For a group of g packages this is done in $O(g \log g)$ time, which makes $O(n \log^2 n)$ in total.

General Structure of the Algorithm. The structure is based on the centroid decomposition. In total, precomputing DBF, *jump* and *CDecomp* and computing semiruns takes $O(n \log n)$ time, whereas transforming semiruns to packages and computing a disjoint representation of packages takes $O(n \log^2 n)$ time.

Theorem 8. *The number of distinct square substrings in an (unrooted) tree with n nodes can be found in $O(n \log^2 n)$ time.*

Algorithm 2. Count-Squares(T)

Compute DBF and *jump* data structure for **T**;   {Facts 1 and 2}

foreach $(T, r) \in CDecomp(\mathbf{T})$ **do**

    Semiruns := *semiruns*(T, r)

    Transform *Semiruns* into a set of packages in T;   {Lemma 6}

    Insert these packages to the set *Packages*

Compute (interval) disjoint representation of *Packages*

return $|sq(\mathbf{T})|$ as the total length of intervals in *Packages*

---

5 Construction of the Basic Tables

The *PREF* and *SUF* tables for ordinary strings are computed by a single simple algorithm, see [6]. This approach fails to generalize for trees, so we develop novel methods, interestingly, totally different for both tables. In order to construct a *PREF* table we generalize the results of Simon [13] originally developed for string pattern matching automata. For the *SUF* table, we use the suffix tree of a tree, originally designed by Kosaraju [10] for tree pattern matching.

5.1 Computation of *PREF*

We compute a slightly modified array $PREF'$ that allows for an overlap of the considered paths. More formally, for a node $v \neq r$, we define $PREF'[v]$ as the lowest node x in the subtree rooted at v such that $val(v, x)$ is a prefix of $val(x)$. Note that having computed $PREF'$, we can obtain $PREF$ by truncating the result so that paths do not overlap. This can be implemented with a single *jump* query.

Observe that $PREF'[v]$ depends only on the path $r \rightsquigarrow v$ and the subtree rooted at v. Hence, instead of a single semideterministic tree with n nodes, we may create a copy of r for each edge going out from r and thus obtain several deterministic trees of total size $O(n)$. For the remainder of this section we assume T is deterministic.

Recall that a border of a string w is a string that is both a prefix and a suffix of w. The *PREF* function for strings is closely related to borders, see [6]. This is inherited by $PREF'$ for deterministic trees, see Figure 3.

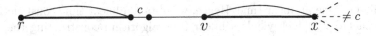

Fig. 3. $PREF'[v] = x$ if and only if $val(v, x)c$ is a border of $val(x)c$ and no edge labeled with c leaves x

For a node x of T and $c \in \Sigma$, let $\pi(x, c)$ (transition function) be a node y such that $val(y)$ is the longest border of $val(x)c$. We say that $\pi(x, c)$ is an *essential transition* if it does not point to the root. Let us define the transition table π and the border table P. For a node x let $\pi[x]$ be the list of pairs (c, y) such that $\pi(x, c) = y$ is an essential transition. For $x \neq r$ we set $P[x]$ as the node y such that $val(y)$ is the longest border of $val(x)$. The following lemma generalizes the results of [13] and gives the crucial properties of essential transitions. The proof is left for the full version of the paper.

Lemma 9. *Let T be a deterministic tree with n nodes. There are no more than $2n - 1$ essential transitions in T. Moreover, the π and P tables can be computed in $O(n)$ time.*

In the algorithm computing the $PREF'$ table, for each x we find all nodes v such that $PREF[v] = x$. This is done by iterating the P table starting from $\pi(x, c)$, see also Fig. 3.

Lemma 10. *For a deterministic tree T, the table $PREF'(T)$ can be computed in linear time.*

5.2 Computation of *SUF*

Let T be a deterministic tree rooted at r and $v \neq r$ be a node of T. We define $SUF'[v]$ as the lowest node x of T such that $val(x)$ is a prefix of $val^R(v)$. Hence, we relax the condition that $LCA(v, x) = r$ and add a requirement that T is deterministic. The technical proof of the following lemma is left for the full version of the paper.

Lemma 11. *Let T be an arbitrary rooted tree with n nodes. The $SUF(T)$ table can be computed in $O(n)$ time from $SUF'(dtr(T))$.*

Observe that *tries* are exactly the deterministic rooted trees. Let $S_1 = \{val(x) : x \in T\}$ and $S_2 = \{val^R(x) : x \in T\}$. Assume that we have constructed a trie $\mathcal{T}$ of all the strings $S_1 \cup S_2$ and that we store pointers to the nodes in $\mathcal{T}$ that correspond to elements of S_1 and S_2. Then for any $v \in T$, $SUF'[v]$ corresponds to the lowest ancestor of $val^R(v)$ in $\mathcal{T}$ which comes from S_1. Such ancestors can be computed for all nodes by a single top-bottom tree traversal, so the SUF' table can be computed in time linear in $\mathcal{T}$.

Unfortunately, the size of $\mathcal{T}$ can be quadratic, so we store its compacted version in which we only have explicit nodes corresponding to $S_1 \cup S_2$ and nodes having at least two children. The trie of S_1 is exactly T, whereas the compacted trie of S_2 is known as a *suffix tree of the tree T*. This notion was introduced in [10] and a linear time construction algorithm for an integer alphabet was given in [12]. The compacted trie $\mathcal{T}$ can therefore be obtained by merging T with its suffix tree, i.e. identifying nodes of the same value. Since T is not compacted, this can easily be done in linear time. Hence, we obtain a linear time construction of the compacted $\mathcal{T}$, which yields a linear time algorithm constructing the SUF' table for T and consequently the following result.

Lemma 12. *The SUF table of a rooted tree can be computed in linear time.*

5.3   Computation of *SHIFT* and *SHIFT*^{R}

Recall that the maximal rotation of w corresponds to the maximal suffix of ww, see [6]. We develop algorithms based on this relation using a concept of *redundant* suffixes. A redundant suffix of w cannot become maximal under any extension of w, regardless of the direction we extend, see Observation 14.

Definition 13. *A suffix u of the string w is* redundant *if for every string z there exists another suffix v of w such that $vz > uz$. Otherwise we call u* nonredundant.

Observation 14. *If u is a redundant suffix of w, then for any string z it holds that uz is a redundant suffix of wz and u is a redundant suffix of zw.*

In the following easy fact and subsequent lemmas we build tools, which allow to focus on a logarithmic number of suffixes, discarding others as redundant.

Fact 15. *If u is a nonredundant suffix of w, then u is a prefix, and therefore a border, of maxSuf(w), the lexicographically maximum suffix of w.*

Lemma 16. *Let i be a position in a string w. Assume i is a square center, i.e. there exists a square factor of w whose second half starts at i. Then the suffix of w starting at i is redundant.*

Proof. Let $w = uxxv$, where $|ux| = i-1$. We need to show that xv is a redundant suffix of w. Let z be an arbitrary string. Consider three suffixes of wz: vz, xvz and $xxvz$. If $vz < xvz$, then $xvz < xxvz$, otherwise $xxvz < vz$. Hence, for each z there exists a suffix of wz greater than xvz, which makes xv redundant.   □

Lemma 17. *If u, v are borders of maxSuf(w) such that $|u| < |v| \le 2|u|$ then u is a redundant suffix of w.*

Proof. Due to Fine & Wilf's periodicity lemma [6] such a pair of borders induces a period of v of length $|v| - |u| \le |u|$. This concludes that there is a square in w centered at the position $|w| - |u| + 1$. Hence, by Fact 16, u is redundant.   □

Above lemmas do not give a full characterization of nonredundant suffixes, hence our algorithm maintains a carefully defined set that might be slightly larger.

Definition 18. *We call a set C a* small candidate set *for a string w if C is a subset of borders of maxSuf(w), contains all nonredundant suffixes of w and $|C| \le \max(1, \log|w| + 1)$. Any small candidate set for w is denoted by Cand(w).*

Lemma 19. *Assume we are given a string w and we are able to compare factors of w in constant time. Then for any $a \in \Sigma$, given small candidate set Cand(w) (Cand(w^R)) we can compute Cand(wa) (resp. Cand$((wa)^R)$) in $O(\log|w|)$ time.*

Proof. We represent the sets *Cand* as sorted lists of lengths of the corresponding suffixes. For *Cand*(wa) we apply the following procedure.

1. $C := \{va : v \in Cand(w)\} \cup \{\varepsilon\}$, where ε is an empty string.

2. Determine the lexicographically maximal element of $\mathcal{C}$, which must be equal to $maxSuf(wa)$ by definitions of redundancy and small candidate set.
3. Remove from $\mathcal{C}$ all elements that are not borders of $maxSuf(wa)$.
4. While there are $u, v \in \mathcal{C}$ such that $|u| < |v| \leq 2|u|$, remove u from $\mathcal{C}$.
5. $Cand(wa) := \mathcal{C}$

All steps can be done in time proportional to the size of $\mathcal{C}$. It follows from Observation 14, Fact 15 and Lemma 17 that the resulting set $Cand(wa)$ is a small candidate set. $Cand((wa)^R)$ is computed in a similar way.   □

Lemma 20. *For a labeled rooted tree T the tables $SHIFT$ and $SHIFT^R$ can be computed in $O(n \log n)$ time.*

Proof. We traverse the tree in DFS order and compute $maxSuf(ww)$ for each prefix path as: $maxSuf(ww) = \max\{yw : y \in Cand(w)\}$. Here we use tree DBF and *jump* queries for lexicographical comparison. If we know $maxSuf(ww)$, maximal cyclic shift of w is computed in constant time.   □

References

1. Bender, M.A., Farach-Colton, M.: The level ancestor problem simplified. Theor. Comput. Sci. 321(1), 5–12 (2004)
2. Bresar, B., Grytczuk, J., Klavzar, S., Niwczyk, S., Peterin, I.: Nonrepetitive colorings of trees. Discrete Mathematics 307(2), 163–172 (2007)
3. Crochemore, M., Ilie, L., Rytter, W.: Repetitions in strings: Algorithms and combinatorics. Theor. Comput. Sci. 410(50), 5227–5235 (2009)
4. Crochemore, M., Iliopoulos, C.S., Kociumaka, T., Kubica, M., Radoszewski, J., Rytter, W., Tyczyński, W., Waleń, T.: The Maximum Number of Squares in a Tree. In: Kärkkäinen, J., Stoye, J. (eds.) CPM 2012. LNCS, vol. 7354, pp. 27–40. Springer, Heidelberg (2012)
5. Crochemore, M., Iliopoulos, C.S., Kubica, M., Radoszewski, J., Rytter, W., Waleń, T.: Extracting Powers and Periods in a String from Its Runs Structure. In: Chavez, E., Lonardi, S. (eds.) SPIRE 2010. LNCS, vol. 6393, pp. 258–269. Springer, Heidelberg (2010)
6. Crochemore, M., Rytter, W.: Jewels of Stringology. World Scientific (2003)
7. Fraenkel, A.S., Simpson, J.: How many squares can a string contain? J. of Combinatorial Theory Series A 82, 112–120 (1998)
8. Gusfield, D., Stoye, J.: Linear time algorithms for finding and representing all the tandem repeats in a string. J. Comput. Syst. Sci. 69(4), 525–546 (2004)
9. Harel, D., Tarjan, R.E.: Fast algorithms for finding nearest common ancestors. SIAM J. Comput. 13(2), 338–355 (1984)
10. Kosaraju, S.R.: Efficient tree pattern matching (preliminary version). In: FOCS, pp. 178–183. IEEE Computer Society (1989)
11. Main, M.G., Lorentz, R.J.: An O(n log n) algorithm for finding all repetitions in a string. J. Algorithms 5(3), 422–432 (1984)
12. Shibuya, T.: Constructing the Suffix Tree of a Tree with a Large Alphabet. In: Aggarwal, A., Pandu Rangan, C. (eds.) ISAAC 1999. LNCS, vol. 1741, pp. 225–236. Springer, Heidelberg (1999)
13. Simon, I.: String Matching Algorithms and Automata. In: Karhumäki, J., Rozenberg, G., Maurer, H.A. (eds.) Results and Trends in Theoretical Computer Science. LNCS, vol. 812, pp. 386–395. Springer, Heidelberg (1994)

A General Method for Improving Insertion-Based Adaptive Sorting[*]

Riku Saikkonen and Eljas Soisalon-Soininen

Aalto University, School of Science and Technology, Department of Computer Science and Engineering, P.O. Box 15400, FI-00076 Aalto, Finland
{rjs,ess}@cs.hut.fi

Abstract. A presortedness measure describes to which extent a sequence of key values to be sorted is already partially sorted. We introduce a new natural measure of presortedness, which is a composition of two existing ones: *Block* that gives the number of already sorted disjoint subsequences of the input, and *Loc* defined as $\prod_{i=2}^{n} d_i$, where d_i denotes the distance between the $(i-1)$th and the ith element of the input in the ordered sequence up to the ith element. We also give a general method for improving insertion-based adaptive sorting, applying it to Splaysort to produce an algorithm that is optimal with respect to the new composite measure. Our experiments are performed for splay-tree sorting which has been reported to be among the most efficient adaptive sorting algorithms. Our experimental results show that, in addition to the theoretical superiority, our method improves standard Splaysort by a large factor when the input contains blocks of reasonable size.

Keywords: Adaptive sorting, Measures of presortedness, Search trees.

1 Introduction

One natural measure of presortedness is *Block* [2], defined as the number of maximal contiguous segments of the input $X = \langle x_1, \ldots, x_n \rangle$ that remain as such in the sorted sequence, or only change the order from descending to ascending[1].

The number of blocks, $Block(X)$, in a sequence X does not care of the order in which the blocks appear in X. In this sense *Block* does not fully demonstrate the level of presortedness in X: for example, sequence $X_1 = \langle 1, 2, 5, 6, 3, 4, \ldots, n-1, n, n-3, n-2 \rangle$ has the same number of blocks as $X_2 = \langle 1, 2, n-1, n, 5, 6, n-5, n-4, \ldots, 7, 8, n-3, n-2, 3, 4 \rangle$, which is obviously much farther from the sorted sequence than X_1.

In this paper we apply the adaptivity measures *Inv* [3] and *Loc* [4] to blocks instead of single elements. The measure *Inv* (number of inversions) gives the number of pairs of any two input elements that are in wrong order, and *Loc* is defined

---

[*] This research was partially supported by the Academy of Finland. A preliminary version of some of the results is published in the doctoral dissertation [1].

[1] The original definition of *Block* only considers ascending blocks. For this article we use the natural extended definition that also includes descending blocks.

K.-M. Chao, T.-s. Hsu, and D.-T. Lee (Eds.): ISAAC 2012, LNCS 7676, pp. 217–226, 2012.
© Springer-Verlag Berlin Heidelberg 2012

as $\prod_{i=2}^{n} d_i$, where d_i is the distance between the $(i-1)$th and the ith element in the sorted prefix containing keys from x_1 to x_i. We can then define composite measures $Inv \circ Block_{rep}(X) = Inv(Block_{rep}(X))$ and $Loc \circ Block_{rep}(X) = Loc(Block_{rep}(X))$, where $Block_{rep}(X)$ denotes the projection of X onto the largest elements (or any fixed representative elements) of the blocks it is composed of.

Such a composite measure is quite natural; it captures the notion that a sequence containing fewer blocks is closer to being sorted, and also Inv- or Loc-adaptivity of the sequence of representatives of blocks. As is intuitive, the composite measures are strictly superior to the measures they are composed of, in the sense that a sorting algorithm optimal with respect to the composite measures is never asymptotically worse than an algorithm only optimal with respect to the measures which are used in defining the composition, but is asymptotically faster for an infinite number of inputs. These properties are easy to show in the framework defined by Petersson and Moffat [4]. It is also straightforward to see that $Loc \circ Block_{rep}$ is strictly superior to $Inv \circ Block_{rep}$, but $Inv \circ Block_{rep}$ is independent with Loc, that is, there is no superiority relation between the two.

Our main result is to show how a general class of adaptive tree-sorting algorithms can be devised such that they are optimal with respect to $M \circ Block_{rep}$, where M is a suitable adaptiveness measure such as Inv or Loc. Especially we present a variation of Splaysort which is optimal with respect to $Loc \circ Block_{rep}$.

Our results are based on a new search-tree structure, called the bulk tree, that stores "bulks" that are blocks of the portion of input currently stored in the tree. An important feature of the tree is that it contains only one element for each bulk, stored as a pair (p, l), where p is a pointer to the first element of the bulk and l its length. Observe that we let the elements of the bulk stay in the input array and do not at all copy them into the bulk tree.

Our previous article [5] introduced the idea of tree sorting by bulks of the input, and the composite measure $Inv \circ Block_{rep}$. But at that time the concept of the bulk tree was not yet invented, and thus we could not produce an $Inv \circ Block_{rep}$-optimal sorting algorithm.

Apart from the optimality results, we include a detailed experimental study that compares the new bulk-tree-based Splaysort with standard Splaysort and others (Splaysort [6] is known to be one of the most versatile adaptive sorting algorithms), using Loc-, $Block$- and $Loc \circ Block_{rep}$-adaptive input data. The main result of our experiments is that our algorithm was much faster than Splaysort or any of the others for all inputs with reasonably large block size. More specifically, whenever $Block(X) \leq 10^6$ (for $|X| \approx 3.4 \cdot 10^7$ in our experiments), Bulk-tree sort was at least 2.3 and up to 9.1 times faster than Splaysort, which used 2.3–5.0 times as many comparisons.

This paper is organized as follows. Section 2 contains the definiton of the bulk tree structure and proves our main result that using bulk trees instead of standard trees indeed yields an $M \circ Block_{rep}$-optimal sorting algorithm, if sorting using the underlying non-bulk tree is M-optimal. In Sect. 3 we report our experimental work, and Sect. 4 gives the conclusion.

2   The Bulk Tree

How can a sorting algorithm be made optimal to $Loc \circ Block_{rep}$? It is not feasible to first find $Block_{rep}(X)$ using a $Block$-optimal algorithm and then sort it with a Loc-optimal algorithm: Since the $Loc \circ Block_{rep}$ measure is superior to $Block$, an algorithm that finds $Block_{rep}(X)$ spends too much time doing it (for some inputs), unless that algorithm by itself is already $Loc \circ Block_{rep}$-optimal. Therefore, a $Loc \circ Block_{rep}$-optimal algorithm cannot be made by simply combining $Block$- and Loc-optimal algorithms.

Here we present a general class of tree-sorting algorithms that are able to process whole blocks as if they were single keys. This method, which we call the *bulk tree*, is applicable to at least Splaysort [6], various AVL-tree-based sorting algorithms (e.g., [5]), and Local Insertion Sort which uses B-tree-based finger trees [7]. We present the general method and apply it to enhance Splaysort, which appears to be one of the most practical adaptive sorting algorithms. The enhanced algorithm will be $Loc \circ Block_{rep}$-optimal, since Splaysort by itself is Loc-optimal [8].

The idea behind the bulk tree is to store each block as a single key in the tree (where a normal tree would store each individual element of the block separately). Then the tree traversal takes time proportional to the number of blocks traversed, independent of the block sizes.

However, before sorting is complete (i.e., when only part of the input sequence has been inserted into the tree), it is not possible to know the actual blocks: later elements in the input sequence may split a previous "block" in parts. For instance, if $X = \langle 7, 8, 1, 2, 5, 6, 3, 4 \rangle$ and only the prefix $X' = \langle 7, 8, 1, 2, 5, 6 \rangle$ has yet been inserted, the blocks of X' are $\langle 7, 8 \rangle$ and $\langle 1, 2, 5, 6 \rangle$. After the sorting is complete, it is seen that X has the blocks $\langle 1, 2 \rangle$, $\langle 3, 4 \rangle$, $\langle 5, 6 \rangle$ and $\langle 7, 8 \rangle$; and the block $\langle 1, 2, 5, 6 \rangle$ of X' has been split in two.

Our bulk tree always stores the blocks of the currently-inserted prefix X' of the input X; we call these blocks *bulks*. Since appending additional keys to a sequence X' can split but not merge its blocks, the bulks represent a subset of the actual blocks of X; more formally:

Lemma 1. *If $X' = \langle x_1, x_2, \ldots, x_{i-1} \rangle$ is a prefix of $X = \langle x_1, x_2, \ldots, x_n \rangle$ so that $1 < i \le n$ is at a block boundary, then $Block_{rep}(X') \subseteq (Block_{rep}(X) \cap X')$.*

2.1   Inserting Bulks

Each element of the bulk tree (each node in a binary search tree) contains a pointer to the first key of the bulk in the input sequence X, the number of keys, and a flag that notes whether the bulk is in ascending or descending order. The first bulk is the longest ascending or descending prefix of X.

A new bulk is inserted into the bulk tree as follows, when the prefix $X' = X[1..i-1]$, $2 \le i \le n$, has already been inserted. Search in the tree for $X[i]$; it belongs either between two existing bulks, or inside an existing bulk which needs to be split in two. An efficient way of searching in the tree is to compare

$X[i]$ only to the minimum key of each bulk, keeping track of the bulk $X[a..b]$ with largest minimum key a (this is in the last visited node which is not the leftmost child of its parent). When the search finishes at a leaf – always at the location where the immediate successor of $X[a..b]$ would be inserted – compare $X[i]$ with b (the maximum key of $X[a..b]$) to find out whether $X[a..b]$ needs to be split.

The new bulk $X[i..j]$ is the longest ascending or descending prefix of $X[i..n]$ with keys between the next-smaller and next-larger keys in the tree (these are found during the tree traversal or, in the bulk-split case, are the keys surrounding the split position). A few optimizations for finding the end j of the bulk are discussed in [1,5].

If no split is needed, the new bulk $X[i..j]$ is simply inserted in the tree in the current position. Otherwise, when the new keys $X[i..j]$ belong inside the existing bulk $X[a..b]$, the latter is split into $X[a..s]$ and $X[s+1..b]$, where $X[s] < X[i] < X[s+1]$. Then $X[a..b]$ in the tree is replaced by one of the three bulks (say $X[a..s]$), and the other two ($X[i..j]$ and $X[s+1..b]$) are inserted in the tree as new elements. (The above assumes that $X[a..b]$ is an ascending bulk; a descending $X[a..b]$ is symmetric.)

We use the name *bulk-tree sort* for the algorithm that sorts an input sequence X by inserting it into the bulk tree and reading the sorted result from the tree. After all of X has been inserted, the bulk tree contains one element for each block in $Block_{rep}(X)$:

Lemma 2. *After every insertion of a new bulk during bulk-tree sort, the bulk tree stores the blocks of the currently-inserted prefix X' of the input sequence X.*

Proof. Assume that the bulk tree currently stores the bulks of the prefix $X[1..i-1]$. The next bulk $X[i..j], 1 < i < j \leq n$ is a block of $X[1..j]$ (and $X[i..j+1]$ is not a block of $X[1..j+1]$ or we would have increased j). An existing bulk (which is a block of $X[1..i-1]$) is split exactly when it forms two blocks in $X[1..j]$. □

2.2   Analysis

The work done in sorting using the bulk tree can be divided in five parts: (i) tree traversal for finding each insertion position; (ii) the insertions and possible rebalancing; (iii) for each new bulk, finding the prefix of the remaining keys that forms the bulk; (iv) splitting previous bulks; (v) $O(|X|)$ time to read the sorted result by traversing the final tree in in-order. We begin with (iii) and (iv), which are independent of the underlying search tree.

Lemma 3. *Assume that n keys that form k bulks are inserted into the bulk tree during sorting. Determining the keys that belong to each new bulk can be done in total time $O(n)$.*

Proof. Except for the first key of each bulk, each key is compared once to the previous one (to see if the sequence continues to be ascending or descending) and to a minimum or maximum value (the next-smaller or next-larger key in the currently-inserted keys). It is actually possible to avoid most of the comparisons to the minimum or maximum: see [5, Theorem 2].   □

Since the bulks are stored only as pointers to the actual keys (which are consecutive in the input array), the only non-constant-time operation involved in splitting a bulk is finding the split position. We obviously cannot use standard binary search for this, since it would use $O(Block(X) \log Block(X))$ comparisons, which is only enough for $Block$-optimality. But using exponential and binary search as defined by Fredman [9] we have:

Lemma 4. *Assume that the keys of sequence $X = \langle x_1, x_2, \ldots, x_n \rangle$ are inserted as k bulks into an initially empty bulk tree T. Then the total time needed for all searches in the bulks of T for finding their split positions is $O(n)$.*

We have now shown that parts (iii), (iv) and (v) all take $O(|X|)$ time. It remains to analyze the tree traversal and insertion costs, which depend on the search tree used as the basis of the bulk tree.

Lemma 5. *Let A be a tree-insertion-based sorting algorithm and A' be the same algorithm augmented to use bulk trees. The total time taken by tree traversals for finding the positions of new bulks, when using A' to sort a sequence X, is $O(t_A(Block_{rep}(X)))$, where $t_A(Block_{rep}(X))$ is the time taken by algorithm A to sort $Block_{rep}(X)$.*

Proof. This would be trivial if the tree always contained exactly the blocks of the input X that are present in the currently-inserted prefix X'. In reality, the tree contains representative bulks for some but not necessarily all of these blocks (Lemmas 1 and 2). But the fact that some of the actual blocks of X' are missing from the tree can only decrease the cost of the tree traversals.   □

We then need to consider rebalancing done after insertions, and the additional tree traversal required in the bulk-split case for inserting a second element in the next-larger position. The latter is slightly different in various underlying trees:

Lemma 6. *Assume that the search tree underlying bulk-tree sort has amortized constant rebalancing cost for a sequence of insertions, and that the next-larger position from the previous insertion can be found in amortized constant time. When sorting X with bulk-tree sort, all insertions, with rebalancing but without tree traversals for finding the positions of new bulks, take total time $O(Block(X))$.*

Proof. There are exactly $Block(X)$ insertions (Lemma 2), so all rebalancing is $O(Block(X))$ by the assumption of amortized constant rebalancing cost.

For traversal in the second insertion of the bulk-split case, we need to analyze the distance between the insertion position of a new bulk and its successor position. All search trees except splay trees insert a new element in a leaf (where the next-larger position is its rightmost child), but in some trees the leaf may propagate upward during rebalancing. (For instance in an AVL tree, a rotation done in the parent of the inserted leaf moves the inserted leaf up.) But this propagation is limited by the (amortized constant) amount of rebalancing that was done, so the distance is $O(1)$ per insertion when amortized over all the insertions.

Splay trees place a newly-inserted element at the root, so the next-larger position can be far away in the worst case. But Cole [8] shows that the amortized cost of traversal to the next-larger position is constant (this is a special case of what is required for *Loc*-optimal sorting: traversal to a distance d should take amortized time $O(\log(d+1))$).   □

We have:

Theorem 1. *Let A be a sorting algorithm that sorts by insertion into a search tree and that can be augmented to use bulk trees. If A is M-optimal for any presortedness measure M, then the augmented algorithm A' will be $M \circ Block_{rep}$-optimal.*

Proof. By Lemmas 3–6 the time bound of A' is $O(|X|+t_A(Block_{rep}(X)))$, where X is the input sequence and $t_A(Block_{rep}(X))$ is the time taken by A to sort $Block_{rep}(X)$. Thus, $t_{A'}(X)$ is $M \circ Block_{rep}$-optimal if $t_A(Block_{rep}(X))$ is M-optimal.   □

Since Splaysort is *Loc*-optimal [8], Bulk-tree sort using splay trees is $Loc \circ Block_{rep}$-optimal (and thus *Loc*-, *Block*- and $Inv \circ Block_{rep}$-optimal due to the superiority relationships of the measures, as well as *Inv*-, *Osc*-, *Rem*-, *Exc*-, *Max*- and *Runs*-optimal using the hierarchy defined by Petersson and Moffat [4]).

2.3   Optimization: Making Small Bulks Larger

Tree insertion-based sorting is known to be slow with inputs that are close to random [5,6,10]. Even though large blocks are not present ($Block(X)$ is large), using the bulk tree can still make sorting these inputs faster: we can enforce a minimum size for the bulks in the bulk tree by adding the keys of a new small bulk to its neighbor instead of creating a new bulk.

We merge a newly-inserted bulk with its neighbors whenever the resulting bulk has at most u elements (u is a constant). It is possible to maintain the invariant that, except for the two bulks at the edges of the tree, every pair of consecutive bulks has at least $u/2$ keys in total, giving an average bulk size of at least $u/4$. When a new bulk is inserted between existing bulks, the new keys are added to either the predecessor or successor bulk if the result has at most u elements. In the bulk-split case, we look at the new bulk, the halves produced by the split and the predecessor and successor bulks, merging any combinations that result in bulks of at most u elements.

To merge bulks that are not adjacent in the input sequence, we need to make copies of their elements (instead of only storing pointers to the input sequence). When a bulk takes part in a merge for the first time, we allocate u elements of temporary space for it, which is enough for all subsequent merges involving this bulk.

The choice of u is arbitrary; a large value increases the cost of the actual insertion (because of merging), but greatly decreases search times (the bulk tree has at most $4|X|/u+2$ bulks). Our experiments indicate that values of u even up to 200 work well; we used $u = 100$ in the ones reported below. Any constant u preserves the optimality properties of bulk-tree sort, since this optimization does not affect bulks with more than u elements.

The optimization needs access to the predecessor and successor bulks at each insertion point. In some search trees they are immediately available; otherwise pointers can be added to the nodes as necessary. (We needed to add a pointer to the predecessor in our implementation of bulk-tree sort with splay trees.)

3 Experiments

We compared our implementation of Bulk-tree sort using splay trees to normal Splaysort and some other sorting algorithms, using randomly generated input data. We use *Block-*, *Loc-* and *Loc ∘ Block$_{rep}$*-adaptive input data, which best capture the theoretical adaptivity of the Bulk-tree sort algorithm.

All of our implementations were written in C, and run under GNU/Linux on a 4-core 64-bit Intel Core2 Quad running at 2.6 GHz (only one CPU core was used, since the algorithms are not concurrent). Each experiment was repeated 10 times using newly generated input; we report averages. The keys were simple 4-byte integers in the range $[1, n]$ with $n = |X| = 2^{25}$. We measured both the number of comparisons and actual running time.

3.1 Other Sorting Algorithms

Since we based Bulk-tree sort on splay trees, the most interesting comparison will be to normal Splaysort, which is known to be efficient especially in the number of comparisons [6,10]. To give a more complete picture, we included the *Inv*-optimal merge-sort-like Splitsort [11], which is not based on tree insertion and is efficient especially in running time [10,12]. In addition, we used the Merge sort and Quicksort implementations from the GNU C library (version 2.11.3) – the latter is an engineered implementation that cites [13].

We also tried our AVL-tree bulk-insertion sort algorithm [5]. It was slower than Bulk-tree sort in all cases (understandably since it is not *Loc ∘ Block$_{rep}$*- or even *Loc*-optimal), so we omit it from the figures shown here.

We reimplemented Splaysort with a space optimization: instead of storing 64-bit pointers in nodes, we stored all nodes in an array so that pointers were 32-bit indices to this array. We also tried the Splaysort implementation from [6]: it performed the same number of comparisons as our own, but was much slower (probably because it was engineered for 32-bit computers in 1995). Our Splitsort implementation used $2n$ pointers of extra space and not the n-pointer space optimization of [11], which we had previously [5] found to be slower.

All tree-based algorithms wrote the sorted result back into the original array (unpacking any bulks into single elements).

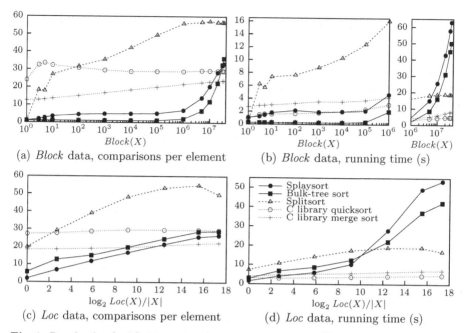

(a) *Block* data, comparisons per element

(b) *Block* data, running time (s)

(c) *Loc* data, comparisons per element

(d) *Loc* data, running time (s)

Fig. 1. Results for **(a–b)** *Block* and **(c–d)** *Loc* data, $n = 2^{25} \approx 3.4 \cdot 10^7$. For clarity, the plot legend is given only in (d), and the plot for (b) has been divided in two horizontally.

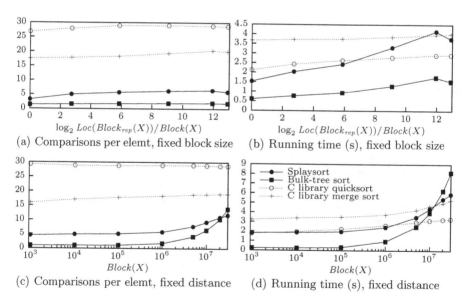

(a) Comparisons per elemt, fixed block size

(b) Running time (s), fixed block size

(c) Comparisons per elemt, fixed distance

(d) Running time (s), fixed distance

Fig. 2. Results for $Loc \circ Block_{rep}$ data, $n = 2^{25} \approx 3.4 \cdot 10^7$. **(a–b)** With a fixed number of blocks $(Block(X) \approx 9.8\text{–}10 \cdot 10^5$, or 8.7 at the leftmost data point). **(c–d)** With fixed distance between blocks $(\log_2 Loc(Block_{rep}(X))/Block(X) \approx 5.8\text{–}5.9$, or 4.5 and 5.6 at the two leftmost data points). For clarity, the plot legend is given only in (d), and Splitsort is not drawn since it was much slower (26–54 compar., 8–16 seconds).

3.2   Results

Figures 1 and 2 show results from our experiments. The x-axis of each figure ranges from almost sorted sequences at the left to almost random ones at the right.

For *Block* data, Figs. 1(a–b), the leftmost data point is fully sorted and the rightmost fully random (i.e., every permutation is equally likely). With this data, Bulk-tree sort was always faster than standard Splaysort, and used fewer comparisons except when $Block(X) > 2 \cdot 10^7$. For fully sorted data, both used the minimum number of comparisons, but Splaysort required 4.3 times as much running time. For the large range $10 \leq Block(X) \leq 10^5$, Splaysort used 3.6–5.0 times as many comparisons and 5.7–9.1 times as much time.

In the *Loc*-adaptive inputs of Figs. 1(c–d), large blocks are never present. The bulk tree is thus at a disadvantage, and Bulk-tree sort performed slightly worse than Splaysort: it used 1.1–1.8 times as many comparisons (2.8 at the leftmost data point) and up to 1.9 times as much time as Splaysort. But when the input was close to random (both here and for *Block* data, i.e., the right-hand side of Figs. 1(b,d)), the optimization of Sect. 2.3 improved running times: then Splaysort used 1.3 times as much time as Bulk-tree sort.

Finally, Fig. 2 shows our results for $Loc \circ Block_{rep}$-adaptive data, for which Splaysort is not even optimal. This type of input data has two parameters: for Figs. 2(a–b) we fixed the block size (*Block*) and altered the distance between blocks ($Loc \circ Block_{rep}$), and then did vice versa for Figs. 2(c–d). Except for close-to-random data ($Block(X) > 10^6$ in Figs. 2(c–d)), Splaysort performed 2.3–4.8 times as many comparisons as Bulk-tree sort and used 2.4–7.7 times as much time.

Summarizing from all the experiments, our main result is that for all inputs with reasonably large blocks ($Block(X) \leq 10^6$), Bulk-tree sort was the fastest algorithm, and standard Splaysort needed 2.3–5.0 times as many comparisons and 2.3–9.1 times as much time as Bulk-tree sort (except for fully sorted sequences where both used the minimum number of comparisons). Even with small blocks, the optimization of Sect. 2.3 improved upon standard Splaysort for close-to-random data, though here the non-adaptive algorithms are much faster.

4   Conclusions

Splaysort is known to be an efficient algorithm, yet in our experiments we were able to improve its running time by a factor of at least 5 for a large range of *Block*- and $Loc \circ Block_{rep}$-adaptive inputs. Our algorithm was consistently more efficient than the others for all inputs that contained reasonably large bulks. The reason is twofold: (*i*) fewer comparisons are required, because the bulk tree allows insertion-based sorting to jump over large bulks while looking only at their endpoints, and (*ii*) the search tree is much smaller when it only stores one element per bulk (especially with our optimization that merges small bulks together), so more of the tree fits in the hardware cache.

The bulk tree can also be used as a general-purpose search tree optimized for bulk insertion. The best bulk-insertion algorithms for traditional search trees require $O(\log m)$ amortized time for rebalancing when inserting a bulk of m keys to a specific position. In the bulk tree, because inserting a bulk only requires inserting one or two new elements, rebalancing is done in amortized $O(1)$ time independent of the bulk size.

Our ongoing research applies the bulk-tree idea to a completely different application: to serve as an index structure for databases stored on solid-state drives. In addition to optimizing bulk operations, our storage method can take advantage of the small physical size of the bulk tree, when a minimum bulk size is enforced.

References

1. Saikkonen, R.: Bulk Updates and Cache Sensitivity in Search Trees. PhD thesis, Helsinki University of Technology (2009)
2. Carlsson, S., Levcopoulos, C., Petersson, O.: Sublinear merging and natural mergesort. Algorithmica 9(6), 629–648 (1993)
3. Mehlhorn, K.: Sorting Presorted Files. In: Weihrauch, K. (ed.) GI-TCS 1979. LNCS, vol. 67, pp. 199–212. Springer, Heidelberg (1979)
4. Petersson, O., Moffat, A.: A framework for adaptive sorting. Discrete Applied Mathematics 59(2), 153–179 (1995)
5. Saikkonen, R., Soisalon-Soininen, E.: Bulk-Insertion Sort: Towards Composite Measures of Presortedness. In: Vahrenhold, J. (ed.) SEA 2009. LNCS, vol. 5526, pp. 269–280. Springer, Heidelberg (2009)
6. Moffat, A., Eddy, G., Petersson, O.: Splaysort: Fast, versatile, practical. Software, Practice and Experience 126(7), 781–797 (1996)
7. Mannila, H.: Measures of presortedness and optimal sorting algorithms. IEEE Transactions on Computers C-34, 318–325 (1985)
8. Cole, R.: On the dynamic finger conjecture for splay trees, part II: The proof. SIAM Journal on Computing 30(1), 44–85 (2000)
9. Fredman, M.L.: Two applications of a probabilistic search technique: Sorting X+Y and building balanced search trees. In: 7th Annual ACM Symposium on Theory of Computing (STOC 1975), pp. 240–244. ACM Press (1975)
10. Elmasry, A., Hammad, A.: An Empirical Study for Inversions-Sensitive Sorting Algorithms. In: Nikoletseas, S.E. (ed.) WEA 2005. LNCS, vol. 3503, pp. 597–601. Springer, Heidelberg (2005)
11. Levcopoulos, C., Petersson, O.: Splitsort – an adaptive sorting algorithm. Information Processing Letters 39, 205–211 (1991)
12. Estivill-Castro, V., Wood, D.: A survey of adaptive sorting algorithms. ACM Computing Surveys 24(4), 441–476 (1992)
13. Bentley, J.L., McIlroy, M.D.: Engineering a sort function. Software, Practice and Experience 23(11), 1249–1265 (1993)

Counting Partitions of Graphs

Pavol Hell[1,*], Miki Hermann[2,**], and Mayssam Mohammadi Nevisi[1,*]

[1] School of Computing Science, Simon Fraser University, Burnaby, Canada
{pavol,maysamm}@sfu.ca
[2] LIX CNRS UMR 7161, École Polytechnique, Palaiseau, France
hermann@lix.polytechnique.fr

Abstract. Recently, there has been much interest in studying certain graph partitions that generalize graph colourings and homomorphisms. They are described by a pattern, usually viewed as a symmetric $\{0, 1, *\}$-matrix M. Existing results focus on recognition algorithms and characterization theorems for graphs that admit such M-partitions, or M-partitions in which vertices of the input graph G have lists of admissible parts. In this paper we study the complexity of counting M-partitions. The complexity of counting problems for graph colourings and homomorphisms have been previously classified, and most turned out to be #P-complete, with only trivial exceptions where the counting problems are easily solvable in polynomial time. By contrast, we exhibit many M-partition problems with interesting non-trivial counting algorithms; moreover these algorithms appear to depend on highly combinatorial tools. In fact, our tools are sufficient to classify the complexity of counting M-partitions for all matrices M of size less than four. It turns out that, among matrices not acccounted for by the existing results on counting homomorphisms, all matrices which do not contain the matrices for independent sets or cliques yield tractable counting problems.

Keywords: partitions, polynomial algorithms, #P-completeness, dichotomy, counting problems.

1 Introduction

It is well known that the number of bipartitions of a graph G can be computed in polynomial time. Indeed, we can first check, in polynomial time, if G is bipartite, and if not, the answer is 0. If G is bipartite, we can find, in polynomial time, the number c of connected components of G. Since each such component admits exactly two bipartitions, the answer in this case is 2^c. Interestingly, the number of bipartitions can also be counted using linear algebra: if each vertex v is associated with a variable x_v over the field $F_2 = \{0, 1\}$, and each edge uv with the equation $x_u + x_v = 1$ in F_2 (i.e., modulo 2), then the number solutions of this system is precisely the number of bipartitions of G. Thus 2-colourings of graphs can

* Supported by NSERC Canada.
** Supported by ANR Blanc International ALCOCLAN.

K.-M. Chao, T.-s. Hsu, and D.-T. Lee (Eds.): ISAAC 2012, LNCS 7676, pp. 227–236, 2012.

be counted in polynomial time. It is also known that the counting problem for
m-colourings of G with $m > 2$ is #P-complete [5]. This is the *dichotomy* of the
counting problems for graph colourings.

A *homomorphism* f of G to H is a mapping $V(G) \rightarrow V(H)$ such that $uv \in$
$E(G)$ implies $f(u)f(v) \in E(H)$. If $H = K_m$, a homomorphism of G to H is an
m-colouring of G. Dichotomy of counting homomorphisms to graphs H has been
established by Dyer and Greenhill [6]. Namely, if each connected component of
H is either a reflexive complete graph, or an irreflexive complete bipartite graph,
then counting homomorphisms to H can be solved by trivial methods, as in the
above example (or, once again, by linear algebra). In all other cases, counting
homomorphisms to H is #P-complete [6]. Other dichotomies for homomorphism
counting problems, in bounded degree graphs, or for homomorphisms with lists,
are discussed in [18]. (In the list version of the problem, the graph G has a list
$L(v) \subseteq V(H)$ for each vertex $v \in V(G)$ and only homomorphisms f that satisfy
$f(v) \in L(v)$, for all $v \in V(G)$, are counted.) In particular, it is proved in [18],
that, as without lists, if each connected component of H is either a reflexive
complete graph, or an irreflexive complete bipartite graph, then counting list
homomorphisms to H can be solved by easy polynomial time methods, and in
all other cases counting list homomorphisms to H is #P-complete [18].

Homomorphisms to H can be viewed as partitions of the input graph G into
parts corresponding to the vertices of H. Specifically, if $x \in V(H)$ has no loop,
the corresponding part P_x is an independent set in G, and if $xy \notin E(H)$, then
there are no edges between the parts P_x and P_y. A further generalization of
homomorphisms allows us to specify that certain parts P_x must be cliques, and
between certain parts P_x and P_y there must be all possible edges.

Throughout the paper, M will always be assumed to be a symmetric m by
m matrix over $0, 1, *$. An *M-partition* of a graph G is a partition $P_1, P_2, \ldots, P_m$
of $V(G)$, such that two distinct vertices in (possibly equal) parts P_i and P_j are
adjacent if $M(i,j) = 1$, and nonadjacent if $M(i,j) = 0$; the entry $M(i,j) =$
$*$ signifies no restriction. Since we admit $i = j$, a part P_i is independent if
$M(i,i) = 0$, and a clique if $M(i,i) = 1$. (We usually refer to P_i as *the i-th
part*.) Note that when M has no 1's, the matrix M corresponds to an adjacency
matrix of a graph H, if we interpret $*$ as adjacent and 0 as non-adjacent; and
in this case an M-partition of G is precisely a homomorphism of G to H. Thus
M-partitions generalize homomorphisms and hence also graph colourings. They
are frequently encountered in the study of perfect graphs. A simple example is
the matrix $M = \begin{pmatrix} 0 & * \\ * & 1 \end{pmatrix}$: in this case a graph G is M-partitionable if and only
if it is a split graph. Other examples of matrices M such that M-partitions are
of interest in the study of perfect graphs can be found in [8]. They include the
existence of a homogeneous set [13] (M has size three, see the next section), the
existence of a clique cutset (M has size three), the existence of a skew cutset
(M has size four), and many other popular problems [3,8,10,15,22]. (If M has a
diagonal $*$, it is usual for the existence problems to focus on M-partitions with
all parts non-empty, otherwise the problems become trivial; this is in particular
the case in the previous three examples.) In any event, we emphasize the fact

that M-partition problems tend to be difficult and interesting even for small matrices M.

We note for future reference the following *complementarity* of M-partitions. Denote by $\overline{M}$ the matrix obtained from M be replacing each 0 by 1 and vice versa. Then an $\overline{M}$-partition of a graph G is precisely an M-partition of the complement $\overline{G}$.

In the literature there are several papers dealing with algorithms and characterizations of graphs admiting M-partitions (or list M-partitions) [4,7,8,10]; these are detailed in a recent survey [14], cf. also a slightly older survey [19], or the book [16]. We focus on the counting problem for M-partitions. Recall that for homomorphism problems, except for trivial cases, counting homomorphisms turned out to be #P-complete [6]. More generally, in constraint satisfaction problems [9,20], the situation is similar, and only the counting problems that can be solved using algebraic methods turned out to be tractable [1]. We contrast these facts by exhibiting several counting problems for M-partitions, where highly combinatorial methods seem to be needed. In the process, we completely classify the complexity of counting the number of M-partitions for all matrices of size less than four.

Given a matrix M, we want to know how hard it is to count the number of M-partitions. As a warm-up, we prove the following classification for two by two matrices $M = \begin{pmatrix} a & c \\ c & b \end{pmatrix}$.

Theorem 1. *If c and exactly one of a, b is $*$, then the problem of counting the number of M-partitions is #P-complete. Otherwise there is a polynomial algorithm to count the number of M-partitions.*

Proof. If, say, $a = c = *$ and $b = 0$, then the number of M-partitions of G is precisely the number of independent sets in G, which is known to be #P-complete [21]. Similarly, if $a = c = *$ and $b = 1$, we are counting the number of cliques in G, which is also #P-complete [21] (or by complementarity).

If M has no 1, the result follows by [6], so we assume that M contains at least one 1, and, by complementarity, also at least one 0. If $c = 0$, the two parts have no edges joining them, and at least one part is a clique. The number of M-partitions can easily be determined once the connected components of G have been computed. (For instance if $a = 1$, $b = *$, and t connected components of G are cliques, then G has t M-partitions. When $a = 1$, $b = 0$, the counting is even easier.) If $c = 1$, the result follows by complementation.

Thus we assume that $c = *$. By symmetry, we may assume without loss of generality that $a = 0$ and $b = 1$. Such an M-partition of G is called a *split partition*. It follows from [8] that the number of split partitions is polynomial, and can be found in polynomial time (see Theorem 3.1 in [8]). Therefore they can also be counted in polynomial time. $\square$

The two matrices $\begin{pmatrix} * & * \\ * & 0 \end{pmatrix}$ and $\begin{pmatrix} * & * \\ * & 1 \end{pmatrix}$ from the first paragraph of the proof will play a role in the sequel, and we will refer to the as the *matrices for independent sets, and cliques.*

2   Decomposition Techniques

The last two by two matrix, corresponding to split partitions, illustrates the fact that there are interesting combinatorial algorithms for counting M-partitions. In this section we examine a few other examples, in this case of three by three matrices, documenting this fact.

In the remainder of the paper, we assume we have the matrix $M = \begin{pmatrix} a & d & e \\ d & b & f \\ e & f & c \end{pmatrix}$.

We begin by discussing some cases related to the example of split partitions at the end of the previous section. In fact, the general technique of Theorem 3.1 from [8] is formulated in the language of so-called *sparse-dense partitions*. If $\mathcal{S}$ and $\mathcal{D}$ are two families of subsets of $V(G)$, and if there exists a constant c such that all intersections $S \cap D$ with $S \in \mathcal{S}, D \in \mathcal{D}$, have at most c vertices, then G with n vertices has at most n^{2c} sparse-dense partitions $V(G) = S \cup D$, with $S \in \mathcal{S}, D \in \mathcal{D}$, and they can be generated in polynomial time [8]. For split partitions, we take $\mathcal{S}$ to be all independent sets, $\mathcal{D}$ all cliques, and $c = 1$. If we take for $\mathcal{S}$ all bipartite induced subgraphs, for $\mathcal{D}$ all cliques, and $c = 2$, we can conclude that any graph G has only a polynomial number of partitions $V(G) = B \cup C$, where B induces a bipartite graph and C induces a clique, and all these partitions can be generated in polynomial time. This result is sufficient to cover a number of polynomial cases.

Theorem 2. *If a, b, c are not all the same and none is $*$, then the number of M-partitions can be counted in polynomial time.*

Proof. Up to symmetry and complementarity we may assume that $a = b = 0, c = 1$. For each sparse-dense partition $V(G) = S \cup D$, we shall test how many partitions of the subgraph induced by S, into the first part and the second part, satisfy the constraints induced by the entries d, e, and f of the matrix M. We impose the constraints due to e and f by introducing lists on the vertices in S. (If, say, $e = 1$, then only vertices completely adjacent to D will have the third part in their lists, and similarly for other values of e and f.) If d is 0 or $*$, this corresponds to counting list homomorphisms to a complete bipartite graph (K_2), which is polynomial by [18], as noted above. When $d = 1$, there are at most two different partitions of S to consider, so we can check these as needed. $\square$

The next result deals with a class of problems related to homogeneous sets and modular decomposition [13]. A *module* (or a *homogeneous set*) in a graph G is a set $S \subseteq V(G)$ such that every vertex not in S is either adjacent to all vertices of S or to none of them. Trivially, each singleton vertex, as well as $V(G)$ and $\emptyset$, are modules. For most applications, these trivial modules are ignored, but we will be counting all modules. In fact, we will show that modules can be counted in polynomial time, even under some additional restrictions.

For our purposes, we will only use the following basic theorem of Gallai [11].

Theorem 3 ([11]). *For any graph G one of the following three cases must occur.*

1. G is disconnected, with components $G_1, G_2, \ldots G_k$.
   Each union of the sets $V(G_i)$ is a module of G, and the other modules of G are precisely all the modules of individual components G_i.
2. The complement of G is disconnected, with components $H_1, H_2, \ldots, H_\ell$.
   Each union of the sets $V(H_j)$ is a module of G, and the other modules of G are precisely all the modules of individual subgraphs $\overline{H_j}$.
3. Both G and its complement are connected. There is a partition $S_1, S_2, \ldots, S_r$ of $V(G)$ (which can be computed in linear time), such that all the modules of G are precisely all the modules of individual subgraphs induced by the sets S_t, $t = 1, \ldots, r$, plus the module $V(G)$.

Based on this theorem, one can recursively decompose any graph into modules; this decomposition produces a tree structure called the *modular decomposition tree* of G. Even though the number of modules of G can be exponential (for instance if $G = K_n$), the modular decomposition tree has polynomial size and can be computed in linear time [12].

We will count modules of G, or modules of G with special properties, recursively, using the theorem. This will suffice to provide a polynomial time counting algorithm for these modules. We note however, that there is a natural linear time algorithm that counts these modules directly on the modular decomposition tree. We will describe this algorithm in the full journal version of our paper.

We will call the sets $V(G_i)$, $V(\overline{H_j})$, and S_t from the theorem the *blocks* of G. According to the theorem, all modules are modules of the blocks, except for modules that are unions of (at least two) blocks. We call these latter modules *cross modules*.

We illustrate the technique on a polynomial time algorithm to count the total number $T(G)$ of non-empty modules of a graph G. (This turns out to be more convenient, and one can add 1 for the empty module at the end of the computation.) We first compute the decompositions (1, 2, or 3) in Theorem 3. In cases 1 and 2, we have $T(G) = 2^t - t - 1 + \sum T(B)$, where t is the number of blocks, and the sum is over all blocks B. In case 3, we have $T(G) = 1 + \sum T(B)$. Indeed the number of cross modules is 1 in the case 3 (only the module $V(G)$ is a cross module), and is $2^t - t - 1$ in the other two cases (subtracting one for the empty set and the individual blocks). Since the sizes of the blocks for the recursive calls sum up to n, this yields a recurrence for the running time whose solution in polynomial in n. This shows that counting the number of modules of G is polynomial.

We note that a number of variants can be counted the same way. Consider, for instance, the number of modules that are independent sets. In case 1, if s of the blocks consist of a single vertex, then the number of cross modules changes to $2^s - s - 1$. In case 2, as well as 3, there are no cross module in this case (unless G has no edges). Moreover, it is easy to see that the number of non-empty independent modules of G inside a block B is precisely the number of non-empty independent modules of the graph induced by B, in all three cases. Thus the number of independent non-empty modules of G is $T(G) = 2^s - s - 1 + \sum T(B)$ in case 1, and $T(G) = \sum T(B)$ in cases 2 and 3. Of course, the number of modules that are cliques can be counted in a similar way, or by looking at the complement.

We can in fact handle all restrictions of this type on the module S, its set of neighbours R, and its set of non-neighbours Q. The above examples are respectively: (i) S, R, Q unrestricted; (ii) S independent, R, Q unrestricted. It turns out that all the remaining combinations of restrictions (independent, clique, or unrestricted) for the sets S, Q, R can be treated in similar ways.

Note that the arguments are written so as to count the number of non-empty modules of the various kinds. Of course by adding 1, we can count all such modules.

Theorem 4. *There is a polynomial time algorithm to count the number of modules satisfying any combination of restrictions where the module itself, its set of neighbours, and its set of non-neighbours are an independent set, a clique, or unrestricted.*

It is easy to see that each restricted kind of module corresponds to an M-partition in which $d = 1, e = 0, f = *$, and a is determined by the constraint on S ($a = 0$ if R is to be independent, $a = 1$ if it is to be a clique, and $a = *$ if S is unrestricted), b is determined by the constraint on R, and c by the constraint on Q.

Corollary 5. *If $d = 1, e = 0, f = *$, then the number of M-partitions with non-empty first part can be counted in polynomial time.*

However, the number of M-partitions must also take into account the partitions with the first part empty. By applying Theorem 1, we conclude that the number of M-partitions with empty first part can also be counted in polynomial time, unless the second and third part form the matrix for independent sets or cliques.

Theorem 6. *If d, e, f are all different, and M does not contain, as a principal submatrix, the matrix for independent sets or cliques, then the number of M-partitions can be counted in polynomial time.*

The last kind of decomposition refers to the matrix M. Namely, if two of d, e, f are 0, then the matrix M can be viewed as consisting of two submatrices and the M-partition problem can be reduced to the corresponding problems for these two matrices.

Theorem 7. *Assume that two of d, e, f are 0 and M does not contain as principal submatrix the matrix for independent sets or cliques. Then counting the number of M-partitions is polynomial.*

3   A Special Polynomial Case

Our final example of a polynomial counting problem deals with the following matrix $M = \begin{pmatrix} 0 & * & * \\ * & 0 & 1 \\ * & 1 & 0 \end{pmatrix}$.

We first consider bipartite input graphs G.

Theorem 8. *The number of M-partitions of bipartite input graphs G can be computed in polynomial time.*

Proof. (*Sketch*) Let G be a bipartite graph with parts X and Y. We again call the three parts of an M-partition A, B, C, in that order. Each part X, Y is an independent set, and thus, must be placed either entirely in $A \cup B$ or entirely in $A \cup C$. The number of M-partitions of G with B or C is empty is easily counted, according to Theorem 1. We will add the number of M-partitions with the vertices of X placed in $A \cup B$, and the vertices of Y placed in $A \cup C$ and the number of M-partitions with the opposite assignment (and subtract 1 when G has no edges, because in that case we are counting the one possible solution that places all vertices to A twice.) Thus consider the M-partitions of G with the vertices of X placed in $A \cup B$, and the vertices of Y placed in $A \cup C$: a vertex of $X \cap A$ has no neighbours in B and a vertex of $Y \cap A$ has no neighbours in C. Thus G has some possible edges between $A \cap X$ and $C \cap Y$, all edges between $C \cap Y = C$ and $B \cap Y = B$, and some possible edges between $B \cap Y$ and $A \cap Y$, and no other edges. Such a partition is called a *split* [2]. (In general graphs splits can have edges inside the parts, in our case of bipartite graphs, the parts are independent sets.) Each M-partition of G gives a split, and each split corresponds to two unique M-partitions of G. Thus it will suffice to count the number of splits. Splits form a recursive structure called a *split decomposition tree* [2], akin to the modular decomposition tree discussed earlier. Even though the number of splits can be exponential, the split decomposition tree has polynomially many vertices, and can be computed in linear time [2]. It can be shown that the number of splits of a graph can be computed in linear time from the split decomposition tree. $\square$

We are ready to prove the main result of this section, Theorem 9.

Theorem 9. *The number of M-partitions of any graph G can be computed in polynomial time.*

Proof. By Theorem 8, we may assume that G is not bipartite. If it does not contain a triangle, then the shortest odd cycle has at least five vertices. In such a case, the number of M-partitions of G is zero. Indeed, the largest complete bipartite subgraph of the cycle has three (consecutive) vertices, and hence at least two adjacent vertices of the cycle must be placed in A in any M-partition of G, which is impossible.

Otherwise, we find a triangle uvw in G, in polynomial time, and then add the numbers of M-partitions of G with the six possible assignments of u, v, w to A, B, C. (Since the parts are independent, u, v, w must be placed in distinct parts.) For each such assignment, we shall first extend the assignment by placing vertices that are forced to certain parts uniquely.

In the first phase, we proceed as follows:

- a vertex with neighbours in two different parts is placed in the third part;
- a vertex with a non-neighbour in B as well as a non-neighbour in C is placed in A;

– a vertex with both a neighbour and a non-neighbour in B (respectively C)
  is placed in A.

It is clear that these are forced assignments in any M-partition of G extending
the given assignment on u, v, w. If at any time these assignments (or those below)
violate the requirements of an M-partition, the corresponding count is zero.

After the first phase, every vertex is either fully adjacent to B and not adjacent
to any vertex of $A \cup C$, forming a set called X, or is fully adjacent to C and not
adjacent to any vertex of $A \cup B$, forming a set called Y. Note that the vertices
of X must be placed in $A \cup C$ and the vertices of Y in $A \cup B$.

In the second phase, we extend the assignment using the following rules. (In
brackets we explain why these rules are forced.)

Assume uv is an edge with $u, v \in X$, and $w \in Y$. (Symmetric rules apply for
$u, v \in Y$, and $w \in X$.)

– If w is adjacent to both u and v then w will be placed in B. (This is forced
  because u and v must be in different parts A and C, and w is adjacent to
  both of them.)
– if w is nonadjacent to both u and v, then w is placed in A. (This is forced
  because u and v must be in different parts A and C, and w is not adjacent
  to either of them, so it is nonadjacent to at least one vertex in C.)
– if w is adjacent to u but not to v, then u is placed in C and v is placed in A.
  (This is forced because if $v \in C$ then $u \in A$ and $w \in B$, which is impossible
  as vw is not an edge of G.)

When there are no more choices remaining, either we have X or Y empty, or the
graph induced by $X \cup Y$ is bipartite. The former case corresponds to partitions
of the remaining vertices into two of the three parts, counted by Theorem 1. The
latter case is counted by Theorem 8.   □

4   Dichotomy

Our main result is the following dichotomy. Notice that when M does not contain
any 1's (or does not contain any 0's), then the dichotomy follows from [6].

Theorem 10. *Suppose M is an m by m matrix with $m < 4$, and assume M
contains both a 0 and a 1.*

*If M contains, as a principal submatrix, the matrix for independent sets, or the
matrix for cliques, then the counting problem for M-partitions is #P-complete.
Otherwise, counting M-partitions is polynomial.*

Proof. We first derive the polynomial cases from our existing results. We assume
throughout that M contains both a 0 and a 1.

For $m = 2$ this follows from Theorem 1. Thus we assume that $m = 3$, say
$M = \begin{pmatrix} a & d & e \\ d & b & f \\ e & f & c \end{pmatrix}$, and M does not contain as a principal submatrix the matrix for

independent sets, or the matrix for cliques. If at least two of d, e, f are 0 (or at least two are 1), then M-partitions are counted by Theorem 7. If d, e, f are all different, then the number of M-partitions are counted by Theorem 6.

Hence, in all the remaining cases, we may assume that $d = e = *$, $f \neq 0$ by symmetry and complementarity. Notice that M contains both a 0 and a 1; hence, $f = 1$ implies that at least one of a, b, c is 0, and $f = *$ implies that at least one of a, b, c is 0 and one is 1. Thus, we note that none of a, b, c is $*$, because M does not contain the forbidden principal submatrices, and we conclude by Theorem 9 ($f = 1$, $a = b = c = 0$) or Theorem 2 (otherwise).

It remains to show that if M contains, as a principal submatrix, the matrix for independent sets or cliques, then counting M-partitions is #P-complete. We shall again use the notation $M' = \left(\begin{smallmatrix} b & f \\ f & c \end{smallmatrix} \right)$. We shall assume that M' is the matrix for independent sets, without loss of generality, say, that $b = f = *$, $c = 0$.

We consider two distinct cases, depending on the value of a. Our proof is completed by the following two lemmas. □

For the purposes of the lemmas we introduce two constructions. The *universal vertex extension* G^* of a graph G is a graph obtained from G by adding a new vertex u, adjacent to all vertices of G. The *isolated vertex extension* G^o of G is a graph obtained from G by adding a new isolated vertex u.

Lemma 11. *If $a \neq 0$, $b = f = *$, $c = 0$, then counting the number of M-partitions is #P-complete.*

Proof. In this case, we reduce from the number of independent sets in graph G using the isolated vertex extension of G. Let $\#I(G)$ denote the number of independent sets of G, and $\#M(G)$ the number of M-partitions of G. As before, we will write A, B, C for the first, second, and third part of an M-partition.

Given an input graph G, we first construct G^o. We count the number of M-partitions of G^o according to the placement of the isolated vertex u. When $a = 1$, we consider the two following cases:

We illustrate the proofs on the case when d, e are different from 1. In this case we can show that $\#M(G^o) = \#I(G) + 2\#M(G)$. This implies that counting the number of M-partitions is #P-complete. In all other cases, $\#I(G)$ can also be reduced in polynomial time to $\#M(G)$. □

Lemma 12. *If $a = 0$, $b = f = *$, $c = 0$, then counting the number of M-partitions is #P-complete.*

In this case the proofs use the universal vertex extension.

References

1. Bulatov, A.A.: Tractable conservative constraint satisfaction problems. In: LICS, pp. 321–330 (2003)
2. Charbit, P., de Montgolfier, F., Raffinot, M.: Linear time split decomposition revisited. SIAM J. Discrete Math. 26, 499–514 (2012)

3. Chvátal, V.: Star-cutsets and perfect graphs. J. Comb. Th. B 39, 189–199 (1985)
4. Cygan, M., Pilipczuk, M., Pilipczuk, M., Wojtaszczyk, J.O.: The stubborn problem is stubborn no more. In: SODA 2011, pp. 1666–1674 (2011)
5. Linial, N.: Hard enumeration problems in geometry and combinatorics. SIAM Journal on Algebraic and Discrete Methods 7, 331–335 (1986)
6. Dyer, M., Greenhill, C.: The complexity of counting graph homomorphisms. In: SODA 1999, pp. 246–255 (1999)
7. de Figueiredo, C.M.H.: The P versus NP-complete dichotomy of some challenging problems in graph theory. Discrete Applied Math. (in press)
8. Feder, T., Hell, P., Klein, S., Motwani, R.: List partitions. SIAM J. Discrete Math. 16, 449–478 (2003)
9. Feder, T., Vardi, M.Y.: The computational structure of monotone monadic SNP and constraint satisfaction. SIAM J. Comput. 28, 57–104 (1999)
10. de Figueiredo, C.M.H., Klein, S., Kohayakawa, Y., Reed, B.: Finding skew partitions efficiently. J. Algorithms 37, 505–521 (2000)
11. Gallai, T.: Transitiv orientierbare Graphen. Acta Mathematica Hungarica 18, 25–66 (1967)
12. Habib, M., Paul, C.: A survey of the algorithmic aspects of modular decomposition. Computer Science Review 4, 41–59 (2010)
13. Golumbic, M.C.: Algorithmic Graph Theory and Perfect Graphs. Academic Press, New York (1980)
14. Hell, P.: Graph partitions with prescribed patterns (to appear)
15. Hell, P., Klein, S., Protti, F., Tito, L.: On generalized split graphs. Electronic Notes in Discrete Math. 7, 98–101 (2001)
16. Hell, P., Nešetřil, J.: On the complexity of H–colouring. J. Combin. Theory B 48, 92–110 (1990)
17. Hell, P., Nešetřil, J.: Graphs and Homomorphisms. Oxford Univ. Press (2004)
18. Hell, P., Nešetřil, J.: Counting list homomorphisms and graphs with bounded degrees. In: Nešetřil, J., Winkler, P. (eds.) Graphs, Morphisms and Statistical Physics. DIMACS Series in Discrete Mathematics and Theoretical Computer Science, vol. 63, pp. 105–112 (2004)
19. Hell, P., Nešetřil, J.: Colouring, constraint satisfaction, and complexity. Computer Science Review 2, 143–163 (2008)
20. Jeavons, P.: On the structure of combinatorial problems. Theoretical Comp. Science 200, 185–204 (1998)
21. Provan, J.S., Ball, M.O.: The complexity of counting cuts and of computing the probability that a graph is connected. SIAM Journal on Computing 12, 777–788 (1983)
22. Tarjan, R.E.: Decomposition by clique separators. Discrete Math. 55, 221–232 (1985)

Constant Unary Constraints and Symmetric Real-Weighted Counting CSPs

Tomoyuki Yamakami

Department of Information Science, University of Fukui
3-9-1 Bunkyo, Fukui 910-8507, Japan

Abstract. In a discussion on the computational complexity of approximately solving Boolean counting constraint satisfaction problems (or #CSPs), we demonstrate the approximability of two constant unary constraints by an arbitrary nonempty set of real-valued constraints. A use of auxiliary free unary constraints has proven to be useful in establishing a complete classification of weighted #CSPs. Using our approximability result, we can clarify the role of such auxiliary free unary constraints by constructing approximation-preserving reductions from #SAT to #CSPs with symmetric real-valued constraints of arbitrary arities.

Keywords: counting constraint satisfaction problem, AP-reducible, T-constructible, constant unary constraint.

1 Roles of Constant Unary Constraints

Constraint satisfaction problems (or *CSPs,* in short) are combinatorial problems that have been ubiquitously found in real-life situations. The importance of these problems have led recent intensive studies from various aspects: for instance, decision CSPs [5,9], optimization CSPs [3,12], and counting CSPs [1,4,7,10]. Driven by theoretical and practical interests, in this paper, we are particularly focused on *counting CSPs* (abbreviated as #CSPs) whose goal is to count the number of variable assignments satisfying all given Boolean constraints defined over a fixed series of Boolean variables. Since, in most real-life applications, all available constraints are pre-determined, we naturally fix a collection of allowed constraints, say, $\mathcal{F}$ and wish to compute solutions of a #CSP whose constraints are all chosen from $\mathcal{F}$. Such a problem is conventionally denoted #CSP($\mathcal{F}$). Dyer, Goldberg, and Jerrum [8] first examined the computational complexity of approximately computing the solutions of unweighted #CSPs using a technical tool of *polynomial-time randomized approximation-preserving reductions* (or *AP-reductions*, hereafter) whose formulation is originated from [6].

In the case of *weighted #CSPs*, constraints are expanded to output more general values than Boolean values, and the goal of each weighted #CSP is to calculate the sum, over all possible Boolean assignments, of products of the output values of given constraints. By further allowing a free use of auxiliary unary constraints besides input constraints, Cai, Lu, and Xia [2] pioneered a study on a classification of the complexity of exactly solving complex-weighted

K.-M. Chao, T.-s. Hsu, and D.-T. Lee (Eds.): ISAAC 2012, LNCS 7676, pp. 237–246, 2012.
© Springer-Verlag Berlin Heidelberg 2012

#CSPs for a given set $\mathcal{F}$ of constraints. Similarly, in the presence of auxiliary unary constraints, a classification of the complexity of approximately solving complex-weighted #CSPs was presented first in [10]. In this classification, the free use of auxiliary unary constraints provide enormous power that makes it possible to obtain a "dichotomy" theorem rather than a "trichotomy" theorem of Dyer et al. [8] for Boolean-valued constraints (or simply, *Boolean constraints*). A key to the proof of their trichotomy theorem is an effective approximation of so-called *constant unary constraints*,[1] $\Delta_0 = [1, 0]$ and $\Delta_1 = [0, 1]$. When $\mathcal{F}$ is composed of real-valued constraints, we can claim that either Δ_0 or Δ_1 is always approximated effectively using $\mathcal{F}$.

Theorem 1. *For any nonempty set $\mathcal{F}$ of real-valued constraints, there exists a constant unary constraint $h \in \{\Delta_0, \Delta_1\}$ for which #CSP$(h, \mathcal{F})$ is AP-equivalent to #CSP$(\mathcal{F})$ (i.e., #CSP$(h, \mathcal{F})$ is AP-reducible to #CSP$(\mathcal{F})$ and vice versa).*

When the values of constraints in $\mathcal{F}$ are all limited to Boolean values, the theorem was already proven in [8] based on basic properties of Boolean arithmetic. For real-valued constraints, however, we cannot rely on those properties and thus we need to develop a quite different argument. An important ingredient of our proof is an efficient estimation of a lower bound of an arbitrary polynomial in the values of given constraints. However, since our constraints can output negative real values, the polynomial may produce arbitrary small values. To deal with those values, we restrict our attention onto *algebraic numbers*.

With Theorem 1 as a technical tool, we will be ale to demonstrate an approximation classification of real-weighted #CSPs when arbitrary free unary constraints are permitted to assist standard inputs. In our proof, the constant unary constraints are quite valuable in "reducing" constraints of high arity to those of low arity. In an exact counting model, both Δ_0 and Δ_1 are naturally available to use; however, in our approximate counting model, Theorem 1 guarantees the availability of only one of them. Even with a help of a single constant unary constraint, it is still possible to reduce the arities of target constraints. Moreover, we can build the reduction with no use of auxiliary unary constraint.

More precisely, let us denote by $\mathcal{U}$ the set of all unary constraints. Given each constraint f, the free use of auxiliary unary constraints makes #SAT (counting satisfiability problem) AP-reducible to #CSP$(f, \mathcal{U})$ unless f is factored into three categories of constraints: the equality, the disequality, and unary constraints [10]. Notice that all constraints factored into basic constraints in those categories form a special set $\mathcal{ED}$. The aforementioned fact establishes the following complete classification of the approximation complexity of weighted #CSPs in the presence of $\mathcal{U}$.

Theorem 2. [10, Theorem 1.1] *Let $\mathcal{F}$ be any set of complex-valued constraints. If $\mathcal{F} \subseteq \mathcal{ED}$, then #CSP$(\mathcal{F}, \mathcal{U})$ is solvable in polynomial time; otherwise, it is AP-reduced from #SAT.*

---

[1] A bracket notation $[x, y]$ denotes a unary function g satisfying $g(0) = x$ and $g(1) = y$. Similarly, $[x, y, z]$ expresses a binary function g for which $g(0, 0) = x$, $g(0, 1) = g(1, 0) = y$, and $g(1, 1) = z$.

The proof of this dichotomy theorem in [10] employed two technical notions of "factorization" and "T-constructibility" of constraints. Because the proof is rather complicated and thus lengthy, it is not immediately clear what roles free auxiliary unary constraints play in the theorem. Toward a complete classification of the approximation complexity of #CSPs *without* any auxiliary unary constraint, it is therefore beneficial to clarify the roles of those extra constraints. With a careful use of Theorem 1, we can precisely locate a point where the free auxiliary unary constraints are imperatively required in establishing the desired dichotomy theorem. For this purpose, we will present a new alternative proof in which the use of auxiliary unary constraints is made only at the very end of the proof. To simplify our proof, we intend to restrict our attention only on symmetric real-weighted #CSPs in the subsequent sections.

Our proof proceeds as follows. In the first step, we must recognize constraints g of the following special forms: $[0, x, z]$ and $[x, y, z]$ with $xyz \neq 0$ and $xz \neq y^2$. Those constraints g become crucial elements of our later analysis because, when auxiliary unary constraints are available for free, $\#\mathrm{CSP}(g, \mathcal{U})$ is known to be computationally at least as hard as #SAT with respect to AP-reducibility [10]. In the second step, we must isolate a set $\mathcal{F}$ of constraints whose corresponding counting problem $\#\mathrm{CSP}(\mathcal{F})$ is AP-reduced from a certain $\#\mathrm{CSP}(g)$ with no use of the auxiliary unary constraints. To be more exact, we wish to establish the following specific AP-reduction.

Theorem 3. *Let $\mathcal{F}$ be any set of symmetric real-valued constraints of arity at least 2. If both $\mathcal{F} \not\subseteq \mathcal{DG} \cup \mathcal{ED}_1^{(+)}$ and $\mathcal{F} \not\subseteq \mathcal{DG}^{(-)} \cup \mathcal{ED}_1 \cup \mathcal{AZ}$ hold, then $\#\mathrm{CSP}(\mathcal{F})$ is AP-reduced from $\#\mathrm{CSP}(g)$, where g is an appropriate constraint of the special form described above.*

Here, the constraint set $\mathcal{DG}$ consists of *degenerate* constraints, $\mathcal{ED}_1$ indicates a "generalized" $\mathcal{ED}$, and $\mathcal{AZ}$ contains specific symmetric constraints having alternating zeros. Two additional sets $\mathcal{DG}^{(-)}$ and $\mathcal{ED}_1^{(+)}$ are naturally induced from $\mathcal{DG}$ and $\mathcal{ED}_1$, respectively. For their precise definitions, refer to Section 4. Although $\#\mathrm{CSP}(\mathcal{DG} \cup \mathcal{ED}_1^{(+)})$ and $\#\mathrm{CSP}(\mathcal{DG}^{(-)} \cup \mathcal{ED}_1 \cup \mathcal{AZ})$ are polynomial-time solvable, they behave quite differently in the presence of auxiliary unary constraints. The counting problem $\#\mathrm{CSP}(\mathcal{DG} \cup \mathcal{ED}_1^{(+)}, \mathcal{U})$ remains solvable in polynomial time; on the contrary, $\#\mathrm{CSP}(\mathcal{DG}^{(-)} \cup \mathcal{ED}_1 \cup \mathcal{AZ}, \mathcal{U})$ is AP-reduced from #SAT. These facts immediately prove Theorem 2 for symmetric real-weighted #CSPs.

Comparison of Proof Techniques: In [8], the approximation of the constant unary constraints by any set of Boolean constraints was proven with a notion of "simulatability." Instead, our proof of Theorem 1 employs a direct arity reduction and applies an estimation result in [10]. While a key proof technique used to prove Theorem 2 is the factorization of constraints, our proof of Theorem 3 (which leads to Theorem 2) makes a heavy use of the constant unary constraints. This fact makes the proof cleaner and more straightforward to follow.

2 Fundamental Notions and Notations

We will explain basic concepts that are necessary to read through the rest of this paper. First, let $\mathbb{N}$ denote the set of all *natural numbers* (i.e., nonnegative integers) and let $\mathbb{R}$ be the set of all *real numbers*. For convenience, define $\mathbb{N}^+ = \mathbb{N} - \{0\}$ and, for each number $n \in \mathbb{N}^+$, $[n]$ stands for the integer set $\{1, 2, \ldots, n\}$.

Because our results rely on Lemma 1(3), we need to limit our attention within *algebraic real numbers*. For this purpose, we introduce a special notation $\mathbb{A}$ to indicate the set of all algebraic real numbers. To simplify our terminology throughout the paper, whenever we refer to "real numbers," we actually mean "algebraic real numbers" as long as there is no confusion from the context.

2.1 Constraints and #CSPs

The term "constraint of arity k" always refers to a function mapping $\{0, 1\}^k$ to $\mathbb{A}$. Assuming a standard lexicographic order on $\{0, 1\}^k$, we conveniently express f as a column-vector consisting of its output values; for instance, if f has arity 2, then it is expressed as $(f(00), f(01), f(10), f(11))$. For a k-ary constraint $f = (f_1, f_2, \ldots, f_{2^k})$ in a vector form, $\|f\|_\infty$ means $\max_{i \in [2^k]}\{|f_i|\}$. A k-ary constraint f is called *symmetric* if, for every input $x \in \{0, 1\}^k$, the value $f(x)$ depends only on the Hamming weight (i.e., the number of 1's in x) of the input x; otherwise, f is called *asymmetric*. When f is a symmetric constraint of arity k, we use another succinct notation $f = [f_0, f_1, \ldots, f_k]$, where each f_i expresses the value of f on inputs of Hamming weight i. In particular, we recognize two special constraints, $\Delta_0 = [1, 0]$ and $\Delta_1 = [0, 1]$, which are called *constant unary constraints*.

Restricted to a set $\mathcal{F}$ of constraints, a *real-weighted (Boolean) #CSP*, conventionally denoted $\#\mathrm{CSP}(\mathcal{F})$, takes a *finite* set Ω composed of elements of the form $\langle h, (x_{i_1}, x_{i_2}, \ldots, x_{i_k}) \rangle$, where $h \in \mathcal{F}$ is a function on k Boolean variables $x_{i_1}, x_{i_2}, \ldots, x_{i_k}$ in $X = \{x_1, x_2, \ldots, x_n\}$ with $i_1, \ldots, i_k \in [n]$, and its goal is to compute the real value $csp_\Omega =_{def} \sum_{x_1, x_2, \ldots, x_n \in \{0,1\}} \prod_{\langle h, x \rangle \in \Omega} h(x_{i_1}, x_{i_2}, \ldots, x_{i_k})$, where x denotes $(x_{i_1}, x_{i_2}, \ldots, x_{i_k})$. To illustrate Ω graphically, we tend to view it as a *labeled bipartite graph* $G = (V_1|V_2, E)$ whose left-hand nodes in V_1 are labeled distinctively by $x_1, x_2, \ldots, x_n$ and right-hand nodes in V_2 are labeled by constraints h in $\mathcal{F}$ such that, for each $\langle h, (x_{i_1}, x_{i_2}, \ldots, x_{i_k}) \rangle$, there are k edges between an associated node labeled h and the nodes labeled $x_{i_1}, x_{i_2}, \ldots, x_{i_k}$. The labels of nodes are formally specified by a *labeling function* $\pi : V_1 \cup V_2 \to X \cup \mathcal{F}$ with $\pi(V_1) = X$ and $\pi(V_2) \subseteq \mathcal{F}$ but we often omit it from the description of G. When Ω is viewed as this bipartite graph, it is called a *constraint frame* [10,11].

To simplify later descriptions, we wish to use the following simple abbreviation rule. For instance, when f is a constraint and both $\mathcal{F}$ and $\mathcal{G}$ are constraint sets, we write $\#\mathrm{CSP}(f, \mathcal{F}, \mathcal{G})$ to mean $\#\mathrm{CSP}(\{f\} \cup \mathcal{F} \cup \mathcal{G})$.

2.2 FP$_\mathbb{A}$ and AP-Reducibility

To connect our results to Theorem 2, we will follow notational conventions used in [10,11]. First, $\mathrm{FP}_\mathbb{A}$ denotes the collection of all $\mathbb{A}$-*valued* functions that can be computed deterministically in polynomial time.

Let F be any function mapping $\{0,1\}^*$ to $\mathbb{A}$ and let Σ be any nonempty finite alphabet. A *randomized approximation scheme* (or RAS, in short) for F is a randomized algorithm that takes a standard input $x \in \Sigma^*$ together with an error tolerance parameter $\varepsilon \in (0,1)$, and outputs values w with probability at least $3/4$ for which $\min\{2^{-\varepsilon}F(x), 2^{\varepsilon}F(x)\} \leq w \leq \max\{2^{-\varepsilon}F(x), 2^{\varepsilon}F(x)\}$.

Given two real-valued functions F and G, a *polynomial-time randomized approximation-preserving reduction* (or AP-reduction) from F to G [6] is a randomized algorithm M that takes a pair $(x, \varepsilon) \in \Sigma^* \times (0,1)$ as input, uses an arbitrary randomized approximation scheme N for G as an oracle, and satisfies the following three conditions: (i) using N, M is an RAS for F; (ii) every oracle call made by M is of the form $(w, \delta) \in \Sigma^* \times (0,1)$ with $1/\delta \leq poly(|x|, 1/\varepsilon)$ and its answer is the outcome of N on (w, δ); and (iii) the running time of M is upper-bounded by a certain polynomial in $(|x|, 1/\varepsilon)$, which is not dependent of the choice of N. If such an AP-reduction exists, then we also say that F is *AP-reducible* to G and we write $F \leq_{AP} G$. If $F \leq_{AP} G$ and $G \leq_{AP} F$, then F and G are said to be *AP-equivalent* and we use the special notation $F \equiv_{AP} G$.

2.3   Effective T-Constructibility

Our goal in the subsequent sections is to prove Theorems 1 and 3. For the desired proofs, we will introduce a fundamental notion of *effective T-constructibility*, whose underlying idea is borrowed from a graph-theoretical formulation of *limited T-constructibility* in [11].

We say that an undirected bipartite graph $G = (V_1|V_2, E)$ (together with a labeling function π) *represents* f if V_1 consists only of k nodes labeled $x_1, \ldots, x_k$, which may have a certain number of dangling[2] edges, and V_2 contains only a node labeled f to whom each node x_i is adjacent. Given a set $\mathcal{G}$ of constraints, a graph $G = (V_1|V_2, E)$ is said to *realize* f by $\mathcal{G}$ if the following four conditions are met simultaneously: (i) $\pi(V_2) \subseteq \mathcal{G}$, (ii) G contains at least k nodes having the labels $x_1, \ldots, x_k$, possibly together with nodes associated with other variables, say, $y_1, \ldots, y_m$; namely, $V_1 = \{x_1, \ldots, x_k, y_1, \ldots, y_m\}$, (iii) only the nodes $x_1, \ldots, x_k$ are allowed to have dangling edges, and (iv) $f(x_1, \ldots, x_k)$ equals $\lambda \sum_{y_1, \ldots, y_m \in \{0,1\}} \prod_{w \in V_2} f_w(z_1, \ldots, z_d)$, where $\lambda \in \mathbb{A} - \{0\}$ and $z_1, \ldots, z_d \in V_1$.

We say that f is *effectively T-constructible* from $\mathcal{G}$ if the following condition holds: for any integer $m \geq 2$ and for any graph G representing f with distinct variables $x_1, \ldots, x_k$, there exists another graph G' such that (i') G' realizes f by $\mathcal{G}$ and (ii') G' maintains the same dangling edges as G does. In this case, we write $f \leq_{e\text{-}con} \mathcal{G}$. When $\mathcal{G}$ is a singleton $\{g\}$, we succinctly write $f \leq_{e\text{-}con} g$.

An infinite series $\Lambda = (g_1, g_2, g_3, \ldots)$ of arity-k constraints is called a *p-convergence series* for a target constraint f of arity k if there exist a constant $\lambda \in (0,1)$ and a deterministic Turing machine (or a DTM, in short) M running in polynomial time such that, for every number $m \in \mathbb{N}$, (1) M takes an input

---

[2] A *dangling edge* is obtained from an edge by deleting exactly one end of the edge. These dangling edges are treated as "normal" edges, and therefore the degree of a node counts dangling edges as well.

of the form 1^m and outputs the complete description of the constraint g_m in a vector form $(z_1, z_2, \ldots, z_{2^k})$ and (2) $\|g_m - f\|_\infty \le \lambda^m$. A p-convergence series $\Lambda = (f_1, f_2, \ldots)$ of arity-k constraints is said to be *effectively T-constructible* from a finite set $\mathcal{G} = \{g_1, g_2, \ldots, g_d\}$ of constraints (denoted $\Lambda \le_{e\text{-con}} \mathcal{G}$) if there exists a polynomial-time DTM M such that, for every number $m \in \mathbb{N}^+$, M takes an input of the form $(1^m, G, (g_1, g_2, \ldots, g_d))$, where G represents f_m with distinct variables $x_1, \ldots, x_k$, and outputs a bipartite graph G_m such that (i') G_m realizes f_m by $\mathcal{G}$ and (ii') G_m maintains the same dangling edges as G does.

Lemma 1. *Let f and g be any constraints. Let $\mathcal{F}$ and $\mathcal{G}$ be any constraint sets.*

1. *It holds that $f \le_{e\text{-con}} f$ and that $f \le_{e\text{-con}} g$ and $g \le_{e\text{-con}} h$ imply $f \le_{e\text{-con}} h$.*
2. *If $f \le_{e\text{-con}} \mathcal{G}$, then $\#\mathrm{CSP}(f, \mathcal{F}) \le_{\mathrm{AP}} \#\mathrm{CSP}(\mathcal{G}, \mathcal{F})$.*
3. *Let Λ be any p-convergence series for f. If $\Lambda \le_{e\text{-con}} \mathcal{G}$ and $\mathcal{G}$ is finite, then $\#\mathrm{CSP}(f) \le_{\mathrm{AP}} \#\mathrm{CSP}(\Lambda, \mathcal{F}) \le_{\mathrm{AP}} \#\mathrm{CSP}(\mathcal{G}, \mathcal{F})$.*

In particular, Lemma 1(3) can be proven by modifying (and slightly generalizing) the proof of [10, Lemma 5.2] given for complex-valued constraints. It is important to note that, since we allow constraints to output negative values, the use of *algebraic real numbers* may be necessary in the proof of Lemma 1(3) because the proof heavily relies on a lower bound estimation of arbitrary polynomials over algebraic numbers.

In the subsequent sections, we will use the following notations. Let f be any constraint of arity $k \in \mathbb{N}^+$. Given any index $i \in [k]$ and any bit $c \in \{0, 1\}$, the notation $f^{x_i = c}$ stands for the function g satisfying that $g(x_1, \ldots, x_{i-1}, x_{i+1}, \ldots, x_k) = f(x_1, \ldots, x_{i-1}, c, x_{i+1}, \ldots, x_k)$. For any two distinct indices $i, j \in [k]$, we denote by $f^{x_i = x_j}$ the function g defined as $g(x_1, \ldots, x_{i-1}, x_{i+1}, \ldots, x_k) = f(x_1, \ldots, x_{i-1}, x_j, x_{i+1}, \ldots, x_k)$. Finally, let $f^{x_i = *}$ express the function g defined by $g(x_1, \ldots, x_{i-1}, x_{i+1}, \ldots, x_k) = \sum_{c \in \{0,1\}} f(x_1, \ldots, x_{i-1}, c, x_{i+1}, \ldots, x_k)$.

3   Approximation of the Constant Unary Constraints

In this section, we will prove our first main theorem—Theorem 1—which states that we can effectively approximate one of the constant unary constraints. The theorem thus suggests that we can freely use such a constant unary constraint for a further application presented in Section 4.

3.1   Notion of Complement Stability

To obtain Theorem 1, we will first introduce two new notions. A k-ary constraint f is said to be *complement invariant* if $f(x_1, \ldots, x_k) = f(x_1 \oplus 1, \ldots, x_k \oplus 1)$ holds for every input tuple $(x_1, \ldots, x_k) \in \{0, 1\}^k$, where the notation $\oplus$ means the *bitwise XOR*. In contrast, we say that f is *complement anti-invariant* if, for every input $(x_1, \ldots, x_k) \in \{0, 1\}^k$, $f(x_1, \ldots, x_k) = -f(x_1 \oplus 1, \ldots, x_k \oplus 1)$ holds. For instance, $f = [1, 1]$ is complement invariant and $f' = [1, 0, -1]$ is complement anti-invariant. In addition, we say that f is *complement stable* if f

is either complement invariant or complement anti-invariant. A constraint set $\mathcal{F}$ is *complement stable* if every constraint in $\mathcal{F}$ is complement stable. Whenever f (resp., $\mathcal{F}$) is not complement stable, we conveniently call it *complement unstable*.

We split Theorem 1 into two separate statements, depending on whether or not a given nonempty set of constraints is complement stable.

Lemma 2. *1. If a nonempty set $\mathcal{F}$ of constraints is complement stable, then $\#\mathrm{CSP}(\Delta_i, \mathcal{F}) \equiv_{\mathrm{AP}} \#\mathrm{CSP}(\mathcal{F})$ holds for every index $i \in \{0, 1\}$.*
  2. For any set $\mathcal{F}$ of constraints, if $\mathcal{F}$ is complement unstable, then there exists an index $i \in \{0, 1\}$ such that $\#\mathrm{CSP}(\Delta_i, \mathcal{F}) \equiv_{\mathrm{AP}} \#\mathrm{CSP}(\mathcal{F})$.

Since Lemma 2(1) can be proven rather easily, in the next subsections, we will concentrate our attention on the proof of Lemma 2(2). First, let $\mathcal{F}$ denote any set of constraints. Obviously, $\#\mathrm{CSP}(\mathcal{F})$ AP-reduces to $\#\mathrm{CSP}(\Delta_i, \mathcal{F})$ for every index $i \in \{0, 1\}$. It therefore suffices to show the other direction (i.e., $\#\mathrm{CSP}(\Delta_i, \mathcal{F}) \leq_{\mathrm{AP}} \#\mathrm{CSP}(\mathcal{F})$) for an appropriately chosen index i. Hereafter, we suppose that $\mathcal{F}$ is complement unstable and we choose a constraint f in $\mathcal{F}$ that is complement unstable. Furthermore, we assume that f has the smallest arity k within $\mathcal{F}$; that is, there is no complement unstable constraint of arity smaller than k. Our proof of Lemma 2(2) proceeds by induction on this index k in Sections 3.2–3.3.

3.2   Basis Case: $k = 1, 2$

In the proof of Lemma 2(2), we will consider the basis case where $k \in \{1, 2\}$.

(1) Assuming $k = 1$, let $f = [x, y]$. Note that $x \neq \pm y$. This is because, if $x = \pm y$, then f has the form $x \cdot [1, \pm 1]$ and thus f is complement stable, a contradiction. Hence, it must hold that $|x| \neq |y|$. To appeal to Lemma 1(3), it is enough to assert that a certain p-convergence series is effectively T-constructible from f. This assertion comes from the following observation.

Claim 1. *Let x and y be any two algebraic real numbers with $|x| > |y|$. A p-convergence series $\Lambda = \{[1, |y|^n/|x|^n] \mid n \in \mathbb{N}^+\}$ for Δ_0 is effectively T-constructible from $[1, |y|/|x|]$. In the case of Δ_1, a similar statement holds if $|x| < |y|$ (in place of $|x| > |y|$).*

Using the above claim, Lemma 1(3) clearly leads to the desired lemma for the case of $k = 1$.

(2) Assuming $k = 2$, for $f = (x, y, z, w)$, it suffices to examine two cases, $x = \pm w$ and $x \neq \pm w$, separately. Here, we omit the details.

3.3   General Case: $k \geq 3$

We will examine the remaining case of $k \geq 3$. As the next lemma indicates, given a complement unstable constraint f of arity k, we can effectively T-construct a complement unstable constraint g of arity less than k.

Lemma 3. *Let $k \geq 3$ and let f be any k-ary constraint. If f is complement unstable, then there exists another constraint g of arity less than k for which $g \leq_{e\text{-}con} f$ and g is also complement unstable.*

Assuming that Lemma 3 is true, we take a complement unstable constraint g of arity $< k$. Instead of working on the originally given set $\mathcal{F}$, let us concentrate on another set $\mathcal{F}' = \mathcal{F} \cup \{g\}$. Since g has arity less than k, the induction hypothesis asserts that a certain constant unary constraint Δ_i ($i \in \{0, 1\}$) satisfies $\#\mathrm{CSP}(\Delta_i, \mathcal{F}') \equiv_{\mathrm{AP}} \#\mathrm{CSP}(\mathcal{F}')$. It follows from both $g \leq_{e\text{-}con} f$ and $f \in \mathcal{F}$ that $\#\mathrm{CSP}(\Delta_i, \mathcal{F}) \leq_{\mathrm{AP}} \#\mathrm{CSP}(g, \mathcal{F}) \leq_{\mathrm{AP}} \#\mathrm{CSP}(f, \mathcal{F}) \leq_{\mathrm{AP}} \#\mathrm{CSP}(\mathcal{F})$. Therefore, the remaining task of ours is to give the proof of Lemma 3.

Let $f = (z_1, z_2, \ldots, z_{2^k})$. Since f is complement unstable, there exists an index $\ell \in [2^k]$ satisfying $z_\ell \neq 0$. For each pair of indices $i, j \in [k]$ with $i < j$, we define a new constraint $g^{(i,j)}$ to be $f^{x_i = x_j}$ and then we set $\mathcal{G} = \{g^{(i,j)} \mid 1 \leq i < j \leq k\}$. If $\mathcal{G}$ contains a complement unstable constraint g, then the lemma immediately follows. In what follows, we assume that every constraint in $\mathcal{G}$ is complement stable.

Let us begin with a simple observation.

Claim 2. *Let $k \geq 3$. For any index $j \in [2^k]$, there exist a constraint $g \in \mathcal{G}$ and $k - 1$ bits $a_1, a_2, \ldots, a_{k-1}$ satisfying $z_j = g(a_1, a_2, \ldots, a_{k-1})$.*

Since f is complement unstable, either of the following two cases must occur. (1) There exists an index $i \in [2^{k-1}]$ satisfying $|z_i| \neq |z_{2^k - i + 1}|$. (2) For every index $i \in [2^{k-1}]$, $|z_i| = |z_{2^k - i + 1}|$ holds, but there are two distinct indices $i, j \in [2^{k-1}]$ for which $z_i = z_{2^k - i + 1} \neq 0$ and $z_j = -z_{2^k - j + 1} \neq 0$.

(1) Let us consider the first case. Choose an index $i \in [2^{k-1}]$ satisfying $|z_i| \neq |z_{2^k - i + 1}|$. Claim 2 ensures the existence of a constraint g in $\mathcal{G}$ such that $z_i = g(a_1, a_2, \ldots, a_{k-1})$ for certain $k - 1$ bits $a_1, a_2, \ldots, a_{k-1}$. This implies that $z_{2^k - i + 1} = g(a_1 \oplus 1, a_2 \oplus 1, \ldots, a_{k-1} \oplus 1)$. By the choice of i, g cannot be complement stable. This is a contradiction against our assumption that $\mathcal{G}$ is complement stable. (2) In the second case, take two indices $i_0, j_0 \in [2^{k-1}]$ satisfying that $z_{i_0} = z_{2^k - i_0 + 1} \neq 0$ and $z_{j_0} = -z_{2^k - j_0 + 1} \neq 0$. We will examine two separate cases.

(i) Assume that a certain constraint g in $\mathcal{G}$ satisfies both $z_{i_0} = g(a_1, a_2, \ldots, a_{k-1})$ and $z_{j_0} = g(b_1, b_2, \ldots, b_{k-1})$ for $2(k - 1)$ bits $a_1, a_2, \ldots, a_{k-1}, b_1, b_2, \ldots, b_{k-1}$. Obviously, g is complement unstable, and this leads to a contradiction.

(ii) Finally, assume that Case (i) does not hold. A close analysis can draw a conclusion that each of the cases, $k = 3$, $k = 4$, and $k \geq 5$, leads to a contradiction.

4   AP-Reductions without Auxiliary Unary Constraints

We will discuss a direct application of Theorem 1. What we wish to show in this section is our second main theorem—Theorem 3—presented in Section 1. To clarify the meaning of this theorem, we need to introduce the following sets of constraints. Recall that all constraints in this paper are assumed to output only *algebraic real values*.

1. Let $\mathcal{DG}$ denote the set of all constraints f that are expressed by products of unary functions. A constraint in $\mathcal{DG}$ is called *degenerate*. When f is symmetric, f must have one of the following three forms: $[x, 0, \ldots, 0]$, $[0, \ldots, 0, x]$, and $y \cdot [1, z, z^2, \ldots, z^k]$ with $yz \neq 0$. By restricting $\mathcal{DG}$, we define $\mathcal{DG}^{(-)}$ as the set of constraints of the forms $[x, 0, \ldots, 0]$, $[0, \ldots, 0, x]$, $y \cdot [1, 1, \ldots, 1]$, and $y \cdot [1, -1, 1, \ldots, -1 \text{ or } 1]$, where $y \neq 0$. Both Δ_0 and Δ_1 belong to $\mathcal{DG}^{(-)}$.
2. The notation $\mathcal{ED}_1$ denotes the set of the following constraints: $[x, \pm x]$, $[x, 0, \ldots, 0, \pm x]$ of arity ≥ 2, and $[0, x, 0]$ with $x \neq 0$. As a natural extension of $\mathcal{ED}_1$, let $\mathcal{ED}_1^{(+)}$ consist of $[x, y]$, $[x, 0, \ldots, 0, y]$ of arity ≥ 2, and $[0, x, 0]$ with $x, y \neq 0$.
3. Let $\mathcal{AZ}$ be composed of all constraints of arity at least 3 having the forms $[0, x, 0, x, \ldots, 0 \text{ or } x]$ and $[x, 0, x, 0, \ldots, x \text{ or } 0]$ with $x \neq 0$.
4. Let $\mathcal{OR}$ denote the set of all constraints of the form $[0, x, y]$ with $x, y > 0$. A constraint $OR = [0, 1, 1]$, for instance, belongs to $\mathcal{OR}$.
5. Let $\mathcal{NAND}$ be the set of all constraints of the form $[x, y, 0]$ with $x, y > 0$. A constraint $NAND = [1, 1, 0]$ is in $\mathcal{NAND}$.
6. Let $\mathcal{B}$ contain all constraints of the form $[x, y, z]$ with $xyz \neq 0$ and $xz \neq y^2$.

Note that, for every constraint g in $\mathcal{OR} \cup \mathcal{B}$, #SAT $\leq_{AP}$ #CSP$(g, \mathcal{U})$ holds [10]. Moreover, using an argument for #CSP$(OR, \mathcal{U}) \equiv_{AP}$ #CSP$(NAND, \mathcal{U})$ in [10], it is possible to prove that $\mathcal{NAND}$ and $\mathcal{OR}$ are similar in approximation complexity; more precisely, for any $f \in \mathcal{NAND}$ (resp., $\mathcal{OR}$) , there exists a constraint $g \in \mathcal{OR}$ (resp., $\mathcal{NAND}$) such that #CSP$(g) \leq_{AP}$ #CSP(f).

Concerning the tractability of #CSPs, when either $\mathcal{F} \subseteq \mathcal{DG} \cup \mathcal{ED}_1^{(+)}$ or $\mathcal{F} \subseteq \mathcal{DG}^{(-)} \cup \mathcal{ED}_1 \cup \mathcal{AZ}$ holds, #CSP$(\mathcal{F})$ can be easily solved.

Proposition 1. *Let $\mathcal{F}$ be any set of symmetric real-valued constraints. If either $\mathcal{F} \subseteq \mathcal{DG} \cup \mathcal{ED}_1^{(+)}$ or $\mathcal{F} \subseteq \mathcal{DG}^{(-)} \cup \mathcal{ED}_1 \cup \mathcal{AZ}$, then #CSP$(\mathcal{F})$ belongs to FP$_\mathbb{A}$.*

Finally, we come to the point of proving Theorem 3. Let us analyze the approximation complexity of #CSP(f) for an arbitrary symmetric constraint f that are not in $\mathcal{DG} \cup \mathcal{ED}_1^{(+)} \cup \mathcal{AZ}$. Note that, when f is in $\mathcal{DG} \cup \mathcal{ED}_1^{(+)} \cup \mathcal{AZ}$, #CSP$(f)$ belongs to FP$_\mathbb{A}$ by Proposition 1. A key claim required for the proof of Theorem 3 is the following assertion.

Lemma 4. *Let f be any symmetric real-valued constraint of arity at least 2. If $f \notin \mathcal{DG} \cup \mathcal{ED}_1^{(+)} \cup \mathcal{AZ}$, then there exists a constraint $g \in \mathcal{OR} \cup \mathcal{B}$ such that #CSP$(g) \leq_{AP}$ #CSP(f).*

The proof of Lemma 4 is relatively lengthy, and thus we exclude it from the current extended abstract. Theorem 3 then follows directly by combining Theorem 1 and Proposition 1 with a help of Lemma 4.

Proof of Theorem 3. Assume that $\mathcal{F} \not\subseteq \mathcal{DG} \cup \mathcal{ED}_1^{(+)}$ and $\mathcal{F} \not\subseteq \mathcal{DG}^{(-)} \cup \mathcal{ED}_1 \cup \mathcal{AZ}$. Note that $\mathcal{F}$ should contain a constraint whose entries are not all zero.

If there exists a constraint f in $\mathcal{F}$ satisfying $f \notin \mathcal{DG} \cup \mathcal{ED}_1^{(+)} \cup D_1$, then we apply Lemma 4 to obtain the theorem. Hereafter, we assume that $\mathcal{F} \subseteq \mathcal{DG} \cup \mathcal{ED}_1^{(+)} \cup \mathcal{AZ}$. From this assumption, we can choose two constraints $f_1, f_2 \in \mathcal{DG} \cup \mathcal{ED}_1^{(+)} \cup \mathcal{AZ}$ in $\mathcal{F}$ for which $f_1 \notin \mathcal{DG}^{(-)} \cup \mathcal{ED}_1 \cup \mathcal{AZ}$ and $f_2 \notin \mathcal{DG} \cup \mathcal{ED}_1^{(+)}$. Note that $f_1 \in \mathcal{DG} - \mathcal{ED}_1$ and $f_2 \in \mathcal{AZ}$. Since $f_2 \in \mathcal{AZ}$, there are two cases, $f_2 = [0,1,0,1,\ldots,0 \text{ or } 1]$ and $f_2 = [1,0,1,0,\ldots,1 \text{ or } 0]$, to study; however, since either Δ_0 or Δ_1 is available by Theorem 1, f_2 can be reduced to either $[1,0,1,0]$ or $[0,1,0,1]$. Here, we will consider the case where $f_2 = [1,0,1,0]$. The other case is similarly handled.

(1) Assume that $f_1 = [x,y]$ with $xy \neq 0$. From $f_1 \notin \mathcal{ED}_1$, it follows that $x \neq \pm y$. Define $g(x_1,x_2,x_3) = f_1(x_1)f_2(x_1,x_2,x_3)$ and $h(x_1,x_2,x_3) = g(x_1,x_2,x_3)g(x_2,x_3,x_1)g(x_3,x_1,x_2)$. A simple calculation shows that h equals $[x^2,0,y^2,0]$. Since $x^2 \neq y^2$, we conclude that $h \notin \mathcal{DG} \cup \mathcal{ED}_1^{(+)} \cup \mathcal{AZ}$. We apply Lemma 4 and then obtain the desired consequence.

(2) Assume that $f_1 = y \cdot [1,z,z^2,\ldots,z^k]$ with $y \neq 0$ and $z \neq \pm 1$. Using either Δ_0 or Δ_1, we can assume that f_1 is of the form $[1,z]$. Thus, this case is reduced to (1).   □

References

1. Cai, J., Lu, P.: Holographic algorithms: from arts to science. J. Comput. System Sci. 77, 41–61 (2011)
2. Cai, J., Lu, P., Xia, M.: Holant problems and counting CSP. In: STOC 2009, pp. 715–724 (2009)
3. Creignou, N.: A dichotomy theorem for maximum generalized satisfiability problems. J. Comput. System Sci. 51, 511–522 (1995)
4. Creignou, N., Hermann, M.: Complexity of generalized satisfiability counting problems. Inform. Comput. 125, 1–12 (1996)
5. Dalmau, V., Ford, D.K.: Generalized Satisfiability with Limited Occurrences per Variable: A Study through Delta-Matroid Parity. In: Rovan, B., Vojtáš, P. (eds.) MFCS 2003. LNCS, vol. 2747, pp. 358–367. Springer, Heidelberg (2003)
6. Dyer, M., Goldberg, L.A., Greenhill, C., Jerrum, M.: The relative complexity of approximating counting problems. Algorithmica 38, 471–500 (2004)
7. Dyer, M., Goldberg, L.A., Jerrum, M.: The complexity of weighted Boolean #CSP. SIAM J. Comput. 38, 1970–1986 (2009)
8. Dyer, M., Goldberg, L.A., Jerrum, M.: An approximation trichotomy for Boolean #CSP. J. Comput. System Sci. 76, 267–277 (2010)
9. Schaefer, T.J.: The complexity of satisfiability problems. In: STOC 1978, pp. 216–226 (1978)
10. Yamakami, T.: Approximate counting for complex-weighted Boolean constraint satisfaction problems. Inform. Comput. 219, 17–38 (2012)
11. Yamakami, T.: A dichotomy theorem for the approximation complexity of complex-weighted bounded-degree Boolean #CSPs. Thoer. Comput. Sci. 447, 120–135 (2012)
12. Yamakami, T.: Optimization, Randomized Approximability, and Boolean Constraint Satisfaction Problems. In: Asano, T., Nakano, S.-i., Okamoto, Y., Watanabe, O. (eds.) ISAAC 2011. LNCS, vol. 7074, pp. 454–463. Springer, Heidelberg (2011)

Interval Scheduling
and Colorful Independent Sets

René van Bevern[1,*], Matthias Mnich[2], Rolf Niedermeier[1],
and Mathias Weller[1,**]

[1] Institut für Softwaretechnik und Theoretische Informatik, TU Berlin, Germany
{rene.vanbevern,rolf.niedermeier,mathias.weller}@tu-berlin.de
[2] Cluster of Excellence, Universität des Saarlandes, Germany
mmnich@mmci.uni-saarland.de

Abstract. The NP-hard INDEPENDENT SET problem is to determine for a given graph G and an integer k whether G contains a set of k pairwise non-adjacent vertices. The problem has numerous applications in scheduling, including resource allocation and steel manufacturing. There, one encounters restricted graph classes such as 2-union graphs, which are edge-wise unions of two interval graphs on the same vertex set, or strip graphs, where additionally one of the two interval graphs is a disjoint union of cliques.

We prove NP-hardness of INDEPENDENT SET on a very restricted subclass of 2-union graphs and identify natural parameterizations to chart the possibilities and limitations of effective polynomial-time preprocessing (kernelization) and fixed-parameter algorithms. Our algorithms benefit from novel formulations of the computational problems in terms of (list-)colored interval graphs.

1 Introduction

Many scheduling problems can be formulated as finding maximum independent sets in certain generalizations of interval graphs [14]. Intuitively, finding a maximum number of pairwise non-adjacent vertices in a graph (this is the INDEPENDENT SET problem) corresponds to scheduling a maximum number of jobs (represented by intervals) without conflicts. In this context, we consider two popular and practically motivated graph models, namely 2-union interval graphs [2] (also called 2-union graphs) and strip graphs [8].

A graph $G = (V, E)$ is a *2-union graph* if it can be represented as the union of two interval graphs $G_1 = (V, E_1)$ and $G_2 = (V, E_2)$ on the same vertex set V, that is, $G = (V, E_1 \cup E_2)$. There are numerous applications of solving (weighted) INDEPENDENT SET on 2-union graphs, including scheduling problems such as resource allocation [2] or coil coating in steel manufacturing [9].

2-UNION INDEPENDENT SET:
Input: Two interval graphs $G_1 = (V, E_1), G_2 = (V, E_2)$ and an integer k.
Question: Does $G = (V, E_1 \cup E_2)$ have an independent set of size k?

---

* Supported by the DFG, projects DAPA, NI 369/12, and AREG, NI 369/9.
** Supported by the DFG, project DARE, NI 369/11.

K.-M. Chao, T.-s. Hsu, and D.-T. Lee (Eds.): ISAAC 2012, LNCS 7676, pp. 247–256, 2012.

We found a helpful natural embedding of 2-UNION INDEPENDENT SET into a more general problem by replacing 2-union graphs with list-colored interval graphs and searching for *colorful* independent sets:[1]

COLORFUL INDEPENDENT SET:

Input: An interval graph $G = (V, E)$, a multicoloring col : $V \to 2^{\{1,\ldots,\gamma\}}$, and an integer k.

Question: Does G have a colorful independent set of size k?

An advantage of this model is that we only have to deal with one interval graph instead of two merged ones. Indeed, the modeling proved very useful when studying INDEPENDENT SET on *strip graphs*, an important subclass of 2-union graphs. We believe that introducing our colorful view on finding independent sets and scheduling is of independent interest and might be useful in further studies. This "colored view on scheduling" leads to a useful reformulation of the classic JOB INTERVAL SELECTION problem [8, 15]. The task is to find a maximum set of jobs that can be executed, where each job has multiple possible execution intervals, each job is executed at most once, and a machine can only execute one job at a time. We state this problem using its classical name, but formulate it in terms of colored interval graphs, where the colors correspond to jobs and intervals of the same color correspond to multiple possible execution times of this job:

JOB INTERVAL SELECTION:

Input: An interval graph $G = (V, E)$, a coloring col : $V \to \{1, 2, \ldots, \gamma\}$ and an integer k.

Question: Does G have a colorful independent set of size k?

Here, the definition of "colorful"[1] degenerates to "no two intervals in the independent set have the same color".

Previous results. For 2-UNION INDEPENDENT SET, the following results are known. The problem remains NP-hard even when the two interval graphs are proper (unit interval) [1]. When restricted to so-called 5-claw-free graphs (which comprises the case that both input interval graphs are proper), Bafna et al. [1] provided a polynomial-time ratio-3.25 approximation. Bar-Yehuda et al. [2] showed that the vertex-weighted optimization version of 2-UNION INDEPENDENT SET has a polynomial-time ratio-4 approximation (indeed, they showed a ratio-2t approximation for the generalization to t-union graphs). Recently, Höhn et al. [9] considered so-called m-composite 2-union graphs (which has applications in coil coating) and developed a dynamic programming algorithm running in polynomial time with the polynomial degree depending on m. This generalizes a result of Jansen [11], who gave such an algorithm for a subclass of m-composite 2-union graphs. Concerning parameterized complexity, Jiang [13] answered an open question of Fellows et al. [6] by proving 2-UNION INDEPENDENT SET to be W[1]-hard for the parameter "solution size k". This W[1]-hardness result holds even when both input interval graphs are proper.

Introduced by Nakajima and Hakimi [15] (using a different notion), JOB INTERVAL SELECTION was shown APX-hard by Spieksma [17], who also provided

---

[1] We call an independent set *colorful* if no two of its vertices share a color.

a ratio-2 greedy approximation algorithm. Chuzhoy et al. [5] improved this ratio to 1.582. Halldórsson and Karlsson [8] introduced the equivalent notion of JOB INTERVAL SELECTION as INDEPENDENT SET on strip graphs, which are 2-union graphs where one of the two input interval graphs is a cluster graph. They showed fixed-parameter tractability for a structural parameter and for the parameter "total number of jobs". In related work, Jansen [12] considered INDEPENDENT SET on unions of cographs and cluster graphs (the latter being disjoint unions of cliques).

New results. The main focus of this work is on initiating a systematic parameterized complexity study (particularly featuring kernelization results) for the three NP-hard problems COLORFUL INDEPENDENT SET, 2-UNION INDEPENDENT SET, and JOB INTERVAL SELECTION (here listed in descending degree of generality). Doing so, we also discuss the relevance and interrelationships of several parameterizations. For COLORFUL INDEPENDENT SET, we provide an $O(2^\gamma \cdot n^3)$-time dynamic-programming algorithm for the parameter "number γ of colors". For 2-UNION INDEPENDENT SET, this result translates to a $O(2^{\#mC_{min}} \cdot n^3)$-time algorithm, where $\#mC_{min}$ denotes the minimum of the numbers of maximal cliques in the two input interval graphs. Moreover, we provide an NP-hardness proof for 2-UNION INDEPENDENT SET, even when restricted to the case that one input graph is a collection of paths on three vertices and the other is a collection of edges and triangles. In contrast, if both input graphs are cluster graphs, we show that, 2-UNION INDEPENDENT SET can be solved in $O(n^{1.5})$ time, improving on the $O(n^3)$ time algorithm [16] implied by the claw-freeness of unions of two cluster graphs. Next, stimulated by Jiang's [13] W[1]-hardness result for the parameter "solution size k", we discuss natural structural parameters that are lower-bounded by or closely related to k. Systematically exploring these parameters, we chart the border between tractability and intractability for 2-UNION INDEPENDENT SET. In particular, we initiate the study of the power of polynomial-time data reduction (known as kernelization in parameterized algorithmics) and show that 2-UNION INDEPENDENT SET has a cubic-vertex problem kernel with respect to the parameter "maximum number of maximal cliques in one of the two input interval graphs". This improves to a quadratic-vertex kernel if both input interval graphs are proper. We remark that parameterizing by the number(s) of maximal cliques allows for generalizing previous results of Halldórsson and Karlsson [8]. Our results for 2-UNION INDEPENDENT SET carry over to the vertex-weighted case.

For JOB INTERVAL SELECTION (or, equivalently, INDEPENDENT SET restricted to strip graphs), our main result refers to polynomial-time preprocessing: while we prove the nonexistence (assuming a standard complexity-theoretic conjecture) of polynomial-size problem kernels even for JOB INTERVAL SELECTION with respect to the combination of the parameters "maximum clique size ω" and "number γ of colors", we also show that, while still NP-hard, JOB INTERVAL SELECTION restricted to proper interval graphs has a problem kernel with $O(k^2 \cdot \omega)$ intervals that can be computed in linear time. Here, notably, $k \leq \gamma$.

Due to the lack of space, most technical details are deferred to a full version.

Preliminaries. When speaking of interval graphs, we state our running times under the assumption that an *interval representation* is given in which the intervals are sorted with respect to their starting or ending points. Given a graph G that allows for such an interval representation, the representation can be computed in $O(n + m)$ time [4]. A graph is a *proper interval* graph if it allows for an interval representation such that for no two intervals v and w it holds that $v \subseteq w$. Every interval graph allows for a total and linear-time computable *clique ordering* $\prec$ of its maximal cliques such that, for each vertex, the maximal cliques containing it occur consecutively [7]. Moreover, all maximal cliques of an interval graph can be listed in linear time.

A problem is *fixed-parameter tractable* (FPT) with respect to a parameter k if there is an algorithm solving any problem instance of size n in $f(k) \cdot n^{O(1)}$ time for some computable function f. A *problem kernelization* is a polynomial-time transformation of a problem instance x with a parameter k into a new instance x' with parameter k' such that $|x'|$ is bounded by a function in k (ideally, a polynomial in k), $k' \leq k$, and (x, k) is a yes-instance if and only if (x', k') is a yes-instance. We call (x', k') the *problem kernel* and $|x'|$ its *size*.

2   Independent Set and 2-Union Graphs

This section mainly investigates the standard and parameterized complexity of 2-Union Independent Set. We start by discussing a complexity dichotomy and thereafter consider various parameterizations of the problem. Finally, we provide parameterized tractability results with respect to number of maximal cliques in the input graphs.

A complexity dichotomy. 2-Union Independent Set is known to be NP-hard [1] and APX-hard for 2-union graphs of maximum degree three [2] and for graphs that are the union of an interval graph with pairwise disjoint edges [17]. Using a reduction from 3-Sat, we can impose further restrictions on NP-hard instances, which are important for showing kernelization lower bounds in Section 3.

Theorem 1. 2-Union Independent Set *is NP-hard, even if one of the input graphs is restricted to be a disjoint union of altogether k edges and triangles and the other is restricted to contain only paths of length two.*

In the context of Theorem 1, note that paths of length two are the simplest graphs that are not cluster graphs. If, in contrast, G would be the union of two cluster graphs, then G is claw-free. Independent Set on claw-free graphs is solvable in $O(n^3)$ time [16]. However, for the union of two cluster graphs, we can provide a $O(n^{1.5})$ time algorithm based on computing a matching of the cliques in the two input cluster graphs.

Proposition 1. 2-Union Independent Set *is solvable in $O(n^{1.5})$ time if both input interval graphs are cluster graphs.*

Theorem 1 and Proposition 1 give rise to a complexity dichotomy, stating that 2-Union Independent Set is polynomial-time solvable if both inputs are restricted to be cluster graphs, and NP-complete otherwise, even in the simplest case of non-cluster graphs. A more detailed investigation of our proof of Theorem 1 together with a result of Impagliazzo et al. [10] yields that, even in the restricted case covered by Theorem 1, there is no algorithm with running time $2^{o(k)} \cdot \text{poly}(n)$ for 2-UNION INDEPENDENT SET unless the Exponential Time Hypothesis[2] fails.

Corollary 1. *Under the prerequisites of Theorem 1, there is no algorithm with running time $2^{o(k)} \cdot \text{poly}(n)$ for* 2-UNION INDEPENDENT SET *unless the Exponential Time Hypothesis[2] fails.*

Parameter identification. We now consider suitable parameters for 2-UNION INDEPENDENT SET. Since we have two input graphs, we often consider the maximum or minimum value of parameters taken over the two input graphs. For example, considering the maximum degrees Δ_1 and Δ_2 of G_1 and G_2, respectively, natural parameters are $\Delta_{\min} := \min\{\Delta_1, \Delta_2\}$ and $\Delta_{\max} := \max\{\Delta_1, \Delta_2\}$. However, Theorem 1 implies that 2-UNION INDEPENDENT SET is NP-hard even if $\Delta_{\max} \leq 2$.

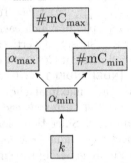

A second view on the parameterized landscape is centered around the fact that 2-UNION INDEPENDENT SET is W[1]-hard with respect to the parameter "solution size k" [13]. We therefore consider parameters that are lower-bounded by k. Unfortunately, the W[1]-hardness proof for parameter k due to Jiang [13] also shows that 2-UNION INDEPENDENT SET is W[1]-hard for the the maximum $\alpha_{\max}$ of the respective independence numbers α_1 and α_2 of G_1 and G_2. In interval graphs, a parameter that is lower bounded by the independence number α is the number of maximal cliques #mC. Indeed, we can show fixed-parameter tractability with respect to #mC$_{\min}$ and #mC$_{\max}$, denoting the minimum, resspectively the maximum, of the numbers of maximal cliques in the two input interval graphs. For the parameter #mC$_{\min}$, we exploit an alternative problem formulation, additionally allowing us to obtain results for the well-known JOB INTERVAL SELECTION problem [15]. An overview of the parameters that are lower-bounded by k is shown above.

Parameterized tractability. In the quest for polynomial-time preprocessing for 2-UNION INDEPENDENT SET, we considered simple-to-implement reduction rules. Surprisingly, a single twin-type reduction rule is sufficient to provide a polynomial-size problem kernel with respect to the parameter #mC$_{\max}$. We reduce the number of vertices having a given "signature" and then bound the number of signatures in a 2-union graph.

Definition 1. *Let (G_1, G_2, k) denote an instance of* 2-UNION INDEPENDENT SET *and let v be a vertex of G_1 and G_2. The* signature sig(v) *of v is the set of all vertex sets C that contain v and form a maximal clique in either G_1 or G_2.*

---

[2] The *Exponential-Time Hypohesis* basically states that there is no $2^{o(n)}$-time algorithm for n-variable 3SAT.

Reduction Rule 1. Let (G_1, G_2, k) denote an instance of 2-UNION INDEPENDENT SET. For each pair of vertices u, v of G_1 and G_2 such that $\text{sig}(v) \subseteq \text{sig}(u)$, delete u from G_1 and G_2.

Theorem 2. *2-UNION INDEPENDENT SET admits a cubic-vertex problem kernel with respect to the parameter "larger number of maximal cliques $\#\text{mC}_{\max}$". A quadratic-vertex problem kernel can be shown if one of the input graphs is a proper interval graph. Both kernels can be computed in $O(n \log^2 n)$ time.*

We can generalize Theorem 2 for the problem of finding an independent set of *weight* at least k: we keep the vertex with highest weight for each signature in the graph. Since each signature is uniquely determined by its first and last maximal cliques in G_1 and G_2 with respect to a clique ordering, there are at most $\#\text{mC}_{\min}^2 \cdot \#\text{mC}_{\max}^2$ different signatures and we obtain a problem kernel with $O(\#\text{mC}_{\min}^2 \cdot \#\text{mC}_{\max}^2)$ vertices for the weighted variant.

In the following, we describe a dynamic programming algorithm that solves 2-UNION INDEPENDENT SET in $O(2^{\#\text{mC}_{\min}} \cdot \#\text{mC}_{\min} \cdot \#\text{mC}_{\max} \cdot n)$ time. To this end, we reformulate the problem in terms of interval graphs in which each vertex has a list out of at most $\#\text{mC}_{\min}$ colors. We call a subset of vertices *colorful* if their color sets are pairwise disjoint. Recall the definition of COLORFUL INDEPENDENT SET in Section 1. We reduce 2-UNION INDEPENDENT SET to COLORFUL INDEPENDENT SET by assigning a color to each maximal clique in G_2 and giving G_1 as input to COLORFUL INDEPENDENT SET such that each vertex has the colors of the maximal cliques of G_2 containing it. Since the color lists generated in this reduction form intervals with respect to a clique ordering of G_2, COLORFUL INDEPENDENT SET can be considered a more general problem than 2-UNION INDEPENDENT SET. Regarding parameters, the numbers $\#\text{mC}_{\min}$ and $\#\text{mC}_{\max}$ of maximal cliques in the input interval graphs translate to the number γ of colors and the number $|\mathcal{C}|$ of maximal cliques in G, respectively.

Given a list-colored interval graph G, the algorithm computes a table T indexed by pairs in $\{0, \ldots, |\mathcal{C}|\} \times 2^{\{1, \ldots, \gamma\}}$ using the clique ordering $\prec$ of G. Let $\mathcal{C}[j]$ denote the j'th element in the ordering $\prec$, and let $G^i = G - \bigcup_{1 \le \ell \le i} \mathcal{C}[\ell]$. We define $T[i, C]$ so that it contains the maximum cardinality of a colorful independent set of G minus the first i maximal cliques (with respect to $\prec$) using only colors in C. For the base case, we set $T[|\mathcal{C}|, C] = 0$ for all $C \subseteq \{1, \ldots, \gamma\}$. Next, observe that for each interval v, there is a unique maximal clique with largest index i_v (according to the ordering $\prec$ of $\mathcal{C}$) containing v. The dynamic programming table can now be filled according to the following recursion:

$$T[i-1, C] = \max\Big\{ T[i, C], \max_{\substack{v \in \mathcal{C}[i] \\ \text{col}(v) \subseteq C}} \{1 + T[i_v, C \setminus \text{col}(v)]\} \Big\}. \tag{1}$$

The cardinality of a maximum colorful independent set of G can be read from $T[0, \{1, \ldots \gamma\}]$. This approach is easily-modifiable to also compute a maximum *weighted* independent set if the input graph is vertex-weighted.

Theorem 3. COLORFUL INDEPENDENT SET *can be solved in* $O(2^\gamma \cdot \gamma \cdot |\mathcal{C}| \cdot n)$ *time*[3], *even if all vertices are integer-weighted.*

In terms of 2-UNION INDEPENDENT SET, Theorem 3 can be stated as follows.

Corollary 2. 2-UNION INDEPENDENT SET *can be solved in* $O(2^{\#\mathrm{mC}_{\min}} \cdot \#\mathrm{mC}_{\max} \cdot \#\mathrm{mC}_{\min} \cdot n)$ *time.*

3   Colorful Independent Sets and Strip Graphs

In Section 2 we reformulated 2-UNION INDEPENDENT SET in terms of finding a maximum colorful independent set in an interval graph and gave a fixed-parameter algorithm for the more general problem COLORFUL INDEPENDENT SET. We now consider the variant of COLORFUL INDEPENDENT SET where each vertex (resp. interval) has only one color instead of a list of colors. This restriction is equivalent to 2-UNION INDEPENDENT SET for input graphs that are the edge-wise union of an interval graph and a cluster graph,[4] a class of graphs called *strip graphs* by Halldórsson and Karlsson [8]. They interpreted each clique in the input cluster graph as an equivalence class of vertices of the input interval graph; we reinterpret these equivalence classes in a natural way: as colors. In the literature, this problem is known as JOB INTERVAL SELECTION [15] (see the definition in Section 1). In our model, colors represent jobs and intervals of the same color in the input graph are possible execution intervals for one job. A solution then shows how to execute at least k jobs.

Jansen [11] showed a polynomial-time algorithm for JOB INTERVAL SELECTION for a constant number γ of colors. The dynamic programming algorithm given by Höhn et al. [9] can be seen as a generalization of this algorithm, since strip graphs are a special case of m-composite graphs. In both cases, the degree of the polynomial depends on γ. Halldórsson and Karlsson [8] gave a fixed-parameter algorithm running in $O(2^Q \cdot n)$ time with Q denoting the "maximum number of live intervals". Omitting the detailed description of the parameter Q, we note that, in instances of the underlying scheduling problem in which there is more than one machine, Q equals the number γ of colors in our interpretation.

Fixed-parameter algorithms for combinations with k. As 2-UNION INDEPENDENT SET is W[1]-hard for the single parameter "solution size k" [13], we combine k with the maximum clique size ω in the input interval graph G, the maximum number ϕ of cliques in G that have a vertex in common, and the number γ of colors in G. These combinations allow for fixed-parameter tractability and kernelization results. In the following, let $C_1, C_2, \ldots$ denote the maximal cliques of G in order of the clique ordering of G. We will reuse the notion of "signatures" (see Definition 1). In the context of colored interval graphs, the *signature* $\mathrm{sig}(v)$ of an interval v is the pair of its color and the set of maximal cliques it is contained in.

---

[3] Assuming that adding, subtracting, and comparing of integers work in $O(1)$ time.

[4] Recall that 2-UNION INDEPENDENT SET is solvable in $O(n^{1.5})$ time if both input graphs are cluster graphs (see Proposition 1).

Fig. 1. Schematic view of the construction of the cross-composition. Circles at the bottom represent the t input instances. Bars at the top represent the newly added intervals spanning over the input instances. Here, each of the $\log t$ rows stands for a new color. A solution (black intervals) for the instance must select one interval in each row, thereby selecting one of the t input instances (x_3 in this example).

The algorithms presented in this section rely on the observation that an optimal solution can be assumed to contain an interval v of the first maximal clique C_1. In the following, assume that there is an interval $v \in C_1$ that is in the sought colorful independent set. Our fixed-parameter algorithms branch on properties of v that allow us to either identify v or remove intervals from G so that an isolated clique containing v is created. These properties are (a) the size of C_1, (b) the last clique containing v, and (c) the color of v. After at most k branchings, we end up with a cluster graph, on which the problem can be solved in polynomial time using Proposition 1. Depending on what property of v we branched on, the exponential components of the running times can be bounded in ω^k, ϕ^k, or γ^k.

Proposition 2. JOB INTERVAL SELECTION *can be solved in* $O(\omega^k \cdot n)$, $O(\phi^k \cdot n^{1.5})$, *and* $O(\gamma^k \cdot n^2 \log^2 n)$ *time.*

Non-existence of polynomial-size kernels for Job Interval Selection. We show that JOB INTERVAL SELECTION is unlikely to admit polynomial-size problem kernels with respect to various parameters. To this end, we employ the technique of "cross-composition" introduced by Bodlaender et al. [3] using a bitmasking approach as standard in previous publications that exclude polynomial-size kernels for other problems. A *cross-composition* is a polynomial-time algorithm that, given t instances x_i with $0 \leq i < t$ of an NP-hard starting problem A, outputs an instance (y, k) of a parameterized problem B such that $k \in \mathrm{poly}(\max_i\{|x_i|\} + \log t)$ and $(y, k) \in B$ if and only if there is some $0 \leq i < t$ with $x_i \in A$. A theorem by Bodlaender et al. [3] states that if a problem B admits such a cross-composition, then there is no polynomial-size kernel for B unless coNP $\subseteq$ NP/poly.

We use an operation on binary-encoded numbers: *shifting* a number i by j bits to the right, denoted by $\mathrm{shift}(i, j) := \lfloor i/2^j \rfloor$. In the following, we present a cross-composition for JOB INTERVAL SELECTION with respect to (ω, γ). For the NP-hard starting problem we use the unparameterized version of JOB INTERVAL SELECTION with the restriction that $k = c$. The NP-hardness of this problem is a direct consequence of Theorem 1. We assume, without loss of generality, that $\log t$ is an integer. The framework of Bodlaender et al. [3] allows us to force the input instances to all have the same value for k and, thus, each instance uses the same color set $\{1, 2, \ldots, k\}$. The steps of the composition are as follows (see Figure 1):

Step 1. Place the t input instances in order of their index on the real line such that no interval of one instance overlaps an interval of another instance.

Step 2. Introduce $\log t$ more colors $k+1, k+2, \ldots, k+\log t$ (the resulting instance then asks for an independent set of size $k + \log t$).

Step 3. For each $1 \leq i \leq \log t$, introduce 2^i new intervals $v_0^i, v_1^i, \ldots, v_{2^i-1}^i$, each with color $k+i$, such that the new interval v_j^i spans over all instances x_ℓ with shift$(\ell, \log t - i) = j$.

It is easy to see that both the number of colors γ and the maximum clique size ω of the constructed instance are at most $\max_i |x_i| + \log t$. In order to show that the presented algorithm constitutes a cross-composition, it remains to prove that the resulting instance has a colorful independent set of size $k + \log t$ if and only if there is an input instance x_i that has a colorful independent set of size k. The presented cross-composition implies the following theorem [3].

Theorem 4. JOB INTERVAL SELECTION *does not admit a polynomial-size problem kernel with respect to the combination of the parameters "number of colors γ" and "maximum clique size ω" unless* coNP $\subseteq$ NP/poly.

Polynomial-size kernel for proper interval graphs. We further restrict JOB INTERVAL SELECTION to proper interval graphs, on which it is still NP-hard, as evident from Section 2. Surprisingly, simple data reduction rules enable us to construct a problem kernel comprising $2\omega k(k-1)$ intervals in this case, sharply contrasting Theorem 4, which excludes a polynomial-size problem kernel with respect to the combined parameter (k, ω) (since $\gamma \geq k$).

Reduction Rule 2. Delete from G every interval that has a color that appears more than $2\omega(k-1)$ times and decrease k by the number of removed colors.

Reduction Rule 3. If G contains more than $2\omega k(k-1)$ intervals, then return a trivial yes-instance.

Reduction Rule 2 can be applied exhaustively in $O(n)$ time. Thereafter executing Reduction Rule 3 immediately yields the following theorem.

Theorem 5. JOB INTERVAL SELECTION *on proper interval graphs admits a problem kernel with at most $2\omega k(k-1)$ intervals that can be computed in $O(n)$ time.*

4   Outlook

Besides hardness results, we also developed encouraging algorithmic results which might find use in practical applications, so future empirical studies seem worthwhile (also see the strong practical results of Höhn et al. [9] with respect to steel manufacturing). As a future challenge, it is interesting to know whether 2-UNION INDEPENDENT SET admits a polynomial-size problem kernel with respect to the parameter $\#mC_{min}$, denoting the smaller number of maximal cliques in one of the input interval graphs. Furthermore, we conjecture that JOB INTERVAL SELECTION with respect to the parameter "solution size k" is fixed-parameter tractable, whereas 2-UNION INDEPENDENT SET is known to be W[1]-hard for this parameter [13].

Acknowledgment. We are grateful to Michael Dom and Hannes Moser for earlier discussions on coil coating.

References

[1] Bafna, V., Narayanan, B.O., Ravi, R.: Nonoverlapping local alignments (weighted independent sets of axis-parallel rectangles). Discrete Appl. Math. 71(1-3), 41–53 (1996)

[2] Bar-Yehuda, R., Halldórsson, M.M., Naor, J., Shachnai, H., Shapira, I.: Scheduling split intervals. SIAM J. Comput. 36(1), 1–15 (2006)

[3] Bodlaender, H.L., Jansen, B.M.P., Kratsch, S.: Cross-composition: A new technique for kernelization lower bounds. In: Proc. 28th STACS. LIPIcs, vol. 9, pp. 165–176. Schloss Dagstuhl–Leibniz-Zentrum für Informatik (2011)

[4] Booth, K.S., Lueker, G.S.: Testing for the consecutive ones property, interval graphs, and graph planarity using PQ-tree algorithms. J. Comput. Syst. Sci. 13(3), 335–379 (1976)

[5] Chuzhoy, J., Ostrovsky, R., Rabani, Y.: Approximation algorithms for the job interval selection problem and related scheduling problems. Math. Oper. Res. 31(4), 730–738 (2006)

[6] Fellows, M.R., Hermelin, D., Rosamond, F.A., Vialette, S.: On the parameterized complexity of multiple-interval graph problems. Theor. Comput. Sci. 410(1), 53–61 (2009)

[7] Fulkerson, D.R., Gross, O.A.: Incidence matrices and interval graphs. Pacific J. Math. 15(3), 835–855 (1965)

[8] Halldórsson, M.M., Karlsson, R.K.: Strip Graphs: Recognition and Scheduling. In: Fomin, F.V. (ed.) WG 2006. LNCS, vol. 4271, pp. 137–146. Springer, Heidelberg (2006)

[9] Höhn, W., König, F.G., Möhring, R.H., Lübbecke, M.E.: Integrated sequencing and scheduling in coil coating. Manage. Sci. 57(4), 647–666 (2011)

[10] Impagliazzo, R., Paturi, R., Zane, F.: Which problems have strongly exponential complexity? J. Comput. Syst. Sci. 63(4), 512–530 (2001)

[11] Jansen, K.: Generalizations of assignments of tasks with interval times. Technical report, Universität Trier (1991)

[12] Jansen, K.: Transfer flow graphs. Discrete Math. 115(1-3), 187–199 (1993)

[13] Jiang, M.: On the parameterized complexity of some optimization problems related to multiple-interval graphs. Theor. Comput. Sci. 411, 4253–4262 (2010)

[14] Kolen, A.W., Lenstra, J.K., Papadimitriou, C.H., Spieksma, F.C.R.: Interval scheduling: A survey. Nav. Res. Log. 54(5), 530–543 (2007)

[15] Nakajima, K., Hakimi, S.L.: Complexity results for scheduling tasks with discrete starting times. J. Algorithms 3(4), 344–361 (1982)

[16] Sbihi, N.: Algorithme de recherche d'un stable de cardinalite maximum dans un graphe sans etoile. Discrete Math. 29(1), 53–76 (1980)

[17] Spieksma, F.C.R.: On the approximability of an interval scheduling problem. J. Sched. 2(5), 215–227 (1999)

More on a Problem of Zarankiewicz

Chinmoy Dutta[1,*] and Jaikumar Radhakrishnan[2]

[1] Northeastern University, Boston, USA
chinmoy@ccs.neu.edu
[2] Tata Institute of Fundamental Research, Mumbai, India
jaikumar@tifr.res.in

Abstract. We show tight necessary and sufficient conditions on the sizes of small bipartite graphs whose union is a larger bipartite graph that has no large bipartite independent set. Our main result is a common generalization of two classical results in graph theory: the theorem of Kővári, Sós and Turán on the minimum number of edges in a bipartite graph that has no large independent set, and the theorem of Hansel (also Katona and Szemerédi, Krichevskii) on the sum of the sizes of bipartite graphs that can be used to construct a graph (non-necessarily bipartite) that has no large independent set. Our results unify the underlying combinatorial principles developed in the proof of tight lower bounds for depth-two superconcentrators.

1 Introduction

Consider a bipartite graph $G = (V, W, E)$, where $|V|, |W| = n$. Suppose every k element subset $S \subseteq V$ is connected to every k element subset $T \subseteq W$ by at least one edge. How many edges must such a graph have? This is the celebrated Zarankiewicz problem.

Definition 1 (Bipartite independent set). *A bipartite independent set of size $k \times k$ in a bipartite graph $G = (V, W, E)$ is a pair of subsets $S \subseteq V$ and $T \subseteq W$ of size k each such that there is no edge connecting S and T, i.e., $(S \times T) \cap E = \emptyset$.*

The Zarankiewicz problem asks for the minimum number of edges in a bipartite graph that does not have any bipartite independent set of size $k \times k$. We may think of an edge as a complete bipartite graph where each side of the bipartition is just a singleton. This motivates the following generalization where we consider bipartite graphs as formed by putting together not just edges, but, more generally, small complete bipartite graphs.

Definition 2. *A bipartite graph $G = (V, W, E)$ is said to be a union of complete bipartite graphs $G_i = (V_i, W_i, E_i = V_i \times W_i)$, $i = 1, 2, \ldots, r$, if each $V_i \subseteq V$, each $W_i \subseteq W$, and $E = E_1 \cup \cdots \cup E_r$.*

* Chinmoy Dutta is supported in part by NSF grant CCF-0845003 and a Microsoft grant to Ravi Sundaram.

K.-M. Chao, T.-s. Hsu, and D.-T. Lee (Eds.): ISAAC 2012, LNCS 7676, pp. 257–266, 2012.
© Springer-Verlag Berlin Heidelberg 2012

Definition 3. *We say that a sequence of positive integers* $(n_1, n_2, \ldots, n_r)$ *is* (n, k)-*strong if there is a bipartite graph* $G = (V, W, E)$ *that is a union of graphs* $G_i = (V_i, W_i, E_i = V_i \times W_i)$, $i = 1, 2, \ldots, r$, *such that*

- $|V|, |W| = n$;
- $|V_i| = |W_i| = n_i$;
- G *has no bipartite independent set of size* $k \times k$.

What conditions must the n_i's satisfy for $(n_1, n_2, \ldots, n_r)$ to be (n, k)-strong? Note that the Zarankiewicz problem is a special case of this question where each n_i is 1 and $\sum_i n_i$ corresponds to the number edges in the final graph G.

Remark. The Zarankiewicz problem is more commonly posed in the following form: What is the maximum number of edges in a bipartite graph with no $k \times k$ bipartite *clique*. By interchanging edges and non-edges, we can ask for the maximum number of non-edges (equivalently the minimum number of edges) such that there is no $k \times k$ bipartite *independent set*. This complementary form is more convenient for our purposes.

The Kővári, Sós and Turán bound
The following classical theorem gives a lower bound on the number of edges in a bipartite graph that has no large independent set.

Theorem 1 (Kővári, Sós and Turán [1]; see, e.g., [2], Page 301, Lemma 2.1.). *If G does not have an independent set of size $k \times k$, then*

$$n \binom{n - \overline{d}}{k} \binom{n}{k}^{-1} \leq k - 1,$$

where $\overline{d}$ is the average degree of G.

The above theorem implies that

$$n \leq (k - 1) \binom{n - \overline{d}}{k}^{-1} \binom{n}{k} \leq (k - 1) \left(\frac{n - k + 1}{n - \overline{d} - k + 1} \right)^k$$

$$= (k - 1) \left(1 + \frac{\overline{d}}{n - \overline{d} - k + 1} \right)^k \leq (k - 1) \exp \left(\frac{\overline{d}k}{n - \overline{d} - k + 1} \right),$$

which yields,

$$\overline{d} \geq \frac{(n - k + 1) \log(n/(k - 1))}{k + \log(n/(k - 1))}.$$

In this paper, we will mainly be interested in $k \in [n^{1/10}, n^{9/10}]$, in which case we obtain

$$|E(G)| = n\overline{d} = \Omega \left(\frac{n^2}{k} \log n \right).$$

For the problem under consideration, this immediately gives the necessary condition

$$\sum_{i=1}^{r} n_i^2 = \Omega\left(\frac{n^2}{k}\log n\right).$$ (1)

It will be convenient to normalize n_i and define $\alpha_i = \frac{n_i}{n/k}$. With this notation, the inequality above can be restated as follows.

$$\sum_{i=1}^{r} \alpha_i^2 = \Omega(k\log n).$$ (2)

The Hansel bound
The same question can also be asked in the context of general graphs. In that case, we have the following classical theorem.

Theorem 2 (Hansel [3], Katona and Szemerédi [4], Krichevskii [5]).
Suppose it is possible to place one copy each of K_{n_i,n_i}, $i = 1, 2, \ldots, r$, in a vertex set of size n such that the resulting graph has no independent set of size k. Then,

$$\sum_{i=1}^{r} n_i \geq n\log\left(\frac{n}{k-1}\right).$$

Although this result pertains to general graphs and is not directly applicable to the bipartite graph setting, it can be used (details omitted as we will use this bound only to motivate our results, not to derive them) to derive a necessary condition for bipartite graphs as well. In particular, normalizing n_i by setting $n_i = \alpha_i \frac{n}{k}$ as before, one can obtain the necessary condition

$$\sum_{i=1}^{r} \alpha_i = \Omega(k\log n).$$ (3)

Note that neither of the two bounds above strictly dominates the other: if all α_i are small (say $\ll 1$), then the first condition derived from the Kővári, Sós and Turán bound is stronger, wheras if all α_i are large ($\gg 1$), then the condition derived from the Hansel bound is stronger.

In our applications, we will meet situations where the α_i's will not be confined to one or the other regime. To get optimal results, one must, therefore, devise a condition appropriate for the entire range of values for the α_i's. Towards this goal, we start by trying to guess the form of this general inequality by asking a dual question: what is a sufficient condition on n_i's (equivalently α_i's) for $(n_1, n_2, \ldots, n_r)$ to be (n, k)-strong? We derive the following (proof omitted).

Theorem 3 (Sufficient condition). *Suppose $k \in [n^{1/10}, n^{9/10}]$, and let $\alpha_i \in [n^{-1/100}, n^{1/100}]$, $i = 1, 2 \ldots, r$. Then, there is a constant $A > 0$ such that if*

$$\sum_{i:\alpha_i \leq 1} \alpha_i^2 + \sum_{i:\alpha_i > 1} \alpha_i \geq Ak\log n,$$

then $(n_1, n_2, \ldots, n_r)$ is (n, k)-strong, where $n_i = \alpha_i(n/k)$.

We might ask if this sufficient condition is also necessary. The Kővári, Sós and Turán bound (Inequality 2) explains the first term in the LHS of the above sufficient condition, and the Hansel bound (Inequality 3) explains the second term. We thus have explanations for both the terms using two classical theorems of graph theory. However, neither of them implies in full generality that the sufficient condition derived above is necessary. In this work, we show that this sufficient condition is indeed also necessary upto constants.

Theorem 4 (Necessary condition). *Suppose $k \in [n^{1/10}, n^{9/10}]$, and let $\alpha_i \in [n^{-1/100}, n^{1/100}]$, $i = 1, 2 \ldots, r$. Then, there is a constant $B > 0$ such that if $(n_1, n_2, \ldots, n_r)$ is (n, k)-strong where $n_i = \alpha_i(n/k)$, then*

$$\sum_{i:\alpha_i \leq 1} \alpha_i^2 + \sum_{i:\alpha_i > 1} \alpha_i \geq Bk \log n.$$

Our proof of Theorem 4 uses refinement of the ideas used in Radhakrishnan and Ta-Shma [6]. In a later section, we also show that our necessary condition leads to a modular proof of their tight lower bound on the size of depth-two superconcentrators.

A tradeoff result for depth-two superconcentrators was shown by Dutta and Radhakrishnan [7]. Their main argument leads one to consider situations where the small bipartite graphs used to build the bigger one are not symmetric, instead of being of the form K_{n_i,n_i}, they are of the form K_{m_i,n_i} (with perhaps $m_i \neq n_i$).

Definition 4. *We say that a sequence $((m_1, n_1), (m_2, n_2), \ldots, (m_r, n_r))$ of pairs of positive integers is (n, k)-strong if there is a bipartite graph $G = (V, W, E)$ that is a union of graphs $G_i = (V_i, W_i, E_i = V_i \times W_i)$, $i = 1, 2, \ldots, r$, such that*

- *$|V|, |W| = n$;*
- *$|V_i| = m_i$ and $|W_i| = n_i$.*
- *G has no bipartite independent set of size $k \times k$.*

We refine our lower bound argument for the symmetric case and provide necessary condition for this asymmetric setting as well.

Theorem 5 (Necessary condition: asymmetric case). *Suppose $\alpha_i, \beta_i \in [n^{-1/100}, n^{1/100}]$, $i = 1, 2 \ldots, r$, and $k \in [n^{1/10}, n^{9/10}]$. Then, there is a constant $C > 0$ such that if the sequence $((m_1, n_1), (m_2, n_2), \ldots, (m_r, n_r))$ is (n, k)-strong where $m_i = \alpha_i(n/k)$ and $n_i = \beta_i(n/k)$, then*

$$\sum_{i \in X} \alpha_i \beta_i + \sum_{i \in \{1, 2, \ldots, r\} \setminus X} (\alpha_i + \beta_i) \mathsf{H}(p_i) \geq Ck \log n$$

for every $X \subseteq \{1, 2, \ldots, r\}$, where $p_i = \frac{\alpha_i}{\alpha_i + \beta_i}$ and $\mathsf{H}(p_i) = -p_i \log(p_i) - (1 - p_i) \log(1 - p_i)$.

As applications of our lower bounds, we provide modular proofs of two known lower bounds for depth-two superconcentrators which are important combinatorial objects useful in both algorithms and complexity (formal definition in

Section 4): the first one is a lower bound on the number of edges in such graphs (Theorem 6 in Section 4) shown in [6] which we reprove here using Theorem 4; the second one is a tradeoff result between the number of edges at different levels of such graphs (Theorem 7 in Section 4) shown in [7] which we reprove using Theorem 5.

2 Building a Bipartite Graph from Smaller Symmetric Bipartite Graphs

Proof of Theorem 4

Let $k \in [n^{\frac{1}{10}}, n^{\frac{9}{10}}]$ and $\alpha_i \in [n^{-1/100}, n^{1/100}]$, $i = 1, 2, \ldots, r$. Suppose we are given a bipartite graph $G = (V, W, E)$ which is a union of complete bipartite graphs $G_1, G_2, \ldots, G_r$ and has no bipartite independent set of size $k \times k$, where $G_i = (V_i, W_i, E_i = V_i \times W_i)$ with $|V_i| = |W_i| = n_i = \alpha_i(n/k)$. We want to show that for some constant $B > 0$,

$$\sum_{i:\alpha_i \leq 1} \alpha_i^2 + \sum_{i:\alpha_i > 1} \alpha_i \geq Bk \log n.$$

We will present the argument for the case when $k = \sqrt{n}$; the proof for other k is similar, and focussing on this k will keep the notation and the constants simple. We will show that if the second term in the LHS of the above inequality is small, say,

$$\mathsf{SecondTerm} = \sum_{i:\alpha_i > 1} \alpha_i \leq \frac{1}{100} k \log n,$$

then the first term must be large, i.e.,

$$\mathsf{FirstTerm} = \sum_{i:\alpha_i \leq 1} \alpha_i^2 \geq \frac{1}{100} k \log n.$$

Assume $\mathsf{SecondTerm} \leq \frac{1}{100} k \log n$. Let us call a G_i for which $\alpha_i > 1$ as *large* and a G_i for which $\alpha_i \leq 1$ as *small*. We start as in [6] by deleting one of the sides of each large G_i independently and uniformly at random from the vertex set of G. For a vertex $v \in V$, let d_v be number of large G_i's such that $v \in V_i$. The probability that v survives at the end of the random deletion is precisely 2^{-d_v}. Now,

$$\sum_v d_v = \sum_{i:\alpha_i > 1} n_i \leq \frac{1}{100} n \log n,$$

where the inequality follows from our assumption that $\mathsf{SecondTerm} \leq \frac{1}{100} k \log n$. That is, the average value of d_v is $\frac{1}{100} \log n$, and by Markov's inequality, at least half of the vertices have their d_v's at most $d = \frac{1}{50} \log n$. We focus on a set V' of $n/2$ such vertices, and if they survive the first deletion, we delete them again with probability $1 - 2^{-(d-d_v)}$, so that every one of these $n/2$ vertices in V' survives with probability exactly $2^{-d} = n^{-1/50}$. Let X be the vertices of V' that survive. Similarly, we define $W' \subseteq W$, and let $Y \subseteq W'$ be the vertices that survive.

Claim. With probability $1 - o(1)$, $|X|, |Y| \geq \frac{n}{4}2^{-d}$.

The claim can be proved as follows. For $v \in V'$, let I_v be the indicator variable for the event that v survives. Then, $\Pr[I_v = 1] = 2^{-d} = n^{-1/50}$ for all $v \in V'$. Furthermore, I_v and $I_{v'}$ are dependent precisely if there is a common large G_i such that both $v, v' \in V_i$. Thus, any one I_v is dependent on at most $\Delta = d_v \times \max\{n_i : \alpha_i > 1\} \leq (1/50)(\log n)n^{1/100}(n/k) = (1/50)n^{51/100}\log n$ such events (recall $k = \sqrt{n}$). We thus have (see Alon-Spencer [8])

$$\mathbb{E}[|X|] = \sum_{v \in V'} I_v = \frac{n}{2}2^{-d} = \frac{1}{2}n^{49/50};$$

$$\mathsf{Var}[|X|] \leq E[|X|]\Delta.$$

By Chebyshev's inequality, the probability that $|X|$ is less than $\frac{\mathbb{E}[|X|]}{2}$ is at most

$$\frac{4\mathsf{Var}[X]}{\mathbb{E}[|X|]^2} \leq \frac{4\Delta}{\mathbb{E}[|X|]} = o(1).$$

A similar calculation can be done for $|Y|$. (End of Claim.)

The crucial consequence of our random deletion process is that no large G_i has any edge between X and Y. Since G does not have any independent set (S, T) of size $k \times k$, the small G_i's must provide the necessary edges to avoid such independent sets between X and Y. Consider an edge (v, w) of a small G_i. The probability that this edge survives in $X \times Y$ is precisely the probability of the event $I_v \wedge I_w$. Note that the two events I_v and I_w are either independent (when v and w do not belong to a common large G_i), or they are mutually exclusive. Thus, the expected number of edges supplied between X and Y by small G_i's is at most

$$\sum_{i:\alpha_i \leq 1} \alpha_i^2(n/k)^2 2^{-2d} = \mathsf{FirstTerm} \times (n/k)^2 2^{-2d},$$

and by Markov's inequality, with probability $1/2$ it is at most twice its expectation. Using the Claim above we conclude that the following three events happen simultaneously: (a) $|X| \geq \frac{n}{4}2^{-d}$, (b) $|Y| \geq \frac{n}{4}2^{-d}$, (c) the number of edges conecting X and Y is at most $2 \times \mathsf{FirstTerm} \times (n/k)^2 2^{-2d}$. Using (1), this number of edges must be at least $\frac{1}{3}\frac{(n2^{-d})^2}{16k}(\frac{49}{50}\log n - 2)$. (Note that $\frac{1}{3}$ suffices as the constant in (1) for the case $|X|, |Y| \geq \frac{n^{49/50}}{4}$ and $k = \sqrt{n}$.) Comparing the upper and lower bounds on the number of edges thus established, we obtain the required inequality

$$\mathsf{FirstTerm} \geq \frac{1}{100}k \log n.$$

3   Building a Bipartite Graph from Smaller Asymmetric Bipartite Graphs

Proof of Theorem 5

Let $k \in [n^{\frac{1}{10}}, n^{\frac{9}{10}}]$ and $\alpha_i, \beta_i \in [n^{-\frac{1}{100}}, n^{\frac{1}{100}}]$, $i = 1, 2, \ldots, r$. Suppose we are given a bipartite graph $G = (V, W, E)$ which is a union of complete bipartite graphs $G_1, G_2, \ldots, G_r$ and has no bipartite independent set of size $k \times k$, where $G_i = (V_i, W_i, E_i = V_i \times W_i)$ with $|V_i| = m_i = \alpha_i(n/k)$ and $|W_i| = n_i = \beta_i(n/k)$. As stated in Theorem 5, we let $p_i = \frac{\alpha_i}{\alpha_i + \beta_i}$ and $H(p_i) = -p_i \log(p_i) - (1 - p_i) \log(1 - p_i)$. We wish to show that there is a constant $C > 0$, such that

$$\sum_{i \in X} \alpha_i \beta_i + \sum_{i \notin X} (\alpha_i + \beta_i) H(p_i) \geq Ck \log n$$

for every $X \subseteq \{1, 2, \ldots, r\}$.

The proof is similar to but more subtle than the proof of Thereom 4 and again we present the argument for the case when $k = \sqrt{n}$. Fix a subset $X \subseteq \{1, 2, \ldots, r\}$. Our plan is to assume that the second term in the LHS of the above inequality is small,

$$\text{SecondTerm} = \sum_{i \notin X} (\alpha_i + \beta_i) H(p_i) \leq \frac{1}{100} k \log n, \tag{4}$$

and from this conclude that the first term must be large,

$$\text{FirstTerm} = \sum_{i \in X} \alpha_i \beta_i \geq \frac{1}{100} k \log n. \tag{5}$$

Assume $\text{SecondTerm} \leq \frac{1}{100} k \log n$. Graphs G_i where $i \in X$ will be called *marked* and graphs G_i where $i \notin X$ will be called *unmarked*. As before, we will delete one of the sides of each unmarked G_i independently at random from the vertex set of G. However, since this time there are different number of vertices on the two sides of G_i, we need to be more careful and choose the deletion probabilities carefully. For every unmarked G_i independently, we delete all the vertices in W_i with probability p_i and all the vertices in V_i with probability $1 - p_i$.

For a vertex $v \in V$, let S_v be the set of $i \notin X$ such that $v \in V_i$. Define d_v to be the quantity $\sum_{i \in S_v} \log(1/p_i)$. The probability that v survives the random deletion process is 2^{-d_v}. Using the fact that $p_i = \frac{\alpha_i}{\alpha_i + \beta_i}$ and plugging the expression for $H(p_i)$ in the assumption (4), we obtain

$$\sum_{i \notin X} (\alpha_i \log(1/p_i) + \beta_i \log(1/(1 - p_i))) \leq \frac{1}{100} k \log n.$$

Multiplying both sides by (n/k), this implies

$$\sum_{i \notin X} m_i \log(1/p_i) \leq \frac{1}{100} n \log n, \tag{6}$$

and

$$\sum_{i \notin X} n_i \log(1/(1 - p_i)) \leq \frac{1}{100} n \log n. \tag{7}$$

Since

$$\sum_{v \in V} d_v = \sum_{i \notin X} m_i \log(1/p_i),$$

the average value of d_v is at most $\frac{1}{100}\log n$, and by Markov's inequality, at least $3n/4$ vertices $v \in V$ have their d_v at most $d = \frac{1}{25}\log n$. Moreover, since $\alpha_i, \beta_i \in [n^{-1/100}, n^{1/100}]$, we have

$$p_i \le \frac{n^{1/100}}{n^{1/100} + n^{-1/100}} \le 1 - \frac{n^{-1/100}}{n^{-1/100} + n^{1/100}} \le \exp\left(-\frac{n^{-1/100}}{n^{1/100} + n^{-1/100}}\right),$$

and thus

$$\frac{1}{p_i} \ge \exp\left(\frac{n^{-1/100}}{n^{1/100} + n^{-1/100}}\right) \ge \exp\left(\frac{1}{2}n^{-1/50}\right).$$

The above implies $\log(1/p_i) \ge \frac{1}{2}n^{-1/50}$, which combined with (6) yields

$$\sum_{i \notin X} m_i \le \frac{1}{50}n^{51/50}\log n.$$

Since

$$\sum_{v \in V} |S_v| = \sum_{i \notin X} m_i,$$

the average value of $|S_v|$ is at most $\frac{1}{50}n^{1/50}\log n$, and again by Markov's inequality, at least $3n/4$ vertices $v \in V$ satisfy $|S_v| \le d' = \frac{4}{50}n^{1/50}\log n$.

We focus on a set V' of $n/2$ vertices $v \in V$ such that $d_v \le d$ and $|S_v| \le d'$. If any vertex $v \in V'$ survives the first deletion, we delete it further with probability $1 - 2^{-(d-d_v)}$, so that the survival probability of each vertex in V' is exactly $2^{-d} = n^{-1/25}$. Let X be the set of vertices in V' that survive. Similarly, we define $W' \subseteq W$, and let Y be the set of vertices in W' that survive.

Claim. With probability $1 - o(1)$, $|X|, |Y| \ge \frac{n}{4}2^{-d}$.

The proof of the claim is exactly like the previous time.

Since no unmarked G_i has any edge between X and Y, the marked G_i's must provide enough edges to avoid all independent sets of size $k \times k$ between X and Y. As in the proof of Theorem 4, we can argue that an edge of a marked G_i survives in $X \times Y$ with probability at most 2^{-2d}. Thus the expected number of edges supplied between X and Y by marked G_i's is at most

$$\sum_{i \in X} m_i n_i 2^{-2d} = \sum_{i \in X} \alpha_i \beta_i (n/k)^2 2^{-2d} = \mathsf{FirstTerm} \times (n/k)^2 2^{-2d},$$

and by Markov's inequality with probability $1/2$ it is at most twice its expectation. Thus the event where both X and Y are of size at least $\frac{n}{4}2^{-d}$ and the number of edges connecting them is at most $2 \times \mathsf{FirstTerm} \times (n/k)^2 2^{-2d}$ occurs with positive probability. From (1), this number of edges must be at least $\frac{1}{3}\frac{(n2^{-d})^2}{16k}(\frac{24}{25}\log n - 2)$. (Note that $\frac{1}{3}$ suffices as the constant in (1) when $|X|, |Y| \ge \frac{n^{24/25}}{4}$ and $k = \sqrt{n}$.) Thus we get

$$\mathsf{FirstTerm} \ge \frac{1}{100}k\log n.$$

4   Depth-two Superconcentrators

Definition 5 (Depth-two superconcentrators). *Let $G = (V, M, W, E)$ be a graph with three sets of vertices V, M and W, where $|V|, |W| = n$, such that all edges in E go from V to M or M to W. Such a graph is called a depth-two n-superconcentrator if for every $k \in \{1, 2, \ldots, n\}$ and every pair of subsets $S \subseteq V$ and $T \subseteq W$, each of size k, there are k vertex disjoint paths from S to T.*

We reprove two known lower bounds for depth-two superconcentrators.

Theorem 6 (Radhakrishnan and Ta-Shma [6]). *If the graph $G(V, M, W, E)$ is a depth-two n-superconcentrator, then $|E(G)| = \Omega(n \frac{(\log n)^2}{\log \log n})$.*

Proof. Assume that the number of edges in a depth-two n-superconcentrator G is at most $(B/100)n\frac{(\log n)^2}{\log \log n}$, where B is the constant in Theorem 4. By increasing the number of edges by a factor at most two, we assume that each vertex in M has the same number of edges coming from V and going to W. For a vertex $v \in M$, let $\deg(v)$ denote the number of edges that come from V to v (equivalently the number of edges that go from v to W). For $k \in [n^{1/4}, n^{3/4}]$, define

$$\mathsf{High}(k) = \{v \in M : \deg(v) \geq \frac{n}{k}(\log n)^2\};$$

$$\mathsf{Medium}(k) = \{v \in M : \frac{n}{k}(\log n)^{-2} \leq \deg(v) < \frac{n}{k}(\log n)^2\};$$

$$\mathsf{Low}(k) = \{v \in M : \deg(v) < \frac{n}{k}(\log n)^{-2}\}.$$

Claim. For each $k \in [n^{1/4}, n^{3/4}]$, the number of edges incident on $\mathsf{Medium}(k)$ is at least $\frac{B}{2}n \log n$.

Fix a $k \in [n^{1/4}, n^{3/4}]$. First observe that $|\mathsf{High}(k)| < k$, for otherwise, the number of edges in G would already exceed $n(\log n)^2$, contradicting our assumption. Thus, every pair of subsets $S \subseteq V$ and $T \subseteq W$ of size k each has a common neighbour in $\mathsf{Medium}(k) \cup \mathsf{Low}(k)$. We are now in a position to move to the setting of Theorem 4. For each vertex $v \in \mathsf{Medium}(k) \cup \mathsf{Low}(k)$, consider the complete bipartite graph between its in-neighbours in V and out-neighbours in W. The analysis above implies that the union of these graphs is a bipartite graph between V and W that has no independent set of size $k \times k$. For $v \in \mathsf{Medium}(k) \cup \mathsf{Low}(k)$, let $\alpha_v = \frac{\deg(v)}{n/k}$. Using Theorem 4, it follows that

$$\sum_{v \in \mathsf{Medium}(k) \cup \mathsf{Low}(k) : \alpha_v \leq 1} \alpha_v^2 + \sum_{v \in \mathsf{Medium}(k) \cup \mathsf{Low}(k) : \alpha_v > 1} \alpha_v \geq Bk \log n. \quad (8)$$

For $\alpha_v \leq 1$, $\alpha_v^2 \leq \alpha_v$ and thus we can replace α_v^2 by α_v when $(\log n)^{-2} \leq \alpha_v \leq 1$ and conclude

$$\sum_{v \in \mathsf{Low}(k)} \alpha_v^2 + \sum_{v \in \mathsf{Medium}(k)} \alpha_v \geq Bk \log n. \quad (9)$$

One of the two terms in the LHS is at least half the RHS. If it is the first term then noting that $\alpha_v < (\log n)^{-2}$ for all $v \in \mathsf{Low}(k)$, we obtain

$$\sum_{v \in \mathsf{Low}} \deg(v) = \frac{n}{k} \sum_{v \in \mathsf{Low}} \alpha_v \geq \frac{n}{k}(\log n)^2 \sum_{v \in \mathsf{Low}} \alpha_v^2 \geq \frac{B}{2}n(\log n)^3.$$

Since the left hand side is precisely the number of edges entering $\mathsf{Low}(k)$, this contradicts our assumption that G has few edges. So, it must be that the second term in the LHS of (9) is at least $\frac{B}{2}k \log n$. Then, the number of edges incident on $\mathsf{Medium}(k)$ is

$$\sum_{v \in \mathsf{Medium}} \deg(v) = \frac{n}{k} \sum_{v \in \mathsf{Medium}} \alpha_v \geq \frac{B}{2}n \log n.$$

This completes the proof of the claim.

Now, consider values of k of the form $n^{1/4}(\log n)^{4i}$ in the range $[n^{1/4}, n^{3/4}]$. Note that there are at least $(\frac{1}{10}) \log n / \log \log n$ such values of k and the sets $\mathsf{Medium}(k)$ for these values of k are disjoint. By the claim above, each such $\mathsf{Medium}(k)$ has at least $\frac{B}{2}n \log n$ edges incident on it, that is G has a total of at least $\frac{B}{20}n \frac{(\log n)^2}{\log \log n}$ edges, again contradicting our assumption.

Theorem 7 (Dutta and Radhakrishnan [7]). *If the graph $G = (V, M, W, E)$ is a depth-two n-superconcentrator with average degree of nodes in V and W being a and b respectively and $a \leq b$, then*

$$a \log\left(\frac{a+b}{a}\right) \log b = \Omega(\log^2 n).$$

The proof of the above theorem is omitted due to lack of space.

References

1. Kővári, T., Sós, V., Turán, P.: On a problem of k. zarankiewicz. Colloquium Mathematicum 3, 50–57 (1954)
2. Bollobás, B.: Extremal Graph Theory. Academic Press (1978)
3. Hansel, G.: Nombre minimal de contacts de fermature nécessaires pour réaliser une fonction booléenne symétrique de n variables. C. R. Acad. Sci. Paris 258, 6037–6040 (1964)
4. Katona, G., Szemerédi, E.: On a problem of graph theory. Studia Sci. Math. Hungar. 2, 23–28 (1967)
5. Krichevskii, R.E.: Complexity of contact circuits realizing a function of logical algebra. Sov. Phys. Dokl. 8, 770–772 (1964)
6. Radhakrishnan, J., Ta-Shma, A.: Bounds for dispersers, extractors and depth-two superconcentrators. SIAM J. Disc. Math. 13(1), 2–24 (2000)
7. Dutta, C., Radhakrishnan, J.: Tradeoffs in Depth-Two Superconcentrators. In: Durand, B., Thomas, W. (eds.) STACS 2006. LNCS, vol. 3884, pp. 372–383. Springer, Heidelberg (2006)
8. Alon, N., Spencer, J.H.: The probabilistic method. Wiley-Interscience (2000)

Efficient Dominating and Edge Dominating Sets for Graphs and Hypergraphs

Andreas Brandstädt[1], Arne Leitert[1], and Dieter Rautenbach[2]

[1] Institut für Informatik, Universität Rostock, Germany
ab@informatik.uni-rostock.de, arne.leitert@uni-rostock.de
[2] Institut für Optimierung und Operations Research, Universität Ulm, Germany
dieter.rautenbach@uni-ulm.de

Abstract. Let $G = (V, E)$ be a graph. A vertex *dominates* itself and all its neighbors, i.e., every vertex $v \in V$ dominates its closed neighborhood $N[v]$. A vertex set D in G is an *efficient dominating* (*e.d.*) set for G if for every vertex $v \in V$, there is exactly one $d \in D$ dominating v. An edge set $M \subseteq E$ is an *efficient edge dominating* (*e.e.d.*) set for G if it is an efficient dominating set in the line graph $L(G)$ of G. The ED problem (EED problem, respectively) asks for the existence of an e.d. set (e.e.d. set, respectively) in the given graph.

We give a unified framework for investigating the complexity of these problems on various classes of graphs. In particular, we solve some open problems and give linear time algorithms for ED and EED on dually chordal graphs.

We extend the two problems to hypergraphs and show that ED remains NP-complete on α-acyclic hypergraphs, and is solvable in polynomial time on hypertrees, while EED is polynomial on α-acyclic hypergraphs and NP-complete on hypertrees.

Keywords: efficient domination, efficient edge domination, graphs and hypergraphs, polynomial time algorithms.

1 Introduction and Basic Notions

Packing and covering problems in graphs and their relationships belong to the most fundamental topics in combinatorics and graph algorithms and have a wide spectrum of applications in computer science, operations research and many other fields. Recently, there has been an increasing interest in problems combining packing and covering properties. Among them, there are the following variants of domination problems:

Let G be a finite simple undirected graph with vertex set V and edge set E. A vertex *dominates* itself and all its neighbors, i.e., every vertex $v \in V$ dominates its closed neighborhood $N[v] = \{u \mid u = v \text{ or } uv \in E\}$. A vertex set D in G is an *efficient dominating* (*e.d.*) set for G if for every vertex $v \in V$, there is exactly one $d \in D$ dominating v (sometimes called *independent perfect dominating set*) [1,2]. An edge set $M \subseteq E$ is an *efficient edge dominating* (*e.e.d.*) set for G if it is an

K.-M. Chao, T.-s. Hsu, and D.-T. Lee (Eds.): ISAAC 2012, LNCS 7676, pp. 267–277, 2012.
© Springer-Verlag Berlin Heidelberg 2012

efficient dominating set in the line graph $L(G)$ of G (sometimes called *dominating induced matching*). The ED problem (EED problem, respectively) asks for the existence of an e.d. set (e.e.d. set, respectively) in the given graph. Note that both problems are NP-complete. The complexity of ED (EED, respectively) (and their variants) with respect to special graph classes was studied in various papers; see e.g. [2,6,20,31,32,35,36,37,38] for ED and [9,12,18,27,33,34] for EED. The main contributions of our paper are:

(i)   a unified framework for the ED and EED problems solving some open questions,

(ii) linear time algorithms for ED and EED on dually chordal graphs, and

(iii) an extension of the two problems to hypergraphs, in particular to α-acyclic hypergraphs and hypertrees: We show that ED remains NP-complete on α-acyclic hypergraphs, and is solvable in polynomial time on hypertrees, while EED is polynomial on α-acyclic hypergraphs and NP-complete on hypertrees.

Our approach has the advantage that it unifies the proofs of various results obtained in numerous papers. Our proofs are typically very short since we extensively use some theoretical background on the relations of the considered graph and hypergraph classes, and in particular closure properties of graph classes with respect to squares of their graphs, and polynomial time algorithms for Maximum Weight Independent Set and Minimum Weight Dominating Set on some graph classes. The consequences are some new cases where the corresponding problems can be efficiently solved.

Due to space limitations, all proofs are omitted; see [11] for a full version.

2   Further Basic Notions

2.1   Basic Notions and Properties of Graphs

Let G be a finite undirected graph without loops and multiple edges. Let V denote its vertex (or node) set and E its edge set; let $|V| = n$ and $|E| = m$. A vertex v is *universal* in G if it is adjacent to all other vertices of G. Let $G^2 = (V, E \cup \{uv \mid \exists w \in V : uw, wv \in E\})$ be the square of G.

A *chordless path* P_k (*chordless cycle* C_k, respectively) has k vertices, say $v_1, \ldots, v_k$, and edges $v_i v_{i+1}$, $1 \le i \le k-1$ (and $v_k v_1$, respectively). A *hole* is a chordless cycle C_k for $k \ge 5$. G is *chordal* if no induced subgraph of G is isomorphic to C_k, $k \ge 4$. See e.g. [10] for the many facets of chordal graphs. A vertex set $U \subseteq V$ is *independent* if for all $x, y \in U$, $xy \notin E$ holds. For a graph G and a vertex weight function on G, let the MAXIMUM WEIGHT INDEPENDENT SET (MWIS) problem be the task of finding an independent vertex set of maximum weight.

Let K_i denote the clique with i vertices. Let $K_4 - e$ or *diamond* be the graph with four vertices and five edges, say vertices a, b, c, d and edges ab, ac, bc, bd, cd; ·its *mid-edge* is the edge bc. Let *gem* denote the graph consisting of five vertices, four of which induce a P_4, and the fifth is universal. Let W_4 denote the graph with five vertices consisting of a C_4 and a universal vertex.

For $U \subseteq V$, let $G[U]$ denote the subgraph induced by U. For a set $\mathcal{F}$ of graphs, a graph G is called $\mathcal{F}$-free if G contains no induced subgraph from $\mathcal{F}$. Thus, it is hole-free if it contains no induced subgraph isomorphic to a hole. A graph G is weakly chordal if G and its complement graph is hole-free. Three pairwise non-adjacent distinct vertices form an asteroidal triple (AT) in G if for each choice of two of them, there is a path between the two avoiding the neighborhood of the third. A graph G is AT-free if G contains no AT.

2.2   Basic Notions and Properties of Hypergraphs

Throughout this paper, a hypergraph $H = (V, \mathcal{E})$ has a finite vertex set V and for all $e \in \mathcal{E}$, $e \subseteq V$ ($\mathcal{E}$ possibly being a multiset). For a graph G, let $\mathcal{N}(G)$ denote the closed neighborhood hypergraph, i.e., $\mathcal{N}(G) = (V, \{N[v] \mid v \in V\})$, and let $\mathcal{C}(G) = (V, \{K \subseteq V \mid K$ is an inclusion-maximal clique$\})$ denote the clique hypergraph of G.

A subset of edges $\mathcal{E}' \subseteq \mathcal{E}$ is an exact cover of H if for all $e, f \in \mathcal{E}'$ with $e \neq f$, $e \cap f = \emptyset$ and $\bigcup \mathcal{E}' = V$. The EXACT COVER problem asks for the existence of an exact cover in a given hypergraph H. It is well known that this problem is NP-complete even for 3-element hyperedges (problem X3C [SP2] in [24]). Thus, the ED problem on a graph G is the same as the Exact Cover problem on $\mathcal{N}(G)$.

For defining the class of dually chordal graphs, whose properties will be contrasted with those of chordal graphs, as well as for extending the ED and the EED problems to hypergraphs, we need some basic definitions: For a hypergraph $H = (V, \mathcal{E})$, let $2sec(H)$ denote its 2-section (also called representative or primal) graph, i.e., the graph having the same vertex set V in which two vertices are adjacent if they are in a common hyperedge. Let $L(H)$ denote the line graph of H, i.e., the graph with the hyperedges $\mathcal{E}$ as its vertex set in which two hyperedges are adjacent in $L(H)$ if they intersect each other.

A hypergraph $H = (V, \mathcal{E})$ has the Helly property if the total intersection of every pairwise intersecting family of hyperedges of $\mathcal{E}$ is nonempty. H is conformal if every clique of the 2-section graph $2sec(H)$ is contained in a hyperedge of $\mathcal{E}$ (see e.g. [5,22]).

The notion of α-acyclicity [22] is one of the most important and most frequently studied hypergraph notions. Among the many equivalent conditions describing α-acyclic hypergraphs, we take the following: For a hypergraph $H = (V, \mathcal{E})$, a tree T with node set $\mathcal{E}$ and edge set E_T is a join tree of H if for all vertices $v \in V$, the set of hyperedges containing v induces a subtree of T. H is α-acyclic if it has a join tree. For a hypergraph $H = (V, \mathcal{E})$ and vertex $v \in V$, let $\mathcal{E}_v := \{e \in \mathcal{E} \mid v \in e\}$. Let $H^* := (\mathcal{E}, \{\mathcal{E}_v \mid v \in V\})$ be the dual hypergraph of H. $H = (V, \mathcal{E})$ is a hypertree if there is a tree T with vertex set V such that for all $e \in \mathcal{E}$, $T[e]$ is connected.

Theorem 1 (Duchet, Flament, Slater, see [10]). *H is a hypertree if and only if H has the Helly property and its line graph $L(H)$ is chordal.*

The following facts are well known:

Lemma 1. *Let H be a hypergraph.*

(i) H *is conformal if and only if H^* has the Helly property.*

(ii) $L(H)$ *is isomorphic to $2sec(H^*)$.*

Thus:

Corollary 1. *H is α-acyclic if and only if H is conformal and its 2-section graph is chordal.*

It is easy to see that the dual $\mathcal{N}(G)^*$ of $\mathcal{N}(G)$ is $\mathcal{N}(G)$ itself, and for any graph G:

$$G^2 \text{ is isomorphic to } L(\mathcal{N}(G)). \tag{1}$$

In [8], the notion of dually chordal graphs was introduced: For a graph $G = (V, E)$ and a vertex $v \in V$, a vertex $u \in N[v]$ is a *maximum neighbor* of v if for all $w \in N[v]$, $N[w] \subseteq N[u]$ holds. (Note that by this definition, a vertex can be its own maximum neighbor.) Let $\sigma = (v_1, \ldots, v_n)$ be a vertex ordering of V. Such an ordering σ is a *maximum neighborhood ordering* of G if for every $i \in \{1, 2, \ldots, n\}$, v_i has a maximum neighbor in $G_i := G[\{v_i, \ldots, v_n\}]$. A graph is *dually chordal* if it has a maximum neighborhood ordering. The following is known:

Theorem 2 ([7,8,21]). *Let G be a graph. Then the following are equivalent:*

(i) G *is a dually chordal graph.*

(ii) $\mathcal{N}(G)$ *is a hypertree.*

(iii) $\mathcal{C}(G)$ *is a hypertree.*

(iv) G *is the 2-section graph of some hypertree.*

Thus, Theorems 1 and 2 together with (1) and the duality properties in Lemma 1 imply:

Corollary 2 ([7,8,21]). *Let G be a graph and H be a hypergraph.*

(i) G *is dually chordal if and only if G^2 is chordal and $\mathcal{N}(G)$ has the Helly property.*

(ii) *If H is α-acyclic then its line graph $L(H)$ is dually chordal.*

(iii) *If H is a hypertree then its 2-section graph $2sec(H)$ is dually chordal.*

3   Efficient (Edge) Domination in General

Recall that a subset $D \subseteq V$ of vertices is an efficient dominating set if for all $v \in V$, there is exactly one $d \in D$ such that $v \in N[d]$. Also a subset $M \subseteq E$ of edges is an efficient edge dominating set in G if for all $e \in E$, there is exactly one $e' \in M$ intersecting the edge e.

Both definitions can be extended to hypergraphs: A subset $D \subseteq V$ is an *efficient dominating set* for a hypergraph H if it is an efficient dominating set for its 2-section graph $2sec(H)$. A subset $M \subseteq \mathcal{E}$ of hyperedges is an *efficient edge dominating set* for H if for all $e \in \mathcal{E}$, there is exactly one $e' \in M$ intersecting the edge e.

Corollary 3. *A vertex set D is an efficient dominating set in H if and only if D is an efficient edge dominating set in H^*.*

The following approach developed in [30] and [36] gives a tool for showing that for various classes of graphs, the ED problem can be solved in polynomial time. For a graph $G = (V, E)$, we define the following vertex weight function: Let $\omega(v) := |N_G[v]|$ (i.e., $\omega(v) := deg(v) + 1$), and for $D \subseteq V$, let $\omega(D) := \Sigma_{d \in D}\, \omega(d)$. Obviously, the following holds:

Proposition 1. *Let $G = (V, E)$ be a graph and $D \subseteq V$.*

   (i) *If D is a dominating vertex set in G then $\omega(D) \geq |V|$.*
   (ii) *If D is an independent vertex set in G^2 then $\omega(D) \leq |V|$.*

Lemma 2. *Let $G = (V, E)$ be a graph and $\omega(v) := |N[v]|$ a vertex weight function for G. Then the following are equivalent for any subset $D \subseteq V$:*

   (i) *D is an efficient dominating set in G*
   (ii) *D is a minimum weight dominating vertex set in G with $\omega(D) = |V|$.*
   (iii) *D is a maximum weight independent vertex set in G^2 with $\omega(D) = |V|$.*

Note that D is not any independent (dominating) set, but a maximum (minimum) weight one. This implies:

Corollary 4. *For every graph class $\mathcal{C}$ for which the MWIS problem is solvable in polynomial time on squares of graphs from $\mathcal{C}$, the ED problem for $\mathcal{C}$ is solvable in polynomial time.*

Corollary 5. *Let H be a hypergraph, $L(H) = (V, E)$ its line graph and $\omega(v) := |N[v]|$ a vertex weight function for $L(H)$ as above. Then the following are equivalent for any subset $D \subseteq V$:*

   (i) *D is an efficient edge dominating set in H*
   (ii) *D is an efficient dominating set in $L(H)$.*
   (iii) *D is a minimum weight dominating vertex set in $L(H)$ with $\omega(D) = |V|$.*
   (iv) *D is a maximum weight independent vertex set in $L(H)^2$ with $\omega(D) = |V|$.*

4   Efficient Domination in Graphs

This section presents results for the ED problem on some graph classes.

Theorem 3 ([35,38]). *The ED problem is $\mathbb{NP}$-complete for bipartite graphs, for chordal graphs as well as for chordal bipartite graphs.*

By Corollary 2 (i), the square of a dually chordal graph is chordal. Thus, based on Lemma 2, ED for dually chordal graphs can be solved in polynomial time by solving the MWIS problem on chordal graphs. However, the MWIS problem is solvable in linear time for chordal graphs with the following algorithm:

Algorithm 1 ([23]).
Input: A chordal graph $G = (V, E)$ with $|V| = n$ and a vertex weight function ω.
Output: A maximum weight independent set $\mathcal{I}$ of G.

(1) Find a perfect elimination ordering $(v_1, \ldots, v_n)$ and set $\mathcal{I} := \emptyset$.
(2) **For** $i := 1$ **To** n
    If $\omega(v_i) > 0$, mark v and set $\omega(u) := \max(\omega(u) - \omega(v_i), 0)$ for all vertices $u \in N(v_i)$.
(3) **For** $i := n$ **DownTo** 1
    If v_i is marked, set $\mathcal{I} := \mathcal{I} \cup \{v_i\}$ and unmark all $u \in N(v_i)$.

By using the following lemmas, the algorithm can be modified in such a way, that it solves the ED problem for dually chordal graphs in linear time.

Lemma 3 ([7]). *A maximum neighborhood ordering of G which simultaneously is a perfect elimination ordering of G^2 can be found in linear time.*

The algorithm given in [7] not only finds a maximum neighborhood ordering $(v_1, \ldots, v_n)$. It also computes the maximum neighbors m_i for each vertex v_i with the property that for all $i < n$ no vertex v_i is its own maximum neighbor $(v_i \neq m_i)$. This is necessary for the following lemma.

Lemma 4. *Let $G = (V, E)$ be a graph with $G^2 = (V, E^2)$ and a maximum neighborhood ordering $(v_1, \ldots, v_n)$ where m_i is the maximum neighbor of v_i with $v_i \neq m_i$ and $1 \leq i < j \leq n$. Then: $v_i v_j \in E^2 \Leftrightarrow m_i v_j \in E$.*

This allows to modify Algorithm 1 in a way, that it is no longer necessary to compute the square of the given dually chordal graph G. Instead, a maximum weight independent set of G^2 can be computed on G in linear time.

Algorithm 2 ([30]).
Input: A dually chordal graph $G = (V, E)$.
Output: An efficient dominating set D (if existing).

(1) $D = \emptyset$.
(2) **For All** $v \in V$
    Set $\omega(v) := |N(v)|$ and $\omega_p(v) := 0$. v is unmarked and not blocked.
(3) Find a maximum neighborhood ordering $(v_1, \ldots, v_n)$ with the corresponding maximum neighbors $(m_1, \ldots, m_n)$ where $v_i \neq m_i$ for $1 \leq i < n$.
(4) **For** $i := 1$ **To** n
    For all $u \in N[v_i]$ set $\omega(v_i) := \omega(v_i) - \omega_p(u)$.
    If $\omega(v_i) > 0$, mark v_i and set $\omega_p(m_i) := \omega_p(m_i) + \omega(v_i)$.
(5) **For** $i := n$ **DownTo** 1
    If v_i is marked and m_i is not blocked, set $D := D \cup \{v_i\}$ and block all $u \in N(v_i)$.
(6) D is an efficient dominating set if and only if $\sum_{v \in D} |N[v]| = |V|$.

Theorem 4. *Algorithm 2 works correctly and runs in linear time.*

Note that strongly chordal graphs are dually chordal [8]. In [35] one of the open problems is the complexity of (weighted) ED for strongly chordal graphs which is solved by Theorem 4 (for the weighted case see [30]).

Theorem 5. *For AT-free graphs, the ED problem is solvable in polynomial time.*

This partially extends the result of [20] showing that the (weighted) ED problem for co-comparability graphs is solvable in polynomial time.

In [35], one of the open problems is the complexity of ED for convex bipartite graphs. This class of graphs is contained in interval bigraphs, and a result of [29] shows that the boolean width of interval bigraphs is at most $2 \log n$, based on a corresponding result for interval graphs [3]. By a result of [4], this leads to a polynomial time algorithm for Minimum Weight Domination on interval bigraphs.

Corollary 6. *For interval bigraphs, the ED problem is solvable in polynomial time.*

This solves the open question from [35] for convex bipartite graphs.

5   Efficient Edge Domination in Graphs

Lemma 5 ([9,12]). *Let G be a graph that has an e.e.d. set M.*

  (i) *M contains exactly one edge of every triangle of G.*
  (ii) *G is K_4-free.*
  (iii) *If xy is the mid-edge of an induced diamond in G then M necessarily contains xy. Thus, in particular, G is W_4-free and gem-free.*

In [33], it was shown that the EED problem is solvable in linear time on chordal graphs. This allows us to solve the EED problem for dually chordal graphs using the following lemma:

Lemma 6. *Let G be a graph that has an e.e.d. set. Then G is chordal if and only if G is dually chordal.*

Corollary 7. *The EED problem can be solved in linear time for dually chordal graphs.*

Efficient edge dominating sets are closely related to maximum induced matchings; it is not hard to see that every efficient edge dominating set is a maximum induced matching but of course not vice versa. However, when the graph has an efficient edge dominating set and is regular then every maximum induced matching is an efficient edge dominating set [17]. On the other hand, the complexity of the two problems differs on some classes such as claw-free graphs where the Maximum Induced Matching (MIM) problem is NP-complete [28] (even on line graphs) while the EED problem is solvable in polynomial time [18]. While every graph has a maximum induced matching, this is not the case for efficient edge dominating sets. Thus, if the graph G has an efficient edge dominating set, this gives also a maximum induced matching but in the other case, the MIM problem is hard for claw-free graphs.

For the MIM problem, there is a long list of results of the following type: If a graph G is in a graph class $\mathcal{C}$ then also $L(G)^2$ is in the same class (see e.g. [15,16]), and if the MWIS problem is solvable in polynomial time for the same class, this leads to polynomial algorithms for the MIM problem on the class $\mathcal{C}$. A very large class of this type are interval-filament graphs [25] which include co-comparability graphs and polygon-circle graphs; the latter include circle graphs, circular-arc graphs, chordal graphs, and outerplanar graphs. AT-free graphs include co-comparability graphs, permutation graphs and trapezoid graphs (see [10]).

Theorem 6 ([15,19]). *Let $G = (V, E)$ be a graph, $L(G)$ its line graph, and $L(G)^2$ its square. Then the following conditions hold:*

(i) *If G is an interval-filament graph then $L(G)^2$ is an interval-filament graph.*
(ii) *If G is an AT-free graph then $L(G)^2$ is an AT-free graph.*

The MWIS problem for interval-filament graphs is solvable in polynomial time [25]; as a consequence, it is mentioned in [15] that the MIM problem is efficiently solvable for interval-filament graphs. In [13] it is shown that the MWIS problem is solvable for AT-free graphs in time $O(n^4)$. For the EED problem, it follows:

Corollary 8. *The EED problem is solvable in polynomial time for interval-filament graphs and for AT-free graphs.*

This generalizes the corresponding result for bipartite permutation graphs in [33]; bipartite permutation graphs are AT-free. It also generalizes a corresponding result for MIM on trapezoid graphs [26] (which are AT-free).

In [18], the complexity of the EED problem for weakly chordal graphs was mentioned as an open problem; in [9], however, it was shown that the EED problem (DIM problem, respectively) is solvable in polynomial time for weakly chordal graphs. It is easy to see that long antiholes (i.e., complements of C_k, $k \geq 6$) have no e.e.d. set, i.e., a hole-free graph having an e.e.d. set is weakly chordal. Thus, for hole-free graphs, the EED problem is solvable in polynomial time [9]. Corollary 5 leads to an easier way of solving the EED problem on weakly chordal graphs:

Corollary 9. *The EED problem is solvable in polynomial time for weakly chordal graphs.*

The next result contrasts to the fact that the MIM and the EED problem are solvable in polynomial time on chordal graphs:

Proposition 2. *The MIM problem is $\mathbb{NP}$-complete for dually chordal graphs.*

6   Some Results for Hypergraphs

Theorem 3 and the fact that every chordal graph is the 2-section graph of an α-acyclic hypergraph (namely, of its clique hypergraph $\mathcal{C}(G)$) implies:

Corollary 10. *The ED problem is NP-complete for α-acyclic hypergraphs.*

This situation is better for hypertrees:

Corollary 11. *For hypertrees, the ED problem is solvable in polynomial time.*

Based on Corollary 3 and the duality of hypertrees and α-acyclic hypergraphs it follows:

Corollary 12. *The EED problem for hypertrees is NP-complete.*

Corollary 13. *For α-acyclic hypergraphs, the EED problem is solvable in polynomial time.*

For the MIM problem we show:

Theorem 7. *The MIM problem is solvable in polynomial time for α-acyclic hypergraphs.*

Theorem 8. *The MIM problem for hypertrees is NP-complete.*

Theorem 9. *The Exact Cover problem is NP-complete for α-acyclic hypergraphs and solvable in polynomial time for hypertrees.*

7   Conclusion

The subsequent scheme summarizes some of our results; NP-c. means NP-complete, pol. (linear) means polynomial-time (linear-time) solvable, and XC means the Exact Cover problem.

	chordal gr.	dually chordal gr.	α-acyclic hypergr.	hypertrees
ED	NP-c. [38]	linear	NP-c.	pol.
EED	linear [33]	linear	pol.	NP-c.
MIM	pol. [14]	NP-c.	pol.	NP-c.
XC			NP-c.	pol.

Acknowledgement. The first author is grateful to J. Mark Keil and Haiko Müller for stimulating discussions and related results.

References

1. Bange, D.W., Barkauskas, A.E., Slater, P.J.: Efficient dominating sets in graphs. In: Ringeisen, R.D., Roberts, F.S. (eds.) Applications of Discrete Math., pp. 189–199. SIAM, Philadelphia (1988)
2. Bange, D.W., Barkauskas, A.E., Host, L.H., Slater, P.J.: Generalized domination and efficient domination in graphs. Discrete Math. 159, 1–11 (1996)

3. Belmonte, R., Vatshelle, M.: Graph Classes with Structured Neighborhoods and Algorithmic Applications. In: Kolman, P., Kratochvíl, J. (eds.) WG 2011. LNCS, vol. 6986, pp. 47–58. Springer, Heidelberg (2011)
4. Bui-Xuan, B.-M., Telle, J.A., Vatshelle, M.: Boolean width of graphs. Theor. Computer Science 412, 5187–5204 (2011)
5. Berge, C.: Graphs and Hypergraphs. North-Holland (1973)
6. Biggs, N.: Perfect codes in graphs. J. of Combinatorial Theory (B) 15, 289–296 (1973)
7. Brandstädt, A., Chepoi, V.D., Dragan, F.F.: The algorithmic use of hypertree structure and maximum neighbourhood orderings. Discrete Applied Math. 82, 43–77 (1998)
8. Brandstädt, A., Dragan, F.F., Chepoi, V.D., Voloshin, V.I.: Dually chordal graphs. SIAM J. Discrete Math. 11, 437–455 (1998)
9. Brandstädt, A., Hundt, C., Nevries, R.: Efficient Edge Domination on Hole-Free Graphs in Polynomial Time. In: López-Ortiz, A. (ed.) LATIN 2010. LNCS, vol. 6034, pp. 650–661. Springer, Heidelberg (2010)
10. Brandstädt, A., Le, V.B., Spinrad, J.P.: Graph Classes: A Survey. SIAM Monographs on Discrete Math. Appl., vol. 3. SIAM, Philadelphia (1999)
11. Brandstädt, A., Leitert, A., Rautenbach, D.: Efficient Dominating and Edge Dominating Sets for Graphs and Hypergraphs. Technical report CoRR, arXiv:1207.0953v2, cs.DM (2012)
12. Brandstädt, A., Mosca, R.: Dominating Induced Matchings for P_7-free Graphs in Linear Time. In: Asano, T., Nakano, S.-i., Okamoto, Y., Watanabe, O. (eds.) ISAAC 2011. LNCS, vol. 7074, pp. 100–109. Springer, Heidelberg (2011)
13. Broersma, H.J., Kloks, T., Kratsch, D., Müller, H.: Independent sets in asteroidal-triple-free graphs. SIAM J. Discrete Math. 12, 276–287 (1999)
14. Cameron, K.: Induced matchings. Discrete Applied Math. 24, 97–102 (1989)
15. Cameron, K.: Induced matchings in intersection graphs. Discrete Mathematics 278, 1–9 (2004)
16. Cameron, K., Sritharan, R., Tang, Y.: Finding a maximum induced matching in weakly chordal graphs. Discrete Math. 266, 133–142 (2003)
17. Cardoso, D.M., Cerdeira, J.O., Delorme, C., Silva, P.C.: Efficient edge domination in regular graphs. Discrete Applied Math. 156, 3060–3065 (2008)
18. Cardoso, D.M., Korpelainen, N., Lozin, V.V.: On the complexity of the dominating induced matching problem in hereditary classes of graphs. Discrete Applied Math. 159, 521–531 (2011)
19. Chang, J.-M.: Induced matchings in asteroidal-triple-free graphs. Discrete Applied Math. 132, 67–78 (2003)
20. Chang, G.J., Pandu Rangan, C., Coorg, S.R.: Weighted independent perfect domination on co-comparability graphs. Discrete Applied Math. 63, 215–222 (1995)
21. Dragan, F.F., Prisacaru, C.F., Chepoi, V.D.: Location problems in graphs and the Helly property. Discrete Mathematics, Moscow 4, 67–73 (1992) (in Russian); The full version appeared as preprint: Dragan, F.F., Prisacaru, C.F., Chepoi, V.D.: r-Domination and p-center problems on graphs: special solution methods and graphs for which this method is usable, Kishinev State University, preprint MoldNIINTI, N. 948–M88 (1987) (in Russian)
22. Fagin, R.: Degrees of Acyclicity for Hypergraphs and Relational Database Schemes. Journal ACM 30, 514–550 (1983)
23. Frank, A.: Some polynomial algorithms for certain graphs and hypergraphs. In: Proceedings of the 5th British Combinatorial Conf. (Aberdeen 1975). Congressus Numerantium, vol. XV, pp. 211–226 (1976)

24. Garey, M.R., Johnson, D.S.: Computers and Intractability – A Guide to the Theory of NP-completeness. Freeman, San Francisco (1979)
25. Gavril, F.: Maximum weight independent sets and cliques in intersection graphs of filaments. Information Processing Letters 73, 181–188 (2000)
26. Golumbic, M.C., Lewenstein, M.: New results on induced matchings. Discrete Applied Math. 101, 157–165 (2000)
27. Grinstead, D.L., Slater, P.L., Sherwani, N.A., Holmes, N.D.: Efficient edge domination problems in graphs. Information Processing Letters 48, 221–228 (1993)
28. Kobler, D., Rotics, U.: Finding maximum induced matchings in subclasses of claw-free and P_5-free graphs, and in graphs with matching and induced matching of equal maximum size. Algorithmica 37, 327–346 (2003)
29. Keil, J.M.: The dominating set problem in interval bigraphs, abstract. In: Proceedings of the 3rd Annual Workshop on Algorithmic Graph Theory, Nipissing University, North Bay, Ontario (2012)
30. Leitert, A.: Das Dominating Induced Matching Problem für azyklische Hypergraphenl. Diploma Thesis, University of Rostock, Germany (2012)
31. Liang, Y.D., Lu, C.L., Tang, C.Y.: Efficient Domination on Permutation Graphs and Trapezoid Graphs. In: Jiang, T., Lee, D.T. (eds.) COCOON 1997. LNCS, vol. 1276, pp. 232–241. Springer, Heidelberg (1997)
32. Lin, Y.-L.: Fast Algorithms for Independent Domination and Efficient Domination in Trapezoid Graphs. In: Chwa, K.-Y., Ibarra, O.H. (eds.) ISAAC 1998. LNCS, vol. 1533, pp. 267–275. Springer, Heidelberg (1998)
33. Lu, C.L., Ko, M.-T.; Tang, C.Y.: Perfect edge domination and efficient edge domination in graphs. Discrete Applied Math. 119, 227–250 (2002)
34. Lu, C.L., Tang, C.Y.: Solving the weighted efficient edge domination problem on bipartite permutation graphs. Discrete Applied Math. 87, 203–211 (1998)
35. Lu, C.L., Tang, C.Y.: Weighted efficient domination problem on some perfect graphs. Discrete Applied Math. 117, 163–182 (2002)
36. Milanič, M.: A hereditary view on efficient domination, extended abstract. In: Proceedings of the 10th Cologne-Twente Workshop, pp. 203–206 (2011); Full version to appear under the title "Hereditary efficiently dominatable graphs"
37. Yen, C.-C.: Algorithmic aspects of perfect domination. Ph.D. Thesis, Institute of Information Science, National Tsing Hua University, Taiwan (1992)
38. Yen, C.-C., Lee, R.C.T.: The weighted perfect domination problem and its variants. Discrete Applied Math. 66, 147–160 (1996)

On the Hyperbolicity of Small-World
and Tree-Like Random Graphs

Wei Chen[1], Wenjie Fang[2], Guangda Hu[3], and Michael W. Mahoney[4]

[1] Microsoft Research Asia
weic@microsoft.com
[2] Ecole Normale Supérieure de Paris
Wenjie.Fang@ens.fr
[3] Princeton University
guangdah@cs.princeton.edu
[4] Stanford University
mmahoney@cs.stanford.edu

Abstract. Hyperbolicity is a property of a graph that may be viewed as being a "soft" version of a tree, and recent empirical and theoretical work has suggested that many graphs arising in Internet and related data applications have hyperbolic properties. Here, we consider Gromov's notion of δ-hyperbolicity, and we establish several positive and negative results for small-world and tree-like random graph models. In particular, we show that small-world random graphs built from underlying grid structures do not have strong improvement in hyperbolicity, even when the rewiring greatly improves decentralized navigation. On the other hand, for a class of tree-like graphs called ringed trees that have constant hyperbolicity, adding random links among the leaves in a manner similar to the small-world graph constructions may easily destroy the hyperbolicity of the graphs, except for a class of random edges added using an exponentially decaying probability function based on the ring distance among the leaves. Our study provides the first significant analytical results on the hyperbolicity of a rich class of random graphs, which shed light on the relationship between hyperbolicity and navigability of random graphs, as well as on the sensitivity of hyperbolic δ to noises in random graphs.

Keywords: Graph hyperbolicity, complex networks, small-world networks, random graphs, decentralized navigation.

1 Introduction

Hyperbolicity, a property of metric spaces that generalizes the idea of Riemannian manifolds with negative curvature, has received considerable attention in both mathematics and computer science. When applied to graphs, one may think of hyperbolicity as characterizing a "soft" version of a tree—trees have hyperbolicity zero, and graphs that "look like" trees in terms of their metric structure have "small" hyperbolicity. Since trees are an important class of graphs and since tree-like graphs arise in numerous applications, the idea of hyperbolicity has received attention in a range of applications. For example, it has found usefulness in the visualization of the Internet, the Web, and other large

K.-M. Chao, T.-s. Hsu, and D.-T. Lee (Eds.): ISAAC 2012, LNCS 7676, pp. 278–288, 2012.

graphs [22,26,31]; it has been applied to questions of compact routing, navigation, and decentralized search in Internet graphs and small-world social networks [11,19,1,20,8]; and it has been applied to a range of other problems such as distance estimation, sensor networks, and traffic flow and congestion minimization [2,13,27,10].

The hyperbolicity of graphs is typically measured by Gromov's hyperbolic δ [12,4] (see Section 2). The hyperbolic δ of a graph measures the "tree-likeness" of the graph in terms of the graph distance metric. It can range from 0 up to the half of the graph diameter, with trees having $\delta = 0$, in contrast of "circle graphs" and "grid graphs" having large δ equal to roughly half of their diameters.

In this paper, we study the δ-hyperbolicity of families of random graphs that intuitively have some sort of tree-like or hierarchical structure. Our motivation comes from two angles. First, although there are a number of empirical studies on the hyperbolicity of real-world and random graphs [2,13,24,23,27,10], there are essentially no systematic analytical study on the hyperbolicity of popular random graphs. Thus, our work is intended to fill this gap. Second, a number of algorithmic studies show that good graph hyperbolicity leads to efficient distance labeling and routing schemes [6,11,9,7,21,8], and the routing infrastructure of the Internet is also empirically shown to be hyperbolic [2]. Thus, it is interesting to further investigate if efficient routing capability implies good graph hyperbolicity.

To achieve our goal, we first provide fine-grained characterization of δ-hyperbolicity of graph families relative to the graph diameter: A family of random graphs is (a) *constantly hyperbolic* if their hyperbolic δ's are constant, regardless of the size or diameter of the graphs; (b) *logarithmically (or polylogarithmically) hyperbolic* if their hyperbolic δ's are in the order of logarithm (or polylogarithm) of the graph diameters; (c) *weakly hyperbolic* if their hyperbolic δ's grow asymptotically slower than the graph diameters; and (d) *not hyperbolic* if their hyperbolic δ's are at the same order as the graph diameters.

We study two families of random graphs. The first family is Kleinberg's grid-based small-world random graphs [16], which build random long-range edges among pairs of nodes with probability inverse proportional to the γ-th power of the grid distance of the pairs. Kleinberg shows that when γ equals to the grid dimension d, decentralized routing can be improved from $\Theta(n)$ in grid to $O(\text{polylog}(n))$, where n is the number of vertices in the graph. Contrary to the improvement in decentralized routing, we show that when $\gamma = d$, with high probability the small-world graph is not polylogarithmically hyperbolic. We further show that when $0 \leq \gamma < d$, the random small-world graphs is not hyperbolic and when $\gamma > 3$ and $d = 1$, the random graphs is not polylogarithmically hyperbolic. Although there still exists a gap between hyperbolic δ and graph diameter at the sweetspot of $\gamma = d$, our results already indicate that long-range edges that enable efficient navigation do not significantly improve the hyperbolicity of the graphs.

The second family of graphs is random *ringed trees*. A ringed tree is a binary tree with nodes in each level of the tree connected by a ring (Figure 1(d)). Ringed trees can be viewed as an idealized version of hierarchical structure with local peer connections, such as the Internet autonomous system (AS) topology. We show that ringed tree is quasi-isometric to the Poincaré disk, the well known hyperbolic space representation, and thus it is constantly hyperbolic. We then study how random additions of long-range links on the leaves of a ringed tree affect the hyperbolicity of random ringed trees. Note that due to the tree base structure, random ringed trees allow efficient routing within $O(\log n)$ steps using tree branches. Our results show that if the random long-range edges between leaves are added according to a probability function that decreases exponentially fast with the ring distance between leaves, then the resulting random graph is logarithmically hyperbolic, but if the probability function decreases only as a power-law with ring distance, or based on another tree distance measure similar to [17], the resulting random graph is not hyperbolic. Furthermore, if we use binary trees instead of ringed trees as base graphs, none of the above versions is hyperbolic. Taken together, our results indicate that δ-hyperbolicity of graphs is quite sensitive to both base graph structures and probabilities of long-range connections.

To summarize, we provide the first significant analytical results on the hyperbolicity properties of important families of random graphs. Our results demonstrate that efficient routing performance does not necessarily mean good graph hyperbolicity (such as logarithmic hyperbolicity).

Related Work. There has been a lot of work on decentralized search subsequent to Kleinberg's original work [16,17], much of which has been summarized in the review [18]. In a parallel with this, there has been empirical and theoretical work on hyperbolicity of real-world complex networks as well as simple random graph models. On the empirical side, [2] showed that measurements of the Internet are negatively curved; [13,24,23] provided empirical evidence that randomized scale-free and Internet graphs are more hyperbolic than other types of random graph models; [27] measured the average δ and related curvature to congestion; and [10] measured treewidth and hyperbolicity properties of the Internet. However, on theoretical analysis of δ-hyperbolicity, the only prior work we are aware of is [28], which proves that with non-zero probability extremely sparse Erdős-Rényi random graphs are not δ-hyperbolic for any positive constant δ.

There are a number of works that connect graph hyperbolicity with efficient distance labeling and routing schemes [6,11,9,7,21,8]. Understanding the relationship between graph hyperbolicity and the ability of efficient routing is one motivation of our research. Our analytical results show, however, that the ability of efficient routing does not necessarily mean low hyperbolicity δ.

Ideas related to hyperbolicity have been applied in numerous other networks applications, e.g., to problems such as distance estimation, sensor networks, and traffic flow and congestion minimization [30,14,15,27,3], as well as large-scale data visualization [22,26,31]. The latter applications typically take important advantage of the idea that data are often hierarchical or tree-like and that there is "more room" in hyperbolic spaces of a given dimension than corresponding Euclidean spaces.

The full version of this conference paper, including detailed proofs and additional results, is available as the technical report [5].

2   Preliminaries on Hyperbolic Spaces and Graphs

We provide basic concepts concerning hyperbolic spaces and graphs used in this paper. For more comprehensive coverage on hyperbolic spaces, see, e.g., [4].

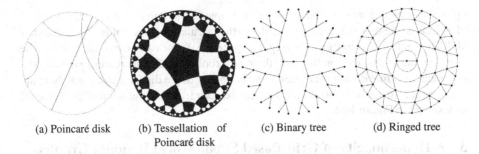

| (a) Poincaré disk | (b) Tessellation of Poincaré disk | (c) Binary tree | (d) Ringed tree |

Fig. 1. Poincaré disk, its tessellation, a binary tree, and a ringed tree

2.1 Gromov's δ-Hyperbolicity

In this paper, we use the following four-point condition originally introduced by Gromov [12] as the hyperbolicity measure of a metric space.

Definition 1 (Gromov's four-point condition). *In a metric space (X, d), given u, v, w, x with $d(u, v) + d(w, x) \geq d(u, x) + d(w, v) \geq d(u, w) + d(v, x)$ in X, we denote $\delta(u, v, w, x) = (d(u, v) + d(w, x) - d(u, x) - d(w, v))/2$. (X, d) is called δ-hyperbolic for some non-negative real number δ if for any four points $u, v, w, x \in X$, $\delta(u, v, w, x) \leq \delta$. Let $\delta(X, d)$ be the smallest possible value of such δ, which can also be defined as $\delta(X, d) = \sup_{u,v,w,x \in X} \delta(u, v, w, x)$.*

An undirected, unweighted and connected graph $G = (V, E)$ can be viewed as a metric space (V, d_G) with the standard graph distance metric d_G. We then apply the four-point condition defined above to define the δ-*hyperbolicity* of graph G, denoted as $\delta(G) = \delta(V, d_G)$. Trees are 0-hyperbolic, and it is often helpful to view graphs with a low hyperbolic δ as tree-like when viewed at large-size scales.

Let $D(G)$ denote the diameter of the graph G. By the triangle inequality, we have $\delta(G) \leq D(G)/2$. We will use the asymptotic difference between the hyperbolicity $\delta(G)$ and the diameter $D(G)$ to characterize the hyperbolicity of the graph G.

Definition 2 (Hyperbolicity of a graph). *For a family of graphs $\mathcal{G}$ with diameter $D(G), G \in \mathcal{G}$ going to infinity, we say that graph family $\mathcal{G}$ is constantly (resp. logarithmically, polylogarithmically, or weakly) hyperbolic, if $\delta(G) = O(1)$ (resp. $O(\log D(G))$, $O((\log D(G))^c)$ for some constant $c > 0$, or $o(D(G))$) when $D(G)$ goes to infinity; and $\mathcal{G}$ is not hyperbolic if $\delta(G) = \Theta(D(G))$, where $G \in \mathcal{G}$.*

The above definition provides more fine-grained characterization of hyperbolicity of graph families than one typically sees in the literature, which only discusses whether or not a graph family is constantly hyperbolic.

2.2 Poincaré Disk

The Poincaré disk is a well-studied hyperbolic metric space. In this paper, we use the Poincaré disk to mainly convey some intuition about hyperbolicity and tree-like behaviors, and thus we defer its technical definition to [5]. Visually, the Poincaré disk is an open disk with unit radius, and a (hyperbolic) line in the Poincaré disk is the segment of a circle in the disk that is perpendicular to the circular boundary of the disk, and thus all lines bend inward towards the origin (Figure 1(a)). For two points maintaining the same Euclidean distance on the disk, their hyperbolic distance increases exponentially fast when they move from the center to the boundary of the disk, meaning that there are "more room" towards the boundary. This can be seen from a tessellation of the Poincaré disk shown in Figure 1(b).

3 δ-Hyperbolicity of Grid-Based Small-World Random Graphs

In this section, we consider the δ-hyperbolicity of random graphs constructed according to the small-world graph model of Kleinberg [16], in which long-range edges are added on top of a base grid, which is a discretization of a low-dimensional Euclidean space. The model starts with n vertices forming a d-dimensional base grid (with wrap-around). More precisely, given positive integers n and d such that $n^{1/d}$ is also an integer, let $B = (V, E)$ be the base grid, with $V = \{(x_1, x_2, \ldots, x_d) \mid x_i \in \{0, 1, \ldots, n^{1/d} - 1\}, i \in [d]\}$, $E = \{((x_1, x_2, \ldots, x_d), (y_1, y_2, \ldots, y_d)) \mid \exists j \in [d], y_j = x_j + 1 \bmod n^{1/d} \text{ or } y_j = x_j - 1 \bmod n^{1/d}, \forall i \neq j, y_i = x_i\}$. Let d_B denote the graph distance metric on the base grid B. We then build a random graph G on top of B, such that G contains all vertices and all edges (referred to as grid edges) of B, and for each node $u \in V$, it has one long-range edge (undirected) connected to some node $v \in V$, with probability proportional to $1/d_B(u, v)^\gamma$, where $\gamma \geq 0$ is a parameter. We refer to the probability space of these random graphs as $KSW(n, d, \gamma)$; and we let $\delta(KSW(n, d, \gamma))$ denote the random variable of the hyperbolic δ of a randomly picked graph G in $KSW(n, d, \gamma)$. Recall that Kleinberg showed that the small-world graphs with $\gamma = d$ allow efficient decentralized routing (with $O(\log^2 n)$ routing hops in expectation), whereas graphs with $\gamma \neq d$ do not allow any efficient decentralized routing (with $\Omega(n^c)$ routing hops for some constant c) [16]; and note that the base grid B has large hyperbolic δ, i.e., $\delta(B) = \Theta(n^{1/d}) = \Theta(D(B))$. Intuitively, the structural reason for the efficient routing performance at $\gamma = d$ is that long-range edges are added "hierarchically" such that each node's long-range edges are nearly uniformly distributed over all "distance scales".

Results and Their Implications. The following theorem summarizes our main technical results on the hyperbolicity of small-world graphs.

Theorem 1. *With probability* $1 - o(1)$ *(when n goes to infinity), we have*

1. $\delta(KSW(n, d, \gamma)) = \Omega((\log n)^{\frac{1}{1.5(d+1)+\varepsilon}})$ *when $d \geq 1$ and $\gamma = d$, for any $\varepsilon > 0$ independent of n;*
2. $\delta(KSW(n, d, \gamma)) = \Omega(\log n)$ *when $d \geq 1$ and $0 \leq \gamma < d$; and*
3. $\delta(KSW(n, d, \gamma)) = \Omega(n^{\frac{\gamma-2}{\gamma-1}-\epsilon})$ *when $d = 1$ and $\gamma > 3$, for any $\epsilon > 0$ independent of n.*

This theorem, together with the results of [16] on the navigability of small-world graphs, have several implications. The first result shows that when $\gamma = d$, with high probability the hyperbolic δ of the small-world graphs is at least $c(\log n)^{\frac{1}{1.5(d+1)}}$ for some constant c. We know that the diameter is $\Theta(\log n)$ in expectation when $\gamma = d$ [25]. Thus the small-world graphs at the sweetspot for efficient routing is not polylogarithmically hyperbolic, i.e., δ is not $O(\log^c \log n)$-hyperbolic for any constant $c > 0$. However, there is still a gap between our lower bound and the upper bound provided by the diameter, and thus it is still open whether small-world graphs are weakly hyperbolic or not hyperbolic. Overall, though, our result indicates no drastic improvement on the hyperbolicity (relative to the improvement of the diameter) for small-world graphs at the sweetspot (where a dramatic improvement was obtained for the efficiency of decentralized routing).

The second result shows that when $\gamma < d$, then $\delta = \Omega(\log n)$. The diameter of the graph in this case is $\Theta(\log n)$ [25]; thus, we see that when $\gamma < d$ the hyperbolic δ is asymptotically the same as the diameter, i.e., although δ decreases as edges are added, small-world graphs in this range are not hyperbolic. The third result concerns the case $\gamma > d$, in which case the random graph degenerates towards the base grid (in the sense that most of the long-range edges are very local), which itself is not hyperbolic. For the general γ, we show that for the case of $d = 1$ the hyperbolic δ is lower bounded by a (low-degree) polynomial of n; this also implies that the graphs in this range are not polylogarithmically hyperbolic. Our polynomial exponent $\frac{\gamma-2}{\gamma-1} - \epsilon$ matches the diameter lower bound proven in [29].

Outline of the Proof of Theorem 1. In our analysis, we use two different techniques, one for the first two results in Theorem 1, and the other for the last result. For the first two results, we further divide the analysis into two cases $d \geq 2$ and $d = 1$.

When $d \geq 2$ and $0 \leq \gamma \leq d$, we first pick an arbitrary square grid with ℓ_0 nodes on each side. We know that when only grid distance is considered, the four corners of the square grid have the Gromov δ value equal to ℓ_0. We will show that, as long as ℓ_0 is not very large (to be exact, $O((\log n)^{\frac{1}{1.5(d+1)+\epsilon}})$ when $\gamma = d$ and $O(\log n)$ when $0 \leq \gamma < d$), the probability that any pair of vertices on this square grid have a shortest path shorter than their grid distance after adding long-range edges is close to zero. Therefore, with high probability, the four corners selected have Gromov δ as desired in the lower bound results.

To prove this result, we study the probability that any pair of vertices u and v at grid distance ℓ are connected with a path that contains at least one long-range edge and has length at most ℓ. We upper bound such ℓ's so that this probability is close to zero. To do so, we first classify such paths into a number of categories, based on the pattern of paths connecting u and v: how it alternates between grid edges and long-range edges, and the direction on each dimension of the grid edges and long-range edges (i.e., whether it is the same direction as from u to v in this dimension, or the opposite direction, or no move in this dimension). We then bound the probability of existing a path in each category and finally bound all such paths in aggregate. The most difficult part of the analysis is the bounding of the probability of existing a path in each category.

For the case of $d = 1$ and $0 \leq \gamma \leq d$, the general idea is similar to the above. The difference is that we do not have a base square to start with. Instead, we find a base ring of length $\Theta(\ell_0)$ using one long-range edges e_0, where ℓ_0 is fixed to be the same as the case of $d \geq 2$. We show that with high probability, (a) such an edge e_0 exists, and (b) the distance of any two vertices on the ring is simply their ring distance. This is enough to show the lower bound on the hyperbolic δ.

For the case of $\gamma > 3$ and $d = 1$, a different technique is used to prove the lower bound on hyperbolic δ. We first show that, in this case, with high probability all long-range edges only connect two vertices with ring distance at most some $\ell_0 = o(\sqrt{n})$. Next, on the one dimensional ring, we first find two vertices A and B at the two opposite ends on the ring. Then we argue that there must be a path $\mathcal{P}^+_{AB}$ that only goes through the clockwise side of ring from A to B, while another path $\mathcal{P}^-_{AB}$ that only goes through the counter-clockwise side of the ring from A to B, and importantly, the shorter length of these two paths are at most $O(\ell_0)$ longer than the distance between A and B. We then pick the middle point C and D of $\mathcal{P}^+_{AB}$ and $\mathcal{P}^-_{AB}$, respectively, and argue that the δ value of the four points A, B, C, and D give the desired lower bound.

Extensions to Other Models. We further study several extensions of the *KSW* model, including base grid without wrap-around, constant number of long-range links per node, and independent linking probabilities of each edge. We show that Theorem 1 still holds in all these models (except the case of $d = 1$ and $\gamma > 3$ for the grid with no wrap-around extension) and their combinations.

4 δ-Hyperbolicity of Ringed Trees

In this section, we consider the δ-hyperbolicity of graphs constructed according to a variant of the small-world graph model, in which long-range edges are added on top of a base binary tree or tree-like low-δ graph. In particular, we consider as based graphs both binary trees (Figure 1(c)) and ringed trees (Figure 1(d)), which contain concentric rings connecting all nodes in the same level of the binary tree, and adding long range links on these base graphs. The ringed tree is formally defined as follows.

Definition 3 (Ringed tree). *A* ringed tree *of level k, denoted $RT(k)$, is a fully binary tree with k levels (counting the root as a level), in which all vertices at the same level are connected by a ring. More precisely, we can use a binary string to represent each vertex in the tree, such that the root (at level 0) is represented by an empty string, and the left child and the right child of a vertex with string σ are represented as $\sigma 0$ and $\sigma 1$, respectively. Then, at each level $i = 1, 2, \ldots, k - 1$, we connect two vertices u and v represented by binary strings σ_u and σ_v if $(\sigma_u + 1) \mod 2^i = \sigma_v$, where the addition treats the binary strings as the integers they represent. As a convention, we say that a level is higher if it has a smaller level number and thus is closer to the root.*

Note that the diameter of the ringed tree $RT(k)$ is $\Theta(\log n)$, where $n = 2^k - 1$ is the number of vertices in $RT(k)$, and we will use $RT(\infty)$ to denote the infinite ringed tree when k in $RT(k)$ goes to infinity. Thus, a ringed tree may be thought of as a soft version of a binary tree. To some extent, a ringed tree can also be viewed as an idealized picture reflecting the hierarchical structure in real networks coupled with local neighborhood connections, such as Internet autonomous system (AS) networks, which has both a hierarchical structure of different level of AS'es, and peer connections based on geographical proximity.

Results and Their Implications. A visual comparison of the ringed tree of Figure 1(d) with the tessellation of Poincaré disk (Figure 1(b)) suggests that the ringed tree can been seen as an approximate tessellation or coarsening of the Poincaré disk, just as a two-dimensional grid can been seen as a coarsening of a two dimensional Euclidean space. Quasi-isometry is a technical concept making it precise what coarsening means. We show that the infinite ringed tree $RT(\infty)$ is indeed quasi-isometric to the Poincaré disk. This also implies that ringed tree $RT(k)$ for any k is constantly hyperbolic (technical definition of quasi-isometry and the above results are included in [5]).

We now consider random ringed trees constructed by adding random edges between two vertices at the outermost level, i.e., level $k-1$, such that the probability connecting two vertices u and v is determined by a function $g(u, v)$. Let V_{k-1} denote the set of vertices at level $k - 1$. Given a real-valued positive function $g(u, v)$, let $RRT(k, g)$ denote a random ringed tree constructed as follows. We start with the ringed tree $RT(k)$, and then for each vertex $v \in V_{k-1}$, we add one long-range edge to a vertex u with probability proportional to $g(u, v)$, that is, with probability $g(u, v)\rho_v^{-1}$ where $\rho_v = \sum_{u \in V_{k-1}} g(u, v)$.

We study three families of functions g, each of which has the characteristic that vertices closer to one another (by some measure) are more likely to be connected by a long-range edge. The first two families use the ring distance $d_R(u, v)$ as the closeness measure: the first family uses an exponential decay function $g_1(u, v) = e^{-\alpha d_R(u,v)}$, and the second family uses a power-law decay function $g_2(u, v) = d_R(u, v)^{-\alpha}$, where $\alpha > 0$. The third family uses the height of the lowest common ancestor of u and v, denoted as $h(u, v)$, as the closeness measure, and the function is $g_3 = 2^{-\alpha h(u,v)}$. Note that this last probability function matches the function used in a tree-based small-world model of Kleinberg [17]. The following theorem summarizes the hyperbolicity behavior of these three families of random ringed trees.

Theorem 2. *Considering the follow families of functions (with u and v as the variables of the function) for random ringed trees $RRT(k, g)$, for any positive integer k and positive real number α, with probability $1 - o(1)$ (when n tends to infinity), we have*

1. $\delta(RRT(k, e^{-\alpha d_R(u,v)})) = O(\log \log n)$;
2. $\delta(RRT(k, d_R(u, v)^{-\alpha})) = \Theta(\log n)$;
3. $\delta(RRT(k, 2^{-\alpha h(u,v)})) = \Theta(\log n)$;

where $n = 2^k - 1$ is the number of vertices in the ringed tree $RT(k)$.

Theorem 2 states that, when the random long-range edges are selected using exponential decay function based on the ring distance measure, the resulting graph is logarithmically hyperbolic, i.e., the constant hyperbolicity of the original base graph is degraded only slightly; but when a power-law decay function based on the ring distance measure or an exponential decay function based on common ancestor measure is used, then hyperbolicity is destroyed and the resulting graph is not hyperbolic. Intuitively, when it is more likely for a long-range edge to connect two far-away vertices, such an edge creates a shortcut for many internal tree nodes so that many shortest paths will go through this shortcut instead of traversing through tree nodes. In Internet routing paths going through such shortcuts are referred to as *valley routes*.

As a comparison, we also study the hyperbolicity of random binary trees $RBT(k, g)$, which are the same as random ringed trees $RRT(k, g)$ except that we remove all ring edges.

Theorem 3. *Considering the follow families of functions (with u and v as the variables of the function) for random binary trees $RBT(k, g)$, for any positive integer k and positive real number α, with probability $1 - o(1)$ (when n tends to infinity), we have*

$$\delta(RBT(k, e^{-\alpha d_R(u,v)})) = \delta(RBT(k, d_R(u,v)^{-\alpha})) = \delta(RBT(k, 2^{-\alpha h(u,v)})) = \Theta(\log n),$$

where $n = 2^k - 1$ is the number of vertices in the binary tree $RBT(k, g)$.

Thus, in this case, the original hyperbolicity of the base graph ($\delta = 0$ for the binary tree) is destroyed. Comparing with Theorem 2, our results above suggest that the "softening" of the hyperbolicity provided by the rings is essential in maintaining good hyperbolicity: with rings, random ringed trees with exponential decay function (depending on the ringed distance) are logarithmically hyperbolic, but without the rings the resulting graphs are not hyperbolic.

Extensions of the Random Ringed Tree Model. We further show that all our results in this section apply to extended models that allow a constant number of long-range edges per node, or independent selection of long-range edges for each node, or both.

5   Discussions and Open Problems

Perhaps the most obvious extension of our results is to close the gap in the bounds on the hyperbolicity in the low-dimensional small-world model when γ is at the "sweetspot", as well as extending the results for large γ to dimensions $d \geq 2$. Also of interest is characterizing in more detail the hyperbolicity properties of other random graph models, in particular those that have substantial heavy-tailed properties. Finally, exact computation of δ by its definition takes $O(n^4)$ time, which is not scalable to large graphs, and thus the design of more efficient exact or approximation algorithms would be of interest.

From a broader perspective, however, our results suggest that δ is a measure of tree-like-ness that can be quite sensitive to noise in graphs, and in particular to randomness as it is implemented in common network generative models. Moreover, our results for the δ hyperbolicity of rewired trees versus rewired low-δ tree-like metrics suggest that, while quite appropriate for continuous negatively-curved manifolds, the usual definition of δ may be somewhat less useful for discrete graphs. Thus, it would be of interest to address questions such as: does there exist a measure other than Gromov's δ that is more appropriate for graph-based data or more robust to noise/randomness as it is used in popular network generation models; is it possible to incorporate in a meaningful way nontrivial randomness in other low δ-hyperbolicity graph families; and can we construct non-trivial random graph families that contain as much randomness as possible while having low δ-hyperbolicity comparing to graph diameter?

References

1. Abraham, I., Balakrishnan, M., Kuhn, F., Malkhi, D., Ramasubramanian, V., Talwar, K.: Reconstructing approximate tree metrics. In: PODC (2007)
2. Baryshnikov, Y.: On the curvature of the Internet. In: Workshop on Stochastic Geometry and Teletraffic, Eindhoven, The Netherlands (April 2002)
3. Baryshnikov, Y., Tucci, G.H.: Asymptotic traffic flow in an hyperbolic network I: Definition and properties of the core. Technical Report Preprint: arXiv:1010.3304 (2010)

4. Bridson, M.R., Haefliger, A.: Metric Spaces of Non-Positive Curvature. Springer (1999)
5. Chen, W., Fang, W., Hu, G., Mahoney, M.W.: On the hyperbolicity of small-world and tree-like random graphs. Technical Report Preprint: arXiv:1201.1717v2 (2012)
6. Chepoi, V., Dragan, F.: A note on distance approximating trees in graphs. European Journal of Combinatorics 21(6), 761–766 (2000)
7. Chepoi, V., Dragan, F.F., Estellon, B., Habib, M., Vaxès, Y.: Diameters, centers, and approximating trees of δ-hyperbolic geodesic spaces and graphs. In: SoCG (2008)
8. Chepoi, V., Dragan, F.F., Estellon, B., Habib, M., Vaxès, Y., Xiang, Y.: Additive spanners and distance and routing labeling schemes for hyperbolic graphs. Algorithmica 62(3-4), 713–732 (2012)
9. Chepoi, V., Estellon, B.: Packing and Covering δ-Hyperbolic Spaces by Balls. In: Charikar, M., Jansen, K., Reingold, O., Rolim, J.D.P. (eds.) APPROX and RANDOM 2007. LNCS, vol. 4627, pp. 59–73. Springer, Heidelberg (2007)
10. de Montgolfier, F., Soto, M., Viennot, L.: Treewidth and hyperbolicity of the internet. In: IEEE Networks Computing and Applications 2011. IEEE (2011)
11. Gavoille, C., Ly, O.: Distance Labeling in Hyperbolic Graphs. In: Deng, X., Du, D.-Z. (eds.) ISAAC 2005. LNCS, vol. 3827, pp. 1071–1079. Springer, Heidelberg (2005)
12. Gromov, M.: Hyperbolic groups. Essays in Group Theory 8, 75–263 (1987)
13. Jonckheere, E., Lohsoonthorn, P.: Hyperbolic geometry approach to multipath routing. In: MED (2002)
14. Jonckheere, E., Lou, M., Bonahon, F., Baryshnikov, Y.: Euclidean versus hyperbolic congestion in idealized versus experimental networks. Technical Report Preprint: arXiv:0911.2538 (2009)
15. Jonckheere, E.A., Lou, M., Hespanha, J., Barooah, P.: Effective resistance of Gromov-hyperbolic graphs: Application to asymptotic sensor network problems. In: CDC (2007)
16. Kleinberg, J.: The small-world phenomenon: an algorithm perspective. In: STOC (2000)
17. Kleinberg, J.: Small-world phenomena and the dynamics of information. In: NIPS (2002)
18. Kleinberg, J.: Complex networks and decentralized search algorithms. In: ICM (2006)
19. Kleinberg, R.: Geographic routing using hyperbolic space. In: Infocom (2007)
20. Krioukov, D., Claffy, K.C., Fall, K., Brady, A.: On compact routing for the Internet. Computer Communication Review 37(3), 41–52 (2007)
21. Krioukov, D., Papadopoulos, F., Kitsak, M., Vahdat, A., Boguñá, M.: Hyperbolic geometry of complex networks. Physical Review E 82, 036106 (2010)
22. Lamping, J., Rao, R.: Laying out and visualizing large trees using a hyperbolic space. In: UIST (1994)
23. Lohsoonthorn, P.: Hyperbolic Geometry of Networks. PhD thesis, University of Southern California (2003)
24. Lou, M.: Traffic pattern in negatively curved network. PhD thesis, University of Southern California (2008)
25. Martel, C.U., Nguyen, V.: Analyzing Kleinberg's (and other) small-world models. In: PODC (2004)
26. Munzner, T., Burchard, P.: Visualizing the structure of the World Wide Web in 3D hyperbolic space. In: Web3D-VRML (1995)
27. Narayan, O., Saniee, I.: The large scale curvature of networks. Technical Report Preprint: arXiv:0907.1478 (2009)

28. Narayan, O., Saniee, I., Tucci, G.H.: Lack of spectral gap and hyperbolicity in asymptotic Erdős-Rényi random graphs. Technical Report Preprint: arXiv:1009.5700 (2010)
29. Nguyen, V., Martel, C.U.: Analyzing and characterizing small-world graphs. In: SODA (2005)
30. Shavitt, Y., Tankel, T.: Hyperbolic embedding of Internet graph for distance estimation and overlay construction. IEEE/ACM Transactions on Networking 16(1), 25–36 (2008)
31. Walter, J.A., Ritter, H.: On interactive visualization of high-dimensional data using the hyperbolic plane. In: KDD (2002)

On the Neighbourhood Helly of Some Graph Classes and Applications to the Enumeration of Minimal Dominating Sets

Mamadou Moustapha Kanté, Vincent Limouzy, Arnaud Mary, and Lhouari Nourine

Clermont-Université, Université Blaise Pascal, LIMOS, CNRS, France
{mamadou.kante,limouzy,mary,nourine}@isima.fr

Abstract. We prove that *line graphs* and *path graphs* have bounded neighbourhood Helly. As a consequence, we obtain output-polynomial time algorithms for enumerating the set of *minimal dominating sets* of line graphs and path graphs. Therefore, there exists an output-polynomial time algorithm that enumerates the set of *minimal edge-dominating sets* of any graph.

1 Introduction

A *hypergraph* $\mathcal{H}$ is a pair $(V(\mathcal{H}), E(\mathcal{H}))$ where $E(\mathcal{H})$, its set of hyperedges, is a family of subsets of $V(\mathcal{H})$, its set of vertices. A hypergraph is called *Sperner* if there is no hyperedge that is contained in another hyperedge. In [1] Berge defined the notion of k-conformality of hypergraphs. A hypergraph $\mathcal{H}$ is called *k-conformal* if $X \subseteq V(\mathcal{H})$ is contained in a hyperedge of $\mathcal{H}$ whenever each subset of X of size at most k is contained in a hyperedge. The *conformality* of a hypergraph $\mathcal{H}$ is defined as the least k such that $\mathcal{H}$ is k-conformal. An interesting property of the conformality notion is that it leads to an output polynomial time algorithm for the *Transversal problem* in Sperner hypergraphs of bounded conformality [14]. A *transversal* in a hypergraph $\mathcal{H}$ is a subset T of $V(\mathcal{H})$ such that T intersects any hyperedge of $\mathcal{H}$. If we denote by $Tr(\mathcal{H})$ the set of (inclusionwise) minimal transversals, the *Transversal Problem* consists in given a hypergraph $\mathcal{H}$ to compute $Tr(\mathcal{H})$. This problem has applications in graph theory, database theory, data mining, ... (see, e.g., [5,6,7,8]). It is an open question whether we can compute $Tr(\mathcal{H})$ in time $O((\|\mathcal{H}\| + |Tr(\mathcal{H})|)^k)$ for some constant k, where $\|\mathcal{H}\|$ is defined as $|V(\mathcal{H})| + |E(\mathcal{H})|$ (an algorithm achieving such a time is called an output-polynomial time algorithm). The best known algorithm for the *Transversal problem* is the one by Fredman and Khachiyan [9] which runs in time $O(N^{\log(N)})$ where $N = \|\mathcal{H}\| + |Tr(\mathcal{H})|$.

In this paper, we are interested in the conformality of the *closed neighbourhood hypergraphs* of graphs. Let us give some preliminary definitions and notations. A graph is a hypergraph where each hyperedge has size two (and are called *edges*). An edge of a graph is written xy (equivalently yx) instead of $\{x, y\}$. We refer to [3] for graph terminologies not defined in this paper. The *neighbourhood* of a vertex

K.-M. Chao, T.-s. Hsu, and D.-T. Lee (Eds.): ISAAC 2012, LNCS 7676, pp. 289–298, 2012.
© Springer-Verlag Berlin Heidelberg 2012

x in a graph G, *i.e.* , $\{y \mid xy \in E(G)\}$, is denoted by $N_G(x)$ and we let $N_G[x]$, the *closed neighbourhood* of x, be $N_G(x) \cup \{x\}$. The *closed neighbourhood hypergraph* $\mathcal{N}(G)$ of a graph G is the hypergraph $(V(G), \{N_G[x] \mid x \in V(G)\})$. A graph is called *$k$-conformal* if $\mathcal{N}(G)$ is k-conformal. The k-conformality of a graph is also known in the literature under the name of *k-neighbourhood Helly* [2,4]. Dually chordal graphs, chordal bipartite graphs, ptolemaic graphs are examples of graphs that have conformality at most 3.

A *cycle of length n* is denoted by C_n. A *claw* is a graph with four vertices isomorphic to the graph $(\{x_1, \ldots, x_4\}, \{x_1 x_2, x_1 x_3, x_1 x_4\})$. A *chordal graph* is a graph without an induced cycle of length greater or equal to 4. The *line graph* of a graph G, denoted by $L(G)$, is the graph with vertex-set $E(G)$ and edge-set $\{ef \mid e, f \in E(G) \text{ and } e \cap f \neq \emptyset\}$.

Let $\mathcal{F}$ be a family of subsets of some ground set. A graph G is an *intersection graph of $\mathcal{F}$* if there exists a bijection between $V(G)$ and $\mathcal{F}$ and such that there exists an edge between x and y if and only if their corresponding images in $\mathcal{F}$ intersect. A *path graph* is an intersection graph of paths in a tree. Path graphs constitute a subclass of chordal graphs [10].

We let $Min(\mathcal{N}(G))$ be the hypergraph obtained from $\mathcal{N}(G)$ by removing those hyperedges that contain a hyperedge. It is clear that $Min(\mathcal{N}(G))$ is Sperner. Notice that the conformality of a hypergraph $\mathcal{H}$ may be different from that of $Min(\mathcal{H})$. In this paper, we prove the following.

Theorem 1. *Line graphs are 6-conformal, path graphs and (C_4, C_5, claw)-free graphs are 3-conformal. Moreover, if we let $\mathcal{ML} := \{Min(\mathcal{N}(G)) \mid G$ is a line graph$\}$ and $\mathcal{MP} := \{Min(\mathcal{N}(G)) \mid G$ is a path graph or a $(C_4, C_5, \text{claw})\}$-free, then $\mathcal{ML}$ and $\mathcal{MP}$ have conformality bounded by 6 and 3 respectively.*

A subset D of the vertex-set of a graph G is called a *dominating set* if every vertex in $V(G) \setminus D$ is adjacent to a vertex in D. We denote by $\mathcal{D}(G)$ the set of (inclusionwise) minimal dominating sets of a graph G. The computation of $\mathcal{D}(G)$ of every graph G, known as the *DOM problem*, in output-polynomial time is a hard task and it was known for a while that $\mathcal{D}(G) = Tr(\mathcal{N}(G))$ for every graph G. Therefore, an output polynomial-time algorithm for the *Transversal problem* is also an output-polynomial time algorithm for the *DOM problem*. The authors have proved in [13] that the other direction also holds, *i.e.* , an output-polynomial time algorithm for *DOM problem* is also an output-polynomial time algorithm for the *Transversal problem*.

Output-polynomial time algorithms for the *DOM problem* are only known for few graph classes (see [12,13] for some of them). As a corollary of Theorem 1, we obtain output-polynomial time algorithms for the *DOM problem* in line graphs, path graphs and (C_4, C_5, claw)-free graphs, and to our knowledge this was not known.

Theorem 2. *1. For every line graph G, one can compute $\mathcal{D}(G)$ in time $O(\|G\|^5 \cdot |\mathcal{D}(G)|^6)$.*

2. For every path graph or (C_4, C_5, claw)-free graph G, one can compute $\mathcal{D}(G)$ in time $O(\|G\|^2 \cdot |\mathcal{D}(G)|^3)$.

A subset F of the edge-set of G is called an *edge-dominating set* if every edge in $E(G) \setminus D$ is incident to an edge in D. We denote by $\mathcal{ED}(G))$ the set of (inclusionwise) minimal edge-dominating sets of a graph G. It was open whether an output-polynomial time algorithm for computing $\mathcal{ED}(G)$ exists. It is well established that D is a dominating set of $L(G)$ if and only if D is an edge-dominating set of G. As a corollary of Theorem 2 we obtain the following theorem.

Theorem 3. *For every graph G, one can compute $\mathcal{ED}(G)$ in time $O(\|L(G)\|^5 \cdot |\mathcal{ED}(G)|^6)$.*

2 Some Remarks on the k-conformality

Definition 1. *Let $\mathcal{H}$ be a hypergraph and let k be a positive integer. A k-bad-set in $\mathcal{H}$ is a subset S of $V(\mathcal{H})$ with $|S| > k$ and such that:*

- *for all subsets $S' \subseteq S$ of size k, there exists $e \in E(\mathcal{H})$ such that $S' \subseteq e$,*
- *for all hyperedges $e \in E(G)$, $S \nsubseteq e$.*

A graph G is said to have a k-bad-set if $\mathcal{N}(G)$ has one.

Remark that a k-bad-set is also a k'-bad-set for every $k' \le k$. The proof of the following is immediate from the definition of k-conformality.

Definition 2. *Let k be a positive integer. A hypergraph is k-conformal if and only if it has no k-bad-set.*

A *minimal k-bad-set* in a hypergraph $\mathcal{H}$ is a k-bad-set in $\mathcal{H}$ of size $k + 1$. Notice that, for every vertex x of a minimal k-bad-set S, there exists a hyperedge e which contains all S, but x.

Proposition 1. *Let k be a positive integer. A hypergraph is k-conformal if and only if it has no minimal k'-bad-set for every $k' \ge k$.*

Proof. The first direction is straigtforward since a k'-bad-set with $k' \ge k$ is also a k-bad-set. Now assume that a hypergraph $\mathcal{H}$ has no minimal k'-bad-set with $k' \ge k$ but has a non minimal k-bad-set S. Assume that S is minimal with respect to inclusion. Then, for every $x \in S$, $S \setminus \{x\}$ is not a k-bad-set and since S is a k-bad-set, there exists $e \in E(\mathcal{H})$ such that $S \setminus \{x\} \subseteq e$. Therefore, S is a minimal k'-bad-set with $k' := |S| - 1 \ge k$. This is a contradiction with the assumption that $\mathcal{H}$ has no k'-bad-set with $k' \ge k$. $\square$

3 Line graphs

For a graph G we denote by $G(F)$, for $F \subseteq E(G)$, the graph with vertex-set $\{x \in V(G) \mid x$ is incident with an edge in $F\}$ and F as edge-set. A *clique* is a graph with pairwise adjacent vertices (it is denoted by K_n if it has n vertices). A *tree* is an acyclic (without induced cycle) connected graph.

A *vertex cover* in a graph G is a subset S of $V(G)$ such that every edge of G intersects S and a *matching* is a subset M of $E(G)$ such that for every $e, f \in M$ we have $e \cap f = \emptyset$. We denote by $\tau(G)$ and $\nu(G)$, respectively, the maximum size of a vertex cover and of a matching. If e is an edge of a graph G, we define the closed neighbourhood of e as $N_{L(G)}[e]$.

Lemma 1. *Let $k \geq 3$ be a positive integer. Let G be a graph and let S be a subset of $E(G)$. If S is a k-bad-set of $L(G)$, then $\nu(G(S)) = 2$.*

Proof. Let $S \subseteq E(G)$ be a k-bad-set of $L(G)$ and assume that $\nu(G(S)) \geq 3$. Let M be a maximum matching of $G(S)$. Let S' be a subset of M with $|S'| = 3$ (such a subset of M exists since $\nu(G(S)) \geq 3$). It is easy to see that no edge of $E(G)$ can be incident to all edges in S' and then any subset of S of size k and containing S' is not included in the closed neighbourhood of an edge of G, contradicting the fact that S is a k-bad-set of $L(G)$. Therefore, $\nu(G(S)) \leq 2$. Assume now that $\nu(G(S)) = 1$. This implies that there exists an edge $e \in S$ incident to all edges in S, which contradicts again the fact that S is a k-bad-set. We can thus conclude that $\nu(G(S)) = 2$.   □

Lemma 2. *Let $k \geq 3$ be a positive integer. Let G be a bipartite graph and let S be a subset of $E(G)$. If S is a k-bad-set of $L(G)$, then $\tau(G(S)) = 2$.*

Proof. By Lemma 1, we have $\nu(G(S)) = 2$. By König's Theorem, we have that $\tau(G(S)) = \nu(G(S)) = 2$.   □

Proposition 2. *Line graphs are 6-conformal.*

Proof. Let G be a graph and assume that $L(G)$ has a 6-bad-set $S \subseteq E(G)$. By Lemma 1, we have $\nu(G(S))) = 2$. Let $\{x_1 x_2, x_3 x_4\}$ be a maximum matching of $G(S)$. By definition of a maximum matching, we know that every edge of S intersects $\{x_1, x_2, x_3, x_4\}$. For $i \in \{1, \ldots, 4\}$, we let P_i be the set $\{e \in S \mid e \cap \{x_1, \ldots, x_4\} = \{x_i\}\}$. One of P_1 or P_2 must be empty. Otherwise let $e_1 \in P_1$ and $e_2 \in P_2$, then $\{e_1, e_2, x_3 x_4, x_1 x_2\}$ would not be in a closed neighbourhood, contradicting the fact that S is a 6-bad-set. Similarly, one of P_3 or P_4 is empty. Therefore, at most two sets among $P_1, \ldots, P_4$ are non empty. We identify two cases.

Case 1. Two sets among $P_1, \ldots, P_4$ are non empty. Assume without loss of generality that they are P_1 and P_3. Let $e_1 \in P_1$ and $e_2 \in P_3$. Let S' be a subset of S of size 6 that contains $\{e, e', x_1 x_2, x_3 x_4\}$. Since S is a 6-bad-set, there exists an edge whose closed neighbourhood contains S', and the only possible one is $x_1 x_3$. Moreover, there must exist an edge e that is neither incident to x_1 nor to x_3, otherwise S would not be a 6-bad-set (the closed neighbourhood of $x_1 x_3$ would contain S). Let again S' be a subset of S of size 6 and that contains $\{x_1 x_2, x_3 x_4, e_1, e_2, e\}$. But, the subset $\{x_1 x_2, x_3 x_4, e_1, e_2, e\}$ of S cannot be contained in the closed neighbourhood of any edge in $E(G)$. So, we can conclude that at most one set among $P_1, \ldots, P_4$ is non empty.

Case 2. One set among $P_1, \ldots, P_4$ is non empty, say P_1 is this set. Let $e \in P_1$. Then, the set $\{x_1x_2, x_3x_4, e\}$ must be included in the closed neighbourhood of some edge. The only two possible such edges are x_1x_3 and x_1x_4. Assume without loss of generality that $x_1x_3 \in E(G)$. Since S is a 6-bad-set of $L(G)$ there exists an edge e in S which is not in the closed neighbourhood of x_1x_3. Since P_2, P_3 and P_4 are empty, that edge must be x_2x_4. Then, $\{e, x_1x_2, x_3x_4, x_2x_4\}$ must be included in a closed neighbourhood of an edge, and that edge is clearly x_1x_4. Again, since S is a 6-bad-set, there must exist an edge not in the neighboorhood of x_1x_4, and the only possible choice is x_2x_3. Therefore, $\{e, x_1x_2, x_3x_4, x_2x_4, x_2x_3\}$ is included in S and since it has size 5, it is included in a closed neighbourhood of an edge. But no such edge exists. So, we can conclude that P_1 is also empty.

From the two cases above, we have that P_i is empty for all $i \in \{1, \ldots, 4\}$, and then $V(G(S)) = \{x_1, x_2, x_3, x_4\}$. Since S is a 6-bad-set, its size must be at least 7, which contradicts the fact that $|V(G(S))| = 4$ since the number of edges in a graph with four vertices is at most 6. This completes the proof. $\qquad\square$

Since the line graph of K_4 is 6-conformal and not 5-conformal, the bound from Proposition 2 is tight. Therefore, line graphs of graphs that contain K_4 as induced subgraphs have conformality 6. By adapting the proof of Proposition 2, we can prove the following.

Proposition 3. *Line graphs of K_4-free graphs are 5-conformal.*

Proof. If we define the sets P_i, $i \in \{1, \ldots, 4\}$ as in the proof of Proposition 2 and follow the same proof with S being a 5-bad-set and not a 6-bad-set, the only way for a line graph to have a 5-bad-set S is when all P_i's are empty. But, in this case $|V(G(S))| = 4$ and since a 5-bad-set must have at least 6 edges, we conclude that the only possible 5-bad-set for a line graph $L(G)$ is that G contains K_4. $\quad\square$

Even if the bound 6 is optimal in the class of line graphs, it is not at all in the class of line graphs of bipartite graphs.

Proposition 4. *Line graphs of bipartite graphs are 4-conformal.*

Proof. Let G be a bipartite graph and let S be a 4-bad-set of $L(G)$. By definition of a 4-bad-set, $|S| \geq 5$. From Lemma 2, $\tau(G(S)) = 2$. Let $\{x, y\}$ be a maximum vertex cover of $G(S)$. Then $xy \notin E(G)$, otherwise the edge xy will be incident to all edges of S, contradicting the fact that S is a 4-bad-set of $L(G)$. We identify three cases.

Case 1. x has only one neighbour x' in $G(S)$. Let S' be a subset of S of size 4 that contains the edge xx'. Since S is a 4-bad-set of $L(G)$, there must exist an edge e of $E(G)$ whose closed neighbourhood contains S'. Since xy is not an edge, e must be yx'. But, in this case the closed neighbourhood of yx' will contain S, contradicting the fact that S is a 4-bad-set of $L(G)$.

Case 2. y has only one neighbour y' in $G(S)$. This case is similar to Case 1 (replace x by y and x' by y').

Case 3. Each of x and y has at least two neighbours in $G(S)$. Let x_1 and x_2 (resp. y_1 and y_2) be two neighbours of x (resp. y) in $G(S)$. Let $S' := \{xx_1, xx_2, yy_1, yy_2\}$ be a subset of S. There must exist an edge e in $E(G)$ whose closed neighbourhood contains S'. One can easily check that the only possible choice for e is xy which is a contradiction with the fact that $xy \notin E(G)$. Since there is no edge whose closed neighbourhood contains S', we have a contradiction with the fact that S is a 4-bad-set of $L(G)$. This concludes the proof. □

The bound in Proposition 4 is tight because the line graph of the cycle C_4 is 4-conformal but not 3-conformal. One easily checks that if a graph contains C_4 as an induced cycle, then its line graph is 4-conformal, but not 3-conformal.

4   Path Graphs

We now prove that path graphs are 3-conformal. A *clique tree* of a graph G is a tree T whose vertices are in bijection with the (inclusionwise) maximal cliques of G and such that those maximal cliques that contain a vertex x induce a subtree of T, which we will denote by T^x. Observe that G is the intersection graph of these subtrees. It is well-known that a graph is chordal if and only if it has a clique-tree [10] and path graphs are exactly those chordal graphs where for every vertex x, T^x is a path. (It is worth noticing that path graphs can be recognised in polynomial time [11].)

A *rooted tree* is a tree with a distinguished vertex called its *root*. In a rooted tree T we define the partial order $\preceq_T$ where $x \preceq_T y$ if and only if the path from the root to x goes through y. For v a vertex of a rooted tree T, we let T_v be the subtree of T rooted at v and induced by the vertices in $\{x \in V(T) \mid x \preceq_T v\}$.

Proposition 5. *Path graphs are 3-conformal.*

Proof. Let G be a path graph and let T be its clique-tree. Assume that G has a minimal $(k-1)$-bad-set $S := \{x_1, x_2, ..., x_k\}$ with $k > 3$. Since S is a minimal $(k-1)$-bad-set, there exists a vertex x such that $\{x_1, x_2, ..., x_{k-1}\} \subseteq N_G[x]$. Let $T^x = (t_1, t_2, \ldots, t_\ell)$. Since, $\{x_1, x_2, ..., x_{k-1}\} \subseteq N_G[x]$, each subtree T^{x_i}, for $i \in \{1, \ldots, k-1\}$, must intersect T^x and $P_{x_i} := T^{x_i} \cap T^x$ forms a sub-path of T^x. Let $s_i := \min\{j \mid t_j \in P_{x_i}\}$ and $e_i := \max\{j \mid t_j \in P_{x_i}\}$ for $i \in \{1, \ldots, k-1\}$. Assume without loss of generality that the vertices $x_1, \ldots, x_{k-1}$ are ordered such that $i < j \implies s_i \leq s_j$. Since S is a $(k-1)$-bad-set, we know that T^{x_k} does not intersect T^x. We let t_r be the unique vertex of T^x such that every path from t_r to any vertex of T^{x_k} intersects T^x only on t_r. We root T at t_r. We identify two cases(see Fig. 1).

Case 1. For all $j \in \{1, \ldots, k-1\}$, T^{x_j} does not contain t_r. Assume first that for all $i \in \{1, \ldots, k-1\}$, $e_i < r$, and let $e_j := \min\{e_i \mid i \in \{1, \ldots, k-1\}\}$. Since

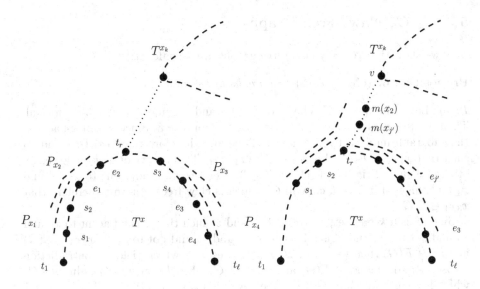

Fig. 1. The first case of the proof of Proposition 5, is illustrated on the left part. Case two is described on the opposite side.

S is a $(k-1)$-bad-set, there exists a vertex z such that $\{x_j, x_k\} \subseteq N_G[z]$. But in this case T^z would intersect T^{x_j} for every $x_j \in S$ which leads to a contradiction. Similarly, if for all $j \in \{1, \ldots, k-1\}$, we have $s_j > r$, any vertex who is adjacent to x_{k-1} and x_k would also be adjacent to $x_1, \ldots, x_{k-2}$, yielding a contradiction. We can therefore assume that $S \setminus \{x_k\} = S_1 \cup S_2$, with $S_1, S_2 \neq \emptyset$ and such that $e_j < r$ for every $x_j \in S_1$ and $s_j > r$ for every $x_j \in S_2$. Let us choose $x_i \in S_1$ and $x_j \in S_2$. Then since S is a $(k-1)$-bad-set, there must exist a vertex z such that $\{x_i, x_j, x_k\} \subseteq N_G[z]$, but no path in T can intersect at the same time the three paths T^{x_i}, T^{x_j} and T^{x_k}, which yields again a contradiction.

Case 2. There is at least one vertex $x_j \in S \setminus \{x_k\}$ such that $t_r \in T^{x_j}$ (ie. $s_j \leq r$ and $e_j \geq r$). Note that in this case, for every vertex $x_i \in S \setminus \{x_k, x_j\}$, we have $t_r \in T^{x_i}$ otherwise, every vertex whose neighborhood contains x_k and x_i would be adjacent to x_j, which is in contradiction with the fact that S is a minimal $(k-1)$-bad-set. Hence, for every $i \in \{1, \ldots, k-1\}$, $s_i \leq r \leq e_i$. Let v be the vertex of T_{t_r} which is the greatest vertex of T^{x_k} (greatest with respect to $\preceq_T$) and let P be the path between t_r and v. Note that for all $i \in \{1, \ldots, k-1\}$, T^{x_i} must intersect P on at least one different vertex from t_r (which implies that $s_i = t_r$ or $e_i = t_r$). Otherwise, for any vertex z such that $\{x_k, x_i\} \in N_G[z]$, T^z would contain t_r and hence S would be included in $N_G[z]$. For every $i \in \{1, \ldots, k-1\}$, let $m(x_i)$ be $\max_{\preceq_T} \{x \in T \mid x \in P \cap T^{x_i}\}$ ($m(x_i)$ is the greatest vertex, with respect to $\preceq_T$, of $P \cap T^{x_i}$). Let j' be such that $m(x_{j'}) := \max_{\preceq_T} \{m(x_i) \mid i \in \{1, \ldots, k-1\}\}$. Then any closed neighborhood that contains x_k and $x_{j'}$ will also contain S, which yields a contradiction. $\qquad\square$

5 (C_4, C_5, claw)-Free Graphs

Now we show that (C_4, C_5, claw)-free graphs are 3-conformal.

Proposition 6. $(C_4, C_5, claw)$-free graphs are 3-conformal.

Proof. Let G be a (C_4, C_5, claw)-free graph and assume it is not 3-conformal. Then, there exists a k-bad-set S with $|S| > k$ for $k \geq 3$. Since S is a k-bad-set, the subgraph induced by S is not a clique and therefore there exist x_1 and x_2 such that $x_1 x_2 \notin E(G)$. Let $x_3 \in S \setminus \{x_1, x_2\}$. Then, $x_3 x_1$ or $x_3 x_2$ is an edge, otherwise since S is k-bad-set for $k \geq 3$, there exists z adjacent to x_1, x_2 and x_3 and this will induce a claw in G (which is claw-free). Assume therefore that $x_3 x_1 \in E(G)$.

Since S is a k-bad-set, there exists z and z' such that z is adjacent to x_1 and x_2 and not to x_3, and z' is adjacent to x_2 and x_3 and not to x_1. If $x_2 x_3 \notin E(G)$ and $zz' \notin E(G)$ then $\{z, x_1, x_3, z', x_2\}$ induces a C_5 which yields a contradiction (G is C_5-free). If $x_2 x_3 \notin E(G)$ and $zz' \in E(G)$, then $\{z, x_1, x_3, z'\}$ induces a C_4 which is a contradiction (G is C_4-free). And if $x_2 x_3 \in E(G)$, then $\{z, x_1, x_3, x_2\}$ induces a C_4 which is again a contradiction. We can therefore conclude that no k-bad-set for $k \geq 3$ exists and hence G is 3-conformal. $\square$

6   Proofs of Theorems

We can now prove Theorems 1, 2 and 3.

Proof (Proof of Theorem 1). The first part of the theorem follows from Propositions 2, 5 and 6. For the second part, one easily checks that if we replace in the arguments "there exists z such that $S' \subseteq N_G[z]$" by "there exists a closed neighbourhood $N_G[z] \supseteq S'$" the same arguments follow. So, the second statement is also true. $\square$

It is clear that Theorems 2 and 3 follow from Theorem 1, and Theorem 4 and Proposition 7 stated below.

Theorem 4 ([14]). *Let $\mathcal{H}$ be a k-conformal Sperner hypergraph. Then one can compute $Tr(\mathcal{H})$ in time $O(\|H\|^{k-1} \cdot |Tr(\mathcal{H})|^k)$.*

Proposition 7 (Folklore). *Let G be a graph and let D be a subset of $E(G)$. Then D is a dominating set of $L(G)$ if and only if D is an edge-dominating set of G.*

7   Conclusion

We have proven that line graphs, path graphs and (C_4, C_5, claw)-free graphs have bounded conformality. A direct consequence, using the result by Boros et al.

in [14] is that we can enumerate minimal dominating sets in output-polynomial time in line graphs, path graphs and (C_4, C_5, claw)-free graphs. Path graphs was one of the maximal subclasses of chordal graphs where no output-polynomial time algorithm for the *DOM problem* was known. *Chordal domination perfect graphs*, which form a subclass of chordal graphs, do not have bounded conformality and therefore we cannot expect using the algorithm by Boros et al. to get an output-polynomial time algorithm for the *DOM problem* in chordal graphs. Notice that chordal claw-free graphs is a maximal subclass of chordal domination perfect graphs and have conformality at most 3 by Proposition 6. We leave open the quest for an output-polynomial time algorithm for the *DOM problem* in chordal graphs, or at least in its other subclasses such as chordal domination perfect graphs.

Acknowledgement A. Mary and L. Nourine are partially supported by the ANR (french National Research Agency) project DAG (ANR-09-DEFIS, 2009-2012). M.M. Kanté and V. Limouzy are supported by the ANR junior project DORSO (2011-2015).

References

1. Berge, C.: Hypergraphs. North Holland Mathematical Library, vol. 445. Elsevier-North Holland, Amsterdam (1989)
2. Brandstädt, A., Bang Le, V., Spinrad, J.P.: Graph Classes: A Survey. SIAM Monographs on Discrete Mathematics and Applications. SIAM (1987)
3. Diestel, R.: Graph Theory, 3rd edn. Springer (2005)
4. Dourado, M.C., Protti, F., Szwarcfiter, J.L.: Complexity aspects of the helly property: Graphs and hypergraphs. Electronic Journal of Combinatorics (2009); Dynamic Survey, DS17
5. Eiter, T., Gottlob, G.: Identifying the minimal transversals of a hypergraph and related problems. SIAM J. Comput. 24(6), 1278–1304 (1995)
6. Eiter, T., Gottlob, G.: Hypergraph Transversal Computation and Related Problems in Logic and AI. In: Flesca, S., Greco, S., Leone, N., Ianni, G. (eds.) JELIA 2002. LNCS (LNAI), vol. 2424, pp. 549–564. Springer, Heidelberg (2002)
7. Eiter, T., Gottlob, G., Makino, K.: New results on monotone dualization and generating hypergraph transversals. SIAM J. Comput. 32(2), 514–537 (2003)
8. Eiter, T., Makino, K., Gottlob, G.: Computational aspects of monotone dualization: A brief survey. Discrete Applied Mathematics 156(11), 2035–2049 (2008)
9. Fredman, M.L., Khachiyan, L.: On the complexity of dualization of monotone disjunctive normal forms. J. Algorithms 21(3), 618–628 (1996)
10. Gavril, F.: The intersection graphs of subtrees in trees are exactly the chordal graphs. J. Combinatorial Theory Ser. (B 16), 47–56 (1974)
11. Gavril, F.: A recognition algorithm for the intersection graphs of paths in trees. Discrete Math. (23), 211–227 (1978)

12. Kanté, M.M., Limouzy, V., Mary, A., Nourine, L.: Enumeration of Minimal Dominating Sets and Variants. In: Owe, O., Steffen, M., Telle, J.A. (eds.) FCT 2011. LNCS, vol. 6914, pp. 298–309. Springer, Heidelberg (2011)
13. Kanté, M.M., Limouzy, V., Mary, A., Nourine, L.: On the enumeration of minimal dominating sets and related notions. Technical report, Clermont-Université, Université Blaise Pascal, LIMOS, CNRS (2012)
14. Khachiyan, L., Boros, E., Elbassioni, K.M., Gurvich, V.: On the dualization of hypergraphs with bounded edge-intersections and other related classes of hypergraphs. Theor. Comput. Sci. 382(2), 139–150 (2007)

Induced Immersions[*]

Rémy Belmonte[1], Pim van 't Hof[1], and Marcin Kamiński[2,**]

[1] Department of Informatics, University of Bergen, Norway
{remy.belmonte,pim.vanthof}@ii.uib.no
[2] Département d'Informatique, Université Libre de Bruxelles, Belgium
Marcin.Kaminski@ulb.ac.be

Abstract. A graph G contains a multigraph H as an induced immersion if H can be obtained from G by a sequence of vertex deletions and lifts. We present a polynomial-time algorithm that decides for any fixed multigraph H whether an input graph G contains H as an induced immersion. We also show that for every multigraph H with maximum degree at most 2, there exists a constant c_H such that every graph with treewidth more than c_H contains H as an induced immersion.

1 Introduction

A recurrent problem in algorithmic graph theory is to decide, given two graphs G and H, whether the structure of H appears as a pattern within the structure of G. The notion of appearing as a pattern gives rise to various graph containment problems depending on which operations are allowed. Maybe the most famous example is the minor relation that has been widely studied, in particular in the seminal Graph Minor series of papers by Robertson and Seymour (see, e.g., [16,17]). A graph G contains a graph H as a *minor* if H can be obtained from G by a sequence of vertex deletions, edge deletions and edge contractions. One of the highlights of the Graph Minor series is the proof that, for every fixed graph H, there exists a cubic-time algorithm that decides whether an input graph G contains H as a minor [16].

If the contraction operation is restricted so that we may only contract an edge if at least one of its endpoints has degree 2, then we obtain the operation known as *vertex dissolution*. If H can be obtained from G by a sequence of vertex deletions, edge deletions and vertex dissolutions, then H is a *topological minor* of G. It is a trivial observation that if G contains H as topological minor, then it also contains H as a minor. Numerous results on topological minors exist in the literature, notable examples being the recent proof that testing for topological minors is FPT by Grohe et al. [12] and the characterization of graphs excluding a fixed graph as a topological minor by Grohe and Marx [13].

Grohe and Marx [13] also show that another containment relation can be decided in FPT time, namely *immersion*. In order to define immersion, we first

---

[*] This work is supported by the Research Council of Norway (197548/F20).
[**] Chargé de recherches du FNRS.

need to define another graph operation. The *lift* (or *split-off*) operation is defined as follows: given three (not necessarily distinct) vertices u, v, w such that $uv, vw \in E(G)$, we delete uv and vw and replace them by a new edge uw, possibly creating a loop or multiple edges. A graph G contains H as an *immersion* if H can be obtained from G by a sequence of vertex deletions, edge deletions, and lifts. It is not hard to verify that if G contains H as a topological minor, then G also contains H as an immersion, as a dissolution can be simulated by a lift and a vertex deletion. The interest in the immersion relation has steadily been growing [1,2,8,9,11,12,18,20].

Our Results. We introduce and study the *induced immersion* relation, which is equivalent to the immersion relation where no edge deletions are allowed (see Figure 1 for an overview of the different containment relations mentioned in this paper). Our main result is that, for any fixed multigraph H, there exists a polynomial-time algorithm deciding whether a simple graph G contains H as an induced immersion. It is interesting to note that it is highly unlikely that a similar result exists for the induced minor and induced topological minor relations, as there exist fixed graphs H for which the problem of deciding whether a graph contains H as an induced minor or as an induced topological minor is NP-complete [7,15]. We complement this result by showing that, for every fixed multigraph H of maximum degree at most 2, there exists a constant c_H such that every graph with treewidth more than c_H contains H as an induced immersion.

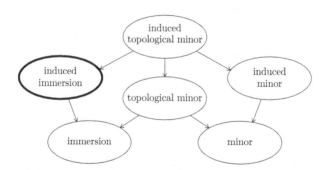

Fig. 1. The relationship between the different containment relations mentioned in this paper. An arrow from relation A to relation B indicates that if G contains H with respect to relation A, then G also contains H with respect to relation B. For example, if G contains H as an induced topological minor, then G also contains H as an induced immersion, since a dissolution can be simulated by a lift and a vertex deletion.

2   Preliminaries

Following the terminology of Diestel [3], we define a *multigraph* to be a pair $G = (V, E)$, where V is a set, and E is a multiset such that every $e \in E$ is a multiset of two elements of V. The elements of V and E are called the *vertices* and *edges* of G, respectively. For convenience, we write uv instead of $\{u, v\}$ to denote an edge between u and v; in particular, uu denotes a loop at u. The *multiplicity* of an edge $uv \in E$, denoted by $\text{mult}_G(uv)$, is the number of times the element $\{u, v\}$ appears in the multiset E. By slight abuse of notation, we define $\text{mult}(uv) = 0$ whenever $uv \notin E$. A *graph* is a multigraph without loops in which the multiplicity of every edge is at most 1. We refer to the textbook by Diestel [3] for graph terminology not defined below, and to the monograph by Downey and Fellows [5] for a background on parameterized complexity.

Let $G = (V, E)$ be a multigraph. For two vertices $u, v \in V$, we say that u is a *neighbor* of v if $\{u, v\} \in E$. Note that if $\{u, u\} \in E$, then u is its own neighbor. The *degree* of a vertex $u \in V$, denoted by $d_G(u)$, is the number of neighbors of u, where a loop at u contributes 2 to the degree of u. A multigraph in which every vertex has degree exactly k is called k-*regular*. Let P be a path in G from u to v that has at least one edge. The *length* of P is its number of edges. The vertices u and v are the *endpoints* of P, and every other vertex of P is an *internal* vertex of P. By abuse of terminology, we will allow the endpoints u and v of a path to be the same vertex, even though strictly speaking such a path is a cycle.

Let G be a multigraph. Let u, v, w be three (not necessarily distinct) vertices in G such that $uv, vw \in E(G)$. The *lift* (or *split-off*) $\{uv, vw\}$ of the edges uv and vw is the operation that deletes uv and vw and replaces them by a new edge uw, thereby increasing the multiplicity of uw by exactly 1. In particular, note that lifting two copies of an edge uv creates a loop either at u or at v, and lifting a loop uu with any incident edge uv simply deletes the loop uu and leaves uv unchanged. When we lift two edges uv and vw, we say that v is the *pivot* of the lift $\{uv, vw\}$. The multigraph obtained from G by applying the lift $\{uv, vw\}$ is denoted by $G \vee \{uv, vw\}$. Similarly, given a sequence of lifts $\mathcal{L} = (\ell_1, \ldots, \ell_q)$, we define $G \vee \mathcal{L} = (((G \vee \ell_1) \vee \ell_2) \cdots \vee \ell_q)$. Let $P = p_1 \cdots p_\ell$ be a path of length at least 2 in G. When we say that we *lift* the path P, we mean that we perform the lifts $\{p_1 p_i, p_i p_{i+1}\}$ for i from 2 up to $\ell - 1$. Note that lifting the path P is equivalent to deleting all the edges of P from G, and adding a new edge with endpoints $p_1 p_\ell$.

Let G and H be two multigraphs. Then G contains H as an *immersion* if H can be obtained from G by a sequence of vertex deletions, edge deletions, and lifts. Equivalently, G contains H as an immersion if there exists a subset S of $|V(H)|$ vertices in G and a bijection ϕ from $V(H)$ to S such that, for each edge $uv \in E(H)$, there exists a path P_{uv} in G from $\phi(u)$ to $\phi(v)$, and all these paths P_{uv}, $uv \in E(H)$, are mutually edge-disjoint [18]. We point out that if $uu \in E(H)$, then the path P_{uu} starts and ends in the same vertex; even though strictly speaking P_{uu} is a cycle, recall that P_{uu} satisfies our definition of a path (from u to u). If H can be obtained from G by a sequence of vertex deletions and lifts, then G contains H as an *induced immersion*.

Observation 1. *Let G and H be two multigraphs. Then G contains H as an induced immersion if and only if there exists a sequence of lifts $\mathcal{L}$ such that $G \vee \mathcal{L}$ contains H as an induced subgraph.*

The definition of induced immersion gives rise to the following definition, which will be used frequently throughout the paper.

Definition 1 (H-model). *Let G and H be two multigraphs such that G contains H as an induced immersion. Given a sequence of lifts $\mathcal{L}$, a set of vertices $S \subseteq V(G)$ and a bijection ϕ from $V(H)$ to S, we say that $(S, \mathcal{L}, \phi)$ is an H-model of G if ϕ is an isomorphism from H to $G'[S]$, where $G' = G \vee \mathcal{L}$.*

Observation 2. *Let G and H be two multigraphs. Then G contains H as an induced immersion if and only if G has an H-model.*

The INDUCED IMMERSION problem takes as input a graph G and a multigraph H, and the task is to decide whether G contains H as an induced immersion.

The *elementary wall* of height r is the graph W_r whose vertex set is $\{(x, y) \mid 0 \leq x \leq 2r + 1, 0 \leq y \leq r\} \setminus \{(0,0), (2r+1, r)\}$ if r is even and $\{(x, y) \mid 0 \leq x \leq 2r+1, 0 \leq y \leq r\} \setminus \{(0,0), (0, r)\}$ if r is odd, and such that there is an edge between any vertices (x, y) and (x', y') if either $|x' - x| = 1$ and $y = y'$, or if $x = x', |y' - y| = 1$ and x and $\max\{y, y'\}$ have the same parity.

Robertson, Seymour and Thomas [19] proved that for $r \geq 1$, every graph with treewidth more than 20^{2r^5} contains the $r \times r$-grid as a minor. Note that every $r \times r$-grid contains an elementary wall of height $\lfloor r/2 \rfloor - 1$ as a minor. Since an elementary wall has maximum degree at most 3, and every minor with maximum degree at most 3 is also a topological minor of the same graph, every $r \times r$-grid also contains $W_{\lfloor r/2 \rfloor - 1}$ as a topological minor. As we pointed out in the introduction, this means that $W_{\lfloor r/2 \rfloor - 1}$ is also contained as an immersion in an $r \times r$-grid. Hence, the aforementioned result of Robertson, Seymour and Thomas implies the following result.

Theorem 1 ([19]). *For $r \geq 1$, every graph with treewidth more than $20^{10(r+1)^5}$ contains W_r as an immersion.*

3   Finding a Fixed Multigraph as an Induced Immersion

The goal of this section is to show that, for every fixed multigraph H, there exists a polynomial-time algorithm that decides whether a graph G contains H as an induced immersion. We first show that if both G and H are given as input, this problem cannot be solved in polynomial time, unless P = NP.

Lemma 1. ($\bigstar$)[1] *The INDUCED IMMERSION problem is NP-complete, even if G is a planar graph of maximum degree at most 4 and H is a cycle.*

Let G and H be two multigraphs such that G contains H as an induced immersion, and let $(S, \mathcal{L}, \phi)$ be an H-model of G. Lemma 3 below shows that we may assume $\mathcal{L}$ to satisfy a property that will be exploited in the algorithm in the proof of Theorem 2. The proof of Lemma 3 relies on the following result, which states that we can "safely" swap certain consecutive pairs of lifts in $\mathcal{L}$.

---

[1] Proofs marked with a star have been omitted due to page restrictions.

Lemma 2. (★) *Let G and H be two multigraphs such that G contains H as an induced immersion, and let $(S, \mathcal{L}, \phi)$ be an H-model of G. Let $\{xy, yz\}$ and $\{uv, vw\}$ be two consecutive lifts in $\mathcal{L}$, occurring at positions $i - 1$ and i in $\mathcal{L}$, respectively, such that $xz \neq uv$ and $xz \neq vw$. Then G has an H-model $(S, \mathcal{L}', \phi)$ where $\mathcal{L}'$ is obtained from $\mathcal{L}$ by swapping $\{xy, yz\}$ and $\{uv, vw\}$.*

Lemma 3. (★) *Let G and H be two multigraphs such that G contains H as an induced immersion, and let $(S, \mathcal{L}, \phi)$ be an H-model of G. For any edge uv of $G[S]$, there exists an H-model $(S, \mathcal{L}', \phi)$ of G such that uv appears in the first lift in $\mathcal{L}'$.*

Informally speaking, the next lemma shows that given an H-model $(S, \mathcal{L}, \phi)$ of a graph G, there exists a short sequence of lifts $\mathcal{L}^*$ whose application removes any unwanted edges between vertices of S. Here, an edge between vertices $\phi(u), \phi(v) \in S$ is *unwanted* if the multiplicity of $\phi(u)\phi(v)$ in G is strictly larger than the multiplicity of uv in H.

Lemma 4. (★) *Let G be a graph and let H be a multigraph such that G contains H as an induced immersion, and let $(S, \mathcal{L}, \phi)$ be an H-model of G. Then there exists a sequence $\mathcal{L}^*$ of lifts that satisfies the following four properties:*

(i) $|\mathcal{L}^*| \leq |E(G[S])|$;
(ii) *for every $\{uv, vw\} \in \mathcal{L}^*$, we have $v \in S$ and $\{u, w\} \cap S \neq \emptyset$;*
(iii) *for every $u, v \in V(H)$, we have $\mathrm{mult}_{G \vee \mathcal{L}^*}(\phi(u)\phi(v)) \leq \mathrm{mult}_H(uv)$;*
(iv) $G \vee \mathcal{L}^*$ *contains H as an induced immersion.*

The next lemma shows that after unwanted edges between vertices of S have been removed by applying $\mathcal{L}^*$, deciding whether G contains H as an induced immersion is equivalent to finding a family of mutually edge-disjoint paths.

Lemma 5. *Let G and H be two multigraphs. Suppose there is a set $S \subseteq V(G)$ and a bijection ϕ from $V(H)$ to S such that $\mathrm{mult}_G(\phi(u)\phi(v)) \leq \mathrm{mult}_H(uv)$ for every $u, v \in V(H)$. Then G has an H-model $(S, \mathcal{L}, \phi)$ for some sequence $\mathcal{L}$ if and only if the following two properties are satisfied:*

(i) *for every $u, v \in V(H)$, there is a set $\mathcal{P}_{uv}$ of $\mathrm{mult}_H(uv)$ paths in G from $\phi(u)$ to $\phi(v)$;*
(ii) *the paths in $\mathcal{P} = \bigcup_{u,v \in V(H)} \mathcal{P}_{uv}$ are mutually edge-disjoint.*

Proof. First suppose G has an H-model $(S, \mathcal{L}, \phi)$ for some sequence $\mathcal{L}$. Then G contains H as an induced immersion by Observation 2, and hence G also contains H as an immersion with respect to S and ϕ. The existence of the family $\mathcal{P}$ then follows from the edge-disjoint paths definition of immersion.

Now suppose that there exists a family $\mathcal{P}$ of paths in G as mentioned in the lemma. Among all such families $\mathcal{P}$, let $\mathcal{P}' = \bigcup_{u,v \in V(H)} \mathcal{P}'_{uv}$ be one that contains the largest number of paths of length 1.

Claim 1. For every $\phi(u), \phi(v) \in V(G)$, each of the $\mathrm{mult}_G(\phi(u)\phi(v))$ edges between $\phi(u)$ and $\phi(v)$ is the unique edge of a path of length 1 in $\mathcal{P}'_{uv}$.

Proof of Claim 1. For contradiction, suppose there exist $\phi(u), \phi(v) \in V(G)$ such that there is an edge e in G between $\phi(u)$ and $\phi(v)$ that is not the unique edge of a path in $\mathcal{P}'_{uv}$. Note that this means that e is not contained in any path of $\mathcal{P}'_{uv}$. We claim that $\mathcal{P}'_{uv}$ contains a path of length at least 2. For contradiction, suppose that each of the $\text{mult}_H(uv)$ paths in $\mathcal{P}_{uv}$ has length 1. By assumption, $\text{mult}_G(\phi(u)\phi(v)) \leq \text{mult}_H(uv)$. Since all the paths in $\mathcal{P}'$ are edge-disjoint and e is not contained in any path of $\mathcal{P}'_{uv}$, the number of paths in $\mathcal{P}'_{uv}$ can be at most $\text{mult}_H(uv) - 1$, yielding the desired contradiction.

We now distinguish two cases, depending on whether or not e belongs to some path in $\mathcal{P}'$. We obtain a contradiction in both cases, which will imply the validity of the claim.

First suppose that there exists a path $P \in \mathcal{P}'$ such that e is an edge of P. Observe that P must have length at least 2, and hence $P \in \mathcal{P}'_{xy}$ for some $\{x, y\} \neq \{u, v\}$. Recall that $\mathcal{P}'_{uv}$ contains a path P' from $\phi(u)$ to $\phi(v)$ of length at least 2. Consider the two paths P' and P. Recall that P is a path between $\phi(x)$ and $\phi(y)$ containing the edge e. Let P_{uv} be the path that has e as its only edge, and let P_{xy} be the path obtained from P by removing e and adding the internal vertices and the edges of P_{uv}. Observe that the paths P_{uv} and P_{xy} use exactly the same vertices and edges as P' and P. Let $\mathcal{P}''_{uv} = \mathcal{P}'_{uv} \setminus \{P'\} \cup \{P_{uv}\}$ and $\mathcal{P}''_{xy} = \mathcal{P}'_{xy} \setminus \{P\} \cup \{P_{xy}\}$, and let $\mathcal{P}''_{ab} = \mathcal{P}'_{ab}$ for every $\{a, b\} \notin \{\{u, v\}, \{x, y\}\}$. Then $\mathcal{P}'' = \bigcup_{u,v \in V(H)} \mathcal{P}''_{uv}$ is a family of paths that satisfies properties (i) and (ii), and contains one more path of length 1 than $\mathcal{P}'$, contradicting the choice of $\mathcal{P}'$.

Now suppose that e is not contained in any path of $\mathcal{P}'$. Let $P' \in \mathcal{P}'_{uv}$ be a path of length at least 2, and let P_{uv} be the path that has e as its unique edge. We define $\mathcal{P}''_{uv} = \mathcal{P}'_{uv} \setminus \{P'\} \cup \{P_{uv}\}$ and $\mathcal{P}''_{xy} = \mathcal{P}'_{xy}$ for every $\{x, y\} \neq \{u, v\}$. Then $\mathcal{P}'' = \bigcup_{u,v \in V(H)} \mathcal{P}''_{uv}$ is a family of paths that satisfies properties (i) and (ii), and contains one more path of length 1 than $\mathcal{P}'$. As in the previous case, this contradicts the choice of $\mathcal{P}'$. This concludes the proof of Claim 1.   ◇

Consider the set $\mathcal{P}'_{uv}$ for some $u, v \in V(H)$. By property (i), $\mathcal{P}'_{uv}$ consists of $\text{mult}_H(uv)$ mutually edge-disjoint paths in G from $\phi(u)$ to $\phi(v)$. By Claim 1, at least $\text{mult}_G(\phi(u)\phi(v))$ many of these paths have length 1. By the definition of multiplicity, there are exactly $\text{mult}_G(\phi(u)\phi(v))$ edges between $\phi(u)$ and $\phi(v)$, which implies that $\mathcal{P}'_{uv}$ contains exactly $\text{mult}_G(\phi(u)\phi(v))$ paths of length 1. This, together with property (ii), implies that no path in $\mathcal{P}' \setminus \mathcal{P}'_{uv}$ contains an edge between $\phi(u)$ and $\phi(v)$. By symmetry, no path in $\mathcal{P}'_{uv}$ contains an edge between $\phi(x)$ and $\phi(y)$ for $\{x, y\} \neq \{u, v\}$. Hence, if we lift each of the $\text{mult}_H(uv) - \text{mult}_G(\phi(u)\phi(v))$ paths in $\mathcal{P}'_{uv}$ of length at least 2, then we create $\text{mult}_H(uv) - \text{mult}_G(\phi(u)\phi(v))$ new edges between $\phi(u)$ and $\phi(v)$, and do not change $\text{mult}_G(\phi(x)\phi(y))$ for any $\{x, y\} \neq \{u, v\}$. Note that the number of edges between $\phi(u)$ and $\phi(v)$ after lifting the paths in $\mathcal{P}'_{uv}$ is exactly $\text{mult}_H(uv)$.

From the above arguments, it is clear that if we lift each of the paths of length at least 2 in $\mathcal{P}'$, then we obtain a graph G' such that for every $u, v \in V(H)$, we have that $\text{mult}_{G'}(\phi(u)\phi(v)) = \text{mult}_H(uv)$. Hence ϕ is an isomorphism from H to

$G'[S]$. This shows that there exists a sequence of lifts $\mathcal{L}$ such that $G \vee \mathcal{L}$ contains H as an induced subgraph, implying that $(S, \mathcal{L}, \phi)$ is an H-model of G.   □

We are now ready to state the main theorem of this section.

Theorem 2. *For every fixed multigraph H, there is a polynomial-time algorithm that decides for any graph G whether G contains H as an induced immersion.*

Proof. Let H be a fixed multigraph, and let G be a graph. Deciding whether G contains H as an induced immersion is equivalent to deciding whether G has an H-model due to Observation 2. We describe an algorithm with running time $O(|V(G)|^{|V(H)|^2+2})$ that finds an H-model $(S, \mathcal{L}, \phi)$ of G, or decides that such a H-model does not exist. Note that this is a polynomial-time algorithm since H is a fixed multigraph.

Suppose G has an H-model $(S, \mathcal{L}, \phi)$. Then there exists a sequence of lifts $\mathcal{L}^*$ that satisfies conditions (i)–(iv) of Lemma 4. Let us determine an upper bound on the number of sequences of lifts $\mathcal{L}$ that satisfy conditions (i) and (ii). First note that the number of possible triples $u, v, w \in V(G)$ such that $\{uv, vw\}$ is a lift satisfying $v \in S$ and $\{u, w\} \cap S \neq \emptyset$ is at most $|V(H)|^2 \cdot |V(G)|$. Consequently, the number of sequences of at most $|E(G[S])| \leq |V(H)|^2$ such lifts is at most $(|V(H)|^2 \cdot |V(G)|)^{|V(H)|^2}$. Hence, there are at most $(|V(H)|^2 \cdot |V(G)|)^{|V(H)|^2}$ sequences of lifts that satisfy conditions (i) and (ii) of Lemma 4, and all these sequences can easily be generated in time $(|V(H)|^2 \cdot |V(G)|)^{|V(H)|^2}$.

For all possible subsets $S \subseteq V(G)$ of size $|V(H)|$ and all possible bijections ϕ from $V(H)$ to S, our algorithm acts as follows. For all possible sequences of lifts $\mathcal{L}$ that satisfy conditions (i) and (ii) of Lemma 4, the algorithm checks whether $\mathcal{L}$ satisfies (iii). If $\mathcal{L}$ does not satisfy (iii), then $\mathcal{L}$ is not a valid candidate for $\mathcal{L}^*$, and we can safely discard it. If $\mathcal{L}$ satisfies condition (iii), then we determine whether $\mathcal{L}$ satisfies (iv) as follows. We use the algorithm of Kawarabayashi, Kobayashi and Reed [14] in order to decide whether conditions (i) and (ii) of Lemma 5 hold for the graphs $G \vee \mathcal{L}$ and H. If so, then Lemma 5 guarantees that $G \vee \mathcal{L}$ contains H as an induced immersion, i.e., $\mathcal{L}$ satisfies condition (iv) of Lemma 4. This implies that G contains H as an induced immersion, so the algorithm outputs "yes". If $\mathcal{L}$ does not satisfy condition (iv), we proceed to the next sequence $\mathcal{L}$. If no sequence $\mathcal{L}$ yields a "yes"-answer, then Lemmas 4 and 5 ensure that G has no H-model $(S, \mathcal{L}, \phi)$ for this particular choice of S and ϕ, so the algorithm chooses the next combination of S and ϕ. If none of the combinations of S and ϕ yields a "yes"-answer, then we know from Lemma 4 that G does not have any H-model, and the algorithm outputs "no".

It remains to analyze the running time of the algorithm. There are at most $|V(G)|^{|V(H)|}$ subsets $S \subseteq V(G)$ of size $|V(H)|$, and for each of these sets S, there are $|V(H)|!$ bijections from $V(H)$ to S. As we saw earlier, there are at most $(|V(H)|^2 \cdot |V(G)|)^{|V(H)|^2}$ different sequences of lifts $\mathcal{L}$ that satisfy conditions (i) and (ii) of Lemma 4. This means the algorithm considers at most $O(|V(G)|^{|V(H)|^2})$ combinations of S, ϕ and $\mathcal{L}$. For each of these combinations, testing whether condition (iii) holds can be done in time $O(|V(H)|^2(|E(H)|+1))$.

It takes $O(|V(G)|^2)$ time [14] to test whether condition (iv) holds, as we ask for edge-disjoint paths between at most $|E(H)|$ pairs of terminals, and $|E(H)|$ is a constant due to the assumption that H is fixed. This yields an overall running time of $O(|V(G)|^{|V(H)|^2+2})$. □

4   Excluding a Fixed Multigraph as an Induced Immersion

In this section, we show that every multigraph with large enough treewidth contains every multigraph of maximum degree at most 2 as an induced immersion. We first show that every multigraph H is contained as an induced immersion in any multigraph G that contains a sufficiently large clique.

Lemma 6. (★) *Let G and H be two multigraphs. If G contains $K_{2(|V(H)|+|E(H)|)}$ as a subgraph, then G contains H as an induced immersion.*

In an elementary wall W of height r, we define the i-th *row* of W to be the set of vertices $\{(j,i) \mid 0 \leq j \leq 2r+1\}$. Similarly, the set $\{(j,i) \mid 0 \leq i \leq r\}$ is the j-th *column* of W. We write $P_i(j,k)$ to denote the unique path in W between (j,i) and (k,i) that contains only vertices of row i.

Lemma 7. (★) *Let G be a multigraph, and let H be a multigraph of maximum degree at most 2. If G contains an elementary wall W as a subgraph such that $G[V(W)]$ contains an independent set of size $4|V(H)|(|V(H)|+2)$, then G contains H as an induced immersion.*

We are now ready to prove the main result of this section.

Theorem 3. *For every multigraph H of maximum degree at most 2, there exists a constant c_H such that every multigraph with treewidth more than c_H contains H as an induced immersion.*

Proof. Let G and H be two multigraphs such that G does not contain H as an induced immersion. Let r be the largest integer such that G contains the elementary wall W_r as an immersion. Then there is a sequence $\mathcal{X}$ of vertex deletions, edge deletions and lifts such that applying $\mathcal{X}$ to G yields W_r. Let W' be the graph obtained from G by applying only the vertex deletions and lifts in $\mathcal{X}$. Then G contains W' as an induced immersion. Note that W' contains W_r as a spanning subgraph. Since G does not contain H as an induced immersion, Lemma 7 implies that W' does not have an independent set of size $4|V(H)|(|V(H)|+2)$. In addition, we know that W' does not contain a clique of size $2(|V(H)|+|E(H)|)$ as a subgraph, as otherwise G would contain H as an induced immersion as a result of Lemma 6. Ramsey's Theorem (cf. [3]) states that a graph that has neither a clique nor an independent set of size more than k must have at most 2^{2k-3} vertices. Hence, we know that W' has at most 2^{2k-3} vertices, where $k \leq \max\{4|V(H)|(|V(H)|+2), 2(|V(H)|+|E(H)|)\}$. Since H has maximum degree at most 2, we have $|E(H)| \leq 2|V(H)|$ and consequently $k \leq 4|V(H)|(|V(H)|+2)$. On the other hand, by the definition of an elementary wall of height r, we know that W' has exactly $2(r+1)^2 - 2$ vertices. Therefore we obtain that $2(r + 1)^2 - 2 \leq 2^{2k-3}$, which implies $r \leq 2^k$. By the definition of r, G does not contain W_{r+1} as an immersion. Hence, by Theorem 1, the treewidth of G is at most $20^{10(r+2)^5} \leq 20^{10(2^k+2)^5}$. We conclude that every multigraph with treewidth more than $20^{10(2^{4|V(H)|(|V(H)|+2)}+2)^5}$ contains H as an induced immersion. $\square$

5   Concluding Remarks

It is not hard to show that every induced immersion of an elementary wall is a planar graph with maximum degree at most 3, and that an elementary wall of height c has treewidth at least c. This implies that we cannot replace "maximum degree at most 2" in Theorem 3 by "maximum degree at most 4". A natural question is whether Theorem 3 holds for every planar multigraph H of maximum degree at most 3.

Our results exhibit some interesting relations between induced immersions and immersions. For example, Lemma 5 readily implies the following result.

Corollary 1. *Let G and H be two multigraphs. If there is a set $S \subseteq V(G)$ such that $G[S]$ is isomorphic to a spanning subgraph of H, then G contains H as an induced immersion if and only if G contains H as an immersion.*

We can also note the following corollary of Lemma 6.

Corollary 2. ($\bigstar$) *Let G and H be two multigraphs. If G contains the graph $K_{2(|V(H)|+|E(H)|)}$ as an immersion, then G contains H as an induced immersion.*

Corollary 2 implies that forbidding a graph as an induced immersion also forbids another graph as an immersion. In particular, the structure theorem for graphs excluding a clique of fixed size as an immersion [20] can also be applied to induced immersions, at the cost of larger constants. Moreover, every class of graphs closed under taking induced immersions has bounded degeneracy [1].

Two very interesting and challenging questions on induced immersions remain. In terms of parameterized complexity [5], Theorem 2 states that INDUCED IMMERSION is in XP when parameterized by the size of H, i.e., $|V(H)|+|E(H)|$. A natural question is whether the problem is fixed-parameter tractable (FPT) when parameterized by the size of H. Very recently, Grohe et al. [12] established fixed-parameter tractability of the closely related IMMERSION problem, thereby resolving a longstanding open question by Downey and Fellows [4]. Interestingly, the next lemma shows that an FPT algorithm for INDUCED IMMERSION would immediately imply an FPT algorithm for IMMERSION, rendering the problem of finding such an FPT algorithm a very challenging one.

Lemma 8. ($\bigstar$) *There exists a parameterized reduction from IMMERSION to INDUCED IMMERSION if both problems are parameterized by the size of H.*

Another celebrated result on immersions, due to Robertson and Seymour [18], states that all graphs are well-quasi ordered with respect to the immersion relation. Does the same hold for induced immersions? If so, then proving such a statement seems to be a formidable task, as it would imply the aforementioned result of Robertson and Seymour. An easier task, recently proposed by Fellows, Hermelin and Rosamond [6], would be to identify interesting classes of graphs that are well-quasi ordered under the induced immersion relation.

References

1. DeVos, M., Dvořák, Z., Fox, J., McDonald, J., Mohar, B., Scheide, D.: Minimum degree condition forcing complete graph immersion (submitted for publication)
2. DeVos, M., Kawarabayashi, K., Mohar, B., Okamura, H.: Immersing small complete graphs. Ars Math. Contemp. 3, 139–146 (2010)
3. Diestel, R.: Graph Theory. Electronic edn. Springer (2005)
4. Downey, R.G., Fellows, M.R.: Fixed-parameter intractability. In: Structure in Complexity Theory Conference, pp. 36–49 (1992)
5. Downey, R.G., Fellows, R.: Parameterized Complexity. Monographs in Computer Science. Springer (1999)
6. Fellows, M.R., Hermelin, D., Rosamond, F.A.: Well quasi orders in subclasses of bounded treewidth graphs and their algorithmic applications. Algorithmica 64, 3–18 (2012)
7. Fellows, M.R., Kratochvíl, J., Middendorf, M., Pfeiffer, F.: The complexity of induced minors and related problems. Algorithmica 13, 266–282 (1995)
8. Fellows, M.R., Langston, M.A.: On well-partial-order theory and its application to combinatorial problems of VLSI design. SIAM J. Disc. Math. 5(1), 117–126 (1992)
9. Ferrara, M., Gould, R., Tansey, G., Whalen, T.: On H-immersions. J. Graph Theory 57, 245–254 (2008)
10. Garey, M.R., Johnson, D.S., Tarjan, R.E.: The planar Hamiltonian circuit problem is NP-complete. SIAM J. Computing 5(4), 704–714 (1976)
11. Giannopoulou, A., Kamiński, M., Thilikos, D.M.: Forbidding Kuratowski graphs as immersions (manuscript)
12. Grohe, M., Kawarabayashi, K., Marx, D., Wollan, P.: Finding topological sugraphs is fixed-parameter tractable. In: STOC 2011, pp. 479–488. ACM (2011)
13. Grohe, M., Marx, D.: Structure theorem and isomorphism test for graphs with excluded topological subgraphs. In: STOC 2012, pp. 173–192. ACM (2012)
14. Kawarabayashi, K., Kobayashi, Y., Reed, B.: The disjoint paths problem in quadratic time. J. Comb. Theory, Ser. B 102, 424–435 (2012)
15. Lévêque, B., Lin, D.Y., Maffray, F., Trotignon, N.: Detecting induced subgraphs. Discrete Applied Math. 157(17), 3540–3551 (2009)
16. Robertson, N., Seymour, P.D.: Graph minors XIII: The disjoint paths problem. J. Comb. Theory, Ser. B 63(1), 65–110 (1995)
17. Robertson, N., Seymour, P.D.: Graph minors XX: Wagner's conjecture. J. Comb. Theory, Ser. B 92(2), 325–357 (2004)
18. Robertson, N., Seymour, P.D.: Graph Minors XXIII: Nash-Williams' immersion conjecture. J. Comb. Theory, Ser. B 100(2), 181–205 (2010)
19. Robertson, N., Seymour, P.D., Thomas, R.: Quickly excluding a planar graph. J. Comb. Theory, Ser. B 62(2), 323–348 (1994)
20. Seymour, P.D., Wollan, P.: The structure of graphs not admitting a fixed immersion (manuscript)

Rectilinear Covering for Imprecise Input Points*

(Extended Abstract)

Hee-Kap Ahn[1], Sang Won Bae[2,**], and Shin-ichi Tanigawa[3]

[1] Department of Computer Science and Engineering, POSTECH, Pohang, Korea
heekap@postech.ac.kr
[2] Department of Computer Science, Kyonggi University, Suwon, Korea
swbae@kgu.ac.kr
[3] Research Institute for Mathematical Science, Kyoto University, Kyoto, Japan
tanigawa@kurims.kyoto-u.ac.jp

Abstract. We consider the rectilinear k-center problem in the presence of impreciseness of input points. We assume that the input is a set S of n unit squares, possibly overlapping each other, each of which is interpreted as a measured point with an identical error bound under the L_∞ metric on $\mathbb{R}^2$. Our goal, in this work, is to analyze the worst situation with respect to the rectilinear k-center for a given set S of unit squares. For the purpose, we are interested in a value $\lambda_k(S)$ that is the minimum side length of k congruent squares by which any possible true point set from S can be covered. We show that, for $k = 1$ or 2, computing $\lambda_k(S)$ is equivalent to the problem of covering the input squares S completely by k squares, and thus one can solve the problem in linear time. However, for $k \geq 3$, this is not the case, and we present an $O(n \log n)$-time algorithm for computing $\lambda_3(S)$. For structural observations, we introduce a new notion on geometric covering, namely the covering-family, which is of independent interest.

1 Introduction

Given a set P of points in the plane and a positive integer k, the *rectilinear k-center problem* is to find k service points $q_1, \ldots, q_k$ in the plane that minimizes the maximum L_∞ (or equivalently L_1) distance from each point of P to its nearest service points. The problem is equivalent to a covering problem of finding k congruent squares of minimum side length whose union covers all points in P.

The rectilinear k-center problem has been studied extensively [3, 6, 11, 15, 16, 18]. When k is a part of input, the problem has been proved to be NP-complete by Megiddo and Supowit [15], and a polynomial-time approximation algorithm with approximation ratio 2 is proposed by Ko et al. [10]. For small fixed k, efficient algorithms are known. Drezner [3] presented $O(n)$-time algorithms for $k = 1$ or 2. For $k = 3$, Sharir and Welzl [18] presented a first linear-time algorithm and Hoffmann [6] a simpler algorithm based on the prune-and-search technique. Indeed, it is known that the rectilinear k-center problem is of LP-type for $k \leq 3$ [18]. For any fixed $k > 3$, exact algorithms

* Work by H.-K. Ahn was supported by NRF grant 2011-0030044 (SRC-GAIA) funded by the government of Korea. Work by S.W.Bae was supported by National Research Foundation of Korea(NRF) grant funded by the Korea government(MEST) (No. 2011-0005512).
** Corresponding author.

K.-M. Chao, T.-s. Hsu, and D.-T. Lee (Eds.): ISAAC 2012, LNCS 7676, pp. 309–318, 2012.
© Springer-Verlag Berlin Heidelberg 2012

are also known: $O(n \log n)$ time for $k = 4$ and $k = 5$ and $O(n^{k-4} \log n)$ time for $k \geq 6$ [16, 18]. In a viewpoint as covering problem, the rectilinear k-center problem is extended to problems of covering a variant geometric objects by k squares. Among those, Hoffmann [5] considered line segments or planar regions as input and showed that, for $k \leq 3$, it is possible in linear time to find k congruent squares of minimum side length that completely cover such input objects.

Recently, how to handle imprecise data in geometric problems has received a remarkable interest. Most of previous results on geometric problems, including the rectilinear k-center problem and its variations, assume that the input points are given precisely. This is, however, not necessarily true in many practical situations. For example, there could be some imprecision arising when we measure real-world objects due to limited precision of measuring devices. This impreciseness of geometric data has been studied lately in computational geometry, and few algorithms that handle imprecise input data have been presented for fundamental geometric problems: computing the axis-aligned bounding box [12], smallest enclosing circle [7, 12], the Hausdorff distance [9], the discrete Fréchet distance [1], Voronoi diagrams [17], planar convex hulls [14], and Delaunay triangulations [2, 8, 13].

In this paper, we consider the rectilinear k-center problem in the presence of impreciseness of input points. In order to properly model the impreciseness, we assume that the input is given a set S of n unit squares, possibly overlapping each other, each of which is interpreted as a measured point with an identical error bound under the L_∞ metric on $\mathbb{R}^2$. We thus assert that for every unit square $s \in S$, there should exist exactly one true point lying in s. As done in most recent research on impreciseness in geometric problems, our goal is to specify the worst situation with respect to the rectilinear k-center for a given set S of unit squares. More precisely, we call any point set $R \subset \mathbb{R}^2$ a *realization* of S if R chooses exactly one point $p \in s$ from each $s \in S$. We are then interested in the maximum value of side lengths of the optimal k squares for all such realizations R of S, denoted by $\lambda_k(S)$. In a worst case, to cover the true point set will need k squares of side length exactly $\lambda_k(S)$.

As our results, we show that, for $k \leq 2$, specifying the worst case of the rectilinear k-center problem for imprecise points is equivalent to the problem of covering the input squares S completely by k squares. This means that the worst situation can be specified in linear time by running any existing algorithm for covering unit squares by one or two squares [5]. Interestingly, for $k \geq 3$, the above relation does not hold any more. In this paper, we mainly focus on the case of $k = 3$, and show that our problem can be solved in $O(n \log n)$ time. For our purpose, we introduce a new notion on geometric covering, namely the *covering-family*, which is defined to be a family of k-tuples of congruent squares such that every realization of S is covered by one of its members.

Due to lack of space, almost all proofs are omitted from this extended abstract but will be presented in a full version.

2 Preliminaries

Let $d(p, q)$ be the L_∞ distance between $p \in \mathbb{R}^2$ and $q \in \mathbb{R}^2$. Let $\mathbf{B}(A)$ denote the axis-aligned bounding box of a subset A of $\mathbb{R}^2$, or a family of subsets of $\mathbb{R}^2$. For a rectangle

A, we denote the length of its horizontal side by $w(A)$ and that of its vertical side by $h(A)$; its top-left, top-right, bottom-left, and bottom-right corner by $\ulcorner(A)$, $\urcorner(A)$, $\llcorner(A)$, and $\lrcorner(A)$; the top, bottom, left, and right side of A by $\top(A)$, $\bot(A)$, $\vdash(A)$, and $\dashv(A)$, respectively.

Given a set P of points in the plane and a positive integer k, the rectilinear k-center problem is to find k points $q_1, \ldots, q_k \in \mathbb{R}^2$ that minimizes $\max_{p \in P} \min_{i=1,\ldots,k} d(p, q_i)$. We shall call a k-tuple of congruent closed squares a k-covering, or simply covering, and the side length of a k-covering denotes that of its member squares. For a covering C, by an abuse of notation, we also mean C by the union of the k squares. As known well, the rectilinear k-center problem is equivalent to finding a k-covering $C = (\sigma_1, \ldots, \sigma_k)$ of minimum side length that contains all the points in P. To see the equivalence, note that the k squares σ_i can be specified by its center at q_i and its side length to be $2 \max_{p \in P} \min_{i=1,\ldots,k} d(p, q_i)$, and vice versa. We define $\rho_k(P)$ to be the side length of such a k-covering C.

The problem naturally extends to various type of input objects. In this paper, we are interested in the case where input objects are given as squares. Let S be a set of n unit closed squares in $\mathbb{R}^2$, possibly overlapping. Then, the value $\rho_k(\cdot)$ can be extended to a set of squares: $\rho_k(S)$ denotes the minimum side length of a k-covering C that completely covers S, that is, $S \subseteq C$. Note that this problem of covering a set S of squares is not equivalent to the original rectilinear k-center problem.

In this paper, we interpret each $s \in S$ as a sampled input point with an identical error bound under the L_∞ metric. We thus assume that the original point lies in s but we do not know its exact position. A *realization* R of S is a set of n points each of which is chosen from a distinct input square in S. Under this assumption, we are interested in the worst situation that the original point set yields the maximum ρ_k value among all possible realizations of S. This can be defined as follows:

$$\lambda_k(S) := \max_{\text{realization } R \text{ of } S} \rho_k(R).$$

The value $\lambda_k(S)$ is well defined since we deal with closed squares only.

Note by definition that for any realization R of S there exists a k-covering C of side length $\lambda_k(S)$ such that $R \subset C$, that is, $\rho_k(R) \leq \lambda_k(S)$. This, however, does not imply that such a k-covering C covers all realizations R of S. A witness for the value $\lambda_k(S)$ would be a family $\mathcal{C}$ of k-coverings such that for any realization R of S, there exists a covering $C \in \mathcal{C}$ with $R \subset C$. We call such a family $\mathcal{C}$ a *covering-family* for S. A covering-family is called *optimal* if all of its coverings are of side length exactly $\lambda_k(S)$. For a k-covering $C \in \mathcal{C}$ in a covering-family for S, a realization R of S is called C-*responsible* if $R \subset C$ but $R \not\subseteq C'$ for any other $C' \in \mathcal{C}$ with $C' \neq C$. A covering-family $\mathcal{C}$ is called *minimal* if for each covering $C \in \mathcal{C}$ there exists a C-responsible realization R. The side length of a covering-family $\mathcal{C}$ denotes the maximum value of those of all coverings in $\mathcal{C}$.

One can find an easy bound $\lambda_k(S) \leq \rho_k(S)$: Since there exists a k-covering C of side length $\rho_k(S)$ that completely covers S, for any realization R of S, we have $R \subset C$. If the equality $\lambda_k(S) = \rho_k(S)$ holds, then the problem of computing $\lambda_k(S)$ and the corresponding optimal covering-family C^* becomes equivalent to the rectilinear k-covering problem for squares S.

Lemma 1. *For any given S and k, the equality $\lambda_k(S) = \rho_k(S)$ holds if and only if there is an optimal and minimal covering-family $\mathcal{C}^*$ such that $|\mathcal{C}^*| = 1$.*

Also, observe that if $k \geq n = |S|$, then $\lambda_k(S)$ becomes trivially zero. In this case, any realization R of S can be covered by a k-covering C of side length zero; that is, $C = R$, so an optimal covering-family $\mathcal{C}$ consists of all possible realizations themselves. We thus assume that $k < n$ throughout the paper. Then, we obtain a non-trivial lower bound on $\lambda_k(S)$.

Lemma 2. *For any positive integer k and any set S of n unit squares with $n > k$, possibly being overlapped, we have $\lambda_k(S) \geq \frac{1}{\lfloor \sqrt{k} \rfloor}$. Moreover, this bound is tight.*

3   General Structure of Optimal Covering-Family

Our goal in the paper is to compute $\lambda_k(S)$ and an optimal and minimal covering-family for given S and small k. In this section, we investigate the general structure of optimal covering-family for any $k > 0$.

Let $\mathcal{C}^*$ be an optimal and minimal covering-family for given S and k. For $s \in S$, an s-*covering-group* is a subfamily $G_s \subseteq \mathcal{C}^*$ such that $s \subseteq \bigcup_{C \in G_s} C$ and $s \not\subseteq \bigcup_{C' \in G_s \setminus \{C\}} C'$ for each $C \in G_s$.

We then claim the following.

Lemma 3 (Grouping Lemma). *For any $s \in S$ and any $C \in \mathcal{C}^*$, C belongs to an s-covering-group G_s for which there exists a realization R_0 of $S \setminus \{s\}$ such that $R_0 \cup \{p\}$ is C-responsible for some $p \in s$ and $R_0 \cup \{p'\} \subset C'$ is covered by some $C' \in G_s$ for any $p' \in s$.*

Note that it is not always true that each covering $C \in \mathcal{C}^*$ belongs to a unique s-covering-group. One can construct an example where a covering belongs to two different covering-groups.

Consider any optimal and minimal covering-family $\mathcal{C}^*$. Unless $\lambda_k(S)$ is larger than the length of the shorter side of $\mathbf{B}(S)$, we can assume that the bounding box $\mathbf{B}(C)$ of each $C \in \mathcal{C}^*$ is included in the bounding box $\mathbf{B}(S)$ of S; that is, $\mathbf{B}(C) \subseteq \mathbf{B}(S)$. The following lemma shows the reverse inclusion relation between them under a reasonable condition.

Lemma 4 (Fitting Lemma). *For given S, let $\mathcal{C}^*$ be any optimal and minimal covering-family and suppose that $\lambda_k(S) \geq 1$. Then, for any covering $C \in \mathcal{C}^*$, it holds that $\mathbf{B}(S) \subseteq \mathbf{B}(C)$.*

4   Optimal Covering-Family for $k \leq 2$

Now, we discuss the case of $k \leq 2$. As aforementioned, we in general have that $\lambda_k(S) \leq \rho_k(S)$ for any S and k. In this section, we show that the equality $\lambda_k(S) = \rho_k(S)$ holds for $k = 1, 2$. By Lemma 1, we conclude that the problem of computing $\lambda_k(S)$ is equivalent to computing minimum covering that completely covers S. Computing $\rho_k(S)$ for $k \leq 2$ and the corresponding k-covering can be done in linear time: for $k = 1$, it can be done by computing the smallest square covering all in S, and Hoffmann [5] reported a linear-time algorithm for $k = 2$. We therefore conclude the following.

Theorem 1. *Let $k = 1$ or 2 and S be a set of $n > k$ unit squares. Then, it holds that $\lambda_k(S) = \rho_k(S)$ and the corresponding covering-family can be computed in $O(n)$ time.*

Sketch of Proof. For $k = 1$, its proof is not difficult by exploiting Lemmas 2 and 4. We focus on proving $\lambda_2(S) = \rho_2(S)$.

Let C^* be any optimal and minimal covering-family for S and $k = 2$. By Lemma 2, we have $\lambda_2(S) \geq 1$ and thus for any 2-covering $C \in C^*$ we have $\mathbf{B}(S) \subseteq \mathbf{B}(C)$ by Lemma 4. Without loss of generality, we assume that the horizontal side of $\mathbf{B}(S)$ is not shorter than its vertical side, that is, $h(\mathbf{B}(S)) \leq w(\mathbf{B}(S))$. Let $C = (\sigma_1, \sigma_2) \in C^*$ be a 2-covering. By a typical translation process, one can assume that either (1) $\ulcorner(\sigma_1) = \ulcorner(\mathbf{B}(S))$ and $\llcorner(\sigma_2) = \llcorner(\mathbf{B}(S))$, or (2) its symmetric configuration; $\llcorner(\sigma_1) = \llcorner(\mathbf{B}(S))$ and $\urcorner(\sigma_2) = \urcorner(\mathbf{B}(S))$. This implies that C^* consists of at most two coverings.

If $|C^*| = 1$, then we are done by Lemma 1. Suppose that $|C^*| = 2$. Then, no covering in C^* completely covers S. Let $C = (\sigma_1, \sigma_2) \in C^*$ be such that $\ulcorner(\sigma_1) = \ulcorner(\mathbf{B}(S))$ and $\llcorner(\sigma_2) = \llcorner(\mathbf{B}(S))$, and $C' = (\sigma_1', \sigma_2') \in C^*$ be the other such that $\llcorner(\sigma_1') = \llcorner(\mathbf{B}(S))$ and $\urcorner(\sigma_2') = \urcorner(\mathbf{B}(S))$. There exists $s \in S$ such that $s \not\subseteq C$ since $S \not\subseteq C$. We claim that at least one corner of s is not contained in C; if every corner of s lies in C though $s \not\subseteq C$, then there exists a point $p \in s$ between two squares σ_1 and σ_2. This leads to a contradiction since we have $p \notin C'$ either.

Without loss of generality, we assume that $\ulcorner(s) \in \sigma_1$ but $\urcorner(s) \notin \sigma_1$. The other cases are symmetric to this configuration. By Lemma 3, C should belong to an s-covering-group $G_s \subseteq C^*$ whose cardinality is more than one. Since $|C^*| = 2$, we have $G_s = C^*$.

Note that the left side $\vdash(\sigma_2')$ of σ_2' should be to the left of the right side $\dashv(\sigma_1)$ of σ_1. (Otherwise, there is a point $p \in s$ such that $p \notin C \cup C'$, a contradiction.) Also, by the definition of s-covering-groups, we have that $s \not\subseteq \sigma_2'$. This implies that $2\lambda_2(S) - 1 < w(\mathbf{B}(S)) \leq 2\lambda_2(S)$. We also observe that $h(\mathbf{B}(S)) > \lambda_2(S)$ since, otherwise, we have $\mathbf{B}(S) \subseteq \sigma_1 \cup \sigma_2 = C$, a contradiction to the minimality of C^*.

Moreover, both σ_1 and σ_2' should touch the square $s_\top \in S$ touching the top side of $\mathbf{B}(S)$. However, since $w(\mathbf{B}(S)) > 2\lambda_2(S) - 1$, either σ_1 or σ_2' cannot completely cover $s_\top$. This implies that C or C' cannot cover $s_\top$ by $h(\mathbf{B}(S)) > \lambda_2(S)$, and hence Lemma 3 further implies that both C and C' cannot cover $s_\top$, as they should form an $s_\top$-covering-group. Now, consider a realization R in which p is chosen from s such that $p \in s \setminus C$ and $p_\top$ is chosen from $s_\top$ such that $p_\top \in s_\top \setminus C'$. Such points p and $p_\top$ exist by the definition of covering-groups. Then, we have that $R \not\subseteq C$ and simultaneously $R \not\subseteq C'$, a contradiction to the assumption that C^* is a covering-family for S. $\square$

5   Optimal Covering-Family for $k = 3$

Now, we discuss the case of $k = 3$. For ease of discussion, we assume a general position on S that the center of each input square in S has a distinct x- and y-coordinate. We call $s \in S$ *extremal* if s touches the boundary of $\mathbf{B}(S)$. Let $s_\top$, $s_\bot$, $s_\vdash$, and $s_\dashv$ be the extremal squares in S, each of which touches the top, bottom, left and right side of $\mathbf{B}(S)$, respectively. Note that they may not be distinct when there is $s \in S$ sharing a corner with $\mathbf{B}(S)$. Also, let $\sigma_\vdash$, $\sigma_\dashv$, $\sigma_\llcorner$, and $\sigma_\lrcorner$ be squares of side length $\lambda_3(S)$ such that $\ulcorner(\sigma_\vdash) = \ulcorner(\mathbf{B}(S))$, $\urcorner(\sigma_\dashv) = \urcorner(\mathbf{B}(S))$, $\llcorner(\sigma_\llcorner) = \llcorner(\mathbf{B}(S))$, and $\lrcorner(\sigma_\lrcorner) = \lrcorner(\mathbf{B}(S))$, respectively.

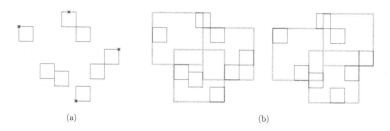

Fig. 1. (a) An example of 8 input squares S and (b) two 3-coverings that form an optimal and minimal covering-family

Unlike the case of $k \leq 2$, the equality $\lambda_3(S) = \rho_3(S)$ does not always hold. We start with introducing an example of input squares S where the strict inequality $\lambda_3(S) < \rho_3(S)$ holds, and thus any optimal and minimal covering-family consists of two or more 3-coverings. See Fig. 1: Let S be the set of 8 squares as shown in Fig. 1(a). Then, an optimal and minimal covering-family $\mathcal{C}^*$ consists of two coverings as in Fig. 1(b), while $\rho_3(S)$ is strictly larger than the side length of $\mathcal{C}^*$.

We first discuss some easy cases where the equality $\lambda_3(S) = \rho_3(S)$ holds. As stated above, in such a case, it is relatively easy to find an optimal covering-family: Lemma 1 states that there exists an optimal and minimal covering-family $\mathcal{C}^*$ with $|\mathcal{C}^*| = 1$, consisting of a 3-covering C such that $S \subseteq C$. For $k = 3$, Hoffmann presented a linear time algorithm to compute $\rho_3(S)$ and the corresponding 3-covering C.

Lemma 5. *If* $\min\{h(\mathbf{B}(S)), w(\mathbf{B}(S))\} \leq \lambda_3(S)$ *or* $\max\{h(\mathbf{B}(S)), w(\mathbf{B}(S))\} \geq 3\lambda_3(S)$, *then it holds that* $\lambda_3(S) = \rho_3(S)$.

In the remaining of the section, we investigate covering-families with good properties and bound the cardinality of such a covering-family, provided that $\lambda_3(S) < h(\mathbf{B}(S)), w(\mathbf{B}(S)) < 3\lambda_3(S)$, and thus $\lambda_3(S) < \rho_3(S)$. These properties and observations based on them at last play a key role to discretize a class of coverings that we should search for, even into a constant size. We then present an algorithm that computes $\lambda_3(S)$ and its corresponding minimal covering-family in $O(n \log n)$ time.

5.1 Properties of Covering-Family for $k = 3$

From now, we focus on properties of optimal and minimal covering-family when $\lambda_3(S) < \rho_3(S)$. By Lemma 5, if the strict inequality holds, then we may assume that $\lambda_3(S) < h(\mathbf{B}(S)), w(\mathbf{B}(S)) < 3\lambda_3(S)$. We start with observing the existence of a covering-family with specific properties.

Lemma 6. *There exists an optimal and minimal covering-family* $\mathcal{C}^*$ *for* S *and* $k = 3$ *such that any 3-covering* $C \in \mathcal{C}^*$ *fulfills the following properties CF1–CF4, including their symmetric analogues, provided that* $\lambda_3(S) < h(\mathbf{B}(S)), w(\mathbf{B}(S)) < 3\lambda_3(S)$:

CF1 $\mathbf{B}(C) = \mathbf{B}(S)$.
CF2 *If a square* $\sigma \in C$ *touches the top side of* $\mathbf{B}(S)$, *then* σ *intersects* $s_\top$.

CF3 *If there is $s \in S$ such that a side of $\sigma \in C$ intersects s but $s \not\subseteq C$, then there is $s' \in S$ with $s' \neq s$ such that the opposite side of σ intersects s' at a point that is avoided by the other two squares in C.*

CF4 *Suppose that C contains $\sigma_\top$ touching the top side of $\mathbf{B}(S)$. If there is $p \in s \in S$ such that $d(p, \top(\mathbf{B}(S))) \leq \lambda_3(S)$ and $p \notin \sigma_\top$, say p lies to the left of $\vdash(\sigma_\top)$, then there is $p' \in s' \in S$ with $s' \neq s$ such that $p' \in s' \cap \sigma_\top$, $d(p, p') > \lambda_3(S)$, and the other two squares in C than σ avoid p'.*

Now, we focus only on optimal and minimal covering-families satisfying **CF1–CF4** as in Lemma 6. Property **CF1**, together with $\min\{h(\mathbf{B}(S)), w(\mathbf{B}(S))\} > \lambda_3(S)$, implies that each $C \in C^*$ includes at least one of the four *cornered squares* $\sigma_\ulcorner, \sigma_\urcorner, \sigma_\llcorner$, and $\sigma_\lrcorner$, since C consists of three squares. Also, each side of $\mathbf{B}(S)$ touches at least one square in C; we call a square $\sigma \in C$ that touches a side of $\mathbf{B}(S)$ a *touching square*, and more specifically, if σ touches the top, bottom, left, or right side of $\mathbf{B}(S)$, we call σ a *top-touching, bottom-touching, left-touching,* or *right-touching* square of C, respectively. Notice by definition that any cornered square is also a touching square. Observe that there can be a square in C that touches none of the sides of $\mathbf{B}(S)$. We call such a square in C the *floating square* of C.

For $C \in C^*$, $\sigma \in C$, and $s \in S$, we say that s is σ-*reliable with respect to* C if $s \cap \sigma \not\subseteq \bigcup_{\sigma' \in C \setminus \{\sigma\}} \sigma'$ and there exists a C-responsible realization R in which a point in $s \cap \sigma$ is chosen from s. Note that by definition there exists a C-responsible realization R for any covering C in any minimal covering-family, and thus that for any $s \in S$ there is at least one $\sigma \in C$ such that s is σ-reliable with respect to C.

We consider the set $U(C) \subseteq S$ of input squares $s \in S$ such that $s \not\subseteq C$. Indeed, it will be shown later that its cardinality $|U(C)|$ is a key for bounding $|C^*|$, thus we would like to bound $|U(C)|$ by a reasonable number.

For a square σ, a side e of σ is said to be *aligned with* $s \in S$ if e intersects a side e' of s in a segment of positive length and either $e = \top(\sigma)$ and $e' = \top(s)$, $e = \bot(\sigma)$ and $e' = \bot(s)$, $e = \vdash(\sigma)$ and $e' = \vdash(s)$, or $e = \dashv(\sigma)$ and $e' = \dashv(s)$.

Lemma 7. *There exists an optimal and minimal covering-family C^* for S and $k = 3$ satisfying properties **CF1–CF4** and **CF5–CF6** in addition, provided that $\lambda_3(S) < h(\mathbf{B}(S)), w(\mathbf{B}(S)) < 3\lambda_3(S)$:*

CF5 *For any $C \in C^*$ and any $\sigma \in C$, at least one e of two opposite sides of σ is aligned with some $s \in S$. Moreover, the side e' of s touching e is not completely contained in the union of the other two squares in C than σ.*

CF6 *For any $C \in C^*$, if $C = (\sigma_\ulcorner, \sigma_\bot, \sigma_\dashv)$ and $\sigma_\lrcorner \notin C$, where $\sigma_\bot$ is bottom-touching and $\sigma_\dashv$ is right-touching, then $\vdash(\sigma_\bot)$ is aligned with some $s \in S$ and $\top(\sigma_\dashv)$ is aligned with some $s' \in S$. This also holds for the other symmetric configurations.*

By Lemma 7, we can assume that every $\sigma \in C$ for any $C \in C^*$ has two adjacent sides that are aligned. This implies that σ always has a corner located at an intersection point between two lines extending two sides of input squares. Note that this observation tells us that the number of possible positions of squares σ involved in C^* is bounded by $O(n^2)$. Property **CF5** is also exploited to discretize a possible search space for the exact value of $\lambda_3(S)$.

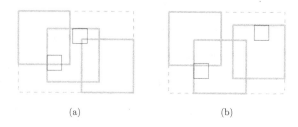

(a)   (b)

Fig. 2. An illustration to (a) a covering in $\mathcal{C}_{\ulcorner\lrcorner}^*$ and (b) a covering in $\mathcal{C}_{\ulcorner}^*$. Those in other subfamilies have symmetric configurations.

Lemma 8. *Let Λ be the set of real numbers λ such that λ is the distance $d(c_1, c_2)$ between a corner c_1 of some $s_1 \in S$ and a corner c_2 of another $s_2 \in S$, or its half $d(c_1, c_2)/2$. Then, $\lambda_3(S) \in \Lambda$, provided that $\lambda_3(S) < h(\mathbf{B}(S)), w(\mathbf{B}(S)) < 3\lambda_3(S)$.*

5.2   Bounds on $|U(C)|$ and $|\mathcal{C}^*|$

We are then able to bound the cardinality of $U(C)$ by a constant for any $C \in \mathcal{C}^*$, provided that $\mathcal{C}^*$ fulfills **CF1–CF6**. In this paper, we focus on obtaining constant bounds on the quantities but not on trying to get tight bounds since it suffices to devise an efficient algorithm later.

Lemma 9. *Suppose that $\mathcal{C}^*$ fulfills properties **CF1–CF6**, and let $C \in \mathcal{C}^*$. If C has a floating square, then $|U(C)| \leq 12$; otherwise, $|U(C)| \leq 14$.*

Now, we bound $|\mathcal{C}^*|$. We consider following six subfamilies of $\mathcal{C}^*$: Let $\mathcal{C}_{\ulcorner\lrcorner}^* := \{C \in \mathcal{C}^* \mid \sigma_{\ulcorner}, \sigma_{\lrcorner} \in C\}$, and $\mathcal{C}_{\llcorner\urcorner}^* := \{C \in \mathcal{C}^* \mid \sigma_{\llcorner}, \sigma_{\urcorner} \in C\}$. Let $\mathcal{C}_{\ulcorner}^* := \{C \in \mathcal{C}^* \setminus (\mathcal{C}_{\ulcorner\lrcorner}^* \cup \mathcal{C}_{\llcorner\urcorner}^*) \mid \sigma_{\ulcorner} \in C\}$. Define $\mathcal{C}_{\urcorner}^*, \mathcal{C}_{\llcorner}^*$, and $\mathcal{C}_{\lrcorner}^*$ analogously. Note that any covering $C \in \mathcal{C}^*$ falls into at least one of the six subfamilies, but the subfamilies do not form a disjoint partition of $\mathcal{C}^*$. More specifically, each $C \in \mathcal{C}^*$ having a floating square belongs to $\mathcal{C}_{\ulcorner\lrcorner}^*$ or $\mathcal{C}_{\llcorner\urcorner}^*$. By **CF5** and **CF6**, each covering $\mathcal{C}^*$ is of shape described in Fig. 2. Then, Lemma 9 gives a constant bound on the number of those $s \in S$ with which a square of C in each subfamily may be aligned. Based on these observations, we obtain following bounds.

Lemma 10. *Let $\mathcal{C}^*$ be any optimal and minimal covering-family for S and $k = 3$ satisfying **CF1–CF6**. Then, $|\mathcal{C}^*|$ and $|\bigcup_{C \in \mathcal{C}^*} U(C)|$ are upper bounded by a constant. More specifically, $|\mathcal{C}^*| \leq 2252$ and $|\bigcup_{C \in \mathcal{C}^*} U(C)| \leq 304$.*

5.3   Algorithm

Now, we describe an algorithm that computes the exact value of $\lambda_3(S)$ and an optimal and minimal covering-family $\mathcal{C}^*$ for a given set S of input unit squares. We first present a decision algorithm that determines if $\lambda \geq \lambda_3(S)$ for an additional input $\lambda \in \mathbb{R}$. In the complete algorithm that computes $\lambda_3(S)$ and the corresponding covering-family $\mathcal{C}^*$, we search Λ for the exact value of $\lambda_3(S)$ based on a sorted matrix search technique and our decision algorithm. As a preprocessing, we compute and maintain the sorted list of input squares in S both in horizontal and vertical directions.

Decision Algorithm. Let $\lambda > 0$ be a given real number. We first compute $\rho_3(S)$ in $O(n)$ time using any existing algorithm such as [5]. If $\lambda \geq \rho_3(S)$, then the answer is yes since $\lambda_3(S) \leq \rho_3(S)$. In the following, we discuss only the case of $\lambda < \rho_3(S)$. To decide whether $\lambda \geq \lambda_3(S)$, we perform the following steps in order: (1) we construct a set $\mathcal{C}_\lambda$ of all possible 3-coverings with side length λ that obey the properties **CF1–CF6**, and then (2) test whether $\mathcal{C}_\lambda$ is a covering-family, that is, whether all realizations of S is covered by one of its members $C \in \mathcal{C}_\lambda$. If the answer is yes, then (3) we extract a minimal covering-family $\mathcal{C}_\lambda^*$ by removing redundant coverings from $\mathcal{C}_\lambda$.

Let us give a sketch of the decision algorithm. In Step (1), we gather all candidate coverings whose union form a covering-family $\mathcal{C}_\lambda$ for S if $\lambda \geq \lambda_3(S)$. This can be done by separately handling six subfamilies $\mathcal{C}_{\llcorner}$, $\mathcal{C}_{\ulcorner}$, $\mathcal{C}_{\vdash}$, $\mathcal{C}_{\urcorner}$, $\mathcal{C}_{\llcorner}$, and $\mathcal{C}_{\lrcorner}$, which are analogous to those defined above Lemma 10. By **CF5–CF6**, together with Lemma 9, it suffices to consider a constant number of coverings to solve the decision problem. Thus, it is shown that the number of coverings in $\mathcal{C}_\lambda$ is bounded by a constant as in Lemma 10. Step (1) can be done in $O(n)$ time after $O(n \log n)$ time preprocessing. Then, Steps (2) and (3) can be performed in constant time since it is done by handling constant sized data, $\mathcal{C}_\lambda$ and $\bigcup_{C \in \mathcal{C}_\lambda} U(C)$.

Lemma 11. *Given a set S of n unit squares and a real number $\lambda \in \mathbb{R}$, one can decide whether $\lambda \geq \lambda_3(S)$ or not in $O(n)$ time, provided that S is sorted both in horizontal and vertical directions. Moreover, a minimal covering-family of side length λ can be found in the same time bound when $\lambda \geq \lambda_3(S)$.*

Computing $\lambda_3(S)$ Let Λ be the set as defined in Lemma 8. We basically search Λ to find $\lambda_3(S) \in \Lambda$ by exploiting our decision algorithm. Since $|\Lambda| = \Theta(n^2)$, however, a straightforward method would take $\Omega(n^2)$ time. We thus adopt the selection algorithm on sorted matrices by Frederickson and Johnson [4]. A matrix is said to be *sorted* if each row and each column of it is in a nondecreasing order.

Let $x_1, \ldots, x_{2n}$ be the x-coordinates of all corners of the n input squares S, in a nondecreasing order. Let M_X be a $2n \times 2n$ matrix such that its (i, j)-entry is defined as $M_X(i, j) := x_i - x_{2n-j+1}$ for each $1 \leq i, j \leq 2n$. Also, let $M_{X/2}$ be another matrix with $M_{X/2}(i, j) := (x_i - x_{2n-j+1})/2$. Let $y_1, \ldots, y_{2n}$ be the y-coordinates of all corners of $s \in S$ in a nondecreasing order. Define M_Y and $M_{Y/2}$ in an analogous way to M_X and $M_{X/2}$.

Observe that the four matrices are all sorted and $\Lambda \subset M_X \cup M_{X/2} \cup M_Y \cup M_{Y/2}$. This fulfills the precondition to apply the technique of Frederickson and Johnson [4], together with our decision algorithm. Subsequently, one can find in $O(n \log n)$ time a smallest value $\lambda_X \in M_X$ such that our decision algorithm reports "YES". Note that we do not explicitly compute the matrices. We perform the same procedure for $M_{X/2}$, M_Y, and $M_{Y/2}$ each, to get three more values $\lambda_{X/2}$, λ_Y, and $\lambda_{Y/2}$, respectively. Then, the smallest one of the four obtained values is exactly $\lambda_3(S)$ by Lemma 8, if $\lambda_3(S) < h(\mathbf{B}(S)), w(\mathbf{B}(S)) < 3\lambda_3(S)$. Otherwise, it holds that $\lambda_3(S) = \rho_3(S)$ by Lemma 5, and we have in this case that $\rho_3(S) \leq \min\{\lambda_X, \lambda_{X/2}, \lambda_Y, \lambda_{Y/2}\}$.

Theorem 2. *Given a set S of n unit squares, the exact value of $\lambda_3(S)$ and a corresponding covering-family $\mathcal{C}^*$ for S of side length $\lambda_3(S)$ with $|\mathcal{C}^*| = O(1)$ can be computed in $O(n \log n)$ time.*

References

1. Ahn, H.K., Knauer, C., Scherfenberg, M., Schlipf, L., Vigneron, A.: Computing the discrete Fréchet distance with imprecise input. Int. J. Comput. Geometry Appl. 22(1), 27–44 (2012)
2. Buchin, K., Löffler, M., Morin, P., Mulzer, W.: Preprocessing imprecise points for Delaunay triangulation: Simplified and extended. Algorithmica 61(3), 674–693 (2011)
3. Drezner, Z.: On the rectangular p-center problem. Naval Res. Logist. 34(2), 229–234 (1987)
4. Frederickson, G., Johnson, D.: Generalized selection and ranking: Sorted matrices. SIAM J. Comput. 13(1), 14–30 (1984)
5. Hoffmann, M.: Covering polygons with few rectangles. In: Proc. 17th Euro. Workshop Comp. Geom. (EuroCG 2001), pp. 39–42 (2001)
6. Hoffmann, M.: A simple linear algorithm for computing rectilinear 3-centers. Comput. Geom. Theory Appl. 31, 150–165 (2005)
7. Jadhav, S., Mukhopadhyay, A., Bhattacharya, B.K.: An optimal algorithm for the intersection radius of a set of convex polygons. J. Algo. 20, 244–267 (1996)
8. Khanban, A.A., Edalat, A.: Computing Delaunay triangulation with imprecise input data. In: Proc. 15th Canadian Conf. Comput. Geom., pp. 94–97 (2003)
9. Knauer, C., Löffler, M., Scherfenberg, M., Wolle, T.: The directed Hausdorff distance between imprecise point sets. Theoretical Comput. Sci. 412(32), 4173–4186 (2011)
10. Ko, M., Lee, R., Chang, J.: An optimal approximation algorithm for the rectilinear m-center problem. Algorithmica 5, 341–352 (1990)
11. Ko, M., Lee, R., Chang, J.: Rectilinear m-center problem. Naval Res. Logist. 37(3), 419–427 (1990)
12. Löffler, M.: Data Imprecision in Computational Geometry. Ph.D. thesis, Utrecht University (2009)
13. Löffler, M., Snoeyink, J.: Delaunay triangulation of imprecise points in linear time after preprocessing. Comput. Geom.: Theory and Appl. 43(3), 234–242 (2010)
14. Löffler, M., van Kreveld, M.J.: Largest and smallest tours and convex hulls for imprecise points. In: Proc. 10th Scandinavian Workshop Algo. Theory, pp. 375–387 (2006)
15. Megiddo, M., Supowit, K.J.: On the complexity of some common geometric location problems. SIAM J. Comput. 13(1), 182–196 (1984)
16. Segal, M.: On piercing sets of axis-parallel rectangles and rings. Int. J. Comput. Geometry Appl. 9(3), 219–234 (1999)
17. Sember, J., Evans, W.: Guaranteed Voronoi diagrams of uncertain sites. In: Proc. 20th Canadian Conf. Comput. (2008)
18. Sharir, M., Welzl, E.: Rectilinear and polygonal p-piercing and p-center problems. In: Proc. 12th Annu. Sympos. Comp. Geom (SoCG 1996), pp. 122–132 (1996)

Robust Nonparametric Data Approximation of Point Sets via Data Reduction*

Stephane Durocher, Alexandre Leblanc, Jason Morrison, and Matthew Skala

University of Manitoba, Winnipeg, Canada
{durocher,mskala}@cs.umanitoba.ca,
{alex_leblanc,jason_morrison}@umanitoba.ca

Abstract. In this paper we present a novel nonparametric method for simplifying piecewise linear curves and we apply this method as a statistical approximation of structure within sequential data in the plane. We consider the problem of minimizing the average length of sequences of consecutive input points that lie on any one side of the simplified curve. Specifically, given a sequence P of n points in the plane that determine a simple polygonal chain consisting of $n - 1$ segments, we describe algorithms for selecting a subsequence $Q \subset P$ (including the first and last points of P) that determines a second polygonal chain to approximate P, such that the number of crossings between the two polygonal chains is maximized, and the cardinality of Q is minimized among all such maximizing subsets of P. Our algorithms have respective running times $O(n^2 \sqrt{\log n})$ when P is monotonic and $O(n^2 \log^{4/3} n)$ when P is any simple polyline.

1 Introduction

Given a simple polygonal chain P (a *polyline*) defined by a sequence of points $(p_1, p_2, \ldots, p_n)$ in the plane, the *polyline approximation problem* is to produce a simplified polyline $Q = (q_1, q_2, \ldots, q_k)$, where $k < n$. The polyline Q represents an approximation of P that optimizes one or more functions of P and Q. For P to be *simple*, the points $p_1, \ldots, p_n$ must be distinct and P cannot intersect itself.

Motivation for studying polyline approximation comes from the fields of computer graphics and cartography, where approximations are used to render vector-based features such as streets, rivers, or coastlines onto a screen or a map at appropriate resolution with acceptable error [1], as well as in problems involving computer animation, pattern matching and geometric hashing (see Alt and Guibas' survey for details [3]). Our present work removes the arbitrary parameter previously required to describe acceptable error between P and Q, and provides an approximation method that is robust to some forms of noise. While previous work uses the Real RAM model, our analysis primarily assumes a Word RAM

---

* Work was supported in part by the Natural Science and Engineering Research Council of Canada (NSERC).

K.-M. Chao, T.-s. Hsu, and D.-T. Lee (Eds.): ISAAC 2012, LNCS 7676, pp. 319–331, 2012.
© Springer-Verlag Berlin Heidelberg 2012

model to ensure clarity in discussions of lower bounds on time complexity. We comment on our results in both models. We further note that the Word RAM model requires that non-integer coordinates (floats and rationals) be contained in a constant number of words and be easily comparable. See the work of Han [8] and of Chan and Pătrașcu [4] for more on this model.

Typical polyline approximation algorithms require that distance between two polylines be measured using a function denoted here by $\zeta(P, Q)$. The specific measure of interest differs depending on the focus of the particular problem or article; however, three measures are popular: Chebyshev distance ζ_C, Hausdorff distance ζ_H, and Fréchet distance ζ_F. In informal terms, the Chebyshev distance is the maximum absolute difference between y-coordinates of P and Q (maximum residual); the symmetric Hausdorff distance is the distance between the most isolated point of P or Q and the other polyline; and the Fréchet distance is more complicated, being the shortest possible maximum distance between two particles each moving forward along P and Q. Alt and Guibas [3] give more formal definitions. We define a new measure of quality or similarity, to be maximized, rather than an error to be minimized. Our crossing measure is a combinatorial description of how well Q approximates P. It is invariant under a variety of geometric transformations of the polylines, and is often robust to uncertainty in the locations of individual points. Specifically, we consider the problem of minimizing the average length of sequences of consecutive input points that lie on any one side of the simplified curve. Given a sequence P of n points in the plane that determine a simple polygonal chain consisting of $n - 1$ segments, we describe algorithms for selecting a subsequence $Q \subset P$ (including the first and last points of P) that determines a second polygonal chain to approximate P, such that the number of crossings between the two polygonal chains is maximized, and the cardinality of Q is minimized among all such maximizing subsets of P.

Our algorithm for minimizing $|Q|$ while optimizing our nonparametric quality measure requires $O(n^2\sqrt{\log n})$ time when P is monotonic in x, or $O(n^2 \log^{4/3} n)$ time when P is a non-monotonic simple polyline on the plane, both in $O(n)$ space. The near-quadratic times are slightly larger in their polylog exponents for the Real RAM model ($O(n^2 \log n)$ and $O(n^2 \log^2 n)$, respectively), and are remarkably similar to the optimal times achieved in the parametric version of the problem using Hausdorff distance [1,5], suggesting the possibility that the problems may have similar complexities.

In Section 3, we define the crossing measure $\chi(Q, P)$ and relate the concepts and properties of $\chi(Q, P)$ to previous work in both polygonal curve simplification and robust approximation. In Section 4, we describe our algorithms to compute approximations of monotonic and non-monotonic simple polylines that maximize $\chi(Q, P)$.

2   Related Work

Previous work on polyline approximation is generally divided into four categories depending on what property is being optimized and what restrictions are placed on Q [3]. Problems can be classified as requiring an approximating polyline Q having the minimum number of segments (minimizing $|Q|$) for a given acceptable error $\zeta(P,Q) \leq \epsilon$, or a Q with minimum error $\zeta(P,Q)$ for a given value of $|Q|$. These are called min-# problems and min-ϵ problems respectively. These two types of problems are each further divided into "restricted" problems, where the points of Q are required to be a subset of those in P and to include the first and last points of P ($q_1 = p_1$ and $q_k = p_n$), and "unrestricted" problems, where the points of Q may be arbitrary points on the plane. Under this classification, the polyline approximation Q we examine is a restricted min-# problem for which a subset of points of P is selected (including p_1 and p_n) where the objective measure is the number of crossings between P and Q and an optimal approximation first maximizes the crossing number (rather than minimizing it), and then has a minimum $|Q|$ given the maximum crossing number.

While the restricted min-# problems find the smallest sized approximation within a given error ϵ, an earlier approach was to find any approximation within the given error. The cartographers Douglas and Peucker [6] developed a heuristic algorithm where an initial (single segment) approximation was evaluated and the furthest point was then added to the approximation. This technique remained inefficient until Hershberger and Snoeyink [9] concluded that it could be applied in $O(n \log^* n)$ time and linear space.

The most relevant previous literature is on restricted min-# problems. Imai and Iri [10] presented an early solution to the restricted polyline approximation problem using $O(n^3)$ time and $O(n)$ space. The version they study optimizes $k = |Q|$ while maintaining that the Hausdorff metric between Q and P is less than the parameter ϵ. Their algorithm was subsequently improved by Melkman and O'Rourke [11] to $O(n^2 \log n)$ time and then by Chan and Chin [5] to $O(n^2)$ time. Subsequently, Agarwal and Varadarajan [2] changed the approach from finding a shortest path in an explicitly constructed graph to an implicit method that runs in $O(f(\delta)n^{4/3+\delta})$ time, where δ is an arbritrarily chosen constant. Agarwal and Varadarajan used the L_1 Manhattan and L_∞ Chebyshev metrics instead of the previous works' Hausdorff metric. Finally, Agarwal et al. [1] study a variety of metrics and give approximations of the min-# problem in $O(n)$ or $O(n \log n)$ time.

3   A Crossing Measure and Its Computation

3.1   Crossing Measure

The crossing measure $\chi(Q, P)$ is defined for a sequence of n distinct points $P = (p_1, p_2, \ldots, p_n)$ and a subsequence of k distinct points $Q \subset P, Q = (q_1, q_2, \ldots, q_k)$ with the same first and last values: $q_1 = p_1$ and $q_k = p_n$. For each p_i let $(x_i, y_i) = p_i \in \mathbb{R}^2$. To understand the crossing measure, it is necessary to make use of the idea of left and right sidedness of a point relative to a directed line segment. A point p_j is on the left side of a segment $S_{i,i+1} = [p_i, p_{i+1}]$ if the signed area of the triangle formed by the points p_i, p_{i+1}, p_j is positive. Correspondingly, p_j is on the right side of the segment if the signed area is negative. The three points are collinear if the area is zero.

Fig. 1. Crossings are indicated with a square and false crossings are marked with a +. Crossings are only counted when the simplifying segment intersects the chain that begins and ends with its own endpoints.

For any endpoint q_j of a segment in Q it is possible to determine the side of P on which q_j lies. Since Q is a polyline using a subset of the points defining P, for every segment $S_{i,i+1}$ there exists a corresponding segment of $S_{\pi(j),\pi(j+1)}$ such that $1 \leq \pi(j) \leq i < i+1 \leq \pi(j+1) \leq n$. The function $\pi : \{1, \ldots, n\} \to \{1, \ldots, k\}$ maps a point $q_j \in Q$ to its corresponding point $p_{\pi(j)} \in P$ such that $p_{\pi(j)} q_j$. The endpoints of $S_{\pi(j),\pi(j+1)}$ are given a side based on $S_{i,i+1}$ and vice versa. Two segments intersect if they share a point. Such a point is interior to both segments if and only if the segments change sides with respect to each other. The intersection is at an endpoint if at least one endpoint is collinear to the other segment [12, p. 566]. The *crossing measure* $\chi(Q, P)$ is the number of times that Q changes sides from properly left to properly right of P due to an intersection between the polylines, as shown in Figure 1. A single crossing can be generated by any of five cases (see Figure 2):

1. A segment of Q intersects P at a point distinct from any endpoints;
2. two consecutive segments of P meet and cross Q at a point interior to a segment of Q;
3. one or more consecutive segments of P are collinear to the interior of a segment of Q with the previous and following segments of P on opposite sides of that segment of Q;
4. two consecutive segments of P share their common point with two consecutive segments of Q and form a crossing; or
5. in a generalization of the previous case, instead of being a single point, the intersection comprises one or more sequential segments of P and possibly Q that are collinear or identical.

In Section 3.2, we discuss how to compute the crossings for the first three cases, which are all cases where crossings involve only one segment of Q. The remaining cases involve more than one segment of Q, because an endpoint of one segment of Q or even some entire segments of Q are coincident with one or more segments of P; those cases are discussed in Section 3.3.

In the case where the x-coordinates of P are monotonic, P describes a function Y of x and Q is an approximation $\hat{Y}$ of that function. The signs of the residuals $r = (r_1, r_2, \ldots, r_n) = Y_P - \hat{Y}$ are computed at the x-coordinates of P and are equivalent to the sidedness described above. The crossing number is the number of proper sign changes in the sequence of residuals. The resulting approximation maximizes the likelihood that adjacent residuals would have different signs,

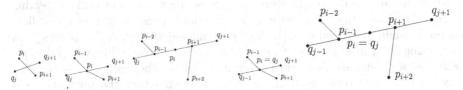

Fig. 2. Examples of the five cases generating a single crossing

while minimizing the number of original data points retained conditional on that number of sign changes. Note that if r was independently and identically selected at random from a distribution with median zero, then any adjacent residuals in the sequence r would have different signs with probability $1/2$.

3.2 Counting Crossings with a Segment

To compute an approximation Q with optimal crossing number for a given P, we consider the optimal number of crossings for segments of P and combine them in a dynamic programming algorithm. Starting from a point p_i we compute optimal crossing numbers for each of the $n - i$ segments that start at point p_i and end at some p_j with $i < j \leq n$. Computing all $n - i$ optimal crossing numbers for a given p_i simultaneously in a single pass is more efficient than computing them for each (p_i, p_j) pair separately. These batched computations are performed for each p_i and the results used to find Q.

To compute a single batch, we will consider the angular order of points in $P_{i+1,n} = \{p_{i+1}, \ldots, p_n\}$ with respect to p_i. Let $\rho_i(j)$ be a function on the indices representing the angular order of segments (p_i, p_j) within this set with respect to vertical, such that $\rho_i(j) = 1$ for all p_j that are closest to the vertical line passing through p_i, and $\rho_i(j) \leq \rho_i(k)$ if and only if the clockwise angle for p_j from the vertical is less than or equal to the corresponding angle for p_k. Note that within axis-aligned quadrants that are centered on p_i, a larger angle corresponds to a smaller slope. This confirms that angular order comparisons are $O(1)$ time computable using basic arithmetic in the Word RAM Model. See Figure 3(a). Using this angular ordering we partition $P_{i+1,n}$ into chains and process the batch of crossing number problems as discussed below.

We define a *chain* with respect to p_i to be a consecutive sequence $P_{\ell,\ell'} \subseteq P_{i+1,n}$ with non-decreasing angular order. That is, either $\rho_i(\ell') \geq \rho_i(j + 1) \geq \cdots \geq \rho_i(\ell)$ or $\rho_i(\ell) \leq \rho_i(j + 1) \leq \cdots \leq \rho_i(\ell')$, with the added constraint that chains cannot cross the vertical ray above p_i. Each segment that does cross is split into two pieces using two artificial points on the ray per crossing segment. The points on the low segment portions have rank $\rho_i = 1$ and the identically placed other points have rank $\rho_i = n + 1$. These points do not increase the complexities by more than a constant factor and are not mentioned further. Processing $P_{i+1,n}$ into its chains is done by first computing the slope and quadrant for each point and storing that information with the points. Then the points are sorted by angular order around p_i and $\rho_i(j)$ is computed as the rank of p_j in the sorted list. This step requires $O(n \log \log n)$ time and linear space in the Word RAM model

[7] with linear extra space to store the slopes, quadrants and ranks. Creating a list of chains is then feasable in $O(n)$ time and space by storing the indices of the beginning and end of each chain encountered while checking points p_j in increasing order from $j = i+1$ to $j = n$. Identifying all chains involves two steps. First, all segments are checked to determine whether they intersect the vertical ray, each in $O(1)$ time. Such an intersection implies that the previous chain should end and the segment that crosses the ray should be a new chain (note that an artificial index of $i + \frac{1}{2}$ can denote the point that crosses the vertical). The second check is to determine whether the most recent pair of points has a different angular rank ordering from the previous pair. If so, the previous chain ended with the previous point and the new chain begins with the current segment. Each chain is oriented from lowest angular index to highest angular index.

Lemma 1. *Consider any chain $P_{\ell,\ell'}$ (w.l.o.g. assume $\ell < \ell'$). With respect to p_i the segment $S_{i,j} : (i < j \leq n)$ can have at most one crossing strictly interior to $P_{\ell,\ell'}$.*

Proof. CASE 1. Suppose $\rho_i(\ell) = \rho_i(j)$ or $\rho_i(j) = \rho_i(\ell')$. If $\ell = n$ then no crossing can exist because at least one end (or all) of $P_{k,\ell}$ is collinear with $S_{i,j}$ and no proper change in sidedness can occur in this chain to generate a crossing.

CASE 2. Suppose $\rho_i(j) \notin [\rho_i(\ell), \rho_i(\ell')]$. These cases have no crossings with the chain because $P_{k,\ell}$ is entirely on one side of $S_{i,j}$. A ray exists between either $\rho_i(j) < \rho_i(\ell)$ or $\rho_i(m) < \rho_i(j)$ that separates $P_{k,l}$ from $S_{i,j}$ and thus no crossings can occur between the segment and the chain.

CASE 3. Suppose $\rho_i(j) \in (\rho_i(\ell), \rho_i(\ell'))$. Assume that the chain causes at least two crossings. Pick the lowest index segment for each of the two crossings that are the fewest segments away from p_i. By definition there are no crossings of segments between these two segments. Label the point with lowest index of these two segments p_λ and the point with greatest index $p_{\lambda'}$. Define a possibly degenerate cone Φ with a base p_i and rays through p_λ and $p_{\lambda'}$. This cone, by definition, separates the segments from $p_{\lambda+1}$ to $p_{\lambda'-1}$ from the remainder of the chain. Since this sub-chain cannot circle p_i entirely there must exist one or more points that have a maximum (or minimum) angular order, which contradicts the definition of the chain. Hence there must be at most one crossing. □

The algorithm for computing the crossing measure on a batch of segments depends on the nature of P. If P is x-monotone, then the chains can be ordered by increasing x-coordinates or equivalently by the greatest index among the points that define them. In this case, a segment $S_{i,j}$ intersects any chain $P_{\ell,\ell'}$ exactly once if its x-coordinates are less than p_j and $\rho_i(j) \in (\rho_i(\ell), \rho_i(\ell'))$ (i.e., Case 3 of Lemma 1).

The algorithm represents each of the $O(n)$ segments $S_{i,j}$ as a blue point $(j, \rho_i(j))$ and each chain $P_{\ell,\ell'}$ as two red points: the start $(\max(\ell, \ell'), \min(\rho_i(\ell), \rho_i(\ell')))$ and end $(\max(\ell, \ell'), \max(\rho_i(\ell), \rho_i(\ell')))$. For every starting red point that is dominated (strictly greater than in both coordinates) by a single blue point a crossing is generated only if the corresponding red point is not dominated. The count of red start points dominated by each blue point is an offline counting query solvable in $O(n\sqrt{\log n})$ time and linear space using the Word RAM model [4, Theorem 2]. The count of red end points dominated by each blue point must then be subtracted from the start domination counts using the same method. Correctness of the result follows from x-monotonicity and the proof of Lemma 1.

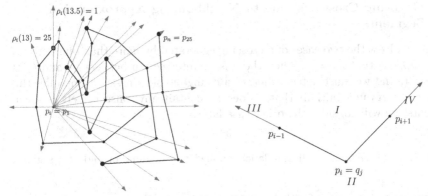

(a) An example of the rays' angular order of vertices in $P_{i+1,n}$ and the resulting chains

(b) Regions around p_j that determine a crossing at p_j

Fig. 3. Angular ordering with chains around p_i and regions local to p_i

The problem becomes more difficult if we assume that P is simple but not necessarily monotonic in x. While chains describe angular order nicely, a non-monotonic P does not imply a consistent ordering of chain boundaries. Thus, queries will be of a specific nature: for a given point p_j, we must determine how many chains are closer to i, have a lower maximum index than j, and are within the angular order (as with the monotonic example). We use the same strategy of two sets of dominance queries as in the monotonic case. The difference is that, instead of using the maximum index on a chain both for the chain's location within the polyline and its distance from p_i, we precompute a distance ranking of chains from p_i as well as the maximum index. Then, the start and end of each chain is represented as a three-dimensional red point.

Chains do not cross, and can only intersect at their endpoints, due to the non-overlapping definition of chains and the simplicity of P. Therefore, to compute the closeness of chains, we sweep a ray from p_i, initially vertical, in increasing ρ_i order (increasing angle). This defines a partial order on chains with respect to their distance from p_i. Using a topological sweep [12, p. 481] it is possible to determine a unique order that preserves this partial ordering of chains. Since there are $O(n)$ chains and changes in neighbours defining the partial order occur only at chain endpoints, there are $O(n)$ edges in the partial order. This further implies that $O(n \log n)$ time is required to determine the events in a sweep and $O(n)$ time to compute the topological ordering. Without loss of generality assume that the chains closest to p_i have a lower topological index. This is a traditional Real RAM approach, and while a more efficient Word RAM approach could be found, it would be unnecessary given that this is a preprocessing step. Computing each batch of domination queries requires $O(n \log^{4/3} n)$ time and linear space using a result of Chan and Pătraşcu [4, Corollary 3.2].

3.3 Counting Crossings Due to Neighbouring Approximation Segments

We now address the two cases of a crossing generated by more than one segment of Q. Suppose that $p_i = q_j$. Then there is an intersection between P and Q at this point, and we must detect whether a change in sidedness accompanies this intersection. Assume initially that P does not contain any consecutively collinear segments; we will consider the other case later.

Table 1. Left, right, and collinear labels applied to beginning or end of a segment at p_j

Categorization	Conditions for End of $S_{\pi(j-1),i}$	Conditions for Beginning of $S_{i,\pi(j+1)}$
collinear (1)	$q_{j-1} = p_{i-1}$ $q_{j+1} = p_{i+1}$	$q_{j-1} = p_{i-1}$ $q_{j+1} = p_{i+1}$
left (2)	$q_{j-1} \in I$ $(q_{j-1} \in III) \wedge (p_{i-2} \in II)$ $(q_{j-1} \in IV) \wedge (p_{i+2} \in II)$	$q_{j+1} \in I$ $(q_{j+1} \in III) \wedge (p_{i-2} \in II)$ $(q_{j+1} \in IV) \wedge (p_{i+2} \in II)$
right (3)	$q_{j-1} \in II$ $(q_{j-1} \in III) \wedge (p_{i-2} \in I)$ $(q_{j-1} \in IV) \wedge (p_{i+2} \in I)$	$q_{j+1} \in II$ $(q_{j+1} \in III) \wedge (p_{i-2} \in I)$ $(q_{j+1} \in IV) \wedge (p_{i+2} \in I)$

We begin with the non-degenerate case where $(p_{i-1}, p_{i+1}, q_{i-1}, q_{i+1})$ are all distinct points (i.e., case 4 in Figure 2). Each of the points q_{j-1} and q_{j+1} can be in one of four locations: in the cone left of (p_{i-1}, p_i, p_{i+1}); in the cone right of (p_{i-1}, p_i, p_{i+1}); on the ray defined by $S_{i,i-1}$; or on the ray defined by $S_{i,i+1}$. These are labelled in Figure 3(b) as regions I through IV, respectively. In Cases III and IV it may also be necessary to consider the location of q_{i-1} or q_{j+1} with respect to $S_{i-2,i-1}$ or $S_{i+1,i+2}$.

For the degenerate case where the points may not be unique, if $p_i = q_j$ and $p_{i+1} \neq q_{j+1}$, then any change in sidedness is handled at p_i and can be detected by verifying the previous side from the polyline. If, however, $p_{i+1} = q_{j+1}$, then any change in sidedness will be counted further along in the approximation.

By examining these points it is possible to assign a sidedness to the end of $S_{\pi(j-1),\pi(j)}$ and the beginning of $S_{\pi(j),\pi(j+1)}$. Note that the sidedness of a point q_{j-1} with respect to $S_{i-2,i-1}$ can be inferred from the sidedness of p_{i-2} with respect to $S_{\pi(j-1),i-1}$, and that property is used in the case of regions

III and *IV*. The assumed lack of consecutive collinear segments requires that $\{p_{i-2}, p_{i+2}\} \in I \cup II$ and thus Table 1 is a complete list of the possible cases when $|P| \geq 5$. Cases involving *III* or *IV* where $i \notin [3, n-2]$ are labelled collinear. We discuss the consequences of this choice later.

A single crossing occurs if and only if the end of $S_{\pi(j-1),i}$ is on the left or right of P, while the beginning of $S_{i,\pi(j+1)}$ is on the opposite side. Furthermore, the end of any approximation $Q_{1,j}$ of $P_{1,i}$ that ends in $S_{\pi(j-1),i}$ inherits the same labelling as the end of $S_{\pi(j-1),i}$. This labelling is consistent with the statement that the approximation last approached the polyline P from the side indicated by the labelling. To maintain this invariant in the labelling of the end of polylines, if $S_{\pi(j-1),i}$ is labelled as collinear then the approximation $Q_{1,j}$ needs to have the same labelling as $Q_{1,j-1}$. As a basis case, the approximations of $P_{1,2}$ and $P_{1,1}$ are the result of the identity operation so they must be collinear. Note that an approximation labelled collinear has no crossings.

The constant number of cases in Table 1, and the constant complexity of the sidedness test, imply that we can compute the number of crossings between a segment and a chain, and therefore the labelling for the segment, in constant time. Let $\eta(Q_{1,j}, S_{\pi(j-1),i})$ represent the number of extra crossings (necessarily 0 or 1) introduced at p_j by joining $Q_{1,j}$ and $S_{\pi(j-1),i}$. We have $\chi(Q_{1,j}, P_{1,i}) = \chi(Q_{1,j-1}, P_{1,\pi(j-1)}) + \chi(S_{\pi(j-1),i}, P_{\pi(j-1),i}) + \eta(Q_{1,j}, S_{\pi(j-1),i})$, which lends itself to computing the optimal approximation incrementally using dynamic programming.

It remains to consider the case of sequential collinear segments (i.e., case 5 in Figure 2). The polyline P' can be simplified into P by merging sequential collinear segments, effectively removing points of P' without changing its shape. When joining two segments where $p'_i = q_j$, p'_{i-1} and p'_{i+1} define the regions as before but there is no longer a guarantee regarding non-collinearity of p'_{i-2} or p'_{i+2} with respect to the other points. The points q_{j-1} and q_{j-2} are now collinear if and only if either of them are entirely collinear to the relevant segments of P. Our check for equality is changed to a check for equality or collinearity. We examine the previous and next points of P' that are not collinear to the two segments $[p'_{i-1}, p'_i]$ and $[p'_i, p'_{i+1}]$. We find such points for every p'_i in a preprocessing step requiring linear time and space, by scanning the polyline for turns and keeping two queues of previous and current collinear points.

4   Finding a Polyline That Maximizes the Crossing Measure

This section describes our dynamic programming approach to computing a polyline Q that is a subset of P that maximizes the crossing measure $\chi(Q, P)$. Our algorithm returns a subset of minimum cardinality k among all such maximizing subsets. We compute $\chi(S_{i,j}, P_{i,j})$ in batches, as described in the previous section. Our algorithm maintains the best known approximations of $P_{1,i}$ for all $i \in [1, n]$ and each of the three possible labellings of the ends. We refer to these paths as $Q_{\sigma,i}$ where σ describes the labelling at i: $\sigma = 1$ for collinear, $\sigma = 2$ for left, or $\sigma = 3$ for right.

To reduce the space complexity we do not explicitly maintain the (potentially exponential-size) set of all approximations $\mathcal{Q}_{\sigma,i}$. Instead, for each approximation corresponding to (σ, i) we maintain: $\chi(\mathcal{Q}_{\sigma,i}, P_{1,i})$ (initially zero); the size of the approximation found, $|\mathcal{Q}_{\sigma,i}|$ (initially $n+1$); the starting index of the last segment added, $\beta_{\sigma,i}$ (initially zero); and the end labelling of the best approximation to which the last segment was connected, $\tau_{\sigma,i}$ (initially zero). The initial values described represent the fact that no approximation is yet known. The algorithm begins by setting the values for the optimal identity approximation for $P_{1,1}$ to the following values (note $\sigma = 1$):

$$\chi(\mathcal{Q}_{1,1}, P_{1,1}) = 0, \qquad |\mathcal{Q}_{1,1}| = 1, \qquad \beta_{1,1} = 1, \qquad \tau_{1,1} = 1.$$

A total of $n - 1$ iterations are performed, one for each $i \in [1, n-1]$, where for each of a batch of segments $S_{i,j} : i < j \leq n$ the algorithm considers a possible approximation ending in that segment. Each iteration begins with the set of approximations $\{\forall \sigma, \mathcal{Q}_{\sigma,\ell} : \ell \leq i\}$ being optimal, with maximal values of $\chi(\mathcal{Q}_{\sigma,\ell}, P_{1,\ell})$ and minimum size $|\mathcal{Q}_{\sigma,\ell}|$ for each of the specified σ and ℓ combinations. The iteration proceeds to calculate the crossing numbers of all segments starting at i and ending at a later index, $\{\chi(S_{i,j}, P_{i,j})|j \in (i,n]\}$, using the method from Section 3. For each of the segments $S_{i,j}$ we compute the sidedness of both the end at j (σ'_j) and the start at i (v'_j). Using v'_j and all values of $\{\sigma : \beta_{\sigma,i} \geq 0\}$ it is possible to compute $\eta(\mathcal{Q}_{\sigma,i}, S_{i,j})$ using just the labellings of the two inputs (see Table 2). It is also possible to determine the labelling of the end of the concatenated polyline $\psi(\sigma, \sigma'_j)$ using the labelling of the end of the previous polyline σ and the end of the additional segment σ'_j (also shown in Table 2).

Table 2. Crossings $\eta(\beta_{\sigma,i}, v'_j)$ due to concatenation, and the end labelling $\psi(\sigma, \sigma'_j)$ of the polyline

$\eta(\beta_{\sigma,i}, v'_j)$		$\beta_{\sigma,i}$ 1 2 3
	1	0 0 0
v'_j	2	0 0 1
	3	0 1 0

$\psi(\sigma, \sigma'_j)$		σ 1 2 3
	1	1 2 3
σ'_j	2	2 2 2
	3	3 3 3

With these values computed, the current value of $\chi(\mathcal{Q}_{\psi(\sigma,\sigma'_j),j}, P_{1,j})$ is compared to $\chi(\mathcal{Q}_{\sigma,i}, P_{1,j}) + \chi(S_{i,j}, P_{i,j}) + \eta(\beta_{\sigma,i}, v'_j)$ and if the new approximation has a greater or equal number of crossings, then we compute:

$$\chi(\mathcal{Q}_{\psi(\sigma,\sigma'_j),j}, P_{1,j}) = \chi(\mathcal{Q}_{\sigma,i}) + \chi(S_{i,j}, P_{i,j}) + \eta(\beta_{\sigma,i}, v'_j),$$
$$|\mathcal{Q}_{\psi(\sigma,\sigma'_j),j}| = |\mathcal{Q}_{\sigma,i}| + 1, \qquad \beta_{\psi(\sigma,\sigma'_j),j} = i, \qquad \tau_{\psi(\sigma,\sigma'_j),j} = \sigma.$$

Correctness of this algorithm follows from the fact that each possible segment ending at $i + 1$ is considered before the $(i + 1)$-st iteration. For each segment and each labelling, at least one optimal polyline with that labelling and leading to the beginning of that segment must have been considered, by the inductive assumption. Since the number of crossings in a polyline depends only on the crossings within the segments and the labelings where the segments meet, the inductive hypothesis is maintained through the $(i+1)$-st iteration. It is also trivially true in the basis case $i = 1$. With the exception of computing the crossing number for all of the segments, the algorithm requires $O(n)$ time and space to update the remaining information in each iteration. The final post-processing step is to determine $\sigma_{max} = \arg\max_\sigma \chi(\mathcal{Q}_{\sigma,n})$, finding the approximation that has the best crossing number. We use the β and τ information to reconstruct $\mathcal{Q}_{\sigma_{max},n}$ in $O(k)$ time.

The algorithm requires $O(n)$ space in each iteration and $O(n \log^{4/3} n)$ time per iteration to compute crossings of each batch of segments dominates the remaining time per iteration. Thus for simple polylines, $\mathcal{Q}_{\sigma_{max},n}$ is computable in $O(n^2 \log^{4/3} n)$ time and $O(n)$ space and for monotonic polylines it is computable in $O(n^2 \sqrt{\log n})$ time and $O(n)$ space.

5   Results

Here we present results of applying our method to approximate x-monotonic data with and without noise in a parameter-free fasion.

Our first point set is given by $p = (x, x^2 + 10 \cdot \sin x)$ for 101 equally spaced points $x \in [-10, 10]$. The maximal-crossing approximation for this point set has 5 points and 7 crossings. We generated a second point set by adding standard normal noise generated in MATLAB with $randn$ to the first point set. The maximal-crossing approximation of the data with standard normal noise has 19 points and a crossing number of 54. We generated a third point set from the first by adding heavy-tailed noise consisting of standard normal noise for 91 data points and standard normal noise multiplied by ten for ten points selected uniformly at random without replacement. The maximal-crossing approximation of the signal contaminated by heavy-tailed noise has 20 points and a crossing number of 50, which is quite comparable to the results obtained with the uncontaminated normal noise. These results are shown in Figure 4.

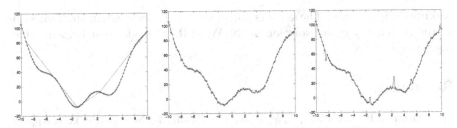

Fig. 4. The optimal crossing path for $p = (x, x^2 + 10 \cdot \sin x)$, without noise, with standard normal noise and with heavy-tailed (mixed gaussian) noise

As can be seen in Figure 4, the crossing-maximization procedure gives a much closer approximation to the signal when there is some nonzero amount of noise present to provide opportunities for crossings. It is reasonable to conclude that, in the case of a clean signal, we could obtain a more useful approximation by artificially adding some noise centered at zero before computing our maximal-crossing polyline. However, to do so requires choosing an appropriate distribution for the added noise, and we wish to keep our procedure free of any parameters.

6   Discussion and Conclusions

The maximal-crossing approximation is robust to small changes of x- or y-coordinates of any p_i when the points are in general position. This robustness can be seen by considering that the crossing number of every approximation depends on the arrangement of lines induced by the line segments, and any point in general position can be moved some distance ϵ without affecting the combinatorial structure of the arrangement. The approximation is also invariant under affine transformations because these too do not modify the combinatorial structure of the arrangement. For x-monotonic polylines, the approximation possesses another useful property: the more a point is an outlier, the less likely it is to be included in the approximation. To see this, consider increasing the y-coordinate of any point p_i to infinity while x-monotonicity remains unchanged. In the limit, this will remove p_i from the approximation. That is, if p_i is initially in the approximation, then once p_i moves sufficiently upward, the two segments of the approximation adjacent to p_i cease to cross any segments of P.

An additional improvement in speed is achievable by bounding sequence lengths. If a parameter m is chosen in advance such that we require that the longest segment considered can span at most $m-2$ vertices, with the appropriate changes, the algorithm can then find the minimum sized approximation conditional on maximum crossing number and having a longest segment of length at most m in $O(nm \log^{4/3} m)$ time for simple polylines or $O(nm\sqrt{\log m})$ time for monotonic polylines, both with linear space. As it seems natural for long segments to be rare in good approximations, setting m to a relatively small value should still lead to good approximations while significantly improving speed.

Acknowledgements. The authors are grateful to Timothy Chan for discussions on the complexity of our solution in the Word RAM model that helped reduce the running times.

References

1. Agarwal, P.K., Har-Peled, S., Mustafa, N.H., Wang, Y.: Near-Linear Time Approximation Algorithms for Curve Simplification. In: Möhring, R.H., Raman, R. (eds.) ESA 2002. LNCS, vol. 2461, pp. 29–202. Springer, Heidelberg (2002)
2. Agarwal, P.K., Varadarajan, K.R.: Efficient algorithms for approximating polygonal chains. Discrete and Computational Geometry 23, 273–291 (2000)

3. Alt, H., Guibas, L.J.: Discrete geometric shapes: Matching, interpolation, and approximation. In: Sack, J.-R., Urrutia, J. (eds.) Handbook of Computational Geometry, pp. 121–153. Elsevier (2000)
4. Chan, T.M., Pătraşcu, M.: Counting inversions, offline orthogonal range counting and related problems. In: SODA, pp. 161–173 (2010)
5. Chan, W.S., Chin, F.: Approximation of Polygonal Curves with Minimum Number of Line Segments. In: Ibaraki, T., Iwama, K., Yamashita, M., Inagaki, Y., Nishizeki, T. (eds.) ISAAC 1992. LNCS, vol. 650, pp. 378–387. Springer, Heidelberg (1992)
6. Douglas, D., Peucker, T.: Algorithms for the reduction of points required to represent a digitised line or its caricature. The Canadian Cartographer 10, 112–122 (1973)
7. Han, Y.: Deterministic sorting in $o(n \log \log n)$ time and linear space. J. of Algorithms 50, 96–105 (2004)
8. Han, Y., Thorup, M.: Integer sorting in $o(n\sqrt{\log \log n})$ expected time and linear space. In: FOCS, pp. 135–144 (2002)
9. Hershberger, J., Snoeyink, J.: Cartographic line simplification and polygon csg formulae in $o(n \log^* n)$ time. Computational Geometry: Theory and Applications 11(3-4), 175–185 (1998)
10. Imai, H., Iri, M.: Polygonal approximation of curve-formulations and algorithms. In: Toussaint, G.T. (ed.) Computational Morphology, pp. 71–86. North-Holland (1988)
11. Melkman, A., O'Rourke, J.: On polygonal chain approximation. In: Toussaint, G.T. (ed.) Computational Morphology, pp. 87–95. North-Holland (1988)
12. Skiena, S.S.: The Algorithm Design Manual, 2nd edn. Springer (2008)

Optimal Point Movement for Covering Circular Regions[*]

Danny Z. Chen[1], Xuehou Tan[2], Haitao Wang[3,**], and Gangshan Wu[4]

[1] Department of Computer Science and Engineering
University of Notre Dame, Notre Dame, IN 46556, USA
dchen@cse.nd.edu
[2] Tokai University, 4-1-1 Kitakaname, Hiratsuka 259-1292, Japan
tan@wing.ncc.u-tokai.ac.jp
[3] Department of Computer Science
Utah State University, Logan, UT 84322, USA
haitao.wang@usu.edu
[4] State Key Lab. for Novel Software Technology
Nanjing University, Hankou Road 22, Nanjing 210093, China
gswu@nju.edu.cn

Abstract. Given n points in a circular region C in the plane, we study the problem of moving these points to the boundary of C to form a regular n-gon such that the maximum of the Euclidean distances traveled by the points is minimized. These problems find applications in mobile sensor barrier coverage of wireless sensor networks. The problem further has two versions: the decision version and optimization version. In this paper, we present an $O(n \log^2 n)$ time algorithm for the decision version and an $O(n \log^3 n)$ time algorithm for the optimization version. The previously best algorithms for these two problem versions take $O(n^{3.5})$ time and $O(n^{3.5} \log n)$ time, respectively. A by-product of our techniques is an algorithm for dynamically maintaining the maximum matching of a circular convex bipartite graph; our algorithm performs each vertex insertion or deletion on the graph in $O(\log^2 n)$ time. This result may be interesting in its own right.

1 Introduction

Given n points in a circular region C in the plane, we study the problem of moving these points to the boundary of C to form a regular n-gon such that the maximum of the Euclidean distances traveled by the points is minimized. Let $|ab|$ denote the Euclidean length of the line segment with two endpoints a and b. Let C be a circular region in the plane. Given a set of n points, $S = \{A_0, A_1, \ldots, A_{n-1}\}$, in C (i.e., in the interior or on the boundary of C), we wish

---

[*] Chen's research was supported in part by NSF under Grant CCF-0916606. Work by Tan was partially supported by Grant-in-Aid (MEXT/JPSP KAKENHI 23500024) for Scientific Research from Japan Society for the Promotion of Science.
[**] Corresponding author.

K.-M. Chao, T.-s. Hsu, and D.-T. Lee (Eds.): ISAAC 2012, LNCS 7676, pp. 332–341, 2012.

to move all n points to n positions $A'_0, A'_1, \ldots, A'_{n-1}$ on the boundary of C to form a regular n-gon, such that the maximum Euclidean distance traveled by the points, i.e., $\max_{0 \leq i \leq n-1}\{|A_i A'_i|\}$, is minimized.

Further, given a value $\lambda \geq 0$, the *decision version* of the problem is to determine whether it is possible to move all points of S to the boundary of C to form a regular n-gon such that the distance traveled by each point is no more than λ. For discrimination, we refer to the original problem as the *optimization version*.

As discussed in [1], the problem finds applications in mobile sensor barrier coverage of wireless sensor networks [1,5,14]. Bhattacharya *et al.* [1] proposed an $O(n^{3.5})$ time algorithm for the decision version and an $O(n^{3.5} \log n)$ time algorithm for the optimization version, where the decision algorithm uses a brute force method and the optimization algorithm is based on a parametric search approach [8,16]. Recently, it was claimed in [18] that these two problem versions were solvable in $O(n^{2.5})$ time and $O(n^{2.5} \log n)$ time, respectively. However, the announced algorithms in [18] contain errors. In this paper, we solve the decision version in $O(n \log^2 n)$ time and the optimization version in $O(n \log^3 n)$ time, which significantly improve the previous results.

A by-product of our techniques that may be interesting in its own right is an algorithm for dynamically maintaining the maximum matchings of *circular convex bipartite graphs*. Our algorithm performs each (online) vertex insertion or deletion on an n-vertex circular convex bipartite graph in (deterministic) $O(\log^2 n)$ time. This matches the performance of the best known dynamic matching algorithm for *convex bipartite graphs* [3]. Note that convex bipartite graphs are a subclass of circular convex bipartite graphs [15].

In fact, the optimization version is equivalent to finding a regular n-gon on the boundary of C such that the bottleneck matching distance between the points in S and the vertices of the n-gon is minimized. The bottleneck matching problems have been studied before, e.g., [4,10,11]. Another related work given by Bremner *et al.* [2] concerns two sets of points on a cycle (neither set of points have to form a regular n-gon) and one wants to rotate one set of points to minimize the matching distance between the two sets.

To distinguish from normal points in the plane, henceforth in the paper, we refer to each point $A_i \in S$ as a *sensor*.

2   The Decision Version

For simplicity, we assume the radius of the circle C is 1. Denote by ∂C the boundary of C. Let λ_C be the maximum distance traveled by the sensors in S in an optimal solution for the optimization problem, i.e., $\lambda_C = \min\{\max_{0 \leq i \leq n-1}\{|A_i A'_i|\}\}$. Since the sensors are all in C, $\lambda_C \leq 2$. Given a value λ, we aim to determine whether $\lambda_C \leq \lambda$. We present an $O(n \log^2 n)$ time algorithm.

We first discuss some concepts. A *bipartite* graph $G = (V_1, V_2, E)$ with $|V_1| = O(n)$ and $|V_2| = O(n)$ is *convex* on the vertex set V_2 if there is a linear ordering on V_2, say, $V_2 = \{v_0, v_1, \ldots, v_{n-1}\}$, such that if any two edges $(v, v_j) \in E$ and $(v, v_k) \in E$ with $v_j, v_k \in V_2$, $v \in V_1$, and $j < k$, then $(v, v_l) \in E$ for all $j \leq l \leq k$.

In other words, for any vertex $v \in V_1$, the subset of vertices in V_2 connected to v forms an interval on the linear ordering of V_2. For any $v \in V_1$, suppose the subset of vertices in V_2 connected to v is $\{v_j, v_{j+1}, \ldots, v_k\}$; then we denote $begin(v, G) = j$ and $end(v, G) = k$. Although E may have $O(n^2)$ edges, it can be represented implicitly by specifying $begin(v, G)$ and $end(v, G)$ for each $v \in V_1$. A *vertex insertion* on G is to insert a vertex v into V_1 with an edge interval $[begin(v, G), end(v, G)]$ and implicitly connect v to every $v_i \in V_2$ with $begin(v, G) \leq i \leq end(v, G)$. Similarly, a *vertex deletion* on G is to delete a vertex v from V_1 as well as all its adjacent edges.

A bipartite graph $G = (V_1, V_2, E)$ is *circular convex* on the vertex set V_2 if there is a circular ordering on V_2 such that for each vertex $v \in V_1$, the subset of vertices in V_2 connected to v forms a circular-arc interval on that ordering. Precisely, suppose such a *clockwise* circular ordering of V_2 is $v_0, v_1, \ldots, v_{n-1}$. For any two edges $(v, v_j) \in E$ and $(v, v_k) \in E$ with $v_j, v_k \in V_2$, $v \in V_1$, and $j < k$, either $(v, v_l) \in E$ for all $j \leq l \leq k$, or $(v, v_l) \in E$ for all $k \leq l \leq n - 1$ and $(v, v_l) \in E$ for all $0 \leq l \leq j$. For each $v \in V_1$, suppose the vertices of V_2 connected to v are from v_j to v_k clockwise on the ordering, then $begin(v, G)$ and $end(v, G)$ are defined to be j and k, respectively. Vertex insertions and deletions on G are defined similarly.

A maximum matching in a convex bipartite graph can be found in $O(n)$ time [12,13,17]. The same time bound holds for a circular convex bipartite graph [15]. Brotal *et al.* [3] designed a data structure for dynamically maintaining the maximum matchings of a convex bipartite graph that can support each vertex insertion or deletion in (deterministic) $O(\log^2 n)$ amortized time. For circular convex bipartite graphs, however, to our best knowledge, we are not aware of any previous work on dynamically maintaining their maximum matchings.

The main idea of our algorithm for determining whether $\lambda_C \leq \lambda$ is as follows. First, we model the problem as finding the maximum matchings in a sequence of $O(n)$ circular convex bipartite graphs, which is further modeled as dynamically maintaining the maximum matching of a circular convex bipartite graph under a sequence of $O(n)$ vertex insertion and deletion operations. Second, we develop an approach for solving the latter problem.

2.1 The Problem Modeling

Our goal is to determine whether $\lambda_C \leq \lambda$. Let P be an arbitrary regular n-gon with its vertices $P_0, P_1, \ldots, P_{n-1}$ ordered clockwise on ∂C. We first consider the following sub-problem: Determine whether we can move all sensors to the vertices of P such that the maximum distance traveled by the sensors is at most λ. Let G_P be the bipartite graph between the sensors $A_0, \ldots, A_{n-1}$ and the vertices of P, such that a sensor A_i is connected to a vertex P_j in G_P if and only if $|A_i P_j| \leq \lambda$. The following lemma is obvious.

Lemma 1. *The bipartite graph G_P is circular convex.*

To solve the above sub-problem, it suffices to compute a maximum matching M in G_P [15]. If M is a perfect matching, then the answer to the sub-problem is "yes"; otherwise, the answer is "no". Thus, the sub-problem can be solved in $O(n)$ time (note that the graph G_P can be constructed implicitly in $O(n)$ time, after $O(n \log n)$ time preprocessing). If the answer to the sub-problem is "yes", then we say that P is *feasible* with respect to the value λ.

If P is feasible, then clearly $\lambda_C \leq \lambda$. If P is not feasible, however, $\lambda_C > \lambda$ does not necessarily hold, because P may not be positioned "right" (i.e., P may not be the regular n-gon in an optimal solution of the optimization version). To further decide whether $\lambda_C \leq \lambda$, we rotate P clockwise on ∂C by an arc distance at most $2\pi/n$. Since the perimeter of C is 2π, the arc distance between any two neighboring vertices of P is $2\pi/n$. A simple yet critical observation is that $\lambda_C \leq \lambda$ holds if and only if during the rotation of P, there is a moment (called a *feasible moment*) at which P becomes feasible with respect to λ. Thus, our task is to determine whether a feasible moment exists during the rotation of P.

Consider the graph G_P. For each sensor A_i, let $E(A_i) = \{P_j, P_{j+1}, \ldots, P_k\}$ be the subset of vertices of P connected to A_i in G_P, where the indices of the vertices of P are taken as module by n. We assume that $E(A_i)$ does not contain all vertices of P (otherwise, it is trivial). Since the arc distance from P_{j-1} to P_j is $2\pi/n$, during the (clockwise) rotation of P, there must be a moment after which P_{j-1} becomes connected to A_i, and we say that P_{j-1} is *added* to $E(A_i)$; similarly, there must be a moment after which P_k becomes disconnected to A_i, and we say that P_k is *removed* from $E(A_i)$. Note that these are the moments when the edges of A_i (and thus the graph G_P) are changed due to the rotation of P. Also, note that during the rotation, all vertices in $E(A_i) \setminus \{P_k\}$ remain connected to A_i and all vertices in $P \setminus \{E(A_i) \cup \{P_{j-1}\}\}$ remain disconnected to A_i. Hence throughout this rotation, there are totally n additions and n removals on the graph G_P. If we sort all these additions and removals based on the time moments when they occur, then we obtain a sequence of $2n$ circular convex bipartite graphs, and determining whether there exists a feasible moment is equivalent to determining whether there is a graph in this sequence that has a perfect matching. With the $O(n)$ time maximum matching algorithm for circular convex bipartite graphs of n vertices in [15], a straightforward solution for determining whether there is a feasible moment would take $O(n^2)$ time.

To do better, we further model the problem as follows. Consider the addition of P_{j-1} to $E(A_i)$. This can be done by deleting the vertex of G_P corresponding to A_i and then inserting a new vertex corresponding to A_i with its edges connecting to the vertices in $\{P_{j-1}\} \cup E(A_i)$. The removal of P_k from $E(A_i)$ can be handled similarly. Thus, each addition or removal on $E(A_i)$ can be transformed to one vertex deletion and one vertex insertion on G_P. If we sort all vertex updates (i.e., insertions and deletions) by the time moments when they occur, then the problem of determining whether there is a feasible moment is transformed to determining whether there exists a perfect matching in a sequence of vertex updates on the graph G_P. In other words, we need to dynamically maintain the maximum matching in a circular convex bipartite graph to support a sequence of $2n$ vertex insertions and $2n$ vertex deletions. This problem is handled in the next subsection, where we treat all vertex updates in an online fashion.

2.2   Dynamic Maximum Matching

Let $G = (V_1, V_2, E)$ with $|V_1| = O(n)$ and $|V_2| = O(n)$ be a circular convex bipartite graph on the vertex set V_2, i.e., the vertices of V_2 connected to each vertex in V_1 form a circular-arc interval on the sequence of the vertex indices of V_2. Suppose $V_2 = \{v_0, v_1, \ldots, v_{n-1}\}$ is ordered clockwise. Recall that a vertex insertion on G is to insert a vertex v into V_1 with an edge interval $[begin(v, G), end(v, G)]$ such that v is (implicitly) connected to all vertices of V_2 from $begin(v, G)$ clockwise to $end(v, G)$. A vertex deletion is to delete a vertex v from V_1 and all its adjacent edges (implicitly). Our task is to design an algorithm for maintaining the maximum matching of G to support such update operations (i.e., vertex insertions and deletions) efficiently. We present an algorithm that takes $O(\log^2 n)$ time in the worst case for each update operation.

Our approach can be viewed as a combination of the data structure given by Brodal, Georgiadis, Hansen, and Katriel [3] for dynamically maintaining the maximum matching in a convex bipartite graph and the linear time algorithm given by Liang and Blum [15] for computing a maximum matching in a circular convex bipartite graph. Denote by $M(G)$ the maximum matching in G. We have the following Theorem 1. Due to the space limit, our algorithm for Theorem 1 is omitted and can be found in the full paper [6].

Theorem 1. *A data structure on a circular convex bipartite graph* $G = (V_1, V_2, E)$ *can be built in* $O(n \log^2 n)$ *time for maintaining its maximum matching* $M(G)$ *so that each online vertex insertion or deletion on* V_1 *can be done in* $O(\log^2 n)$ *time in the worst case. After each update operation, the value* $|M(G)|$ *can be obtained in* $O(1)$ *time and the maximum matching* $M(G)$ *can be reported in* $O(|M(G)|)$ *time.*

Recall that we have reduced the problem of determining whether $\lambda_C \leq \lambda$ to dynamically maintaining the maximum matching in a circular convex bipartite graph with a sequence of $2n$ vertex insertions and $2n$ vertex decisions. We determine whether $\lambda_C \leq \lambda$ as follows. After each update operation, we check whether $|M(G)| = n$, and if this is true, then we report $\lambda_C \leq \lambda$ and halt the algorithm. If all $4n$ updates have been processed but it is always $|M(G)| < n$, then we report $\lambda_C > \lambda$. Based on Theorem 1, we have the result below.

Theorem 2. *Given a value* λ, *we can determine whether* $\lambda_C \leq \lambda$ *in* $O(n \log^2 n)$ *time for the decision version.*

3   The Optimization Version

In this section, we present an $O(n \log^3 n)$ time algorithm for the optimization version. The main task is to compute the value λ_C.

Let o be the center of C. For simplicity of discussion, we assume that no sensor lies at o. Denote by X_i and Y_i the two points on ∂C which are closest and farthest to each sensor A_i, respectively. Clearly, X_i and Y_i are the two intersection points of ∂C with the line passing through A_i and the center o of C (see Figure 1(a)). The lemma below has been proved in [18].

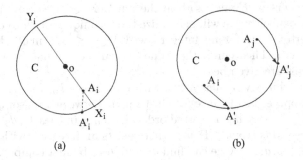

Fig. 1. (a) The points X_i and Y_i on ∂C for A_i; (b) $|A_i A_i'| = |A_j A_j'|$

Lemma 2. [18] *Suppose an optimal solution for the optimization problem is achieved with $\lambda_C = |A_i A_i'|$ for some $i \in \{0, \ldots, n-1\}$. Then either A_i' is the point X_i, or there is another sensor A_j $(j \neq i)$ such that $\lambda_C = |A_j A_j'|$ also holds. In the latter case, any slight rotation of the regular n-gon that achieves λ_C in either direction causes the value of λ_C to increase (i.e., it makes one of the two distances $|A_i A_i'|$ and $|A_j A_j'|$ increase and the other one decrease).*

The points on ∂C satisfying the conditions specified in Lemma 2 may be considered as those defining candidate values for λ_C, i.e., they can be considered as some vertices of possible regular n-gons on ∂C in an optimal solution. The points X_h of all sensors A_h $(0 \leq h \leq n-1)$ can be easily determined. Define $D_1 = \cup_{h=0}^{n-1} \{|A_h X_h|\}$, which can be computed in $O(n)$ time. But, the challenging task is to handle all the pairs (A_i, A_j) $(i \neq j)$ such that the distance from A_i to a vertex of a regular n-gon is equal to the distance from A_j to another vertex of that n-gon and a slight rotation of the n-gon in either direction monotonically increases one of these two distances but decreases the other. We refer to such distances as the *critical equal distances*. Denote by D_2 the set of all critical equal distances. Let $D = D_1 \cup D_2$. By Lemma 2, $\lambda_C \in D$. Thus, if D is somehow available, then λ_C can be determined by using our algorithm in Theorem 2 in a binary search process. Since D_1 is readily available, the key is to deal with D_2 efficiently. An easy observation is $\max_{0 \leq h \leq n-1} |A_h X_h| \leq \lambda_C$. We can use the algorithm in Theorem 2 to check whether $\lambda_C \leq \max_{0 \leq h \leq n-1} |A_h X_h|$, after which we know whether $\lambda_C = \max_{0 \leq h \leq n-1} |A_h X_h|$. Below, we assume $\max_{0 \leq h \leq n-1} |A_h X_h| < \lambda_C$ (otherwise, we are done). Thus, we only need to focus on finding λ_C from the set D_2.

It has been shown in [18] that $|D_2| = O(n^3)$. Of course, we need to avoid computing D_2. To do so, first we determine a subset D_2' of D_2 such that $\lambda_C \in D_2'$ but with $|D_2'| = O(n^2)$. Furthermore, we do not compute D_2' explicitly. Specifically, our idea is as follows. We show that the elements of D_2' are the y-coordinates of a subset of intersection points among a set F of $O(n)$ functional curves in the plane such that each curve is x-monotone and any two such curves intersect in at most one point at which the two curves cross each other. (Such a set of curves is sometimes referred to as *pseudolines* in the literature.) Let $\mathcal{A}_F$

be the arrangement of F and $|\mathcal{A}_F|$ be the number of vertices of $\mathcal{A}_F$. Without computing $\mathcal{A}_F$ explicitly, we will generalize the techniques in [9] to compute the k-th highest vertex of $\mathcal{A}_F$ for any integer k with $1 \leq k \leq |\mathcal{A}_F|$ in $O(n \log^2 n)$ time. Consequently, with Theorem 2, the value λ_C can be computed in $O(n \log^3 n)$ time. The details are given below.

Let P be an arbitrary regular n-gon with its vertices P_0, P_1, ..., P_{n-1} clockwise on ∂C. Suppose the distances of all the pairs between a sensor and a vertex of P are $d_1 \leq d_2 \leq \cdots \leq d_{n^2}$ in sorted order. Let $d_0 = 0$. Clearly, $d_0 < \lambda_C \leq d_{n^2}$ (the case of $\lambda_C = 0$ is trivial). Hence, there exists an integer k with $0 \leq k < n^2$ such that $\lambda_C \in (d_k, d_{k+1}]$. One can find d_k and d_{k+1} by first computing all these n^2 distances explicitly and then utilizing our algorithm in Theorem 2 in a binary search process. But that would take $\Omega(n^2)$ time. The following lemma gives a faster procedure without having to compute these n^2 distances explicitly.

Lemma 3. *The two distances d_k and d_{k+1} can be obtained in $O(n \log^3 n)$ time.*

Proof. We apply a technique, called *binary search in sorted arrays*, as follows. Given M arrays A_i, $1 \leq i \leq M$, each containing $O(N)$ elements in sorted order, the task is to find a certain unknown element $\delta \in A = \cup_{i=1}^{M} A_i$. Further, there is a "black-box" decision procedure Π available, such that given any value a, Π reports $a \leq \delta$ or $a > \delta$ in $O(T)$ time. An algorithm is given in [7] that can find the sought element δ in $O((M + T) \log(NM))$ time. We use this technique to find d_k and d_{k+1}, as follows.

Consider a sensor A_i. Let $S(A_i)$ be the set of distances between A_i and all vertices of P. In $O(\log n)$ time, we can implicitly partition $S(A_i)$ into two sorted arrays in the following way. By binary search, we can determine an index j such that X_i lies on the arc of ∂C from P_j to P_{j+1} clockwise (the indices are taken as module by n). Recall that X_i is the point on ∂C closest to A_i. If a vertex of P is on X_i, then define j to be the index of that vertex. Similarly, we can determine an index h such that Y_i (i.e., the farthest point on ∂C to A_i) lies on the arc from P_h to P_{h+1} clockwise. If a vertex of P is on Y_i, then define h to be the index of that vertex. Both j and h can be determined in $O(\log n)$ time, after which we implicitly partition $S(A_i)$ into two sorted arrays: One array consists of all distances from A_i to $P_j, P_{j-1}, \ldots, P_{h+1}$, and the other consists of all distances from A_i to $P_{j+1}, P_{j+2}, \ldots, P_h$ (again, all indices are taken as module by n). Note that both arrays are sorted increasingly and each element of them can be obtained in $O(1)$ time by using its index in the corresponding array.

Thus, we obtain $2n$ sorted arrays (represented implicitly) for all n sensors in $O(n \log n)$ time, and each array has no more than n elements. Therefore, by using the technique of binary search in sorted arrays, with our algorithm in Theorem 2 as the black-box decision procedure, both d_k and d_{k+1} can be found in $O(n \log^3 n)$ time. The lemma thus follows.

By applying Lemma 3, we have $\lambda_C \in (d_k, d_{k+1}]$. Below, for simplicity of discussion, we assume $\lambda_C \neq d_{k+1}$. Thus $\lambda_C \in (d_k, d_{k+1})$. Since $\max_{0 \leq h \leq n-1} |A_h X_h| < \lambda_C$, we redefine $d_k := \max\{d_k, \max_{0 \leq h \leq n-1} |A_h X_h|\}$. We still have $\lambda_C \in (d_k, d_{k+1})$. Let D_2' be the set of all critical equal distances in the range (d_k, d_{k+1}).

Then $\lambda_C \in D_2'$. We show below that $|D_2'| = O(n^2)$ and λ_C can be found in $O(n \log^3 n)$ time without computing D_2' explicitly.

Suppose we rotate the regular n-gon $P = (P_0, P_1, \ldots, P_{n-1})$ on ∂C clockwise by an arc distance $2\pi/n$ (this is the arc distance between any two adjacent vertices of P). Let $A_i(P_h(t))$ denote the distance function from a sensor A_i to a vertex P_h of P with the time parameter t during the rotation. Clearly, the function $A_i(P_h(t))$ increases or decreases monotonically, unless the interval of ∂C in which P_h moves contains the point X_i or Y_i; if that interval contains X_i or Y_i, then we can further divide the interval into two sub-intervals at X_i or Y_i, such that $A_i(P_h(t))$ is monotone in each sub-interval. The functions $A_i(P_h(t))$, for all P_h's of P, can thus be put into two sets S_{i1} and S_{i2} such that all functions in S_{i1} monotonically increase and all functions in S_{i2} monotonically decrease. Let $m = |S_{i1}|$. Then $m \le n$. Denote by $d_1^i < d_2^i < \cdots < d_m^i$ the sorted sequence of the initial values of the functions in S_{i1}. Also, let $d_0^i = 0$ and $d_{m+1}^i = 2$ (recall that the radius of C is 1). It is easy to see that the range (d_k, d_{k+1}) obtained in Lemma 3 is contained in $[d_j^i, d_{j+1}^i]$ for some $0 \le j \le m$. The same discussion can be made for the distance functions in the set S_{i2} as well.

Since we rotate P by only an arc distance $2\pi/n$, during the rotation of P, each sensor A_i can have at most two distance functions (i.e., one decreasing and one increasing) whose values may vary in the range (d_k, d_{k+1}). We can easily identify these at most $2n$ distance functions for the n sensors in $O(n \log n)$ time. Denote by F' the set of all such distance functions. Clearly, all critical equal distances in the range (d_k, d_{k+1}) can be generated by the functions in F' during the rotation of P. Because every such distance function either increases or decreases monotonically during the rotation of P, each pair of one increasing function and one decreasing function can generate at most one critical equal distance during the rotation. (Note that by Lemma 2, a critical equal distance cannot be generated by two increasing functions or two decreasing functions.) Since $|F'| \le 2n$, the total number of critical equal distances in (d_k, d_{k+1}) is bounded by $O(n^2)$, i.e., $|D_2'| = O(n^2)$. For convenience of discussion, since we are concerned only with the critical equal distances in (d_k, d_{k+1}), for each function in F', we restrict it to the range (d_k, d_{k+1}) only.

Let the time t be the x-coordinate and the function values be the y-coordinates of the plane. Then each function in F' defines a curve segment that lies in the strip of the plane between the two horizontal lines $y = d_k$ and $y = d_{k+1}$. We refer to a function in F' and its curve segment interchangeably, i.e., F' is also a set of curve segments. Clearly, a critical equal distance generated by an increasing function and a decreasing function is the y-coordinate of the intersection point of the two corresponding curve segments. Note that every function in F' has a simple mathematical description. Below, we simply assume that each function in F' is of $O(1)$ complexity. Thus, many operations on them can each be performed in $O(1)$ time, e.g., computing the intersection of a decreasing function and an increasing function.

The set D_2' can be computed explicitly in $O(n^2)$ time, after which λ_C can be easily found. Below, we develop a faster solution without computing D_2'

explicitly, by utilizing the property that each element of D_2' is the y-coordinate of the intersection point of a decreasing function and an increasing function in F' and generalizing the techniques in [9].

A slope selection algorithm for a set of points in the plane was given in [9]. We will extend this approach to help solve our problem. To this end, we need the following lemma, whose proof can be found in the full paper [6].

Lemma 4. *For any two increasing (resp., decreasing) functions in F', if the curve segments defined by them are not identical to each other, then the two curve segments intersect in at most one point and they cross each other at their intersection point (if any).*

We further extend every curve segment in F' into an x-monotone curve, as follows. For each increasing (resp., decreasing) curve segment, we extend it by attaching two half-lines with slope 1 (resp., -1) at the two endpoints of that curve segment, respectively, such that the resulting new curve is still monotonically increasing (resp., decreasing). Denote the resulting new curve set by F. Obviously, an increasing curve and a decreasing curve in F intersect once and they cross each other at their intersection point. For any two different increasing (resp., decreasing) curves in F, by Lemma 4 and the way we extend the corresponding curve segments, they can intersect in at most one point and cross each other at their intersection point (if any). In other words, F can be viewed as a set of pseudolines. Let $\mathcal{A}_F$ be the arrangement of F. Observe that the elements in D_2' are the y-coordinates of a subset of the vertices of $\mathcal{A}_F$. Since $\lambda_C \in D_2'$, λ_C is the y-coordinate of a vertex of $\mathcal{A}_F$. Denote by $|\mathcal{A}_F|$ the number of vertices in $\mathcal{A}_F$. Of course, we do not want to compute the vertices of $\mathcal{A}_F$ explicitly. By generalizing some techniques in [9], we have the following lemma, whose proof can be found in the full paper [6].

Lemma 5. *The value $|\mathcal{A}_F|$ can be computed in $O(n \log n)$ time. Given an integer k with $1 \le k \le |\mathcal{A}_F|$, the k-th highest vertex of $\mathcal{A}_F$ can be found in $O(n \log^2 n)$ time.*

Recall that λ_C is the y-coordinate of a vertex of $\mathcal{A}_F$. Our algorithm for computing λ_C works as follows. First, compute $|\mathcal{A}_F|$. Next, find the $(|\mathcal{A}_F|/2)$-th highest vertex of $\mathcal{A}_F$, and denote its y-coordinate by λ_m. Determine whether $\lambda_C \le \lambda_m$ by the algorithm in Theorem 2, after which one half of the vertices of $\mathcal{A}_F$ can be pruned away. We apply the above procedure recursively on the remaining vertices of $\mathcal{A}_F$, until λ_C is found. Since there are $O(\log n)$ recursive calls to this procedure, each of which takes $O(n \log^2 n)$, the total time for computing λ_C is $O(n \log^3 n)$.

Theorem 3. *The optimization version is solvable in $O(n \log^3 n)$ time.*

References

1. Bhattacharya, B., Burmester, B., Hu, Y., Kranakis, E., Shi, Q., Wiese, A.: Optimal movement of mobile sensors for barrier coverage of a planar region. Theoretical Computer Science 410(52), 5515–5528 (2009)

2. Bremner, D., Chan, T., Demaine, E., Erickson, J., Hurtado, F., Iacono, J., Langerman, S., Taslakian, P.: Necklaces, convolutions, and X + Y. In: Proc. of the 14th conference on Annual European Symposium on Algorithms, pp. 160–171 (2006)
3. Brodal, G.S., Georgiadis, L., Hansen, K.A., Katriel, I.: Dynamic Matchings in Convex Bipartite Graphs. In: Kučera, L., Kučera, A. (eds.) MFCS 2007. LNCS, vol. 4708, pp. 406–417. Springer, Heidelberg (2007)
4. Chang, M., Tang, C., Lee, R.: Solving the Euclidean bottleneck matching problem by k-relative neighborhood graphs. Algorithmica 8, 177–194 (1992)
5. Chen, A., Kumar, S., Lai, T.: Designing localized algorithms for barrier coverage. In: Proc. of the 13th Annual ACM International Conference on Mobile Computing and Networking, pp. 63–73 (2007)
6. Chen, D., Tan, X., Wang, H., Wu, G.: Optimal point movement for covering circular regions. arXiv:1107.1012v1 (2012)
7. Chen, D., Wang, C., Wang, H.: Representing a functional curve by curves with fewer peaks. Discrete and Computational Geometry 46(2), 334–360 (2011)
8. Cole, R.: Slowing down sorting networks to obtain faster sorting algorithms. Journal of the ACM 34(1), 200–208 (1987)
9. Cole, R., Salowe, J., Steiger, W., Szemerédi, E.: An optimal-time algorithm for slope selection. SIAM Journal on Computing 18(4), 792–810 (1989)
10. Efrat, A., Itai, A., Katz, M.: Geometry helps in bottleneck matching and related problems. Algorithmica 31(1), 1–28 (2001)
11. Efrat, A., Katz, M.: Computing Euclidean bottleneck matchings in higher dimensions. Information Processing Letters 75, 169–174 (2000)
12. Gabow, H., Tarjan, R.: A linear-time algorithm for a special case of disjoint set union. Journal of Computer and System Sciences 30, 209–221 (1985)
13. Lipski Jr., W., Preparata, F.P.: Efficient algorithms for finding maximum matchings in convex bipartite graphs and related problems. Acta Informatica 15(4), 329–346 (1981)
14. Kumar, S., Lai, T., Arora, A.: Barrier coverage with wireless sensors. Wireless Networks 13(6), 817–834 (2007)
15. Liang, Y., Blum, N.: Circular convex bipartite graphs: Maximum matching and Hamiltonian circuits. Information Processing Letters 56, 215–219 (1995)
16. Megiddo, N.: Applying parallel computation algorithms in the design of serial algorithms. Journal of the ACM 30(4), 852–865 (1983)
17. Steiner, G., Yeomans, J.: A linear time algorithm for maximum matchings in convex, bipartite graphs. Computers and Mathematics with Applications 31(2), 91–96 (1996)
18. Tan, X., Wu, G.: New Algorithms for Barrier Coverage with Mobile Sensors. In: Lee, D.-T., Chen, D.Z., Ying, S. (eds.) FAW 2010. LNCS, vol. 6213, pp. 327–338. Springer, Heidelberg (2010)

Solving Circular Integral Block Decomposition in Polynomial Time

Yunlong Liu[1] and Xiaodong Wu[1,2]

[1] Electrical and Computer Engineering, The University of Iowa
[2] Department of Radiation Oncology, The University of Iowa

Abstract. The circular integral block decomposition (CIBD) problem seeks an optimal set of circular blocks that stack up to approximate a given reference integral function defined on a circular interval. This problem models the radiation dose delivery in Dynamic Rotating-Shield Brachytherapy (D-RSBT). The challenge lies in the circularity of the problem domain and the maximum length constraint of the circular blocks. We give an efficient polynomial time algorithm for solving the CIBD problem. The key idea is based on several new observations, enabling us to formulate the CIBD problem as the convex cost integer dual network flow. Implementation results show that our CIBD algorithm runs fast and produces promising D-RSBT treatment plans.

1 Introduction

In this paper, we study an interesting geometric optimization problem: the *Circular Integral Block Decomposition* (CIBD). Consider two integer parameters $w > 0$ and $H > 0$, and a nonnegative integral function t defined on a circular interval $\mathcal{C} = [0, n-1]$. We define a *circular window function* $f_k(x)$, with

$$f_k(x) = \begin{cases} h_k, & \text{if } x \in \mathcal{I}_k \subset \mathcal{C}, \\ 0, & \text{otherwise,} \end{cases} \quad \text{where } h_k > 0 \text{ is an integer constant and the size}$$

of the circular interval $|\mathcal{I}_k| \leq w$. Intuitively, the CIBD problem seeks to find a set of circular window functions $f_k(x)$ that "approximates" the given function t by "tiling" them up and the total height of the window functions $\sum h_k \leq H$.

As shown by Fig. 1, a function t is defined on a circular interval with size $n = 4$, namely $x \in \{0, 1, 2, 3\}$ (CCW) and $t(x) = \{4, 5, 2, 4\}$. The function is then perfectly decomposed to a set of 4 circular window functions, denoted as $\mathbf{B} = \{\langle 0, 2, 2\rangle, \langle 1, 0, 1\rangle, \langle 2, 0, 1\rangle, \langle 3, 2, 2\rangle\}$, where each triplet $\langle a_k, b_k, h_k\rangle$ represents a circular window function, with

$$f_k(x) = \begin{cases} h_k, & \text{if } a_k < b_k, x \in \mathcal{I}_k = [a_k, b_k - 1] \\ h_k, & \text{if } a_k \geq b_k, x \in \mathcal{I}_k = [a_k, n-1] \cup [0, b_k - 1] \\ 0, & \text{otherwise} \end{cases} \quad (1)$$

We thus also call the circular window function a *block*. The maximal window size w among those blocks is 3 and the total height of all window functions is

K.-M. Chao, T.-s. Hsu, and D.-T. Lee (Eds.): ISAAC 2012, LNCS 7676, pp. 342–351, 2012.
© Springer-Verlag Berlin Heidelberg 2012

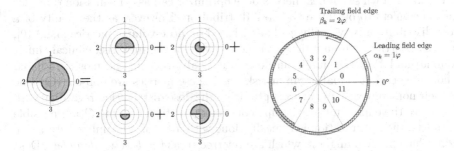

Fig. 1. Illustration of a CIBD example   **Fig. 2.** D-RSBT model

$\sum h_k = 6$. The function attained by tiling up all the window functions in **B** is defined as

$$\mathcal{F}_{\mathbf{B}}(x) = \sum_{f_k \in \mathbf{B}, x \in \mathcal{I}_k} h_k \qquad (2)$$

and $\forall x \in [0, n-1], \mathcal{F}_{\mathbf{B}}(x) = t(x)$ in the example shown in Fig. 1.

In general, due to the constraint $\sum h_k \leq H$, a perfect decomposition of t may not be found. Formally, the CIBD problem is defined as the following optimization problem:

$$\min \mathcal{E}(\mathbf{B}) = \sum_{x=0}^{n-1} (\mathcal{F}_{\mathbf{B}}(x) - t(x))^2$$

$$\text{s.t.} \quad 1 \leq b_k - a_k \leq w \text{ or } 1 \leq b_k + n - a_k \leq w, \qquad k \in [1, |\mathbf{B}|] \qquad (3a)$$

$$\sum_{k=1}^{|\mathbf{B}|} h_k \leq H \qquad (3b)$$

$$a_k, b_k \in [0, n-1], h_k > 0, a_k, b_k, h_k \in \mathbb{Z}, \qquad k \in [1, |\mathbf{B}|] \qquad (3c)$$

1.1 Application Background

The CIBD problems first presented itself during the development of a new brachytherapy technique called *Dynamic Rotating-Shield Brachytherapy* (D-RSBT). This is an intensity-modulation technology for delivering radiation doses in during brachytherapy.

In D-RSBT, the radiation source is partially-covered by a multi-layered radiation-attenuating shield, forming directed apertures called "beamlets" by rotating the field edges [7] (See Fig. 2). The leading and trailing field edge can rotate independently, stopping at n discrete positions distributed evenly along the circle. Each beamlet can be defined by the direction of the leading and trailing field edges with their rotation angles related to reference angle $0°$.

For any known set of beamlets, a dose optimizer can assign emission times for those beamlets to optimize the dose distribution. However, as the quality of a dose distribution is typically evaluated based on dose-volume metrics [3,4,9,10], such as D_{90}(the minimum dose that covers 90% of the high risk clinical tumor volume) and D_{2cc}(the minimum dose that is absorbed in the most irradiated 2 cm^3 of each individual organ at risk), and these metrics are non-convex. Due to their non-convexity in nature, optimizing the dose distribution regarding with respect to the emission times is time-consuming. Instead of using $\mathcal{O}(n^2)$ possible beamlets, the optimization is typically done with a set of n beamlets with a fine azimuthal emission angle φ, which are referred to as the *baseline beamlets*. Dose optimization with the baseline beamlets yields high-quality dose distributions, but the delivery is typically impractical as the total emission time is prohibitively long. The output of dose optimization, in fact, defines an integral function t assigning each baseline beamlet an integral emission time. Unlike Intensity Modulated RadiationTherapy (IMRT), the field edges can rotate at a sufficiently fast speed (e.g. 1 rotation per second) so that the set up time for D-RSBT is negligible compared to the emission time. Therefore, the total D-RSBT emission time dominants the time cost for the whole delivery. RSBT is time-critical since the process must occur rapidly in order to ensure effective utilization of clinical resources, as the patient is under general, epidural throughout the process. An additional sequencing step is needed to make a compromise between the delivery time and the dose quality. To reduce the delivery time, we can combine several consecutive baseline beamlets into a larger deliverable beamlet B_k, denoted by $\langle a_k, b_k, h_k \rangle$ with the leading field edge pointing to $\alpha_k = a_k\varphi$ and the trailing field edge pointing to $\beta_k = b_k\varphi$ with an emission time h_k. The delivery time is then the total sum of h_k of all the deliverable beamlets used. Given a delivery time threshold H, the sequencing problem is to find a set $\mathcal{B}$ of deliverable beamlets whose total delivery time is no larger than H and well approximates the dose distribution output by the dose optimization with minimum dose errors, that is, $\sum_{x=0}^{n-1}(\mathcal{F}_\mathcal{B}(x) - t(x))^2$ is minimized. In addition, due to the physical constraint of the shielding device (Fig. 2), there is a maximum opening w of the deliverable beamlets. Hence, the D-RSBT sequencing problem can be modeled as a CIBD problem.

1.2   Related Works

Although the CIBD problem arises with D-RSBT, it is similar to the Coupled Path Planning (CPP) problems [2] encountered with IMRT. The CIBD problem, however is different from CPP since it is defined on a circular interval with the maximum window constraint; while the CPP problem is defined on a linear interval. The circularity of the problem domain and the maximum window constraint complicates the CIBD problem.

The CIBD problem is also closely related to the CTP$_0$ problem, in which the energy function $\mathcal{E}_0(\mathbf{y}) = \sum_{(u,v) \in E} V_{uv}(x_v - x_u)$ is minimized subject to $\mathbf{y} \in \mathbb{Z}^V$, where V_{uv} are convex functions [6]. CTP$_0$ can be solved by the algorithm proposed by Ahuja *et al.* [1] with time $\mathcal{O}(nm \log(n^2/m) \log(nK))$, which is the best known algorithm on this problem. The basic differences between the CIBD problem and the CTP_0 problem are that: (i) CIBD is not $L^\natural$-convex due to the maximal window constraints and the circular domain constraint [5]; (ii) the number of functions V_{uv} is bounded by $\mathcal{O}(n)$.

1.3 Our Contributions

By fully exploiting the problem structures, we solved the challenges brought up by the maximal window constraint and the circularity of the CIBD problem by formulating them as convex cost integer dual network flow problems. Thus, we are able to solve the CIBD problem in $\mathcal{O}(n^2 \log nH)$ time. Due to the space constraints, some proofs may be found in the journal version of this paper.

2 Canonical Blocksets and Admissible Function Pairs

Note that the CIBD problem is defined on a circular interval $\mathcal{C} = [0, n-1]$, and a window function (a block) is defined on a sub-interval $[a_k, b_k] \subset \mathcal{C}$ with $a_k, b_k \in [0, n-1]$. Without loss of generality, we unify the representation of a block, that is, a block is a feasible one if and only if $\langle a_k, b_k, h_k \rangle$ with $b_k > a_k \geq 0, (b_k - a_k) \leq w, a_k < n$ and $h_k > 0$. Thus, we will have $a_k \in [0, n-1]$ and $b_k \in [0, n+w-1]$.

Definition 1. *A blockset* $\mathbf{B}$ *is feasible if and only if every* $B_k \in \mathbf{B}$ *is feasible.*

Definition 2. *Two blockset* $\mathbf{B}$ *and* $\mathbf{B}'$ *are equivalent if and only if* $\mathcal{F}_{\mathbf{B}} = \mathcal{F}_{\mathbf{B}'}$ *and* $H_{\mathbf{B}} = H_{\mathbf{B}'}$, *where* $\mathcal{F}_{\mathbf{B}} = \mathcal{F}_{\mathbf{B}'}$ *stands for a function equivalence:* $\forall x \in [0, n-1], \mathcal{F}_{\mathbf{B}}(x) = \mathcal{F}_{\mathbf{B}'}(x)$; *and* $H_{\mathbf{B}} = \sum_k h_k$ *stands for the total height of blocks in a blockset* $\mathbf{B}$.

Definition 3. *A feasible blockset* $\mathbf{B} = \{\langle a_k, b_k, h_k \rangle | k \in [1, K]\}$ *is canonical if and only if* $\mathbf{B}$ *satisfies the following properties:*

CB1. $\forall k \in [1, K-1], a_k \leq a_{k+1}, b_k \leq b_{k+1}$;
CB2. $b_K - n \leq b_1$;

Lemma 1. *For any feasible blockset* $\mathbf{B}$, *there exists a canonical blockset* $\bar{\mathbf{B}} = \{\langle \bar{a}_k, \bar{b}_k, \bar{h}_k \rangle | k \in [1, \bar{K}]\}$ *such that* $\bar{\mathbf{B}}$ *and* $\mathbf{B}$ *are equivalent.*

According to Lemma 1, the CIBD problem can be solved by considering canonical blocksets only.

For each canonical blockset $\mathbf{B} = \{B_1, B_2, \ldots, B_K\}$, a pair of functions $(\mathcal{L}, \mathcal{R})$ is defined, as follows:

$$\mathcal{L}(x) = \sum_{B_k \in \mathbf{B}, a_k \leq x} h_k, \ \forall x \in [0, n-1] \tag{4a}$$

$$\mathcal{R}(x) = \sum_{B_k \in \mathbf{B}, b_k \leq x} h_k, \ \forall x \in [0, n+w-1] \tag{4b}$$

$$F_{(\mathcal{L},\mathcal{R})}(x) = \begin{cases} \mathcal{L}(x) - \mathcal{R}(x) + \mathcal{L}(n-1) - \mathcal{R}(n+x), \forall x < w \\ \mathcal{L}(x) - \mathcal{R}(x), \forall x \geq w \end{cases} \tag{5}$$

Notice that $\mathcal{L}(n-1) = \sum_{k, a_k \leq n-1} h_k = \sum_{k=1}^{K} h_k = H_{\mathbf{B}}$.

Lemma 2. *If $(\mathcal{L}, \mathcal{R})$ is defined with a canonical blockset $\mathbf{B}$, then $F_{(\mathcal{L},\mathcal{R})}(x) = F_{\mathbf{B}}(x)$ for any $x \in [0, n+w-1]$.*

Definition 4. *A function pair $(\mathcal{L}, \mathcal{R})$ with $\mathcal{L} : [0, n-1] \to \mathbb{Z}$ and $\mathcal{R} : [0, n+w-1] \to \mathbb{Z}$ is admissible if and only if $(\mathcal{L}, \mathcal{R})$ satisfies the following properties:*

AD1: *$\mathcal{L}$ and $\mathcal{R}$ are non-negative, $\mathcal{R}(0) = 0$;*
AD2: *$\mathcal{L}$ and $\mathcal{R}$ are monotonically non-decreasing, i.e. $\forall x \in [0, n-2], \mathcal{L}(x) \leq \mathcal{L}(x+1); \forall x \in [0, n+w-2], \mathcal{R}(x) \leq \mathcal{R}(x+1)$;*
AD3: *$\forall x \in [0, n-1], \mathcal{L}(x) \geq \mathcal{R}(x); \forall x \in [n, n+w-1], \mathcal{L}(n-1) \geq \mathcal{R}(x)$; particularly, $\mathcal{R}(n+w-1) = \mathcal{L}(n-1)$;*
AD4: *$\forall x \in [0, n-1], \mathcal{L}(x) \leq \mathcal{R}(x+w)$;*
AD5: *$\forall x \in [0, n-1], \mathcal{L}(x) \geq \mathcal{R}(x+1)$;*
AD6: *$\forall x \geq b_1 + n, \mathcal{R}(x) = \mathcal{L}(n-1)$, where $b_1 = \min \arg(\mathcal{R}(x) > 0)$.*

Theorem 1. *For any canonical blockset $\mathbf{B}$, we can find an admissible function pair $(\mathcal{L}, \mathcal{R})$ with $F_{\mathbf{B}}(x) = F_{(\mathcal{L},\mathcal{R})}(x)$, $H_{\mathbf{B}} = \mathcal{L}(n-1)$, and vice versa.*

Then, according to Theorem 1, the objective of the CIBD problem can be formulated as:

$$\min \mathcal{E}(\mathcal{L}, \mathcal{R}) = \sum_{x=0}^{w-1} (\mathcal{L}(x) - \mathcal{R}(x) + \mathcal{L}(n-1) - \mathcal{R}(n+x) - t(x))^2$$
$$+ \sum_{x=w}^{n-1} (\mathcal{L}(x) - \mathcal{R}(x) - t(x))^2 \tag{6}$$

However, not all properties can be expressed with linear constraints defined with $(\mathcal{L}, \mathcal{R})$ since b_1 in (AD6) remains unknown until $(\mathcal{L}(x), \mathcal{R}(x))$ is known. Moreover, Equation (6) is not submodular since not all off-diagonal coefficients in the constraint matrix are nonpositive with more than 2 variables in a single term of the quadratic objective function(Prop 2.6 [8]). Further the lack of submodularity makes this problem hard to solve.

We introduce the following transformation for admissible function pairs $(\mathcal{L}, \mathcal{R})$:

$$\bar{\mathcal{R}}(x) = \begin{cases} \mathcal{R}(n+x) - \mathcal{L}(n-1), \forall x \in [0, b_1 - 1] \\ \mathcal{R}(x), \forall x \in [b_1, n-1] \end{cases} \tag{7}$$

The CIBD problem is then formulated, as follows.

$$\min \mathcal{E}(\mathcal{L}, \bar{\mathcal{R}}) = \sum_{x=0}^{n-1} (\mathcal{L}(x) - \bar{\mathcal{R}}(x) - t(x))^2$$

s.t. $\mathcal{L}(x) \leq \mathcal{L}(x+1),\ \forall x \in [0, n-2]$ (8a)

$\bar{\mathcal{R}}(x) \leq \bar{\mathcal{R}}(x+1),\ \forall x \in [0, n-2]$ (8b)

$\mathcal{L}(x) \leq \bar{\mathcal{R}}(x+w),\ \forall x \in [0, n-w-1]$ (8c)

$\mathcal{L}(x) \leq \bar{\mathcal{R}}(x+w-n) + \mathcal{L}(n-1),\ \forall x \in [n-w, n-1]$ (8d)

$\bar{\mathcal{R}}(x) \leq \mathcal{L}(x-1),\ \forall x \in [1, n-1]$ (8e)

$\bar{\mathcal{R}}(0) \leq 0, \mathcal{L}(0) \geq 0, \mathcal{L}(n-1) \leq H$ (8f)

Lemma 3. *For any admissible function pair* $(\mathcal{L}, \mathcal{R})$, $(\mathcal{L}, \bar{\mathcal{R}})$ *is feasible to Equation* (8) *with* $\mathcal{E}(\mathcal{L}, \mathcal{R}) = \mathcal{E}(\mathcal{L}, \bar{\mathcal{R}})$; *and for any feasible solution* $(\mathcal{L}, \bar{\mathcal{R}})$ *to Equation* (8), *there exist an admissible function pair* $(\mathcal{L}, \mathcal{R})$ *such that* $\mathcal{E}(\mathcal{L}, \mathcal{R}) = \mathcal{E}(\mathcal{L}, \bar{\mathcal{R}})$.

Proof. (*Sketch.*) To proof Lemma 3, first, we can show the one-to-one correspondence between admissible function pairs $(\mathcal{L}, \mathcal{R})$ and feasible solutions $(\mathcal{L}, \bar{\mathcal{R}})$ to Equation (8).

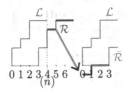

Fig. 3. Visual illustration of mapping from $(\mathcal{L}, \mathcal{R})$ to $(\mathcal{L}, \bar{\mathcal{R}})$

According to Equation (7), any admissible function pairs $(\mathcal{L}, \mathcal{R})$ can be uniquely mapped to a solution $(\mathcal{L}, \bar{\mathcal{R}})$ by shifting $\mathcal{R}(x \in [n, n+w-1])$ n units leftwards and $\mathcal{L}(n-1)$ units downwards, as illustrated by Fig. 3. Notice that, the domain of $\bar{\mathcal{R}}$ is reduced from $[0, n+w-1]$ to $[0, n-1]$ compared to $\mathcal{R}$ as the shifting operation overlapped the intervals $[0, w-1]$ and $[n, n+w-1]$. According to (AD6), $\forall x \in [0, b_1], \mathcal{R}(x) = 0$, and $\forall x \in [b_1, w-1], \mathcal{R}(n+x) = \mathcal{L}(n-1)$; i.e. $\forall x \in [0, w-1]$, either $\mathcal{R}(x) = 0$ or $\mathcal{R}(n+x) = \mathcal{L}(n-1)$ (or both). That serves as the key for making a unique mapping from $(\mathcal{L}, \bar{\mathcal{R}})$ back to $(\mathcal{L}, \mathcal{R})$ with the following equation:

$$\mathcal{R}(x) = \begin{cases} \bar{\mathcal{R}}(x), & x \in [w, n-1] \text{ or } (x < w, \bar{\mathcal{R}}(x) \geq 0) \\ 0, & x < w, \bar{\mathcal{R}}(x) < 0 \\ \mathcal{L}(n-1), & x \geq n, \bar{\mathcal{R}}(x-n) \geq 0 \\ \mathcal{L}(n-1) + \bar{\mathcal{R}}(x-n), & x \geq n, \bar{\mathcal{R}}(x-n) < 0 \end{cases} \quad (9)$$

Together with Equation (9), Equations (8a) and (8b) are used to enforce the non-decreasing property AD2; Equations (8c) and (8d) enforce the maximal window constraint AD4; Equation (8e) encodes AD5 which excludes infeasible blocks with 0 width; AD6 is enforced by $\bar{\mathcal{R}}(0) \leq 0$ based on Equations (7) and (9); the non-negativity AD1 can be inferred from $\mathcal{L}(0) \geq 0$, Equations (8a) and (8b); and $\mathcal{L}(n-1) \leq H$ is used to enforce the constraint on total height of blocks. AD3 is inferred by AD2 and AD5.

The optimization problem in Equation (8) is similar to the problem addressed by Ahuja's algorithm [1], except that the constraint matrices in Ahuja's problems are network matrices (i.e. every column contains two non-zero entries, one of them is $+1$, and the other is -1). The constraint matrices in Equation (8) are not network matrices unless $\mathcal{L}(n-1)$ is known. Thus, by enumerating $\mathcal{L}(n-1)$, Equation (8) can be solved in $\mathcal{O}(n^2 H \log(nH))$ time. However, we can do much better by discovering the following property of the problem.

Theorem 2. *If there exist some feasible solution to Equation* (8), *i.e. dom* $\mathcal{E} \neq \emptyset$, *and* $H \leq \sum_{x=0}^{n-1} t(x)$, *then there exist a solution* $\mathbf{y}^* = (\mathcal{L}^*, \bar{\mathcal{R}}^*)$ *such that* $\mathcal{L}^*(n-1) = H$ *and* $\forall \mathbf{y} \in \text{dom } \mathcal{E}, \mathcal{E}(\mathbf{y}^*) \leq \mathcal{E}(\mathbf{y})$.

Proof. (*Sketch.*) Theorem 2 can be proved in a constructive way, i.e. suppose there exist some other optimal solution $\mathbf{y}' = (\mathcal{L}', \bar{\mathcal{R}}')$ such that $H' = \mathcal{L}'(n-1) \leq H$ and $\forall \mathbf{y} \in \text{dom } \mathcal{E}, \mathcal{E}(\mathbf{y}') \leq \mathcal{E}(\mathbf{y})$, then we can find another solution $\mathbf{y}^* = (\mathcal{L}^*, \bar{\mathcal{R}}^*)$ with $\mathcal{L}^*(n-1) = H$ and $\mathcal{E}(\mathbf{y}^*) \leq \mathcal{E}(\mathbf{y}')$.

The construction of $\mathbf{y}^*$ differs in two different cases. For the first case, if $H \leq \sum_{x=0}^{n-1}(\mathcal{L}'(x) - \bar{\mathcal{R}}'(x))$, set $\mathbf{y}^* = (\mathcal{L}' + \delta, \bar{\mathcal{R}}' + \delta)$, where

$$\delta(x) = \begin{cases} \min\{-\bar{\mathcal{R}}'(0), H - H'\} & x = 0 \\ \min\{\mathcal{L}'(x-1) - \bar{\mathcal{R}}'(x) + \delta(x-1), H - H'\}, & x > 0 \end{cases} \quad (10)$$

Essentially, the function δ is applied to $\mathbf{y}'$ in order to make the new solution $\mathbf{y}^* = (\mathcal{L}^*, \bar{\mathcal{R}}^*)$ satisfy $\mathcal{L}^*(n-1) = H$ without changing the objective value while preserving all the constraints.

For the second case, where $H > \sum_{x=0}^{n-1}(\mathcal{L}'(x) - \bar{\mathcal{R}}'(x))$, let $\mathbf{y}'' = (\mathcal{L}'', \bar{\mathcal{R}}'') = (\mathcal{L}' + \delta, \bar{\mathcal{R}}' + \delta)$, where δ is the same as defined in Equation (10). As same as the previous case $\mathbf{y}''$ is feasible to Equation (8) and $\mathcal{E}(\mathbf{y}'') = \mathcal{E}(\mathbf{y}')$, however, $\mathcal{L}''(n-1) < H$. But, $\mathbf{y}''$ has its specialties: $\bar{\mathcal{R}}''(0) = 0$ and $\forall x \in [0, n-1], \bar{\mathcal{R}}''(x) = \mathcal{L}''(x-1)$ (define $\mathcal{L}''(-1) = 0$). By enforcing these two specialties into Equation (8), Equation (8b)-(8e) becomes redundant, and Equation (8f) can be rewrote to $\mathcal{L}(n-1) \leq H, \mathcal{L}(0) = 0$. We denote the problem by further relaxing the constraint $\mathcal{L}(0) = 0$ as CIBD''. Assume $\mathcal{L}°(n-1) = H$, finding the solution $\mathbf{y}° = \mathcal{L}°$ with $\mathcal{E}''(\mathcal{L}°) = 0$ to CIBD'' can be done in linear time (the objective function of CIBD'' is defined as $\mathcal{E}''(\mathcal{L}) = \sum_{x=0}^{n-1}(\mathcal{L}(x) - \mathcal{L}(x-1) - t(x))^2$, the $\bar{\mathcal{R}}$ part of the solution is omit since it can be determined by $\mathcal{L}$).

Then, $\mathbf{y}^* = (\mathcal{L}^*, \bar{\mathcal{R}}^*)$ can be assigned with $\forall x \in [0, n-1], \mathcal{L}^*(x) = (\mathcal{L}'' \wedge \mathcal{L}°)(x), \bar{\mathcal{R}}^*(x) = (\mathcal{L}'' \wedge \mathcal{L}°)(x-1)$. According to the L$^\natural$-convexity of CIBD'', $\mathcal{E}(\mathbf{y}^*) = \mathcal{E}'(\mathcal{L}'' \vee \mathcal{L}°) \leq \mathcal{E}'(\mathcal{L}'') = \mathcal{E}(\mathbf{y}')$. By further showing $(\mathcal{L}'' \vee \mathcal{L}°)(-1) = \mathcal{L}''(-1) = 0, (\mathcal{L}'' \vee \mathcal{L}°)(n-1) = \mathcal{L}°(n-1) = H$, we can show that $\mathbf{y}^*$ is feasible to Equation (8) and it is also a global optimizer.

According to Theorem 2, whenever $H \leq \sum_{x=0}^{n-1} t(x)$, Equation (8) can be solved by setting $\mathcal{L}(n-1) = H$. And setting $\mathcal{L}(n-1) = H$ makes Equation (8) an instance of convex cost integer dual network flow problem, which can be solved in time $\mathcal{O}(n^2 \log(nH))$ for this case [1]. If $H > \sum_{x=0}^{n-1} t(x)$, it can be intuitively solved in linear time.

3 ˙ Experimental Results

Although Ahuja's algorithm has the best know theoretical complexity, Kolmogorov *et al.* demonstrate that their algorithm runs better in practice [6]. We implemented our CIBD algorithm using the C++ based on Kolmogorov's framework [6] with a specialized local search step and the total time complexity is $\mathcal{O}(n^3 \log n \log H)$. For the combinations of parameters n and H, 100 computer-generated testcases were used to test the efficiency of our algorithm on an Intel Xeon workstation (Xeon 2.66 GHz, 16 GB memory, Linux 2.6.37 64bit). Fig. 4(a) and 4(b) show the impact of parameters n and H on the running time, respectively. Based on the experimental results, the running time quadratically increases with n but is not noticeably impacted by H.

We also validated our algorithm with 5 clinical cases from 5 different patients. An example of a DVH (Dose-Volumn Histogram) plot for one of the 5 cases is shown in Fig. 5. In a DVH plot, each point on the curve represents the volume of the structure (y-axis) receiving greater than or equal to that dose (x-axis). The baseline plans were generated by a simulated annealing dose optimizer developed by ourselves and used as the inputs for the CIBD algorithm. The delivery plans were evaluated with HRCTV (High Risk Clinical Tumor Volume) D_{90} and the delivery time in minutes per fraction. All the testcases were finished within 1 second. The plan quality comparisons are shown in Table 1.

Our implementation also features the tradeoff between the delivery time and D_{90} such that the users may choose the best tradeoff by selecting different time budgets or quality goals.

Table 1. Clinical verification results for 5 patient cases

Case #	Baseline plans		CIBD plans	
	D_{90}	time	D_{90}	time
#1	101 Gy_{10}	151 min/fx	100 Gy_{10}	20 min/fx
#2	107 Gy_{10}	234 min/fx	91 Gy_{10}	20 min/fx
#3	92 Gy_{10}	260 min/fx	81 Gy_{10}	19 min/fx
#4	91 Gy_{10}	369 min/fx	84 Gy_{10}	30 min/fx
#5	95 Gy_{10}	205 min/fx	95 Gy_{10}	20 min/fx

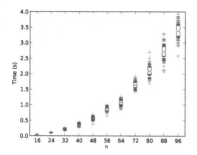 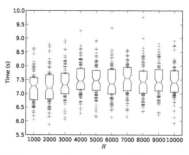

(a) Impact of parameter n on running time, $H = 10000$

(b) Impact of parameter H on running time, $n = 128$

Fig. 4. Running time analyses with phantom testcases

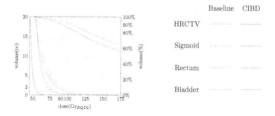

Fig. 5. DVH plot for case #2

References

1. Ahuja, R.K., Hochbaum, D.S., Orlin, J.B.: Solving the convex cost integer dual network flow problem. Management Science 49(7), 950–964 (2003)
2. Chen, D., Luan, S., Wang, C.: Coupled path planning, region optimization, and applications in intensity-modulated radiation therapy. Algorithmica 60, 152–174 (2011)
3. Dimopoulos, J., Lang, S., Kirisits, C., et al.: Dose-Volume histogram parameters and local tumor control in magnetic resonance Image-Guided cervical cancer brachytherapy. International Journal of Radiation Oncology* Biology* Physics 75(1), 56–63 (2009)
4. Haie-Meder, C., Pötter, R., Van Limbergen, E., et al.: Recommendations from gynaecological (GYN) GEC-ESTRO working group[star, open] (I). Radiotherapy and Oncology 74(3), 235–245 (2005)
5. Hochbaum, D., Levin, A.: Optimizing over consecutive 1's and circular 1's constraints. SIAM Journal on Optimization 17(2), 311–330 (2006)
6. Kolmogorov, V., Shioura, A.: New algorithms for convex cost tension problem with application to computer vision. Discrete Optimization 6(4), 378–393 (2009)
7. Liu, Y., Flynn, R., Kim, Y., Wu, X.: Dynamic-shield intensity modulated brachytherapy (IMBT) for cervical cancer. International Journal of Radiation Oncology* Biology* Physics 81(2), S201 (2011)
8. Murota, K.: Discrete Convex Analysis. Siam Monographs on Discrete Mathematics and Applications. Society for Industrial and Applied Mathematics (2003)

9. Pötter, R., Dimopoulos, J., Georg, P., Lang, S., et al.: Clinical impact of MRI assisted dose volume adaptation and dose escalation in brachytherapy of locally advanced cervix cancer. Radiotherapy and Oncology 83(2), 148–155 (2007)
10. Pötter, R., Haie-Meder, C., Limbergen, E.V., et al.: Recommendations from gynaecological (GYN) GEC ESTRO working group (II). Radiotherapy and Oncology 78(1), 67–77 (2006)

The Canadian Traveller Problem Revisited

Yamming Huang and Chung-Shou Liao[*]

Department of Industrial Engineering and Engineering Management,
National Tsing Hua University, Hsinchu 30013, Taiwan
csliao@ie.nthu.edu.tw

Abstract. This study investigates a generalization of the CANADIAN
TRAVELLER PROBLEM (CTP), which finds real applications in dynamic
navigation systems used to avoid traffic congestion. Given a road network
$G = (V, E)$ in which there is a source s and a destination t in V, every
edge e in E is associated with two possible distances: original $d(e)$ and
jam $d^+(e)$. A traveller only finds out which one of the two distances of
an edge upon reaching an end vertex incident to the edge. The objective
is to derive an adaptive strategy for travelling from s to t so that the
competitive ratio, which compares the distance traversed with that of
the static s, t-shortest path in hindsight, is minimized. This problem
was defined by Papadimitriou and Yannakakis. They proved that it is
PSPACE-complete to obtain an algorithm with a bounded competitive
ratio. In this paper, we propose tight lower bounds of the problem when
the number of "traffic jams" is a given constant k; and we introduce a
simple deterministic algorithm with a $\min\{r, 2k + 1\}$-ratio, which meets
the proposed lower bound, where r is the worst-case performance ratio.
We also consider the uniform jam cost model, i.e., for every edge e,
$d^+(e) = d(e) + c$, for a constant c. Finally, we discuss an extension to the
metric TRAVELLING SALESMAN PROBLEM (TSP) and propose a touring
strategy within an $O(\sqrt{k})$-competitive ratio.

Keywords: Canadian traveller problem, competitive ratio, travelling
salesman problem.

1 Introduction

Consider a road map $G = (V, E)$ represented by a set V of vertices connected
by edges, where each edge $e \in E$ is associated with the time it takes for the
traveller to traverse the edge. From an online perspective, the traveller is aware
of the entire structure of the road network in advance; however, some edges may
be blocked by accidents during the trip, but the problem would only become
evident when the traveller reaches an end vertex incident to the blocked edge.
This problem, called the CANADIAN TRAVELLER PROBLEM (CTP), was defined
by Papadimitriou and Yannakakis [9] in 1991. The objective is to design an
efficient routing policy from a source to a destination under this condition of

[*] Supported by the National Science Council of Taiwan under Grants NSC100-2221-
E-007-108-MY3 and NSC100-3113-P-002-012.

K.-M. Chao, T.-s. Hsu, and D.-T. Lee (Eds.): ISAAC 2012, LNCS 7676, pp. 352–361, 2012.

uncertainty. The CTP is actually a two-player game between a traveller and a malicious adversary who sets up road blockages in order to maximize the gap between the performance of the online strategy and that of the offline optimum (with the blocked edges removed). The criterion for measuring the quality of an online strategy is usually the competitive ratio of the algorithm [3,10]. The competitive ratio can be defined as follows: for any instance, the total cost of the online algorithm is at most its ratio times that of the optimal offline approach (under complete information). We will provide the formal definition later in the paper. Papadimitriou and Yannakakis [9] showed that it is PSPACE-complete for the CTP to devise a strategy that guarantees a bounded competitive ratio.

For several years, there has been no significant progress in the development of approximation algorithms for this problem. Bor-Noy and Schieber [1] explored several variations of the CTP from the worst-case scenario perspective, where the objective is to find a static (offline) algorithm that minimizes the maximum travel cost [2]. They considered the k-CTP in which the number of blockages is bounded from below by k. Note that for an arbitrary k, the problem of designing a strategy that guarantees a given travel time remains PSPACE-complete, as shown in [1,9]. In addition, Bor-Noy and Schieber discussed the Recoverable k-CTP, where each blocked edge is associated with a recovery time, which is not very long relative to the traversed time, to reopen. They also studied the stochastic model, where an independent blockage probability for each edge is given in advance. This model, which tries to minimize the expected ratio to the offline optimum, is known to be $\sharp$P-hard [9]. Subsequently, Karger and Nikolova [8] investigated the stochastic CTP in special graph classes and developed exact algorithms by applying techniques from the theory of Markov Decision Processes.

Recently, Westphal [13] proved that there are no deterministic online algorithms within a $(2k + 1)$-competitive ratio for the k-CTP. The author designed a simple *reposition* algorithm that satisfies the lower bound, and also proposed a lower bound of $k + 1$ for the competitive ratio of any randomized online algorithms. Xu *et al.* [14] developed two deterministic adaptive policies: a *greedy* strategy and a *comparison* strategy that incorporates the concept of reposition. The latter strategy also achieves the tight lower bound.

In this paper, we study a natural generalization of CTP, called the DOUBLE-VALUED GRAPH, which was initiated by Papadimitriou and Yannakakis [9]. Given a graph $G = (V, E)$ with a source s and a destination t in V, each edge e in E is associated with two possible distances: original $d(e)$ and jam $d^+(e)$, where $d, d^+ : E \to R^+$ and $d(e) < d^+(e)$, for each $e \in E$. A traveller only learns about the distance cost $(d(e)$ or $d^+(e))$ of an edge e on arrival at one of its end vertices. The goal is to develop an adaptive strategy for traversing the graph from s to t under incomplete information about traffic conditions so that the competitive ratio is minimized. This problem is also PSPACE-complete, as shown by a reduction from quantified SAT (QSAT) [9]. Here, for a graph G without any jammed edges, i.e., $d^+(e) = d(e)$, for any $e \in E$, the length of the s, t-shortest path is denoted by $d(s, t)$. On the other hand, consider the s, t-shortest path P with k jammed edges that have the maximum jam costs. More precisely, select an edge e^* in P with the maximum jam cost and add it to an edge subset E', where $e^* = \arg\max_{e \in P}\{d^+(e) - d(e)\}$, and repeat the argument until $|E'| = k$, if any. The distance of such a path from s to t is $d(s,t) + \sum_{e \in E'}(d^+(e) - d(e))$, denoted by $d^{+k}(s,t)$.

Our Contribution. The main results of this study are detailed below.

1. We provide tight lower bounds for deterministic and randomized algorithms for DOUBLE-VALUED GRAPH in terms of k and $r = \frac{d^{+k}(s,t)}{d(s,t)}$ when the number of traffic jams is up to a given constant k.
2. We present a simple deterministic adaptive strategy with a $\min\{r, 2k+1\}$-competitive ratio that meets the proposed lower bound. The algorithm can also be applied directly to the Recoverable k-CTP that assumes the blocked edges are not found to be blocked again.
3. We also study the uniform jam cost model of this problem, i.e., for every edge e, $d^+(e) = d(e) + c$, for a constant c [5], and derive a tight lower bound with an additive ratio.
4. Finally, the main contribution of this study is that we investigate an extension of the k-CTP to the metric TRAVELLING SALESMAN PROBLEM (TSP). The goal is to design a tour of a set of vertices whereby the traveller visits each vertex and returns to the origin under the same uncertainty as the k-CTP, such that the competitive ratio is minimized. We propose a touring strategy within an $O(\sqrt{k})$-competitive ratio.

2   Preliminaries

We consider the DOUBLE-VALUED GRAPH problem with at most k traffic jams. Given a connected graph $G = (V, E)$ with a source s and a destination t, we denote the sequence of traffic jams in E learned by an online algorithm A during the trip as $S_i^A = (e_1, e_2, \ldots, e_i)$, where $1 \leq i \leq k$. Let $E_i^A = \{e_1, e_2, \ldots, e_i\} \subseteq E$, $1 \leq i \leq k$, consist of these jammed edges, and let E_k be the set of all jammed edges. In addition, let $d : E \to R^+$ be the original distance function. The (traffic) jam distance function is $d^+ : E \to R^+$; that is, for each edge $e = (u, v) \in E$, $d(u,v) < d^+(u,v)$. Moreover, in the online problem, let $d_{E_i^A}(s,t)$ denote the travel cost from s to t derived by an adaptive algorithm A that learns about traffic jam information E_i during the trip; and let $d_{E_k}(s,t)$ be the offline optimum from s to t under complete information E_k. We have the following property immediately, where $E_1 \subseteq E_2 \subseteq \ldots \subseteq E_k$.

$$d(s,t) \leq d_{E_1}(s,t) \leq \cdots \leq d_{E_k}(s,t). \tag{1}$$

We refer to [3,10] and formally define the competitive ratio as follows: an online algorithm A is c^A-*competitive* for the DOUBLE-VALUED GRAPH problem if

$$d_{E_i^A}(s,t) \leq c^A \cdot d_{E_k}(s,t) + \varepsilon, \quad 1 \leq i \leq k.$$

where c^A and ε are constants.

Similar to the proof in [13], we propose tight lower bounds for DOUBLE-VALUED GRAPHS when we use deterministic and randomized algorithms. Due to space limitations, we present the following theorems without proofs. For further details, readers may refer to [7].

Lemma 1. *For the* DOUBLE-VALUED GRAPH *problem, there is no deterministic online algorithm within a* $\min\{r, 2k+1\}$*-competitive ratio, when the number of jammed edges is up to* k.

Next, given the independent probabilities of traffic congestion on all the edges, we consider randomized strategies to solve this online problem.

Lemma 2. *For the* DOUBLE-VALUED GRAPH *problem, there is no randomized online algorithm with a competitive ratio less than* $\min\{r, k+1\}$ *when the number of jammed edges is up to* k.

Based on the proofs of Lemmas 1 and 2, we introduce two strategies: the *greedy algorithm* and the *reposition algorithm* [13,14], which will be used later. We denote them as GA and RA respectively.

Greedy Algorithm (GA): Starting at a vertex v (including the source s), the traveller selects the shortest path from v to t by using Dijkstra's algorithm [6] in a greedy manner based on the current information E_i; that is, the distance cost of the path is $d_{E_i}(v,t)$. Note that if all the k jammed edges are known at the outset, the cost of the path derived by GA from the source s is the same as that of the offline optimum, $d_{E_k}(s,t)$.

Reposition Algorithm (RA): The traveller begins at the source s and follows the s,t-path with the cost $d(s,t)$. When the traveller learns about a jammed edge on the path to t, he/she returns to s and takes the s,t-path with the cost $d_{E_i}(s,t)$ based on the current information E_i. The traveller repeats this strategy until he/she arrives at t.

3 Double-Valued Graph

In this section, we present a simple algorithm called GR, which combines GA with RA to solve the DOUBLE-VALUED GRAPH problem. Its ratio meets the proposed lower bound. We also study the uniform jam cost model. For convenience, let $e_i = (v_i, v_{i'})$, $1 \leq i \leq k$, be a jammed edge learned by the traveller; and let v_i be the first end vertex of the jammed edge e_i the traveller visits during the trip.

Note that the ratio r has to be updated while the traveller is learning about a new jammed edge. In addition, because $2(k-i)+1$ decreases during the trip, $r \leq \frac{d^{+k}(s,t)}{d(s,t)}$ once the traveller follows a path derived by GA.

Algorithm 1. Greedy&Reposition Algorithm (GR)

Input : $G = (V, E)$, $d : E \to R^+$, $d^+ : E \to R^+$ and a constant k;
Output : A route from s to t;
1: Initialize $i = 0$;
2: **while** the traveller does not arrive at t **do**
3:      Let $r = \frac{d^{+(k-i)}(v_i,t)}{d_{E_i}(s,t)}$;   ▷ $v_0 = s$ and $E_0 = \emptyset$
4:      **if** $r \leq 2(k - i) + 1$ **then**
5:          the traveller traverses a path from v_i to t derived by GA;
6:      **else**
7:          **if** $d_{E_{i-1}^{GR}}(s,t) + d_{E_i}(v_i,t) \leq (i+1) \cdot d_{E_i}(s,t)$ **then**
8:              the traveller moves from v_i to t via GA unless he finds a jammed edge;
9:          **else**
10:              the traveller moves from v_i to t via RA unless he finds a jammed edge;
11:          **end if**
12:      **end if**
13:      Let $i = i + 1$ and let the new jammed edge be $e_i = (v_i, v_{i'})$ during the trip;
14: **end while**

Lemma 3. *If the ratio* $r = \frac{d^{+(k-i)}(v_i,t)}{d_{E_i}(s,t)} > 2(k - i) + 1$, *for each* i *during the whole trip, then the total distance cost of GR satisfies the following property, where* $1 \leq i \leq k$, *provided there is a set of jammed edges* E_i .

$$
d_{E_i^{GR}}(s,t) \leq \begin{cases} (i+1) \cdot d_{E_i}(s,t), & \text{if the traveller uses } GA \text{ at } v_i; \\ (2i+1) \cdot d_{E_i}(s,t), & \text{if the traveller uses } RA \text{ at } v_i. \end{cases}
$$

The above lemma shows the competitive ratio without following a path derived by GA irrespective of whether the traveller finds a jammed edge. We prove that the GR algorithm is $\min\{r, 2k + 1\}$-competitive.

Theorem 1. *For the* DOUBLE-VALUED GRAPH *problem with at most* k *traffic jams, the competitive ratio of GR is at most* $\min\{r, 2k + 1\}$, *where* $r = \frac{d^{+k}(s,t)}{d(s,t)}$ *and* r *might decrease during the trip.*

Regarding the time complexity analysis, the number of iterations in the while loop, i.e., the number of updates for the ratio r, is at most k. Each of the three strategies can apply Dijkstra's algorithm [6] to devise a path from s or v_i to t, for some i. In addition, RA just takes the original s, v_i-path when the traveller needs to return to s from v_i. Thus, for a given constant k, the running time is a constant factor times $D(n)$, where n is the order of a graph G and $D(n)$ is the running time of Dijkstra's algorithm.

The GR algorithm can be extended to the MULTIPLE-VALUED GRAPH problem in which each edge is associated with more than two possible distances. We regard the largest distance of each edge e as $d^+(e)$ and GR performs in a similar way to travel from s to t. Hence, the competitive ratio remains the same.

Corollary 1. *The* MULTIPLE-VALUED GRAPH *problem can be approximated within a competitive ratio* $\min\{r, 2k+1\}$ *when the number of traffic jams is up to a given constant* k.

In addition, we consider the Recoverable k-CTP in which each blocked edge e is associated with a recovery time $r(e)$ to reopen. In this online problem, it is assumed that the blocked edges will not be blocked again. An instance I of the Recoverable k-CTP can be transformed into an instance I' of the DOUBLE-VALUED GRAPH problem by letting $r(e)$ be represented in terms of the distance and letting $d^+(e) = d(e) + r(e)$ for every edge $e \in E$ in I'.

Corollary 2. *The Recoverable k-CTP can be approximated within a competitive ratio* $\min\{r, 2k+1\}$, *when the number of blockages is bounded from below by* k.

Su et al. [11,12] considered the Recoverable k-CTP and proposed two policies: a *waiting* strategy and a *greedy* strategy. For example, if $r(e) \leq d(e)$, i.e., $d^+(e) \leq 2d(e)$ in the DOUBLE-VALUED GRAPH problem, then $r = \frac{d^{+k}(s,t)}{d(s,t)} \leq 2$ and GR will follow the path derived by GA initially. Thus, we have the competitive ratio $r \leq 2$. Besides, if $r(e) \leq d(s,t)$, it implies that $r = \frac{d^{+k}(s,t)}{d(s,t)} \leq k+1$. Therefore, GR will follow a path derived by GA and obtain the competitive ratio $r \leq k+1$. The results show that the GR approach is at least as good as the previous result in [11,12].

3.1   The Uniform Jam Cost Model

We refer to [5] and suppose the jam cost of each edge is a constant c. We propose a tight lower bound below.

Lemma 4. *For the uniform jam cost model of the* DOUBLE-VALUED GRAPH *problem with at most k traffic jams, there is no deterministic online algorithm within a kc additive ratio; that is, given a uniform jam cost c, the derived solution cannot be better than* $d_{E_k}(s,t) + kc$.

By Lemma 4, no deterministic algorithm can derive a better additive ratio than kc. Actually, the traveller can use a very straightforward algorithm to achieve the lower bound: following the s,t-path with the cost $d(s,t)$ irrespective of whether the traveller finds jammed edges. However, it is possible to improve the ratio slightly under some conditions.

For instance, if $c \geq 2\delta \cdot d(s,t)$ for some $\delta > 1$, we let a threshold be $\frac{c}{2\epsilon}$, for a constant $1 \leq \epsilon < \delta$. Assume there is a nonempty subset of jammed edges E_i such that $d_{E_i}(s,t) \leq \frac{c}{2\epsilon}$ for some $i \leq k$, when the traveller learns about jammed edges. Based on this assumption, when $d_{E_j}(s,t) \leq \frac{c}{2\epsilon}$, the traveller will use RA; however, when $d_{E_j}(s,t) > \frac{c}{2\epsilon}$, the traveller will use GA until he/she arrives at t. Thus, we let e_ℓ be the first jammed edge such that $d_{E_\ell}(s,t) > \frac{c}{2\epsilon}$, if any. That is,

$$\begin{cases} d_{E_j}(s,t) \leq \frac{c}{2\epsilon}, & \text{if } 1 \leq j < \ell; \\ d_{E_j}(s,t) > \frac{c}{2\epsilon}, & \text{if } \ell \leq j \leq k. \end{cases}$$

Then, the total travel cost can be formulated as follows:

$$2 \cdot d(s,t) + \ldots + 2 \cdot d_{E_{\ell-2}}(s,t) + d_{E_{\ell-1}}(s,t) + (k - \ell + 1)c$$
$$\leq 2(\ell-1) \cdot \frac{c}{2\epsilon} + d_{E_k}(s,t) + (k - \ell + 1)c$$
$$\leq d_{E_k}(s,t) + kc - (\ell-1)(1 - \frac{1}{\epsilon})c.$$

If there is no such ℓ, the traveller uses RA until he/she arrives at t. The above equation implies that the total travel cost is at most $2k(\frac{c}{2\epsilon}) + d_{E_k}(s,t) = \frac{kc}{\epsilon} + d_{E_k}(s,t)$.

4 k-CTP for Metric TSP

In this section, we extend the k-CTP to the metric travelling salesman problem (TSP). We refer to [14] and present several natural assumptions in the following. Given a complete graph $G = (V,E)$ of order n, i.e., $|V| = n$, the graph is still connected even if blocked edges are removed. Otherwise, the traveller would not be able to visit all the vertices in G. Moreover, we assume that the upper bound of the number of blockages is $k < n - 1$ for the same reason. In addition, the traveller will only find a blockage upon reaching an end vertex of the blocked edge, and the state of the edge will not change again after the traveller learns about the blockage. The objective is to design a tour that enables the traveller to visit every vertex and return to the origin under the uncertainty, such that the competitive ratio is minimized. Note that the problem allows the traveller to visit a vertex more than once because the blocked edges may cause the assumption to occur.

The rationale behind our approach is that the traveller will visit as many vertices as possible via alternative routes based on a tour derived by Christofides' algorithm [4], denoted by $P : s = v_1 - v_2 - \ldots - v_n - s$. Assume the traveller takes the tour P, which is supposed to visit all the vertices in V in m^* rounds. Let V_m be the set of unvisited vertices in the mth round, $1 \leq m \leq m^* + 1$, and $V_1 = V \setminus \{s\}$. Then, we have $V \supset V_1 \supseteq V_2 \supseteq \ldots \supseteq V_{m^*} \supset V_{m^*+1} = \emptyset$. In addition, we denote a path over V_m as $P_m : v_{m,0} - v_{m,1} - \ldots - v_{m,|V_m|}$ when the traveller tries to visit the vertices in V_m in the mth round by using the Cyclic Routing (CR) algorithm (see Algorithm 2). Note that $v_{m,0}$ is the last vertex visited in the $(m-1)^{th}$ round when $m > 1$, i.e., $v_{m,0} \notin V_m$.

As mentioned earlier, we let E_i^{CR} consist of the blocked edges revealed by the CR algorithm, $1 \leq i \leq k$. We decompose E_i^{CR} into m^* subsets called $E_{m,i}^{CR}$. Each subset is learned by the traveller in the mth round, $1 \leq m \leq m^*$, i.e., $E_i = \{e_1, \ldots, e_i\} = \bigcup_{m=1}^{m^*} E_{m,i}^{CR}$. For convenience, we use E'_m instead of $E_{m,i}^{CR}$; and we denote a path p from u to v as $p : u \sim v$. Note that in the mth round, the traveller may traverse the path P_m in a either clockwise direction or counterclockwise direction $1 \leq m \leq m^*$.

Algorithm 2. Cyclic Routing Algorithm (CR)

Input : $G = (V, E)$ of order n, $d : E \to R^+$ and a constant k;
Output : A tour P^* that visits every vertex in V;
1: Use Christofides' algorithm to find a tour $P : s = v_1 - v_2 - \ldots - v_n - s$ that visits every vertex exactly once from s and return to s, and initialize $P^* = \emptyset$, $m = 1$, $V_m = V \setminus \{s\}$, and $v_{m,0} = s$;
2: **while** $V_m \neq \emptyset$ **do**
3:     Let a path over V_m be $P_m : v_{m,0} - v_{m,1} - \ldots - v_{m,|V_m|}$;
4:     **if** $m = 1$ or $v_{m,0} = v_{m-1,|V_{m-1}|}$ **then**
5:         Perform **procedure** DETOUR in the same direction as that in the $(m-1)^{th}$ round or in a clockwise direction when $m = 1$;
6:         **if** $V_{m+1} = V_m$ **then**
7:             Perform **procedure** DETOUR in the opposite direction;
8:         **end if**
9:     **else**
10:         Perform **procedure** DETOUR in the opposite direction to that in the $(m-1)^{th}$ round;
11:     **end if**
12:     $m = m + 1$;
13: **end while**
14: The traveller returns to s directly or finds a previously visited vertex u such that $v_{m,0} - u - s$ is not blocked and traverse the alternative route;
15: $P^* = P^* \cup \{v_{m,0} \sim s\}$;
16: **return** P^*;

Lemma 5. *The traveller can visit at least one vertex in V_m by using CR in the mth round, $1 \leq m \leq m^*$; that is, $|V_1| > |V_2| > \ldots > |V_m| > |V_{m+1}| > \ldots > |V_{m^*}|$.*

Based on the above key lemma, the tour P^* derived by the CR algorithm traverses all the vertices in V until $V_{m^*+1} = \emptyset$; that is, P^* is a feasible solution. The next lemmas follow immediately.

Lemma 6. $E'_i \cap E'_j = \emptyset$ *for any* $1 \leq i < j \leq m^*$.

Lemma 7. *The number of new blockages learned in the mth round is not less than the number of unvisited vertices in the $(m+1)^{th}$ round, i.e., $|E'_m| \geq |V_{m+1}|$, $1 \leq m \leq m^*$.*

We develop the CR strategy based on a tour P derived by Christofides' algorithm for the original metric TSP. It combines the minimum spanning tree of G with the minimum weight perfect matching on the vertices with odd degree in the tree to obtain a Hamiltonian tour with a $\frac{3}{2}$-approximation ratio if G satisfies the triangle inequality property. Let OPT be the offline optimum of the k-CTP for the metric TSP. Obviously, OPT cannot be less than the optimum of the original TSP.

```
1: procedure DETOUR
2:     Let V_{m+1} = V_m and E'_m = ∅;
3:     for i = 1 to |V_m| do
4:         if v_{m,i-1} − v_{m,i} is not blocked or there is an internal visited vertex u in
           v_{m,i-1} ∼ v_{m,i} of P such that v_{m,i-1} − u − v_{m,i} is not blocked then
5:             V_{m+1} = V_{m+1} \ {v_{m,i}} and P* = P* ∪ {v_{m,i-1} ∼ v_{m,i}};
6:         else
7:             Let v_{m,i} = v_{m,i-1};
8:         end if
9:         E'_m = E'_m ∪ {e}, for each new blockage e the traveller just learned;
10:    end for
11: end procedure
```

Lemma 8. *When the traveller uses CR to traverse P_m in the mth round, $1 \leq m \leq m^*$, the travel cost is not larger than $3OPT$; and for the $(m^* + 1)^{th}$ round, the travel cost is OPT at most.*

Theorem 2. *The k-CTP for the metric TSP can be approximated within an $O(\sqrt{k})$-competitive ratio when the number of blockages is bounded from below by k.*

The next corollary shows the tightness of the analysis of the competitive ratio and provides a tight example that attains the ratio.

Corollary 3. *The analysis of an $O(\sqrt{k})$-competitive ratio of the CR algorithm is tight.*

5 Concluding Remarks

In this paper, we have studied the DOUBLE-VALUED GRAPH problem, which is a generalization of k-CTP when the number of traffic jams is up to k. We have presented tight lower bounds and a simple adaptive algorithm that can satisfy the lower bound. In addition, we have derived a lower bound with an additive ratio for the uniform jam cost model. We have also extended this problem to the metric TSP and proposed an online touring $O(\sqrt{k})$-competitive algorithm. It would be worthwhile investigating these online route planning problems because they find real applications in dynamic navigation systems designed to avoid traffic congestion. We conclude the study with two observations: First, compared with the *larger* lower bound of deterministic algorithms, it would be very interesting to develop a randomized online algorithm that can yield a better competitive ratio. On the other hand, a traveller could learn about a blockage or traffic congestion in advance from road sensor networks; for example, a GPS navigation system could indicate traffic conditions as the traveller approaches within a distance ℓ of an end vertex of a blocked edge (or a jammed edge) for a given constant ℓ. The question is how much the earlier information could improve online route planning. We will consider these issues in our future research.

References

1. Bar-Noy, A., Schieber, B.: The Canadian traveller problem. In: Proc. of the 2nd ACM-SIAM Symposium on Discrete Algorithms (SODA), pp. 261–270 (1991)
2. Ben-David, S., Borodin, A.: A new measure for the study of online algorithms. Algorithmica 11(1), 73–91 (1994)
3. Borodin, A., El-Yaniv, R.: Online computation and competitive analysis. Cambridge Univeristy Press, Cambridge (1998)
4. Christofides, N.: Worst-case analysis of a new heuristic for the traveling salesman problem. Techical report, Graduate School of Industrial Administration, Carnegie-Mellon University (1976)
5. Chuzhoy, J.: Routing in undirected graphs with constant congestion. In: Proc. of the 44th ACM Symposium on Theory of Computing, STOC (2012)
6. Dijkstra, E.W.: A note on two problems in connexion with graphs. Numerische Mathematik 1(1), 269–271 (1959)
7. Huang, Y., Liao, C.S.: The Canadian traveller problem revisited (2012) (manuscripts), http://acolab.ie.nthu.edu.tw/draft/ctp.pdf
8. Karger, D., Nikolova, E.: Exact algorithms for the Canadian traveller problem on paths and trees. Technical report, MIT Computer Science & Artificial Intelligence Lab (2008), http://hdl.handle.net/1721.1/40093
9. Papadimitriou, C.H., Yannakakis, M.: Shortest paths without a map. Theoretical Computer Science 84(1), 127–150 (1991)
10. Sleator, D., Tarjan, R.E.: Amortized efficiency of list update and paging rules. Communications of the ACM 28, 202–208 (1985)
11. Su, B., Xu, Y.F.: Online recoverable Canadian traveller problem. In: Proc. of the International Conference on Management Science and Engineering, pp. 633–639 (2004)
12. Su, B., Xu, Y., Xiao, P., Tian, L.: A Risk-Reward Competitive Analysis for the Recoverable Canadian Traveller Problem. In: Yang, B., Du, D.-Z., Wang, C.A. (eds.) COCOA 2008. LNCS, vol. 5165, pp. 417–426. Springer, Heidelberg (2008)
13. Westphal, S.: A note on the k-Canadian traveller problem. Information Processing Letters 106, 87–89 (2008)
14. Xu, Y.F., Hu, M.L., Su, B., Zhu, B.H., Zhu, Z.J.: The Canadian traveller problem and its competitive analysis. Journal of Combinatorial Optimization 18, 195–205 (2009)
15. Zhang, H., Xu, Y.: The k-Canadian Travelers Problem with Communication. In: Atallah, M., Li, X.-Y., Zhu, B. (eds.) FAW-AAIM 2011. LNCS, vol. 6681, pp. 17–28. Springer, Heidelberg (2011)

Vehicle Scheduling on a Graph Revisited*

Wei Yu[1,2], Mordecai Golin[2], and Guochuan Zhang[1]

[1] College of Computer Science, Zhejiang University, Hangzhou, 310027, China
{yuwei2006831,zgc}@zju.edu.cn
[2] Department of Computer Science and Engineering, Hong Kong University of
Science and Technology, Clear Water Bay, Kowloon, Hong Kong
golin@cse.ust.hk

Abstract. We consider a generalization of the well-known Traveling
Salesman Problem, called the Vehicle Scheduling Problem (VSP), in
which each city is associated with a release time and a service time.
The salesman has to visit each city at or after its release time. Our
main results are three-fold. First, we devise an approximation algorithm
for VSP with performance ratio less than 5/2 when the number of distinct release times is fixed, improving the previous algorithm proposed
by Nagamochi et al. [12]. Then we analyze a natural class of algorithms
and show that no performance ratio better than 5/2 is possible unless
the Metric TSP can be approximated with a ratio strictly less than 3/2,
which is a well-known longstanding open question. Finally, we consider a
special case of VSP, that has a heavy edge, and present an approximation
algorithm with performance ratio less than 5/2 as well.

1 Introduction

The Traveling Salesman Problem (TSP) is one of the most intensively studied
problems in combinatorial optimization. The input to the problem is a complete
graph in which vertices represent cities and edge lengths indicate the travel times
between the cities. The goal is to find a route for a salesman to visit all the cities
and return to his starting point such that (i) each city is visited exactly once,
and (ii) the total travel time of the salesman is minimum. This is a fundamental
problem that finds applications in most areas of discrete optimization. Unfortunately, it turns out that TSP is NP-hard [7] and one cannot give an efficient
algorithm to produce the optimal route for each instance of TSP unless $P = NP$.
For this reason, researchers concentrate on designing approximation algorithms
for TSP. However, as shown in Sahni and Gonzalez [13], if the travel times are arbitrary, for any constant $c > 0$ it is NP-hard to approximate TSP within ratio c.
If the travel times obey the triangle inequality (called Metric TSP), Christofides
[5] presents an elegant 3/2-approximation algorithm, which is currently the best
available, though the problem is still NP-hard [7].

Given a graph G, we can naturally induce a complete graph G_c (called the
closure of G) with edge lengths satisfying the triangle inequality. In G_c, the

---

* Research supported in part by NSFC (10971192).

K.-M. Chao, T.-s. Hsu, and D.-T. Lee (Eds.): ISAAC 2012, LNCS 7676, pp. 362–371, 2012.
© Springer-Verlag Berlin Heidelberg 2012

vertex set is the same as G and the length of edge (u, v) is equal to the length of the shortest path between u and v in G. In this work, we focus on the *Vehicle Scheduling Problem (VSP)*, a natural generalization of Metric TSP in which each city has a release time, the earliest possible time at which the salesman can visit the city, and a service time indicating the duration for the visit. The problem also specifies a home city at which the salesman starts and finishes the tour.

The literature contains a large body of research devoted to some special cases of VSP, in which the complete graph is the closure of a line (or a cycle, a tree) network. For VSP on a line, Tsitsiklis [15] proves the NP-hardness. Karuno et al. [10] give a 3/2-approximation algorithm when the home is one of the end vertices of the line. Gaur et al. [8] develop a 5/3-approximation algorithm when the home is located at an arbitrary point, which is further improved to a PTAS by Karuno, Nagamochi [11] and Augustine, Seiden [2], independently. Bhattacharya et al. [4], Yu and Liu [16] devise faster 3/2-approximation algorithms for VSP on a line as well as its variants. For VSP on a cycle and a tree, Bhattacharya et al. [4] design approximation algorithms with performance ratios 9/5 and 11/6, respectively, which are improved to 12/7 and 9/5 by Bao and Liu [3] recently.

For the general case, as far as we know, there is only a 5/2-approximation algorithm proposed by Nagamochi et al. [12]. The algorithm simply works as follows: the salesman waits at home till all cities are available to visit (at the largest release time), and then visits the vertices along a tour generated by Christofides' Algorithm. It leaves a natural question if one can do better by serving some cities that are released earlier. Along this line, in this work, we achieve the following nontrivial results.

When the number of distinct release times is fixed, we are able to approximate VSP with a ratio better than 5/2. In contrast, for an arbitrary number of distinct release times, we carefully analyze a class of algorithms, called partition-based algorithms, that group cities based on the release times and employ approximation algorithms for the metric TSP to serve the cities in a group. We show that the performance ratio for such algorithms cannot be better than 5/2 unless metric TSP can be approximated strictly better than 3/2. This implies that in order to break the barrier of 5/2 it is necessary to introduce some new approaches. Finally, we deal with a special case of VSP, that has a heavy edge. Such a problem admits an approximation algorithm with performance ratio better than 5/2.

The remainder of the paper is organized as follows. Section 2 describes the problem formally, and introduces some notations. In Section 3 we present a recursive algorithm and analyze its performance ratio. A negative result is showed in Section 4 for partition-based algorithms which generalize the algorithm in Section 3. In Section 5 we discuss the case with a heavy edge. Section 6 concludes the paper.

2   Problem Formulation and Notations

VSP is formally described below. Given a complete graph $G = (V, E)$ with vertex set $V = \{0, 1, 2, \ldots, n\}$ and edge set E, each edge has a non-negative length satisfying the triangle inequality. Each vertex $u \in V \setminus \{0\}$ corresponds to exactly one customer. Except for vertex 0 we use the terms *vertices* and *customers* interchangeably if no confusion is caused. A vehicle (or salesman), initially located at vertex 0 (called its home), travels along the graph to serve all the customers. The travel time $t_{u,v}$ of edge (u,v) for the vehicle is equal to the length of this edge. It takes a time duration of $p(u)$ (called the service time) to serve customer u, but this service cannot start before its release time $r(u)$. The goal is to find a route for the vehicle to serve all the customers and return home as soon as possible.

For a customer set A, a schedule S on A with starting vertex u and ending vertex v is an order of vertices on $A \cup \{u,v\}$ in which u is the first element and v is the last. The makespan of S, denoted by $C_{\max}(S)$, is the least time needed for the vehicle to start from u, serve all the customers in A following S and arrive at v. If we refer to a schedule without mentioning its starting vertex (or ending vertex), then the starting point (or ending point) is automatically the vertex 0. Therefore, the goal of VSP is to find a schedule on $V \setminus \{0\}$ to minimize the makespan.

As a subroutine we will often be required to solve (approximately) the Shortest Hamiltonian Path Problem (SHPP), in which we are required to find a shortest Hamiltonian path with one (or two) specified ending vertices, ignoring release and service times. SHPP is clearly NP-hard. Hoogeveen [9] presents an algorithm, which we refer to as Hoogeveen's algorithm, to find an approximate Hamiltonian path. We will always denote a Hamiltonian path on a vertex set $V' \cup \{l\}$ (resp. $V' \cup \{l',l\}$) with one specified ending vertex l (resp. two specified ending vertices $\{l',l\}$) by an order of the vertices in $V' \cup \{l\}$ (resp. $V' \cup \{l',l\}$) such that vertex l is the last element in this order.

Given an instance I of VSP, let $0 \le r_1 < r_2 < \cdots < r_m$ be the m distinct release times of the customers. Clearly, $m \le n$ (if the release times are all distinct then $m = n$). Denote by V_i the set of vertices released at time r_i. Set $V_{i,j} = \cup_{k=i}^{j} V_k$. We write $V_{\le i}, V_{\ge i}$, respectively, for $V_{1,i}, V_{i,m}$. $P_{\le i}$ denotes the total service time of vertices in $V_{\le i}$. $P_{\ge i}$ are defined similarly for $V_{\ge i}$. We simply write P for $P_{\le m}$. For $V_{\le i} \cup \{0\}$, we denote respectively by $F_{\le i}, H_{\le i}$ the minimum spanning tree and the optimal Hamiltonian path with one ending vertex 0 on the specified vertex set. We similarly define $F_{\ge i}, H_{\ge i}$ for $V_{\ge i} \cup \{0\}$. Denote by T the optimal tour on V (ignoring release and service times). Finally, $OPT(I)$ is the optimal value of I, i.e., the minimum possible makespan of a schedule on $V \setminus \{0\}$, which is usually replaced by OPT if the instance is explicit in the context.

For simplicity, the notation for a tour (path, tree) also represents its length, i.e., the sum of travel times along it. Then we can show the following useful facts and lower bounds on the optimal value of I.

Lemma 1. *(i)* $F_{\le i} \le H_{\le i} \le T$; *(ii)* $OPT \ge P+T$; *(iii)* $OPT \ge r_i+P_{\ge i}+H_{\ge i}$.

Proof. The inequalities (i) and (ii) are obvious. For the correctness of (iii), we only need to note the following facts. No vertex in $V_{\ge i}$ is available before time r_i and the vehicle is required to return home. Thus, in any feasible schedule, after time r_i the vehicle has to serve all the vertices in $V_{\geq i}$ and travel along a path going through all the vertices in $V_{\geq i} \cup \{0\}$ with ending vertex 0.

Observe that (i) still holds by substituting $F_{\geq i}, H_{\geq i}$ for $F_{\leq i}, H_{\leq i}$, respectively.

Due to the page limit, the proofs of all the following lemmas and Theorem 4 are omitted and will appear in the full version of this paper.

3   A Recursive Algorithm

In this section, we present an approximation algorithm for VSP and analyze its performance ratio. Consider an instance I with m distinct release times. Since the salesman starts from vertex 0 to serve the customers and returns home eventually, we can get a schedule directly from a Hamiltonian path on $V_{\leq m}$ with two specified ending vertices $\{0, 0\}$, which is actually a tour on V. Moreover, we observe that for each $1 \leq i \leq m - 1$, a schedule on $V_{\leq m}$ can be obtained by appending a schedule on $V_{\geq i+1}$ with some starting vertex u to a schedule on $V_{\leq i}$ that ends at u.

Since the vehicle has to return home, the schedule on $V_{\geq i+1}$ can be generated by computing a Hamiltonian path $\tilde{H}_{\geq i+1}$ with one specified ending vertex 0. Let l be the other ending vertex of $\tilde{H}_{\geq i+1}$ in the generated schedule. Next we obtain a schedule on $V_{\leq i}$ with ending vertex l. Such a schedule can be found recursively by replacing $V_{\leq m}$ and 0 with $V_{\leq i}$ and l, respectively.

In summary, we have the following procedure to compute a best available schedule on $V_{\leq i}$ with ending vertex l for all pairs (i, l) such that either (1) $i = m$ and $l = 0$ or (2) $1 \leq i \leq m - 1$ and $l \in V_{\geq i+1}$.

Algorithm Recur(i, l)

Step 1. If $i > 1$, for each $j = 1, 2, \ldots, i - 1$, run Hoogeveen's algorithm to compute a Hamiltonian path $\tilde{H}_{j+1,i}$ on $V_{j+1,i} \cup \{l\}$ with one specified ending vertex l. Let l' be the other ending vertex of $\tilde{H}_{j+1,i}$. By a recursive call Recur(j, l'), we obtain a schedule on $V_{\leq j}$ with ending vertex l' and makespan t_j. After that follow $\tilde{H}_{j+1,i}$ to serve the vertices in $V_{j+1,i}$. Let S_j be the resulting schedule with ending vertex l.

Step 2. Run Hoogeveen's algorithm to compute a Hamiltonian path $\tilde{H}_{\leq i}$ on $V_{\leq i} \cup \{0, l\}$ with two specified ending vertices $\{0, l\}$. Follow $\tilde{H}_{\leq i}$ to serve all the vertices in $V_{\leq i}$. The obtained schedule with ending vertex l is denoted by S_i.

Step 3. Choose one among $S_1, S_2, \ldots, S_i$ with minimum makespan as the output.

For any instance I with m distinct release times, we run Recur$(m, 0)$ to obtain an approximate solution S^* which is concatenated by a series of paths. Since each call of Recur(i, l) takes a polynomial time by employing all Recur(j, l') with $j < i$ and $l' \in V_{\geq j+1}$ and there are at most $O(mn)$ recursive calls to compute, S^* can be found in polynomial time. Next we analyze the performance of this algorithm. First we present some bounds on the length of the constructed path.

Lemma 2. *(i)* $\tilde{H}_{j+1,i} \leq \frac{3}{2}H_{\geq j+1}$; *(ii)* $\tilde{H}_{\leq j} \leq F_{\geq 1} + \frac{1}{2}T \leq \frac{3}{2}T$.

Remark 1. Note that Hoogeveen's algorithm has a performance ratio of 5/3 for a shortest Hamiltonian path with two specified ending vertices [9], which is improved by An et al. [1] who propose a better algorithm with an upper bound of $(1 + \sqrt{5})/2$. However, as shown in Lemma 2(ii), if comparing the length of the path produced by Hoogeveen's algorithm to the optimal **tour** length the ratio is at most 3/2. Note that this ratio is currently the best available since a smaller ratio implies an improvement for the Metric TSP. Similarly, the ratio 3/2 in Lemma 2(i) is also the best available unless there exists an approximation algorithm for SHPP with one specified ending vertex whose performance ratio is less than 3/2.

We can bound the makespan of S_j and t_j as follows.

Lemma 3. *(i)* $C_{\max}(S_j) \leq \max\{t_j, r_i\} + P_{\geq j+1} + \frac{3}{2}H_{\geq j+1}$, *for* $j = 1, \ldots, i-1$; *(ii)* $C_{\max}(S_i) \leq r_i + P_{\leq i} + \tilde{H}_{\leq i}$; *(iii)* $t_j \leq r_j + P_{\leq j} + \tilde{H}_{\leq j}$.

In the following, by an induction on i we prove that each recursive call $\text{Recur}(i, l)$ returns a schedule on $V_{\leq i}$ with ending vertex l whose makespan is at most $\rho_i OPT$ for some ρ_i. Using Lemmas 2 and 3, for $i = 1, 2$, we have

Lemma 4. $\rho_1 \leq \frac{3}{2}$.

Lemma 5. $\rho_2 \leq \frac{2}{5}\rho_1 + \frac{3}{2}$.

Now we move to the general case that $i \geq 3$.

Lemma 6. *For any* $0 < \lambda \leq 1$, *if both* $r_{j+1} \geq \lambda r_i$ *and* $t_j \leq r_i$ *hold, we have* $C_{\max}(S_j) \leq g(\lambda)OPT$, *where*

$$g(\lambda) = \begin{cases} \frac{5}{2} - \frac{3}{2}\lambda, & \text{if } 0 < \lambda < \frac{2}{3} \\ \frac{3}{2}, & \text{if } \lambda \geq \frac{2}{3}. \end{cases}$$

Lemma 7. *For any* $0 < \lambda \leq 1$, *if* $r_{i-1} \leq \lambda r_i$ *holds, then*

$$\min\{C_{\max}(S_{i-1}), C_{\max}(S_i)\} \leq \frac{21 - 6\lambda}{10 - 4\lambda}OPT.$$

Lemma 8. *For* $i \geq 3$, $\rho_i \leq \frac{4}{19}\rho_{i-2} + \frac{75}{38}$.

Using Lemmas 4, 5, 8 we can establish the following theorem.

Theorem 1. *There exists a polynomial time algorithm that produces a* ρ_m-*approximate solution for any instance I with m distinct release times, where*

$$\rho_m \leq \begin{cases} \frac{5}{2} - \left(\frac{4}{19}\right)^{(m-1)/2}, & m \text{ is odd,} \\ \frac{5}{2} - \frac{\sqrt{19}}{5}\left(\frac{4}{19}\right)^{(m-1)/2}, & m \text{ is even.} \end{cases}$$

Proof. For any instance I with m distinct release times we run $\mathrm{Recur}(m, 0)$ to obtain a ρ_m-approximate solution.

If m is an odd number, by Lemmas 8, 4 we have

$$\rho_m \leq \frac{4}{19}\rho_{m-2} + \frac{75}{38} \leq \left(\frac{4}{19}\right)^2 \rho_{m-4} + \frac{75}{38}\left(\frac{4}{19} + 1\right) \leq \cdots \cdots$$

$$\leq \left(\frac{4}{19}\right)^{(m-1)/2} \rho_1 + \frac{75}{38}\left(\left(\frac{4}{19}\right)^{(m-3)/2} + \cdots + \frac{4}{19} + 1\right)$$

$$\leq \frac{5}{2} - \left(\frac{4}{19}\right)^{(m-1)/2} .$$

Similarly, if m is an even number, using Lemmas 8, 5 we can prove the other inequality in the theorem.

By the above theorem, for VSP with a fixed number of distinct release times, we have an approximation algorithm with performance ratio less than $5/2$. The following theorem reveals the relation between algorithms for VSP and those for VSP with a fixed number of distinct release times.

Theorem 2. *For any fixed $m \geq 1$, given an algorithm with performance ratio at most $\rho(m)$ for VSP with m distinct release times, one can build a new algorithm for VSP with performance ratio at most $\rho(m) + \frac{1}{m}$.*

Proof. Given an instance I of VSP, let $r_{\max}$ be the maximum release time, we round the release times such that $r'(u) = \lfloor \frac{mr(u)}{r_{\max}} \rfloor \frac{r_{\max}}{m}$ to get an instance I' with m distinct release times. Run the algorithm for I' to obtain a solution S' with makespan at most $\rho(m)OPT(I')$. By delaying the serving of each vertex for $\frac{r_{\max}}{m}$ time units in S' we obtain a solution of I with makespan at most

$$\rho(m)OPT(I') + \frac{r_{\max}}{m} \leq \left(\rho(m) + \frac{1}{m}\right)OPT(I).$$

By the above theorem, we can obtain an approximation algorithm for VSP with performance ratio less than $5/2$ if there is an algorithm for VSP with two distinct release times whose performance ratio is less than 2.

4   The Bottleneck of Partition-Based Algorithms

We generalize the algorithm in the previous section to a natural class of algorithms, called partition-based algorithms, and figure out the bottleneck of using this class of algorithms to obtain an algorithm for VSP with performance ratio less than $5/2$.

Recall that in the algorithm by Nagamochi et al.[12], for any instance I with m release times, the vehicle waits at the home until the largest release time r_m and then serves all the customers along a shortest possible approximate tour. To improve the algorithm, the vehicle may serve some customers before r_m instead of waiting at the home. However, the problem arises that we have to carefully choose these customers first served. Like our algorithm in the last section, a natural choice is $V_{\leq i}$, which motivates us to introduce partition-based schedules. For any $l \in V_{\geq i+1} \cup \{0\}$, let $H(i, l)$ be a Hamiltonian path on $V_{\leq i} \cup \{0, l\}$ with two specified ending vertices $\{0, l\}$. Note that such a path must be computed by a polynomial time oracle (Hoogeveen's algorithm is one of such oracles), so that the following schedule can be achieved polynomially.

Definition 1. *A schedule S is called a partition-based schedule, if in S there exists some vertex $l \in V_{\geq i+1} \cup \{0\}$ with $1 \leq i \leq m$ such that all the customers in $V_{\leq i}$ are served before l along with $H(i, l)$, while all the customers in $V_{\geq i+1}$ are served after l. A partition-based algorithm is a polynomial time algorithm that always produces partition-based schedules.*

Next we derive a lower bound of any partition-based algorithm. This bound holds even for a special case of VSP, called Traveling Salesman Problem with release times(TSPR), in which all the service times are zero. We achieve this bound by considering an oracle for a special kind of metric graphs, called bounded metric graphs (see Trevisan [14] and Engebretsen, Karpinski [6]), in which all the edge lengths lie in $\{1, 2, \ldots, B\}$ for some fixed integer B independent of the number of the vertices. Defining

$\gamma_0 = \inf\{\gamma \mid$ there is an oracle that produces for any bounded metric graph a
     Hamiltonian path with two specified vertices of length at most γ
     times that of the optimal tour on the graph $\}$,

we can show that

Theorem 3. *The performance ratio of any partition-based algorithm on TSPR is at least $1 + \gamma_0$.*

Proof. Given any partition-based algorithm, let BL be the oracle to compute schedules. Suppose G_n is a series of worst-case bounded metric graphs for BL with vertex set $V_n = \{0, 1, \ldots, n\}$, then there exist two vertices u_n, v_n such that BL returns for the input (G_n, u_n, v_n) a Hamiltonian path from u_n to v_n with length at least γ_0 times that of the optimal tour, say $T(n)$. Without loss of generality we assume that $T(n) = (0, 1, 2, \ldots, n)$. Since G_n is a bounded metric graph, we have $T(n) \geq n \geq \frac{n}{B} t_{i,j}$, for all i, j, which implies for any two specified vertices $\{u, v\}$, BL will output a Hamiltonian path of length at least $\left(\gamma_0 - \frac{2B}{n}\right) T(n)$, since a Hamiltonian path with ending vertices $\{u, v\}$ can be transformed into a Hamiltonian path with ending vertices $\{u_n, v_n\}$ by adding two edges (u, u_n), (v, v_n).

Next we construct an instance of TSPR defined on a complete graph G, which is obtained from G_n by adding $i - 1$ copies of vertex i for $i = 1, 2, \ldots, n$ and n copies of vertex 0. For $i = 1, 2, \ldots, n$, the release times of vertex i and its $i - 1$ copies are $r_i, r_{i-1}, \ldots, r_1$, respectively, where r_i is the length of the path from vertex 0 to vertex i along $T(n)$. The release times of the n copies of vertex 0 are $r_n, r_{n-1}, \ldots, r_1$, respectively. It can be seen that the optimal solution simply travels along $T(n)$ to visit all the vertices with makespan $T(n)$.

Now consider a partition-based schedule $S(i, l)$, where l represents vertex l as well as its copies, we have $H(i, l) \geq \left(\gamma_0 - \frac{2B}{n}\right) T(n)$. Note that the optimal path on $V_{\geq i+1} \cup \{0\}$ with two specified ending vertices $\{0, l\}$ is of length at least $T(n) - r_i - t_{i,i+1} - t_{i+1,l}$ $(t_{n+1,j} = t_{0,j})$, otherwise contradicts the optimality of $T(n)$. Moreover, since there is a copy of vertex 0 with release time r_i, we have

$$
\begin{aligned}
C_{\max}(S(i, l)) &\geq r_i + H(i, l) + T(n) - r_i - t_{i,i+1} - t_{i+1,l} \\
&\geq (1 + \gamma_0) T(n) - \frac{2B}{n} T(n) - t_{i,i+1} - t_{i+1,l} \\
&\geq (1 + \gamma_0) T(n) - \frac{4B}{n} T(n).
\end{aligned}
$$

Since i and l are arbitrary, the theorem follows by letting n tend to infinity.

Remark 2. It can be verified that the above proof still holds even if we modify the definition of γ_0 by replacing bounded metric graphs with a more general class of graphs $\mathbf{G}_0$, in which the ratio of the maximum edge length to the length of the optimal tour approaches 0 when the number of vertices tends to infinity.

By Theorem 3 and Remark 2, to obtain a partition-based algorithm on TSPR with performance ratio less than $\frac{5}{2}$, it is necessary to show the value of γ_0 defined with respect to $\mathbf{G}_0$ is less than $\frac{3}{2}$, which asks to improve the currently best-known approximation algorithm for TSP defined on graphs in $\mathbf{G}_0$.

5   VSP with a Heavy Edge

In the last section, we have shown that there is a bottleneck to obtain a partition-based algorithm for TSPR on graphs in $\mathbf{G}_0$. In this section we discuss VSP with a heavy edge, which is a generalization of TSPR on graphs not in $\mathbf{G}_0$, and give a simple approximation algorithm with performance ratio less than $\frac{5}{2}$. To characterize the class of graphs not in $\mathbf{G}_0$, we introduce heavy edges.

Definition 2. *Given some $c_0 > 0$, an edge (v, w) of an instance of VSP is called a heavy edge if $p(v) + p(w) + t_{v,w} \geq 2c_0(P + T)$.*

VSP with a heavy edge consists of only VSP instances with heavy edges. Now we describe an algorithm, called HeavyEdge, for VSP with a heavy edge. Given any instance I, let (v, w) be the heavy edge with $p(v) + p(w) + t_{v,w} \geq 2c_0(P + T)$. By the triangle inequality $\max\{p(v) + t_{0,v}, p(w) + t_{0,w}\} \geq \frac{1}{2}(p(v) + p(w) + t_{v,w}) \geq c_0(P + T)$. Without loss of generality we assume that $p(v) + t_{0,v} \geq c_0(P + T)$. The algorithm generates two schedules. The first one is S_1 in which the vehicle first travels to vertex v and then serves all the customers along a Hamiltonian path $\tilde{H}$ with specified vertices $\{v, 0\}$ obtained by Hoogeveen's algorithm. The construction of the other schedule S_2 is identical to Nagamochi et al. [12] in which the vehicle waits at the home until the largest release time r_m and then serves all the customers by following an approximate tour $\tilde{T}$ on V computed by Christofides' Algorithm. The best one of S_1, S_2 is taken as the approximate solution.

Next we analyze the performance ratio of Algorithm HeavyEdge.

Theorem 4. *Algorithm HeavyEdge is a $\rho(c_0)$-approximation algorithm for VSP with a heavy edge, where*

$$\rho(c_0) = \left(\frac{6}{\sqrt{(6+c_0)^2 + 12c_0} + c_0} + \frac{3}{2} \right) < \frac{5}{2}.$$

By this theorem, a ρ-approximation algorithm, say A_0, for TSP defined on graphs in $\mathbf{G}_0$ with $\rho < \frac{3}{2}$ is sufficient to obtain an algorithm for VSP with performance ratio less than $\frac{5}{2}$. And this algorithm works in the following way: given any instance I, if the complete graph of I is in $\mathbf{G}_0$, run Nagamochi's algorithm with subroutine A_0 instead of Christofides' algorithm; otherwise, run Algorithm HeavyEdge.

6   Concluding Remarks

In this paper, we derive an approximation algorithm for VSP whose performance ratio is less than $5/2$ when the number of distinct release times is fixed. And we show that the performance ratio of any algorithm belonging to a natural class of algorithms, called partition-based algorithms, is at least $5/2$, unless TSP defined on graphs in $\mathbf{G}_0$ can be approximated within ratio $3/2$. Moreover, for VSP with a heavy edge, we develop an approximation algorithm with performance ratio less than $5/2$.

To get a better algorithm for VSP, we have the following possibilities. If we only concentrate on partition-based algorithms and Algorithm HeavyEdge in Section 5, it is necessary and sufficient to show an approximation algorithm with performance ratio less than $3/2$ for TSP defined on graphs in $\mathbf{G}_0$. Otherwise, we need to consider non-partition-based algorithm besides Algorithm HeavyEdge. Finally, as shown by Theorem 2 finding a better algorithm for VSP with a fixed number of distinct release times can be helpful. Particularly, we can break the barrier of $5/2$ by designing an algorithm for VSP with two distinct release times with performance ratio less than 2.

References

1. An, H.-C., Kleinberg, R., Shmoys, D.B.: Improving Christofides' Algorithm for the s-t path TSP. In: the Proceedings of the 44th Annual ACM Symposium on Theory of Computing, pp. 875–886 (2012)
2. Augustine, J.E., Seiden, S.: Linear time approximation schemes for vehicle scheduling problems. Theoretical Computer Science 324, 147–160 (2004)

3. Bao, X., Liu, Z.: Approximation algorithms for single vehicle scheduling problems with release and service times on a tree or cycle. Theoretical Computer Science 434, 1–10 (2012)
4. Bhattacharya, B., Carmi, P., Hu, Y., Shi, Q.: Single Vehicle Scheduling Problems on Path/Tree/Cycle Networks with Release and Handling Times. In: Hong, S.-H., Nagamochi, H., Fukunaga, T. (eds.) ISAAC 2008. LNCS, vol. 5369, pp. 800–811. Springer, Heidelberg (2008)
5. Christofides, N.: Worst-case analysis of a new heuristic for the traveling salesman problem. Technical Report, Graduate School of Industrial Administration, Carnegie-Mellon University, Pittsburgh, PA (1976)
6. Engebretsen, L., Karpinski, M.: TSP with bounded metrics. Journal of Computer and System Sciences 72, 509–546 (2006)
7. Garey, M.R., Johnson, D.S.: Computers and Intractability: A Guide to the Theory of NP-completeness. Freeman, San Francisco (1979)
8. Gaur, D.R., Gupta, A., Krishnamurti, R.: A $\frac{5}{3}$-approximation algorithm for scheduling vehicles on a path with release and handling times. Information Processing Letters 86, 87–91 (2003)
9. Hoogeveen, J.A.: Analysis of Christofide's heuristic: some paths are more difficult than cycles. Operations Research Letters 10, 291–295 (1991)
10. Karuno, Y., Nagamochi, H., Ibaraki, T.: A 1.5-approximation for single vehicle scheduling problem on a line with release and handling times. In: Proceedings of ISCIE/ASME 1998 Japan-USA Symposium on Flexible Automation, vol. 3, pp. 1363–1368 (1998)
11. Karuno, Y., Nagamochi, H.: An approximability result of the multi-vehicle scheduling problem on a path with release and handling times. Theoretical Computer Science 312, 267–280 (2004)
12. Nagamochi, H., Mochizuki, K., Ibaraki, T.: Complexity of the single vehicle scheduling problems on graphs. Information Systems and Operations Research 35, 256–276 (1997)
13. Sahni, S., Gonzalez, T.: P-complete approximation problems. Journal of the Association for Computing Machinery 23, 555–565 (1976)
14. Trevisan, L.: When Hamming meets Euclid: the approximability of geometric TSP and Steiner tree. SIAM Journal on Computing 30, 475–485 (2000)
15. Tsitsiklis, J.N.: Special cases of traveling salesman and repairman problems with time windows. Networks 22, 263–282 (1992)
16. Yu, W., Liu, Z.: Single-vehicle scheduling problems with release and service times on a line. Networks 57, 128–134 (2011)

A 4.31-Approximation for the Geometric Unique Coverage Problem on Unit Disks[*]

Takehiro Ito[1], Shin-ichi Nakano[2], Yoshio Okamoto[3],
Yota Otachi[4], Ryuhei Uehara[4], Takeaki Uno[5], and Yushi Uno[6]

[1] Tohoku University, Aoba-yama 6-6-05, Sendai, 980-8579, Japan
[2] Gunma University, Kiryu, 376-8515, Japan
[3] University of Electro-Communications,
Chofugaoka 1-5-1, Chofu, Tokyo 182-8585, Japan
[4] Japan Advanced Institute of Science and Technology,
Asahidai 1-1, Nomi, Ishikawa 923-1292, Japan
[5] National Institute of Informatics,
2-1-2 Hitotsubashi, Chiyoda-ku, Tokyo 101-8430, Japan
[6] Osaka Prefecture University, 1-1 Gakuen-cho, Naka-ku, Sakai, 599-8531, Japan
takehiro@ecei.tohoku.ac.jp, nakano@cs.gunma-u.ac.jp, okamotoy@uec.ac.jp,
{otachi,uehara}@jaist.ac.jp, uno@nii.ac.jp, uno@mi.s.osakafu-u.ac.jp

Abstract. We give an improved approximation algorithm for the unique unit-disk coverage problem: Given a set of points and a set of unit disks, both in the plane, we wish to find a subset of disks that maximizes the number of points contained in exactly one disk in the subset. Erlebach and van Leeuwen (2008) introduced this problem as the geometric version of the unique coverage problem, and gave a polynomial-time 18-approximation algorithm. In this paper, we improve this approximation ratio 18 to $2 + 4/\sqrt{3} + \varepsilon$ ($< 4.3095 + \varepsilon$) for any fixed constant $\varepsilon > 0$. Our algorithm runs in polynomial time which depends exponentially on $1/\varepsilon$. The algorithm can be generalized to the budgeted unique unit-disk coverage problem in which each point has a profit, each disk has a cost, and we wish to maximize the total profit of the uniquely covered points under the condition that the total cost is at most a given bound.

1 Introduction

Motivated by applications from wireless networks, Erlebach and van Leeuwen [2] study the following problem. Let $\mathcal{P}$ be a set of points and $\mathcal{D}$ a set of unit disks, both in the plane $\mathbb{R}^2$. For a subset $\mathcal{C} \subseteq \mathcal{D}$ of unit disks, we say that a point $p \in \mathcal{P}$ is *uniquely covered* by $\mathcal{C}$ if there is exactly one disk $D \in \mathcal{C}$ containing p. In the (maximum) *unique unit-disk coverage problem*, we are given a pair $\langle \mathcal{P}, \mathcal{D} \rangle$ of a set $\mathcal{P}$ of points and a set $\mathcal{D}$ of unit disks as input, and we are asked to find a subset $\mathcal{C} \subseteq \mathcal{D}$ such that the number of points in $\mathcal{P}$ uniquely covered by $\mathcal{C}$

---

[*] This work is partially supported by Grant-in-Aid for Scientific Research, and by the Funding Program for World-Leading Innovative R&D on Science and Technology, Japan.

K.-M. Chao, T.-s. Hsu, and D.-T. Lee (Eds.): ISAAC 2012, LNCS 7676, pp. 372–381, 2012.

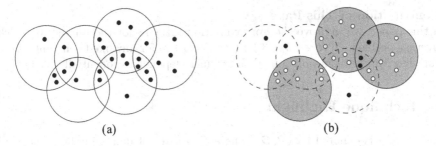

Fig. 1. (a) An instance $\langle \mathcal{P}, \mathcal{D} \rangle$ of the unique unit-disk coverage problem, and (b) an optimal solution $\mathcal{C}^*$ to $\langle \mathcal{P}, \mathcal{D} \rangle$, where each disk in $\mathcal{C}^*$ is hatched and each uniquely covered point is drawn as a small white circle

is maximized. For example, Fig. 1(b) illustrates an optimal solution $\mathcal{C}^*$ for the instance in Fig. 1(a).

In the context of wireless networks, as described by Erlebach and van Leeuwen [2], each point corresponds to a customer location, and the center of each disk corresponds to a place where the provider can build a base station. If several base stations cover a certain customer location, then the resulting interference might cause this customer to receive no service at all. Ideally, each customer should be serviced by exactly one base station, and service should be provided to as many customers as possible. This situation corresponds to the unique unit-disk coverage problem.

[Past Work and Motivation]

Demaine et al. [1] formulated the non-geometric unique coverage problem in a more general setting. They gave a polynomial-time $O(\log n)$-approximation algorithm[1] for the non-geometric unique coverage problem, where n is the number of elements (in the geometric version, n corresponds to the number of points). Guruswami and Trevisan [3] studied the same problem and its generalization, which they called 1-in-k SAT. The appearance of the unique coverage problem is not restricted to wireless networks. The previous papers [1,3] provide a connection with unlimited-supply single-minded envy-free pricing. The maximum cut problem can also be modeled as the unique coverage problem [1,3].

Erlebach and van Leeuwen [2] studied geometric versions of the unique coverage problem. They showed that the unique unit-disk coverage problem is strongly NP-hard, and gave a polynomial-time 18-approximation algorithm. They also consider the problem on unit squares, and gave a polynomial-time $(4 + \varepsilon)$-approximation algorithm for any $\varepsilon > 0$. Later, van Leeuwen [6] gave a proof that the unit-square version is strongly NP-hard, and improved the approximation ratio for unit squares to $2 + \varepsilon$. Recently, we exhibit a polynomial-time approximation scheme (PTAS) for the unique unit-square coverage problem [5].

---

[1] Throughout the paper, we say that an algorithm for a maximization problem is α-*approximation* if it returns a solution with the objective value APX such that OPT $\leq \alpha$APX, where OPT is the optimal objective value, and hence $\alpha \geq 1$.

[Contribution of This Paper]
In this paper, we improve the approximation ratio 18 for the unique unit-disk coverage problem [2] to $2 + 4/\sqrt{3} + \varepsilon$ ($< 4.3095 + \varepsilon$) for any fixed constant $\varepsilon > 0$. Our algorithm runs in polynomial time which depends exponentially on $1/\varepsilon$.

2 Technique Highlight

An instance is denoted by $\langle \mathcal{P}, \mathcal{D} \rangle$, where $\mathcal{P}$ is a set of points in the plane, and $\mathcal{D}$ is a set of unit disks in the plane. A unit disk in this paper means a closed disk with radius $1/2$, and hence has the boundary. Without loss of generality, we assume that any two points in $\mathcal{P}$ (resp., any two centers of disks in $\mathcal{D}$) have distinct x-coordinates and distinct y-coordinates [6]. We also assume that no two disks in $\mathcal{D}$ touch, and no point in $\mathcal{P}$ lies on the boundary of any disk in $\mathcal{D}$ [6]. For brevity, the x-coordinate of the center of a disk is referred to as the x-coordinate of the disk. The same applies to the y-coordinate too.

[Our Approach]
We use the following two general techniques. (1) The shifting technique by Hochbaum and Maass [4]: This subdivides the whole plane into some smaller pieces, and ignores some points so that the combination of approximate solutions to smaller pieces will yield an approximate solution to the whole plane. (2) A classification of disks: Namely, for each instance on a smaller piece, we partition the set of disks into a few classes so that the instance on a restricted set of disks can be handled in polynomial time. Taking the best solution in those classes yields a constant-factor approximation.

More specifically, our algorithm in this work exploits the techniques above in the following way. (1) Our smaller pieces are stripes, which consists of some number of horizontal ribbons such that each ribbon is of height $h = \sqrt{3}/4$ and the gap between ribbons is of height $b = 1/2$. At this step, we lose the approximation ratio of $1 + b/h = 1 + 2/\sqrt{3}$, as shown later in Lemma 1. (2) We classify the disks intersecting a stripe into two classes. The first class consists of the disks whose centers lie outside the ribbons in the stripe, and the second class consists of the disks whose centers lie inside the ribbons. It is important to notice that we will not solve the classified instances exactly, but rather we design a PTAS for each of them. Namely, we provide a polynomial-time algorithm for each of the classified instances with approximation ratio $1 + \varepsilon'$, where $\varepsilon' > 0$ is a fixed constant. Note that the polynomial running times depend exponentially on $1/\varepsilon'$. Then, since we have two classes, we only lose the approximation ratio of $2(1 + \varepsilon')$ at this step (Lemma 2). Thus, choosing ε' appropriately, we can achieve the overall approximation ratio of $(1 + 2/\sqrt{3}) \times 2(1 + \varepsilon') = 2 + 4/\sqrt{3} + \varepsilon$.

[Comparison with the Unit-Square Case]
The PTAS in this paper for each of the classified instances uses an idea similar to our PTAS for unit squares [5]. However, there is a big difference, as explained below, that makes us unable to give a PTAS for the original instance on unit disks. Look at a horizontal ribbon. For the unit-square case, the intersection of

the ribbon and a unit square is a rectangle. Then, its boundary is an x-monotone curve. The monotonicity enables us to provide a PTAS. However, for the unit-disk case, if we look at the intersection of the ribbon and a unit disk, then its boundary is not necessarily x-monotone. To make it x-monotone, we need to give a gap between ribbons; this is why we classified the disks into two classes, as mentioned above. It should be noted that, by this disk classification, we can get the x-monotonicity only for the disks whose centers lie outside the ribbons. To obtain the approximation ratio of $2 + 4/\sqrt{3} + \varepsilon$, we need to construct a PTAS for the classified instance in which the centers of disks lie inside the ribbons. We thus develop several new techniques to deal with such disks.

3   Main Result and Outline

The following is the main result of the paper.

Theorem 1. *For any fixed constant $\varepsilon > 0$, there is a polynomial-time $(2 + 4/\sqrt{3} + \varepsilon)$-approximation algorithm for the unique unit-disk coverage problem.*

In the remainder of the paper, we give a polynomial-time $2(1 + \varepsilon')(1 + 2/\sqrt{3})$-approximation algorithm for the unique unit-disk coverage problem, where ε' is a fixed positive constant such that $2\varepsilon'(1 + 2/\sqrt{3}) = \varepsilon$. (However, due to the page limitation, we omit proofs from this extended abstract.)

[Restricting the Problem to a Stripe]
A rectangle is *axis-parallel* if its boundary consists of horizontal and vertical line segments. Let R_W be an (unbounded) axis-parallel rectangle of width W and height ∞ which properly contains all points in $\mathcal{P}$ and all unit disks in $\mathcal{D}$. We fix the origin of the coordinate system on the left vertical boundary of R_W. For two positive real numbers h, b and a non-negative real number $q \in [0, h+b)$, we define a *stripe* $R_W(q, h, b)$ as follows: $R_W(q, h, b) = \{[0, W] \times [q+i(h+b), q+(i+1)h+ib) \mid i \in \mathbb{Z}\}$, that is, $R_W(q, h, b)$ is a set of rectangles with width W and height h; each rectangle in $R_W(q, h, b)$ is called a *ribbon*. It should be noted that the upper boundary of each ribbon is open, while the lower boundary is closed. We denote by $\mathcal{P} \cap R_W(q, h, b)$ the set of all points in $\mathcal{P}$ contained in $R_W(q, h, b)$. We have the following lemma, by applying the well-known shifting technique [2,4].

Lemma 1. *Suppose that there is a polynomial-time α-approximation algorithm for the unique unit-disk coverage problem on $\langle \mathcal{P} \cap R_W(q, h, b), \mathcal{D} \rangle$ for arbitrary constant q and fixed constants h, b. Then, there is a polynomial-time $\alpha(1 + b/h)$-approximation algorithm for the unique unit-disk coverage problem on $\langle \mathcal{P}, \mathcal{D} \rangle$.*

For the sake of further simplification, we assume without loss of generality that no ribbon has a point of $\mathcal{P}$ or the center of a disk of $\mathcal{D}$ on its boundary (of the closure).

[Approximating the Problem on a Stripe]
In the rest of the paper, we fix a stripe $R_W(q, h, b)$ for $h = \sqrt{3}/4$, $b = 1/2$ and some real number $q \in [0, h + b)$. Then, using Lemma 1, one can obtain

a polynomial-time $\alpha(1 + 2/\sqrt{3})$-approximation algorithm for the problem on $\langle \mathcal{P}, \mathcal{D} \rangle$. Therefore, to complete the proof of Theorem 1, we give a polynomial-time $2(1 + \varepsilon')$-approximation algorithm for the problem on $\langle \mathcal{P} \cap R_W(q, h, b), \mathcal{D} \rangle$ for any fixed constant $\varepsilon' > 0$.

We first partition the disk set $\mathcal{D}$ into two subsets $\mathcal{D}_O$ and $\mathcal{D}_I$ under the stripe $R_W(q, h, b)$. Let $\mathcal{D}_O \subseteq \mathcal{D}$ be the set of unit disks whose centers are not contained in the stripe $R_W(q, h, b)$. Let $\mathcal{D}_I = \mathcal{D} \setminus \mathcal{D}_O$, then $\mathcal{D}_I$ is the set of unit disks whose centers are contained in $R_W(q, h, b)$. Let $\mathcal{P}_q = \mathcal{P} \cap R_W(q, h, b)$. In Sections 4 and 5, we will show that each of the problems on $\langle \mathcal{P}_q, \mathcal{D}_O \rangle$ and $\langle \mathcal{P}_q, \mathcal{D}_I \rangle$ admits a polynomial-time $(1 + \varepsilon')$-approximation algorithm for any fixed constant $\varepsilon' > 0$, respectively. We choose a better solution from $\langle \mathcal{P}_q, \mathcal{D}_O \rangle$ and $\langle \mathcal{P}_q, \mathcal{D}_I \rangle$ as our approximate solution to $\langle \mathcal{P}_q, \mathcal{D} \rangle$. The following lemma shows that this choice gives rise to a $2(1 + \varepsilon')$-approximation for the problem on $\langle \mathcal{P}_q, \mathcal{D} \rangle$.

Lemma 2. *Let* $\langle \mathcal{P}, \mathcal{D} \rangle$ *be an instance of the unique unit-disk coverage problem, and let* $\mathcal{D}_1$ *and* $\mathcal{D}_2$ *partition* $\mathcal{D}$ *(i.e.,* $\mathcal{D}_1 \cup \mathcal{D}_2 = \mathcal{D}$ *and* $\mathcal{D}_1 \cap \mathcal{D}_2 = \emptyset$*). Let* $\mathcal{C}_1 \subseteq \mathcal{D}_1$ *and* $\mathcal{C}_2 \subseteq \mathcal{D}_2$ *be* β-approximate solutions to the instances $\langle \mathcal{P}, \mathcal{D}_1 \rangle$ *and* $\langle \mathcal{P}, \mathcal{D}_2 \rangle$, *respectively. Then, the better of* $\mathcal{C}_1$ *and* $\mathcal{C}_2$ *is a* 2β-approximate solution to $\langle \mathcal{P}, \mathcal{D} \rangle$.

We may assume without loss of generality that each ribbon in $R_W(q, h, b)$ contains at least one point in $\mathcal{P}$. (We can simply ignore the ribbons containing no points.) We thus deal with only a polynomial number of ribbons. Let $R_1, R_2, \ldots, R_t$ be the ribbons in $R_W(q, h, b)$ ordered from bottom to top.

4 PTAS for the Problem on $\langle \mathcal{P}_q, \mathcal{D}_O \rangle$

In this section, we give a PTAS for the problem on $\langle \mathcal{P}_q, \mathcal{D}_O \rangle$.

Lemma 3. *For any fixed constant* $\varepsilon' > 0$, *there is a polynomial-time* $(1 + \varepsilon')$-*approximation algorithm for the unique unit-disk coverage problem on* $\langle \mathcal{P}_q, \mathcal{D}_O \rangle$.

Let $k = \lceil 1/\varepsilon' \rceil$. A proof of Lemma 3 is given by the following two lemmas.

Lemma 4. *Suppose that we can obtain an optimal solution to* $\langle \mathcal{P}_q \cap G, \mathcal{D}_O \rangle$ *in polynomial time for every set* G *consisting of at most* k *ribbons. Then, we can obtain a* $(1 + \varepsilon')$-*approximate solution to* $\langle \mathcal{P}_q, \mathcal{D}_O \rangle$ *in polynomial time.*

Lemma 5. *We can obtain an optimal solution to* $\langle \mathcal{P}_q \cap G, \mathcal{D}_O \rangle$ *in polynomial time for every set* G *consisting of at most* k *ribbons.*

The proof of Lemma 5 is one of the cruxes in this paper. We give a constructive proof, namely, we give such an algorithm.

[Basic Ideas]
Our algorithm employs a dynamic programming approach based on the line-sweep paradigm. Namely, we look at points and disks from left to right, and extend the uniquely covered region sequentially. However, adding one disk D at the rightmost position can influence a lot of disks that were already chosen, and

can change the situation drastically (we say that D *influences* a disk D' if the region uniquely covered by D' changes after the addition of D). We therefore need to keep track of the disks that are possibly influenced by a newly added disk. Unless the number of those disks is bounded by some constant (or the logarithm of the input size), this approach cannot lead to a polynomial-time algorithm. Unfortunately, new disks may influence a super-constant (or super-logarithmic) number of disks.

Instead of adding a disk at the rightmost position, we add a disk D such that the number of disks that were already chosen and influenced by D can be bounded by a constant. Lemmas 6 and 7 state that we can do this for any set of disks, as long as a trivial condition for the disk set to be an optimal solution is satisfied. Furthermore, such a disk can be found in polynomial time.

[Basic Definitions]

We may assume without loss of generality that the set G consists of consecutive ribbons forming a *group*; otherwise we can simply solve the problem for each group, because those groups have pairwise distance more than one. Suppose that G consists of k consecutive ribbons $R_{j+1}, R_{j+2}, \ldots, R_{j+k}$ in $R_W(q, h, b)$, ordered from bottom to top, for some integer j. If a disk in $\mathcal{D}_O$ can cover points in $\mathcal{P}_q \cap G$, then its center lies between R_{j+i} and R_{j+i+1} for some $i \in \{0, \ldots, k\}$. For notational convenience, we assume $j = 0$ without loss of generality. Note that the two ribbons R_0 and R_{k+1} are not in G.

For each $i \in \{0, \ldots, k\}$, we denote by $\mathcal{D}_{i,i+1}$ the set of all disks in $\mathcal{D}_O$ with their centers lying between R_i and R_{i+1}. Then, each disk in $\mathcal{D}_{i,i+1}$ intersects R_i and R_{i+1}. Note that $\mathcal{D}_{0,1}, \mathcal{D}_{1,2}, \ldots, \mathcal{D}_{k,k+1}$ form a partition of the disks in $\mathcal{D}_O$ intersecting G.

For a disk set $\mathcal{C} \subseteq \mathcal{D}$, let $A_0(\mathcal{C})$, $A_1(\mathcal{C})$, $A_2(\mathcal{C})$ and $A_{\geq 3}(\mathcal{C})$ be the areas covered by no disk, exactly one disk, exactly two disks, and three or more disks in $\mathcal{C}$, respectively. Then, each point contained in $A_1(\mathcal{C})$ is uniquely covered by $\mathcal{C}$.

[Properties on Disk Subsets of $\mathcal{D}_{i,i+1}$]

We first deal with disks only in a set $\mathcal{C} \subseteq \mathcal{D}_{i,i+1}$ and the region uniquely covered by them. Of course, disks in $\mathcal{D}_{i-1,i} \cup \mathcal{D}_{i+1,i+2}$ may influence disks in $\mathcal{C}$; this difficulty will be discussed later. We sometimes denote by $R_{i,i+1}$ the set of two consecutive ribbons R_i and R_{i+1}.

Upper and Lower Envelopes. Let $\mathcal{C} \subseteq \mathcal{D}_{i,i+1}$ be a disk set. Since any two unit disks have distinct x-coordinates and distinct y-coordinates, we can partition the boundary of the closure of $A_1(\mathcal{C})$ into two types: the boundary between $A_0(\mathcal{C})$ and $A_1(\mathcal{C})$; and that between $A_1(\mathcal{C})$ and $A_2(\mathcal{C})$. We call the former type of the boundaries above the lower boundary of R_{i+1} (and below the upper boundary of R_i) the *upper* (resp., *lower*) *envelope* of $\mathcal{C}$. (See Fig. 2.) We say that a disk D *forms* the boundary of an area A if a part of the boundary of D is a part of that of A. Let $UE(\mathcal{C})$ and $LE(\mathcal{C})$ be the sequences of disks that form the upper and lower envelopes of $\mathcal{C}$, from right to left, respectively. Note that a disk $D \in \mathcal{C}$ may appear in both $UE(\mathcal{C})$ and $LE(\mathcal{C})$.

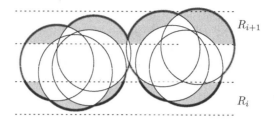

Fig. 2. A set $\mathcal{C}$ of disks in $\mathcal{D}_{i,i+1}$, together with $A_1(\mathcal{C}) \cap R_{i,i+1}$ (gray), the upper envelope (red), the lower envelope (blue) and the other part of the boundary between $A_0(\mathcal{C})$ and $A_1(\mathcal{C})$ (green). The dotted lines show the boundaries of R_i and R_{i+1}

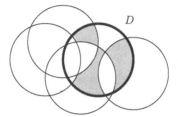

Fig. 3. The gray region shows $\overline{\Delta}(\mathcal{C}, D)$ for the thick disk D

Consider an arbitrary optimal solution $\mathcal{C}^* \subseteq \mathcal{D}_{i,i+1}$ to $\langle \mathcal{P}_q \cap R_{i,i+1}, \mathcal{D}_{i,i+1} \rangle$. If there is a disk $D \in \mathcal{C}^*$ that is not part of $A_1(\mathcal{C}^*)$, we can simply remove it from $\mathcal{C}^*$ without losing the optimality. Thus, hereafter we deal with a disk set $\mathcal{C} \subseteq \mathcal{D}_{i,i+1}$ such that every disk D in $\mathcal{C}$ forms the upper or lower boundaries of $\mathcal{C}$, that is, $D \in UE(\mathcal{C})$ or $D \in LE(\mathcal{C})$ holds. This property enables us to sweep the ribbons $R_{i,i+1}$, roughly speaking from left to right, and to extend the upper and lower envelopes sequentially.

Top Disks and the Key Lemma. When we add a "new" disk D to the current disk set $\mathcal{C} \setminus \{D\}$, we need to know the symmetric difference between $A_1(\mathcal{C})$ and $A_1(\mathcal{C} \setminus \{D\})$: the area $A_1(\mathcal{C}) \setminus A_1(\mathcal{C} \setminus \{D\}) \subseteq A_1(\mathcal{C})$ is the uniquely covered area obtained newly by adding the disk D, and the area $A_1(\mathcal{C} \setminus \{D\}) \setminus A_1(\mathcal{C}) \subseteq A_2(\mathcal{C})$ is the non-uniquely covered area due to D. However, it suffices to know the area $A_1(\mathcal{C} \setminus \{D\}) \setminus A_1(\mathcal{C})$ and its boundary, because the boundary of $A_1(\mathcal{C}) \setminus A_1(\mathcal{C} \setminus \{D\})$ is formed only by D and disks forming the boundary of $A_1(\mathcal{C} \setminus \{D\}) \setminus A_1(\mathcal{C})$.

For a disk D in a set $\mathcal{C} \subseteq \mathcal{D}$, let $\overline{\Delta}(\mathcal{C}, D)$ be the area $A_1(\mathcal{C} \setminus \{D\}) \setminus A_1(\mathcal{C})$, and $\Delta(\mathcal{C}, D)$ be the set of all disks in $\mathcal{C}$ that form the boundary of $\overline{\Delta}(\mathcal{C}, D)$. (See Fig. 3.) Clearly, every disk in $\Delta(\mathcal{C}, D)$ has non-empty intersection with D. As we mentioned, $\Delta(\mathcal{C}, D)$ may contain a super-constant (or super-logarithmic) number of disks if we simply choose the rightmost disk D in $\mathcal{C}$. We will show that, for any disk set $\mathcal{C} \subseteq \mathcal{D}_{i,i+1}$, there always exists a disk $D \in \mathcal{C}$ such that $\Delta(\mathcal{C}, D)$ contains at most 16 disks, called top disks, and D itself is a top disk.

For a disk set $\mathcal{C} \subseteq \mathcal{D}_{i,i+1}$, a disk $D \in \mathcal{C}$ is called a *top disk* of $\mathcal{C}$ if one of the following conditions (i)–(iv) holds:

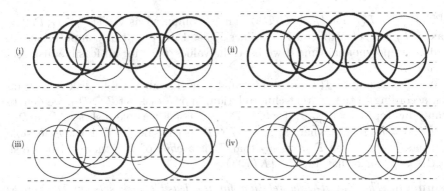

Fig. 4. An example of top disks. The (blue) thick disks are top disks, and the numbers correspond to the conditions in the definition.

(i) D is one of the first six disks of $UE(\mathcal{C})$;
(ii) D is one of the first six disks of $LE(\mathcal{C})$;
(iii) D is one of the first two disks of $UE(LE(\mathcal{C}) \setminus UE(\mathcal{C}))$; and
(iv) D is one of the first two disks of $LE(UE(\mathcal{C}) \setminus LE(\mathcal{C}))$.

An example is given in Fig. 4. Remember that the disks in $UE(\mathcal{C})$ and $LE(\mathcal{C})$ are ordered from right to left. We denote by $\mathsf{Top}(\mathcal{C})$ the set of top disks of C. Note that a disk may satisfy more than one of the conditions above. A disk set $\mathcal{T} \subseteq \mathcal{D}_{i,i+1}$ is *feasible on* $\mathcal{D}_{i,i+1}$ if $\mathsf{Top}(\mathcal{T}) = \mathcal{T}$. For a feasible disk set $\mathcal{T}$ on $\mathcal{D}_{i,i+1}$, we denote by $\mathfrak{C}_{i,i+1}(\mathcal{T})$ the set of all disk sets whose top disks are equal to $\mathcal{T}$, that is, $\mathfrak{C}_{i,i+1}(\mathcal{T}) = \{\mathcal{C} \subseteq \mathcal{D}_{i,i+1} \mid \mathsf{Top}(\mathcal{C}) = \mathcal{T}\}$. A top disk D in a feasible set $\mathcal{T}$ is said to be *stable in* $\mathcal{T}$ if $\Delta(\mathcal{C}, D)$ consists only of top disks in $\mathcal{T}$ for any disk set $\mathcal{C} \in \mathfrak{C}_{i,i+1}(\mathcal{T})$. Indeed, stable top disks will be crucial to our algorithm: if a top disk D is stable in a feasible set $\mathcal{T} \subseteq \mathcal{D}_{i,i+1}$, then $\Delta(\mathcal{C}, D)$ contains at most 16 top disks in $\mathcal{T}$ for any disk set $\mathcal{C} \in \mathfrak{C}_{i,i+1}(\mathcal{T})$; and hence we can compute $\Delta(\mathcal{C}, D)$ in polynomial time. Therefore, below is the key lemma.

Lemma 6. *For any feasible disk set* $\mathcal{T}$ *on* $\mathcal{D}_{i,i+1}$, *at least one top disk* $K(\mathcal{T})$ *is stable in* $\mathcal{T}$. *Moreover,* $K(\mathcal{T})$ *can be found in polynomial time.*

[Properties on Disk Subsets of $\mathcal{D}_O$]
A disk set $\mathcal{T} \subseteq \mathcal{D}_O$ is *feasible on* $\mathcal{D}_O$ if $\mathsf{Top}(\mathcal{T} \cap \mathcal{D}_{i,i+1}) = \mathcal{T} \cap \mathcal{D}_{i,i+1}$ for each $i \in \{0, \ldots, k\}$. For a feasible disk set $\mathcal{T}$ on $\mathcal{D}_O$ and $i \in \{0, \ldots, k\}$, let $\mathcal{T}_{i,i+1} = \mathcal{T} \cap \mathcal{D}_{i,i+1}$, and let $\mathfrak{C}(\mathcal{T}) = \{\mathcal{C} \subseteq \mathcal{D}_O \mid \mathsf{Top}(\mathcal{C} \cap \mathcal{D}_{i,i+1}) = \mathcal{T}_{i,i+1} \text{ for each } i \in \{0, \ldots, k\}\}$. We say that $\mathcal{T}_{i,i+1}$ is *safe for* $\mathcal{T}$ if $\Delta(\mathcal{C}, K(\mathcal{T}_{i,i+1})) \subset \mathcal{T}$ for any disk set $\mathcal{C} \in \mathfrak{C}(\mathcal{T})$, where $K(\mathcal{T}_{i,i+1})$ is the top disk in $\mathcal{T}_{i,i+1}$ selected by Lemma 6.

Lemma 7. *For any feasible disk set* $\mathcal{T}$ *on* $\mathcal{D}_O$, *there exists an index* $s \in \{0, \ldots, k\}$ *such that* $\mathcal{T}_{s,s+1}$ *is safe for* $\mathcal{T}$.

[Algorithm for the Problem on $\langle \mathcal{P}_q \cap G, \mathcal{D}_O \rangle$]
For a feasible disk set $\mathcal{T}$ on $\mathcal{D}_O$, let $f(\mathcal{T})$ be the maximum number of points in $\mathcal{P}_q \cap G$ uniquely covered by a disk set in $\mathfrak{C}(\mathcal{T})$. Then, the optimal value

$\mathrm{OPT}(\mathcal{P}_q \cap G, \mathcal{D}_O)$ for $\langle \mathcal{P}_q \cap G, \mathcal{D}_O \rangle$ can be computed as $\mathrm{OPT}(\mathcal{P}_q \cap G, \mathcal{D}_O) = \max\{f(\mathcal{T}) \mid \mathcal{T}$ is feasible on $\mathcal{D}_O\}$. Since $|\mathcal{T}| < 16(k+1)$, this computation can be done in polynomial time if we have the values $f(\mathcal{T})$ for all feasible disk sets $\mathcal{T}$ on $\mathcal{D}_O$.

We thus compute $f(\mathcal{T})$ in polynomial time for all feasible disk sets $\mathcal{T}$ on $\mathcal{D}_O$, according to the "parent-child relation." For a disk set $\mathcal{C} \subseteq \mathcal{D}_O$, we denote simply by $\mathsf{Top}(\mathcal{C}) = \bigcup_{0 \le i \le k} \mathsf{Top}(\mathcal{C} \cap \mathcal{D}_{i,i+1})$. For a feasible disk set $\mathcal{T}$ on $\mathcal{D}_O$, let $K(\mathcal{T}) = K(\mathcal{T}_{s,s+1})$ where $\mathcal{T}_{s,s+1} = \mathcal{T} \cap \mathcal{D}_{s,s+1}$ is safe for $\mathcal{T}$. For two feasible disk sets $\mathcal{T}$ and $\mathcal{T}'$ on $\mathcal{D}_O$, we say that $\mathcal{T}'$ is a *child* of $\mathcal{T}$ if there exists a disk set $\mathcal{C} \in \mathfrak{C}(\mathcal{T})$ such that $\mathsf{Top}(\mathcal{C} \setminus \{K(\mathcal{T})\}) = \mathcal{T}'$.

Lemma 8. *The parent-child relation for the feasible disk sets on $\mathcal{D}_O$ can be constructed in polynomial time. The parent-child relation is acyclic.*

We finally give our algorithm to solve the problem on $\langle \mathcal{P}_q \cap G, \mathcal{D}_O \rangle$.

For each $i \in \{0, \ldots, k\}$, let $\mathcal{T}_{i,i+1}^0$ be the disk set consisting of the first 16 disks in $\mathcal{D}_{i,i+1}$ having the smallest x-coordinates. Let $\mathcal{T}^0 = \bigcup_{0 \le i \le k} \mathcal{T}_{i,i+1}^0$, then $|\mathcal{T}^0| \le 16(k+1)$. As the initialization, we first compute $f(\mathcal{T})$ for all feasible sets $\mathcal{T}$ on $\mathcal{T}^0$. Since $|\mathcal{T}^0|$ is a constant, the total number of feasible sets $\mathcal{T}$ on $\mathcal{T}^0$ is also a constant. Therefore, this initialization can be done in polynomial time.

We then compute $f(\mathcal{T})$ for a feasible disk set $\mathcal{T}$ on $\mathcal{D}_O$ from $f(\mathcal{T}')$ for all children $\mathcal{T}'$ of $\mathcal{T}$. Since the parent-child relation is acyclic, we can find a feasible disk set $\mathcal{T}$ such that $f(\mathcal{T}')$ are already computed for all children $\mathcal{T}'$ of $\mathcal{T}$. For a disk set $\mathcal{C} \subseteq \mathcal{D}_O$ and a disk $D \in \mathcal{C}$, we denote by $z(\mathcal{C}, D)$ the difference of uniquely covered points in $\mathcal{P}_q \cap G$ caused by adding D to $\mathcal{C} \setminus \{D\}$, that is, the number of points in $\mathcal{P}_q \cap G$ that are included in $D \cap A_1(\mathcal{C})$ minus the number of points in $\mathcal{P}_q \cap G$ that are included in $D \cap A_1(\mathcal{C} \setminus \{D\})$. Since $K(\mathcal{T}) = K(\mathcal{T}_{s,s+1})$ and $\mathcal{T}_{s,s+1} = \mathcal{T} \cap \mathcal{D}_{s,s+1}$ is safe for $\mathcal{T}$, we have $z(\mathcal{T}, K(\mathcal{T})) = z(\mathcal{C}, K(\mathcal{T}))$ for all disk sets $\mathcal{C} \in \mathfrak{C}(\mathcal{T})$. Therefore, we can correctly update $f(\mathcal{T})$ by $f(\mathcal{T}) := \max\{f(\mathcal{T}') \mid \mathcal{T}'$ is a child of $\mathcal{T}\} + z(\mathcal{T}, K(\mathcal{T}))$. This way, the algorithm correctly solves the problem on $\langle \mathcal{P}_q \cap G, \mathcal{D}_O \rangle$ in polynomial time.

This completes the proof of Lemma 5.   □

5   PTAS for the Problem on $\langle \mathcal{P}_q, \mathcal{D}_I \rangle$

We finally give the following lemma, which completes the proof of Theorem 1.

Lemma 9. *For any fixed constant $\varepsilon' > 0$, there is a polynomial-time $(1 + \varepsilon')$-approximation algorithm for the unique unit-disk coverage problem on $\langle \mathcal{P}_q, \mathcal{D}_I \rangle$.*

Remember that the upper boundary of each ribbon R_i in the stripe $R_W(q, h, b)$ is open. Therefore, the ribbons in $R_W(q, h, b)$ have pairwise distance strictly greater than $b = 1/2$. Since $\mathcal{D}_I$ consists of unit disks (with radius $1/2$) whose centers are contained in ribbons, no disk in $\mathcal{D}_I$ can cover points in two distinct ribbons. Therefore, we can independently solve the problem on $\langle \mathcal{P}_q \cap R_i, \mathcal{D}_I \rangle$ for each ribbon R_i in $R_W(q, h, b)$. Thus, if there is a PTAS for the problem on $\langle \mathcal{P}_q \cap R_i, \mathcal{D}_I \rangle$, then we can obtain a PTAS for the problem on $\langle \mathcal{P}_q, \mathcal{D}_I \rangle$;

we combine the approximate solutions to $\langle \mathcal{P}_q \cap R_i, \mathcal{D}_I \rangle$, and output it as our approximate solution to $\langle \mathcal{P}_q, \mathcal{D}_I \rangle$.

We now give a PTAS for the problem on $\langle \mathcal{P}_q \cap R_i, \mathcal{D}_I \rangle$ for each ribbon R_i. We first vertically divide R_i into rectangles, called *cells*, so that the diagonal of each cell is of length exactly $1/2$. Let W_c be the width of each cell, that is, $W_c = 1/4$ since $h = \sqrt{3}/4$. We may assume that, in each cell, the left boundary is closed and the right boundary is open. Let $r = 4$, then $rW_c = 1$.

Let $k = \lceil 1/\varepsilon' \rceil$. Similarly as in the PTAS for $\langle \mathcal{P}_q, \mathcal{D}_O \rangle$, we remove r consecutive cells from every $r(1+k)$ consecutive cells, and obtain the "sub-ribbon" consisting of "groups," each of which contains at most rk consecutive cells. Then, these groups have pairwise distance more than one, and hence no unit disk (with radius $1/2$) can cover points in two distinct groups. Therefore, we can independently solve the problem on $\langle \mathcal{P}_q \cap G, \mathcal{D}_I \rangle$ for each group G in the sub-ribbon. The similar arguments in Lemma 4 establishes that the problem on $\langle \mathcal{P}_q \cap R_i, \mathcal{D}_I \rangle$ admits a PTAS if there is a polynomial-time algorithm which optimally solves the problem on $\langle \mathcal{P}_q \cap G, \mathcal{D}_I \rangle$ for each group G. Therefore, the following lemma completes the proof of Lemma 9.

Lemma 10. *There is a polynomial-time algorithm which optimally solves the problem on $\langle \mathcal{P}_q \cap G, \mathcal{D}_I \rangle$ for a group G consisting of at most rk consecutive cells.*

6   Concluding Remark

Consider the *budgeted* unique unit-disk coverage problem, in which we are given a budget B, each point in $\mathcal{P}$ has a profit, each disk in $\mathcal{D}$ has a cost, and we wish to find $\mathcal{C} \subseteq \mathcal{D}$ that maximizes the total profit of the uniquely covered points by $\mathcal{C}$ under the condition that the total cost of $\mathcal{C}$ is at most B. The generality of our approach enables us to give a polynomial-time $(2+4/\sqrt{3}+\varepsilon)$-approximation algorithm, for any fixed constant $\varepsilon > 0$, for the budgeted version, too.

References

1. Demaine, E.D., Hajiaghayi, M.T., Feige, U., Salavatipour, M.R.: Combination can be hard: approximability of the unique coverage problem. SIAM J. on Computing 38, 1464–1483 (2008)
2. Erlebach, T., van Leeuwen, E.J.: Approximating geometric coverage problems. In: SODA 2008, pp. 1267–1276 (2008)
3. Guruswami, V., Trevisan, L.: The Complexity of Making Unique Choices: Approximating 1-in-k SAT. In: Chekuri, C., Jansen, K., Rolim, J.D.P., Trevisan, L. (eds.) APPROX and RANDOM 2005. LNCS, vol. 3624, pp. 99–110. Springer, Heidelberg (2005)
4. Hochbaum, D.S., Maass, W.: Approximation schemes for covering and packing problems in image processing and VLSI. J. ACM 32, 130–136 (1985)
5. Ito, T., Nakano, S.-I., Okamoto, Y., Otachi, Y., Uehara, R., Uno, T,, Uno, Y.: A Polynomial-Time Approximation Scheme for the Geometric Unique Coverage Problem on Unit Squares. In: Fomin, F.V., Kaski, P. (eds.) SWAT 2012. LNCS, vol. 7357, pp. 24–35. Springer, Heidelberg (2012)
6. van Leeuwen, E.J.: Optimization and approximation on systems of geometric objects. Ph.D. Thesis, University of Amsterdam (2009)

The Minimum Vulnerability Problem

Sepehr Assadi[1], Ehsan Emamjomeh-Zadeh[1], Ashkan Norouzi-Fard[1],
Sadra Yazdanbod[1], and Hamid Zarrabi-Zadeh[1,2,*]

[1] Department of Computer Engineering,
Sharif University of Technology, Tehran, Iran
{s_asadi,emamjomeh,noroozifard,yazdanbod}@ce.sharif.edu,
zarrabi@sharif.edu
[2] Institute for Research in Fundamental Sciences (IPM), Tehran, Iran

Abstract. We revisit the problem of finding k paths with a minimum
number of shared edges between two vertices of a graph. An edge is
called *shared* if it is used in more than one of the k paths. We provide a $\lfloor k/2 \rfloor$-approximation algorithm for this problem, improving the
best previous approximation factor of $k - 1$. We also provide the first
approximation algorithm for the problem with a sublinear approximation factor of $O(n^{3/4})$, where n is the number of vertices in the input
graph. For sparse graphs, such as bounded-degree and planar graphs, we
show that the approximation factor of our algorithm can be improved to
$O(\sqrt{n})$. While the problem is NP-hard, and even hard to approximate
to within an $O(\log n)$ factor, we show that the problem is polynomially
solvable when k is a constant. This settles an open problem posed by
Omran *et al.* regarding the complexity of the problem for small values of k. We present most of our results in a more general form where
each edge of the graph has a sharing cost and a sharing capacity, and
there is vulnerability parameter r that determines the number of times
an edge can be used among different paths before it is counted as a
shared/vulnerable edge.

1 Introduction

In this paper, we investigate a family of NP-Hard network design problems.
Our study is motivated by the *minimum shared edges* (MSE) problem, formally
defined as follows:

Problem 1 (Minimum Shared Edges). Given a directed graph $G = (V, E)$, an
integer $k > 0$, and two distinct vertices s and t in V, find k paths from s to t
minimizing the number of shared edges. An edge is called *shared* if it is used in
more than one of the k paths.

The minimum shared edges problem arises in a number of transportation and
communication network design problems. As an example, consider a VIP who
wishes to travel safely between two places of a network (see [10]). To achieve a

---

* This author's research was partially supported by IPM under grant No: CS1391-4-04.

K.-M. Chao, T.-s. Hsu, and D.-T. Lee (Eds.): ISAAC 2012, LNCS 7676, pp. 382–391, 2012.

minimum level of security assurance, the usual strategy is to pre-select k paths, and then, choose one of the k paths at random just before the actual trip. To bound the probability of being attacked by an adversary (who knows the strategy and the paths) to at most $1/k$, we need to put guards on high-risk edges, i.e., those edges shared among more than one of the pre-selected paths. To reduce the guarding cost, the obvious objective is to find paths with a minimum number of shared edges. A similar problem arises in the context of communication network design, e.g., in designing reliable client-server networks [13], reliable multicast communications [11], and distributed communication protocols [3].

In this work, we obtain results for a generalized version of the minimum shared edges problem. More precisely, we generalize MSE (Problem 1) in three directions. Firstly, we assign a cost c_e to each edge e, which represents the cost of guarding the edge. This weighted version is closer to the practical applications, in which guarding edges have different costs, depending on, say, the length of the edges. Secondly, we make the problem capacitated by assigning to each edge an upper bound specifying the maximum number of times an edge can be used among the k paths. Thirdly, we generalize the problem by adding a parameter r that specifies a threshold on the number of times an edge can be used before it becomes vulnerable, and needs to be guarded. The generalized problem, which we call *minimum vulnerability*, is formally defined as follows:

Problem 2 (Minimum Vulnerability). Given a directed graph $G = (V, E)$ with nonnegative edge costs c_e and maximum edge capacities U_e assigned to the edges $e \in E$, two distinct vertices $s, t \in V$, and two integers r and k with $0 \leqslant r < k$, find k paths from s to t so as to minimize the total cost of r-vulnerable edges. An edge is called r-*vulnerable* if it is used in more than r of the k paths.

Clearly, the minimum 1-vulnerability problem (i.e., when $r = 1$) is equivalent to the weighted capacitated MSE problem. Furthermore, the minimum 0-vulnerability problem is equivalent to the classic *minimum edge-cost flow* (MECF) problem, in which we are given a graph $G = (V, E)$ with nonnegative edge costs and capacities, and the goal is to find a min-cost subset $A \subseteq E$ so that the flow from s to t in (V, A) is at least a given value k. The MECF problem is one of the fundamental NP-hard problems in network design (see Garey and Johnson [4]). It includes several other interesting problems as special case, such as the Steiner tree problem [4] and some of its generalizations [5,8].

Previous Work. The best previous approximation algorithm for the MSE problem has an approximation factor of $k - 1$ [10], which is based on a k-approximation algorithm for the MECF problem, proposed by Krumke *et al.* [9]. Both the MSE and MECF problems are known to be hard to approximate to within a factor of $2^{\log^{1-\varepsilon} n}$, for any constant $\varepsilon > 0$ [2,10].

A restriction of the minimum vulnerability problem to the case where no r-vulnerable edge ($r > 0$) is allowed is equivalent to the well-known *disjoint paths* problem, which can be solved polynomially using a standard maximum flow algorithm (e.g., [6]). A closely related problem studied in the literature [13,14] is the *minimum sharability* problem in which the cost of sharing each edge is equal to the number of times the edge is shared (i.e., the flow of the edge minus one) times the cost of the edge. This sharability problem can be solved efficiently using minimum-cost flow algorithms. Another related problem is the *fixed-charge flow* problem in which each edge has a fixed building cost as well as a per-unit flow cost, and the objective is to select a subset of edges to route a flow of size k between two nodes s and t such that the total cost of building the network and sending the flow is minimized. The best current approximation factor for this problem is $\beta(G) + 1 + \varepsilon$ where $\beta(G)$ is the size of a maximum s-t cut in the graph [1].

Our Results. In this paper, we study the minimum vulnerability problem as a generalization of the MSE and MECF problems, and obtain several results, a summary of which is listed below.

- We present a primal-dual algorithm for the minimum r-vulnerability problem that achieves an approximation factor of $\lfloor \frac{k}{r+1} \rfloor$. This improves, in particular, the best previous approximation factor of $k - 1$ for the MSE problem to $\lfloor k/2 \rfloor$. It also yields an alternative k-approximation algorithm for the MECF problem.

- We show that for any $r \geqslant 0$ and $\varepsilon > 0$, the minimum r-vulnerability problem is hard to approximate to within a factor of $2^{\log^{1-\varepsilon} n}$ unless $NP \subseteq DTIME(n^{\text{polylog}\, n})$. This eliminates the possibility of obtaining a polylogarithmic approximation factor for the minimum vulnerability problem.

- Despite the fact that the minimum vulnerability problem is NP-hard (and even hard to approximate), we show that for any constant k and any $r > 0$, the minimum r-vulnerability problem can be solved exactly in polynomial time. This settles an open problem posed by Omran *et al.* [10] regarding the complexity of the MSE problem for small values of k. Our result indeed shows that the hardness of the minimum r-vulnerability problem, for any $r > 0$, crucially relies on the number of paths in the problem instance.

- For the MSE problem, we present an approximation algorithm that achieves an approximation guarantee of $O(n^{3/4})$, where n is the number of vertices in the graph. This improves upon the trivial factor-n approximation available for the problem, and is the first algorithm for the problem with a sublinear approximation factor. When the input graph is sparse—which is the case in most real-world applications, e.g., in road-map networks with bounded vertex-degrees—we show that the approximation factor of our algorithm can be further improved to $O(\sqrt{n})$.

Our results are mainly based on a clever use of max-flow min-cut duality. In Section 2, we use a primal-dual method to pick a bounded-cost set of edges, out of which the final vulnerable edges are selected. In Section 3, we find an ordered set of min-cuts that leads to an exact solution to the minimum r-vulnerability problem for any fixed k via a dynamic programming approach. In Section 4, we use a combination of the primal-dual method and a shortest path algorithm to obtain the first sublinear approximation factor for the MSE problem.

2   A Primal-Dual Algorithm

In this section, we present a primal-dual[1] algorithm for the minimum r-vulnerability problem with an approximation factor of $\lfloor \frac{k}{r+1} \rfloor$.

A s-t *cut* is defined as a minimal set of edges whose removal disconnects t from s. Let S be the set of all s-t cuts of size less than $\lceil k/r \rceil$ in G. For the special case of $r = 0$, we define S to be the set of all s-t cuts in G. An obvious constraint is that in any feasible solution, at least one edge from each cut $C \in S$ must be r-vulnerable. If not, at most $(\lceil k/r \rceil - 1) \times r < k$ paths can pass through C, making the solution infeasible. Let x_e be a 0/1 variable which is set to 1 if edge e is r-vulnerable in our solution, and is set to 0 otherwise. The minimum vulnerability problem with no capacity bounds (i.e., when $U_e = \infty$ for all edges) can be expressed as the following integer program:

$$\min \quad \sum_{e \in E} c_e x_e \qquad \qquad \text{(IP)}$$

$$\text{s.t.} \quad \sum_{e \in C} x_e \geqslant 1 \quad \forall C \in S$$

$$x_e \in \{0, 1\} \quad \forall e \in E$$

We relax the integer program to a linear program by replacing the constraint $x_e \in \{0, 1\}$ with $x_e \geqslant 0$. The following is the dual of the resulting linear program:

$$\max \quad \sum_{C \in S} y_C$$

$$\text{s.t.} \quad \sum_{C \ni e} y_C \leqslant c_e \quad \forall e \in E$$

$$y_C \geqslant 0 \quad \forall C \in S$$

Our primal-dual algorithm is presented in Algorithm 1. We start with a feasible dual solution $y = 0$, and an empty set of vulnerable edges R, that represents an infeasible primal solution. We initialize the capacity u_e of each edge to r, allowing each edge to pass at most r paths initially. We then iteratively improve the feasibility of the primal solution by choosing a s-t cut C whose capacity is less than k, and increase its corresponding variable y_C, until a dual constraint $\sum_{C \ni e} y_C \leqslant c_e$ becomes tight for some edge e. We then add e to the set of vulnerable edges, and set its capacity to U_e. The loop is terminated when all s-t cuts have capacity at least k, admitting a s-t flow f of value k, which is returned as the final solution.

Let OPT be the cost of an optimal solution for the minimum vulnerability problem, let Z_{IP} be the optimal value of the objective function of (IP), and APX be the cost of the solution returned by our algorithm. Obviously, $Z_{\text{IP}} \leqslant$ OPT, because every feasible solution to the capacitated problem is also a feasible solution for the uncapacitated one. We further prove the following.

---

[1] Readers not familiar with the primal-dual framework are referred to the textbooks on approximation algorithms, e.g., [12].

Algorithm 1. PRIMAL-DUAL

1: $y \leftarrow 0$, $R \leftarrow \emptyset$
2: set $u_e \leftarrow r$ for all $e \in E$
3: **while** there exists a s-t cut C of capacity less than k in G **do**
4:     increase y_C until $\sum_{C \ni e} y_C = c_e$ for some edge e
5:     $R \leftarrow R \cup \{e\}$, $u_e \leftarrow U_e$
6: find an integral s-t flow f of value k in graph G with edge capacities u_e
7: **return** f

Lemma 3. APX $\leqslant \lfloor \frac{k}{r+1} \rfloor$ OPT.

Proof. Let T be the set of edges carrying a flow more than r in f. Clearly, $T \subseteq R$. Now,

$$\text{APX} = \sum_{e \in T} c_e$$

$$= \sum_{e \in T} \sum_{C \ni e} y_C \qquad \text{(by line 4 of algorithm)}$$

$$= \sum_{C \in S} y_C \times |\{e \in T \cap C\}|$$

$$\leqslant \left\lfloor \frac{k}{r+1} \right\rfloor \sum_{C \in S} y_C \qquad (*)$$

$$\leqslant \left\lfloor \frac{k}{r+1} \right\rfloor Z_{\text{IP}} \qquad \text{(by weak duality)}$$

where the inequality (*) holds, because at most $\lfloor \frac{k}{r+1} \rfloor$ edges of each cut C can have a flow more than r in f. The lemma follows by the fact $Z_{\text{IP}} \leqslant$ OPT. $\square$

Theorem 4. *There is a $\lfloor \frac{k}{r+1} \rfloor$-approximation algorithm for the minimum vulnerability problem that runs in $O(nm^2 \log(n^2/m))$ time on a graph with n vertices and m edges.*

Proof. The approximation factor of Algorithm 1 follows from Lemma 3. The main loop iterates at most m times. At each iteration, we need to compute a min-cut, which can be done in $O(nm \log(n^2/m))$ time [7]. Line 4 involves comparing at most k values, taking $O(k) = O(n)$ time. The total time is therefore $O(nm^2 \log(n^2/m))$. $\square$

3   An Exact Algorithm for Fixed k

The minimum vulnerability problem is not only NP-hard, but is also hard to approximate to within a factor of $2^{\log^{1-\varepsilon} n}$, for any $\varepsilon > 0$ (the proof is omitted

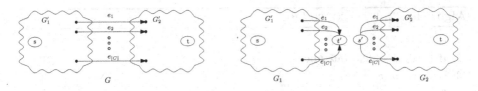

Fig. 1. A graph G with a s-t cut C is divided into two graphs G_1 and G_2

in this version). Despite this fact, we show in this section that if k is a constant, then the minimum r-vulnerability problem, for any $r > 0$, can be solved exactly in polynomial time.

Given a directed graph G and a s-t cut C, we say that an edge $e = (u, v) \notin C$ is *before* C if there is a path from s to u not using any edge of C, and we say that e is *after* C if a path exists from v to t with no edge from C. Note that an edge cannot be both before and after C because C is a s-t cut. Given two s-t cuts C_1 and C_2, we write $C_1 \leqslant C_2$ if each edge of C_1 is either before or in C_2.

Consider an instance of the minimum r-vulnerability problem. We call a capacity function $u : E \to \mathbb{Z}^+$ *proper* if there exists a s-t flow f of value k such that $f(e) \leqslant u(e)$ for all edges $e \in E$. A proper capacity function is *minimal* if decreasing the capacity of any edge e with $u(e) > r$ makes u improper.

Lemma 5. *Given a minimal capacity function u, a sequence $C_1 \leqslant \cdots \leqslant C_\gamma$ of s-t cuts can be found such that $\sum_{e \in C_i} u(e) = k$ for all $1 \leqslant i \leqslant \gamma$, and that each edge $e \in E$ with $u(e) > r$ lies in at least one of the γ cuts.*

Proof. Pick an arbitrary edge $e \in E$ with $u(e) > r$ such that its head is not t and its tail is not s. If no such e exists, we are done. There must exist a s-t cut C containing e such that $\sum_{e \in C} u(e) = k$ by the minimality of u. We construct a graph G_1 from G by removing all edges after C, and then, merging all heads of the edges in C into a new vertex t' (see Figure 1). Similarly, we construct G_2 from G by removing all edges before C, and merging all tails of the edges in C into a new vertex s'. For any set P of k paths from s to t, all edges of P are either in G_1 or G_2. Therefore, the problem of finding k paths from s to t in G can be reduced into two subproblems: finding k paths from s to t' in G_1 and finding k paths from s' to t in G_2. By induction, there exists a sequence of cuts for each of G_1 and G_2 as stated in the lemma. Therefore, the sequence of cuts in G_1, followed by C and then the sequence of cuts in G_2 yields the desired cut sequence. □

Theorem 6. *If k is a constant, then the minimum r-vulnerability problem can be solved exactly in polynomial time for any $r > 0$.*

Proof. We define a *state* as a pair (C, θ), where C is a s-t cut, and θ is a $|C|$-tuple with $\sum_{e \in C} \theta_e = k$ and $\theta_e \leqslant U_e$. A set of k paths from s to t, represented by a s-t flow f of value k, is *suitable* for the state (C, θ) if $f(e) \leqslant \theta_e$ for all $e \in C$, and $f(e) \leqslant U_e$ for all $e \in E \setminus C$.

Algorithm 2. FIND-COST(C, θ)

1: $\mathrm{cost}_C \leftarrow \sum_{e \in C, \theta_e > r} c_e$
2: **if** (C, θ) is a final state **then**
3:     **return** cost_C
4: **for each** edge $e \in C$ **do**
5:     **if** $\theta_e > U_e$ **then**
6:         **return** ∞
7: ans $\leftarrow \infty$
8: **for each** cut C' with $C \leqslant C'$ **do**
9:     **for each** $\theta' = (\theta'_1, \theta'_2, \ldots, \theta'_{|C'|})$ with $\sum_{i=1}^{|C'|} \theta'_i = k$ **do**
10:        **if** (C', θ') is immediately after (C, θ) **then**
11:            ans $\leftarrow \min\{\text{ans, FIND-COST}(C', \theta') + \mathrm{cost}_C\}$
12: **return** ans

The *answer* to the state (C, θ) is defined as the minimum total cost of r-vulnerable edges either in C or after it, over all suitable sets of k paths. Obviously, if there exists a suitable set of k paths for a state (C, θ) such that no edge after C is r-vulnerable, then the answer to this state is equal to the total cost of r-vulnerable edges in C. We call such states the *final states*.

A state (C', θ') is *immediately after* (C, θ), if $C \leqslant C'$, and there exists a set of k paths suitable for both (C, θ) and (C', θ'), with no r-vulnerable edge between C and C' (i.e., after C and before C'). Given two states (C, θ) and (C', θ'), we define the following capacity function:

$$
w(e) = \begin{cases}
\theta_e & \text{if } e \in C \\
\theta'_e & \text{if } e \in C' \\
\min\{U_e, r\} & \text{if } e \text{ between } C \text{ and } C' \\
U_e & \text{otherwise}
\end{cases}
$$

Observe that the maximum flow in graph G with capacity function w is at least k if and only if (C', θ') is immediately after (C, θ). We use this observation to check whether a state is immediately after another one.

Algorithm 2 computes the answer to the state (C, θ) recursively, based on the answers to the states immediately after it. The algorithm works as follows. Given a state (C, θ), we find all states (C', θ') immediately after (C, θ), solve the problem recursively for (C', θ'), add the cost of the current cut (i.e., sum of costs of edges $e \in C$ with $\theta_e > r$), and then return the minimum of all these values. The final answer to the problem is the minimum answer to the states (C, θ) such that there exists a suitable set of k paths with no r-vulnerable edge before C.

The correctness of the algorithm follows from Lemma 5, since all possible sequences of s-t cuts that can be the sequence of cuts in an optimal solution are examined by the algorithm. We can use dynamic programming to store the answers to the states, and avoid recomputing. Note that the number of s-t cuts of size less than k is $O(m^{k-1})$ (where $m = |E|$), and the number of solutions to $\sum_{i=1}^{|C|} \theta_i = k$ is at most $\binom{2k-1}{k-1} = O(1)$, implying that the number of states is $O(m^{k-1})$. On the other hand, checking if a state is final, and checking if a state is immediately after another one can be both done in $O(n^3)$ time using a standard max flow algorithm. Therefore, the total running time of the algorithm is $O(m^{2(k-1)} n^3)$.   □

4   A Sublinear Approximation Factor

In this section, we present an approximation algorithm for the MSE problem (Problem 1) with a sublinear approximation factor. Our algorithm is a combination of the primal-dual method presented in Section 2 and a simple shortest path algorithm. The pseudo-code is presented in Algorithm 3.

Algorithm 3. SUBLINEAR-APPROX

1: let P_1 be the output of Algorithm 1, having w shared edges
2: let P_2 be a shortest s-t path of length ℓ
3: **return** P_1 if $w < \ell$ else P_2

The feasibility of the returned solution is clear, as we can route all the k paths through a shortest s-t path. Let P^* be an optimal solution to the MSE problem with a minimum number of used edges (i.e., edges carrying non-zero flow). Let $\mathcal{D}$ be the graph induced by P^*, and m^* be the number of its edges. Denote by OPT the number of shared edges in any optimal solution. We assume, w.l.o.g., that OPT $\neq 0$.

Lemma 7. $\mathcal{D}$ *is a DAG.*

Proof. Suppose that there is a cycle in $\mathcal{D}$. Reduce the capacity of each edge in the cycle by the minimum amount of flow along the edges of the cycle. This results in decreasing the number of edges in $\mathcal{D}$ by at least one without increasing the number of shared edges, contradicting the minimality of the number of edges in $\mathcal{D}$.   □

Lemma 8. $\frac{k\ell - m^*}{k} \leqslant$ OPT, *where ℓ is the length of a shortest s-t path.*

Proof. Let f be a s-t flow of value k in $\mathcal{D}$. We have:

$$\sum_{e \in E} \max \{0, f(e) - 1\} \leqslant k\text{OPT}, \tag{1}$$

where the the left-hand side counts the number of shared edges in f. On the other hand, any shared edge can be in at most k paths, so the inequality above holds. Furthermore:

$$\sum_{e \in E} \max \{0, f(e) - 1\} = \sum_{e \in E} f(e) - m^*. \tag{2}$$

The length of every s-t path is at least ℓ. Therefore, the total number of edges used in the k paths is at least $k\ell$. Combined with (1) and (2) we get:

$$k\ell - m^* \leqslant \sum_{e \in E} f(e) - m^* \leqslant k\text{OPT}.$$

$\square$

Lemma 9. *Let $G = (V, E)$ be a DAG with n vertices such that for all, but k vertices, in-degree equals out-degree. Then G has $O(kn\sqrt{n} + k^2)$ edges.*

(Proof can be found in the full version.)

Theorem 10. *Algorithm 3 is an $O(\min\{n^{\frac{3}{4}}, m^{\frac{1}{2}}\})$-approximation algorithm for the MSE problem.*

Proof. Let $\alpha = \ell/\text{OPT}$ be the approximation factor achievable by just returning the shortest path. Algorithm 3 has therefore an approximation factor of $\min\{k, \alpha\}$. We consider two cases.

CASE 1. $k\ell \geqslant 2m^*$. By Lemma 8, $\alpha \leqslant \frac{k\ell}{k\ell - m^*}$. Therefore,

$$\alpha \leqslant \frac{k\ell}{k\ell - m^*} \leqslant \frac{2m^*}{2m^* - m^*} = 2.$$

Hence, $\min\{k, \alpha\} \leqslant 2$ in this case.

CASE 2. $k\ell < 2m^*$. In this case

$$k\alpha \leqslant k\alpha\text{OPT} = k\ell < 2m^* \leqslant 2m.$$

Therefore, $\min\{k, \alpha\} < \sqrt{2m}$, implying an approximation factor of $O(\sqrt{m})$.

We can put a second upper bound on the approximation factor. Consider the DAG $\mathcal{D}$ defined earlier in this section. In this graph, except for the endpoints of shared edges and s and t, all other vertices have equal in/out degrees, because each path enters a vertex with an edge, and exits it with another edge. By Lemma 9, the number of edges in $\mathcal{D}$ is $O(n\sqrt{n}\text{OPT} + \text{OPT}^2)$. Observe that if $\text{OPT} \geqslant \sqrt{n}$, a shortest path is a $\sqrt{n}$-approximation to MSE, because $\ell \leqslant n$. If $\text{OPT} < \sqrt{n}$, then m^* is upper-bounded by $O(n\sqrt{n}\text{OPT})$, resulting in:

$$k\ell < 2m^* < cn\sqrt{n}\text{OPT} \implies k\alpha\text{OPT} < cn\sqrt{n}\text{OPT}$$

$$\implies k\alpha < cn\sqrt{n} \implies \min\{k, \alpha\} < \sqrt{cn\sqrt{n}}$$

Combined with the previous result we get $\min\{k, \alpha\} = O(\min\{n^{\frac{3}{4}}, m^{\frac{1}{2}}\})$. $\square$

Corollary 11. *For sparse graphs with $m = O(n)$, such as planar graphs and bounded-degree graphs, Algorithm 3 yields an $O(\sqrt{n})$ approximation factor.*

5   Conclusion

In this paper, we introduced the minimum vulnerability problem which is an extension of the two previously-known problems, MECF and MSE. We obtained a $\lfloor \frac{k}{r+1} \rfloor$-approximation algorithm for the problem in general form, and the first sublinear approximation factor for the MSE problem. While the problem is hard to approximate, we showed that the minimum $(r > 0)$-vulnerability problem can be solved exactly in polynomial time for any fixed k. We leave this question open whether a same poly-time algorithm can be obtained for the case of $r = 0$, i.e., for the MECF problem. Another open problem is whether better approximation factors can be obtained for MSE, and in general, for the minimum vulnerability problem.

References

1. Carr, R.D., Fleischer, L.K., Leung, V.J., Phillips, C.A.: Strengthening integrality gaps for capacitated network design and covering problems. In: Proc. 11th ACM-SIAM Sympos. Discrete Algorithms, pp. 106–115 (2000)
2. Even, G., Kortsarz, G., Slany, W.: On network design problems: fixed cost flows and the covering steiner problem. ACM Trans. Algorithms 1(1), 74–101 (2005)
3. Franklin, M.K.: Complexity and security of distributed protocols. PhD thesis, Dept. of Computer Science, Columbia University (1994)
4. Garey, M., Johnson, D.S.: Computers and intractability: A guide to the theory of NP-completeness. W.H. Freeman (1979)
5. Garg, N., Ravi, R., Konjevod, G.: A polylogarithmic approximation algorithm for the group Steiner tree problem. J. Algorithms 37(1) (2000)
6. Goldberg, A.V., Rao, S.: Beyond the flow decomposition barrier. J. ACM 45(5), 783–797 (1998)
7. Goldberg, A.V., Tarjan, R.E.: A new approach to the maximum-flow problem. J. ACM 35(4), 921–940 (1988)
8. Konjevod, G., Ravi, R., Srinivasan, A.: Approximation algorithms for the covering steiner problem. Random Structures & Algorithms 20(3), 465–482 (2002)
9. Krumke, S.O., Noltemeier, H., Schwarz, S., Wirth, H.-C., Ravi, R.: Flow improvement and network flows with fixed costs. In: Proc. Internat. Conf. Oper. Res.: OR 1998, pp. 158–167 (1998)
10. Omran, M.T., Sack, J.-R., Zarrabi-Zadeh, H.: Finding Paths with Minimum Shared Edges. In: Fu, B., Du, D.-Z. (eds.) COCOON 2011. LNCS, vol. 6842, pp. 567–578. Springer, Heidelberg (2011)
11. Wang, J., Yang, M., Yang, B., Zheng, S.Q.: Dual-homing based scalable partial multicast protection. IEEE Trans. Comput. 55(9), 1130–1141 (2006)
12. Williamson, D.P., Shmoys, D.B.: The design of approximation algorithms. Cambridge University Press (2011)
13. Yang, B., Yang, M., Wang, J., Zheng, S.Q.: Minimum cost paths subject to minimum vulnerability for reliable communications. In: Proc. 8th Internat. Symp. Parallel Architectures, Algorithms and Networks, ISPAN 2005, pp. 334–339. IEEE Computer Society (2005)
14. Zheng, S.Q., Wang, J., Yang, B., Yang, M.: Minimum-cost multiple paths subject to minimum link and node sharing in a network. IEEE/ACM Trans. Networking 18(5), 1436–1449 (2010)

A Strongly Polynomial Time Algorithm for the Shortest Path Problem on Coherent Planar Periodic Graphs

Norie Fu

Department of Computer Science, University of Tokyo
f_norie@is.s.u-tokyo.ac.jp

Abstract. A *periodic graph* is an infinite graph obtained by copying a finite graph to each room of $\mathbb{Z}^d$-lattice and connecting them regularly. Höfting and Wanke formulated the shortest path problem on periodic graphs as an integer programming and showed that it is NP-hard, together with a pseudopolynomial time algorithm for bounded d. Using Iwano and Steiglitz's result, the time complexity can be shown to be weakly polynomial on planar periodic graphs with $d = 2$. In this paper, we show a strongly polynomial time algorithm for the shortest path problem on *coherent* planar periodic graphs with $d = 2$. The coherence is a combinatorial property of periodic graphs introduced in this paper, which naturally holds for planar regular tilings, etc. We show that the coherence is the necessary and sufficient condition that an optimal solution to the integer programming is still optimal if the connectedness constraint is removed. Using the theory of toric ideal, we further show that the *incidence-transit matrix* describing the linear constraints in the integer linear programming for planar cases with $d = 2$ form a new class of unimodular matrices, which itself is of theoretical interest and leads to the strongly polynomial time algorithm.

1 Introduction

A *periodic graph* is an infinite graph which has a finite description by a finite directed graph with vector labels on its edges, called a *static graph*. The periodic graph is obtained as follows: Copy the vertices of the static graph to each cell of $\mathbb{Z}^d$-lattice and make a directed edge from the vertex s in the cell $\boldsymbol{z}$ to the vertex t in the cell $\boldsymbol{z}'$ if an edge from s to t with edge label $\boldsymbol{z}' - \boldsymbol{z}$ is in the static graph. Periodic graphs naturally arise as a model of various periodic structures; crystal structure [5], VLSI circuits [10], systems of uniform recurrence equations [12] and so on. Thus fundamental problems on them are widely investigated; connectivity by Cohen and Megiddo [4], planarity by Iwano and Steiglitz [11], for instance.

The shortest path problem on the periodic graph is the problem to find a path from s in the cell $\boldsymbol{z}'$ to t in the cell $\boldsymbol{z}' + \boldsymbol{z}$ with the minimum number of edges, for given s, t and $\boldsymbol{z} \in \mathbb{Z}^d$. By Höfting and Wanke [9], the shortest path problem on general periodic graph is formulated as an integer programming obtained by adding the connectedness constraint to the integer linear programming defined

K.-M. Chao, T.-s. Hsu, and D.-T. Lee (Eds.): ISAAC 2012, LNCS 7676, pp. 392–401, 2012.
© Springer-Verlag Berlin Heidelberg 2012

by the *incidence-transit matrix* encoding the static graph. They showed the NP-hardness of this problem and a pseudopolynomial time algorithm on periodic graphs with bounded d, which solves the problem by enumerating pseudopolynomial number of possible decompositions of the shortest path into a path and cycles and then solving an integer linear programming for each decomposition.

For efficient use of the nanotechnology to construct an intended configuration of atoms on a physical crystal surface by repeated swaps of atoms [1], an efficient algorithm for the shortest path problem on planar periodic graph is necessary. Motivated by this application, we are interested in the next question: *Is it possible to construct a strongly polynomial time algorithm for the shortest path problem on planar periodic graph?* Using the result of Iwano and Steiglitz [11], the complexity of Höfting and Wanke's algorithm on a planar periodic graph turns out to be weakly polynomial, where the size of the vector z appears, however, we still need to solve strongly polynomial number of integer linear programmings.

In this paper we show an $O(n^8)$ time algorithm on a planar periodic graph with $d = 2$, which outputs the shortest path if the given planar periodic graph is *coherent* and otherwise outputs "no", where n is the number or the vertices and m is the number of the edges of the static graph. Here the coherence is a combinatorial property of periodic graphs introduced in this paper, which makes it possible to compute the shortest path by solving only one integer linear programming defined by the incidence-transit matrix and then converting it into the shortest path. The time complexity for testing coherence and this convert are both $O(n^8)$ on planar periodic graphs. We also show that there are infinite number of coherent periodic graphs by establishing that the periodic graphs with the well-studied property ℓ_1-*rigidity* [7] are coherent. A catalog of ℓ_1-rigid planar periodic graphs is provided by Deza, Grishukhin and Shtogrin [6].

Furthermore we show that the incidence-transit matrix of a planar periodic graph with $d = 2$ forms a new class of unimodular matrices. By the theory of integer and linear programming [15], this means that the integer linear programming defined by the incidence-transit matrix can be solved by computing its LP relaxation. Using Iwano and Steiglitz's theory mentioned above, we can also show that this LP relaxation can be solved in $O(n^{6.5} \log n)$ time using Orlin's dual version [14] of Tardos' strongly polynomial algorithm for the linear programming defined by a matrix with bounded entries [18]. Note that the incidence-transit matrix of a planar periodic graph is not always totally unimodular and we cannot use Seymour's decomposition theorem for totally unimodular matrices [16] to prove this result. We utilize the theory of computational algebra, especially the theory on toric ideal [17] for the proof.

The time complexity of our algorithm is large, however, we can use fast algorithms such as the simplex method for the LP relaxation. The bottleneck of the algorithm is thus the convert into the shortest path, but the computations for the convert into are also necessary in the execution of Höfting and Wanke's algorithm. A more sophisticated way or a new theory to deal with the connectedness condition may be necessary to overcome this bottleneck.

2 Preliminaries

Definition 1. *The pair* $\mathcal{G} = (\mathcal{V}, \mathcal{E})$ *of a vertex set* $\mathcal{V} = \{1, \ldots, n\}$ *and a set of directed edges with vector labels* $\mathcal{E} = \{e^{(1)}, \ldots, e^{(m)}\} \subseteq \{((i,j), \boldsymbol{g}) : i, j \in \mathcal{V}, \boldsymbol{g} \in \mathbb{Z}^d\}$ *is called a static graph. For a static graph* $\mathcal{G} = (\mathcal{V}, \mathcal{E})$, *the periodic graph* $G = (V, E)$ *generated by* $\mathcal{G}$ *is defined by* $V = \mathcal{V} \times \mathbb{Z}^d$, $E = \{((i, \boldsymbol{h}), (j, \boldsymbol{h} + \boldsymbol{g})) : \boldsymbol{h} \in \mathbb{Z}^d, ((i,j), \boldsymbol{g}) \in \mathcal{E}\}$.

For an edge $e = ((i,j), \boldsymbol{g}) \in \mathcal{E}$, $\boldsymbol{g}$ is called the *transit vector* of e, denoted by $\mathrm{tran}(e)$. By definition, periodic graphs are generally directed graphs. However, if for all $((i,j), \boldsymbol{g}) \in \mathcal{E}$ its *reversed edge* $((j,i), -\boldsymbol{g})$ is also in $\mathcal{E}$, the periodic graph can be regarded as an undirected graph. An example is shown in Figure 1.

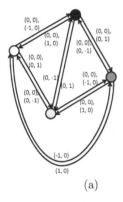
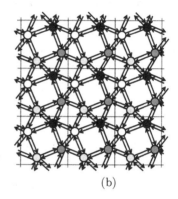

(a)   (b)

Fig. 1. (a) is a static graph $\mathcal{G}$ with $d = 2$ and (b) is the periodic graph G generated by $\mathcal{G}$. The multiple edges with different edge labels on $\mathcal{G}$ are represented by an edge with multiple edge labels. The periodic graph G can be regarded as undirected, since for every directed edge there exists an oppositely-oriented edge of it. The oppositely-oriented edges on G are induced by the reversed edges on the static graph $\mathcal{G}$.

Note that different two static graphs can generate isomorphic periodic graphs. In the rest, we assume that G is undirected and connected, and $d = 2$. By $\mathcal{G} = (\mathcal{V}, \mathcal{E})$ we denote a static graph with $\mathcal{V} = \{1, \ldots, n\}$ and $\mathcal{E} = \{e^{(1)}, \ldots, e^{(m)}\}$ and by $G = (V, E)$ the corresponding periodic graph.

A *semi-walk* is an alternating sequence of vertices and edges $v^{(1)}, f^{(1)}, \ldots, v^{(k-1)}, f^{(k-1)}, v^{(k)}$, where $f^{(i)}$ is an edge from v_i to v_{i+1} or else $f^{(i)}$ is an edge from v_{i+1} to v_i. If $v^{(1)} = v^{(k)}$, the semi-walk is *closed* and otherwise the semi-walk is *open*. The semi-walk is a *semi-path* if it has no cycle as a proper subsequence. A closed semi-walk is called *cycle* and said to be *simple* if it is a semi-path. In the former case edge $f^{(i)}$ is a *forward edge* of the semi-walk, and in the latter case edge $f^{(i)}$ is a *backward edge*. A semi-walk is a *directed walk* (or simply *walk*) if it has no backward edges. For a semi-walk $\mathcal{P}$ on $\mathcal{G}$ with the forward edges $\mathcal{F}$ and the backward edges $\mathcal{B}$, we call $\sum_{e \in \mathcal{F}} \mathrm{tran}(e) - \sum_{e' \in \mathcal{B}} \mathrm{tran}(e')$ the *transit vector* of $\mathcal{P}$, denoted by $\mathrm{tran}(\mathcal{P})$. We denote the number of edges in $\mathcal{P}$ by $\mathrm{len}(\mathcal{P})$.

Denote by $\boldsymbol{u}^{(k)} \in \mathbb{Z}^{(n+2)}$ the k-th unit vector and by g_k the k-th entry of a vector $\boldsymbol{g}$. For $e = ((i,j), \boldsymbol{g}) \in \mathcal{E}$, let $\boldsymbol{a}(e)$ be the column vector $(\boldsymbol{u}^{(i)} - \boldsymbol{u}^{(j)}) + (g_1 \boldsymbol{u}^{(n+1)} + g_2 \boldsymbol{u}^{(n+2)})$. When $i \neq j$, $\boldsymbol{a}(e) = (\ldots, 1, \ldots, -1, \ldots, g_1, g_2)^T$ and otherwise $\boldsymbol{a}(e) = (0, \ldots, 0, g_1, g_2)^T$.

Definition 2. *The incidence-transit matrix of $\mathcal{G}$, denoted by $A_{\mathcal{G}}$, is the matrix* $(\boldsymbol{a}(e^{(1)}), \ldots, \boldsymbol{a}(e^{(m)}))$.

For a multiset $\mathcal{S}$ on the set $\{s_1, \ldots, s_m\}$, we call the vector $\boldsymbol{v} = (v_1, \ldots, v_m)^T$ the *characteristic vector of S* if v_i is the multiplicity of s_i in S. We also refer to the characteristic vector of a semi-walk by regarding it as a multiset of edges.

Theorem 1 ([9]). *Let $s, t \in \mathcal{V}$ and $\boldsymbol{z} \in \mathbb{Z}^2$. Then the characteristic vector of a shortest walk $\mathcal{P}$ on $\mathcal{G}$ from s to t with* $\mathrm{tran}(\mathcal{P}) = \boldsymbol{z}$ *can be computed by the following mathematical programming denoted by $\mathrm{IP}_{\mathcal{G}}(s, t, \boldsymbol{z})$: $\min \mathbf{1} \cdot \boldsymbol{x}$ subject to (1)$A_{\mathcal{G}} \boldsymbol{x} = (\boldsymbol{u}^{(s)} - \boldsymbol{u}^{(t)}) + (z_1 \boldsymbol{u}^{(n+1)} + z_2 \boldsymbol{u}^{(n+2)})$, (2)$\boldsymbol{x} \geq \mathbf{0}$, (3)$\boldsymbol{x} \in \mathbb{Z}^m$ and (4)$\{e^{(i)} \in \mathcal{E} : x_i > 0\}$ is weakly connected and is incident to s and t on $\mathcal{G}$.*

For a fixed $\boldsymbol{z}' \in \mathbb{Z}^2$, there is a one-to-one correspondence between a walk connecting $(s, \boldsymbol{z}')$ and $(t, \boldsymbol{z}' + \boldsymbol{z})$ on G and a walk from s to t on $\mathcal{G}$ with transit vector $\boldsymbol{z}$. Thus $\mathrm{IP}_{\mathcal{G}}(s, t, \boldsymbol{z})$ also computes a shortest path from $(s, \boldsymbol{z}')$ to $(t, \boldsymbol{z}' + \boldsymbol{z})$ on G. By Höfting and Wanke, a pseudopolynomial time algorithm for $\mathrm{IP}_{\mathcal{G}}(s, t, \boldsymbol{z})$ is proposed [9]. By $\mathrm{ILP}_{\mathcal{G}}(s, t, \boldsymbol{z})$ (resp. $\mathrm{LP}_{\mathcal{G}}(s, t, \boldsymbol{z})$) let us denote the integer linear programming (resp. the linear programming) obtained by removing the constraint (4) (resp. the constraint (3) and (4)) from $\mathrm{IP}_{\mathcal{G}}(s, t, \boldsymbol{z})$. The next proposition is easy to verify.

Proposition 1. *Let $\boldsymbol{p}, \boldsymbol{q} \in \mathbb{N}^m$, $s, t \in \mathcal{V}$ and $\boldsymbol{z} \in \mathbb{Z}^2$. A multiset $\mathcal{E}'$ with its characteristic vector $\boldsymbol{p} + \boldsymbol{q}$ satisfies 1, 2 and 3 if and only if $A_{\mathcal{G}}(\boldsymbol{p} - \boldsymbol{q}) = \boldsymbol{u}^{(s)} - \boldsymbol{u}^{(t)} + (z_1 \boldsymbol{u}^{(n+1)} + z_2 \boldsymbol{u}^{(n+2)})$.*

1. *$\mathcal{E}'$ is the union of vertex-disjoint weakly connected multisets corresponding to semi-walks $\mathcal{P}^{(1)}, \ldots, \mathcal{P}^{(k)}$ ($k \geq 1$) on $\mathcal{G}$ such that $\sum_{i=1}^{k} \mathrm{tran}(\mathcal{P}^{(i)}) = \boldsymbol{z}$.*
2. *$\boldsymbol{p}$ is the characteristic vector of the multiset of the forward edges in $\mathcal{E}'$ and $\boldsymbol{q}$ is that of the backward edges in $\mathcal{E}'$.*
3. *$\mathcal{P}^{(1)}$ is a semi-walk from s to t and $\mathcal{P}^{(2)}, \ldots, \mathcal{P}^{(k)}$ are closed semi-walks.*

A *drawing* of a graph is a function Γ which maps each vertex v to a distinct point $\Gamma(v) \in \mathbb{R}^2$ and each directed edge (u, v) to a simple open curve $\Gamma(u, v)$ with endpoints $\Gamma(u)$ and $\Gamma(v)$ so that $\Gamma(u, v) = \Gamma(v, u)$. If the curves $\Gamma(u, v)$ and $\Gamma(u', v')$ do not intersect for any $(u, v) \neq (u', v')$, then Γ is called a *planar drawing*. For a planar drawing Γ, a point $K \in \mathbb{R}^2$ is called a *vertex accumulation point*, abbreviated by VAP, if for every $\varepsilon > 0$ there exist an infinite vertex set $U \subset V$ such that $\Gamma(U) \subseteq \{K' \in \mathbb{R}^2 : \|K - K'\|_{\ell_2} < \varepsilon\}$. A graph is *VAP-free planar* if it admits a planar drawing with no VAP. Proposition 2 and Proposition 3 hold by Theorem 5.1 and Theorem 6.1 in [11], respectively.

Proposition 2. *If G is VAP-free planar, then there exists a static graph $\mathcal{G}'$ of G such that $A_{\mathcal{G}'}$ is a $\{0, \pm1\}$-matrix. Such a static graph can be computed in $O(m)$ time from a given static graph of G.*

Proposition 3. *If G is VAP-free planar, then m is $O(n)$.*

The next follows as a corollary of Corollary 4.9 in [9] and Proposition 2.

Corollary 1. *On a connected VAP-free planar periodic graph with $d = 2$, there exists a weakly polynomial time algorithm for $\mathrm{IP}_\mathcal{G}(s, t, z)$.*

Theorem 2. *A connected periodic graph with $d = 2$ is planar if and only if it is VAP-free planar.*

Theorem 2 can be derived using Ayala, Domínguez, Márquez and Quintero's theory on planar infinite graphs using the *end*, which is an equivalence relation on the rays on infinite graphs [2]. Due to the page limitation, we omit the proof.

3   Coherence of Periodic Graphs

In this section, we use the notations $\mathrm{IP}_\mathcal{G}^*(s, t, z), \mathrm{ILP}_\mathcal{G}^*(s, t, z)$ to describe the optimal value $\mathbf{1} \cdot \boldsymbol{x}$ to $\mathrm{IP}_\mathcal{G}(s, t, z), \mathrm{ILP}_\mathcal{G}(s, t, z)$, respectively.

Definition 3. *The periodic graph G generated by a static graph $\mathcal{G}$ is coherent if for any $s, t \in \mathcal{V}$ and $z \in \mathbb{Z}^2$, $\mathrm{IP}_\mathcal{G}^*(s, s, z) = \mathrm{IP}_\mathcal{G}^*(t, t, z)$.*

First we present a pseudopolynomial time algorithm to compute an optimal solution to $\mathrm{IP}_\mathcal{G}(s, t, z)$ from a given optimal solution to $\mathrm{ILP}_\mathcal{G}(s, t, z)$. Let $\mathcal{P}$ be a walk on $\mathcal{G}$. Then, $\mathcal{P}$ decomposes into a path and simple cycles on $\mathcal{G}$. Note that when $\mathcal{P}$ is a cycle, the path is an empty path. Suppose $\mathcal{P}$ decomposes into a path $\mathcal{P}'$ and y_i simple cycles $\mathcal{C}_i$ starting at the vertex s_i with $\mathrm{tran}(\mathcal{C}_i) = z^{(i)}$ and $\mathrm{len}(\mathcal{C}_i) = c_i$ where i runs in $\{1, \dots, k\}$. Then the system $(\mathcal{P}', (y_1, s_1, z^{(1)}, c_1), \dots, (y_k, s_k, z^{(k)}, c_k))$ is called a *complete path decomposition* of $\mathcal{P}$ [9]. By Proposition 1, the edge multiset corresponding to the solution to $\mathrm{ILP}_\mathcal{G}(s, t, z)$ decomposes into one (possibly empty) walk and cycles on $\mathcal{G}$ such that each of them is vertex-disjoint and corresponds to a shortest path on G. Each of them also have its complete path decomposition. We use the same notation $(\mathcal{P}', (y_1, s_1, z^{(1)}, c_1), \dots, (y_k, s_k, z^{(k)}, c_k))$ to describe the sum of the complete path decompositions of the walk and cycles in the solution to $\mathrm{ILP}_\mathcal{G}(s, t, z)$ and by abusing the term call it the complete path decomposition of the solution. By $\mathcal{Z}_\mathcal{G}$, let us denote the maximum value in $\{\|\mathrm{tran}(e)\|_{\ell_\infty} : e \in \mathcal{E}\}$.

Lemma 1. *On a coherent periodic graph an optimal solution to $\mathrm{IP}_\mathcal{G}(s, t, z)$ can be computed in $O(n^7 m \mathcal{Z}_\mathcal{G}^4)$ time if an optimal solution to $\mathrm{ILP}_\mathcal{G}(s, t, z)$ is given.*

Proof. Let $\boldsymbol{x}^*$ be an optimal solution to $\mathrm{ILP}_\mathcal{G}(s, t, z)$ and $\mathcal{E}' := \{e^{(i)} : x_i^* > 0\}$. The complete path decomposition $(\mathcal{P}', (y_1, s_1, z^{(1)}, c_1), \dots, (y_k, s_k, z^{(k)}, c_k))$ of $\boldsymbol{x}^*$ can be computed in $O(n + m)$ time by computing the depth-first forest of $\mathcal{E}'$ and decomposing each components to a path and simple cycles. For each $i = 1, \dots, k$ let $\boldsymbol{x}^{(i)}$ be an optimal solution to $\mathrm{IP}_\mathcal{G}(s, s, z^{(i)})$ and $\boldsymbol{x}'$ be the characteristic vector of $\mathcal{P}'$. Let $\boldsymbol{x} := \boldsymbol{x}' + \sum_{i=1}^{k} \boldsymbol{x}^{(i)}$. Since G is coherent, for each $i = 1, \dots, k$, $\mathrm{IP}_\mathcal{G}(s, s, z^{(i)}) = c_i$. Hence $\mathbf{1} \cdot \boldsymbol{x} = \mathbf{1} \cdot \boldsymbol{x}^*$ and $\boldsymbol{x}$ is an optimal solution to $\text{ILP}_{\mathcal{G}}(s,t,z)$. Since the edge set $\{e^{(i)} : x_i > 0\}$ is weakly-connected, x is also an optimal solution to $\text{IP}_{\mathcal{G}}(s,t,z)$. Since $(s_i, z^{(i)}) \neq (s_j, z^{(j)})$ for all $i \neq j$ by definition of complete path decomposition, by Fact 4.1.2 in [9] $k \leq n \cdot (2 \mathcal{Z}_{\mathcal{G}} n + 1)^2$. Since each cycle in the complete path decomposition is simple, $c_i < n$ and each $x^{(i)}$ can be computed in $O(n^4 m \mathcal{Z}_{\mathcal{G}}^2)$ time by Fact 4.1.3 in [9]. $\qquad \square$

Paying attention to the fact that each cycle in the complete path decomposition in the proof of Lemma 1, the next lemma can be easily established.

Lemma 2. *Let $C_1, \dots, C_k$ be all simple cycles on $\mathcal{G}$ such that each C_i corresponds to a shortest path on G generated by $\mathcal{G}$. G is coherent if and only if for all $s \in V$ and $i \in \{1, \dots, k\}$, $\text{IP}_{\mathcal{G}}(s, s, \text{tran}(C_i)) = \text{len}(C_i)$.*

Theorem 3. *It can be computed whether a periodic graph generated by a given static graph $\mathcal{G}$ is coherent or not in $O(n^7 m \mathcal{Z}_{\mathcal{G}}^4)$ time.*

Theorem 3 can be shown by using the breadth-first search in Fact 4.1 of [9], the fact that the length of a simple cycle is at most n and Lemma 2.

Theorem 4. *G is coherent if and only if for all $s, t \in V$ and $z \in \mathbb{Z}^2$, $\text{IP}_{\mathcal{G}}^*(s, t, z) = \text{ILP}_{\mathcal{G}}^*(s, t, z)$.*

Proof. By the proof of Lemma 1, the sufficiency of the coherence obviously holds. If the latter condition holds, then for any $s, t \in V$ and any $z \in \mathbb{Z}^2$, $\text{IP}_{\mathcal{G}}^*(s, s, z) = \text{ILP}_{\mathcal{G}}^*(s, s, z) = \text{ILP}_{\mathcal{G}}^*(t, t, z) = \text{IP}_{\mathcal{G}}^*(t, t, z)$. $\qquad \square$

Finally we show that periodic graphs with a well-known property ℓ_1-*rigidity* are coherent. An infinite graph with vertex set U is ℓ_1-*embeddable* if for some $\lambda, d \in \mathbb{N}$, there exists a map $\phi : U \to \mathbb{Z}^d$ such that for any $s_1, s_2 \in U$, the length of the shortest path between s_1 and s_2 is $\frac{1}{\lambda} \|\phi(s_1) - \phi(s_2)\|_{\ell_1}$. An ℓ_1-embeddable infinite graph is ℓ_1-*rigid* if it has a unique ℓ_1-embedding into $\mathbb{Z}^d$, up to the symmetry of $\mathbb{Z}^d$. For sufficient conditions for ℓ_1-rigidity, see Corollary 2 in [3]. In chapter 9 of [6], a catalog of ℓ_1-rigid periodic graphs is given.

Theorem 5. *If a periodic graph is ℓ_1-rigid, then it is coherent.*

Proof. Let ϕ be the ℓ_1-embedding of G. For all $s, t \in V$ and any $z \in \mathbb{Z}^2$, $\text{IP}_{\mathcal{G}}^*(s, s, z) = \text{IP}_{\mathcal{G}}^*(t, t, z)$ since $\phi((s, z)) - \phi((s, 0)) = \phi((t, z)) - \phi((t, 0))$ by Proposition 2 in [8]. $\qquad \square$

4 Unimodularity of the Incidence-Transit Matrices of Planar Periodic Graphs

The objective of this section is to show the following theorem:

Theorem 6. *The incidence-transit matrix $A_{\mathcal{G}}$ of a connected planar periodic graph with $d = 2$ is unimodular, i.e., every $r \times r$ minor of $A_{\mathcal{G}}$ has determinant 1, 0 or -1 where $r = \text{rank} A_{\mathcal{G}}$.*

To show Theorem 6, we need some preparations shown below.

Theorem 7 ([11]). *For an edge* $f := ((s, \boldsymbol{y}), (t, \boldsymbol{z}))$ *on* G *and a vector* $\boldsymbol{g} \in \mathbb{Z}^2$, *denote the edge* $((s, \boldsymbol{y} + \boldsymbol{g}), (t, \boldsymbol{z} + \boldsymbol{g}))$ *by* $f_{\boldsymbol{g}}$. *If* G *is VAP-free planar, then* G *has a planar drawing* Γ *such that for any edge* f *on* G *and* $\boldsymbol{g} \in \mathbb{Z}^2$, $\Gamma(f_{\boldsymbol{g}})$ *is the translation* $\Gamma(f) + \boldsymbol{g}$ *of the simple open curve* $\Gamma(f)$.

By Theorem 2, Theorem 7 holds on any planar periodic graphs. In the rest, whenever we refer to a planar drawing, it is the one in Theorem 7.

For a given $\boldsymbol{p} = (p_1, \ldots, p_m) \in \mathbb{N}^m$, we denote the monomial $x_1^{p_1} \ldots x_m^{p_m}$ by $\boldsymbol{x}^{\boldsymbol{p}}$. When $\boldsymbol{x}^{\boldsymbol{p}}$ divides $\boldsymbol{x}^{\boldsymbol{q}}$, we denote $\boldsymbol{x}^{\boldsymbol{p}} \mid \boldsymbol{x}^{\boldsymbol{q}}$.

Definition 4. *Let* $A \in \mathbb{Z}^{n' \times m'}$ *be a matrix. The toric ideal* I_A *of* A *is the ideal generated by the set* $ker(A) = \{\boldsymbol{x}^{\boldsymbol{p}} - \boldsymbol{x}^{\boldsymbol{q}} : Ap = Aq, \text{ and } \boldsymbol{p}, \boldsymbol{q} \in \mathbb{N}^{m'}\}$, *i.e., the set of polynomials* $\{g_1 b_1 + \cdots + g_l b_l : g_i \in k[x_1, \ldots, x_{m'}], b_i \in ker(A), l \in \mathbb{N}\}$.

Definition 5. *A binomial* $\boldsymbol{x}^{\boldsymbol{p}} - \boldsymbol{x}^{\boldsymbol{q}} \in I_A$ *is primitive if there exists no other binomial* $\boldsymbol{x}^{\boldsymbol{p}'} - \boldsymbol{x}^{\boldsymbol{q}'} \in I_A$ *such that* $\boldsymbol{x}^{\boldsymbol{p}'} \mid \boldsymbol{x}^{\boldsymbol{p}}$ *and* $\boldsymbol{x}^{\boldsymbol{q}'} \mid \boldsymbol{x}^{\boldsymbol{q}}$.

Theorem 8 ([17]). *If every primitive binomial* $\boldsymbol{x}^{\boldsymbol{p}} - \boldsymbol{x}^{\boldsymbol{q}}$ *in the toric ideal* I_A *is square-free, i.e., each* p_i *and* q_i *are at most 1, then the matrix* A *is unimodular.*

In the rest of this section, we prove the next theorem.

Theorem 9. *If* G *is planar, then every primitive binomial in* $I_{A_{\mathcal{G}}}$ *is square-free.*

By Theorem 8, Theorem 6 follows as a corollary of Theorem 9.

The i-th variable x_i of $I_{A_{\mathcal{G}}}$ corresponds to the i-th column of $A_{\mathcal{G}}$, which corresponds to the i-th edge $e^{(i)}$ in $\mathcal{E}$. A binomial $\boldsymbol{x}^{\boldsymbol{p}} - \boldsymbol{x}^{\boldsymbol{q}} \in I_{A_{\mathcal{G}}}$ induces the multiset of edges in $\mathcal{E}$ such that its characteristic vector is $\boldsymbol{p} + \boldsymbol{q}$.

Proposition 4. *Let* $\boldsymbol{x}^{\boldsymbol{p}} - \boldsymbol{x}^{\boldsymbol{q}}$ *be a binomial in* $I_{A_{\mathcal{G}}}$. *The edge multiset* $\mathcal{E}'$ *induced by* $\boldsymbol{x}^{\boldsymbol{p}} - \boldsymbol{x}^{\boldsymbol{q}}$ *corresponds to a set of vertex-disjoint closed semi-walks* $\mathcal{P}^{(1)}, \ldots, \mathcal{P}^{(k)}$ *with* $\sum_{i=1}^{k} \text{tran}(\mathcal{P}^{(i)}) = \boldsymbol{0}$ *and* $k \geq 1$. *Furthermore, if* $\boldsymbol{x}^{\boldsymbol{p}} - \boldsymbol{x}^{\boldsymbol{q}}$ *is primitive, then* $\mathcal{E}'$ *has no closed semi-walk* $\mathcal{C}$ *with* $\text{tran}(\mathcal{C}) = 0$ *as a proper multisubset.*

Proposition 4 can be easily shown using Proposition 1 and the definition of primitive binomials. Note that in Proposition 4 if the binomial is primitive and $k = 1$ then $\mathcal{P}^{(1)}$ corresponds to a closed semi-path on G and if the binomial is primitive and $k \geq 2$ then $\mathcal{P}^{(1)}, \ldots, \mathcal{P}^{(k)}$ correspond to k open semi-paths on G.

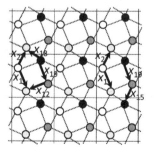

Fig. 2. The closed semi-path on the left is induced by b_1 and the set of two open semi-paths on the right is induced by b_2

Example 1. Let $\mathcal{G}, G$ be the static graph and the periodic graph in Figure 1. $b_1 := x_2 x_{12} x_{19} - x_1 x_{18}$ and $b_2 := x_2 x_{15} x_{19} - x_1$ are both primitive binomials in $I_{A_{\mathcal{G}}}$. As shown in Figure 2, b_1 induces a closed semi-path on G, and b_2 induces two open semi-paths on G.

Lemma 3. *Let $b = x^p - x^q \in I_{A_\mathcal{G}}$ be a binomial inducing a closed semi-path on G. If G is planar and b is not square-free, then there exists a square-free binomial $b' = x^{p'} - x^{q'} \in I_{A_\mathcal{G}}$ inducing another closed semi-path on G and satisfying $x^{p'} \mid x^p$ and $x^{q'} \mid x^q$.*

Proof (Sketch). We construct such b' from the given binomial b, by making a sequence of binomials $b = b^{(0)}, b^{(1)}, \ldots, b^{(k)} = b'$ with $b^{(i)} = x^{p^{(i)}} - x^{q^{(i)}}$ such that $x^{p^{(i+1)}} \mid x^{p^{(i)}}$ and $x^{q^{(i+1)}} \mid x^{q^{(i)}}$ for $i = 0, \ldots, k-1$. Let C (resp. $\mathcal{C}$) be the closed semi-path on G (resp. the closed semi-walk on $\mathcal{G}$) induced by $b^{(0)}$. Since $b^{(0)}$ is not square-free, two directed edges f, f' with $f \neq f'$ in C correspond to the same edge of $\mathcal{G}$. Via the drawing Γ of G, C corresponds to a Jordan curve J on the plane. Since f and f' correspond to the same edge of $\mathcal{G}$, there exists $z \neq 0 \in \mathbb{Z}^2$ such that $\Gamma(f') = \Gamma(f) + z$. Let $F := \Gamma(f')$. The Jordan curve J and its copy $J' := J + z$ has F as their intersection. Note that by definition of Γ, the copy J' corresponds to another simple closed semi-path on G induced by $\mathcal{C}$. It is not hard to prove that J and J' must have another intersection F' because J passes through both of the exterior and the interior of J'.

Let K (resp. K') be a point in F (resp. F'). Let J_L (resp. J'_L) be the curve obtained by following J (resp. J') from K to K' counterclockwise and J_R (resp. J'_R) be the curve obtained by following J (resp. J') from K to K' clockwise. Then, it is easy to show that one of the follows holds: (a) $(J_L + z) \subset J'_L$ and $J'_R \subset (J_R + z)$, or (b) $J'_L \subset (J_L + z)$ and $(J_R + z) \subset J'_R$. If (a) holds then let $J'' := J_L \cup J'_R$ and let $J'' := J'_L \cup J_R$ if (b) holds. Since G is a planar graph, J'' induces a closed semi-path C'' on G via Γ. The closed semi-walk $\mathcal{C}''$ on $\mathcal{G}$ corresponding to C'' is a proper submultiset of $\mathcal{C}$, regarding $\mathcal{C}$ and $\mathcal{C}''$ as multisets of edges. Thus there exists a binomial $b^{(1)} := x^{p^{(1)}} - x^{q^{(1)}} \in I_{A_\mathcal{G}}$ inducing C'' such that $x^{p^{(1)}} \mid x^{p^{(0)}}$, $x^{q^{(1)}} \mid x^{q^{(0)}}$ and $|b^{(1)}| < |b^{(0)}|$. Here, for $b^{(i)}$, by $|b^{(i)}|$ we denote the sum $\sum_{j=1}^m p_j + \sum_{j=1}^m q_j$.

If one of the terms of $b^{(1)}$ is not square-free, we can obtain the next binomial $b^{(2)} \in I_{A_\mathcal{G}}$ inducing a simple closed semi-walk on G such that $x^{p^{(2)}} \mid x^{p^{(1)}}$, $x^{q^{(2)}} \mid x^{q^{(1)}}$ and $|b^{(2)}| < |b^{(1)}|$, by applying the above construction to $b^{(1)}$ again. Since $|b^{(0)}|$ is finite, the repeated construction must stop with finite k steps and the obtained binomial $b^{(k)} = x^{p^{(k)}} - x^{q^{(k)}}$ inducing a closed semi-path on G is square-free. Since $x^{p^{(i+1)}} \mid x^{p^{(i)}}$ and and $x^{q^{(i+1)}} \mid x^{q^{(i)}}$ for each i, $x^{p^{(k)}} \mid x^{p^{(0)}} = x^p$ and $x^{q^{(k)}} \mid x^{q^{(0)}} = x^q$. Thus $b^{(k)}$ is the objective binomial. $\square$

Corollary 2. *If G is planar, then a primitive binomial $b \in I_{A_\mathcal{G}}$ inducing a closed semi-path on G must be square-free.*

A proof of Proposition 5 can be found in Maehara's concise proof of the Jordan curve theorem using Brouwer's fix point theorem [13].

Proposition 5. *Consider a rectangular $ABCD$. Let F (resp. F') be a continuous curve in $ABCD$ from A to C (resp. from B to D). Then, F and F' have an intersection in the boundary or the interior of $ABCD$.*

Lemma 4. *Let* $b \in I_{A_G}$ *be a primitive binomial inducing a set of vertex-disjoint closed semi-walks* $\mathcal{P}^{(1)}, \ldots, \mathcal{P}^{(k)}$ $(k \geq 2)$ *on* $\mathcal{G}$ *and* $z^{(i)}$ *be the transit vector of* $\mathcal{P}^{(i)}$. *If* G *is planar, then all* $z^{(1)}, \ldots, z^{(k)}$ *are parallel to each other.*

Proof (Sketch). We prove the lemma by contradiction, assuming that $z^{(1)}$ and $z^{(2)}$ are not parallel without loss of generality. Since b is primitive, $z^{(1)}$ and $z^{(2)}$ are not $\mathbf{0}$. For $i = 1, 2$, let $P_\infty^{(i)}$ be the doubly infinite path on G corresponding to the infinite walk on $\mathcal{G}$ obtained by combining $\mathcal{P}^{(i)}$ infinite times. By the definition of the drawing Γ, for all $\lambda \in \mathbb{Z}$, $\Gamma(P_\infty^{(i)}) = \Gamma(P_\infty^{(i)}) + \lambda z^{(i)}$ and thus $\Gamma(P_\infty^{(i)}) = \bigcup_{\lambda \in \mathbb{Z}} \left(\left(\Gamma(P_\infty^{(i)}) \cap \left([0, z_1^{(i)}] \times [0, z_2^{(i)}] \right) \right) + \lambda z^{(i)} \right)$. Hence there exists a straight line $L^{(i)}$ with slope $z^{(i)}$ and a constant $\varepsilon^{(i)}$ such that $\Gamma(P_\infty^{(i)})$ is contained in the relative interior of the strip $S^{(i)}$ consisting of the points whose distance from $L^{(i)}$ is equal to or less than $\varepsilon^{(i)}$. The region $S = S^{(1)} \cap S^{(2)}$ is a parallelogram. Using Proposition 5, it is easy to prove that the curves $\Gamma(P_\infty^{(1)}) \cap S$ and $\Gamma(P_\infty^{(2)}) \cap S$ have an intersection K^*. Since Γ is a planar drawing, K^* corresponds to a vertex. Thus the doubly infinite paths $P_\infty^{(1)}$ and $P_\infty^{(2)}$ have a common vertex, contradicting to the assumption that $\mathcal{P}^{(1)}$ and $\mathcal{P}^{(2)}$ are vertex-disjoint.   □

Lemma 5. *Let* $x^p - x^q \in I_{A_G}$ *be a primitive binomial inducing a set of vertex-disjoint closed semi-walks* $\mathcal{P}^{(1)}, \ldots, \mathcal{P}^{(k)}$ $(k \geq 2)$ *on* $\mathcal{G}$. *If* G *is planar, then* $x^p - x^q$ *is square-free.*

Lemma 5 follows from Lemma 3 and Lemma 4, without further use of planarity. Theorem 9 follows by combining Proposition 4, Corollary 2 and Lemma 5.

5   The Strongly Polynomial Time Algorithm for the Shortest Path Problem

Lemma 6. *On planar periodic graphs, there exists an* $O(n^{6.5} \log n)$ *time algorithm to compute an optimal solution to* $\mathrm{ILP}_\mathcal{G}(s, t, z)$.

Proof. By Proposition 2, Proposition 3 and Theorem 2, if G is planar then an optimal solution to $\mathrm{LP}_\mathcal{G}(s, t, z)$ can be computed in $O(n^{6.5} \log n)$ arithmetic steps using Orlin's strongly polynomial time algorithm for the linear programming defined by a matrix with bounded entries [14]. The optimal solution to the linear programming defined by a unimodular matrix is integral [15]. Thus by Theorem 6, the optimal solution is also an optimal solution to $\mathrm{ILP}_\mathcal{G}(s, t, z)$.   □

Theorem 10. *There is an* $O(n^8)$ *time algorithm on a planar periodic graph* G *which outputs "no" if* G *is not coherent and otherwise outputs an optimal solution to* $\mathrm{IP}_\mathcal{G}(s, t, z)$ *for given* $s, t \in \mathcal{V}$ *and* $z \in \mathbb{Z}^2$.

Proof. By Proposition 2, Theorem 2 and Theorem 3, it can be determined whether a given planar periodic graph is coherent or not in $O(n^8)$ time. Suppose G is found to be coherent. An optimal solution to $\mathrm{ILP}_\mathcal{G}(s, t, z)$ can be found in $O(n^{6.5} \log n)$ time by Lemma 6. By Proposition 3 and Lemma 1, an optimal solution to $\mathrm{IP}_\mathcal{G}(s, t, z)$ can be computed in $O(n^8)$ time.   □

Acknowledgement. The author appreciate constructive comments and tolerant support of Prof. Imai. She is grateful to Prof. Avis for pointing out the incompleteness in the prototype of the proof. She would like to thank Prof. Ohsugi and Dr. Shibuta for fruitful discussions. This work is supported by the Grant-in-Aid for JSPS Fellows.

References

1. Abe, M., Sugimoto, Y., Namikawa, T., Morita, K., Oyabu, N., Morita, S.: Drift-compensated data acquisition performed at room temperature with frequecy modulation atomic force microscopy. Applied Physics Letters 90, 203103 (2007)
2. Ayala, R., Domínguez, E., Márquez, A., Quintero, A.: On the graphs which are the edge of a plane tiling. Mathematica Scandinavica 77, 5–16 (1995)
3. Chepoi, V., Deza, M., Grishukhin, V.: Clin d'oeil on L_1-embeddable planar graphs. Discrete Applied Mathematics 80(1), 3–19 (1997)
4. Cohen, E., Megiddo, N.: Recognizing properties of periodic graphs. Applied Geometry and Discrete Mathematics 4, 135–146 (1991)
5. Delgado-Friedrichs, O., O'Keeffe, M.: Crystal nets as graphs: Terminology and definitions. Journal of Solid State Chemistry 178, 2480–2485 (2005)
6. Deza, M., Grishukhin, V., Shtogrin, M.: Scale-Isometric Polytopal Graphs in Hypercubes and Cubic Lattices, ch. 9. World Scientific Publishing Company (2004)
7. Deza, M., Laurent, M.: Geometry of Cuts and Metrics, ch. 21. Springer (1997)
8. Fu, N., Hashikura, A., Imai, H.: Geometrical treatment of periodic graphs with coordinate system using axis-fiber and an application to a motion planning. In: Proceedings of the Ninth International Symposium on Voronoi Diagrams in Science and Engineering, pp. 115–121 (2012)
9. Höfting, F., Wanke, E.: Minimum cost paths in periodic graphs. SIAM Journal on Computing 24(5), 1051–1067 (1995)
10. Iwano, K., Steiglitz, K.: Optimization of one-bit full adders embedded in regular structures. IEEE Transaction on Acoustics, Speech and Signal Processing 34, 1289–1300 (1986)
11. Iwano, K., Steiglitz, K.: Planarity testing of doubly periodic infinite graphs. Networks 18, 205–222 (1988)
12. Karp, R., Miller, R., Winograd, A.: The organization of computations for uniform recurrence equiations. Journal of the ACM 14, 563–590 (1967)
13. Maehara, R.: The Jordan curve theorem via the Brouwer fixed point theorem. The American Mathematical Monthly 91(10), 641–643 (1984)
14. Orlin, J.B.: A dual version of Tardos's algorithm for linear programming. Operations Research Letters 5(5), 221–226 (1985)
15. Schrijver, A.: Theory of linear and integer programming. John Wiley & Sons, Inc. (1986)
16. Seymour, P.D.: Decomposition of regular matroids. Journal of Combinatorial Theory (B) 28, 305–359 (1980)
17. Sturmfels, B.: Gröbner Basis and Convex Polytopes. American Mathematical Society (1995)
18. Tardos, É.: A strongly polynomial algorithm to solve combinatorial linear programs. Operation Research 34, 250–256 (1986)

Cubic Augmentation of Planar Graphs

Tanja Hartmann*, Jonathan Rollin, and Ignaz Rutter

Karlsruhe Institute of Technology (KIT)
{firstname.lastname}@kit.edu

Abstract. In this paper we study the problem of augmenting a planar graph such that it becomes 3-regular and remains planar. We show that it is NP-hard to decide whether such an augmentation exists. On the other hand, we give an efficient algorithm for the variant of the problem where the input graph has a fixed planar (topological) embedding that has to be preserved by the augmentation. We further generalize this algorithm to test efficiently whether a 3-regular planar augmentation exists that additionally makes the input graph connected or biconnected.

1 Introduction

An *augmentation* of a graph $G = (V, E)$ is a set $W \subseteq E^c$ of edges of the complement graph. The *augmented graph* $G' = (V, E \cup W)$ is denoted by $G + W$. We study several problems where the task is to augment a given planar graph to be 3-regular while preserving planarity. The problem of augmenting a graph such that the resulting graph has some additional properties is well-studied and has applications in network planning [4]. Often the goal is to increase the connectivity of the graph while adding few edges. Nagamochi and Ibaraki [9] study the problem making a graph biconnected by adding few edges. Watanabe and Nakamura [13] give an $O(c \min\{c, n\} n^4 (cn + m))$ algorithm for minimizing the number of edges to make a graph c-edge-connected. The problem of biconnecting a graph at minimum cost is NP-hard, even if all weights are in $\{1, 2\}$ [9]. Motivated by graph drawing algorithms that require biconnected input graphs, Kant and Bodlaender [8] initiated the study of augmenting the connectivity of planar graphs, while preserving planarity. They show that minimizing the number of edges for the biconnected case is NP-hard and give efficient 2-approximation algorithms for both variants. Rutter and Wolff [11] give a corresponding NP-hardness result for planar 2-edge connectivity and study the complexity of geometric augmentation problems, where the input graph is a plane geometric graph and additional edges have to be drawn as straight-line segments. On plane geometric graph augmentation see also [7]. Abellanas et al. [1], Tóth [12] and Al-Jubeh et al. [2] give upper bounds on the number of edges required to make a plane straight-line graph c-connected for $c = 2, 3$.

We study the problem of augmenting a graph to be 3-regular while preserving planarity. In doing so, we additionally seek to raise the connectivity as much as possible. Specifically, we study the following problems.

Problem: PLANAR 3-REGULAR AUGMENTATION (PRA)
Instance: Planar graph $G = (V, E)$
Task: Find an augmentation W such that $G + W$ is 3-regular and planar.

* Partially supported by the DFG under grant WA 654/15 within the Priority Programme "Algorithm Engineering".

K.-M. Chao, T.-s. Hsu, and D.-T. Lee (Eds.): ISAAC 2012, LNCS 7676, pp. 402–412, 2012.
© Springer-Verlag Berlin Heidelberg 2012

Problem: FIXED-EMBEDDING PLANAR 3-REGULAR AUGMENTATION (FERA)
Instance: Planar graph $G = (V, E)$ with a fixed planar (topological) embedding
Task: Find an augmentation W such that $G + W$ is 3-regular, planar, and W can be added
in a planar way to the fixed embedding of G.

Moreover, we study *c-connected* FERA, for $c = 1, 2$, where the goal is to find a solution
to FERA, such that the resulting graph additionally is c-connected.

Contribution and Outline. Using a modified version of an NP-hardness reduction by
Rutter and Wolff [11], we show that PRA is NP-hard; the proof is in the full paper [6].

Theorem 1. PRA *is NP-complete, even if the input graph is biconnected.*

Our main result is an efficient algorithm for FERA and c-connected FERA for $c = 1, 2$. We note that Pilz [10] has simultaneously and independently studied the planar 3-regular augmentation problem. He showed that it is NP-hard and posed the question on the complexity if the embedding is fixed. Our hardness proof strengthens his result (to biconnected input graphs) and our algorithmic results answer his open question. In the full paper [6] we further show that for $c = 3$ c-connected FERA is again NP-hard.

We introduce basic notions used throughout the paper in Section 2. We present our results on FERA in Section 3. The problem is equivalent to finding a *node assignment* that assigns the vertices with degree less than 3 to the faces of the graph, such that for each face f an augmentation exists that can be embedded in f in a planar way and raises the degrees of all its assigned vertices to 3. We completely characterize these assignments and show that their existence can be tested efficiently. We strengthen our characterizations to the case where the graph should become c-connected for $c = 1, 2$ in Section 4 and show that our algorithm can be extended to incorporate these constraints. Proofs omitted due to space constraints can be found in the full version [6].

2 Preliminaries

A graph $G = (V, E)$ is *3-regular* if all vertices have degree 3. It is a *maxdeg-3 graph* if all vertices have at most degree 3. For a vertex set V, we denote by $V^{⓪}, V^{①}$ and $V^{②}$ the set of vertices with degree 0, 1 and 2, respectively. For convenience, we use $V^{⊖} = V^{⓪} \cup V^{①} \cup V^{②}$ to denote the set of vertices with degree less than 3. Clearly, an augmentation W such that $G + W$ is 3-regular must contain $3 - i$ edges incident to a vertex $V^{⊖}$. We say that a vertex $v \in V^{⊖}$ has $3 - i$ *(free) valencies* and that an edge of an augmentation incident to v *satisfies* a valency of v. Two valencies are adjacent if their vertices are adjacent.

Recall that a graph G is *connected* if it contains a path between any pair of vertices, and it is *c-(edge)-connected* if it is connected and removing any set of at most $c - 1$ vertices (edges) leaves G connected. A 2-connected graph is also called *biconnected*. We note that the notions of c-connectivity and c-edge-connectivity coincide on maxdeg-3 graphs. Hence a maxdeg-3 graph is biconnected if and only if it is connected and does not contain a *bridge*, i.e., an edge whose removal disconnects the graph.

A graph is *planar* if it admits a *planar embedding* into the Euclidean plane, where each vertex (edge) is mapped to a distinct point (Jordan curve between its endpoints) such that curves representing distinct edges do not cross. A planar embedding of a graph subdivides the Euclidean plane into *faces*. When we seek a planar augmentation preserving a fixed embedding, we require that the additional edges can be embedded into these faces in a planar way.

3   Planar 3-Regular Augmentation with Fixed Embedding

In this section we study the problem FERA of deciding for a graph $G = (V, E)$ with fixed planar embedding, whether there exists an augmentation W such that $G + W$ is 3-regular and the edges in W can be embedded into the faces of G in a planar way.

An augmentation W is *valid* only if the endpoints of each edge in W share a common face in G. We assume that a valid augmentation is associated with a (not necessarily planar) embedding of its edges into the faces of G such that each edge is embedded into a face shared by its endpoint. A valid augmentation is *planar* if the edges can be further embedded in a planar way into the faces of G.

Let F denote the set of faces of G and recall that $V^\oplus$ is the set of vertices with free valencies. A *node assignment* is a mapping $A \colon V^\oplus \to F$ such that each $v \in V^\oplus$ is incident to $A(v)$. Each valid 3-regular augmentation W induces a node assignment by assigning each vertex v to the face where its incident edges in W are embedded: this is well-defined since vertices in $V^\oplus \cup V^\oplus$ are incident to a single face. A node assignment A is *realizable* if it is induced by a valid augmentation W. If W is also planar, A is further *realizable in a planar way*. We call the corresponding augmentation a *realization*. A realizable node assignment can be found efficiently by computing a matching in the subgraph of G^c that contains edges only between vertices that share a common face. The existence of such a matching is a necessary condition for the existence of a planar realization. The main result of this section is that this condition is also sufficient.

Both valid augmentations and node assignments are local by nature, and can be considered independently for distinct faces. Let A be a node assignment and let f be a face. We denote by V_f the vertices that are assigned to f. We say that A is *realizable for f* if there exists an augmentation $W_f \subseteq \binom{V_f}{2}$ such that in $G + W_f$ all vertices of V_f have degree 3. It is *realizable for f in a planar way* if additionally W_f can be embedded in f without crossings. We call the corresponding augmentations *(planar) realizations for f*. The following lemma is obtained by glueing (planar) realizations for all faces.

Lemma 1. *A node assignment is realizable (in a planar way) for a graph G if and only if it is realizable (in a planar way) for each face f of G.*

Note that a node assignment induces a unique corresponding assignment of free valencies, and we also refer to the node assignment as assigning free valencies to faces. In the spirit of the notation $G + W$ we use $f + W_f$ to denote the graph $G + W_f$, where the edges in W_f are embedded into the face f. If W_f consists of a single edge e, we write $f + e$. For a fixed node assignment A we sometimes consider an augmentation W_f that realizes A for f only in parts by allowing that some vertices assigned to f have still a degree less than 3 in $f + W_f$. We then seek an augmentation W_f' such that $W_f \cup W_f'$ forms a realization of A for f. We interpret A as a node assignment for $f + W_f$ that assigns to f all vertices that were originally assigned to f by A and do not yet have degree 3 in $f + W_f$. Observe that in doing so, we still assign to the faces of G but when considering free valencies and adjacencies, we consider $G + W_f$.

3.1 (Planarly) Realizable Assignments for a Face

Throughout this section we consider an embedded graph G together with a fixed node assignment A and a fixed face f of G. The goal of this section is to characterize when A is realizable (in a planar way) for f. We first collect some necessary conditions for a realizable assignment.

Condition 1 (parity). *The number of free valencies assigned to f is even.*

Furthermore, we list certain *indicator sets* of vertices assigned to f that demand additional valencies outside the set to which they can be matched, as otherwise an augmentation is impossible. Note that these sets may overlap.

(1) **Joker:** A vertex in $V^{\textcircled{2}}$ whose neighbors are not assigned to f demands *one* valency.
(2) **Pair:** Two adjacent vertices in $V^{\textcircled{2}}$ demand *two* valencies.
(3) **Leaf:** A vertex in $V^{\textcircled{1}}$ whose neighbor has degree 3 demands *two* valencies from two *distinct vertices*.
(4) **Branch:** A vertex in $V^{\textcircled{1}}$ and an adjacent vertex in $V^{\textcircled{2}}$ demand *three* valencies from at least *two distinct vertices* with at most one valency adjacent to the vertex in $V^{\textcircled{2}}$.
(5) **Island:** A vertex in $V^{\textcircled{1}}$ demands *three* valencies from *distinct vertices*.
(6) **Stick:** Two adjacent vertices of degree 1 demand *four* valencies of which *at most two* belong to the *same vertex*.
(7) Two vertices in $V^{\textcircled{2}}$ demand *four* valencies; at most two from the *same vertex*.
(8) **3-cycle:** A cycle of three vertices in $V^{\textcircled{2}}$ demands *three* valencies.

Condition 2 (matching). *The demands of all indicator sets formed by vertices assigned to f are satisfied.*

Each indicator set contains at most three vertices and provides at least the number of valencies it demands; only sets of type (7) provide more. The demand of a joker is implicitly satisfied by the parity condition. We call an indicator set with maximum demand *maximum indicator set*, and we denote its demand by $k_{\max}$. Note that $k_{\max} \leq 4$. We observe that inserting edges does not increase $k_{\max}$; see Observation 1 in the full paper [6]. Lemma 2 reveals the special role of maximum indicator sets. While the necessity of the parity and the matching condition is obvious, Theorem 2 states that they are also sufficient for a node assignment to be realizable for f.

Lemma 2. *Let S be a maximum indicator set in f. Then A satisfies the matching condition for f if and only if the demand of S is satisfied.*

Theorem 2. *A is realizable for f $\Leftrightarrow$ A satisfies the parity and matching condition for f.*

Sketch of proof. The case that A assigns less than seven vertices to f is handled by a case distinction; see Lemma 3 in [6]. Assume A assigns at least seven vertices to f and satisfies the parity and the matching condition for f. Suppose there exists a partial augmentation W_1 of f such that A still assigns $k \geq 6$ vertices to $f + W_1$ and each assigned vertex is in $V^{\textcircled{2}}$. Consider the graph H^c that consists of the vertices assigned to $f + W_1$ and contains an edge if and only if the endpoints are not adjacent in $f + W_1$. Since each vertex assigned to $f + W_1$ is in $V^{\textcircled{2}}$, it has at most two adjacencies in $f + W_1$ and at least $k - 1 - 2 \geq k/2$ (for $k \geq 6$) adjacencies in H^c. Thus, by a theorem of Dirac [3], a Hamiltonian cycle exists in H^c, which induces a perfect matching W_2 of the degree-2 vertices in $f + W_1$. Hence $W_1 \cup W_2$ is a 3-regular augmentation for f. The proof proceeds by showing that such a partial augmentation W_1 always exists, distinguishing cases on the number of assigned vertices in $V^\textcircled{\tiny 2}$ and $V^\textcircled{\tiny 1}$. □

Given a node assignment A that satisfies the parity and the matching condition for a face f, the following rule picks an edge that can be inserted into f. Lemma 4 states that afterwards the remaining assignment still satisfies the parity and the matching condition. Iteratively applying Rule 1 hence yields a (not necessarily planar) realization.

Rule 1.   *1. If $k_{max} \geq 3$ let S denote a maximum indicator set. Choose a vertex u of lowest degree in S and connect this to an arbitrary assigned vertex $v \notin S$.*
2. If $k_{max} = 2$ and u is a leaf, choose $S = \{u\}$, and connect u to an assigned vertex v.
3. If $k_{max} = 2$ and there is no leaf, let S denote a path xuy of assigned vertices in $V^\textcircled{\tiny 2}$. Connect u to an arbitrary assigned vertex $v \notin S$.
4. If $k_{max} = 2$ and there is neither a leaf nor a path of three assigned vertices in $V^\textcircled{\tiny 2}$, let S denote a pair uw. Connect u to an arbitrary assigned vertex $v \notin S$.
5. If $k_{max} = 1$, choose $S = \{u\}$, where u is a joker, and connect u to another joker v.

Lemma 4. *Assume A satisfies the parity and matching condition for f and let e denote an edge chosen according to Rule 1. Then A satisfies the same conditions for $f + e$.*

Sketch of proof. We use Theorem 2 and rather show that there exists a realization of A for f that contains e. Since A satisfies the parity and the matching condition for f, there exists a realization W_f for f. If $e \in W_f$, we are done. Otherwise, recall that besides the edge $e = uv$ the rule also determines a set S (which except for subrule 3 is a maximum indicator set). We remove from W_f all edges incident to S and, if afterwards v still has degree 3, we remove an additional arbitrary edge incident to v. Then we insert the edge e, and it remains to show that the small and quite restricted problem of having assigned to f only the vertices that have free valencies after these steps admits a solution. To this end, we show that the demand of any maximum indicator set of this small instance is satisfied. This is done by a case distinction on k_{max} of the small instance, using the fact that (in most cases) S was a maximum indicator set in f. □

Our next goal is to extend this characterization and the construction of the assignment to the planar case. Consider a path of degree-2 vertices that are incident to two distinct faces f and f' but are all assigned to f. Then a *planar* realization for f may not connect any two vertices of the path. Hence the following sets of vertices demand additional valencies, which gives a new condition.

(1) A path π of $k > 2$ assigned degree-2 vertices that are incident to two distinct faces (end vertices not adjacent) demands either k further valencies or at least one valency from a different connected component.

(2) A cycle π of $k > 3$ assigned degree-2 vertices that are incident to two distinct faces demands either k further valencies or at least two valencies from two distinct connected components different from π.

Condition 3 (planarity). *The demand of each path of $k > 2$ and each cycle of $k > 3$ degree-2 vertices that are incident to two faces and that are assigned to f, is satisfied.*

Obviously, the planarity condition is satisfied if and only if the demand of a longest such path or cycle is satisfied. We prove for a node assignment A and a face f that the parity, matching, and planarity condition together are necessary and sufficient for the existence of a planar realization for a face f. To construct a corresponding realization we give a refined selection rule that iteratively chooses edges that can be embedded in f, such that the resulting augmentation is a planar realization of A for f. The new rule considers the demands of both maximum paths and cycles and maximum indicator sets, and at each moment picks a set with highest demand. If an indicator set is chosen, essentially Rule 1 is applied. However, we exploit the freedom to choose the endpoint v of $e = uv$ arbitrarily, and choose v either from a different connected component incident to f (if possible) or by a right-first (or left-first) search along the boundary of f. This guarantees that even if inserting the edge uv splits f into two faces f_1 and f_2, one of them is incident to all vertices that are assigned to f. Slightly overloading notation, we denote this face by $f + e$ and consider all remaining valencies assigned to it. We show in Lemma 5 that A then satisfies all three conditions for $f + e$ again.

Rule 2. *Phase 1: Different connected components assign valencies to f.*

1. *If there exists a path (or cycle) of more than $k_{\max}$ assigned degree-2 vertices, let u denote the middle vertex $v_{\lceil k/2 \rceil}$ of the longest such path (or cycle) $\pi = v_1, \ldots v_k$. Connect u to an arbitrary assigned vertex v in another component.*

2. *If all paths (or cycles) of assigned degree-2 vertices have length at most $k_{\max}$, apply Rule 1, choosing the vertex v in another component.*

Phase 2: All assigned valencies are on the same connected component. Consider only paths of assigned degree-2 vertices that are incident to two distinct faces:

1. *If there exists a path that is longer than $k_{\max}$, let u denote the right endvertex v_k of the longest path $\pi = v_1, \ldots v_k$. Choose v as the first assigned vertex found by a right-first search along the boundary of f, starting from u.*

2. *If all paths have length at most $k_{\max}$, apply Rule 1, choosing v as follows: Let v_1, v_2 denote the first assigned vertices not adjacent to u found by a left- and right-first search along the boundary of f, starting at u. If S is a branch and one of v_1, v_2 has degree 2, choose it as v. In all other cases choose $v = v_1$.*

Lemma 5. *Assume A satisfies the parity, matching, and planarity condition for f and let e be an edge chosen according to Rule 2. Then A satisfies all conditions also for $f + e$.*

Sketch of proof. Clearly, the parity condition is always preserved. In both phases, when applying subrule 1, after connecting one vertex of a path of length k to a vertex of a different connected component, the remainder of this path still provides $k - 1$ free valencies. This is enough to satisfy both the planarity condition and the matching condition for any other set of vertices disjoint from this path. Hence the matching condition and the planarity condition are preserved.

In subrule 2 of both phases, the matching condition follows directly from the correctness of Rule 1 (Lemma 4). For the planarity condition observe that since $k_{\max} \leq 4$ is at least as large as the longest path or cycle in the planarity condition, all these paths and cycles are relatively short. The matching condition and the fact that the planarity condition holds for f then imply that a sufficient number of valencies is provided to ensure the planarity condition for $f + e$. □

Given a node assignment A and a face f satisfying the parity, matching, and planarity condition, iteratively picking edges according to Rule 2 hence yields a planar realization of A for f. Applying this to every face yields the following theorem.

Theorem 3. *There exists a planar realization W of A if and only if A satisfies for each face the parity, matching, and planarity condition; W can be computed in $O(n)$ time.*

3.2 Globally Realizable Node Assignments and Planarity

In this section we show how to compute a node assignment that is realizable in a planar way if one exists. By Theorem 3, this is equivalent to finding a node assignment satisfying for each face the parity, matching, and planarity condition. In a first step, we show that the planarity condition can be neglected as an assignment satisfying the other two conditions can always be modified to additionally satisfy the planarity condition.

Lemma 6. *Given a node assignment A that satisfies the parity and matching condition for all faces, a node assignment A' that additionally satisfies the planarity condition can be computed in $O(n)$ time.*

Sketch of proof. To produce A' from A we traverse all faces. Let f be a face and let $\pi = v_1, \ldots, v_k$ be a path (or a cycle) in $V^\oslash$ that is incident to f and another face f'. Further assume that π violates the planarity condition for f. Let $u = v_1$ and choose $v = v_{\lceil (k+2)/2 \rceil}$. We reassign u and v to f'. Note that the edge uv can be embedded in a planar way into f' such that all remaining vertices assigned to f' again share a common face. Hence reassigning preserves all conditions of f'. We argue that it also ensures the planarity condition for f. Namely, the remaining vertices of π provide $k - 2$ free valencies. Since π violated the planarity condition all other paths in $V^\oslash$ incident to two faces demand at most $k - 1$ free valencies; the missing valency exists due to the parity condition. Similarly, since we reassigned the middle vertex, the two subpaths of π remaining assigned to f satisfy each others demands, up to one valency given by the parity condition. A similar argument holds for the matching condition since a maximum indicator set is either contained in π or it is disjoint, but then π provides enough free valencies.   □

Lemma 6 and Theorem 3 together imply the following characterization.

Theorem 4. *G admits a planar 3-regular augmentation if and only if it admits a node assignment that satisfies for all faces the parity and matching condition.*

To find a node assignment satisfying the parity and matching condition, we compute a (generalized) perfect matching in the following (multi-)graph $G_A = (V^\oslash, E')$, called *assignment graph*. It is defined on $V^\oslash$, and the demand of a vertex in $V^\oslash$ is $3 - i$ for $i = 0, 1, 2$. For a face f let $V_f^\oslash \subseteq V^\oslash$ denote the vertices incident to f. For each face f of G, G_A contains the edge set $E_f = \binom{V_f^\oslash}{2} \setminus E$, connecting non-adjacent vertices in $V^\oslash$ that share the face f. We seek a perfect (generalized) matching M of G_A satisfying exactly the demands of all vertices. The interpretation is that we assign a vertex v to a face f if and only if M contains an edge incident to v that belongs to E_f. It is not hard to see that for each face f the edges in $M \cap E_f$ are a (non-planar) realization of this assignment, implying the parity condition and the matching condition; the converse holds too.

Lemma 7. *A perfect matching of G_A corresponds to a node assignment that satisfies the parity and matching condition for all faces, and vice versa.*

Since testing whether the assignment graph admits a perfect matching can be done in $O(n^{2.5})$ time [5], this immediately implies the following theorem.

Theorem 5. FERA *can be solved in* $O(n^{2.5})$ *time.*

4 C-Connected FERA

In this section, we extend our results to testing for augmentations that additionally make the input graph connected or biconnected. We start with the connected case. Observe that an augmentation makes G connected if and only if in each face all incident connected components are connected by the augmentation. We characterize the node assignments admitting such *connected realizations* and modify the assignment graph from the previous section to yield such assignments.

Let $G = (V, E)$ be a planar graph with a fixed planar embedding, let f be a face of G, and let z_f denote the number of connected components incident to f. Obviously, an augmentation connecting all these components must contain at least a spanning tree on these components, which consists of $z_f - 1$ edges. Thus the following *connectivity condition* is necessary for a node assignment to admit a connected realization for f.

Condition 4 (connectivity).
(1) If $z_f > 1$, *each connected component incident to* f *must have at least one vertex assigned to* f.
(2) The number of valencies assigned to f *must be at least* $2z_f - 2$.

It is not difficult to see that this condition is also sufficient (both in the planar and in the non-planar case) since both Rule 1 and Rule 2 gives us freedom to choose the second vertex v arbitrarily. We employ this degree of freedom to find a connected augmentation by choosing v in a connected component distinct from the one of u, which is always possible due to the connectivity condition.

Theorem 6. *There exists a connected realization* W *of* A *if and only if* A *satisfies the parity, matching, and connectivity condition for all faces. Moreover,* W *can be chosen in a planar way if and only if* A *additionally satisfies the planarity condition for all faces. Corresponding realizations can be computed in* $O(n)$ *time.*

The following corollary follows from Theorem 3 by showing that the reassignment which establishes the planarity condition preserves the connectivity condition.

Corollary 1. *Given a node assignment* A *that satisfies the parity, matching and connectivity condition for all faces, a node assignment* A' *that additionally satisfies the planarity condition can be computed in* $O(n)$ *time.*

Corollary 1 and Theorem 6 together imply the following characterization.

Theorem 7. G *admits a connected planar 3-regular augmentation iff it admits a node assignment that satisfies the parity, matching and connectivity condition for all faces.*

We describe a modified assignment graph, the *connectivity assignment graph* G'_A, whose construction is such that there is a correspondence between the perfect matchings of G'_A and node assignments satisfying the parity, matching and connectivity condition.

To construct the connectivity assignment graph a more detailed look at the faces and how vertices are assigned, is necessary. A *triangle* is a cycle of three degree-2 vertices in G. An *empty triangle* is a triangle that is incident to a face that does not contain any further vertices. The set V_{in} (for inside) contains all vertices from $V^{①} \cup V^{①}$, all degree-2 vertices incident to bridges (they are all incident to only a single face), and all vertices of empty triangles (although technically they are incident to two faces, no augmentation edges can be embedded on the empty side of the triangle). We call the set of remaining vertices V_b (for boundary). We construct a preliminary assignment $\widetilde{A}$ that assigns the vertices in the set V_{in} of G whose assignment is basically unique. The remaining degree of freedom is to assign vertices in V_b to one of their incident faces. The connectivity assignment graph G'_A again has an edge set E'_f for each face f of G. Again the interpretation will be that a perfect matching M of G induces a node assignment by assigning to f all vertices that are incident to edges in $M \cap E'_f$.

If a face f is incident to a single connected component, we use for E'_f the ordinary assignment graph; the connectivity condition is trivial in this case. Now let f be a face with $z_f > 1$ incident connected components. For each component C incident to f that does not contain a vertex that is preassigned to f, we add a dummy vertex $v_{C,f}$ with demand 1 and connect it to all degree-2 vertices of C incident to f; this ensures connectivity condition (1). Let c_f denote the number of these dummy vertices, and note

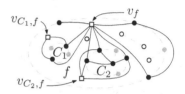

Fig. 1. Graph (dashed lines; preassigned vertices are empty) and its connectivity assignment graph (solid lines, dummy vertices as boxes)

that there are exactly c_f valencies assigned to f due to these dummy vertices. Let $\widetilde{a}_f$ denote the number of free valencies assigned by $\widetilde{A}$, and let $\widetilde{d}_f$ denote the number of valencies a maximum indicator set in f with respect to $\widetilde{A}$ misses. To ensure that the necessary valencies for the matching condition are present, we need that at least $\widetilde{d}_f - c_f$ vertices of V_b are assigned to f. For connectivity condition (2) we need at least $2z_f - 2 - \widetilde{a}_f - c_f$ such vertices assigned to f. We thus create a dummy vertex v_f whose demand is set to $s_f = \max\{2z_f - 2 - \widetilde{a}_f - c_f, \widetilde{d}_f - c_f, 0\}$, possibly increasing this demand by 1 to guarantee the parity condition. Finally, we wish to allow an arbitrary even number of vertices in V_b to be assigned to f. Since some valencies are already taken by dummy vertices, we do not just add to E'_f edges between non-adjacent vertices of V_b incident to f but for all such pairs. The valencies assigned by $\widetilde{A}$ and the dummy vertices satisfy the demand of any indicator set. Fig. 1 shows an example; for clarity edges connecting vertices in V_b are omitted in f and the outer face.

Lemma 8. *A perfect matching of G'_A (together with $\widetilde{A}$) induces a node assignment that satisfies parity, matching, and connectivity condition for all faces, and vice versa.*

Together with the previous observations this directly implies an algorithm for finding connected 3-regular augmentations.

Theorem 8. *Connected* FERA *can be solved in $O(n^{2.5})$ time.*

Biconnected case. For the biconnected case we consider the *bridge forest* B_f of a face f, whose nodes and edges correspond to the 2-edge connected components and bridges of G incident to f. A 2-edge connected component is a *leaf component* if its node in B_f has degree 1. To find an augmentation for a face f with z_f incident connected components, such that no bridges remain in f, the following condition is necessary.

Condition 5 (biconnectivity). *(1) If $z_f > 1$, each connected component incident to f must have at least two valencies assigned to f, and (2) each leaf component of f must assign at least one valency to f.*

Again condition 5 is also sufficient, and with similar techniques as in the connected case, a biconnectivity assignment graph can be constructed, whose perfect matchings correspond to node assignments satisfying the parity, matching, and biconnectivity condition. Again reassigning some valencies further ensures the planarity condition (see the full paper [6] for a proof).

Theorem 9. *Biconnected* FERA *can be solved in $O(n^{2.5})$ time.*

5   Conclusion

In this paper we have given efficient algorithms for deciding whether a given planar graph with a fixed embedding admits a 3-regular planar augmentation. We note that the running time of $O(n^{2.5})$ is due to the potentially quadratic size of our assignment graphs. Recently, we succeeded in constructing equivalent assignment graphs with only $O(n)$ edges. This immediately improves the running time of all our algorithms to $O(n^{1.5})$.

References

1. Abellanas, M., García, A., Hurtado, F., Tejel, J., Urrutia, J.: Augmenting the connectivity of geometric graphs. Comp. Geom. Theor. Appl. 40(3), 220–230 (2008)
2. Al-Jubeh, M., Ishaque, M., Rédei, K., Souvaine, D.L., Tóth, C.D.: Tri-Edge-Connectivity Augmentation for Planar Straight Line Graphs. In: Dong, Y., Du, D.-Z., Ibarra, O. (eds.) ISAAC 2009. LNCS, vol. 5878, pp. 902–912. Springer, Heidelberg (2009)
3. Dirac, G.A.: Some theorems on abstract graphs. Proceedings of the London Mathematical Society s3-2, 69–81 (1952)
4. Eswaran, K.P., Tarjan, R.E.: Augmentation problems. SIAM Journal on Computing 5(4), 653–665 (1976)
5. Gabow, H.N.: An efficient reduction technique for degree-constrained subgraph and bidirected network flow problems. In: STOC 1983, pp. 448–456. ACM (1983)
6. Hartmann, T., Rollin, J., Rutter, I.: Cubic augmentation of planar graphs. arXiv e-print (2012), http://arxiv.org/abs/1209.3865
7. Hurtado, F., Tóth, C.D.: Plane geometric graph augmentation: a generic perspective. In: Pach, J. (ed.) Thirty Essays on Geometric Graph Theory, vol. 29 (2012)
8. Kant, G., Bodlaender, H.L.: Planar Graph Augmentation Problems. In: Dehne, F., Sack, J.-R., Santoro, N. (eds.) WADS 1991. LNCS, vol. 519, pp. 286–298. Springer, Heidelberg (1991)
9. Nagamochi, H., Ibaraki, T.: Graph connectivity and its augmentation: applications of ma orderings. Discrete Applied Mathematics (1-3), 447–472 (2002)

10. Pilz, A.: Augmentability to cubic graphs. In: Proceedings of the 28th European Workshop on Computational Geometry (EuroCG 2012), pp. 29–32 (2012)
11. Rutter, I., Wolff, A.: Augmenting the connectivity of planar and geometric graphs. Journal of Graph Algorithms and Applications 16(2), 599–628 (2012)
12. Tóth, C.D.: Connectivity augmentation in plane straight line graphs. Electronic Notes in Discrete Mathematics 31, 49–52 (2008)
13. Watanabe, T., Nakamura, A.: Edge-connectivity augmentation problems. Journal of Computer and System Sciences 35(1), 96–144 (1987)

On the Number of Upward Planar Orientations
of Maximal Planar Graphs

Fabrizio Frati[1], Joachim Gudmundsson[1], and Emo Welzl[2,*]

[1] School of Information Technologies – The University of Sydney
brillo@it.usyd.edu.au,
joachim.gudmundsson@sydney.edu.au
[2] Institute of Theoretical Computer Science – ETH Zurich
emo@inf.ethz.ch

Abstract. We consider the problem of determining the maximum and the minimum number of upward planar orientations a maximal planar graph can have. We show that n-vertex maximal planar graphs have at least $\Omega(n \cdot 1.189^n)$ and at most $O(n \cdot 4^n)$ upward planar orientations. Moreover, there exist n-vertex maximal planar graphs having as few as $O(n \cdot 2^n)$ upward planar orientations and n-vertex maximal planar graphs having $\Omega(2.599^n)$ upward planar orientations.

1 Introduction

A drawing of a graph G in the plane is *upward* if every edge is represented by a y-monotone Jordan curve, and it is *planar* if no two such curves meet other than at common endpoints. An upward drawing induces an orientation of the edges of G (where each edge is oriented from the vertex with smaller y-coordinate to the one with larger y-coordinate) – this provides a directed graph $\mathbf{G}$. An orientation $\mathbf{G}$ of a planar graph G is *upward planar*, if there exists an upward planar drawing of G that induces $\mathbf{G}$. In this paper we study the number of possible upward planar orientations of an n-vertex planar graph; in fact, we concentrate on *maximal planar graphs* (also called *triangulations*, since all faces in any planar drawing are triangles).

Upward planarity is a natural generalization of planarity to directed graphs. When dealing with the visualization of directed graphs, one usually requires an *upward drawing*, i.e., a drawing such that each edge monotonically increases in the y-direction. As a consequence, there has been a lot of work on testing whether a directed graph admits an upward planar drawing (a directed graph that admits such a drawing is called *upward planar graph*) and on constructing upward planar drawings of directed graphs (see, e.g., [1,6]). Remarkable results in the area are that every upward planar graph is a subgraph of a *planar st-graph* and that every upward planar graph admits a *straight-line* upward planar drawing [4,9]. Here we study the minimum and the maximum number of upward planar orientations of maximal planar graphs providing the following results.

Theorem 1. *Every n-vertex maximal planar graph has at most $O(n \cdot 4^n)$ upward planar orientations. Moreover, there exists an n-vertex maximal planar graph that has*
$\Omega((23 + 3\sqrt{57})^{n/4}) = \Omega(2.599^n)$ *upward planar orientations.*

* Support from EuroCores/EuroGiga/ComPoSe SNF 20GG21 134318/1 is acknowledged.

K.-M. Chao, T.-s. Hsu, and D.-T. Lee (Eds.): ISAAC 2012, LNCS 7676, pp. 413–422, 2012.

Theorem 2. *Every n-vertex maximal planar graph has at least $\Omega(n \cdot 2^{n/4}) = \Omega(n \cdot 1.189^n)$ upward planar orientations. Moreover, there exists an n-vertex maximal planar graph that has $O(n \cdot 2^n)$ upward planar orientations.*

The proof of the upper bound in Theorem 1 relies on a "canonical ordering" for maximal upward planar graphs and on a counting argument that uses such a canonical ordering (Sect. 3). The proofs of the lower bound in Theorem 1 and of the upper bound in Theorem 2 are constructive, as they show maximal planar graphs with the claimed number of upward planar orientations (Sect. 4). The proof of the lower bound in Theorem 2 exploits the decomposition of a maximal planar graph G into outerplanar levels in order to construct, level by level, many upward planar orientations of G (Sect. 5).

2 Preliminaries

A *planar drawing* of a graph maps each vertex to a distinct point in the plane and each edge to a Jordan curve between its endpoints so that no two edges cross. A planar drawing partitions the plane into topologically connected regions, called *faces*. The bounded faces are called *internal*, while the unbounded face is the *outer face*. A *planar graph* is a graph admitting a planar drawing. A planar graph is *maximal* if no edge can be added to it while maintaining its planarity. Any two drawings of the same maximal planar graph determine the same faces; however, their outer faces might be different. An *internally-triangulated* planar graph is a planar graph with a fixed outer face whose every internal face is delimited by three edges.

Let G be a directed graph. A vertex v of G is a *source (sink)* if v has no incoming (outgoing) edges. A *monotone path* $P = (v_1, v_2, \ldots, v_k)$ is such that edge (v_i, v_{i+1}) is directed from v_i to v_{i+1}, for $i = 1, 2, \ldots, k - 1$. An *upward drawing* of a directed graph is such that each edge is represented by a curve monotonically increasing in the y-direction. An *upward planar graph* is a directed graph that admits an upward planar drawing. An *orientation* of a graph G is an assignment of directions to the edges of G. An orientation is *acyclic* if it contains no directed cycle. An orientation is *bimodal* if, in any planar embedding of G, the edges incident to each vertex v of G can be partitioned into two sets of consecutive edges, one containing all the edges outgoing v and one containing all the edges incoming v. An orientation of a graph G is *upward planar* if the resulting directed graph is upward planar. Observe that an upward planar orientation is both acyclic and bimodal. Two upward planar orientations G^1 and G^2 of a graph G are *distinct* if there exists an edge (u, v) in G which is directed from u to v in G^1 and from v to u in G^2. We say that a maximal planar graph G has a *fixed orientation for its outer face* if all the considered upward planar orientations of G lead to upward planar drawings in which the outer face is the same and its incident edges are oriented in the same way. If G is an internally-triangulated upward planar graph and (u, v, z) is a cycle delimiting an internal face of G, we say that (u, v, z) is a (u, z)-*monotone face* if u and z are the source and the sink of (u, v, z), respectively. If G is an internally-triangulated upward planar graph whose outer face is delimited by two monotone paths P_1 and P_2 connecting the unique source s of G and the unique sink t of G, the *leftmost path* (the *rightmost path*) of G is the one of P_1 and P_2 whose edges flow in clockwise (resp. counter-clockwise) direction along the outer face of G.

3   Upper Bound for Theorem 1

In this section we show a proof for the upper bound of Theorem 1. The proof is based on two ingredients. First, we show a "canonical" construction for upward planar orientations. That is, we prove that every internally-triangulated upward planar graph can be constructed starting from its leftmost path and repeatedly adding a single vertex or a single edge to the current graph while maintaining strong monotonicity properties. Such a construction is equivalent to a construction presented by Mehlhorn in [11]; still, for sake of completeness, we explicitly state it and prove its correctness in Lemma 1. Second, we use an inductive argument (and the canonical construction) to count the number of upward planar orientations of an internally-triangulated planar graph.

Lemma 1. *Let G be an internally-triangulated upward planar graph whose outer face is delimited by two monotone paths connecting the unique source s of G to the unique sink t of G. Then, there exists a sequence $G_1, G_2, \ldots, G_k$ of upward planar graphs such that: (1) G_1 coincides with the leftmost path of G; (2) G_k coincides with G; (3) for $1 \le i \le k$, all the vertices and edges in G and not in G_i lie in the outer face of G_i; (4) for $1 \le i \le k$, the rightmost path of G_i is a monotone path connecting s and t, and (5) for $2 \le i \le k$, G_i is obtained from G_{i-1} by*

- *either adding a vertex in the outer face of G_{i-1} and connecting it to two vertices in the rightmost path of G_{i-1},*
- *or adding an edge in the outer face of G_{i-1}.*

Proof: Properties (1)–(5) of the lemma are clearly satisfied if $i = 1$, given that G_1 coincides with the leftmost path of G. Next, assume that G_{i-1} satisfies Properties (1)–(5). We show how to construct G_i so that it also satisfies Properties (1)–(5). Denote by $(s = v_1, v_2, \ldots, v_m = t)$ the rightmost path of G_{i-1}.

Case 1: Suppose that, for some $1 \le j \le m - 2$, there exists an edge (v_j, v_{j+2}) in G such that: (a) (v_j, v_{j+2}) is not in G_{i-1}; and (b) (v_j, v_{j+1}, v_{j+2}) is a face of G. Then, let $G_i = G_{i-1} \cup \{(v_j, v_{j+2})\}$. Observe that G_i trivially satisfies Properties (1)–(2). Further, G_i satisfies Property (3) since G_{i-1} satisfies Property (3) and since (v_j, v_{j+1}, v_{j+2}) is a face of G. Moreover, G_i satisfies Property (4), since (a) its rightmost path is $(v_1, v_2, \ldots, v_j, v_{j+2}, v_{j+3}, \ldots, v_m)$, (b) G_{i-1} satisfies Property (4), and (c) edge (v_j, v_{j+2}) is outgoing v_j, as otherwise (v_j, v_{j+1}, v_{j+2}) would be a directed cycle in G. Finally, G_i satisfies Property (5) given that (v_j, v_{j+2}) is in the outer face of G_{i-1}, since G_{i-1} satisfies Property (3).

Case 2: Consider, for any $1 \le j \le m - 2$, a vertex u_j in G such that: (a) u_j is not in G_{i-1}; and (b) (v_j, v_{j+1}, u_j) is a (v_j, v_{j+1})-monotone face of G. The existence of such a vertex u_j implies that $G_i = G_{i-1} \cup \{u_j, (v_j, u_j), (u_j, v_{j+1})\}$ satisfies Properties (1)–(5). Namely, G_i trivially satisfies Properties (1)–(2). Further, G_i satisfies Property (3) since G_{i-1} satisfies Property (3) and since (v_j, v_{j+1}, u_j) is a face of G. Moreover, G_i satisfies Property (4), since (a) its rightmost path is $(v_1, v_2, \ldots, v_j, u_j, v_{j+1}, \ldots, v_m)$, (b) G_{i-1} satisfies Property (4), and (c) edges (v_j, u_j) and (u_j, v_{j+1}) are outgoing v_j and u_j, respectively. Finally, G_i satisfies Property (5) given that u_j is in the outer face of G_{i-1}, since G_{i-1} satisfies Property (3).

It remains to prove that, if Case 1 does not apply, then such a vertex u_j always exists. Consider the vertex u_1 of G that forms a face with edge (v_1, v_2) such that (v_1, v_2, u_1) is not a face of G_{i-1}. Since v_1 is the source of G, edge (v_1, u_1) is outgoing v_1. Hence, if edge (u_1, v_2) is outgoing u_1, then u_1 is the desired vertex. Otherwise, edge (v_2, u_1) is outgoing v_2. Since Case 1 does not apply, then $(v_2, u_1) \neq (v_2, v_3)$. Suppose that, for some $1 \leq j \leq m - 2$, a vertex u_j has been found such that (v_j, v_{j+1}, u_j) is not a face of G_{i-1} and is a (v_j, u_j)-monotone face of G, and such that $(v_{j+1}, u_j) \neq (v_{j+1}, v_{j+2})$. Then, consider the vertex u_{j+1} that forms a face with edge (v_{j+1}, v_{j+2}) such that $(v_{j+1}, v_{j+2}, u_{j+1})$ is not a face of G_{i-1}. Since $(v_{j+1}, u_{j+1}) \neq (v_{j+1}, v_j)$ and since (v_j, v_{j+1}, u_j) is a (v_j, u_j)-monotone face of G, edge (v_{j+1}, u_{j+1}) is outgoing v_{j+1}, as otherwise the upward planarity of G would be violated. Hence, if edge (u_{j+1}, v_{j+2}) is outgoing u_{j+1}, then u_{j+1} is the desired vertex, otherwise $(v_{j+1}, v_{j+2}, u_{j+1})$ is a (v_{j+1}, u_{j+1})-monotone face of G, with $(v_{j+2}, u_{j+1}) \neq (v_{j+2}, v_{j+3})$. Assuming that, for each $1 \leq j \leq m - 2$, vertex u_j is not the desired vertex, consider the vertex u_{m-1} that forms a face with edge (v_{m-1}, v_m) such that (v_{m-1}, v_m, u_{m-1}) is not a face of G_{i-1} and such that edge (v_{m-1}, u_{m-1}) is outgoing v_{m-1}. Since v_m is a sink of G, edge (u_{m-1}, v_m) is outgoing u_{m-1}, hence u_{m-1} is the desired vertex.   □

Now let G be an internally-triangulated planar graph. Let s and t be two consecutive vertices on the outer face of G. Let $P = (s = v_0, v_1, \ldots, v_k = t)$ be the path connecting s and t along the outer face of G different from edge (s, t). Let p be any integer such that $0 \leq p \leq k - 1$ and such that no internal edge (v_j, v_{j+2}) of G is such that (v_j, v_{j+1}, v_{j+2}) is an internal face of G, for any $0 \leq j \leq p - 2$. Let l be the number of internal vertices of G. Denote by $N(l, k, p)$ the number of upward planar orientations G of G such that: (1) The outer face of G is delimited by two monotone paths P and (s, t) connecting s and t; and (2) for $0 \leq j \leq p$, the internal face of G having (v_j, v_{j+1}) as an incident edge is not (v_j, v_{j+1})-monotone. We have the following:

Lemma 2. $N(l, k, p) \leq 4^l 2^{k-p}$.

Proof: The proof is by induction on l and, secondarily, on k. If $l = 0$, then, in any upward planar orientation G of G, every internal edge $(v_j, v_{j'})$ of G, with $j' \geq j$, is oriented from v_j to $v_{j'}$, as otherwise $(v_j, v_{j+1}, \ldots, v_{j'})$ would be a directed cycle. Hence G has a unique upward planar orientation. Then, the bound follows from $2^{k-p} > 1$, given that $k > p$. Suppose next that $l > 0$. By Lemma 1, either (Case 1) there exists an edge (v_j, v_{j+2}) such that (v_j, v_{j+1}, v_{j+2}) is a (v_j, v_{j+2})-monotone face of G or (Case 2) there exists a vertex u_j, for some $p \leq j \leq k - 1$, such that (v_j, v_{j+1}, u_j) is a (v_j, v_{j+1})-monotone face of G and Case 1 does not apply.

We discuss Case 1. By assumption, $j \geq p - 1$. Since (v_j, v_{j+2}) is outgoing v_j, the number of upward planar orientations of G satisfying Properties (1) and (2) is equal to the number of upward planar orientations of $G' = G \setminus \{v_{j+1}, (v_j, v_{j+1}), (v_{j+1}, v_{j+2})\}$ satisfying Properties (1) and (2), where $p' = p - 1$ if $j = p - 1$, and $p' = p$ otherwise. Observe that G' has $l' = l$ internal vertices and that the path delimiting the outer face of G' different from edge (s, t) has $k' = k - 1$ edges. No internal edge of G' connects two vertices among $(v_0, v_1, \ldots, v_{p'})$ given that no internal edge of G connects two vertices among $(v_0, v_1, \ldots, v_p)$. By induction on k, we have that G' (and hence G) has at most $N(l', k', p') \leq 4^l 2^{k-p}$ upward planar orientations satisfying Properties (1) and (2).

We now discuss Case 2. We partition the upward planar orientations of G satisfying Properties (1) and (2) into sets $S(p), S(p+1), \ldots, S(k-1)$ where, for each $p \leq j \leq k-1$, each upward planar orientation G of G in $S(j)$ is such that j is the minimum index for which there exists a vertex u_j for which (v_j, v_{j+1}, u_j) is a (v_j, v_{j+1})-monotone face of G. Observe that the number of upward planar orientations G of G satisfying Properties (1) and (2) is at most $|S(p)| + |S(p+1)| + \ldots + |S(k-1)|$. For any $p \leq j \leq k-1$, the number of upward planar orientations of G in $S(j)$ is equal to the number of upward planar orientations of $G' = G \setminus \{(v_j, v_{j+1})\}$ satisfying Properties (1) and (2), where $p' = j$. Namely, by assumption, in any of such orientations there exists no edge (v_h, v_{h+2}) such that (v_h, v_{h+1}, v_{h+2}) is an internal face of G, for any $0 \leq h \leq j-2$, and there exists no vertex u_h such that (v_h, v_{h+1}, u_h) is a (v_h, v_{h+1})-monotone face of G, for any $h < j$. Observe that G' has $l' = l - 1$ internal vertices and that the path delimiting the outer face of G' different from edge (s, t) has $k' = k + 1$ edges. By induction on l, $|S(j)| \leq N(l', k', p') \leq 4^{l-1} 2^{(k+1)-j}$. Thus the total number of upward planar orientations of G satisfying Properties (1) and (2) is at most $\sum_{j=p}^{k-1} |S(j)| \leq \sum_{j=p}^{k-1} 4^{l-1} 2^{(k+1)-j} = 4^{l-1} 2^{k-p+1} \sum_{j=0}^{k-p-1} 2^{-j} < 4^l 2^{k-p}$. $\square$

Lemma 2 implies the upper bound of Theorem 1. Namely, consider any n-vertex maximal planar graph G. Arbitrarily choose the outer face of G and fix an orientation for it (this can be done in $O(n)$ ways). By Lemma 2 with $l = n - 3$, $k = 2$, and $p = 0$, we have that G has at most 4^{n-2} upward planar orientations with a fixed orientation for its outer face, hence G has at most $O(n \cdot 4^n)$ upward planar orientations.

4   Upper Bound for Theorem 2 and Lower Bound for Theorem 1

In this section we show two classes of maximal planar graphs, the first one providing the upper bound in Theorem 2, the second one providing the lower bound in Theorem 1.

The class of maximal planar graphs providing the upper bound for Theorem 2 is the one of *planar 3-trees*. A planar 3-tree is inductively defined as follows (see Fig. 1(a)). A 3-cycle is the only planar 3-tree with 3 vertices; every planar 3-tree G_n with n vertices can be obtained from a planar 3-tree G_{n-1} with $n - 1$ vertices by inserting a vertex w inside an internal face (u, v, z) of G_{n-1} and connecting w with u, v, and z. We prove the following statement: Every n-vertex planar 3-tree with a fixed orientation for the outer face has 2^{n-3} upward planar orientations. The proof is by induction on n.

If $n = 3$ the statement holds since the outer face of G has a fixed orientation. Suppose that $n > 3$. Consider any n-vertex planar 3-tree G_n and let w be a degree-3 vertex of G_n whose removal yields an $(n-1)$-vertex planar 3-tree G_{n-1}. Let u, v, and z be the neighbors of w in G_n. Consider any upward planar orientation G_{n-1} of G_{n-1}. Assume w.l.o.g. that u is the source and z is the sink of 3-cycle (u, v, z). We claim that, in any upward planar orientation G_n of G_n in which the orientation of G_{n-1} is G_{n-1}, edge (u, w) is outgoing u. See Fig. 1(b)–(c). If u is the source of G_{n-1}, then u is the source of G_n, hence (u, w) is outgoing u in G_n. Otherwise, u contains an incoming edge (x, u). In order for G_n to be bimodal, edge (u, w) has to be oriented from u to w, given that edges $(x, u), (u, v), (u, w)$, and (u, z) appear in this circular order around u. An analogous argument proves that edge (w, z) is incoming z in G_n. On the other hand, the two orientations of (v, w) lead to two distinct upward planar orientations G_n^1

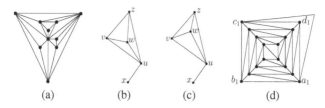

Fig. 1. (a) A planar 3-tree. (b)–(c) Orientations of the edges incident to a vertex w. (d) Graph G in the proof of the lower bound in Theorem 1.

and G_n^2 of G_n in which the orientation of G_{n-1} is $\boldsymbol{G}_{n-1}$. By induction G_{n-1} has 2^{n-4} upward planar orientations with a fixed orientation for its outer face, hence G_n has 2^{n-3} upward planar orientations with a fixed orientation for its outer face, thus proving the claim. Finally, every maximal planar graph has $O(n)$ choices for its outer face and, for each choice, the cycle delimiting the outer face has $O(1)$ orientations (observe that a planar 3-tree G_n can be constructed by iteratively adding a degree-3 internal vertex for every choice of the outer face of G_n). It follows that G has $O(n \cdot 2^n)$ upward planar orientations, thus proving the upper bound in Theorem 2.

Next, we are going to prove the lower bound in Theorem 1.

Let G be the n-vertex maximal planar graph, where n is a multiple of 4, defined as follows (see Fig. 1(d)). For $i = 1, 2, \ldots, \frac{n}{4}$, G contains a cycle $C_i = (a_i, b_i, c_i, d_i)$; for $i = 1, 2, \ldots, \frac{n}{4} - 1$, G contains edges (a_i, a_{i+1}), (b_i, b_{i+1}), (c_i, c_{i+1}), (d_i, d_{i+1}), (a_i, b_{i+1}), (b_i, c_{i+1}), (c_i, d_{i+1}), and (d_i, a_{i+1}); finally, G contains edges (a_1, c_1) and $(a_{n/4}, c_{n/4})$. The outer face of G is delimited by cycle (a_1, b_1, c_1), that is oriented so that a_1 and c_1 are its source and sink, respectively. We are going to prove that G has $\Omega((23 + 3\sqrt{57})^{n/4})$ upward planar orientations.

An orientation $\boldsymbol{C_i}$ of C_i is of *type A* if $\boldsymbol{C_i}$ has one source and one sink, and such vertices are not adjacent in C_i; further, $\boldsymbol{C_i}$ is of *type B* if $\boldsymbol{C_i}$ has one source and one sink, and such vertices are adjacent in C_i; finally, $\boldsymbol{C_i}$ is of *type C* if $\boldsymbol{C_i}$ has two sources and two sinks. Denote by $C_{i,i+1}$ the subgraph of G induced by the vertices of C_i and by the vertices of C_{i+1}, for any $1 \le i \le \frac{n}{4} - 1$ (minus edge (a_1, c_1) if $i = 1$ and minus edge $(a_{n/4}, c_{n/4})$ if $i = n/4 - 1$). Denote by $N(j, k)$ the number of upward planar orientations of $C_{i,i+1}$ such that each edge of C_i has a fixed orientation of type j and the orientation of C_{i+1} is any orientation of type k, for each $j, k \in \{A, B, C\}$. The following inequalities hold: (1) $N(A, A) \ge 20$; (2) $N(A, B) \ge 30$; (3) $N(A, C) \ge 12$; (4) $N(B, A) \ge 14$; (5) $N(B, B) \ge 24$; (6) $N(B, C) \ge 10$; (7) $N(C, A) \ge 4$; (8) $N(C, B) \ge 8$; and (9) $N(C, C) \ge 4$. The proof of such inequalities can be conducted by exhibiting, for each $j, k \in \{A, B, C\}$, $N(j, k)$ upward planar orientations in which C_i has an orientation of type j, C_{i+1} has an orientation of type k, and C_i has exactly the same sources and sinks in each of the $N(j, k)$ orientations. Figures illustrating such orientations are shown in the extended version of the paper.

For each $1 \le i \le \frac{n}{4} - 1$, let C_i^* be the subgraph of G induced by the vertices in $C_{\frac{n}{4}-i}, C_{\frac{n}{4}-i+1}, \ldots, C_{\frac{n}{4}-1}$ (excluding edge (a_1, c_1) when $i = \frac{n}{4} - 1$). Observe that $C_{\frac{n}{4}-1}^*$ coincides with G minus vertices $a_{n/4}$, $b_{n/4}$, $c_{n/4}$, and $d_{n/4}$, and minus edge (a_1, c_1). Denote by $A(i)$, $B(i)$, and $C(i)$ the number of upward planar orientations of C_i^* such that $C_{\frac{n}{4}-i}$ has a fixed orientation of type A, B, and C, respectively. Clearly, we have $A(1) = B(1) = C(1) = 1$. From the lower bounds on $N(j, k)$ we immediately get the following: (a) $A(i) \geq 20A(i-1) + 30B(i-1) + 12C(i-1)$, (b) $B(i) \geq 14A(i-1) + 24B(i-1) + 10C(i-1)$, and (c) $C(i) \geq 4A(i-1) + 8B(i-1) + 4C(i-1)$.

We are going to prove that (d) $A(i) \geq at^{i-1}$, (e) $B(i) \geq bt^{i-1}$, and (f) $C(i) \geq ct^{i-1}$, for some constants $0 < a, b, c \leq 1$ and $t > 0$ to be determined later. Observe that (d), (e), and (f) hold true if $i = 1$ (given that $a, b, c \leq 1$). Suppose that (d), (e), and (f) hold · true for $i - 1$. We compute for which values of a, b, c, and t they hold true for i.

From (a), (b), and (c) we have that (d), (e), and (f) hold true if (g) $20at^{i-1} + 30bt^{i-1} + 12ct^{i-1} = at^i$, (h) $14at^{i-1} + 24bt^{i-1} + 10ct^{i-1} = bt^i$, and (i) $4at^{i-1} + 8bt^{i-1} + 4ct^{i-1} = ct^i$ hold true. Simplifying (g), (h), and (i), we get (j) $(20-t)a + 30b + 12c = 0$, (k) $14a + (24 - t)b + 10c = 0$, and (l) $4a + 8b + (4 - t)c = 0$.

Whenever equations (j), (k), and (l) are linearly independent, the only solution to such equalities is $a = b = c = 0$. However, when the determinant of the matrix associated with (j), (k), and (l) is equal to zero, one of such equations is a linear combination of the others. Simple calculations show that this happens if $t = 2, t = 23 - 3\sqrt{57}$, and $t = 23 + 3\sqrt{57}$. Focusing on the value $t = 23 + 3\sqrt{57}$, we get that (j) and (k) become (m) $(-3 - 3\sqrt{57})a + 30b + 12c = 0$ and (n) $14a + (1 - 3\sqrt{57})b + 10c = 0$. Solving (n) with respect to c we get (o) $c = -\frac{14a+(1-3\sqrt{57})b}{10}$. Plugging such a value in (m) and solving with respect to b we get (p) $b = \frac{33+5\sqrt{57}}{48+6\sqrt{57}}a$. From (p) and (o), we get (q) $c = \frac{15+\sqrt{57}}{48+6\sqrt{57}}a$. Hence, we have that $(a, b, c, t) = (1, \frac{33+5\sqrt{57}}{48+6\sqrt{57}}, \frac{15+\sqrt{57}}{48+6\sqrt{57}}, 23 + 3\sqrt{57})$ solves (j), (k), and (l). Observe that $a, b, c \leq 1$. Thus, it holds $A(i) \geq (23 + 3\sqrt{57})^{i-1}$, and hence graph $C_{\frac{n}{4}-1}^*$ admits $A(\frac{n}{4} - 1) \geq (23 + 3\sqrt{57})^{\frac{n}{4}-2}$ upward planar orientations with a fixed outer face of Type A having a_1 as a source and c_1 as a sink. In order to conclude the proof of the lower bound in Theorem 1, it suffices to observe that, for every upward planar orientation $C_{\frac{n}{4}-1}^*$ of $C_{\frac{n}{4}-1}^*$ with a fixed outer face of Type A having a_1 as a source and c_1 as a sink, it is possible to suitably orient edge (a_1, c_1) together with the edges incident to the vertices in G that do not belong to $C_{\frac{n}{4}-1}^*$ (that is, $a_{n/4}$, $b_{n/4}, c_{n/4}$, and $d_{n/4}$) in such a way that the resulting orientation of G is upward planar.

5   Lower Bound for Theorem 2

In this section we show a proof for the lower bound of Theorem 2.

Let G be any n-vertex maximal planar graph. For $i = 0, \ldots, t$, denote by G_i the subgraph of G induced by the vertices at graph-theoretic distance i from the outer face of G. Observe that the *outerplanarity* of G is $t + 1$. Also, for $i = 0, \ldots, t$, denote by G_i^* the subgraph of G induced by the vertices at graph-theoretic distance less than or equal to i from the outer face of G. Observe that $G_0^* = G_0$ and $G_t^* = G$. Let $n_i = |G_i|$.

In the following, whenever we say that we construct a certain number of upward planar drawings of a graph, we always mean that such upward planar drawings correspond to distinct upward planar orientations of the graph with a fixed outer face.

In order to prove the lower bound of Theorem 2, we exhibit two algorithms, Algorithm A and Algorithm B, that construct a set $S_A(t)$ and a set $S_B(t)$ of upward planar drawings of $G_t^* = G$, respectively. Both algorithms construct upward planar drawings one *outerplanar level* at a time, i.e., upward planar drawings of G_{i+1}^* are constructed

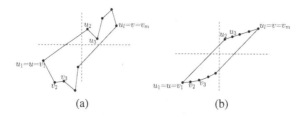

Fig. 2. (a) An x-monotone face. (b) An (x, y)-monotone face.

by plugging upward planar drawings of G_{i+1} into upward planar drawings of G_i^* (and drawing the edges connecting the vertices in G_i to the vertices in G_{i+1}). In fact, for each $0 \leq i \leq t$, Algorithm A and Algorithm B construct two sets $S_A(i)$ and $S_B(i)$ of upward planar drawings of G_i^*, respectively. Algorithm A and Algorithm B both maintain strong invariants on the geometry of the faces of G_i^* in any constructed upward planar drawing. Call *new face* any face that is delimited exclusively by edges of G_i.

Algorithm A maintains the invariant that, in every upward planar drawing of G_i^* in $S_A(i)$, all the new faces of G_i^* are *x-monotone*: Denote by u and v two vertices incident to any new face f of G_i^*, and denote by $(u = v_1 = u_1, u_2, \ldots, u_l = v = v_m, v_{m-1}, \ldots, v_2)$ the clockwise order of the vertices on the border of f; a drawing of f is *x-monotone* if it is either *positively x-monotone* or *negatively x-monotone*, where f is positively x-monotone if (see Fig. 2(a)): (1) $x(v_1) < x(v_2) < \ldots < x(v_{m-1}) < x(u_2) < x(u_3) < \ldots < x(u_{l-1}) < x(u_l)$ and (2) $y(v_1), y(v_2), \ldots, y(v_{m-1}) < y(u_2), y(u_3), \ldots, y(u_{l-1}), y(u_l)$; moreover, f is negatively x-monotone if: (1) $x(u_1) < x(u_2) < x(u_3) < \ldots < x(u_{l-1}) < x(v_2) < \ldots < x(v_{m-1}) < x(v_m)$ and (2) $y(v_2), y(v_3), \ldots, y(v_{m-1}), y(v_m) < y(u_1), y(u_2), y(u_3), \ldots, y(u_{l-1}), y(u_l)$.

Algorithm B maintains the invariant that, in every upward planar drawing of G_i^* in $S_B(i)$, all the new faces of G_i^* are (x, y)-*monotone*: Denote by u and v two vertices incident to any new face f of G_i^*, and denote by $(u = v_1 = u_1, u_2, \ldots, u_l = v = v_m, v_{m-1}, \ldots, v_2)$ the clockwise order of the vertices on the border of f; a drawing of f is (x, y)-*monotone* if it positively (x, y)-*monotone* or negatively (x, y)-*monotone*, where a drawing of f is positively (x, y)-monotone if it is positively x-monotone and further (see Fig. 2(b)): (3) $y(v_1) < y(v_2) < \ldots < y(v_{m-1}) < y(v_m)$ and (4) $y(u_1) < y(u_2) < \ldots < y(u_{l-1}) < y(u_l)$; a drawing of f is negatively (x, y)-monotone if it is negatively x-monotone and if: (3) $y(v_1) > y(v_2) > \ldots > y(v_{m-1}) > y(v_m)$ and (4) $y(u_1) > y(u_2) > \ldots > y(u_{l-1}) > y(u_l)$.

Algorithm A constructs the upward planar drawings in $S_A(i+1)$ by plugging upward planar drawings of G_{i+1} into each of the upward planar drawings of G_i^* that are in $S_A(i)$. Analogously, Algorithm B constructs the upward planar drawings in $S_B(i + 1)$ by plugging upward planar drawings of G_{i+1} into each of the upward planar drawings of G_i^* that are in $S_B(i)$.

Algorithms A and B construct the upward planar drawings in $S_A(t)$ and the upward planar drawings in $S_B(t)$, respectively, so that (at least) one of $S_A(t)$ and $S_B(t)$ contains "many" upward planar drawings of $G_t^* = G$. The overall idea behind this argument is as follows. Graph G_i, for $1 \leq i \leq t$, consists of several connected components $G_i^1, G_i^2, \ldots, G_i^{k(i)}$. For each $1 \leq i \leq t$ and for each $1 \leq j \leq k(i)$, if G_i^j is a tree, then denote by h_i^j the number of vertices of G_i^j, otherwise denote by h_i^j the number of bridges of G_i^j (where a *bridge* is a maximal biconnected component of G_i^j consisting of a single edge). Further, let $h_i = \sum_{j=1}^{k(i)} h_i^j$ and let $h = \sum_{i=1}^{t} h_i$. If h is "small", then $S_A(t)$ contains many upward planar drawings of G_t^*, while if h is "large", then $S_B(t)$ contains many upward planar drawings of G_t^*. This is formalized as follows.

Lemma 3. *Consider any upward planar drawing Γ_i^* of G_i^* such that all the new faces of G_i^* are x-monotone. Then, there exist $2^{(n_{i+1}-h_{i+1})/3}$ upward planar drawings of G_{i+1}^* such that, for each of such drawings Γ_{i+1}^*, all the new faces of G_{i+1}^* are x-monotone in Γ_{i+1}^* and such that the restriction of Γ_{i+1}^* to G_i^* coincides with Γ_i^*.*

Lemma 4. *Consider any upward planar drawing Γ_i^* of G_i^* such that all the new faces of G_i^* are (x,y)-monotone. Then, there exist $2^{h_{i+1}}$ upward planar drawings of G_{i+1}^* such that, for each of such drawings Γ_{i+1}^*, all the new faces of G_{i+1}^* are (x,y)-monotone in Γ_{i+1}^* and such that the restriction of Γ_{i+1}^* to G_i^* coincides with Γ_i^*.*

Algorithm A initializes $S_A(0)$ with one upward planar drawing, corresponding to the fixed orientation of G_0. For $0 \le i \le t - 1$, Algorithm A constructs $S_A(i + 1)$ by considering each upward planar drawing Γ_i^* in $S_A(i)$, whose new faces are x-monotone, and by inserting $2^{(n_{i+1}-h_{i+1})/3}$ upward planar drawings of G_{i+1}^* into $S_A(i+1)$ so that, for each of such drawings Γ_{i+1}^*, all the new faces of G_{i+1}^* are x-monotone in Γ_{i+1}^* and the restriction of Γ_{i+1}^* to G_i^* coincides with Γ_i^* (this can be done by Lemma 3).

Algorithm B initializes $S_B(0)$ with one upward planar drawing, corresponding to the fixed orientation of G_0. For $0 \le i \le t - 1$, Algorithm B constructs $S_B(i + 1)$ by considering each upward planar drawing Γ_i^* in $S_B(i)$, whose new faces are (x,y)-monotone, and by inserting $2^{h_{i+1}}$ upward planar drawings of G_{i+1}^* into $S_B(i + 1)$ so that, for each of such drawings Γ_{i+1}^*, all the new faces of G_{i+1}^* are (x,y)-monotone in Γ_{i+1}^* and the restriction of Γ_{i+1}^* to G_i^* coincides with Γ_i^* (this can be done by Lemma 4).

By Lemma 3, we have $|S_A(t)| = 2^{((n_1-h_1)+(n_2-h_2)+...+(n_t-h_t))/3}$. Moreover, by Lemma 4, we have $|S_B(t)| = 2^{h_1+h_2+...+h_t}$. Observe that $n_1+n_2+...+n_t = n-3$ (the only vertices of G that do not belong to any graph G_i, with $1 \le i \le t$, are the vertices in G_0). Hence, we have $|S_A(t)| = 2^{(n-h-3)/3}$ and $|S_B(t)| = 2^h$. Thus, if $h \le n/4$, we have that $|S_A(t)| \in \Omega(2^{n/4})$, while if $h > n/4$, we have that $|S_B(t)| \in \Omega(2^{n/4})$. Thus, one of $S_A(t)$ and $S_B(t)$ contains $\Omega(2^{n/4})$ upward planar drawings of distinct upward planar orientations of G with fixed outer face. Since there are $\Omega(n)$ choices for the outer face of G, the lower bound in Theorem 2 follows.

6   Conclusions

In this paper we considered the problem of determining the maximum and the minimum number of upward planar orientations a maximal planar graph can have. Tightening the bounds we provided in this paper is an interesting open problem. In particular, we suspect that a suitable combination of Algorithms A and B presented in Section 5 would lead to improve the lower bound in Theorem 2. Deep techniques that might be helpful for improving the bounds we presented in this paper are provided by the large body of literature on *bipolar orientations* of biconnected planar graphs (see, e.g., [3,5]).

Extending our results to general planar graphs is, in our opinion, worth research efforts. In particular, we pose the following question: Is it true that if an n-vertex planar graph G has x upward planar orientations then an n-vertex maximal planar graph exists containing G as a subgraph and having at least x upward planar orientations?

Any upward planar orientation of a graph is acyclic, thus our investigations relate to counting the number of acyclic orientations of a graph, which has a rich body of literature, with links to chromatic polynomials. Surprisingly, we did not find any work dealing with the asymptotic number of acyclic orientations of n-vertex planar graphs. Still, there is a general "trick" to derive an upper bound, due to Fredman (see [7] and [10]): Fixing the out-degree of each vertex of a planar graph G completely determines the orientation itself. This easily gives an upper bound of $\prod_v (\deg_v + 1)$ (where the sum is over all the vertices v of the graph and $\deg_v$ represented the degree of v). With the use of appropriate inequalities and the fact that planar graphs have average degree at most 6, this entails an upper bound of $O(7^n)$ for the number of acyclic orientations of a planar graph and an upper bound of $O(n^2 5^n)$ for the number of acyclic orientations with unique source and sink of a planar graph. Also observe that, if we count acyclic orientations of planar graphs with a unique source and multiple sinks, then this number is upper bounded by the number of spanning trees, see [8], which is at most $O(5.3^n)$, see [2]. It is also easy to see that there exist maximal planar graphs, namely planar 3-trees, having $\Theta(4^n)$ acyclic orientations. Deepening the study of the number of acyclic orientations a planar graph can have seems to us an interesting research direction.

References

1. Bertolazzi, P., Di Battista, G., Liotta, G., Mannino, C.: Upward drawings of triconnected digraphs. Algorithmica 12(6), 476–497 (1994)
2. Buchin, K., Schulz, A.: On the Number of Spanning Trees a Planar Graph Can Have. In: de Berg, M., Meyer, U. (eds.) ESA 2010, Part I. LNCS, vol. 6346, pp. 110–121. Springer, Heidelberg (2010)
3. de Fraysseix, H., de Mendez, P.O., Rosenstiehl, P.: Bipolar orientations revisited. Discr. Appl. Math. 56(2-3), 157–179 (1995)
4. Di Battista, G., Tamassia, R.: Algorithms for plane representations of acyclic digraphs. Theor. Comput. Sci. 61, 175–198 (1988)
5. Fusy, E., Poulalhon, D., Schaeffer, G.: Bijective counting of plane bipolar orientations. Elec. Notes Discr. Math. 29, 283–287 (2007)
6. Garg, A., Tamassia, R.: On the computational complexity of upward and rectilinear planarity testing. SIAM J. Comput. 31(2), 601–625 (2001)
7. Graham, R.L., Yao, A.C., Yao, F.F.: Information bounds are weak in the shortest distance problem. J. ACM 27(3), 428–444 (1980)
8. Kahale, N., Schulman, L.J.: Bounds on the chromatic polynomial and on the number of acyclic orientations of a graph. Combinatorica 16(3), 383–397 (1996)
9. Kelly, D.: Fundamentals of planar ordered sets. Disc. Math. 63(2-3), 197–216 (1987)
10. Manber, U., Tompa, M.: The effect of number of hamiltonian paths on the complexity of a vertex-coloring problem. SIAM J. Comput. 13(1), 109–115 (1984)
11. Mehlhorn, K.: Data Structures and Algorithms: Multi-dimensional Searching and Computational Geometry, vol. 3. Springer (1984)

Universal Point Subsets for Planar Graphs

Patrizio Angelini[1], Carla Binucci[2], William Evans[3], Ferran Hurtado[4],
Giuseppe Liotta[2], Tamara Mchedlidze[5], Henk Meijer[6], and Yoshio Okamoto[7]

[1] Roma Tre University, Italy
[2] University of Perugia, Italy
[3] University of British Columbia, Canada
[4] Universitat Politécnica de Catalunya, Spain
[5] Karlsruhe Institute of Technology, Germany
[6] Roosevelt Academy, Netherlands
[7] University of Electro-Communications, Japan

Abstract. A set S of k points in the plane is a *universal point subset* for a class $\mathcal{G}$ of planar graphs if every graph belonging to $\mathcal{G}$ admits a planar straight-line drawing such that k of its vertices are represented by the points of S. In this paper we study the following main problem: For a given class of graphs, what is the maximum k such that there exists a universal point subset of size k? We provide a $\lceil \sqrt{n} \, \rceil$ lower bound on k for the class of planar graphs with n vertices. In addition, we consider the value $F(n, \mathcal{G})$ such that *every* set of $F(n, \mathcal{G})$ points in general position is a universal subset for all graphs with n vertices belonging to the family $\mathcal{G}$, and we establish upper and lower bounds for $F(n, \mathcal{G})$ for different families of planar graphs, including 4-connected planar graphs and nested-triangles graphs.

1 Introduction

A classic result in graph theory states that every planar graph $G = (V, E)$ can be drawn without crossings on the plane using some set S of points as vertices, and straight-line segments with endpoints in S to represent the edges [7,16,21]. However, not every set S with $n \doteq |S| = |V|$ is suitable for such a representation; for example, the drawing is impossible if G is a maximal planar graph with $n > 3$ vertices and S is in convex position, because in this case $|E| = 3n - 6$ while at most $2n - 3$ segments can be drawn between points of S without crossings. In fact, Cabello [4] proved that deciding whether there is a planar straight-line drawing of $G = (V, E)$ using a point set S with $|S| = |V|$ is an NP-complete problem.

As the number of combinatorially different sets of n points is finite [8], it is obvious that there exist some adequate yet huge sets of points U, such that given any planar graph G with n vertices, some n-subset of U admits a planar straight-line drawing of G. The challenge though, is to find sets U with that property, yet as small as possible. We define next this problem more precisely.

A set U of k points in the plane is a *universal point set* if every planar graph with n vertices admits a planar straight-line drawing whose vertices are a subset of the points of U. From the literature it is known that if U is a universal point set for planar graphs

K.-M. Chao, T.-s. Hsu, and D.-T. Lee (Eds.): ISAAC 2012, LNCS 7676, pp. 423–432, 2012.
© Springer-Verlag Berlin Heidelberg 2012

then $1.235n \leq |U| \leq 8n^2/9$. Indeed, Kurowski [10] proved that the size of U requires at least $1.235n$ points, while de Fraysseix, Pach, and Pollack [5], Schnyder [15] and Brandenburg [3] showed that a $O(n) \times O(n)$ grid of points is a universal point set.

This topic has been a very active area of research since it was introduced, and several variations have been considered. For example, one can restrict the family of graphs to be represented. In this sense, Gritzman, Mohar, Pach and Pollack [9] proved that every set of n distinct points in the plane in general position (no three collinear) is universal for the class of outerplanar graphs with n vertices.

In this paper we introduce and study the notion of a *universal point subset*. A set S of k points is a *universal point subset* for a class $\mathcal{G}$ of planar graphs if every graph in $\mathcal{G}$ admits a planar straight-line drawing such that k of its vertices are represented by the points of S.

In Section 2 we prove that a particular, very flat convex chain of $\lceil \sqrt{n} \rceil$ points is a universal point subset for the class of (maximal) planar graphs with n vertices.

For a certain subfamily of 4-connected planar graphs we have been able to obtain a bound that is stronger in a particular sense, namely that *every* set of $\lceil \frac{\lg n}{4} \rceil$ points in general position is a universal point subset for all the graphs with n vertices in this family. Inspired by this result, we consider in Section 3 the value $F(n, \mathcal{G})$ such that every set of $F(n, \mathcal{G})$ points in general position is a universal subset for the planar graphs with n vertices belonging to $\mathcal{G}$. It is trivial to prove that every set of 1, 2, or 3 points in general position is a universal point subset for every planar graph and every value of n (Tutte's algorithm [19,20]). On the other hand there exists a set of 4 points in general position that is not a universal point subset for planar graphs having $n = 5$ vertices.

We show lower and upper bounds for $F(n, \mathcal{G})$ for different families of planar graphs. In particular, we show that every set of 4 points in general position is a universal point subset for all planar graphs with at least 6 vertices and, on the other hand, we show that there exists a set of $2\lceil \frac{n}{3} \rceil + 2$ points in convex position that is not a universal subset for the class of planar graphs. In other words, we prove that $4 \leq F(n, \mathcal{G}) \leq 2\lceil \frac{n}{3} \rceil + 1$, for all $n \geq 6$, when $\mathcal{G}$ is the class of all planar graphs. In addition, we improve the lower bound and the upper bound for some subfamilies of planar graphs; specifically, we study the case that $\mathcal{G}$ is the class of 4-connected planar graphs whose outer face is a quadrilateral, and the case that $\mathcal{G}$ is the class of nested-triangles graphs.

We conclude in Section 4 with some remarks and open problems. In that section we also briefly discuss the relationships between our problem and the related *allocation problem* that has been the subject of recent studies (see, e.g. [11,13]).

Definitions and Notation

Point Sets. A set S of points is in *general position* if no three points are collinear. The *convex hull* $CH(S)$ of S is the point set obtained as a convex combination of the points of S. If no point is in the convex hull of the others, then S is in *convex position*. A set S in convex position is *one-sided* if it can be rotated in such a way that the leftmost and rightmost points are consecutive in the convex hull.

Graphs. We denote by (u, v) both an undirected and a directed edge, in the latter case meaning the edge is directed from u to v. Also, we use the term *triangle* to denote both a 3-cycle and its drawing. A graph $G = (V, E)$ is *planar* if it has a drawing Γ without edge crossings. Drawing Γ splits the plane into connected regions called *faces*; the unbounded region is the *outer* face and the other faces are the *internal* faces. The cyclic ordering of edges around each vertex of Γ together with a choice of the outer face is a *planar embedding* of G. A *plane graph* is a graph with a fixed planar embedding. A planar (plane) graph is *maximal* if each face of the graph is a triangle, thus no edge can be added to it without violating planarity. A graph G is k-*connected* if it does not contain a set of $k - 1$ vertices whose removal disconnects it.

Let $G = (V, E)$ be a maximal plane graph with outer face (v_1, v_2, v_n). A *canonical ordering* [5] of G is an order $\sigma = (v_1, v_2, v_3, \ldots, v_n)$ of its vertices satisfying the following properties: (1) The subgraph G_{i-1} induced by $v_1, v_2, \ldots, v_{i-1}$ is 2-connected and the boundary of the outer face of G_{i-1} is a cycle C_{i-1} containing edge (v_1, v_2); (2) vertex v_i is in the outer face of G_i and its neighbors in G_{i-1} form a (non-trivial) subpath of path $C_{i-1} - (v_1, v_2)$.

Let G be a planar graph with a planar drawing Γ. Let t_1 and t_2 be two disjoint triangles of G. We say that t_2 is *nested in* t_1, and write $t_2 < t_1$, if t_2 is in the bounded region of the plane delimited by t_1. A *nested-triangles graph* G with n vertices (n is a multiple of 3) is a 3-connected graph admitting a planar drawing Γ in which $n/3$ disjoint triangles $t_1, t_2, \ldots, t_{n/3}$ exist such that $t_1 > t_2 > \cdots > t_{n/3}$.

2   A Universal Point Subset for Planar Graphs

In this section we provide a universal point subset of size $\lceil \sqrt{n} \rceil$ for (maximal) planar graphs with n vertices. Note that considering maximal planar graphs is not a limitation, since any planar graph is a subgraph of a maximal planar graph.

Let G be a maximal planar graph with n vertices. Let $\sigma = (v_1, v_2, \ldots, v_n)$ be a canonical ordering of the vertices of G for some planar embedding of G. Let G_i be the subgraph of G induced by the first i vertices in σ and let C_i be the outer face of G_i. Bose et al. [2] define the *frame* G^σ of G with respect to σ to be a directed subgraph of G with edges: (v_1, v_2) and, for every v_i ($i \geq 3$), $(v_{a(i)}, v_i)$ and $(v_i, v_{b(i)})$ where $v_{a(i)}$ is the first and $v_{b(i)}$ the last vertex that are adjacent to v_i on path $C_{i-1} - (v_1, v_2)$.

Let $<_\sigma$ be the partial order on the vertices of G where $u <_\sigma v$ if and only if G^σ contains a path from u to v. Notice that $v_{a(i)}$ is the smallest vertex and $v_{b(i)}$ is the largest vertex according to $<_\sigma$ that are adjacent to v_i in G and precede v_i in σ. A sequence of numbers $(x_1, x_2, \ldots, x_n)$ *obeys* the partial order $<_\sigma$ if $x_a < x_b$ for all $v_a <_\sigma v_b$.

Lemma 1. *Given a canonical ordering* $\sigma = (v_1, v_2, \ldots, v_n)$ *of the vertices of a maximal planar graph G and a sequence of x-coordinates* $(x_1, x_2, \ldots, x_n)$ *that obeys the partial order* $<_\sigma$ *with* $x_i \in [1, n]$, *for any sequence of y-coordinates* $(y_1, y_2, \ldots, y_n)$ *satisfying* $y_1 = y_2 = 0$ *and* $y_i > \frac{n-1}{\Delta} y_{i-1}$ *for* $i \geq 3$, *where* $0 < \Delta \leq \min_{v_a <_\sigma v_b} x_b - x_a$, *the drawing of G with v_i at point* (x_i, y_i) *for all $i \in [n]$ is a plane drawing.*

Proof: Suppose that the drawing of G_{i-1} with v_j at point (x_j, y_j) for $j \in [i - 1]$ is a plane drawing, and furthermore that C_{i-1} is an x-monotone chain. Clearly, this holds for $i - 1 = 2$. If the vertex v_i at point (x_i, y_i) lies in the intersection of the half-planes above the lines defined by consecutive vertices on the chain, then v_i can connect to any subsequence of chain vertices without intersecting the drawing of G_{i-1}. By adding v_i at (x_i, y_i), we obtain a plane drawing of G_i since $(v_1, v_2, \ldots, v_n)$ is a canonical ordering. Since the sequence $(x_1, x_2, \ldots, x_n)$ obeys the partial order $<_\sigma$, C_i is x-monotone.

It remains to show that (x_i, y_i) is above the lines through every pair of adjacent vertices in C_{i-1}. Let v_a precede v_b on the chain C_{i-1}. Since $v_a <_\sigma v_b$, $x_a < x_b$. The point (x_i, y_i) lies above the line through (x_a, y_a) and (x_b, y_b) if $y_i(x_b - x_a) > y_a(x_b - x_i) + y_b(x_i - x_a)$. By choosing $y_i > \frac{n-1}{\Delta} y_{i-1}$ this inequality holds for any $x_i \in [1, n]$, since $x_b - x_a \geq \Delta$, $y_a, y_b \leq y_{i-1}$, and $x_a, x_b \in [1, n]$.   ■

Let $\mathcal{U}_k = \{((2n)^{-ni}, (2n)^{ni}) \mid i \in [k]\}$ be a nearly vertical set of k points in convex position. Observe that $\mathcal{U}_k$ is a one-sided convex set.

Lemma 2. *If a maximal planar graph G has a canonical ordering σ so that $<_\sigma$ has an anti-chain of size k, then G admits a planar straight-line drawing with k of its vertices placed on $\mathcal{U}_k$.*

Proof: Let $v_1, v_2, \ldots, v_n$ be the vertices of G in canonical order σ. Let $A = \{v_{i_1}, v_{i_2}, \ldots, v_{i_k}\}$ be an anti-chain in $<_\sigma$ with $i_1 < i_2 < \cdots < i_k$. Note that $i_1 > 2$ (unless $k = 1$) since v_1 and v_2 cannot be part of an anti-chain of size greater than one: $v_1 <_\sigma v$ for all $v \neq v_1$ and $v <_\sigma v_2$ for all $v \neq v_2$. Let A^* be the set of vertices less than (according to $<_\sigma$) some vertex in A. We create a sequence of x-coordinates $(x_1, x_2, \ldots, x_n)$ that obeys the partial order $<_\sigma$ with each x_i an integer in $[1, n]$ for $v_i \notin A$ and $x_{i_j} = |S| + 1 + (2n)^{-nj}$ for all $j \in [k]$. This is easy to achieve using a topological sort of A^* and a topological sort of $V \setminus (A^* \cup A)$.

We create a sequence of y-coordinates $(y_1, y_2, \ldots, y_n)$ with $y_1 = y_2 = 0$ and $y_i = (2n)^{jn + (i - i_j)}$ for $i_j \leq i < i_{j+1}$ where, for convenience, we have assumed $i_0 = 1$ and $i_{k+1} = n + 1$. This assigns the jth vertex in the anti-chain a y-coordinate of the form $(2n)^{jn}$, and it assigns vertices not in the anti-chain, that are between the jth and $(j+1)$th anti-chain vertices (in the canonical ordering σ), y-coordinates between $(2n)^{jn}$ and $(2n)^{(j+1)n}$. Since no two vertices in A are related by $<_\sigma$, the minimum of $x_b - x_a$ for $v_a <_\sigma v_b$ is at least $1 - (2n)^{-n} > 1/2 = \Delta$. Thus the sequence $(y_1, y_2, \ldots, y_n)$ satisfies the conditions of Lemma 1. By that lemma, there is a plane drawing of G with these x- and y-coordinates. Shifting this drawing by $-|A^*| - 1$ in the x-coordinate places the anti-chain A on the points $\mathcal{U}_k$.   ■

Lemma 3. *Let π be a maximal chain of $<_\sigma$. Then, the subgraph of G induced by the vertices of π is outerplanar.*

Proof: Since π is a maximal chain, it corresponds to a directed path, P, in G^σ from v_1 to v_2. Let C be the undirected cycle in G composed of P and the edge (v_1, v_2). We prove that all the chords of C in G lie inside it with respect to the embedding used to derive the canonical ordering σ. Assume, for a contradiction, that C has a chord (u, v) outside C, where u occurs before v on P. Let $P' = (u, w, \ldots, z, v)$ be the subpath of P from u to v. Suppose that u precedes v in σ. Since (u, v) is an outside chord, the vertices in $P' - v$ precede v in σ. However, the fact that both u and z precede v in σ and $u <_\sigma z$ contradicts the fact that (z, v) is an edge of G^σ. Indeed, G^σ contains only one directed edge to v from a vertex that precedes v in σ. The edge is from the first vertex in $<_\sigma$ among the neighbors of v in G that precede v in σ. Since $u <_\sigma z$ and since u

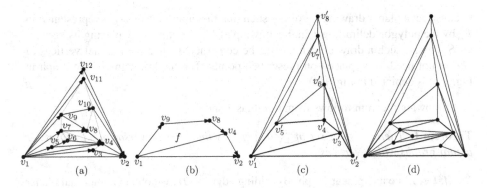

Fig. 1. (a) A maximal planar graph G. The edges of G^σ are directed and black, while edges not in G^σ are gray. $\pi = (v_1, v_9, v_8, v_4, v_2)$ is a maximal chain of $<_\sigma$. (b) A drawing of the subgraph $G(\pi)$ of G on a one-sided convex point set. (c) Extending the drawing of $G(\pi)$. (d) The final drawing after filling in the faces of $G(\pi)$.

precedes v in σ, vertex z cannot be the first neighbor. Hence, edge (z, v) cannot be in G^σ. Similarly, if v precedes u in σ, both v and w precede u, and $w <_\sigma v$ contradicts the fact that (u, w) is an edge of G^σ. ∎

Lemma 4. *If a maximal planar graph G has a canonical ordering σ such that $<_\sigma$ has a maximal chain π of size k, then G admits a planar straight-line drawing with k of its vertices placed on any one-sided convex set of size k.*

Proof: Consider any one-sided convex point set S of size k. Assume, without loss of generality up to a rotation of the coordinate system, that such points are ordered based on their x-coordinate and that the leftmost and the rightmost points are also the bottommost ones. By Lemma 3, the subgraph $G(\pi)$ of G induced by the vertices of π is outerplanar. Hence, such a subgraph can be drawn [1] on the points of S in such a way that the vertices of π are assigned increasing x-coordinates according to the order they appear on π. Figure 1(a) illustrates a maximal planar graph G, together with the frame G^σ associated with a partial order $<_\sigma$ of G. Figure 1(b) illustrates a drawing of the subgraph $G(\pi)$ of G on a one-sided convex point set.

Further, consider the planar graph G' obtained from G by removing every vertex that is internal to some face of $G(\pi)$. Since $G(\pi)$ is outerplanar, there exists a canonical ordering σ' of the vertices of G' such that the k vertices of $G(\pi)$ appear in the first k positions of σ', that is, $G(\pi) = G'_k$. Since the drawing of $G(\pi) = G'_k$ obtained by placing its vertices on S is such that C'_k is an x-monotone chain, such a drawing can be extended to a planar drawing of G' by applying an algorithm that is analogous to the one given by de Fraysseix, et al. [5] to construct polynomial area drawings of planar graphs. Namely, place each vertex v'_j of G', with $j > k$, in such a way that G'_j is an x-monotone chain. Note that this is always possible, since vertices v'_j, with $j > k$, do not need to be placed on prescribed points (neither a point of the prescribed point set nor an integer grid point, as it happens in [5]). See Fig. 1(c).

Finally, for each face f of $G(\pi)$, consider the subgraph G_f of G induced by the vertices of f and by the vertices that are internal to f. Then, apply Tutte's algorithm [19,20] to construct a planar drawing Γ_f of G_f such that the outer face of G_f is represented in Γ_f by the polygon delimiting f in the drawing of $G(\pi)$ obtained by placing its vertices on S. Again, such a drawing can always be constructed since the internal vertices of G_f do not need to be placed on prescribed points. The final drawing of the graph in Fig. 1(a) is depicted in Fig. 1(d). ∎

The following theorem follows from Lemmas 2 and 4.

Theorem 1. *There exists a set of $\lceil \sqrt{n} \rceil$ points that is a universal point subset for planar graphs with n vertices.*

Proof: Let H be any planar graph. By adding edges to H, we obtain a maximal planar graph G. Let σ be a canonical ordering of G for some planar embedding of G. Let G^σ be a frame of G and let $<_\sigma$ be the corresponding partial order. By Dilworth's theorem [6], there exists in $<_\sigma$ either a chain of $\lceil \sqrt{n} \rceil$ vertices or an anti-chain of $\lceil \sqrt{n} \rceil$ vertices. In either case, by Lemma 2 or 4, G admits an embedding-preserving planar straight-line drawing with $k = \lceil \sqrt{n} \rceil$ of its vertices placed on the points of the one-sided convex point set $\mathcal{U}_k = \{((2n)^{-ni}, (2n)^{ni}) \mid i \in [k]\}$. Removing the added edges gives a drawing of H. ∎

3   Universalizing the Size of Universal Point Subsets

Let $\mathcal{G}$ be a class of planar graphs. We define $F(n, \mathcal{G})$ as the maximum value such that *every* set of $F(n, \mathcal{G})$ points in general position is a universal point subset for the graphs in $\mathcal{G}$ with n vertices. When $\mathcal{G}$ coincides with the class of all planar graphs, we simply denote this value by $F(n)$. In this section we give lower and upper bounds for $F(n, \mathcal{G})$ for some classes of planar graphs.

3.1   Planar Graphs

When $\mathcal{G}$ coincides with the class of all planar graphs, we show that $F(5) = 3, F(6) = 4$, and $4 \leq F(n) \leq 2\lceil \frac{n}{3} \rceil + 1$ for all the other values of n.

First observe that there exists a set of 4 points in general position that is not a universal point subset for planar graphs with $n = 5$ vertices. Indeed, consider the set S of 4 points at the corners of a unit square. The claim follows from the fact that the outer face of every maximal planar graph with 5 vertices can use at most one point of S, as otherwise it could not contain all the remaining points of S in its interior. This, together with the fact that, by Tutte's theorem [19,20], every set of 3 points in general position is a universal point subset for planar graphs, implies that $F(5) = 3$. Then, we consider planar graphs with $n = 6$. Again, with the same argument as the one used for $n = 5$, we can prove that a set of 5 points composed of the corners of a regular pentagon is not a universal point subset for all planar graphs of size 6, which means $F(6) \leq 4$. In the following lemma we prove that also this bound is tight (that is, $F(6) = 4$). First, we note that all maximal planar graphs with six vertices are those depicted in Fig. 2(a-d).

Lemma 5. $F(6) = 4$.

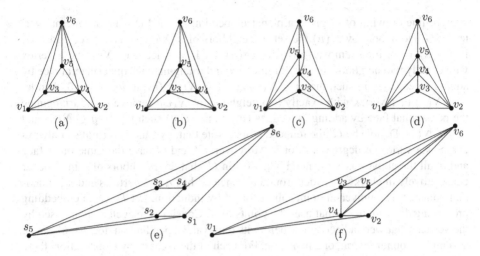

Fig. 2. (a-d) All maximal six-vertex planar graphs. (e) The common structure used to draw them on a set of four points in convex position. (f) The corresponding drawing of the graph in (a).

Proof: Let S be a set of 4 points in general position. If S is not in convex position, we map the vertices of the outer face of G to the three points on the convex hull $CH(S)$ of S. For all cases of Fig. 2(a-d), the remaining vertices are drawn inside $CH(S)$, by using the fourth point of S to place one of them.

If the points of S are in convex position, let s_1, s_2, s_3, and s_4 be the points of $CH(S)$ in clockwise order. Let s_5 and s_6 be points not in S so that triangle $s_1s_5s_6$ contains $S \setminus \{s_1\}$ in its interior; s_5 sees s_3, s_4, s_2, and s_1 in clockwise order; and s_6 sees s_3, s_4, s_2, and s_1 in counterclockwise order. It is straightforward to confirm that such points exist since S is in convex position. Add segments s_1s_2, s_2s_4, s_4s_3, s_5s_6, s_5s_3, s_5s_2, s_5s_1, s_6s_3, s_6s_4, and s_6s_1. See Fig. 2(e). For each case of Fig. 2(a-d), we map the outer vertices of G to the points s_1, s_5, s_6, we map the internal vertices of G to the points s_2, s_3, s_4 and we insert the remaining edges of G. As an example, the drawing of the graph in Fig. 2(a) is shown in Fig. 2(f). ∎

Finally, we consider the general case, namely planar graphs with $n > 6$ vertices. We first observe that, by using an argument similar to the one used to prove $F(5) \leq 3$ and $F(6) \leq 4$, we can prove that $F(n) \leq n - 2$. Indeed, a set of $n - 1$ points composed of the corners of a regular $(n-1)$-gon is not a universal point subset for all maximal planar graphs of size n, as at most one point in this point set can be used to place vertices of the outer face. However, as shown in the following theorem, in the general case we can prove a better upper bound. On the other hand, the lower bound of 4 is obtained by extending the result for planar graphs with six vertices.

Theorem 2. *If $n > 6$ then $4 \leq F(n) \leq 2\lceil \frac{n}{3} \rceil + 1$.*

Proof: We first prove the lower bound. Let S be a set of 4 points in general position. Consider a maximal planar graph G with $n \geq 6$ vertices. The proof is by induction on the number of vertices of G. In the base case G has $n = 6$ vertices, and the statement follows from Lemma 5. For $n > 6$, we can use Read's algorithm [14] to produce a straight-line drawing of G given a planar embedding of G. Let u be an internal vertex of G and denote by $N(u)$ the set of neighbors of u. We can have three cases: (i) if $\deg(u) = 3$, then remove u; (ii) if $\deg(u) = 4$, then let $v \in N(u)$ be a vertex with exactly two neighbors in $N(u)$: remove u and triangulate the quadrilateral face by adding an edge (v, x) such that $x \in N(u)$ and $x \notin N(v)$; (iii) if $\deg(u) = 5$, then let $v \in N(u)$ be a vertex with exactly two neighbors in $N(u)$: remove u and triangulate the pentagonal face by adding two edges (v, x) and (v, y) such that $x, y \in N(u)$ and $x, y \notin N(v)$. Denote by G' the reduced graph. Note that, by Euler's formula, G always contains a vertex of degree 3, 4, or 5. Moreover, G' and G have the same outer face and in all the above cases the neighbors of v in G' that are neighbors of u in G occur consecutively in clockwise order around v. Since G' has $n - 1$ vertices and it is maximal planar, S is a universal point subset for G' by induction. Let Γ' be an embedding preserving drawing of G' that uses the points of S. Consider the cycle C composed by the vertices that were adjacent to u before its removal. Depending on $\deg(u)$, C can be a triangle, a quadrilateral, or a pentagon. For each of these cases, by construction, there exists a point p inside C sufficiently close to v on which we can draw u and obtain an embedding preserving drawing of G [14].

Now we prove the upper bound. Let G be a graph with $n > 6$ vertices that contains $\lfloor n/3 \rfloor$ nested triangles. Let S be a set of $n - (\lfloor \frac{n}{3} \rfloor - 2) \leq 2\lceil \frac{n}{3} \rceil + 2$ points in convex position. All but $\lfloor n/3 \rfloor - 2$ vertices of G must map to points of S. Thus, there are at least two nested triangles t_1 and t_2 of G that have all three vertices mapped on points of S. However, since S is in convex position, t_1 cannot include t_2 and vice-versa.   ∎

3.2   4-Connected Planar Graphs

Next, we consider the value of $F(n, \mathcal{G})$ when $\mathcal{G}$ coincides with the class of 4-connected planar graphs whose outer face is a 4-cycle. For this class we can prove a stronger lower bound than for planar graphs, namely $F(n, \mathcal{G}) \geq \frac{\lg n}{4}$.

Theorem 3. *Let $\mathcal{G}$ be the class of 4-connected planar graphs whose outer face has size at least 4. Then, $\frac{\lg n}{4} \leq F(n, \mathcal{G}) \leq n - 2$.*

Proof: The upper bound comes analogously to $F(n) \leq n - 2$ for planar graphs. In fact, consider a 4-connected planar graph with $n > 8$ vertices and whose outer face is a 4-cycle, and consider a point set S composed of the corners of a regular $(n - 1)$-gon. Again, at least two points not in S have to be used to place the vertices of the outer face.

We prove the lower bound. Let S be any set of $\lceil \frac{\lg n}{4} \rceil$ points in general position. Let G be an internally 4-connected plane graph with n vertices and outer face of size at least 4. Thomassen [17] showed that G is the dual of a rectangular subdivision of a rectangle. Tóth [18] showed that there exists a horizontal or vertical line (called a *stabber*) that intersects at least $\frac{\lg n}{4}$ rectangles in this rectangular subdivision. Find such a stabber that intersects rectangles $r_1, r_2, ..., r_k$ ($k \geq \frac{\lg n}{4}$). Add points to S to create a set S' of k points. Choose axes so that no two points of S' have the same x-coordinate. Let $p_1, p_2, ..., p_k$ be the points of S' sorted by x-coordinate. Place the vertex corresponding to rectangle r_i at p_i for $i \in [k]$. So S' supports the x-monotone drawing of a path in G, and the vertices corresponding to p_1 and p_k lie on the outer face of G. This path divides G into two subgraphs G_1 and G_2, one on each side of the path. So we can use the construction by de Fraysseix, et al. [5] twice, once for G_1 and once for G_2.   ∎

3.3 Nested-Triangles Graphs

Finally, we consider the value of $F(n, \mathcal{G})$ when $\mathcal{G}$ coincides with the class of nested-triangles graphs. We prove that for this class $F(n, \mathcal{G}) \geq \frac{n}{3}$, almost matching the upper bound.

Theorem 4. *Let $\mathcal{G}$ be the class of nested-triangles graphs. Then,* $\frac{n}{3} \leq F(n, \mathcal{G}) \leq 2\lceil \frac{n}{3} \rceil + 1$.

Proof: The upper bound is the same proved in Theorem 2. As for the lower bound, let S be any set of $\frac{n}{3}$ points in general position and choose the coordinate axes in such a way that no two points have the same y-coordinate. Let G be a nested-triangles graph with a given planar embedding. Let v be a vertex of the triangle t representing the outer face of G and let p be the point of S having the largest y-coordinate. Map v to p and represent t as a triangle that encloses all remaining points of S. Remove the outer face of G and repeat the argument on $S \setminus \{p\}$; at every step, the sides of the triangles that represent the outer face are drawn parallel to one another in such a way that the inclusion of the triangular faces is respected and no two edges cross. ∎

4 Final Remarks and Open Problems

We remark that in this paper we assumed the points to be in general position. This is coherent with most of the literature in combinatorial and computational geometry, where geometric graphs (i.e. planar straight line drawings) are defined on point sets in general position [12]. However, one might also consider point sets allowing collinearities. In this scenario, some of our results should definitely be reformulated. For example, it is easy to see that a point set of four collinear points cannot be a universal point subset for the class of maximal planar graphs with six vertices. For a class $\mathcal{G}$ of planar graphs one could define $\overline{F}(n, \mathcal{G})$ as the maximum value such that *every* set of $\overline{F}(n, \mathcal{G})$ distinct points, whether it contains collinearities or not, is a universal point subset for the graphs in $\mathcal{G}$ with n vertices, and analogously define $\overline{F}(n)$ for planar graphs. Note that, allowing collinearities makes it possible to relate the problem of determining the value of $\overline{F}(n)$ with the *allocation problem* for planar graphs [11,13]. In this problem, the input is an n-vertex planar graph G and a point set X of size n, possibly with collinearities, and the goal is to construct a planar drawing Γ of G such that as many vertices of G as possible are placed in Γ on points of X. In particular, by exploiting this relationship, a slightly sublinear upper bound can be proved for $\overline{F}(n)$ via a construction from [13] that makes heavy use of collinearity.

We conclude with a few open problems that we find particularly interesting. (i) Narrow the gaps between the upper and lower bounds of Section 3. (ii) Prove/disprove a sublinear upper bound for $F(n)$ when points are in general position. (iii) Does the $\lim_{n \to \infty} F(n) = \infty$ hold? (iv) Is there any universal subset for the set of all planar graphs with n vertices that consists of more than $\lceil \sqrt{n} \rceil$ points?

Acknowledgments. Work on this problem began at the BICI Workshop on Graph Drawing, held in Bertinoro, Italy, in March 2012. We thank all the participants for many fruitful discussions. Research supported in part by the MIUR project "AlgoDEEP" prot. 2008TFBWL4 and by the ESF project 10-EuroGIGA-OP-003 GraDR "Graph Drawings and Representations". William Evans is partially supported by NSERC of Canada. Ferran Hurtado is partially supported by projects MICINN MTM2009-07242, Gen. Cat. DGR2009SGR1040, and ESF EUROCORES programme EuroGIGA, CRP ComPoSe: MICINN Project EUI-EURC-2011-4306. Yoshio Okamoto is partially supported by Grand-in-Aid for Scientific Research from Ministry of Education, Science and Culture, Japan and Japan Society for the Promotion of Science.

References

1. Bose, P.: On embedding an outer-planar graph in a point set. Comp. Geom. 23(3), 303–312 (2002)
2. Bose, P., Dujmovic, V., Hurtado, F., Langerman, S., Morin, P., Wood, D.R.: A polynomial bound for untangling geometric planar graphs. Discrete & Comp. Geom. 42(4), 570–585 (2009)
3. Brandenburg, F.-J.: Drawing planar graphs on $\frac{8}{9}n^2$ area. Electronic Notes in Discrete Mathematics 31, 37–40 (2008)
4. Cabello, S.: Planar embeddability of the vertices of a graph using a fixed point set is NP-hard. J. Graph Algor. and Applic. 10(2), 353–363 (2006)
5. de Fraysseix, H., Pach, J., Pollack, R.: How to draw a planar graph on a grid. Combinatorica 10(1), 41–51 (1990)
6. Dilworth, R.: A decomposition theorem for partially ordered sets. Annals of Mathematics 51(1), 161–166 (1950)
7. Fáry, I.: On straight-line representation of planar graphs. Acta Sci. Math. (Szeged) 11, 229–233 (1948)
8. Goodman, J.E., Pollack, R.: Allowable sequences and order types in discrete and computational geometry. In: New Trends in Discrete and Comp. Geom., pp. 103–134 (1993)
9. Gritzmann, P., Mohar, B., Pach, J., Pollack, R.: Embedding a planar triangulation with vertices at specified points. Amer. Math. Monthly 98(2), 165–166 (1991)
10. Kurowski, M.: A 1.235 lower bound on the number of points needed to draw all n-vertex planar graphs. Inf. Process. Lett. 92(2), 95–98 (2004)
11. Olaverri, A.G., Hurtado, F., Huemer, C., Tejel, J., Valtr, P.: On triconnected and cubic plane graphs on given point sets. Comput. Geom. 42(9), 913–922 (2009)
12. Pach, J., Agarwal, P.K.: Geometric graphs. In: Comb. Geom., pp. 223–239. Wiley (1995)
13. Ravsky, A., Verbitsky, O.: On Collinear Sets in Straight-Line Drawings. In: Kolman, P., Kratochvíl, J. (eds.) WG 2011. LNCS, vol. 6986, pp. 295–306. Springer, Heidelberg (2011)
14. Read, R.: A new method for drawing a planar graph given the cyclic order of the edges at each vertex. Congressus Numeration 56, 31–44 (1987)
15. Schnyder, W.: Embedding planar graphs on the grid. In: Johnson, D.S. (ed.) SODA, pp. 138–148. SIAM (1990)
16. Stein, S.K.: Convex maps. Proc. of the Amer. Math. Society 2(3), 464–466 (1951)
17. Thomassen, C.: Interval representations of planar graphs. J. of Comb. Theory, Series B 40(1), 9–20 (1986)
18. Tóth, C.D.: Axis-aligned subdivisions with low stabbing numbers. SIAM J. Discrete Math. 22(3), 1187–1204 (2008)
19. Tutte, W.T.: Convex representations of graphs. Proc. London Math. Soc. 10, 304–320 (1960)
20. Tutte, W.T.: How to draw a graph. Proc. London Math. Soc. 13, 743–768 (1963)
21. Wagner, K.: Bemerkungen zum vierfarbenproblem. Jahresbericht. German. Math.-Verein. 46, 26–32 (1936)

Abstract Flows over Time: A First Step towards Solving Dynamic Packing Problems

Jan-Philipp W. Kappmeier, Jannik Matuschke, and Britta Peis

TU Berlin, Institut für Mathematik, Straße des 17. Juni 136, 10623 Berlin, Germany
{kappmeier,matuschke,peis}@math.tu-berlin.de

Abstract. *Flows over time* [4] generalize classical network flows by introducing a notion of time. Each arc is equipped with a transit time that specifies how long flow takes to traverse it, while flow rates may vary over time within the given edge capacities. In this paper, we extend this concept of a dynamic optimization problem to the more general setting of *abstract flows* [8]. In this model, the underlying network is replaced by an abstract system of linearly ordered sets, called "paths" satisfying a simple switching property: Whenever two paths P and Q intersect, there must be another path that is contained in the beginning of P and the end of Q.

We show that a maximum abstract flow over time can be obtained by solving a weighted abstract flow problem and constructing a temporally repeated flow from its solution. In the course of the proof, we also show that the relatively modest switching property of abstract networks already captures many essential properties of classical networks.

1 Introduction

Time plays a crucial role in many applications of combinatorial optimization, e.g., in the context of transportation, communication, or productional planning. Therefore, extending classical problem formulations by a temporal dimension is of particular interest. So far the most prominent example in this direction is the concept of *flows over time* – also called "dynamic flows" in the literature – which was first introduced and investigated by Ford and Fulkerson [4]. A key challenge in the context of flows over time is that an explicit specification of all flow values at each time step leads to an output that is exponential in the input size. Ford and Fulkerson resolved this issue by showing that the maximum flow over time problem allows for a so-called *temporally repeated* solution, which can be obtained by solving a single static flow problem. Since then, numerous results on different variants of flow over time problems have emerged. Outstanding results include [2,10,12], see [16] for a general survey.

Network flows can be interpreted as a special case of packing problems: we try to pack the capacitated edges of the graph by assigning flow values to the source-sink-paths. Given the impact of Ford and Fulkerson's result, which spawned a whole theory of flows over time, one now might ask how the concept of time can

K.-M. Chao, T.-s. Hsu, and D.-T. Lee (Eds.): ISAAC 2012, LNCS 7676, pp. 433–443, 2012.

be extended to other packing problems. A first natural candidate are generalizations of static network flows, as, e.g., *abstract flows*. The notion of abstract flows goes back to Hoffman [8], who observed that Ford and Fulkerson's original proof of the max flow/min cut theorem [3] does not use the underlying network structure directly but only exploits one particular property of the path system, the so-called *switching property*. Hoffman succeeded in showing that packing problems defined on general set systems (called *abstract networks*) with this switching property are totally dual integral (TDI). These structural results were later complemented by the combinatorial primal-dual algorithms of Martens and McCormick [14,13]. Inspired by Hoffman's work, further abstractions based on uncrossing axioms have been proposed and corresponding TDI results have been established, e.g., lattice polyhedra [9] or switchdec polyhedra [5], see [15] for a survey. In light of these generalizations, abstract flows appear to serve as an ideal first stepstone in our endeavour towards dynamic formulations of more general packing integer programs.

Our Contribution. In this paper, we introduce and investigate *abstract flows over time* and show how a temporally repeated abstract flow and a corresponding minimum cut can be computed by solving a single static weighted abstract flow problem. This immediately leads to the max flow/min cut theorem for abstract flows over time as our main result. Although our construction resembles that of Ford and Fulkerson's original result [4] on (non-abstract) flows over time, the proof turns out to be considerably more involved and we will need to take a detour via a relaxed version of abstract flows over time that also considers storage of flow at intermediate elements. However, our results also imply that this relaxation is not proper and there always is an optimal solution that does not wait at intermediate nodes. In the course of our proof, we also establish some interesting structural properties of abstract networks, showing that the relatively modest switching property of abstract path systems already captures many essential properties of classical networks.

Structure of This Paper. In the remainder of this section, we introduce Hoffman's model of abstract flows in detail. In Section 2, we show how to conduct a time expansion on this model and point out differences to the time expanded network for classical network flows by Ford and Fulkerson [4]. In Section 3, we will show how to construct the temporally repeated abstract flow and a corresponding minimum abstract cut of same value. In order to validate feasibility of this cut, we will prove the necessary properties on the structure of abstract networks in Section 4. Using these results, we can finally show in Section 5 that the cut actually intersects all temporal paths, completing the proof of our main theorem.

Introduction to Abstract Flows

An *abstract path system* consists of a ground set E of *elements* and a family of *paths* $\mathcal{P} \subseteq 2^E$. For every $P \in \mathcal{P}$ there is an order $<_P$ of the elements in P.

A path system is an *abstract network*, if the *switching property* is fulfilled: For every $P, Q \in \mathcal{P}$ and every $e \in P \cap Q$, there is a path

$$P \times_e Q \subseteq \{p \in P : \ p \leq_P e\} \cup \{q \in Q : \ q \geq_Q e\}.$$

For the sake of convenience, we define $[P, e] := \{p \in P : p \leq_P e\}$, $[e, P] := \{p \in P : p \geq_P e\}$, $(P, e) := \{p \in P : p <_P e\}$, and $(e, P) := \{p \in P : p >_P e\}$.

Given an abstract network with capacities $c \in \mathbb{R}_+^E$ for all elements, the *maximum abstract flow problem* asks for an assignment of flow values $x \in \mathbb{R}_+^{\mathcal{P}}$ to the paths such as to maximize the total flow value while not violating the capacity of any element. The problem can be generalized further by introducing a weight function $r \in \mathbb{R}_+^{\mathcal{P}}$ that specifies the "reward" per unit of flow sent along each path. It is easy to see that allowing general weight functions renders the problem *NP*-hard. Thus, the choice of weight functions is restricted to *supermodular* functions, i.e., we require $r(P \times_e Q) + r(Q \times_e P) \geq r(P) + r(Q)$ for every $P, Q \in \mathcal{P}$ and $e \in P \cap Q$.

The dual of the maximum weighted abstract flow problem is the *minimum weighted abstract cut problem*, which assigns a value $y(e)$ to every element $e \in E$ so as to cover every path according to its weight. The two problems can be stated as follows.

$$
\begin{array}{lll}
\max & \displaystyle\sum_{P \in \mathcal{P}} r(P) x(P) & \quad\quad \min \ \displaystyle\sum_{e \in E} c(e) y(e) \\[2.5em]
\text{s.t.} & \displaystyle\sum_{P \in \mathcal{P}: e \in P} x(P) \leq c(e) \quad \forall e \in E & \quad\quad \text{s.t.} \ \displaystyle\sum_{e \in P} y(e) \geq r(P) \ \ \forall P \in \mathcal{P} \\[2.5em]
& x(P) \geq 0 \quad\quad \forall P \in \mathcal{P} & \quad\quad\quad\quad\ \ y(e) \geq 0 \quad\quad \forall e \in E
\end{array}
$$

Hoffman [8] showed that for every integral supermodular weight function, the abstract cut LP is totally dual integral. This implies a generalized version of Ford and Fulkerson's max flow/min cut result in two ways: On the one hand, the switching property represents a significant abstraction, allowing for more general structures. On the other hand, supermodular weight functions lead to weighted cuts, i.e., elements can appear multiple times in the cut. We will later see a useful example of such weights in the context of temporally repeated flows, which also yields an intuitive interpretation of these cut values.

Hoffman's structural result was extended by McCormick [14], who presented a combinatorial algorithm that solves the unweighted version ($r \equiv 1$) of the maximum abstract flow problem in time polynomial in $|E|$, if the abstract network is given by a separation oracle for the abstract cut LP (in the unweighted case, this is equivalent to deciding whether a given set of elements contains a path or not). Later, Martens and McCormick [13] extended this result and presented an algorithm that also solves the weighted case.

While these results indicate that the switching property is the essential force behind max flow/min cut and similar total dual integrality results for flow based problems, we want to close this section by pointing out an example that shows how abstract networks actually may differ from classical networks. In classical networks, if two paths P and Q both intersect a third path R, then there either is a path from the beginning of P to the end of Q or the other way around. The following example shows that this is not true in abstract networks, even in cases where the switching property preserves the order of intersecting abstract paths.

Example. Consider the abstract network $(E, \mathcal{P})$ with $E = \{1, 2, 3, 4, a, b, c, d\}$ and $\mathcal{P} = \{(1, 2, 3, 4), (a, 2, c), (b, 3, d), (1, c), (1, d), (a, 4), (b, 4)\}$. Although both $(a, 2, c)$ and $(b, 3, d)$ intersect the path $(1, 2, 3, 4)$, there is neither a path that starts with a and ends with d nor one that starts with b and ends with c.

2   Time Expansion of Abstract Networks

Time plays an important role in many application areas of network flows. Flow rates can vary over time, and flow also takes time to travel within the network. One concept to capture these temporal effects is the so-called time expanded network introduced by Ford and Fulkerson [4]. The basic idea is to introduce multiple copies of the nodes in the network, one for each point in time. Then arcs connect copies of vertices according to their travel time. We extend this concept to the world of abstract flows by introducing the time expansion of an abstract network. In the spirit of Ford and Fulkerson's idea, we will introduce multiple copies of the abstract network. In contrast to the classical case however, not copies of individual arcs but of whole paths will be introduced.

The *time expansion* of an abstract network consists of a (static) abstract network with capacities $c \in \mathbb{R}_+^E$, *transit times* $\tau \in \mathbb{Z}_+^E$ and a *time horizon* $T \in \mathbb{Z}_+$. The time from 0 to T is discretized into T intervals $[0, 1), \ldots, [T - 1, T)$ which we identify with the set of their starting times $\mathcal{T} := \{0, \ldots, T - 1\}$. For each interval, a copy of the ground set E is introduced, i.e., the time expanded ground set is $E_T := E \times \mathcal{T}$.

A *temporal path* is denoted by P_t, where P is a path of the underlying static abstract network and $t \in \mathcal{T}$ specifies the starting time of the path. Flow sent along the temporal path P_t enters element e at time $t + \sum_{p \in (P, e)} \tau(e)$, which is the time it needs for traversing all preceeding elements plus the initial offset of the path. Accordingly, we identify P_t with the set of its temporal elemtents by defining

$$P_t := \left\{ (e, \theta) \in E_T : e \in P, \ \theta = t + \sum_{p \in (P, e)} \tau(p) \right\}.$$

The *arrival time* of the temporal path P_t is $t + \sum_{e \in P} \tau(e)$, i.e., the time at which the flow arrives the end of the path. Since all flow is supposed to arrive its destination within the time horizon, we only allow copies of paths with a maximum arrival time of $T - 1$, which is the final element of $\mathcal{T}$. Thus, the set of temporal paths is defined by

$$\mathcal{P}_T := \left\{ P_t : \ P \in \mathcal{P}, \ t \in \mathcal{T}, \ t + \sum_{p \in P} \tau(p) < T \right\}.$$

We now can define the maximum abstract flow over time problem in analogy to the (static) maximum abstract flow problem. An *abstract flow over time* is an assignment $x : \mathcal{P}_T \to \mathbb{R}_+$ of non-negative flow values to all temporal paths. It is feasible if and only if the capacity of every element at every point in time is respected. The maximum abstract flow over time problem asks for an abstract flow over time that maximizes the total value of the flow:

$$\max \quad \sum_{P_t \in \mathcal{P}_T} x(P_t)$$

$$\text{s.t.} \quad \sum_{P_t \in \mathcal{P}_T:\ (e,\theta)\in P_t} x(P_t) \le c(e) \quad \forall e \in E,\ \theta \in \mathcal{T}$$

$$x(P_t) \ge 0 \qquad \forall P_t \in \mathcal{P}_T.$$

In analogy to the static case, the maximum value of an abstract flow over time can be bounded by an *abstract cut over time*, i.e., a subset $C \subseteq E_T$ of the time expanded ground set such that for each $P_t \in \mathcal{P}_T$ the set $P_t \cap C$ is nonempty. It is not hard to observe that the sum of capacities of the elements in such a cut is an upper bound on the value of a flow over time (see [11] for a formal proof).

Lemma 1. *Let x be an abstract flow over time and let C be an abstract cut over time. Then $\sum_{P_t \in \mathcal{P}_T} x(P_t) \le \sum_{(e,\theta)\in C} c(e)$.*

Remark. (Time expansion of an abstract network vs. time expanded network) While the time expansion of abstract networks as defined above is similar to the notion of a *time expanded network* as defined by Ford and Fulkerson [4] for classical network flows, the two definitions are not quite identical. Time expanded networks are based on the arc formulation of network flows. They are constructed by introducing copies of both the nodes and arcs of the underlying static network and adjusting the end points of the arcs according to their transit times. By construction, the resulting structure is guaranteed to be a network again. Unfortunately, there is no correspondence to the arc formulation for abstract flows – their definition is inherently tied to the path system, which does not allow for local concepts such as flow conservation at a particular element. Our model of time expansion therefore introduces copies of each path as a whole. In contrast to time expanded networks, the time expansion of an abstract network is *not* an abstract network in general, as can be seen in the following example.

Example. Let $E = \{s, a, b, t\}$ and $\mathcal{P} = \{P, Q, R, S\}$ with $P = (s, a, b, t)$, $Q = (s, b, a, t)$, $R = (s, a, t)$, and $S = (a, b, t)$. It is easy to verify that $\mathcal{P}$ in fact fulfills the switching property. Now assume all elements have unit transit times, i.e., $\tau \equiv 1$. The temporal paths P_0 and Q_1 intersect in the element $(b, 2)$. However, there is no temporal path in $\mathcal{P}_T$ that can be constructed from the elements $\{(s, 1), (b, 2), (t, 4)\}$, as there is a "time gap" between $(b, 2)$ and $(t, 4)$. Thus, the time expansion violates the switching property.

In view of this example, it is not even clear whether max flow/min cut results are still valid in the context of abstract flows over time or how far existing algorithms for abstract flow problems can be applied to the time expansion of the abstract network. Fortunately, the proof of our main result in the following sections will dissipate these concerns.

Theorem 2 (Abstract max flow/min cut over time). *The value of a maximum abstract flow over time equals the capacity of a minimum abstract cut over time. Both a maximum flow and a minimum cut over time can be computed by solving a single (static) maximum weighted abstract flow problem.*

Our proof of Theorem 2 involves constructing an abstract cut over time. In order to show feasibility of this cut, we will have to introduce the possibility of waiting at intermediate elements as an important device in our proof (see Section 4). Storage of flow at intermediate nodes plays an interesting role in the field of flows over time: While in some settings, such as the maximum flow over time problem or the NP-hard minimum cost flow over time problem, there always exist optimal solutions that do not wait at intermediate nodes [4,1], this is not true in other settings: e.g., for multi-commodity flows over time, the decision of allowing flow storage at intermediate nodes has an influence on the value of the solution and also on the complexity [7,6]. In the context of abstract flows over time, our results imply that the possibility of waiting has no influence on the problem, as we prove in Section 5 that the temporally repeated solution constructed in Section 3 is optimal even if waiting is allowed.

Theorem 3. *If waiting at intermediate elements is allowed, there still is a maximum abstract flow over time that does not wait at intermediate elements.*

3   Constructing a Maximum Abstract Flow over Time

The number of paths created by applying the time expansion is linear in T and thus exponential in the size of the input. Hence, even encoding a solution in the straightforward way results in an exponentially sized output. Ford and Fulkerson [4] resolved this problem for the classical (non-abstract) flow over time problem by introducing so-called temporally repeated flows, i.e., a flow over time constructed by temporally repeating a static flow pattern.

A *temporally repeated abstract flow* is an abstract flow over time x that is constructed from a static abstract flow $\tilde{x}$ by setting $x(P_t) := \tilde{x}(P)$ for $P \in \mathcal{P}$ and $0 \leq t < T - \sum_{e \in P} \tau(e)$. In other words, the static flow on each path is repeatedly sent as long as possible before the time horizon is reached. It is easy to check that feasibility of the underlying static flow implies feasibility of the temporally repeated flow (see [11] for a formal proof).

Lemma 4. *A temporally repeated abstract flow derived from a feasible abstract flow is a feasible abstract flow over time.*

In order to construct a *maximum* temporally repeated abstract flow, we first observe that flow can be sent along path $P \in \mathcal{P}$ up to time $r(P) := T - \sum_{e \in P} \tau(e)$, i.e., the flow value $\tilde{x}(P)$ is repeated $r(P)$ times. Thus, the total flow value of the temporally repeated flow x resulting from the static flow $\tilde{x}$ is $\sum_{P \in \mathcal{P}} r(P)\tilde{x}(P)$ and a maximum temporally repeated flow corresponds to a static abstract flow that is maximum with respect to the weights $r(P)$. It is not hard to see that the weight function defined in this way is supermodular (see [11] for a formal proof).

Observation 5. The weight function $r(P) := T - \sum_{e \in P} \tau(e)$ is supermodular.

Thus, we can solve the weighted abstract flow problem defined by these weights using the algorithm from [13], yielding a (static) abstract flow $\tilde{x}^*$ of maximum weight and the corresponding temporally repeated flow x^*. We will show that the value of x^* is not only maximum among the temporally repeated abstract flows but also among *all* abstract flows over time. To this end, we now construct an abstract cut over time whose capacity matches the flow value of x^*. Let $\tilde{y}$ be an optimal solution to the dual of the static weighted abstract flow problem with the weights $r(P)$ used to construct the temporally repeated flow. Note that by [8], we can assume $\tilde{y}$ to be integral. We will interpret the values $\tilde{y}(e)$ as the number of time steps for which element e is contained in the cut. We define the time at which $e \in E$ enters the cut by setting

$$\alpha(e) := \min_{P \in \mathcal{P}} \sum_{e \in P} (\tau(e) + \tilde{y}(e))$$

and define

$$C := \{(e, \theta) \in E_T : \ \alpha(e) \leq \theta < \alpha(e) + \tilde{y}(e)\} .$$

Theorem 6. C *is a feasible abstract cut over time.*

The proof of Theorem 6 involves some additional results on the structure of abstract networks, which we will elaborate on in the following sections. Using LP duality, Theorem 6 immediately leads to the following corollary, which implies Theorem 2 (see [11] for a formal proof).

Corollary 7. *The temporally repeated abstract flow x^* is a maximum abstract flow over time, and C is a minimum abstract cut over time whose capacity is equal to the flow value.*

4   Waiting at Intermediate Elements and the Structure of Abstract Networks

In order to prove that the set C constructed in the preceeding section actually covers all temporal paths, we need to show that we can ensure w.l.o.g. that the switching operation $\times$. preserves the order of the intersecting paths. We start by showing a weaker version of this statement, asserting that we can always choose the path resulting from an application of $\times$. in such a way that the two subpaths used for its construction are not mixed.

Lemma 8. *Let $P, Q \in \mathcal{P}$, $e \in P \cap Q$, then there is a path $R \subseteq [P, e] \cup [e, Q]$ such that $a \in R \cap [P, e]$ and $b \in R \setminus [P, e]$ implies $a <_R b$.*

Proof. Let $P, Q \in \mathcal{P}$ and $e \in P \cap Q$. Let R to be a path contained in $[P, e] \cup [e, Q]$ such that $|R \setminus [P, e]|$ is minimal. By contradiction assume there is $a \in R \cap [P, e]$ and $b \in R \setminus [P, e]$ with $b <_R a$. Let $R' := P \times_a R$. Observe that $R' \subset [P, e] \cup [e, Q]$ and $R' \setminus [P, e] \subset R \setminus [P, e]$ as $a \notin R'$, contradicting the choice of R. □

As a result of Lemma 8, the following assumption is without loss of generality.

Assumption A. If $a \in P \times_e Q \cap [P, e]$ and $b \in P \times_e Q \setminus [P, e]$, then $a <_{P \times_e Q} b$.

In order to show that $\times$. actually preserves the internal order of P and Q, we will – temporally – extend our model of time expansion by allowing flow to deliberately delay its traversal at intermediate elements.

Waiting at Intermediate Elements. A *temporal path with intermediate waiting* is denoted by P_σ, where $P \in \mathcal{P}$ is a path of the underlying static abstract network and $\sigma : P \to \mathcal{T}$ specifies the waiting time $\sigma(e)$ before traversing element $e \in P$. Flow sent along P_σ enters element e at time $\gamma(P_\sigma, e) := \sum_{p \in (P, e)} (\sigma(p) + \tau(p)) + \sigma(e)$ which is the time it needs for traversing all preceeding elements and the time it spends waiting at those elements and at e itself. Accordingly, we identify P_σ with the set of its temporal elemtents by defining

$$P_\sigma := \{(e, \theta) \in E_T : e \in P, \ \theta = \gamma(P_\sigma, e)\}.$$

The set of all temporal paths with intermediate waiting is denoted by

$$\mathcal{P}_T^* := \left\{ P_\sigma : \ P \in \mathcal{P}, \ \sigma \in \mathcal{T}^P, \ \sum_{e \in P} (\sigma(e) + \tau(e)) < T \right\}.$$

We will identify $P_t \in \mathcal{P}_T$ with $P_{(t,0,\ldots,0)} \in \mathcal{P}_T^*$. Note that the maximum abstract flow over time problem with waiting at intermediate elements is a relaxation of the maximum abstract flow over time without waiting, and the temporally repeated abstract flow x^* defined in Section 3 is a feasible solution to this relaxation. We will show that C actually covers all paths in $\mathcal{P}_T^*$, and thus x^* is optimal even if waiting is allowed. This implies that the relaxation is not proper, i.e., the possibility of waiting does not have any effect on the value of the optimal solution.

However, the extension of the model allows us to delete certain paths from the network. Observe that if Q is a strict subset of P, and $<_Q$ is identical to the restriction of $<_P$ to Q, then there always is an optimal abstract flow over time that does not use any copy of P (since it can wait at intermediate elements and use Q instead). Thus we can safely erase P from the base network in this case (without violating the switching axiom as Q can always replace P as switching choice). Hence, if we allow waiting at intermediate elements, the following assumption is without loss of generality.

Assumption B. If $Q \subset P$ then there are $a, b \in Q$ with $a <_P b$ and $b <_Q a$.

In the remainder of this section, we will show that Assumption B implies the following lemma. As a corollary, we can assume w.l.o.g. the switching operation to preserve order (see [11] for a proof of the corollary).

Lemma 9. *There are no paths $P, Q \in \mathcal{P}$ such that $Q \subset P$.*

Corollary 10. *Let $R := P \times_e Q$. If $a, b \in R \cap [P, e]$ and $a <_P b$ then $a <_R b$. If $a, b \in R \setminus [P, e]$ and $a <_Q b$ then $a <_R b$.*

Proof of Lemma 9. By contradiction assume there are $P, Q \in \mathcal{P}$ with $Q \subset P$. Let P^* be such that $|P^*|$ is minimal among all possible choices of such a P.

For $Q \subset P^*$ define $b(Q) \in Q$ to be the maximal element w.r.t. $<_Q$ such that $p <_{P^*} b(Q)$ for all $p \in (Q, b(Q))$, i.e., until element $b(Q)$ the order of Q is identical to that of P. By Assumption B, $b(Q)$ cannot be the last element of Q. So let $a(Q) \in Q$ be the successor of $b(Q)$ in Q. Note that this implies $a <_{P^*} b$ by definition of $b(Q)$. Among all paths $Q \subset P^*$, choose Q^* such that $b^* := b(Q^*)$ is maximal w.r.t. $<_{P^*}$. Let $a^* := a(Q^*)$.

Let $R := Q^* \times_{b^*} P^*$. Note that $a^* \notin R$, as $a^* >_{Q^*} b^*$, and therefore $R \subset P^*$. We now claim that $<_R$ is identical to $<_{Q^*}$ on the (Q^*, b^*)-part of R.

Claim. For all $c, d \in R \cap (Q^*, b^*)$ with $c <_{Q^*} d$, we have $c <_R d$.

Proof. If $c <_{Q^*} d$ but $d <_R c$, let $R' := R \times_d Q^*$. Note that $c \notin R'$ and by Assumption A, we have chosen R such that $[R, d] \subset Q^*$. Thus $R' \subset Q^* \subset P^*$ which contradicts the choice of P^*.   □

By definition of $b(Q^*)$, the order $<_{Q^*}$ is identical to $<_{P^*}$ on (Q^*, b^*) and thus $<_R$ is identical to $<_{P^*}$ on the (Q^*, b^*)-part of R. This implies that $a(R), b(R)$ cannot be both in the (Q^*, b^*)-part of R. Thus, $a(R) \in [b^*, P^*]$, which by $a(R) <_{P^*} b(R)$ implies that $b(R) \in (b^*, P^*)$. However this means $b(R) >_{P^*} b^*$ contradicting our choice of Q^* maximizing b^*.   □

5 Proof of Theorem 6

We will show that C not only covers all paths in $\mathcal{P}_T$ but even those paths that use waiting at intermediate elements, implying optimality of the constructed temporally repeated abstract flow for the relaxation of the problem. We are thus allowed to use the results from Section 4 in the proof, which is only sketched here (a complete proof can be found in [11]).

Theorem 6a. $C \cap P_\sigma \neq \emptyset$ for every $P_\sigma \in \mathcal{P}_T^*$.

Proof (sketch). By contradiction assume there is a path that is not covered by C. Among all uncovered paths choose $P_\sigma \in \mathcal{P}_T$ such that $\sum_{e \in P}(\tau(e) + \tilde{y}(e))$ is minimal. We will show that there is an uncovered path R whose length is strictly shorter, yielding a contradiction.

Let $\bar{e} \in P$ be maximal w.r.t. $<_P$ among all elements on P with $\gamma(P_\sigma, \bar{e}) \geq \alpha(\bar{e})$ By multiple careful applications of the switching operation, we can show that $\alpha(\bar{e}) < \sum_{e \in (P, \bar{e})}(\tau(e) + \tilde{y}(e))$.

Now let $Q \in \mathcal{P}$ be a path with $\sum_{e \in (Q, \bar{e})}(\tau(e) + \tilde{y}(e)) = \alpha(\bar{e})$. We consider the path $R := Q \times_{\bar{e}} P$. Let $t := \sum_{e \in [Q, \bar{e}]} \tilde{y}(e) + \sum_{e \in [Q, \bar{e}] \setminus R} \tau(e)$. Our results from Section 4 ensure that R inherited its internal order from Q and P. Thus, our choice of t guarantees that R_t arrives at elements from $[Q, \bar{e}]$ after they left the cut, but arrives at the elements from $(\bar{e}, P)$ before they enter the cut. This implies that R_t is not covered by C. However, observe that

$$\sum_{e \in R}(\tau(e) + \tilde{y}(e)) \leq \alpha(\bar{e}) + \sum_{e \in (\bar{e}, P)}(\tau(e) + \tilde{y}(e)) < \sum_{e \in P}(\tau(e) + \tilde{y}(e)),$$

contradicting the choice of P.   □

6 Conclusion

We presented abstract flows over time, an extension of flows over time that can be viewed as a first approach towards more general dynamic packing IPs. Our main result shows that the max flow/min cut result of Ford and Fulkerson still is valid in Hoffman's setting of abstract flows, emphasizing the robustness of the concept. At their heart, our proofs relied exclusively on the switching axiom for abstract networks, showing how this abstraction actually captures the essence of total dual integrality in network-based packing problems.

Acknowledgements. This work was supported by Deutsche Forschungsgemeinschaft (DFG) as part of the Priority Program "Algorithm Engineering" (1307), by DFG Research Center MATHEON "Mathematics for key technologies" in Berlin, and the Berlin Mathematical School.

References

1. Fleischer, L., Skutella, M.: Minimum cost flows over time without intermediate storage. In: Proceedings of the Fourteenth Annual ACM-SIAM Symposium on Discrete Algorithms, pp. 66–75 (2003)
2. Fleischer, L., Tardos, E.: Efficient continuous-time dynamic network flow algorithms. Operations Research Letters 23(3-5), 71–80 (1998)
3. Ford, L., Fulkerson, D.: Maximal flow through a network (1954)
4. Ford, L., Fulkerson, D.: Flows in networks. Princeton University Press (1962)
5. Gaillard, A.: Switchdec polyhedra. Discrete Appl. Math. 76(1), 141–163 (1997)
6. Groß, M., Skutella, M.: Maximum Multicommodity Flows over Time without Intermediate Storage. In: Epstein, L., Ferragina, P. (eds.) ESA 2012. LNCS, vol. 7501, pp. 539–550. Springer, Heidelberg (2012)
7. Hall, A., Hippler, S., Skutella, M.: Multicommodity flows over time: Efficient algorithms and complexity. Theoretical Computer Science 379(3), 387–404 (2007)
8. Hoffman, A.: A generalization of max flow-min cut. Mathematical Programming 6(1), 352–359 (1974)
9. Hoffman, A., Schwartz, D.: On lattice polyhedra. In: Proceedings of the 5th Hungarian Coll. on Combinatorics, pp. 593–598. North Holland (1978)
10. Hoppe, B., Tardos, E.: The quickest transshipment problem. Mathematics of Operations Research 25(1), 36–62 (2000)
11. Kappmeier, J.-P., Matuschke, J., Peis, B.: Abstract flows over time: A first step towards solving dynamic packing problems. Preprint 001-2012, TU Berlin
12. Klinz, B., Woeginger, G.: Minimum-cost dynamic flows: The series-parallel case. Networks 43(3), 153–162 (2004)
13. Martens, M., McCormick, S.T.: A Polynomial Algorithm for Weighted Abstract Flow. In: Lodi, A., Panconesi, A., Rinaldi, G. (eds.) IPCO 2008. LNCS, vol. 5035, pp. 97–111. Springer, Heidelberg (2008)

14. McCormick, S.T.: A polynomial algorithm for abstract maximum flow. In: Proceedings of the Seventh Annual ACM-SIAM Symposium on Discrete Algorithms, pp. 490–497 (1996)
15. Schrijver, A.: Total dual integrality from directed graphs, crossing families and sub- and supermodular functions. Progress in combinatorial optimization, pp. 315–361 (1984)
16. Skutella, M.: An introduction to network flows over time. In: Research Trends in Combinatorial Optimization, pp. 451–482 (2009)

Extending Partial Representations
of Subclasses of Chordal Graphs

Pavel Klavík[1,*], Jan Kratochvíl[1,*], Yota Otachi[2], and Toshiki Saitoh[3]

[1] Department of Applied Mathematics, Faculty of Mathematics and Physics,
Charles University, Malostranské náměstí 25, 118 00 Prague, Czech Republic
{klavik,honza}@kam.mff.cuni.cz
[2] School of Information Science, Japan Advanced Institute of Science and
Technology. Asahidai 1-1, Nomi, Ishikawa 923-1292, Japan
otachi@jaist.ac.jp
[3] Graduate School of Engineering, Kobe University, Rokkodai 1-1,
Nada, Kobe, 657-8501, Japan
saitoh@eedept.kobe-u.ac.jp

Abstract. Chordal graphs are intersection graphs of subtrees in a tree.
We investigate complexity of the partial representation extension problem for chordal graphs. A partial representation specifies a tree T' and
some pre-drawn subtrees. It asks whether it is possible to construct a
representation inside a modified tree T which extends the partial representation (keeps the pre-drawn subtrees unchanged).

We consider four modifications of T' and get vastly different problems.
In some cases, the problem is interesting even if just T' is given and no
subtree is pre-drawn. Also, we consider three well-known subclasses of
chordal graphs: Proper interval graphs, interval graphs and path graphs.
We give an almost complete complexity characterization.

In addition, we study parametrized complexity by the number of pre-
drawn subtrees, the number of components and the size of the tree T'.
We describe an interesting relation with integer partition problems. The
problem 3-PARTITION is used in the NP-completeness reductions. The BIN-
PACKING problem is closely related to the extension of interval graphs when
space in T' is limited, and we obtain "equivalency" with BINPACKING.

1 Introduction

Geometric representations of graphs and graph drawing are well-studied topics
in graph theory. We study *intersection representations* of graphs where the goal
is to assign geometrical objects to the vertices of the graph and encode edges by
intersections of the objects. An intersection-defined class restricts the geometrical
objects and contains all graphs representable by these restricted objects; for
example, interval graphs are intersection graphs of closed intervals of the real
line. Intersection-defined classes have many interesting properties and appear
naturally in numerous applications; for details see for example [8].

For a fixed class, its recognition problem asks whether an input graph belongs
to this class; in other words, whether it has an intersection representation of this

* Supported by ESF Eurogiga project GraDR as GAČR GIG/11/E023.

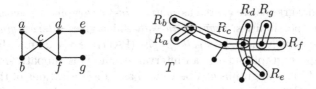

Fig. 1. An example of a chordal graph with one of its representations

class. The complexity of recognition is for many classes well-understood; for example interval graphs can be recognized in linear-time [2,4].

A recent paper [12] introduced the following new problem called *partial representation extension*. Given a graph and a partial representation (a representation of an induced subgraph), it asks whether it is possible to extend this representation to the entire graph. This problems falls into the paradigm of extending partial solutions, an approach that has been studied frequently in other circumstances. Often it proves to be much harder than building a solution from scratch, for example for graph coloring. Surprisingly, a very natural problem of partial representation extension was only considered recently.

The paper [12] gives an $\mathcal{O}(n^2)$-algorithm for interval graphs and an $\mathcal{O}(nm)$-algorithm for proper interval graphs. Also, several other papers consider this problem. Interval graphs can be extended in time $\mathcal{O}(n + m)$ [1]; proper interval graphs in time $\mathcal{O}(n + m)$ and unit interval graphs in time $\mathcal{O}(n^2)$ [11]; function and permutation graphs in polynomial time [10].

In this paper, we follow this recent trend and investigate the complexity of partial representation extension for chordal graphs. Our mostly negative results are interesting since chordal graphs are the first class for which the partial representation problem becomes harder than the original recognition problem. Also, we investigate three well-known subclasses proper interval graphs, interval graphs and path graphs, for which the complexity results are more rich. We believe that better understanding of these simpler cases will give tools to attack chordal graphs and beyond (for example, from the point of parametrized complexity).

1.1 Chordal Graphs and Their Subclasses

A graph is *chordal* if it does not contain an induced cycle of the length four or more, i.e., each "long" cycle is triangulated. The class of chordal graphs, denoted by CHOR, is well-studied and has many wonderful properties. Chordal graphs are closed under induced subgraphs and admit so-called *perfect elimination schemes*, closely related to optimal ways for Gaussian elimination for sparse matrices. Chordal graphs are perfect and many hard combinatorial problems are easy to solve on chordal graphs: maximum clique, maximum independent set, k-coloring, etc. Chordal graphs can be recognized in time $\mathcal{O}(n + m)$ [14].

Chordal graphs have the following intersection representations [7]. For every chordal graph G there exists a tree T and a collection $\{R_v : v \in V(G)\}$ of subtrees of T such that $R_u \cap R_v \neq \emptyset$ if and only if $uv \in E(G)$. For an example of a chordal graph and one of its intersection representations, see Figure 1.

When chordal graphs are viewed as *subtrees-in-tree* graphs, it is natural to consider two other possibilities: *subpaths-in-path* which gives *interval graphs* (INT),
and *subpaths-in-tree* which gives *path graphs* (PATH). For example the graph in
Figure 1 is a path graph but not an interval graph. This subpaths-in-path representations of interval graphs can be viewed as a discretizations of the real-line
representations. Interval graphs can be recognized in time $\mathcal{O}(n+m)$ [2] and path
graphs in time $\mathcal{O}(nm)$ [15].

In addition, we consider proper interval graphs (PINT). An interval graphs
is a proper interval graph if it has a representation $\mathcal{R}$ for which $R_u \subseteq R_v$
implies $R_u = R_v$; so no interval is a proper subset of another. From point of
our results, PINT behaves very similar to INT, but there are subtle differences
which we consider interesting. Also, partial representation extension of PINT
is surprisingly very closely related to partial representation extension of unit
interval graphs considered in [11]; in details discussed below. Proper interval
graphs can be recognized in time $\mathcal{O}(n + m)$ [13,3].

1.2 Partial Representation Extension

For a class $\mathcal{C}$, we denote the recognition problem by RECOG($\mathcal{C}$). For an input graph G, it asks whether it belongs to $\mathcal{C}$, and moreover we may certify it
by a representation. The partial representation extension problem denoted by
REPEXT($\mathcal{C}$) asks whether a part of the representation given by the input can be
extended to a representation of the whole graph.

A partial representation $\mathcal{R}'$ of G is a representation of an induced subgraph
G'. The vertices of G' are called *pre-drawn*. A representation $\mathcal{R}$ extends $\mathcal{R}'$ if
$R_v = R'_v$ for every $v \in V(G')$. The meta-problem we deal with is the following.

Problem: REPEXT($\mathcal{C}$) (Partial Representation Extension of $\mathcal{C}$)
  Input: A graph G with a partial representation $\mathcal{R}'$.
  Output: Does G have a representation $\mathcal{R}$ that extends $\mathcal{R}'$?

In this paper, we study complexity of the partial representation extension problems for CHOR, PATH and INT in the setting of subtree-representations. Here the
partial representation $\mathcal{R}'$ fixes subtrees belonging to G' and also specifies some
tree T' in which these subtrees are placed. The representation $\mathcal{R}$ is placed in a
tree T which is created by some modification of T'. We consider four possible
modifications and get different extension problems:

- FIXED – the tree can not be modified at all, $T = T'$.
- SUB – the tree can only be subdivided; T is a subdivision of T'.[1]
- ADD – we can add branches to the tree; T' is a subgraph of T.
- BOTH – we can both add branches and subdivide; a subgraph of T is a
  subdivision of T', or in other words T' is a topological minor of T.

---

[1] Let an edge $xy \in E(T')$ be subdivided (with a vertex z added in the middle). Then
also pre-drawn subtrees containing both x and y are modified and contain z as well.
So technically in the case of subdivision, it is not true that $R'_u = R_u$ for every
pre-drawn interval but from the topological point of view the partial representation
is extended.

We denote the problems by REPEXT($\mathcal{C}$, type). Constructing a representation in a specified tree T' is interesting even if no subtree is pre-drawn, i.e., G' is empty; this problem is denoted by RECOG*($\mathcal{C}$, type). Clearly, hardness of the RECOG* problem implies the hardness of the corresponding REPEXT problem.

Concerning chordal graphs, the types ADD and BOTH allow to construct an arbitrary tree T, so the RECOG* problem is equivalent to the standard RECOG problem. For interval graphs, the types ADD and SUB behave differently. The type ADD allows to extend the ends of the paths. The type SUB allows to expand the middle of the path but if the endpoint of the path is contained in some pre-drawn subpath, it remains there even after subdivision. The type BOTH is equivalent to the RECOG and REPEXT problems for the real line.

1.3 Our Results

We consider the complexity of the RECOG* and REPEXT problems for all four classes and all four types. Our results are displayed in the table in Figure 2.

- All NP-complete results are reduced from the 3-PARTITION problem. The reductions are very similar and the basic case is REPEXT(PINT, FIXED).
- Polynomial cases for INT and PINT are based on the known algorithm for recognition and extension. But since the space in T is limited, we adapt the algorithm for the specific problems.

Also, we study parametrized complexity of these problems with respect to three parameters: The number of pre-drawn subtrees k, the number of components c and the size t of the tree T'. In some cases, the parametrization does not help and the problem is NP-complete even if the value of the parameter is zero or one. In other cases, the problems are fixed-parameter tractable (FPT), W[1]-hard or in XP.

The main result concerning parametrization is the following. The BINPACKING problem is a well-known problem concerning integer partitions; more details in Section 3. For two problems A and B, we denote by $A \leq B$ a polynomial reduction

		PINT	INT	PATH	CHOR
FIXED	RECOG*	$\mathcal{O}(n+m)$	$\mathcal{O}(n+m)$	NP-complete	NP-complete
	REPEXT	NP-complete	NP-complete	NP-complete	NP-complete
SUB	RECOG*	$\mathcal{O}(n+m)$ [15, 3]	$\mathcal{O}(n+m)$ [2]	NP-complete	NP-complete
	REPEXT	$\mathcal{O}(n+m)$	$\mathcal{O}(n+m)$	NP-complete	NP-complete
ADD	RECOG*	$\mathcal{O}(n+m)$ [15, 3]	$\mathcal{O}(n+m)$ [2]	$\mathcal{O}(nm)$ [17]	$\mathcal{O}(n+m)$ [16]
	REPEXT	$\mathcal{O}(n+m)$	NP-complete	NP-complete	NP-complete
BOTH	RECOG*	$\mathcal{O}(n+m)$ [15, 3]	$\mathcal{O}(n+m)$ [2]	$\mathcal{O}(nm)$ [17]	$\mathcal{O}(n+m)$ [16]
	REPEXT	$\mathcal{O}(n+m)$ [13]	$\mathcal{O}(n+m)$ [1]	**open**	NP-complete

Fig. 2. Table of the complexity of different problems for all four considered classes. Results without references are new results of this paper.

and by $A \leq_{wtt} B$ a weak truth-table reduction (roughly, we may use a number of B-oraculum questions bounded by a computable function to solve A):

Theorem 1. BinPacking $\leq$ RepExt(PINT, Fixed) $\leq_{wtt}$ BinPacking *where the weak truth-table reduction needs to solve* 2^k *instances of* BinPacking.

We note that due to space limitations, this paper contains only sketches of the techniques and the proofs. Refer to the full version attached in the end.

2 Preliminaries

Notation. We reserve n for the number of vertices and m for the number of edges of the main considered graph G. The set of vertices is denoted by $V(G)$ and the set of edges by $E(G)$. For a vertex $v \in V(G)$, we let $N(v) = \{x : vx \in E(G)\}$ denote the *open neighborhood*, and $N[v] = N(v) \cup \{v\}$ the *closed neighborhood*.

Basic Concepts. A component C is called a *located component* if it has at least one vertex pre-drawn, and called *unlocated* otherwise. For interval graphs, the located components have to be ordered from left to right, otherwise the partial representation is clearly not extendible.

Let u and v be two vertices of G such that $N[u] = N[v]$. These two vertices are called *indistinguishable* since they can be represented exactly the same, having $R_u = R_v$ (a common property of all intersection representations). We assume that all input graphs are pruned, having only one vertex per group of indistinguishable vertices. It can be done in time $\mathcal{O}(n + m)$. We need to be a little careful since we cannot prune pre-drawn vertices.

To every maximal clique K, there is a subtree R_K contained exactly in R_u's of $u \in K$ (due to the Helly property). Moreover, these subtrees are for different maximal cliques pairwise disjoint. So for example, if $|T|$ is smaller than the number of maximal cliques of G, the graph is clearly not representable in T.

3 Interval Graphs

In this section, we deal with classes PINT and INT. The results obtained here are used as tools for PATH and CHOR graphs in Section 4.

3.1 Polynomial Cases

First we deal with all polynomial cases. Also, we describe several concepts as minimum span, useful in the rest of the paper.

Non-Fixed Type Recognition. The only limitation for recognition of interval graphs inside a given path is the length of the path. In all three types Sub, Add and Both, we can produce a path as long as necessary. Every possible representation can be realized in a tree T with at least $2n$ vertices. Thus the problems are equivalent to the standard recognition on the real line. For PINT, it can be solved in time $\mathcal{O}(n + m)$, for example [13,3]. Similarly for INT, it can be solved in time $\mathcal{O}(n + m)$ [2].

Both Type Extension. Similarly, the path T' can be extended as much as necessary which makes the problem equivalent to the partial representation extension problems on the real line, both solvable in time $\mathcal{O}(n + m)$ [11,1].

Sub Type Extension. It is possible to modify the above algorithms for partial representation extension of INT and PINT. Since we do not want to go in details of these algorithms, we instead reduce to the BOTH type extensions which we can solve in time $\mathcal{O}(n + m)$ (as discussed above):

Theorem 2. *The problems* REPEXT(PINT, SUB) *and* REPEXT(INT, SUB) *can be solved in time* $\mathcal{O}(n + m)$.

Proof (Sketch). We extend the paths at the ends by one and add two pre-drawn intervals $v_{\leftarrow}$ and $v_{\rightarrow}$ in such a way that the entire graph G has to live in between $R_{v_{\leftarrow}}$ and $R_{v_{\rightarrow}}$. Thus only subdivision involves G. $\qquad\square$

General Properties. A basic tool for proper interval graphs is uniqueness of the ordering of the intervals from left to right. For each component, there exists a unique ordering $<$ up to complete reversal (and possibly reordering of groups of indistinguishable intervals which are pruned) [5]. For each partial representation, the pre-drawn intervals appear in some specific order. This order has to be compatible with $<$ or its reversal (possibly both), otherwise the component is clearly not extendible.

For types FIXED and ADD, the space is limited and we need to save it. For a component C, we denote by $\mathrm{minspan}(C)$ the size of the smallest possible representation, taking as little vertices of T as possible. We can compute this value and construct such a representation:

Lemma 1. *For every component C (both located, or unlocated) of* PINT, *the value* $\mathrm{minspan}(C)$ *can be computed in time* $\mathcal{O}(n + m)$ *(together with a realizing representation).*

Proof (Sketch). Process the component according to $<$ and leave the gaps between intervals as small as possible. If the component is located, the pre-drawn intervals are fixed and for the rest of the intervals we minimize the gaps. $\qquad\square$

For INT, let $\mathrm{cl}(C)$ denote the number of maximal cliques of C.

Lemma 2. *For an unlocated component C of* INT, $\mathrm{minspan}(C) = \mathrm{cl}(C)$. *We can find a smallest representation in time* $\mathcal{O}(n + m)$.

Proof (Sketch). We identify maximal cliques [14] and construct the representation of size $\mathrm{cl}(C)$ using PQ-trees [2], both in time $\mathcal{O}(n + m)$. Clearly $\mathrm{minspan}(C) \geq \mathrm{cl}(C)$. $\qquad\square$

Fixed Type Recognition. We just need to use the values minspan we already know how to compute.

Proposition 1. *Both* RECOG*(PINT, FIXED) *and* RECOG*(INT, FIXED) *can be solved in time* $\mathcal{O}(n + m)$.

Proof. We process components $C_1, \ldots, C_c$ one-by-one and place them from left to right on T'. If $\sum_{i=1}^{c} \text{minspan}(C_i) \leq |T'|$, we can place the components using the smallest representation from Lemma 1 for PINT, resp. Lemma 2 for INT. Otherwise, the path is too small and the representation cannot be constructed.
□

Add Type Extension, PINT. Again, we approach this problem using minimum spans and Lemma 1.

Theorem 3. *The problem* REPEXT(PINT, ADD) *can be solved in time* $\mathcal{O}(n+m)$.

Proof (Sketch). We place unlocated components far to the left (since the path can be arbitrary stretched). Now, we process components in the ordering $C_1 < \cdots < C_c$ from left to right. We try to place each component as far to the left as possible, while on the right of the previously placed component. We have two possible representations and we choose the one more to the left.
□

3.2 NP-Complete Cases

Our reductions are from 3-PARTITION. The input of 3-PARTITION consists of integers k, M and $A_1, \ldots, A_{3k}$ such that $\frac{M}{4} < A_i < \frac{M}{2}$ for each A_i and $\sum A_i = kM$. It asks whether it is possible to partition A_i's into k triples such that the sets A_i of each triple sum to exactly M. This problem is known to be strongly NP-complete (even with all integers of a polynomial size) [6].

Theorem 4. *The problems* REPEXT(PINT, FIXED) *and* REPEXT(INT, FIXED) *are* NP-*complete.*

Proof (Sketch). We describe just the case of INT. For a given input of 3-PARTITION, we construct a graph G as follows. It consists of *split gadgets* making k gaps of size M and *take gadgets* representing sets A_i, all as separate components. The split gadgets are single pre-drawn vertices $v_0, \ldots, v_k$, pre-drawn as depicted in Figure 3. The take gadgets corresponding to A_i is just a path P_{A_i}. Notice that $\text{minspan}(P_k) = k$. The take gadgets are distributed into k gaps which gives a solution to the partition problem.
□

Corollary 1. *The problem* REPEXT(INT, ADD) *is* NP-*complete.*

Proof. Use the above reduction with one additional pre-drawn interval v attached to everything in G. We put $R_v = \{p_0, \ldots, p_{(M+1)k}\}$, so it contains the whole tree T'. Now since the representation of each take gadget has to intersect R_v, it has to be placed inside of the k gaps as before.
□

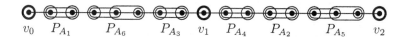

$v_0 \quad P_{A_1} \qquad P_{A_6} \qquad P_{A_3} \quad v_1 \quad P_{A_4} \qquad P_{A_2} \qquad P_{A_5} \quad v_2$

Fig. 3. An example of the reduction for the input set of 3-PARTITION: $k = 2$, $M = 7$, $A_1 = A_2 = A_3 = A_4 = 2$ and $A_5 = A_6 = 3$. The partial representation (depicted in bold) is extended, giving a solution to 3-PARTITION.

3.3 Parametrized Complexity

In this subsection, we study parametrized complexity of the above NP-complete problems. The parameters are the number c of components, the number k of pre-drawn intervals and the size t of the tree.

By Number of Components. For INT, it clearly does not help, we can add a vertex attached to everything. For PINT, the hardness lies in c:

Proposition 2. *The problem* REPEXT(PINT, FIXED) *is fixed-parameter tractable in the number of components c, solvable in time $\mathcal{O}((n + m)c!)$.*

Proof. There are $c!$ possible orderings of the components from left to right, we test each (some orderings may be excluded by pre-drawn intervals), using similar approach as in the proof of Theorem 3 (only here we deal with unlocated components as well). □

By Number of Pre-drawn Intervals. In the reduction in Theorem 4, we need to have k pre-drawn intervals. One could ask, whether the problems becomes simpler with a small number of pre-drawn intervals. We answer this negatively, for PINT the problem is in XP and W[1]-hard with respect to k. This is closely related a well-known integer partition problem called BINPACKING:

Problem: BINPACKING
    Input: Integers k, ℓ, V and $A_1, \ldots, A_\ell$.
  Output: Does there exist a k-partition $\mathcal{P}_1, \ldots, \mathcal{P}_k$ of $A_1, \ldots, A_\ell$ such that $\sum_{A_i \in \mathcal{P}_j} A_i \leq V$ for every $\mathcal{P}_j$.

When the sizes are encoded in unary, BINPACKING can be solved in time $t^{\mathcal{O}(k)}$ using dynamic programming where t is the total size of all items. And it is W[1]-hard with respect to the parameter k [9]. Similar holds for REPEXT(PINT, FIXED):

Proof (Sketch of Theorem 1). Modify the reduction in Theorem 4 to solve BINPACKING. For the other implication, we first deal with located components using Lemma 1, there are at most 2^k possible smallest representations (for each component, we have two possible orderings). For unlocated components, we have $k + 1$ gaps in which we can place them. This involves integer partiton of minimum spans which we can solve using a generalization of BINPACKING. □

Corollary 2. *The problem* REPEXT(PINT, FIXED) *is* W[1]*-hard and belongs to* XP *with respect to the parameter k, solvable in time $n^{\mathcal{O}(k)}$, where k is the number of pre-drawn intervals.*

Proposition 3. *The problems* REPEXT(INT, FIXED) *and* REPEXT(INT, ADD) *are* W[1]*-hard with respect to the parameter k.*

By Size of the Path. When the tree is fixed, it is easy to construct an FPT algorithm by a brute-force testing:

Proposition 4. *For t the size of T', the problems* REPEXT(PINT, FIXED) *and* REPEXT(INT, FIXED) *can be solved in time* $\mathcal{O}(n + m + f(t))$ *where* $f(t) = t^{2t^2}$.

Proof. We can have at most t^2 different subpaths and at most t^2 different vertices in G, so we test all possible representations.   □

4   Path and Chordal Graphs

We present several results concerning PATH and CHOR classes. We use many results from Section 3 as basic tools here.

4.1 Polynomial Cases

The recognition problem for types ADD and BOTH is equivalent to standard recognition without any additional tree T'. For path graphs, the current fastest algorithm is by Schäffer [15] in time $\mathcal{O}(nm)$. For chordal graphs, there is a beautiful simple algorithm in time $\mathcal{O}(n + m)$ by Rose et al. [14].

4.2 NP-Complete Cases

In all cases, we modify the reduction in Theorem 4.

Proposition 5. *Both* RECOG*(PATH, FIXED) *and* RECOG*(CHOR, FIXED) *are* NP-*complete.*

Proof (Sketch). As the split gadget, we consider a subdivided star. The tree T is a path $P_{(M+1)k}$ with three paths of length two attached to every vertex $p_{(M+1)i}$. The split gadgets can be represented only in the attached stars to T, thus splitting T into k gaps of size M. The take gadgets are the same.   □

Theorem 5. *The problems* RECOG*(PATH, SUB) *and* RECOG*(CHOR, SUB) *are* NP-*complete.*

Proof (Sketch). Subdivision can produce more vertices of degree two. We modify both gadgets so they require a certain number of vertices of degree at least three. Each split gadget is a larger tree than before. For the take gadget, we consider a caterpillar-like structure of length A_i.   □

Theorem 6. *Even with only a single subtree pre-drawn, i.e, $|G'| = 1$, the problems* REPEXT(CHOR, ADD) *and* REPEXT(CHOR, BOTH) *are* NP-*complete.*

Proof (Sketch). Modify the above reductions by adding one pre-drawn vertex v attached to everything, such that $R_v = T'$. Every intersection has to happen inside of T'.   □

To use the same technique for path graphs, we need the input partial tree T' to be a path. The complexity of REPEXT(PATH, BOTH) remains open.

Proposition 6. *The problem* REPEXT(PATH, ADD) *is* NP-*complete.*

Proof. Modify the reduction in Theorem 4 as above, with $R_v = T'$.   □

5   Conclusions

In this paper, we have considered different problems concerning extending partial representations of chordal graphs and their three subclasses. One of the main goals of this paper is to stimulate future research in this area. Therefore, we conclude with two open problems.

The first problems concerns the only open case in the table in Figure 2.

Problem 1. What is complexity of REPEXT(PATH, BOTH)?

Concerning parametrized complexity, we believe it is useful to first attack problems related to interval graphs. This allows to develop tools for more complicated chordal graphs. A generalization of Theorem 1 and Corollary 2 for INT seems to be particularly interesting. The PQ-tree approach seems to be a good starting point.

Problem 2. Does REPEXT(INT, FIXED) belong to XP with respect to k where k is the number of pre-drawn intervals?

References

1. Bläsius, T., Rutter, I.: Simultaneous PQ-ordering with applications to constrained embedding problems. CoRR abs/1112.0245 (2011)
2. Booth, K.S., Lueker, G.S.: Testing for the consecutive ones property, interval graphs, and planarity using pq-tree algorithms. Journal of Computational Systems Science 13, 335–379 (1976)
3. Corneil, D.G., Kim, H., Natarajan, S., Olariu, S., Sprague, A.P.: Simple linear time recognition of unit interval graphs. Information Processing Letters 55(2), 99–104 (1995)
4. Corneil, D.G., Olariu, S., Stewart, L.: The LBFS structure and recognition of interval graphs. SIAM Journal on Discrete Mathematics 23(4), 1905–1953 (2009)
5. Deng, X., Hell, P., Huang, J.: Linear-time representation algorithms for proper circular-arc graphs and proper interval graphs. SIAM J. Comput. 25(2), 390–403 (1996)
6. Garey, M.R., Johnson, D.S.: Complexity results for multiprocessor scheduling under resource constraints. SIAM Journal on Computing 4(4), 397–411 (1975)
7. Gavril, F.: The intersection graphs of subtrees in trees are exactly the chordal graphs. Journal of Combinatorial Theory, Series B 16(1), 47–56 (1974)
8. Golumbic, M.C.: Algorithmic Graph Theory and Perfect Graphs. North-Holland Publishing Co. (2004)
9. Jansen, K., Kratsch, S., Marx, D., Schlotter, I.: Bin Packing with Fixed Number of Bins Revisited. In: Kaplan, H. (ed.) SWAT 2010. LNCS, vol. 6139, pp. 260–272. Springer, Heidelberg (2010)
10. Klavík, P., Kratochvíl, J., Krawczyk, T., Walczak, B.: Extending Partial Representations of Function Graphs and Permutation Graphs. In: Epstein, L., Ferragina, P. (eds.) ESA 2012. LNCS, vol. 7501, pp. 671–682. Springer, Heidelberg (2012)
11. Klavík, P., Kratochvíl, J., Otachi, Y., Ignaz, R., Saitoh, T., Saumell, M., Vyskočil, T.: Extending partial representations of proper and unit interval graphs (in preparation, 2012)

12. Klavík, P., Kratochvíl, J., Vyskočil, T.: Extending Partial Representations of Interval Graphs. In: Ogihara, M., Tarui, J. (eds.) TAMC 2011. LNCS, vol. 6648, pp. 276–285. Springer, Heidelberg (2011)
13. Looges, P.J., Olariu, S.: Optimal greedy algorithms for indifference graphs. Comput. Math. Appl. 25, 15–25 (1993)
14. Rose, D.J., Tarjan, R.E., Lueker, G.S.: Algorithmic aspects of vertex elimination on graphs. SIAM Journal on Computing 5(2), 266–283 (1976)
15. Schäffer, A.A.: A faster algorithm to recognize undirected path graphs. Discrete Appl. Math. 43, 261–295 (1993)

Isomorphism for Graphs of Bounded Connected-Path-Distance-Width

Yota Otachi

School of Information Science, Japan Advanced Institute of Science and Technology
Asahidai 1-1, Nomi, Ishikawa 923-1292, Japan

Abstract. We show that GRAPH ISOMORPHISM problem (GI) can be solved in $\mathcal{O}(n^2)$ time for graphs of bounded connected-path-distance-width, and more generally, in $\mathcal{O}(n^{c+1})$ time for graphs of bounded c-connected-path-distance-width, where n is the number of vertices. These results extend the result of Yamazaki, Bodlaender, de Fluiter, and Thilikos [Isomorphism for graphs of bounded distance width. Algorithmica 24, 105–127 (1999)], who showed the fixed-parameter tractability of GI parameterized by rooted-path-distance-width.

Keywords: Graph isomorphism, Fixed-parameter tractability, Connected-path-distance-width, Treewidth.

1 Introduction

GRAPH ISOMORPHISM problem (GI, for short) is a practically and theoretically important graph problem of determining whether two graphs have the same structure. More precisely, GI asks the existence of a bijective correspondence between the vertex sets of two graphs in which the edge relation is preserved. Despite intensive efforts, GI for general graphs is not known to be polynomial-time solvable nor NP-complete (see [10,15,21]). It is known that if GI is NP-complete, then polynomial hierarchy collapses to its second level [22].

For some restricted classes of graphs, tractability of GI is known (see [14] and the references therein). It is notable that, quite recently, Grohe and Marx [11] have shown that GI for graphs excluding a graph H as a topological subgraph can be solved in $\mathcal{O}(n^{f(H)})$ time, where n is the number of vertices of the input graphs and f is a computable function. For some graph parameters, such as maximum degree [17], genus [8,18], eigenvalue multiplicity [1,7], and treewidth [2], GI is known to be in P. However, the running times of the algorithms, except for one for bounded eigenvalue multiplicity [7], exponentially depend on the parameters. For instance, Bodlaender's algorithm for GI parameterized by treewidth runs in $\mathcal{O}(n^{k+4.5})$ time, where k is the treewidth [2]. To give an $\mathcal{O}(f(k) \cdot n^c)$ time algorithm for GI parameterized by treewidth is an important open problem in parameterized complexity theory [3].

Some partial answers for the open problem are known. Yamazaki, Bodlaender, de Fluiter, and Thilikos [24] studied GI for graphs of bounded distance-width that form subsets of the class of bounded treewidth graphs, and presented some polynomial-time algorithms. Their algorithm for graphs of rooted-tree-distance-width at most k runs in $\mathcal{O}(f(k) \cdot n^3)$ time. They also presented an algorithm for graphs of rooted-path-distance-width at most k with running time $\mathcal{O}(f(k) \cdot n^2)$. Recently, Kratsch and Schweitzer [16] have investigated GI for graphs of bounded feedback vertex set number that also form a subset of the class of bounded treewidth graphs. They presented an $\mathcal{O}(f(k) \cdot n^2)$-time algorithm for deciding GI for graphs of feedback vertex number at most k. The results of these groups are incomparable by the definition of their graph parameters.

In this paper, we introduce $(c\text{-})$*connected-path-distance-width* of graphs, and present an $\mathcal{O}(f(k) \cdot n^2)$-time algorithm for deciding GI for graphs of connected-path-distance-width at most k and an $\mathcal{O}(f(k) \cdot n^{c+1})$-time algorithm for graphs of c-connected-path-distance-width at most k. Thus we show that GI is fixed-parameter tractable when parameterized by these graph parameters. By the definitions of the graph parameters, we can show that our result is an extension of one of the results by Yamazaki et al. [24].

The rest of this paper is organized as follows. In Section 2, we introduce basic concepts of graphs and define graph parameters. We also briefly review the theory of parameterized complexity there. In Section 3, we show hardness and fixed-parameter tractability of newly introduced graph parameters. In Section 4, using the results in the previous section, we present the algorithms solving GI for graphs of bounded connected-path-distance-width and for graphs of bounded c-connected-path-distance-width. In Section 5, we discuss relationships among graph parameters, and show that our results properly extend a known result. In Section 6, we conclude the paper with some open problems.

Due to the space limitation, the proofs in Section 5 are omitted.

2   Preliminaries

The graphs considered are undirected, simple, and connected. We denote by $V(G)$ and $E(G)$ the vertex set and the edge set of a graph G, respectively. For $S \subseteq G$, we denote by $G[S]$ the subgraph of G induced by S. If $G[S]$ is connected, then we say that S is *connected* in G.

Let G be a connected graph. The *distance* between $u \in V(G)$ and $v \in V(G)$ in G, denoted $d_G(u, v)$, is the length of a shortest u–v path in G. We define the *distance* between $S \subseteq V(G)$ and $v \in V(G)$ in G as $d_G(S, v) = \min_{u \in S} d_G(u, v)$. The *diameter* of G is $\mathrm{diam}(G) = \max_{u,v \in V(G)} d_G(u, v)$, and the *radius* of G is $\mathrm{rad}(G) = \min_{u \in V(G)} \max_{v \in V(G)} d_G(u, v)$. A vertex $v \in V(G)$ is a *center* of G if $\max_{v \in V(G)} d_G(u, v) = \mathrm{rad}(G)$. It is not difficult to see that $\mathrm{diam}(G) \leq |V(G)| - 1$ and $\mathrm{rad}(G) \leq \lfloor |V(G)|/2 \rfloor$. The *(open) neighborhood* of a vertex v in G, denoted $N_G(v)$, is the set of vertices adjacent to v; that is $N_G(v) = \{u \mid \{u, v\} \in E(G)\}$. The *closed neighborhood* of v in G, denoted $N_G[v]$, is the set $\{v\} \cup N_G(v)$. The *open neighborhood* of a vertex set $S \subseteq V(G)$ in G, denoted $N_G(S)$, is the set of vertices not in S and adjacent to some vertex $v \in S$; that is $N_G(S) = \bigcup_{v \in S} N_G(v) \setminus S$.

Path-distance-width and other graph parameters Path-distance-width is a graph parameter to measure how close a graph is to a path [24,23], and is a restricted variant of bandwidth (and thus, of pathwidth and of treewidth). It is known that

$$\mathsf{tw}(G) \leq \mathsf{pw}(G) \leq \mathsf{bw}(G) < 2 \cdot \mathsf{pdw}(G)$$

for any connected graph G [12,24], where tw is treewidth, pw is pathwidth, bw is bandwidth, and pdw is path-distance-width. A sequence $(L_0, \ldots, L_t)$ of subsets of vertices is a *distance structure* of a graph G if $\bigcup_{0 \leq i \leq t} L_i = V(G)$ and $L_i = \{v \in V(G) \mid d_G(L_0, v) = i\}$ for $0 \leq i \leq t$. Each L_i is a *level* and especially L_0 is the *initial set*. The *width* of $(L_0, \ldots, L_t)$, denoted $\mathsf{pdw}_{L_0}(G)$, is $\max_{0 \leq i \leq t} |L_i|$. The *path-distance-width* of G, denoted $\mathsf{pdw}(G)$, is defined as

$$\mathsf{pdw}(G) = \min\{\mathsf{pdw}_S(G) \mid S \subseteq V(G)\}.$$

The distance structure with a given initial set can be computed in linear time.

Lemma 2.1 ([24]). *Given a vertex set $S \subseteq V(G)$ of a graph G, the distance structure of G with initial set S can be constructed in $\mathcal{O}(|E(G)|)$ time. As a consequence, $\mathsf{pdw}_S(G)$ can be computed in the same time complexity.*

A distance structure of G is *connected* if its initial set is connected in G. The *connected-path-distance-width* of G, denoted $\mathsf{cpdw}(G)$, is the minimum width over all its connected distance structures; that is,

$$\mathsf{cpdw}(G) = \min\{\mathsf{pdw}_S(G) \mid S \subseteq V(G),\ G[S] \text{ is connected}\}.$$

As a generalization of connected-path-distance-width, we define *c-connected-path-distance-width* of a graph G, denoted $c\text{-}\mathsf{cpdw}(G)$, in which an initial set induces a graph with at most c connected components. For instance, $1\text{-}\mathsf{cpdw}(G) = \mathsf{cpdw}(G)$, and $2\text{-}\mathsf{cpdw}(G)$ is defined as follows:

$$2\text{-}\mathsf{cpdw}(G) = \min\{\mathsf{pdw}_S(G) \mid S \subseteq V(G),\ G[S] \text{ has at most two components}\}.$$

If the initial set of a distance structure of G is a set that consists of only one vertex, then we say that it is a *rooted distance structure* of G. The *rooted-path-distance-width* of G, denoted $\mathsf{rpdw}(G)$, is the minimum width over all its rooted distance structures; that is,

$$\mathsf{rpdw}(G) = \min\{\mathsf{pdw}_{\{v\}}(G) \mid v \in V(G)\}.$$

The following relations follow immediately from the definitions.

Proposition 2.2. *For any connected graph G,*

$$\mathsf{rpdw}(G) \geq \mathsf{cpdw}(G) \geq 2\text{-}\mathsf{cpdw}(G) \geq \mathsf{pdw}(G) \geq |V(G)|/(\mathrm{diam}(G) + 1).$$

Parameterized complexity A *parameterized problem* is a subset $L \subseteq \Sigma^* \times \mathbf{N}$, where Σ is a fixed alphabet, Σ^* is the set of all finite length strings over Σ, and $\mathbf{N}$ is the set of natural numbers. An input (x, k) to a parameterized problem consists of two parts: the main part x of the input and the parameter k. In this paper, the main part of the input is a graph or a pair of graphs. A parameterized problem L is *fixed-parameter tractable* if there exists an algorithm that decides whether $(x, k) \in L$ in $f(k) \cdot n^c$ time, where f is a computable function, c is a fixed constant, and $n = |x| + k$. The class of fixed-parameter tractable problems is FPT. The class of parameterized problems that can be solved in $\mathcal{O}(n^{f(k)})$ time is XP. As like NP-hardness in classical computational complexity theory, we have the concept of W[t]-hardness, where t is a positive integer. We only mention here that if FPT = W[1], then 3-SAT can be solved in time $2^{o(n)}$ (see [4]), and that

$$\text{FPT} \subseteq \text{W}[1] \subseteq \text{W}[2] \subseteq \cdots \subseteq \text{W}[t] \subseteq \text{XP}.$$

See standard textbooks [6,9,19] on parameterized complexity and fixed-parameter algorithms.

3   Hardness and Fixed-Parameter Tractability of cpdw

In this section, we present the main tool for developing the fixed-parameter tractable algorithm for GI parameterized by cpdw. We first show NP-hardness of the problem of determining cpdw. Next we show that the parameterized problem of cpdw parameterized by itself is fixed-parameter tractable. Indeed, we show a slightly stronger fact: all connected set with width at most k in a graph can be listed in $\mathcal{O}(f(k) \cdot n^2)$ time. This stronger fact is necessary for our GI algorithm.

3.1   NP-Hardness

It is known that rpdw can be determined in linear time, while determining pdw is NP-hard [24]. Here we show that determining cpdw is NP-hard even for co-bipartite graphs. A graph G is *cobipartite* if its vertices can be partitioned into two cliques, where a *clique* is a set of pairwise adjacent vertices. The following fact was shown in our work on approximability of pdw for AT-free graphs.

Theorem 3.1 ([20]). *Given a cobipartite graph G, it is NP-complete to decide whether* $\text{pdw}(G) \leq |V(G)|/3$.

Using the fact above, we can show the analogous result for cpdw.

Theorem 3.2. *Given a cobipartite graph G, it is NP-complete to decide whether* $c\text{-cpdw}(G) \leq |V(G)|/3$ *for any $c \geq 1$.*

Proof. Let G be a cobipartite graph. We shall show that $c\text{-cpdw}(G) \leq |V(G)|/3$ if and only if $\text{pdw}(G) \leq |V(G)|/3$. The only-if part is trivial. For the if part assume that $\text{pdw}_S(G) \leq |V(G)|/3$ for some S. Let (X, Y) be a partition of $V(G)$ such that both X and Y are cliques. If S intersects both X and Y, then every vertex is included in $N_G[S]$. Hence the distance structure with S as its initial set has at most two levels. Therefore, it follows $\mathsf{pdw}_S(G) \geq |V(G)|/2$, a contradiction. Now we have $S \subseteq X$ or $S \subseteq Y$, and thus S is connected in G. □

Corollary 3.3. *Given a cobipartite graph G, it is NP-complete to decide whether* $\mathsf{cpdw}(G) \leq |V(G)|/3$.

3.2 Fixed-Parameter Tractability

A vertex set $S \subseteq V(G)$ is *feasible* with G and k if S is connected in G and $\mathsf{pdw}_S(G) \leq k$. In other words, a feasible set with G and k is a certificate for the fact $\mathsf{cpdw}(G) \leq k$. Similarly, a *c-feasible set* with G and k is a set $S \subseteq V(G)$ such that $\mathsf{pdw}_S(G) \leq k$ and $G[S]$ has at most c connected components. To show the fixed-parameter tractability of cpdw parameterized by itself, it suffices to present a fixed-parameter tractable algorithm that finds a feasible set or decides nonexistence of such a set. However, to show the fixed-parameter tractability of GI parameterized by cpdw, we have to present a fixed-parameter tractable algorithm that finds all the feasible sets.

The following two simple facts are crucial for developing our fixed-parameter tractable algorithm.

Lemma 3.4. *Let G be a graph and $S \subseteq V(G)$. If there exists $S' \subseteq V(G)$ such that $S \subseteq S'$ and $\mathsf{pdw}_{S'}(G) \leq k$, then $|S| \leq k$ and $|N[S]| \leq 2k$.*

Proof. The fact $|S| \leq k$ is trivially true. Let $(L_0, \dots, L_t)$ be the distance structure with $L_0 = S'$. Since $N[S'] = L_0 \cup L_1$, it follows that $N[S] \subseteq L_0 \cup L_1$. As $\mathsf{pdw}_{S'}(G) = \max_i |L_i| \leq k$, the lemma holds. □

Lemma 3.5. *Let G be a graph. If $S \subsetneq S' \subseteq V(G)$ and S' is connected in G, then there is a vertex $u \in S' \setminus S$ such that $u \in N_G(S)$.*

Proof. If there is no such vertex u, then S' is not connected. □

Now we are ready to present the main theorem in this section.

Theorem 3.6. *All the feasible sets with a graph G and an integer k can be enumerated in $\mathcal{O}(n^2(2k)!/(k!))$ time, where $n = |V(G)|$.*

Proof. First observe that if $\mathsf{pdw}(G) \leq k$, then $\deg_G(v) \leq 3k - 1$ for each $v \in V(G)$. This is because the vertices in $N[v]$ can be distributed to at most three consecutive levels. Hence if $|E(G)| > n(3k - 1)/2$, then we can conclude that there is no feasible sets with G and k. This check can be done in linear time. In the following, we assume that $|E(G)| \leq n(3k - 1)/2$.

The algorithm and its correctness Now we describe our algorithm. See Algorithm 1. For each $v \in V(G)$, we enumerate all the feasible sets that include v. Lines 1–3 in Algorithm 1 correspond to this loop. Procedure ENUM(S) enumerate all the feasible sets that contain S. If $|S| > k$ or $|N[S]| > 2k$, then there is no feasible superset of S with G and k by Lemma 3.4 (Lines 5–7). If S itself is feasible, then the algorithm outputs S (Lines 8–10). Now it finds feasible sets that contain S as a proper subset (Lines 11–13). By Lemma 3.5, if S is a proper subset of a feasible set S', then there is a vertex $u \in S' \setminus S$ such that $u \in N_G(S)$. Thus for each $u \in N_G(S)$ we compute $\mathrm{ENUM}(S \cup \{u\})$. Clearly, every feasible set is listed by the algorithm.[1]

Algorithm 1. Enumerate all the feasible sets with G and k

1: **for all** $v \in V(G)$ **do**
2:     $\mathrm{ENUM}(\{v\})$
3: **end for**

4: **procedure** $\mathrm{ENUM}(S)$
5:     **if** $|S| > k$ **or** $|N[S]| > 2k$ **then**
6:         **return**
7:     **end if**
8:     **if** $\mathsf{pdw}_S(G) \leq k$ **then**
9:         Output S
10:     **end if**
11:     **for all** $u \in N_G(S)$ **do**
12:         $\mathrm{ENUM}(S \cup \{u\})$
13:     **end for**
14: **end procedure**

The running time of the algorithm In the for loop, we have n branches corresponding to the calls of $\mathrm{ENUM}(\{v\})$ for all $v \in V(G)$. Each call of $\mathrm{ENUM}(S)$ has $|N_G(S)|$ branches corresponding to the calls of $\mathrm{ENUM}(S \cup \{u\})$ for each $u \in N_G(S)$ if $|S| \leq k$ and $|N[S]| \leq 2k$. Therefore, the search-tree has depth k in which the root has n children, and each node in the depth $d \in \{1, \ldots, k-1\}$ has at most $2k - d$ children. Thus it has at most $n(2k-1)(2k-2)\cdots(k+1) = n(2k-1)!/(k!)$ leaves, and hence it has less than $2n(2k-1)!/(k!)$ nodes. Since the algorithm takes $\mathcal{O}(|E(G)|) = \mathcal{O}(nk)$ time to check whether $\mathsf{pdw}_S(G) \leq k$ for each node of the search-tree, the total running time is $\mathcal{O}(n^2(2k)!/(k!))$. $\square$

By the theorem above, the following fact, which can be of independent interest, immediately follows.

Corollary 3.7. *Given a graph G and an integer k, the problem of deciding whether $\mathsf{cpdw}(G) \leq k$ can be solved in $\mathcal{O}(n^2(2k)!/(k!))$ time.* $\square$

Now we generalize the algorithm so that it works for c-feasible sets. The modification is very simple. We just start with each set with at most c vertices in the outer loop.

---

[1] The algorithm may output a feasible set more than once. This fact does not affect the correctness of our algorithm.

Theorem 3.8. *All the c-feasible sets with a graph G and an integer k can be enumerated in $\mathcal{O}(n^{c+1}(2k)!/(k!))$ time, where $n = |V(G)|$.*

Proof. We first guess $c' \le c$ vertices that belong to different components. This needs $\mathcal{O}(n^c)$ first branches instead of $\mathcal{O}(n)$ first branches in Algorithm 1. The other steps are the same. See Algorithm 2. □

Corollary 3.9. *Given a graph G and an integer k, the problem of deciding whether c-cpdw$(G) \le k$ can be solved in $\mathcal{O}(n^{c+1}(2k)!/(k!))$ time.* □

Algorithm 2. Enumerate all the c-feasible sets with G and k

1: **for all** $S \subseteq V(G)$ such that $|S| \le c$ **do**
2:    ENUM(S)
3: **end for**

---

4   Fixed-Parameter Tractable Algorithms for GI

Using the algorithms in the previous section, we present the main algorithm of this paper here. Two distance structures $(L_0, \ldots, L_t)$ of G and $(L'_0, \ldots, L'_t)$ of H are *isomorphic* if there exists an isomorphism η from G to H such that $\eta(L_i) = L'_i$ for $0 \le i \le t$. It is easy to see that two graphs G and H are isomorphic if and only if there exist isomorphic distance structures of them.

Yamazaki et al. [24] presented a procedure SUB-RPDW that takes two distance structures of graphs as input and decides whether the distance structures are isomorphic in $\mathcal{O}(k!^2 k^2 \cdot n)$ time.

Theorem 4.1. *Isomorphism for graphs of connected-path-distance-width at most k can be decided in $\mathcal{O}((2k)!k!k \cdot n^2)$ time, where n is the number of vertices.*

Proof. Assume that $|V(G)| = |V(H)| = n$ and $|E(G)| = |E(H)| \le n(3k-1)/2$. The algorithm first finds a feasible set S with G and k by Algorithm 1. It next enumerates all the feasible sets S' with H and k, and for each S' determines whether the distance structures with initial sets S and S' are isomorphic by using SUB-RPDW. This can be done by slightly modifying Algorithm 1. See Algorithm 3 for the full description. The correctness of Algorithm 3 follows from the correctness of Algorithm 1 and SUB-RPDW.

Now we analyze the running time. The first step takes $\mathcal{O}(n^2(2k)!/(k!))$ time. The second step takes $\mathcal{O}(|E(G)|) = \mathcal{O}(nk)$ time. In the for loop (Lines 3–8), the algorithm traverses its search tree of less than $2n(2k-1)!/(k!)$ nodes. At each node, the algorithm takes $\mathcal{O}(nk)$ time for constructing the distance structure L' of H, and $\mathcal{O}(k!^2 k^2 \cdot n)$ time for calling SUB-RPDW(L, L'). Therefore, the total running time is $\mathcal{O}((2k)!k!k \cdot n^2)$. □

Algorithm 3. Isomorphism for graphs G and H of cpdw at most k

---
1: Find a minimal feasible set S with G and k.
2: Let $L = (L_0, \ldots, L_t)$ be the distance structure of G with $L_0 = S$.
3: **for all** minimal feasible set S' with H and k **do**
4:    Let $L' = (L'_0, \ldots, L'_{t'})$ be the distance structure of H with $L'_0 = S'$.
5:    **if** SUB-RPDW(L, L') = true **then**
6:        **return** yes
7:    **end if**
8: **end for**
9: **return** no

Generalizing the algorithm above for c-cpdw can be done by using c-feasible sets instead of feasible sets.

Theorem 4.2. *Isomorphism for graphs of c-connected-path-distance-width at most k can be decided in $\mathcal{O}((2k)!k!k \cdot n^{c+1})$ time.* $\square$

The running time $\mathcal{O}((2k)!k!k \cdot n^{c+1})$ is unacceptable if c can be large. However, if we fix c to be a constant, then we have a fixed-parameter tractable algorithm for a larger class. For example, we can decide GI for graphs of bounded 2-cpdw in $\mathcal{O}(f(k) \cdot n^3)$ time.

5   Relationships among Graph Parameters

To clarify the significance of our results, we discuss the relationships among some graph parameters related to parameterized complexity of GI.

By the definition, $\mathsf{rtdw}(G) \leq \mathsf{rpdw}(G)$ and $2\text{-}\mathsf{cpdw}(G) \leq \mathsf{cpdw}(G) \leq \mathsf{rpdw}(G)$ for any graph G, where $\mathsf{rtdw}(G)$ is rooted-tree-distance-width. Here we obtain two nontrivial relations. First we show that rpdw of a graph is bounded by a function that depends only on its cpdw. This relation implies that, although our algorithm is significantly faster than theirs in the worst case, the algorithm by Yamazaki et al. [24] for GI parameterized by rpdw also runs in $\mathcal{O}(f(k) \cdot n^2)$ time for GI parameterized by cpdw. Next we show that there is no such relation between rtdw and $2\text{-}\mathsf{cpdw}$, that is, there is a family of graphs that has constant $2\text{-}\mathsf{cpdw}$ and unbounded rtdw. Thus we show that graphs of bounded $2\text{-}\mathsf{cpdw}$ form a new subclass of graphs of bounded treewidth that admits a fixed-parameter tractable algorithm for GI.

5.1   Upper Bounding rpdw by cpdw

The next two propositions show that $\mathsf{rpdw}(G) \leq (\mathsf{cpdw}(G)/2 + 1)\mathsf{cpdw}(G)$, and that this bound is tight up to a constant factor.

Proposition 5.1. *If $\mathsf{cpdw}(G) = k$, then $\mathsf{rpdw}(G) \leq (\lfloor k/2 \rfloor + 1)k$.*

Proposition 5.2. *For any integer $k \geq 2$, there is a graph G such that $\mathsf{cpdw}(G) \leq 2k$ and $\mathsf{rpdw}(G) \geq k(k+1)/2$.*

5.2 The Classes of Bounded 2-cpdw and of Unbounded rtdw

Now we show that the classes of bounded 2-cpdw and of bounded rtdw are incomparable. Observe that $\mathsf{rtdw}(K_{1,n}) = 1$ since the star $K_{1,n}$ is a tree, while $2\text{-}\mathsf{cpdw}(K_{1,n}) \geq \mathsf{pdw}(K_{1,n}) = \lceil n/2 \rceil$. Thus any function only depends on rtdw cannot be an upper bound of 2-cpdw. In what follows, we show that any function only depends on 2-cpdw cannot be an upper bound of rtdw as well.

Yamazaki et al. [24] showed that a class of trees has bounded pdw and unbounded rpdw. We can modify each tree in the class so that the new class consists of all the modified graphs has bounded 2-cpdw and unbounded rtdw.

Proposition 5.3. *For any positive integer $k \geq 3$, there is a graph G such that $2\text{-}\mathsf{cpdw}(G) \leq 6$ and $\mathsf{rtdw}(G) \geq k$.*

6 Concluding Remarks

We show that new subclasses of the class of bounded treewidth graphs admit fixed-parameter tractable algorithms for GI. However, the original open problem is still unsettled.

Open Problem . Is GI parameterized by treewidth fixed-parameter tractable?

It would be also interesting to study GI parameterized by pathwidth, tree-distance-width, and path-distance-width.

There is another aspect of study on GI for bounded treewidth graphs. Das, Torán, and Wagner [5] studied space complexity of the problem. They showed that GI for graphs of bounded tree-distance-width is L-complete, and that GI for graphs of bounded treewidth is in LogCFL. Whether the latter result can be improved to L is unknown. Recently, in his blog post, Kintali [13] has announced that GI for graphs of treewidth 4 is $\oplus$L-hard, which means that GI for graphs of bounded treewidth (at least 4) is not in L unless $L = \oplus L$.

References

1. Babai, L., Grigoryev, D., Mount, D.: Isomorphism of graphs with bounded eigenvalue multiplicity. In: 14th Annual ACM Symposium on Theory of Computing (STOC 1982), pp. 310–324 (1982)
2. Bodlaender, H.L.: Polynomial algorithms for graph isomorphism and chromatic index on partial k-trees. J. Algorithms 11, 631–643 (1990)
3. Bodlaender, H.L., Demaine, E.D., Fellows, M.R., Guo, J., Hermelin, D., Lokshtanov, D., Müller, M., Raman, V., van Rooij, J., Rosamond, F.A.: Open problems in parameterized and exact computation — IWPEC 2008. Tech. Rep. UU-CS-2008-017, Department of Information and Computing Sciences, Utrecht University (2008)
4. Cai, L., Juedes, D.: On the existence of subexponential parameterized algorithms. J. Comput. System Sci. 67, 789–807 (2003)

5. Das, B., Torán, J., Wagner, F.: Restricted space algorithms for isomorphism on bounded treewidth graphs. In: 27th International Symposium on Theoretical Aspects of Computer Science (STACS 2010). LIPIcs, vol. 5, pp. 227–238 (2010)
6. Downey, R.G., Fellows, M.R.: Parameterized Complexity. Springer (1998)
7. Evdokimov, S., Ponomarenko, I.: Isomorphism of coloured graphs with slowly increasing multiplicity of jordan blocks. Combinatorica 19, 321–333 (1999)
8. Filotti, I.S., Mayer, J.N.: A polynomial-time algorithm for determining the isomorphism of graphs of fixed genus. In: 12th Annual ACM Symposium on Theory of Computing (STOC 1980), pp. 236–243 (1980)
9. Flum, J., Grohe, M.: Parameterized Complexity Theory. Springer (2006)
10. Garey, M.R., Johnson, D.S.: Computers and Intractability: A Guide to the Theory of NP-Completeness. Freeman (1979)
11. Grohe, M., Marx, D.: Structure theorem and isomorphism test for graphs with excluded topological subgraphs. In: 44th Annual ACM Symposium on Theory of Computing (STOC 2012), pp. 173–192 (2012)
12. Kaplan, H., Shamir, R.: Pathwidth, bandwidth, and completion problems to proper interval graphs with small cliques. SIAM J. Comput. 25, 540–561 (1996)
13. Kintali, S.: Hardness of graph isomorphism of bounded treewidth graphs, http://kintali.wordpress.com/2011/11/16/
14. Köbler, J.: On Graph Isomorphism for Restricted Graph Classes. In: Beckmann, A., Berger, U., Löwe, B., Tucker, J.V. (eds.) CiE 2006. LNCS, vol. 3988, pp. 241–256. Springer, Heidelberg (2006)
15. Köbler, J., Schöning, U., Torán, J.: The Graph Isomorphism Problem: Its Structural Complexity. Birkhauser Verlag (1993)
16. Kratsch, S., Schweitzer, P.: Isomorphism for Graphs of Bounded Feedback Vertex Set Number. In: Kaplan, H. (ed.) SWAT 2010. LNCS, vol. 6139, pp. 81–92. Springer, Heidelberg (2010)
17. Luks, E.M.: Isomorphism of graphs of bounded valence can be tested in polynomial time. J. Comput. System Sci. 25, 42–65 (1982)
18. Miller, G.: Isomorphism testing for graphs of bounded genus. In: 12th Annual ACM Symposium on Theory of Computing (STOC 1980), pp. 225–235 (1980)
19. Niedermeier, R.: Invitation to Fixed Parameter Algorithms. Oxford University Press (2006)
20. Otachi, Y., Saitoh, T., Yamanaka, K., Kijima, S., Okamoto, Y., Ono, H., Uno, Y., Yamazaki, K.: Approximability of the Path-Distance-Width for AT-free Graphs. In: Kolman, P., Kratochvíl, J. (eds.) WG 2011. LNCS, vol. 6986, pp. 271–282. Springer, Heidelberg (2011)
21. Read, R.C., Corneil, D.G.: The graph isomorphism disease. J. Graph Theory 1, 339–363 (1977)
22. Schöning, U.: Graph isomorphism is in the low hierarchy. J. Comput. System Sci. 37, 312–323 (1988)
23. Yamazaki, K.: On approximation intractability of the path-distance-width problem. Discrete Appl. Math. 110, 317–325 (2001)
24. Yamazaki, K., Bodlaender, H.L., de Fluiter, B., Thilikos, D.M.: Isomorphism for graphs of bounded distance width. Algorithmica 24, 105–127 (1999)

Algorithmic Aspects of the Intersection and Overlap Numbers of a Graph

Danny Hermelin[1], Romeo Rizzi[2], and Stéphane Vialette[3],[*]

[1] Max-Planck-Institut für Informatik, Saarbrücken, Germany
hermelin@mpi-inf.mpg.de
[2] Dipartimento di Matematica ed Informatica, Universit degli Studi di Udine, Italy
rrizzi@dimi.uniud.it
[3] LIGM CNRS UMR 8049, Université Paris-Est, France ·
vialette@univ-mlv.fr

Abstract. The intersection number of a graph G is the minimum size of a set S such that G is an intersection graph of some family of subsets $\mathcal{F} \subseteq 2^S$. The overlap number of G is defined similarly, except that G is required to be an overlap graph of $\mathcal{F}$. Computing the overlap number of a graph has been stated as an open problem in [B. Rosgen and L. Stewart, 2010, arXiv:1008.2170v2] and [D.W. Cranston, *et al.*, J. Graph Theory., 2011]. In this paper we show two algorithmic aspects concerning both these graph invariants. On the one hand, we show that the corresponding optimization problems associated with these numbers are both **APX**-hard, where for the intersection number our results hold even for biconnected graphs of maximum degree 7, strengthening the previously known hardness result. On the other hand, we show that the recognition problem for any specific intersection graph class (*e.g.* interval, unit disc, ...) is easy when restricted to graphs with fixed intersection number.

1 Introduction

An *intersection graph* is a graph that represents the pattern of intersections of a family of sets. Any undirected graph G may be represented as an intersection graph: For each vertex of G, form a set consisting of the edges incident to this vertex; then two such sets have a nonempty intersection if and only if the corresponding vertices share an edge. Erdős, Goodman, and Pósa [5] provided a construction that is more efficient in which the total number of set elements is at most $n^2/4$, where n is the number of vertices in the graph. Many important graph families can be described as intersection graphs of more restricted types of set families, in particular sets corresponding to geometric objects. Examples of such graph classes are *interval graphs* (intersection graphs of intervals on the real line), *circle graphs* (intersection graphs of chords in a circle), and *unit disc graphs* (intersection graphs of unit discs in the plane).

---

[*] Partially supported by ANR project BIRDS JCJC SIMI 2-2010.

K.-M. Chao, T.-s. Hsu, and D.-T. Lee (Eds.): ISAAC 2012, LNCS 7676, pp. 465–474, 2012.

The *intersection number* of a graph G, denoted $i(G)$, is defined to be the minimum cardinality of a (ground) set S such that G is an intersection graph of a family of subsets $\mathcal{F} \subseteq 2^S$ of S. In [5], it was shown that $i(G)$ also equals the minimum number of complete subgraphs needed to cover the edges of G. This latter number is known as the *edge-clique cover* number of G, and is denoted $\theta(G)$. (The best general reference is [14].) Computing $\theta(G)$ (and hence $i(G)$) is **NP**-hard [11,16], even when restricted to planar graphs [3] or graphs with maximum degree 6 [9]. It is polynomial-time solvable for chordal graphs [13], graphs with maximum degree 5 [9], line graphs [16], and circular-arc graphs [10]. By way of contrast, it is not approximable within ratio n^ε for some $\varepsilon > 0$ unless $\mathbf{P} = \mathbf{NP}$ [12], and so far nothing better than a polynomial ratio of $O(n^2 \frac{(\log \log n)^2}{(\log n)^3})$ is known [2]. As for its parameterized complexity, computing $\theta(G)$ is fixed-parameter tractable under the standard parameterization [6]. Computing $\theta(G)$ is very widely applicable to discover underlying structure in complex real-world networks [7], and [15] gives a bioinformatic application for this problem.

The overlap model for graph representations arose much later and is not as well studied [4]. The *overlap graph* of a family of sets $\mathcal{F} = \{S_1, S_2, \ldots, S_n\}$, denoted $O(\mathcal{F})$, is the graph having $\mathcal{F}$ as vertex set with S_i adjacent to S_j if and only if S_i and S_j intersect and neither set is contained in the other, *i.e.*, $S_i \cap S_j \neq \emptyset$, $S_i \setminus S_j \neq \emptyset$, and $S_j \setminus S_i \neq \emptyset$. Notice that some graph classes can play it both ways: A graph is an intersection graph of chords in a circle if and only if it is has an overlap representation using intervals on a line. The *overlap number* of a graph G, denoted $\varphi(G)$, is the minimum size of the ground set in any overlap representation of G. Extending an overlap representation and finding a minimum overlap representation with limited containment have been shown to be **NP**-hard problems [19]. However, the general problem of computing the overlap number of a graph has been stated as an open problem in [19] and [4]. The following upper bounds for the overlap number of a n-vertex graph are known [18,19]: $n+1$ for trees, $2n$ for chordal graphs, $\frac{10}{3} n - 6$ for planar graphs, and $\lfloor n^2/4 \rfloor + n$ for general graphs. In [4], it is shown among other results that an optimal overlap representation of a tree can be produced in linear-time, and its size is the number of vertices in the largest subtree in which the neighbor of any leaf has degree 2.

The results in this paper are twofold. In the first part of the paper, we consider the INTERSECTION NUMBER and OVERLAP NUMBER problems, the optimization problems that ask to respectively determine the intersection and overlap number of a given input graph. We show that both problems are **APX**-hard. While for INTERSECTION NUMBER this was already known for general graphs [12], our result proves this is the case also for graphs of maximum degree 7. Moreover, this result is used to show the **APX**-hardness of OVERLAP NUMBER. In the second part of the paper, we show that for any intersection graph class $\mathcal{G}$, *i.e.* any graph class defined by specifying the allowed intersection model, the recognition problem associated with $\mathcal{G}$ is linear-time solvable when restricted to graphs with bounded fixed intersection number.

2   Notations

Let G be a graph. We write $\mathbf{V}(G)$ for the set of vertices and $\mathbf{E}(G)$ for the set of edges of G. The *neighborhood* of a vertex u, denoted $N_G(u)$ or (when the graph is unambiguous) $N(u)$, is the set of adjacent vertices to u. We let $N[v]$ denote the set $\{v\} \cup N(v)$. The *degree* of a vertex $u \in \mathbf{V}(G)$, denoted $d(u)$, is the number of vertices adjacent to u. The *maximum degree* of G, denoted by $\Delta(G)$, is the maximum degree of its vertices. A *biconnected graph* is a connected graph that is not broken into disconnected pieces by deleting any single vertex (and its incident edges). An *edge-clique cover* of G is any family $\mathcal{E} = \{Q_1, Q_2, \ldots, Q_k\}$ of complete subgraphs of G such that every edge of G is in at least one of $Q_1, Q_2, \ldots, Q_k$. The minimum cardinality of an edge-clique cover of G is denoted $\theta(G)$, and we write EDGE-CLIQUE COVER for the combinatorial problem of computing $\theta(G)$.

The *Cartesian product* $G \times H$ of graphs G and H is the graph such that the vertex set of $G \times H$ is the Cartesian product $\mathbf{V}(G) \times \mathbf{V}(H)$, and any two vertices (u, u') and (v, v') are adjacent in $G \times H$ if and only if either $u = v$ and u' is adjacent with v' in H, or $u' = v'$ and u is adjacent with v in G. A *column* of $G \times H$ is the set of vertices $\{(u, u') : u \in \mathbf{V}(G)\}$ for some vertex $u' \in \mathbf{V}(H)$, and a *row* of $G \times H$ is the set of vertices $\{(u, u') : u' \in \mathbf{V}(H)\}$ for some vertex $u \in \mathbf{V}(G)$. Observe that each row induces a copy of H, and each column induces a copy of G (see Figure 1). This terminology is consistent with a representation of $G \times H$ by the points of the $|\mathbf{V}(G)| \times |\mathbf{V}(H)|$ grid. (See Figure 1 for an illustration.)

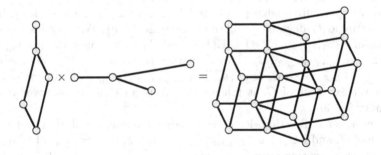

Fig. 1. The cartesian products of two graphs

3   Hardness of Approximating $i(G)$

Since $i(G) = \theta(G)$ [5], we prove hardness of approximation for bounded degree graphs in terms of edge-clique covers. Notice that this result will be the main ingredient of upcoming Proposition 2.

Proposition 1. EDGE-CLIQUE COVER *is* **APX**-*hard for biconnected graphs with maximum degree 7.*

Proof. The reduction is from VERTEX COVER in cubic graphs, which is known to be **APX**-hard [17,1]. Let G be a cubic n-vertex graph. We construct a new graph H as follows. We represent each vertex $u \in \mathbf{V}(G)$ by a triangle T_u with vertices u_0, u_1 and u_2 in the new graph H. These n triangles are all vertex disjoint in H, and each of them can offer a different edge for three connections. Let us be more specific. For each vertex $u \in \mathbf{V}(G)$ with incident edges $e_u[0]$, $e_u[1]$ and $e_u[2]$ (the order is arbitrary), the edge $\{u_i, u_{i+1 \pmod 3}\} \in T_u$, $0 \le i \le 2$, is devoted to the edge $e_u[i] \in \mathbf{E}(G)$. Now, to represent an edge $\{u, v\} \in \mathbf{E}(G)$ in H, where $\{u_i, u_{i+1 \pmod 3}\}$, $0 \le i \le 2$, is the edge of the triangle T_u devoted to representing $\{u, v\}$, and $\{v_j, v_{j+1 \pmod 3}\}$, $0 \le j \le 2$, is the edge of the triangle T_v devoted to representing $\{u, v\}$, we introduce two new vertices $A_{u,v}$ and $B_{u,v}$ and the 6 edges $\{A_{u,v}, u_i\}$, $\{A_{u,v}, v_j\}$, $\{B_{u,v}, u_{i+1 \pmod 3}\}$, $\{B_{u,v}, v_{j+1 \pmod 3}\}$, and $\{u_i, v_j\}$. What is left is to add m non-incident edges to H (one additional edge for each edge of G): For each edge $\{u, v\} \in \mathbf{E}(G)$ in H, where $\{u_i, u_{i+1 \pmod 3}\}$, $0 \le i \le 2$, is the edge of the triangle T_u devoted to representing $\{u, v\}$, and $\{v_j, v_{j+1 \pmod 3}\}$, $0 \le j \le 2$, is the edge of the triangle T_v devoted to representing $\{u, v\}$, we add the edge $\{u_i, v_{j+1 \pmod 3}\}$ or $\{u_{i+1 \pmod 3}, v_j\}$ (the choice is made so that these m additional edges form a matching). We refer the reader to Figure 2 for an illustration. Clearly $|\mathbf{V}(H)| = \frac{9}{2}n$ and $|\mathbf{E}(H)| = \frac{27}{2}n$. Moreover, it follows from the construction that H is a biconnected graph with maximum degree 7. We claim that G has a vertex cover of size k if and only if $\theta(H) \le k + 3m$, where $m = \frac{3}{2}n$ is the number of edges of G.

Suppose G has a vertex cover $V' \subseteq \mathbf{V}(G)$ of size k. Construct an edge-clique cover $\mathcal{E}$ of H as follows. For each $u \in V'$, add T_u to $\mathcal{E}$. For each edge $\{u, v\} \in \mathbf{E}(G)$, let $\{u_i, u_{i+1 \pmod 3}\}$, $i \in \{0, 1, 2\}$, be the edge of triangle T_u devoted to representing edge $\{u, v\}$, and $\{v_j, v_{j+1 \pmod 3}\}$, $j \in \{0, 1, 2\}$, be the edge of triangle T_v devoted to representing edge $\{u, v\}$. Without loss of generality, assume $\{u_i, v_{j+1 \pmod 3}\} \in \mathbf{E}(H)$. Add the two cliques $\{u_i, v_j, A_{u,v}\}$ and $\{u_{i+1 \pmod 3}, v_{j+1 \pmod 3}, B_{u,v}\}$ to $\mathcal{E}$. Furthermore, if $u \in V'$, add the clique $\{u_i, v_j, v_{j+1 \pmod 3}\}$ to $\mathcal{E}$, and $\{u_i, u_{i+1 \pmod 3}, v_{j+1 \pmod 3}\}$ otherwise. Since V' is a vertex cover of G, it follows that $\mathcal{E}$ is an edge-clique cover of H of cardinality $k + 3m$.

For the reverse direction, let $\mathcal{E}$ be an edge-clique cover of H. Let $\{u, v\}$ be any edge of G, and let $\{u_i, u_{i+1 \pmod 3}\}$, $i \in \{0, 1, 2\}$, be the edge of triangle T_u devoted to representing edge $\{u, v\}$, and $\{v_j, v_{j+1 \pmod 3}\}$, $j \in \{0, 1, 2\}$, be the edge of triangle T_v devoted to representing edge $\{u, v\}$. Without loss of generality, assume $\{u_i, v_{j+1 \pmod 3}\} \in \mathbf{E}(H)$. If we let $H_{u,v}$ stand for be the subgraph of H induced by the subset $\{u_i, u_{i+1 \pmod 3}, A_{u,v}, B_{u,v}, v_j, v_{j+1 \pmod 3}\}$, we make the easy observations (see Figure 2) that (i) 4 cliques are needed to cover the edges of $H_{u,v}$, and (ii) 3 cliques are needed to cover the edges of $H_{u,v}$ if $\{u_i, u_{i+1 \pmod 3}\}$ or $\{v_j, v_{j+1 \pmod 3}\}$ (possibly both) is removed. Therefore, $|\mathcal{E}| = 3m + k$ for some non-negative integer $k \le m$. But each triangle T_w, $w \in \mathbf{V}(G)$, can be covered by 1 clique, and hence there is no loss of generality in assuming $k \le n$. Furthermore, there is no loss go generality in assuming that $\mathcal{E}$ satisfies the following property: for every edge $\{u, v\} \in \mathbf{E}(G)$, either T_u or T_v (possibly both) is in $\mathcal{E}$. Let $V' \subseteq$

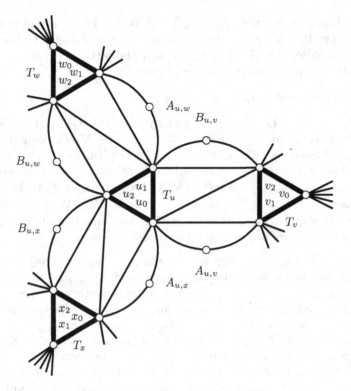

Fig. 2. The three edge-gadgets used in the proof of Proposition 1 for vertex $u \in \mathbf{V}(G)$ with edges $e_u[0] = \{u, v\}$, $e_u[1] = \{u, w\}$, and $e_u[2] = \{u, x\}$

$\mathbf{V}(G)$ be the subset defined as follows: $u \in V'$ if $T_u \in \mathcal{E}$. According to the above, if $|\mathcal{E}| = 3m + k$ for some non-negative integer $k \leq n$, then $|V'| = k$ is a vertex cover of G. $\square$

4   Hardness of Approximating $\varphi(G)$

This section is devoted to proving that there exists a constant $c > 1$ such that computing the overlap number of a graph is hard to approximate to within c.

Proposition 2. OVERLAP NUMBER *is* **APX**-*hard.*

Proof. According to Proposition 1, there exists a constant $c > 0$ such that $\theta(G)$ cannot be approximated to within c (unless $\mathbf{P} = \mathbf{NP}$). We shall prove that a $\sqrt{c}$-approximation algorithm for OVERLAP NUMBER would yield a c-approximation algorithm for EDGE-CLIQUE COVER

Let G be a n-vertex graph with maximum degree 7 for which we are asked to c-approximate $\theta(G)$. Without loss of generality, we assume that G has no isolated vertices and is biconnected (see Proposition 1). Let m be the smallest integer such that $m \geq n$ and $\frac{m}{m-1} < \sqrt{c}$, and let K_m be the complete graph on m vertices. Let $H = K_m \times G$ be the Cartesian product of K_m by G. For the sake of simplicity, write $\mathbf{V}(K_m) = \{u_1, u_2, \ldots, u_m\}$ and $\mathbf{V}(G) = \{v_1, v_2, \ldots, v_n\}$. We have divided the proof into a sequence of claims.

Claim. $\varphi(H) \leq n + m\,\theta(G)$.

Proof (Of Claim 4). Let $k = \theta(G)$ and let $\mathcal{E} = \{Q_1, Q_2, \ldots, Q_k\}$ be a size-k edge-clique cover of G. To every node $(u_i, v_j) \in \mathbf{V}(H)$, we associate the set $S_{(u_i,v_j)}$ defined as follows: $S_{(u_i,v_j)} = \{v_j\} \cup \{(u_i, p) : v_j \in Q_p\}$. Consider the family $\mathcal{F} = \{S_{(u_i,v_j)} : (u_i, v_j) \in \mathbf{V}(H)\}$ defined over the ground set $X = \bigcup_{(u_i,v_j) \in \mathbf{V}(H)} S_{(u_i,v_j)} = \mathbf{V}(G) \cup (\mathbf{V}(K_m) \times [k])$, where $[k]$ is the set of the first k integers $\{1, 2, \ldots, k\}$. Notice that $|X| = n + km$. We prove that $O(\mathcal{F})$ and H are isomorphic graphs, thereby proving the claim. Indeed, let $S_{(u_i,v_j)}$ and $S_{(u_r,v_s)}$ be two subsets of $\mathcal{F}$. We need to consider 3 cases.

- If $u_i \neq u_r$ and $v_j \neq v_s$, then (u_i, v_j) and (u_r, v_s) are not adjacent vertices in H. It can be easily verified that $S_{(u_i,v_j)}$ and $S_{(u_r,v_s)}$ are disjoint subsets, and hence $S_{(u_i,v_j)}$ and $S_{(u_r,v_s)}$ are not adjacent vertices in $O(\mathcal{F})$.
- If $u_i \neq u_r$ and $v_j = v_s$, then (u_i, v_j) and (u_r, v_s) are adjacent vertices in H since K_m is a clique. Firstly, $v_j \in S_{(u_i,v_j)}$ and $v_j \in S_{(u_r,v_s)}$ since $v_j = v_s$, and hence $S_{(u_i,v_j)} \cap S_{(u_r,v_s)} \neq \emptyset$. Secondly, both $v_j \in S_{(u_i,v_j)} \setminus S_{(u_r,v_s)}$ and $v_s \in S_{(u_r,v_s)} \setminus S_{(u_i,v_j)}$ are non-empty (*i.e.*, the two sets have some private element) since $u_i \neq u_r$ and v_j is not an isolated vertex of G. Therefore, $S_{(u_i,v_j)}$ and $S_{(u_r,v_s)}$ overlap, and hence $S_{(u_i,v_j)}$ and $S_{(u_r,v_s)}$ are adjacent vertices in $O(\mathcal{F})$.
- If $u_i = u_r$ and $v_j \neq v_s$, then (u_i, v_j) and (u_r, v_s) are adjacent vertices in H if and only if $\{v_i, v_j\} \in \mathbf{E}(G)$. We have $v_j \in S_{(u_i,v_j)} \setminus S_{(u_r,v_s)}$ and $v_s \in S_{(u_r,v_s)} \setminus S_{(u_i,v_j)}$ (*i.e.*, the two sets have some private element) Therefore, the two sets overlap if and only if v_j and v_j belong to a same Q_p for some $1 \leq p \leq k$, which amounts to saying that $\{v_i, v_j\} \in \mathbf{E}(G)$. Hence, $S_{(u_i,v_j)}$ and $S_{(u_r,v_s)}$ are adjacent vertices in $O(\mathcal{F})$ if and only if $\{v_i, v_j\} \in \mathbf{E}(G)$.

$\square$

Now we proceed to the reduction in the reverse direction. For this we will need the following technical result.

Claim. Let $(\mathcal{F} = \{S_{(u_i,v_j)} : (u_i, v_j) \in \mathbf{V}(H)\}, X)$ be an overlap representation of H. If $S_{(u_r,v_s)} \subset S_{(u_i,v_j)}$ for some vertices (u_i, v_j) and (u_r, v_s) of H, then $S_{(u_p,v_q)} \subset S_{(u_i,v_j)}$ for every vertex (u_p, v_q) of H which is not adjacent to vertex (u_i, v_j).

Proof (Of claim 4). First, if $S_{(u_r,v_s)} \subset S_{(u_i,v_j)}$ then vertices (u_r, v_s) and (u_i, v_j) are not adjacent in H since $(\mathcal{F}, X)$ is an overlap representation of H. Now, let (u_p, v_q) be any vertex of H distinct from (u_r, v_s) that is not adjacent to (u_i, v_j). Also, let H' be the graph obtained from H by deleting every vertex in the close neighborhood of vertex (u_i, v_j). Notice that, since (u_r, v_s) and (u_p, v_q) are not adjacent to (u_i, v_j) in H, they are both vertices of H'. We claim that there exists a path between vertices (u_r, v_s) and (u_p, v_q) in H'. Indeed, since G is biconnected there exists a path in G between vertices v_s and v_q that does not go through vertex v_j, and hence there exists a path in H' between vertices (u_r, v_s) and (u_r, v_q). If $u_r = u_p$ we are done. Otherwise, since each column of H' is a clique then the two vertices (u_r, v_q) and (u_p, v_q) are connected by an edge in H'.

To prove the claim it is now enough to show that $S_{(u_p,v_q)} \subset S_{(u_i,v_j)}$ for any vertex (u_p, v_q) of H that is adjacent to (u_r, v_s) but not to (u_i, v_j). The proof follows from an easy contradiction. Suppose $S_{(u_p,v_q)} \not\subset S_{(u_i,v_j)}$. Since $S_{(u_p,v_q)} \neq \emptyset$ (H does not contain any isolated vertex), then there exists $x \in X$ such that $x \in S_{(u_p,v_q)}$ and $x \notin S_{(u_i,v_j)}$, and hence $S_{(u_p,v_q)} \setminus S_{(u_i,v_j)} \neq \emptyset$. Furthermore, (u_p, v_q) and (u_r, v_s) are adjacent vertices in H, and hence (since $S_{(u_p,v_q)}$ and $S_{(u_r,v_s)}$ have to overlap) there exist $x', x'' \in X$ such that (i) $x' \in S_{(u_p,v_q)}$ and $x' \in S_{(u_r,v_s)}$, and (ii) $x'' \notin S_{(u_p,v_q)}$ and $x'' \in S_{(u_r,v_s)}$. But $S_{(u_r,v_s)} \subset S_{(u_i,v_j)}$, and hence $x' \in S_{(u_i,v_j)}$ and $x'' \in S_{(u_i,v_j)}$. Then it follows that $S_{(u_i,v_j)} \setminus S_{(u_p,v_q)} \neq \emptyset$ and $S_{(u_p,v_q)} \cap S_{(u_i,v_j)} \neq \emptyset$, and hence $S_{(u_p,v_q)}$ and $S_{(u_i,v_j)}$ overlap. This is the sought contradiction since this would result in (u_p, v_q) and (u_i, v_j) being adjacent in H. $\square$

Claim. $\theta(G) \leq \frac{\varphi(H)-n-1}{m-1} + 7.$

Proof (Of claim 4). Let $(\mathcal{F} = \{S_{(u_i,v_j)} : (u_i, v_j) \in \mathbf{V}(H)\}, X)$ be an overlap representation of H. Suppose that there exists some subset $S_{(u_i,v_j)} \in \mathcal{F}$ that strictly contains at least one set of $\mathcal{F}$. Then, according to Claim 4, $S_{(u_i,v_j)}$ contains all subsets $S_{(u_r,v_s)} \in \mathcal{F}$ such that $u_i \neq u_r$ and $v_j \neq v_s$ (i.e., $S_{(u_i,v_j)}$ contains all those subsets of $\mathcal{F}$ that are associated to vertices of H that are not in the same row nor column of vertex (u_i, v_j)). Furthermore, if there exist subsets $S_{(u_r,v_s)}, S_{(u_p,v_q)} \in \mathcal{F}$ distinct from $S_{(u_i,v_j)}$ such that $S_{(u_r,v_s)} \subset S_{(u_p,v_q)}$, then $u_i = u_p$ or $v_j = v_q$ (i.e., vertex (u_p, v_q) is on the same row or on the same column of vertex (u_i, u_j)). Indeed, assuming $u_i \neq u_p$ and $v_j \neq v_q$, Claim 4 would yield to $S_{(u_i,v_j)} \subset S_{(u_p,v_q)}$ and $S_{(u_i,v_j)} \subset S_{(u_p,v_q)}$, a contradiction. Now, let H' be the graph obtained from H by deleting all vertices (u_r, v_s) such that $u_r = u_i$ or $v_s = v_j$ (i.e., deleting all vertices that are in the same row or column of vertex (u_i, v_j)). Also, let $\mathcal{F}' \subseteq \mathcal{F}$ be those subsets of $\mathcal{F}$ that correspond to vertices of H', and $X' \subseteq X$ be the union of the subsets in $\mathcal{F}'$ (X' is the ground set of $\mathcal{F}'$). Notice that $\mathcal{F}'$ is an overlap representation of H' where no subset being a subset of another, and that $|X'| \leq |S_{(u_i,v_j)}|$ since every subset of $\mathcal{F}'$ is strictly contained in $S_{(u_i,v_j)}$. Moreover, if we let G' stand for the graph obtained from G by deleting vertex v_j we have $H' = K_{m-1} \times G'$. We now claim that $\theta(G') \leq \frac{|X|-n-1}{m-1}$. Indeed, consider the "*edge-multi-coloring*" procedure of H' defined by assigning to every edge $e = \{(u_r, v_s), (u_p, v_q)\}$ of H' the "*colors*" col$(e) = S_{(u_r,v_s)} \cap S_{(u_p,v_q)}$. Since $\mathcal{F}'$ is an overlap representation of H', it follows that at least one color is assigned to every edge of H'. Furthermore, since no subset being a subset of another in $\mathcal{F}'$, it follows that for every color c, $\{e \in \mathbf{E}(H') : c \in \text{col}(e)\}$ induces a clique in G', and hence H' can be covered with at most $|X|$ cliques. But the maximal cliques of H' are either columns (there are $n - 1$ of these and at least $n - 1$ vertical edges must have received a different color), or are contained in a single row and correspond to maximal clique of G. Therefore, $m-1$ disjoint copies of G' can be covered with at most $|X|-n-1$ cliques. This proves $\theta(G') \leq \frac{|X|-n-1}{m-1}$. What is left is to prove $\theta(G) \leq \frac{\varphi(H)-n-1}{m-1}+7$. This follows from $\theta(G') \leq \frac{|X|-n-1}{m-1}$ and $\theta(G) \leq \theta(G') + \Delta(G)$.   □

Suppose, aiming at a contradiction, that there exists a $\sqrt{c}$-approximation algorithm B for EDGE-CLIQUE COVER. Then, we have $\mathsf{B}(H) \leq \sqrt{c}\,\varphi(H)$. Combining this inequality with Claim 4 yield $\mathsf{B}(H) \leq \sqrt{c}\,(n+m\,\theta(G))$. We now apply the constructive proof of Claim 4 to obtain an approximate $\mathsf{A}(G)$ of $\theta(G)$. We have $\mathsf{A}(G) \leq \frac{\mathsf{B}(H)-n-1}{m-1}+7 = \frac{\mathsf{B}(H)}{m-1} - \frac{n+1}{m-1}+7 \leq \frac{\sqrt{c}\,(n+m\,\theta(G))}{m-1} - \frac{n+1}{m-1}+7 \leq \frac{n\sqrt{c}}{m-1} + \frac{(\sqrt{c})^2\,\theta(G)}{m-1} - \frac{n+1}{m-1}+7 \leq (\sqrt{c})^2\,\theta(G) + \frac{n\sqrt{c}}{m-1} - \frac{n+1}{m-1}+7 = c\,\theta(G) + O(1)$. The constant makes no problem since $\theta(G)$ is bound to grow with n since we assume $\Delta(G)$ is bounded.   □

5   Recognizing Intersection Graph Classes

A central algorithmic problem corresponding to an intersection graph class $\mathcal{G}$ is the so called $\mathcal{G}$-RECOGNITION problem: *Given a graph G, is $G \in \mathcal{G}$?* In this section, we show that this problem is linear-time solvable for graphs G with fixed intersection number $i(G)$. Our proof here uses the fact that $i(G) = \theta(G)$ [5]. For ease of presentation, we assume all graphs have no isolated vertices.

Proposition 3. *Let $\mathcal{G}$ be any intersection graph class, and let $k \in \mathbb{N}$. The $\mathcal{G}$-RECOGNITION problem can be solved in linear-time when restricted to graphs with intersection number at most k.*

The proof of proposition 3 heavily relies on the notion of well quasi orders. A quasi order (*i.e.*, a binary reflexive transitive relation) is a *well quasi order* (or *wqo* for short) if it does not contain infinitely descending sequences nor infinite antichains. For example, the standard order $\leq$ of the natural numbers $\mathbb{N}$ is a well quasi order. Another less obvious example is given by considering vectors in $\mathbb{N}^k$. For two vectors $\overrightarrow{x}, \overrightarrow{y} \in \mathbb{N}^k$, let us write $\overrightarrow{x} \leq \overrightarrow{y}$ if $x_i \leq y_i$ for $1 \leq i \leq k$. The following lemma follows directly from a classical result known as Higman's Lemma [8].

Lemma 1. *The set $\mathbb{N}^K$ is well quasi ordered by $\leq$ for any fixed $K \in \mathbb{N}$.*

Let G be a graph with an edge-clique cover $\mathcal{E} = \{Q_1, Q_2, \ldots, Q_k\}$ (here, and in what follows, we allow $\mathcal{E}$ to be a multiset). A *characteristic vector of $\mathcal{E}$ for G* is a vector $\overrightarrow{c} \in \mathbb{N}^{2^k}$ that is indexed by subsets $S \subseteq \{1, \ldots, k\}$ such that $\overrightarrow{c}[S] = |\bigcap_{i \in S} Q_i|$. In other words, $\overrightarrow{c}$ contains the number of vertices in the intersection of any subset of cliques in $\mathcal{E}$. We say that a vector $\overrightarrow{c} \in \mathbb{N}^{2^k}$ is a characteristic vector of G if $\overrightarrow{c}$ is characteristic vector of some $\mathcal{E}$, $|\mathcal{E}| = k$, for G. Notice that there can be several characteristic vectors for a graph. The following lemma shows the connection between characteristic vectors and the induced subgraph order.

Lemma 2. *Let H and G be two graphs, and let $\vec{c}(G) \in \mathbb{N}^{2^{\theta(G)}}$ be a characteristic vector for G. Then H is an induced subgraph of G if and only if $\vec{c}(H) \leq \vec{c}(G)$ for some characteristic vector $\vec{c}(H) \in \mathbb{N}^{2^{\theta(G)}}$ for H.*

Proof. Let $\mathcal{E}_G := \{Q_1, Q_2, \ldots, Q_k\}$ denote the edge clique cover of G corresponding to $\vec{c}(G)$, and let $\vec{c}(H)$ be some characteristic vector for H of dimension 2^k with $\vec{c}(H) \leq \vec{c}(G)$. Let $\mathcal{E}_H := \{P_1, P_2, \ldots, P_k\}$ denote an edge clique cover for H corresponding to $\vec{c}(H)$. We map H to G by mapping each vertex in P_i to a vertex in Q_i for all $i \in \{1, \ldots, k\}$. Since $\vec{c}(H)[\{i\}] \leq \vec{c}[\{i\}]$ for each i, an injective mapping f obeying this property can easily be constructed. Furthermore, it can easily be seen that $\{u, v\} \in \mathbf{E}(H) \Rightarrow \{f(u), f(v)\} \in \mathbf{E}(G)$ for any pair $u, v \in \mathbf{V}(H)$, since two vertices in any graph are adjacent if and only if they belong together in some clique of the graph. This implies one direction of the lemma. The converse direction can be obtained by similar arguments. $\square$

We next show two applications of Lemma 2. The first application allows us to show that graphs of bounded intersection number are wqo by the induced subgraph order.

Lemma 3. *Let $k \in \mathbb{N}$. The set of all graphs G with $\theta(G) \leq k$ is wqo by the induced subgraph order.*

For a fixed graph H, the H-INDUCED SUBGRAPH problem asks to determine whether H is an induced subgraph of an input graph G. The second application of Lemma 2 is that H-INDUCED SUBGRAPH can be solved in linear-time for any fixed H, when its input is restricted to graphs of bounded edge clique cover number. The proof of this lemma uses the fact that for any fixed k, there is a linear-time algorithm for constructing an edge clique cover of size k for an input graph G with $\theta(G) \leq k$ [6].

Lemma 4. *Let H be an arbitrary graph, and let $k \in \mathbb{N}$. There is a linear time algorithm for H-INDUCED SUBGRAPH when restricted to graphs G with $\theta(G) \leq k$.*

We are now ready to give the proof of Proposition 3.

Proof (of proposition 3). Let $\mathcal{G}$ be any intersection graph class, and let $\overline{\mathcal{G}}$ denote the set of all finite graphs not in $\mathcal{G}$. Also, let $\mathcal{H}$ denote the set of all minimal graphs in $\overline{\mathcal{G}}$ w.r.t. the induced subgraph order. That is, $\mathcal{H} = \{H \in \overline{\mathcal{G}} : \nexists H' \in \overline{\mathcal{G}}$ such that H' is an induced subgraph of $H\}$. Observe that $\mathcal{G}$ is closed under induced subgraphs (*i.e.*, $H \in \mathcal{G}$ whenever H is an induced subgraph of some graph $G \in \mathcal{G}$). This implies that a graph G belongs to $\mathcal{G}$ if and only if no graph $H \in \mathcal{H}$ is an induced subgraph of G. Now by Lemma 3, the set $\mathcal{H}$ is finite, and its size depends only on $\mathcal{G}$. Thus our recognition algorithm for $\mathcal{G}$ has the set of graphs $\mathcal{H}$ "hard-wired" into it, and on given input graph G, it simply checks whether any $H \in \mathcal{H}$ is an induced subgraph of G, determining that $G \notin \mathcal{G}$ if and only if any of these checks turns out positive. The running-time of this algorithm is linear by Lemma 4, and since the number and sizes of graphs in $\mathcal{H}$ is constant w.r.t. the size of G. $\square$

References

1. Alimonti, P., Kann, V.: Some APX-completeness results for cubic graphs. TCS 237(1-2), 123–134 (2000)
2. Ausiello, G., Crescenzi, P., Gambosi, G., Kann, V., Marchetti-Spaccamela, A., Protasi, M.: Complexity and Approximation: Combinatorial optimization problems and their approximability properties. Springer (1999)
3. Chang, M.-S., Müller, H.: On the Tree-Degree of Graphs. In: Brandstädt, A., Le, V.B. (eds.) WG 2001. LNCS, vol. 2204, pp. 44–54. Springer, Heidelberg (2001)
4. Cranston, D.W., Korula, N., LeSaulnier, T.D., Milans, K., Stocker, C., Vandenbussche, J., West, D.B.: Overlap number of graphs. J. Graph Theory 70(1), 10–28 (2012)
5. Erdős, P., Goodman, A.W., Pósa, L.: The intersection of a graph by set intersections. Canad. J. Math. 18, 106–112 (1966)
6. Gramm, J., Guo, J., Hüffner, F., Niedermeier, R.: Data reduction and exact algorithms for clique cover. ACM J. of Experimental Algo. 13, 2.2:1–2.2:15 (2008)
7. Guillaume, J.-L., Latapy, M.: Bipartite structure of all complex networks. IPL 90(5), 215–221 (2004)
8. Higman, G.: Ordering by divisibility in abstract algebras. Proc. London Math. Society III 2(7), 326–336 (1952)
9. Hoover, D.N.: Complexity of graph covering problems for graphs of low degree. J. Comb. Math. and Comb. Comp. 11, 187–200 (1992)
10. Hsu, W.-L., Tsai, K.-H.: Linear time algorithms on circular-arc graphs. IPL 40(3), 123–129 (1991)
11. Kou, L.T., Stockmeyer, L.J., Wong, C.K.: Covering graphs by cliques with regard to keyword conflicts and intersection graphs. Comm. ACM 21, 135–139 (1978)
12. Lund, C., Yannakakis, M.: The Approximation of Maximum Subgraph Problems. In: Shamir, E., Abiteboul, S. (eds.) ICALP 1994. LNCS, vol. 820, pp. 40–51. Springer, Heidelberg (1994)
13. Ma, S., Wallis, W.D., Wu, J.: Clique covering of chordal graphs. Utilitas Mathematica 36, 151–152 (1989)
14. McKee, T.A., McMorris, F.R.: Topics in intersection graph theory. SIAM Monographs on Discrete Mathematics and Applications (1999)
15. Nor, I., Hermelin, D., Charlat, S., Engelstadter, J., Reuter, M., Duron, O., Sagot, M.-F.: Mod/Resc Parsimony Inference. In: Amir, A., Parida, L. (eds.) CPM 2010. LNCS, vol. 6129, pp. 202–213. Springer, Heidelberg (2010)
16. Orlin, J.B.: Contentment in graph theory: Covering graphs with cliques. Indagationes Mathematicae (Proc.) 80(5), 406–424 (1977)
17. Papadimitriou, C.H., Yannakakis, M.: Optimization, approximation and complexity classes. J. Comp. Sys. Sc. 43, 425–440 (1991)
18. Rosgen, B.: Set representations of graphs. Master's thesis, Univ. Alberta (2005)
19. Rosgen, B., Stewart, L.: The overlap number of a graph (2010) (submitted, arXiv:1008.2170v2)

Linear Layouts in Submodular Systems

Hiroshi Nagamochi

Graduate School of Informatics, Kyoto University, Japan
nag@amp.i.kyoto-u.ac.jp

Abstract. Linear layout of graphs/digraphs is one of the classical and important optimization problems that have many practical applications. Recently Tamaki proposed an $O(mn^{k+1})$-time and $O(n^k)$-space algorithm for testing whether the pathwidth (or vertex separation) of a given digraph with n vertices and m edges is at most k. In this paper, we show that linear layout of digraphs with an objective function such as cutwidth, minimum linear arrangement, vertex separation (or pathwidth) and sum cut can be formulated as a linear layout problem on a submodular system (V, f) and then propose a simple framework of search tree algorithms for finding a linear layout (a sequence of V) with a bounded width that minimizes a given cost function. According to our framework, we obtain an $O(kmn^{2k})$-time and $O(n + m)$-space algorithm for testing whether the pathwidth of a given digraph is at most k.

1 Introduction

Let $G = (V, E)$ stand for an undirected or directed graph with a set V of n vertices and a set E of m edges. Linear layout of graphs is a problem of finding a linear arrangement (a sequence of V) $\sigma = (v_1, \ldots, v_n)$ of the vertex set V of G so that a prescribed cost function $cost(\sigma)$ is minimized. The problem is one of the classical and important optimization problems that have many practical applications (e.g., see [4]). From practical point of views, there have been introduced several different choices of cost functions, among which the following ones can be described by vertex/edge-cut functions of digraphs (where we regard an undirected graph as a symmetric digraph i.e., treat each undirected edge uv as two oppositely directed edges (u, v) and (v, u)):

- CUTWIDTH: $cost_{\mathrm{CW}}(\sigma) = \max\{d_G^+(\{v_1, \ldots, v_i\}) \mid 1 \le i \le n - 1\}$, where $d_G^+(X)$ denotes the number of directed edges with a tail in X and a head in $V - X$;
- MINIMUM LINEAR ARRANGEMENT: $cost_{\mathrm{MLA}}(\sigma) = \sum\{d_G^+(\{v_1, \ldots, v_i\}) \mid 1 \le i \le n - 1\}$;
- VERTEX SEPARATION (or PATHWIDTH): $cost_{\mathrm{VS}}(\sigma) = \max\{\Gamma_G^+(\{v_1, \ldots, v_i\}) \mid 1 \le i \le n - 1\}$, where $\Gamma_G^+(X)$ denotes the number of out-neighbors of a subset X (the vertices $v \in X$ that have directed edges from v to a vertex in $V - X$); and
- SUM CUT: $cost_{\mathrm{SC}}(\sigma) = \sum\{\Gamma_G^+(\{v_1, \ldots, v_i\}) \mid 1 \le i \le n - 1\}$.

K.-M. Chao, T.-s. Hsu, and D.-T. Lee (Eds.): ISAAC 2012, LNCS 7676, pp. 475–484, 2012.

In these functions, directed edges with the backward direction in a sequence σ are ignored. The *cutwitdh* (resp., *vertex separation*) of G is defined to be the minimum of $cost_{CW}(\sigma)$ (resp., $cost_{VS}(\sigma)$) over all sequences σ of V. The vertex separation of a digraph G is equal to the "pathwidth" of G (e.g., [12]), which is a width of a path-decomposition of G.

Bodlaender et al. [3] showed that a class of linear layout problems including the above four can be solved (i) in $O^*(2^n)$ time and $O^*(2^n)$ space by a dynamic programming; and (ii) in $O^*(4^n)$ time and polynomial space by a search tree algorithm (where the O^*-notation suppresses factors that are polynomial in n).

When a problem is parameterized by value k of its cost function, it is known that CUTWIDTH and VERTEX SEPARATION admit faster exact algorithms. We let CUTWIDTH(k) (resp., VERTEX SEPARATION(k)) stand for the problem of testing whether a given graph/digraph G has a sequence σ of V such that $cost_{CW}(\sigma) \leq k$ (resp., $cost_{VS}(\sigma) \leq k$) or not. Gurari and Sudborough [8] presented an $O(n^k)$-time and exponential-space dynamic programming algorithm for CUTWIDTH(k) in undirected graphs, and Makedon and Sudborough [13] later improved the time bound to $O(n^{k-1})$. For CUTWIDTH(k) in undirected graphs with a fixed k, Fellows and Langston [6] obtained an $O(n^2)$-time algorithm, and the time bound is improved to linear by Abrahamson and Fellows [1] and Thilikos et al. [20]. For VERTEX SEPARATION(k) in undirected graphs with a fixed k, Fellows and Langston [6] designed an $O(n^3)$-time algorithm, and afterwards Bodlaender [2] gave a linear time algorithm. For undirected graphs, it is known that the graph minor theorem by Robertson and Seymour [17] implies polynomial-time algorithms for problems CUTWIDTH(k) and VERTEX SEPARATION(k) with fixed k and that the theorem, however, cannot be applied to the directed case (e.g., see [19,20])

Recently Tamaki [19] proposed an $O(mn^{k+1})$-time and $O(n^k)$-space algorithm for testing whether the pathwidth (or vertex separation) of a given digraph with n vertices and m edges is at most k. Although it remains open whether VERTEX SEPARATION(k) in digraphs is fixed-parameter tractable or not, it is the first nontrivial step toward design of efficient exact algorithms for computing graph parameters of digraphs. His algorithm is a search tree algorithm equipped with a pruning procedure that tries to discard one of two partial sequences with the same length by a dominance relationship. It is proven that the number of all partial sequences with the same length during an execution is always $O(n^k)$, which ensures the claimed time and space complexities of the algorithm. More interestingly, although the submodularity of function Γ_G^+ is used to derive the upper bound $O(n^k)$, the mechanism of the algorithm is self-contained in the sense that it never relies on any other optimization mechanism such as submodular minimization and dynamic programming to attain the nontrivial upper bound. In fact, recently Nagamochi [14] proved that the new mechanism can be conversely used to solve the submodular minimization problem, the most representative optimization problem.

From these observations, it would be natural to find a way of applying submodular minimization to the pathwidth problem in digraphs. Our research group has implemented Tamaki's algorithm to investigate the distribution of pathwidth of chemical graphs, and it turned out that the $O(n^k)$-space algorithm easily uses up the memory allowed for graphs with over 100 vertices [9]. This is another motivation for us to develop a more space-efficient algorithm for the problem.

In this paper, we show that linear layout of digraphs with an objective function such as cutwidth, minimum linear arrangement, vertex separation (or pathwidth) and sum cut can be formulated as a linear layout problem on a submodular system (V, f), and then propose a simple framework of search tree algorithms for finding a linear layout (a sequence of V) with a bounded width that minimizes a given cost function. When a cost function is given as $\sum_{1 \leq i \leq n-1} f(\{v_1, \ldots, v_i\})$, the linear layout problem on a submodular system (V, f) has been introduced by Iwata et al. [11], and they proposed a $(2 - 2/(n+1))$-approximation algorithm to the problem when f is a monotone submodular function.

The paper is organized as follows. Section 2 reviews basic results on submodular functions and introduces a layout problem in submodular systems. Section 3 presents a key property on sequences in submodular systems, based on which a search tree algorithm is designed. Section 4 analyzes the time complexity of the algorithm applied to the problem of testing whether the cutwidth/pathwidth of a given digraph is at most k. Finally Section 5 makes concluding remarks.

2   Preliminaries

Submodular Systems. Let V denote a given finite set with $n \geq 1$ elements. A set function f on V is called *submodular* if $f(X) + f(Y) \geq f(X \cap Y) + f(X \cup Y)$ for every pair of subsets $X, Y \subseteq V$. There are numerous examples of submodular set functions such as cut function of digraphs and hypergraphs, matroid rank function, and entropy function. The problem of finding a subset X that minimizes $f(X)$ over a submodular set function f is one of the most fundamental and important issues in optimization. Grotschel, Lovasz, and Schrijver gave the first polynomial time algorithm for minimizing a submodular set function [7]. Schrijver [18] and Iwata, Fleischer, and Fujishige [10] independently developed strongly polynomial time combinatorial algorithms for the submodular minimization. Currently an $O(n^6 + n^5\theta)$-time minimization algorithm is obtained by Orlin [15], where $n = |V|$ and θ is the time to evaluate $f(X)$ of a specified subset X.

For two disjoint subsets $S, T \subseteq V$, an (S, T)-*separator* is defined to be a subset X such that $S \subseteq X \subseteq V - T$, and let $f_{\min}(S, T)$ denote the minimum $f(X)$ of an (S, T)-separator X, where such a set X is called a *minimum* (S, T)-separator. We denote (S, T) with $S = \{s\}$ and $T = \{t\}$ by (s, t).

We here remark that the problem of finding a subset X with minimum $f(X)$ in a submodular system (V, f) is essentially equivalent to that of finding a minimum (S, T)-separator in a submodular system.

Sequences. For two integers $i \leq j$, the set of all integers h with $i \leq h \leq j$ is denoted by $[i, j]$. A sequence σ consisting of some elements in a finite set V is called *non-duplicating* if each element of V occurs at most once in σ. We denote by Σ_i the set of all non-duplicating sequences of exactly i elements in V, where Σ_0 contains only the null sequence (the sequence of length zero). We denote $\cup_{0 \le i \le n} \Sigma_i$ by Σ. Let $\sigma \in \Sigma$ be a sequence. We denote by $V(\sigma)$ the set of elements constituting σ and by $|\sigma| = |V(\sigma)|$ the length of σ. Let $\sigma(i)$ denote the ith element in a sequence σ, and let σ_i be the sequence that consists of the first i elements of σ, i.e., $\sigma_i = (\sigma(1), \sigma(2), \ldots, \sigma(i))$. Given two disjoint subsets $S, T \subseteq V$, a sequence σ is called an (S, T)-sequence if $V(\sigma_{|S|}) = S$ and $V - V(\sigma_{|V-T|}) = T$. We let $\overline{X}$ denote $V - X$.

For two sequences $\alpha, \beta \in \Sigma$ such that $V(\alpha) \cap V(\beta) = \emptyset$, we denote by $\alpha\beta$ the sequence $\sigma \in \Sigma_{|\alpha|+|\beta|}$ obtained by appending β to α so that $\sigma(i) = \alpha(i)$ for $i \le |\alpha|$ and $\sigma(i) = \beta(i - |\alpha|)$ otherwise.

For a subset $X \subseteq V$, let $\sigma[X]$ denote the sequence $\sigma' \in \Sigma_{|V(\sigma) \cap X|}$ such that $V(\sigma') = V(\sigma) \cap X$ and for every two elements $u, v \in V(\sigma')$, u precedes v in σ' if and only if u precedes v in σ.

Linear Layouts. We consider a cost function *cost* on sequences $\sigma \in \Sigma$. A cost function *cost* is called *non-decreasing* if $cost(\sigma)$ is determined only by $\{f(\sigma_1), f(\sigma_2), \ldots, f(\sigma_{\ell-1})\}$ ($\ell = |\sigma|$) and $cost(\sigma)$ does not decrease when $f(\sigma_i)$ for some i increases, where we regard $\{f(\sigma_1), \ldots, f(\sigma_{\ell-1})\}$ as a multiset consisting of exactly ℓ numbers. For example, the following three functions are all non-decreasing:

$$f_{\min}(\sigma) = \min\{f(\sigma_i) \mid 1 \le i \le \ell - 1\},$$

$$f_{\max}(\sigma) = \max\{f(\sigma_i) \mid 1 \le i \le \ell - 1\},$$

$$f_{\text{sum}}(\sigma) = \sum\{f(\sigma_i) \mid 1 \le i \le \ell - 1\}.$$

We call $f_{\max}(\sigma)$ the f-*width* of σ.

For a subset X of a digraph $G = (V, E)$, let $\Gamma_G^-(X)$ denote the number of in-neighbors of a subset X (the vertices $v \in V - X$ that have directed edges from v to a vertex in X), and let $d_G^-(X) = d_G^+(V - X)$. Observe that $cost_{\text{CW}} = f_{\max}$ and $cost_{\text{MLA}} = f_{\text{sum}}$ for the edge-cut function $f = d_G^+$, and $cost_{\text{VS}} = f_{\max}$ and $cost_{\text{SC}} = f_{\text{sum}}$ for the vertex-cut function $f = \Gamma_G^+$.

We are ready formulate a general form of the problems studied in this paper:

Linear Layouts in Submodular Systems. Given a nonnegative submodular system (V, f) with $f(\emptyset) = f(V) = 0$ ($n = |V|$), a positive real $k > 0$ and a non-decreasing cost function *cost*, find a sequence $\sigma \in \Sigma_n$ with f-width at most k that minimizes $cost(\sigma)$ among all sequences with f-width at most k.

Note that there is a chance that a sequence $\tau \in \Sigma_n$ with f-width greater than k attains $cost(\tau)$ smaller than the minimum $cost(\sigma)$ of the above problem when *cost* is not given by $f_{\max}$. However our main result (Theorem 1) still suggests that for the problem of minimizing $cost_{\text{MLA}}$ or $cost_{\text{SC}}$, $f_{\max}$ is a useful measure to parameterize these problems, since values of these cost functions in strongly connected digraphs are not less than n are inadequate to measure the computational tractability.

Precedent Constraint. In some application of arrangement of elements such as scheduling problems (e.g., see section 11.2 in [4]), an output sequence is required to meet a precedent relation among elements such that an element u precedes another element v, denoted by $u \prec v$. The set of such ordered pairs (u, v) can be given by a poset P on V, where P is represented by a set of directed edges (u, v) such that $u \prec v$ and there is no element w with $u \prec w$ and $w \prec v$ (in general P is not necessarily equal to a given digraph G itself). We can naturally include the side constraint as a penalty function into a given submodular system (V, f). Define the DAG (V, P), and let p be the submodular function on V by defining $p(X) = (k+1)d_P^-(X)$ for each subset $X \subseteq V$, where $d_P^-(X)$ denotes the the number of directed edges of (V, P) with a tail in $V - X$ and a head in X. Clearly $(V, f' = f + g)$ remains a submodular system, and any sequence σ of V with $f'_{\max}(\sigma) \leq k$ satisfies $f_{\max}(\sigma_i) = f'_{\max}(\sigma_i) \leq k$ for $i = 1, 2, \ldots, n-1$, which indicates that there is no edge $(u, v) \in P$ such that $i > j$ for $\sigma(i) = u$ and $\sigma(j) = v$, i.e., the given precedent constraint is met.

Main Result. In this paper, we prove the next.

Theorem 1. *Given a submodular system (V, f), a real k, and a non-decreasing function cost, a minimum cost sequence σ with f-width at most k (if any) can be obtained by solving submodular minimization $O(n^{2\Delta(k)+2})$ times using $O(|V|)$ work space except for storage of f, where $\Delta(k)$ denotes the number of distinct values in $\{f(X) \leq k \mid \emptyset \subsetneq X \subsetneq V\}$.*

In particular, when f is integer-valued and k is a positive integer, it holds $\Delta(k) \leq k - \min_{\emptyset \subsetneq X \subsetneq V} f(X)$.

3   Algorithm

This section proves Theorem 1 by presenting a search tree algorithm that solves the problem. All we need to design our new algorithm is the following observation.

Lemma 1. *For a submodular system (V, f), let τ be an (S, T)-sequence $\tau \in \Sigma_n$ of V. For a minimum (S, T)-separator A in (V, f), let $\sigma = \tau[A]\tau[\overline{A}] \in \Sigma_n$, and ψ be a bijection on $[|S| + 1, n - |T|]$ such that $\psi(i)$ is the index j such that $\sigma(i) = \tau(j)$. Then*

$$f(\sigma_i) \leq f(\tau_{\psi(i)}) \text{ for all } i \in [|S| + 1, n - |T|]. \tag{1}$$

Proof. Fix $i \in [|S|+1, n-|T|]$, and let $j = \psi(i)$. Since $V(\tau_j) \cup A$ and $V(\tau_j) \cap A$ are (S, T)-separators, we have $f(A) = f_{\min}(S, T) \leq \min\{f(V(\tau_j) \cup A), f(V(\tau_j) \cap A)\}$. Hence by the submodularity of f, it holds $f(A) + f(\tau_j) \geq f(V(\tau_j) \cap A) + f(V(\tau_j) \cup A)$, from which we have $f(\tau_j) \geq \max\{f(V(\tau_j) \cap A), f(V(\tau_j) \cap A)\}$.

We first consider the case where $|S| + 1 \leq i \leq |A|$. In this case it holds $V(\sigma_i) = V(\tau_j) \cap A$ and we have $f(\sigma_i) = f(V(\tau_j) \cap A) \leq f(\tau_j)$. On the other hand $(|A| + 1 \leq i \leq n - |T|)$, it holds $V(\sigma_i) = V(\tau_j) \cup A$ and we have $f(\sigma_i) = f(V(\tau_j) \cup A) \leq f(\tau_j)$. ∎

Note that (1) implies that $cost(\sigma[V - S - T]) \leq cost(\tau[V - S - T])$ for any non-decreasing cost function $cost$.

Fix a nonnegative submodular system (V, f) with $f(\emptyset) = f(V) = 0$ and a real number k, an instance of our problem is specified by an ordered pair (S, T) of disjoint subsets $S, T \subseteq V$, to which we wish to find an (S, T)-sequence σ such that the f-width of the subsequence $\sigma[V-S-T]$ is at most k and $cost(\sigma[V-S-T])$ is minimized among all such sequences σ. Such an (S, T)-sequence σ is called a *solution* to the instance (S, T).

To find a solution to a given instance (S, T) by a search tree algorithm, we introduce branch operations based on Lemma 1.

For every two elements $s, t \in V$ in a given submodular system (V, f), we first genetare an instance $(S = \{s\}, T = \{t\})$. There are at most n^2 such instances. Let $f^* = \min\{f(X) \mid \emptyset \subsetneq X \subsetneq V\}$.

An instance (S, T) with $|V-S-T| \leq 1$ is trivial since it has a unique solution (if any). Let $|V- S- T| \geq 2$ Compute $f_{\min}(S, T)$ invoking submodular minimization on f. Assume that $f_{\min}(S, T) \leq k$, since otherwise there is no (S, T)-sequence σ such that the f-width of $\sigma[V - S - T]$ is at most k.

Case 1. $f_{\min}(S, T) = k$: In this case, we can reduce (S, T) into trivial one. Choose an arbitrary element $u \in V - S- T$ such that $f(S \cup \{u\}) = k$ (if no such element $u \in V - S - T$ exits then the instance (S, T) has no solution either). By Lemma 1, a solution to (S, T) can be obtained by combining solutions to $(S, T' = V - S- \{u\})$ and $(S' = S \cup \{u\}, T)$. Since $(S, T' = V - S- \{u\})$ has a unique solution, this reduces the current instance (S, T) to $(S' = S \cup \{u\}, T)$, where $f_{\min}(S', T) = k$ still holds. Hence we can apply the above procedure until the instance becomes trivial (or we find out infeasibility of (S, T)).

Case 2. $f_{\min}(S, T) < k$: We further test whether there is a minimum (S, T)-separator A with $S \subsetneq A \subsetneq V - T$ (this can be done by computing $f_{\min}(S \cup \{u\}, T \cup \{v\})$ for all pairs $u, v \in V - S - T$, thus $O(|V - S - T|^2)$ times of submodular minimization).

Case 2a. A minimum (S, T)-separator A with $S \subsetneq A \subsetneq V - T$ exists: We split the current instance into two instances $(S, T' = V - A)$ and $(S' = A, T)$. By Lemma 1, a solution to (S, T) can be obtained by combining solutions to $(S, T' = V - A)$ and $(S' = A, T)$.

Case 2b. No minimum (S, T)-separator A with $S \subsetneq A \subsetneq V - T$ exists; i.e., only S or $V - T$ is a minimum (S, T)-separator:

(i) Exactly one of S and $V - T$, say S is a minimum (S, T)-separator: We branch into $|V - S- T|$ instances $I_u = (S_u = S \cup \{u\}, T)$, $u \in V- S- T$, and select an (S, T)-sequence with minimum $cost$ among solutions to I_u, $u \in V-S-T$ as a solution to (S, T). Note that $f_{\min}(S_u, T) > f_{\min}(S, T)$.

(ii) Both of S and $V - T$ are minimum (S, T)-separators: We branch into $|V- S- T|(|V- S- T| - 1)$ instances $I_{uv} = (S_u = S \cup \{u\}, T_v = T \cup \{v\})$, $u, v \in V- S- T$, and select an (S, T)-sequence with minimum $cost$ among solutions to I_{uv}, $u, v \in V- S- T$ as a solution to (S, T). Note that $f_{\min}(S_u, T_v) > f_{\min}(S, T)$.

The above branching rules give our search tree algorithm.

We now analyze the time and space complexities of our algorithm.

For each instance (S, T), we solve submodular minimization $O(|V - S - T|^2)$ times to generate a set of instances in Case 2. Let $\Delta(a, b)$ denote the number of distinct values in $\{f(X) \mid a \leq f(X) < b, X \subseteq V\}$. It is not difficult to see that the number of instances in the search tree rooted at an instance (S, T) is at most $|V - S - T|^{2\Delta(f_{\min}(S,T),k)}$ since the number of branches is $|V - S - T|$ and the depth of the rooted tree is $\Delta(f_{\min}(S, T), k)$. Hence we have the next (the proof is omitted due to space limitation).

Lemma 2. *From an instance (S, T), at most $|V - S - T|^{2\Delta(f_{\min}(S,T),k)}$ instances that invoke submodular minimization will be generated.*

It always holds $f_{\min}(S, T) \geq f^*$ for any generated instances (S, T). By Lemma 2, our algorithm generates from each instance $(S = \{s\}, T = \{t\})$, at most $n^{2\Delta(f^*,k)} = n^{2(\Delta(k)-1)}$ instances that invokes submodular minimization, where $\Delta(k) = |\{f(X) \leq k \mid \emptyset \subsetneq X \subsetneq V\}|$. Since there are at most n^2 pairs of (s, t) and each instance invokes at most n^2 submodular minimization, the number of times for solving submodular minimizations is at most $n^2 n^{2\Delta(k)-2} n^2 = n^{2\Delta(k)+2}$. This proves Theorem 1.

4 Digraph Case

In this section, we consider layout of a digraph $G = (V, E)$ with cost functions $cost_{\mathrm{CW}}$, $cost_{\mathrm{MLA}}$, $cost_{\mathrm{VS}}$ and $cost_{\mathrm{SC}}$, and analyze upper bounds on the time and space complexities of our algorithm applied to these problems using flow technique. We consider the problem of finding a minimum cost of an (S, T)-sequence σ with f-width at most k.

4.1 Cutwidth and Minimum Linear Arrangement

We here show how to find a minimum cost layout of a digraph under a fixed cutwidth. First consider the case where there is no precedent constraint, i.e., we set $f = d_G^+$; we assume that G is connected and $m \geq n - 1$. Let λ denote the edge-connectivity of G, i.e., $\lambda = \min_{\emptyset \subsetneq X \subsetneq V} d^+(X)$. In this case, $f_{\min}(S, T)$ and a minimum (S, T)-separator can be obtained by computing a maximum (s', t')-flow φ in a directed network G' obtained from G by contracting S and T into single vertices s' and t', where the capacity of each directed edge is 1. From a maximum (s', t')-flow φ, we can find a minimum (S, T)-separator A with $S \subsetneq A \subsetneq V - T$ in G in linear time (if any) by constructing a DAG representation of all minimum (S, T)-separators in linear time [16] without newly solving $O(n^2)$ minimization problems. Hence for each instance (S, T), we need to solve a single maximum flow problem, which takes $O(k(m + n))$ time and $O(n + m)$ space [5], where we do not need to find any minimum (S, T)-separator when the flow value exceeds k. Since the total number of instances to be generated is at most $n^2 n^{2\Delta(k)-2} \leq n^2 n^{2(k-\lambda+1)-2}$, the entire time complexity is $O(kmn^{2(k-\lambda+1)})$.

Theorem 2. *Given a digraph $G = (V, E)$ with n vertices and m edges and an integer $k \geq 1$, whether there is a sequence of V with pathwidth at most k can be tested in $O(kmn^{2(k-\lambda+1)})$ time and $O(n + m)$ space. When such a sequence exists, a sequence $\sigma \in \Sigma_n$ with pathwidth at most k that minimizes $cost_{CW}$ can be found in the same time and space complexities.*

We next consider the case where a precedent constraint is imposed as a poset $P \subseteq V \times V$ i.e., we set $f = d_G^+ + (k+1)d_P^-$ (note that f-width at most k is equal to d_G^+-width in any sequences). In this case, let $\overline{P} = \{(v, u) \mid (u, v) \in P\}$, and augment G by adding all edges $(v, u) \in \overline{P}$ to obtain a directed network, where the capacity of each directed edge in E is 1 and we treat each $(v, u) \in \overline{P}$ as $k + 1$ multiple edges with capacity 1. Hence the number m' of edges in the augmented multigraph is at most $m + (k+1)|P|$. For a given (S, T), we can obtain $f_{min}(S, T)$ and a minimum (S, T)-separator in a similar way; we compute a maximum (s', t')-flow in the directed network after contracting S and T into single vertices s' and t', taking $O(km'n^{2(k-\lambda+1)}) = O(k(m+n+k|P|)n^{2(k-\lambda+1)})$ time and $O(n + m + |P|)$ space.

Theorem 3. *Given a digraph $G = (V, E)$ with n vertices and m edges, a poset $P \subseteq V \times V$ and an integer $k \geq 1$, whether there is a sequence of V with pathwidth at most k which meets the precedent constraint by P can be tested in $O(k(m + n+k|P|)n^{2(k-\lambda+1)})$ time and $O(n+m+|P|)$ space. When such a sequence exists, a sequence $\sigma \in \Sigma_n$ with pathwidth at most k that minimizes $cost_{CW}$ under the precedent constraint by P can be found in the same time and space complexities.*

For the layout of digraphs with sum cut $cost_{CW}$, the same statements of Theorems 2 and 3 hold by replacing $cost_{CW}$ with $cost_{MLA}$.

4.2   Vertex Separation and Sum Cut

We here show how to find a minimum cost layout of a digraph under a fixed pathwidth (or vertex separation). We consider the case where a precedent constraint where a precedent constraint is imposed as a poset $P \subseteq V \times V$ i.e., we set $f = \Gamma_G^+ + (k+1)d_P^-$ (note that f-width at most k is equal to Γ_G^+-width in any sequences).

In this case, we can compute a minimum (S, T)-separator by computing a maximum flow applying the standard technique of converting vertex-cuts into edge-cuts (however $\min_{\emptyset \subsetneq X \subsetneq V} \Gamma^+(X) \leq \min_{v \in V} \Gamma^+(V - \{v\}) \leq 1$ is not the vertex-connectivity of G). For this, we construct a digraph $G_P = (V' \cup V'', A_E \cup A_V \cup A_{\overline{P}})$ as follows. Let $\overline{P} = \{(v, u) \mid (u, v) \in P\}$. Replace each vertex $v \in V$ with two copies v' and v'' with a new directed edge (v', v''), and let $A_V = \{(v', v'') \mid v \in V\}$. For each directed edge $(u, v) \in E$, we set a directed edge (u'', v') in G_P, and let $A_E = \{(u'', v') \mid (u, v) \in E\}$. For each directed edge $(v, u) \in \overline{P}$, we set a directed edge (v'', u''), and let $A_{\overline{P}} = \{(v'', u'') \mid (v, u) \in \overline{P}\}$, where we treat each edge (v'', u'') in G_P as $k+1$ multiple edges. The number m' of edges in the multigraph G_P is at most $m + n + (k+1)|P|$. The next lemma verifies that we can obtain a minimum (S, T)-separator A with $\Gamma^+(A) \leq k$ (if any) by computing a minimum $(\hat{S}, \hat{T})$-separator in G_P for $\hat{S} = \{u', u'' \mid u \in S\}$ and $\hat{T} = \{u'' \mid u \in T\}$(the proof is omitted due to space limitation).

Lemma 3. *For the vertex-cut function Γ_G^+ of a digraph $G = (V, E)$, and the penalty function $p = (k + 1)d_P^-$ defined by a poset P on V, let $f = \Gamma_G^+ + p$ be a set function on V. Let g be the edge-cut function $d_{G_P}^+$ of $G_P = (V' \cup V'', A_E \cup A_V \cup A_{\overline{P}})$ defined from (G, P, k) in the above. Given two disjoint subsets $S, T \subseteq V$, let $\hat{S} = \{u', u'' \mid u \in S\}$ and $\hat{T} = \{u'' \mid u \in T\}$. Then $f_{\min}(S, T) > k$ if and only if $g_{\min}(\hat{S}, \hat{T}) > k$; and if $f_{\min}(S, T) \leq k$, then $g_{\min}(\hat{S}, \hat{T}) = f_{\min}(S, T)$.*

Since a minimum $(\hat{S}, \hat{T})$-separator in G_P can be obtained by computing a maximum (s', t')-flow after contracting $\hat{S}$ and $\hat{T}$ into single vertices s' and t'. The single maximum flow problem can be solved in $O(km') = O(k(m+n+|P|))$ time and $O(n+m') = O(m+n+|P|)$ space analogously with the case of cutwidth. Hence the time bound is $O(k(m+n+|P|)n^2n^{2\Delta(k)-2}) = O(k(m+n+|P|)n^2n^{2(k+1)-2})$. In particular, when no precedent constraint is imposed, we can assume that G is strongly connected (otherwise a solution is easily obtained) and we can set $\Delta(k) \leq k$ and $|P| = 0$ in these bounds. Therefore we obtain the following results.

Theorem 4. *Given a digraph $G = (V, E)$ with n vertices and m edges and an integer $k \geq 1$, whether there is a sequence of V with pathwidth at most k can be tested in $O(kmn^{2k})$ time and $O(n + m)$ space. When such a sequence exists, a sequence $\sigma \in \Sigma_n$ with pathwidth at most k that minimizes $cost_{VS}$ can be found in the same time and space complexities.*

Theorem 5. *Given a digraph $G = (V, E)$ with n vertices and m edges, a poset $P \subseteq V \times V$ and an integer $k \geq 1$, whether there is a sequence of V with pathwidth at most k which meets the precedent constraint by P can be tested in $O(k(m + n + k|P|)n^{2k+2})$ time and $O(n + m + |P|)$ space. When such a sequence exists, a sequence $\sigma \in \Sigma_n$ with pathwidth at most k that minimizes $cost_{VS}$ under the precedent constraint by P can be found in the same time and space complexities.*

For the layout of digraphs with sum cut $cost_{SC}$, the same statements of Theorems 4 and 5 hold by replacing $cost_{VS}$ with $cost_{SC}$.

5   Concluding Remarks

In this paper, we introduced a linear layout in submodular systems (V, f), which includes several linear layout problems in graphs/digraphs, defining non-decreasing cost functions and f-width. We proposed a framework for search tree algorithms of finding a minimum cost layout with a bounded f-width. Our result in contrast to Tamak's algorithm has a similar trade-off between the $O^*(2^n)$-time and space algorithm and the $O^*(4^n)$-time and polynomial-space algorithms; reducing the space complexity to polynomial one increases the time complexity up to the square of it (the work complexity). Theorem 1 would indicate that f-width is a useful parameter to investigate the tractability of layout problems with cost functions whose value is as large as n.

Acknowledgment. The author would like to thank Prof. Hisao Tamaki for useful discussions.

References

1. K. Abrahamson, M. Fellows. Finite automata, bounded treewidth and well-quasiordering. in: Contemp. Math., 147: Amer. Math. Soc., Providence, RI, 539–563, 1993.
2. Bodlaender, H.L.: A linear-time algorithm for finding tree-decompositions of small treewidth. SIAM J. Comput. 25(6), 1305–1317 (1996)
3. Bodlaender, H.L., Fomin, F.V., Koster, A.M.C.A., Kratsch, D., Thilikos, D.M.: A note on exact algorithms for vertex ordering problems on graphs. Theory Comput. Syst. 50, 420–432 (2012)
4. Díaz, J., Petit, J., Serna, M.: A survey of graph layout problems. ACM Computing Surveys (CSUR) 34(3), 313–356 (2002)
5. Even, S., Tarjan, R.E.: Network flow and testing graph connectivity. SIAM J. Comput. 4, 507–518 (1975)
6. Fellows, M.R., Langston, M.A.: Layout permutation problems and well-partially ordered sets. In: Advanced Research in VLSI, pp. 315–327. MIT Press, Cambridge (1988)
7. Grötschel, M., Lovász, L., Schrijver, A.: The ellipsoid algorithm and its consequences in combinatorial optimization. Combinatorica 1, 499–513 (1981)
8. Gurari, E., Sudborough, I.H.: Improved dynamic programming algorithms for the bandwidth minimization and the mincut linear arrangement problem. J. Algorithm 5, 531–546 (1984)
9. Ikeda, M., Nagamochi, H.: A method for computing the pathwidth of chemical graphs. In: The 15th Japan-Korea Joint Workshop on Algorithms and Computation, Tokyo, Japan, July 10-11, pp. 140–145 (2012)
10. Iwata, S., Fleischer, L., Fujishige, S.: A combinatorial, strongly polynomial-time algorithm for minimizing submodular functions. J. ACM 48, 761–777 (2001)
11. Iwata, S., Tetali, P., Tripathi, P.: Approximating Minimum Linear Ordering Problems. In: Gupta, A., Jansen, K., Rolim, J., Servedio, R. (eds.) APPROX 2012 and RANDOM 2012. LNCS, vol. 7408, pp. 206–217. Springer, Heidelberg (2012)
12. Kinnersley, N.G.: The vertex separation number of a graph equals its path-width. Inf. Proc. Lett. 42, 345–350 (1992)
13. Makedon, F., Sudborough, I.H.: On minimizing width in linear layouts. Dis. Appl. Math. 23(3), 243–265 (1989)
14. Nagamochi, H.: Submodular Minimization via Pathwidth. In: Agrawal, M., Cooper, S.B., Li, A. (eds.) TAMC 2012. LNCS, vol. 7287, pp. 584–593. Springer, Heidelberg (2012)
15. Orlin, J.B.: A faster strongly polynomial time algorithm for submodular function minimization. Math. Program., Ser. A 118, 237–251 (2009)
16. Picard, J.-C., Queyranne, M.: On the structure of all minimum cuts in a network and applications. Math. Prog. Study 13, 8–16 (1980)
17. Robertson, N., Seymour, P.: Graph Minors. XX. Wagner's conjecture. J. Combin. Theory Ser. B 92(2), 325–335 (2004)
18. Schrijver, A.: A combinatorial algorithm minimizing submodular functions in strongly polynomial time. J. Combin. Theory Ser. B 80, 346–355 (2000)
19. Tamaki, H.: A Polynomial Time Algorithm for Bounded Directed Pathwidth. In: Kolman, P., Kratochvíl, J. (eds.) WG 2011. LNCS, vol. 6986, pp. 331–342. Springer, Heidelberg (2011)
20. Thilikos, D.M., Serna, M., Bodlaender, H.L.: Cutwidth I: A linear time fixed parameter algorithm. J. of Algorithms 56, 1–24 (2005)

Segmental Mapping and Distance for Rooted Labeled Ordered Trees[*]

Tomohiro Kan[1], Shoichi Higuchi[1], and Kouichi Hirata[2]

[1] Graduage School of Computer Science and Systems Engineering
[2] Department of Artificial Intelligence & Biomedical Informatics R&D Center
Kyushu Institute of Technology Kawazu 680-4, Iizuka 820-8502, Japan
{kan,syou_hig,hirata}@dumbo.ai.kyutech.ac.jp

Abstract. In this paper, as a variation of a Tai mapping between trees, we introduce a *segmental* mapping to preserve the parent-children relationship as possible. Then, we show that the segmental mapping provides a new hierarchy for the classes of Tai mappings in addition to a well-known one. Also we show that the *segmental distance* as the minimum cost of segmental mappings is a metric. Finally, we design the algorithm to compute the segmental distance in quadratic time and space.

1 Introduction

Comparing tree-structured data such as HTML and XML data for web mining or DNA and glycan data for bioinformatics is one of the important tasks for data mining. In this paper, we formulate such data as *rooted labeled ordered trees* (*trees*, for short) and then focus on distance measures between trees.

The most famous distance measure between trees is the *edit distance* [5]. The edit distance is formulated as the minimum cost to transform from a tree to another tree by applying *edit operations* of a *substitution*, a *deletion* and an *insertion* to trees. It is known that the edit distance is closely related to the notion of a *Tai mapping* (TAI) [5], which is a one-to-one node correspondence between trees preserving ancestor and sibling relations. The minimum cost of possible mappings coincides with the edit distance [5]. After introducing the edit distance, the time complexity to compute it has been improved as $O(n^3)$ time [2], where n is the maximum number of nodes in given two trees.

While the edit distance is the standard measure for comparing trees, it is too general for several applications. Therefore, more structural sensitive variations of the edit distance such as the *top-down* (or *degree*-1) distance [1,4], the *degree*-2 distance [10], the *accordant* distance [3], the *isolated-subtree* (or *constrained*) distance [7,9] and the *bottom-up* distance [6] are required for these applications. Such variations are formulated as the minimum cost of restricted mappings such

---

[*] This work is partially supported by Grand-in-Aid for Scientific Research 22240010, 24240021 and 24300060 from the Ministry of Education, Culture, Sports, Science and Technology, Japan.

K.-M. Chao, T.-s. Hsu, and D.-T. Lee (Eds.): ISAAC 2012, LNCS 7676, pp. 485–494, 2012.

as *top-down* (TOP), *degree*-2 (DG2), *accordant* (ACC), *isolated-subtree* (ISST) and *bottom-up* (BOT) mappings, respectively, and computed in $O(n^2)$ time[1].

It is known that these mappings provide the hierarchy described in Figure 1 (left) [3,7] as a Hasse diagram. This diagram claims that if $M \in A$ then $M \in B$ for a mapping M, a lower class A and an upper class B in Figure 1 (left).

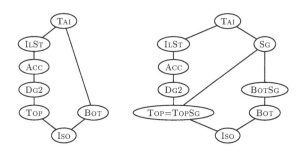

Fig. 1. A mapping hierarchy [3,7] (left) and a new mapping hierarchy (right)

In the above mappings, the parent-children relationship is just preserved by both top-down and bottom-up mappings, which are too restricted. On the other hand, it is sometimes important in several applications such as the function determination of glycan data, the parse trees of programs, the trace patterns of procedure calls and the change detection of XML documents (*cf.*, [1,3,6]).

As the generalization of top-down and bottom-up mappings, we introduce a *segmental mapping* (SG) preserving the parent-children relationship as possible. The segmental mapping requires that, for every pair of nodes in a mapping, if the mapping contains a pair of the ancestors of the nodes, then it always contains the pair of the parents of the nodes. Also we formulate *top-down* and *bottom-up segmental mappings* (TOPSG and BOTSG) that are segmental mappings always containing the pair of the roots and the pair of leaves as descendants, respectively.

In this paper, first we show that SG, TOPSG and BOTSG provide a new hierarchy in Figure 1 (right). Next, we show that the *segmental distance* and the *bottom-up segmental distance* as the minimum cost of SG and BOTSG are metrics. Finally, we design the algorithm to compute the segmental distance in $O(n^2)$ time and space.

2   Preliminaries

A *tree* is a connected graph without cycles. For a tree $T = (V, E)$, we denote V and E by $V(T)$ and $E(T)$, respectively. Also the *size* of T is $|V|$ and denoted by $|T|$. We sometime denote $v \in V(T)$ by $v \in T$. We denote an empty tree by $\emptyset$.

---

[1] While Valiente [6] has first introduced a bottom-up distance, his distance does not allow the substitution. Then, his distance is an indel distance, which runs in $O(n)$ time, rather than an edit distance, which runs in $O(n^2)$ time. See [8].

A *rooted tree* is a tree with one node r chosen as its *root*. We denote the root of a rooted tree T by $r(T)$. For each node v in a rooted tree with the root r, let $UP_r(v)$ be the unique path from v to r. The *parent* of $v(\neq r)$, which we denote by $par(v)$, is its adjacent node on $UP_r(v)$ and the *ancestors* of $v(\neq r)$ are the nodes on $UP_r(v) - \{v\}$. We denote the set of all ancestors of v by $anc(v)$. We say that u is a *child* of v if v is the parent of u, and u is a *descendant* of v if v is an ancestor of u. In this paper, we use the ancestor orders $<$ and $\leq$, that is, $u < v$ if v is an ancestor of u and $u \leq v$ if $u < v$ or $u = v$. We say that w is the *least common ancestor* of u and v, denoted by $u \sqcup v$, if $u \leq w$, $v \leq w$ and there exists no w' such that $w' \leq w$, $u \leq w'$ and $v \leq w'$.

A *leaf* is a node having no children. We denote the set of all leaves in T by $lv(T)$. The *degree* of a node $v \in V(T)$, denoted by $deg(v)$, is the number of children of v. A (*complete*) *subtree* of T rooted by v, denoted by $T(v)$, is a tree consisting of v and all of the descendants of v.

We say that a rooted tree is *ordered* if a left-to-right order among siblings is given. For a rooted ordered tree T, a node v in T and its children $v_1, \ldots, v_i$, the *preorder traversal* of $T(v)$ is obtained by visiting v and then recursively visiting $T(v_k)$ $(1 \leq k \leq i)$ in order. Similarly, the *postorder traversal* of $T(v)$ is obtained by first visiting $T(v_k)$ $(1 \leq k \leq i)$ and then visiting v. The *preorder* (*resp.*, *postorder*) *number* of $v \in T$ is the number of nodes preceding v in the preorder (*resp.* postorder) traversal of T and denote it by $pre(v)$ (*resp.*, $post(v)$). The nodes *to the left of* $v \in T$ is the set of nodes $u \in T$ satisfying that (1) $pre(u) \leq pre(v)$ and (2) $post(u) \leq post(v)$. If u is to the left of v, then v is to the *right* of u. We denote that u is to the left of v by $u \preceq v$.

We say that a rooted tree is *labeled* if each node is assigned a symbol from a fixed finite alphabet Σ. For a node v, we denote the label of v by $l(v)$, and sometimes identify v with $l(v)$. In this paper, we call a rooted labeled ordered tree a tree simply. A(n *ordered*) *forest* is a sequence of trees. We denote a forest consisting of trees $T_1, \ldots, T_m$ by $[T_1, \ldots, T_m]$.

Definition 1 (Edit operations). We define *edit operations* of a tree T as follows. See Figure 2.

1. *Substitution*: Change the label of the node v in T. ·
2. *Deletion*: Delete a non-root node v in T with parent v', making the children of v become the children of v'. The children are inserted in the place of v as a subsequence in the left-to-right order of the children of v'.
3. *Insertion*: The complement of deletion. Insert a node v as a child of v' in T making v the parent of a consecutive subsequence of the children of v'.

Let $\varepsilon \notin \Sigma$ denote a special *blank* symbol and define $\Sigma_\varepsilon = \Sigma \cup \{\varepsilon\}$. Then, we represent each edit operation by $(l_1 \mapsto l_2)$, where $(l_1, l_2) \in (\Sigma_\varepsilon \times \Sigma_\varepsilon - \{(\varepsilon, \varepsilon)\})$. The operation is a substitution if $l_1 \neq \varepsilon$ and $l_2 \neq \varepsilon$, a deletion if $l_2 = \varepsilon$, and an insertion if $l_1 = \varepsilon$. For nodes v and w, we also denote $(l(v) \mapsto l(w))$ by $(v \mapsto w)$.

We define a *cost function* $\gamma : (\Sigma_\varepsilon \times \Sigma_\varepsilon - \{(\varepsilon, \varepsilon)\}) \mapsto \mathbf{R}$ on pairs of labels. We often constrain a cost function γ to be a *metric*, that is, $\gamma(l_1, l_2) \geq 0$, $\gamma(l_1, l_2) = 0$ iff $l_1 = l_2$, $\gamma(l_1, l_2) = \gamma(l_2, l_1)$ and $\gamma(l_1, l_3) \leq \gamma(l_1, l_2) + \gamma(l_2, l_3)$.

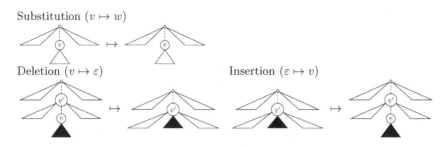

Fig. 2. Edit operations for trees

Definition 2 (Edit distance). For a cost function γ, the *cost* of an edit operation $e = l_1 \mapsto l_2$ is given by $\gamma(e) = \gamma(l_1, l_2)$. The *cost* of a sequence $E = e_1, \ldots, e_k$ of edit operations is given by $\gamma(E) = \sum_{i=1}^{k} \gamma(e_i)$. Then, an *edit distance* $\tau(T_1, T_2)$ between trees T_1 and T_2 is defined as follows:

$$\tau(T_1, T_2) = \min \left\{ \gamma(E) \,\middle|\, \begin{array}{l} E \text{ is a sequence of edit operations} \\ \text{transforming } T_1 \text{ to } T_2 \end{array} \right\}.$$

Definition 3 (Mapping). Let T_1 and T_2 be trees. We say that a triple (M, T_1, T_2) (or simply M when there is no confusion) is a *Tai mapping* (a *mapping*, for short) between T_1 and T_2, which we denote by $M \in \text{TAI}$, if $M \subseteq V(T_1) \times V(T_2)$ and every pair (v_1, w_1) and (v_2, w_2) in M satisfies the following three conditions.

1. $v_1 = v_2$ iff $w_1 = w_2$.   2. $v_1 \leq v_2$ iff $w_1 \leq w_2$.   3. $v_1 \preceq v_2$ iff $w_1 \preceq w_2$.

Let M be a mapping between T_1 and T_2. Let I and J be the sets of nodes in T_1 and T_2 but not in M. Then, the *cost* of M is given as follows.

$$\gamma(M) = \sum_{(v,w)\in M} \gamma(v \mapsto w) + \sum_{v\in I} \gamma(v \mapsto \varepsilon) + \sum_{w\in J} \gamma(\varepsilon \mapsto w).$$

Theorem 1 (Tai [5]). $\tau(T_1, T_2) = \min\{\gamma(M) \mid M \in \text{TAI}\}$.

Trees T_1 and T_2 are *isomorphic*, denoted by $T_1 \equiv T_2$, if there exists a mapping M between T_1 and T_2 such that $\gamma(M) = 0$, which we denote by $M \in \text{ISO}$.

Definition 4 (Variations). Let T_1 and T_2 be trees and $M \subseteq V(T_1) \times V(T_2)$ a mapping between T_1 and T_2. Also we denote $M - \{(r(T_1), r(T_2))\}$ by M^-.

1. We say that M is an *isolated-subtree mapping* [7] (or a *constrained mapping* [9]), denoted by $M \in \text{ILST}$, if M satisfies the following condition.
   $$\forall (v_1, w_1), (v_2, w_2), (v_3, w_3) \in M \left(v_3 < v_1 \sqcup v_2 \iff w_3 < w_1 \sqcup w_2 \right).$$
2. We say that M is an *accordant mapping* [3], denoted by $M \in \text{ACC}$, if M satisfies the following condition.
   $$\forall (v_1, w_1), (v_2, w_2), (v_3, w_3) \in M \left(\begin{array}{c} v_1 \sqcup v_2 = v_1 \sqcup v_3 \\ \iff w_1 \sqcup w_2 = w_1 \sqcup w_3 \end{array} \right).$$

3. We say that M is a *degree-2 mapping* [10], denoted by $M \in \mathrm{DG2}$, if M satisfies the following condition.

$$\forall (v_1, w_1), (v_2, w_2) \in M^{-}\Big((v_1 \sqcup v_2, w_1 \sqcup w_2) \in M\Big).$$

4. We say that M is a *top-down mapping* [1,4], denoted by $M \in \mathrm{TOP}$, if M satisfies the following condition.

$$\forall (v, w) \in M^{-}\Big((par(v), par(w)) \in M\Big).$$

5. We say that M is a *bottom-up mapping* [3,6,8][2], denoted by $M \in \mathrm{BOT}$, if M satisfies the following condition.

$$\forall (v, w) \in M \begin{pmatrix} \forall v' \in T_1(v) \exists w' \in T_2(w)\Big((v', w') \in M\Big) \\ \wedge \forall w' \in T_2(w) \exists v' \in T_1(v)\Big((v', w') \in M\Big) \end{pmatrix}.$$

Also we define the *top-down distance* $\tau_{\mathrm{T}}(T_1, T_2)$ as $\min\{\gamma(M) \mid M \in \mathrm{TOP}\}$.

Example 1. Consider the mappings M_i $(1 \le i \le 6)$ in Figure 3. Then, it holds that $M_1 \in \mathrm{TOP}$; $M_2 \notin \mathrm{TOP}$ but $M_2 \in \mathrm{DG2}$; $M_3 \notin \mathrm{DG2}$ but $M_3 \in \mathrm{ACC}$; $M_4 \notin \mathrm{ACC}$ but $M_4 \in \mathrm{ILST}$; $M_5 \notin \mathrm{ILST}$ but $M_5 \in \mathrm{TAI}$. Also it holds that $M_6 \in \mathrm{BOT}$ but $M_6 \notin \mathrm{ILST}$. Furthermore, it holds that $M_i \notin \mathrm{BOT}$ $(1 \le i \le 5)$.

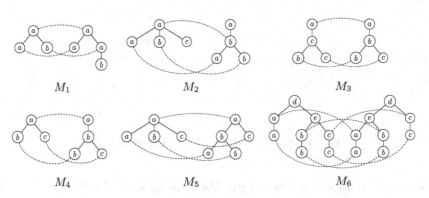

Fig. 3. Mappings M_i $(1 \le i \le 6)$ in Example 1

3 Segmental Mapping and Distance

In this section, we introduce a *segmental mapping* and a *segmental distance*.

Definition 5 (Segmental mapping). Let T_1 and T_2 be trees and $M \subseteq V(T_1) \times V(T_2)$ a mapping between T_1 and T_2.

---

[2] While Valiente [6] has introduced a bottom-up mapping that requires an isolated-subtree mapping, his algorithm computes one that is not an isolated-subtree distance. Then, we adopt the revised definition of a bottom-up mapping [3,8].

1. We say that M is a *segmental* mapping, denoted by $M \in \mathrm{SG}$, if M satisfies the following condition.

$$\forall(v,w) \in M^- \left(\begin{array}{l} \left(\big((v',w') \in M\big) \wedge \big(v' \in anc(v)\big) \wedge \big(w' \in anc(w)\big) \right) \\ \implies (par(v), par(w)) \in M \end{array} \right).$$

2. We say that M is a *top-down segmental* mapping, denoted by $M \in \mathrm{TopSG}$, if M is a segmental mapping such that $(r(T_1), r(T_2)) \in M$.

3. We say that M is a *bottom-up segmental* mapping, denoted by $M \in \mathrm{BotSG}$, if M is a segmental mapping satisfying the following condition.

$$\forall(v,w) \in M \left(\begin{array}{l} \exists(v',w') \in M \left(\begin{array}{l} \big(v \in anc(v')\big) \wedge \big(w \in anc(w')\big) \\ \wedge \big(v' \in lv(T_1)\big) \wedge \big(w' \in lv(T_2)\big) \end{array} \right) \\ \vee \left(\big(v \in lv(T_1)\big) \wedge \big(w \in lv(T_1)\big) \right) \end{array} \right).$$

Example 2. Consider the mappings M_i ($7 \le i \le 9$) in Figure 4. For M_7, it holds that $M_7 \in \mathrm{TOP}$, $M_7 \in \mathrm{TopSG}$, $M_7 \in \mathrm{BotSG}$ and $M_7 \in \mathrm{SG}$ but $M_7 \notin \mathrm{BOT}$. For M_8, it holds that $M_8 \in \mathrm{BotSG}$ and $M_8 \in \mathrm{SG}$ but $M_8 \notin \mathrm{TOP}$ and $M_8 \notin \mathrm{TopSG}$. For M_9, it holds that $M_9 \in \mathrm{SG}$ but $M_9 \notin \mathrm{BotSG}$, $M_9 \notin \mathrm{TopSG}$ and $M_9 \notin \mathrm{TOP}$. Also it holds that $M_9 \notin \mathrm{ILST}$. Furthermore, for M_3 and M_6 in Example 1, it holds that $M_3 \in \mathrm{ILST}$ but $M_3 \notin \mathrm{SG}$; $M_6 \in \mathrm{BotSG}$ and $M_6 \in \mathrm{SG}$ but $M_6 \notin \mathrm{ILST}$.

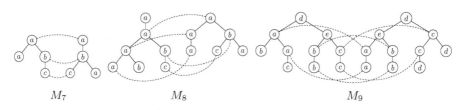

M_7 M_8 M_9

Fig. 4. Mappings M_i ($7 \le i \le 9$) in Example 2

Theorem 2 (Mapping hierarchy). *The mapping hierarchy illustrated in Figure 1 (right) in Section 1 holds. that is:*

1. $\mathrm{TOP} = \mathrm{TopSG} \subset \mathrm{SG} \subset \mathrm{TAI}$ *and* $\mathrm{BOT} \subset \mathrm{BotSG} \subset \mathrm{SG} \subset \mathrm{TAI}$.
2. $A \not\subseteq B$ *and* $B \not\subseteq A$ *for* $A \in \{\mathrm{BotSG}, \mathrm{SG}\}$ *and* $B \in \{\mathrm{TOP}, \mathrm{DG2}, \mathrm{ACC}, \mathrm{ILST}\}$.

Proof. The formula in Definition 5 implies that $\mathrm{TOP} = \mathrm{TopSG}$. Other inclusion, properness and incomparability follow from Definition 5 and Example 2.   □

For segmental mappings M_i ($i = 1, 2$) between T_i and T_{i+1}, we define the *composition* $M_1 \circ M_2$ as $\{(u,w) \mid \exists v \in T_2 \text{ s.t. } (u,v) \in M_1 \text{ and } (v,w) \in M_2\}$. Then, we can show the following lemma from Definition 5 as similar as [9].

Lemma 1. 1. *$M_1 \circ M_2$ is a segmental mapping between T_1 and T_3.*
2. *For a cost function γ that is a metric, $\gamma(M_1 \circ M_2) \le \gamma(M_1) + \gamma(M_2)$.*

Definition 6 (Segmental distance). A *segmental distance* $\delta(T_1, T_2)$ and a *bottom-up segmental distance* $\delta_\perp(T_1, T_2)$ between T_1 and T_2 are defined as:

$$\delta(T_1, T_2) = \min\{\gamma(M) \mid M \in \text{SG}\}, \quad \delta_\perp(T_1, T_2) = \min\{\gamma(M) \mid M \in \text{BOTSG}\}.$$

Theorem 3. *Both δ and $\delta_\perp$ are metrics.*

Proof. It is sufficient to show the triangle inequality for δ. Let M_1 (*resp.*, M_2) be the minimum cost segmental mapping between T_1 and T_2 (*resp.*, between T_2 and T_3). By Lemma 1, it holds that $\delta(T_1, T_3) \leq \gamma(M_1 \circ M_2) \leq \gamma(M_1) + \gamma(M_2) = \delta(T_1, T_2) + \delta(T_2, T_3)$, so δ is a metric. Similarly, $\delta_\perp$ is also a metric   □

4 Computing Segmental Distance

In this section, we identify a node in T_1 (*resp.*, T_2) with its postorder number i ($1 \leq i \leq |T_1|$) (*resp.*, j ($1 \leq j \leq |T_2|$)) of T_1 (*resp.*, T_2), where 0 denotes the empty tree. We denote the postorder number of the leftmost leaf of $T_1(i)$ (*resp.*, $T_2(j)$) by $ll(i)$ (*resp.*, $ll(j)$). Also let $F_1(i)$ (*resp.*, $F_2(j)$) denote the forest obtained by deleting i (*resp.*, j) from $T_1(i)$ (*resp.*, $T_2(j)$). Let $n = \max\{|T_1|, |T_2|\}$.

Let M be a segmental mapping between T_1 and T_2. Then, there exists at least one pair $(i, j) \in M$ such that $(i', j') \notin M$ for every ancestor i' of i in T_1 and every ancestor j' of j in T_2. We call such a pair a *maximal pair* of M and denote the set of all maximal pairs of M by P_M. Also, for every $(i, j) \in P_M$, we can obtain the subset $M_{(i,j)} \subseteq M$ such that $M_{(i,j)} = \{(i', j') \in M \mid i' \in T_1(i), j' \in T_2(j)\}$. We denote the set of nodes that are not descendants of every i (*resp.*, j) such that $(i, j) \in P_M$ by R_M^1 (*resp.*, R_M^2). Then, the following equation is straightforward.

$$\gamma(M) = \sum_{(i,j) \in P_M} \gamma(M_{(i,j)}) + \sum_{v \in R_M^1} \gamma(v \mapsto \varepsilon) + \sum_{w \in R_M^2} \gamma(\varepsilon \mapsto w). \tag{1}$$

Lemma 2. *For every $(i, j) \in P_M$, $M_{(i,j)}$ is a top-down mapping between $T_1(i)$ and $T_2(j)$. Hence, it holds that $\gamma(M_{(i,j)}) \geq \tau_\top(T_1(i), T_2(j))$.*

Proof. For $(i', j') \in M_{(i,j)}$, it holds that $i' \leq i$ in T_1 and $j' \leq j$ in T_2. Since $M_{(i,j)}$ is a segmental mapping, there exists a sequence $(i'_1, j'_1), \ldots, (i'_a, j'_a)$ of pairs in $M_{(i,j)}$ such that $i'_1 = i$, $j'_1 = j$, $i'_a = i'$, $j'_a = j'$, $i'_b = par(i'_{b+1})$ and $j'_b = par(j'_{b+1})$ for $1 \leq b \leq a - 1$. This implies that $M_{(i,j)}$ is a top-down mapping.   □

Lemma 3. *Let M^* be the minimum cost segmental mapping between T_1 and T_2. Then, the following equation holds.*

$$\delta(T_1, T_2) = \sum_{(i,j) \in P_{M^*}} \tau_\top(T_1(i), T_2(j)) + \sum_{v \in R_{M^*}^1} \gamma(v \mapsto \varepsilon) + \sum_{w \in R_{M^*}^2} \gamma(\varepsilon \mapsto w). \tag{2}$$

Proof. Since $\gamma(M^*) = \delta(T_1, T_2)$ and the minimality of $\gamma(M^*)$ implies that $\gamma(M_{(i,j)}^*) = \tau_\top(T_1(i), T_2(j))$, the equation (2) follows from the equation (1).   □

procedure SegDist(T_1, T_2, γ)

/* T_1, T_2 : trees, γ : cost function */

1 **for** $i = 1$ **to** $|T_1|$ **do**

2   **for** $j = 1$ **to** $|T_2|$ **do**

3      $TD[i,j] \leftarrow$ TopDownPair(i, j, γ);

4 $D[0,0] \leftarrow 0$;

5 **for** $i = 1$ **to** $|T_1|$ **do**

6    $D[i,0] \leftarrow D[i-1,0] + \gamma(i \mapsto \varepsilon)$;

7 **for** $j = 1$ **to** $|T_2|$ **do**

8    $D[0,j] \leftarrow D[0,j-1] + \gamma(\varepsilon \mapsto j)$;

9 **for** $i = 1$ **to** $|T_1|$ **do**

10   **for** $j = 1$ **to** $|T_2|$ **do**

11      $D[i,j] \leftarrow \min \left\{ \begin{array}{l} D[i-1,j] + \gamma(i \mapsto \varepsilon), \ D[i,j-1] + \gamma(\varepsilon \mapsto j), \\ D[ll(i)-1, ll(j)-1] + TD[i,j] \end{array} \right\}$;

12 **output** $D[|T_1|, |T_2|]$;

procedure TopDownPair(i, j, γ)

/* $i \in T_1$, $F_1(i) = [T_1(i_1), \ldots, T_1(i_m)]$, where $i_0 = 0$ */

/* $j \in T_2$, $F_2(j) = [T_2(j_1), \ldots, T_2(j_n)]$, where $j_0 = 0$ */

13 $F[0,0] \leftarrow 0$;

14 **for** $k = 1$ **to** m **do**

15    $F[i_k, 0] \leftarrow F[i_{k-1}, 0] + |T_1(i_k)| \times \gamma(i_k \mapsto \varepsilon)$;

16 **for** $l = 1$ **to** n **do**

17    $F[0, j_l] \leftarrow F[0, j_{l-1}] + |T_2(j_l)| \times \gamma(\varepsilon \mapsto j_l)$;

18 **for** $k = 1$ **to** m **do**

19   **for** $l = 1$ **to** n **do**

20      $F[i_k, j_l] \leftarrow \min \left\{ \begin{array}{l} F[i_{k-1}, j_l] + |T_1(i_k)| \times \gamma(i_k \mapsto \varepsilon), \\ F[i_k, j_{l-1}] + |T_2(j_l)| \times \gamma(\varepsilon \mapsto j_l), \\ F[i_{k-1}, j_{l-1}] + TD[i_k, j_l] \end{array} \right\}$;

21 **output** $F[i_m, j_n] + \gamma(i \mapsto j)$;

Algorithm 1. SegDist

The equation (2) claims that we can compute the segmental distance $\delta(T_1, T_2)$ by first computing the top-down distance $\tau_T(T_1(i), T_2(j))$ for every pair $(i, j) \in T_1 \times T_2$ and then combining pairs such that the total cost of a mapping is minimum, which we achieve in $O(n^4)$ time by using a naive method [1,4]. In this paper, we design an $O(n^2)$ time algorithm SegDist in Algorithm 1.

Lemma 4. *For $i \in T_1$ and $j \in T_2$, the algorithm* TopDownPair(i, j, γ) *computes the top-down distance $\tau_T(T_1(i), T_2(j))$ in $O(deg(i) \times deg(j))$ time.*

Proof. Let $i_1, \ldots, i_m$ be the children of i in T_1 and $j_1, \ldots, j_n$ the children of j in T_2, that is, let $F_1(i) = [T_1(i_1), \ldots, T_1(i_m)]$ and $F_2(j) = [T_2(j_1), \ldots, T_2(j_n)]$. Also let $I = \{i_1, \ldots, i_m\}$ and $J = \{j_1, \ldots, j_n\}$. Furthermore, since the for-loop of lines 1 and 2 in SegDist executes in postorder traversal, we can suppose that $TD[i_a, j_b] (= \tau_\top(T_1(i_a), T_2(j_b)))$ has been already computed for $1 \leq i_a < i$ $(1 \leq a \leq m)$ and $1 \leq j_b < j$ $(1 \leq b \leq n)$ when computing $\tau_\top(T_1(i), T_2(j))$.

Since $\gamma(i \mapsto j)$ in line 21 is the cost of the pair (i, j), which is contained from every top-down mapping between $T_1(i)$ and $T_2(j)$, we can obtain the top-down distance $\tau_\top(T_1(i), T_2(j))$ by adding $\gamma(i \mapsto j)$ to the combination of I and J providing the minimum cost. As the same discussion of [9], we can regard such a combination as the string edit distance between $i_1 \cdots i_m$ and $j_1 \cdots j_n$ under the cost function c such that $c(i_a, \varepsilon) = |T_1(i_a)| \times \gamma(i_a \mapsto \varepsilon) = \tau_\top(T_1(i_a), \emptyset)$, $c(\varepsilon, j_b) = |T_2(j_b)| \times \gamma(\varepsilon \mapsto j_b) = \tau_\top(\emptyset, T_2(j_a))$ and $c(i_a, j_b) = \tau_\top(T_1(i_a), T_2(j_b))$ for $1 \leq a \leq m$ and $1 \leq b \leq n$, each of which is a formula in line 20. It is obvious that the algorithm TopDownPair(i, j, γ) runs in $O(deg(i) \times deg(j))$ time. □

Theorem 4. *The algorithm* SegDist *computes the segmental distance* $\delta(T_1, T_2)$ *between* T_1 *and* T_2 *in* $O(n^2)$ *time and space.*

Proof. Let $F_1[i]$ (*resp.*, $F_2[j]$) be the forest of T_1 (*resp.*, T_2) constructing the nodes from 1 to i (*resp.*, from 1 to j) in postorder of T_1 (*resp.*, T_2). By the definition of ll, $ll(i)-1$ and $ll(j)-1$ are the left siblings of i in $F_1[i]$ and j in $F_2[j]$, that is, $F_1[i] = [\ldots, T_1(ll(i) - 1), T_1(i)]$ and $F_2[j] = [\ldots, T_2(ll(j) - 1), T_2(j)]$.

Suppose that $D[k, l]$ is the segmental distance between $F_1[k]$ and $F_2[l]$ for $1 \leq k \leq i$ and $1 \leq l \leq j$, and consider the segmental distance between $F_1[i]$ and $F_2[j]$. If j is inserted, then $D[i, j]$ is the sum of the segmental distance $D[i, j-1]$ between $F_1[i]$ and $F_2[j - 1]$ and the cost $\gamma(\varepsilon \mapsto j)$ of the insertion of j. If i is deleted, then $D[i, j]$ is the sum of the segmental distance $D[i - 1, j]$ between $F_1[i-1]$ and $F_2[j]$ and the cost $\gamma(i \mapsto \varepsilon)$ of the deletion of i. If i is substituted to j, then, by Lemma 3, $D[i, j]$ is the sum of the segmental distance $D[ll(i)-1, ll(j)-1]$ between $F_1[ll(i)-1]$ and $F_2[ll(j)-1]$ and the top-down distance $TD[i, j]$ between $T_1(i)$ and $T_2(j)$. Hence, $\delta(T_1, T_2)$ is given as $D[|T_1|, |T_2|]$.

The algorithm SegDist uses $O(|T_1| \times |T_2|)$ space. Also, by Lemma 4, the time complexity of the algorithm SegDist is given as $\sum_{i=1}^{|T_1|} \sum_{j=1}^{|T_2|} O(deg(i) \times$ $deg(j)) + O(|T_1|) + O(|T_2|) + O(|T_1| \times |T_2|) \leq O \left(\sum_{i=1}^{|T_1|} deg(i) \times \sum_{j=1}^{|T_2|} deg(j) \right) +$ $O(|T_1| \times |T_2|) \leq O(|T_1| \times |T_2|)$. □

Furthermore, we can design the algorithm to compute the bottom-up segmental distance $\delta_\perp(T_1, T_2)$ in $O(n^2)$ time and space, by adding the routine of determining that a current top-down mapping contains a pair of leaves when the third statement of $F[i_{k-1}, j_{l-1}] + TD[i_k, j_l]$ in line 20 is executed to SegDist.

Figure 5 illustrates distributions and the correlation diagrams to the edit distance τ of segmental, top-down and bottom-up distances for N-glycan data provided from KEGG[3]. Hence, the segmental distance preserves the parent-children relationship more than the top-down and the bottom-up distances nearer to τ.

---

[3] Kyoto Encyclopedia of Genes and Genomes, http://www.kegg.jp/. The number of N-glycan data is 2142, the average number of nodes is 11.09, the average number of labels is 5.43 and the average depth and degree are 5.38 and 2.07, respectively.

494   T. Kan, S. Higuchi, and K. Hirata

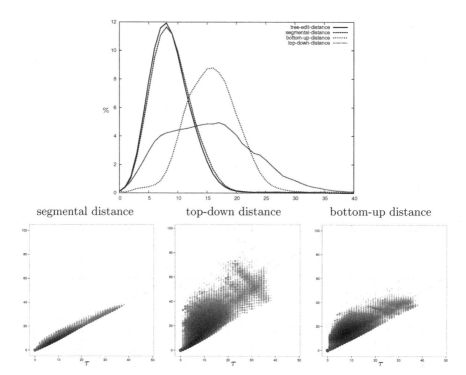

Fig. 5. The distributions (upper) and the correlation diagrams (lower) of segmental, top-down and bottom-up distances to an edit distance τ for N-glycan data

References

1. Chawathe, S.S.: Comparing hierarchical data in external memory. In: Proc. VLDB 1999, pp. 90–101 (1999)
2. Demaine, E.D., Mozes, S., Rossman, B., Weimann, O.: An optimal decomposition algorithm for tree edit distance. ACM Trans. Algorithms 6 (2009)
3. Kuboyama, T.: Matching and learning in trees. Ph.D thesis, University of Tokyo (2007), http://tk.cc.gakushuin.ac.jp/doc/kuboyama2007phd.pdf
4. Selkow, S.M.: The tree-to-tree editing problem. Inform. Process. Lett. 6, 184–186 (1977)
5. Tai, K.-C.: The tree-to-tree correction problem. J. ACM 26, 422–433 (1979)
6. Valiente, G.: An efficient bottom-up distance between trees. In: Proc. SPIRE 2001, pp. 212–219 (2001)
7. Wang, J.T.L., Zhang, K.: Finding similar consensus between trees: An algorithm and a distance hierarchy. Pattern Recog. 34, 127–137 (2001)
8. Yamamoto, Y., Hirata, K., Kuboyama, T.: A bottom-up edit distance between rooted labeled trees. In: Proc. LLLL 2011, pp. 26–33 (2011)
9. Zhang, K.: Algorithms for the constrained editing distance between ordered labeled trees and related problems. Pattern Recog. 28, 463–474 (1995)
10. Zhang, K., Wang, J.T.L., Shasha, D.: On the editing distance between undirected acyclic graph. Int. J. Found. Comput. Sci. 7, 43–58 (1995)

Detecting Induced Minors in AT-Free Graphs*

Petr A. Golovach[1], Dieter Kratsch[2], and Daniël Paulusma[3]

[1] Department of Informatics, Bergen University, PB 7803, 5020 Bergen, Norway
petr.golovach@ii.uib.no
[2] Laboratoire d'Informatique Théorique et Appliquée,
Université de Lorraine, 57045 Metz Cedex 01, France
kratsch@univ-metz.fr
[3] School of Engineering and Computing Sciences, Durham University,
South Road, Durham DH1 3LE, UK
daniel.paulusma@durham.ac.uk

Abstract. The problem INDUCED MINOR is to test whether a graph
G can be modified into a graph H by a sequence of vertex deletions
and edge contractions. We prove that INDUCED MINOR is polynomial-
time solvable when G is AT-free, and H is fixed, i.e., not part of the
input. Our result can be considered to be optimal in some sense as we
also prove that INDUCED MINOR is W[1]-hard on AT-free graphs, when
parameterized by $|V_H|$. In order to obtain it we prove that the SET-
RESTRICTED k-DISJOINT PATHS problem can be solved in polynomial
time on AT-free graphs for any fixed k. We also use the latter result to
prove that the SET-RESTRICTED k-DISJOINT CONNECTED SUBGRAPHS
problem is polynomial-time solvable on AT-free graphs for any fixed k.

1 Introduction

In this paper we study graph containment problems. Whether or not a graph
contains some other graph depends on the notion of containment used. In the literature several natural definitions have been studied such as containing a graph
as a contraction, dissolution, immersion, (induced) minor, (induced) topological
minor, (induced) subgraph, or (induced) spanning subgraph. We focus on the
containment relation "induced minor". A graph G contains a graph H as an
induced minor if G can be modified into a graph H by a sequence of vertex
deletions and edge contractions. Here, the operation *edge contraction* removes
the end-vertices u and v of an edge from G and replaces them by a new vertex
adjacent to precisely those vertices to which u or v were adjacent. The corresponding decision problem asking whether H is an induced minor of G is called
INDUCED MINOR. This problem is NP-complete even when G and H are trees of
bounded diameter or trees, the vertices of which have degree at most 3 except
for at most one vertex, as shown by Matoušek and Thomas [14]. It is therefore
natural to fix the graph H and to consider only the graph G to be part of the
input. We denote this variant as H-INDUCED MINOR.

* Supported by EPSRC (EP/G043434/1) and ERC (267959).

K.-M. Chao, T.-s. Hsu, and D.-T. Lee (Eds.): ISAAC 2012, LNCS 7676, pp. 495–505, 2012.
© Springer-Verlag Berlin Heidelberg 2012

The computational complexity classification of H-INDUCED MINOR is far from being settled, although both polynomial-time and NP-complete cases are known. In contrast, the two related problems H-MINOR and H-TOPOLOGICAL MINOR, which are to test whether a graph G contains a graph H as a minor or topological minor, respectively, can be solved in cubic time for any fixed graph H, as shown by Robertson and Seymour [16] and Grohe et al. [9], respectively. Fellows et al. [5] showed that there exists a graph H for which H-INDUCED MINOR is NP-complete. This specific graph H has 68 vertices and is still the smallest H for which H-INDUCED MINOR is known to be NP-complete. The question whether H-INDUCED MINOR is polynomial-time solvable for any fixed tree H was posed as an open problem at the AMS-IMS-SIAM Joint Summer Research Conference on Graph Minors in 1991. So far this question could only be answered for trees on at most 7 vertices except for one case [6].

Due to the notorious difficulty of solving H-INDUCED MINOR for general graphs, the input has been restricted to special graph classes. Fellows et al. [5] showed that for every fixed graph H, the H-INDUCED MINOR problem can be solved in linear time on planar graphs. Van 't Hof et al. [10] extended this result by proving that for every fixed planar graph H, the H-INDUCED MINOR problem is polynomial-time solvable on any minor-closed graph class not containing all graphs. Belmonte et al. [1] showed that for every fixed graph H, the H-INDUCED MINOR problem is polynomial-time solvable for chordal graphs, whereas for claw-free graphs a number of partial results, which only include polynomial-time solvable cases, are known [7].

We consider H-INDUCED MINOR restricted to the class of *asteroidal triple-free* graphs, also known as *AT-free* graphs. An *asteroidal triple* is a set of three mutually non-adjacent vertices such that each two of them are joined by a path that avoids the neighborhood of the third, and AT-free graphs are exactly those graphs that contain no such triple. AT-free graphs, defined fifty years ago by Lekkerkerker and Boland [13], are well studied in the literature and contain many well-known classes, e.g., cobipartite graphs, cocomparability graphs, cographs, interval graphs, permutation graphs, and trapezoid graphs (cf. [3]). All these graph classes have geometric intersection models being extremely useful when designing polynomial-time algorithms for hard problems. No such model is available for AT-free graphs. Recently, Golovach et al. [8] showed that the H-INDUCED TOPOLOGICAL MINOR problem is polynomial-time solvable on AT-free graphs for every fixed H. They also showed that this problem is W[1]-hard when parameterized by $|V_H|$.

Our Results. We show that H-INDUCED MINOR can be solved in polynomial time on AT-free graphs for any fixed graph H. Consequently, on AT-free graphs, all four problems H-MINOR, H-INDUCED MINOR, H-TOPOLOGICAL MINOR and H-INDUCED TOPOLOGICAL MINOR are polynomial-time solvable for any fixed graph H. In addition, we prove that INDUCED MINOR is W[1]-hard when parameterized by $|V_H|$. Our proof also implies the NP-completeness of INDUCED MINOR for AT-free graphs, which was not known before.

The celebrated result by Robertson and Seymour that H-MINOR is FPT on general graphs [16] is closely connected to the fact that k-DISJOINT PATHS is FPT with parameter k. To solve H-INDUCED TOPOLOGICAL MINOR on AT-free graphs, Golovach et al. [8] considered the variant k-INDUCED DISJOINT PATHS, in which the paths must not only be vertex-disjoint but also mutually induced, i.e., edges between vertices of any two distinct paths are forbidden. Here we must consider another variant, which was introduced by Belmonte et al. [1]. A *terminal pair* in a graph $G = (V, E)$ is a specified pair of vertices s and t called *terminals*, and the *domain* of a terminal pair (s, t) is a specified subset $U \subseteq V$ containing both s and t. We say that two paths, each of which is between some terminal pair, are *vertex-disjoint* if they have no common vertices except possibly the vertices of the terminal pairs. This leads to the following decision problem, which is NP-complete on general graphs even when $k = 2$ [1].

SET-RESTRICTED k-DISJOINT PATHS
Instance: a graph G, terminal pairs $(s_1, t_1), \ldots, (s_k, t_k)$, and domains $U_1, \ldots, U_k$.
Question: does G contain k mutually vertex-disjoint paths $P_1, \ldots, P_k$ such that
  P_i is a path from s_i to t_i using only vertices from U_i for $i = 1, \ldots, k$?

Note that the domains $U_1, \ldots, U_k$ are not necessarily pairwise disjoint. If we let every domain contain all vertices of G, we obtain exactly the DISJOINT PATHS problem. We give an algorithm that solves SET-RESTRICTED k-DISJOINT PATHS in polynomial time on AT-free graphs for any fixed integer k. We then use this algorithm as a subroutine in our polynomial-time algorithm that solves H-INDUCED MINOR on AT-free graphs for any fixed graph H. We emphasize that we can not apply the algorithm for k-INDUCED DISJOINT PATHS on AT-free graphs [8] as a subroutine to solve H-INDUCED MINOR on AT-free graphs. Also, the techniques used in that algorithm are quite different from the techniques we use here to solve SET-RESTRICTED k-DISJOINT PATHS on AT-free graphs. Moreover, when k is in the input, k-INDUCED DISJOINT PATHS and SET-RESTRICTED k-DISJOINT PATHS have a different complexity for AT-free graphs. Golovach et al. [8] proved that in that case k-INDUCED DISJOINT PATHS is polynomial-time solvable for AT-free graphs, whereas k-DISJOINT PATHS, and consequently SET-RESTRICTED k-DISJOINT PATHS, is already NP-complete for interval graphs [15], a subclass of AT-free graphs.

We use our algorithm for solving SET-RESTRICTED k-DISJOINT PATHS to obtain two additional results on AT-free graphs. The first result is that we can solve the problem SET-RESTRICTED k-DISJOINT CONNECTED SUBGRAPHS in polynomial time on AT-free graphs for any fixed integer k. A *terminal set* in a graph $G = (V, E)$ is a specified subset $S_i \subseteq V$.

SET-RESTRICTED k-DISJOINT CONNECTED SUBGRAPHS
Instance: a graph G, terminal sets $S_1, \ldots, S_k$, and domains $U_1, \ldots, U_k$.
Question: does G have k pairwise vertex-disjoint connected subgraphs $G_1, \ldots, G_k$,
  such that $S_i \subseteq V_{G_i} \subseteq U_i$, for $1 \leq i \leq k$?

If $|S_i| = 2$ for all $1 \leq i \leq k$, then we obtain the SET-RESTRICTED k-DISJOINT PATHS problem. If $U_i = V_G$ then we obtain the k-DISJOINT CONNECTED SUBGRAPHS problem. The latter problem has been introduced by Robertson and

Seymour [16] and is NP-complete on general graphs even when $k = 2$ and $\min\{|Z_1|, |Z_2|\} = 2$ [11]. The second result is that we can solve the problem H-CONTRACTIBILITY in polynomial time on AT-free graphs for any fixed triangle-free graph H. This problem is to test whether a graph G can be modified into a graph H by a sequence of contractions only. For general graphs, its complexity classification is still open but among other things it is known that the problem is already NP-complete when H is the 4-vertex path or the 4-vertex cycle [2].

2   Preliminaries

We only consider finite undirected graphs without loops and multiple edges. Let G be a graph. We denote the vertex set of G by V_G and the edge set by E_G. The subgraph of G induced by a subset $U \subseteq V_G$ is denoted by $G[U]$. We say that $U \subseteq V_G$ is *connected* if $G[U]$ is a connected graph. The graph $G - U$ is the graph obtained from G by removing all vertices in U. If $U = \{u\}$, we also write $G - u$. The *open neighborhood* of a vertex $u \in V_G$ is defined as $N_G(u) = \{v \mid uv \in E_G\}$, and its *closed neighborhood* is defined as $N_G[u] = N_G(u) \cup \{u\}$. For $U \subseteq V_G$, $N_G[U] = \cup_{u \in U} N_G[u]$. The degree of a vertex $u \in V_G$ is denoted $d_G(u) = |N_G(u)|$. The *distance* $\mathrm{dist}_G(u, v)$ between a pair of vertices u and v of G is the number of edges of a shortest path between them. Two sets $U, U' \subseteq V_G$ are called *adjacent* if there exist vertices $u \in U$ and $u' \in U'$ such that $uu' \in E_G$. A set $U \subseteq V_G$ *dominates* a vertex w if $w \in N_G[U]$, and U *dominates* a set $W \subseteq V_G$ if U dominates each vertex of W. In these two cases, we also say that $G[U]$ *dominates* w or W, respectively. A set $U \subseteq V_G$ is a *dominating set* of G if U dominates V_G.

The graph $P = u_1 \cdots u_k$ denotes the *path* with vertices $u_1, \ldots, u_k$ and edges $u_i u_{i+1}$ for $i = 1, \ldots, k - 1$. We also say that P is a (u_1, u_k)-*path*. For a path P with some specified end-vertex s, we write $x \prec_s y$ if $x \in V_P$ lies in P between s and $y \in V_P$; in this definition, we allow that $x = s$ or $x = y$. A pair of vertices $\{x, y\}$ is a *dominating pair* if the vertex set of every (x, y)-path is a dominating set of G. Corneil et al. [3,4] proved the following structural theorem.

Theorem 1 ([3,4]). *Every connected AT-free graph has a dominating pair and such a pair can be found in linear time.*

Using these results, Kloks et al. [12] gave the following tool for constructing dynamic programming algorithms on AT-free graphs. For a vertex u of a graph G, we call the sets $L_i(u) = \{v \in V_G \mid \mathrm{dist}_G(u, v) = i\}$ the *BFS-levels* of G. Note that the BFS-levels of a vertex can be determined in linear time by the Breadth-First Search algorithm (BFS).

Theorem 2 ([12]). *Every connected AT-free graph contains a dominating path $P = u_0 \cdots u_\ell$ that can be found in linear time such that i) ℓ is the number of BFS-levels of u_0, ii) $u_i \in L_i(u_0)$ for $i = 1, \ldots, \ell$, and iii) each $z \in L_i(u_0)$ is adjacent to u_{i-1} or to u_i for all $1 \leq i \leq \ell$.*

3   Set-Restricted Disjoint Paths

We show that SET-RESTRICTED k-DISJOINT PATHS can be solved in polynomial time on AT-free graphs for any fixed integer k. We need some extra terminology. Let G be a graph, and let $W \subseteq V_G$. Consider an induced path P in G. Then $V_P \cap W$ and $V_P \setminus W$ induce a collection of subpaths of P called W-*segments*, or *segments* if no confusion is possible. Segments induced by $V_P \cap W$ are said to lie *inside* W, whereas segments induced by $V_P \setminus W$ lie *outside* W. We need the following three lemmas (two proofs are omitted due to space restrictions).

Lemma 1. *Let P be an induced path in an AT-free graph G. Let $U \subseteq V_G$ be connected. Then P has at most three segments inside $N_G[U]$.*

Lemma 2. *Let P be an induced path in an AT-free graph G. Let $U \subseteq V_G$ be connected. Then every segment of P outside $N_G[U]$ that contains no end-vertex of P has at most two vertices.*

The next lemma directly follows from the condition on the path P to be induced.

Lemma 3. *Let u be a vertex of an induced path P in a graph G. Then P has one segment inside $N_G[u]$ and this segment has at most three vertices.*

Let G be a graph with terminal pairs $(s_1, t_1), \ldots (s_k, t_k)$ and corresponding domains $U_1, \ldots, U_k$. Let $\{P_1, \ldots, P_k\}$ be a set of mutually vertex-disjoint paths, such that P_i is a path from s_i to t_i using only vertices from U_i for $i = 1, \ldots, k$. We say that $\{P_1, \ldots, P_k\}$ is a *solution*. A solution $\{P_1, \ldots, P_k\}$ is *minimal* if no P_i can be replaced by a shorter (s_i, t_i)-path P_i' that uses only vertices of U_i in such a way that $P_1, \ldots, P_{i-1}, P_i', P_{i+1}, \ldots, P_k$ are mutually vertex-disjoint. Clearly, every yes-instance of SET-RESTRICTED k-DISJOINT PATHS has a minimal solution. We also observe that any path in a minimal solution is induced. We need Lemma 4 (proof omitted).

Lemma 4. *Let G be a graph with terminal pairs $(s_1, t_1), \ldots (s_k, t_k)$ and corresponding domains $U_1, \ldots, U_k$. Let $u \in U_i$ for some $1 \leq i \leq k$, and let $\{P_1, \ldots, P_k\}$ be a minimal solution with $u \notin \bigcup_{j=1}^{k} V_{P_j}$. Then P_i has at most two segments inside $N_G[u]$. Moreover, if P_i has one segment inside $N_G[u]$, then P_i has at most three vertices. If P_i has two segments Q_1 and Q_2 inside $N_G[u]$, then Q_1 and Q_2 each has precisely one vertex, and the segment Q' outside $N_G[u]$ that lies between Q_1 and Q_2 in P_i also has one vertex.*

We apply dynamic programming to prove that SET-RESTRICTED k-DISJOINT PATHS is polynomial-time solvable on AT-free graphs for every fixed integer k. Our algorithm solves the decision problem, but can easily be modified to produce the desired paths if they exist. It is based on the following idea. We find a shortest dominating path $u_0 \ldots u_\ell$ in G as described in Theorem 2. For $0 \leq i \leq \ell$, we trace the segments of (s_j, t_j)-paths inside $N_G[\{u_0, \ldots, u_i\}]$ by extending the segments inside $N_G[\{u_0, \ldots, u_{i-1}\}]$ in $N_G[u_i] \setminus N_G[\{u_0, \ldots, u_{i-1}\}]$. Note that if some path is traced from the middle, then we have to extend the corresponding segment in two directions, i.e., we have to trace two paths. The paths inside $N_G[u_i] \setminus N_G[\{u_0, \ldots, u_{i-1}\}]$ are constructed recursively, as by Lemmas 3 and 4 we can reduce the number of domains by distinguishing whether u_i is used by one of the paths or not. Hence, it is convenient for us to generalize as follows:

SET-RESTRICTED r-GROUP DISJOINT PATHS
Instance: A graph H, positive integers $p_1, \ldots, p_r$, terminal pairs (s_i^j, t_i^j) for $i \in \{1, \ldots, r\}$ and $j \in \{1, \ldots, p_i\}$, and domains $U_1, \ldots, U_r$.
Question: Does H contain mutually vertex-disjoint paths P_i^j, where $i \in \{1, \ldots, r\}$ and $j \in \{1, \ldots, p_i\}$, such that P_i^j is a path from s_i^j to t_i^j using only vertices from U_i for $i = 1, \ldots, r$?

Note that if $p_1 = \ldots = p_r = 1$, then we have SET-RESTRICTED r-DISJOINT PATHS. We say that for each $1 \leq i \leq r$, the pairs $(s_i^1, t_i^1), \ldots, (s_i^{p_i}, t_i^{p_i})$ (or corresponding paths) form a *group*. We are going to solve SET-RESTRICTED r-GROUP DISJOINT PATHS for induced subgraphs H of G and $r \leq k$ recursively to obtain a solution that can be extended to a solution of SET-RESTRICTED k-DISJOINT PATHS in such a way that $P_i^1, \ldots, P_i^{p_i}$ are disjoint subpaths of the (s_i, t_i)-path P_i in the solution of SET-RESTRICTED k-DISJOINT PATHS. Hence, we are interested only in some special solutions of SET-RESTRICTED r-GROUP DISJOINT PATHS.

For $r = 1$, SET-RESTRICTED r-GROUP DISJOINT PATHS is the p_1-DISJOINT PATHS problem in $H[U_1]$. By the celebrated result of Robertson and Seymour [16], we immediately get the following lemma.

Lemma 5. *For $r = 1$ and any fixed positive integer p_1, SET-RESTRICTED r-GROUP DISJOINT PATHS can be solved in $O(n^3)$ time on n-vertex graphs.*

Now we are ready to describe our algorithm for SET-RESTRICTED r-GROUP DISJOINT PATHS. First, we recursively apply the following preprocessing rules.

Rule 1. If H has a vertex $u \notin \cup_{i=1}^r U_i$, then we delete it and solve the problem on $H - u$.

Rule 2. If there are $i \in \{1, \ldots, r\}$ and $j \in \{1, \ldots, p_i\}$ such that s_i^j and t_i^j are in different components of $H[U_i]$, then stop and return No.

Rule 3. If H has components $H_1, \ldots, H_q$ and $q > 1$, then solve the problem for each component H_h for the pairs of terminals (s_i^j, t_i^j) such that $s_i^j, t_i^j \in V_{H_h}$ and the corresponding domains. We return Yes if we get a solution for each component H_h, and we return No otherwise.

Rule 4. If $r = 1$, then solve the problem by Lemma 5.

From now we assume that $r \geq 2$ and H is connected. Let $p = p_1 + \ldots + p_r$.

By Theorem 2, we can find a vertex $u_0 \in V_H$ and a dominating path $P = u_0 \ldots u_\ell$ in H with the property that for $i \in \{1, \ldots, \ell\}$, $u_i \in L_i$ and for any $z \in L_i$, z is adjacent to u_{i-1} or u_i, where $L_0, \ldots L_\ell$ are the BFS-levels of u_0. For $i \in \{0, \ldots, \ell\}$, let $W_i = N_H[\{u_0, \ldots, u_i\}]$, $W_{-1} = \emptyset$, and $S_i = N_G[u_i] \setminus W_{i-1}$. To simplify notations, we assume that for $i > \ell$, $S_i = \emptyset$, and $S_{-1} = \emptyset$. Notice that by the choice of P, there are no edges $xy \in E_H$ with $x \in S_j$ and $y \in N_H[\{u_0, \ldots, u_i\}]$ if $j - i > 2$.

Our dynamic programming algorithm keeps a table for each $i \in \{0, \ldots, \ell\}$, $X_i \subseteq S_{i+1}$ and $Y_i \subseteq S_{i+2}$, where $|X_i| \leq 4p$, $|Y_i| \leq 4p$, and an integer $next_i \in \{0, \ldots, r\}$. The table stores information about segments of (s_j^h, t_j^h)-paths inside W_i. Recall that each path can have more than one segment inside W_i, but in this case by Lemma 1, there are at most three such segments, and by Lemma 2, the number of vertices of the segments outside W_i, that join the segments inside, is bounded. We keep information about these vertices in X_i, Y_i. If $next_i = 0$, then no path in the partial solution includes u_{i+1}, and if $next_i = j > 0$, then only $(s_j^h, \tilde{t}_j^h)$, $(\tilde{s}_j^h, t_j^h)$, $(\tilde{s}_j^h, \tilde{t}_j^h)$-paths can use u_{i+1} (if $i = \ell$, then we assume that $next_i = 0$). For each $i, X_i, Y_i, next_i$, the table stores a collection of records $\mathcal{R}(i, X_i, Y_i, next_i)$ with the elements

$$\{(State_j^h, R_j^h)|1 \leq j \leq r, 1 \leq h \leq p_i\},$$

where R_j^h are ordered multisets of size at most two without common vertices except (possibly) terminals $s_1, \ldots, s_k, t_1, \ldots, t_k$ of the original instance of SET-RESTRICTED k-DISJOINT PATHS, $R_j^h \subseteq U_j$, and where each $State_j^h$ can have one of the following five values:

Not initialized, Started from s, Started from t, Started from middle, Completed.

These records correspond to a partial solution of SET-RESTRICTED r-GROUP DISJOINT PATHS for $H_i = H[W_i \cup X_i \cup Y_i]$ with the following properties.

- If $State_j^h = Not\ initialized$, then (s_j^h, t_j^h)-paths have no vertices in H_i in the partial solution and $R_j^h = \emptyset$.
- If $State_j^h = Started\ from\ s$, then $s_j^h \in W_i$, $t_j^h \notin V_{H_i}$ and R_j^h contains one vertex. Let $R_j = (\tilde{t}_j^h)$. Then $\tilde{t}_j^h \in S_{i-1} \cup S_i$ and the partial solution contains an $(s_j^h, \tilde{t}_j^h)$-path.
- If $State_j^h = Started\ from\ t$, then $s_j^h \notin V_{H_i}$, $t_j^h \in W_i$ and R_j^h contains one vertex. Let $R_j^h = (\tilde{s}_j^h)$. Then $\tilde{s}_j^h \in S_{i-1} \cup S_i$ and the partial solution contains an $(\tilde{s}_j^h, t_j^h)$-path.
- If $State_j^h = Started\ from\ middle$, then $s_j^h, t_j^h \notin V_{H_i}$ and R_j^h contains two vertices. Let $R_j^h = (\tilde{s}_j^h, \tilde{t}_j^h)$ (it can happen that $\tilde{t}_j^h = \tilde{s}_j^h$). Then $\tilde{s}_j^h, \tilde{t}_j^h \in S_{i-1} \cup S_i$ and the partial solution contains an $(\tilde{s}_j^h, \tilde{t}_j^h)$-path.
- If $State_j^h = Completed$, then $s_j^h, t_j^h \in W_i$, $R_j^h = \emptyset$, and it is assumed that the partial solution contains an (s_j^h, t_j^h)-path.

We consequently construct the tables for $i = 0, \ldots, \ell$. The algorithm returns Yes if $\mathcal{R}(\ell, X_\ell, Y_\ell, next_\ell)$ for $X_\ell = Y_\ell = \emptyset$ contains the record $\{(State_j^h, R_j^h)|1 \leq j \leq r, 1 \leq h \leq p_i\}$, where each $State_j^h =$ (Completed). The details and the proof of the main theorem have been omitted.

Theorem 3. SET-RESTRICTED k-DISJOINT PATHS *can be solved in* $O(n^{f(k)})$ *time for n-vertex AT-free graphs for some function $f(k)$ that only depends on k.*

4 Induced Minors

In this section we consider the H-INDUCED MINOR problem. It is convenient for us to represent this problem in the following way. An H-*witness structure* of G is a collection of $|V_H|$ non-empty mutually disjoint sets $W(x) \subseteq V_G$, one set for each $x \in V_H$, called H-*witness sets*, such that

(i) each $W(x)$ is a connected set; and
(ii) for all $x, y \in V_H$ with $x \neq y$, sets $W(x)$ and $W(y)$ are adjacent in G if and only if x and y are adjacent in H.

Observe that H is an induced minor of G if and only if G has an H-witness structure.

Theorem 4. H-INDUCED MINOR *can be solved in polynomial time on AT-free graphs for any fixed graph* H.

Proof. Suppose that H is an induced minor of G. Then G has an H-witness structure, i.e., sets $W(x) \subseteq V_G$ for $x \in V_H$. For each $x \in V_H$, $G[W(x)]$ is a connected AT-free graph. Hence, by Theorem 1, $G[W(x)]$ has a dominating pair (u_x, v_x).

For each $x \in V_H$, we guess the pair (u_x, v_x) (it can happen that $u_x = v_x$), and guess at most six vertices of a shortest (u_x, v_x)-path P_x in $G[W(x)]$ as follows: if P_x has at most five vertices, then we guess all vertices of P_x, and if P_x has at least six vertices, then we guess the first three vertices u_1^x, u_2^x, u_3^x and the last three vertices v_1^x, v_2^x, v_3^x such that $u_x = u_1^x$, $v_x = v_3^x$ and $u_1^x \prec_{u_x} u_2^x \prec_{u_x} u_3^x \prec_{u_x} v_1^x \prec_{u_x} v_2^x \prec_{u_x} v_3^x$ in P_x. Observe that P_x is an induced path. We denote by X_1, X_2 the partition of V_H (one of the sets can be empty), where for $x \in X_1$, all at most five vertices of P_x were chosen, and for $x \in X_2$, we have the vertices $u_1^x, u_2^x, u_3^x, v_1^x, v_2^x, v_3^x$. Further, for each edge $xy \in E_H$, we guess adjacent vertices $s_{xy}, s_{yx} \in V_G$, where $s_{xy} \in W(x)$ and $s_{yx} \in W(y)$. Notice that the vertices s_{xy} are not necessarily distinct, and some of them can coincide with the vertices chosen to represent P_x. Let $S(x) = \{s_{xy} | xy \in E_H\}$. All the guesses should be consistent with the witness structure, i.e., vertices included in distinct $W(x)$ should be distinct, and if $xy \notin E_H$, then the vertices included in $W(x)$ and $W(y)$ should be non-adjacent in G.

For $x \in X_1$, we check whether the guessed path P_x dominates $S(x)$, and if it is so, then we let $W'(x) = V_{P_x} \cup S(x)$. Otherwise we discard our choice.

Recall that we already selected some vertices, and that we cannot use these vertices and also not their neighbors in case non-adjacencies in H forbid this. Hence, for each $x \in X_2$, we obtain the set

$$U_x = V_G \setminus \big((\cup_{y \in X_1, xy \in E_H} W'(y)) \cup (\cup_{y \in X_1, xy \notin E_H} N_G[W'(y)]) \cup$$
$$\cup (\cup_{y \in X_2, xy \in E_H} (S(y) \cup \{u_1^y, u_2^y, u_3^y, v_1^y, v_2^y, v_3^y\})) \cup$$
$$\cup (\cup_{y \in X_2 \setminus \{x\}, xy \notin E_H} N_G[S(y) \cup \{u_1^y, u_2^y, u_3^y, v_1^y, v_2^y, v_3^y\}] \cup$$
$$\cup N_G[\{u_1^x, u_2^x, v_2^x, v_3^x\}]) \big) \cup \{u_1^x, u_2^x, u_3^x, v_1^x, v_2^x, v_3^x\}.$$

Then for each $x \in X_2$, we check whether $S'(x) = S(x) \setminus N_G[\{u_1^x, u_2^x, v_2^x, v_3^x\}]$ is included in one component of $G[U_x]$. If it is not so, then we discard our choice, since we cannot have a path with the first vertices u_1^x, u_2^x, u_3^x and the last vertices v_1^x, v_2^x, v_3^x that dominates $S(x)$. Otherwise we denote by U_x' the set of vertices of the component of $G[U_x]$ that contains $S'(x)$. Notice that (u_1^x, v_3^x) is a dominating pair in $G[U_x']$ for $x \in X_2$. To show it, consider a dominating pair (u, v) in $G[U_x']$. Any (u, v)-path P dominates u_1^x and v_3^x. It follows that one vertex of the pair is in $\{u_1^x, u_2^x\}$ and another is in $\{v_2^x, v_3^x\}$. It remains to observe that if u_2^x (v_2^x respectively) is in the pair, then it can be replaced by u_1^x (v_3^x respectively). We solve SET-RESTRICTED $|X_2|$-DISJOINT PATHS for the pairs of terminals (u_1^x, v_3^x) with domains U_x' for $x \in X_2$. If we get a No-answer, then we discard our guess since there are no P_x that satisfy our choices. Otherwise, let P_x' be the (u_1^x, v_3^x)-path in the obtained solution for $x \in X_2$. We let $W'(x) = P_x' \cup S(x)$.

We claim that the sets $W'(x)$ compose an H-witness structure. To show it, observe first that by the construction of these sets, $W'(x)$ are disjoint. If $xy \in E_H$, then as $s_{xy} \in W'(x)$ and $s_{yx} \in W'(y)$, $W'(x)$ and $W'(y)$ are adjacent. It remains to prove that if $xy \notin E_G$, then $W'(x)$ and $W'(y)$ are not adjacent. To obtain a contradiction, assume that $W'(x)$ and $W'(y)$ are adjacent for some $x, y \in V_H$, i.e., there is $uv \in E_G$ with $u \in W'(x)$ and $v \in W'(y)$, where $xy \notin E_H$. By the construction of $W'(x), W'(y)$, $x, y \in X_2$. Moreover, $u \notin N_G[\{u_1^x, u_2^x, v_2^x, v_3^x\}]$ or $v \notin N_G[\{u_1^y, u_2^y, v_2^y, v_3^y\}]$. If $u \notin N_G[\{u_1^x, u_2^x, v_2^x, v_3^x\}]$, then we consider u_1^x, v_3^x, u_1^y and observe that these vertices compose an asteroidal triple. Clearly, the (u_1^x, v_3^x)-path P_x' avoids $N_G[u_1^y]$, because $N_G[u_1^y] \cap U_x = \emptyset$. Because $u \notin N_G[\{u_1^x, u_2^x, v_2^x, v_3^x\}]$, u is either in P_x' or adjacent to a vertex in P_x' and v is either in P_y' or adjacent to a vertex in P_y', $G[W'(x) \cup W'[y]] - N_G[u_1^x]$ and $G[W'(x) \cup W'[y]] - N_G[v_3^x]$ are connected. Hence, there are (u_1^x, u_1^y) and (v_3^x, u_1^y)-paths that avoid $N_G[v_3^x]$ and $N_G[u_1^x]$ respectively. By symmetry, we conclude that if $v \notin N_G[\{u_1^y, u_2^y, v_2^y, v_3^y\}]$, then u_1^y, v_3^y, u_1^x is an asteroidal triple. This contradiction proves our claim. To complete the proof, note that we guess at most $6|V_H| + 2|E_H|$ vertices of G, and we can consider all possible choices in time $n^{O(|V_H|+|E_H|)}$, where $n = |V_G|$. If for one of the choices we get an H-witness structure, then H is an induced minor of G, otherwise we return No. As we can solve SET-RESTRICTED $|X_2|$-DISJOINT PATHS in time $n^{f(|V_H|)}$ by Theorem 3, the claim follows. $\qquad\square$

A graph is *cobipartite* if its vertex set can be partitioned into two cliques. Such a graph is AT-free. Hence, the next theorem complements Theorem 4. It is proven by a reduction from the CLIQUE problem; the details have been omitted.

Theorem 5. *The H-INDUCED MINOR problem is* NP-*complete for cobipartite graphs, and* W[1]-*hard for cobipartite graphs when parameterized by* $|V_H|$.

5   Concluding Remarks

We have presented a polynomial-time algorithm that solves SET-RESTRICTED k-DISJOINT PATHS on AT-free graphs for any fixed integer k, and applied this algorithm to solve H-INDUCED MINOR in polynomial time on this graph class for any fixed graph H. We give (without proofs) two further applications of our algorithm for SET-RESTRICTED k-DISJOINT PATHS.

Theorem 6. SET-RESTRICTED k-DISJOINT CONNECTED SUBGRAPHS *can be solved in polynomial time on AT-free graphs for any fixed integer k.*

Theorem 7. H-CONTRACTIBILITY *can be solved in polynomial time on AT-free graphs for any fixed triangle-free graph H.*

The *join* of two vertex-disjoint graphs $G_1 = (V_1, E_1)$ and $G_2 = (V_2, E_2)$ is the graph $G_1 \bowtie G_2 = (V_1 \cup V_2, E_1 \cup E_2 \cup \{uv \mid u \in V_1, v \in V_2\})$. A graph G contains a graph H as an induced minor if and only if $K_1 \bowtie G$ contains $K_1 \bowtie H$ as a contraction [10]. This fact together with Theorem 5 yields Corollary 1.

Corollary 1. H-CONTRACTIBILITY *is* NP-*complete for cobipartite graphs, and* W[1]-*hard for cobipartite graphs when parameterized by* $|V_H|$.

Determining the complexity classification of H-CONTRACTIBILITY on AT-free graphs when H is a fixed graph that is not triangle-free is an open problem.

References

1. Belmonte, R., Golovach, P.A., Heggernes, P., van 't Hof, P., Kamiński, M., Paulusma, D.: Finding Contractions and Induced Minors in Chordal Graphs via Disjoint Paths. In: Asano, T., Nakano, S.-i., Okamoto, Y., Watanabe, O. (eds.) ISAAC 2011. LNCS, vol. 7074, pp. 110–119. Springer, Heidelberg (2011)
2. Brouwer, A.E., Veldman, H.J.: Contractibility and NP-completeness. Journal of Graph Theory 11, 71–79 (1987)
3. Corneil, D.G., Olariu, S., Stewart, L.: Asteroidal triple-free graphs. SIAM Journal on Discrete Mathematics 10, 299–430 (1997)
4. Corneil, D.G., Olariu, S., Stewart, L.: Linear time algorithms for dominating pairs in asteroidal triple-free graphs. SIAM Journal on Computing 28, 1284–1297 (1999)
5. Fellows, M.R., Kratochvíl, J., Middendorf, M., Pfeiffer, F.: The complexity of induced minors and related problems. Algorithmica 13, 266–282 (1995)
6. Fiala, J., Kamiński, M., Paulusma, D.: Detecting induced star-like minors in polynomial time (preprint)
7. Fiala, J., Kamiński, M., Paulusma, D.: A note on contracting claw-free graphs (preprint)
8. Golovach, P.A., Paulusma, D., van Leeuwen, E.J.: Induced Disjoint Paths in AT-Free Graphs. In: Fomin, F.V., Kaski, P. (eds.) SWAT 2012. LNCS, vol. 7357, pp. 153–164. Springer, Heidelberg (2012)
9. Grohe, M., Kawarabayashi, K., Marx, D., Wollan, P.: Finding topological subgraphs is fixed-parameter tractable. In: Proceedings of STOC 2011, pp. 479–488 (2011)
10. van' t Hof, P., Kaminski, M., Paulusma, D., Szeider, S., Thilikos, D.M.: On graph contractions and induced minors. Discrete Applied Mathematics 160, 799–809 (2012)

11. van' t Hof, P., Paulusma, D., Woeginger, G.J.: Partitioning graphs in connected parts. Theoretical Computer Science 410, 4834–4843 (2009)
12. Kloks, T., Kratsch, D., Müller, H.: Approximating the bandwidth for AT-free graphs. Journal of Algorithms 32, 41–57 (1999)
13. Lekkerkerker, C.G., Boland, J.Ch.: Representation of a finite graph by a set of intervals on the real line. Fundamenta Mathematicae 51, 45–64 (1962)
14. Matoušek, J., Thomas, R.: On the complexity of finding iso- and other morphisms for partial k-trees. Discrete Mathematics 108, 343–364 (1992)
15. Natarajan, S., Sprague, A.P.: Disjoint paths in circular arc graphs. Nordic Journal of Computing 3, 256–270 (1996)
16. Robertson, N., Seymour, P.D.: Graph minors. XIII. The disjoint paths problem. Journal of Combinatorial Theory B 63, 65–110 (1995)

Degree-Constrained Orientations
of Embedded Graphs

Yann Disser[1] and Jannik Matuschke[2]

[1] ETH Zurich, Institute of Theoretical Computer Science
ydisser@inf.ethz.ch
[2] TU Berlin, Institut für Mathematik
matuschke@math.tu-berlin.de

Abstract. We investigate the problem of orienting the edges of an embedded graph in such a way that the in-degrees of both the nodes and faces meet given values. We show that the number of feasible solutions is bounded by 2^{2g}, where g is the genus of the embedding, and all solutions can be determined within time $\mathcal{O}(2^{2g}|E|^2 + |E|^3)$. In particular, for planar graphs the solution is unique if it exists, and in general the problem of finding a feasible orientation is fixed-parameter tractable in g. In sharp contrast to these results, we show that the problem becomes NP-complete even for a fixed genus if only upper and lower bounds on the in-degrees are specified instead of exact values.

1 Introduction

Graph orientation is an area of combinatorial optimization that deals with the problem of assigning directions to the edges of an undirected graph, subject to certain problem-specific requirements. Besides yielding useful structural insights, e.g., with respect to connectivity of graphs [14] and hypergraphs [11], research in graph orientation is motivated by applications in areas such as graph drawing [2,5] or efficient data structures for planar graphs [3].

A particularly well-studied class of orientation problems are degree-constrained problems, i.e., where the in-degree of each vertex in the resulting orientation has to lie within certain bounds. Hakimi [12] and Frank [10] provided good characterizations[1] for the existence of such orientations. In this paper, we answer a question raised by András Frank [9], asking for a good characterization for the following problem: Given an embedding of a graph in the plane, is there an orientation of the edges that meets prescribed in-degrees both in the primal and the dual graph at the same time? We show that if such an orientation exists, it is unique and can be computed by combining a feasible orientation for the primal graph with a feasible orientation for the dual graph. Our result generalizes to graph embeddings of higher genus, showing that the number of

---

[1] A good characterization of a decision problem in the sense of Edmonds [6] is a description of polynomially verifiable certificates for both yes- and no-instances of the problem.

K.-M. Chao, T.-s. Hsu, and D.-T. Lee (Eds.): ISAAC 2012, LNCS 7676, pp. 506–516, 2012.
© Springer-Verlag Berlin Heidelberg 2012

feasible orientations is bounded by a function of the genus, and the set of all solutions can be computed efficiently as long as the genus is fixed. We also show that the problem becomes NP-complete as soon as upper and lower bounds on the in-degrees are specified instead of exact values.

Related Work. Research in graph orientation has a long history that revealed many interesting structural insights and applications. E.g., a classical result by Robbins [14] states that an undirected graph is 2-edge-connected if and only if it has an orientation that is strongly connected. This result was translated to hypergraphs by Frank et al. [11]. Graph orientation is also closely connected to graph drawing. For example, Eades and Wormald [5] showed hardness of a fixed edge-length graph drawing problem using an orientation problem on a planar graph as an important device in their reduction. More recently, Biedl et al. [2] provided a 13/8-approximation algorithm for finding a balanced acyclic orientation, with implications for orthogonal graph drawing.

Regarding degree-constrained orientation problems, Hakimi [12] gave a good characterization for the existence of orientations that match given prescribed in-degrees exactly and also for the existence of orientations that fulfill either lower or upper bounds on the in-degrees. Frank and Gyárfás [10] observed that the results for lower and upper bounds can easily be combined in a constructive way to find orientations that fulfill upper and lower bounds at the same time. Asahiro et al. [1] consider an optimization version of the degree-constrained orientation problem where a penalty function on the violated degree-bounds is to be minimized. They find that the problem is solvable in polynomial time if the penalty function is convex, but APX-hard in case of concave penalty functions.

Orientations of planar graphs received special attention by the research community because they revealed several interesting properties. Based on the insight that every planar graph allows for an orientation with maximum in-degree 3, Chrobak and Eppstein [3] designed a highly efficient data structure for adjacency queries in planar graphs. In a distinct line of research, Felsner [7] showed that the set of orientations fulfilling a prescribed in-degree in a planar graph carries the structure of a distributive lattice.

Contribution and Structure of the Paper. In this paper, we consider an extension of the degree-constrained problem, which we call *primal-dual orientation problem*. The input to this problem is an embedding of a graph in a surface and we require in-degree prescriptions not only to be met for every vertex but also for every face of the embedding (in this context, the in-degree of a face refers to the number of edges on its boundary oriented in counter-clockwise direction). This variant of the problem was first proposed by András Frank for the special case of plane graphs [9] in conjunction with the question for a good characterization of the existence of such an orientation.

Before we present our results, we give a short introduction to orientations and embedded graphs in Section 2. Section 3 then deals with the primal-dual orientation problem with fixed in-degrees and contains two different proofs that yield the answer to Frank's question. Subsection 3.1 comprises a combinatorial proof for the uniqueness of the solution in plane graphs, also reducing the problem to solving the original degree-constrained orientation problem once in the primal and once in the dual graph. In Subsection 3.2, an alternative proof based on a simple linear algebra argument also yields a bound on the number of feasible orientations in embeddings of higher genus—showing that the problem is fixed-parameter tractable in terms of the genus. In Section 4, we show that if we accept bounds on the in-degrees instead of exact values, the problem becomes NP-complete. In Section 5, we point out an open question, which will be subject of future research.

2   Preliminaries

We give a short introduction on graph embeddings and orientations of those embeddings. Throughout this paper we will assume all graphs to be connected but not necessarily simple, i.e., loops and multi-edges are allowed. While the connectedness assumption is very common in the context of graph embeddings, all results presented here can be extended to non-connected graphs by temporarily introducing additional edges (and adjusting the in-degree specifications accordingly) so as to render the graph connected.

Embedded Graphs. An *embedding* of a graph is a mapping of its vertices and edges onto a closed surface (e.g., a sphere or a torus) such that edges meet only at common vertices. This mapping partitions the surface into several regions, called *faces*. The *dual* of an embedded graph is the graph that is obtained by the following procedure: For every face in the embedding, introduce a vertex in the dual graph. For every edge of primal graph, introduce an edge in the dual graph that connects the faces that are adjacent to the original edge. The genus g of the embedding is determined by Euler's formula: If E is the set of edges, V is the set of vertices and V^* is the set of faces, then $|V| + |V^*| - |E| = 2 - 2g$.

If $g = 0$, i.e., the graph is embedded in a sphere, the embedding is called *planar* (as embeddings in spheres and planes are combinatorially equivalent). Planar embeddings have several features that make them particularly interesting. In this work, we will make use of the following fact, called *cycle-cut duality* [15], which holds (exclusively) in planar embeddings: A set of edges is a simple cycle in the primal if and only if it is a simple cut[2] in the dual and vice versa.

Orientations of Primal and Dual Graphs. An orientation of a graph is an assignment of directions to the edges, i.e., for every edge we specify one of the two endpoints of the edge as its head and the other as its tail. By convention, we orient the edges in the dual graph in such a way that they cross their primal "alter egos" from right to left (cf. Figure 1). Thus every orientation of the primal graph induces an orientation of the dual graph and vice versa. Given an orientation D, we denote the set of edges whose head is the vertex v by $\delta_D^-(v)$

---

[2] A simple cut is a cut whose edge set is minimal w.r.t. inclusion. In a connected graph, a cut is simple if and only if it splits the graph into two connected components.

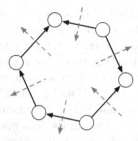

Fig. 1. Induced orientations of the edges in the dual graph. An edge in the dual graph crosses its corresponding edge in the primal graph from right to left.

and the set of edges whose tail is v by $\delta_D^+(v)$. In accordance with our convention for dual orientations, we let $\delta_D^-(f)$ be the set of edges whose left face is f, and $\delta_D^+(f)$ be the set of edges whose right face is f.

We mention that our convention for primal and dual orientations extends cycle/cut duality in the sense that a directed simple cycle in the primal is a directed simple cut in the dual and vice versa.[3]

3   Orientations with Fixed In-degrees

We consider the problem of finding an orientation that meets given fixed in-degrees for both the vertices and faces of the embedded graph, called the primal-dual orientation problem. We start this section by stating a formal description of the problem.

Problem 1. *(Primal-dual orientation problem)*

Given: *an embedded graph* $G = (V, E)$,
   two functions $\alpha : V \to \mathbb{N}_0$, $\alpha^* : V^* \to \mathbb{N}_0$
Task: *Find an orientation* D *of the edges* E *such that* $|\delta_D^-(v)| = \alpha(v)$ *for all* $v \in V$ *and* $|\delta_D^-(f)| = \alpha^*(f)$ *for all* $f \in V^*$, *or prove that there is none.*

Primal and Dual Feasibility. The following notation will be useful throughout the proofs in this section. Given an instance of the primal-dual orientation problem, we call an orientation D

- *primally feasible* if $|\delta_D^-(v)| = \alpha(v)$ for all $v \in V$.
- *dually feasible* if $|\delta_D^-(f)| = \alpha^*(f)$ for all $f \in V^*$.
- *totally feasible* if it is primally and dually feasible.

---

[3] A cut or cycle is directed if all its edges are oriented in the same direction.

The primal-dual orientation problem thus asks for a totally feasible orientation. It is clear that the existence of both primally feasible solutions and dually feasible solutions is necessary for the existence of such a totally feasible orientation. However, it can easily be checked that this is not sufficient: For example, consider a planar graph with two vertices and two parallel edges connecting them, and let $\alpha(v) = 1$ and $\alpha^*(f) = 1$ for all $v \in V$ and $f \in V^*$. While orienting both edges in opposite directions in the primal graph is primally feasible, orienting them in the same direction (which is orienting them in oposite directions in the dual graph) is dually feasible. However, none of the orientations is totally feasible.

In this section, we will present two approaches for obtaining necessary and sufficient conditions for the existence of totally feasible solutions.

3.1   A Combinatorial Approach for Planar Embeddings

In this section we want to provide a combinatorial argument for the uniqueness of a feasible solution to the primal-dual orientation problem in the planar case. We show how to construct a totally feasible solution from an orientation that is feasible in the primal graph and an orientation that is feasible in the dual graph.

Rigid Edges. Hakimi [12] showed that a primally feasible orientation exists if and only if $\sum_{v \in V} \alpha(v) = |E|$ and $\sum_{v \in S} \alpha(v) \geq |E[S]|$ for all $S \subseteq V$, where $E[S]$ is the set of edges with both endpoints in S. The necessity follows from the fact that every edge in $E[S]$ contributes to the in-degree of a node in S, independent of its orientation.

Now consider a subset $S \subseteq V$ with $\sum_{v \in S} \alpha(v) = |E[S]|$. All edges that have one end point in S and one end point in $V \setminus S$ must be oriented from S to $V \setminus S$ in all primally feasible orientations. We call edges whose orientation is fixed in this way *primally rigid*[4] and denote the set of all primally rigid edges by R. Analogously, we define the set of *dually rigid* edges R^* as those that are fixed for all dually feasible orientations due to a tight set $S^* \subseteq V^*$ of faces with $\sum_{f \in S^*} \alpha^*(f) = |E[S^*]|$. It is easy to check that an edge is primally rigid if and only if it is on a directed cut in the primal graph with respect to any primally feasible orientation. Likewise, an edge is dually rigid if it is on a directed cut in the dual graph with respect to any dually feasible orientation. Furthermore, note that the set of edges on directed cuts is invariant for all feasible solutions.

Our main result in this section follows from this characterization of rigid edges and the duality of cycles and cuts in planar graphs.

Theorem 1. *In case of a planar embedding, there exists a totally feasible orientation if and only if the following three conditions are fulfilled.*

(1) There exists both a primally feasible orientation D and a dually feasible orientation D^*.
(2) The edge set can be partitioned into primally and dually rigid edges ($E = R \dot\cup R^*$).

---

[4] The term "rigid" for edges on a directed cut of an orientation is taken from [7].

(3) The orientation obtained by orienting all primally rigid edges in the same direction as they are oriented in D and all dually rigid edges in the same orientation as they are oriented in D^* is totally feasible.

If it exists, the solution is unique.

Proof. The sufficiency of the conditions is trivial, as the third condition requires the existence of a totally feasible orientation. In order to show necessity, assume there exists a totally feasible orientation D_0. As D_0 is both primally and dually feasible, it fulfills Condition (1) of the theorem. An edge is primally rigid if and only if it is on a directed cut (w.r.t. D_0) in the primal graph. It is dually rigid, if and only if it is on a directed cut in the dual graph. Thus, by cycle/cut duality of planar graphs, an edge is dually rigid if and only if it is on a directed cycle in the primal graph. As every edge in the primal graph is either on a directed cut or on a directed cycle, the sets of primally and dually rigid edges comprise a partition of E, proving Condition (2). Now, let D be a primally feasible orientation and D^* be a dually feasible orientation. As D_0 equals D on all primally rigid edges and equals D^* on all dually rigid edges, the construction described in Condition (3) yields D_0 and is feasible. As all edges are either primally of dually rigid, they must have the same orientation in all totally feasible solutions, and D_0 is unique.   □

Note that the totally feasible solution constructed in the third condition does not depend on the choice of D and D^*. Theorem 1 also yields a polynomial time algorithm to solve the problem for planar embeddings.

Corollary 2. *The primal-dual orientation problem in planar embeddings can be solved in time $\mathcal{O}(|E|^2)$.*

Proof. By Theorem 1, the problem can be solved by computing a primally feasible solution and a dually feasible solution and identifying the corresponding rigid edges. A primally feasible orientation can be found in time $\mathcal{O}(|V||E|)$ by using a simple push/relabel type algorithm [8]. Now applying the same result to the dual gives a total time of $\mathcal{O}((|V| + |V^*|) \cdot |E|) = \mathcal{O}(|E|^2)$ for determining the two orientations. Identifying directed cuts is equivalent to identifying strongly connected components, which can be done in time $\mathcal{O}(|E|)$.   □

3.2   A Linear Algebra Analysis for General Embeddings

The primal-dual orientation problem can be formulated as a system of linear equalities over binary variables. To this end, we fix an arbitrary orientation D of the graph and introduce for every edge $e \in E$ a decision variable $x(e)$ that determines whether the orientation of the edge should be reversed (if it is 1) or not (if it is 0) in order to become totally feasible. The vector $x \in \{0, 1\}^E$ yields a feasible orientation if and only if it satisfies the following system of equalities:

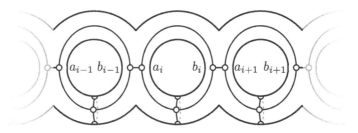

Fig. 2. Construction of an instance with 2^{2g} feasible orientations, showing the tightness of the bound in Theorem 3. The base graph consists of two cycles of length 3 intersecting in a common vertex and is embedded in a torus. Examples of genus g are obtained by introducing g copies of the base graph.

$$\sum_{e \in \delta_D^+(v)} x(e) - \sum_{e \in \delta_D^-(v)} x(e) = \alpha(v) - |\delta_D^-(v)| \quad \forall v \in V$$

$$\sum_{e \in \delta_D^+(f)} x(e) - \sum_{e \in \delta_D^-(f)} x(e) = \alpha^*(f) - |\delta_D^-(f)| \quad \forall f \in V^*$$

The matrix corresponding to the first set of equalities is the incidence matrix of the primal graph, while the matrix corresponding to the second type of equalities is the incidence matrix of the dual graph (both graphs directed according to the orientation D). As we assume the graph to be connected, we know that the rank of the former matrix is $|V| - 1$, while the rank of the latter matrix is $|V^*| - 1$. Using the fact that the boundary of a face is a closed walk in the primal graph, it is easy to see that the rows of the first matrix are orthogonal to the rows of the second matrix. This implies that all feasible solutions are contained in a subspace of $\mathbb{R}^E$ of dimension $|E| - |V| - |V^*| + 2 = 2g$.

Theorem 3. *There are at most 2^{2g} distinct solutions to the primal-dual orientation problem. The set of all totally feasible orientations can be determined in time $\mathcal{O}(2^{2g}|E|^2 + |E|^3)$. The bound on the number of orientations is tight, i.e., there are embedded graphs of genus g that allow for 2^{2g} distinct orientations.*

Proof. By basis augmentation, there is a set $A \subseteq E$ of $2g$ edges such that adding equalities $x(e) = a(e)$ with $a(e) \in \{0,1\}$ for all $e \in A$ results in a system with full rank, i.e., it has at most one solution. If for some $a \in \{0,1\}^A$ the unique solution exists and is a 0-1-vector, it corresponds to the unique totally feasible orientation that orients the edges of A according to the values $a(e)$. Otherwise, there is no such totally feasible orientation. Thus, solving the equality system for all $|\{0,1\}^A| = 2^{2g}$ possible values of a yields all possible solutions to the primal-dual orientation problem. This takes time $\mathcal{O}(|E|^3)$ for inverting the $|E| \times |E|$-matrix and $\mathcal{O}(2^{2g}|E|^2)$ for multiplying the 2^{2g} distinct right hand side vectors.

To see that the bound on the number of orientations is tight, consider the example depicted in Figure 2. It is constructed from a base graph consisting of two cycles of length 3 sharing a common vertex. The base graph is embedded in a torus, thus featuring only a single face f. When setting $\alpha^*(f) = |E| = 6$, any orientation is dually feasible as all dual edges are self-loops. We set the in-degree specification to 2 for the vertex at the intersection of the cycles and to 1 for the other vertices. Now, an orientation of the base graph is primally feasible, if and only if the edges of each cycle are all oriented in the same direction. As the two cycles can be oriented independently, the base graph has 4 feasible orientations. Examples of higher genus can be obtained by introducing g copies of the embedding described above. The graphs are joined via an edge from node b_i to a_{i+1} for $i \in \{1, \ldots, g - 1\}$. The resulting embedding has $5g$ vertices and $7g - 1$ edges and still has only a single face. We increase the in-degree specifications of each base graph by setting $\alpha(a_{i+1}) = 2$ for $i \in \{1, \ldots, g - 1\}$, so that the new edges joining the copies have to be oriented from copy i to copy $i + 1$. The in-degree specification of the face is set to $|E| = 7g - 1$. Now each copy of the base graph still has its 4 feasible orientations, so in total there are 4^g feasible orientations.[5]   □

4   Orientations with Upper and Lower Bounds

A generalization of the primal-dual orientation problem asks for an orientation that fulfills upper and lower bounds on the in-degrees of vertices and faces instead of attaining fixed values. We show that this problem becomes NP-complete, even when restricted to instances with embeddings of a fixed genus (e.g., planar graphs).

Problem 2. *(Bounded primal-dual orientation problem)*

Given: *an embedded graph $G = (V, E)$,*
 two pairs of functions $\alpha, \beta : V \to \mathbb{N}_0$ and $\alpha^*, \beta^* : V^* \to \mathbb{N}_0$
Task: *Find an orientation D of the edges E such that $\alpha(v) \leq |\delta_D^-(v)| \leq \beta(v)$*
 for all $v \in V$ and $\alpha^*(f) \leq |\delta_D^-(f)| \leq \beta^*(f)$ for all $f \in V^*$, or prove that
 there is none.

Theorem 4. *The bounded primal-dual orientation problem is NP-complete for graphs with any fixed genus.*

Proof sketch. An orientation that solves the bounded primal-dual orientation problem can easily be verified in polynomial time. Hence, it remains to show that the problem is NP-hard. It is sufficient to do this for planar graphs. We use a reduction from planar 3-SAT, which is known to be an NP-hard problem [13]. We construct an instance of the bounded primal-dual orientation problem that

---

[5] Note that while the primal graph in the construction described above could also be embedded in a plane, this can be avoided by introducing additional vertices and edges.

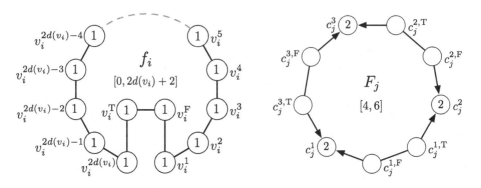

Fig. 3. Illustration of the variable gadget (left) and the clause gadget (right). The degree (number of occurrences) of variable v_i is denoted by $d(v_i)$. The labels of the nodes correspond to their prescribed in-degrees, and the intervals on the faces correspond to a range of permitted in-degrees. The orientation of the edge (v_i^T, v_i^F) corresponds to a truth assignment to variable v_i. The orientation of the edge $(c_j^{l,T}, c_j^{l,F})$ corresponds to a truth assignment to literal l in clause C_j.

has a solution if and only if the instance of planar 3-SAT has a solution. The construction consists of three main devices: For each variable, there is a *variable gadget*, for each clause, there is a *clause gadget*, and whenever a clause contains a variable, the corresponding gadgets are connected by an *edge gadget*. Figures 3 and 4 illustrate the construction. For a complete proof, please refer to the full version of this paper [4].   □

Corollary 5. *The bounded primal-dual orientation problem is NP-complete even when restricted to instances with $\alpha = \beta$ or $\alpha^* = \beta^*$.*

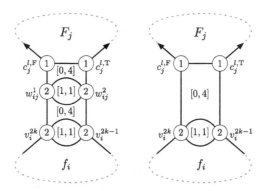

Fig. 4. Illustration of the edge gadget for an edge connecting variable v_i with clause C_j. The gadget on the left is used when v_i appears in a positive literal in C_j, and the one on the right is used when v_i appears in a negative literal.

Proof. This follows from the fact that the construction in the proof of Theorem 4 has $\alpha = \beta$. By duality, the reduction can also be achieved by an instance with $\alpha^* = \beta^*$. □

5 Conclusion

We have shown that the primal-dual orientation problem in an embedded graph of genus g has at most 2^{2g} feasible solutions and the set of all solutions can be computed in time $\mathcal{O}(2^{2g}|E|^2 + |E|^3)$. In particular, the solution is unique if the embedding is planar. However, the problem becomes NP-hard immediately, if only upper and lower bounds on the in-degrees are specified.

While these results give a relatively clear characterization of the complexity of the primal-dual orientation problem, we still want to point out an open question resulting from our research: The algorithm proposed in the proof of Theorem 3 has a running time that is exponential in the genus of the embedding. Is it possible to devise an algorithm that finds a totally feasible orientation in time polynomial in the size of the graph and the genus of the embedding? In the case of the bounded primal-dual orientation problem, our results imply that parameterization by the genus did not have any effect on the complexity of the problem. This might be an indication that the primal-dual orientation problem with exact degree specifications is not only fixed-parameter tractable but actually solvable in polynomial time.

Acknowledgements. We thank Kristóf Bérczi and Júlia Pap for providing many helpful suggestions. This work has been supported by the Berlin Mathematical School and by Deutsche Forschungsgemeinschaft (DFG) as part of the Priority Program "Algorithm Engineering" (1307).

References

1. Asahiro, Y., Jansson, J., Miyano, E., Ono, H.: Upper and Lower Degree Bounded Graph Orientation with Minimum Penalty. In: Proceedings of the 18th Computing: The Australasian Theory Symposium (CATS 2012). CRPIT, vol. 128, pp. 139–146 (2012)
2. Biedl, T., Chan, T., Ganjali, Y., Hajiaghayi, M.T., Wood, D.R.: Balanced vertex-orderings of graphs. Discrete Applied Mathematics 148(1), 27–48 (2005)
3. Chrobak, M., Eppstein, D.: Planar orientations with low out-degree and compaction of adjacency matrices. Theoretical Computer Science 86(2), 243–266 (1991)
4. Disser, Y., Matuschke, J.: Degree-constrained orientations of embedded graphs. Technical Report 032, TU Berlin (2011)
5. Eades, P., Wormald, N.C.: Fixed edge-length graph drawing is NP-hard. Discrete Applied Mathematics 28(2), 111–134 (1990)
6. Edmonds, J.: Minimum partition of a matroid into independent subsets. Journal of Research National Bureau of Standards Section B 69, 67–72 (1965)
7. Felsner, S.: Lattice structures from planar graphs. Journal of Combinatorics 11(1), 15 (2004)

8. Frank, A.: Connections in combinatorial optimization. Oxford University Press (2011)
9. Frank, A.: Personal communication (February 2010)
10. Frank, A., Gyárfás, A.: How to orient the edges of a graph. Colloquia Mathematica Societatis Janos Bolyai 18, 353–364 (1976)
11. Frank, A., Király, T., Király, Z.: On the orientation of graphs and hypergraphs. Discrete Applied Mathematics 131(2), 385–400 (2003)
12. Hakimi, S.L.: On the degrees of the vertices of a directed graph. Journal of the Franklin Institute 279(4), 290–308 (1965)
13. Lichtenstein, D.: Planar formulae and their uses. SIAM Journal on Computing 11(2), 329–343 (1982)
14. Robbins, H.: A theorem on graphs, with an application to a problem of traffic control. The American Mathematical Monthly 46(5), 281–283 (1939)
15. Whitney, H.: Non-separable and planar graphs. Transactions of the American Mathematical Society 34(2), 339–362 (1932)

Interval Graph Representation
with Given Interval and Intersection Lengths

Johannes Köbler[1], Sebastian Kuhnert[1,*], and Osamu Watanabe[2]

[1] Humboldt-Universität zu Berlin, Inst. für Informatik
[2] Tokyo Institute of Technology, Dept. of Mathematical and Computing Sciences

Abstract. We consider the problem of finding interval representations of graphs that additionally respect given interval lengths and/or pairwise intersection lengths, which are represented as weight functions on the vertices and edges, respectively. Pe'er and Shamir proved that the problem is NP-complete if only the former are given [SIAM J. Discr. Math. 10.4, 1997]. We give both a linear-time and a logspace algorithm for the case when both are given, and both an $\mathcal{O}(n \cdot m)$ time and a logspace algorithm when only the latter are given. We also show that the resulting interval systems are unique up to isomorphism.

Complementing their hardness result, Pe'er and Shamir give a polynomial-time algorithm for the case that the input graph has a unique interval ordering of its maxcliques. For such graphs, their algorithm computes an interval representation that respects a given set of distance inequalities between the interval endpoints (if it exists). We observe that deciding if such a representation exists is NL-complete.

1 Introduction

Algorithmic aspects of interval graphs have been the subject of ongoing research for several decades, stimulated by their numerous applications; see e.g. [Gol04].

The *interval representation problem* asks, given a graph G, if G is an interval graph, and if so, to compute an interval representation for it. Booth and Lueker [BL76] solve this problem in linear time, introducing the widely used concept of PQ-trees to efficiently encode all possible orderings of the maximal cliques. Hsu and Ma [HM99] give a simpler linear-time algorithm that relies on modular decomposition instead. Corneil, Olariu, and Stewart [COS09] show a further simplification, avoiding ordering the maximal cliques, by using lexicographic breadth first search. Klein gave a parallel AC^2 algorithm [Kle96]. Köbler et al. [KKLV11] show that the interval representation problem is complete for logspace.

In this paper, we consider the problems whether a graph with a weight function ℓ on its vertices and/or a weight function s on its edges admits ℓ-respecting interval representations (where for each vertex v, its weight $\ell(v)$ prescribes the length of its interval), s-respecting interval representations (where for each

* Supported by DFG grant KO 1053/7–1.

K.-M. Chao, T.-s. Hsu, and D.-T. Lee (Eds.): ISAAC 2012, LNCS 7676, pp. 517–526, 2012.
© Springer-Verlag Berlin Heidelberg 2012

edge $\{u, v\}$, its weight $s(\{u, v\})$ prescribes the length of the intersection of the intervals of u and v), and (ℓ, s)-respecting interval representations (which are required to fulfill both these restrictions). Pe'er and Shamir showed that it is NP-complete to decide if a graph G admits an ℓ-respecting interval representation [PS97]. The problem of finding s-respecting interval representations was introduced in [Yam07].

Our Results. We show how to construct (ℓ, s)-respecting interval representations in linear time or alternatively in logspace, and s-respecting interval representations in $\mathcal{O}(n \cdot m)$ time or alternatively in logspace. Since computing ℓ-respecting interval representations is NP-hard, our result illustrates that the information on interval intersections is quite helpful.

The first step towards our algorithms is to show that all interval representations of the appropriate type have the same inclusion and overlap relationships, and that these relations can be computed efficiently when G, ℓ (and s) are given as input. This is described in Section 3.

To obtain our results on (ℓ, s)-respecting interval representations (which are in Section 4), we first focus on graphs with overlap-connected representations. We show that these representations are unique up to reflection and can be computed efficiently (if they exist). For graphs with several overlap components we arrange these components into a tree, and combine their (ℓ, s)-respecting interval representations into one for the whole graph. We also show that all (ℓ, s)-respecting interval representations are isomorphic.

In Section 5 we show how to compute s-respecting interval representations efficiently. To obtain our result, we repeatedly use our algorithm for computing an (ℓ, s)-respecting interval representation as a subroutine. We prove that the lengths of the pairwise intersections already determine the interval lengths (up to insertion of points that are only present in a single interval). The resulting s-respecting interval representation is minimal, i.e., it contains no superfluous points. We also show that all minimal s-respecting interval representations are isomorphic.

In Section 6, we consider the variant of the interval representation problem for which Pe'er and Shamir gave a polynomial time algorithm [PS97]: On the one hand, the input graph is required to have a unique interval ordering of its inclusion-maximal cliques (up to reflection); on the other hand, general lower and upper bounds on distances between interval endpoints are allowed. We observe that this variant is in fact NL-complete. That is, it is unlikely that this generalization of the ℓ-respecting interval representation problem is solvable in deterministic logspace even for the restricted input graphs.

2   Preliminaries

We say that two sets A and B *overlap* and write $A \between B$, if $A \cap B \neq \varnothing$, $A \setminus B \neq \varnothing$, and $B \setminus A \neq \varnothing$. The cardinality of a finite set A is denoted by $\|A\|$.

For a graph $G = (V, E)$, the set of neighbors of a vertex $v \in V$ is denoted by $N(v)$. G is an *interval graph* if there is a system $\mathcal{I}$ of nonempty intervals over $\mathbb{N}$

(we allow $\mathcal{I}$ to be a multiset) and a bijection $\rho\colon V \to \mathcal{I}$ such that $\{u,v\} \in E \Leftrightarrow$ $\rho(u) \cap \rho(v) \neq \varnothing$. In this case, ρ is called an *interval representation* of G and $\mathcal{I}$ is called an *interval model* of G. The latter is also denoted by $\rho(G)$.

We write $[l, r]$ to denote the interval $\{i \in \mathbb{N} \mid l \leq i \leq r\}$. With the *length* of an interval we denote the number of points in it.[1] For an interval model $\mathcal{I}$ we always suppose $\bigcup_{I \in \mathcal{I}} I = [1, k]$ for some k, i.e., we disallow shifting and gaps between connected components. $\mathcal{I}$ can be regarded as hypergraph with nodes $[1, k]$ and hyperedges $\mathcal{I}$. Two interval models $\mathcal{I}$ and $\mathcal{I}'$ with points $[1, k]$ are *isomorphic* if they are isomorphic as hypergraphs, i.e., if there is a permutation $\pi\colon [1, k] \to$ $[1, k]$ of the points that induces a bijection between the intervals of $\mathcal{I}$ and $\mathcal{I}'$ (preserving multiplicities). We call two interval representations ρ_1 and ρ_2 of a graph G isomorphic if $\rho_1(G)$ and $\rho_2(G)$ are isomorphic. The *slots* of $\mathcal{I}$ are the equivalence classes on $[1, k]$ w.r.t. containment in the intervals in $\mathcal{I}$. That is, two vertices are in the same slot, if all hyperedges contain either both or none of them.

For functions $\ell\colon V \to \mathbb{N}$ and $s\colon E \to \mathbb{N}$, an interval representation $\rho\colon V \to \mathcal{I}$ of $G = (V, E)$ is called ℓ-*respecting* if $\|\rho(v)\| = \ell(v)$ for all $v \in V$, s-*respecting* if $\|\rho(u) \cap \rho(v)\| = s(\{u, v\})$ for all $\{u, v\} \in E$, and (ℓ, s)-*respecting* if both conditions hold. An s-respecting interval representation ρ of G is called *minimal* if there is no s-respecting interval representation ρ' of G that uses fewer points, i.e., that satisfies $\|\bigcup_{v \in V} \rho'(v)\| < \|\bigcup_{v \in V} \rho(v)\|$.

As usual, L is the class of all languages decidable by Turing machines with a read-only input tape using only $\mathcal{O}(\log N)$ space on the working tapes, where N is the input size. FL is the class of all functions computable by such machines that additionally have a write-only output tape. Note that FL is closed under composition: To compute $f(g(x))$ for $f, g \in$ FL, simulate the Turing machine for f and keep track of the position of its input head. Every time this simulation needs a character from f's input tape, simulate the Turing machine for g on input x until it outputs the required character. Note also that g can first output a copy of its input x and afterwards compute additional information to be used by f. This construction can be iterated a constant number of times, still preserving the logarithmic space bound. We will utilize this closure property in our logspace algorithms by employing pre- and post-processing steps.

This closure property can also be used to generalize our logspace results to the case where the prescribed lengths are rational: Bring all lengths to a common denominator and use the resulting numerators. This transformation is possible in logspace as iterative integer multiplication is in DLOGTIME-uniform TC^0 [HAB02].

3 Deriving Structural Information

Let $G = (V, E)$ be a graph, let $n = \|V\|$ and $m = \|E\|$, and let $\ell\colon V \to \mathbb{N}$ and $s\colon E \to \mathbb{N}$ specify the desired interval and intersection lengths. For convenience,

---

[1] This does not coincide with the usual notion of length $r - l$. However, if we use the real interval $(l - 0.5, r + 0.5)$, then both measures coincide.

we write $s(u, v)$ instead of $s(\{u, v\})$ for $\{u, v\} \in E$; for $\{u, v\} \notin E$ we let $s(u, v) = 0$. Using this convention, we define two relations $R_{\ell,s}, R_s \subseteq V^2$:

$$(u, v) \in R_{\ell,s} \Leftrightarrow \{u, v\} \in E \wedge \ell(u) > s(u, v)$$
$$(u, v) \in R_s \;\; \Leftrightarrow \{u, v\} \in E \wedge \exists w \in V \setminus \{u, v\} : s(w, u) > \min\{s(w, v), s(u, v)\}$$

By the following lemma, these relations characterize a structural property that *all* (ℓ, s)-respecting (resp., minimal s-respecting) interval representations of G have in common.

Lemma 1.
(a) Let $\rho \colon V \to \mathcal{I}$ be any (ℓ, s)-respecting interval representation of G, and let $\{u, v\} \in E$. Then $\rho(u) \setminus \rho(v) \neq \varnothing$ if and only if $(u, v) \in R_{\ell,s}$.
(b) Let $\rho \colon V \to \mathcal{I}$ be any minimal s-respecting interval representation of G, and let $\{u, v\} \in E$. Then $\rho(u) \setminus \rho(v) \neq \varnothing$ if and only if $(u, v) \in R_s$.

Proof. Part (a) follows directly from the definitions.
We now show part (b). By definition, $(u, v) \in R_s$ means that there is a $w \in V$ such that $s(w, u) > s(w, v)$ or $s(w, u) > s(u, v)$. Either way, there must be a point $p \in \rho(w) \cap (\rho(u) \setminus \rho(v))$, implying $\rho(u) \setminus \rho(v) \neq \varnothing$.
For the backward direction, consider a point $p \in \rho(u) \setminus \rho(v)$. By minimality of ρ, there is a vertex $w \in V \setminus \{u\}$ with $p \in \rho(w)$. Note that $w \neq v$ by choice of p. If $\rho(w) \supset \rho(u) \cap \rho(v)$, it follows that $s(w, u) > s(u, v)$. Otherwise $\rho(w) \cap \rho(u) \supsetneq \rho(w) \cap \rho(v)$ and thus $s(w, u) > s(w, v)$. $\square$

Lemma 2. $R_{\ell,s}$ and R_s can be enumerated in time $\mathcal{O}(m)$ and $\mathcal{O}(n \cdot m)$, respectively, and both can be enumerated in logspace.

Proof. The logspace part is obvious. To enumerate $R_{\ell,s}$ in linear time, loop over all edges $\{u, v\} \in E$ (considering both orientations) and output (u, v) if $\ell(u) > s(u, v)$. To enumerate R_s, loop over all edges $\{w, u\} \in E$ (again, considering both orientations) and all nodes $v \in V \setminus \{w, u\}$, and output (u, v) if $s(w, u) > \min\{s(w, v), s(u, v)\}$. $\square$

We write $u \between_{\ell,s} v$ if $(u, v) \in R_{\ell,s} \wedge (v, u) \in R_{\ell,s}$, and $u \subseteq_{\ell,s} v$ if $\{u, v\} \in E \wedge (u, v) \notin R_{\ell,s}$. The relations $\between_s$ and $\subseteq_s$ are defined analogously using R_s. By Lemma 1, these relations describe the situation in any appropriate representation of G, e.g. we have $u \between_{\ell,s} v \Leftrightarrow \rho(u) \between \rho(v)$ in any (ℓ, s)-respecting interval representation ρ of G, and $u \between_s v \Leftrightarrow \rho'(u) \between \rho'(v)$ in any minimal s-respecting interval representation ρ' of G.

Lemma 3. Let $\rho \colon V \to \mathcal{I}$ be any s-respecting interval representation of G. For any three vertices $v, w_1, w_2 \in V$ such that $\rho(w_1) \between \rho(v) \between \rho(w_2)$, the intervals $\rho(w_1)$ and $\rho(w_2)$ overlap $\rho(v)$ from the same side if and only if $s(w_1, w_2) > \min\{s(w_1, v), s(w_2, v)\}$.

Note that this condition can be decided both in constant time and in logspace.

Proof. If $\rho(w_1)$ and $\rho(w_2)$ overlap $\rho(v)$ from the same side, then $\rho(w_1)$ and $\rho(w_2)$ contain at least one common point outside $\rho(v)$, making their intersection larger than the minimum of $\|\rho(w_1) \cap \rho(v)\|$ and $\|\rho(w_2) \cap \rho(v)\|$.

Now suppose to the contrary that $\rho(w_1)$ and $\rho(w_2)$ overlap $\rho(v)$ from different sides. In this case $(\rho(w_1) \cap \rho(v)) \setminus \rho(w_2)$ and $(\rho(w_2) \cap \rho(v)) \setminus \rho(w_1)$ are both non-empty, implying that $\|\rho(w_1) \cap \rho(w_2)\|$ is smaller than both $\|\rho(w_1) \cap \rho(v)\|$ and $\|\rho(w_2) \cap \rho(v)\|$. □

4   Given Interval and Intersection Lengths

Let $G = (V, E)$ be a graph, and let $\ell \colon V \to \mathbb{N}$ and $s \colon E \to \mathbb{N}$ specify the desired interval and intersection lengths, respectively. In this section, we give linear-time and logspace algorithms that construct an (ℓ, s)-respecting interval representation of G, or detect that such a representation does not exist.

We define $E_{\ell,s} = \{\{u, v\} \in E \mid u \, \wp_{\ell,s} \, v\}$ and $G_{\ell,s} = (V, E_{\ell,s})$ and call the connected components of $G_{\ell,s}$ the *overlap components* of G. As a first step, we consider overlap-connected graphs.

Lemma 4. *Given $G = (V, E)$, ℓ and s, such that $G_{\ell,s}$ is connected, it is possible in linear time (resp., in logspace) to compute an (ℓ, s)-respecting interval representation $\rho \colon V \to \mathcal{I}$ of G, or to detect that none exists. Moreover, if existent, ρ is unique up to reflection.*

Proof. Let $v_1, v_2, \ldots, v_N$ be a walk in $G_{\ell,s}$ that visits every vertex at least once; such a walk can be constructed in linear time using depth first search or in logspace using Reingold's universal exploration sequences [Rei08]. The following algorithm computes an interval representation $\rho \colon V \to \mathcal{I}$ of G by moving along this walk (which we assume has been computed in a pre-processing step). It computes an interval I_i for v_i at each step and outputs $\rho(v_i) = I_i$, if there is no $j < i$ with $v_j = v_i$. Define $I_1 = [1, \ell(v_1)]$ and $I_2 = [\ell(v_1) - s(v_1, v_2) + 1, \ell(v_1) - s(v_1, v_2) + \ell(v_2)]$. Note that after I_1 has been placed, there are only two possibilities for I_2 that respect (ℓ, s); see Fig. 1 for an illustration. After that, all further intervals are completely determined because of Lemma 3, and can be computed from the walk, ℓ and s, remembering only the two previous intervals.

In a post-processing step, check that ρ is (ℓ, s)-respecting. Additionally, shift the resulting intervals such that 1 becomes the smallest point.

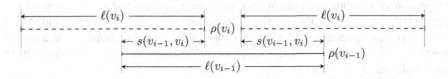

Fig. 1. Proof of Lemma 4: If $v_i \, \wp_{\ell,s} \, v_{i-1}$, and if $\rho(v_{i-1})$ is already determined, there remain only the two dashed possibilities for $\rho(v_i)$

The uniqueness up to reflection follows from the fact that the only arbitrary decision (except shifting) was to place $\rho(v_2)$ right of $\rho(v_1)$.   □

The next step is to generalize Lemma 4 to the case that $G_{\ell,s}$ is not connected. We can assume that there are no vertices v and v' such that both $v \subseteq_{\ell,s} v'$ and $v' \subseteq_{\ell,s} v$ hold; otherwise compute an (ℓ, s)-respecting interval representation for $G \setminus \{v'\}$ and extend it by $v' \mapsto \rho(v)$ afterwards. Let $\mathcal{C} = \{G_1, \ldots, G_k\}$ be the connected components of $G_{\ell,s}$. We write $G_i \leq_{\ell,s} G_j$ if $i = j$ or if there are vertices u in G_i and v in G_j such that $v \subseteq_{\ell,s} u$. The latter implies that, for any (ℓ, s)-respecting interval representation ρ of G, the interval $\bigcup_{u \in G_i} \rho(u)$ is contained in some slot $S \subseteq \rho(v)$ of $\rho(G_j)$, because otherwise there would be an overlap-path from $\rho(G_i)$ to $\rho(G_j)$. Thus $\leq_{\ell,s}$ is a partial order on the overlap components of G. If G is connected, $(\mathcal{C}, \leq_{\ell,s})$ is also connected; by removing reflexive and transitive edges, we obtain a rooted tree $T_{\ell,s}$, which we call the *overlap component tree* of G.

Theorem 5. *Given $G = (V, E)$, ℓ and s, it is possible in linear time (resp., logspace) to compute an (ℓ, s)-respecting interval representation $\rho \colon V \to \mathcal{I}$ of G, or to detect that none exists. Moreover, ρ is unique up to isomorphism.*

Proof. We assume that G is connected, otherwise consider its connected components separately and concatenate their representations afterwards.

The algorithm works as follows: As pre-processing steps, compute the connected components $G_1, \ldots, G_k$ of $G_{\ell,s}$, an (ℓ, s)-respecting interval representation for each of them, and the overlap component tree $T_{\ell,s}$. The main part of the algorithm constructs an (ℓ, s)-respecting interval representation of G by combining appropriately shifted copies of the representations of the overlap components. This is done in a depth-first traversal of the overlap component tree. The representation of the root component is not shifted. The representations of the other components are shifted to the appropriate slot of their parent component; if several child components are contained in the same slot, they are placed beside each other in the order in which they are encountered. It remains to check that the result is indeed an (ℓ, s)-respecting interval representation of G.

If G admits an (ℓ, s)-respecting interval representation, then this algorithm will find it: The representations of the components are unique up to reflection by Lemma 4, implying that they have the same length in all representations; and in every (ℓ, s)-respecting interval representation of G, each overlap component must be placed in the appropriate slot of its parent overlap component. In the construction of the representation, the only arbitrary choices are the precise placement of overlap components within their containing slot, the order of the connected components of G, and whether the representations of the individual overlap components are reflected. All these choices can be transformed into one another by isomorphisms of the resulting interval system, so ρ is unique up to isomorphism.

To finish the proof, we show that the algorithm can be implemented in linear time or logspace. Connected components can be found in linear time using depth first search, and in logspace using Reingold's connectivity algorithm [Rei08]. The

(ℓ, s)-respecting representations of the components of $G_{\ell,s}$ can be computed using Lemma 4. The construction of the overlap component tree $T_{\ell,s}$ can easily be implemented in logspace. To obtain it in linear time, compute $\leq_{\ell,s}$ by iterating over the edges of G, and remove reflexive and transitive arcs; see [HMR93, Proposition 3.6] for how the latter is possible in linear time. Computing the offsets for shifting is clearly possible in linear time, and also in logspace if during the tree traversal (see e.g. [Lin92] for how to do this in logspace) a current shift-offset is maintained. □

5 Given Intersection Lengths

Let $G = (V, E)$ be a graph and let $s\colon E \to \mathcal{I}$ prescribe the desired intersection lengths. In this section, we reduce finding a minimal s-respecting interval representation of G to finding (ℓ, s)-respecting interval representations. In particular, we show that the lengths of the intervals in a minimal s-respecting representation are determined by G and s, and can be computed efficiently.

Note that we need minimality here, in contrast to the case of (ℓ, s)-respecting representations. The reason is that adding a point to an interval of an (ℓ, s)-respecting representation always destroys this property, while in an s-respecting representation, we can always duplicate points that are contained in a single interval.

Lemma 6. *Let $G = (V, E)$ be an interval graph with length function $s\colon E \to \mathbb{N}$, and let $\rho\colon V \to \mathcal{I}$ be an arbitrary minimal s-respecting interval representation of G. Then the interval lengths $\ell(v) = \|\rho(v)\|$ do not depend on the choice of ρ and can be computed from G and s in logspace; or in $\mathcal{O}(n + m)$ time, if R_s is given as additional input.*

Proof. We first describe the algorithm. For each $v \in V$, consider these cases:
1. If $N(v) = \varnothing$, set $\ell(v) := 1$.
2. If $\exists w \in N(v) : v \subseteq_s w$, then set $\ell(v) := s(v, w)$.
3. Else, if $\exists w_1, w_2 \in N(v)$ such that $v \between_s w_1 \between_s w_2 \between_s v$ and $s(w_1, w_2) < \min\{s(w_1, v), s(w_2, v)\}$, then set $\ell(v) := s(w_1, v) + s(w_2, v) - s(w_1, w_2)$.
4. Otherwise, consider the subgraph $G[N(v)]$ and define $\ell_v\colon N(v) \to \mathbb{N}$ by $\ell_v(w) = s(w, v)$ for all $w \in N(v)$. Additionally, define $s_v\colon \left(E \cap \binom{N(v)}{2}\right) \to \mathbb{N}$ by $s_v(w_1, w_2) = \min\{s(w_1, v), s(w_2, v)\}$ if w_1 and w_2 overlap v from the same side, and $s_v(w_1, w_2) = s(w_1, w_2)$ otherwise. Compute an (ℓ_v, s_v)-respecting interval representation $\rho_v\colon N(v) \to \mathcal{I}_v$ of $G[N(v)]$, and set $\ell(v) := \|\bigcup_{I \in \mathcal{I}_v} I\|$.

Next, we show that the computed ℓ satisfies $\ell(v) = \|\rho(v)\|$ for each $v \in V$. For an isolated vertex v, as considered in case 1, we have $\|\rho(v)\| = 1$ by minimality of ρ, so $\ell(v) = 1$ is correct. By Lemma 1(b) and the definitions of $\between_s$ and $\subseteq_s$, we have $u \between_s v \Leftrightarrow \rho(u) \between \rho(v)$ and $u \subseteq_s v \Leftrightarrow \rho(u) \subseteq \rho(v)$. In case 2, this immediately implies $\ell(v) = s(v, w) = \|\rho(v) \cap \rho(w)\| = \|\rho(v)\|$.

In case 3, $\rho(w_1)$ and $\rho(w_2)$ cover $\rho(v)$, overlapping it from different sides (the latter is true by Lemma 3), so we have the situation depicted in Fig. 2.

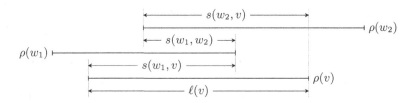

Fig. 2. Proof of Lemma 6, case 3: $\rho(w_1)$ and $\rho(w_2)$ cover $\rho(v)$, overlapping it from different sides

Thus, $\ell(v) = s(w_1, v) + s(w_2, v) - s(w_1, w_2) = \|\rho(w_1) \cap \rho(v)\| + \|\rho(w_2) \cap \rho(v)\| - \|\rho(w_1) \cap \rho(w_2)\| = \|(\rho(w_1) \cup \rho(w_2)) \cap \rho(v)\| = \|\rho(v)\|$.

In case 4, the definitions of ℓ_v and s_v truncate the intervals of the vertices in $N(v)$ to include only their intersections with $\rho(v)$. We have $\|\rho_v(u)\| = \|\rho(u)\|$ for all $u \subseteq_s v$, and $\|\rho_v(w)\| = \|\rho(w) \cap \rho(v)\|$ for all $w \between_s v$. So truncating $\rho(G[N(v)])$ gives an (ℓ_v, s_v)-respecting model $\rho_v(G[N(v)])$ of $G[N(v)]$. By Theorem 5, this model is unique up to isomorphism; in particular, its length is uniquely determined, implying $\|\rho(v)\| \geq \ell(v)$. By minimality of ρ, both values are equal.

It is obvious that this algorithm can be implemented in logspace. To see that it is also possible in linear time, observe that in case 3, Lemma 3 allows us to partition the $\between_s$-neighbors of v into two sets W_1 and W_2, where neighbors that overlap from the same side are in the same set, and that we can require $w_1 \in W_1$ and $w_2 \in W_2$. For the linear-time implementation of case 4, observe that each vertex u of G can occur in at most three of the auxiliary graphs: Suppose to the contrary that there are vertices v_1, v_2, v_3, v_4 such that for each $i \in [1, 4]$, $u \in N(v_i)$ and case 4 is reached for v_i. The latter implies that no $\rho(v_i) = [v_i^-, v_i^+]$ is contained in any other interval, and that none of them is covered by two overlapping intervals. Because case 2 does not hold, there are no containments, so we can assume $v_1^- < v_2^- < v_3^- < v_4^-$ and $v_1^+ < v_2^+ < v_3^+ < v_4^+$. As case 3 holds neither, it follows that $v_1^+ < v_3^-$ and $v_2^+ < v_4^-$. Now let $\rho(u) = [u^-, u^+]$. As u is a neighbor of all v_i, we know $u^- \leq v_1^+$ and $v_4^- \leq u^+$. But this implies that $\rho(u)$ either covers $\rho(v_2)$ alone or together with $\rho(v_1)$, contradicting that case 4 is reached for v_2.   □

The following is a consequence of Theorem 5 and Lemmas 2 and 6.

Corollary 7. *Given $G = (V, E)$ and s, it is possible in $\mathcal{O}(n \cdot m)$ time (resp., in logspace) to compute a minimal s-respecting interval representation $\rho \colon V \to \mathcal{I}$ of G, or to detect that none exists. Moreover, ρ is unique up to isomorphism.*

6   Interval Graphs with Unique Maxclique Ordering

As mentioned before, deciding if a graph has an ℓ-respecting interval representation is NP-complete [PS97]. However, if the input graph G is required to have a unique interval ordering of its inclusion-maximal cliques (up to reflection), even the more general problem DCIG (short for *distance constrained interval graph*) becomes tractable: Additionally to G, a system of difference inequalities of the form $x_i - x_j \geq c$ is given, where the variables are the left and right endpoints of the intervals (strict inequalities are allowed, too). The problem is to decide if G has an interval model that satisfies these inequalities. Pe'er and Shamir show that DCIG is linear-time equivalent to the problem NEGCYCLE, i.e., deciding if a digraph has a negative cycle [PS97]. Based on the following facts, we observe that this problem is NL-complete.

Fact 8 NEGCYCLE *is* NL-*complete*.

Proof. The problem is in NL, because one can check if a nondeterministically chosen path is a negative cycle, storing only the first vertex, the number of steps taken so far and the accumulated weight. To prove the hardness, we reduce from the NL-complete problem s-t-CON to decide if there is a directed path from s to t in a given digraph: Let all arcs have weight 1, except (t, s), which is introduced if not yet present, and assigned the weight $-n$. $\square$

Fact 9 *The linear-time reductions between* NEGCYCLE *and* DCIG *for interval graphs with unique maxclique ordering can be implemented in logspace.*

Proof idea. For most steps of the reductions in [PS97] this is obvious, only computing the unique maxclique ordering requires the algorithm from [KKLV11]. $\square$

We remark that the reduction from NEGCYCLE to DCIG generates only lower and upper bounds on interval lengths, so NL-hardness holds for this special case, too.

7   Conclusion

We have shown how to compute (ℓ, s)- and s-respecting interval representations, giving a linear-time algorithm for the former, an $\mathcal{O}(n \cdot m)$ time algorithm for the latter, and logspace algorithms for both. We remark that deciding whether a graph admits an (ℓ, s)- or s-respecting interval representation is L-complete: In the reduction proving that recognizing interval graphs is L-hard [KKLV11, Theorem 7.7], all generated *yes*-instances are paths; so these graphs have (ℓ, s)- and s-respecting interval representations if we let $\ell(v) = 2$ and $s(e) = 1$ for all vertices u and edges e.

We also have shown that (ℓ, s)- and minimal s-respecting interval representations are unique up to isomorphism. This implies that any algorithm that computes canonical interval representations of interval hypergraphs can be used to obtain *canonical* (ℓ, s)- and s-respecting interval representations. The algorithm given in [KKLV11, Theorem 4.6] solves this in logspace, and it can also be done in linear time using the PQ-tree algorithms of [BL76].

Open Questions. The bottleneck in our $\mathcal{O}(n \cdot m)$ time algorithm for computing s-respecting interval representations is the enumeration of R_s (see Lemma 2). Can this also be implemented in linear, or at least $\mathcal{O}(n^2)$, time? Does the complexity of computing (ℓ, s)- and s-respecting interval representations increase, when the interval and intersection lengths are restricted only for some vertices? Our techniques are not directly applicable in this case, as the algorithm of Lemma 4 relies on the uniqueness of the representation, which is not necessarily preserved in the modified scenario.

Acknowledgement. We thank Oleg Verbitsky for interesting discussions about these results.

References

[BL76]   Booth, K.S., Lueker, G.S.: Testing for the consecutive ones property, interval graphs, and graph planarity using PQ-tree algorithms. J. Comput. Syst. Sci. 13(3), 335–379 (1976)

[COS09]   Corneil, D.G., Olariu, S., Stewart, L.: The LBFS Structure and Recognition of Interval Graphs. SIAM J. Discr. Math. 23(4), 1905–1953 (2009)

[Gol04]   Golumbic, M.C.: Algorithmic graph theory and perfect graphs, 2nd edn. Annals of Discrete Mathematics 57. Elsevier, Amsterdam (2004)

[HAB02]   Hesse, W., Allender, E., Barrington, D.A.M.: Uniform constant-depth threshold circuits for division and iterated multiplication. J. Comput. Syst. Sci. 65(4), 695–716 (2002)

[HM99]   Hsu, W.-L., Ma, T.-H.: Fast and simple algorithms for recognizing chordal comparability graphs and interval graphs. SIAM J. Comput. 28(3), 1004–1020 (1999)

[HMR93]   Habib, M., Morvan, M., Rampon, J.-X.: On the calculation of transitive reduction—closure of orders. Discrete Math. 111(1-3), 289–303 (1993)

[KKLV11] Köbler, J., Kuhnert, S., Laubner, B., Verbitsky, O.: Interval graphs: Canonical representations in logspace. SIAM J. Comput. 40(5), 1292–1315 (2011)

[Kle96]   Klein, P.N.: Efficient parallel algorithms for chordal graphs. SIAM J. Comput. 25(4), 797–827 (1996)

[Lin92]   Lindell, S.: A logspace algorithm for tree canonization. extended abstract. In: Proc. 24th STOC, pp. 400–404 (1992)

[PS97]   Pe'er, I., Shamir, R.: Realizing interval graphs with size and distance constraints. SIAM J. Discr. Math. 10(4), 662–687 (1997)

[Rei08]   Reingold, O.: Undirected connectivity in log-space. J. ACM 55(4), 17:1–17:24 (2008)

[Yam07]   Yamamoto, N.: Weighted interval graphs and their representations. Master's Thesis. Tokyo Inst. of Technology (2007) (in Japanese)

Finger Search in the Implicit Model

Gerth Stølting Brodal, Jesper Sindahl Nielsen, and Jakob Truelsen

MADALGO*, Department of Computer Science, Aarhus University, Denmark
{gerth,jasn,jakobt}@madalgo.au.dk

Abstract. We address the problem of creating a dictionary with the finger search property in the strict implicit model, where no information is stored between operations, except the array of elements. We show that for any implicit dictionary supporting finger searches in $q(t) = \Omega(\log t)$ time, the time to move the finger to another element is $\Omega(q^{-1}(\log n))$, where t is the rank distance between the query element and the finger. We present an optimal implicit static structure matching this lower bound. We furthermore present a near optimal implicit dynamic structure supporting **search**, **change-finger**, **insert**, and **delete** in times $\mathcal{O}(q(t))$, $\mathcal{O}(q^{-1}(\log n)\log n)$, $\mathcal{O}(\log n)$, and $\mathcal{O}(\log n)$, respectively, for any $q(t) = \Omega(\log t)$. Finally we show that the **search** operation must take $\Omega(\log n)$ time for the special case where the finger is always changed to the element returned by the last query.

1 Introduction

We consider the problem of creating an implicit dictionary [4] that supports finger search. A dictionary is a data structure storing a set of elements with distinct comparable keys such that an element can be located efficiently given its key. It may also support predecessor and successor queries where given a query k it must return the element with the greatest key less than k or the element with smallest key greater than k. A dynamic dictionary also supports insertion the and deletion of elements.

A dictionary has the finger search property if the time for searching is dependent on the rank distance t between a specific element f, called the finger, and the query key k. In the static case $\mathcal{O}(\log t)$ search can be achieved by exponential search on a sorted array of elements starting at the finger. Dynamic finger search data structures have been widely studied, e.g. some of the famous dynamic structures that support finger searches are splay trees, randomized skip lists and level linked (2-4)-trees. These all support finger search in $\mathcal{O}(\log t)$ time, respectively in the amortized, expected and worst case sense. For an overview of data structures that support finger search see [3].

We consider two variants of finger search structures. The first variant is the *finger search dictionary* where the **search** operation also changes the finger to the returned element. The second variant is the *change finger dictionary*

---

* Center for Massive Data Algorithmics, a Center of the Danish National Research Foundation.

K.-M. Chao, T.-s. Hsu, and D.-T. Lee (Eds.): ISAAC 2012, LNCS 7676, pp. 527–536, 2012.

where the `change-finger` operation is separate from the `search` operation. We consider the two problems in the strict implicit model where we are only allowed to explicitly store the elements and the number of elements n between operations as defined in [1,6]. Note that the static sorted array solution does not fit into this model, since we are not allowed to use additional space to store the index of f between operations. Other papers allow $\mathcal{O}(1)$ additional words [4,5,8]. We call this the weak implicit model. In both models almost all structure has to be encoded in the order of the elements. In either model the only allowed operations on elements are comparisons and swaps. As there is no agreement on the exact definition of the implicit model, it is interesting to study the limits of the strict model. We show that for a static dictionary in the strict model, if we want a `search` time of $\mathcal{O}(\log t)$, then `change-finger` must take time $\Omega(n^\varepsilon)$, while in the weak model a sorted array achieves $\mathcal{O}(1)$ `change-finger` time.

Much effort has gone into finding a worst case optimal implicit dictionary. Among the first [7] gave a dictionary supporting insert, delete and search in $\mathcal{O}(\log^2 n)$ time. In [5] an implicit B-tree is presented, and finally in [4] a worst case optimal and cache oblivious dictionary is presented. To prove our dynamic upper bounds we use the movable implicit dictionary presented in [2], supporting `insert`, `delete`, `predecessor`, `successor`, `move-left` and `move-right`. The operation `move-right` moves the dictionary laid out in cells i through j to $i+1$ through $j+1$ and `move-left` moves the dictionary the other direction.

Preliminaries. A common implicit data structure technique is the pair encoding of bits. When we have two distinct consecutive elements x and y, then they encode a 1 if $x \leq y$ and 0 otherwise. The running time of the `search` operation is hereafter denoted by $q(t, n)$. Throughout the paper we require that $q(t, n)$ is non decreasing in both t and n, $q(t, n) \geq \log t$ and that $q(0, n) < \log \frac{n}{2}$. We define $Z_q(n) = \min\{t \in \mathbb{N} \mid q(t, n) \geq \log \frac{n}{2}\}$, i.e. $Z_q(n)$ is the smallest rank distance t, such that $q(t, n) > \log \frac{n}{2}$. Note that $Z_q(n) \leq \frac{n}{2}$ (since by assumption $q(t, n) \geq \log t$), and if q is a function of only t, then Z_q is essentially equivalent to $q^{-1}(\log \frac{n}{2})$. As an example $q(t, n) = \frac{1}{\varepsilon} \log t$, gives $Z_q(n) = \lceil (\frac{n}{2})^\varepsilon \rceil$, for $0 < \varepsilon \leq 1$. We require that for a given q, $Z_q(n)$ can be evaluated in constant time, and that $Z_q(n+1) - Z_q(n)$ is bounded by a fixed constant for all n.

We will use set notation on a data structure when appropriate, e.g. $|X|$ will denote the number of elements in the structure X and $e \in X$ will denote that the element e is in the structure X. Given two data structures or sets X and Y, we say that $X \prec Y \Leftrightarrow \forall (x, y) \in X \times Y : x < y$. We use $d(e_1, e_2)$ to denote the rank distance between two elements, that is the difference of the index of e_1 and e_2 in the sorted key order of all elements in the structure. At any time f will denote the current finger element and t the rank distance between this and the current search key.

Our Results. In Section 2 we present a static *change-finger implicit dictionary* supporting `predecessor` in time $\mathcal{O}(q(t, n))$, and `change-finger` in time $\mathcal{O}(Z_q(n) + \log n)$, for any function $q(t, n)$. Note that by choosing $q(t, n) = \frac{1}{\varepsilon} \log t$, we get a **search** time of $\mathcal{O}(\log t)$ and a change finger time of $\mathcal{O}(n^\varepsilon)$ for any $0 < \varepsilon \le 1$.

In Section 3 we prove our lower bounds. First we prove (Lemma 1) that for any algorithm A on a strict implicit data structure of size n that runs in time at most τ, whose arguments are keys or elements from the structure, there exists a set $\mathcal{X}_{A,n}$ of at most $\mathcal{O}(2^\tau)$ array entries, such that A touches only array entries from $\mathcal{X}_{A,n}$, no matter the arguments to A or the content of the data structure. We use this to show that for any *change-finger implicit dictionary* with a search time of $q(t, n)$, **change-finger** will take time $\Omega(Z_q(n) + \log n)$ for some t (Theorem 1). We prove that for any *change-finger implicit dictionary* **search** will take time at least $\log t$ (Theorem 2). A similar argument applies for **predecessor** and **successor**. This means that the requirement $q(t, n) \ge \log t$ is necessary. We show that for any *finger-search implicit dictionary* **search** must take at least $\log n$ time as a function of both t and n, i.e. it is impossible to create any meaningful finger-search dictionary in the strict implicit model (Theorem 3).

By Theorem 1 and 2 the static data structure presented in Section 2 is optimal w.r.t. **search** and **change-finger** time trade off, for any function $q(t, n)$ as defined above. In the special case where the restriction $q(0, n) < \log \frac{n}{2}$ does not hold [4] provides the optimal trade off.

Finally in Section 4 we outline a construction for creating a dynamic *change-finger implicit dictionary*, supporting **insert** and **delete** in time $\mathcal{O}(\log n)$, **predecessor** and **successor** in time $\mathcal{O}(q(t, n))$ and **change-finger** in time $\mathcal{O}(Z_q(n) \log n)$. Note that by setting $q(t, n) = \frac{2}{\varepsilon} \log t$, we get a **search** time of $\mathcal{O}(\log t)$ and a **change-finger** time of $\mathcal{O}(n^{\varepsilon/2} \log n) = \mathcal{O}(n^\varepsilon)$ for any $0 < \varepsilon \le 1$, which is asymptotically optimal in the strict model. It remains an open problem if one can get better bounds in the dynamic case by using $\mathcal{O}(1)$ additional words.

2 Static Finger Search

In this section we present a simple *change-finger implicit dictionary*, achieving an optimal trade off between the time for **search** and **changer-finger**.

Given some function $q(t, n)$, as defined in Section 1, we are aiming for a **search** time of $\mathcal{O}(q(t, n))$. Let $\Delta = Z_q(n)$. Note that we are allowed to use $\mathcal{O}(\log n)$ time searching for elements with rank-distance $t \ge \Delta$ from the finger, since $q(t, n) = \Omega(\log n)$ for $t \ge \Delta$.

Intuitively, we start with a sorted list of elements. We cut the $2\Delta + 1$ elements closest to f (f being in the center), from this list, and swap them with the first $2\Delta + 1$ elements, such that the finger element is at position $\Delta + 1$. The elements that were cut out form the *proximity structure* P, the rest of the elements are in the *overflow structure* O (see Figure 2). A **search** for x is performed by first doing an exponential search for x in the proximity structure, and if x is not found there, by doing binary searches for it in the remaining sorted sequences.

The proximity structure consists of sorted lists $XS \prec S \prec \{f\} \prec L \prec XL$. The list S contains the up to Δ elements smaller then f that are closest to f w.r.t.

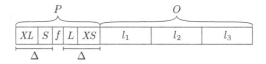

Fig. 1. Memory layout of the static dictonary

rank distance. The list L contains the up to Δ closest to f, but larger than f. Both are sorted in ascending order. XL contains a possibly empty sorted sequence of elements larger than elements from L, and XS contains a possibly empty sorted sequence of elements smaller than elements from S. Here $|XL| + |S| = \Delta = |L| + |XS|$, $|S| = \min\{\Delta, \operatorname{rank}(f) - 1\}$ and $|L| = \min\{\Delta, n - \operatorname{rank}(f)\}$. The overflow structure consists of three sorted sequences $l_2 \prec l_1 \prec \{f\} \prec l_3$, each possibly empty.

To perform a **change-finger** operation, we first revert the array back to one sorted list and the index of f is found by doing a binary search. Once f is found there are 4 cases to consider, as illustrated in Figure 2. Note that in each case, at most $2|P|$ elements have to be moved. Furthermore the elements can be moved such that at most $\mathcal{O}(|P|)$ swaps are needed. In particular case 2 and 4 can be solved by a constant number of list reversals.

For reverting to a sorted array and for doing **search**, we need to compute the lengths of all sorted sequences. These lengths uniquely determine the case used for construction, and the construction can thus be undone. To find $|S|$ a binary search for the split point between XL and S, is done within the first Δ elements of P. This is possible since $S \prec \{f\} \prec XL$. Similarly $|L|$ and $|XS|$ can be found. The separation between l_2 and l_3, can be found by doing a binary search for f in O, since $l_1 \cup l_2 \prec \{f\} \prec l_3$. Finally if $|l_3| < |O|$, the separation between l_1 and l_2 can be found by a binary search, comparing candidates against the largest element from l_2, since $l_2 \prec l_1$.

When performing the **search** operation for some key k, we first determine if $k < f$. If this is the case, an exponential search for k in S is performed. We

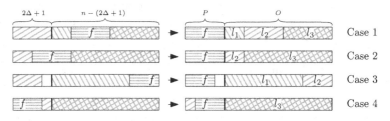

Fig. 2. Cases for the **change-finger** operation. The left side is the sorted array. In all cases the horizontally marked segment contains the new finger element and must be moved to the beginning. In the final two cases, there are not enough elements around f so P is padded with what was already there. The emphasized bar in the array is the $2\Delta + 1$ break point between the proximity structure and the overflow structure.

can detect if we have crossed the boundary to XL, since $S \prec \{f\} \prec XL$. If the element is found it can be returned. If $k > f$ we do an identical search in L. Otherwise the element is neither located in S nor L, and therefore $d(k, f) > \Delta$. All lengths are reconstructed as above, and the element is searched for using binary search in X_l and l_3 if $k > f$ and, otherwise in X_s, $l1$ and l_2.

Analysis. The `change-finger` operation first computes the lengths of all lists in $\mathcal{O}(\log n)$ time. The case used for constructing the current layout is then identified and reversed in $\mathcal{O}(\Delta)$ time. We locate the new finger f' by binary search in $\mathcal{O}(\log n)$ time and afterwards the $\mathcal{O}(\Delta)$ elements closest to f' are moved to P. We get $\mathcal{O}(\Delta + \log n)$ time for `change-finger`.

For searches there are two cases to consider. If $t \leq \Delta$, it will be located by the exponential search in P in $\mathcal{O}(\log t) = \mathcal{O}(q(t, n))$ time, since by assumption $q(t, n) \geq \log t$. Otherwise the lengths of the sorted sequences will be recovered in $\mathcal{O}(\log n)$ time, and a constant number of binary searches will be performed in $\mathcal{O}(\log n)$ time total. Since $t \geq \Delta \Rightarrow q(t, n) \geq \log \frac{n}{2}$, we again get a search time of $\mathcal{O}(q(t, n))$.

3 Lower Bounds

To prove our lower bounds we use an abstracted version of the strict implicit model. The strict model requires that *nothing* but the elements and the number of elements are stored between operations, and that during computation elements can only be used for comparison. With these assumptions a decision tree can be formed for a given n, where nodes correspond to element comparisons and loads and leaves contain the answers. Note that in the weak model a node could probe a cell containing an integer, giving it a degree of n, which prevents any of our lower bound arguments.

Lemma 1. *Let A be an operation on an implicit data structure of length n, running in time τ worst case, that takes any number of keys as arguments. Then there exists a set $\mathcal{X}_{A,n}$ of size 2^τ, such that executing A with any arguments will touch only cells from $\mathcal{X}_{A,n}$ no matter the content of the data structure.*

Proof. Before loading any elements from the data structure, A can reach only a single state which gives rise to a root in a decision tree. When A is in some node s, the next execution step may load some cell in the data structure, and transition into another fixed node, or A may compare two previously loaded elements or arguments, and given the result of this comparison transition into one of two distinct nodes. It follows that the total number of nodes A can enter within its τ steps is $\sum_{i=0}^{\tau-1} 2^i < 2^\tau$. Now each node can access at most one cell, so it follows that at most 2^τ different cells can be probed by any execution of A within τ steps. □

Observe that no matter how many times an operation that take at most τ time is performed they will only be able to reach the same set of cells, since the decision tree is the same for all invocations.

Theorem 1. *For any* change-finger *implicit dictionary with a search time of $q(t,n)$ as defined in Section 1,* change-finger *requires $\Omega(Z_q(n) + \log n)$ time.*

Proof. Let $e_1 \ldots e_n$ be a set of elements in sorted order with respect to the keys $k_1 \ldots k_n$. Let $t = Z_q(n) - 1$. By definition $q(t + 1, n) \geq \log \frac{n}{2} > q(t,n)$. Consider the following sequence of operations:

> **for** $i = 0 \ldots \frac{n}{t}$:
> change-finger(k_{it})
> **for** $j = 0 \ldots t - 1$: search(k_{it+j})

Since the rank distance of any query element is at most t from the current finger and q is non-decreasing each search operation takes time at most $q(t,n)$. By Lemma 1 there exists a set $\mathcal{X}$ of size $2^{q(t,n)}$ such that all queries only touch cells in $\mathcal{X}$. We note that $|\mathcal{X}| \leq 2^{q(t,n)} \leq 2^{\log(n/2)} = \frac{n}{2}$.

Since all n elements were returned by the query set, the change-finger operations, must have copied at least $n - |\mathcal{X}| \geq \frac{n}{2}$ elements into $\mathcal{X}$. We performed $\frac{n}{t}$ change-finger operations, thus on average the change-finger operations must have moved at least $\frac{t}{2} = \Omega(Z_q(n))$ elements into $\mathcal{X}$.

For the $\log n$ term in the lower bound, we consider the sequence of operations change-finger(k_i) followed by search(k_i) for i between 1 and n. Since the rank distance of any search is 0 and $q(0,n) < \log \frac{n}{2}$ (by assumption), we know from Lemma 1 that there exists a set $\mathcal{X}_s$ of size at most $2^{\log(n/2)}$, such that search only touches cells from $\mathcal{X}_s$. Assume that change-finger runs in time $c(n)$, then from Lemma 1 we get a set $\mathcal{X}_c$ of size at most $2^{c(n)}$ such that change-finger only touches cells from $\mathcal{X}_c$. Since every element is returned, the cell initially containing the element must be touched by either change-finger or search at some point, thus $|\mathcal{X}_c| + |\mathcal{X}_s| \geq n$. We see that $2^{c(n)} \geq |\mathcal{X}_c| \geq n - |\mathcal{X}_s| \geq n - 2^{\log(n/2)} = 2^{\log(n/2)}$, i.e. $c(n) \geq \log \frac{n}{2}$. □

Theorem 2. *For a change-finger implicit dictionary with* search *time $q'(t,n)$, where q' is none decreasing in both t and n, it holds that $q'(t,n) \geq \log t$.*

Proof. Let $e_1 \ldots e_n$ be a set of elements with keys $k_1 \ldots k_n$ in sorted order. Let $t \leq n$ be given. First perform change-finger(k_1), then for i between 1 and t perform search(k_i). From Lemma 1 we know there exists a set $\mathcal{X}$ of size at most $2^{q'(t,n)}$, such that any of the search operations touch only cells from $\mathcal{X}$ (since any element searched for has rank distance at most t from the finger). The search operations return t distinct elements so $t \leq |\mathcal{X}| \leq 2^{q'(t,n)}$, and $q'(t,n) \geq \log t$. □

Theorem 3. *For finger-search implicit dictionary, the* finger-search *operation requires at least $g(t,n) \geq \log n$ time for any rank distance $t > 0$ where $g(t,n)$ is non decreasing in both t and n.*

Proof. Let $e_1 \ldots e_n$ be a set of elements with keys $k_1 \ldots k_n$ in sorted order. First perform finger-search(k_1), then perform finger-search(k_i) for i between 1 and n. Now for all queries except the first, the rank distance $t \leq 1$ and by

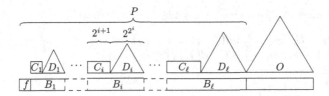

Fig. 3. Memory layout

Lemma 1 there exists a set of memory cells $\mathcal{X}$ of size $2^{g(1,n)}$ such that all these queries only touch cells in $\mathcal{X}$. Since all elements are returned by the queries we have $|\mathcal{X}| = n$, so $g(1,n) \geq \log n$, since this holds for $t = 1$ it holds for all t.   □

We can conclude that it is not possible to achieve any form of meaningful finger-search in the strict implicit model. The static *change-finger implicit dictionary* from Section 2 is by Theorem 1 optimal within a constant factor, with respect to the `search` to `change-finger` time trade off, assuming the running time of `change-finger` depends only on the size of the structure.

4   A Dynamic Structure

For any function $q(t, n)$, as defined in the introduction, we present a dynamic *change-finger implicit dictionary* that supports `change-finger, search, insert` and `delete` in $\mathcal{O}(\Delta \log n), \mathcal{O}(q(t,n)), \mathcal{O}(\log n)$ and $\mathcal{O}(\log n)$ time respectively, where $\Delta = Z_q(n)$ and n is the number of elements when the operation was started.

The data structure consists of two parts: a *proximity structure* P which contains the elements near f and an *overflow structure* O which contains elements further from f w.r.t. rank distance. We partition P into several smaller structures $B_1, \ldots, B_\ell$. Elements in B_i are closer to f than elements in B_{i+1}. The overflow structure O is an *implicit movable dictionary* [2] that supports `move-left` and `move-right` as described in the Section 1. See Figure 3 for the layout of the data structure. During a `change-finger` operation the proximity structure is rebuilt such that $B_1, \ldots, B_\ell$ correspond to the new finger, and the remaining elements are put in O.

The total size of P is $2\Delta + 1$. The i'th block B_i consists of a counter C_i and an implicit movable dictionary D_i. The counter C_i contains a pair encoded number c_i, where c_i is the number of elements in D_i smaller than f. The sizes within B_i are $|C_i| = 2^{i+1}$ and $|D_i| = 2^{2^i}$, except in the final block B_ℓ where they might be smaller (B_ℓ might be empty). In particular we define:

$$\ell = \min \left\{ \ell' \in \mathbb{N} \,\Big|\, \sum_{i=0}^{\ell'} \left(2^{i+1} + 2^{2^i} \right) > 2\Delta \right\}.$$

We will maintain the following invariants for the structure:

I.1 $\forall i < j, e_1 \in B_i, e_2 \in B_j : d(f, e_1) < d(f, e_2)$
I.2 $\forall e_1 \in B_1 \cup \cdots \cup B_\ell, e_2 \in O : d(f, e_1) \leq d(f, e_2)$
I.3 $|P| = 2\Delta + 1$
I.4 $|C_i| \leq 2^{i+1}$
I.5 $|D_i| > 0 \Rightarrow |C_i| = 2^{i+1}$
I.6 $|D_\ell| < 2^{2^\ell}$ and $\forall i < \ell : |D_i| = 2^{2^i}$
I.7 $|D_i| > 0 \Rightarrow c_i = |\{e \in D_i \mid e < f\}|$

We observe that the above invariants imply:

O.1 $\forall i < \ell : |B_i| = 2^{i+1} + 2^{2^i}$ (From I.5 and I.6)
O.2 $|B_\ell| < 2^{\ell+1} + 2^{2^\ell}$ (From I.4 and I.6)
O.3 $d(e, f) \leq 2^{2^k - 1} \leq \Delta \Rightarrow e \in B_j$ for some $j \leq k$ (From I.1 – I.6)

4.1   Block Operations

The following operations operate on a single block and are internal helper functions for the operations described in Section 4.2.

block_delete(k, B_i): Removes the element e with key k from the block B_i. This element *must* be located in B_i. First we scan C_i to find e. If it is not found it must be in D_i, so we delete it from D_i. If $e < f$ we decrement c_i. In the case where $e \in C_i$ and D_i is nonempty, an arbitrary element g is deleted from D_i and if $g < f$ we decrement c_i. We then overwrite e with g, and fix C_i to encode the new number c_i. In the final case where $e \in C_i$ and D_i is empty, we overwrite e with the last element from C_i.

block_insert(e, B_i): Inserts e into block B_i. If $|C_i| < 2^{i+1}$, e is inserted into C_i and we return. Else we insert e into D_i. If D_i was empty we set $c_i = 0$. In either case if $e < f$ we increment c_i.

block_search(k, B_i): Searches for an element e with key k in the block B_i. We scan C_i for e, if it is found we return it. Otherwise if D_i is nonempty we perform a search on it, to find e and we return it. If the element is not found nil is returned.

block_predecessor(k, B_i): Finds the predecessor element for the key k in B_i. Do a linear scan through C_i and find the element l_1 with largest key less than k. Afterwards do a predecessor search for key k on D_i, call the result l_2. Return $\max(l_1, l_2)$, or that no element in B_i has key less than k.

4.2   Operations

In order to maintain correct sizes of P and O as the entire structure expands or contracts a rebalance operation is called in the end of every insert and delete operation. This is an internal operation that does not require I.3 to be valid before invocation.

rebalance(): Balance B_ℓ such that the number of elements in P less than f is as close to the number of elements greater than f as possible. We start by evaluating $\Delta = Z_q(n)$, the new desired proximity size. Let s be the number of elements in B_ℓ less than f which can be computed as $c_\ell + |\{e \in C_\ell \mid e < f\}|$. While $2\Delta + 1 > |P|$ we move elements from O to P. We move the predecessor of f from O to B_ℓ if $O \prec \{f\} \vee (s < \frac{|B_\ell|}{2} \wedge \neg(\{f\} \prec O))$ and otherwise we move the successor of f to O. While $2\Delta + 1 < |P|$ we move elements from B_ℓ to O. We move the largest element from B_ℓ to O if $s < \frac{B_\ell}{2}$. Otherwise we move the smallest element.

change-finger(k): To change the finger of the structure to k, we first insert every element of $B_\ell \dots B_1$ into O. We then remove the element e with key k from O, and place it at index 1 as the new f, and finish by performing rebalance.

insert(e): Assume $e > f$. The case $e < f$ can be handled similarly. Find the first block B_i where e is smaller than the largest element l_i from B_i (which can be found using a predecessor search) or $l_i < f$. Now if $l_i > f$ for all blocks $j \geq i$, block_delete the largest element and block_insert it into B_{j+1}. In the other case where $l_i < f$ for all blocks $j \geq i$, block_delete the smallest element and block_insert it into B_{j+1}. The final element that does not have a block to go into, will be put into O, then we put e into B_i. In the special case where e did not fit in any block, we insert e into O. In all cases we perform rebalance.

delete(k): We perform a block_search on all blocks and a search in O to find out which structure the element e with key k is located in. If it is in O we just delete it from O. Otherwise assume $k < f$ (the case $k > f$ can be handled similarly), and assume that e is in B_i, then block_delete e from B_i. For each $j > i$ we block_delete the predecessor of f in B_j, and insert it into B_{j-1} (in the case where there is no predecessor, we block_delete the successor of f instead). We also delete the predecessor of f from O and insert it in B_ℓ. The special case where $k = f$, is handled similarly to $k < f$, we note that after this the predecessor of f will be the new finger element. In all cases we perform a rebalance.

search(k), predecessor(k) and successor(k), all follow the same general pattern. For each block B_i starting from B_1, we compute the largest and the smallest element in the block. If k is between these two elements we return the result of block_search, block_predecessor or block_successor respectively on B_i, otherwise we continue with the next block. In case k is not within the bounds of any block, we return the result of search(k), predecessor(k) or successor(k) respectively on O.

4.3   Analysis

By the invariants, we see that every C_i and D_i except the last, have fixed size. Since O is a movable dictionary it can be moved right or left as this final C_i or D_i expands or contracts. Thus the structure can be maintained in a contiguous memory layout.

The correctness of the operations follows from the fact that I.1 and I.2, implies that elements in B_j or O are further away from f than elements from B_i where $i < j$. We now argue that search runs in time $\mathcal{O}(q(t, n))$. Let e be the element we are searching for. If e is located in some B_i then at least half the elements in B_{i-1} will be between f and e by I.1. We know from O.1 that $t = d(f, e) \geq \frac{|B_{i-1}|}{2} \geq 2^{2^{i-1}-1}$. The time spent searching is $\mathcal{O}(\sum_{j=1}^{i} \log |B_j|) = \mathcal{O}(2^i) = \mathcal{O}(\log t) = \mathcal{O}(q(t, n))$. If on the other hand e is in O, then by I.3 there are $2\Delta + 1$ elements in P, of these at least half are between f and e by I.2, so $t \geq \Delta$, and the time used for searching is $\mathcal{O}(\log n + \sum_{j=1}^{k} \log |B_j|) = \mathcal{O}(\log n) = \mathcal{O}(q(t, n))$. The last equality follows by the definition of Z_q. The same arguments work for predecessor and successor.

Before the change-finger operation the number of elements in the proximity structure by I.3 is $2\Delta + 1$. During the operation all these elements are inserted into O, and the same number of elements are extracted again by rebalance. Each of these operations are just insert or delete on a movable dictionary or a block taking time $\mathcal{O}(\log n)$. In total we use time $\mathcal{O}(\Delta \log n)$.

Finally to see that both Insert and Delete run in $\mathcal{O}(\log n)$ time, notice that in the proximity structure doing a constant number of queries in every block is asymptotically bounded by the time to do the queries in the last block. This is because their sizes increase double-exponentially. Since the size of the last block is bounded by n we can guarantee $\mathcal{O}(\log n)$ time for doing a constant number of queries on every block (this includes predecessor/successor queries). In the worst case, we need to insert an element in the first block of the proximity structure, and "bubble" elements all the way through the proximity structure and finally insert an element in the overflow structure. This will take $\mathcal{O}(\log n)$ time. At this point we might have to rebalance the structure, but this merely requires deleting and inserting a constant number of elements from one structure to the other, since we assumed $Z_q(n)$ and $Z_q(n + 1)$ differ by at most a constant. Deletion works in a similar manner.

References

1. Borodin, A., Fich, F.E., Meyer auf der Heide, F., Upfal, E., Wigderson, A.: A Trade-off Between Search and Update Time for the Implicit Dictionary Problem. In: Kott, L. (ed.) ICALP 1986. LNCS, vol. 226, pp. 50–59. Springer, Heidelberg (1986)
2. Brodal, G.S., Kejlberg-Rasmussen, C., Truelsen, J.: A Cache-Oblivious Implicit Dictionary with the Working Set Property. In: Cheong, O., Chwa, K.-Y., Park, K. (eds.) ISAAC 2010, Part II. LNCS, vol. 6507, pp. 37–48. Springer, Heidelberg (2010)
3. Brodal, G.S.: Finger search trees. In: Mehta, D., Sahni, S. (eds.) Handbook of Data Structures and Applications, ch. 11. CRC Press (2005)
4. Franceschini, G., Grossi, R.: Optimal Worst-Case Operations for Implicit Cache-Oblivious Search Trees. In: Dehne, F., Sack, J.-R., Smid, M. (eds.) WADS 2003. LNCS, vol. 2748, pp. 114–126. Springer, Heidelberg (2003)
5. Franceschini, G., Grossi, R., Munro, J.I., Pagli, L.: Implicit B-Trees: New results for the dictionary problem. In: Proc. 43rd FOCS, pp. 145–154. IEEE (2002)
6. Frederickson, G.N.: Implicit data structures for the dictionary problem. JACM 30(1), 80–94 (1983)
7. Munro, J.I.: An implicit data structure supporting insertion, deletion, and search in $\mathcal{O}(\log^2 n)$ time. JCSS 33(1), 66–74 (1986)
8. Munro, J.I., Suwanda, H.: Implicit data structures for fast search and update. JCSS 21(2), 236–250 (1980)

A Framework for Succinct Labeled Ordinal Trees over Large Alphabets*

Meng He[1], J. Ian Munro[2], and Gelin Zhou[2]

[1] Faculty of Computer Science, Dalhousie University, Canada
mhe@cs.dal.ca
[2] David R. Cheriton School of Computer Science, University of Waterloo, Canada
{imunro,g5zhou}@uwaterloo.ca

Abstract. We consider succinct representations of labeled ordinal trees that support a rich set of operations. Our new representations support a much broader collection of operations than previous work [10,8,1]. In our approach, labels of nodes are stored in a *preorder label sequence*, which can be compressed using any succinct index for strings that supports rank$_\alpha$ and select$_\alpha$ operations. In other words, we present a framework for succinct representations of labeled ordinal trees that allows alphabets to be large. This answers an open problem presented by Geary et al. [10]. We further extend our work and present the first succinct representation of dynamic labeled ordinal trees that supports several label-based operations including finding the level ancestor with a given label.

1   Introduction

We address the issue of succinct representations of ordinal (or ordered) trees with satellite data over a large alphabet. Much of this is motivated by the needs of large text-dominated databases that store and manipulate XML documents, which can be essentially modeled as ordinal trees in which each node is assigned a tag drawn from a tag set.

Our representations support a much broader collection of operations than previous work [10,8,1], particularly those operations that aim at XML-style document retrieval, such as queries written in the XML path language (XPath). Our data structures are succinct, occupying space close to the information-theoretic lower bound, which are essential to systems and applications that deal with very large data sets.

Our approach is based on "tree extraction", that is, constructing subtrees consisting of nodes with appropriate labels (and their parents). The basic idea of tree extraction was introduced by He et al. [14,15], where it was used to answer queries that are generalizations of geometric queries such as range counting from planar point sets to trees. Here we follow a different approach for a completely different class of operations that originate from text databases. Most of these operations are not required in [14,15]. In our data structures, an input tree is

---

* This work was supported by NSERC and the Canada Research Chairs Program.

K.-M. Chao, T.-s. Hsu, and D.-T. Lee (Eds.): ISAAC 2012, LNCS 7676, pp. 537–547, 2012.
© Springer-Verlag Berlin Heidelberg 2012

split according to labels of nodes, such that we can maintain structural information and labels jointly in a space-efficiently way. Previous solutions to the same problem are all based on different ideas [10,8,1].

In this paper, we consider the operations listed in Table 1, in which DFUDS denotes *depth-first unary degree sequence* order as defined by Benoit et al. [3]. We list only the labeled versions of these operations. The unlabeled versions simply include all nodes. In other words, the support for unlabeled versions can be reduced to the support for labeled versions by setting the alphabet size to 1. For simplicity, we refer to the labeled versions of operations as α-operations. We call a node that has label α an α-node (hence α-children, α-ancestor, etc). In addition, we define the α-rank of node x in a list to be the number of α-nodes to the left of x in the list.

Table 1. Operations considered in this paper. Here we give only the definitions of the labeled versions of operations.

Operation	Description
$\text{depth}_\alpha(x)$	α-depth of x, i.e., number of α-nodes from x to the root
$\text{parent}_\alpha(x)$	closest α-ancestor of x
$\text{level\_anc}_\alpha(x,i)$	α-ancestor y of x satisfying $\text{depth}_\alpha(x) - \text{depth}_\alpha(y) = i$
$\text{deg}_\alpha(x)$	number of α-children of x
$\text{child\_rank}_\alpha(x)$	α-rank of x in the list of children of $\text{parent}(x)$
$\text{child\_select}_\alpha(x,i)$	i-th α-child of x
$\text{nbdesc}_\alpha(x)$	number of α-nodes in the subtree rooted at x
$\text{pre\_rank}_\alpha(x)/\text{pre\_select}_\alpha(i)$	α-rank of x in preorder/i-th α-node in preorder
$\text{post\_rank}_\alpha(x)/\text{post\_select}_\alpha(i)$	α-rank of x in postorder/i-th α-node in postorder
$\text{height}_\alpha(x)$	α-height of x, i.e., maximum number of α-nodes from x to its leaf descendant
$\text{LCA}_\alpha(x,y)$	lowest common α-ancestor of nodes x and y
$\text{dfuds\_rank}_\alpha(x)/\text{dfuds\_select}_\alpha(i)$	α-rank of x in DFUDS order/i-th α-node in DFUDS order
$\text{leaf\_lmost}_\alpha(x)/\text{leaf\_rmost}_\alpha(x)$	leftmost/rightmost α-leaf in the subtree rooted at x
$\text{leaf\_rank}_\alpha(x)$	number of α-leaves to the left of x in preorder
$\text{leaf\_select}_\alpha(i)$	i-th α-leaf in preorder
$\text{nbleaf}_\alpha(x)$	number of α-leaves in the subtree rooted at node x
$\text{insert}_\alpha(x)$	insert an α-node x as an internal node or a leaf
$\text{delete}(x)$	delete non-root node x

Geary et al. [10] presented data structures to support in constant time the first group of α-operations in Table 1 and their unlabeled versions. The overall space cost of their data structures is $n(\lg \sigma + 2) + O(\frac{n\sigma \lg \lg \lg n}{\lg \lg n})$ bits, which is much more than the information-theoretic lower bound of $n(\lg \sigma + 2) - O(\lg n)$ bits when $\sigma = \Omega(\lg \lg n)$. Ferragina et al. [8] and Barbay et al. [1] designed data structures for labeled trees that use space close to the information-theoretic lower bound, but supporting a more restricted set of α-operations. Ferragina et al.'s [8] xbw-based representation supports only $\text{child\_select}_\alpha(x,i)$ and $\text{deg}_\alpha(x)$ [1], while

---

[1] It also supports SubPathSearch queries, which return the number of nodes whose upward paths start with a given query string.

Barbay et al.'s [1] data structure supports only $\mathtt{pre\_rank}_\alpha(x)$, $\mathtt{pre\_select}_\alpha(i)$ and $\mathtt{nbdesc}_\alpha(x)$. These results are for static labeled trees; to the best of our knowledge, there is no succinct data structure for dynamic labeled ordinal trees with efficient query and update time.

Our results for static and dynamic labeled ordinal trees are summarized in Theorems 1 to 3. First, as a preliminary result, we improve the succinct representation of labeled ordinal trees of Geary et al. [10]. As shown in Theorem 1, the improved representation supports more operations while occupying less space, where PLS_T is the preorder label sequence of T, and H_k is the k-th order empirical entropy [16], which is bounded above by $\lg \sigma$. However, this data structure is succinct only if the size of alphabet is very small, i.e., $\sigma = o(\lg \lg n)$.

$\mathtt{deepest}_\alpha$ and $\mathtt{min\_depth}_\alpha$ are auxiliary α-operations used in Section 3: $\mathtt{deepest}_\alpha(i,j)$ returns a node (there could be a tie) with preorder rank in $[i,j]$ that has the maximum α-depth, and $\mathtt{min\_depth}_\alpha(i,j)$ returns an α-node with preorder rank in $[i,j]$ that has the minimum depth (i.e., is closest to the root).

Theorem 1. *Let T be a static ordinal tree on n nodes, each having a label drawn from an alphabet of size $\sigma = o(\lg \lg n)$. Under the word RAM with word size $w = \Omega(\lg n)$, for any $k = o(\log_\sigma n)$, there exists a data structure that encodes T using $n(H_k(PLS_T) + 2) + O(\frac{n(k \lg \sigma + \lg \lg n)}{\log_\sigma n}) + O(\frac{n\sigma \lg \lg \lg n}{\lg \lg n})$ bits of space, supporting the first two groups of α-operations in Table 1 and their unlabeled versions, plus two additional α-operations $\mathtt{deepest}_\alpha$ and $\mathtt{min\_depth}_\alpha$, in constant time.*

Theorem 2 is the main result in this paper. To achieve this result, we present a framework for succinct representations of labeled ordinal trees, in which an α-operation is reduced to a constant number of well-supported operations on simpler data structures such as bit vectors, preorder label sequences, and unlabeled and 0/1-labeled ordinal trees, where a 0/1-labeled ordinal tree is an ordinal tree over the alphabet $\{0,1\}$. This creative reduction allows us to deal with large alphabets, and to compress labels of nodes into entropy bounds. Thus our framework answers an open problem proposed by Geary et al. [10].

Theorem 2. *Let T be a static ordinal tree on n nodes, each having a label drawn from an alphabet Σ of size σ. Under the word RAM with word size $w = \Omega(\lg n)$,*

(a) for $\sigma = O(\mathtt{polylog}(n))$, T can be encoded using $n(H_0(PLS_T) + 9) + o(n)$ bits of space to support the first two groups of α-operations in Table 1 in constant time;

(b) for general Σ, T can be encoded using $nH_0(PLS_T) + O(n)$ bits of space to support the first two groups of α-operations in Table 1 in $O(\lg \lg \sigma)$ time;

(c) for general Σ and $k = o(\log_\sigma n)$, T can be encoded using $nH_k(PLS_T) + \lg \sigma \cdot o(n) + O(\frac{n \lg \sigma}{\lg \lg \lg \sigma})$ bits to support the first two groups of α-operations in Table 1 in $O(\lg \lg \sigma (\lg \lg \lg \sigma)^2)$ time.

In addition, these data structures support the unlabeled versions of these α-operations in constant time.

Theorem 3 further extends our work to the dynamic case. Here we only list the result with the fastest query time. One can make use of the dynamic strings in [12,17] when worst-case update time is desired.

Theorem 3. *Let T be a dynamic ordinal tree on n nodes, each having a label drawn from an alphabet Σ of size σ. Under the word RAM with word size $w = \Omega(\lg n)$, T can be represented using $n(H_0(PLS_T) + 5) + O(\frac{n(1+H_0(PLS_T))}{\lg^{1-\epsilon} n} + \sigma(\lg \sigma + \lg^{1+\epsilon} n))$ bits of space, for any constant $0 < \epsilon < 1$, such that* $\mathtt{depth}_\alpha$, $\mathtt{parent}_\alpha$, $\mathtt{nbdesc}_\alpha$, $\mathtt{LCA}_\alpha$, $\mathtt{pre\_rank}_\alpha$, $\mathtt{pre\_select}_\alpha$, $\mathtt{post\_rank}_\alpha$, $\mathtt{post\_select}_\alpha$, *and the* $\mathtt{leaf}$ *α-operations can be supported in $O(\frac{\lg n}{\lg \lg n})$ time,* $\mathtt{level\_anc}_\alpha$ *can be supported in $O(\lg n)$ time, and* $\mathtt{insert}_\alpha$ *and* $\mathtt{delete}$ *can be supported in $O(\frac{\lg n}{\lg \lg n})$ amortized time.*

The rest of this paper is organized as follows. In Section 2 we review the data structures and the techniques used in this paper. In Section 3 we describe the construction of our data structures for static trees over large alphabets, i.e., the proof of Theorem 2. In Section 4, we sketch our data structures for dynamic trees, i.e., the proof of Theorem 3. Due to space limitations, we omit the proof of Theorem 1.

2   Preliminaries

2.1   Bit Vectors, Strings and the Related Operations

In this subsection, we review bit vectors, strings, and the operations performed on them. Bit vector plays a central role in many succinct data structures. For a bit vector $B[1..n]$, $\mathtt{rank}_0(i)$ and $\mathtt{rank}_1(i)$ return the numbers of 0-bits and 1-bits in $B[1..i]$, respectively. $\mathtt{select}_0(i)$ and $\mathtt{select}_1(i)$ return the positions of the i-th 0-bit and the i-th 1-bit in B, respectively. Bit vectors can be generalized to strings, in which characters are drawn from an alphabet Σ of size σ. For a string $S[1..n]$ and $\alpha \in \Sigma$, $\mathtt{rank}_\alpha(i)$ returns the number of α's in $S[1..i]$, and $\mathtt{select}_\alpha(i)$ returns the position of the i-th α. Another operation studied by researchers is the random access to any substring of length $O(\log_\sigma n)$.

2.2   Tree Extraction

Tree extraction [14,15], based on the deletion operation of tree edit distance [4], is a technique used to decompose a tree by deleting nodes, moving their children into their positions in the sibling order. A crucial fact is that the ancestor-descendant and preorder/postorder relationships between the remaining nodes are preserved. To support the α-operations related to children, we develop a new space-efficient approach based on tree extraction that is very different from the strategy used in [14,15], so that the parent-child relationship is preserved.

Let $V(T)$ be the set of nodes in T. For any set $X \subseteq V(T)$ that contains the root of T, we denote by T_X the ordinal tree obtained by deleting all the nodes that are not in X from T, where the nodes are deleted in level order. T_X is called the X-*extraction* of T. It is easy to see that there is a natural one-to-one correspondence between the nodes in X and the nodes in T_X. Lemma 1 captures an essential property of tree extraction. The proof is omitted here.

Lemma 1. *For any two sets of nodes* $X, X' \subseteq V(T)$, *the nodes in* $X \cap X'$ *have the same relative positions in the preorder and the postorder traversal sequences of* T_X *and* $T_{X'}$.

3 Static Trees over Large Alphabets: Theorem 2

For each possible subscript $\alpha \in \Sigma$, we could support the first two groups of operations in Table 1 by relabeling T into a 0/1-labeled tree and indexing the relabeled tree, where a node is relabeled 1 if and only if it is an α-node in T. However, we would have to store σ trees that have $n\sigma$ nodes in total if we simply apply this idea for each $\alpha \in \Sigma$, which we could not afford.

Instead of storing all the n nodes for each $\alpha \in \Sigma$, we store only the nodes that are closely relevant to α, i.e., the α-nodes and their parents, and the ancestor-descendant relationship between these nodes. We apply tree extraction to summarize the information, where the tree constructed for label α is denoted by T_α.

For $\alpha \in \Sigma$, we create a new root r_α, and make the original root of T be the only child of r_α. The structure of T_α is obtained by computing the X_α-extraction of the augmented tree rooted at r_α, where X_α is the union of r_α, the α-nodes in T, and the parents of the α-nodes. The natural one-to-one mapping between the nodes in X_α and T_α determines the labels of the nodes in T_α. The root of T_α is always labeled 0. A non-root node in T_α is labeled 1 if its corresponding node in T is an α-node, and 0 otherwise. Thus, the number of 1-nodes in T_α is equal to the number of α-nodes in T. Let n_α denote both values.

To clarify notation, the nodes in T are denoted by lowercase letters, while the nodes in T_α are denoted by lowercase letters plus prime symbols. To illustrate the one-to-one mapping, we denote by x' a node in T_α if and only if its corresponding node in T is denoted by x. The root of T_α, which corresponds to r_α, is denoted by r'_α. We show how to convert the corresponding nodes in T and T_α using the preorder label sequence of T in Subsection 3.2.

Since the structure of T_α is different from T, we need also store the structure of T and the labels of the nodes in T so that we can perform conversions between the nodes in T and T_α. In addition, to support the leaf α-operations, we store for each $\alpha \in \Sigma$ a bit vector $L_\alpha[1..n_\alpha]$ in which the i-th bit is one if and only if the i-th 1-node in preorder of T_α corresponds to a leaf in T.

Following this approach, our succinct representation consists of four components: (a) the structure of T; (b) PLS_T, the preorder label sequence of T; (c) a 0/1-labeled tree T_α for each $\alpha \in \Sigma$; (d) and a bit vector L_α for each $\alpha \in \Sigma$. The unlabeled versions of the first two groups of operations in Table 1 are directly supported by the data structure that maintains the structure of T. For α-operations, our basic idea is to reduce an α-operation to a constant number of well-supported operations on T, PLS_T, T_α's, and L_α's, for which we summarize the previous work in Subsection 3.5. In the following subsections, we describe our algorithms in terms of T, PLS_T, T_α's and L_α's. For each operation, we specify the component on which it performs as the first parameter. If such a component is not specified in context, then this operation is performed on T.

3.1   pre_rank$_\alpha$, pre_select$_\alpha$ and nbdesc$_\alpha$

By the definitions, we have pre_rank$_\alpha(x)$ = rank$_\alpha(PLS_T,$ pre_rank$(x))$ and pre_select$_\alpha(i)$ = pre_select(select$_\alpha(PLS_T, i))$. We make use of them to find the α-predecessor and the α-successor of node x, which are defined to be the last α-node preceding x and the first α-node succeeding x in preorder (both can be x itself).

We support nbdesc$_\alpha(x)$ in the same way as [1]. The descendants of x form a consecutive substring in PLS_T, which starts at index pre_rank(x) and ends at index pre_rank(x) + nbdesc$(x) - 1$. We can compute the number of α-nodes lying in this range using rank$_\alpha$ on PLS_T. Providing that x has an α-descendant, we can further compute the first and the last α-descendant of node x in preorder, which are the α-successor of x and the α-predecessor of the node that has preorder rank pre_rank(x) + nbdesc$(x) - 1$, respectively. Let these α-nodes be u and v. For simplicity, we call $[u, v]$ the α-boundary of the subtree rooted at x.

3.2   Conversion between the Nodes in T and T_α

The conversion between node x in T and the corresponding node x' in T_α plays an important role in supporting α-operations. If x is an α-node, then x' must exist in T_α. By Lemma 1, the conversion can be done using $x' =$ pre_select$_1(T_\alpha,$ rank$_\alpha(PLS_T, x))$, and $x =$ pre_select$_\alpha($pre_rank$_1(T_\alpha, x'))$.

The other case in which x is not an α-node is more complex, since the node in T_α that corresponds to x may or may not exist. By the definition of T_α, x must have an α-child if x' exists in T_α. In addition, every α-child of x in T must appear as a 1-child of x' in T_α. Using this, we can compute x from x': We first find the first 1-child of x', say $y' =$ child_select$_1(T_\alpha, x', 1)$, and compute node y in T that corresponds to y' using the equation in the first paragraph. Then x must be the parent of y in T.

Algorithm 1. Compute x' when x is not an α-node

---

1 **if** x *has no α-descendant* **then return** NULL;
2 $[u, v] \leftarrow$ *the α-boundary of the subtree rooted at x*;
3 **if** $x \neq$ LCA(u, v) **then**
4 | $y \leftarrow$ *the child of x that is an ancestor of* LCA(u, v);
5 | **if** y *is an α-node* **then return** parent(T_α, y');
6 | **else return** NULL;
7 **else**
8 | $w' =$ LCA(T_α, u', v');
9 | **if** w' *corresponds to x* **then return** w';
10 | **else return** NULL;

Algorithm 1 shows how to compute x' from x when x is not an α-node, and it returns NULL if no such x' exists. We first verify whether x has at least one α-descendant in line 1 using $\mathtt{nbdesc}_\alpha$. x' does not exist in T_α if x has no α-descendant. Otherwise, we compute the α-boundary, $[u, v]$, of the subtree rooted at x in line 2. We have two cases, depending on whether x is the lowest common ancestor of u and v. If $x \neq \mathtt{LCA}(u, v)$, then x must have a child y that is a common ancestor of u and v. All α-descendants of x must also be descendants of y, or an α-descendant of x may precede u or succeed v in preorder. Thus, we need only check if y is an α-node, as shown in lines 5 and 6. Now suppose $x = \mathtt{LCA}(u, v)$. We claim that x' must be the lowest common ancestor of u' and v' if x' exists in T_α. Thus, we need only compute $w' = \mathtt{LCA}(T_\alpha, u', v')$ in line 8, and verify if w' corresponds to x in lines 9 and 10.

3.3 $\mathtt{parent}_\alpha$, $\mathtt{level\_anc}_\alpha$, $\mathtt{LCA}_\alpha$ and $\mathtt{depth}_\alpha$

We first show how to compute $\mathtt{parent}_\alpha(x)$ in Algorithm 2. The case in which x is an α-node is solved in lines 2 to 3. We simply compute $y' = \mathtt{parent}_1(T_\alpha, x')$ and return y. Suppose x is not an α-node. We compute u, the α-predecessor of x in preorder, in line 4. We claim that x has no α-parent if x has no α-predecessor in preorder, since the ancestors of x precede x in preorder. If such u exists, we take a look at $v = \mathtt{LCA}(u, x)$ in line 6. We further claim that there is no α-node on the path between v and x (excluding v), because u would not be the α-predecessor if such an α-node exists. In addition, we know that v has at least one α-descendant because of the existence of u. We return v if v is an α-node. Otherwise, we compute the first α-descendant, w, of v. It is clear that there is no α-node on the path between w and v (excluding w). Thus, the α-parent of w, being computed in line 9, must be the α-parent of both v and x.

Algorithm 2. $\mathtt{parent}_\alpha(x)$

1 **if** x *is an α-node* **then**
2 $\quad$ $y' \leftarrow \mathtt{parent}_1(T_\alpha, x')$;
3 $\quad$ **return** y;

4 $u \leftarrow$ *the α-predecessor of x in preorder of T*;
5 **if** x *has no α-predecessor* **then return** NULL;
6 $v \leftarrow \mathtt{LCA}(u, x)$;
7 **if** v *is an α-node* **then return** v;
8 $w \leftarrow$ *the first α-descendant of v in preorder*;
9 $y' \leftarrow \mathtt{parent}_1(T_\alpha, w')$;
10 **return** y;

Then we make use of $\mathtt{parent}_\alpha(x)$ to support $\mathtt{level\_anc}_\alpha(x, i)$: We first compute $y = \mathtt{parent}_\alpha(x)$, where y must be an α-node or NULL. We return y if $y = \mathtt{NULL}$ or $i = 1$. Otherwise, we compute $z' = \mathtt{level\_anc}_1(T_\alpha, y', i - 1)$ and return z.

LCA_α and $depth_\alpha$ can also be easily supported using $parent_\alpha$. $LCA_\alpha(x, y)$ is equal to $LCA(x, y)$ if the lowest common ancestor of x and y is an α-node; otherwise it is equal to $parent_\alpha(LCA(x, y))$. To compute $depth_\alpha(x)$, let $y = x$ if x is an α-node, or $y = parent_\alpha(x)$ if x is not. It is then clear that $depth_\alpha(x) = depth_\alpha(y) = depth_1(T_\alpha, y')$, since each α-ancestor of y in T corresponds to a 1-ancestor of y' in T_α.

3.4   child_rank$_\alpha$, child_select$_\alpha$ and deg$_\alpha$

We can support $child\_select_\alpha(x, i)$ and $deg_\alpha(x)$ using the techniques shown in Subsection 3.2. We first try to find x', the node in T_α that corresponds to x. If such x' does not exist, then x must have no α-child. Thus, we return NULL for $child\_select_\alpha(x, i)$ and return 0 for $deg_\alpha(x)$. Otherwise, we compute $y' = child\_select_1(T_\alpha, x', i)$ and return y for $child\_select_\alpha(x, i)$, as well as return $deg_1(T_\alpha, x')$ for $deg_\alpha(x)$.

Algorithm 3. child_rank$_\alpha(x)$

---

1 if x is an α-node then return child_rank$_1(T_\alpha, x')$;
2 $u \leftarrow parent(x)$;
3 if u has no α-child then return 0;
4 $v \leftarrow$ the α-predecessor of x in preorder;
5 if x has no α-predecessor or pre_rank$(v) \leq$ pre_rank(u) then return 0;
6 compute u' and v', the nodes in T_α that correspond to u and v;
7 $w' \leftarrow$ the child of u' that is an ancestor of v';
8 return child_rank$_1(T_\alpha, w')$;

The algorithm to support $child\_rank_\alpha(x)$ is shown as Algorithm 3. The case in which x is an α-node is easy to handle, as shown in line 1. We consider only the case in which the label of x is not α. In lines 2 to 3, we compute node u that is the parent of x, and verify if u has an α-child using deg_α. We return 0 if u has no α-child. Otherwise, we compute the α-predecessor, v, of x in preorder. If such v does not exist, or v is not a proper descendant of u, then x has no α-sibling preceding it and we can return 0, since an sibling preceding x occurs before x in preorder. Suppose v exists as a proper descendant of u. We can find u' and v', the nodes in T_α that correspond to u and v, since both u and v are α-nodes. In addition, we find the child, w', of u' that is an ancestor of v'. We claim that each α-child of u in T that precedes x corresponds to a 1-child of u' in T_α that precedes w'. Otherwise, v would not be the α-predecessor. Finally, we return $child\_rank_1(T_\alpha, w')$ as the answer.

3.5   Completing the Proof of Theorem 2

Due to space limitations, the support for the other static α-operations is omitted here. In the current state, we have σ 0/1-labeled trees and σ bit vectors. To reduce redundancy, we merge T_α's into a single tree $\mathcal{T}$, and merge L_α's into a single bit vector $\mathcal{L}$. We list the characters in Σ as $\alpha_1, \cdots, \alpha_\sigma$. Initially, $\mathcal{T}$ contains a root node $\mathcal{R}$ only, on which the label is 0. Then, for $i = 1$ to σ, we append r'_{α_i}, the root of T_{α_i}, to the list of children of $\mathcal{R}$. Let n_α be the number of α-nodes in T. For $\alpha \in \Sigma$, T_α has at most $2n_\alpha + 1$ nodes, since each α-node adds a 1-node and at most one 0-node into T_α. In addition, the T_α that corresponds to the label of the root of T has at most $2n_\alpha$ nodes, since the root does not add an 0-node to T_α. Hence, $\mathcal{T}$ has at most $2n + \sigma$ nodes in total. By the construction of $\mathcal{T}$, the preorder/postorder traversal sequence of T_α occurs as a substring in the preorder/postorder traversal sequence of $\mathcal{T}$. Also, the DFUDS traversal sequence of T_α with r'_α removed occurs as a substring in the DFUDS traversal sequence of $\mathcal{T}$.

In addition, we append L_{α_i} to $\mathcal{L}$, which is initially empty, for $i = 1$ to σ. The length of $\mathcal{L}$ is clearly n. It is not hard to verify that the reductions described in early subsections can still be performed on the merged tree $\mathcal{T}$ and the merged bit vector $\mathcal{L}$. The following lemma generalizes the discussion.

Lemma 2. *Let T be an ordinal tree on n nodes, each having a label drawn from an alphabet Σ of size σ. Suppose that there exist*

- *a data structure $\mathcal{D}_1$ that represents a unlabeled ordinal tree on n nodes using $\mathcal{S}_1(n)$ bits and supports the unlabeled versions of the first two groups of α-operations in Table 1;*
- *a data structure $\mathcal{D}_2$ that represents a string S using $\mathcal{S}_2(S)$ bits and supports rank$_\alpha$ and select$_\alpha$ for $\alpha \in \Sigma$;*
- *a data structure $\mathcal{D}_3$ that represents a 0/1-labeled ordinal tree on n nodes using $\mathcal{S}_3(n)$ bits and supports the first two groups of α-operations in Table 1 and their unlabeled versions, plus two additional α-operations deepest$_\alpha$ and min_depth$_\alpha$;*
- *and a data structure $\mathcal{D}_4$ that represents a bit vector of length n using $\mathcal{S}_4(n)$ bits and supports rank$_\alpha$ and select$_\alpha$ for $\alpha \in \{0,1\}$.*

Then there exists a data structure that encodes T using $\mathcal{S}_1(n) + \mathcal{S}_2(PLS_T) + \mathcal{S}_3(2n + \sigma) + \mathcal{S}_4(n)$ bits of space, supporting the first two groups of α-operations in Table 1 and their unlabeled versions using a constant number of operations mentioned above on $\mathcal{D}_1$, $\mathcal{D}_2$, $\mathcal{D}_3$ and $\mathcal{D}_4$.

Proof. The unlabeled versions are supported by $\mathcal{D}_1$ directly. The reductions for the α-operations are shown in Subsections 3.1 to 3.4, and more details are omitted due to space limitations. Here we consider the space cost only. We maintain the structure of T, PLS_T, $\mathcal{T}$, and $\mathcal{L}$ using $\mathcal{D}_1$, $\mathcal{D}_2$, $\mathcal{D}_3$, and $\mathcal{D}_4$, respectively. The overall space cost is $\mathcal{S}_1(n) + \mathcal{S}_2(PLS_T) + \mathcal{S}_3(2n + \sigma) + \mathcal{S}_4(n)$ bits. $\square$

With the following three lemmas the proof of Theorem 2 follows.

Lemma 3 ([9,11,2]). *Let S be a string of length n over an alphabet of size σ. Under the word RAM with word size $w = \Omega(\lg n)$,*

(a) for $\sigma = O(\texttt{polylog}(n))$, S can be represented using $nH_0(S) + o(n)$ bits of space to support rank$_\alpha$ and select$_\alpha$ in $O(1)$ time;

(b) for general Σ, S can be represented using $nH_0(S) + O(n)$ bits of space to support rank_α and select_α in $O(\lg \lg \sigma)$ time;

(c) for any $k = o(\log_\sigma n)$, S can be represented using $nH_k(S) + \lg \sigma \cdot o(n) + O(\frac{n \lg \sigma}{\lg \lg \lg \sigma})$ bits to support rank_α and select_α in $O(\lg \lg \sigma (\lg \lg \lg \sigma)^2)$ time.

Lemma 4 ([5]). *A bit vector of length n can be represented in $n + o(n)$ bits to support rank_α and select_α in constant time, for $\alpha \in \{0,1\}$.*

Lemma 5 ([13,6,7,17]). *Let T be an ordinal tree on n nodes. T can be represented using $2n + o(n)$ bits such that the unlabeled versions of the first two groups of α-operations in Table 1 can be supported in constant time.*

Proof (Theorem 2). Applying Lemma 5, one of Lemma 3 (a,b,c), Theorem 1, and Lemma 4 for $\mathcal{D}_1$, $\mathcal{D}_2$, $\mathcal{D}_3$, and $\mathcal{D}_4$, respectively, we obtain the conclusion. □

4 Dynamic Trees that Support Level-Ancestor Operations : Theorem 3

Our succinct representation of dynamic trees also consists of four components: the structure of T, PLS_T, T_α's and L_α's. The construction of T_α is different in this scenario in order to facilitate update operations. For each $\alpha \in \Sigma$, we still add a new root r_α to T, and compute the X_α-extraction of the augmented tree rooted at r_α. However, X_α contains r_α and the α-nodes in T only, and we do not assign labels to the nodes in T_α. Finally, we still merge T_α's into a single tree $\mathcal{T}$, and merge L_α's into a single bit vector $\mathcal{L}$. $\mathcal{T}$ has exactly $n + \sigma + 1$ nodes. We omit the details of supporting operations due to space constraints.

References

1. Barbay, J., Golynski, A., Munro, J.I., Rao, S.S.: Adaptive searching in succinctly encoded binary relations and tree-structured documents. Theor. Comput. Sci. 387(3), 284–297 (2007)
2. Barbay, J., He, M., Munro, J.I., Rao, S.S.: Succinct indexes for strings, binary relations and multilabeled trees. ACM Transactions on Algorithms 7(4), 52 (2011)
3. Benoit, D., Demaine, E.D., Munro, J.I., Raman, R., Raman, V., Rao, S.S.: Representing trees of higher degree. Algorithmica 43(4), 275–292 (2005)
4. Bille, P.: A survey on tree edit distance and related problems. Theor. Comput. Sci. 337(1-3), 217–239 (2005)
5. Clark, D.R., Munro, J.I.: Efficient suffix trees on secondary storage (extended abstract). In: SODA, pp. 383–391 (1996)
6. Farzan, A., Munro, J.I.: A Uniform Approach Towards Succinct Representation of Trees. In: Gudmundsson, J. (ed.) SWAT 2008. LNCS, vol. 5124, pp. 173–184. Springer, Heidelberg (2008)
7. Farzan, A., Raman, R., Rao, S.S.: Universal Succinct Representations of Trees? In: Albers, S., Marchetti-Spaccamela, A., Matias, Y., Nikoletseas, S., Thomas, W. (eds.) ICALP 2009, Part I. LNCS, vol. 5555, pp. 451–462. Springer, Heidelberg (2009)

8. Ferragina, P., Luccio, F., Manzini, G., Muthukrishnan, S.: Compressing and indexing labeled trees, with applications. J. ACM 57(1) (2009)
9. Ferragina, P., Manzini, G., Mäkinen, V., Navarro, G.: Compressed representations of sequences and full-text indexes. ACM Transactions on Algorithms 3(2) (2007)
10. Geary, R.F., Raman, R., Raman, V.: Succinct ordinal trees with level-ancestor queries. ACM Transactions on Algorithms 2(4), 510–534 (2006)
11. Golynski, A., Munro, J.I., Rao, S.S.: Rank/select operations on large alphabets: a tool for text indexing. In: SODA, pp. 368–373 (2006)
12. He, M., Munro, J.I.: Succinct Representations of Dynamic Strings. In: Chavez, E., Lonardi, S. (eds.) SPIRE 2010. LNCS, vol. 6393, pp. 334–346. Springer, Heidelberg (2010)
13. He, M., Munro, J.I., Rao, S.S.: Succinct Ordinal Trees Based on Tree Covering. In: Arge, L., Cachin, C., Jurdziński, T., Tarlecki, A. (eds.) ICALP 2007. LNCS, vol. 4596, pp. 509–520. Springer, Heidelberg (2007)
14. He, M., Munro, J.I., Zhou, G.: Path Queries in Weighted Trees. In: Asano, T., Nakano, S.-i., Okamoto, Y., Watanabe, O. (eds.) ISAAC 2011. LNCS, vol. 7074, pp. 140–149. Springer, Heidelberg (2011)
15. He, M., Munro, J.I., Zhou, G.: Succinct Data Structures for Path Queries. In: Epstein, L., Ferragina, P. (eds.) ESA 2012. LNCS, vol. 7501, pp. 575–586. Springer, Heidelberg (2012)
16. Manzini, G.: An analysis of the burrows-wheeler transform. J. ACM 48(3), 407–430 (2001)
17. Sadakane, K., Navarro, G.: Fully-functional succinct trees. In: SODA, pp. 134–149 (2010)

A Space-Efficient Framework
for Dynamic Point Location

Meng He[1], Patrick K. Nicholson[2], and Norbert Zeh[1]

[1] Faculty of Computer Science, Dalhousie University, Canada
{mhe,nzeh}@cs.dal.ca
[2] Cheriton School of Computer Science, University of Waterloo, Canada
p3nichol@uwaterloo.ca

Abstract. Let G be a planar subdivision with n vertices. A succinct geometric index for G is a data structure that occupies $o(n)$ bits beyond the space required to store the coordinates of the vertices of G, while supporting efficient queries. We describe a general framework for converting dynamic data structures for planar point location into succinct geometric indexes, provided that the subdivision G to be maintained has bounded face size. Using this framework, we obtain several succinct geometric indexes for dynamic planar point location on G with query times matching the currently best (non-succinct) data structures and polylogarithmic update times.

1 Introduction

Many fundamental problems in computational geometry involve constructing data structures over large geometric data sets to support efficient queries on the data. Such queries include point location, ray shooting, nearest neighbour searching, as well as a plethora of range searching variants (see [5,1]). Data structures supporting these types of queries provide the building blocks of software in many important application areas, such as geographic information systems, network traffic monitoring, database systems, computer aided design (e.g., very-large-scale integration), and numerous graphical applications.

One of the most heavily studied queries of this type is that of planar point location: Given an n-vertex planar subdivision, the goal is to support queries of the form, "Which region of the subdivision contains the query point?" Starting with Kirkpatrick's work [14], many linear-space data structures have been proposed to support planar point location queries in $O(\lg n)$ time[1], which is the asymptotically optimal number of point-line comparisons. Further research has focused on determining the exact number of point-line comparisons [12,17], developing data structures that bound the query time based on the entropy of the query distribution [13], and exploiting word-RAM parallelism [8,7].

Recently, Bose et al. [6] presented a space-efficient framework for planar point location in the word-RAM model. In this framework, the coordinates of the

---

[1] We use $\lg n$ to denote $\lceil \log_2 n \rceil$.

K.-M. Chao, T.-s. Hsu, and D.-T. Lee (Eds.): ISAAC 2012, LNCS 7676, pp. 548–557, 2012.
© Springer-Verlag Berlin Heidelberg 2012

Table 1. Previous results for linear-space data structures for dynamic planar point location. In the model column, "PM" denotes the pointer machine, and "RAM" the word-RAM model. In the restrictions column, "General" denotes an arbitrary planar subdivision, and "Horizontal" denotes point location among horizontal segments (vertical ray shooting). The letters "a" and "p" in the columns showing query, insertion, and deletion bounds indicate amortized bounds and high-probability bounds (i.e., probability $1 - O(1)/n^c$, for some $c \geq 1$), respectively. The value ε is any positive constant.

Source	Model	Restrictions	Query	Insert	Delete
CJ92 [9]	PM	General	$O(\lg^2 n)$	$O(\lg n)$	$O(\lg n)$
BJM94 [4]	PM	General	$O(\lg n \lg \lg n)$	$O(\lg n \lg \lg n)^{\mathrm{a}}$	$O(\lg^2 n)^{\mathrm{a}}$
ABG06 [3]	PM	General	$O(\lg n)$	$O(\lg^{1+\varepsilon} n)^{\mathrm{a}}$	$O(\lg^{2+\varepsilon})^{\mathrm{a}}$
ABG06 [3]	RAM	General	$O(\lg n)$	$O(\lg n \lg^{1+\varepsilon} \lg n)^{\mathrm{a,p}}$	$O(\frac{\lg^{2+\varepsilon} n}{\lg \lg n})^{\mathrm{a,p}}$
GK09 [11]	RAM	Horizontal	$O(\lg n)$	$O(\lg n)$	$O(\lg n)$
N10 [15]	RAM	Horizontal	$O(\frac{\lg n}{\lg \lg n})$	$O(\lg^{1+\varepsilon} n)^{\mathrm{a}}$	$O(\lg^{1+\varepsilon} n)^{\mathrm{a}}$

n vertices of the subdivision are permuted and stored along with an auxiliary data structure called a *succinct geometric index*. The succinct geometric index occupies only $o(n)$ *bits*, which is asymptotically negligible compared to the space occupied by the coordinates. With only this index and access to the permuted sequence, they showed how to match the efficiency of many of the previous data structures for planar point location. Specifically, they presented several succinct geometric indexes that can answer point location queries in $O(\lg n)$ time; $O(H + 1)$ time, where H is the entropy of the query distribution; $O(\min\{\lg n/ \lg \lg n, \sqrt{\lg U}\})$ time, if the coordinates are integers in the range $[1, U]$; and, finally, $\lg n + 2\sqrt{\lg n} + O(\lg^{1/4} n)$ point-line comparisons. Furthermore, they showed how to make their data structure *implicit*. Their implicit data structure uses only $O(1)$ words of space beyond the permuted sequence of coordinates for the vertices and supports point location queries in $O(\lg^2 n)$ time.

But what about the *dynamic case*, where we are allowed to modify the structure of the planar subdivision? Many data structures have been proposed to solve this problem, though the best choice depends on whether we desire fast queries, fast updates, worst-case behaviour, deterministic behaviour, or are operating on a restricted class of subdivisions. Table 1 summarizes the skyline results for dynamic planar point location. Note that we consider only linear-space data structures that are fully dynamic, i.e., support insertions and deletions. So far, no point location data structures have been proposed that are dynamic *and* use only $o(n)$ bits of memory beyond the space required for the coordinates of the vertices. Developing such structures is the focus of this paper.

1.1   Our Results

Our main result is a framework for creating succinct geometric indexes for dynamic point location in planar subdivisions with *bounded face size*: i.e., the maximum number of vertices defining a face is a fixed constant that does not depend on n. This framework allows us to convert any existing linear-space data structure for dynamic planar point location into a succinct geometric index, subject to the constraint that the subdivision to be maintained have bounded face size[2]. The query times of our data structures can be set to match any of the data structures for general subdivisions from Table 1, since our framework introduces only an additive $o(\lg n)$ term to the query cost. Updates are supported in polylogarithmic time, where the exact update time depends on the choice of the underlying data structure. Our *update operations* allow for insertion and deletion of vertices and edges into/from G, with some restrictions that we describe in Section 3. We note that the types of update operations are similar to previous work [9], and complete in the sense that they allow the assembly and disassembly of any planar subdivision of bounded face size. All our results hold in the word-RAM model with word size $\Theta(\lg n)$ bits. The following theorem, which is a consequence of our main theorem (Theorem 3 on page 556), summarizes our contributions.

Theorem 1. *Let G be an n-vertex planar subdivision with bounded face size and each of whose vertices has coordinates occupying $M = O(\lg n)$ bits. For any constant $\varepsilon > 0$, there exists a data structure for dynamic planar point location in G that occupies $nM + o(n)$ bits and*

- *Supports queries in $O(\lg n)$ time and updates in $O(\lg^{3+\varepsilon} n)$ amortized time with high probability (using [3]),*
- *Supports queries in $O(\lg n \lg \lg n)$ time and updates in $O(\lg^{2+\varepsilon} n)$ amortized time (using [4]), or*
- *Supports queries in $O(\lg^2 n)$ time and updates in $O(\lg^{2+\varepsilon} n)$ worst-case time (using [9]).*

Techniques and Overview: Our point location framework is based on the two-level decomposition used in Bose et al.'s framework for obtaining succinct indexes for *static* planar point location [6]. This framework uses a two-level partition of the subdivision using planar separators. The main challenge in obtaining a *dynamic* framework based on these ideas is to maintain the separator decomposition under updates of the subdivision. Aleksandrov and Djidjev [2] introduced the P-tree, a linear-space data structure for maintaining planar graph partitions under updates of the graph. Our main technical contribution is to develop a succinct version of this data structure that requires only $o(n)$ bits of space beyond the space required to store the coordinates of the vertices of the graph. Obtaining this structure requires a non-trivial combination of the original P-tree data structure with the labelling scheme by Bose et al. [6]. While our motivation to develop this data structure was its importance as part of our point location framework, we expect it to be of independent interest, as graph partitions find applications in a wide range of algorithms.

---

[2] Previous results for general subdivisions, e.g., [9], do not have this constraint.

2   Definitions and Preliminaries

We require the following definitions, closely following Aleksandrov and Djid-jev [2], but making some slight modifications. For a graph G, let $V(G)$ and $E(G)$ denote the vertex and edge sets of G, respectively. A *planar graph* G is any graph that can be embedded (drawn) in the plane so that its edges intersect only at their endpoints. A *straight-line embedding* of G represents each edge as a line segment. A *planar subdivision* is a straight-line embedding of a 2-edge-connected planar graph. The *faces* of the subdivision are the connected components of $\mathbb{R}^2 \setminus (V(G) \cup E(G))$. We denote the set of faces by $F(G)$, the number of faces by N, and the number of vertices by n.

A *region* R is any set of faces, and we use $G(R)$ to denote the subgraph spanned by the edges on their boundaries. An edge $e \in E(G(R))$ is a *boundary edge* of R if only one of the faces in $F(G)$ incident to e is in $G(R)$. All other edges of $G(R)$ are referred to as *interior* edges. The boundary of R, denoted ∂R, is the subgraph induced by the boundary edges of R. We say a region is *connected* if the dual of $G(R)$ is connected.

A *partition* $\mathcal{R} = \{R_1, ..., R_r\}$ of G is a set of regions such that each face in $F(G)$ appears in exactly one region in $\mathcal{R}$. A partition is *weakly connected* if every region is either connected or adjacent to (i.e., shares a boundary edge with) at most two other regions, in which case these two adjacent regions are connected. The boundary of partition $\mathcal{R}$, denoted $\partial \mathcal{R}$, is the the union of the boundaries ∂R_i, $1 \leq i \leq r$, of its regions.

Let G be a planar graph with N faces, and $\varepsilon > 0$. A partition $\mathcal{R}$ is an ε-partition of G if no region of $\mathcal{R}$ contains more than εN faces. In the remainder of this paper we deal with graphs whose face size is bounded by a constant. We note that the number of vertices, n, is $\Theta(N)$ in this case.

We now review some preliminary lemmas that we will use extensively. The following lemma is an extension of the result of [10], and is used throughout our data structures to save space.

Lemma 1 ([6], Lemma 4.1). *Given a planar subdivision of n vertices, for a sufficiently large n, there exists an algorithm that can encode it as a permutation of its point set in $O(n)$ time and such that the subdivision can be decoded from this permutation in $O(n)$ time.*

For a sequence S of length n, let $S[i]$ denote the ith symbol in S. We use $\text{rank}_b(S, i)$ to denote the frequency of symbol b in the prefix $S[1], ..., S[i]$. Similarly, we use $\text{select}_b(S, i)$ to denote the index j containing the ith occurrence of symbol b in S. We make use of the following lemma, which can be used to support rank and select operations on a *bit sequence*, while also compressing the sequence if it sparse.

Lemma 2 ([16]). *Given a sequence S of n bits, with m one bits 1, there exists a data structure that represents S using $m \lg(n/m) + 1.92m + o(m)$ bits and supports rank and select operation in $O(\lg(n/m) + (\lg^4 m)/\lg n)$ and $O((\lg^4 m)/\lg n)$ time, respectively. Construction of the data structure takes $O(n)$ time.*

3   P-Trees

The *P-tree*, introduced by Aleksandrov and Djidjev [2], is a dynamic data structure for maintaining ε-partitions of a planar graph under the following operations, assuming the face size is bounded by a constant d:

- $\mathtt{insert\_vertex}(v, e)$: Create a new vertex v and replace the edge $e = (u, w)$ with two new edges $e_1 = (u, v)$ and $e_2 = (v, w)$.
- $\mathtt{insert\_edge}(u, w, f)$: Insert a new edge $e = (u, w)$ across face f. Vertices u and w have to be on the boundary of face f.
- $\mathtt{delete\_vertex}(v)$: Assuming v is a degree-2 vertex with neighbours u and w, delete v and replace the edges (u, v) and (v, w) with a single edge (u, w).
- $\mathtt{delete\_edge}(e)$: Delete the edge $e = (u, w)$, assuming both u and w are of degree greater than two.
- $\mathtt{list\_partition}(\varepsilon)$: Return an ε-partition of G.

As noted by Aleksandrov and Djidjev [2], these operations can be used to transform any planar graph into any other planar graph, as long as neither contains a vertex of degree less than two. Next we give a brief summary of this data structure. For full details, refer to [2].

The P-tree represents a hierarchy of graphs $G_0, G_1, \ldots, G_\ell$, where $G_0 = G$ and $|G_\ell| = O(1)$. For $1 \le i \le \ell$, G_i is obtained from G_{i-1} by computing a weakly connected h/N_{i-1}-partition $\mathcal{R}_{i-1}$ of G_{i-1}, where N_{i-1} is the number of faces of G_{i-1} and h is an appropriate constant. The faces of G_i represent the regions in this partition. Every edge of G_i represents a maximal list of boundary edges on the boundaries of the two adjacent regions. Every face f of G_i has the faces of G_{i-1} in the corresponding region of $\mathcal{R}_{i-1}$ as its children in the P-tree and stores a pointer to the list of edges on its boundary. Every edge of G_i stores pointers to the edges of G_{i-1} it represents. By following these pointers recursively, every face f of G_i represents a collection of faces of G, a region $R(f)$, and every edge of G_i represents a collection of edges of G. The edges of G corresponding to the edges on the boundary of a face f of G_i are exactly the boundary edges of the region $R(f)$. We define the *cost* of an edge of G_i as the number of edges in G it represents. For a P-tree T, we use $\mathcal{R}(T, i)$ to denote the partition $\{R(f_1), \ldots, R(f_r)\}$, where $f_1, \ldots, f_r$ are the faces of G_i, that is, the faces represented by nodes in T at distance i from the leaf level.

We call a node z of a P-tree T with children $z_1, z_2, \ldots, z_q$ *balanced* if it satisfies three properties:

(B1) $z, z_1, z_2, \ldots, z_q$ have at most h children each, for an appropriate constant h.

(B2) If $\#\mathtt{c}(z)$ and $\#\mathtt{g}(z)$ respectively denote the number of children and the number of grandchildren of z, then $\#\mathtt{c}(z)/\#\mathtt{g}(z) \le c/h$, for an appropriate constant $c \le h/2$.

(B3) Let k be the level of z in the P-tree, i.e., its distance from the leaf level. Then the total cost of all edges on z_i's boundary, for every $1 \le i \le q$, is at most $dh^{(k-1)/2}$.

We note that all nodes of the P-tree are balanced after applying the described construction algorithm. Furthermore, using these balancing conditions, it is easy to prove the following lemma.

Lemma 3. *Let $1 \le k \le \ell$ and $\varepsilon_k = h^k/N$, where N is the number of faces in G. The partition $\mathcal{R}(T, k)$ is an ε_k-partition of G with boundary size not exceeding $d\sqrt{(Nc^{2k})/\varepsilon_k}$.*

By Lemma 3, a list_partition(ε) query amounts to finding the right level in the P-tree and reporting the regions corresponding to the nodes at this level and their boundaries.

The update operations supported by the P-tree may create or destroy a constant number of leaves (faces of G) and change the boundary of a constant number of leaves. For example, insert_edge increases the number of leaves by one. This may affect the balancing of the ancestors of these leaves. In order to rebalance the tree, the path from each such leaf to the root is traversed and every unbalanced node z is rebalanced using the following lemma.[3]

Lemma 4. *Let G be a planar graph with N faces, assume the edges of G have associated costs, and assume the total cost of the edges bounding each face is bounded by some parameter b. Then there exists a weakly connected h/N-partition $\mathcal{R} = \{R_1, ..., R_r\}$ such that the cost of each region's boundary is at most $b\sqrt{h}$ and $r \le cN/h$, for some constant c. Furthermore, $\mathcal{R}$ can be constructed in $O(N \lg N)$ time.*

We apply this lemma to every unbalanced node z we encounter. Since we rebalance nodes in a bottom-up fashion, z can violate condition (B1) only if one of its children has too many children. While rebalancing, we maintain the invariant that every node, balanced or not, has at most $2ch$ children and every node below the current node is balanced. Thus, once we are done processing the root, the entire tree is balanced.

Now consider an unbalanced level-k node z with at most h children $z_1, \ldots, z_q$ that each satisfy the conditions for z to be balanced, except one which may have between h and $2ch$ children or boundary cost greater than $dh^{(k-1)/2}$. Moreover, z may violate condition (B2). To rebalance z, we consider the constant-size subgraph of G_{k-2} consisting of z's grandchildren and their boundaries, and partition it into subgraphs with at most h faces each. Each region in this partition becomes a new face of G_{k-1}, and these faces become the new children of z. By Lemma 4, this restores condition (B2), the part of condition (B1) bounding the number of children of each child of z and, since each grandchild of z has boundary cost at most $b \le dh^{(k-2)/2}$, condition (B3). Since z has at most h children before rebalancing, one of which has up to $2ch$ children, while all others have at most h children, z has less than $h^2 + 2ch \le 2h^2$ grandchildren. Thus,

---

[3] In [2, Theorem 1], this result was stated incorrectly, but apparently the remainder of the paper applied it correctly. The running time stated in the lemma can be reduced to $O(N)$, but it would have no effect on our data structure and would require a significantly more tedious analysis.

by Lemma 4, we partition the subgraph defined by these grandchildren into at most $2h^2 \cdot c/h \leq 2ch$ regions, each of which becomes a child of z. Thus, z satisfies the upper bound on its number of children necessary to proceed to rebalancing its parent. Since each rebalancing operation works with a graph of constant size and the height of the tree is $O(\lg N)$, each update takes $O(\lg N)$ time.

Based on Lemmas 3 and 4, we get the following corollary:

Corollary 1. *Let $h \geq c^\alpha$, for some $\alpha > 2$, and let $k = \log_h \lg^\lambda N$, for some $\lambda > 0$. The partition $\mathcal{R}(T, k)$ is a $(\lg^\lambda N/N)$-partition with boundary size that does not exceed $dN/\lg^{\lambda/2 - \lambda/\alpha} N$. Furthermore, the number of regions in $\mathcal{R}(T, k)$ does not exceed $N/\lg^{\lambda - \lambda/\alpha} N$.*

4   Succinct P-Trees

In this section we introduce the *succinct P-tree*, which uses only $nM + o(n)$ bits of space to represent an n-vertex planar graph whose vertices have M-bit coordinates in the plane. The price we pay for this space reduction is a polylogarithmic slowdown in update time.

The main idea of the succinct P-tree is to prune all nodes below level $\log_h \lg^\lambda N$ in the tree, for some $\lambda > 0$, referred to as the *pruning level*. We call the nodes at the pruning level *pruned nodes* and the nodes above the pruning level *internal nodes*. By Corollary 1, there are $r \leq N/\lg^{\lambda - \lambda/\alpha} N$ pruned nodes. We refer to the regions defined by the pruned nodes as *pruned regions*.

Suppose we apply Lemma 1 to a pruned subgraph $G(R(z))$, obtaining a permutation of the coordinates of its vertices that uniquely identifies the structure of $G(R(z))$. Storing this permuted sequence of coordinates in the nodes at the pruning level allows us to perform operations as in the original P-tree, at the cost of decoding and re-encoding a subgraph of size $O(\lg^\lambda N)$ during each update. However, explicitly storing this permutation in each pruned node causes the coordinates of the boundary vertices to be duplicated in several pruned regions. To avoid this, we adapt the labelling scheme of Bose et al. [6] to the dynamic setting. The details are as follows.

Data Structures of Pruned Nodes: Let π_z denote the permutation of vertices obtained by applying Lemma 1 to $G(R(z))$, for a fixed pruned node $z \in T$. We denote the ith vertex in the permutation as $\pi_z(i)$. We separate the vertices into two categories: vertices that are endpoints of edges on the boundary of the pruned region are *boundary vertices*; all other vertices are *interior vertices*. Every interior vertex belongs to exactly one pruned region. Every boundary vertex belongs to more than one pruned region. Each pruned node z now stores the following data structures, where $\partial R(z)$ and n_z denote the boundary of and the number of vertices in $R(z)$, respectively.

- A binary sequence B_z, where $B_z[j] = 0$ if $\pi_z(j)$ is an interior vertex, and $B_z[j] = 1$ if $\pi_z(j)$ is a boundary vertex of $R(z)$. We represent B_z using the data structure of Lemma 2.

- An array I_z that stores the coordinates for interior vertices. Entry $I_z[j]$ stores the coordinates for the vertex $\pi_z(\text{select}_0(B_z, j))$, for $1 \leq j \leq \text{rank}_0(B_z, n_z)$.
- An array X_z that stores pointers to *records* external to node z. For $1 \leq j \leq \text{rank}_1(B_z, n_z)$, the record pointed to by entry $X_z[j]$ stores the coordinates of vertex $\pi_z(\text{select}_1(B_z, j))$: the jth boundary vertex in $R(z)$.
- An array E_z that stores pointers to records representing the edges on the boundary of $\partial R(z)$. The ordering of E_z is any canonical ordering based on the permutation π_z. For example, we can order the edges lexicographically by the positions of their endpoints in π_z. Let $\mathcal{C}$ be the record representing the jth boundary edge e of $R(z)$ in this ordering, pointed to by entry $E_z[j]$. $\mathcal{C}$ stores the indices of e's endpoints in arrays X_z, as well as a pointer to z. Furthermore, $\mathcal{C}$ stores the symmetric information about the other pruned region $R(z')$ that has e on its boundary.

Each internal node stores the same information as in a standard P-tree. Using Corollary 1 and Lemma 2, we can bound the space occupied by the succinct P-tree data structure as follows.

Lemma 5. *The succinct P-tree occupies* $nM + O(n/\lg^{\lambda/2-\beta-1} n)$ *bits, for any constant* $\beta > 0$.

We next state the following lemma about supporting updates. Intuitively, the idea is to simulate the operation of a standard P-tree. Above the pruning level, each update operation proceeds identically as if it were run on a standard P-tree and, thus, takes $O(\lg n)$ time. In order to implement the portion of the update operation that operates on nodes below the pruning level, we reconstruct the affected pruned region from its succinct representation and build a P-tree from it. This takes $O(\lg^\lambda n \lg \lg n)$ time. Since each update affects only a constant number of pruned regions, the lemma follows.

Lemma 6. *A succinct P-tree supports the operations* insert_vertex, insert_edge, delete_vertex, *and* delete_edge *in* $O(\lg^\lambda n \lg \lg n)$ *time.*

Combining Lemmas 5 and 6, and setting $\lambda > 2$ leads us to our main theorem of this section.

Theorem 2. *Let G be a planar subdivision with n vertices and N faces, where each face has at most d vertices, for some constant $d \geq 3$. Each vertex is assumed to store M-bit coordinates. Let ε be any positive constant. There exists a data structure representing G in $nM + o(n)$ bits of space that can perform the operations* insert_vertex, delete_vertex, insert_edge, *and* delete_edge *in $O(\lg^{2+\varepsilon} n)$ time. The operation* list_partition(ε') *can be performed in time proportional to the partition's size, for $\lg^{2+\varepsilon} N/N \leq \varepsilon' \leq 1$. The boundary size of the partition returned by* list_partition(ε') *does not exceed $d\sqrt{(N^{1+\delta})/(\varepsilon'^{1-\delta})}$, where $\delta > 0$ is an arbitrarily small constant but depends on our choice of ε.*

5 Dynamic Planar Point Location

As an application of Theorem 2, we develop a succinct geometric index for dynamic planar point location. To do this, we add extra data structures to the pruned nodes of the succinct P-tree. These extra data structures are analogous to the *subregion* level data structures of the two-level index of Bose et al. [6].

Let γ be a positive constant in the range $(0, 2]$. We refer to the regions in $\mathcal{R}(T, \log_h \lg^\gamma n)$ as γ-*subregions*. Since $\lambda > 2$, each γ-subregion is contained in a pruned region and thus is not accessible without decoding this pruned region. The next lemma states that we can augment the succinct P-tree to provide efficient access to γ-subregions.

Lemma 7. *The succinct P-tree can be augmented to support extraction of the graph structure of an arbitrary γ-subregion (within a specified pruned region) in $O(\lg^\gamma(n)\mathtt{polyloglog}(n))$ time, where $\gamma \in (0, 2]$ is fixed at construction time. The space bound becomes $nM + O((n \lg \lg n)/\lg^{\gamma/2 - \gamma/\alpha} n)$, which is $o(n)$ for $\gamma \in (0, 2]$ and $\alpha > 2$, and the update costs remain as stated in Theorem 2.*

We now sketch how to use γ-subregion extraction to efficiently support point location queries, resulting in our main theorem:

Theorem 3. *Let $\mathcal{D}$ be a dynamic point location data structure that uses $O(n)$ words of space to store an n-vertex subdivision and supports queries and updates on this subdivision in $Q(n)$ and $U(n)$ time, respectively. We assume $Q(n) + U(n) = O(\mathtt{polylog}(n))$. Let G be a planar subdivision, each of whose faces has a constant number of vertices, and assume the coordinates of each vertex can be stored in M bits. Let $\varepsilon > 0$ be any positive constant, and choose any constant $\gamma \in (0, 2]$. There exists a data structure for dynamic planar point location that occupies $nM + o(n)$ bits of space, supports queries in $O(\lg^\gamma(n)\mathtt{polyloglog}(n) + Q(n))$ time, and supports the operations* insert_vertex, delete_vertex, insert_edge, *and* delete_edge *in $O(U(n)\lg^{1 + \varepsilon/2} n + \lg^{2 + \varepsilon} n)$ time.*

Proof (Sketch). We maintain G in a succinct P-tree T, augmented as in Lemma 7. The planar subdivision defined by the boundaries of the pruned regions of G is stored in the data structure $\mathcal{D}$. This is the *first level* point location structure. Inside each pruned node z, we also store another instance of $\mathcal{D}$, denoted $\mathcal{D}_z$. This *second-level* point location structure stores the planar subdivision defined by the boundaries of the γ-subregions contained in $R(z)$. Each face f of this subdivision corresponds to a connected component of a γ-subregion S_i. We store its index i with f. To answer a query for a point p, we first query $\mathcal{D}$ to identify the pruned region $R(z)$ that contains p. Next we query $\mathcal{D}_z$ to identify the γ-subregion S_i that contains p. Finally, we extract S_i and perform a brute-force search to find the face of S_i that contains p. The queries to $\mathcal{D}$ and $\mathcal{D}_z$ require $Q(n)$ time, and the γ-subregion extraction step requires $O(\lg^\gamma(n)\mathtt{polyloglog}(n))$ time, by Lemma 7. The total query time is therefore $O(Q(n) + \lg^\gamma(n)\mathtt{polyloglog}(n))$. The key idea of achieving the claimed update time is to use condition (B3) to bound the number of edges and vertices in D that are changed by an update. □

By choosing the data structures from Table 1 as $\mathcal{D}$ in the previous theorem, and setting γ appropriately, we get the result of Theorem 1.

References

1. Agarwal, P., Erickson, J.: Geometric range searching and its relatives. Contemporary Mathematics 223, 1–56 (1999)
2. Aleksandrov, L.G., Djidjev, H.N.: A Dynamic Algorithm for Maintaining Graph Partitions. In: Halldórsson, M.M. (ed.) SWAT 2000. LNCS, vol. 1851, p. 71. Springer, Heidelberg (2000)
3. Arge, L., Brodal, G., Georgiadis, L.: Improved dynamic planar point location. In: Proceedings of the 47th Annual IEEE Symposium on Foundations of Computer Science, pp. 305–314. IEEE Computer Society (2006)
4. Baumgarten, N., Jung, H., Mehlhorn, K.: Dynamic point location in general subdivisions. Journal of Algorithms 17(3), 342–380 (1994)
5. de Berg, M., Cheong, O., van Kreveld, M., Overmars, M.: Computational Geometry: Algorithms and Applications, 3rd edn. Springer, Santa Clara (2008)
6. Bose, P., Chen, E., He, M., Maheshwari, A., Morin, P.: Succinct geometric indexes supporting point location queries. ACM Trans. on Algorithms 8(2), 10 (2012)
7. Chan, T.M.: Persistent predecessor search and orthogonal point location on the word ram. In: SODA, pp. 1131–1145 (2011)
8. Chan, T.M., Patrascu, M.: Transdichotomous Results in Computational Geometry, I: Point Location in Sublogarithmic Time. SIAM J. Comput. 39(2), 703–729 (2009)
9. Cheng, S., Janardan, R.: New results on dynamic planar point location. SIAM Journal on Computing 21, 972 (1992)
10. Denny, M., Sohler, C.: Encoding a triangulation as a permutation of its point set. In: Proc. CCCG (1997)
11. Giora, Y., Kaplan, H.: Optimal dynamic vertical ray shooting in rectilinear planar subdivisions. ACM Transactions on Algorithms (TALG) 5(3), 28 (2009)
12. Goodrich, M., Orletsky, M., Ramaiyer, K.: Methods for achieving fast query times in point location data structures. In: Proc. SODA, pp. 757–766. SIAM (1997)
13. Iacono, J.: A static optimality transformation with applications to planar point location. In: Symposium on Computational Geometry, pp. 21–26 (2011)
14. Kirkpatrick, D.: Optimal search in planar subdivisions. SIAM J. Comput. 12(1), 28–35 (1983)
15. Nekrich, Y.: Searching in dynamic catalogs on a tree. Arxiv preprint arXiv:1007.3415 (2010)
16. Okanohara, D., Sadakane, K.: Practical entropy-compressed rank/select dictionary. In: ALENEX (2007)
17. Seidel, R., Adamy, U.: On the exact worst case query complexity of planar point location. Journal of Algorithms 37(1), 189–217 (2000)

Selection in the Presence of Memory Faults, with Applications to In-place Resilient Sorting*

Tsvi Kopelowitz and Nimrod Talmon

Weizmann Institute of Science, Rehovot, Israel
{kopelot,nimrodtalmon77}@gmail.com

Abstract. The selection problem, where one wishes to locate the kth smallest element in an unsorted array of size n, is one of the basic problems studied in computer science. The main focus of this work is designing algorithms for solving the selection problem in the presence of memory faults.

Specifically, the computational model assumed here is a faulty variant of the RAM model (abbreviated as $FRAM$), which was introduced by Finocchi and Italiano [FI04]. In this model, the content of memory cells might get corrupted adversarially during the execution, and the algorithm cannot distinguish between corrupted cells and uncorrupted cells. The model assumes a constant number of reliable memory cells that never become corrupted, and an upper bound δ on the number of corruptions that may occur, which is given as an auxiliary input to the algorithm.

The main contribution of this work is a deterministic resilient selection algorithm with optimal $O(n)$ worst-case running time. Interestingly, the running time does not depend on the number of faults, and the algorithm does not need to know δ. As part of the solution, several techniques that allow to sometimes use non-tail recursion algorithms in the FRAM model are developed. Notice that using recursive algorithms in this model is problematic, as the stack might be too large to fit in reliable memory.

The aforementioned resilient selection algorithm can be used to improve the complexity bounds for resilient k-d trees developed by Gieseke, Moruz and Vahrenhold [GMV10]. Specifically, the time complexity for constructing a k-d tree is improved from $O(n \log^2 n + \delta^2)$ to $O(n \log n)$.

Besides the deterministic algorithm, a randomized resilient selection algorithm is developed, which is simpler than the deterministic one, and has $O(n + \alpha)$ expected time complexity and $O(1)$ space complexity (i.e., is in-place). This algorithm is used to develop the first resilient sorting algorithm that is in-place and achieves optimal $O(n \log n + \alpha\delta)$ expected running time.

1 Introduction

Computing devices are becoming smaller and faster. As a result, the likelihood of soft memory errors (which are not caused by permanent failures) is increased.

---

* A full version appears at http://arxiv.org/abs/1204.5229. This work was supported in part by The Israel Science Foundation (grant #452/08), by a US-Israel BSF grant #2010418, and by the Citi Foundation.

In fact, a recent practical survey [Sem04] concludes that a few thousands of soft errors per billion hours per megabit is fairly typical, which would imply roughly one soft error every five hours on a modern PC with 24 gigabytes of memory [CDK11]. The causes of these soft errors vary and include cosmic rays [Bau05], alpha particles [MW79], or hardware failures [LHSC10].

To deal with these faults, the faulty RAM (FRAM) model has been proposed by Finocchi and Italiano [FI04], and has received some attention [BFF+07, BJM09, BJMM09, CFFS11, FGI09a, FGI09b, GMV10, JMM07]. In this model, an upper bound on the number of corruptions is given to the algorithm, and is denoted by δ, while the *actual* number of faults is denoted by α ($\alpha \leq \delta$). Memory cells may become corrupted at any time during an algorithm's execution and the algorithm cannot distinguish between corrupted cells and uncorrupted cells. The same memory cell may become corrupted multiple times during a single execution of an algorithm. In addition, the model assumes the existence of $O(1)$ reliable memory cells, which are needed, for example, to reliably store the code itself. A cell is assumed to contain $\Theta(\log n)$ bits, where n is the size of the input, as is usual in the RAM model.

One of the interesting aspects of developing algorithms in the FRAM model is that the notion of correctness is not always clear. Usually, correctness is defined with respect to the subset of uncorrupted memory cells and in a worst-case sense, implying that for an algorithm to be correct, it must be correct in the presence of any faulty environment, including an adversarial environment. In this model, an algorithm that is always correct (which is problem dependent) is called *resilient*.

The main focus of this work is on the *selection problem* (sometimes called the k-order statistic problem) in the FRAM model, where one wishes to locate the kth smallest element in an unsorted array of size n, in the presence of memory faults. The following main theorem is proved in Section 3.

Theorem 1. *There exists a deterministic resilient selection algorithm with time complexity $O(n)$.*

Interestingly, the running time does not depend on the number of faults. Moreover, the algorithm does not need to know δ explicitly. The selection problem is a classic problem in computer science. Along with searching and sorting, it is one of the basic problems studied in the field, taught already at undergraduate level (e.g., [CLRS09]). There are numerous applications for the selection problem, thus devising efficient algorithms is of practical interest.

When considering the selection problem in the FRAM model, the first difficulty is to define correctness[1]. To this end, the correctness definition used here allows to return an element, which may even be corrupted, whose rank is between $k - \alpha$ to $k + \alpha$ in the input array. Notice that when $\alpha = 0$ this definition coincides with the non-faulty definition (for a formal definition see Section 2).

---

[1] The common notion of considering only the non-corrupted elements is somewhat misleading in the selection problem. This is because of the difficulty of not being able to distinguish between corrupted and uncorrupted data.

Besides the deterministic algorithm, a randomized and in-place counterpart is developed as well. Specifically, a randomized and in-place resilient selection algorithm with expected time complexity $O(n+\alpha)$ is developed. The randomized selection algorithm is simpler than to the deterministic one, and is likely to beat the deterministic algorithm in practice. Details are left for the full paper.

The selection algorithm presented here can be used to improve the complexity bounds for resilient k-d trees developed by Gieseke et al. [GMV10]. There, a deterministic resilient algorithm for constructing a k-d tree with $O(n \log^2 n + \delta^2)$ time complexity is shown. This can be improved to $O(n \log n)$ by using the deterministic resilient selection algorithm developed here. Details are left for the full paper.

The problem of sorting in the FRAM model is also revisited, as an application of the resilient selection algorithm. Finocchi et al. [FGI09a], already developed a resilient Mergesort algorithm, sorting an array of size n in $O(n \log n + \alpha\delta)$ time, where the uncorrupted subset of the array is guaranteed to be sorted. They also proved that this bound is tight. A new in-place randomized sorting algorithm which resembles Quicksort and runs in $O(n \log n + \alpha\delta)$ expected time is presented. This sorting algorithm uses the randomized selection algorithm as a black box. Details are left for the full paper.

In the (non-faulty) RAM model the recursion stack needs to reliably store the local variables, as well as the frame pointer and the program counter. Corruptions of this data can cause the algorithm to behave unexpectedly, and in general the recursion stack cannot fit in reliable memory. Some new techniques for implementing a specific recursion stack which suffices for solving the selection problem are developed in Section 4, and are used to develop the resilient deterministic selection algorithm presented in Section 3. It is likely that these techniques can be used to help implement recursive algorithms for other problems in the FRAM model, as these techniques are somewhat general and can be used due to the following four points: (i) *Easily Inverted Size Function*: When performing a recursive call, the function which determines the size of the input to the recursive call is easily inverted, while needing only $O(1)$ bits to maintain the data needed to perform the inversion. (ii) *Bounded Depth*: The depth of the recursion is bounded by $O(\log n)$ and so using $O(1)$ bits per level can fit in reliable memory. (iii) *Verification*: A linear verification procedure is used such that once a recursive call is returned, if the procedure accepts, then the algorithm may proceed even if some errors did occur in the recursive call. (iv) *Amortization*: If the verification procedure fails, then the number of errors which caused the failure is linear in the amount of time spent on the recursive call. This allows to amortize the cost of work on the number of errors, up to a multiplicative factor.

The only previous work done in the FRAM model for non-tail recursion was done by Caminiti et al. [CFFS11] where they developed a recursive algorithm for solving dynamic programming. However, the recursion inherited in the problem of dynamic programming is simpler compared to the recursion treated in the selection problem, due to the structural behavior of the dynamic programming table (the recursions depend on positioning within the table, and not on the actual data). Moreover, their solution only works with high probability (due to using fingerprints for the verification procedure).

2 Preliminaries

2.1 Definitions

Let X be an array of size n of elements taken from a totally ordered set. Let X^0 denote the state of X at the beginning of the execution of an algorithm A executed on X. Let $\alpha \leq \delta$ be the number of corruptions that occurred during such execution.

Definition 1. *Let X be an array and let e be an element. The* rank *of e in X is defined as $rank_X(e) = |\{i : X[i] \leq e\}|$. The α-rank of k in X is defined as $\alpha\text{-}rank_X(k) = \{e : rank_X(e) \in [k - \alpha, k + \alpha]\}$.*

Notice that the α-rank of k in X is an interval containing the elements whose rank in X is not smaller than $k - \alpha$ and not larger than $k + \alpha$. In particular, if $\alpha \geq n$, this interval is equal to $[-\infty, \infty]$. Moreover, if $\alpha = 0$, then this interval is equal to the k-order statistic, thus coincides with the non-faulty definition.

Definition 2. *A resilient k-selection algorithm is an algorithm that is given an array X of size n and an integer k, and returns an element $e \in \alpha\text{-}rank_{X^0}(k)$, where $\alpha \leq \delta$ is the number of faults that occurred during the execution of the algorithm.*

2.2 Basic Procedures

Lemma 1. *There exists a* resilient ranking procedure *with time complexity $O(n)$, that is given an array X of size n and an element e, and returns an integer k such that $e \in \alpha\text{-}rank_{X^0}(k)$.*

Proof. A resilient ranking procedure can be implemented by scanning X while counting the number of elements smaller or equal to e, denoted by k. If $\alpha = 0$, then $k = rank_X(e)$. If $\alpha > 0$, then $e \in \alpha\text{-}rank_{X^0}(k)$, because each corruption can change at most one memory cell, changing the rank of e in X by at most 1. □

Lemma 2. *There exists a* resilient partition procedure *with time complexity $O(n)$ and space complexity $O(1)$, that is given an array X of size n and an element e, and reorders X such that the uncorrupted elements smaller (larger) than e are placed before (after) e, and returns an element k such that $e \in \alpha\text{-}rank_{X^0}(k)$.*

Proof. A resilient partition procedure can be implemented by scanning X while counting the number of elements smaller or equal to e, denoted by k, such that whenever an element smaller than e is encountered it is swapped with the element at position $k + 1$. □

Notice that both procedures compute an integer k such that $e \in \alpha\text{-rank}_{X^0}(k)$. Let $rank^c_X(e)$ denote the value k computed by either procedure, such that whenever the notation $rank^c_X(e)$ will be used, it will be understood from the context which procedure is used. Notice that if $\alpha = 0$, then $rank^c_X(e) = rank_X(e)$.

3 Deterministic Resilient Selection Algorithm

The following deterministic resilient selection algorithm is similar in nature to the non-resilient algorithm by Blum, Floyd, Pratt, Rivest, and Tarjan [BFP+73], but several major modifications are introduced in order to make it resilient. The algorithm is presented in a recursive form, but the recursion is implemented in a very specific way, as explained in Section 4.

Generally, a recursive computation can be thought of as a traversal on a recursion tree T, where the computation begins at the root. Each internal node $u \in T$ performs several recursive calls, which can be partitioned into two types: the *first* type and the *second* type. Each node performs at least one call of each type, and the calls may be interleaved. The idea is for each node u, to locate the k^{th}_u smallest element in the array X_u of size n_u. However, due to corruptions, this cannot be guaranteed, therefore a weaker guarantee is used, as explained later.

3.1 Algorithm Description

The root of the recursion tree is a call to *Determinstic-Select(X, k, −∞, ∞)*. The computation of an inner node u has two phases.

First Phase. The goal of the first phase is to find a *good* pivot, specifically, a pivot whose rank is in the range $[f_u, n_u - f_u]$, where $f_u = \lfloor \frac{3n_u}{10} \rfloor - \lfloor \frac{n_u}{11} \rfloor - 6.$[2] Finding a pivot is done by computing the median of each group of five consecutive elements in X, followed by a recursive call of the first type, to compute the median of these medians. The process is repeated until a *good* pivot is found.

Second Phase. The goal of the second phase is to find a *good* element. Specifically, an element whose rank is in $[k_u \pm n_v]$ where v is a second type child of u. This will be shown to be sufficient[3]. This is done by making a recursive call of the second type, which considers only the relevant sub-array with the updated

---

[2] The exact choice of f_u (which is a function of n_u, the size of the node u) relates to the recursion implementation as explained in Section 4. The idea is to always partition the array at a predetermined ratio, in order to provide more structure to the recursion, and this is what allows for the recursion size function to be easily invertible, as mentioned in Section 1. Notice that the $\lfloor \frac{n_u}{11} \rfloor$ could be picked to be $\lfloor \epsilon \cdot n_u \rfloor$ for any constant $\epsilon < \frac{1}{10}$, because this is needed for the running time of the algorithm, as explained in the proof of Theorem 2.

[3] The exact choice of $[k_u \pm n_v]$ relates to the proof by induction for the correctness of the algorithm. The idea is that as long as less then n_v corruptions occurred during the computation of v, the rank of the element located by v is guaranteed, by induction, to be in these bounds.

·

order statistic. Notice that, unlike the non-faulty selection algorithm, here the appropriate sub-array might be padded with more elements, so that the size of the sub-array is $n_u - f_u$. This is important for the recursion implementation, as explained in Section 4. If the returned value from the recursive call is not in the accepted range, the entire computation of the node repeats, starting from the first phase. Once a *good* element is found, it is returned to the caller.

Algorithm 1. Deterministic-Select(X, n, k, lb, ub)

---

1 # The algorithm uses the recursion implementation from Lemma 3
2 **repeat**
3 　# Let f denote $\lfloor \frac{3n}{10} \rfloor - \lfloor \frac{n}{11} \rfloor - 6$
4 　**begin** First Phase
5 　　**repeat**
6 　　　$X_m \leftarrow []$
7 　　　**for** $i \in [1..\lceil n/5 \rceil]$ **do**
8 　　　　$X_m[i] \leftarrow$ median of $X[5i, 5i + min(4, n - 5i)]$
9 　　　$x_p \leftarrow$ Deterministic-Select(X_m, $\lceil |X_m|/2 \rceil$, lb, ub)
10 　　　partition X around x_p # using the algorithm from Lemma 2
11 　　　# Let p denote $rank^c_X(x_p)$
12 　　**until** $p \in [f, n - f]$
13 　**begin** Second Phase
14 　　**if** $p = k$ **then return** $e = min(max(x_p, lb), ub)$
15 　　**else if** $p > k$ **then** $e \leftarrow$ Deterministic-Select($X[1, n - f]$, k, lb, x_p)
16 　　**else if** $p < k$ **then** $e \leftarrow$ Deterministic-Select($X[f, n]$, $k - f$, x_p, ub)
17 **until** $rank^c_X(e) \in [k \pm n_v]$ # v is a second type child of the node
18 **return** $e = min(max(e, lb), ub)$

Let α_u be the number of corruptions that occurred in u's sub-tree. Each node uses two boundary values lb_u and ub_u which are used similarly to the bounds used in the randomized resilient algorithm.

The recursive calls are made with the parameters X_u, n_u, k_u, lb_u, ub_u, and each recursive call returns an element x. In Section 4, a recursion implementation with the following properties is described.

Lemma 3. *There exists a recursion implementation for the resilient determin-istic selection algorithm with the following properties:*

1. *The position of X_u, n_u, the return value, and program counter are reliable.*[4]
2. *If $\alpha_u \leq n_u$, then lb_u, ub_u, k_u are reliable.*[5]
3. *The time overhead induced by the implementation is $O(n_u)$ per call.*

---

[4] This means that these variables are correct, as long as no more than δ faults occurred.
[5] This means that these variables are correct, as long as no more than n_u faults occurred.

The proof of the Lemma is given in Section 4.

3.2   Analysis

Let u be a node. Let $V = (v_1, \ldots, v_{|V|})$ be u's children. v_1 is always a first type node, and $v_{|V|}$ is always a second type node. Every second type child, except $v_{|V|}$, is followed by a first type child, therefore there cannot be two adjacent second type children. Let α_u denote the number of corruptions that occur in u's sub-tree and let α_u^{local} denote the number of corruptions that occur only in u's data. Let $\alpha_u^{v_i}$ denote the number of corruptions that occur in u's data between the execution of v_i and the execution of v_{i+1} (or until u finishes its computation, if v_i is the last child of u) and let α_v^0 denote the number of corruptions that occur in u's data before the execution of v_1. It follows that, $\alpha_u = \alpha_u^{local} + \sum_{v=0}^{|V|} \alpha_{v_i} = \sum_{v=1}^{|V|} (\alpha_u^{v_i} + \alpha_{v_i})$. Let X_u^0 denote the state of X_u at the beginning of u's computation. Let $X_u^{v_i}$ denote the state of X_u at the moment of the call to v_i.

The following Lemmas are used to prove the correctness and the running time of *Deterministic-Select* in Thm. 2. The proofs of the Lemmas are left for the full paper.

Lemma 4. *If* $\alpha_u \leq n_u$, *then* $e_u \in \alpha_u\text{-}rank_{X_u^0}(k_u)$.

Lemma 5. *Let* $w = v_i$ *be a first type child of* u. *If* v_{i+1} *is not a second type node, then* $\alpha_u^w + \alpha_w \geq \Omega(n_u)$.

Lemma 6. *Let* $w = v_i$ *be a second type child of* u. *If* w *is not the last child of* u, *then* $\alpha_u^w + \alpha_w \geq \Omega(n_u)$.

Theorem 2. Deterministic-Select *is a deterministic resilient selection algorithm with time complexity* $O(n + \alpha)$.

Proof. First, *Deterministic-Select* is shown to be resilient. Let u be the root of the recursion tree, T. If $\delta \leq n = n_u$, then by Lemma 4, $e \in \alpha\text{-}rank_{X^0}(k)$, as needed. Otherwise, if $\delta \geq n$, then there are two cases to consider. If $\alpha \leq n$, then by Lemma 4, $e \in \alpha\text{-}rank_{X^0}(k)$, as before. Otherwise, if $\alpha \geq n$, then by definition, $[-\infty, \infty] = n\text{-}rank_{X^0}(k) = \alpha\text{-}rank_{X^0}(k)$. Therefore, for any element e, $e \in \alpha\text{-}rank_{X^0}(k)$. In particular, the element returned is correct.

With regard to the time complexity, consider a non-faulty execution (i.e., $\alpha = 0$). The time complexity $T(n) = T(\lceil n/5 \rceil) + T(\lceil 7n/10 \rceil + \lceil n/11 \rceil + 6) + O(n) = O(n)$ follows, because $\lceil n/5 \rceil + \lceil 7n/10 \rceil + \lceil n/11 \rceil < n$.

If $\alpha > 0$, then there might be some repetitions. Lemma 5 and Lemma 6 show that enough corruptions can be charged for the time spent in those repetitions. In particular, the $\Omega(n_u)$ corruptions that cause a first type child repetition pay for the $O(n_u)$ computation time of the child, and the $\Omega(n_u)$ corruptions that cause a second type child repetition pay for the $O(n_u)$ computation time of the child, and for the $O(n_u)$ computation time of the first type child that precedes it. In both cases there is $O(1)$ amortized cost per corruption. Therefore, the overall time complexity is $O(n + \alpha)$. $\qquad\square$

Theorem 3. *There exists a deterministic resilient selection algorithm with time complexity $O(n)$.*

The idea is to maintain a counter, which is a lower bound on the number of corruptions that occurred. The full proof is left for the full paper.

4 Recursion Implementation

In this section, an abstract recursion stack for *Deterministic-Select* is developed. The data structures used by this abstract stack are described, followed by the implementation of the operations on it. This leads to the proof of Lemma 3 at the end of this section.

4.1 Data Structures

Two stacks, one reliable and the other one faulty, together with a constant number of reliable memory cells, are used to implement the recursion for the algorithm *Deterministic-Select*. An execution path in the recursion tree, T, starts from the root and ends at the current node. In each stack, the entire execution path is stored in a contiguous region in memory, where the root is at the beginning, and the current node is at the end.

Reliable Stack. The reliable stack stores only 9 bits of information per node. The height of T is $O(\log n)$, therefore it can be stored in a constant number of reliable memory cells. For each inner node $u \in T$, the reliable stack stores 1 bit to distinguish between a first type child and a second type child. Let ρ_x^y denote the remainder of the division of x by y. For a node of the first type, $\rho_{n_u}^5$ is stored. For a node of the second type, $\rho_{n_u}^{10/3}$ and $\rho_{n_u}^{11}$ are stored. Notice that the $O(1)$ reliable memory cells are used down to the bit level.

Faulty Stack. The faulty stack stores $O(n_u)$ words of information per node. For each node $u \in T$, the faulty stack stores the elements of X_u, as well as k_u, lb_u, and ub_u. The elements of X_u are stored using 1 copy per element, while k_u, lb_u, and ub_u are stored using $2n_u + 1$ copies per variable.

Global Variables. Each one of the following global variables is stored using a reliable memory cell: The current array size, the reliable stack's frame pointer, the faulty stack's frame pointer, the program counter, and the return value.

Notice that at a given moment in an execution only one value per each global variable needs to be stored.

4.2 Operations

Two operations are implemented by the recursion implementation. A push operation corresponds to a recursive call, and a pop operation corresponds to returning from a recursive call.

Push. When a node u calls its child v, the following is done. The information of whether v is a first type child or a second type child of u is written to the reliable stack, as well as the relevant remainders (i.e., $\rho_{n_u}^5$ or $\rho_{n_u}^{10/3}$ and $\rho_{n_u}^{11}$), and the reliable stack's frame pointer is incremented by 9 bits. Then, the relevant sub-array is pushed to the faulty stack, followed by the values lb_v, ub_v, and k_v. If v is a first type child, then n_v is updated to $\lceil n_u/5 \rceil$. If v is a second type child, then n_v is updated to $n_u - f_u$. The faulty stack's frame pointer is updated accordingly, and the program counter is set to line 1. Then, the computation continues to v.

Pop. When v finishes its computation, the following is done. First, the reliable stack's frame pointer is decremented by 9 bits, and the information of whether v is a first type or a second type child of u is read, as well as the remainder (i.e., $\rho_{n_u}^5$ or $\rho_{n_u}^{10/3}$ and $\rho_{n_u}^{11}$).

If v is a first type child, then n_u is updated to $5(n_v - 1) + \rho_{n_u}^5$. If v is a second type child, then n_u is updated to $(110/87) \cdot (n_v - \rho_{n_u}^{10/3}/(10/3) + \rho_{n_u}^{11}/11 - 6)$. Notice that this function is the inverse function of $n_u - f_u$, which is the function used to update n when calling a second type child, as explained before. The faulty stack's frame pointer is decremented by $n_u + 3(2n_u + 1)$ words.

The $2n_u + 1$ copies of lb_u, ub_u, and k_u are read, and the computed majority of their copies are stored in reliable memory and used as the values for lb_u, ub_u, and k_u. Then, the computation returns to u, either to line 8 or to line 18, depending on the type of u.

4.3   Proof of Lemma 3

Proof. The frame pointers, the return value, and the program counter were shown to be reliable, as well as the location of the array X_u and its size n_u. lb_u, ub_u, and k_u are stored using $2n_u + 1$ copies each, therefore, if $\alpha_u \leq n_u$, then these parameters are reliable. The time overhead induced by the frame pointers, return value, program counter, location of the array X_u and its size n_u is a constant. The time overhead induced by lb_u, ub_u, and k_u is $O(n_u)$. Therefore, the time overhead of the recursive implementation is $O(n_u)$.   □

References

[Bau05]   Baumann, R.C.: Radiation-induced soft errors in advanced semiconductor technologies. IEEE Transactions on Device and Materials Reliability 5(3), 305–316 (2005)

[BFF+07]   Brodal, G.S., Fagerberg, R., Finocchi, I., Grandoni, F., Italiano, G.F., Jørgensen, A.G., Moruz, G., Mølhave, T.: Optimal Resilient Dynamic Dictionaries. In: Arge, L., Hoffmann, M., Welzl, E. (eds.) ESA 2007. LNCS, vol. 4698, pp. 347–358. Springer, Heidelberg (2007)

[BFP+73]   Blum, M., Floyd, R.W., Pratt, V.R., Rivest, R.L., Tarjan, R.E.: Time bounds for selection. J. Comput. Syst. Sci. 7(4), 448–461 (1973)

[BJM09]   Brodal, G.S., Jørgensen, A.G., Mølhave, T.: Fault Tolerant External Memory Algorithms. In: Dehne, F., Gavrilova, M., Sack, J.-R., Tóth, C.D. (eds.) WADS 2009. LNCS, vol. 5664, pp. 411–422. Springer, Heidelberg (2009)

[BJMM09]   Brodal, G.S., Jørgensen, A.G., Moruz, G., Mølhave, T.: Counting in the Presence of Memory Faults. In: Dong, Y., Du, D.-Z., Ibarra, O. (eds.) ISAAC 2009. LNCS, vol. 5878, pp. 842–851. Springer, Heidelberg (2009)

[CDK11]   Christiano, P., Demaine, E.D., Kishore, S.: Lossless Fault-Tolerant Data Structures with Additive Overhead. In: Dehne, F., Iacono, J., Sack, J.-R. (eds.) WADS 2011. LNCS, vol. 6844, pp. 243–254. Springer, Heidelberg (2011)

[CFFS11]   Caminiti, S., Finocchi, I., Fusco, E.G., Silvestri, F.: Dynamic programming in faulty memory hierarchies (cache-obliviously). In: Proceedings of FSTTCS, pp. 433–444 (2011)

[CLRS09]   Cormen, T.H., Leiserson, C.E., Rivest, R.L., Stein, C.: Introduction to Algorithms, 3rd edn. The MIT Press (2009)

[FGI09a]   Finocchi, I., Grandoni, F., Italiano, G.F.: Optimal resilient sorting and searching in the presence of memory faults. Theor. Comput. Sci. 410(44), 4457–4470 (2009)

[FGI09b]   Finocchi, I., Grandoni, F., Italiano, G.F.: Resilient dictionaries. ACM Transactions on Algorithms 6(1) (2009)

[FI04]   Finocchi, I., Italiano, G.F.: Sorting and searching in the presence of memory faults (without redundancy). In: Proceedings of STOC, pp. 101–110 (2004)

[GMV10]   Gieseke, F., Moruz, G., Vahrenhold, J.: Resilient k-d trees: K-means in space revisited. In: Proceedings of ICDM, pp. 815–820 (2010)

[JMM07]   Jørgensen, A.G., Moruz, G., Mølhave, T.: Priority Queues Resilient to Memory Faults. In: Dehne, F., Sack, J.-R., Zeh, N. (eds.) WADS 2007. LNCS, vol. 4619, pp. 127–138. Springer, Heidelberg (2007)

[LHSC10]   Li, X., Huang, M.C., Shen, K., Chu, L.: A realistic evaluation of memory hardware errors and software system susceptibility. In: Proceedings of USENIX, p. 6 (2010)

[MW79]   May, T.C., Woods, M.H.: Alpha-particle-induced soft errors in dynamic memories. IEEE Transactions on Electron Devices 26(1), 2–9 (1979)

[Sem04]   Tezzaron Semiconductor. Soft errors in electronic memory - a white paper (2004),
           http://www.tezzaron.com/about/papers/soft_errors_1_1_secure.pdf

An Improved Algorithm for Static 3D Dominance Reporting in the Pointer Machine

Christos Makris and Konstantinos Tsakalidis

Computer Engineering and Informatics Department,
University of Patras, 26500 Patras, Greece
{makri,tsakalid}@ceid.upatras.gr

Abstract. We present an efficient algorithm for the pointer machine model that preprocesses a set of n three-dimensional points in $O(n \log n)$ worst case time to construct an $O(n)$ space data structure that supports three-dimensional dominance reporting queries in $O(\log n + t)$ worst case time, when t points are reported. Previous results achieved either $O(n^2)$ worst case or $O(n \log n)$ expected preprocessing time. The novelty of our approach is that we employ persistent data structures and exploit geometric observations of previous works, in order to achieve a drastic reduction in the worst case preprocessing time.

Keywords: computational geometry, dominance reporting, persistent data structures, pointer machine.

1 Introduction

We study static dominance reporting in three dimensions for the pointer machine model of computation. Dominance reporting is an important special case of the orthogonal range reporting problem [2,3]. Let P be a set of n points in $\mathcal{R}^3$. A point $p=(p_x, p_y, p_z)$ *dominates* another point $q=(q_x, q_y, q_z)$, if $p_x > q_x, p_y > q_y$ and $p_z > q_z$ hold. A *three-dimensional (3d) dominance reporting query* reports all the points that dominate a given query point. We study the problem of efficiently constructing a static data structure that supports 3d dominance reporting queries, using minimal space.

Previous Work. The first static pointer-based data structures for 3d dominance reporting were presented by Chazelle and Edelsbrunner [8]. They present two linear space data structures with $O(\log n(1+t))$ and $O(\log^2 n + t)$ worst case query time when t points are reported, and with $O(n^2)$ and $O(n \log^2 n)$ worst case preprocessing time, respectively. Notice that by employing the algorithm of Nekrich [12] for 3d layers of maxima, the worst case preprocessing time of the first data structure is improved to $O(n \log n)$. Makris and Tsakalidis [11] improved the query time to $O(\log n \log \log n + t)$ with a linear space data structure that is preprocessed in $O(n \log^2 n)$ worst case time. Afshani [1] further improved the query time to optimal $O(\log n + t)$ with a linear space data structure. However, the preprocessing algorithm uses shallow cuttings and

K.-M. Chao, T.-s. Hsu, and D.-T. Lee (Eds.): ISAAC 2012, LNCS 7676, pp. 568–577, 2012.

Table 1. Linear space pointer-based data structures for 3d dominance reporting with input size n and output size t. [†] Expected bound.

	Space	Query	Preprocessing
[8]	n	$\log n(1+t)$	$n \log n$
[8]	n	$\log^2 n + t$	$n \log^2 n$
[11]	n	$\log n \log \log n + t$	$n \log^2 n$
[1]	n	$\log n + t$	polynomial(n)
[1]	n	$\log n + t$	$n \log n$[†]
New	n	$\log n + t$	$n \log n$

hence it needs $O(n \log n)$ expected or polynomial worst case time. We should also note that Saxena [14] presents a simplified structure for the RAM model with $O(\log n + t)$ and $O(n \log n)$ worst case query and preprocessing time, respectively, however using $O(n \log n)$ space. Table 1 summarizes the results.

Our Results. In Section 3 we present a static data structure for the pointer machine model that supports 3d dominance reporting queries in $O(\log n + t)$ worst case time and is preprocessed in $O(n \log n)$ worst case time, using $O(n)$ space. The preprocessing time improves the previous deterministic polynomial time and randomized $O(n \log n)$ time preprocessing algorithms. The result can be considered of significance, since efficient preprocessing time is important when the static data structure is employed as a subroutine for solving the off-line version of the problem in three and higher dimensional spaces. See the relevant discussion in the conclusion of the paper.

In order to achieve the result, we follow the approach of Chazelle and Edelsbrunner [8], who in general compute the layers of maxima of the pointset and preprocess each layer in a data structure that supports reporting the points of the layer that dominate a given query point. The implementation of the data structure for each layer follows in principle the hive graph approach [7] that reduces the problem to point location queries on isothetic planar subdivisions. Nonetheless, we replace this approach with an implementation based on simple partially persistent data structures that moreover allows for simultaneous point location queries to a logarithmic number of layers, by use of colored data structures. Finally, we modify the filtering search approach of [8] to support 3d dominance reporting queries efficiently on a set of general points.

2 Preliminaries

3d Layers of Maxima. Let P be a set of n points in $\mathcal{R}^3$. A point is called *maximal* in P, if it is not dominated by any other point in P. Let *layer* P_1 denote the set of maximal points in P. Layer P_i is defined as the set of maximal points among the points in the set $P \setminus \cup_{j=1}^{i-1} P_j$, where the layers P_1 to P_{i-1} have been iteratively removed. The *three-dimensional layers of maxima problem* asks to assign integer i to every point in P that belongs to layer P_i. The $O(n)$ space algorithm of Nekrich [12] solves the problem in the pointer machine in $O(n \log n)$ worst case time.

3d Maxima Dominance Reporting. Given a set of n three-dimensional maximal points, a linear space data structure that supports the reporting of the t maximal points that dominate a given query point in $O(\log n + t)$ worst case time, can be constructed in $O(n \log n)$ worst case time [8, Lemma 5]. The set of weighted two-dimensional points P, obtained by projecting the n maximal points to the xy-plane and assigning their z-coordinate as weight, satisfies the *appearance property*. Namely, for any two points p and q in P with weights w_p and w_q respectively, when p dominates q then $w_q > w_p$ holds [8].

The data structure consists of an isothetic planar subdivision called *map $M(P)$* and a directed acyclic planar *graph $G(P)$*. The map $M(P)$ associates every point $p = (p_x, p_y) \in P$ of weight w with a horizontal segment ph and a vertical segment pv, for which p is called the *anchor point*. In particular, point $h = (h_x, p_y)$ (resp. $v = (p_x, v_y)$) has the x-coordinate $h_x \leq p_x$ (resp. y-coordinate $v_y \leq p_y$) of the point with maximum x-coordinate (resp. y-coordinate) among the points $p' \in P$ with weight $w' > w$, where $p'_y > p_y$ (resp. $p'_x > p_x$) holds. Refer to Figure 1. The graph $G(P) = (V, E)$ encodes the adjacency relationship of the anchor points in $M(P)$ with respect to the intersections of their segments. In particular, there exists a node $u \in V$ for every anchor point $p(u) \in M(P)$, and there exists a directed edge $(u, u') \in E$ from node u to node u', if and only if $p(u)_x = h_x$ (resp. $p(u)_y = v_y$), where $p(u')h$ (resp. $p(u')v$) is the horizontal (resp. vertical) segment in $M(P)$ with anchor point $p(u')$.

Given a three-dimensional query point $q = (q_x, q_y, q_z)$, a *ray intersection query* on $M(P)$ determines the two-dimensional anchor points whose associated segments intersect the upwards vertical and rightwards horizontal rays that emanate from the xy-projection (q_x, q_y) of the query point toward the non-decreasing y- and x-directions, respectively. A crucial observation is that the appearance property allows for reporting all maximal points that dominate q in time linear in the output size, by accessing the determined anchor points a by non-decreasing y- and x-coordinate respectively, and by traversing $G(P)$ from the corresponding nodes $a(u)$, if the weight of a is more than q_z.

Persistent Data Structures. Ordinary dynamic data structures are *ephemeral*, meaning that updates create a new version of the data structure without maintaining previous versions. A *persistent* data structure remembers all versions as updates are performed to it. The data structure is called *partially persistent*, when the previous versions can only be queried and only the last version can be updated. In this case the versions form a list, called the *version list*. Brodal [4] presents a method that makes any pointer-based ephemeral data structure partially persistent with $O(1)$ worst case time overhead per access step and $O(1)$ worst case time and space overhead per update step, given that any node of the underlying graph has in-degree bounded by a constant.

Sarnak and Tarjan [13] utilize *partially persistent red-black trees* in order to solve the point location problem for planar polygonal subdivisions. In a similar manner, we apply the persistence method [4] on the lazy red-black trees presented by Driscoll et al. [9]. Since they induce $O(1)$ worst case extra rebalancing cost per update operation, we obtain partially persistent red-black trees that store a set of elements from a total order, and support the following operations in $O(\log m)$ worst case time and $O(m)$ space, after m updates: Predecessor(value x, version i) and Successor(value x, version i) that access the element stored in the i-th version of the red-black tree with the largest value smaller or equal to x and the smallest value larger or equal to x, respectively, Insert(element e) and Delete(element e) that inserts and deletes element e to the latest version of the red-black tree, respectively, creating a new version.

Interval trees store a set of n intervals on the one-dimensional axis, and support the operations: Stab(value x) that returns all stored intervals $[l, r]$ such that $l \leq x \leq r$ holds, Insert(interval $[l, r]$) and Delete(interval $[l, r]$) that inserts and deletes interval $[l, r]$ to the interval tree, respectively. By applying the persistence method [4] to binary interval trees, we obtain *partially persistent interval trees* that report the t stabbed intervals at any version in $O(\log m + t)$ worst case time, and support updates to the latest version in $O(\log m)$ worst case time, using $O(m)$ space.

3   3d Dominance Reporting

In this Section we present a pointer-based data structure that preprocesses a set of n three-dimensional points in $O(n \log n)$ worst case time and supports reporting the t points that dominate a given query point in $O(\log n + t)$ worst case time, using $O(n)$ space. The main intuition behind our approach is to redesign the 3d maxima dominance reporting structure of [8] by employing persistent data structures, and to redesign their filtering search approach for 3d dominance reporting when dealing with a general set of points.

3.1   Maximal Input Points

We first design a data structure that stores a set P of n maximal points and supports reporting the t maximal points that dominate a given query point. We follow the approach of [8], namely we compute the map $M(P)$ and the graph $G(P)$. To implement the dominance query, we perform a ray intersection query on $M(P)$ that determines the anchor points within the output. We then visit the respective nodes in $G(P)$ and report the remaining output points. The novelty in our construction is that instead of using a hive graph to implement the map $M(P)$ [8], we preprocess it by employing persistent and sweep like techniques. In particular, to support the upwards ray shooting queries, we use a persistent sorted list S that stores y-coordinates as elements and is implemented as a persistent leaf-oriented red-black tree, whose leaves are threaded as a linked list by non-decreasing y-coordinate [13]. The versions of S are stored in a linked version list.

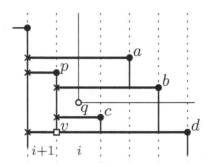

Fig. 1. The input horizontal and vertical segments of the planar subdivision for one maximal layer are shown by thick lines and their anchor points by black dots. A version of the persistent red-black tree for the layer corresponds to the x-range between two consecutive vertical dotted lines. Version i contains elements a, b, c, d. Version $i+1$ is created by the insertion of the input vertical segment pv. In particular, elements b and c are removed (shown by crosses) and element p is inserted. A query for q accesses first element b in the i-th version.

Preprocessing Operation. To preprocess the construction, we sweep the anchor points in $M(P)$ by non-increasing x-coordinate. For every anchor point p with associated vertical segment pv in $M(P)$, we insert p_y to S and remove the y-coordinates that are strictly smaller than p_y and strictly larger than v_y. This operation creates a new version of S that we associate with p_x and store in the version list. Notice that by definition v_y already belongs to S and is in fact the predecessor of p_y in the new version. In this way, the versions of S divide the x-plane into disjoint x-ranges, defined by two consecutive vertical segments of $M(P)$. The x-range of every version is in turn divided into disjoint y-ranges, defined by horizontal segments that span the x-range at the y-coordinates stored in the version. See Figure 1 for an example.

Query Operation. To answer an upward ray shooting query for a query point $q = (q_x, q_y)$ with weight q_z, we perform a binary search for the predecessor version of q_x in the version list of S, and search in the returned version for the successor element e of q_y. If the anchor point of e has z-coordinate more than q_z, we return it to $G(P)$ and access recursively the next larger y-coordinate in the version of S by traversing its leaf linked list. The query returns the version of S with the x-range that contains q_x, and the largest y-coordinate in the y-range of the version's x-range that contains q_y. Thus, the anchor points whose horizontal segments intersect the upward vertical ray that emanates from q are accessed by non-decreasing y-coordinate. To support rightwards ray shooting queries we preprocess and query $M(P)$ symmetrically.

Lemma 1. *There exists an $O(n)$ space data structure for the pointer machine that preprocesses a set of n maximal three-dimensional points in $O(n \log n)$ worst case time and supports reporting the t points that dominate a given query point in $O(\log n + t)$ worst case time.*

Proof. $M(P)$ and $G(P)$ are preprocessed in $O(n \log n)$ time and occupy $O(n)$ space [8, Lemma 5]. All points are inserted to the persistent sorted list S, and each point is inserted at most once. In particular, let p be the point associated with the horizontal segment ph. Point h has the x-coordinate of the point p', where $p'_x < p_x, p'_y > p_y$ and $p'_z > p_z$. When we insert p' to S, p_y is deleted from S and p is never inserted again. Thus the number of versions of S is $O(n)$. Thus, the space of S and its version list is $O(n)$. An update takes $O(\log n)$ time, which yields $O(n \log n)$ total worst case preprocessing time.

Querying the version list and searching in a version of S for element e costs $O(\log n)$ time. The anchor points accessed by the traversal of the linked list occur by non-decreasing y-coordinate, and by the appearance property [8, Lemma 1] they also occur by non-increasing weight. Thus the t' anchor points within the output are accessed in $O(t'+1)$ time, and traversing $G(P)$ takes $O(t+t')$ time. Since $t' \leq t$, the total worst case query time is $O(\log n + t)$.

$\square$

3.2 General Input Points

Although similar to [8, Lemma 5], the data structure of Lemma 1 allows for supporting general 3d dominance reporting queries more efficiently, by using techniques that permit processing of multiple layers simultaneously. In particular, let P be a set of n three-dimensional points. We first compute the layers of maxima of P [12]. To achieve $O(k+t)$ query time for a parameter k, we follow an approach similar to the filtering search approach of [8]. We divide the layers into groups of at most k consecutive layers in a top-down manner, where k is a parameter chosen so that the time complexity for reporting all the t_i output points in the i-th group of layers is $O(k+t_i)$. To perform a query, we start from the topmost group and check if there exists an output point in any of its layers. If there exists no such point the search stops, otherwise it reports all the output points in the current group, and continues to the next group. Since the $O(k)$ cost in each of the queried groups, except from the last one, can be charged to the output, as each maximal layer has at least one output point, the total worst case query time cost is $O(k+t)$.

It remains to show how to perform a 3d dominance query simultaneously to all the at most k layers of a group. To support upwards ray intersection queries, we design a data structure that supports interval stabbing queries on a set of vertical colored intervals, and combine it with our previous structure for 3d maxima dominance reporting. In particular, we color the points of each layer with the same color, such that every layer in the group has a different color. We use a persistent interval tree T that stores colored vertical intervals, and we preprocess the points of every color i to a persistent red-black tree S_i as described before. We model every two consecutive y-coordinates in every version of S_i as an interval of color i in T. In particular, we implement T as a static binary base tree built on the y-coordinates of all points in the group. We associate every internal node with the median y-coordinate of the y-coordinates stored in its subtree. An interval $[\ell, r]$ is associated with the highest node u whose y-coordinate stabs $[\ell, r]$.

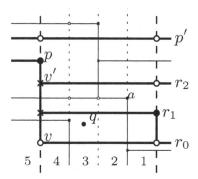

Fig. 2. The input horizontal and vertical segments of the planar subdivisions for two maximal layers are shown; the background layer with thin lines and its anchor points with small squares, and the foreground layer with thick lines and its anchor points with black dots. A version of the persistent interval tree corresponds to the x-range between two consecutive vertical lines. Versions $1, 2, 3, 4$ contain the vertical intervals $[r_{2y}, p'_y], [r_{1y}, r_{2y}], [r_{0y}, r_{1y}]$ with foreground color. Their upper endpoints are shown with white dots. Version 5 is created by the insertion of the input vertical segment pv with foreground color to version 4. In particular, intervals $[r_{1y}, r_{2y}], [r_{0y}, r_{1y}]$ are removed (their upper endpoints shown by crosses) and interval $[r_{2y}, p'_y]$ is replaced by intervals $[p_y, p'_y]$ and $[v_y, p_y]$ with foreground color.

Node u stores respectively the lower and upper endpoint ℓ and r of its associated intervals in two sorted list L_u and R_u, which are implemented as persistent red-black trees. Moreover, for every interval $[\ell, r]$ of color i stored in T, we maintain a pointer $\vec{r}$ from the occurence of the upper endpoint r at a particular version j of T to the occurence of the y-coordinate r_y at the version of S_i that in a temporal sense is the latest version that precedes or equals j. The structure T and each one of the structures S_i separately have indegree bounded by a constant, and thus they can be rendered partially persistent [4]. We store the versions of T in a linked version list.

Preprocessing Operation. To preprocess the construction, we sweep all the points in the group by non-increasing x-coordinate. For every anchor point p of color i associated with the vertical segment pv in $M(P)$, we update S_i as described for Lemma 1. Then, for every removed y-coordinate from S_i that corresponds to anchor point r, we stab T with r_y to remove from T the interval with color i that has r as its upper endpoint. We perform a stabbing query to T with p_y, remove the stabbed interval $[v'_y, p'_y]$ with color i from T, and insert the vertical intervals $[p_y, p'_y]$ and $[v_y, p_y]$ with color i to T. These update operations create a new version of S_i and of T respectively, that we associate with p_x and store to the corresponding version lists. We set a new pointer $\vec{p}$ from the occurence of upper endpoint p_y in the new version of T to the inserted occurrence of y-coordinate p_y in the new version of S_i. Due to the workings of the persistence method [4], the new version of S_i may contain new copies of some elements p'_y, besides element p_y. To update the pointer $\overrightarrow{p}'$ in the new version of T to correctly point to the new copy, during the sweep we maintain for every anchor point p a temporary *indirection pointer* to the position of pointer $\overrightarrow{p}$ in the latest version of T. The indirection pointer is traversed, whenever a new copy of element p_y is created in S_i. It is created when the sweep algorithm accesses anchor point p for the first time, and it is removed when element p_y is removed from S_i.

Geometrically the correctness of the above procedure is derived from the fact that the vertical intervals of color i in every version of T divide the y-range of the version into disjoint y-ranges. In particular, the removed elements r from S_i correspond to the upper endpoints of the vertical intervals of color i in the old version of T that stab the y-range $[v_y, p_y]$. In fact, v_y' is equal to the upper endpoint of the highest removed vertical interval, and v_y is equal to the lower endpoint of the lowest removed vertical interval. We ensure that the vertical intervals of color i cover the y-range of the new version of T, by replacing the remaining vertical interval $[v_y', p_y']$ with intervals $[p_y, p_y']$ and $[v_y, p_y]$ with color i. See Figure 2 for an example.

Query Operation. To support the upwards ray intersection query for a query point (q_x, q_y, q_z) simultaneously on all layers of a group, we perform a binary search for the predecessor of q_x on the version list of T, and stab the returned version with q_y. For every returned interval $[\ell, r]$ of color i, we traverse the pointer $\overrightarrow{r}$ to the occurrence of r_y in the pointed version of S_i. We report the points in layer i that dominate q, by traversing the linked list of the accessed version of S_i starting from r_y, as described before.

When the preprocessing algorithm creates a new version of T associated with p_x, it also creates a new version of S_i for at most one color i. Since the rest of the persistent red-black trees $S_{i'}$ remain unchanged, the x-coordinates associated with their latest versions are predecessors of p_x for all $i' \neq i$. Therefore the version of T that precedes the query parameter q_x contains the pointers $\overrightarrow{p}$ that point to the versions of S_i that precede q_x for all colors i. Moreover, since the intervals of a particular color in T do not overlap, at most one interval of color i is stabbed, and thus a layer is visited at most once. To support rightwards ray shooting queries we preprocess and query the construction symmetrically.

Theorem 1. *There exists an $O(n)$ space data structure for the pointer machine that preprocesses a set of n three-dimensional points in $O(n \log n)$ worst case time and supports reporting the t points that dominate a given query point in $O(\log n + t)$ worst case time.*

Proof. Computing the layers of maxima, $M(P)$ and $G(P)$ takes $O(n \log n)$ time and $O(n)$ space [12,8]. Let the j-th group of layers P_j contain n_j points, where $j \in [1, \lceil n/k \rceil]$. Since the total number of employed intervals is $O(n_j)$ we get that the size of T is also $O(n_j)$. An update operation to T takes $O(\log n_j)$ worst case time, and incurs $O(1)$ worst case structural changes to its secondary structures. Moreover, an update operation to persistent lazy red-black trees also incurs $O(1)$ structural changes in the worst case. Therefore creating, updating and removing the temporary indirection pointers incurs no extra cost. Since there are $O(n_j)$

versions, the j-th group is preprocessed in $O(n_j \log n_j)$ worst case time, using $O(n_j)$ space [4]. Since a point can participate in only one layer, we get that $\Sigma_{j=1}^{\lceil n/k \rceil} n_j = n$. Thus the persistent interval trees for all groups of layers are preprocessed in $O(n \log n)$ worst case time and occupy $O(n)$ space. Moreover, by Lemma 1 it follows that the persistent sorted lists for all groups of layers take $O(n \log n)$ preprocessing time and $O(n)$ space.

Querying a version of the persistent interval tree for the group of layers P_j takes $O(\log n_j + k')$ time, where $k' \leq k$ is the number of stabbed intervals in the group. Since at most one interval is returned for every color, k' is also the number of layers in P_j that contain at least one output point. By following the pointers from the upper endpoint of the stabbed interval of color i to the version of S_i, and using a similar procedure for the rightward ray intersection query, we locate the t_i output points in the layer of color i in extra $O(t_i+1)$ time. Hence, the overall time for querying a group of layers is $O(\log n_j + k' + t')$, where $t' = \Sigma_{i=1}^{k'} t_i$ is the output points in the group. Since when a point is dominated by a point in group j, it is also dominated by at least one point in each of the other groups j' such that $j' \leq j$, and since we stop at the bottommost group that can produce useful output, it follows that the time to report all t output points is $O(\log n + t)$ time, when parameter $k = \lceil \log n \rceil$. □

4   Conclusion

We have presented a static algorithm for the 3-dimensional dominance reporting problem in the pointer machine that improves upon the preprocessing cost of the previous optimal solution [1]. It would be interesting to exploit this contribution in order to improve upon the static offline four-dimensional variant of the problem in the pointer machine, also known as rectangle enclosure reporting [10,6]. In particular our idea of performing simultaneously dominance reporting in groups of maximal layers, empowered with the fast maximal layers computation algorithm of [12] could possibly exploit techniques described in [6], designed for the rectangle enclosure reporting problem in the RAM model, in order to lead to an optimal solution for the problem in the pointer machine model.

Moreover we plan to investigate the applicability of our technique in the I/O model of computation (either the classical two level, or the cache oblivious model) in order to obtain an algorithm optimal in query, space and preprocessing time complexity. Finally, it should be noted that our technique of replacing point location with persistence-based sweep-like techniques is not directly applicable in the RAM model of computation. There, utilizing the existing partially persistent van Emde Boas trees [5] does not achieve worst case efficient preprocessing cost, since their update time is efficient in the expected and not in the worst case sense.

References

1. Afshani, P.: On Dominance Reporting in 3D. In: Halperin, D., Mehlhorn, K. (eds.) ESA 2008. LNCS, vol. 5193, pp. 41–51. Springer, Heidelberg (2008)

2. Agarwal, P.K.: Range searching. In: Handbook of Discrete and Computational Geometry, pp. 575–598. CRC Press, Inc. (1997)
3. Agarwal, P.K., Erickson, J.: Geometric range searching and its relatives. In: Advances in Discrete and Computational Geometry, pp. 1–56. American Mathematical Society (1999)
4. Brodal, G.S.: Partially persistent data structures of bounded degree with constant update time. Nord. J. Comput. 3(3), 238–255 (1996)
5. Chan, T.M.: Persistent predecessor search and orthogonal point location on the word ram. In: Proceedings of the Twenty-Second Annual ACM-SIAM Symposium on Discrete Algorithms, SODA 2011, pp. 1131–1145. SIAM (2011)
6. Chan, T.M., Larsen, K.G., Patrascu, M.: Orthogonal range searching on the RAM, revisited. In: Hurtado, F., van Kreveld, M.J. (eds.) Symposium on Computational Geometry, pp. 1–10. ACM (2011)
7. Chazelle, B.: Filtering search: A new approach to query-answering. SIAM J. Comput. 15(3), 703–724 (1986)
8. Chazelle, B., Edelsbrunner, H.: Linear space data structures for two types of range search. Discrete & Computational Geometry 2, 113–126 (1987)
9. Driscoll, J.R., Sarnak, N., Sleator, D.D., Tarjan, R.E.: Making data structures persistent. J. Comput. Syst. Sci. 38, 86–124 (1989)
10. Lee, D.-T., Preparata, F.P.: An improved algorithm for the rectangle enclosure problem. J. Algorithms 3(3), 218–224 (1982)
11. Makris, C., Tsakalidis, A.K.: Algorithms for three-dimensional dominance searching in linear space. Inf. Process. Lett. 66(6), 277–283 (1998)
12. Nekrich, Y.: A Fast Algorithm for Three-Dimensional Layers of Maxima Problem. In: Dehne, F., Iacono, J., Sack, J.-R. (eds.) WADS 2011. LNCS, vol. 6844, pp. 607–618. Springer, Heidelberg (2011)
13. Sarnak, N., Tarjan, R.E.: Planar point location using persistent search trees. Commun. ACM 29(7), 669–679 (1986)
14. Saxena, S.: Dominance made simple. Inf. Process. Lett. 109(9), 419–421 (2009)

The Multi-Service Center Problem[*]

Hung-I Yu[1] and Cheng-Chung Li[2]

[1] Institute of Information Science, Academia Sinica,
Nankang, Taipei 115, Taiwan
herbert@iis.sinica.edu.tw
[2] Intel-NTU Connected Context Computing Center,
National Taiwan University, Taipei 106, Taiwan
f92922087@ntu.edu.tw

Abstract. We propose a new type of multiple facilities location problem, called the p-service center problem. In this problem, we are to locate p facilities in the graph, each of which provides distinct service required by all vertices. For each vertex, its p-service distance is the summation of its weighted distances to the p facilities. The objective is to minimize the maximum value among the p-service distances of all vertices.

In this paper, we show that the p-service center problem on a general graph is NP-hard, and propose a polynomial-time approximation algorithm. Moreover, we study the basic case $p = 2$ on paths and trees, and provide linear and near-linear time algorithms.

Keywords: location theory, p-service center, general graphs, paths, trees.

1 Introduction

Network location problems have received much attention from researchers in the fields of transportation and communication over four decades [1,2,3,5,6,8,9]. Traditionally, network location problems consider only identical facilities while determining the optimal location of multiple facilities, and each client will thus be served by its closest facility. For example, the well-known *p-center* problem is to place p identical facilities in the network such that the maximum distance from each client to its closest facility is minimized.

However, while providing large-scale and integrated services, it is possible that each single facility cannot afford to provide all kinds of services due to the complexity and cost. Take the freight traffic of logistics centers as an example. We want to setup several logistics centers in the transportation network to provide many kinds of goods, but various factors make it difficult to store all kinds of goods in each single logistics center, such as places of production, storage cost, market demand, and so on. It is more efficient for each logistics center to store only one or several kinds of goods, and together they still meet all

---

[*] Research supported by the National Science Council, National Taiwan University and Intel Corporation under Grants No. 100-2911-I-002-001, 101-2221-E-005-019, 101-2221-E-005-026, 101-2811-E-005-005, and 101R7501.

K.-M. Chao, T.-s. Hsu, and D.-T. Lee (Eds.): ISAAC 2012, LNCS 7676, pp. 578–587, 2012.

kinds of requirements from everywhere in the network. Obviously, the optimal placement of such distinct but cooperative facilities will be very different from that of identical facilities. We are interested in this new kind of network location problems, and call them the *multi-service location* problems.

In this paper, we define the first problem of this kind, the *p-service center* problem. There are p facilities to be placed in a network of n clients, each of which provides one distinct kind of service. Each client has its own need for each kind of service, and how well the client is served is measured by its *p-service distance*, the total transportation cost from the p facilities to the client. The objective of the p-service center problem is to find a placement of the p facilities minimizing the maximum value among the p-service distances of all clients.

We study the p-service center problem and obtain approximate and exact solutions for different cases. When the underlying graph is a general graph, we show that the p-service center problem for general p is NP-hard, and propose an approximation algorithm of factor p/c, where c is an arbitrary integer constant. This is based on a procedure that can solve the p-service center problem in polynomial time when p is a fixed value. We further consider the case that $p = 2$ and provide efficient solutions on paths and trees. When the underlying graph is a path, we solve this problem in linear time. For a tree, we propose an $O(n \log n)$-time algorithm for the weighted case and a linear-time algorithm for the unweighted case. Due to page limits, the proofs are omitted.

The rest of this paper is organized as follows. Section 2 gives formal definitions and basic properties. In Section 3, we study p-service center problem on general graphs. Then, in Section 4, we discuss the 2-service center problem on paths and trees. Finally, in Section 5, we conclude the paper.

2   Notation and Preliminaries

Let $p \geq 1$ be an integer, and $G = (V, E, W)$ be an undirected connected graph, where V is the vertex set, E is the edge set, and W is an n-by-p matrix of positive weights $[w_{i,j}]_{n \times p}$. Let $V = \{1, 2, \ldots, n\}$ and $m = |E|$. Each vertex $i \in V$ is associated with p positive weights $w_{i,1}, w_{i,2}, \ldots, w_{i,p}$, which correspond to the p elements in the i-th row of W. Each edge $e \in E$ has a nonnegative length and is assumed to be rectifiable. Thus, we will refer to interior points on an edge by their distances (along the edge) from the two nodes of the edge. Let G also denote the set of all points of the graph. For any two points $a, b \in G$, let $P(a, b)$ be the shortest path between a and b, and $d(a, b)$ be the distance of length of $P(a, b)$. Suppose that the matrix of shortest distances between vertices of G is given. A *p-location* is a tuple $(x_1, x_2, \ldots, x_p)$ of p points in G (not necessarily distinct). For an arbitrary p-location $X = (x_1, x_2, \ldots, x_p)$ and a vertex $i \in V$, the *p-service distance* from i to X is defined to be $D(i, X) = \sum_{1 \leq j \leq p} w_{i,j} \times d(i, x_j)$, and the *p-service cost* of X is defined to be $F(X) = \max_{i \in V} D(i, X)$. The *p-service center* problem is to find a p-location X that minimizes $F(X)$. Note that, in the case that $p = 1$, the p-service center problem is equivalent to the p-center problem.

A *p-family* is a tuple $(H_1, H_2, \ldots, H_p)$ of p subgraphs of G (not necessarily distinct). Given a p-location $X = (x_1, x_2, \ldots, x_p)$ and a p-family $(H_1, H_2, \ldots, H_p)$, we write $X \in (H_1, H_2, \ldots, H_p)$ to mean that $x_j \in H_j$ for $1 \le j \le p$, and $X \notin (H_1, H_2, \ldots, H_p)$ if there exists at least one index j such that $x_j \notin H_j$. For a p-family $(H_1, H_2, \ldots, H_p)$, a p-location $X \in (H_1, H_2, \ldots, H_p)$ is called an $(H_1, H_2, \ldots, H_p)$-*optimal solution* if $F(X) = \min_{X' \in (H_1, H_2, \ldots, H_p)} F(X')$. For simplicity, let H^p denote the p-family $(H, H, \ldots, H)$ for any subgraph $H \subseteq G$. A G^p-optimal solution is thus an optimal solution to the p-service center problem.

3   The p-Service Center Problem on a General Graph

In this section, we study the complexity results of the p-service center problem on a general graph. In Subsection 3.1, we first show that this problem is NP-hard for general p. As a procedure required by the approximation algorithm, in Subsection 3.2, we develop a polynomial-time algorithm for solving this problem exactly when p is a constant. Finally, in Subsection 3.3, we give a factor p/c approximation algorithm for general p, where c is an arbitrary integer constant.

3.1   NP-Hardness

We show that finding a p-service center in a general graph is an NP-hard optimization problem by a reduction from 3-CNF-SAT [4]. A 3-CNF-SAT instance $(\mathbf{B}, \mathbf{C})$ consists of a set of M clauses $\mathbf{C} = \{\mathbf{c_1}, \mathbf{c_2}, \ldots, \mathbf{c_M}\}$ and a set of N Boolean variables $\mathbf{B} = \{\mathbf{b_1}, \mathbf{b_2}, \ldots, \mathbf{b_N}\}$. A literal is either a Boolean variable $\mathbf{b_\ell}$ or its negation $\overline{\mathbf{b}}_\ell$. Each clause $\mathbf{c_k}$ is the disjunction of three distinct literals. The instance is satisfiable if there exists a truth assignment to $\mathbf{B}$ such that all clauses in $\mathbf{C}$ are **True** under the assignment. The 3-CNF-SAT problem is to determine whether a given instance is satisfiable, which is known to be NP-complete [4]. Without loss of generality, we assume that for each variable at least one clause contains its literal, and no clause contains both $\mathbf{b_\ell}$ and $\overline{\mathbf{b}}_\ell$ for any ℓ.

Given a 3-CNF-SAT instance $(\mathbf{B}, \mathbf{C})$, we construct a graph $G = (V, E, W)$ as an instance of the p-service center problem, where $p = N$. (In the remaining of this subsection, we replace p by N in order to avoid confusion.) First, we create vertices to represent literals and clauses. For each variable $\mathbf{b_\ell} \in \mathbf{B}$, two *Boolean vertices* $b_\ell^\mathbf{T}$ and $b_\ell^\mathbf{F}$ are created to represent **True** and **False** of $\mathbf{b_\ell}$, respectively. For each clause $\mathbf{c_k} \in \mathbf{C}$, a *clause vertex* c_k is created to represent $\mathbf{c_k}$. Then, we create edges between Boolean vertices and clause vertices. For each clause $\mathbf{c_k} \in \mathbf{C}$, if $\mathbf{c_k}$ contains the literal $\mathbf{b_\ell}$, an edge is introduced to connect c_k and $b_\ell^\mathbf{T}$; on the other hand, if $\mathbf{c_k}$ contains $\overline{\mathbf{b}}_\ell$, c_k and $b_\ell^\mathbf{F}$ are connected. By assumption, each clause vertex connects to three distinct Boolean vertices. Furthermore, we create two types of auxiliary vertices: N Type-1 vertices $a_1^\mathrm{I}, a_2^\mathrm{I}, \cdots, a_N^\mathrm{I}$ and $N+1$ Type-2 vertices $a_0^\mathrm{II}, a_1^\mathrm{II}, \cdots, a_N^\mathrm{II}$. For each Type-1 vertex a_ℓ^I, $1 \le \ell \le N$, two edges are introduced for connecting a_ℓ^I to $b_\ell^\mathbf{T}$ and $b_\ell^\mathbf{F}$. On the other hand, we connect each Type-2 vertex a_ℓ^II to all Boolean vertices, $0 \le \ell \le N$.

The N weights of each vertex are set as follows. For each Boolean vertex $b_\ell^\mathbf{T}$, let $w_{b_\ell^\mathbf{T}, j} = 1$ for $1 \le j \le N$. Similarly, let $w_{b_\ell^\mathbf{F}, j} = 1$ for all j. For each clause vertex c_k, let $w_{c_k, j} = 10/9$ if $\mathbf{c_k}$ contains either $\mathbf{b_j}$ or $\overline{\mathbf{b}}_j$, and $w_{c_k, j} = 1$

otherwise. For each Type-1 vertex a_ℓ^I, $1 \leq \ell \leq N$, let $w_{a_{\ell,j}^I} = 3$ if $j = \ell$ and $w_{a_{\ell,j}^I} = 1$ otherwise; for each Type-2 vertex a_ℓ^{II}, $0 \leq \ell \leq N$, let $w_{a_{\ell,j}^{II}} = 3$ for all j. Finally, the lengths of all edges are set to 1. It is easy to see that the transformation can be done in time polynomial to N and M.

Given an N-location $X = (x_1, x_2, \cdots, x_N) \in G^N$, a point x_j in X is said to be *assigned to* the Boolean variable $\mathbf{b}_\ell$ if x_j is located at either $b_\ell^{\mathbf{T}}$ or $b_\ell^{\mathbf{F}}$. The N-location X is said to be *legally-assigned* if x_j is assigned to $\mathbf{b_j}$ for $1 \leq j \leq N$. In other words, X represents a truth assignment to $\mathbf{B}$. From the setting of auxiliary vertices, the following lemma can be obtained by pigeonhole principle.

Lemma 1. *For any N-location $X = (x_1, x_2, \cdots, x_N) \in G^N$, $F(X) \geq 3N$. If X is not legally-assigned, $F(X) > 3N$.*

Then, the connectivity of clause vertices guarantees the following.

Lemma 2. *The 3-CNF-SAT instance $(\mathbf{B}, \mathbf{C})$ is satisfiable if and only if any G^N-optimal solution X^* has its N-service cost $F(X^*) = 3N$.*

Lemma 2 enables us to decide the satisfiability of $(\mathbf{B}, \mathbf{C})$ by finding an N-service center X^* in G. Moreover, if satisfiable, X^* must be legally-assigned by Lemma 1, and the satisfying truth assignment can be directly obtained from the location of X^*. Therefore, we have the following result.

Theorem 1. *The p-service center problem on a general graph is NP-hard.*

3.2   Exact Algorithm for Fixed p

Given a general graph $G = (V, E, W)$, in this subsection, we propose a polynomial algorithm for solving the p-service center problem on G when p is a constant. Let $e = (i, i')$ be an arbitrary edge in G. A point $x \in e$ is called a *bottleneck point* of e if there exists a vertex k such that $d(k, i) + d(i, x) = d(k, i') + d(i', x)$. By definition, there are at most n bottleneck points on e (including i and i'), which partition e into $O(n)$ segments. Similarly, every edge in G can be partitioned into $O(n)$ segments by its bottleneck points. Let $\mathcal{I}$ denote the set of all segments in G, and $\mathcal{F}$ denote the collection of p-families $\{(I_1, I_2, \ldots, I_p) \mid I_j \in \mathcal{I}$ for $1 \leq j \leq p\}$. The p-service center problem on G can be solved by computing the $(I_1, I_2, \ldots, I_p)$-optimal solution for each p-family $(I_1, I_2, \ldots, I_p) \in \mathcal{F}$ and then choosing a solution with minimum p-service cost as the G^p-optimal solution.

Let $(I_1, I_2, \ldots, I_p)$ be a fixed p-family in $\mathcal{F}$. In the following, we discuss how to compute the $(I_1, I_2, \ldots, I_p)$-optimal solution. For $1 \leq j \leq p$, let q_j and r_j be the endpoints of the segment I_j, and $l_j = d(q_j, r_j)$ be the length of I_j. For ease of description, each I_j is regarded as an interval $[0, l_j]$ on the real line, and the points along I_j are identified by the values of a real number $0 \leq x \leq l_j$, where $x = 0$ represents q_j and $x = l_j$ represents r_j. For any vertex $i \in V$, $d(i, x)$ is a linear function on I_j, since by definition I_j does not contain any bottleneck point in its interior. Let $s_{i,j}$ be the slope of $d(i, x)$ on I_j, where $s_{i,j} = +1$ if $d(i, l_j) > d(i, 0)$ and $s_{i,j} = -1$ otherwise. We then have that $d(i, x) = d(i, 0) + s_{i,j}x$ for $x \in I_j$.

For $1 \leq j \leq p$, let $x_j \in I_j$ be a variable. The tuple $X = (x_1, x_2, \ldots, x_p)$ thus represents a variable p-location in $(I_1, I_2, \ldots, I_p)$. For each $i \in V$, the p-service distance from i to X can be formulated as $D(i, X) = \sum_{1 \leq j \leq p} w_{i,j} \times d(i, x_j) = \sum_{1 \leq j \leq p} s_{i,j} w_{i,j} x_j + \sum_{1 \leq j \leq p} w_{i,j} d(i, 0)$. Together with the variables x_j, we add an auxiliary variable z for simulating $F(X)$, and construct a linear program:

Minimize z

$$\text{Subject to } z \geq \sum_{1 \leq j \leq p} s_{i,j} w_{i,j} x_j + \sum_{1 \leq j \leq p} w_{i,j} d(i, 0), \text{for } i \in V,$$

$$0 \leq x_j \leq l_j, \text{for } 1 \leq j \leq p,$$

$$z \geq 0.$$

$$(1)$$

In this linear program, the constraints ensure that $z \geq D(i, X)$ for $i \in V$, which implies that $z \geq \max_{i \in V} D(i, X) = F(X)$. The objective of minimizing z ensures that $z = F(X)$ for any $X \in (I_1, I_2, \ldots, I_p)$, and then plays the role of finding the solution that minimizes $F(X)$. Therefore, solving this linear program gives us the $(I_1, I_2, \ldots, I_p)$-optimal solution. Since p is a constant, this linear program can be solved in $O(n)$ time by applying Megiddo's linear-time algorithm for constant-variable linear programming [7].

Thus, the G^p-optimal solution can be obtained by constructing and solving the linear programs for all p-families in $\mathcal{F}$ in $O(n|\mathcal{F}|)$ time and then choosing a solution with minimum objective value in $|\mathcal{F}|$ time. Since $|\mathcal{I}| = m \times O(n) = O(mn)$ and $|\mathcal{F}| = |\mathcal{I}|^p$ by definition, we have the following result.

Theorem 2. *The p-service center problem on a general graph can be solved in $O(m^p n^{p+1})$ time when p is a constant.*

3.3   Approximation Algorithm for General p

Using the algorithm developed in Subsection 3.2, we propose an algorithm to find an approximate p-service center in a given graph $G = (V, E, W)$ for general p. Let c be an arbitrary integer constant. Suppose that $p > c$, otherwise the problem can be directly solved by Theorem 2. Furthermore, we assume that p is divisible by c for brevity's sake. Let $p_c = p/c$. We partition the weight matrix W into p_c submatrices $W_0, W_1, \cdots, W_{p_c-1}$, where W_k is an n-by-c matrix corresponding to the $(ck + 1)$st to the $(ck + c)$th columns of W. For each $0 \leq k \leq p_c - 1$, we duplicate the vertices and edges of G, and construct an instance $G_k = (V, E, W_k)$ of the c-service center problem. Since c is a constant, for each k a $(G_k)^c$-optimal solution $X_k = (x_{k,1}, x_{k,2}, \cdots, x_{k,c})$ can be computed by Theorem 2. We then create a p-location X by concatenating $X_0, X_1, \cdots, X_{p_c-1}$, that is, $X = (x_{0,1}, x_{0,2}, \cdots, x_{0,c}, x_{1,1}, x_{1,2}, \cdots, x_{1,c}, \cdots)$. The following lemma can be obtained for the solution X.

Lemma 3. *$F(X) \leq p_c \times F(X^*)$, where X^* is an arbitrary G^p-optimal solution.*

The p_c instances have total $p_c n$ vertices, $p_c m$ edges and pn weights, which can be constructed in $O(p_c m + pn)$ time. Computing optimal solutions for them by Theorem 2 takes $O(p_c m^c n^{c+1})$ time. Therefore, we have the following result.

Theorem 3. *Let $c \geq 1$ be an arbitrary integer constant. There is a (p/c)-approximation algorithm for the p-service center problem on a general graph, which runs in $O((p/c)m^c n^{c+1})$ time.*

4 The 2-Service Center Problem on a Path and a Tree

In this section, we focus on the basic case $p = 2$ of the p-service center problem. In Subsection 4.1, we obtain an $O(n)$-time algorithm for the 2-service center problem on a path. In Subsections 4.2 and 4.3, we consider the problem on a tree, and propose an $O(n \log n)$-time algorithm for a weighted tree and an $O(n)$-time algorithm for an unweighted tree (all weights are equal to 1), respectively.

4.1 Algorithm on a Path

In this subsection, we assume the underlying graph G is a path $P = (V, E, W)$ and propose an $O(n)$-time algorithm, improved from the $O(n^5)$-time result which follows from directly applying Theorem 2.

Without loss of generality, suppose that $E = \{(1, 2), (2, 3), \ldots, (n - 1, n)\}$. For $i \in V$, let $d_i = d(1, i)$. For convenience, the whole path P is regarded as an interval $[0, d_n]$ on the real line, and the points along P are identified by the values of a real number $0 \leq x \leq d_n$, where $x = d_i$ represents the vertex i for $i \in V$. For any vertex $i \in V$, it is easy to see that $d(i, x) = \max\{d_i - x, x - d_i\}$.

Again, we use linear programming to find a P^2-optimal solution. Let x_1 and x_2 be two points in P. For each $i \in V$, the 2-service distance from i to the tuple $X = (x_1, x_2)$ is $D(i, X) = w_{i,1} \times \max\{d_i - x_1, x_1 - d_i\} + w_{i,2} \times \max\{d_i - x_2, x_2 - d_i\}$. To ensure that $z \geq D(i, X)$ for each $i \in V$, we use four constraints $z \geq \pm w_{i,1}(d_i - x_1) \pm w_{i,2}(d_i - x_2)$ in the linear program. Similar to Subsection 3.2, we have that $z \geq \max_{i \in V} D(i, X) = F(X)$. Thus, solving the linear program in $O(n)$ time by [7] gives a P^2-optimal solution, and the theorem follows.

Theorem 4. *The 2-service center problem on a path can be solved in $O(n)$ time.*

4.2 Algorithm on a Tree

In this subsection, we assume the underlying graph G is a tree $T = (V, E, W)$ and propose an $O(n \log n)$-time algorithm for finding a T^2-optimal solution in T. The approach is to perform two mutually dependent prune-and-search procedures on two specified subtrees T_1 and T_2, which iteratively pick two paths in T_1 and T_2, compute the local optimal solution of the paths, and then prune T_1 and T_2.

The algorithm itself is fairly simple, but establishing its theoretical basis is quite lengthy. In the following, we begin from the exploration of the relationship between the local optimal solution of paths and the T^2-optimal solution. Based on the relationship, we describe the algorithm and analyze its time complexity.

H.-I Yu and C.-C. Li

Properties. Let x be an arbitrary point in T. Removing x will break T into several partial trees, called the *open x-branches*, each of which consists of a subtree B of T and a partial edge connecting x and B. The union of x and any open x-branch is called an *x-branch*. For any point $y \neq x$ in T, let $B^-(x, y)$ and $B(x, y)$ denote the open x-branch and x-branch that contain y, respectively.

For any two subgraphs $H_1, H_2 \subseteq T$, we say H_1 *intersects* H_2 if there exists a vertex i such that $i \in H_1$ and $i \in H_2$. For any 2-location $X = (x_1, x_2) \in T^2$, a vertex $i \in V$ is called a *critical vertex* of X if $D(i, X) = F(X)$. Let $CR(X)$ denote the set of critical vertices of X.

Let c^* be the point that minimizes $F((c^*, c^*))$, and C^* denote the 2-location (c^*, c^*). Note that c^* corresponds to the weighted 1-center of the tree $T' = (V, E, W')$, where W' is an n-by-1 matrix $[w'_i = w_{i,1} + w_{i,2}]_{n \times 1}$. The properties of the weighted 1-center ensure that c^* is unique [6] and $CR(C^*)$ intersects at least 2 open c^*-branches [5]. Furthermore, we obtain the following lemma.

Lemma 4. *If $CR(C^*)$ intersects 3 or more open c^*-branches, C^* is the only T^2-optimal solution.*

In the following, suppose that $CR(C^*)$ intersects exactly 2 open c^*-branches. Let $\delta_{\max} = \max_{i \in CR(C^*)} w_{i,1}/w_{i,2}$, and i_1 be an arbitrary vertex in $CR(C^*)$ such that $w_{i_1,1}/w_{i_1,2} = \delta_{\max}$. Let T_1 denote the c^*-branch that contains i_1 and T_2 denote the other c^*-branch that intersects $CR(C^*)$. We can simplify the finding of a T^2-optimal solution by the following lemma.

Lemma 5. *A (T_1, T_2)-optimal solution is also a T^2-optimal solution.*

By Lemma 5, we can concentrate on the finding of a (T_1, T_2)-optimal solution. A (T_1, T_2)-optimal solution $X = (x_1, x_2)$ is called a (T_1, T_2)-*extreme solution* if, for any other (T_1, T_2)-optimal solution $X' = (x'_1, x'_2)$, $P(x_1, x_2) \nsubseteq P(x'_1, x'_2)$. The following lemma shows the uniqueness of the (T_1, T_2)-extreme solution.

Lemma 6. *There exists exactly one (T_1, T_2)-extreme solution.*

Let $X^* = (x_1^*, x_2^*)$ denote the unique (T_1, T_2)-extreme solution. Our objective is to find X^* as the T^2-optimal solution. This can be done by building relationships between X^* and the local optimal solution of specified paths in T. For any leaf $i \in T_1(T_2)$, the path $P(c^*, i)$ is called a T_1-*arm* (T_2-*arm*). Consider an arbitrary T_1-arm P_1 and an arbitrary T_2-arm P_2. A (P_1, P_2)-optimal solution $Y = (y_1, y_2)$ is called a (P_1, P_2)-*extreme solution* if, for any other (P_1, P_2)-optimal solution $Y' = (y'_1, y'_2)$, $P(y_1, y_2) \nsubseteq P(y'_1, y'_2)$. We have the following lemma.

Lemma 7. *There exists exactly one (P_1, P_2)-extreme solution.*

Let $Y^* = (y_1^*, y_2^*)$ denote the unique (P_1, P_2)-extreme solution. If $X^* \in (P_1, P_2)$, we obviously have that $Y^* = X^*$. In the following, by assuming that $X^* \notin (P_1, P_2)$, we develop several lemmas that describe the relationship between X^* and Y^*. Note that the assumption $X^* \notin (P_1, P_2)$ implies that $x_1^* \neq x_2^*$, otherwise $x_1^* = x_2^* = c^*$ makes $X^* \in (P_1, P_2)$.

Let b_1 denote the point closest to x_1^* on P_1 and b_2 denote the point closest to x_2^* on P_2, which are called the *boundary points* of P_1 and P_2, respectively. Let $P_1^{in} = P(c^*, b_1) - \{x_1^*\}$ and $P_2^{in} = P(c^*, b_2) - \{x_2^*\}$. For $j = 1, 2$, if $x_j^* \notin P_j$, b_j is a vertex and belongs to P_j^{in} by definition. If otherwise $x_j^* \in P_j$, $b_j = x_j^*$ and hence $b_j \notin P_j^{in}$. The following lemma tells us that the (P_1, P_2)-extreme solution Y^* always appears in between the boundary points and includes a boundary point.

Lemma 8. *If* $X^* \notin (P_1, P_2)$, *either* $Y^* \in (P_1^{in}, \{b_2\})$ *or* $Y^* \in (\{b_1\}, P_2^{in})$.

For any 2-location $Y \in (T_1, T_2)$, let $CR_1(Y) = CR(Y) \bigcap T_1$ and $CR_2(Y) = CR(Y) \bigcap T_2$. The following two lemmas summarize the above discussion and play the key role in the finding of X^*.

Lemma 9. *If* $X^* \notin (P_1, P_2)$, *there always exists an index* $j \in \{1, 2\}$ *for* Y^*, *such that* $y_j^* = b_j$ *satisfies the following conditions:*

(A) There is an open y_j^*-branch B^- that does not intersect P_j,
(B) $x_j^* \in B^-$,
(C) $CR_j(Y^*) \subset B^-$.

Lemma 10. *Let* P_1 *and* P_2 *be two arbitrary T_1-arm and T_2-arm, respectively, and* $Y^* = (y_1^*, y_2^*)$ *be the (P_1, P_2)-extreme solution. If there exists an index j that satisfies the following conditions:*

(A) y_j^* is a vertex,
(B) There is an open y_j^*-branch B^- that does not intersect P_j,
(C) $CR_j(Y^*) \subset B^-$,

then all open y_j^*-branches other than B^- do not contain x_j^*.

Lemmas 9 and 10 in pairs provide a basic idea for finding X^*. For any given P_1 and P_2, if $X^* \notin (P_1, P_2)$, by Lemma 9, there must exist some open branch B^- satisfying the conditions of Lemma 10. Thus, we can redirect one of the arms toward B^- so that the next Y^* better approaches X^*. Iteratively applying the arguments will eventually make $Y^* = X^*$.

Algorithm. Now, we are ready to propose the algorithm for finding a 2-service center in T. The algorithm is based on the idea mentioned above, but directs the arms in a simple and efficient way. In the following, we give its high level description and then analyze its time complexity.

At first, we compute c^* and check whether (c^*, c^*) satisfies Lemma 4. If yes, (c^*, c^*) serves as a T^2-optimal solution. Otherwise, by Lemmas 5 and 6, we can identify T_1 and T_2 so that the problem is reduced to the finding of the (T_1, T_2)-extreme solution X^*. The finding of X^* is based on performing the idea of redirection iteratively, and further applies the prune-and-search strategy to both T_1-arms and T_2-arms.

Some notation is required for discussion. For each leaf $i \in T$, let $ord(i)$ denote the order of i visited by a depth-first traversal starting from c^* to all leaves.

Given a set L of leaves, let $med(L)$ be the median of the set $\{ord(i) \mid i \in L\}$. Also, we identify and prune the arms by their ending leaves. A leaf $i \in T$ is called a *candidate* if $P(c^*, i)$ might contain x_1^* or x_2^*. Let L_1 and L_2 denotes the sets of candidates in T_1 and T_2, which initially consist of all leaves in T_1 and T_2, respectively.

The prune-and-search process is performed as follows. At each iteration, we pick the T_1-arm $P_1 = P(c^*, l_1)$ and the T_2-arm $P_2 = P(c^*, l_2)$, where $l_1 \in L_1$ is the leaf with $ord(l_1) = med(L_1)$ and $l_2 \in L_2$ is the leaf with $ord(l_2) = med(L_2)$. Then, we compute the (P_1, P_2)-extreme solution Y^* and determine whether Y^* satisfies the conditions in Lemma 10.

When Y^* satisfies the conditions, Lemma 10 can be applied for pruning candidates. If it is y_1^* and the open y_1^*-branch B_1^- that satisfy the conditions, Lemma 10 shows that x_1^* is not contained in all open y_1^*-branches except B_1^-. It follows that all leaves outside B_1^- are no longer candidates. Thus, we remove these leaves from L_1. Similarly, if it is y_2^* and the open y_2^*-branch B_2^- that satisfy the conditions, we remove from L_2 all leaves not in B_2^-.

The prune-and-search process iterates until Y^* violates the conditions. Since Lemma 9 guarantees that Y^* always satisfies the conditions if $X^* \notin (P_1, P_2)$, the violation implies that $X^* \in (P_1, P_2)$ and $Y^* = X^*$, thereby solves the problem.

The time complexity of the algorithm is analyzed as follows. Computing c^* can be done in $O(n)$ time by Megiddo's algorithm for finding the weighted 1-center on a tree [6]. Then, identifying T_1 and T_2 and assigning the order labels to their leaves also takes $O(n)$ time.

For the prune-and-search process, we first show that it performs at most $O(\log n)$ iterations. By Lemma 9, at each iteration before termination, Y^* satisfies the conditions in Lemma 10 for either $j = 1$ or 2, which means that B^- does not intersect P_j. Since the leaf labels are assigned by depth-first traversal, either $ord(l_j) > ord(i)\, \forall i \in B^-$ or $ord(l_j) < ord(i)\, \forall i \in B^-$. By the chosen rule of l_j, B^- contains at most half of the leaves in L_j. Removing all leaves outside B^- thus reduces the size of L_j by a factor of at least 2. It follows that the pruning is done at most $O(\log n)$ times for $j = 1$ and at most $O(\log n)$ times for $j = 2$.

For each fixed iteration, picking arms and verifying conditions for Y^* take $O(n)$ time. By a simple extension from the algorithm in Subsection 4.1, we can obtain the following result to find Y^* in $O(n)$ time.

Lemma 11. *Given an arbitrary T_1-arm P_1 and an arbitrary T_2-arm P_2, the (P_1, P_2)-extreme solution Y^* can be computed in $O(n)$ time.*

Consequently, each iteration of the loop takes $O(n)$ time in total, and the algorithm requires $O(n) + O(n) \times O(\log n) = O(n \log n)$ time.

Theorem 5. *The 2-service center problem on a weighted tree can be solved in $O(n \log n)$ time.*

4.3   Algorithm on an Unweighted Tree

The 2-service center problem on an unweighted tree, where $w_{i,j} = 1$ for $i \in V$ and $j = 1, 2$, is far more simple. Suppose that (c^*, c^*) does not satisfy Lemma 4 so that T_1 and T_2 can be identified. Let l_1^* be a leaf in T_1 farthest from c^* and l_2^* be a leaf in T_2 farthest from c^*. We can directly find the arms that containing X^* as shown below, and then solving the problem by Lemma 11.

Lemma 12. *The* (T_1, T_2)-*extreme solution* $X^* \in (P(c^*, l_1^*), P(c^*, l_2^*))$.

Theorem 6. *The 2-service center problem on an unweighted tree can be solved in* $O(n)$ *time.*

5 Concluding Remarks

In this paper, we have proposed a whole new topic in location theory. The multi-service location problems have very different properties from traditional ones. In traditional location problems, facilities serve only their nearby vertices, and naturally partition the graph into relatively independent regions. On the other hand, in multi-service location problems, every vertex receives services from all facilities, so the placement of facilities are mutually related. Thus, new observation and approach are required for better solving this kind of problems.

As a good starting point, we obtained several algorithms for the p-service center problem. However, some of these results are yet to be improved. For the problem on general graphs, the p/c approximation factor does not sound good and is possible for further refinement. For the 2-service center problem on trees, there is still a gap of factor $O(\log n)$ (in terms of time complexity) between the weighted and unweighted cases. It would be challenging to close the gap.

References

1. Ben-Moshe, B., Bhattacharya, B., Shi, Q.: An Optimal Algorithm for the Continuous/Discrete Weighted 2-Center Problem in Trees. In: Correa, J.R., Hevia, A., Kiwi, M. (eds.) LATIN 2006. LNCS, vol. 3887, pp. 166–177. Springer, Heidelberg (2006)
2. Cole, R.: Slowing down sorting networks to obtain faster sorting algorithms. J. ACM 34, 200–208 (1987)
3. Frederickson, G.: Parametric Search and Locating Supply Centers in Trees. In: Dehne, F., Sack, J.-R., Santoro, N. (eds.) WADS 1991. LNCS, vol. 519, pp. 299–319. Springer, Heidelberg (1991), 10.1007/BFb0028271
4. Garey, M.R., Johnson, D.S.: Computers and Intractability: A Guide to the Theory of NP-Completeness. W.H. Freeman & Co., New York (1979)
5. Kariv, O., Hakimi, S.L.: An algorithmic approach to network location problems. I: The p-centers. SIAM Journal on Applied Mathematics 37(3), 513–538 (1979)
6. Megiddo, N.: Linear-time algorithms for linear programming in R^3 and related problems. SIAM Journal on Computing 12(4), 759–776 (1983)
7. Megiddo, N.: Linear programming in linear time when the dimension is fixed. J. ACM 31, 114–127 (1984)
8. Plesník, J.: On the computational complexity of centers locating in a graph. Apl. Mat. 25, 445–452 (1980)
9. Plesník, J.: A heuristic for the p-center problems in graphs. Discrete Applied Mathematics 17(3), 263–268 (1987)

Computing Minmax Regret 1-Median on a Tree Network with Positive/Negative Vertex Weights

Binay Bhattacharya[*], Tsunehiko Kameda[*], and Zhao Song

School of Computing Science, Simon Fraser University, Canada
{binay,tiko,zhaos}@sfu.ca

Abstract. In a facility location problem, if the vertex weights are uncertain one may look for a "robust" solution that minimizes "regret." The most efficient previously known algorithm for finding the minmax regret 1-median on trees with positive and negative vertex weights takes $O(n^2)$ time. In this paper, we improve it to $O(n \log^2 n)$.

1 Introduction

Deciding where to locate facilities to minimize the communication or travel costs is known as the *facility location problem*. For a recent review of this subject, the reader is referred to [11]. In the 1-median problem, the objective function is formulated as the sum of the distances from the customer sites weighted by the weights of the sites. In the *minmax regret* version of this problem, there is uncertainty in the weights of the vertices and/or edge lengths, and only their ranges are known [10,13]. In this paper, our model assumes that the edge lengths are nonnegative and fixed, and uncertainty is only in the weights of the vertices.[1] A particular *realization* (assignment of a weight to each vertex) is called a *scenario*. Intuitively, the minmax regret 1-median problem can be understood as a 2-person game as follows. The first player picks a location x to place a facility. The opponent's move is to pick a scenario s. The payoff to the second player is the cost of x minus the cost of the median, both under s, and he wants to pick the scenario s that maximizes his payoff. Our objective (as the first player) is to select x that minimizes this payoff in the worst case (i.e., over all scenarios).

The problem of finding the minmax regret median in a network, and a tree in particular, has attracted great research interest in recent years. Many researchers have worked on this problem. Kouvelis et al. [13] discussed the problem of finding the minmax regret 1-median on a tree and proposed an $O(n^4)$ solution, where n is the number of vertices. Chen and Lin [10] improved it to $O(n^3)$. Averbakh and Berman then found a simple $O(n^2)$ algorithm [1] and improved it later to $O(n \log^2 n)$ [2]. Yu et al. [15] and Brodal et al. [6] independently proposed an $O(n \log n)$ implementation of the algorithm in [2]. More recently, Bhattacharya

---

[*] Supported in part by the NSERC of Canada.
[1] Theorem 1 in [10] shows that for trees it is sufficient to set all edge lengths at their maximum values.

K.-M. Chao, T.-s. Hsu, and D.-T. Lee (Eds.): ISAAC 2012, LNCS 7676, pp. 588–597, 2012.

and Kameda [5] came up with the optimal $O(n)$ time algorithm. When the vertices can have negative weights, Burkard and Dollani have an $O(n^2)$ time algorithm [7]. A vertex with a negative weight has an interesting interpretation as an *obnoxious* site [8]. In this paper, we present an $O(n \log^2 n)$ time algorithm for trees where each vertex may have a positive or negative weight from an interval.

The paper is organized as follows. In Sec. 2, we introduce basic terms that are used throughout the paper, and review some properties of the median in a tree. Sec. 3 computes (classical) medians and their costs. We then present our main result in Sec. 4 that shows the minmax 1-median can be computed in $O(n \log^2 n)$ time. Sec. 5 concludes the paper.

2 Preliminaries

2.1 Definitions

Let $T = (V, E)$ be a tree network with vertex set V and edge set E, where $|V| = n$. We also use T to denote the set of all points (vertices and points on edges) on T. Each vertex $v \in V$ is associated with an interval of integer weights $W(v) = [\underline{w}_v, \overline{w}_v]$, and each edge $e \in E$ is associated with a non-negative length (or distance). For any two points $x, y \in T$, let $d(x, y)$ denote the distance between x and y on T, and $\pi[x, y]$, $\pi(x, y)$, and $\pi[x, y)$ denote the closed, open, and half-open path from x to y, respectively. For an interior point x of an edge $e = (u, v)$, we assume $d(x, u)$ is a prorated fraction of the length of e. Let $\mathcal{S}$ be the Cartesian product of all $W(v)$, $v \in V$:

$$\mathcal{S} \triangleq \prod_{v \in V} [\underline{w}_v, \overline{w}_v].$$

Under a *scenario* $s \in \mathcal{S}$, we define the *cost* of a point $x \in T$ by

$$F^s(x) \triangleq \sum_{v \in V} d(v, x) w_v^s,$$

A point that minimizes the above cost is called a *weighted 1-median*, or just a *median* for short. We call

$$R^s(x) \triangleq F^s(x) - F^s(m(s)) \tag{1}$$

the *regret* of x, where $m(s)$ denotes a median under s. We finally define the *maximum regret* of x by

$$R^*(x) \triangleq \max_{s \in \mathcal{S}} R^s(x). \tag{2}$$

Note that $R^*(x)$ is the maximum payoff with respect to x that we mentioned in the Introduction. We seek location $x^* \in T$, called the *minimum regret median*, that minimizes $R^*(x)$. Suppose $s = \hat{s}(x)$ maximizes (1) for a given $x \in T$. We call $\hat{s}(x)$ and $m(\hat{s}(x))$ the *worst case scenario* and the *worst case alternative* for x, respectively. We thus have

$$R^*(x) = R^{\hat{s}(x)}(x). \tag{3}$$

2.2   Basic Properties of Trees

Lemma 1. [12] *In a tree there is always a median at a vertex.*   ☐

The above lemma is proved in [12], assuming that the vertex weights are positive, but it holds without this assumption. However, the minmax regret median may not be at a vertex [13]. If there exists a vertex u that is a median for all the scenarios in $\mathcal{S}$, then clearly u is the min-regret location. In such a case, the problem instance is said to be *degenerate* [5].

Lemma 2. [5] *In a general network, we have $R^*(x^*) = 0$ if and only if the problem instance is degenerate.*   ☐

Let $e = (u, u')$ be any edge of T, and let T_u (resp. $T_{u'}$) denote the subtree containing u (resp. u') but not e. Let $s(u)$ be a scenario such that $w_v^s = \overline{w}_v$ for each vertex $v \in T_u$, and $w_v^s = \underline{w}_v$ for each vertex $v \in T_{u'}$. Such a scenario s is called a *bipartite* scenario, and u is the *front* of s, denoted by $f(s)$ [5]. Let $\mathcal{S}^* \subset \mathcal{S}$ be the set of all bipartite scenarios. Under $s(u)$, we call T_u the *max-weighted component* and $T_{u'}$ the *min-weighted component*. Chen and Lin showed

Lemma 3. (Theorem 1(a) in [10])

$$\forall x \in T : R^*(x) = \max_{s \in \mathcal{S}^*} R^s(x). \tag{4}$$

☐

Let $\tilde{\mathcal{S}} \subset \mathcal{S}^*$ be the set of bipartite scenarios such that the median is in the max-weighted component. The proof of Lemma 3, together with Lemma 2, implies

Lemma 4.

$$\forall x \in T : R^*(x) = \max_{s \in \tilde{\mathcal{S}}} R^s(x). \tag{5}$$

☐

Thus, we only consider scenarios in $\tilde{\mathcal{S}}$ in the rest of this paper. We sometimes refer to $R^*(x)$ as the *upper envelope* of $\{R^s(x) \mid s \in \tilde{\mathcal{S}}\}$.

2.3   Weight and Cost Arrays

We pick an arbitrary vertex as the root r, and from now on consider T as a rooted tree. For $v \in V$ such that $v \neq r$, we denote the *parent* of v by $p(v)$. Let $T(v)$ denote the subtree of T rooted at v, and call $T^c(v) = T \setminus T(v)$ the *complement* of $T(v)$. We define two arrays for *subtree weights*, $\overline{W}_t[\cdot]$ and $\underline{W}_t[\cdot]$, and two arrays for *complement weights*, $\overline{W}_c[\cdot]$ and $\underline{W}_c[\cdot]$, as follows.

$$\overline{W}_t[v] \triangleq \sum_{u \in T(v) \cap V} \overline{w}_u, \quad \underline{W}_t[v] \triangleq \sum_{u \in T(v) \cap V} \underline{w}_u,$$

$$\overline{W}_c[v] \triangleq \sum_{u \in T^c(v) \cap V} \overline{w}_u, \quad \underline{W}_c[v] \triangleq \sum_{u \in T^c(v) \cap V} \underline{w}_u.$$

We can compute $\overline{W}_t[\cdot]$ and $\underline{W}_t[\cdot]$ easily, using the post-order depth first search in $O(n)$ time. Once they have been computed, to compute $\overline{W}_c[v]$, for example, we simply use the relation [7]

$$\overline{W}_c[v] = \overline{W}_t[r] - \overline{W}_t[v].$$

Note that $\overline{W}_c[r] = \underline{W}_c[r] = 0$. We now define the *subtree costs* (with subscript t) and *complement costs* (with subscript c) as follows:

$$\overline{C}_t[v] \triangleq \sum_{u \in T(v) \cap V} d(u,v)\overline{w}_u, \quad \underline{C}_t[v] \triangleq \sum_{u \in T(v) \cap V} d(u,v)\underline{w}_u,$$

$$\overline{C}_c[v] \triangleq \sum_{u \in T^c(v) \cap V} d(u,v)\overline{w}_u, \quad \underline{C}_c[v] \triangleq \sum_{u \in T^c(v) \cap V} d(u,v)\underline{w}_u.$$

Arrays $\overline{C}_t[\cdot]$, $\underline{C}_t[\cdot]$, $\overline{C}_c[\cdot]$, and $\underline{C}_c[\cdot]$ can be computed in $O(n)$ time [7]. For compact representation, we also introduce the following notation:

$$C_t^{\pm}[u] \triangleq \overline{C}_t[u] - \underline{C}_t[u], \quad C_c^{\pm}[u] \triangleq \overline{C}_c[u] - \underline{C}_c[u],$$
$$W_t^{\pm}[u] \triangleq \overline{W}_t[u] - \underline{W}_t[u], \quad W_c^{\pm}[u] \triangleq \overline{W}_c[u] - \underline{W}_c[u].$$

3 Medians and Their Costs

3.1 Computing $m(s)$ and $F^s(m(s))$ for all $s \in \tilde{S}$

Let $s(u)$ (resp. $s^c(u)$) denote the scenario under which all the vertices of $T(u)$ (resp. $T^c(u)$) have the maximum weights, and the rest of the vertices have the minimum weights. Fig. 1(a) (resp. (b)) illustrates scenario $s(u)$ (resp. $s^c(u)$),

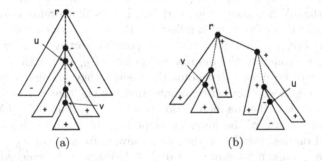

Fig. 1. (a) Scenario $s(u)$; (b) Scenario $s^c(u)$

where a vertex with the maximum (resp. minimum) weight is indicated by a $+$ (resp. $-$). Let us consider the contributions to the cost of vertex v from different parts of tree T.

Case (a) $[F^{s(u)}(v)$. Fig. 1(a)]:

1. From $T(v)$: $\overline{C}_t[v]$.
2. From $T(u) \backslash T(v)$ (if $v \neq u$): $\overline{C}_c[v] - \{\overline{C}_c[u] + d(u,v)\overline{W}_c[u]\}$.
3. From $T^c(u)$: $\underline{C}_c[u] + d(u,v)\underline{W}_c[u]$.

It is clear that, using arrays $\overline{C}_*[\cdot]$, $\underline{C}_*[\cdot]$ ($* \in \{t,c\}$), $\overline{W}_c[\cdot]$, and $\underline{W}_c[\cdot]$, the above three quantities can be computed in constant time. Adding all of them, we obtain

$$F^{s(u)}(v) = \{\overline{C}_t[v] + \overline{C}_c[v]\} - \{C_c^{\pm}[u] + d(u,v)W_c^{\pm}[u]\}. \tag{6}$$

We introduce the notation $g_v(u) \triangleq F^{s(u)}(v)$, with the intention of treating it as a function of u with parameter v. For two vertices $v_1, v_2 \in T(u)$, the difference in their costs under $s(u)$ is

$$g_{v_2}(u) - g_{v_1}(u) = \{\overline{C}_t[v_2] + \overline{C}_c[v_2]\} - \{\overline{C}_t[v_1] + \overline{C}_c[v_1]\} - \{d(u,v_2) - d(u,v_1)\}W_c^{\pm}[u].$$

Let w be the lowest common ancestor of v_1 and v_2. Then $d(u,v_i) = d(w,v_i) + d(w,u)$, so that the common part $d(w,u)$ cancels in the above equation, and

$$g_{v_2}(u) - g_{v_1}(u) = \{\overline{C}_t[v_2] + \overline{C}_c[v_2]\} - \{\overline{C}_t[v_1] + \overline{C}_c[v_1]\} - \{d(w,v_2) - d(w,v_1)\}W_c^{\pm}[u]. \tag{7}$$

Note that the only term in (7) that depends on u is $W_c^{\pm}[u]$. If we define $X(u) \triangleq W_c^{\pm}[u]$, then (7) is a linear function of $X(u)$. Clearly, $X(u)$ has $O(n)$ distinct values. Let us define function $g'_v(\cdot)$ by $g'_v(X(u)) \triangleq g_v(u)$.

We compute X_{min} (resp. X_{max}) by finding the vertex u_{min} (resp. u_{max}) that minimizes (resp. maximizes) $X(u)$. Clearly, we have $u_{min} = r$ and u_{max} is a leaf. Thus the range of $X(\triangleq X(u))$ is $I = [X_{min}, X_{max}]$. In the next paragraph we show how to find the lower envelope of $\{g'_v(X) \mid v \in T(u)\}$. Using the lower envelope, for each $u \in V$, we can identify one or more vertices v such that $g'_v(X(u))$ takes the minimum over all vertices, which implies that v is a median under $s(u)$.

To find the lower envelope of $\{g'_v(X) \mid v \in T(u)\}$, we perform "sweeping," pretending that X is a continuous variable. Let us first evaluate $g_v(u_{min})$ for all v and order them from the smallest to the largest, which takes $O(n \log n)$ time. Let $L = \langle v_1, v_2, \ldots, v_l \rangle$ be the corresponding sorted list of vertices. For each $v_i \in L$, we compute the intersection points of $g'_{v_i}(X)$ with $g'_{v_{i-1}}(X)$ and $g'_{v_{i+1}}(X)$. It can be done in constant time, since they are linear functions of X. From among those intersection points, find the one that corresponds to the smallest X, and let it be the intersection point of $g'_{v_j}(X)$ and $g'_{v_{j+1}}(X)$. If $j = 1$, then the first interval of the lower envelope is from X_{min} to this intersection point. If $j > 1$ on the other hand, then we remove v_j from L, since $g'_{v_j}(X(u))$ will never be the minimum for any value of $X \in I$. When v_j is removed, v_{j-1} and v_{j+1} become adjacent in L, so we compute their intersection point, and repeat the above process. It is easy to see that the whole process of identifying all the segments of the lower envelope can be carried out in $O(n \log n)$ time.

Case (b) $[F^{s^c(u)}(v)$. Fig. 1(b)]: A scenario of the type $s^c(u)$ is shown in Fig. 1(b). In this case, we have

$$F^{s^c(u)}(v) = \overline{C}_t[v] + \overline{C}_c[v] - \{C_t^{\pm}[u] + d(u,v)W_t^{\pm}[u]\}, \tag{8}$$

which is essentially the same as (6). Formula (8) is also valid if v and u are in the same subtree of T. Therefore, we can find the vertex v that minimizes it as above, thus obtaining $m(s^c(u))$.

Lemma 5. *We can compute $\{m(s) \mid s \in \tilde{S}\}$ and $\{F^s(m(s)) \mid s \in \tilde{S}\}$ in $O(n \log n)$ time.*

Proof. We showed how to compute $m(s)$ above. We can plug $v = m(s)$ in (6) or (8) to compute $F^{s(u)}(v)$ or $F^{s^c(u)}(v)$, respectively, which takes constant time. $\square$

4   Optimal Facility Location

We assume that the given $T(V, E)$ is a balanced binary tree, rooted at vertex r, and having n vertices, hence of height $O(\log n)$. We will discuss a general tree later in Sec. 4.3. For subtree $T(v)$ of T, let $N_n(T(v))$ denote the number of vertices in $T(v)$, and let $S(T(v)) \subseteq \tilde{S}$ be the set of scenarios whose max weighted components are totally contained in $T(v)$.

4.1   Algorithm

For $x \notin T(v)$, let us introduce

$$R^s_{T(v)}(x) \triangleq \sum_{u \in T(v) \cap V} d(u, x) w^s_u - F^s(m(s))$$

$$R^*_{T(v)}(x) \triangleq \max_{s \in S(T(v))} R^s_{T(v)}(x), \tag{9}$$

ignoring the cost contributions from the vertices in $T^c(v)$. In the *preprocessing* phase, for each $v \in V$, we compute the upper envelope, i.e., $R^*_{T(v)}(x)$, which is valid for $x \notin T(v)$.

Lemma 6. *Let $u, v \in V$ be the two child vertices of a non-root vertex y in T. If $R^*_{T(u)}(x)$ and $R^*_{T(v)}(x)$ are available, then we can compute $R^*_{T(y)}(x)$ for $x \in (y, p(y))$ in $O(N_n(T(y)))$ time.*

Proof. Note that $R^*_{T(u)}(x)$ consists of $O(N_n(T(u)))$ line segments for $x \notin T(u)$. In order to compute $R^*_{T(y)}(x)$, we merge $R^*_{T(u)}(x)$, $R^*_{T(v)}(x)$ and $R^{s(y)}_{T(y)}(x)$ for points $x \notin T(y)$. Note that $d(x, y) \underline{w}_y$ must be added to $R^*_{T(u)}(x)$ and $R^*_{T(v)}(x)$ before computing the upper envelope of these three functions. It is easy to see that all this can be done in $O(N_n(T(y)))$ time. $\square$

Note that the subtrees at each level of T have disjoint sets of descendant leaves. By Lemma 6, therefore, the total time need to their upper envelopes is $O(n)$. Since there are $O(\log n)$ levels in T, we have

Lemma 7. *The total preprocessing time for computing $\{R^*_{T(v)}(x) \mid v \in V\}$ for a balanced binary tree T is $O(n \log n)$.* $\square$

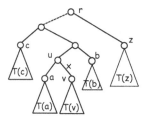

Fig. 2. Finding median for point $x \in e = (u, v)$

From now on we assume that $R^*_{T(v)}(x)$ is stored at v in terms of its bending points. Let $e = (u, v) \in E$, where v is farther than u from the root r of T. We now show how to compute $R^*(x)$ (global upper envelope) for $x \in e$, with the help of Fig. 2, which shows the path from v to r and the $O(\log n)$ subtrees hanging from it, $T(v), T(a), T(b), \ldots, T(z)$. Assume that $R^*_{T(v)}(x), R^*_{T(a)}(x), \ldots, R^*_{T(z)}(x)$ are available (by Lemma 7). Note that $R^*(x)$ is the upper envelope of the following functions:

1. The scenarios in $S(T(v))$:

$$R^*_{T(v)}(x) + \{\underline{C}_c[v] - d(v, x)\underline{W}_c[v]\}. \tag{10}$$

2. For $k = a, b, \ldots, z$:

$$R^*_{T(k)}(x) + \{\underline{C}_t(v) + d(x, v)\underline{W}_t(v)\}$$
$$+\{\underline{C}_c(v) - d(v, x)\underline{W}_c(v)\} - \{\underline{C}_t(k) + d(x, k)\underline{W}_t(k)\}. \tag{11}$$

3. Scenario $s^c(k')$ for $k = a, b, \ldots, z$, where k' is the sibling of k:

$$\{\underline{C}_t(v) + d(x, v)\underline{W}_t(v)\} + \{\underline{C}_c(v) - d(v, x)\underline{W}_c(v)\}$$
$$+\{C_c^{\pm}[k'] + d(x, p(k'))W_c^{\pm}[k']\} - F^{s^c(k')}(m(s^c(k'))), \tag{12}$$

where $p(k')$ is the parent of k'. The first (resp. second (in brackets)) term in (11) is the contribution from $T(k)$ (resp. $T(v)$) for the scenarios in $S(T(k))$, and (12) is the contribution from $T \setminus \{T(v) \cup T(k)\}$.

Lemma 8. *Given $x \in e$, where $e \in E$, we can evaluate $R^*(x)$ in $O(\log^2 n)$ time.*

Proof. For any scenario $s \in \tilde{S}$, $R^s(x)$ is linear over the entire e, which implies that $R^*(x)$ is convex over e. Note that each of the functions of x listed above consists of a sequence of linear segments over e, so that it can be represented by a sequence of bending points. Thus, for any x, evaluating each such function takes $O(\log n)$ time by binary search, and the total for all of them is $O(\log^2 n)$. □

When we evaluate $R^*(x)$ at $x \in e$, we can identify the scenario s such that $R^s(x) = R^*(x)$ and the slope of $F^s(x)$ at x. This slope (positive or negative) indicates which side of x, the optimal location on that edge lies. Thus we can use binary search to find the optimal location on e in $O(\log n)$ steps. Thus the globally optimal location x^* among all the edges can be found in $O(n \log^3 n)$ time. In the next subsection, we improve this to $O(n \log^2 n)$.

4.2   Efficient Implementation

In the proof of Lemma 8, we found the dominating scenario at x, performing binary search separately for $O(\log n)$ upper envelopes, resulting in $O(\log^2 n)$ total time. Here we perform binary search "simultaneously," and use the result individually, reducing the total time down to $O(\log n)$. This method is reminiscent of *fractional cascading* [9]. Fig. 3(a) illustrates the idea. The three lines, labeled u,

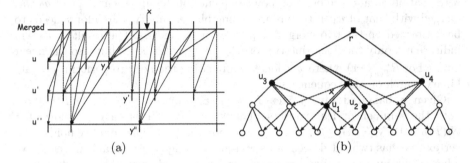

(a)   (b)

Fig. 3. "Cascading" pointers: (a) Principle; (b) Rightward pointers (rightward arrows) and leftward pointers (leftward arrows)

u', and u'', represent the sequence of the linear intervals of three precomputed upper envelopes at vertices u, u', and u'', respectively, namely, the bending points of $R^*_{T(u)}(x)$, $R^*_{T(u')}(x)$, and $R^*_{T(u'')}(x)$, respectively. In the "Merged" sequence at the top, all bending points are merged together. In general, as many as $O(\log n)$ sequences may be merged, instead of just three in this illustration. The downward arrows and dashed lines represent pointers. From the left end of each interval (which is a bending point) in the "Merged" sequence, there are three pointers (k pointers if k sequences are merged.) For example, the first marker to the left of point x has a pointer pointing to y (resp. y', y'') in the sequence of u (resp. u', u''). This means that y (resp. y', y'') is the start of the linear interval of $R^*_{T(u)}(x)$ (resp. $R^*_{T(u')}(x)$, $R^*_{T(u'')}(x)$) that contains point x. Thus, once the interval in which x lies in the "Merged" sequence is determined by binary search, the intervals in which x lies in the sequences of u, u', and u'' can be found in constant time by tracing the appropriate pointers. The pointer to the sequence of u'' is shown by a dashed line, because the marker itself represents the bending point y'' on u''.

Fig. 3(b) shows which upper envelopes should be merged. Let u' be any vertex that lies on the leftmost (resp. rightmost) path from the root of T and u' be its sibling. Let v be the left (resp. right) child of any vertex belonging to $T(u')$. Then we merge $R^*_{T(v)}(x)$ with $R^*_{T(u)}(x)$. In Fig. 3(b), for example, the bending point sequences at the seven vertices that are reachable by following leftward arrows from u_4 are merged with the bending point sequence at u_4. Binary search for x needs to be performed only on the merged sequences at u_4 and u_3. In general, we need to perform binary search at at most two vertices, to evaluate the functions (10) and (11).

In Fig. 3(b), vertex u_1 is the root of the subtree just "below" the edge on which x lies. The vertices on the path from x to the root r are indicated by squares, and the roots (i.e., u_2, u_3, and u_4) of the subtrees hanging from this path are indicated by black circles. In terms of Fig. 2, we have $u_1 = v$, $u_2 = a$, $u_3 = c$, and $u_4 = z$. We need to examine the upper envelopes stored at these four vertices. The bending point sequence of the upper envelope of u_1 was merged with that of u_3, and the bending point sequence of the upper envelope of u_2 was merged with that of u_4 in the preprocessing phase. Binary search for x needs to be performed on the two merged sequences at u_3 and u_4. The result at u_3 will indicate not only the linear interval of $R^*_{T(u_3)}(x)$ that x is on, but also the linear interval of $R^*_{T(u_1)}(x)$ where x is located, obviating the need to perform separate binary search on the sequence at u_1.

Given a balanced binary tree $T = (V, E)$, for $u \in V$, let $N_l(T(u))$ denote the number of leaf vertices of subtree $T(u)$. It is easy to show that $N_l(T(p(u)))/N_l(T(u)) \geq 3/2$ for any non-root vertex u. The equality holds for a vertex that has two left descendants and just one right descendant. In the above scheme, the number of bending points in the merged sequence at node v is at most

$$N_l(T(v)) + \{N_l(T(v)) + 1\}/(3/2) + \{N_l(T(v)) + 1\}/(3/2)^2 + \cdots$$
$$< 3\{N_l(T(v)) + 1\}. \tag{13}$$

Note that $sib(v)$ may have one more descendant leaf than v.

Lemma 9. *The total number of pointers at all levels is $O(n \log^2 n)$.*

Proof. Since each bending point in a merged sequence has $O(\log n)$ pointers, and the sum of $N_l(T(v))$ for all nodes v at any level of T is $O(n)$, the total number of pointers stored at a level of T is $O(n \log n)$. This should be multiplied by the number of levels in T, i.e., $O(\log n)$, resulting in $O(n \log^2 n)$.   □

Theorem 1. *The minmax regret 1-median for a balanced binary tree with positive/negative vertex weights can be found in $O(n \log^2 n)$ time.*   □

4.3   General Trees

Given a tree network, it is is not binary, then we first convert it into a binary tree, introducing $O(n)$ new vertices with 0 weight and $O(n)$ new edges of 0 length [14]. Using *spine decomposition* [3,4], we can convert the binary tree into a structure that has properties of a balanced binary tree, introducing $O(n)$ new vertices and $O(n)$ new edges. Then we can apply our algorithm described above to this structure.

5   Conclusion

We have presented an $O(n \log^2 n)$ time algorithm for computing the minmax regret 1-median for trees with positive/negative vertex weights. This is an improvement over the best previously known time complexity of $O(n^2)$ [8].

References

1. Averbakh, I., Berman, O.: Minmax regret median location on a network under uncertainty. INFORMS Journal of Computing 12(2), 104–110 (2000)
2. Averbakh, I., Berman, O.: An improved algorithm for the minmax regret median problem on a tree. Networks 41, 97–103 (2003)
3. Benkoczi, R.: Cardinality constrained facility location problems in trees. Ph.D. thesis, School of Computing Science, Simon Fraser University, Canada (2004)
4. Benkoczi, R., Bhattacharya, B., Chrobak, M., Larmore, L.L., Rytter, W.: Faster Algorithms for k-Medians in Trees. In: Rovan, B., Vojtáš, P. (eds.) MFCS 2003. LNCS, vol. 2747, pp. 218–227. Springer, Heidelberg (2003)
5. Bhattacharya, B., Kameda, T.: A Linear Time Algorithm for Computing Minmax Regret 1-Median on a Tree. In: Gudmundsson, J., Mestre, J., Viglas, T. (eds.) COCOON 2012. LNCS, vol. 7434, pp. 1–12. Springer, Heidelberg (2012)
6. Brodal, G.S., Georgiadis, L., Katriel, I.: An $O(n \log n)$ version of the Averbakh–Berman algorithm for the robust median of a tree. Operations Research Letters 36, 14–18 (2008)
7. Burkard, R.E., Dollani, H.: Robust location problems with pos/neg-weights on a tree. Tech. Rep. Diskrete Optimierung Bericht Nr. 148, Karl-Franzens-Universiät Graz & Technische Universiät Graz (1999)
8. Burkard, R., Krarup, J.: A linear algorithm for the pos/neg-weighted 1-median problem on a cactus. Computing 60, 193–215 (1998)
9. Chazelle, B., Guibas, L.J.: Fractional cascading: I. A data structuring technique. Algorithmica 1, 133–162 (1986)
10. Chen, B., Lin, C.S.: Minmax-regret robust 1-median location on a tree. Networks 31, 93–103 (1998)
11. Hale, T.S., Moberg, C.R.: Location science research: A review. Annals of Operations Research 123, 21–35 (2003)
12. Kariv, O., Hakimi, S.: An algorithmic approach to network location problems, part 2: The p-median. SIAM J. Appl. Math. 37, 539–560 (1979)
13. Kouvelis, P., Vairaktarakis, G., Yu, G.: Robust 1-median location on a tree in the presence of demand and transportation cost uncertainty. Tech. Rep. Working Paper 93/94-3-4, Department of Management Science, The University of Texas, Austin (1993)
14. Tamir, A.: An $O(pn^2)$ algorithm for the p-median and the related problems in tree graphs. Operations Research Letters 19, 59–64 (1996)
15. Yu, H.I., Lin, T.C., Wang, B.F.: Improved algorithms for the minmax-regret 1-center and 1-median problem. ACM Transactions on Algorithms 4(3), 1–1 (2008)

Fence Patrolling by Mobile Agents
with Distinct Speeds

Akitoshi Kawamura* and Yusuke Kobayashi**

University of Tokyo, Tokyo, Japan
kawamura@is.s.u-tokyo.ac.jp,
kobayashi@mist.i.u-tokyo.ac.jp

Abstract. Suppose we want to patrol a fence (line segment) using k mobile agents with speeds v_1, ..., v_k so that every point on the fence is visited by an agent at least once in every unit time period. Czyzowicz et al. conjectured that the maximum length of the fence that can be patrolled is $(v_1 + \cdots + v_k)/2$, which is achieved by the simple strategy where each agent i moves back and forth in a segment of length $v_i/2$. We disprove this conjecture by a counterexample involving $k = 6$ agents. We also show that the conjecture is true for $k \leq 3$.

1 Introduction

In *patrolling problems*, a set of mobile agents move around a given area to protect or surveil it. We want to make sure that each point in the area is visited frequently. Although there are many ways to measure frequency, our objective in this paper is to find a strategy such that each point is visited by an agent at least once in every fixed time period T. Patrolling problems have been well-studied in robotics, motivated by practical situations such as tracking chemical spills and surveilling an environment (see e.g. [3,6]). Most of the previous studies suggest heuristic patrolling strategies and analyze their performance theoretically or experimentally [1,2,5,6,8]. Recently, theoretical optimality of patrolling strategies has been studied for some cases [4,7]. Czyzowicz et al. [4] consider two problems called *boundary patrolling* and *fence patrolling*. In the boundary patrolling problem, k mobile agents patrol the boundary of a planar object represented by a cycle; in the fence patrolling problem, they patrol a fence represented by a segment. For the boundary patrolling problem, they show that a simple partition-based strategy is optimal in some restricted cases but not in general. For the fence patrolling problem, they showed that a similar partition-based strategy is optimal for $k = 2$, and conjectured that it is true for every k. In this paper, we prove that this conjecture holds also for $k = 3$ but not in general.

---

* Supported by Grant-in-Aid for Scientific Research, Japan.
** Supported by Grant-in-Aid for Scientific Research and by the Global COE Program "The research and training center for new development in mathematics", MEXT, Japan.

K.-M. Chao, T.-s. Hsu, and D.-T. Lee (Eds.): ISAAC 2012, LNCS 7676, pp. 598–608, 2012.
© Springer-Verlag Berlin Heidelberg 2012

Formal Description of the Fence Patrolling Problem. We are given a segment of length l, which is identified with the interval $[0, l]$. The set of mobile agents $a_1, a_2, \ldots, a_k$ are moving along the segment, and they are allowed to move in both directions. The speed of each agent a_i may vary during its motion, but its absolute value is bounded by the predefined maximum speed v_i. The position of agent a_i at time $t \in [0, \infty)$ is denoted by $a_i(t) \in [0, l]$. Thus, for any $t > 0$ and $\epsilon > 0$,

$$|a_i(t) - a_i(t + \epsilon)| \le v_i \cdot \epsilon. \tag{1}$$

We say that the segment $[0, l]$ is *patrolled by* $a_1, \ldots, a_k$ *with idle time* $T > 0$ if for any $x \in [0, l]$ and $t^* \in [T, +\infty)$, the pair $(x; t^*)$ is covered by some a_i, where a_i is said to *cover* $(x; t^*)$ if $a_i(t) = x$ for some $t \in [t^* - T, t^*)$. In the original setting [4], the definition of coverage merely required $t \in [t^* - T, t^*]$ instead of $t \in [t^* - T, t^*)$, but this makes no difference, because each $a_i(t)$ satisfying (1) is a continuous function.

The Partition-Based Strategy. A simple strategy for fence patrolling is as follows:

1. Partition the segment $[0, l]$ into k segments such that the length of the ith segment is $\frac{lv_i}{v_1 + \cdots + v_k}$.
2. Each mobile agent a_i patrols the ith segment by alternately visiting both endpoints with maximum speed.

Performance of a patrolling strategy is measured by the length of the longest fence that can be patrolled with fixed idle time $T > 0$ (or equivalently, the minimum idle time that is needed to patrol a fence of fixed length). Since each mobile agent a_i can patrol a segment of length $\frac{v_i T}{2}$, the partition-based strategy can patrol a segment of length $l = \sum_{i=1}^{k} \frac{v_i T}{2}$. Czyzowicz et al. [4] showed that this is optimal when $k = 2$. They also conjectured that it is true for any k, that is, a segment of length $l > \sum_{i=1}^{k} \frac{v_i T}{2}$ cannot be patrolled with idle time T.

In this paper, we disprove this conjecture by demonstrating $k = 6$ agents that patrol a fence of length greater than $\sum_{i=1}^{k} \frac{v_i T}{2}$ (Theorem 1). On the other hand, we show that the partition-based strategy is optimal when $k = 3$ (Theorem 4). We also show that the partition-based strategy is optimal when all agents have the same speed (Theorem 2), but not when there are two distinct speeds (Theorem 1).

The Weighted Setting. We also consider the *weighted fence patrolling problem* in which the idle time $T_i > 0$ depends on the agent a_i. Here, T_i is called the *weight* of a_i, and can be interpreted as a power of influence. For the weighted problem, we say that the segment $[0, l]$ is *patrolled by* $a_1, \ldots, a_k$ if for any $x \in [0, l]$ and $t^* \in [\max_i(T_i), +\infty)$, the pair $(x; t^*)$ is covered by some a_i, where a_i is said to cover $(x; t^*)$ if $a_i(t) = x$ for some $t \in [t^* - T_i, t^*)$.

As in the unweighted case, we can consider the *partition-based strategy*: partition the segment $[0, l]$ into k segments so that the length of the ith segment is $\frac{lv_i T_i}{v_1 T_1 + \cdots + v_k T_k}$, and let each agent a_i patrol the ith segment. For the weighted case, we show that the partition-based strategy is optimal when there are two agents with different speeds (Theorem 3), but not when there are three (Theorem 1).

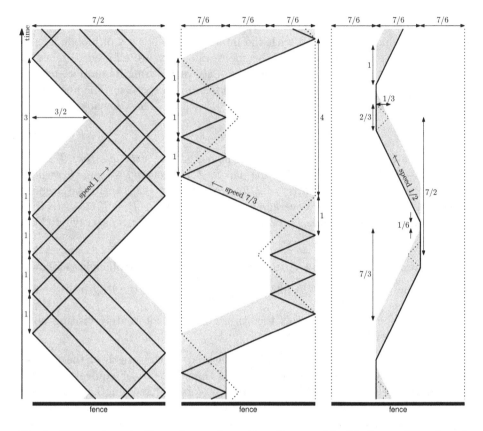

Fig. 1. Six agents patrolling a longer fence than they would with the partition-based strategy. The four agents on the left diagram can also be seen as one agent of weight 4.

Organization of the Paper. In Section 2, we present our counterexamples to the conjecture of Czyzowicz et al. [4]. Section 3 is about the case where all agents have one speed. After discussing the case of $k = 2$ in Section 4, we show that the partition-based strategy is optimal for (unweighted) fence patrolling with $k = 3$ in Section 5.

2   The Partition-Based Strategy Is Not Always Optimal

Fig. 1 shows six (unweighted) agents with speeds 1, 1, 1, 1, 7/3, 1/2 who patrol a fence of length 7/2 with idle time $T = 1$. The fence is placed horizontally and time flows upwards. The regions covered by each agent are shown shaded (i.e., the agent itself moves along the lower edge of each shaded band of height 1). The four agents with speed 1, shown in the diagram on the left, visit the two endpoints alternately. Another agent with speed 7/3 in the middle diagram almost covers the remaining regions (whose boundaries are shown in dotted lines), but misses some small triangles. They are covered by the last agent with speed 1/2 in the

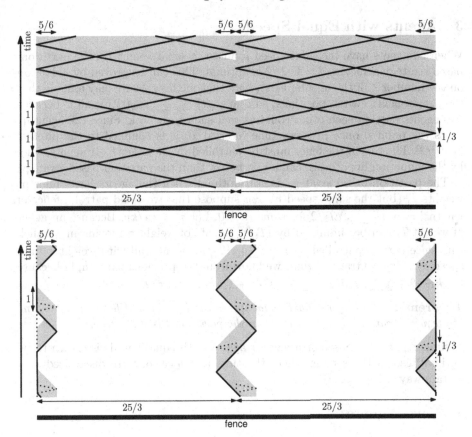

Fig. 2. Six agents with speed 5 (top) and three agents with speed 1 (bottom) patrolling a longer fence than they would with the partition-based strategy

diagram on the right. This disproves [4, Conjecture 1], since the partition-based strategy with these agents would only patrol the length $(1 + 1 + 1 + 1 + 7/3 + 1/2)/2 = 41/12 < 7/2$. Note that the first four agents can also be seen as one agent of weight 4, so that there are three agents altogether with different weights.

Another counterexample involving more agents but perhaps simpler is shown in Fig. 2, where six agents with speed 5 and three with speed 1 patrol a fence of length $50/3$ with idle time $T = 1$. Thus,

Theorem 1. *There are settings of agents' speeds and weights for which the partition-based strategy is not the one that patrols the longest possible fence. Such an example can consist of*

- *three agents (with different weights and speeds),*
- *six agents of equal weights, or*
- *nine agents of equal weights but having two distinct speeds.*

3   Agents with Equal Speeds

When all agents have the same speed v (and the same weight T), the partition-based strategy achieving $l = kvT/2$ is optimal. This can be proved by induction on the number k of the agents: First, note that in this case we may assume that the agents never switch positions, so that $a_1(t) \leq \cdots \leq a_k(t)$ for all t, because two agents passing each other could as well each turn back. Since the agent a_1 visits the point 0 once in every time interval T, it is confined to the interval $[0, vT/2]$. The rest of the fence must be patrolled by the other $k-1$ agents, who, by the induction hypothesis, cannot do better than the partition-based strategy.

The partition-based strategy is still optimal when the agents have different weights T_i (but the same speed v). For suppose that we could patrol a fence of length $l = \alpha + \sum_{i=1}^{k} vT_i/2$ for some $\alpha > 0$. Let $\tau = 2\alpha/kv$. Because an agent of weight T_i can be simulated by $\lceil T_i/\tau \rceil$ agents of weight τ moving in parallel, this fence can be patrolled by $\kappa = \sum_{i=1}^{k} \lceil T_i/\tau \rceil$ agents, all with weight τ (and speed v). This contradicts what we know from the previous paragraph, because $l = k\tau v/2 + \sum_{i=1}^{k} T_i v/2 = \sum_{i=1}^{k} (T_i/\tau + 1)v\tau/2 > \kappa v\tau/2$.

Theorem 2. *If all agents have equal speeds (and possibly different weights), no strategy can patrol a longer fence than the partition-based strategy.*

A similar partition-based strategy for agents with equal speeds is shown to be optimal also in the setting where the time and locations are discretized in a certain way [7, Section III].

4   Two Agents

In this section, we show that the partition-based strategy is optimal when the number of agents is $k = 2$. Although this is shown in [4] for two agents with equal weight, our proof is simpler even for this special case. Some ideas in the proof will be used for the case of three unweighted agents in Section 5.

Theorem 3. *For two agents (of possibly distinct weights), no strategy patrols a longer fence than the partition-based strategy.*

Proof. To derive a contradiction, assume that the segment $[0, l]$ can be patrolled by agents a_1 and a_2 for some $l > v_1T_1/2 + v_2T_2/2$. Let $l_1 = v_1T_1 l/(v_1T_1 + v_2T_2)$ and $l_2 = v_2T_2 l/(v_1T_1 + v_2T_2)$. Then, $l = l_1 + l_2$ and $l_i > v_iT_i/2$ for $i = 1, 2$. Without loss of generality, we may assume that $v_1 \geq v_2$.

For any time $t \geq 0$, each agent must visit one of the endpoints 0 and l some time after t. To see this, note that clearly one of the agents visits 0 after time t, say at $t_0 > t$. Then $(l; t_0 + l/v_1)$ cannot be covered by this agent, and thus has to be covered by the other agent.

We may therefore assume that the slower agent a_2 visits 0 at some time $t_2 > \max\{T_1, T_2\}$. Since $(l_2; t_2 + \frac{l_2}{v_2})$ cannot be covered by a_2, the other agent a_1 must visit l_2 at some time $t_1 \in [t_2 + \frac{l_2}{v_2} - T_1, t_2 + \frac{l_2}{v_2})$. Then neither a_1 nor a_2 can cover $(l; t_1 + \frac{l_1}{v_1})$, which is a contradiction.   □

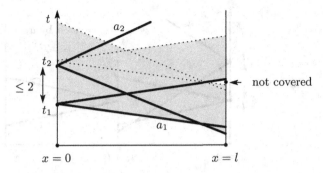

Fig. 3. Proof of Lemma 2

5   Three Agents of Equal Weights

In this section, we show that Czyzowicz et al.'s conjecture is true for three agents:

Theorem 4. *For three agents with equal weights, no strategy patrols a longer fence than the partition-based strategy.*

By scaling all speeds, we may assume that $T = 2$. For a contradiction, suppose that agents a_1, a_2, a_3 with speeds $v_1 \geq v_2 \geq v_3$ can patrol $[0, l]$, where $l > v_1 + v_2 + v_3$. For $i = 1, 2, 3$ let $l_i = \frac{v_i l}{v_1 + v_2 + v_3}$, so that $l = l_1 + l_2 + l_3$ and $l_i > v_i$.

5.1   Some Observations

In this subsection, we give some properties of the case $k = 3$.

Lemma 1. *For any $t^* \geq 0$, at least two different agents visit 0 after the time t^*, and at least two different agents visit l after the time t^*.*

Proof. Assume that a_1 is the only agent that visits 0 after time t^*. This forces it to stay (after time $t^* + 1$) in the part $[0, l_1]$, so the remaining part $[l_1, l]$ of length $l_2 + l_3$ has to be patrolled by a_2 and a_3, contradicting Theorem 3.

Similarly, neither a_2 nor a_3 can be the only agent that visits 0 after t^*. The same argument applies to the other endpoint l.   □

Lemma 2. *For any $t^* \geq 0$ and for each $i = 1, 2, 3$, the agent a_i visits at least one of 0 and l after the time t^*.*

Proof. Assume that a_3 does not visit 0 after t^*. By Lemma 1, both a_1 and a_2 visit 0 infinitely often after t^*. Thus, $a_1(t_1) = a_2(t_2) = 0$ for some $t_1, t_2 > t^* + 1$ with $t_1 \leq t_2 \leq t_1 + 2$ (see Fig. 3). The pair $(l; t_1 + \frac{l}{v_1})$ is not covered by a_1, because $\left(t_1 + \frac{l}{v_1}\right) - \left(t_1 - \frac{l}{v_1}\right) > 2$. It is not covered by a_2 either, because

$$\left(t_1 + \frac{l}{v_1}\right) - \left(t_2 - \frac{l}{v_2}\right) > (t_1 - t_2) + 2 + \left(\frac{v_2}{v_1} + \frac{v_1}{v_2}\right) \geq 2.$$

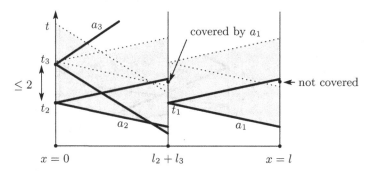

Fig. 4. Proof of Lemma 3

Hence, it must be covered by a_3, which means that a_3 visits l after the time t^*. The same argument can be applied to a_1 and a_2. □

Lemma 3. *Suppose that $a_2(t_2) = a_3(t_3) = 0$ (resp. $= l$) for some $t_2, t_3 > 2$. Then $a_1(t_1) = 0$ (resp. $= l$) for some $t_1 \in [t_2, t_3]$ (or $t_1 \in [t_3, t_2]$).*

Proof. Assume that $a_2(t_2) = a_3(t_3) = 0$ for some $t_3 \geq t_2 > 2$ such that $a_1(t_1) \neq 0$ for any $t_1 \in [t_2, t_3]$. By choosing t_2 and t_3 that minimize $t_3 - t_2$, we may assume that $t_3 - t_2 \leq 2$ (see Fig. 4).

The pair $(l_2 + l_3; t_2 + \frac{l_2 + l_3}{v_2})$ is not covered by a_2, because $(t_2 + \frac{l_2 + l_3}{v_2}) - (t_2 - \frac{l_2 + l_3}{v_2}) > 2$. It is not covered by a_3 either, because

$$\left(t_2 + \frac{l_2 + l_3}{v_2}\right) - \left(t_3 - \frac{l_2 + l_3}{v_3}\right) > (t_2 - t_3) + 2 + \left(\frac{v_3}{v_2} + \frac{v_2}{v_3}\right) \geq 2.$$

Hence, it must be covered by a_1, which means that $a_1(t_1) = l_2 + l_3$ for some t_1 with $t_2 + \frac{l_2 + l_3}{v_2} - 2 \leq t_1 < t_2 + \frac{l_2 + l_3}{v_2}$. Since $v_1 \geq v_2 \geq v_3$, $(l; t_1 + \frac{l_1}{v_1})$ is covered by none of a_1, a_2, and a_3, which is a contradiction.

The same argument can be applied to the case of $t_2 \geq t_3$ and to the case of $a_2(t_2) = a_3(t_3) = l$. □

By Lemmas 2 and 3, for any $t^* \geq 2$, there exist $t_2 \geq t_1 > t^*$ such that either $a_1(t_1) = a_2(t_2) = 0$ and $a_3(t) \neq 0$ for any $t \in [t_1, t_2]$ or $a_1(t_1) = a_2(t_2) = l$ and $a_3(t) \neq l$ for any $t \in [t_1, t_2]$. Let t^* be a sufficiently large number (e.g. $t^* = 10$). By choosing t_1 and t_2 with a minimal interval $[t_1, t_2]$, without loss of generality, we may assume that $t^* < t_1 \leq t_2 \leq t_1 + 2$ and $a_1(t_1) = a_2(t_2) = 0$.

We consider the cases of $v_1 \geq 2v_2 + v_3$ and $v_1 \leq 2v_2 + v_3$ in Sections 5.2 and 5.3, respectively.

5.2   Case of $v_1 \geq 2v_2 + v_3$

As in Section 5.1, let t_1 and t_2 be such that $t^* < t_1 \leq t_2 \leq t_1 + 2$ and $a_1(t_1) = a_2(t_2) = 0$. The pair $(l_1 + l_2 - l_3; t_1 + \frac{l_1 + l_2 - l_3}{v_1})$ is not covered by a_1, because

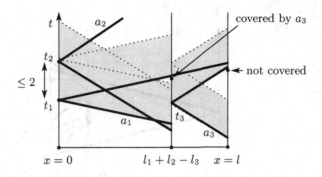

Fig. 5. Case of $v_1 \geq 2v_2 + v_3$

$\left(t_1 + \frac{l_1+l_2-l_3}{v_1}\right) - \left(t_1 - \frac{l_1+l_2-l_3}{v_1}\right) > 2$. It is not covered by a_2 either, because

$$\left(t_1 + \frac{l_1 + l_2 - l_3}{v_1}\right) - \left(t_2 - \frac{l_1 + l_2 - l_3}{v_2}\right) > (t_1 - t_2) + 2 + \frac{v_1 - v_3}{v_2} \geq 2.$$

Hence, it must be covered by a_3, which means that $a_3(t_3) = l_1 + l_2 - l_3$ for some t_3 with $t_1 + \frac{l_1+l_2-l_3}{v_1} - 2 \leq t_3 < t_1 + \frac{l_1+l_2-l_3}{v_1}$ (see Fig. 5).

If $t_3 + \frac{2l_3}{v_3} \leq t_1 + \frac{l}{v_1}$, then $(l; t_3 + \frac{2l_3}{v_3})$ is covered by none of a_1, a_2, a_3 (see Fig. 5). Otherwise, $(l; t_1 + \frac{l}{v_1})$ is covered by none of a_1, a_2, a_3 (not by a_3 because $(t_1 + \frac{l}{v_1}) - (t_3 - \frac{2l_3}{v_3}) > t_3 - (t_3 - 2) = 2$).

5.3   Case of $v_1 \leq 2v_2 + v_3$

Again, let t_1 and t_2 be such that $t^* < t_1 \leq t_2 \leq t_1 + 2$ and $a_1(t_1) = a_2(t_2) = 0$.

Lemma 4. $a_2(t) \neq l$ for any $t \in [t_1 - \frac{l}{v_1}, t_1 + \frac{l}{v_1}]$.

Proof. By the same argument as in Lemma 2, $(l; t_1 + \frac{l}{v_1})$ is covered by a_3. Since a_1 cannot visit l during $[t_1 - \frac{l}{v_1}, t_1 + \frac{l}{v_1}]$, neither can a_2, because of Lemma 3.   □

Lemma 5. $a_3(t) \neq l_1 + l_2$ for any $t \in [t_1 - \frac{l_2+l_3}{v_1}, t_1 + \frac{l_2+l_3}{v_1}]$ (see Fig. 6).

Proof. Assume that $a_3(t) = l_1 + l_2$ for some $t \in [t_1 - \frac{l_2+l_3}{v_1}, t_1 + \frac{l_2+l_3}{v_1}]$. Then, since $[t - 1, t + 1] \subseteq [t_1 - \frac{l}{v_1}, t_1 + \frac{l}{v_1}]$, $(l; t + 1)$ is covered by neither a_1 nor a_3. Furthermore, by Lemma 4, it is not covered by a_2, which is a contradiction.   □

Lemma 6. $a_3(t) = l_1 + l_2$ for some t with $t_1 + \frac{l_2+l_3}{v_1} < t < t_1 + \frac{l_1+l_2}{v_1}$ (see Fig. 6).

Proof. The pair $(l_1 + l_2; t_1 + \frac{l_1+l_2}{v_1})$ is not covered by a_1, because $\left(t_1 + \frac{l_1+l_2}{v_1}\right) - \left(t_1 - \frac{l_1+l_2}{v_1}\right) > 2$. It is not covered by a_2 either, because

$$\left(t_1 + \frac{l_1 + l_2}{v_1}\right) - \left(t_2 - \frac{l_1 + l_2}{v_2}\right) > (t_1 - t_2) + 2 + \left(\frac{v_2}{v_1} + \frac{v_1}{v_2}\right) \geq 2.$$

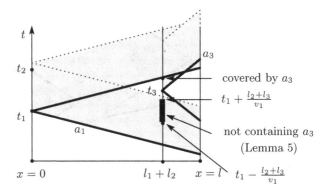

Fig. 6. Lemmas 5 and 6

Hence, it must be covered by a_3, which means that $a_3(t) = l_1 + l_2$ for some t with $t_1 + \frac{l_1+l_2}{v_1} - 2 \leq t < t_1 + \frac{l_1+l_2}{v_1}$.

Since $\left(t_1 + \frac{l_1+l_2}{v_1} - 2\right) - \left(t_1 - \frac{l_2+l_3}{v_1}\right) > \frac{2l_2+l_3}{v_1} - 1 > 0$ by the assumption $v_1 \leq 2v_2 + v_3$, we have $t_1 + \frac{l_1+l_2}{v_1} - 2 > t_1 - \frac{l_2+l_3}{v_1}$. Hence, by Lemma 5, $a_3(t) = l_1 + l_2$ for some t with $t_1 + \frac{l_2+l_3}{v_1} < t < t_1 + \frac{l_1+l_2}{v_1}$.   □

In what follows, let t_3 be the minimum number such that $a_3(t_3) = l_1 + l_2$ and $t_1 + \frac{l_2+l_3}{v_1} < t_3 < t_1 + \frac{l_1+l_2}{v_1}$ (see Fig. 6).

Lemma 7. $a_3(t) = l$ for some $t \in [t_1 + \frac{l}{v_1} - 2, t_3 - \frac{l_3}{v_3}]$.

Proof. By Lemma 4, $(l; t_1 + \frac{l}{v_1})$ is covered by neither a_1 nor a_2, which means that $a_3(t) = l$ for some t with $t_1 + \frac{l}{v_1} - 2 \leq t < t_1 + \frac{l}{v_1}$. On the other hand, since $a_3(t_3) = l_1 + l_2$, we have $a_3(t) \neq l$ for any t with $t_3 - \frac{l_3}{v_3} < t < t_3 + \frac{l_3}{v_3}$. By combining them, we have that $a_3(t) = l$ for some $t \in [t_1 + \frac{l}{v_1} - 2, t_3 - \frac{l_3}{v_3}]$.   □

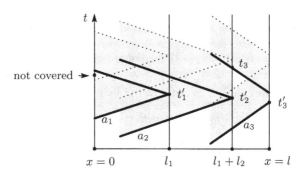

Fig. 7. Construction of t_3', t_2', and t_1'

Let t'_3 be the maximum number such that $a_3(t'_3) = l$ and $t'_3 \in [t_1 + \frac{l}{v_1} - 2, t_3 - \frac{l_3}{v_3}]$. Then, the pair $(l_1 + l_2; t_3)$ is not covered by a_3, because $t_3 > (t'_3 - \frac{l_3}{v_3}) + 2$ and $t'_3 + \frac{l_3}{v_3} > t_1 - \frac{l_2 + l_3}{v_1}$. It is not covered by a_1 either, because $t_3 > t_1 + \frac{l_2 + l_3}{v_1} > (t_1 - \frac{l_1 + l_2}{v_1}) + 2$ by $v_1 \le 2v_2 + v_3$. Hence, it is covered by a_2, which means that $a_2(t'_2) = l_1 + l_2$ for some t'_2 with $t_3 - 2 \le t'_2 < t_3$ (see Fig. 7).

Since $(l_1; t'_2 + \frac{l_2}{v_2})$ is covered by neither a_2 nor a_3, it is covered by a_1, which means that $a_1(t'_1) = l_1$ for some t'_1 with $t'_2 + \frac{l_2}{v_2} - 2 \le t'_1 < t'_2 + \frac{l_2}{v_2}$. In this case, $(0; t'_1 + \frac{l_1}{v_1})$ is covered by none of a_1, a_2, and a_3, which is a contradiction.

6 Conclusion

We have proved:

- The partition-based strategy is optimal when all agents have the same speed (Theorem 2), but not when there are two distinct speeds (Fig. 2).
- The partition-based strategy is optimal when there are two agents with different speeds and weights (Theorem 3), but not when there are three (Fig. 1).
- The partition-based strategy is optimal when there are three agents with the same weight (Theorem 4), but not when there are six (Fig. 1).

The third result settles a conjecture of Czyzowicz et al. [4]. It remains open whether the partition-based strategy is optimal for four and five (unweighted) agents.

Our examples in Theorem 1 outperform the partition-based strategy only slightly. Thus we propose a revised conjecture: there is some constant $c < 1$ such that a fence of length exceeding $c \sum_{i=1}^{k} v_i$ can never be patrolled.

Acknowledgements. The authors thank Yoshio Okamoto for suggesting this research. They also thank Kohei Shimane and Yushi Uno for helpful comments.

References

1. Almeida, A., Ramalho, G., Santana, H., Tedesco, P., Menezes, T., Corruble, V., Chevaleyre, Y.: Recent advances on multi-agent patrolling. In: Proc. 17th Brazilian Symposium on Artificial Intelligence, pp. 474–483 (2004)
2. Chevaleyre, Y.: Theoretical analysis of the multi-agent patrolling problem. In: Proc. IEEE/WIC/ACM Int. Conf. Intelligent Agent Technology, pp. 302–308 (2004)
3. Clark, J., Fierro, R.: Mobile robotic sensors for perimeter detection and tracking. ISA Transactions 46, 3–13 (2007)
4. Czyzowicz, J., Gąsieniec, L., Kosowski, A., Kranakis, E.: Boundary Patrolling by Mobile Agents with Distinct Maximal Speeds. In: Demetrescu, C., Halldórsson, M.M. (eds.) ESA 2011. LNCS, vol. 6942, pp. 701–712. Springer, Heidelberg (2011)
5. Elmaliach, Y., Agmon, N., Kaminka, G.A.: Multi-robot area patrol under frequency constraints. Ann. Math. Artif. Intell. 57, 293–320 (2009)

6. Elmaliach, Y., Shiloni, A., Kaminka, G.A.: A realistic model of frequency-based multi-robot polyline patrolling. In: Proc. 7th Int. Conf. Autonomous Agents and Multiagent Systems, pp. 63–70 (2008)
7. Pasqualetti, F., Franchi, A., Bullo, F.: On optimal cooperative patrolling. In: Proc. 49th IEEE Conference on Decision and Control, pp. 7153–7158 (2010)
8. Yanovski, V., Wagner, I.A., Bruckstein, A.M.: A distributed ant algorithm for efficiently patrolling a network. Algorithmica 37, 165–186 (2003)

Weak Visibility Queries of Line Segments in Simple Polygons

Danny Z. Chen[1,*] and Haitao Wang[2,**]

[1] Department of Computer Science and Engineering
University of Notre Dame, Notre Dame, IN 46556, USA
dchen@cse.nd.edu
[2] Department of Computer Science
Utah State University, Logan, UT 84322, USA
haitao.wang@usu.edu

Abstract. Given a simple polygon in the plane, we study the problem of computing the weak visibility polygon from any query line segment in the polygon. This is a basic problem in computational geometry and has been studied extensively before. In this paper, we present new algorithms and data structures that improve the previous results.

1 Introduction

Given a simple polygon $\mathcal{P}$ of n vertices in the plane, two points in $\mathcal{P}$ are *visible* to each other if the line segment joining them lies in $\mathcal{P}$. For a line segment s in $\mathcal{P}$, a point p is *weakly visible* (or *visible* for short) to s if s has at least one point that is visible to p. The *weak visibility polygon* (or *visibility polygon* for short) of s, denoted by $Vis(s)$, is the set of all points in $\mathcal{P}$ that are visible to s. The *weak visibility query problem* is to build a data structure for $\mathcal{P}$ such that $Vis(s)$ can be computed efficiently for any query line segment s in $\mathcal{P}$.

This problem has been studied before. Bose *et al.* [4] built a data structure of $O(n^3)$ size in $O(n^3 \log n)$ time that can compute $Vis(s)$ in $O(k \log n)$ time for any query, where k is the size of $Vis(s)$. Throughout this paper, we always let k be the size of $Vis(s)$ for any query line segment s. Bygi and Ghodsi [5] gave an improved data structure with the same size and preprocessing time as that in [4] but its query time is $O(k + \log n)$. Aronov *et al.* [1] proposed a smaller data structure of $O(n^2)$ size with $O(n^2 \log n)$ preprocessing time and $O(k \log^2 n)$ query time. Table 1 gives a summary. If the problem is to compute $Vis(s)$ for a single segment (not queries), then there is an $O(n)$ time algorithm [10].

In this paper, we present two new data structures whose performances are also given in Table 1. Let K be the size of the visibility graph of $\mathcal{P}$ [12], which is proportional to the number of pairs of vertices of $\mathcal{P}$ that are mutually visible. Note that $K = O(n^2)$. Our first data structure can be built in $O(K)$ preprocessing time, its size is $O(K)$, and it can compute $Vis(s)$ in $O(k \log n)$ time for any

---

* Chen's research was supported in part by NSF under Grant CCF-0916606.
** Corresponding author.

K.-M. Chao, T.-s. Hsu, and D.-T. Lee (Eds.): ISAAC 2012, LNCS 7676, pp. 609–618, 2012.
© Springer-Verlag Berlin Heidelberg 2012

Table 1. A summary of the data structures. The value k is the size of $Vis(s)$ for any query s and $K = O(n^2)$ is the size of the visibility graph of $\mathcal{P}$

Data Structure	Preprocessing Time	Size	Query Time
[4]	$O(n^3 \log n)$	$O(n^3)$	$O(k \log n)$
[5]	$O(n^3 \log n)$	$O(n^3)$	$O(k + \log n)$
[1]	$O(n^2 \log n)$	$O(n^2)$	$O(k \log^2 n)$
Our Result 1	$O(K)$	$O(K)$	$O(k \log n)$
Our Result 2	$O(n^3)$	$O(n^3)$	$O(k + \log n)$

query. Comparing with the data structure in [1], our data structure reduces the query time by a logarithmic factor and uses less preprocessing time and space.

The preprocessing time and size of our second data structure are both $O(n^3)$, and each query time is $O(k + \log n)$. Comparing with the result in [5], our data structure has less preprocessing time. Our techniques explore many geometric observations on the problem that may be useful elsewhere. For example, we prove a tight combinatorial bound for the "zone" in a line segment arrangement contained in a simple polygon, which is interesting in its own right.

2   Preliminaries

In this section, we review some geometric structures and discuss an algorithmic scheme that will be used by the query algorithms of both our data structures given in Sections 3 and 4. For simplicity of discussion, we assume no three vertices of $\mathcal{P}$ are collinear; we also assume for any query segment s, s is not collinear with any vertex of $\mathcal{P}$ and each endpoint of s is not collinear with any two vertices of $\mathcal{P}$. As in [1,4], our approaches can be easily extended to the general situation.

Denote by $\partial \mathcal{P}$ the boundary of $\mathcal{P}$. The *visibility graph* of $\mathcal{P}$ is a graph whose vertex set consists of all vertices of $\mathcal{P}$ and whose edge set consists of edges defined by all visible pairs of vertices of $\mathcal{P}$. Note that two adjacent vertices on $\partial \mathcal{P}$ are considered *visible* to each other. In this paper, we always use K to denote the size of the visibility graph of $\mathcal{P}$. Clearly, $K = O(n^2)$. The visibility graph can be computed in $O(K)$ time [12].

We discuss the *visibility decomposition* of $\mathcal{P}$ [1,4]. Consider a point p in $\mathcal{P}$ and a vertex v of $\mathcal{P}$. Suppose the line segment $\overline{pv}$ is in $\mathcal{P}$, i.e., p is visible to v. We extend $\overline{pv}$ along the direction from p to v and suppose we stay inside $\mathcal{P}$ (when this happens, v must be a reflex vertex). Let w be the point on the boundary $\mathcal{P}$ that is hit first by our above extension of $\overline{pv}$ (e.g., see Fig. 1). We call the line segment $\overline{vw}$ the *window* of p. The point p is called the *defining point* of the window and the vertex v is called the *anchor vertex* of the window. It is commonly known that the boundary of the visibility polygon of the point p consists of parts of $\partial \mathcal{P}$ and the windows of p [1,4]. If the point p is a vertex of $\mathcal{P}$, then the window $\overline{vw}$ is called a *critical constraint* of $\mathcal{P}$ and p is called the *defining vertex* of the critical constraint. For example, in Fig. 2, the two critical

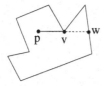
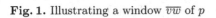

Fig. 1. Illustrating a window $\overline{vw}$ of p

Fig. 2. Illustrating the two critical constraints $\overline{vp_v}$ and $\overline{up_u}$ defined by the two mutually visible vertices u and v

constraints $\overline{up_u}$ and $\overline{vp_v}$ are both defined by u and v; for $\overline{up_u}$, its anchor vertex is u and its defining vertex is v, and for $\overline{vp_v}$, its anchor vertex is v and defining vertex is u. It is easy to see that the total number of critical constraints is $O(K)$ because each critical constraint corresponds to a visible vertex pair of $\mathcal{P}$ and a visible vertex pair corresponds to at most two critical constraints.

As in [1,4], we can represent the visibility polygon $Vis(s)$ of a segment s by a cyclic list of the vertices and edges of $\mathcal{P}$ in the order in which they appear on the boundary of $Vis(s)$, and we call the above list the *combinatorial representation* of $Vis(s)$ [1]. With the combinatorial representation, $Vis(s)$ can be explicitly determined in linear time in terms of the size of $Vis(s)$. Our query algorithms given later always report the combinatorial representation of $Vis(s)$.

The critical constraints of $\mathcal{P}$ partition $\mathcal{P}$ into cells, called the *visibility decomposition* of $\mathcal{P}$ and denoted by $VD(\mathcal{P})$. The visibility decomposition $VD(\mathcal{P})$ has a property that for any two points p and q in the same cell of $VD(\mathcal{P})$, the two visibility polygons $Vis(p)$ and $Vis(q)$ have the same combinatorial representation. Also, the combinatorial representations of the visibility polygons of two adjacent cells in $VD(\mathcal{P})$ have only $O(1)$ differences. The visibility decomposition has been used for computing visibility polygons of query points (not line segments) [1,4].

Consider a query segment s in $\mathcal{P}$. In the following, we discuss an algorithmic scheme for computing $Vis(s)$. Denote by a and b the two endpoints of s. Suppose we move a point p on s from a to b. We want to capture the combinatorial representation changes of $Vis(p)$ of the point p during its movement. Initially, p is at a and we have $Vis(p) = Vis(a)$. As p moves, the combinatorial representation of $Vis(p)$ changes if and only if p crosses a critical constraint of $\mathcal{P}$ [1,4]. $Vis(s)$ is the union of all such visibility polygons as p moves from a to b. Therefore, to compute $Vis(s)$, as in [1,4], we use the following approach. Initially, let $Vis(s) = Vis(p) = Vis(a)$. As p moves from a to b, when p crosses a critical constraint, either p sees one more vertex/edge, or p sees one less vertex/edge. If p sees one more vertex/edge, then we update $Vis(s)$ in constant time by inserting the new vertex/edge to the appropriate position of the combinatorial representation of $Vis(s)$. Otherwise, we do nothing (because although a vertex/edge is not visible to p any more, it is visible to s and thus should be kept; refer to [4] for details).

The above algorithm has two remaining issues. The first one is how to compute $Vis(a)$ of the point a. The second issue is how to determine the next critical constraint that will be crossed by p as p moves. Each of our two data structures

Fig. 3. Illustrating the principle child w of v in $T(p)$

given in Sections 3 and 4 does some preprocessing such that the corresponding query algorithm can resolve the above two issues efficiently.

3   The First Data Structure

Our goal is to compute $Vis(s)$ for any query segment s. Again, let $s = \overline{ab}$. As discussed before, we need to resolve two issues. The first issue is to compute $Vis(a)$. For this, we use a ray-shooting data structure [7,13], which, with $O(n)$ time preprocessing, can answer a ray-shooting query in $O(\log n)$ time. As discussed in [1], by using the ray-shooting data structure, with $O(n)$ time preprocessing, we can compute $Vis(a)$ in $O(|Vis(a)|\log n)$ time, where $|Vis(a)|$ is the size of $Vis(a)$.

The second issue is how to determine the next critical constraint of $\mathcal{P}$ that will be crossed by the point p as p moves from a to b. Suppose at the moment we know $Vis(p)$ (initially, $Vis(p) = Vis(a)$). Let β be the critical constraint that is crossed next by p. To determine β, we sketch an observation given in [1].

Denote by $T(p)$ the shortest path tree rooted at p, which is the union of the shortest paths in $\mathcal{P}$ from p to all vertices of $\mathcal{P}$. A vertex of $\mathcal{P}$ is in $Vis(p)$ if and only if it is a child of p in $T(p)$. For a node v of the tree $T(p)$ with $v \neq p$, suppose u is the parent of v; define the *principal child* of v to be the vertex w among the children of v such that the angle formed by $\overrightarrow{vw}$ and $\overrightarrow{uv}$ is the smallest among all such angles (see Fig. 3, where $u = p$). In other words, if we go from p to any child of v along the shortest path and we turn to the left (resp., right), then w is the first child of v that is hit by rotating counterclockwise (resp., clockwise) the line containing $\overline{uv}$ around v. To determine the next critical constraint β, the following observation was shown in [1]. Two children of p in $T(p)$ are *consecutive* if there are no other children of p between them in the cyclic order.

Observation 1. [1] *The next critical constraint β is defined by two vertices of $\mathcal{P}$ that are either two consecutive children of p or one, say v, is a child of p and the other is the principal child of v.*

Based on Observation 1, Aronov *et al.* [1] maintain $T(p)$ as p moves and used the balanced triangulation of $\mathcal{P}$ to determine the principal children. Here, we use a different approach, although we still need Observation 1. Our data structure consists of the following: the ray-shooting data structure [7,13], the visibility graph of $\mathcal{P}$, denoted by G, and a priority queue Q.

The ray-shooting data structure is used to compute $Vis(a)$ in $O(|Vis(a)|\log n)$ time. The visibility graph G is used to determine the principal children of all children of p, as follows. First of all, since we already know $Vis(p)$, we have all p's children in $T(p)$, sorted cyclically around p. Note that we do not store the entire tree $T(p)$. For any child v of p, the vertices of $\mathcal{P}$ that are visible to v are those adjacent vertices of v in the visibility graph G. (For simplicity, vertices of G also refer to the corresponding vertices of $\mathcal{P}$, and vice versa.) Denote by $adj(v)$ the set of the vertices of $\mathcal{P}$ visible to v. Let w be the principal child of v that we are looking for. Clearly, $w \in adj(v)$. Consider the ray $\rho(v)$ originating from v with the direction from p to v. By the definition of principle children, w is the vertex hit first by the ray $\rho(v)$ if we rotate $\rho(v)$ along the direction that is consistent with the turning direction of the shortest path from p to the children of v in $T(p)$. It is easy to see that once we know the above direction, we can obtain w in $O(\log n)$ time by binary search with the help of the visibility graph G. To determine the above direction, we only need to look at the relationship between the line containing $\rho(v)$ and the two edges of $\mathcal{P}$ adjacent to v. Specifically, assume the line containing $\rho(v)$ has the same direction as $\rho(v)$. If the two adjacent edges of v lie to the left of the above line (e.g., see Fig. 3), then we should rotate $\rho(v)$ counterclockwise to determine w; otherwise we rotate $\rho(v)$ clockwise. Hence, we can determine the principle child of v in $O(\log n)$ time. Initially, $p = a$ and we determine the principle children of all children of a in $T(a)$ in $O(|Vis(a)|\log n)$ time since a has $O(|Vis(a)|)$ children.

We use the priority queue Q to store the critical constraints stated in Observation 1 that intersect the line segment s, where the key of each such critical constraint is the position of its intersection with s. Initially when $p = a$, we compute the critical constraints defined by all pairs of consecutive children of p in $T(p)$. Similarly, for each child v of $T(p)$, we compute the critical constraint defined by v and its principal child. Note that the total number of the above critical constraints are $O(|Vis(a)|)$. For each such critical constraint, we check whether it intersects s. If so, we insert it into Q; otherwise, we do nothing. Then, the first critical constraint in Q is the next critical constraint that p will cross as it moves. In general, after p crosses a critical constraint, p either sees one more vertex or sees one less vertex. In either case, there are only a constant number of insertion or deletion operations on Q. Specifically, consider the case where p sees one more vertex u (and an adjacent edge of u). By the implementation given in [4], we can update the combinatorial representation of $Vis(p)$ in constant time (i.e., insert u and the adjacent edge to the appropriate positions of the cyclic list of $Vis(p)$). After this, u becomes a child of p in the new tree $T(p)$, and we can determine p's two other children, say, u_1 and u_2, which are neighbors of u, in constant time. Then, for u_1, we check whether the critical constraint defined by u and u_1 intersects s, and if so, we insert it into Q. For u_2, we do the same thing. Further, we compute the principal child of u in $T(p)$, in $O(\log n)$ time, by the approach discussed before. For the other case where p sees one less vertex after it crosses the critical constraint, we do similar processing. After p arrives at the other endpoint b of s, we obtain the combinatorial representation of $Vis(s)$.

We claim that the above algorithm takes $O(k \log n)$ time (recall $k = |Vis(s)|$). Indeed, the initialization takes $O(|Vis(a)| \log n)$ time. Clearly, $|Vis(a)| = O(k)$ since each vertex of $\mathcal{P}$ that is in $Vis(a)$ also appears in $Vis(s)$. If we consider every time p crosses a critical constraint as an *event*, each event takes $O(\log n)$ time. It has been shown in [1] that the total number of events as p moves from a to b is $O(k)$. Hence, the overall running time for computing $Vis(s)$ is $O(k \log n)$.

For the preprocessing, the ray-shooting data structure needs $O(n)$ time and space to build. Computing the visibility graph G takes $O(K)$ time and space. Further, in our query algorithm, the space used in the priority queue Q is always bounded by $O(k)$. We conclude this section with the following theorem.

Theorem 1. *For any simple polygon $\mathcal{P}$, we can build a data structure of size $O(K)$ in $O(K)$ time that can compute $Vis(s)$ in $O(|Vis(s)| \log n)$ time for each query segment s, where $K = O(n^2)$ is the size of the visibility graph of $\mathcal{P}$.*

4   The Second Data Structure

In general, the preprocessing of our second data structure is very similar to that in [4], and we make it faster by using better tools. Our improvement on the query algorithm is due to many new observations, e.g., a combinatorial bound of the "zone" of the line segment arrangements in simple polygons. For completeness, we briefly discuss the approach in [4].

The preprocessing in [4] has several steps, whose running time is $O(n^3 \log n)$ and is dominated by the first two steps. The other steps together take $O(n^3)$ time. We show below that the first two steps can be implemented in $O(n^3)$ time.

The preprocessing in [4] first computes the visibility decomposition $VD(\mathcal{P})$ of $\mathcal{P}$. Although there may be $\Omega(n^2)$ critical constraints in $\mathcal{P}$, it has been shown [4] that any line segment in $\mathcal{P}$ can intersect only $O(n)$ critical constraints, which implies the size of $VD(\mathcal{P})$ is $O(n^3)$ instead of $O(n^4)$. As discussed in [4], all critical constraints of $\mathcal{P}$ can be easily computed in $O(n^2)$ time. After that, to compute $VD(\mathcal{P})$, the authors in [4] used the algorithm in [2], which takes $O(n^3 \log n)$ time. In fact, a faster algorithm is available for computing $VD(\mathcal{P})$. Chazelle and Edelsbrunner's algorithm [6] can compute the planar subdivision induced by a set of m line segments in $O(m \log m + I)$ time, where I is the number of intersections of all lines. In our problem, we have $O(n^2)$ critical constraints each of which is a line segment and the boundary of $\mathcal{P}$ has n edges. Therefore, by using the algorithm in [6], we can compute $VD(\mathcal{P})$ in $O(n^3)$ time.

The second step of the preprocessing in [4] is to build a point location data structure on $VD(\mathcal{P})$ in $O(n^3 \log n)$ time. By the approaches in [9] or [14], we can build a point location data structure in $O(n^3)$ time.

The remaining steps of our preprocessing algorithm are the same as those in [4], which together take $O(n^3)$ time. Hence, the total preprocessing time is $O(n^3)$. With the preprocessing, for each query point q in $\mathcal{P}$, we can compute the visibility polygon $Vis(q)$ of q in $O(|Vis(q)| + \log n)$ time.

For a query segment $s = \overline{ab}$, the query algorithm in [4] first computes $Vis(a)$. Then, again, let a point p move on s from a to b. The algorithm maintains $Vis(p)$

as p moves, and initially $Vis(p) = Vis(a)$. Again, whenever p crosses a critical constraint, the combinatorial representation of $Vis(p)$ changes. Unlike our first data structure in Section 3, here we have $VD(\mathcal{P})$ explicitly. Therefore, we can determine the next critical constraint much easier. Specifically, the algorithm in [4] uses the following approach. Suppose p is currently in a cell of $VD(\mathcal{P})$; then the next critical constraint crossed by p must be on the boundary of the cell. Since each cell is convex, we can determine the above critical constraint in $O(\log n)$ time. The algorithm stops when p arrives at b. The total running time of the query algorithm is $O(k \log n)$, where $k = |Vis(s)|$.

We propose a new and simpler query algorithm. We follow the previous query algorithmic scheme. The only difference is when we determine the next critical constraint that will be crossed by p, we simply check each edge in the boundary of the current cell that contains p, and the running time is linear in terms of the number of edges of the cell. Therefore, the total running time of finding all critical constraints crossed by p as it moves on s is proportional to the total number of edges on all faces of $VD(\mathcal{P})$ that intersect s, and denote by $F(s)$ the set of such faces of $VD(\mathcal{P})$. Let $E(s)$ denote the set of edges of the faces in $F(s)$. The total running time of finding all critical constraints crossed by p is $O(|E(s)|)$. Note that the running time of the overall query algorithm is the sum of the time for computing $Vis(a)$ and the time for finding all critical constraints crossed by p. Since $Vis(a)$ can be found in $O(|Vis(a)| + \log n)$, the running time of the query algorithm is $O(\log n + |Vis(a)| + |E(s)|)$. Recall that $|Vis(a)| = O(k)$. In the following Lemma 1, we will prove that $|E(s)| = O(k)$. Consequently, the query algorithm takes $O(\log n + k)$ time and Theorem 2 thus follows.

Lemma 1. *The size of the set $E(s)$ is $O(k)$.*

Theorem 2. *For any simple polygon $\mathcal{P}$, we can build a data structure of size $O(n^3)$ in $O(n^3)$ time that can compute $Vis(s)$ in $O(|Vis(s)| + \log n)$ time for each query segment s.*

It remains to prove Lemma 1. Note that each edge of $E(s)$ lies either on $\partial \mathcal{P}$ or on a critical constraint. We partition the set $E(s)$ into two subsets $E_1(s)$ and $E_2(s)$. For each edge of $E(s)$, if it lies on $\partial \mathcal{P}$, then it is in $E_1(s)$; otherwise, it is in $E_2(s)$. We will show that both $|E_1(s)| = O(k)$ and $E_2(s) = O(k)$ hold.

Denote by $C(s)$ the set of all critical constraints each of which contains at least one edge of $E(s)$. Due to the space limit, the proof of Lemma 2 is omitted and can be found in our full paper.

Lemma 2. *The size of the set $C(s)$ is $O(k)$.*

In the next lemma, we bound the size of the subset $E_1(s)$.

Lemma 3. *The size of the set $E_1(s)$ is $O(k)$.*

Proof. Denote by $V(s)$ the set of vertices of $\mathcal{P}$ visible to s. Clearly, $|V(s)| \le k$. Consider an edge e in $E_1(s)$. To prove the lemma, we will charge e either to a vertex of $V(s)$ or to a critical constraint of $C(s)$. We will also show that each vertex of $V(s)$ will be charged at most twice and each critical constraint of $C(s)$ will be charged at most four times. Consequently, due to $|V(s)| \leq k$ and $|C(s)| = O(k)$ (by Lemma 2), the lemma follows.

By the definition of $E_1(s)$, e is on an edge of $\mathcal{P}$. If e has an endpoint that is a vertex of $\mathcal{P}$, say u, then clearly, u is visible to s. We charge e to u. Otherwise, both endpoints of e are endpoints of critical constraints, and we charge e to an arbitrary one of the above two critical constraints.

For each vertex of $\mathcal{P}$, it has two adjacent edges in $\mathcal{P}$, and therefore, it has at most two adjacent edges in $E_1(s)$. Hence, each vertex of $V(s)$ can be charged at most twice. On the other hand, each critical constraint has two endpoints, and each endpoint is adjacent to at most two edges in $E_1(s)$. Therefore, each critical constraint of $C(s)$ can be charged at most four times.

To prove Lemma 1, it remains to show $|E_2(s)| = O(k)$. To this end, we discuss a more general problem, in the following.

Assume we have a set S of line segments in $\mathcal{P}$ such that the endpoints of each segment are on $\partial\mathcal{P}$. Let $\mathcal{A}$ be the arrangement formed by the line segments in S and the edges of $\partial\mathcal{P}$. For any line segment s in $\mathcal{P}$ (the endpoints of s do not need to be on $\partial\mathcal{P}$), the *zone* of s is defined to be the set of all faces of $\mathcal{A}$ that s intersects. Denote by $Z(s)$ the zone of s. For each edge of a face in $\mathcal{A}$, it either lies on a line segment of S or lies on $\partial\mathcal{P}$; if it is the former case, we call the edge the *S-edge*. We define the *complexity* of $Z(s)$ as the number of S-edges of the faces in $Z(s)$ (namely, the edges on $\partial\mathcal{P}$ are not considered), denoted by Λ. Our goal is to find a good upper bound for Λ. By using the zone theorem for the general line segment arrangement [8], we can easily obtain $\Lambda = O(|S|\alpha(|S|))$, where $\alpha(\cdot)$ is the functional inverse of Ackermann's function [11].

Denote by S_s the set of line segments in S that intersect $Z(s)$, i.e., each segment in S_s contains at least one S-edge of $Z(s)$. Let $m = |S_s|$ (note that $m \leq |S|$). By using the property that each segment in S has both endpoints on $\partial\mathcal{P}$, we show that $\Lambda = O(m)$ in the following Theorem 3, which we call the *zone theorem*. The proof of Theorem 3 is given in Section 4.1.

Theorem 3. *The complexity of $Z(s)$ is $O(m)$.*

Now consider our original problem for proving $|E_2(s)| = O(k)$. By using the zone theorem, we have the following corollary.

Corollary 1. *The size of the set $E_2(s)$ is $O(k)$.*

Proof. The set $E_2(s)$ consists of all edges of $E(s)$ that lie on critical constraints. Consider the arrangement formed by all critical constraints of $\mathcal{P}$ and $\partial\mathcal{P}$. The complexity of the zone $Z(s)$ of the query segment s in the arrangement is exactly $|E_2(s)|$. Let $C'(s)$ be the set of critical constraints of $\mathcal{P}$ each of which contains at least one edge in $E_2(s)$. Then, by the zone theorem (Theorem 3), we have $|E_2(s)| = O(|C'(s)|)$. Note that $C'(s) \subseteq C(s)$. Due to $|C(s)| = O(k)$ (Lemma 2), we have $|E_2(s)| = O(k)$. The corollary thus follows.

Lemma 3 and Corollary 1 together lead to Lemma 1.

4.1 Proving the Zone Theorem (i.e., Theorem 3)

This subsection is entirely devoted to proving the zone theorem, i.e., Theorem 3. All notations here follow those defined before.

We partition the set S_s into two subsets: S_s^1 and S_s^2. For each segment in S_s, if it does not intersect the interior of s, then it is in S_s^1; otherwise, it is in S_s^2. Let $m_1 = |S_s^1|$ and $m_2 = |S_s^2|$. Hence, $m = m_1 + m_2$. Consider the arrangement formed by the line segments in S_s^1 and ∂P. Since no segment in S_s^1 intersects the interior of s, s must be contained in a face of the above arrangement and we denote by F_s the face. For each edge of F_s, if it lies on a segment of S, we also call it an S-edge. Note that edges of F_s that are not S-edges are on ∂P.

Lemma 4. *The number of S-edges of the face F_s is $O(m_1)$; the shortest path in P between any two points in F_s is contained in F_s.*

Proof. For each segment s' in S_s^1, since both endpoints of s' are on ∂P, s' partitions P into two simple polygons and one of them contains s, which we denote by $P(s')$. Note that the face F_s is the common intersection of $P(s')$'s for all s' in S_s^1. To prove the lemma, it is sufficient to show that each segment s' in S_s^1 has only one (maximal) continuous portion on the boundary of F_s, as follows.

For any two points p and q in P, denote by $\pi(p,q)$ the shortest path between p and q in P. Note that since P is a simple polygon, $\pi(p,q)$ is unique. We claim that for any two points p and q in the face F_s, $\pi(p,q)$ is contained in F_s. Indeed, suppose to the contrary that $\pi(p,q)$ is not contained in F_s. Then, $\pi(p,q)$ must cross the boundary of F_s. Since $\pi(p,q)$ cannot cross the boundary of P, $\pi(p,q)$ must cross an S-edge of F_s, and we assume s' is the segment in S_s^1 that contains the above S-edge. That implies $\pi(p,q)$ is also not contained in the polygon $P(s')$. Recall that the line segment s' partitions P into two simple polygons and one of them is $P(s')$. It is easy to show that for any two points in $P(s')$, their shortest path in P must be contained in $P(s')$. Therefore, we obtain a contradiction.

Now assume to the contrary that a segment s' in S_s^1 have two different continuous portions on the boundary of F_s. Let p and q be two points on the two portions of s', respectively. Thus, both p and q are in F_s. Since these are two discontinuous portions of s' on the boundary of F_s, the line segment $\overline{pq}$ is not contained in F_s. Since $\overline{pq}$ is on s', the shortest path $\pi(p,q)$ is $\overline{pq}$. But this means $\pi(p,q)$ is not contained in F_s, which incurs contradiction with our previous claim that $\pi(p,q)$ should be contained in F_s. Hence, we obtain that each segment s' in S_s^1 has only one continuous portion on the boundary of F_s, and consequently, the number of S-edges of the face F_s is $O(m_1)$.

Lemma 5 below shows a property of the face F_s.

Lemma 5. *For any line segment s' in P with two endpoints on ∂P, s' has at most one (maximal) continuous portion intersecting F_s; consequently, s' intersects at most two edges of F_s.*

Proof. Assume to the contrary that s' has two continuous portions intersecting F_s. Let p and q be two points on the two portions of s', respectively. Thus, both

p and q are in F_s. Clearly, the line segment $\overline{pq}$ is not contained in F_s. Since $\overline{pq}$ is on s', $\overline{pq}$ is the shortest path $\pi(p, q)$ between p and q in $\mathcal{P}$. But this means $\pi(p, q)$ is not contained in F_s, which incurs a contradiction with Lemma 4.

For each S-edge of $Z(s)$, it lies either on a segment in S_s^1 or on a segment in S_s^2; we call it an S_s^1-edge if it lies on a segment in S_s^1 and an S_s^2-edge otherwise. Due to $m = m_1 + m_2$, our zone theorem is an immediate consequence of the following Lemma 6. Note that we can obtain the zone $Z(s)$ of s by adding the segments in S_s^2 to F_s. To prove Lemma 6, we use induction on m_2, i.e., $|S_s^2|$. The approach is very similar to that in [3] used for line arrangements. Here, although we have line segments, the property that each line segment has both endpoints on $\partial \mathcal{P}$ makes the approach in [3] applicable with some modifications. The proof of Lemma 6 can be found in our full paper.

Lemma 6. *The zone $Z(s)$ has $O(m_2)$ S_s^2-edges and $O(m_1 + m_2)$ S_s^1-edges.*

References

1. Aronov, B., Guibas, L., Teichmann, M., Zhang, L.: Visibility queries and maintenance in simple polygons. Discrete and Computational Geometry 27(4), 461–483 (2002)
2. Bentley, J., Ottmann, T.: Algorithms for reporting and counting geometric intersections. IEEE Transactions on Computers 28(9), 643–647 (1979)
3. de Berg, M., Cheong, O., van Kreveld, M., Overmars, M.: Computational Geometry — Algorithms and Applications, 3rd edn. Springer, Berlin (2008)
4. Bose, P., Lubiw, A., Munro, J.: Efficient visibility queries in simple polygons. Computational Geometry: Theory and Applications 23(3), 313–335 (2002)
5. Bygi, M., Ghodsi, M.: Weak visibility queries in simple polygons. In: Proc. of the 23rd Canadian Conference on Computational Geometry, CCCG (2011)
6. Chazelle, B., Edelsbrunner, H.: An optimal algorithm for intersecting line segments in the plane. Journal of the ACM 39(1), 1–54 (1992)
7. Chazelle, B., Edelsbrunner, H., Grigni, M., Guibas, L., Hershberger, J., Sharir, J., Snoeyink, J.: Ray shooting in polygons using geodesic triangulations. Algorithmica 12(1), 54–68 (1994)
8. Edelsbrunner, H., Guibas, L., Pach, J., Pollack, R., Seidel, R., Sharir, M.: Arrangements of curves in the plane topology, combinatorics, and algorithms. Theoretical Computer Science 92(2), 319–336 (1992)
9. Edelsbrunner, H., Guibas, L., Stolfi, J.: Optimal point location in a monotone subdivision. SIAM Journal on Computing 15(2), 317–340 (1986)
10. Guibas, L., Hershberger, J., Leven, D., Sharir, M., Tarjan, R.: Linear-time algorithms for visibility and shortest path problems inside triangulated simple polygons. Algorithmica 2(1-4), 209–233 (1987)
11. Hart, S., Sharir, M.: Nonlinearity of Davenport–Schinzel sequences and of generalized path compression schemes. Combinatorica 6(2), 151–177 (1986)
12. Hershberger, J.: An optimal visibility graph algorithm for triangulated simple polygons. Algorithmica 4, 141–155 (1989)
13. Hershberger, J., Suri, S.: A pedestrian approach to ray shooting: Shoot a ray, take a walk. Journal of Algorithms 18(3), 403–431 (1995)
14. Kirkpatrick, D.: Optimal search in planar subdivisions. SIAM Journal on Computing 12(1), 28–35 (1983)

Beyond Homothetic Polygons: Recognition and Maximum Clique*

Konstanty Junosza-Szaniawski[1], Jan Kratochvíl[2],
Martin Pergel[3], and Paweł Rzążewski[1]

[1] Warsaw University of Technology, Faculty of Mathematics and Information Science,
Koszykowa 75, 00-662 Warszawa, Poland
{k.szaniawski,p.rzazewski}@mini.pw.edu.pl
[2] Department of Applied Mathematics, and Institute for Theoretical Computer
Science, Charles University, Malostranské nám. 25, 118 00 Praha 1, Czech Republic
honza@kam.ms.mff.cuni.cz
[3] Department of Software and Computer Science Education, Charles University,
Malostranské nám. 25, 118 00 Praha 1, Czech Republic
perm@kam.mff.cuni.cz

Abstract. We study the CLIQUE problem in classes of intersection graphs of convex sets in the plane. The problem is known to be NP-complete in convex-sets intersection graphs and straight-line-segments intersection graphs, but solvable in polynomial time in intersection graphs of homothetic triangles. We extend the latter result by showing that for every convex polygon P with k sides, every n-vertex graph which is an intersection graph of homothetic copies of P contains at most n^{2k} inclusion-wise maximal cliques. We actually prove this result for a more general class of graphs, so called k_{DIR}-CONV, which are intersection graphs of convex polygons whose all sides are parallel to at most k directions. We further provide lower bounds on the numbers of maximal cliques, discuss the complexity of recognizing these classes of graphs and present relationship with other classes of convex-sets intersection graphs.

1 Introduction

Geometric representations of graphs, and intersection graphs in particular, are widely studied both for their practical applications and motivations, and for their interesting theoretical and structural properties. It is often the case that optimization problems NP-hard for general graphs can be solved, or at least approximated, in polynomial time on such graphs. Classical examples are the STABLE SET, CLIQUE or COLORING problems for interval graphs, one of the oldest intersection defined classes of graphs [4]. The former two problems remain polynomially solvable in circle and polygon-circle graphs, while the last one already becomes NP-complete. For definitions and more results about these issues, the interested reader is referred to [5] or [18].

---

* This work was supported by a Czech research grant GAČR GIG/11/E023.

K.-M. Chao, T.-s. Hsu, and D.-T. Lee (Eds.): ISAAC 2012, LNCS 7676, pp. 619–628, 2012.
© Springer-Verlag Berlin Heidelberg 2012

In this paper we want to investigate subclasses of the class of intersection graphs of convex sets in the plane, denoted by CONV, and the computational complexity of the problem of finding a maximum clique in such graphs. This has been inspired by a few starting points. First, the CLIQUE problem was shown polynomial time solvable for intersection graphs of homothetic triangles in the plane by Kaufmann et al. [8]. (It has been shown that these graphs are equivalent to the so called max-tolerance graphs, and as such found direct application in DNA sequencing.) Secondly, the CLIQUE problem is known to be NP-complete in CONV graphs [12], and so it is interesting to inspect the boundary between easy and hard instances more closely. Straight line segments are the simplest convex sets, and it is thus natural to ask how difficult is CLIQUE in intersection graphs of segments in the plane (this class is denoted by SEG). Kratochvíl and Nešetřil posed this problem in [14] after they observed that if the number of different directions of the segments is bounded by a constant, say k, a maximum clique can be found in time $O(n^{k+1})$ (this class of graphs is denoted by k-DIR). This question has been answered very recently by Cabello et al. [2] who showed that CLIQUE is NP-complete in SEG graphs.

In [15], Kratochvíl and Pergel initiated a study of P_{hom} graphs, defined as intersection graphs of convex polygons homothetic to a single polygon P. They announced that for every convex polygon P, recognition of P_{hom} graphs is NP-hard, and asked in Problem 3.1 if P_{hom} graphs can have superpolynomial number of maximal cliques. Our main result shows that for every convex k-gon P, every P_{hom} graph with n vertices contains at most n^{2k} maximal cliques, and hence CLIQUE is solvable in polynomial time on P_{hom} graphs for every fixed polygon P. For the sake of completeness, we will also present the proof of NP-hardness of P_{hom} recognition.

In [19], E.J. and J. van Leeuwens considered a more general class of graphs based on affine transformations of one (or more) master objects, called $\mathcal{P}$-intersection graphs, where $\mathcal{P} = (S, T)$ is a signature consisting of a set of master objects S and a set of transformations T. They prove that if all objects in the signature are described by rational numbers, such graphs have representations of polynomial size and the recognition problem is in NP. As a corollary, recognition of P_{hom} graphs is in NP (and hence NP-complete) for every rational polygon P. In [16], van Leeuwens and T. Müller prove tight bounds on the maximum sizes of representations (in terms of coordinate sizes) of $P_{translate}$ and P_{hom} graphs.

In proving the main result of our paper, the polynomial bound on the number of maximal cliques, we go beyond the homothetic intersection graphs. We observe that in any representation by polygons homothetic to a master one (or to one of a finite set of master polygons), the sides of the polygons are parallel to a bounded number of directions in the plane. So we relax the requirement on homothetic relation of the polygons in the representation and we simply consider consider a finite set of directions and look after graphs that have intersection representations by convex polygons with sides parallel to (some of) these directions (see Figure 1). We prove that every such graph has at most n^{2k} maximal cliques, where k is the number of chosen directions. We find this fact worth emphasizing, as it also covers van Leeuwens' $\mathcal{P}$-intersection graphs for transformations without rotations. So we further investigate the class of k_{DIR}-CONV graphs (i.e. intersection graphs of convex polygons whose all sides are parallel to at most k directions), discuss the complexity of its recognition and relationship to other relevant graph classes (SEG, k-DIR, and P_{hom}).

In the last section we pay a closer attention to maximal cliques in P_{hom} graphs for specific polygons P. We show that under certain conditions on the shape of P (parallel opposite sides or certain angles of consecutive sides), one can improve the upper bounds. We also show that, for every fixed polygon but parallelograms, there exists a P_{hom} graph with $\Omega(n^3)$ maximal cliques (by a modification of a construction for triangles from [8]). It is worth noting that also for the max-coordinate results of [16], parallelograms play an exceptional role.

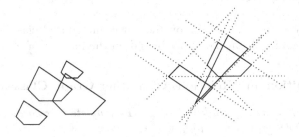

Fig. 1. Homothetic pentagons (left) and polygons with 5 directions of sides (right)

2   Preliminaries and Basic Definitions

In the paper we deal with intersection graphs of subsets of the Euclidean plane $\mathbb{R}^2$. The following concepts are standard and we only briefly overview them to make the paper self-contained. For a collection of sets R the *intersection graph* of R is denoted by $IG(R)$; its vertices are in 1-1 correspondence with the sets and two vertices are adjacent if and only if the corresponding sets are non-disjoint. In such a case the collection R is called an *(intersection) representation* of G, and the set corresponding to a vertex $v \in V(G)$ is denoted by R_v.

The intersection graphs of straight-line segments are called the SEG graphs, of convex sets the CONV graphs, and k-DIR is used for SEG graphs having a representation with all the segments being parallel to at most k directions (thus 1-DIR are exactly the interval graphs). For a fixed set (in most cases a convex polygon) P, the class of intersection graphs of sets homothetic to P is denoted by P_{hom} (two sets are homothetic if one can be obtained from the other one by scaling and/or translating). If P is a disk, we get disk-intersection graphs, a well studied class of graphs. Pseudodisk intersection graphs are intersection graphs of collections of sets that are pairwise in the pseudodisk relationship, i.e., both differences $A \setminus B$ and $B \setminus A$ are arc-connected. It is well known that P_{hom} graphs are pseudodisk intersection graphs for every convex set P.

Now we introduce the main character of the paper. Let $\mathcal{L}$ be the set of all distinct lines in $\mathbb{R}^2$ that contain the point $(0,0)$. For a k-tuple of lines $L = \{\ell_1, .., \ell_k\} \in \binom{\mathcal{L}}{k}$, we denote by $\mathcal{P}(L)$ the family of all polygons P such that every side of P is parallel to some $\ell \in L$ and we denote by $k_{\text{DIR}(L)}$-CONV the class of intersection graphs of polygons of $\mathcal{P}(L)$. Moreover, we define k_{DIR}-CONV $= \bigcup_{L \in \binom{\mathcal{L}}{k}} k_{\text{DIR}(L)}$-CONV. Fig. 1 shows examples of representations of the same graph by intersections of homothetic pentagons and as a 5_{DIR}-CONV graph. The following property of disjoint convex polygons is well known.

Lemma 1 (Folklore). *Let $L \subseteq \mathcal{L}$. Any two disjoint convex polygons in $\mathcal{P}(L)$ can be separated by a line parallel to some $\ell \in L$.*

3 k_{DIR}-CONV Graphs

In this section we present results on the newly introduced class of graphs, including the main result on the upper bound on the number of maximal cliques.

3.1 Recognition and Relations with other Graph Classes

Theorem 1. *For every fixed $k \geq 2$, it is NP-complete to recognize*
1. *$k_{DIR(L)}$-CONV graphs for any $L \in \binom{\mathcal{L}}{k}$,*
2. *k_{DIR}-CONV graphs.*

Proof. As 2_{DIR}-CONV graphs are exactly graphs of boxicity at most 2, they are NP-complete to recognize [10].

For $k > 2$, the class of k_{DIR}-CONV graphs contains the class of 3-DIR graphs and simultaneously is contained in CONV, thus to show NP-hardness we may apply the reduction from [10]. This is a unified reduction that shows NP-hardness of k-DIR graphs (for any $k \geq 3$) and CONV graphs in one step (reducing from satisfiability and showing that the obtained graph is in 3-DIR if the initial formula is satisfiable, but is not in CONV if it is not). Since for $k = 3$ all triples of directions are equivalent under an affine transformation of the plane, this shows that for $k > 2$, recognition of k_{DIR}-CONV, and also of $k_{\text{DIR}(L)}$-CONV for any $L \in \binom{\mathcal{L}}{k}$, are NP-hard.

Similarly to [13] we show that the recognition problem for both $k_{\text{DIR}(L)}$-CONV and k_{DIR}-CONV is in NP. We have to establish a polynomially-large certificate. As this certificate we take a combinatorial description of the arrangement, i.e., we guess a description saying in what order individual sides of individual polygons get intersected. We also need the information about particular corners of individual polygons. For individual sides of polygons we also need the information what direction they follow (for the direction it is sufficient to keep the index of the direction, i.e. a number in $\{1, .., k\}$). To make the situation formally easier, instead of segments we consider a description of the whole underlying lines. Note that a corner of a polygon and the intersection of boundaries appear here as intersection of two lines.

Now we have to verify the realizability of such an arrangement (again, in the same way as [13]). To do this we construct a linear program consisting of inequalities describing the ordering of the intersections along each side of each polygon. For a line p described by the equation $y = a_p x + b_p$ the intersection with q precedes the intersection with r ("from the left to the right") if $\frac{b_q - b_p}{a_p - a_q} < \frac{b_r - b_p}{a_p - a_r}$. In the case of prescribed directions (a_p, a_q, a_r) we have a linear program (whose variables are the b's) and this linear program can be solved in a polynomial time. This shows NP-membership for $k_{\mathrm{DIR}(L)}$-CONV.

For k_{DIR}-CONV we follow the argument of [13], which argues that the directions (obtained as solutions of this linear program) are of polynomial size and thus they may be also a part of a polynomial certificate. □

We observe that for every $k \geq 2$, each k-DIR graph (see [13] for more information about this class of graphs) is also a k_{DIR}-CONV graph. To show that, let us consider a segment representation of some $G \in k$-DIR. It is enough to extend every segment from a segment representation of G to a very narrow parallelogram in such a way that no new intersections appear. Moreover, one can show that this inclusion is proper, i.e. k_{DIR}-CONV $\not\subseteq k$-DIR.

3.2   The Number of Maximal Cliques

Theorem 2. *The number of maximal cliques in any n-vertex graph in k_{DIR}-CONV is at most n^{2k}.*

Proof. Let G be an n-vertex graph in k_{DIR}-CONV and let R be its representation by convex sets with sides parallel to k directions. Let $L = \{\ell_1, .., \ell_k\} \in \binom{\mathcal{L}}{k}$ be the set of lines determined by the sides of the polygons (without loss of generality we may assume that the representation determines exactly k directions). Let Q be an arbitrary maximal clique in G.

For $i \in \{1, .., k\}$, let w_i be a normal vector of ℓ_i (there are two possible orientations for w_i). Let $\overline{W} = \{w_i, -w_i : i \in \{1, .., k\}\}$. For every $w \in \overline{W}$, let P_w denote the last polygon in $\{R_v : v \in Q\}$ encountered by a sweeping line perpendicular to w, in the direction of w. Let $\widetilde{Q} = \{v \in V(G) : R_v \cap P_w \neq \emptyset$ for all $w \in \overline{W}\}$.

Since Q is a clique, obviously $Q \subseteq \widetilde{Q}$. Suppose that there exists $x \in Q$ and $y \in \widetilde{Q}$ such that $xy \notin E(G)$ (and therefore $R_x \cap R_y = \emptyset$). From Lemma 1 we know that there exists a line ℓ separating R_x and R_y, which is perpendicular to some $w \in \overline{W}$. Since both w and $-w$ belong to $\overline{W}$, without loss of generality we may assume that R_x is above ℓ and R_y is below ℓ with respect to direction w. Since P_w was the last polygon found by the sweeping line in the direction w, it is clearly above ℓ and therefore R_y does not intersect it – a contradiction. Therefore every $y \in \widetilde{Q}$ is adjacent to every $x \in Q$. Since Q is a maximal clique, we obtain that $Q = \widetilde{Q}$.

Notice that we can choose the set $\widetilde{Q}$ in at most n^{2k} ways – there are $2k$ vectors w in $\overline{W}$ and for each of them we choose one polygon P_w out of n. □

Theorem 3. *For any $k \geq 2$, the maximum number of maximal cliques over all n-vertex graphs in k_{DIR}-CONV is $\Omega(n^{k(1-\epsilon)})$, for any $\epsilon > 0$.*

Proof (Sketch). Because every k-DIR graph is k_{DIR}-CONV, it is enough to give the construction for k-DIR graphs. For each direction we take $\frac{n}{k}$ segments and position them in such a way that every pair of non-parallel segments intersect. Every maximal clique contains exactly one segment from each direction, so the number of maximal cliques is $(\frac{n}{k})^k = \Omega(n^{k-\log k}) = \Omega(n^{k(1-\epsilon)})$ for any $\epsilon > 0$.   □

4 P_{hom} Graphs

In this section we deal with intersection graphs of homothetic polygons. We first present the proof of the NP-hardness result that was announced in [15] and relationships with other subclasses of CONV. Then we present improved upper bounds on the number of maximal cliques for special shapes of P and cubic lower bounds.

4.1   Recognition and Relations with other Graph Classes

Theorem 4. *For every convex polygon P, the recognition problem of P_{hom} graphs is NP-hard.*

Proof. As the first step we refer to [8] which proves the NP-hardness for homothetic triangles.

For polygons with more corners, we use the same reduction as is used in [6]. The reduction establishes a graph from an instance of a special version of NAE-SAT problem, in which each clause contains three literals and each variable occurs at most four times. The formula is satisfiable iff the graph corresponding to it is a pseudodisk intersection graph, and if the formula is satisfiable, the graph is even a disk intersection one. We show that the same holds also for P_{hom} graphs.

Given a formula, we establish a graph such that the variables get replaced by variable-gadgets, clauses get replaced by clause-gadgets. The variable gadget is a cycle (from which individual pairs of paths depart). The truth assignment is reflected by the orientation of the cycle (in the representation the cycle is represented either "clockwise" or "counter-clockwise"). Each occurrence of a variable is represented by a pair of paths forming a ladder. In any representation by pseudodisks, one (path) is always to the left, the other to the right. This orientation carries the truth-assignment from variable-gadget to the clause-gadget. The ladders are further connected by cross-over gadgets that allow the ladders either to cross or just to touch, depending on what is needed in the construction (while still keeping the left-right order within the ladders). All these gadgets are the same as in [6].

First we observe use the fact that the graph constructed from a nonsatisfiable formula is not a pseudodisk intersection graph, and hence not a P_{hom} one.

The counterpart (showing that graph corresponding to satisfiable formulas are in P_{hom}) is rather technical and we just sketch it. For a given polygon P, we first find a bounding box. The bounding box is (for our purpose) a parallelogram such that each of its sides contains exactly one corner of P, and none of the corners of the bounding box belongs to P. The rest is just a technical analysis that the appropriate gadgets can be represented using the fact that the shape contains at least the convex combination of the corners on the sides of the parallelogram and that the sides of that parallelogram contain no other points from the shape.

Note that for the analysis we may consider the parallelogram to be a rectangle (if it is not, we apply a linear transformation on the whole arrangement to get a rectangle, we find a representation with rectangles and we apply the inverse transformation).

To show representability of clause-gadgets, it is necessary to draw appropriate pictures which we omit due to the page-limit. They can be found in [17]. $\square$

The NP-membership of the recognition of P_{hom} graphs has been shown in [19]. We quickly review an argument based on an approach of [13]. We proceed exactly in the same way as for k_{DIR}-CONV graphs, but we extend the linear program with equations controlling the ratios of side-lengths (for individual polygons). For intersections of a line p with neighboring sides q and r or a polygon A instead of the inequality checking the ordering of the intersections we add the following equation: $\frac{b_r - b_p}{a_p - a_r} - \frac{b_q - b_p}{a_p - a_q} = k_p \cdot s_A$, where s_A (a variable) represents the size of a polygon A. Note that the denominators are again constants, again we obtain a linear program. If the shape is fixed, we are done. But even if the polygon P is not given, we can regard the directions of its sides as variables and use the same trick as in the proof of Theorem 1. Thus we obtain the following strengthening, which partially solves Problem 6.3 in [16].

Theorem 5. *For every fixed k, the problem of deciding whether there exists a convex k-gon P such that an input graph is in P_{hom} is NP-complete.*

We conclude this subsection by pointing out the relationships of P_{hom} with the classes k_{DIR}-CONV and k-DIR. One can show that k_{DIR}-CONV $\not\subseteq P_{hom}$ for any k and P. Moreover, k-DIR $\not\subseteq P_{hom}$ and $P_{hom} \not\subseteq k$-DIR for any P and $k \geq 2$.

4.2   The Number of Maximal Cliques

By considering directions of the sides we observe that for any convex p-gon P with q pairs of parallel sides, every P_{hom} graph is also a $(p-q)_{DIR}$-CONV graph. From Theorem 2 then follows that there are at most $n^{2(p-q)}$ maximal cliques in any P_{hom} graph, for such a P. We can improve this bound, but we need one more definition.

Definition 1. *For a polygon P, let F_1, F_2 and F_3 be three consecutive sides and let ℓ_1, ℓ_2 and ℓ_3 denote the lines containing them, respectively. Let h be a half-space with boundary ℓ_2, which does not contain P. A side F_2 is weak if ℓ_1 and ℓ_3 do not intersect in h and there is no side parallel to F_2.*

For a polygon P and a vector w let ℓ be a supporting line of P perpendicular to w such that P is above ℓ in the direction of w, by $F(P, w)$ we denote the side of P contained in ℓ.

Lemma 2. *Let P_1 and P_2 be two homothetic polygons. Let w be a vector perpendicular to a side of P_1 (and therefore to a side of P_2). Assume P_1 and P_2 intersect and that P_2 is below P_1 in the direction of w (i.e. a sweeping line in direction of w encounters P_1 after P_2). If the side $F(P_1, w)$ is weak, then P_2 intersects it.*

Analogously as in Theorem 2 we can show the following theorem. The improvement is based on Lemma 2 and the observation that for each weak side we need only one normal vector instead of two opposite ones.

Theorem 6. *Let P be a convex p-gon with q pairs of parallel sides and s weak sides, which is not a parallelogram. Any P_{hom} graph contains at most $n^{2(p-q)-s}$ maximal cliques.*

Finally we list explicit upper bounds for special shapes P, accompanied also by lower bounds.

Theorem 7. *Let $f_P(n)$ denote the maximum number of maximal cliques over all n-vertex P_{hom} graphs for a polygon P. Then:*
1. $f_P(n) = \Theta(n^2)$ *if P is a parallelogram,*
2. $f_P(n) = \Theta(n^3)$ *if P is a triangle,*
3. $f_P(n) = O(n^4)$ *if P is a trapezoid,*
4. $f_P(n) = O(n^6)$ *if P is an arbitrary tetragon,*
5. $f_P(n) = \Omega(n^3)$ *if P is not a parallelogram.*

Proof (Sketch). 1. The proof for the upper bound is similar to the proof of Theorem 6. The difference is that in the case of parallelograms it is enough to consider just two normal vectors, one for each pair of parallel sides.

For the lower bound it is enough to consider the family of squares $\{S(\frac{i}{n/2-1}, 1 - \frac{i}{n/2-1}), S(\frac{1}{2} + \frac{i}{n/2-1}, \frac{3}{2} - \frac{i}{n/2-1}): i \in \{0, .., \frac{n}{2} - 1\}\}$, where $S(x, y)$ denotes the unit square with left-bottom corner at the point (x, y). This construction can be generalized for an arbitrary parallelogram.

2. For the upper bound it is easy to notice that every side of a triangle is weak. Therefore, from Theorem 6 we obtain $f_P(n) \leq n^{2(3-0)-3} = n^3$. The lower bound is obtained with the construction presented by Kaufmann et al. [8].

3. Suppose that P, but not a parallelogram. In every such trapezoid both non-parallel sides are weak. Therefore, $f_P(n) \leq n^{2(4-1)-2} = n^4$.

4. Suppose P is not a trapezoid. Notice that if a side of P is not weak, then the sum of the angles adjacent to this side is greater than π. Since the sum of all angles in a tetragon is 2π, for every pair of opposite sides, at least one side is weak. Therefore, $f_P(n) \leq n^{2(4-0)-2} = n^6$.

5. For a line ℓ, let $D(\ell)$ be the side of P parallel to ℓ (if there is such) or the corner of P at the largest distance from ℓ.

Choose a side of P and call it F_1, let ℓ_1 be the line containing F_1 and $P_1 = D(\ell_1)$. Let F_2, F_3 be sides of P adjacent to P_1 and let ℓ_2, ℓ_3 be lines containing F_2, F_3, respectively and let $P_2 = D(\ell_2)$ and $P_3 = D(\ell_3)$.

Let h, r, t, v be four copies of P. By $F_i^h, F_i^r, F_i^t, F_i^v, P_i^h, P_i^r, P_i^t, P_i^v$ we denote the sides or corners in polygons h, r, t, v corresponding to F_i, P_i in polygon P for $i \in \{1, 2, 3\}$, respectively (see Figure 2). We can adjust the sizes and positions of h, r, t, v in such a way that (i) t and h are touching and F_1^t intersects with P_1^h, (ii) t and v are touching and F_1^t intersects with P_1^v, (iii) h and v are touching and F_2^h intersects with P_2^v, (iv) r and t are touching and F_3^r intersects with P_3^t, (v) r and h intersect, (vi) r and v intersect.

For every polygon $x \in \{h, r, t, v\}$ we make $\frac{n}{4}$ copies $x_1, .. x_{\frac{n}{4}}$ and move them slightly comparing to the position of h, r, t, v in such a way that: (i) t_i and v_j intersect iff $i \geq j$, (ii) h_i and v_j intersect iff $i \leq j$, (iii) h_j and t_j intersect iff $i \leq j$, (iv) r_i and t_j intersect iff $i \geq j$, (v) r_i and v_j intersect for all $i, j \in \{1, .., \frac{n}{4}\}$, (vi) r_i and h_j intersect for all $i, j \in \{1, .., \frac{n}{4}\}$. For any α, β, γ such that $1 \leq \alpha \leq \beta \leq \gamma \leq \frac{n}{4}$ the set $\{h_1, \ldots, h_\alpha, v_\alpha, \ldots, v_\beta, t_\beta, \ldots, t_\gamma, r_\gamma, \ldots, r_{\frac{n}{4}}\}$ is a maximal clique in G. Hence there are $\binom{\frac{n}{4}}{3} = \Omega(n^3)$ maximal cliques in total. □

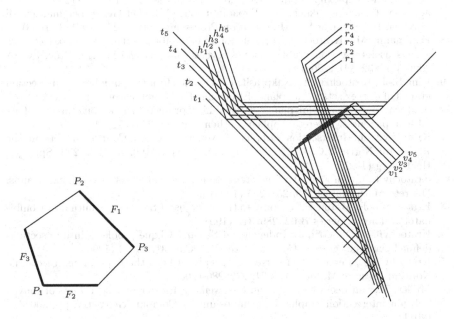

Fig. 2. Idea of the construction (for $n = 20$)

5 Conclusion

In this paper we have shown that the number of maximal cliques in any k_{DIR}-CONV graph (and therefore any P_{hom} graph for a k-gon P) is at most n^{2k}. Since maximal cliques can be enumerated with polynomial delay [7], the

CLIQUE problem can be solved in time $O(n^{f(k)})$ for any $G \in k_{\text{DIR}}$-CONV, and therefore is in XP when parameterized by k. It is interesting to know if this problem is in FPT.

Acknowledgement. We thank Michael Kaufmann for valuable discussions on the topic.

References

1. Badent, M., Binucci, C., Di Giacomo, E., Didimo, W., Felsner, S., Giordano, F., Kratochvíl, J., Palladino, P., Patrignani, M., Trotta, F.: Homothetic Triangle Contact Representations of Planar Graphs. In: Proc. CCCG 2007, pp. 233–236 (2007)
2. Cabello, S., Cardinal, J., Langerman, S.: The Clique Problem in Ray Intersection Graphs. In: Epstein, L., Ferragina, P. (eds.) ESA 2012. LNCS, vol. 7501, pp. 241–252. Springer, Heidelberg (2012)
3. Corneil, D.G., Olariu, O., Stewart, L.: The LBFS Structure and Recognition of Interval Graphs. SIAM J. Discrete Math. 23, 1905–1953 (2009)
4. Gilmore, P.C., Hoffman, A.J.: A characterization of comparability graphs and of interval graphs. Can. J. Math. 16, 539–548 (1964)
5. Golumbic, M.: Algorithmic Graph Theory and Perfect Graphs. Acad. Press (1980)
6. Hliněný, P., Kratochvíl, J.: Representing graphs by disks and balls (a survey of recognition-complexity results). Discrete Mathematics 229, 101–124 (2001)
7. Johnson, D.S., Yannakakis, M., Papadimitriou, C.H.: On GeneraLemmating All Maximal Independent Sets. Information Processing Letters 27, 119–123 (1988)
8. Kaufmann, M., Kratochvíl, J., Lehmann, K., Subramanian, A.: Max-tolerance graphs as intersection graphs: cliques, cycles, and recognition. In: Proc. SODA 2006, pp. 832–841 (2006)
9. Kim, S.-J., Kostochka, A., Nakprasit, K.: On the chromatic number of intersection graphs of convex sets in the plane. Electronic J. of Combinatorics 11, #R52 (2004)
10. Kratochvíl, J.: A special planar satisfiability problem and a consequence of its NP-completeness. Discrete Applied Mathematics 52, 233–252 (1994)
11. Kratochvíl, J.: Intersection Graphs of Noncrossing Arc-Connected Sets in the Plane. In: North, S.C. (ed.) GD 1996. LNCS, vol. 1190, pp. 257–270. Springer, Heidelberg (1997)
12. Kratochvíl, J., Kuběna, A.: On intersection representations of co-planar graphs. Discrete Mathematics 178, 251–255 (1998)
13. Kratochvíl, J., Matoušek, J.: Intersection Graphs of Segments. Journal of Combinatorial Theory, Series B 62, 289–315 (1994)
14. Kratochvíl, J., Nešetřil, J.: Independent Set and Clique problems in intersection-defined classes of graphs. Comm. Math. Uni. Car. 31, 85–93 (1990)
15. Kratochvíl, J., Pergel, M.: Intersection graphs of homothetic polygons. Electronic Notes in Discrete Mathematics 31, 277–280 (2008)
16. Müller, T., van Leeuwen, E.J., van Leeuwen, J.: Integer representations of convex polygon intersection graphs. In: Symposium on Comput. Geometry, pp. 300–307 (2011)
17. Pergel, M.: Special graph classes and algorithms on them. Ph.D.-thesis, Charles University (2008)
18. Spinrad, J.: Efficient Graph Representations. Fields Institute Monographs 19. American Mathematical Society, Providence (2003)
19. van Leeuwen, E.J., van Leeuwen, J.: Convex Polygon Intersection Graphs. In: Brandes, U., Cornelsen, S. (eds.) GD 2010. LNCS, vol. 6502, pp. 377–388. Springer, Heidelberg (2011)

Area Bounds of Rectilinear Polygons Realized by Angle Sequences*

Sang Won Bae[1], Yoshio Okamoto[2], and Chan-Su Shin[3]

[1] Dept. of Computer Science, Kyonggi University, Korea
swbae@kgu.ac.kr
[2] Dept. of Communication Engineering and Informatics,
University of Electro-Communications, Japan
okamotoy@uec.ac.jp
[3] Dept. of Digital Information Engineering,
Hankuk University of Foreign Studies, Korea
cssin@hufs.ac.kr

Abstract. Given a sequence S of angles at n vertices of a rectilinear polygon, S directly defines (or realizes) a set of rectilinear polygons in the integer grid. Among such realizations, we consider the one $P(S)$ with minimum area. Let $\delta(n)$ be the minimum of the area of $P(S)$ over all angle sequences S of length n, and $\Delta(n)$ be the maximum. In this paper, we provide the explicit formula for $\delta(n)$ and $\Delta(n)$.

1 Introduction

A rectilinear polygon is a simple polygon whose edges are either horizontal or vertical segments. The vertices are ordered in counterclockwise order. An edge meets its adjacent edges only at right angles, 90° or 270°. If we associate each vertex with its angle, then we can represent a rectilinear polygon of n vertices as an *angle sequence* or a *turn sequence* S of L and R of length n, where L means "left turn" at convex vertices, i.e., 90° angle and R means "right turn" at reflex vertices, i.e., 270° angle. For example, if $S := $ LLLL, then the rectilinear polygon induced from S is a rectangle consisting of four convex vertices. It is well known in [5] that $n = 2r + 4$ where r is the number of reflex vertices. Thus the number of L's in S is four more than the number of R's in S.

Observe that the converse also holds. Namely, if a sequence S of L and R has even length and four more L's than R's, then there exists a rectilinear polygon that has S as its turn sequence. Thus, it makes sense to call such a sequence a turn sequence without referring to a polygon. Then the following natural question arises.

* Work by S.W. Bae was supported by National Research Foundation of Korea(NRF) grant funded by Korea government(MEST)(No. 2011-0005512). Work by Y. Okamoto was supported by Grand-in-Aid for Scientific Research from Ministry of Education, Science and Culture, Japan and Japan Society for the Promotion of Science. Work by C.-S. Shin was supported by research grant funded by Hankuk University of Foreign Studies.

K.-M. Chao, T.-s. Hsu, and D.-T. Lee (Eds.): ISAAC 2012, LNCS 7676, pp. 629–638, 2012.
© Springer-Verlag Berlin Heidelberg 2012

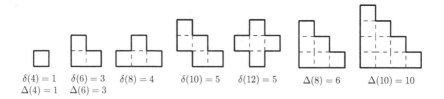

$\delta(4) = 1$ $\delta(6) = 3$ $\delta(8) = 4$ $\delta(10) = 5$ $\delta(12) = 5$ $\Delta(8) = 6$ $\Delta(10) = 10$
$\Delta(4) = 1$ $\Delta(6) = 3$

Fig. 1. The values of $\delta(n)$ and $\Delta(n)$ for small n

Given a turn sequence S of length n, we can *realize* (or *draw*) a rectilinear polygon on the integer grid such that the vertices of the polygon have integer coordinates and S is its turn sequence. There are infinitely many drawings for a fixed S. In this paper, we consider only a drawing $P(S)$ with minimum area that realizes S. We denote the area of $P(S)$ by area($P(S)$). Let us define $\delta(n)$ to be the minimum of area($P(S)$) over all turn sequences S of length n, and $\Delta(n)$ the maximum of area($P(S)$) over all turn sequences S of length n. A natural question here is to determine $\delta(n)$ and $\Delta(n)$. In this paper, we give a complete answer to this question; for small n, we can easily guess what $\delta(n)$ and $\Delta(n)$ are, as listed in Fig. 1, which give us hints for the exact values of $\delta(n)$ and $\Delta(n)$.

Theorem 1. *For* $n (= 2r + 4) \geq 4$,

$$\delta(n) = \begin{cases} \frac{n}{2} - 1 \ (= r + 1) & \text{if } n \equiv 4 \bmod 8, \\ \frac{n}{2} \ (= r + 2) & \text{otherwise.} \end{cases}$$

Theorem 2. *For* $n (= 2r+4) \geq 4$, $\Delta(n) = \frac{1}{8}(n-2)(n+4) \ (= \frac{1}{2}(r+1)(r+2))$.

Reconstruction of a polygon from partial information is a hot topic in computational geometry. There are several variations depending on the information that we may use. One variation dealt with the problem of reconstructing a simple polygon from a sequence of angles defined by all the visible vertices [3,4]. An algorithm proposed by [4] runs in $O(n^3 \log n)$ time, which was improved to the worst-case optimal $O(n^2)$ time [3]. Another variation considers reconstructing rectilinear polygons from a set of points [2,6], i.e., coordinates of the vertices, instead of angles, which could be obtained by laser scanning devices [2].

To the best of authors' knowledge, the variation in this paper has not been investigated in the literature. As a related problem, Bajuelos et al. [1] studied the minimum and maximum area of the drawing of rectilinear polygons in the integer grid without collinear edges, i.e., any grid line contains at most one edge of the drawing. They claimed that the minimum area is at least $2r + 1 = n - 3$ and the maximum area is at most $r^2 + 3 = (\frac{n}{2} - 2)^2 + 3$, but no formal proofs were provided. The main difference with our problem is the collinearity restriction, which makes the problem quite different. In this paper, we want to draw a polygon with the "minimum area", thus we do allow the collinearity.

We omit the proofs of most lemmas in this version due to the space limitation. The readers will find the proofs in the journal version.

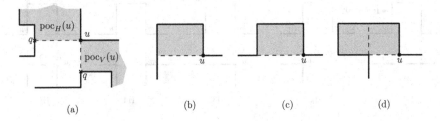

Fig. 2. (a) Horizontal and vertical pockets of a reflex vertex u. (b) A C-type canonical pocket. (c) An H-type canonical pocket. (d) A pocket containing a canonical pocket.

2 Preliminaries

A *turn sequence* is an even-length sequence of L and R such that the number of L's is four more than the number of R's. Let S be a turn sequence of length n. Let $P(S)$ be a rectilinear polygon on the integer grid with minimum area that realizes S. Note that the ordered LR-sequence of the angles at the vertices of P in counterclockwise order is identical to S, provided that the starting vertex is the same. As noted already, L corresponds to a convex vertex of P, and R corresponds to a reflex vertx of P. From now on, we simply use P instead of $P(S)$ if there is no ambiguity. A cell of the grid is a unit square.

We can easily prove that $\delta(n)$ for $n > 4$ is at least the half of the number of convex vertices, i.e., $\delta(n) \geq (r+4)/2 = (n+4)/4$ since a grid cell contained in P can be surrounded by at most two convex vertices when $n > 4$. By a similar argument, a grid cell in P can be surrounded by at most three edges of P when $n > 4$, so $\delta(n) \geq \frac{1}{3}\mathrm{peri}(P)$, where $\mathrm{peri}(P)$ is the perimeter of the polygon P. Thus we have a rough bound on $\delta(n)$ that

$$\delta(n) \geq \max\left\{\frac{1}{4}n + 1, \frac{1}{3}\mathrm{peri}(P)\right\} \qquad \text{if } n > 4.$$

This bound on $\delta(n)$ is not tight. We will show the tight bound of $\delta(n)$ in Section 3.

As an upper bound on $\Delta(n)$, we have $\Delta(n) \leq (r+1)^2$ since we can always realize any turn sequence S of length n in $(r+1) \times (r+1)$ grid [7]. This directly implies that a smallest bounding rectangle R containing P has a dimension $(r+1) \times (r+1)$ at most. Consider a grid cell of $R \setminus P$ which is incident to a reflex vertex of P. It is not difficult to show that the cell can have at most two reflex vertices at its corners. Thus there must be at least $r/2$ such cells in $R \setminus P$, which means $\mathrm{area}(P) \leq \mathrm{area}(R) - r/2 = (r+1)^2 - r/2$. Thus $\Delta(n) \leq (r+1)^2 - r/2$. However, this is still far from what we guess from Fig. 1, $\Delta(n) \leq \frac{1}{2}(r+1)(r+2)$, which is the minimum area of a certain class of monotone polygons. We show the tight bound on $\Delta(n)$ in Section 4.

A reflex vertex u has two incident edges, one horizontal and the other vertical. Draw a ray from u along the direction of the horizontal incident edge toward the interior of P, then it first hits the boundary of P at some point q. The line segment uq is completely contained in P. We cut P along uq, then P is divided

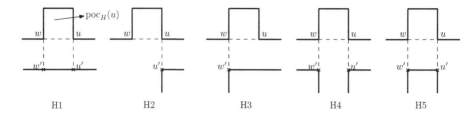

Fig. 3. Five subtypes of H-type canonical pockets. Symmetric ones are omitted.

into two smaller polygons. See Fig. 2(a). We call the one including the vertical edge incident to u the *horizontal pocket* for u, denoted by $\text{poc}_H(u)$. Similarly, we define the *vertical* pocket for u, denoted by $\text{poc}_V(u)$, by shooting the ray along the vertical edge incident to u.

A pocket is said to be *canonical* if the pocket is a rectangle and contains no other pocket. Fig. 2(b)-(d) shows three types of pockets. The last type is not canonical since it contains another pocket inside. We call the first two types *C-type* (Cup-type) and *H-type* (Hat-type), respectively. Namely, a canonical pocket is C-type if the cutting line segment is incident to only one reflex vertex, and H-type if incident to two reflex vertices.

We need to mention that the area of the canonical pocket should be one in the minimum-area drawing $P(S)$, that is, it occupies one grid cell; otherwise we can push the edge incident to u of the pocket so that its area becomes one while keeping the other drawn parts of the polygon untouched. We now show the lemma on the canonical pocket of P as follows.

Lemma 1. *For any turn sequence S of length > 4 and any minimum-area drawing P that realizes S, there are at least two canonical pockets in P.*

We will look into H-type pockets more. Consider a horizontal H-type canonical pocket $\text{poc}_H(u)$ at some reflex vertex u. According to the configuration around $\text{poc}_H(u)$, we classify the five cases as in Fig. 3. Let w be the reflex vertex which defines the canonical pocket together with u. Define u' and w' as the first points on the boundary of P just below u and w, respectively. The points would not be convex vertices, so they would be either reflex vertices or points in the interior of edges. Then we can easily verify that if we ignore the symmetric configurations, then there are only five cases as in Fig. 3. Remember that a canonical pocket has area one in P. The following lemma allows us to consider only the first four cases (subtypes).

Lemma 2. *For any $n > 4$, P has at least one canonical pocket which is not of H5-type ones.*

3   Bounds on $\delta(n)$

3.1   Proving $\delta(n) \geq n/2 - 1$ and Tightness for $n \equiv 4 \bmod 8$

We first prove a lower bound $\delta(n) \geq n/2 - 1$.

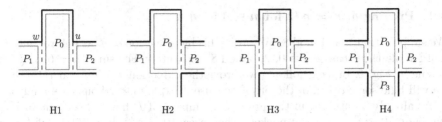

H1   H2   H3   H4

Fig. 4. The proof of Lemma 3

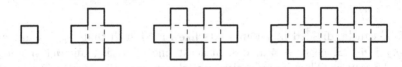

Fig. 5. Tightness for $n \equiv 4 \bmod 8$. Four drawings for $n = 4, 12, 20, 28$. For other n, P is constructed by attaching a part of 8 vertices whose area 4 to the right.

Lemma 3. *For any $n \geq 4$, $\delta(n) \geq n/2 - 1$.*

Proof. We prove the lemma by induction on n. When $n = 4$, P is a unit square, thus it holds. For general n, if P has a C-type canonical pocket, say $\mathrm{poc}_H(u)$, then we divide P into two smaller polygons P_0 and P_1; $P_0 := \mathrm{poc}_H(u)$ and $P_1 := P \backslash \mathrm{poc}_H(u)$. Let n_i denote the number of vertices of P_i. Then $\mathrm{area}(P_0) = 1$ and $\mathrm{area}(P_1) \geq n_1/2 - 1$ by induction hypothesis. Since $n_0 + n_1 - 2 = n$ and $n_0 = 4$, $\mathrm{area}(P) = \mathrm{area}(P_0) + \mathrm{area}(P_1) \geq 1 + (n_1/2 - 1) = n_1/2 = n/2 - 1$. We now suppose that P has no C-type canonical pockets. Then by Lemma 1 P has at least two H-type canonical pockets. Furthermore, we have four subtypes for H-type pockets by Lemma 2.

For H-type, we partition P into three or four parts as in Fig. 4, where P_0 contains the H-type canonical pocket and $\mathrm{area}(P_0) \geq 2$. For H1-type and H2-type, we partition P into three small polygons P_0, P_1, P_2 as in Fig. 4. Since $n_1 + n_2 = n$, $\mathrm{area}(P) = \mathrm{area}(P_0) + \mathrm{area}(P_1) + \mathrm{area}(P_2) \geq 2 + (n_1/2 - 1) + (n_2/2 - 1) = n/2$, so it is done. For H3-type, $n_1 + n_2 = n - 2$. So we have $\mathrm{area}(P) \geq 2 + (n_1 + n_2)/2 - 2 \geq n/2 - 1$. For H4-type, P is partitioned into four parts such that $n_1 + n_2 + n_3 = n$, thus $\mathrm{area}(P) \geq 2 + (n_1 + n_2 + n_3)/2 - 3 = n/2 - 1$. Therefore, $\mathrm{area}(P) \geq n/2$ for H1-type and H2-type and $\mathrm{area}(P) \geq n/2 - 1$ for H3-type and H4-type, which completes the proof. □

This bound is tight when $n \equiv 4 \bmod 8$ because we can construct P of any such n having the area of $n/2 - 1$ as in Fig. 5. So we have the following result.

Lemma 4. *For $n \geq 4$ and $n \equiv 4 \bmod 8$, $\delta(n) = n/2 - 1$.*

3.2   Proving $\delta(n) = n/2$ for $n \not\equiv 4 \bmod 8$

We suppose that $n \not\equiv 4 \bmod 8$. We prove this by induction on n as before, but we will handle three cases, $n \equiv 0$, 2, 6 mod 8, separately. For small $n = 6, 8, 10$, it is true by Fig. 1. In what follows, we prove the induction step for general n, but we will hide some detail of the proof because it exploits a tedious case analysis.

As already mentioned in the proof of Lemma 3, if P has at least one of H1-type and H2-type canonical pockets, then area$(P) \geq n/2$. In addition, if P has an H4-type canonical pocket, it is also true that area$(P) \geq n/2$ as follows.

Lemma 5. *For $n \not\equiv 4 \bmod 8$, if P has an H4-type canonical pocket, then* area$(P) \geq n/2$.

Proof. P can be divided into four parts like Fig. 4 such that $n_0 = 4$ and $n = n_1 + n_2 + n_3$. Since $n \not\equiv 4 \bmod 8$, at least one of n_1, n_2, and n_3 should be $\not\equiv 4 \bmod 8$, say n_1. Then by induction hypothesis, area$(P) \geq 2 + n_1/2 + (n_2/2 - 1) + (n_3/2 - 1) = (n_1 + n_2 + n_3)/2 = n/2$. □

It thus suffices to consider two types only, C-type and H3-type in the induction step. We have the following result.

Lemma 6. *For $n \not\equiv 4 \bmod 8$, $\delta(n) = n/2$.*

As a result, Lemma 6 completes the proof of Theorem 1.

4   Bounds on $\Delta(n)$

Let S be a turn sequence with r right turns (or reflex angles). We also use S_r to emphasize the number of reflex angles. We first consider the special case that S_r is realized as monotone polygons in both axes, and then present an algorithm to give a realization $Q(S_r)$ for arbitrary sequence S_r such that area$(Q(S_r)) \leq \frac{1}{2}(r+1)(r+2)$.

4.1   Realizing Monotone Sequences

A rectilinear polygon is *monotone* in an axis if any line orthogonal to the axis intersects the polygon in at most one connected component. An XY-monotone polygon is a monotone polygon in both axes. We first need the following two observations on the monotonicity.

Observation 1. *If a sequence S is realized as an XY-monotone polygon, then all the polygons realized by S are XY-monotone.*

By the above observation, a turn sequence S is said to be XY-*monotone* if a realization by S is XY-monotone.

Observation 2. *A sequence S is XY-monotone if and only if S does not contain a pattern RR.*

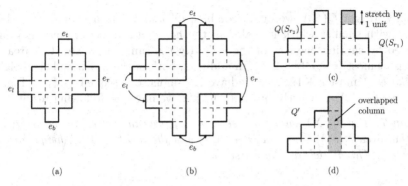

Fig. 6. (a) Four extreme edges for XY-monotone polygon. (b) Split S_r into four "staircase" sequences. (c) Scale $Q(S_{r_1})$. (d) Combine $Q(S_{r_1})$ and $Q(S_{r_2})$ into Q'.

In what follows, we suppose that S_r is XY-monotone. See Fig. 6(a). By the monotone property, any XY-monotone rectilinear polygon Q with r reflex vertices must have four *extreme edges*, leftmost, rightmost, topmost, and bottommost edges, which all have a form of LL. Then the boundary of Q contains at most four staircase chains between the extreme edges. We can split Q into at most four staircase polygons. For instance, see Fig. 6(b). The turn sequences corresponding to such staircase polygons have a form of $LL(RL)^k LL$ for some $k > 0$. So we can now split S_r into (at most) four sequences S_{r_1}, S_{r_2}, S_{r_3}, and S_{r_4} from the original sequence S_r such that they are realized as four staircase polygons. Then we know that $r = r_1 + r_2 + r_3 + r_4$.

We first get a realization $Q(S_{r_i})$ for each $1 \le i \le 4$, and then combine four realizations to get the final realization $Q(S_r)$. To get $Q(S_{r_i})$ in the minimum area, we simply draw it in the staircase shape as in Fig. 6(b). Its area is exactly $1 + 2 + \cdots + (r_i + 1) = \frac{1}{2}(r_i + 1)(r_i + 2)$. Furthermore, its width and height are both $r_i + 1$. We next combine them together into $Q(S_r)$ such that area$(Q(S_r)) \le \frac{1}{2}(r + 1)(r + 2)$ as follows.

We first combine $Q(S_{r_1})$ and $Q(S_{r_2})$. We assume that $r_1 \le r_2$. Let e and e' be the leftmost and the second leftmost vertical edges in $Q(S_{r_1})$, respectively. In fact, these edges are incident to e_t, and they are one unit apart from each other. We scale $Q(S_{r_1})$ in vertical direction so that its height becomes $r_2 + 1$ by stretching e and e' at the same time, as in Fig. 6(c). Then their length increases by $r_2 - r_1$, so the number of cells added newly is also $r_2 - r_1$. The total number of cells of the scaled $Q(S_{r_1})$, i.e., its area is at most $1 + 2 + \cdots + r_1 + (r_1 + 1) + (r_2 - r_1)$. We next overlay the scaled $Q(S_{r_1})$ onto $Q(S_{r_2})$ such that the leftmost column of the cells in $Q(S_{r_1})$ overlaps with the rightmost column in $Q(S_{r_2})$. This combined drawing Q' has the same height as $Q(S_{r_1})$ and $Q(S_{r_2})$, i.e., $r_2 + 1$. The number of cells in Q' is at most the sum of the number of cells in $Q(S_{r_2})$ and $Q(S_{r_1})$ minus the number of cells in the overlapped column, that is, $(1 + \cdots + r_2 + 1) + ((1 + \cdots + r_1 + 1) + (r_2 - r_1)) - (r_2 + 1) = (1 + \cdots + r_2 + 1) + (1 + \cdots + r_1) \le (1 + 2 + \cdots + r_1 + r_2 + 1) = (r_1 + r_2 + 1)(r_1 + r_2 + 2)/2$. In the same way, we get a combined drawing Q'' of $Q(S_{r_3})$ and $Q(S_{r_4})$ such that area$(Q'') \le (r_3 + r_4 + 1)(r_3 + r_4 + 2)/2$.

Finally, to get $Q(S_r)$, we again combine Q' and Q'' in a similar way that the one with smaller width is scaled in the horizontal direction and they are overlayed by sharing one row of the cells. By the same argument, the area of $Q(S_r)$ is at most $(r+1)(r+2)/2$. In addition, both of the width and the height of $Q(S_r)$ is at most $r+1$. Since we have XY-monotone sequences S_r in Fig. 1 whose area($P(S_r)$) is exactly $(r+1)(r+2)/2$, we have the following result.

Lemma 7. *For any XY-monotone sequence S_r, we can have a realization $Q(S_r)$ such that area($Q(S_r)$) $\leq (r+1)(r+2)/2$, and the width and the height are at most $r+1$. This is tight in the worst case.*

4.2 Realizing Arbitrary Sequences

Let S_r be an arbitrary turn sequence with r reflex angles. We now present an algorithm to give a realization $Q(S_r)$ such that area($Q(S_r)$) $\leq \frac{1}{2}(r+1)(r+2)$.

The Outline of the Algorithm. By Lemma 7, if S_r is XY-monotone, then we are done, so assume that S_r is not monotone in at least one axis. Then, by Observation 2, there must be at least one RR in S_r. Furthermore, we can find RRL in S_r. We now represent $S_r := S_{r'} \, \text{RRL} \, S_{r''}$ for $r', r'' \geq 0$ such that $r = r' + r'' + 2$. Define a shorter sequence S_{r-1} by replacing RRL by R, i.e., $S_{r-1} := S_{r'} \, \text{R} \, S_{r''}$. We repeat this replacement process until we get S_{r_0} such that it no longer has RR. By Observation 2, S_{r_0} is XY-monotone. As did in Section 4.1, we split S_{r_0} into (at most) four XY-monotone sequences, and realize them separately as staircase rectilinear polygons. Then we have four initial staircase realizations. We reconstruct the original sequence by inserting RRL back into one of the initial realizations one by one in the reverse order of the replacement process. As the last step, we combine the four realizations into $Q(S_r)$ by the same method as done in Section 4.1.

Realization from a Staircase Subsequence. The realization from four initial staircase polygons can be separately handled in the symmetric way, so we consider the realization only from a fixed staircase polygon as in Fig. 7(a).

Let S_a be the sequence corresponding to the northwest staircase polygon $Q(S_a)$. Let S_b be the original sequence which we have to reconstruct from S_a. Clearly $b \geq a > 0$. S_a has a form of $\text{LL}(\text{RL})^a\text{LL}$, thus $Q(S_a)$ has a reflex vertices $u_1, u_2, \ldots, u_a$ which correspond to R in $(\text{RL})^a$. The replacement process of RRL by R implies that S_b has a form of $\text{LL}(\text{R}S_{r_1}\text{L})(\text{R}S_{r_2}\text{L}) \cdots (\text{R}S_{r_a}\text{L})\text{LL}$ such that $b = a + r_1 + \cdots + r_a$ for $r_i \geq 0$. We now have to reconstruct $Q(S_b)$ from $Q(S_a)$ by realizing each S_{r_i} for $1 \leq i \leq a$. We will realize these S_{r_i} one by one in the order of $i = 1, 2, \ldots, a$.

To realize S_{r_i}, we need some free space to accommodate the edges for RRL in S_{r_i}. For this, as in Fig. 7(a), we will make a room in the cell incident to u_i by inserting r_i rows and r_i columns in total; one row and one column are inserted in the cell whenever a pattern RRL is reconstructed.

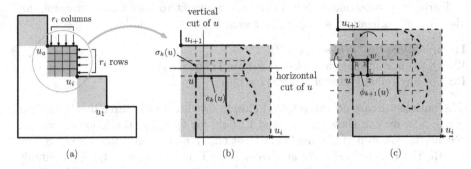

Fig. 7. (a) The initial realization $Q(S_a)$. Realize each S_{r_i} by inserting r_i rows and r_i columns in the shaded cell incident to u_i. (b) Two cuts of u, $e_k(u)$, and $\sigma_k(u)$ in $Q(S_k)$. (c) Reconstruct the pattern RRL in $Q(S_k)$ by inserting one row and one column.

Suppose that the current realization is $Q(S_k)$ for some $a + (r_1 + \cdots + r_{i-1}) < k \leq a + (r_1 + \cdots + r_i)$. Refer to Fig. 7(b) for the notations. For a reflex vertex u in $Q(S_k)$, we denote by $e_k(u)$ an edge incident to u which comes next u in counterclockwise order, and by $\sigma_k(u)$ a unit cell in $Q(S_k)$ whose side is contained in $e_k(u)$ and one corner coincides with u. We call $\sigma_k(u)$ the *ground cell* of u in $Q(S_k)$. A horizontal line (resp., vertical line) through the center of $\sigma_k(u)$ is called a *horizontal cut* of u (resp., *vertical cut* of u) in $Q(S_k)$. The *cut number* $\mathrm{cn}_k(u)$ is the number of the cells in $Q(S_k)$ intersected by at least one of two cuts of u.

We are now going to reconstruct the next pattern RRL into $Q(S_k)$, which results in $Q(S_{k+1})$. Let u be a reflex vertex in $Q(S_k)$ which will be replaced by the next RRL. For drawing the edges for the pattern, we need a free space in $\sigma_k(u)$. Actually, as in Fig. 7(c), one free row and one free column of the grid cells are sufficient. We first insert one empty row just above $e_k(u)$, which is done by stretching all the edges of $Q(S_k)$ intersected by the horizontal cut of u by one unit. We next insert one empty column by stretching the edges of $Q(S_k)$ intersected by the vertical cut of u. Then $\sigma_k(u)$ changes to a square region of side length two and its interior is empty. As the final step, we simply draw the edges for RRL along the added row and column. After this drawing step, we get $Q(S_{k+1})$. Note that the unit cell surrounded by RRL belongs to the exterior of $Q(S_{k+1})$, specially called the *exterior cell* of u in $Q(S_{k+1})$, and denoted by $\phi_{k+1}(u)$. In fact, the exterior cell of u in $Q(S_{k+1})$ was the ground cell of u in $Q(S_k)$. As a result, the reflex vertex u (corresponding to the old R) is deleted in $Q(S_{k+1})$, but two new reflex vertices v, w and one convex vertex z (corresponding to the new RRL) are created in $Q(S_{k+1})$.

How many new cells were inserted in this single reconstruction step? There were two insertions for free row and column. The insertion of the free column (resp., the free row) actually "copies" the cells intersected by the vertical cut (resp., horizontal cut) of u in $Q(S_k)$. In other words, the number of copied cells for the free column and row is the cut number of u in $Q(S_k)$ plus two, i.e., $\mathrm{cn}_k(u) + 2$ cells. But $\phi_{k+1}(u)$ is excluded from the interior of $Q(S_{k+1})$, thus the actual number of inserted cells is at most $\mathrm{cn}_k(u) + 1$.

During the subsequent reconstructions, we want to keep three invariants for the current realization, say $Q(S_k)$ for each $a \leq k \leq b$, as follows.

I1: $\mathrm{area}(Q(S_k)) \leq \frac{1}{2}(k+1)(k+2)$.
I2: $Q(S_k)$ is contained in a $(k+1) \times (k+1)$ square grid.
I3: For any reflex vertex u in $Q(S_k)$, $\mathrm{cn}_k(u) \leq k+1$.

The initial staircase polygons $Q(S_a)$ clearly satisfies all the three invariants. For any $k > a$, we can show that **I1** and **I2** hold for $Q(S_k)$ in the following way. A reconstruction step of RRL inserts at most $\mathrm{cn}_{k-1}(u)+1$ cells more into $Q(S_{k-1})$. By **I3**, the inserted cells are at most $k+1$. Thus the area of $Q(S_k)$ is simply no more than $\mathrm{area}(Q(S_{k-1})) + k + 1 \leq \frac{1}{2}k(k+1) + k + 1 = \frac{1}{2}(k+1)(k+2)$, so **I1** holds. The invariant **I2** holds because one row and one column are inserted exactly whenever a pattern RRL is reconstructed. Hence what remains is to check if **I3** holds for $Q(S_k)$. We omit the proof in this version.

Combining Four Realizations. Let us return to the original problem to realize S_r in the area of $\frac{1}{2}(r+1)(r+2)$. We now have four pieces of the realizations which are reconstructed from four XY-monotone staircase polygons. Since each realization satisfies **I1** and **I2**, we can combine them as $Q(S_r)$ by the similar method explained in Section 4.1, and prove that its area is at most $\frac{1}{2}(r+1)(r+2)$. Therefore, we can show that the upper bound $\Delta(n) \leq \frac{1}{2}(r+1)(r+2)$ and its equality holds for the sequences shown in Fig. 1, which completes the proof of Theorem 2.

References

1. Bajuelos, A.L., Tomás, A.P., Marques, F.: Partitioning Orthogonal Polygons by Extension of All Edges Incident to Reflex Vertices: Lower and Upper Bounds on the Number of Pieces. In: Laganá, A., Gavrilova, M.L., Kumar, V., Mun, Y., Tan, C.J.K., Gervasi, O. (eds.) ICCSA 2004. LNCS, vol. 3045, pp. 127–136. Springer, Heidelberg (2004)
2. Biedl, T., Durocher, S., Snoeyink, J.: Reconstructing polygons from scanner data. Theoretical Computer Science 412, 4161–4172 (2011)
3. Chen, D.Z., Wang, H.: An improved algorithm for reconstructing a simple polygon from its visibility angles. Computational Geometry: Theory and Applications 45, 254–257 (2012)
4. Disser, Y., Mihalák, M., Widmayer, P.: Reconstructing a simple polygon from its angles. Computational Geometry: Theory and Applications 44, 418–426 (2011)
5. O'Rourke, J.: An alternate proof of the rectilinear art gallery theorem. Journal of Geometry 21, 118–130 (1983)
6. O'Rourke, J.: Uniqueness of orthogonal connect-the-dots. In: Toussaint, G.T. (ed.) Computational Morphology, pp. 97–104 (1988)
7. Tomás, A.P., Bajuelos, A.L.: Generating Random Orthogonal Polygons. In: Conejo, R., Urretavizcaya, M., Pérez-de-la-Cruz, J.-L. (eds.) CAEPIA/TTIA 2003. LNCS (LNAI), vol. 3040, pp. 364–373. Springer, Heidelberg (2004)

A Time-Efficient Output-Sensitive Quantum Algorithm for Boolean Matrix Multiplication

François Le Gall

Department of Computer Science, The University of Tokyo
legall@is.s.u-tokyo.ac.jp

Abstract. This paper presents a quantum algorithm that computes the product of two $n \times n$ Boolean matrices in $\tilde{O}(n\sqrt{\ell} + \ell\sqrt{n})$ time, where ℓ is the number of non-zero entries in the product. This improves the previous output-sensitive quantum algorithms for Boolean matrix multiplication in the time complexity setting by Buhrman and Špalek (SODA'06) and Le Gall (SODA'12). We also show that our approach cannot be further improved unless a breakthrough is made: we prove that any significant improvement would imply the existence of an algorithm based on quantum search that multiplies two $n \times n$ Boolean matrices in $O(n^{5/2-\varepsilon})$ time, for some constant $\varepsilon > 0$.

1 Introduction

Multiplying two Boolean matrices, where addition is interpreted as a logical OR and multiplication as a logical AND, is a fundamental problem that have found applications in many areas of computer science (for instance, computing the transitive closure of a graph [7,8,15] or solving all-pairs path problems [5,9,17,18]). The product of two $n \times n$ Boolean matrices can be trivially computed in time $O(n^3)$. The best known algorithm is obtained by seeing the input matrices as integer matrices, computing the product, and converting the product matrix to a Boolean matrix. Using the algorithm by Coppersmith and Winograd [4] for multiplying integer matrices (and more generally for multiplying matrices over any ring), or its recent improvements by Stothers [19] and Vassilevska Williams [20], this gives a classical algorithm for Boolean matrix multiplication with time complexity $O(n^{2.38})$.

This algebraic approach has nevertheless many disadvantages, the main being that the huge constants involved in the complexities make these algorithms impractical. Indeed, in the classical setting, much attention has focused on algorithms that do not use reductions to matrix multiplication over rings, but instead are based on search or on combinatorial arguments. Such algorithms are often called *combinatorial algorithms*, and the main open problem in this field is to understand whether a $O(n^{3-\varepsilon})$-time combinatorial algorithm, for some constant $\varepsilon > 0$, exists for Boolean matrix multiplication. Unfortunately, there have been little progress on this question. The best known combinatorial classical algorithm for Boolean matrix multiplication, by Bansal and Williams [2], has time complexity $O(n^3 / \log^{2.25}(n))$.

K.-M. Chao, T.-s. Hsu, and D.-T. Lee (Eds.): ISAAC 2012, LNCS 7676, pp. 639–648, 2012.

In the quantum setting, there exists a straightforward $\tilde{O}(n^{5/2})$-time[1] algorithm that computes the product of two $n \times n$ Boolean matrices A and B: for each pair of indexes $i, j \in \{1, 2, \ldots, n\}$, check if there exists an index $k \in \{1, \ldots, n\}$ such that $A[i, k] = B[k, j] = 1$ in time $\tilde{O}(\sqrt{n})$ using Grover's quantum search algorithm [10]. Buhrman and Špalek [3] observed that a similar approach leads to a quantum algorithm that computes the product AB in $\tilde{O}(n^{3/2}\sqrt{\ell})$ time, where ℓ denotes the number on non-zero entries in AB. Since the parameter $\ell \in \{0, \ldots, n^2\}$ represents the sparsity of the output matrix, such an algorithm will be referred as *output-sensitive*. Classical output-sensitive algorithms for Boolean matrix multiplication have also been constructed recently: Amossen and Pagh [1] constructed an algorithm with time complexity $\tilde{O}(n^{1.724}\ell^{0.408} + n^{4/3}\ell^{2/3} + n^2)$, while Lingas [14] constructed an algorithm with time complexity $\tilde{O}(n^2\ell^{0.188})$. The above $\tilde{O}(n^{3/2}\sqrt{\ell})$-time quantum algorithm beats both of them when $\ell \leq n^{1.602}$. Note that these two classical algorithms are based on the approach by Coppersmith and Winograd [4] and are thus not combinatorial.

Le Gall [13] has shown recently that there exists an output-sensitive quantum algorithm that computes the product of two $n \times n$ Boolean matrices with time complexity $O(n^{3/2})$ if $1 \leq \ell \leq n^{2/3}$ and $O(n\ell^{3/4})$ if $n^{2/3} \leq \ell \leq n^2$. This algorithm, which improves the quantum algorithm by Buhrman and Špalek [3], was constructed by combining ideas from works by Vassilevska Williams and Williams [21] and Lingas [14].

Several developments concerning the quantum query complexity of this problem, where the complexity under consideration is the number of queries to the entries of the input matrices A and B, have also happened. Output-sensitive quantum algorithms for Boolean matrix multiplication in the query complexity setting were first proposed in [21], and then improved in [13]. Very recently, Jeffery, Kothari and Magniez [11] significantly improved those results: they showed that the quantum query complexity of computing the product of two $n \times n$ Boolean matrices with ℓ non-zero entries is $\tilde{O}(n\sqrt{\ell})$, and gave a matching (up to polylogarithmic factors) lower bound $\Omega(n\sqrt{\ell})$. The quantum query complexity of Boolean matrix multiplication may thus be considered as settled.

Can the quantum time complexity of Boolean matrix multiplication can be further improved as well? The most fundamental question is of course whether there exists a quantum algorithm that uses only quantum search or similar techniques with time complexity $O(n^{5/2-\varepsilon})$, for some constant $\varepsilon > 0$, when $\ell \approx n^2$. This question is especially motivated by its apparently deep connection to the design of subcubic-time classical combinatorial algorithms for Boolean matrix multiplication: a $O(n^{5/2-\varepsilon})$-time quantum algorithm would correspond to an amortized cost of $O(n^{1/2-\varepsilon})$ per entry of the product, which may provides us with a new approach to develop a subcubic-time classical combinatorial algorithm, i.e., an algorithm with amortized cost of $O(n^{1-\varepsilon'})$ per entry of the product. Studying quantum algorithms for Boolean matrix multiplication in the time complexity setting can then, besides its own interest, be considered as a way to

---

[1] In this paper the notation $\tilde{O}$ suppresses $\text{poly}(\log n)$ factors.

gain new insight about the optimal value of the exponent of matrix multiplication in the general case (i.e., for dense output matrices). In comparison, when the output matrix is dense, the classical and the quantum query complexities of matrix multiplication are both trivially equal to $\Theta(n^2)$.

1.1 Statement of Our Results

In this paper we build on the recent approach by Jeffery, Kothari and Magniez [11] to construct a new time-efficient output-sensitive quantum algorithm for Boolean matrix multiplication. Our main result is stated in the following theorem.

Theorem 1. *There exists a quantum algorithm that computes the product of two $n \times n$ Boolean matrices with time complexity $\tilde{O}(n\sqrt{\ell} + \ell\sqrt{n})$, where ℓ denotes the number of non-zero entries in the product.*

Our algorithm improves the quantum algorithm by Le Gall [13] for any value of ℓ other than $\ell \approx n^2$ (we obtain the same upper bound $\tilde{O}(n^{2.5})$ for $\ell \approx n^2$). It also beats the classical algorithms by Amossen and Pagh [1] and Lingas [14] mentioned earlier, which are based on the algebraic approach, for any value $\ell \leq n^{1.847}$ (i.e., whenever $\ell\sqrt{n} \leq n^2\ell^{0.188}$).

As will be explained in more details below, for $\ell \leq n$ our result can be seen as a time-efficient version of the quantum algorithm constructed for the query complexity setting in [11]. The query complexity lower bound $\Omega(n\sqrt{\ell})$ proved in [11] shows that the time complexity of our algorithm is optimal, up to a possible polylogarithmic factor, for $\ell \leq n$. The most interesting part of our results is perhaps the upper bound $\tilde{O}(\ell\sqrt{n})$ we obtain for $\ell \geq n$, which corresponds to the case where the output matrix is reasonably dense and differs from the query complexity upper bounds obtained in [11]. We also show that, for values $\ell \geq n$, no quantum algorithm based on search can perform better than ours unless there exists a quantum algorithm based on search that computes the product of two arbitrary $n \times n$ Boolean matrices with time complexity significantly better that $n^{5/2}$. The formal statement follows.

Theorem 2. *Let δ be any function such that $\delta(n) > 0$ for all $n \in \mathbb{N}^+$. Suppose that, for some value $\lambda \geq n$, there exists a quantum algorithm $\mathcal{Q}$ that, given as input any $n \times n$ Boolean matrices A and B such that the number of non-zero entries in the product AB is at most λ, computes AB in $O\left(\lambda\sqrt{n} \cdot n^{-\delta(n)}\right)$ time. Then there exists an algorithm using $\mathcal{Q}$ as a black-box that computes the product of two $n \times n$ Boolean matrices with overall time complexity $\tilde{O}(n^{5/2-\delta(n)} + n)$.*

The reduction stated in Theorem 2 is actually classical and combinatorial: the whole algorithm uses only classical combinatorial operations and calls to $\mathcal{Q}$. Thus Theorem 2 implies that, if for a given value $\ell \geq n$ the complexity of Theorem 1 can be improved to $O\left(\ell\sqrt{n}/n^\varepsilon\right)$, for some constant $\varepsilon > 0$, using techniques similar to ours (i.e., based on quantum search), then there exists an algorithm based on quantum search (and classical combinatorial operations) that computes the product of two $n \times n$ Boolean matrices with time complexity $\tilde{O}(n^{5/2-\varepsilon} + n)$.

1.2   Overview of Our Techniques

The main tool used to obtain our improvements is the new approach by Jeffery, Kothari and Magniez [11] to find collisions in the graph associated with the multiplication of two $n \times n$ Boolean matrices. More precisely, it was shown in [11] how to find up to t collisions in this graph, on a quantum computer, using $\tilde{O}(\sqrt{nt} + \sqrt{\ell})$ queries, where ℓ is the number of non-zero entries in the product. We construct (in Section 3) a time-efficient version of this algorithm that finds one collision in $\tilde{O}(\sqrt{n} + \sqrt{\ell})$ time. We then use this algorithm to design a quantum algorithm that computes the matrix product in time $\tilde{O}(n\sqrt{\ell})$ when $\ell = O(n)$, which proves Theorem 1 for $\ell = O(n)$. Our key technique is the introduction of a small data structure that is still powerful enough to enable time-efficient access to exactly all the information about the graph needed by the quantum searches. More precisely, while the size of the graph considered is $O(n^2)$, we show that the size of this data structure can be kept much smaller — roughly speaking, the idea is to keep a record of the non-edges of the graph. Moreover, the data structure is carefully chosen so that constructing it, at the beginning of the algorithm, can be done very efficiently (in $\tilde{O}(n)$ time), and updating it during the execution of the algorithm can be done at a cost less than the running time of the quantum searches.

We then prove that the ability of finding up to n non-zero entries of the matrix product is enough by showing (in Section 4) a classical reduction, for $\ell > n$, from the problem of computing the product of two $n \times n$ Boolean matrices with at most ℓ non-zero entries in the product to the problem of computing ℓ/n separate products of two Boolean matrices, each product having at most $O(n)$ non-zero entries. The idea is to randomly permute the rows and columns of the input matrices in order to make the output matrix homogeneous (in the sense that the non-zero entries are distributed almost uniformly), in which case we can decompose the input matrices into smaller blocks and ensure that each product of two smaller blocks contains, with non-negligible probability, at most $O(n)$ non-zero entries. This approach is inspired by a technique introduced by Lingas [14] and then generalized in [12,13]. The main difference is that here we focus on the number of non-zero entries in the product of each pair of blocks, while [12,13,14] focused mainly on the size of the blocks. The upper bounds of Theorem 1 for $\ell \geq n$ follow directly from our reduction, and a stronger version of this reduction leads to the proof of Theorem 2.

2   Preliminaries

In this paper we suppose that the reader is familiar with quantum computation, and especially with quantum search and its variants. We present below the model we are considering for accessing the input matrices on a quantum computer, and computing their product. This model is the same as the one used in [3,13].

Let A and B be two $n \times n$ Boolean matrices, for any positive integer n (the model presented below can be generalized to deal with rectangular matrices in a straightforward way). We suppose that these matrices can be accessed directly by a quantum algorithm. More precisely, we have an oracle O_A that, for any $i, j \in \{1, \ldots, n\}$, any $a \in \{0, 1\}$ and any $z \in \{0, 1\}^*$, performs the unitary mapping $O_A \colon |i\rangle|j\rangle|a\rangle|z\rangle \mapsto |i\rangle|j\rangle|a \oplus A[i, j]\rangle|z\rangle$, where $\oplus$ denotes the bit parity (i.e., the logical XOR). We have a similar oracle O_B for B. Since we are interested in time complexity, we will count all the computational steps of the algorithm and assign a cost of one for each call to O_A or O_B, which corresponds to the cases where quantum access to the inputs A and B can be done at unit cost, for example in a random access model working in quantum superposition (we refer to [16] for an extensive treatment of such quantum random access memories).

Let $C = AB$ denote the product of the two matrices A and B. Given any indices $i, j \in \{1, \ldots, n\}$ such that $C[i, j] = 1$, a witness for this non-zero entry is defined as an index $k \in \{1, \ldots, n\}$ such that $A[i, k] = B[k, j] = 1$. We define a quantum algorithm for Boolean matrix multiplication as follows.

Definition 1. *A quantum algorithm for Boolean matrix multiplication is a quantum algorithm that, when given access to oracles O_A and O_B corresponding to Boolean matrices A and B, outputs with probability at least $2/3$ all the non-zero entries of the product AB along with one witness for each non-zero entry.*

The complexity of several algorithms in this paper will be stated using an upper bound λ on the number ℓ of non-zero entries in the product AB. The same complexity, up to a logarithmic factor, can actually be obtained even if no nontrivial upper bound on ℓ is known a priori. The idea is, similarly to what was done in [21,13], to try successively $\lambda = 2$ (and find up to 2 non-zero entries), $\lambda = 4$ (and find up to 4 non-zero entries), ... and stop when no new non-zero entry is found. The complexity of this approach is, up to a logarithmic factor, the complexity of the last iteration (in which the value of λ is $\lambda = 2^{\lceil \log_2 \ell \rceil + 1}$ if ℓ is a power of two, and $\lambda = 2^{\lceil \log_2 \ell \rceil}$ otherwise). In this paper we will then assume, without loss of generality, that a value λ such that $\ell \leq \lambda \leq 2\ell$ is always available.

3   Finding Up to $O(n)$ Non-zero Entries

Let A and B be the two $n \times n$ Boolean matrices of which we want to compute the product. In this section we define, following [11], a graph collision problem and use it to show how to compute up to $O(n)$ non-zero entries of AB.

Let $G = (I, J, E)$ be a bipartite undirected graph over two disjoint sets I and J, each of size n. The edge set E is then a subset of $I \times J$. When there is no ambiguity it will be convenient to write $I = \{1, \ldots, n\}$ and $J = \{1, \ldots, n\}$. We now define the concept of a collision for the graph G.

Definition 2. *For any index $k \in \{1, \ldots, n\}$, a k-collision for G is an edge $(i, j) \in E$ such that $A[i, k] = B[k, j] = 1$. A collision for G is an edge $(i, j) \in E$ that is a k-collision for some index $k \in \{1, \ldots, n\}$.*

We suppose that the graph G is given by a data structure $\mathcal{M}$ that contains the following information:

- for each vertex u in I, the degree of u;
- for each vertex u in I, a list of all the vertices of J not connected to u.

The size of $\mathcal{M}$ is at most $\tilde{O}(n^2)$, but the key idea is that its size will be much smaller when G is "close to" a complete bipartite graph. Using adequate data structures to implement $\mathcal{M}$ (e.g., using self-balancing binary search trees), we can perform the following four access operations in poly($\log n$) time.

get-degree(u): get the degree of a vertex $u \in I$
check-connection(u,v): check if the vertices $u \in I$ and $v \in J$ are connected
get-vert$_I(r,d)$: get the r-th smallest vertex in I that has degree at most d
get-vert$_J(r,u)$: get the r-th smallest vertex in J not connected to $u \in I$

For the latter two access operations, the order on the vertices refer to the usual order $\leq$ obtained when seeing vertices in I and J as integers in $\{1, \ldots, n\}$. We assume that these two access operations output an error message when the query is not well-defined (i.e, when r is too large).

Similarly, we can update $\mathcal{M}$ in poly($\log n$) time to take in consideration the removal of one edge (u,v) from E (i.e., update the degree of u and update the list of vertices not connected to u). This low complexity will be crucial since our main algorithm (in Proposition 2 below) will remove successively edges from E.

Let L be an integer such that $0 \leq L \leq n^2$. We will define our graph collision problem, denoted GRAPH COLLISION(n,L), as the problem of finding a collision for G under the promise that $|E| \geq n^2 - L$, i.e., there are at most L missing edges in G. The formal definition is as follows.

GRAPH COLLISION(n,L)   [here $n \geq 1$ and $0 \leq L \leq n^2$]
INPUT: two $n \times n$ Boolean matrices A and B
       a bipartite graph $G = (I \cup J, E)$, with $|I| = |J| = n$, given by $\mathcal{M}$
       an index $k \in \{1, \ldots, n\}$
PROMISE: $|E| \geq n^2 - L$
OUTPUT: one k-collision if such a collision exists

The following proposition shows that there exists a time-efficient quantum algorithm solving this problem. The algorithm is similar to the query-efficient quantum algorithm given in [11], but uses the data structure $\mathcal{M}$ in order to keep the time complexity low.

Proposition 1. *There exists a quantum algorithm running in time $\tilde{O}(\sqrt{L}+\sqrt{n})$ that solves, with high probability, the problem GRAPH COLLISION(n,L).*

Proof. We will say that a vertex $i \in I$ is marked if $A[i,k] = 1$, and that a vertex $j \in J$ is marked if $B[k,j] = 1$. Our goal is thus to find a pair $(i,j) \in E$ of marked vertices. The algorithm is as follows.

We first use the minimum finding quantum algorithm from [6] to find the marked vertex u of largest degree in I, in $\tilde{O}(\sqrt{n})$ time using get-degree($\cdot$) to obtain the order of a vertex from the data structure $\mathcal{M}$. Let d denote the degree of u, let I' denote the set of vertices in I with degree at most d, and let S denote the set of vertices in J connected to u. We then search for one marked vertex in S, using Grover's algorithm [10] with check-connection$(u, \cdot)$, in $\tilde{O}(\sqrt{n})$ time. If we find one, then this gives us a k-collision and we end the algorithm. Otherwise we proceed as follows. Note that, since each vertex in I' has at most d neighbors, by considering the number of missing edges we obtain:

$$|I'| \cdot (n - d) \leq n^2 - |E| \leq L.$$

Also note that $|J\backslash S| = n - d$. We do a quantum search on $I' \times (J\backslash S)$ to find one pair of connected marked vertices in time $\tilde{O}(\sqrt{|I'| \cdot |J\backslash S|}) = \tilde{O}(\sqrt{L})$, using get-vert$_I(\cdot, d)$ to access the vertices in I' and get-vert$_J(\cdot, u)$ to access the vertices in $J\backslash S$. □

We now show how an efficient quantum algorithm that computes up to $O(n)$ non-zero entries of the product of two $n \times n$ matrices can be constructed using Proposition 1.

Proposition 2. *Let λ be a known value such that $\lambda = O(n)$. Then there exists a quantum algorithm that, given any $n \times n$ Boolean matrices A and B such that the number of non-zero entries in the product AB is at most λ, computes AB in time $\tilde{O}(n\sqrt{\lambda})$.*

Proof. Let A and B be two $n \times n$ Boolean matrices such that the product AB has at most λ non-zero entries.

We associate with this matrix multiplication the bipartite graph $G = (V, E)$, where $V = I \cup J$ with $I = J = \{1, \ldots, n\}$, and define the edge set as $E = I \times J$. The two components I and J of G are then fully connected: there is no missing edge. It is easy to see that computing the product of A and B is equivalent to computing all the collisions, since a pair (i, j) is a collision if and only if the entry in the i-th row and the j-th column of the product AB is 1.

To find all the collisions, we will basically repeat the following approach: for a given k, search for a new k-collision in G and remove the corresponding edge from E by updating the data structure $\mathcal{M}$ corresponding to G. Since we know that there are at most λ non-zero entries in the matrix product AB, at most λ collisions will be found (and then removed). We are thus precisely interested in finding collisions when $|E| \geq n^2 - \lambda$, i.e., when there are at most λ missing edges in G. We can then use the algorithm of Proposition 1. The main subtlety is that we cannot simply try all the indexes k successively since the cost would be too high. Instead, we will search for good indexes in a quantum way, as described in the next paragraph.

We partition the set of potential witnesses $K = \{1, \ldots, n\}$ into $m = \max(\lambda, n)$ subsets $K_1, \ldots, K_m$, each of size at most $\lceil n/m \rceil$. Starting with $s = 1$, we repeatedly search for a pair (i, j) that is a k-collision for some $k \in K_s$. This is done by doing a Grover search over K_s that invokes the algorithm of Proposition 1. Each time a new collision (i, j) is found (which is a k-collision for some $k \in K_s$), we immediately remove the edge (i, j) from E by updating the data structure $\mathcal{M}$. When no other collision is found, we move to K_{s+1}. We end the algorithm when the last set K_m has been processed.

This algorithm will find, with high probability, all the collisions in the initial graph, and thus all the non-zero entries of AB. Let us examine its time complexity. We first discuss the complexity of creating the data structure $\mathcal{M}$ (remember that updating $\mathcal{M}$ to take in consideration the removal of one edge from E has polylogarithmic cost). Initially $|E| = n^2$, so each vertex of I has the same degree n. Moreover, for each vertex $u \in I$, there is no vertex in J not connected to u. The cost for creating $\mathcal{M}$ is thus $\tilde{O}(n)$ time. Next, we discuss the cost of the quantum search. Let λ_s denote the number of collisions found when examining the set K_s. Note that the search for collisions (the Grover search that invokes the algorithm of Proposition 1) is done $\lambda_s + 1$ times when examining K_s (we need one additional search to decide that there is no other collision). Moreover, we have $\lambda_1 + \cdots + \lambda_m \leq \lambda$. The time complexity of the search is thus

$$\tilde{O}\left(\sum_{s=1}^{m} \sqrt{|K_s|} \times (\sqrt{\lambda} + \sqrt{n}) \times (\lambda_s + 1)\right) = \tilde{O}\left(\sqrt{n}\lambda + n\sqrt{\lambda}\right) = \tilde{O}\left(n\sqrt{\lambda}\right).$$

The overall time complexity of the algorithm is thus $\tilde{O}(n\sqrt{\lambda}+n) = \tilde{O}(n\sqrt{\lambda})$. □

4   Reduction to Several Matrix Multiplications

Suppose that we have a randomized (or quantum) algorithm $\mathcal{A}$ that, given any $m \times n$ Boolean matrix A and any $n \times m$ Boolean matrix B such that the number of non-zero entries in the product AB is known to be at most L, computes AB with time complexity $T(m, n, L)$. For the sake of simplicity, we will make the following assumptions on $\mathcal{A}$:

(1) the time complexity of $\mathcal{A}$ does not exceed $T(m, n, L)$ even if the input matrices do not satisfy the promise (i.e., if there are more than L non-zero entries in the product);
(2) the algorithm $\mathcal{A}$ never outputs that a zero entry of the product is non-zero;
(3) if the matrix product has at most L non-zero entries, then with probability at least $1 - 1/n^3$ all these entries are found.

These assumptions can be done without loss of generality when considering quantum algorithms for Boolean matrix multiplication as defined in Section 2. Assumption (1) can be guaranteed simply by supposing that the algorithm systematically stops after $T(m, n, L)$ steps. Assumption (2) can be guaranteed since a witness is output for each potential non-zero entry found (the witness can be used to immediately check the result). Assumption (3) can be guaranted by repeating the original algorithm (which has success probability at least $2/3$) a logarithmic number of times.

The goal of this section is to show the following proposition.

Proposition 3. *Let L be a known value such that $L \geq n$. Then, for any value $r \in \{1, \ldots, n\}$, there exists an algorithm that, given any $n \times n$ Boolean matrices A*

and B such that the number of non-zero entries in the product AB is at most L, uses algorithm $\mathcal{A}$ to compute with high probability the product AB in time

$$\tilde{O}\left(r^2 \times T\left(\lceil n/r\rceil, n, \frac{100(n + L/r)}{r}\right) + n\right).$$

The proof of Proposition 3 is omitted due to space constraints, but can be found in the full version of the paper. The idea is to show a classical (combinatorial) reduction from the computation of AB to the computation of $\tilde{O}(r^2)$ distinct Boolean matrix products, each having at most $\frac{100}{r}(n + L/r)$ non-zero entries in the product.

5   Proofs of Theorems 1 and 2

In this section we give the proofs of Theorems 1 and 2.

Proof (of Theorem 1). Let A and B be two $n \times n$ Boolean matrices such that the product AB has ℓ non-zero entries. Remember that, as discussed in Section 2, an integer $\lambda \in \{1, \ldots, n^2\}$ such that $\ell \leq \lambda \leq 2\ell$ is known.

If $\lambda \leq n$ then the product AB can be computed in time $\tilde{O}(n\sqrt{\lambda}) = \tilde{O}(n\sqrt{\ell})$ by the algorithm of Proposition 2. Now consider the case $n \leq \lambda \leq n^2$. By Proposition 3 (with the value $r = \lceil\sqrt{\lambda/n}\rceil$), the product of A and B can be computed with complexity $\tilde{O}\left(\frac{\lambda}{n} \times T(n, n, \Delta) + n\right)$, where $\Delta = O(n)$. Combined with Proposition 2, this gives a quantum algorithm that computes the product AB in $\tilde{O}(\lambda\sqrt{n}) = \tilde{O}(\ell\sqrt{n})$ time.   □

Proof (of Theorem 2). Suppose the existence of a quantum algorithm that computes in time $O\left(\lambda\sqrt{n} \cdot n^{-\delta(n)}\right)$ the product of any two $n \times n$ Boolean matrices such that the number of non-zero entries in their product is at most λ. Let c be a positive constant. Using Proposition 3 with the values $r = \lceil cn/\sqrt{\lambda}\rceil$ and $L = n^2$, we obtain a quantum algorithm that computes the product of two $n \times n$ Boolean matrices in time

$$\tilde{O}\left(\frac{n^2}{\lambda} \times T\left(n, n, \frac{100n}{\lceil cn/\sqrt{\lambda}\rceil} + \frac{100n^2}{\lceil cn/\sqrt{\lambda}\rceil^2}\right) + n\right).$$

By choosing the constant c large enough, we can rewrite this upper bound as $\tilde{O}\left(\frac{n^2}{\lambda} \times T(n, n, \lambda) + n\right) = \tilde{O}\left(n^{5/2 - \delta(n)} + n\right)$.   □

Acknowledgments. The author is grateful to Stacey Jeffery, Robin Kothari and Frédéric Magniez for helpful discussions and comments, and for communicating to him preliminary versions of their works. He also acknowledges support from the JSPS and the MEXT, under the grant-in-aids Nos. 24700005, 24106009 and 24240001.

References

1. Amossen, R.R., Pagh, R.: Faster join-projects and sparse matrix multiplications. In: Proceedings of Database Theory, ICDT, pp. 121–126 (2009)
2. Bansal, N., Williams, R.: Regularity lemmas and combinatorial algorithms. In: Proceedings of FOCS 2009, pp. 745–754 (2009)
3. Buhrman, H., Špalek, R.: Quantum verification of matrix products. In: Proceedings of SODA 2006, pp. 880–889 (2006)
4. Coppersmith, D., Winograd, S.: Matrix multiplication via arithmetic progressions. Journal of Symbolic Computation 9(3), 251–280 (1990)
5. Dor, D., Halperin, S., Zwick, U.: All-pairs almost shortest paths. SIAM Journal on Computing 29(5), 1740–1759 (2000)
6. Dürr, C., Høyer, P.: A quantum algorithm for finding the minimum. arXiv:quant-ph/9607014v2 (1996)
7. Fischer, M.J., Meyer, A.R.: Boolean matrix multiplication and transitive closure. In: Proceedings of the 12th Annual Symposium on Switching and Automata Theory, pp. 129–131 (1971)
8. Furman, M.E.: Application of a method of fast multiplication of matrices in the problem of finding the transitive closure of a graph. Soviet Mathematics Doklady (English Translation) 11(5), 1252 (1970)
9. Galil, Z., Margalit, O.: All pairs shortest distances for graphs with small integer length edges. Information and Computation 134(2), 103–139 (1997)
10. Grover, L.K.: A fast quantum mechanical algorithm for database search. In: Proceedings of STOC 1996, pp. 212–219 (1996)
11. Jeffery, S., Kothari, R., Magniez, F.: Improving Quantum Query Complexity of Boolean Matrix Multiplication Using Graph Collision. In: Czumaj, A., Mehlhorn, K., Pitts, A., Wattenhofer, R. (eds.) Automata, Languages, and Programming. LNCS, vol. 7391, pp. 522–532. Springer, Heidelberg (2012)
12. Jeffery, S., Magniez, F.: Improving quantum query complexity of Boolean matrix multiplication using graph collision. arXiv:1112.5855v1 (December 2011)
13. Le Gall, F.: Improved output-sensitive quantum algorithms for Boolean matrix multiplication. In: Proceedings of SODA 2012, pp. 1464–1476 (2012)
14. Lingas, A.: A fast output-sensitive algorithm for boolean matrix multiplication. Algorithmica 61(1), 36–50 (2011)
15. Munro, J.I.: Efficient determination of the transitive closure of a directed graph. Information Processing Letters 1(2), 56–58 (1971)
16. Nielsen, M., Chuang, I.: Quantum Computation and Quantum Information. Cambridge University Press (2000)
17. Seidel, R.: On the all-pairs-shortest-path problem in unweighted undirected graphs. Journal of Computer and System Sciences 51(3), 400–403 (1995)
18. Shoshan, A., Zwick, U.: All pairs shortest paths in undirected graphs with integer weights. In: Proceedings of FOCS 1999, pp. 605–615 (1999)
19. Stothers, A.: On the Complexity of Matrix Multiplication. PhD thesis, University of Edinburgh (2010)
20. Vassilevska Williams, V.: Multiplying matrices faster than Coppersmith-Winograd. In: Proceedings of STOC 2012, pp. 887–898 (2012)
21. Vassilevska Williams, V., Williams, R.: Subcubic equivalences between path, matrix and triangle problems. In: Proceedings of FOCS 2010, pp. 645–654 (2010)

On Almost Disjunct Matrices for Group Testing

Arya Mazumdar*

Massachusetts Institute of Technology
Cambridge, MA 02139, USA
aryam@mit.edu

Abstract. In a *group testing* scheme, a set of tests is designed to identify a small number t of defective items among a large set (of size N) of items. In the non-adaptive scenario the set of tests has to be designed in one-shot. In this setting, designing a testing scheme is equivalent to the construction of a *disjunct matrix*, an $M \times N$ matrix where the union of supports of any t columns does not contain the support of any other column. In principle, one wants to have such a matrix with minimum possible number M of rows (tests). One of the main ways of constructing disjunct matrices relies on *constant weight error-correcting codes* and their *minimum distance*. In this paper, we consider a relaxed definition of a disjunct matrix known as *almost disjunct matrix*. This concept is also studied under the name of *weakly separated design* in the literature. The relaxed definition allows one to come up with group testing schemes where a close-to-one fraction of all possible sets of defective items are identifiable. Our main contribution is twofold. First, we go beyond the minimum distance analysis and connect the *average distance* of a constant weight code to the parameters of an almost disjunct matrix constructed from it. Next we show as a consequence an explicit construction of almost disjunct matrices based on our average distance analysis. The parameters of our construction can be varied to cover a large range of relations for t and N. As an example of parameters, consider any absolute constant $\epsilon > 0$ and t proportional to $N^\delta, \delta > 0$. With our method it is possible to explicitly construct a group testing scheme that identifies $(1 - \epsilon)$ proportion of all possible defective sets of size t using only $O\left(t^{3/2}\sqrt{\log(N/\epsilon)}\right)$ tests (as opposed to $O(t^2 \log N)$ required to identify all defective sets).

1 Introduction

Combinatorial group testing is an old and well-studied problem. In the most general form it is assumed that there is a set of N elements among which at most t are *defective*, i.e., special. This set of defective items is called the *defective set* or *configuration*. To find the defective set, one might test all the elements individually for *defects*, requiring N tests. Intuitively, that would be a waste of resource if $t \ll N$. On the other hand, to identify the defective configuration it is required to ask at least $\log \sum_{i=0}^{t} \binom{N}{i} \approx t \log \frac{N}{t}$ yes-no questions. The main objective is to identify the defective configuration with a number of tests that is as close to this minimum as possible. In the group testing

---

* This work was supported in part by the US Air Force Office of Scientific Research under Grant No. FA9550-11-1-0183, and by the National Science Foundation under Grant No. CCF-1017772.

K.-M. Chao, T.-s. Hsu, and D.-T. Lee (Eds.): ISAAC 2012, LNCS 7676, pp. 649–658, 2012.

problem, a *group* of elements are tested together and if this particular group contains any defective element the test result is positive. Based on the test results of this kind one *identifies* (with an efficient algorithm) the defective set with minimum possible number of tests. The schemes (grouping of elements) can be adaptive, where the design of one test may depend on the results of preceding tests. For a comprehensive survey of adaptive group testing schemes we refer the reader to [7].

In this paper we are interested in non-adaptive group testing schemes: here all the tests are designed together. If the number of designed tests is M, then a non-adaptive group testing scheme is equivalent to the design of a so-called binary *test matrix* of size $M \times N$ where the (i, j)th entry is 1 if the ith test includes the jth element; it is 0 otherwise. As the test results, we see the Boolean OR of the columns corresponding to the defective entries. Extensive research has been performed to find out the minimum number of required tests M in terms of the number of elements N and the maximum number of defective elements t. The best known lower bound says that it is necessary to have $M = \Omega(\frac{t^2}{\log t} \log N)$ tests [8, 10]. The existence of non-adaptive group testing schemes with $M = O(t^2 \log N)$ is also known for quite some time [7, 15]. Evidently, there is a gap by the factor of $O(\log t)$ in these upper and lower bounds. It is generally believed that it is hard to close the gap. In contrast, for the adaptive setting, schemes have been constructed with the optimal number, $O(t \log N)$, of tests [7, 14].

A construction of group testing schemes from error-correcting code matrices and using code concatenation [19] appeared in the seminal paper by Kautz and Singleton [16]. In [16], the authors concatenate a q-ary ($q > 2$) Reed-Solomon code with a unit weight code to use the resulting codewords as the columns of the testing matrix. Recently in [24], an explicit construction of a scheme with $M = O(t^2 \log N)$ tests is provided. The construction of [24] is based on the idea of [16]: instead of the Reed-Solomon code, they take a low-rate code that achieves the Gilbert-Varshamov bound of coding theory [19, 25]. Papers, such as [9,29], also consider construction of non-adaptive group testing schemes.

In this paper we explicitly construct a non-adaptive scheme that requires a number of test proportional to $t^{3/2}$. However, we needed to relax the requirement of identifications of defective elements in a way that makes it amenable for our analysis. This relaxed requirement schemes were considered under the name of *weakly separated designs* in [20] and [30]. Our definition of this relaxation appeared previously in the paper [18]. We (and [18, 20, 30]) aim for a scheme that successfully identifies a large fraction of all possible defective configurations. Non-adaptive group testing has found applications in multiple different areas, such as, multi-user communication [2,28], DNA screening [23], pattern finding [17] etc. It can be observed that in many of these applications it would have been still useful to have a scheme that identifies almost all different defective configurations if not all possible defective configurations. It is known (see, [30]) that with this relaxation it might be possible to reduce the number of tests to be proportional to $t \log N$. However this result is not constructive. The above relaxation and weakly separated designs form a parallel of similar works in compressive sensing (see, [3,21]) where recovery of almost all sparse signals from a generic random model is considered. In the literature, other relaxed versions of the group testing problem have been studied as well. For example, in [13] it is assumed that recovering a large fraction of defective elements is sufficient.

The constructions of [16,24] and many others are based on so-called *constant weight error-correcting codes*, a set of binary vectors of same Hamming weight (number of ones). The group-testing recovery property relies on the pairwise *minimum distance* between the vectors of the code [16]. In this work, we go beyond this minimum distance analysis and relate the group-testing parameters to the *average distance* of the constant weight code. This allows us to connect weakly separated designs to error-correcting codes in a general way. Previously the connection between distances of the code and weakly separated designs was only known for the very specific family of *maximum distance separable* codes [18], where much more than the average distance is known.

Based on the newfound connection, we construct an explicit (constructible deterministically in polynomial time) scheme of non-adaptive group testing that can identify all except an $\epsilon > 0$ fraction of all defective sets of size at most t. To be specific, we show that it is possible to explicitly construct a group testing scheme that identifies $(1-\epsilon)$ proportion of all possible defective sets of size t using only $8et^{3/2}\log N \frac{\sqrt{\log \frac{2(N-t)}{\epsilon}}}{\log t - \log\log \frac{2(N-t)}{\epsilon}}$ tests for any $\epsilon > 2(N-t)e^{-t}$. It can be seen that, with the relaxation in requirement, the number of tests is brought down to be proportional to $t^{3/2}$ from t^2. This allows us to operate with a number of tests that was previously not possible in explicit constructions of non-adaptive group testing. For a large range of values of t, namely t being proportional to any positive power of N, i.e., $t \sim N^\delta$, and constant ϵ our scheme has number of tests only about $\frac{8e}{\delta}t^{3/2}\sqrt{\log(N/\epsilon)}$. Our construction scheme is same as that of [16,24], however relies on a finer analysis on the distance properties of a linear code. This construction is only an example of the consequence of the our main result, and by no mean optimal. Better constructions might be possible using our technique.

In Section 2, we provide the necessary definitions and discuss the main result: we state the connection between the parameters of a weakly separated design and the average distance of a constant weight code. In Section 3 we discuss our construction scheme. The proofs of our claims can be found in Sections 2.3 and 3.

The author has recently been made aware of a parallel and independent unpublished work [12] that claims to construct weakly separated designs with $O(t\,\mathrm{poly}(\log N))$ tests. However the technique of this paper is completely different, and the main emphasis here is to show for the first time the relation between the disjunct-property and average distance of codes which leaves a lot of scope for future explorations.

2   Almost Disjunct Matrices from Codes

It is easy to see that, if an $M \times N$ binary matrix gives a non-adaptive group testing scheme that identify up to t defective elements, then, $\sum_{i=0}^{t}\binom{N}{i} \leq 2^M$. This means that for any group testing scheme, $M \geq \log\sum_{i=0}^{t}\binom{N}{i} \geq t\log\frac{N}{t}$, which is a loose bound. Consider the case when one is interested in a scheme that identifies all possible except an ϵ fraction of the different defective sets. Then it is required that,

$$M \geq \log\left((1-\epsilon)\binom{N}{t}\right) \geq t\log\frac{N}{t} + \log(1-\epsilon). \tag{1}$$

It is shown in [20, 30] that (1) is tight.

2.1 Disjunct Matrices

The *support* of a vector x is the set of coordinates where the vector has nonzero entries. It is denoted by $\mathrm{supp}(x)$. We use the usual set terminology, where a set A contains B if $B \subseteq A$.

Definition 1. *An $M \times N$ binary matrix A is called t-disjunct if the support of any column is not contained in the union of the supports of any other t columns.*

It is not very difficult to see that a t-disjunct matrix gives a group testing scheme that identifies any defective set up to size t. On the other hand any group testing scheme that identifies any defective set up to size t must be a $(t - 1)$-disjunct matrix [7]. To a great advantage, disjunct matrices allow for a simple identification algorithm that runs in time $O(Nt)$. Below we define *relaxed* disjunct matrices. This definition appeared very closely in [20, 30] and independently exactly in [18].

Definition 2. *For any $\epsilon > 0$, an $M \times N$ matrix A is called type-1 (t, ϵ)-disjunct if the set of t-tuple of columns (of size $\binom{N}{t}$) has a subset $\mathcal{B}$ of size at least $(1 - \epsilon)\binom{N}{t}$ with the following property: for all $J \in \mathcal{B}$, $\cup_{\kappa \in J} \mathrm{supp}(\kappa)$ does not contain support of any column $\nu \notin J$.*

In other words, the union of supports of a randomly and uniformly chosen set of t columns from a type-1 (t, ϵ)-disjunct matrix does not contain the support of any other column with probability at least $1 - \epsilon$. It is easy to see the following fact.

Proposition 1. *A type-1 (t, ϵ)-disjunct matrix gives a group testing scheme that can identify all but at most a fraction $\epsilon > 0$ of all defective configurations of size at most t.*

The definition of disjunct matrix can be restated as follows: a matrix is t-disjunct if any $t + 1$ columns indexed by $i_1, \ldots, i_{t+1}$ of the matrix form a sub matrix which must have a row that has exactly one 1 in the i_jth position and zeros in the other positions, for $j = 1, \ldots, t+1$. Recall that, a *permutation matrix* is a square binary $\{0, 1\}$-matrix with exactly one 1 in each row and each column. Hence, for a t-disjunct matrix, any $t + 1$ columns form a sub-matrix that must contain $t + 1$ rows such that a $(t + 1) \times (t + 1)$ permutation matrix is formed of these rows and columns. A statistical relaxation of the above definition gives the following.

Definition 3. *For any $\epsilon > 0$, an $M \times N$ matrix A is called type-2 (t, ϵ)-disjunct if the set of $(t+1)$-tuples of columns (of size $\binom{N}{t+1}$) has a subset $\mathcal{B}$ of size at least $(1-\epsilon)\binom{N}{t+1}$ with the following property: the $M \times (t + 1)$ matrix formed by any element $J \in \mathcal{B}$ must contain $t + 1$ rows that form a $(t + 1) \times (t + 1)$ permutation matrix.*

In other words, with probability at least $1 - \epsilon$, any randomly and uniformly chosen $t + 1$ columns from a type-2 (t, ϵ)-disjunct matrix form a sub-matrix that must has $t + 1$ rows such that a $(t + 1) \times (t + 1)$ permutation matrix can be formed. It is clear that for $\epsilon = 0$, the type-1 and type-2 (t, ϵ)-disjunct matrices are same (i.e., t-disjunct). In the rest of the paper, we concentrate on the design of an $M \times N$ matrix A that is type-2 (t, ϵ)-disjunct. Our technique can be easily extended to the construction of type-1 disjunct matrices.

2.2   Constant Weight Codes and Disjunct Matrices

A binary (M, N, d) code $\mathcal{C}$ is a set of size N consisting of $\{0, 1\}$-vectors of length M. Here d is the largest integer such that any two vectors (codewords) of $\mathcal{C}$ are at least Hamming distance d apart. d is called the *minimum distance* (or *distance*) of $\mathcal{C}$. If all the codewords of $\mathcal{C}$ have Hamming weight w, then it is called a constant weight code. In that case we write $\mathcal{C}$ is an (M, N, d, w) constant weight binary code.

Constant weight codes can give constructions of group testing schemes. One just arranges the codewords as the columns of the test matrix. Kautz and Singleton proved the following in [16].

Proposition 2. *An (M, N, d, w) constant weight binary code provides a t-disjunct matrix where,* $t = \left\lfloor \frac{w-1}{w-d/2} \right\rfloor$.

Proof. The intersection of supports of any two columns has size at most $w - d/2$. Hence if $w > t(w - d/2)$, support of any column will not be contained in the union of supports of any t other columns.

We extend Prop. 2 to have one of our main theorems. However, to do that we need to define the *average distance* D of a code $\mathcal{C}$:

$$D(\mathcal{C}) = \frac{1}{|\mathcal{C}| - 1} \min_{x \in \mathcal{C}} \sum_{y \in \mathcal{C} \setminus \{x\}} d_H(x, y). \tag{2}$$

Here $d_H(x, y)$ denotes the Hamming distance between x and y.

Theorem 1. *Suppose, we have a constant weight binary code $\mathcal{C}$ of size N, minimum distance d and average distance D such that every codeword has length M and weight w. The test matrix obtained from the code is type-2 (t, ϵ)-disjunct for the largest t such that,* $\alpha\sqrt{t \ln \frac{2(t+1)}{\epsilon}} \leq \frac{w-1-t(w-D/2)}{w-d/2}$ *holds. Here α is any absolute constant greater than or equal to $\sqrt{2}(1 + t/(N-1))$.*

The proof of this theorem is deferred until after the following remarks.

Remark: By a simple change in the proof of the Theorem 1, it is possible to see that the test matrix is type-1 (t, ϵ)-disjunct if, $\alpha\sqrt{t \ln \frac{2(N-t)}{\epsilon}} \leq \frac{w-1-t(w-D/2)}{w-d/2}$, for an absolute constant α.

One can compare the results of Prop. 2 and Theorem 1 to see the improvement achieved as we relax the definition of disjunct matrices. This will lead to the final improvement on the parameters of Porat-Rothschild construction [24], as we will see in Section 3.

2.3   Proof of Theorem 1

This section is dedicated to the proof of Theorem 1. Suppose, we have a constant weight binary code $\mathcal{C}$ of size N and minimum distance d such that every codeword has length

M and weight w. Let the average distance of the code be D. Note that this code is fixed: we will prove a property of this code by probabilistic method .

Let us now chose $(t+1)$ codewords randomly and uniformly from all possible $\binom{N}{t+1}$ choices. Let the randomly chosen codewords are $\{c_1, c_2, \ldots, c_{t+1}\}$. In what follows, we adapt the proof of Prop. 2 in a probabilistic setting.

Define the random variables for $i = 1, \ldots, t+1$, $Z^i = \sum_{j=1, j\neq i}^{t+1} \left(w - \frac{d_H(c_i, c_j)}{2} \right)$. Clearly, Z^i is the maximum possible size of the portion of the support of c_i that is common to at least one of $c_j, j = 1, \ldots, t+1, j \neq i$. Note that the size of support of c_i is w. Hence, as we have seen in the proof of Prop. 2, if Z^i is less than w for all $i = 1, \ldots, t+1$, then the $M \times (t+1)$ matrix formed by the $t+1$ codewords must contain $t+1$ rows such that a $(t+1) \times (t+1)$ permutation matrix can be formed. Therefore, we aim to find the probability $\Pr(\exists i \in \{1, \ldots, t+1\} : Z^i \geq w)$ and show it to be bounded above by ϵ under the condition of the theorem.

As the variable Z^is are identically distributed, we see that, $\Pr(\exists i \in \{1, \ldots, t+1\} : Z^i \geq w) \leq (t+1)\Pr(Z^1 \geq w)$. In the following, we will find an upper bound on $\Pr(Z^1 \geq w)$.

Define, $Z_i = \mathbb{E}\left(\sum_{j=2}^{t+1} \left(w - \frac{d_H(c_1, c_j)}{2} \right) \mid d_H(c_1, c_k), k = 2, 3, \ldots, i \right)$. Clearly, $Z_1 = \mathbb{E}\left(\sum_{j=2}^{t+1} \left(w - \frac{d_H(c_1, c_j)}{2} \right) \right)$, and $Z_{t+1} = \sum_{j=2}^{t+1} \left(w - \frac{d_H(c_1, c_j)}{2} \right) = Z^1$.

We have the following three lemmas whose proofs are omitted here.

Lemma 1. $Z_1 \leq t(w - D/2)$.

Lemma 2. *The sequence of random variables $Z_i, i = 1, \ldots, t+1$, forms a martingale.*

The statement is true by construction. Once we have proved that the sequence is a martingale, we show that it is a bounded-difference martingale.

Lemma 3. *For any $i = 2, \ldots, t+1$, $|Z_i - Z_{i-1}| \leq (w - d/2)\left(1 + \frac{t-i+1}{N-i} \right)$.*

Now using Azuma's inequality for martingale with bounded difference [22], we have,

$$\Pr\left(|Z_{t+1} - Z_1| > \nu \right) \leq 2\exp\left(-\frac{\nu^2}{2(w - d/2)^2 \sum_{i=2}^{t+1} c_i^2} \right),$$

where, $c_i = 1 + \frac{t-i+1}{N-i}$. This implies,

$$\Pr\left(|Z_{t+1}| > \nu + t(w - D/2) \right) \leq 2\exp\left(-\frac{\nu^2}{2(w - d/2)^2 \sum_{i=2}^{t+1} c_i^2} \right).$$

Setting, $\nu = w - 1 - t(w - D/2)$, we have,

$$\Pr\left(Z^1 > w - 1 \right) \leq 2\exp\left(-\frac{(w - 1 - t(w - D/2))^2}{2(w - d/2)^2 \sum_{i=2}^{t+1} c_i^2} \right).$$

Now, $\sum_{i=2}^{t+1} c_i^2 \leq t\left(1 + \frac{t-1}{N-2} \right)^2$. Hence, $\Pr(\exists i \in \{1, \ldots, t+1\} : Z^i \geq w) \leq 2(t+1)\exp\left(-\frac{(w-1-t(w-D/2))^2}{2t(w-d/2)^2\left(1 + \frac{t-1}{N-2} \right)^2} \right) < \epsilon$, when, $d/2 \geq w - \frac{w-1-t(w-D/2)}{\alpha\sqrt{t \ln \frac{2(t+1)}{\epsilon}}}$, and α is a constant greater than $\sqrt{2}\left(1 + \frac{t-1}{N-2} \right)$.

3 Construction

As we have seen in Section 2, constant weight codes can be used to produce disjunct matrices. Kautz and Singleton [16] gives a construction of constant weight codes that results in good disjunct matrices. In their construction, they start with a Reed-Solomon (RS) code, a q-ary error-correcting code of length $q - 1$. For a detailed discussion of RS codes we refer the reader to the standard textbooks of coding theory [19, 25]. Next they replace the q-ary symbols in the codewords by unit weight binary vectors of length q. The mapping from q-ary symbols to length-q unit weight binary vectors is bijective: i.e., it is $0 \to 100 \ldots 0; 1 \to 010 \ldots 0; \ldots; q - 1 \to 0 \ldots 01$. We refer to this mapping as ϕ. As a result, one obtains a set of binary vectors of length $q(q - 1)$ and constant weight q. The size of the resulting binary code is same as the size of the RS code, and the distance of the binary code is twice that of the distance of the RS code.

3.1 Consequence of Theorem 1 in Kautz-Singleton Construction

For a q-ary RS code of size N and length $q-1$, the minimum distance is $q-1-\log_q N + 1 = q-\log_q N$. Hence, the Kautz-Singleton construction is a constant-weight code with length $M = q(q - 1)$, weight $w = q - 1$, size N and distance $2(q - \log_q N)$. Therefore, from Prop. 2, we have a t-disjunct matrix with,

$$t = \frac{q - 1 - 1}{q - 1 - q + \log_q N} = \frac{q - 2}{\log_q N - 1} \approx \frac{q \log q}{\log N} \approx \frac{\sqrt{M} \log M}{2 \log N}.$$

On the other hand, note that, the average distance of the RS code is $\frac{N}{N-1}(q-1)(1-1/q)$. Hence the average distance of the resulting constant weight code from Kautz-Singleton construction will be $D = \frac{2N(q-1)^2}{q(N-1)}$. Now, substituting these values in Theorem 1, we have a type-1 (t, ϵ) disjunct matrix, where,

$$\alpha \sqrt{t \ln \frac{2(N - t)}{\epsilon}} \leq \frac{(q - t) \log q}{\log N} \approx \frac{(\sqrt{M} - t) \log M}{2 \log N}.$$

Suppose $t \leq \sqrt{M}/2$. Then, $M(\ln M)^2 \geq 4\alpha^2 t(\ln N)^2 \ln \frac{2(N-t)}{\epsilon}$. This restricts t to be about $O(\sqrt{M})$. Hence, Theorem 1 does not obtain any meaningful asymptotic improvement from the Kautz-Singleton construction except in special cases.

Example: Consider a 4096-ary Reed-Solomon code of length 4095 and size $N = 4096^3 \approx 6.8 \times 10^{10}$ (number of elements). From the above discussion we see that, with number of tests $M = 4096 \cdot 4095 \approx 1.6 \times 10^7$, the resulting matrix is type-1 $(2700, 2^{-4})$-disjunct. Although the number of defectives t seems quite large here, it is very small compared to N. On the other hand the straight-forward Kautz-Singleton construction guarantees that for the same dimension of a matrix, we can have a 2047-disjunct matrix. Roughly speaking, in this example it is possible to identify, by the merit of Theorem 1, 31.9% more defective items, but the tests are successful in 93.75% of the cases. It can be inferred that the improvement suggested in Theorem 1 appear only for very large values of N. However, in the asymptotic limits Kautz-Singleton construction is not optimal, as shown by the next construction.

There are two places where the Kautz-Singleton construction can be improved: 1) instead of Reed-Solomon code one can use any other q-ary code of different length, and 2) instead of the mapping ϕ any binary constant weight code of size q might have been used. For a general discussion we refer the reader to [7, §7.4]. In the recent work [24], the mapping ϕ is kept the same, while the RS code has been changed to a q-ary code that achieve the Gilbert-Varshamov bound [19, 25].

In our construction of disjunct matrices we follow the footsteps of [16, 24]. However, we exploit some property of the resulting scheme (namely, the average distance) and do a finer analysis that was absent from the previous works such as [24].

3.2 q-ary Code Construction

We choose $q > (2e\alpha\sqrt{a} + 1)t$, for some constant $a > 0$ and α being the constant of Theorem 1, to be a power of a prime number. Next, we construct a *linear* q-ary code of size N, length M_q and minimum distance d_q that achieves the Gilbert-Varshamov (GV) bound [19, 25], i.e.,

$$\frac{\log_q N}{M_q} \geq 1 - h_q\left(\frac{d_q}{M_q}\right) - o(1), \tag{3}$$

where h_q is the q-ary entropy function defined by, $h_q(x) = x\log_q\frac{q-1}{x} + (1-x)\log_q\frac{1}{1-x}$.

Porat and Rothschild [24] show that it is possible to construct in time $O(M_q N)$ a q-ary code that achieves the GV bound. To have such construction, they exploit the following well-known fact: a q-ary linear code with random generator matrix achieves the GV bound with high probability [25]. To have an explicit construction of such codes, a derandomization method known as the method of conditional expectation [1] is used. In this method, the entries of the generator matrix of the code are chosen one-by-one so that the minimum distance of the resulting code does not go below the value prescribed by (3). For a detail description of the procedure, see [24].

Next, the q-ary symbols of the codewords of the above code are replaced by binary vectors according to the map ϕ. The N binary vectors of length $M = qM_q$ are used as the rows of the test matrix. This construction with proper parameters gives us a disjunct matrix with the following property.

Theorem 2. *Suppose $\epsilon > 2(t + 1)e^{-at}$ for some constant $a > 1$. It is possible to explicitly construct a type-2 (t, ϵ)-disjunct matrix of size $M \times N$ where*

$$M = O\left(t^{3/2}\ln N\frac{\sqrt{\ln\frac{2(t+1)}{\epsilon}}}{\ln t - \ln\ln\frac{2(t+1)}{\epsilon} + \ln(4a)}\right). \tag{4}$$

The proof of this theorem is omitted.

Note that the implicit constant in Theorem 2 is proportional to $\sqrt{a}$. We have not particularly tried to optimize the constant, and got the value of about $8e\sqrt{a}$.

Remark: As in the case of Theorem 1, with a simple change in the proof, it is easy to see that one can construct a test matrix that is type-1 (t, ϵ)-disjunct if the condition of (4) is satisfied with $(t + 1)$ replaced by $(N - t)$ in the expression.

It is clear from Prop. 1 that a type-1 (t, ϵ) disjunct matrix is equivalent to a group testing scheme. Hence, as a consequence of Theorem 2 (specifically, the remark above), we will be able to construct a testing scheme with $O\left(t^{3/2} \log N \frac{\sqrt{\log \frac{2(N-t)}{\epsilon}}}{\log t - \log \log \frac{2(N-t)}{\epsilon}}\right)$ tests. Whenever the defect-model is such that all the possible defective sets of size t are equally likely and there are no more than t defective elements, the above group testing scheme will be successful with probability at lease $1 - \epsilon$.

Note that, if t is proportional to any positive power of N, then $\log N$ and $\log t$ are of same order. Hence it will be possible to have the above testing scheme with $O(t^{3/2} \sqrt{\log(N/\epsilon)})$ tests, for any $\epsilon > 2(N - t)e^{-t}$.

4 Conclusion

In this work we show that it is possible to construct non-adaptive group testing schemes with small number of tests that identify a uniformly chosen random defective configuration with high probability. To construct a t-disjunct matrix one starts with the simple relation between the minimum distance d of a constant w-weight code and t. This is an example of a scenario where a pairwise property (i.e., distance) of the elements of a set is translated into a property of t-tuples.

Our method of analysis provides a general way to prove that a property holds for almost all t-tuples of elements from a set based on the mean pairwise statistics of the set. Our method will be useful in many areas of applied combinatorics, such as digital fingerprinting or design of key-distribution schemes, where such a translation is evident. For example, with our method new results can be obtained for the cases of cover-free codes [11, 16, 27], traceability and frameproof codes [5, 26]. This is the subject of our ongoing work.

References

1. Alon, N., Spencer, J.: The Probabilistic Method. Wiley & Sons (2000)
2. Berger, T., Mehravari, N., Towsley, D., Wolf, J.: Random multiple-access communications and group testing. IEEE Transactions on Communications 32(7), 769–779 (1984)
3. Calderbank, R., Howard, S., Jafarpour, S.: Construction of a large class of deterministic sensing Matrices that satisfy a statistical isometry property. IEEE Journal of Selected Topics in Signal Processing 4(4), 358–374 (2010)
4. Cheraghchi, M.: Improved Constructions for Non-adaptive Threshold Group Testing. In: Abramsky, S., Gavoille, C., Kirchner, C., Meyer auf der Heide, F., Spirakis, P.G. (eds.) ICALP 2010. LNCS, vol. 6198, pp. 552–564. Springer, Heidelberg (2010)
5. Chor, B., Fiat, A., Naor, M.: Tracing Traitors. In: Desmedt, Y.G. (ed.) CRYPTO 1994. LNCS, vol. 839, pp. 257–270. Springer, Heidelberg (1994)
6. Doob, J.L.: Stochastic Processes. John Wiley & Sons, New York (1953)
7. Du, D., Hwang, F.: Combinatorial Group Testing and Applications. World Scientific Publishing (2000)
8. Dyachkov, A., Rykov, V.: Bounds on the length of disjunctive codes. Problemy Peredachi Informatsii 18, 7–13 (1982)
9. D'yachkov, A., Rykov, V., Macula, A.: New constructions of superimposed codes. IEEE Transactions on Information Theory 46(1) (2000)

10. Dyachkov, A., Rykov, V., Rashad, A.: Superimposed distance codes. Problems of Control and Information Theory 18(4), 237–250 (1989)
11. Dyachkov, A., Vilenkin, P., Torney, D., Macula, A.: Families of finite sets in which no intersection of l sets is covered by the union of s other. Journal of Combinatorial Theory, Series A 99(2), 195–218 (2002)
12. Gilbert, A., Hemenway, B., Rudra, A., Strauss, M., Wootters, M.: Recovering simple signals (manuscript, 2012)
13. Gilbert, A., Iwen, M., Strauss, M.: Group testing and sparse signal recovery. In: Proc. 42nd Asilomar Conference on Signals, Systems and Computers (2008)
14. Hwang, F.: A method for detecting all defective members in a population by group testing. Journal of American Statistical Association 67, 605–608 (1972)
15. Hwang, F., Sos, V.: Non-adaptive hypergeometric group testing. Studia Scient. Math. Hungarica. 22, 257–263 (1987)
16. Kautz, W., Singleton, R.: Nonrandom binary superimposed codes. IEEE Transaction on Information Theory 10(4), 185–191 (1964)
17. Macula, A., Popyack, L.: A group testing method for finding patterns in data. Discrete Applied Mathematics 144(1-2), 149–157 (2004)
18. Macula, A., Rykov, V., Yekhanin, S.: Trivial two-stage group testing for complexes using almost disjunct matrices. Discrete Applied Mathematics 137(1), 97–107 (2004)
19. Macwilliams, F., Sloane, N.: The Theory of Error-Correcting Codes. North-Holland (1977)
20. Malyutov, M.: The separating property of random matrices. Mathematical Notes 23(1), 84–91 (1978)
21. Mazumdar, A., Barg, A.: Sparse-Recovery Properties of Statistical RIP Matrices. In: Proc. 49th Allerton Conference on Communication, Control and Computing, Monticello, IL, September 28–30 (2011)
22. McDiarmid, C.: On the method of bounded differences. In: Surveys in Combinatorics, Cambridge. London Math. Soc. Lectures Notes, pp. 148–188 (1989)
23. Ngo, H., Du, D.: A survey on combinatorial group testing algorithms with applications to DNA library screening. In: Discrete Mathematical Problems with Medical Applications. DIMACS Series Discrete Mathematics and Theoretical Computer Science, vol. 55, pp. 171–182 (1999)
24. Porat, E., Rothschild, A.: Explicit Non-adaptive Combinatorial Group Testing Schemes. In: Aceto, L., Damgård, I., Goldberg, L.A., Halldórsson, M.M., Ingólfsdóttir, A., Walukiewicz, I. (eds.) ICALP 2008, Part I. LNCS, vol. 5125, pp. 748–759. Springer, Heidelberg (2008)
25. R. Roth, *Introduction to Coding Theory*, Cambridge, 2006.
26. Staddon, J.N., Stinson, D.R., Wei, R.: Combinatorial properties of frameproof and traceability codes. IEEE Transaction on Information Theory 47(3), 1042–1049 (2001)
27. Stinson, D.R., Wei, R., Zhu, L.: Some new bounds for cover-free families. Journal of Combinatorial Theory, Series A 90(1), 224–234 (2000)
28. Wolf, J.: Born again group testing: multiaccess communications. IEEE Transaction on Information Theory 31, 185–191 (1985)
29. Yekhanin, S.: Some new constructions of optimal superimposed designs. In: Proc. of International Conference on Algebraic and Combinatorial Coding Theory (ACCT), pp. 232–235 (1998)
30. Zhigljavsky, A.: Probabilistic existence theorems in group testing. Journal of Statistical Planning and Inference 115(1), 1–43 (2003)

Parameterized Clique on Scale-Free Networks

Tobias Friedrich and Anton Krohmer

Friedrich-Schiller-Universität Jena, Germany

Abstract. Finding cliques in graphs is a classical problem which is in general NP-hard and parameterized intractable. However, in typical applications like social networks or protein-protein interaction networks, the considered graphs are scale-free, i.e., their degree sequence follows a power law. Their specific structure can be algorithmically exploited and makes it possible to solve clique much more efficiently. We prove that on inhomogeneous random graphs with n nodes and power law exponent γ, cliques of size k can be found in time $\mathcal{O}(n^2)$ for $\gamma \geq 3$ and in time $\mathcal{O}(n \exp(k^4))$ for $2 < \gamma < 3$.

1 Introduction

The clique problem numbers among the most studied problems in theoretical computer science. Its decision version calls for determining whether a given graph with n vertices contains a clique of size k, i.e., a complete subgraph on k vertices. It is one of Karp's original NP-complete problems [16] and is complete for the class W[1], the parameterized analog of NP [8]. Its optimization variant is a classical example of a problem that is NP-hard to approximate within a factor of $n^{1-\varepsilon}$ for any $\varepsilon > 0$ [11, 31]. Also, on Erdős-Rényi random graphs, the problem is believed to be intractable in general, which is even used for cryptographic schemes [15]. For all functions p, Rossman [27] presented an average-case lower bound of $\omega(n^{k/4})$ on the size of monotone circuits for solving k-CLIQUE.

The term "clique" was first used 1949 by Luce and Perry [19], to describe a group of mutual friends in a social network. Since then, social networks, and likewise the study thereof, increased tremendously. There exist numerous models, most of them having in common a so-called *scale-free* behavior. This means that there is a constant γ such that the fraction of nodes that have degree d is proportional to $d^{-\gamma}$. Besides social networks, many other real-world networks are scale-free, too. Examples are the internet, citation graphs, co-author graphs, protein-protein interaction networks and power supply networks [7, 20, 22].

It is therefore natural to study the clique problem on scale-free graphs. This is not only of theoretical interest, as this question occurs in different application domains. One example for this is bioinformatics. Here, cliques in protein-protein interaction networks are sought in order to identify clusters of proteins that interact tightly with each other [29]. Another bioinformatics example is the clustering of large scale gene expression data using cliques [4]. A different direction is internet marketing, where it is e.g. valuable to find large cliques on Facebook.

K.-M. Chao, T.-s. Hsu, and D.-T. Lee (Eds.): ISAAC 2012, LNCS 7676, pp. 659–668, 2012.
© Springer-Verlag Berlin Heidelberg 2012

We are interested in the complexity of the clique problem on *inhomogeneous random graphs*, where each node i has a weight w_i and an edge $\{i, j\}$ is present independently with probability proportional to $w_i w_j$. This is a generalization of several scale-free random graph models like Chung-Lu [1, 2, 6], Norros-Reittu [23], and generalized random graphs [30].

Our Results

The behavior of scale-free networks depends significantly on the exponent γ of the power law degree distribution. If $\gamma > 3$, the expected maximal size of a clique is constant [5, 13]. This implies that large cliques are very unlikely, but does not imply a fast algorithm that always answers correctly. The difficulty is certifying a negative answer. We prove the following theorem.

Theorem 1. *The k-CLIQUE problem can be solved in expected time $\mathcal{O}(n^2)$ on inhomogeneous random graphs with power law exponent $\gamma \geq 3$.*

All our algorithms are deterministic and always return the correct answer. Note that the above theorem implies that k-CLIQUE, which is NP-complete in general, in this setting becomes avgP, which is the average-case analog of P [18].

  On the other hand, many scale-free networks (e.g. the internet) have a power law exponent γ with $2 < \gamma < 3$ [21]. In this case, the expected maximal size of a clique diverges [5, 13]. and there exists a giant component of polynomial size, the *core*. The core is a subgraph that has a diameter of $\mathcal{O}(\log \log n)$ and contains a dense Erdős-Rényi graph [6, 30]. As this is a known hard problem, we cannot expect similarly good results as for $\gamma \geq 3$. We prove the following theorem.

Theorem 2. *The k-CLIQUE problem can be solved in time $\mathcal{O}(n \exp(k^4))$ with overwhelming[1] probability on inhomogeneous random graphs with power law exponent $2 < \gamma < 3$.*

While in general k-CLIQUE is not believed to be parameterized tractable, i.e. in FPT, the above theorem shows that in this setting k-CLIQUE is typically parameterized tractable, i.e. in typFPT, which is an average-case analog of FPT as defined in [9].

Related Work

Much previous research on cliques in random graphs focuses on Erdős-Rényi random graphs [12]. Rossman [26, 27, 28] provides lower bounds for solving the problem where he uses bounded-depth Boolean circuits and unbounded-depth monotone circuits as the computation model. Using a greedy approach, a clique of size $\log n$ can be found in a $G(n, \frac{1}{2})$ [10], whereas Jerrum [14] showed that one cannot use the Metropolis algorithm to find cliques of size $(1+\varepsilon) \log n$ in $G(n, \frac{1}{2})$,

---

[1] We use the terms *high probability* for probability $1 - o(1)$, *negligible probability* for probability $1/f(n)$, and *overwhelming probability* for probability $1 - 1/f(n)$, where $f(n)$ is any superpolynomially increasing function.

and Peinado [24, 25] proved that several randomized algorithms are also bound to fail on this problem. Kučera [17] shows that a planted clique (i.e., a clique which is explicitly added after drawing a random graph) of size $\Omega(\sqrt{n \log n})$ is easy to find in the $G(n, \frac{1}{2})$, and Alon et al. [3] further improve this bound to $\Omega(\sqrt{n})$.

The inspiration for our work was Fountoulakis et al. [9]. They introduce an average-case analog of FPT and show that the k-CLIQUE problem on $G(n, p)$ can be solved for all edge probabilities $p(n)$ in expected FPT-time and in FPT-time with high probability. Janson et al. [13] also showed that on Norros-Reittu random graphs [23] a simple algorithm with access to the weights of the model can find a $(1 - o(1))$-approximation of maximum clique in polynomial time with high probability.

2 Preliminaries

In order to achieve high general validity, we use the inhomogeneous random graph model of van der Hofstad [30], which generalizes the models of Chung-Lu [1, 2, 6] and Norros-Reittu [23] as well as the generalized random graphs. The model has two adjustable parameters: the exponent of the scale-free network γ and the average degree a. Depending on these two parameters, each node i has a weight w_i. This determines the edge probability $p_{ij} := \Pr[\{i, j\} \in E]$, which should be set proportional to $w_i w_j$.

Weights w_i. A simple way to fix the weights would be for example $w_i = a(n/i)^{\frac{1}{\gamma-1}}$. However, we aim for a more general setting and proceed differently. Given the weights w_i, we can use the empirical distribution function $F_n(w) = \frac{1}{n} \sum_{i=1}^{n} \mathbb{1}[w_i \geq w]$. This gives us $F_n(w) = \Pr[W \geq w]$, where W is a random variable chosen uniformly from the weights $w_1, \ldots, w_n$. Instead of fixing w_i, it is now easier to start from $F_n(w)$ and assume the following.

Definition 1 (Power-Law Weights). *We say that an empirical distribution function $F_n(w)$ follows the power law with exponent γ, if there exist two positive constants α_1, α_2 such that*

$$\alpha_1 w^{-\gamma+1} \leq F_n(w) \leq \alpha_2 w^{-\gamma+1}.$$

Then, we require that weights $w_1, \ldots, w_n$ have the empirical distribution function $F_n(w)$. Following van der Hofstad [30], we moreover require that the empirical distribution function F_n satisfies the following properties.

Definition 2 (Regularity Conditions for Vertex Weights).

(1) **Weak convergence of vertex weights.** *There exists a function F such that $\lim_{n \to \infty} F_n(x) = F(x)$.*

(2) **Convergence of average vertex weight.** *Let W_n and W have distribution functions F_n and F, respectively. Then, $\lim_{n \to \infty} \mathrm{E}[W_n] = \mathrm{E}[W]$ holds. Furthermore, $\mathrm{E}[W] > 0$.*

The regularity of F_n guarantees that the intuition $F_n(w) = \Pr[W \geq w]$ indeed holds. Furthermore, it guarantees that the average degree in the inhomogeneous random graphs converges, and that the largest weight is asymptotically bounded by $o(n)$, i.e. $\max_{i \in \{1,\ldots,n\}} w_i = o(n)$. Both assumptions are sufficient to generate a scale-free network. [30]

Edge Probability p_{ij}. Other inhomogeneous random graph models use the following definitions:

$$p_{ij} = \min\left\{ \frac{w_i w_j}{\sum_{k=1}^{n} w_k}, 1 \right\} \qquad \text{(Chung-Lu)}$$

$$p_{ij} = \frac{w_i w_j}{\sum_{k=1}^{n} w_k + w_i w_j} \qquad \text{(Generalized Random Graph)}$$

$$p_{ij} = 1 - \exp\left\{ -\frac{w_i w_j}{\sum_{k=1}^{n} w_k} \right\} \qquad \text{(Norros-Reittu)}$$

We use a more general approach and only assume the following.

Definition 3. *We call p_{ij} the edge probability between nodes i and j of the inhomogeneous random graph, if it is 0 for $i = j$, and otherwise fulfills*

$$p_{ij} = \mathcal{O}\left(\frac{w_i w_j}{n} \right) \qquad and \qquad p_{ij} = \Omega\left(\frac{w_i w_j}{n + w_i w_j} \right).$$

In order to see that this is a generalization of all aforementioned scale-free random graph models, we observe that when $\gamma \geq 2$, $w_{\min} = \Theta(1)$ and $w_{\min}^{-\gamma+2} > w_{\max}^{-\gamma+2}$ and compute

$$\sum_{k=1}^{n} w_k = n \cdot \mathrm{E}[W] = n \cdot \int_{w_{\min}}^{w_{\max}} F_n(w)\,\mathrm{d}w = \Theta(n \cdot w_{\min}^{-\gamma+2}) = \Theta(n). \qquad (1)$$

It is useful to observe that the expected degree of each vertex can be asymptotically upper bounded by its weight:

$$\mathrm{E}[\deg(i, G)] = \sum_{j=1}^{n} p_{ij} = \mathcal{O}\left(\frac{w_i}{n} \sum_{j=1}^{n} w_j \right) = \mathcal{O}(w_i) \qquad (2)$$

This bound is in fact tight, as shown by the following lemma. The proof is omitted due to space limitations.

Lemma 1. $\mathrm{E}[\deg(i, G)] = \Theta(w_i)$ *for $\gamma \geq 2$.*

Notation. We use $\mathcal{G}_{SF}(\gamma)$ to refer to the probability space of inhomogeneous random graphs that were created as described above, and G to represent a graph drawn from $\mathcal{G}_{SF}(\gamma)$. By $\deg(v, G)$ we refer to the degree of a node v in a graph G. We expect the nodes to be ordered from smallest to greatest weight. Finally, we use the induced subgraph $G_i := G[i, \ldots, n]$ that describes an inhomogeneous random graph $G \leftarrow \mathcal{G}_{SF}$ where nodes $1, \ldots, i-1$ have been removed from the vertex set.

3   Analysis for Power Law Exponent $\gamma \geq 3$

In this section, we describe an algorithm to solve the clique problem in $\mathcal{O}(n^2)$ on average whenever the scale-free network exhibits an exponent of $\gamma \geq 3$.

Greedy Algorithm. We exploit the scale-free structure by processing low-degree nodes first. The algorithm repeats the following steps: Choose a node v with minimum degree. If there is a $(k-1)$-subset of neighbors of v that is a clique, return the resulting k-clique. Otherwise, remove v from the graph. This implies that when the algorithm reaches the high-degree nodes, the graph is almost empty, which means that those nodes are of small degree, too. We use adjacency lists to store the edges. A careful implementation then allows finding a node with minimum degree in expected amortized constant time: An array of length n stores in each position i all nodes of degree i. By memorizing an index in the array, one can extract the currently smallest degree node in constant time. As removing a node from the graph means removing the node from the array and all adjacency lists of its neighbors, the overall runtime for updating the graph is proportional to the sum of all degrees. By equation (2), this is $\mathcal{O}(n)$. We show that the greedy algorithm has an expected runtime of $\mathcal{O}(n^2)$.

Weight Algorithm. The difficulty in the analysis of the greedy algorithm is that the node with the smallest degree may not have the smallest weight, and vice versa. We therefore take a slight detour and analyze another approach, the weight algorithm. We prove in the following Lemma 2 that it is at most $\mathcal{O}(n)$ faster than the greedy algorithm. The weight algorithm works like the greedy algorithm, the only difference being that instead of taking the node with smallest degree, it chooses a node v with minimum weight w_v. This makes it more practical for bounding the runtime. As we only use the weight algorithm for our analysis, it does not matter that the weights are not available for real networks.

Lemma 2. *On all inputs, the greedy algorithm is at most a factor of $\mathcal{O}(n)$ slower than the weight algorithm.*

Proof. Consider a graph $G = (V, E)$ and let nodes $v_1, \ldots, v_n$ be ordered as they are processed by the greedy algorithm. Let $d_{\max}$ be the largest degree that occurs during the greedy algorithm, and let t be an iteration in which this happens. Then, using some constant c, we can upper bound the runtime of this algorithm by $T_{\text{greedy}} = c \cdot n \cdot \binom{d_{\max}}{k-1} \cdot (k-1)^2$. Therefore, by definition, the subgraph $G_t = G[v_t, \ldots, v_n]$ of graph G has minimum degree $d_{\max}$, and the weight algorithm needs at least $\binom{d_{\max}}{k-1} \cdot (k-1)^2$ time for processing a node from this subgraph. $\square$

We now examine the expected degree of a node i in the weight algorithm, when the processed nodes $1, \ldots, i-1$ were already removed from the graph.

Lemma 3. *Let $G = (V, E)$ be a random graph drawn from $\mathcal{G}_{\mathcal{SF}}(\gamma)$, where $\gamma \geq 3$, and let $G_i = G[i, \ldots, n]$. Then, we have*

$$\mathrm{E}\left[\deg(i, G_i)\right] = \mathcal{O}(1).$$

Proof. We use the indicator variable $\mathbb{1}[\{i,j\} \in E]$, which attains value 1 if $\{i,j\}$ is an edge and 0 otherwise. Whenever possible, we hide constants in $\mathcal{O}(1)$.

$$\mathrm{E}\left[\deg(i, G_i)\right] = \sum_{j=i+1}^{n} \mathrm{E}\left[\mathbb{1}[\{i,j\} \in E]\right] = \sum_{j=i+1}^{n} p_{ij} = \mathcal{O}\left(\frac{w_i}{n} \sum_{j=1}^{n} w_j \cdot \mathbb{1}[w_j > w_i]\right)$$

We now use the random variable W as described in equation (1), yielding

$$\mathrm{E}\left[\deg(i, G_i)\right] = \mathcal{O}\left(w_i \cdot \mathrm{E}\left[W \cdot \mathbb{1}[W \geq w_i]\right]\right)$$
$$= \mathcal{O}\left(w_i \cdot \mathrm{E}\left[W \mid W \geq w_i\right] \cdot \Pr[W \geq w_i]\right)$$
$$= \mathcal{O}\left(w_i \cdot F_n(w_i) \cdot \int_{w_1}^{w_n} \Pr[W \geq w \mid W \geq w_i] \, dw\right).$$

To determine the probability that a weight W drawn uniformly at random admits $W \geq w$ given that $W \geq w_i$ holds, we distinguish two cases:

(1) $w \leq w_i$: Since we know that $W \geq w_i$, it is also larger than w.
(2) $w > w_i$: The conditional probability simplifies to $\frac{\Pr[W \geq w]}{\Pr[W \geq w_i]} = \frac{F_n(w)}{F_n(w_i)}$.

This simplifies the integral and yields

$$\mathrm{E}\left[\deg(i, G_i)\right] = \mathcal{O}\left(w_i \cdot F_n(w_i) \cdot \left(\int_{w_1}^{w_i} 1 \, dw + \int_{w_i}^{w_n} \frac{F_n(w)}{F_n(w_i)} \, dw\right)\right)$$
$$= \mathcal{O}\left(w_i(w_i - w_1) \cdot F_n(w_i) + \frac{w_i}{-\gamma + 2}\left[w^{-\gamma+2}\right]_{w_i}^{w_n}\right)$$
$$= \mathcal{O}(w_i^{-\gamma+3}).$$

For $\gamma \geq 3$, this term is constant, as the weights w_i are in $\Omega(1)$.   □

For proving Theorem 1, it remains to show that the processed nodes are very unlikely to have high degrees during the course of the weight algorithm.

Proof of Theorem 1. Let T_{greedy} and T_{weight} be the running time of the greedy and the weight algorithms, respectively. The runtime for processing node i in the weight algorithm is denoted by T_i.

The number of $(k-1)$-subsets of vertices a node with x neighbors allows is $\binom{x}{k-1} \leq 2^x$, and the time needed to check whether a subset is a clique is $(k-1)^2 \leq x^2$. For the expected degree of node i we write μ_i. Thus, we can write

$$\mathrm{E}[T_{\mathrm{greedy}}] = \mathcal{O}(n \cdot \mathrm{E}\left[T_{\mathrm{weight}}\right]) = \mathcal{O}\left(n \sum_{i=1}^{n} \mathrm{E}\left[T_i\right]\right)$$
$$= \mathcal{O}\left(n \sum_{i=1}^{n} \sum_{x=1}^{n} \binom{x}{k-1} \cdot (k-1)^2 \cdot \Pr\left[\deg(i, G_i) = x\right]\right)$$
$$= \mathcal{O}\left(n \sum_{i=1}^{n} \sum_{x=1}^{n} 2^x x^2 \cdot \Pr\left[\deg(i, G_i) \geq \left(1 + \left(\frac{x}{\mu_i} - 1\right)\right)\mu_i\right]\right).$$

By Lemma 3, μ_i is constant. Applying a Chernoff bound gives

$$\mathrm{E}[T_{\text{greedy}}] = \mathcal{O}\left(n\sum_{i=1}^{n}\sum_{x=1}^{n} 2^x x^2 \cdot \left(\exp\left(\frac{x}{\mu_i} - 1\right)\middle/\left(\frac{x}{\mu_i}\right)^{x/\mu_i}\right)^{\mu_i}\right)$$

$$= \mathcal{O}\left(n\sum_{i=1}^{n}\sum_{x=1}^{\infty} (2e\mu_i)^x x^{2-x}\right).$$

Since the inner sum converges to a constant, $\mathrm{E}[T_{\text{greedy}}] = \mathcal{O}(n^2)$. $\square$

4 Analysis for Power Law Exponent $\gamma \in (2, 3)$

Using the greedy algorithm of the previous section for this case would imply a superpolynomial runtime of $\text{poly}(n)^k$ since the neighborhood sizes in the algorithm increase with n. This only yields the result that k-CLIQUE is in expectation in the parameterized class XP [8]. We therefore have to apply a third algorithm to prove a better result. Instead of hoping that there will be few edges and therefore cliques, we hope that the probability that the core contains a k-clique is high. However, this approach is only feasible for small values of k. This shows that k-CLIQUE is parameterized tractable for the parameter k with high probability.

Partitioning Algorithm. To find a k-clique, the partitioning algorithm first removes all nodes with degree below $\sqrt{n/\log\log n}$. The obtained subgraph $G' = (V', E')$ is arbitrarily partitioned into components of size k. Each component is then individually checked to determine if it is a clique. If no clique is found, the algorithm searches exhaustively all k-subsets of V. It is easy to see that this algorithm is correct.

We now want to prove that when k is small, the exhaustive search is triggered only with negligible probability. For this, we first show that there are polynomially many nodes in V' and that their mutual edge probabilities are $\geq 1/\log n$. We then use this to prove that one of the partitions is likely to be a clique. Note that a slightly larger threshold like $\sqrt{n}$ would yield an edge probability of $1 - o(1)$, but the core V' is then empty if γ is close to 3. The chosen threshold $\sqrt{n/\log\log n}$ is therefore more suitable for our analysis. As in the previous section, the analysis would be much easier if, when choosing V', the algorithm was allowed to choose the nodes according to their weight, but unfortunately it only has access to their degree in the given graph. We do know, however, that the weight and the expected degree of a node are equal up to a constant. Using that, we can prove the two following lemmas:

Lemma 4 (Partitioning algorithm keeps polynomially many nodes). *Let $\gamma \in (2, 3)$, and $G = (V, E)$ be a scale-free graph drawn from $\mathcal{G}_{SF}(\gamma)$. Then,*

$$\Pr\left[\exists i > n - n^{\frac{3-\gamma}{2}} : \deg(i, G) < \sqrt{n/\log\log n}\right] \leq \exp\left(-\Theta(\sqrt{n})\right).$$

Lemma 5 (Partitioning algorithm keeps only high-weight nodes). *Let* $\gamma \in (2,3)$, *and* $G = (V, E)$ *be a scale-free graph drawn from* $\mathcal{G}_{SF}(\gamma)$. *Then,*

$$\Pr\left[\exists i \in V' \colon w_i < \sqrt{\alpha_1 n / \log n}\right] \leq \exp\left(-\Theta(n^{\frac{1}{3}})\right).$$

Both proofs will be given in the full version of the paper. It remains to show that the partitioning algorithm needs more than $k^2 n$ time only with negligible probability, if k is small enough. The idea is that the core V' of the scale-free network has larger edge probabilities than a dense Erdős-Rényi random graph, which is known to allow finding cliques fast [9].

Proof of Theorem 2. If $k > \log^{\frac{1}{3}} n$, then $n < e^{k^3}$, which implies that the exhaustive search of the partitioning algorithm runs in time $n^k < e^{k^4}$ and proves the claim. We can therefore assume $k \leq \log^{\frac{1}{3}} n$.

As excluding unlikely events does not affect small failure probabilities, we can condition on the statements of Lemmas 4 and 5 and assume that there are more than $n^{\frac{3-\gamma}{2}}$ nodes in V' and all nodes in V' have weight $\geq \sqrt{\alpha_1 n / \log n}$. This implies that $w_i w_j \geq \frac{\alpha_1 n}{\log n}$ and the edge probability between nodes $i, j \in V'$ is

$$p_{ij} = \Omega\left(\frac{w_i w_j}{n + w_i w_j}\right) = \Omega\left(\frac{1}{\log n}\right).$$

By choosing $g(n) = \frac{\log \log n - \Theta(1)}{\log n}$ suitably, we can write the edge probability as $p := p_{ij} \geq n^{-g(n)}$.

A k-partition is a k-clique with probability $\geq p^{\binom{k}{2}}$. The probability of not finding a clique before the exhaustive search is thus

$$\leq \left(1 - p^{\binom{k}{2}}\right)^{\left\lfloor n^{\frac{3-\gamma}{2}}/k \right\rfloor} \leq \exp\left(-\left\lfloor \frac{n^{\frac{3-\gamma}{2}}}{k} \right\rfloor p^{\binom{k}{2}}\right)$$

$$\leq \exp\left(-\frac{n^{\frac{3-\gamma}{2}}}{2k} p^{\binom{k}{2}}\right) = \exp\left(-\frac{n^{\frac{3-\gamma}{2} - g(n)\binom{k}{2}}}{2k}\right),$$

since we have $k \leq \log^{\frac{1}{3}} n \leq n^{\frac{3-\gamma}{2}}/2$ for large n and therefore $\lfloor p_1(n)/k \rfloor \geq p_1(n)/k - 1 \geq p_1(n)/(2k)$.

Since $k \leq \log^{\frac{1}{3}} n$, we can also assume that $k \leq \sqrt{(3-\gamma)/(10\, g(n))}$ holds for large n. Similarly, we have that $k \leq n^{\frac{3-\gamma}{20}}/2$. Then, we obtain $\binom{k}{2} g(n) \leq (3 - \gamma)/10$ and thus $\frac{3-\gamma}{2} - g(n)\binom{k}{2} \geq \frac{3-\gamma}{10}$. Therefore $n^{\frac{3-\gamma}{2} - g(n)\binom{k}{2}} \geq n^{\frac{3-\gamma}{10}}$. Hence, as $2k < n^{\frac{3-\gamma}{20}}$, it follows that the probability of doing the exhaustive search is $\leq \exp\left(-n^{\frac{3-\gamma}{20}}\right)$.   □

5   Conclusion

Social networks are becoming ubiquitous. There is a significant body of research on the structural properties of such networks, but very little on how this can be exploited algorithmically. We have shown that for scale-free networks with n nodes and power-law exponent γ, the notoriously hard k-CLIQUE problem becomes parameterized tractable for $2 < \gamma < 3$ (runtime $\mathcal{O}(n \exp(k^4))$) and even polynomial time solvable for $\gamma \geq 3$ (runtime $\mathcal{O}(n^2)$). In the future, we plan to improve the latter runtime bound for the greedy algorithm, as well as examine other NP-hard combinatorial problems (e.g. from bioinformatics) on scale-free networks.

Acknowledgements. We thank Jiong Guo and Danny Hermelin for helpful discussions. This work is part of the project "Average-Case Analysis of Parameterized Problems and Algorithms (ACAPA)" of the German Research Foundation (DFG).

References

[1] Aiello, W., Chung, F., Lu, L.: A random graph model for massive graphs. In: 32nd Symp. Theory of Computing (STOC), pp. 171–180 (2000)

[2] Aiello, W., Chung, F., Lu, L.: A random graph model for power law graphs. Experimental Mathematics 10(1), 53–66 (2001)

[3] Alon, N., Krivelevich, M., Sudakov, B.: Finding a large hidden clique in a random graph. In: 9th Symp. Discrete Algorithms (SODA), pp. 594–598 (1998)

[4] Ben-Dor, A., Shamir, R., Yakhini, Z.: Clustering gene expression patterns. Journal of Computational Biology 6(3-4), 281–297 (1999)

[5] Bianconi, G., Marsili, M.: Number of cliques in random scale-free network ensembles. Physica D: Nonlinear Phenomena 224(1-2), 1–6 (2006)

[6] Chung, F., Lu, L.: Connected components in random graphs with given expected degree sequences. Annals of Combinatorics 6(2), 125–145 (2002)

[7] Clauset, A., Shalizi, C.R., Newman, M.E.J.: Power-law distributions in empirical data. SIAM Review 51(4), 661–703 (2009)

[8] Fellows, M.R., Downey, R.G.: Parameterized Complexity. Springer (1998)

[9] Fountoulakis, N., Friedrich, T., Hermelin, D.: On the average-case complexity of parameterized clique (unpublished manuscript)

[10] Grimmett, G.R., McDiarmid, C.J.H.: On colouring random graphs. Mathematical Proceedings of the Cambridge Philosophical Society 77(02), 313–324 (1975)

[11] Håstad, J.: Clique is hard to approximate within $n^{1-\varepsilon}$. Acta Mathematica 182(1), 105–142 (1999)

[12] Janson, S., Łuczak, T., Rucinsky, A.: Random Graphs. John Wiley & Sons (2000)

[13] Janson, S., Łuczak, T., Norros, I.: Large cliques in a power-law random graph. J. Appl. Probab. 47(4), 1124–1135 (2010)

[14] Jerrum, M.: Large cliques elude the metropolis process. Random Structures & Algorithms 3(4), 347–360 (1992)

[15] Juels, A., Peinado, M.: Hiding cliques for cryptographic security. Designs, Codes And Cryptography 20(3), 269–280 (2000)

[16] Karp, R.M.: Reducibility among combinatorial problems. In: Complexity of Computer Computations, pp. 85–103. Plenum Press (1972)

[17] Kučera, L.: Expected complexity of graph partitioning problems. Discrete Applied Mathematics 57(2-3), 193–212 (1995)

[18] Levin, L.A.: Average case complete problems. SIAM Journal on Computing 15(1), 285–286 (1986)

[19] Luce, R., Perry, A.: A method of matrix analysis of group structure. Psychometrika 14(2), 95–116 (1949)

[20] Mitzenmacher, M.: A brief history of generative models for power law and lognormal distributions. Internet Mathematics 1(2), 226–251 (2004)

[21] Newman, M.E.J.: The structure and function of complex networks. SIAM Review 45(2), 167–256 (2003)

[22] Newman, M.E.J.: Power laws, Pareto distributions and Zipf's law. Contemporary Physics 46(5), 323–351 (2005)

[23] Norros, I., Reittu, H.: On a conditionally Poissonian graph process. Advances in Applied Probability 38(1), 59–75 (2006)

[24] Peinado, M.: Hard graphs for the randomized Boppana-Halldórsson algorithm for maxclique. Nordic Journal of Computing 1(4), 493–515 (1994)

[25] Peinado, M.: Go with the winners algorithms for cliques in random graphs. In: 12th Intl. Symp. Algorithms and Computation (ISAAC), pp. 525–536 (2001)

[26] Rossman, B.: On the constant-depth complexity of k-clique. In: 40th Symp. Theory of Computing (STOC), pp. 721–730 (2008)

[27] Rossman, B.: The monotone complexity of k-clique on random graphs. In: 51st Symp. Foundations of Computer Science (FOCS), pp. 193–201 (2010a)

[28] Rossman, B.: Average-case complexity of detecting cliques. PhD thesis, Massachusetts Institute of Technology (2010b)

[29] Spirin, V., Mirny, L.A.: Protein complexes and functional modules in molecular networks. Proceedings of the National Academy of Sciences 100(21), 12123–12128 (2003)

[30] van der Hofstad, R.: Random graphs and complex networks (2009), http://www.win.tue.nl/~rhofstad/NotesRGCN.pdf

[31] Zuckerman, D.: Linear degree extractors and the inapproximability of max clique and chromatic number. Theory of Computing 3(1), 103–128 (2007)

Multi-unit Auctions with Budgets and Non-uniform Valuations

H.F. Ting and Xiangzhong Xiang

Department of Computer Science, The University of Hong Kong, Hong Kong
{hfting,xzxiang}@cs.hku.hk

Abstract. This paper proposes and studies a model of multi-unit auction that allows each bidder to specify a budget and a quantity constraint. The budget tells the auctioneer how much a bidder is willing to pay and the quantity constraint specifies the maximum number of items he wants. Unlike previous studies, which assume uniform valuation (i.e., the value of an item to a bidder is fixed, no matter how many items are allocated to him), we assume that the total value of the items allocated to a bidder stops increasing, or may even start to decrease, if the number of items exceeds his acceptable quantity. We give an auction mechanism for this model and prove that when the budgets and the quantity constraints are publicly known, then our mechanism is Pareto Optimal and Incentive Compatible. On the other hand, we show that if the quantity constraints are private, then no mechanism can be both Pareto Optimal and Incentive Compatible, even if the budgets are public. We also study the revenue generated by our mechanism.

1 Introduction

In this paper, we study the allocational efficiency of multi-unit auction, in which individual bidders may bid for more than one items. A famous example of such auction is the Dutch flower auction. It has also found applications in auctioning goods such as Treasury bills and radio spectrum licences. A more recent example is Google's auction for TV Ads [11], in which a bidder selects shows, times, and days he wishes to advertise on, and then bids for time slots for showing his advertisement. Ausubel [4] was the first to study the allocational efficiency of such auction. He proposed an interesting ascending auction mechanism, now commonly known as Ausubel's Clinching Auction, and proved that under a reasonable assumption, namely the bidders have downward sloping demand curves, his mechanism yields exactly the VCG prices and is thus incentive compatible. It should be noted that Ausubel's Clinching Auction has been used in the FCC spectrum auctions [3,4,10].

In [7], Dobzinski, Lavi and Nisan observed that bidders usually have budgets on their payments to the auction, but these budgets constraints were not adequately considered in previous studies on multi-unit auction. As a consequence, when implementing the auctions from these studies, a mismatch between theory and practice emerged immediately. Hence, they proposed and studied the following model of *multi-unit auction with budget constraints*:

K.-M. Chao, T.-s. Hsu, and D.-T. Lee (Eds.): ISAAC 2012, LNCS 7676, pp. 669–678, 2012.
© Springer-Verlag Berlin Heidelberg 2012

There is a set of identical items for bidding. For every bidder i, if x items are allocated to i, these x items have a total value of $V_i(x)$ to i. It is assumed that each item has a uniform value v_i to i, and $V_i(x) = x \cdot v_i$. Bidder i also has a budget limit b_i, and if i is charged a price of p_i for the x items allocated to him, the utility for i is $V_i(x) - p_i$ if p_i is no more than i's budget b_i, and $-\infty$ otherwise. Every bidder wants to maximize his utility.

Dobzinski *et al.* [8] concentrated on two properties related to allocational efficiency, namely *Incentive-compatible*, which encourages bidders to reveal their true values and budgets, and *Pareto-optimal*, which ensures that it is impossible to strictly improve the utility of some bidder without hurting those of others. They proposed an adaptive version of Ausubel's Clinching auction mechanism for their model, and proved that when the budgets of all bidders are publicly known (or equivalently, when every bidder always tells the auctioneer its true budget), this mechanism is pareto-optimal and incentive-compatible. Furthermore, they proved that the mechanism is the unique mechanism which simultaneously satisfies pareto-optimality and incentive-compatibility. On the other hand, they proved that when bidders' budgets are private, there is no incentive-compatible auction that always produces a Pareto-optimal allocation.

We observe that although the introduction of budget constraints makes the multi-unit auction model more realistic, its uniform valuation assumption is not universally applicable. Recall that we assume $V_i(x) = x \cdot v_i$ for every integer $x \geq 0$, no matter how large x is. However, in some applications, the marginal value gained by a new item acquired will be diminished. For example, consider an auction for used cars. Suppose someone wants to buy two cars for personal use. Then the marginal value of the third car allocated to him is 0, or even worse, may be negative. This motivates us to generalize the uniform valuation assumption as follows:

Every bidder i has an *acceptable quantity* q_i and a value per item v_i. If $x \leq q_i$ items are allocated to i, the total value of these items is $V_i(x) = x \cdot v_i$, as before. However, the total value stops increasing, or may even start decreasing, when the number of items allocated to i reaches his acceptable quantity q_i.

In this paper, we study the multi-unit auction with budget constraints under the above non-uniform valuation assumption. We prove that if the acceptable quantity of every bidder is private, then even when the budgets are public, there is no auction mechanism for our model that is both incentive-compatible and Pareto-optimal. On the other hand, we extend Ausubel's Clinching auction mechanism for our model and prove that when both budgets and acceptable quantities are publicly known, the mechanism is both incentive compatible and Pareto-optimal.

We also consider the revenue of the adaptive clinching auction. The revenue is compared to that of a non-discriminatory monopoly that knows the budgets, demands and values of all bidders, and determines a single unit-price at which all items will be sold. Results show that the adaptive clinching auction extracts a large fraction of the optimal monopoly revenue generally. Because of the page limits, we move the discussion into the full version of this paper (http://www.cs.hku.hk/~xzxiang/papers/mulauc.pdf).

Related Works. There are many work related to the multi-auction with budget constraints model. In [9], Fiat *et al.* extended the model from the case of multiple identical items to a combinatorial setting where items are distinct and different bidders may be interested in different items. They gave an auction mechanism which is incentive compatible and achieves Pareto-optimality, under the assumption that the sets of interests of the bidders are public knowledge. In [5], Bhattacharya *et al.* showed that for one infinitely divisible good, a bidder cannot improve her utility by under reporting her budget. This leads to a randomized, truthful in expectation mechanism (of one infinitely divisible good) with private budgets and private valuations. Bhattacharya *et al.* [6] also considered revenue optimal Bayesian incentive compatible auctions with budgets.

It should be noted that there are also studies on the budget constraints for the single-unit auction. Aggarwal *et al.* [1], and independently Ashlagi *et al.* [2] gave incentive compatible auctions for this model with budgets constraints.

2   Preliminaries

In the auction, there are m identical items and n bidders. Each bidder i has a non-uniform valuation function $V_i(x)$ for x items. When x is no greater than bidder i's acceptable quantity constraint q_i, $V_i(x)$ is $x \cdot v_i$; when x is no less than q_i, $V_i(x)$ is a non-increasing function of x. In other words, the marginal value of each item for bidder i is v_i if $x \le q_i$ and non-positive if $x > q_i$. Each bidder also has a budget b_i on his payment.

An allocation of the auction is a vector of quantities $(x_1, \ldots, x_n)$ and a vector of payments $(p_1, \ldots, p_n)$. We define feasible allocation as follows:

Definition 21. *An allocation $\{(x_i, p_i)\}$ is a feasible allocation if it satisfies the following properties:*

1. *(Feasibility) $x_i \in N$ and $\sum_i x_i \le m$.*
2. *(Indivisible Rationality) $p_i \le V_i(x_i)$ and $\sum p_i \ge 0$.*
3. *(Budget Limit) $p_i \le b_i$.*

We define the utility of bidder i in an allocation $\{(x_i, p_i)\}$ as $u_i = V_i(x) - p_i$, if $p_i \le b_i$; otherwise, u_i is defined as $-\infty$. In the auction, every bidder always aims at getting as high utility as possible. Wlog, we assume the sum of all quantity constraints is large enough, i.e. $\sum_i q_i \ge m$.

Definition 22. *A mechanism is* incentive compatible *if for every $v_1, \ldots, v_n$, $b_1, \ldots, b_n$ and $q_1, \ldots, q_n$ and every possible manipulation v_i', b_i' and q_i', we have that $u_i = V_i(x_i) - p_i \ge V_i(x_i') - p_i' = u_i'$, where (x_i, p_i) are the allocation and payment of i for input (v_i, b_i, q_i) and (x_i', p_i') are the allocation and payment of i*

for input (v_i', b_i', q_i'). *A mechanism is* incentive compatible *for the case of publicly known budgets and quantity constraints if for any* v_i' *and fixed* $b_i' = b_i$, $q_i' = q_i$, *the above definition holds.*

Definition 23. *An allocation* $\{(x_i, p_i)\}$ *is* Pareto-optimal *if there is no other allocation* $\{(x_i', p_i')\}$ *such that* $u_i' = V_i(x_i') - p_i' \geq V_i(x_i) - p_i = u_i$ *for all i, and* $\sum_i p_i' \geq \sum_i p_i$, *with at least one of the inequalities strict.*

In our multi-unit auction model, Pareto optimality is equivalent to a "no trade" condition: no bidder can resell the items he received to other bidders and make a profit. Rename all bidders such that $v_1 \geq v_2 \geq \ldots \geq v_n$.

Lemma 24. *An allocation* $\{(x_k, p_k)\}$ *is Pareto-optimal if and only if (a) for any k,* $x_k \leq q_k$, *and (b)* $\sum_{1 \leq k \leq n} x_k = m$, *and (c) for any* $1 \leq i, j \leq n$ *such that* $v_i > v_j$ *and* $x_j > 0$, *we have either* $x_i = q_i$ *or* $b_i - p_i < v_j$.

Proof. The proof could be found in the full version of this paper.

3   The Adaptive Clinching Auction

In this section, we revise the adaptive clinching auction in [7] for model, and prove its allocation is individual rational (IR), incentive compatible (IC) and Pareto optimal (PO), when budgets and quantity constraints are publicly known.

The auction keeps for each bidder i the current number of items x_i which is already allocated to i, the current payment p_i, the remaining budget $B_i = b_i - p_i$ and the remaining quantity constraint $Q_i = q_i - x_i$. The auction also keeps the global unit-price p and the remaining number of available items M (M is initially equal to m and p is zero). The price p ascends gradually as long as the total demand of all bidders at price p is strictly larger than the total supply M. We define the demand of bidder i at price p by:

$$D_i(p) = \begin{cases} \min(\lfloor \frac{B_i}{p} \rfloor, Q_i) & p < v_i \\ 0 & otherwise \end{cases}$$

It is obvious that the above demand functions are not continuous. In particular, there are change points: when the price reaches the remaining budget B_i, and when the price reaches the value v_i. The former point is identified by using $D_i^+(p) = \lim_{t \to p^+} D_i(t)$ as, for $p = B_i$ and $Q_i > 0$, we have $D_i(p) > 0$ and $D_i^+(p) = 0$. The latter point is identified by using: $D_i^-(p) = \lim_{t \to p^-} D_i(t)$ as, for $p = v_i \leq B_i$ and $Q_i > 0$, we have $D_i(p) = 0$ and $D_i^-(p) > 0$.

Below we describe the Adaptive Clinching Auction:

1. Initialize all variables appropriately.
2. While $\sum_i D_i^+(p) > M$,
   (a) If $\exists i$ such that $D_{-i}^+(p) = \sum_{j \neq i} D_j^+(p) < M$, allocate $M - D_{-i}^+(p)$ items to bidder i for a unit price p. Update all running variables and repeat.
   (b) Otherwise increase the price p, recompute the demands, and repeat.

3. Otherwise $(\sum_i D_i^-(p) \geq M \geq \sum_i D_i^+(p))$:
    (a) For each bidder i with $D_i^+(p) > 0$, allocate $D_i^+(p)$ units to bidder i for a unit-price p and update all variables.
    (b) While $M > 0$ and there exists a bidder i with $D_i(p) > 0$, allocate $\min\{D_i(p), M\}$ units to bidder i for a unit-price p and update all variables.
    (c) While $M > 0$ and there exists a bidder i with $D_i^-(p) > 0$, allocate $\min\{D_i^-(p), M\}$ units to bidder i for a unit-price p.
    (d) Terminate.

Recall that bidder i has budget b_i and quantity constraint q_i. For simplicity, we introduce some new notations:

- $P(t)$: The current price in the auction at time t.
- $x_i(t)$: The number of all items that have already been allocated to bidder i at time t.
- $p_i(t)$: The total payment that bidder i has already paid at time t.

Assume bidder i gets some items just before time t. Define marginal utility $mu_i(t)$ as the utility of the latest item bidder i gets. Note that when $x_i(t) > q_i$, the marginal value of that item is non-positive which implies its marginal utility is negative. So:

$$mu_i(t) = \begin{cases} v_i - P(t) & x_i(t) \leq q_i \text{ and } p_i(t) \leq b_i \\ negative & otherwise \end{cases}$$

Fact 31. *If one bidder i gets some items in step 3c just before time t, then $P(t) = v_i$.*

Proof. As i is allocated some items in step 3c, we know $M > 0$ and $D_i^-(P) > 0$ at the beginning of 3c. (We use P to denote $P(t)$ for simplicity as price is unchanged it step 3.) So it is also true that $M > 0$ at the end of step 3b. We can claim that $D_i(P)$ must be 0 at the end of step 3b. Otherwise, as $M > 0$ and $D_i(P) > 0$, the iteration in 3b would not end. Now we know $D_i(P) = 0$ and $D_i^-(P) > 0$ at the beginning of 3c. This can only happen when $P = v_i$.

Lemma 32. $mu_i(t) \geq 0$ if and only if $P(t) \leq v_i$.

Proof. If $mu_i(t) \geq 0$, it is trivial that $v_i - P(t) \geq 0$. We only need to prove that if $P(t) \leq v_i$ then $mu_i(t) \geq 0$, which is equivalent to that if $P(t) \leq v_i$ then $x_i(t) \leq q_i$ and $p_i(t) \leq b_i$. Use x to denote the number of items newly allocated to bidder i just before time t. We consider two cases:

Case 1: x items are allocated to bidder i in step 2a, 3a or 3b.

- If this happens in 2a, $x = M - \sum_{j \neq i} D_j^+(P)$. As $\sum_i D_i^+(P) > M$, $x < D_i^+(P) \leq D_i(P)$.
- In 3a, $x = D_i^+(P) \leq D_i(P)$.
- In 3b, $x \leq D_i(P)$.

Above all, i gets at most $D_i(P)$ new items just before time t. Assume i gets these x items just after time $t - \epsilon$. Then: $x_i(t) = x_i(t - \epsilon) + x \leq x_i(t - \epsilon) + D_i(P) = x_i(t - \epsilon) + \min\left(\lfloor \frac{b_i - p_i(t-\epsilon)}{P} \rfloor, q_i - x_i(t - \epsilon)\right) \leq x_i(t - \epsilon) + (q_i - x_i(t - \epsilon)) = q_i$, and $p_i(t) = p_i(t - \epsilon) + x \cdot P \leq p_i(t - \epsilon) + D_i(P) \cdot P = p_i(t - \epsilon) + \min\left(\lfloor \frac{b_i - p_i(t-\epsilon)}{P} \rfloor, q_i - x_i(t - \epsilon)\right) \cdot P \leq p_i(t - \epsilon) + \left(\frac{b_i - p_i(t-\epsilon)}{P}\right) \cdot P = b_i$. Note that $x_i(t)$ and $p_i(t)$ are monotonically increasing and $x_i(0) = 0, p_i(0) = 0$.

Case 2: x items are allocated to i in 3c. From Fact 31, this can only happen when $P(t) = v_i$. As $D_i^-(P) = \lim_{\gamma \to v_i^-} D_i(\gamma) = \lim_{\gamma \to v_i^-} \min(\lfloor \frac{B_i}{\gamma} \rfloor, Q_i) = \min(\lfloor \frac{B_i}{v_i} \rfloor, Q_i) = \min(\lfloor \frac{B_i}{P} \rfloor, Q_i)$ and $x \leq D_i^-(P)$, then $x \leq \min(\lfloor \frac{b_i - p_i(t-\epsilon)}{P} \rfloor, q_i - x_i(t - \epsilon))$. Similar to case 1, we can prove that $x_i(t) \leq q_i$ and $p_i(t) \leq b_i$.

Lemma 33. *The adaptive clinching auction produces a feasible allocation and makes sure that all bidders' quantity constraints are not exceeded.*

Proof. For a truthful bidder i, the auction can guarantee he only gets items at price $p \leq v_i$. From lemma 32, the marginal utility of each clinched item is non-negative. So at the end of the auction, the bidder's budget and quantity constraint are not exceeded, i.e., $x_i \leq q_i$ and $p_i \leq b_i$. As i's payment p_i is the sum of the price of each clinched item, $p_i \leq x_i \cdot v_i = V_i(x_i)$. On the other hand, as $p_i \geq 0$ for all i, $\sum_i p_i \geq 0$.

The above lemma implies Individual Rationality and that each truthful bidder obtains a non-negative utility.

Lemma 34. *The adaptive clinching auction satisfies Incentive Compatibility (IC), i.e. a truthful bidder cannot increase his utility by declaring any value different from his true value.*

Proof. Observe that declaring a value in the auction means to decide the price at which the bidder completely drops from the auction. By lemma 32, any item clinched after price $p = v_i$ has strictly negative utility. So declaring $\tilde{v}_i > v_i$ can only decrease the total utility. Similarly, as any item clinched before or at price $p = v_i$ has non-negative utility, declaring $\tilde{v}_i < v_i$ can only decrease the total utility.

Define $D(p) = \sum_i D_i(p)$ and define $D^+(p)$ and $D^-(p)$ similarly. It is obvious that these functions are monotonically decreasing. For any continuity point of $D(p)$, $D(p) = D^+(p) = D^-(p)$. We consider discontinuity point.

Fact 35. *If p^* is a discontinuity point of $D(p)$ and $D^+(p) > M$ for any $p < p^*$, then $D^-(p^*) > M$.*

Proof. $D(p)$ is discontinued at point p^*. However, we can find a point $p' < p^*$ such that $D(p)$ is continual in the interval $[p', p^*)$. As $D(p) = \sum_i D_i(p)$, $D_i(p)$ should also be continual in $[p', p^*)$ for any i. We compare $D_i^-(p^*)$ and $D_i^+(p')$:

$$D_i^-(p^*) = \lim_{\gamma \to p^{*-}} D_i(\gamma) = \lim_{\gamma \to p^{*-}} D_i(p' + (\gamma - p')) = \lim_{\gamma \to p^{*-}} D_i(p') = D_i(p')$$

$$D_i^+(p') = \lim_{\gamma \to p'+} D_i(\gamma) = \lim_{\gamma \to p'+} D_i(p' + (\gamma - p')) = \lim_{\gamma \to p'+} D_i(p') = D_i(p')$$

The third equalities are true as $p' < p^*$ and $D_i(p)$ is continual in the interval $[p', p^*)$. So we know for all i, $D_i^-(p^*) = D_i^+(p')$. Then $D^-(p^*) = D^+(p')$. As $p' < p^*$, we get $D^+(p') > M$, which implies $D^-(p^*) > M$.

Lemma 36. *The adaptive clinching auction always allocates all items.*

Proof. Assume the auction enters step 3 at price p^*. $D^+(p^*) \leq M$ and p^* must be a discontinuity point of $D(p)$.

Firstly, we would prove that $D^-(p^*) > M$ at the beginning of step 3. For any $p < p^*$, at the beginning of step 2a we had $D^+(p) > M$. In step 2a, for any bidder i who is allocated x items at price p, $D_i^+(p)$ would decrease by x as $B_i \leftarrow B_i - x \cdot p$ and $Q_i \leftarrow Q_i - x$. M also decreases by x. So $D^+(p)$ would decrease by x and $D^+(p) > M$ still holds at the end of step 2a. After p is increased in step 2b, we have $D^+(p) > M$ or $D^+(p) \leq M$. If $D^+(p) \leq M$, the auction enter step 3 and the current price p is just the discontinuity point p^* of $D(p)$. Using the above Fact, $D^-(p^*) > M$ at the beginning of step 3.

In step 3a, for any bidder i who is allocated x items at price p^*, $D_i^-(p^*)$ would decrease by x as $B_i \leftarrow B_i - x \cdot p^*$ and $Q_i \leftarrow Q_i - x$. The remaining number of available items M also decreases by x. So $D^-(p^*)$ would decrease by x and $D^-(p^*) > M$ still holds at the end of step 3a. Similarly, $D^-(p^*) > M$ at the end of step 3b. In step 3c, while $M > 0$, allocate $\min\{D_i^-(p^*), M\}$ items to bidder with $D_i^-(p^*) > 0$. Because $D^-(p^*) > M$, M must be 0 at the end of step 3c, which implies all items are allocated.

Lemma 37. *The adaptive clinching auction satisfies Pareto optimality (PO).*

Proof. We would prove the lemma by checking the three conditions in lemma 24. First, $\sum_i x_i = m$ could be induced easily by lemma 36. All items would be allocated. Secondly, lemma 33 shows all x_i are no more than q_i. It remains to check condition (c). Use $B_i' = b_i - p_i$ and $Q_i' = q_i - x_i$ to denote the remaining budget and quantity constraint of bidder i at the end of the auction respectively. Fix an arbitrary bidder j who received at least one item. For any bidder i with $v_i > v_j$, we need to prove that $B_i' < v_j$ or $Q_i' = 0$ at the end of the auction. Consider the last price p at which j received his last item.

Firstly, assume p is not the price that ended the auction. In this case, j received the last item in step 2a. As $D_j^+(p) \neq 0$, $p < v_j$. After that, $M \leftarrow M - (M - D_{-j}^+(p))$. So $M = D_{-j}^+(p)$. Since all items were allocated and j did not get any items any more, we get that each bidder $i \neq j$ received after the the step exactly $D_i^+(p)$ items, his demand at the price p, at some price no less than p. Consider two cases: (1) $D_i^+(p) > 0$. Then $D_i^+(p) = \min(\lim_{t \to p+} \lfloor \frac{B_i}{t} \rfloor, Q_i)$. If $Q_i \leq \lim_{t \to p+} \lfloor \frac{B_i}{t} \rfloor$, then $Q_i' = 0$ at the end of the auction. Otherwise, the remaining budget of i at the end of auction is: $B_i' \leq B_i - \lim_{t \to p+} \lfloor \frac{B_i}{t} \rfloor \cdot p \leq p < v_j$. (2) $D_i^+(p) = 0$. As $v_i > v_j > p$, we get $B_i \leq p$ or $Q_i = 0$. If $B_i \leq p$, then $B_i' = B_i < v_j$. If $Q_i = 0$, then $Q_i' = 0$. So it is always true that $B_i' < v_j$ or $Q_i' = 0$ at the end of the auction.

Now assume p is the price that ended the auction. We can get that bidder j received his last item at price p in step 3 and $v_j \geq p$. We consider four cases about bidder i's demand at the beginning of step 3: (1) $D_i^+(p) > 0$. He could receive all his demand $D_i^+(p)$ in step 3a. Similar to the above argument, $Q_i' = 0$ or $B_i' \leq p$. If $v_j > p$, we can conclude $Q_i' = 0$ or $B_i' \leq p < v_j$. Otherwise $v_j = p$. Bidder j must get his last item in step 3c. In the special case $Q_i' > 0$ and $B_i' = p$, i would get another item in step 3b ($D_i(p) = 1$). After that, $B_i' \leftarrow 0$. So it is always true that $B_i' < v_j$ or $Q_i' = 0$ at the end. (2) $D_i^+(p) = 0$ and $D_i(p) > 0$. This happens when $p = B_i$ and $Q_i > 0$. If i got one item (in step 3b), $B_i' = B_i - \lfloor \frac{B_i}{p} \rfloor \cdot p < p \leq v_j$. Otherwise, M must be 0 at the beginning of step 3c and j got his last item in step 3a or 3b. So $v_j > p = B_i'$. (3) $D_i(p) = 0$ and $D_i^-(p) > 0$. This only happens when $p = v_i$, contradicting with $v_i > v_j \geq p$. (4) $D_i^-(p) = 0$. Then $p > v_i$ or $B_i < p$ or $Q_i = 0$. $p > v_i$ cannot happen and $B_i < p$ implies $B_i' < p$.

4   Impossibility Results

For multi-unit auctions with budgets and quantity constraints, we have designed a mechanism which is individual rational, incentive compatible with respect to value per item and Pareto-optimal. However, for any bidder i, we have made several assumptions: (a) b_i is public; (b) q_i is public; (c) bidder must have uniform value per item when allocated no more than q_i items.

In this section, we talk about the possibility of removing these assumptions. In [7], there is no quantity constraint and it has been proved that there is no truthful, rational and Pareto-optimal auction with private budgets and valuations. We will prove other assumptions are also necessary by contradiction.

Focus on multi-unit auction with two bidders 1, 2 and two items. Bidder 1's quantity constraint is one while bidder 2's is two. Both v_1 and v_2 are positive.

Lemma 41. *In any individual rational, incentive compatible and Pareto-optimal clinching auction: if bidder 2 wins two items then the payment of bidder 1 p_1 is 0.*

Proof. Consider the special case when $v_1 = 0$ and $v_2 > 0$. By Pareto-optimality of the auction, both of the two items have to be allocated to bidder 2. Otherwise we could increase bidder 2's utility by allocating more items to bidder 2 without changing any payment. Because of incentive compatibility, bidder 2 should pay $p_2 = 0$. Otherwise, he would lower his reported value and still attain these two items at a lower price. As the auctioneer rationality, $p_1 + p_2 \geq 0$. So $p_1 \geq 0$. As bidder rationality, $0 - p_1 \geq 0$ (bidder 1 wins no item). We get that $p_1 = 0$.

Now consider the case when $v_1 > 0$ and $v_2 > 0$. In any instance where bidder 1 wins no item it must be that $p_1 = 0$. This follows since by incentive compatibility one's payment cannot depend on his valuation. When bidder 1 reports a valuation of zero, he wins no item and p_1 is zero; when bidder 1 reports v_1 and wins no item, his payment should still be zero.

Lemma 42. *In any individual rational, incentive compatible and Pareto-optimal clinching auction: if bidder 2 wins only one item then his payment p_2 is 0.*

Proof. Consider the special case when $v_1 > 0$ and $v_2 = \epsilon$ (ϵ is positive and small enough). As in the previous proof, by Pareto-optimality, the auction has to assign one item to bidder 1 and the other one to bidder 2. By incentive compatibility, $p_1 = 0$ otherwise bidder 1 would lower his reported valuation and still win one item. As $\epsilon - p_2 \geq 0$ and $p_1 + p_2 \geq 0$, $p_2 = 0$.

Now consider the case when $v_1 > 0$ and $v_2 > 0$. In any instance where bidder 2 wins one item, it must be that $p_2 = 0$, this follows since by incentive compatibility p_2 cannot depend on v_2 and p_2 is zero when bidder 2 reports a valuation of ϵ.

Theorem 43. *There is no individual rational, incentive compatible and Pareto-optimal auction with public budgets, private valuations and quantity constraints.*

Proof. We prove the theorem by contradiction. Assume there is such an auction called A. Recall the uniqueness result of [7]:

Theorem 44 (Theorem 5.1 in [7]). *Let A' be a truthful, individual rational and Pareto-optimal auction with 2 bidders with known budgets b_1, b_2 that are generic. Then if $v_1 \neq v_2$ the allocation produced by A' is identical to that produced by the clinch auction of [7].*

Consider the case of two bidder 1, 2 and two items. Let $q_1 = 2$ and $q_2 = 2$. Additionally, fix $v_1 = 5$, $v_2 = 6$, $b_1 = 4$ and $b_2 = 5$. There is no demand constraint in [7]. However, in this case q_1, q_2 are equal to the total number of available items. Then the allocation of A must coincide with the result of the clinching auction in [7]. Bidder 2 wins one item at price of 2; bidder 1 wins the other item at price of 3. $x_1 = 1$, $p_1 = 3$; $x_2 = 1$, $p_2 = 2$.

Consider a similar case where $q_1 = 1$ instead of $q_1 = 2$. We will argue that A will not allocate any item to bidder 1 and $p_1 = 0$. As A is Pareto-optimal, both items will be allocated and bidder 1 wins at most one item; otherwise we could increase bidder 2's utility by allocating one more item to bidder 2 without changing any payment. All possible allocations are as follows: **Case 1:** $x_1 = 1$, $x_2 = 1$. By lemma 41, $p_2 = 0$. In this case, bidder 2 can buy one item from bidder 1 at price v_1 and both of two bidders are better off ($x_1' = 0$, $x_2' = 2$, $p_1' = p_1 - v_1$, $p_2' = v_1$; $u_1' = v_1 - p_1 = u_1$, $u_2' = v_2 + (v_2 - v_1) > v_2 = u_2$). This contradicts with the Pareto-optimality of A and this case is impossible. **Case 2:** $x_1 = 0$, $x_2 = 2$. By lemma 42, $p_1 = 0$. So case 2 is the only possible allocation, bidder 1 will not be allocated any item and his payment is 0.

Now assume that bidder 1's true quantity constraint is $q_1 = 1$. If he reports q_1 truthfully, $x_1 = 0$, $p_1 = 0$ and his utility $u_1 = 0$. If he lies and reports q_1 as 2, A would allocate him one item, $x_1 = 1$, $p_1 = 3$ and $u_1 = v_1 - p_1 = 2$. Bidder 1 has incentive to lie in order to increase his utility. We can conclude that in any individual rational, incentive compatible and Pareto-optimal multi-unit auction A, bidder 1 has incentive to lie.

Throughout the paper, we assume that bidder i has uniform value per item v_i when allocated $x \leq q_i$ items. A more general model is that bidder i has valuation

$v_{i,x}$ for the x-th item when $x \leq q_i$ (if $v_{i,x}$ is non-increasing, we say bidder i has diminishing marginal valuations).

Corollary 45. *There is no individual rational, incentive compatible and Pareto-optimal multi-unit auction with public budgets, public quantity constraints and private general valuations (even for diminishing marginal valuations).*

Proof. This follows directly from Theorem 43. Consider the case where $v_{1,1} = 5$, $v_{1,2} = 0$, $v_{2,1} = v_{2,2} = 6$; $b_1 = 4$, $b_2 = 5$; $q_1 = q_2 = 2$. (Note that bidders have diminishing marginal valuations in this case.) As shown above, in any individual rational, incentive compatible and Pareto-optimal multi-unit auction, bidder 1 has incentive to lie and report the value of the second item $v_{1,2}$ as 5 instead of 0.

References

1. Aggarwal, G., Muthukrishnan, S., Pál, D., Pál, M.: General auction mechanism for search advertising. In: Proceedings of the 18th International Conference on World Wide Web, pp. 241–250. ACM (2009)
2. Ashlagi, I., Braverman, M., Hassidim, A., Lavi, R., Tennenholtz, M.: Position auctions with budgets: Existence and uniqueness (2010)
3. Ausubel, L.M.: An efficient ascending-bid auction for multiple objects. The American Economic Review 94(5), 1452–1475 (2004)
4. Ausubel, L.M., Milgrom, P.: Ascending auctions with package bidding. Frontiers of Theoretical Economics 1(1), 1–42 (2002)
5. Bhattacharya, S., Conitzer, V., Munagala, K., Xia, L.: Incentive compatible budget elicitation in multi-unit auctions. In: Proceedings of the Twenty-First Annual ACM-SIAM Symposium on Discrete Algorithms, pp. 554–572. Society for Industrial and Applied Mathematics (2010)
6. Bhattacharya, S., Goel, G., Gollapudi, S., Munagala, K.: Budget constrained auctions with heterogeneous items. In: Proceedings of the 42nd ACM Symposium on Theory of Computing, pp. 379–388. ACM (2010)
7. Dobzinski, S., Lavi, R., Nisan, N.: Multi-unit auctions with budget limits. In: 2008 49th Annual IEEE Symposium on Foundations of Computer Science, pp. 260–269. IEEE (2008)
8. Dobzinski, S., Lavi, R., Nisan, N.: Multi-unit auctions with budget limits. Games and Economic Behavior (2011)
9. Fiat, A., Leonardi, S., Saia, J., Sankowski, P.: Single valued combinatorial auctions with budgets. In: Proceedings of the 12th ACM Conference on Electronic Commerce, pp. 223–232. ACM (2011)
10. Milgrom, P.: Putting auction theory to work: The simultaneous ascending auction. Journal of Political Economy 108(2), 245–272 (1999)
11. Nisan, N., Bayer, J., Chandra, D., Franji, T., Gardner, R., Matias, Y., Rhodes, N., Seltzer, M., Tom, D., Varian, H., et al.: Googles auction for tv ads. In: Automata, Languages and Programming, pp. 309–327 (2009)

Efficient Computation of Power Indices for Weighted Majority Games

Takeaki Uno

National Institute of Informatics
2-1-2, Hitotsubashi, Chiyoda-ku, Tokyo 101-8430, Japan
uno@nii.jp

Abstract. Power indices of weighted majority games are measures of the effects of parties on the voting in a council. Among the many kinds of power indices, the Banzhaf index, the Shapley–Shubik index, and the Deegan–Packel index have been studied well. For computing these power indices, dynamic programming algorithms have been proposed. The time complexities of these algorithms are $O(n^2 q)$, $O(n^3 q)$, and $O(n^4 q)$, respectively. We propose new algorithms for the problems whose time complexities are $O(nq)$, $O(n^2 q)$, and $O(n^2 q)$, respectively.

1 Introduction

Power indices are measures for evaluating the power of a party in a council. Let $\{p_1, \ldots, p_n\}$ be a set of players, and $w_1, \ldots, w_n$ be the respective weights. We consider a council composed of parties each of which corresponds to a player. Suppose that the weight of a party is the number of members of that party. For each proposal, each party (player) votes "yes" or "no". If the sum of the weights of the "yes" votes is larger than a constant q, the proposal is accepted. Constant q is called a *quota*. A *weighted majority game* is a game dealing with this situation, composed of these players, their weights, and the quota. Because each player has a distinct weight for their vote, the effect of each player on the voting is different. Many kinds of power indices have been proposed for measuring the differences.

Among the power indices proposed for weighted majority games, the Banzhaf index, the Shapley–Shubik index, and the Deegan–Packel index have been studied well. When the number of player is large, computing exact values of these indices by a straightforward exponential time algorithms requires very long time. When all the weights are integers, these indices can be computed in polynomial time in n and q by using dynamic programming algorithms[4,6,7]. The computation time of the Banzhaf index, the Shapley–Shubik index, and the Deegan–Packel index for one player is $O(nq)$[4,6], $O(n^2 q)$[6,4], and $O(n^3 q)$[7], respectively. For computing indices for all players, these algorithms take $O(n^2 q)$, $O(n^3 q)$, and $O(n^4 q)$ time, respectively.

In recent studies, power indices have been considered for many kinds of problem, such as network flow games[2] and spanning tree games[1], and applicable to auctions, management science, and so on[9]. In these studies, the number of players can be large. Moreover, basic problems such as computing power indices are often solved many times repeatedly, or solved as a sub-problem of another large complicated problem. Algorithms with a small time complexity with respect to the number of players becomes more important.

K.-M. Chao, T.-s. Hsu, and D.-T. Lee (Eds.): ISAAC 2012, LNCS 7676, pp. 679–689, 2012.

In this paper, we propose new algorithms for computing the abovementioned three power indices for all players. The framework of our algorithms is to compute the power indices of two players by existing dynamic programming algorithms in two directions, and then to compute the indices of other players by combining the results. As a result, the Banzhaf and the Shapley–Shubik indices for all players can be computed with the same time complexity as computing these indices for just one player, that is, $O(nq)$ time and $O(n^2q)$ time, respectively. These reduce the time complexities of the existing algorithms by a factor of n. Our algorithm for the Deegan–Packel index is based on a new dynamic programming running in $O(n^2q)$ time for one player, and also $O(n^2q)$ time for all players.

These algorithms are described in the following sections. Section 2 explains our algorithm for the Banzhaf index, which is the simplest algorithm among our three proposed algorithms; in this section, we explain the common basic idea of our algorithms. In Section 3, we explain a way to save unnecessary computation. Sections 4 and 5 describe our algorithms for the Shapley–Shubik index and the Deegan–Packel index, respectively. We conclude the paper in Section 6.

2   Algorithm for the Banzhaf Index

We begin with some definitions. A set of players is called a *coalition*. For a coalition S, we define the weight of S by $\sum_{p_i \in S} w_i$ and denote it by $w(S)$. In particular, we define $w(\emptyset) = 0$. S is called a *winner* if $w(S) \geq q$, and a *loser* otherwise. We assume that $w_i > 0$ for any i. Let P_i be the set of players from p_1 through p_i, i.e., $P_i = \{p_1, p_2, \ldots, p_i\}$, and let $\bar{P}_i$ be the set of players from p_i through p_n, i.e., $\bar{P}_i = \{p_i, p_{i+1}, \ldots, p_n\}$. We define P_0 and $\bar{P}_{n+1}$ by the emptyset. For a function $g(p_i, a, b, \ldots)$ mapping a combination of numbers to an integer, we define $V(g(p_i))$ by the set of the values of g for all possible values of parameters $p_i, a, b, \ldots$. As previous papers, we assume that any arithmetic operation with respect to (large) combinatorial numbers can be done in $O(1)$ time.

Let S be a winner including player p_i. If S becomes a loser when p_i exits from S, then p_i is considered to have a power in S. Assuming that every coalition occurs randomly at the same probability, the probability that p_i belongs to the coalition with having a power in the coalition is

$$| \{S | S \subseteq \{p_1, \ldots, p_n\}, p_i \in S, w(S) \geq q, w(S \setminus \{p_i\}) < q\} | \ / \ 2^n.$$

This probability is the Banzhaf index[3] of player p_i, denoted by $BZ(p_i)$.

A dynamic programming algorithm for computing $BZ(p_n)$ was proposed in [4,6], that computes the index by computing all the values of a function, denoted herein as f. For any player p_i and any $y, 0 \leq y \leq q - 1$, f is defined by the number of coalitions $S \subseteq P_i$ satisfying $w(S) = y$, i.e.,

$$f(i, y) = | \{S | S \subseteq P_i, w(S) = y\} |.$$

Note that f is defined for p_0, so that $f(0, y) = 1$ if $y = 0$ and 0 otherwise. Since

$$BZ(p_n) \times 2^n = |\{S|S \subseteq P_n, p_n \in S, w(S) \geq q, w(S \setminus \{p_n\}) < q\}|$$
$$= |\{S|S \subseteq P_{n-1}, q - w_n \leq w(S) \leq q - 1\}|$$
$$= \sum_{y=q-w_n}^{q-1} f(n-1, y),$$

we can compute $BZ(p_n)$ from $V(f(n-1))$ in $O(q)$ time. To compute the values of f, the following property is used.

Property 1. For any i, y such that $1 \leq i \leq n$ and $0 \leq y \leq q - 1$,

$$f(i, y) = \begin{cases} f(i-1, y) + f(i-1, y - w_i) & \text{if } y \geq w_i \\ f(i-1, y) & \text{if } y < w_i \end{cases}.$$

$\square$

This property implies that $V(f(i))$ is computed from $V(f(i-1))$ in $O(q)$ time. Since each value of $V(f(1))$ can be computed directly in $O(1)$ time, $V(f(n-1))$ can be computed in $O(nq)$ time. This is the basic idea of the existing dynamic programming algorithms[6,4]. For player $p_i, i < n$, we exchange the indices of players p_i and p_n, and compute $BZ(p_n)$. Therefore, the time complexity of the existing dynamic programming algorithm is $O(n^2 q)$ for computing the indices for all players.

This dynamic programming algorithm might seem to do no unnecessary operations to compute $BZ(p_n)$. Hence, reducing time complexity of the dynamic programming algorithm for computing $BZ(p_n)$ would seem to be difficult. However, the algorithm does quite similar operations for computing $BZ(p_n)$ as done to compute $BZ(p_i)$. Here, we can see possibility for improvement. In the following, we introduce a new function $b(i, y)$, that is a fuction symmetric to $f(i, y)$ such that $b(i, y)$ stands for players i to n where $f(i, y)$ stands for 1 to i. We also define a function h, and propose an algorithm for computing $BZ(p_1), BZ(p_2), \ldots, BZ(p_n)$ using these functions, instead of solving n similar dynamic programming problems.

The functions are defined as follows:

$$b(i, y) = |\{S|S \subseteq \bar{P}_i, w(S) = y\}|, \quad \text{and} \quad h(i, z) = \sum_{y=0}^{z} b(i, y).$$

For $y < 0$, we define $h(i, y) = 0$. The function f is used to solve the dynamic programming problem in the forward direction, and b is used to solve the problem in the backward direction. Since f and b are symmetric, the following equation holds for any $1 \leq i \leq n$ according to Property 1.

$$b(i, y) = \begin{cases} b(i+1, y) + b(i+1, y - w_i) & \text{if } y \geq w_i \\ b(i+1, y) & \text{if } y < w_i \end{cases}.$$

This shows that $V(b(i))$ can be computed from $V(b(i+1))$ in $O(q)$ time in the same way as f. Also, $V(h(i))$ can be computed from $V(b(i))$ in $O(q)$ time.

Now we explain how to compute $BZ(p_i)$ using f, b and h. For each coalition S not including p_i, consider the partition of S into $S \cap P_{i-1}$ and $S \cap \bar{P}_{i+1}$. Thus, the condition $q - w_i \leq w(S) \leq q - 1$ is equivalent to

$$q - w_i \leq w(S \cap P_{i-1}) + w(S \cap \bar{P}_{i+1}) \leq q - 1.$$

Hence,

$$BZ(p_i) \times 2^n = |\{S | p_i \notin S, q - w_i \leq w(S) \leq q - 1\}|$$

$$= |\{S | p_i \notin S, q - w_i \leq w(S \cap P_i) + w(S \cap \bar{P}_i) \leq q - 1\}|$$

$$= |\{(S_1, S_2) | S_1 \subseteq P_{i-1}, S_2 \subseteq \bar{P}_{i+1}, q - w_i \leq w(S_1) + w(S_2) \leq q - 1\}|$$

$$= \sum_{z=0}^{q-1} |\{S_1 | S_1 \subseteq P_{i-1}, \quad w(S_1) = z\}| \times \sum_{y=\max\{q-w_i-z,0\}}^{q-1-z} |\{S_2 | S_2 \subseteq \bar{P}_{i+1}, w(S_2) = y\}|.$$

In the last line, the left side and right side can be computed by using f and b, respectively. Specifically, for any $1 \leq i \leq n$,

$$BZ(p_i) \times 2^n = \sum_{z=0}^{q-1} f(i-1,z) \times \sum_{y=\max\{q-w_i-z,0\}}^{q-1-z} b(i+1,y)$$

$$= \sum_{z=0}^{q-1} f(i-1,z) \times (h(i, q-1-z) - h(i, \max\{q - w_i - z, 0\} - 1)).$$

The algorithm computes all values for $f(1,0), \ldots, f(n,q)$ and $b(1,0), \ldots, b(n,q)$, and then compute $h(1,0), \ldots, h(n,q)$ from b. We have the following theorem.

Theorem 1. *The Banzhaf indices of all n players with quota q can be computed in $O(nq)$ time and $O(nq)$ space if the weights of players are all integers.* $\square$

3   Reducing Space Complexity and Unnecessary Operations

The algorithm described in the previous section is quite basic. Hence, slight modifications can reduce its computation time and the space complexity. Note that the modifications do not reduce the time complexity.

First, we will explain the technique for reducing the space complexity. Transforming the statement of Property 1 gives the following equation. For $1 \leq i \leq n$,

$$f(i-1,y) = \begin{cases} f(i,y) & \text{if } 0 \leq y \leq w_i \\ f(i,y) - f(i-1, y - w_i) & \text{if } w_i \leq y \leq q - 1. \end{cases}$$

From this, we can compute $V(f(i))$ from $V(f(i+1))$ in $O(q)$ time. First, we compute $V(f(n-1))$. After computing $BZ(p_n)$, we compute $V(f(i-1))$ and $V(b(i+1))$ for $i = 1, ..., n - 1$, and compute $BZ(p_i)$, in decreasing order of i. After computing $V(f(i-2))$ and $V(b(i))$, we delete $V(f(i-1))$ and $V(b(i+1))$ from the memory since they will never be referred. As a result of this deletion, when we compute $BZ(p_i)$, only $V(f(i-1))$ and $V(b(i+1))$ are in memory. Hence, the space complexity is reduced to $O(q)$.

Theorem 2. *The Banzhaf indices of all n players with quota q can be computed in $O(nq)$ time and $O(q)$ space if the weights of players are all integers.* $\square$

4   Algorithm for Shapley–Shubik Index

Suppose that, at first, a coalition S is the empty set, and players join the coalition one by one, in an order. We suppose that the coalition changes from a loser to a winner when player p_i participates in S. Then, we can naturally consider that p_i has power. Let Π_n be the set of permutations with length n. We note that $|\Pi_n| = n!$. If the order of the coalition building is uniformly random, then the probability that p_i has power is

$$| \{(j_1, \ldots, j_n) \in \Pi_n | q \leq w(\{p_{j_1}, p_{j_2}, \ldots, p_i\}) < q + w_i\} | \; / \; n!.$$

This is the definition of the Shapley–Shubik index[8] of player p_i, denoted by $SS(p_i)$.

For computing the Shapley–Shubik index, a dynamic programming algorithm has previously been proposed [6,4] that computes the indices by computing the values of a function defined in a way similar to that of the Banzhaf index. We will again use f to denote this function central to the computation of the index. For $0 \leq i \leq n$, $0 \leq y \leq q - 1$, and $0 \leq k \leq i$, function f is defined as $f(i, k, y) = | \{S | S \subseteq P_i, w(S) = y, |S| = k\} |$. Note that $f(0, k, y)$ is 1 if $k, y = 0$, and is 0 otherwise. For any player p_i, the number of permutations $(j_1, \ldots, j_n)$ satisfying $\{p_{j_1}, p_{j_2} \ldots, p_i\} = S$ is $(|S| - 1)!(n - |S|)!$, and thus,

$$SS(p_i) \times n! = | \{(j_1, \ldots, j_n) \in \Pi_n | q \leq w(\{p_{j_1}, p_{j_2}, \ldots, p_i\}) < q + w_i\} |$$

$$= \sum_{S | p_i \notin S, q - w_i \leq w(S) \leq q-1} |S|!(n - |S| - 1)!$$

$$= \sum_{y=q-w_i}^{q-1} \sum_{k=0}^{i-1} \sum_{S | p_i \notin S, w(S)=y, |S|=k} k!(n - k - 1)!.$$

Therefore, $SS(p_n)$ can be computed from $V(f(n - 1))$. To compute f, the following property similar to Property 1 is used.

Property 2. For any i, k, and y satisfying $1 \leq i \leq n, 1 \leq k \leq i$ and $0 \leq y \leq q - 1$,

$$f(i, k, y) = \begin{cases} f(i - 1, k, y) + f(i - 1, k - 1, y - w_i) & \text{if } y \geq w_i \\ f(i - 1, k, y) & \text{if } y < w_i \end{cases}. \qquad \Box$$

This property implies that $V(f(i))$ can be computed from $V(f(i - 1))$ in $O(nq)$ time. Since each element of $V(f(1))$ can be computed directly, $V(f(n - 1))$ and $SS(p_n)$ can be computed in $O(n^2q)$ time. The indices of the other players can be computed in the same way by exchanging the indices. Thus, computing $SS(p_i)$ for all players takes $O(n^3q)$ time. This is the framework of the algorithm in [4,6].

This algorithm also seems to have no unnecessary operations involved in the dynamic programming. However, as in the case of the Banzhaf index, computing $SS(p_i)$ for all players involves unnecessary operations in that n similar dynamic programming problems are solved.

Although the idea for the algorithm for the Banzhaf index described in Section 2 might seem to be applicable to the Shapley–Shubik index, applying it directly does not lead to an efficient algorithm. If b is defined in the same way for the Banzhaf index as follows,

$$b(i, k, y) = |\{S | S \subseteq \bar{P}_i, w(S) = y, |S| = k\}|,$$

then $SS(p_i)$ is given by

$$SS(p_i) \times n! = \sum_{z=0}^{q-1} \sum_{k=0}^{i-1} \sum_{y=\max\{0, q-w_i-z\}}^{q-1-z} \sum_{l=0}^{n-(i-1)} f(i-1, k, z) \times b(i+1, l, y) \times (k+l)!(n-k-l-1)!.$$

The right-hand side of the equation includes four summations. Using function h defined in the previous section, the third summation can be eliminated, however three summations still remain. Hence, we need $O(n^2 q)$ time to compute $SS(p_i)$ for each player. This does not decrease the time complexity.

Our improved algorithm instead uses the following definitions for b and h:

$$b(i, k, y) = \sum_{S, S \subseteq \bar{P}_i, w(S) = y} (|S| + k)!(n - |S| - k - 1)!, \text{ and}$$

$$h(i, k, y) = \begin{cases} \sum_{z=0}^{y} b(i, k, z) & \text{if } 0 \le y \le q - 1 \\ 0 & \text{if } y < 0 \end{cases}.$$

Note that $b(n + 1, k, y)$ is $k!(n - k - 1)!$ if $y = 0$ and 0 otherwise.

Lemma 1. *For i and k such that $1 \le i \le n$ and $0 \le k \le i$,*

$$b(i, k, y) = \begin{cases} b(i+1, k, y) + b(i+1, k+1, y - w_i) & \text{if } y \ge w_i \\ b(i+1, k, y) & \text{if } y < w_i \end{cases}.$$

Proof. The equation holds when $i = n$, and thus we consider the case $i < n$. For any $S \subseteq \bar{P}_i$, if $w(S) < w_i$ then $p_i \notin S$. Hence, $b(i, k, y) = b(i+1, k, y)$ holds if $y < w_i$. If $y \ge w_i$, then

$$b(i, k, y) = \sum_{S | S \subseteq \bar{P}_i, w(S) = y} (|S| + k)!(n - |S| - k - 1)!$$

$$= \sum_{S | S \subseteq \bar{P}_i, w(S) = y, p_i \notin S} (|S| + k)!(n - |S| - k - 1)!$$

$$+ \sum_{S | S \subseteq \bar{P}_i, w(S) = y, p_i \in S} (|S| + k)!(n - |S| - k - 1)!$$

$$= b(i+1, k, y) + \sum_{S' | S' \subseteq \bar{P}_{i+1}, w(S') = y - w_i} (|S'| + k + 1)!(n - |S'| - k - 2)!$$

$$= b(i+1, k, y) + b(i+1, k+1, y - w_i). \qquad \Box$$

Lemma 2. *For any i such that $1 \le i \le n$,*

$$SS(p_i) \times n! = \sum_{y=0}^{q-1} \sum_{k=0}^{i-1} (f(i-1, k, y) \times (h(i, k, q-1-y) - h(i, k, \max\{q - w_i - y, 0\} - 1)))..$$

Proof. We consider a partition of a coalition S, given by $S \cap P_{i-1}$ and $S \cap \bar{P}_{i+1}$. The condition $w(S) = y$ is equivalent to $w(S \cap P_{i-1}) + w(S \cap \bar{P}_{i+1}) = y$. Hence,

$$SS(p_i) \times n! = \sum_{S|p_i \notin S, q-w_i \le w(S) \le q-1} |S|!(n-|S|-1)!$$

$$= \sum_{S_1 \subseteq P_{i-1}} \left(\sum_{S_2 \subseteq \bar{P}_{i+1}, q-w_i \le w(S_1)+w(S_2) \le q-1} (|S_1|+|S_2|)!(n-|S_1|-|S_2|-1)! \right)$$

$$= \sum_{y=0}^{q-1} \sum_{k=0}^{i-1} \left(\sum_{S_1 \subseteq P_{i-1}, w(S_1)=y, |S_1|=k} \left(\sum_{y=\max\{q-w_i-y,0\}}^{q-1-y} \sum_{S_2 \subseteq \bar{P}_{i+1}, w(S_2)=y} (k+|S_2|)!(n-k-|S_2|-1)! \right) \right)$$

$$= \sum_{y=0}^{q-1} \sum_{k=0}^{i-1} \left(|\{S_1|S_1 \subseteq P_{i-1}, w(S_1)=y, |S_1|=k\}| \times \sum_{y=\max\{q-w_i-y,0\}}^{q-1-y} b(i+1,k,y) \right)$$

$$= \sum_{y=0}^{q-1} \sum_{k=0}^{i-1} \left(f(i-1,k,y) \times \sum_{y=\max\{q-w_i-y,0\}}^{q-1-y} b(i+1,k,y) \right)$$

$$= \sum_{y=0}^{q-1} \sum_{k=0}^{i-1} f(i-1,k,y) \times (h(i,k,q-1-y) - h(i,k,\max\{q-w_i-y,0\}-1)). \qquad \square$$

Using the equations of the above lemmas, we can compute f, b and h iteratively in $O(nq)$ time, and thereby $SS(p_i)$ is obtained in $O(nq)$ time by combining f and h. Using the technique described in Section 3, we can compute the values of f in the backward direction, hence the required memory space is $O(qn)$.

Theorem 3. *The Shapley–Shubik indices of all n players with quota q can be computed in $O(n^2 q)$ time and $O(nq)$ space if the weights of players are all integers.* $\qquad \square$

5 Algorithm for Deegan–Packel Index

A winner is called a *minimal winner* if the removal of any player from the winner caused it to become a loser. If a player belongs to a minimal winner S, then the player can be considered to have power. This power can be considered to be proportional to $|S|$, hence we define the player's power to be $1/|S|$. The definition of the Deegan–Packel index $DP(p_i)[5]$ of player p_i is the expected value of the power of p_i under the assumption that every minimal winner occurs with equal probability.

For computing the Deegan–Packel index, a dynamic programming algorithm similar to those for computing the Banzhaf and Shapley–Shubik indices has been proposed [7]. The function used has parameters player, weight, size, and minimum weight player of the coalition. Thus, the algorithm takes $O(n^3 q)$ time for computing the index of one player, and $O(n^4 q)$ time for computing indices for all players.

In this section, to compute $DP(p_i)$, we use a new function f having only three parameters. Using this new function, we can construct a dynamic programming algorithm that computes $DP(p_i)$ in $O(n^2 q)$ time. As in the cases of the Banzhaf and Shapley–Shubik indices described in the previous sections, computing $DP(p_i)$ for all players can be done in $O(n^2 q)$ time.

Assume that the indices are assigned to players in decreasing order of their weights, i.e., $w_i \ge w_j$ for any $1 \le i < j \le n$. This order enables the minimal winners to be characterized in a useful way. Let $d(S)$ be the coalition obtained from S by removing the player with maximum index among players in S, and let X be the set of all minimal winners. Then,

$$S \in X \Leftrightarrow w(S) \geq q \text{ and } w(d(S)) < q.$$

Thus,

$$DP(p_i) \times |X| = \sum_{S|S \subseteq P_n, p_i \in S, w(S) \geq q, w(d(S)) < q} \frac{1}{|S|}.$$

For any i, y, and k such that $0 \leq i \leq n$, $0 \leq y \leq q$, and $0 \leq k \leq i$, we define

$$f(i, k, y) = |\{S|S \subseteq P_i, w(S) = y, |S| = k\}|,$$

and for any i and y such that $1 \leq i \leq n$ and $0 \leq y \leq q - 1$, we define

$$b(i, k, y) = \sum_{S, S \subseteq \bar{P}_i, p_i \in S, w(S) \geq y, w(d(S)) < y} \frac{1}{|S| + k}.$$

Note that $b(i, k, y)$ is 0 if either $k = 0$ or $y = 0$ holds. The definition of f is the same as in the previous section.

Property 3. For any $1 \leq i \leq n, 0 \leq k \leq i$ and $w_i \leq y \leq q - 1$,

$$b(i, k, y) = \begin{cases} b(i+1, k, y - w_i + w_{i+1}) + b(i+1, k+1, y - w_i) & \text{if } y \geq w_i \\ 1/(k+1) & \text{if } 1 \leq y \leq w_i \\ 0 & \text{if } y = 0 \end{cases}.$$

Proof. The equation holds when $i = n$ or $y \leq w_i$. For other cases,

$$b(i, k, y) = \sum_{S|S \subseteq \bar{P}_i, p_i \in S, w(S) \geq y, w(d(S)) < y} \frac{1}{|S| + k}$$

$$= \sum_{S|S \subseteq \bar{P}_{i+1}, w(S) \geq y - w_i, w(d(S)) < y - w_i, p_{i+1} \notin S} \frac{1}{|S| + k + 1}$$

$$+ \sum_{S|S \subseteq \bar{P}_{i+1}, w(S) \geq y - w_i, w(d(S)) < y - w_i, p_{i+1} \in S} \frac{1}{|S| + k + 1}$$

$$= \sum_{S|S \subseteq \bar{P}_{i+1}, p_{i+1} \in S, w(S) \geq y - w_i + w_{i+1}, w(d(S)) < y - w_i + w_{i+1}} \frac{1}{|S| + k}$$

$$+ b(i+1, k+1, y - w_i)$$

$$= b(i+1, k, y - w_i + w_{i+1}) + b(i+1, k+1, y - w_i). \qquad \square$$

Lemma 3.

$$DP(p_i) \times |X| = \sum_{y=0}^{q-1} \sum_{k=0}^{i-1} (f(i-1, k, y) \times b(i, k, q - y))$$

Proof. As in Sections 2 and 4, consider a partition of a coalition S including p_i into $S \cap P_{i-1}$ and $S \cap \bar{P}_i$. Then, the condition $w(S) \geq q$ and $w(d(S)) < q$ is equivalent to

$$q \leq w(S \cap P_{i-1}) + w(S \cap \bar{P}_i) \text{ and } w(S \cap P_{i-1}) + w(d(S \cap \bar{P}_i)) < q,$$

since $S \cap \bar{P}_i$ is always non-empty. Hence,

$$DP(p_i) \times |X| = \sum_{S|S \subseteq P_n, p_i \in S, w(S) \geq q, w(d(S)) < q} \frac{1}{|S|}$$

$$= \sum_{(S_1, S_2)|S_1 \subseteq P_{i-1}, S_2 \subseteq \bar{P}_i, p_i \in S_2, w(S_1) + w(S_2) \geq q, w(S_1) + w(d(S_2)) < q} \frac{1}{|S_1| + |S_2|}$$

$$= \sum_{S_1 \subseteq P_{i-1}} \left(\sum_{S_2 \subseteq \bar{P}_i, p_i \in S_2, q \leq w(S_1) + w(S_2), w(S_1) + w(d(S_2)) < q} \frac{1}{|S_1| + |S_2|} \right)$$

$$= \sum_{y=0}^{q-1} \sum_{k=0}^{i-1} \left(\sum_{S_1 \subseteq P_{i-1}, w(S_1) = y, |S_1| = k} \sum_{S_2 \subseteq \bar{P}_i, p_i \in S_2, q \leq y + w(S_2), y + w(d(S_2)) < q} \frac{1}{|S_1| + |S_2|} \right)$$

$$= \sum_{y=0}^{q-1} \sum_{k=0}^{i-1} \left(|\{S_1 | S_1 \subseteq P_{i-1}, w(S_1) = y, |S_1| = k\}| \times \sum_{S_2 \subseteq \bar{P}_i, p_i \in S_2, q \leq y + w(S_2), y + w(d(S_2)) < q} \frac{1}{|S_1| + |S_2|} \right).$$

Substituting f and b into the last line of the above equation gives

$$\sum_{y=0}^{q-1} \sum_{k=0}^{i-1} (f(i-1, k, y) \times b(i, k, q-y)).$$

$\square$

If $i = n$ or $y < w_i$, then $b(i, k, y)$ can be computed in constant time.

Theorem 4. *The Deegan–Packel indices of all n players with quota q can be computed in $O(n^2 q)$ time and $O(nq)$ space if the weights of players are all integers.* $\square$

Table 1. Results of the computational experiments

#players		25	50	100	200	400	800	1,600	3,200	6,400
Banzhaf	uniform	0.002	0.01	0.03	0.09	0.43	2.6	25	197	1,460
time (sec.)	scale free	0.002	0.01	0.03	0.09	0.43	2.6	23	175	1,350
	power law x^2	0.002	0.003	0.03	0.08	0.37	2	15	122	960
memory	uniform	14	14	14	15	18	25	50	141	484
	scale free	14	14	14	15	18	27	57	167	597
	power law x^2	14	14	14	15	18	26	54	152	532
Shapley–Shubik	uniform	0.03	0.12	1.3	15.1	211	3,350			
time (sec.)	scale free	0.016	0.12	1.1	10.7	126	1,950			
	power law x^2	0.021	0.12	0.9	8.8	90	1,170			
memory (MB)	uniform	17	27	76	401	2,741	21,815			
	scale free	17	26	69	305	1,717	11,626			
	power law x^2	17	26	62	227	976	5,023			
Deegan–Packel	uniform	0.015	0.13	1.2	14.5	208	3,480			
time (sec.)	scale free	0.019	0.13	1.1	11.4	134	2,050			
	power law x^2	0.024	0.12	1	9.2	93	1,190			
memory (MB)	uniform	17	27	84	430	2,948	23,363			
	scale free	17	26	72	317	1,806	12,298			
	power law x^2	17	26	63	230	986	5,040			

6 Implementation and Experiments

The above-described algorithms are implemented in C code and are available at the author's Web site (http://research.nii.ac.jp/ũno/codes.html), along with the GMP library for multi-precision integers, used for handling large integers coming from combinatorial numbers. This section shows results of some computational experiments of the implementation, specifically, computation time, memory usage, and the ratio of empty cells, that is, the combinations of i and y (and k) such that $f(i, y) = 0$ ($f(i, k, y) = 0$). All experiments were done on a 3.2 GHz Core i7-960 with a Linux operating system with 24GB of RAM memory. Note that none of the implementations used multiple cores. The instances are generated randomly with the uniform distribution and the power law distribution. In all the instances, the average weight of a player is 100, and the quota is set to $50n$; thus, the time complexities of our algorithms will be $O(n^2)$ and $O(n^3)$. The instances are generated by giving unit weight to a player randomly: the ith player receives unit weight with probability proportional to $1/n$, $1/i$ (scale free, power law), or $1/i^2$ (power law). The instances and the instance generator are also available at the author's Web site.

Table 1 shows the results for computation time. The computation time is almost $O(n^3)$ for the Banzhaf index, and $O(n^{3.5})$ to $O(n^4)$ for the Shapley–Shubik and the Deegan–Packel indices, respectively. Since multi-precision integers need much time for usual arithmetic operations, the computation time is above the time complexity we stated. Since the dynamic programs involve arithmetic operations exactly proportional to $50n^2$ and $50n^3$, we can see that each multi-precision integer requires $O(n^{1/2})$ to $O(n)$ time for one arithmetic operation, on average. In the worst case, each value of f and b can require n-bits, and products of two n bit numbers requires $O(n^2)$ time; thus, the worst case computation time is longer by a factor of n^2. In contrast our implementations need much less time. Computation times for scale free instances and power law instances are less because many players have the same small weights, and thus have the same power indices, thereby can be omitted. Table 2 shows the computation time relative to that for computing the index of just one player. Although our algorithms compute the indices for all players, the factor of the increase is less than four in most cases, with the most significant exceptions being for the Shapley–Shubik and the Deegan–Packel indices for uniform random instances. The high factor of increase in these cases are because, although the values of f require few bits, the values of b require many bits because they involve products of combinatorial numbers or rational numbers. The memory usage is slightly smaller than $O(n^2)$ for the Banzhaf index, and $O(n^{2.5})$ to $O(n^3)$ for the Shapley–Shubik and the Deegan–Packel indices. This implies

Table 2. Comparison of computation time for computing the index for one player

	uniform					scale free					power law $1/x^2$				
	50	100	200	400	800	50	100	200	400	800	50	100	200	400	800
Banzhaf	2.5	2.5	3.33	3.02	3.37	3.33	4.28	2.30	2.86	2.95	0.75	2.3	2.16	2.60	2.53
Shapley–Shubik	2.66	4.67	6.83	11.72	-	2.92	3.09	3.50	4.06	-	2.22	2.50	2.78	2.90	-
Deegan–Packel	2.65	4.25	6.44	11.01	-	2.92	3.09	3.71	4.28	-	2.4	2.83	2.93	2.92	-

that the multi-precision integers use $o(n)$ bits, less than $O(n)$. The proportion of empty cells was not especially large in any instance. Those were between 1% and 70%, and therefore even if we use lists of non-zero cells to save space for zero-cells, we can not increase the performance significantly.

Acknowledgment. This research is supported by Funding Program for World-Leading Innovative R&D on Science and Technology, Japan.

References

1. Aziz, H., Lachish, O., Paterson, M., Savani, R.: Power Indices in Spanning Connectivity Games. In: Goldberg, A.V., Zhou, Y. (eds.) AAIM 2009. LNCS, vol. 5564, pp. 55–67. Springer, Heidelberg (2009)
2. Bachrach, Y., Rosenschein, J.S.: Computing the Banzhaf Power Index in Network Flow Games. In: AAMAS 2007 (2007)
3. Banzhaf III, J.F.: Weighted Voting doesn't work. Rutgers Law Review 19, 317–343 (1965)
4. Brams, S.J., Affuso, P.J.: Power and size: a new paradox. Theory and Decision 7, 29–56 (1975)
5. Deegan, J., Packel, E.W.: A New Index of Power for Simple n-person Games. International Journal of Game Theory 7, 113–123 (1978)
6. Lucas, W.F.: Measuring Power in Weighted Voting Systems. In: Brams, S.J., Lucas, W.F., Straffin, P.D. (eds.) Political and Related Models, pp. 183–238. Springer (1983)
7. Matsui, T., Matsui, Y.: A Survey of Algorithms for Calculating Power Indices of Weighted Majority Games. Journal of the Operations Research Society of Japan 43, 71–86 (2000)
8. Shapley, L.S., Shubik, M.: A Method for Evaluating the Distribution of Power in a Committee System. American Political Science Review 48, 787–792 (1954)
9. Shapley, L.S.: The Shapley value: essays in honor of Lloyd S. Shapley. Cambridge University Press (1988)

Revenue Maximization in a Bayesian Double Auction Market

Xiaotie Deng, Paul Goldberg⋆, Bo Tang, and Jinshan Zhang

Dept. of Computer Science, University of Liverpool, United Kingdom
{Xiaotie.Deng,P.W.Goldberg,Bo.Tang,Jinshan.Zhang}@liv.ac.uk

Abstract. We study double auction market design where the market maker wants to maximize its total revenue by buying low from the sellers and selling high to the buyers. We consider a Bayesian setting where buyers and sellers have independent probability distributions on the values of products on the market.

For the simplest setting where each seller has one kind of item that can be sold in whole to a buyer, and each buyer's value can be represented by a single parameter, i.e., single-parameter setting, we develop a maximum mechanism for the market maker to maximize its own revenue.

For the more general case where the product may be different, we consider various models in terms of supplies and demands constraints. For each of them, we develop a polynomial time computable truthful mechanism for the market maker to achieve a revenue at least a constant α times the revenue of any other truthful mechanism.

1 Introduction

We consider a double auction market maker who collects valuations from buyers and sellers about a certain product to decide on the prices each seller gets and each buyer pays. The buyers may want to buy many units and the sellers may have many units to part with. The buyers and sellers may have different valuations of the product, and there is public knowledge of the probability distributions of the valuations (but each valuation, sampled from its distribution, is known only to its own buyer or seller). For simplicity, we assume that the probability distributions are independent. For the sellers and buyers, they know their own private values exactly. The market maker purchases the products from the sellers and sell them to the buyers. Our goal is to design a market mechanism that maximizes the revenue of the market maker. In other words, the market maker is to buy the same amount of products from the sellers as the amount sold to the buyers with the objective of maximizing the difference of its collected payment from the buyers and the total amount paid to the sellers. When in addition we assume public knowledge of distributions of buyers' private values from the previous sales, we call it a revenue maximization Bayesian double auction market maker.

⋆ Supported by EPSRC grant EP/G069239/1 "Efficient Decentralised Approaches in Algorithmic Game Theory".

There have been many double auction institutions, each of which may be suitable for one type of market environment [9]. Ours is motivated by the growing use of discriminative pricing models over the Internet such as one that is studied in [7] for the prior-free market environment. A possible realistic setting for applications of our model could be Google's ad exchange where Google could play a market maker for advertisers and webpage owners [12]. One may also use it for a market model of Groupon. Our use of the Bayesian model is justified by the repeated uses of a commercial system by registered users. It allows the market maker to gain Bayesian information of the users' valuations of the products being sold. Therefore, the Bayesian model adequately describes the knowledge of the market maker, buyers and sellers for the optimal mechanism design.

Our Results. We provide optimal or constant approximate mechanisms for various settings for double auction design. There are important parameters in the market design issues. The problem can be one or multi dimensional (meaning, one product or multiple different types of products). The buyers can have demand constraints or not, and sellers are supply constrained or not. Players' values are drawn from a continuous or discrete distribution. Our results are summarized in the following table.

Table 1. Results

	Dimension	Demand	Supply	Distribution	Results
Sec. 3	Single	Arbitrary	Arbitrary	Continuous	Optimal
Sec. 4	Multi	Arbitrary	Arbitrary	Continuous	1/4-Approx
Sec. 4	Multi	Arbitrary	Arbitrary	Discrete	1/4-Approx
Sec. 5	Multi	Unlimited	Arbitrary	Discrete	Optimal
Sec. 5	Multi	Arbitrary	Unlimited	Discrete	Optimal

For the demand column, "Arbitrary" refers to the case where buyers can buy at most d_i items where d_i can be an arbitrary number and "Unlimited" means $d_i = +\infty$. The supply column is similar.

In the Bayesian Mechanism Design problems, there are two computational processes involved. The first one is, given the distribution, to design an optimal or approximate mechanism which can be viewed as a function mapping bidders' profiles to allocation and payment outcomes. Since the function maps potentially exponentially many profiles to outcomes, a succinct representation of the function is also an important part in the Bayesian mechanism design. The second process is the implementation of the mechanism, i.e., given a bid profile, we run the mechanism to compute the outcome. Our results imply that all mechanisms described in the table can be represented in polynomial size and be found and implemented in polynomial time.

Related Works. Auction design play an important role in economics in general and especially in electronic commerce [11]. Of particular interest, a number of research works focus on maximizing the auctioneer's revenue, referred as the

optimal auction design problem. Myerson, in his seminal paper [13], characterized the optimal auction for the single-item setting in the Bayesian model. Recently, efforts have been made on extending Myerson's results to border settings
[8,15,17].

Unlike Myerson's optimal auction result, finding the optimal solution is not
easy for multi-dimensional settings. Recent research interest has turned toward
approximate mechanisms [1,5]. Cai et al. [4] presented a characterization of a
rather general multi-dimensional setting and proposed an efficient mechanism for
the special case where no bidders are demand constrained. Using similar ideas,
Alaei et al. [2] present a general framework for reducing multi-agent service
problems to single-agent ones.

The double auction design problem becomes more complicated since the market maker acts as the middle man to bring buyers and sellers together. A guide
to the literature in micro-economics on this topic can be found in [9]. The profit
maximization problem for the single buyer/single seller setting has been studied
by Myerson and Satterthwaite [14]. Our optimal double auction is a direct extension of their work and, to our best knowledge, fills a clear gap in the economic
theory of double auctions. Deshmukh et al. [7], studied the revenue maximization problem for double auctions when the auctioneer has no prior knowledge
about bids. Their prior-free model is essentially different from ours. More auction
mechanism design problems were studied by many researchers in recent years,
but as far as we know, not in the context of optimal double auction design in
the Bayesian setting. While our setting assumes the existence of a monopoly
platform, Rochet and Tirole [16] and Armstrong [3] introduced several different
models for two-sided markets and studied platform competition.

2   Preliminaries

Throughout the paper we consider Bayesian incentive compatible mechanisms
only. Informally, a mechanism is Bayesian incentive compatible if it is optimal
for each buyer and each seller to bid its true value of the items. We will formally
define this concept later. As a consequence, we should consider their bids to be
their true valuations and restrict our discussion to mechanisms that result in
less or equal utility if one deviates to report a false value.

Therefore, we will use the notation v_{ij} to represent the ith buyer's (true) bid
for one of the jth seller's items and w_j for the jth seller's (true) bid. We will
drop the "(true)" subsequently as deviations of bids from the true valuations
will be marked. The ith buyer's bid can be denoted by a vector v_i and bids of all
buyers can be denoted by v or sometimes $(v_i; v_{-i})$ where v_{-i} is the joint profile
of all other bidders. Similarly, we use w and $(w_j; w_{-j})$ for the sellers' bid. [1]

In our model, all players' bids are assumed to be distributed independently
according to publicly known distributions, V for buyers, W for sellers. Note
that we also assume that V and W should be bounded, i.e. $v_{ij} \in [\underline{v}_{ij}, \overline{v}_{ij}]$ and
$w_j \in [\underline{w}_j, \overline{w}_j]$.

---

[1] We use semi-colon to separate the profile of a special player with others and use
comma to separate the buyers' profiles with sellers'.

The outcome of a mechanism M consists of four random variables (x, p, y, q) where x and p are the allocation function and payment functions for buyers, y and q for sellers. That is, buyer i receives item j with probability $x_{ij}(v, w)$ and pays $p_i(v, w)$; seller j sells her item with probability $y_j(v, w)$ and gets a payment $q_j(v, w)$. Thus, the expected revenue of the mechanism is $R(M) = \mathrm{E}_{v,w}[\sum_i p_i(v, w) - \sum_j q_j(v, w)]$ where $\mathrm{E}_{v,w}$ is short for $\mathrm{E}_{v \sim V, w \sim W}$.

In general, a buyer may buy more than one item from the mechanism. We assume buyers' valuation functions are additive, i.e. $v_i(S) = \sum_{j \in S} v_{ij}$. For each buyer i, let d_i denote the demand constraint for buyer i, i.e. buyer i cannot buy more than d_i items. Similarly, let k_j be the supply constraint for seller j, i.e. seller j cannot sell more than k_j items. By the Birkhoff-von Neumann theorem [10][8][6], it suffices to satisfy $\sum_j x_{ij} \le d_i$ and $y_j = \sum_i x_{ij} \le k_j$.

Let $U_i(v, w) = \sum_j x_{ij}(v, w) v_{ij} - p_i(v, w)$ be the expected utility of buyer i when the profile of all players is (v, w) and $T_j(v, w) = q_j(v, w) - y_j(v, w) w_j$ be the expected utility of seller j. We proceed to formally define the concepts of Bayesian Incentive Compatibility of mechanisms and ex-interim Individual Rationality of the buyers and sellers:

Definition 1. *A double auction mechanism M is said to be* Bayesian Incentive Compatible (BIC) *iff the following inequalities hold for all i, j, v, w.*

$$
\begin{aligned}
\mathrm{E}_{v_{-i}, w}[U_i(v, w)] &\ge \mathrm{E}_{v_{-i}, w}[U_i((v_i'; v_{-i}), w)] \\
\mathrm{E}_{v, w_{-j}}[T_j(v, w)] &\ge \mathrm{E}_{v, w_{-j}}[T_j(v, (w_j'; w_{-j}))]
\end{aligned}
\tag{1}
$$

We note that, if $U_i(v, w) \ge U_i((v_i'; v_{-i}), w)$ and $T_j(v, w) \ge T_j(v, (w_j'; w_{-j}))$ for all v, w, v_i', w_j', we say M is Incentive Compatible.

Definition 2. *A double auction mechanism M is said to be* ex-interim Individual Rational (IR) *iff the following inequalities hold for all i, j, v, w.*

$$
\begin{aligned}
\mathrm{E}_{v_{-i}, w}[U_i(v, w)] &\ge 0 \\
\mathrm{E}_{v, w_{-j}}[T_j(v, w)] &\ge 0
\end{aligned}
\tag{2}
$$

Similarly, we note that, if $U_i(v, w) \ge 0$ and $T_j(v, w) \ge 0$ for all v, w, we say M is ex-post Individual Rational.

Finally, we present the formal definition of approximate mechanism.

Definition 3 (α-approximate Mechanism[17]). *Given a set $\mathbb{M}$ of feasible mechanisms, we say mechanism $M \in \mathbb{M}$ is an α-approximate mechanism in $\mathbb{M}$ iff for each mechanism $M' \in \mathbb{M}$, for any set of buyer and sellers $\alpha \cdot R(M') \le R(M)$. A mechanism is* optimal *in $\mathbb{M}$ if it is an 1-approximate mechanism in $\mathbb{M}$.*

3 Optimal Single-Dimensional Double Auction

In this section, we consider the single-dimensional double auction design problem where all sellers sell identical items, that is for all $j, j' \in [m]$, $v_{ij} = v_{ij'}$. Moreover, as shown in Table 1, in this section we assume the bidders' bids are drawn from continuous distributions. Let f_i, F_i be the probability density function (PDF) and cumulative distribution function (CDF) for buyer i's value, g_j, G_j be the PDF and CDF for seller j's value.

Our mechanism can be viewed as a generalization of the classical Myerson's Optimal Auction [13]. It is well known that Myerson's approach is powerful and extensive in the single-dimensional setting. We strengthen this by showing that a similar optimal double auction can be found in this single-dimensional setting. In addition, in Section 4 this optimal mechanism will be used to construct a constant approximate mechanism for a multi-dimensional setting.

Recall that Myerson's virtual value function is defined as $c_i(v_i) = v_i - \frac{1-F_i(v_i)}{f_i(v_i)}$ for each buyer. In the double auction, we define the virtual value functions for buyers and sellers as $c_i(v_i) = v_i - \frac{1-F_i(v_i)}{f_i(v_i)}$ and $r_j(w_j) = w_j + \frac{G_j(w_j)}{g_j(w_j)}$. If $c_i(v_i)$ is not an increasing function of v_i or r_j is not decreasing, by Myerson's ironing technique, we can use the ironed virtual value function $\bar{c}_i$ and $\bar{r}_j$. W.l.o.g, we assume the buyers are sorted in decreasing order with respect to $\bar{c}_i(v_i)$ and all sellers are in increasing order with respect to $\bar{r}_j(w_j)$. Let $D = \max_{i,j}\{\min\{\sum_{s=1}^i d_s, \sum_{t=1}^j k_j\} | \bar{c}_i(v_i) > \bar{r}_j(w_j)\}$. Thus, we can define the optimal auction in the spirit of maximizing virtual surplus.

$$x_i(v,w) = \begin{cases} d_i & \text{if } \sum_{s \leq i} d_s \leq D \\ D - \sum_{s < i} d_s & \text{if } \sum_{s < i} d_s < D < \sum_{s \leq i} d_s \\ 0 & \text{otherwise} \end{cases}$$

$$y_j(v,w) = \begin{cases} k_j & \text{if } \sum_{t \leq j} k_t \leq D \\ D - \sum_{s < j} k_s & \text{if } \sum_{t < j} k_t < D < \sum_{t \leq j} k_t \\ 0 & \text{otherwise} \end{cases}$$

$$p_i(v,w) = x_i(v,w)v_i - \int_{\underline{v_i}}^{v_i} x_i((s;v_{-i}),w)ds$$

$$q_j(v,w) = y_j(v,w)w_j + \int_{w_j}^{\overline{w_j}} y_j(v,(t;w_{-j}))dt$$

Theorem 1. *The above mechanism is an optimal (revenue) mechanism for the single-dimensional double auction setting. Under the assumption that the integration and convex hull of f, g can be computed in polynomial time, the mechanism can be found and implemented. Moreover, the mechanism is deterministic, incentive compatible and ex-post Individual Rational.*

4   Approximate Multi-dimensional Double Auction

In this section, we provide a general framework for approximately reducing the double auction design problem for multiple buyers and sellers to single pair of buyer and seller sub-problems. As an application, we apply the framework to construct a 1/4-approximate mechanism for the multi-dimensional setting. Our approach is inspired by the work of Alaei [1] which provide a general framework for the one sided auction.

Recall that all bids are drawn from public known distributions and our goal is to maximize the expected revenue for the auctioneer. It should be emphasized that, in this section, we assume the buyers' values for different items are independent, i.e. v_{ij} and $v_{ij'}$ are independent.

First of all, we introduce the concept of Primary Mechanism which can be viewed as a mechanism between one buyer and one seller.

Definition 4 (Primary Mechanism/Primary Benchmark).
A primary mechanism denoted by M_{ij} for buyer i and seller j is a single buyer and single seller mechanism which allows specifying an upper bound on the ex-ante expected probability $\bar{k}_{ij}$ of allocating jth item to buyer i. A primary benchmark denoted by $\bar{R}_{ij}$ is a concave function such that the optimal revenue of any primary mechanism M_{ij} subject to $\bar{k}_{ij}$ is upper bounded by $\bar{R}_{ij}(\bar{k}_{ij})$.

Intuitively, for any allocation rule, define the ex-ante probability of assigning jth seller's items to buyer i as $\bar{k}_{ij} = \mathrm{E}_{v_i,w_j}[x_{ij}(v_i, w_j)]$. Then we can relax the supply constraints $\sum_i x_{ij}(v, w) \le k_j$ and demand constraints $\sum_j x_{ij}(v, w) \le d_i$ to the ex-ante probability constraints, $\sum_i \bar{k}_{ij} \le k_j$ and $\sum_j \bar{k}_{ij} \le d_i$. Then we compute the optimal ex-ante probability by convex programming. Obviously, the optimal solution of the relaxed problem must be an upper bound for any original solution. Unfortunately, the solution solved by convex programming may not be a feasible solution of the original problem. To solve this problem, Alaei introduced the following rounding process to round the relaxed solution to a feasible one.

Lemma 1 (γ-Conservative Magician (Theorem 2 in [1])). *In the Magician problem, a magician is presented with a series of boxes one by one. He has k magic wands that can be used to open the boxes. On each box is written a probability q_i. If a wand is used on a box, it opens, but with at most probability q_i the wand breaks. Given $\sum_i q_i \le k$ and any $\gamma \le 1 - \frac{1}{\sqrt{k+3}}$, a γ-conservative magician guarantees that each box is opened with an ex-ante expected probability at least γ.*

Using above lemma, we describe our mechanism for multi-dimensional double auction problem. Recall that in the classical auction setting, all items are sold by the auctioneer. However, in the double auction setting, items are sold by different sellers and more efforts should be taken to handle the truthfulness issue of sellers. We extend Alaei's rounding mechanism from one-dimension (considering buyers one by one) to two-dimension (considering each pair of buyer and seller sequentially) as follows.

Mechanism (Modified γ-Pre-Rounding Mechanism)

(I) Solve the following convex program and let $\bar{k}_{ij}$ denote an optimal assignment for it.

$$\text{Maximize:} \quad \sum_{i \in [n], j \in [m]} \bar{R}_{ij}(x_{ij}) \qquad\qquad\qquad (CP)$$

$$\text{Subject to:} \quad \sum_{j \in [m]} x_{ij} \le d_i \qquad\qquad \text{for all } i \in [n]$$

$$\sum_{i \in [n]} x_{ij} \le k_j \qquad\qquad \text{for all } j \in [m]$$

$$x_{ij} \ge 0 \qquad\qquad \text{for all } i \in [n], \, j \in [m]$$

(II) For each buyer i, create an instance of γ-conservative magician with d_i wands (this will be referred to as the buyer i's magician). For each item j create an instance of γ-conservative magician with k_j wands (this will be referred to as the seller j's magician).

(III) For each pair of buyer and seller (i,j):
   (a) Write $\bar{k}_{ij}$ on a box and present it to the buyer i's magician and the seller j's magician.
   (b) If both of them open the box, run $M_{ij}(\bar{k}_{ij})$ on buyer i and seller j otherwise consider next pair.
   (c) If the mechanism buys an item from seller j and sells it to buyer i, then break the wands of buyer i's magician and seller j's magician.

Theorem 2 (Modified γ-Pre-Rounding Mechanism). *Suppose for each buyer and seller pair (i,j), we have an α-approximate primary mechanism M_{ij} and a corresponding primary benchmark $\bar{R}_{ij}$. Then for any $\gamma \in [0, 1 - \frac{1}{\sqrt{k^*+3}}]$ where $k^* = \min_{i,j}\{d_i, k_j\}$, the Modified γ-Pre-Rounding Mechanism is a $\gamma^2 \cdot \alpha$-approximation mechanism.*

Proof. The proof is similar to the one in [1]. First, we prove that the expected revenue of any mechanism is upper bounded by $\sum_i \sum_j \bar{R}_{ij}(\bar{k}_{ij})$. For any mechanism $M = (x, p, y, q)$, let $k_{ij} = E_{v,w} x_{ij}(v, w)$. Due to the feasibility of M, k_{ij} must be a feasible solution of the convex programming (CP). So we have,

$$R(M) = \sum_i \sum_j R_{ij}(k_{ij}) \le \sum_i \sum_j \bar{R}_{ij}(k_{ij}) \le \sum_i \sum_j \bar{R}_{ij}(\bar{k}_{ij})$$

Then it suffices to show that for each pair (i,j), our mechanism can gain the revenue $\bar{R}_{ij}(\bar{k}_{ij})$ with probability at least $\gamma^2 \cdot \alpha$, i.e. each box will be opened with probability at least γ^2. This can be deduced from Lemma 1 easily. $\square$

Then we consider the multi-dimensional double auction design problem and present a constant approximate mechanism. For each buyer and seller pair i, j, we use the mechanism in Section 3 for one-dimensional cases to be the primary mechanism M_{ij} and the expected revenue of M_{ij} to be the primary benchmark $\bar{R}_{ij}$.

Theorem 3. *Assume that all bidders' bids are drawn from continuous distributions. A 1/4 approximate double auction for the multi-dimensional setting can be found and implemented in polynomial time.*

For the discrete distribution case, the optimal mechanism for single buyer and single seller can be computed by Linear Programming. So we have the similar result.

Theorem 4. *Assume that all bidders' bids are drawn from discrete distributions. A 1/4 approximate double auction for the multi-dimensional setting can be found and implemented in polynomial time.*

5 Optimal Mechanism for Discrete Distributions

In this section, we consider the multi-dimensional double auction when all the bidders' value distributions are discrete. Unlike Section 4, we consider two special cases of the problem. One is the case where all buyers have unlimited demand, i.e., $d_i = +\infty$ for all buyer i and the other one is the case where all sellers have unlimited supply, i.e. $k_j = +\infty$ for all seller j. In this section, we focus on the previous case. The mechanism and the proof of the latter case are similar.

Recall that, in the multi-dimensional setting, the auctioneer collects each buyer's bid, denoted by a vector $v_i = (v_{i1}, \ldots, v_{im})$ drawn from a public known distribution V_i and seller's bid denoted by w_j drawn from W_j. Throughout this section, V_i and W_j are discrete distributions and we use f_i and g_j to denote their probability mass function, i.e. $f_i(t) = \Pr[v_i = t]$ and $g_j(t) = \Pr[w_j = t]$. It should be emphasized that, unlike Section 4, we do not need to assume that the buyer's bids for each item should be independent, i.e. v_{ij} and $v_{ij'}$ can be correlated in this section. We also add a dummy buyer 0 with only one type v_0 for buyers and seller 0 with w_0 for sellers.

Our approach is motivated by the recent results of Cai et al. [4] and Aleai et al. [2] which require a reduced form of x, y, p, q denoted by $\bar{x}$, $\bar{y}$, $\bar{p}$ and $\bar{q}$ respectively, defined as follows:

$$\bar{x}_{ij}(v_i, w_j) = \mathrm{E}_{v_{-i}, w_{-j}}[x_{ij}(v, w)] \quad \bar{y}_j(v_i, w_j) = \mathrm{E}_{v_{-i}, w_{-j}}[y_j(v, w)]$$
$$\bar{p}_i(v_i, w_j) = \mathrm{E}_{v_{-i}, w_{-j}}[p_i(v, w)] \quad \bar{q}_j(v_i, w_j) = \mathrm{E}_{v_{-i}, w_{-j}}[q_j(v, w)]$$

Now we are ready to convert an optimization problem of x, p, y, q to a problem of $\bar{x}, \bar{p}, \bar{y}, \bar{q}$ which can be represented by a Linear Program with polynomial size in T, n and m where T is the maximum among all $|V_i|$ and $|W_j|$.

Then BIC constraints (1) and IR constraints (2) can be rewritten as

$$\mathrm{E}_{w_j}[\textstyle\sum_j \bar{x}_{ij}(v_i, w_j)v_{ij} - \bar{p}_i(v_i, w_j)] \geq \mathrm{E}_{w_j}[\textstyle\sum_j \bar{x}_{ij}(v_i', w_j)v_{ij} - \bar{p}_i(v_i', w_j)]$$
$$\mathrm{E}_{v_i}[\bar{q}_j(v_i, w_j) - \bar{y}_j(v_i, w_j)w_j)] \qquad \geq \mathrm{E}_{v_i}[\bar{q}_j(v_i, w_j') - \bar{y}_j(v_i, w_j')w_j] \tag{3}$$
$$\mathrm{E}_{w_j}[\textstyle\sum_j \bar{x}_{ij}(v_i, w_j)v_{ij} - \bar{p}_i(v_i, w_j)] \geq 0$$
$$\mathrm{E}_{v_i}[\bar{q}_j(v_i, w_j) - \bar{y}_j(v_i, w_j)w_j)] \qquad \geq 0$$

Finally, all mechanism should satisfy the supply constraints, i.e., for each item j and profiles $v, w, y_j(v, w) = \sum_i x_{ij}(v, w) \leq k_j$. Note that there is no demand constraint on buyers. With loss of generality, we assume that $k_j = 1$ for all j. Otherwise, we can normalize x by setting $x'_{ij}(v, w) = x_{ij}(v, w)/k_j$ and refine v, w by setting $v'_{ij} = k_j v_{ij}$ and $w'_j = k_j w_j$ such that $k'_j = 1$ for all item j.

For the single-item setting of classical auction, i.e. $m = 1$ and seller's value for his item is always 0, Alaei et al. [2] prove a sufficient and necessary condition for the supply constraint. We generalize their result to a multi-dimensional double auction setting.

Lemma 2. *Given a reduced form $\bar{x}$, there exists an ex-post implementation x such that $x_{ij}(v, w) \geq 0, \sum_i x_{ij}(v, w) \leq 1$ and $\bar{x}_{ij}(v_i, w_j) = E_{v_{-i}, w_{-j}}[x_{ij}(v, w)]$ iff there exists (s, z) such that, for each seller j and $w_j \in W_j$*

$$s_0^{(j)}(v_0, w_j, 0) = 1$$
$$s_i^{(j)}(v_i, w_j, i) = \sum_{k=0}^{i-1} \sum_{v_k \in V_k} z_{ki}^{(j)}(v_k, v_i, w_j) \qquad \forall i, v_i \in V_i$$
$$s_k^{(j)}(v_k, w_j, i) = s_k^{(j)}(v_k, w_j, i-1) - \sum_{v_i \in V_i} z_{ki}^{(j)}(v_k, v_i, w_j) \quad \forall i, k < i, v_k \in V_k$$
$$z_{ki}^{(j)}(v_k, v_i, w_j) \leq s_k^{(j)}(v_k, w_j, i-1) f_i(v_i) \qquad \forall i, k < i, v_i \in V_i, v_k \in V_k$$
$$\bar{x}_{ij}(v_i, w_j) f_i(v_i) = s_i^{(j)}(v_i, w_j, n) \qquad \forall i, v_i \in V_i$$

$$\tag{4}$$

Moreover, given any feasible reduced allocation rule $\bar{x}$, the ex-post of $\bar{x}$ can be found efficiently.

Finally, we convert the problem of multi-dimensional double auction design problem to a Linear Program with reduced form which can be solved in polynomial time in m, n, T.

Theorem 5. *Assume all bidders' bids are drawn from discrete distributions and all bidders are without demand constraints. An optimal double auction for multi-dimensional setting can be found and implemented in polynomial time.*

Theorem 6. *Assume that all bidders' bids are drawn from discrete distributions and all sellers are without supply constraints. An optimal double auction for multi-dimensional setting can be found and implemented in polynomial time.*

6   Conclusion

In this paper, we present several optimal or approximately-optimal auctions for a double auction market. Double auction platforms have started to gain importance in electronic commerce. One possible example is the ad exchange market proposed to bring advertisers and web publishers together [12]. There is other potential in setting up electronic platforms for sellers and buyers of other types of resources in the context of cloud computing.

Our results on the one hand show the power of recent significant progress in one-sided markets, and on the other hand raise new challenges in the development of mathematical and algorithmic tools for market design.

References

1. Alaei, S.: Bayesian combinatorial auctions: Expanding single buyer mechanisms to many buyers. In: Procs. of 52nd IEEE FOCS Symposium, pp. 512–521 (2011)
2. Alaei, S., Fu, H., Haghpanah, N., Hartline, J., Malekian, A.: Bayesian optimal auctions via multi- to single-agent reduction. In: Proceedings of the 14th ACM Conference on Electronic Commerce, pp. 17–17 (2012)
3. Armstrong, M.: Competition in two-sided markets. The RAND Journal of Economics 37(3), 668–691 (2006)
4. Cai, Y., Daskalakis, C., Weinberg, S.M.: An algorithmic characterization of multidimensional mechanisms. In: Procs. of the 44th Annual ACM STOC Symposium, pp. 459–478 (2012)
5. Chawla, S., Hartline, J.D., Malec, D.L., Ivan, B.S.: Multi-parameter mechanism design and sequential posted pricing. In: Proceedings of the 42nd ACM Symposium on Theory of Computing, STOC 2010, New York, NY, USA, pp. 311–320 (2010)
6. Daskalakis, C., Weinberg, S.M.: Symmetries and optimal multi-dimensional mechanism design. In: Proceedings of the 13th ACM Conference on Electronic Commerce, EC 2012, pp. 370–387. ACM, New York (2012)
7. Deshmukh, K., Goldberg, A.V., Hartline, J.D., Karlin, A.R.: Truthful and Competitive Double Auctions. In: Möhring, R.H., Raman, R. (eds.) ESA 2002. LNCS, vol. 2461, pp. 361–373. Springer, Heidelberg (2002)
8. Dobzinski, S., Fu, H., Kleinberg, R.D.: Optimal auctions with correlated bidders are easy. In: Proceedings of the 43rd Annual ACM Symposium on Theory of Computing, STOC 2011, pp. 129–138. ACM, New York (2011)
9. Friedman, D.: The double auction market institution: A survey. The Double Auction Market Institutions Theories and Evidence 14, 3–25 (1993)
10. Johnson, D.M., Dulmage, A.L., Mendelsohn, N.S.: On an algorithm of G. Birkhoff concerning doubly stochastic matrices. Canadian Mathematical Bulletin (1960)
11. Klemperer, P.: The economic theory of auctions. Edward Elgar Publishing (2000)
12. Muthukrishnan, S.: Ad Exchanges: Research Issues. In: Leonardi, S. (ed.) WINE 2009. LNCS, vol. 5929, pp. 1–12. Springer, Heidelberg (2009)
13. Myerson, R.: Optimal auction design. Mathematics of Operations Research 6(1), 58–73 (1981)
14. Myerson, R.B., Satterthwaite, M.A.: Efficient mechanisms for bilateral trading. Journal of Economic Theory 29(2), 265–281 (1983)
15. Papadimitriou, C.H., Pierrakos, G.: On optimal single-item auctions. In: Procs. of the 43rd Annual ACM STOC Symposium, pp. 119–128 (2011)
16. Rochet, J.-C., Tirole, J.: Platform competition in two-sided markets. Journal of the European Economic Association 1(4), 990–1029 (2003)
17. Ronen, A.: On approximating optimal auctions. In: Procs. of the 3rd ACM Conference on Electronic Commerce, EC 2001, pp. 11–17 (2001)

Author Index

Ahn, Hee-Kap   54, 309
Akutsu, Tatsuya   146
Angelini, Patrizio   423
Assadi, Sepehr   382

Bae, Sang Won   309, 629
Bampis, Evripidis   106
Belmonte, Rémy   299
Bhattacharya, Binay   588
Binucci, Carla   423
Brandstädt, Andreas   267
Brodal, Gerth Stølting   527
Brodnik, Andrej   156
Burcea, Mihai   44

Chan, Timothy M.   2
Cheilaris, Panagiotis   4
Chen, Danny Z.   332, 609
Chen, Jian-Jia   75
Chen, Wei   278

Demaine, Erik D.   3
Deng, Xiaotie   690
Dey, Sandeep Kumar   187
Disser, Yann   506
Dorrigiv, Reza   136
Durocher, Stephane   319
Dutta, Chinmoy   257

Emamjomeh-Zadeh, Ehsan   382
Evans, William   423

Fang, Wenjie   278
Frati, Fabrizio   413
Friedrich, Tobias   659
Fu, Norie   392

Ganguly, Sumit   64
Gargano, Luisa   4
Goldberg, Paul   690
Golin, Mordecai   362
Golovach, Petr A.   14, 495
Grgurovič, Marko   156
Gudmundsson, Joachim   413
Guo, Jiong   126

Hartmann, Tanja   95, 402
He, Meng   136, 537, 548
Hell, Pavol   227
Hermann, Miki   227
Hermelin, Danny   465
Higuchi, Shoichi   485
Hirata, Kouichi   485
Hopcroft, John E.   1
Hu, Guangda   278
Huang, Yamming   352
Hurtado, Ferran   423

Ito, Takehiro   34, 372

Junosza-Szaniawski, Konstanty   619

Kameda, Tsunehiko   588
Kamiński, Marcin   299
Kan, Tomohiro   485
Kanté, Mamadou Moustapha   289
Kao, Mong-Jen   75
Kappmeier, Jan-Philipp W.   433
Kawamura, Akitoshi   598
Kawamura, Kazuto   34
Kim, Hyo-Sil   54
Kim, Sang-Sub   54
Kita, Nanao   85
Klavík, Pavel   444
Kobayashi, Yusuke   598
Köbler, Johannes   517
Kociumaka, Tomasz   207
Kopelowitz, Tsvi   558
Kratochvíl, Jan   444, 619
Kratsch, Dieter   495
Kreveld, Marc van   166
Krohmer, Anton   659
Kuhnert, Sebastian   517
Kuo, Ching-Chen   24

Leblanc, Alexandre   319
Le Gall, François   639
Leitert, Arne   267
Letsios, Dimitrios   106
Li, Cheng-Chung   578
Liao, Chung-Shou   352

Limouzy, Vincent   289
Liotta, Giuseppe   423
Liu, Yunlong   342
Löffler, Maarten   166
Lu, Hsueh-I   24
Lucarelli, Giorgio   106

Mahoney, Michael W.   278
Makris, Christos   568
Mary, Arnaud   289
Matuschke, Jannik   433, 506
Mazumdar, Arya   649
Mchedlidze, Tamara   423
Meijer, Henk   423
Mnich, Matthias   247
Morrison, Jason   319
Munro, J. Ian   537

Nagamochi, Hiroshi   475
Nakano, Shin-ichi   372
Nevisi, Mayssam Mohammadi   227
Nicholson, Patrick K.   548
Niedermeier, Rolf   247
Nielsen, Jesper Sindahl   527
Norouzi-Fard, Ashkan   382
Nourine, Lhouari   289

Okamoto, Yoshio   372, 423, 629
Ono, Hirotaka   34
Otachi, Yota   372, 444, 455

Pach, János   166
Pachocki, Jakub   207
Paluch, Katarzyna   116
Papadopoulou, Evanthia   177, 187
Paulusma, Daniël   14, 495
Peis, Britta   433
Pergel, Martin   619

Radhakrishnan, Jaikumar   257
Radoszewski, Jakub   207
Rautenbach, Dieter   267
Rescigno, Adele A.   4
Rizzi, Romeo   465
Rollin, Jonathan   402
Rutter, Ignaz   75, 402
Rytter, Wojciech   207
Rząźewski, Paweł   619

Saikkonen, Riku   217
Saitoh, Toshiki   444
Sakai, Yoshifumi   197

Shin, Chan-Su   629
Shrestha, Yash Raj   126
Skala, Matthew   319
Smorodinsky, Shakhar   4
Soisalon-Soininen, Eljas   217
Son, Wanbin   54
Song, Jian   14
Song, Zhao   588

Talmon, Nimrod   558
Tamura, Takeyuki   146
Tan, Xuehou   332
Tang, Bo   690
Tanigawa, Shin-ichi   309
Ting, H.F.   669
Truelsen, Jakob   527
Tsakalidis, Konstantinos   568

Uehara, Ryuhei   372
Uno, Takeaki   372, 679
Uno, Yushi   372

van Bevern, René   247
van 't Hof, Pim   299
Vialette, Stéphane   465

Wagner, Dorothea   75, 95
Waleń, Tomasz   207
Wang, Haitao   332, 609
Watanabe, Osamu   517
Weller, Mathias   247
Welzl, Emo   413
Wong, Prudence W.H.   44
Wu, Gangshan   332
Wu, Xiaodong   342

Xiang, Xiangzhong   669

Yamakami, Tomoyuki   237
Yazdanbod, Sadra   382
Yu, Hung-I   578
Yu, Wei   362
Yung, Fencol C.C.   44

Zarrabi-Zadeh, Hamid   382
Zavershynskyi, Maksym   177
Zeh, Norbert   136, 548
Zhang, Guochuan   362
Zhang, Jinshan   690
Zhou, Gelin   537
Zhou, Xiao   34